International Conference on Strongly Correlated Electron Systems

(SCES 2011)

Journal of Physics: Conference Series Volume 391

Cambridge, United Kingdom
29 August - 3 September 2011

ISBN: 978-1-62276-855-4
ISSN: 1742-6588

Printed from e-media with permission by:

Curran Associates, Inc.
57 Morehouse Lane
Red Hook, NY 12571

Some format issues inherent in the e-media version may also appear in this print version.

Printed by Curran Associates, Inc. (2013)

For permission requests, please contact the Institute of Physics at the address below.

Institute of Physics
Dirac House, Temple Back
Bristol BS1 6BE UK

Phone: 44 1 17 929 7481
Fax: 44 1 17 920 0979

techtracking@iop.org

Additional copies of this publication are available from:

Curran Associates, Inc.
57 Morehouse Lane
Red Hook, NY 12571 USA
Phone: 845-758-0400
Fax: 845-758-2634
Email: curran@proceedings.com
Web: www.proceedings.com

TABLE OF CONTENTS

F-ELECTRON COMPOUNDS

D-ELECTRON COMPOUNDS

ORGANICS AND S,P-ELECTRON COMPOUNDS

CUPRATES AND PNICTIDES

ATOMIC GASES

LOW-DIMENSIONAL SYSTEM

TOPOLOGICAL INSULATORS

THEORY

ENERGY

Preface

The Strongly Correlated Electron Systems Conference (SCES) 2011, was held from 29 August–3 September 2011, in Cambridge, UK. SCES'2011 was dedicated to 100 years of superconductivity and covered a range of topics in the area of strongly correlated systems. The correlated electronic and magnetic materials featured include f-electron based heavy fermion intermetallics and d-electron based transition metal compounds.

The meeting welcomed to Cambridge 657 participants from 23 countries, who presented 127 talks (including 16 plenary, 57 invited, and 54 contributed) and 736 posters in 40 sessions over five full days of meetings.

This proceedings volume contains papers reporting on the science presented at the meeting. This work deepens our understanding of the rich physical phenomena that arise from correlation effects. Strongly correlated systems are known for their remarkable array of emergent phenomena: the traditional subjects of superconductivity, magnetism and metal-insulator transitions have been joined by non-Fermi liquid phenomena, topologically protected quantum states, atomic and photonic gases, and quantum phase transitions. These are some of the most challenging and interesting phenomena in science.

As well as the science driver, there is underlying interest in energy-dense materials, which make use of 'small' electrons packed to the highest possible density. These are by definition 'strongly correlated'. For example: good photovoltaics must be efficient optical absorbers, which means that photons will generate tightly bound electron-hole pairs (excitons) that must then be ionised at a heterointerface and transported to contacts; efficient solid state refrigeration depends on substantial entropy changes in a unit cell, with large local electrical or magnetic moments; efficient lighting is in a real sense the inverse of photovoltaics; the limit of an efficient battery is a supercapacitor employing mixed valent ions; fuel cells and solar to fuel conversion require us to understand electrochemistry on the scale of a single atom; and we already know that the only prospect for effective high temperature superconductivity involves strongly correlated materials. Even novel IT technologies are now seen to have value not just for novel function but also for efficiency.

While strongly correlated electron systems continue to excite researchers and the public alike due to the fundamental science issues involved, it seems increasingly likely that support for the science will be leveraged by its impact on energy and sustainability.

The conference owes its success to the large number of devoted workers for the cause, which includes the organising and programme committees and a considerable number of workers on the ground who contributed to the smooth running of the meeting.

The conference received major sponsorship from CamCool Research Limited, the International Institute for Complex Adaptive Matter, from the European Science Foundation through the program INTELBIOMAT, and the Cambridge Central Asia Forum.

On behalf of Conference Chairs PB Littlewood and GG Lonzarich; Secretary S Saxena; Treasurer M Sutherland; Local Organising Committee Chair SE Sebastian; Programme Committee Chairs E Artacho, FM Grosche, Z Hadzibabic

Journal of Physics: Conference Series **391** (2012) 011001 doi:10.1088/1742-6596/391/1/011001

Programme Committee

E. Artacho, Cambridge (chair)
M. Grosche, Cambridge (chair)
Z. Hadzibabic, Cambridge (chair)
H. Alloul, Paris
E. Baggio-Saitovich, Rio Di Janeiro
E. Bauer, Vienna
N. Berloff, Cambridge
D. Bonn, Vancouver
J. van den Brink, Dresden
R. Budhani, Delhi
P. Chandra, Piscataway
S-W. Cheong, Rutgers
P. Coleman, Piscataway
M. Vojta, Cologne
D. Cox, Davis
H. Ding, IOP, China
M. Ellerby, London
Z. Fisk, Irvine
J. Flouquet, Grenoble
A. Galatanu, Romania
P. Gegenwart, Gottingen
L. Greene, Urbana
H. Hwang, Tokyo
A. P. Mackenzie, St.Andrews
N. Mathur, Cambridge
K. Miyake, Osaka
A Navrotsky, Davis
M. Norman, Argonne
Y. Onuki, Osaka
C. Panagopoulos, Singapore
S. Ramakrishnan, Mumbai
A. Ramirez, Santa Cruz
F. Rivadulla, Compostela
S. E. Sebastian, Cambridge
V. Sechovsky, Prague
S. Simon, Oxford
D. Snoke, Pittsburgh
J.C. Gomez-Sal, Santander
V. Tripathi, Mumbai
A. Vasiliev, Moscow

Local Committee

S. E. Sebastian (chair)
N. Mathur
M. Parish
M. Atature
C. Barnes
N. Berloff
M. Blamire
J. Baumberg
A. Cheetham
R. Needs
E. Pugh
M. Carpenter
R. Cowburn
J. McManus Driscoll
A. Ferrari
C. Grey
Z. Hadzibabic
J. Keeling
D. Khmelnitskii
M. Koehl
W. Milne
S. Redfern
D. Ritchie
J. Robertson
B. Simons

National Advisory Committee

G. Aeppli, London
A. Ardavan, Oxford
P. Attfield, Edinburgh
A. Boothroyd, Oxford
A. Coldea, Oxford
L. Eaves, Nottingham
D. Edwards, London
M. Ellerby, London
V. Falko, Lancaster
R. Friend, Cambridge
C. Frost, Rutherford
G. Gehring, Shefield
S. Hayden, Bristol
N. Hussey, Bristol
A. Huxley, Edinburgh
H. Wilhelm, Didcot
M. Pepper, Cambridge
T. Perring, Didcot
J. Saunders, London
A. Schofield, Birmingham
N. Shannon, Bristol
M. Skolnick, Sheffield
S. Thompson, York

Journal of Physics: Conference Series **391** (2012) 011001 doi:10.1088/1742-6596/391/1/011001

International Advisory Committee

E. Abrahams, UCLA
G. Aeppli, London
J. W. Allen, Ann Arbor
P. W. Anderson, Princeton
M. Aronson, Stony Brook
Y. K. Bang, Kwangju and Pohang
M. Barma, Mumbai
G. Baskaran, Chennai
E. Bauer, Vienna
G. Boebinger, Tallahassee
R. Budhani, Delhi
P. Canfield, Ames
M. Continentino, Rio di Janiero
S. Coppersmith, Madison
B. Coqblin, Paris
A. Chubukov, Madison
C. Di Castro, Rome
M. Eremets, Mainz
M. Fiebig, Bonn
Z. FiskIrvine
J. Flouquet, Grenoble
P. Fulde, Dresden
A. Geim, Manchester
J.C. Gomez-Sal, Santander
A. Kavokin, Southampton
J. Goodenough, Austin
H. Hosono, Tokyo
S. Julian, Toronto
G. Kotliar, Piscataway
D. Khmelnitskii, Cambridge, UK
K. Kugel, Moscow
C. Lacroix, Grenoble
P. A. Lee, Cambridge, USA
C.T. Liang, Taipei
P. Majumdar, Allahabad
Y. Maeno, Kyoto
J. Mannhart, Augsburg
M. B. Maple, San Diego
Y. Matsuda, Kyoto
R. Moessner, Dresden
A. Millis, New York
J. Mydosh, Leiden
S. Nakatsuji, Tokyo
G. Oomi, Kyushu
R. Osborn, Chicago
S. Ovchnikov, Krasnoyarsk
C. Panagopoulos, Singapore & Heraklion
S. Paschen, Vienna
C. Pfleiderer, Munich
P. Phillips, Urbana
D. Pines, Davis
T. V. Ramakrishnan, Bangalore
A.K. Raychaudhuri, Calcutta
M. Reifers, Kosice
P. Riseborough, Philadelphia
M. L Saboungi, Orleans
E. V. Sampathkumaran, Mumbai
J. Sarrao, Los Alamos
J. Schilling, St. Louise
A. Schofield, Birmingham
V. Sechovsky, Prague
T. Senthil, Cambridge, USA
J. G. Sereni, Bariloche
K. Shimizu, Osaka
Q. Si, Houston
M. Sigrist, Zurich
A. Simoni, Trento
D. Singh, Oak Ridge
A. Sood, Bangalore
J. Spalek, Krakow
F. Steglich, Dresden
G. R. Stewart, Gainesville
H. Takagi, Tokyo
L. Taillefer, Sherbrooke
J. D. Thompson, Los Alamos
Y. Tokura, Tokyo
K. Ueda, Tokyo
C. M. Varma, Riverside
T. Vojta, Rolla
N.L. Wang, Beijing
T. Xiang, Beijing
L. Yu, Beijing
F. C. Zhang, Hong Kong
G. Zwicknagl, Braunschweig

Operational Team

Anson Cheung (co-ordinator)
Beng Tan (co-ordinator)
Cheng Liu (co-ordinator)
Swee K. Goh (co-ordinator)
Adam Halski
Edd Cavanna
Felix Nissen
Gareth Conduit
Gerie Lonzarich
Greg Lever
Hannah Price
Haruka Taniguchi
Hyeong Jin Kim
Jack Gillett
Jo Wensley
Jonathan Silver
Lara Sibley
Leona Hope
Lina Klintberg
Marianne Bauer
Matt Burgess
Muhammad Ahsan Zeb
Nick Bristowe
Oleksandr Poplavskyy
Paul Nahai-Williamson
Peter Logg
Prajakti Kalra
Richard Brierley
Robert Hay
Seb Haines
Sitikantha Das
Stephen Rowley
Sven Friedemann
Yang Zou
Yiqian Xu
Zhuo Feng

Peer review statement

All papers published in this volume of *Journal of Physics: Conference Series* have been peer reviewed through processes administered by the proceedings Editors. Reviews were conducted by expert referees to the professional and scientific standards expected of a proceedings journal published by IOP Publishing.

Magnetic properties of $Ce_{1-x}La_xRu_2Si_2$ compounds

D. Abe[1], Y. Amakai[1], S. Murayama[1], N. Momono[1], H. Takano[1], T. Nakano[2], K. Matsubayashi[3], Y. Uwatoko[3]

[1] Muroran Institute of Technology, Muroran 050-8585, Japan
[2] Faculty of Engineering, Niigata University, Niigata 950-2181, Japan
[3] Institute for Solid State Physics, the University of Tokyo, Kashiwa 277-8581, Japan

E-mail: murayama@mmm.muroran-it.ac.jp

Abstract. We have measured the susceptibility χ and resistivity ρ on single-crystalline $Ce_{1-x}La_xRu_2Si_2$ for $0 \le x \le 0.8$ in directions parallel (//) and perpendicular (⊥) to the c-axis. The χ// for $0.1 \le x \le 0.7$ indicates a Curie-Weiss form at $T > 20$ K and a typical AF maximum at T_N. A significant crossover between T_N and the magnitude of the Weiss temperature $|\Theta|$ occurs at about $x = 0.2$. The $\rho\perp$ shows a gap-type upturn at T_N for $0.1 \le x \le 0.185$ with rather a downward bend in ρ// at around T_N. The upturn in $\rho\perp$ at T_N disappears for $x = 0.2$ with a sharp maximum in ρ// at T_N. A sharp peak in $\rho\perp$ takes place at T_N for $0.3 \le x \le 0.7$ with the sharp maximum in ρ// at T_N for $0.4 \le x \le 0.6$. These results can be interpreted as the evolution from the itinerant SDW in the heavy-fermion band for $0.1 \le x < 0.2$ to the AF ordering with the local moment for $x > 0.2$ up to $x = 0.7$.

1. Introduction

The heavy-fermion compounds $Ce_{1-x}La_xRu_2Si_2$ with a body-centered tetragonal $ThCr_2Si_2$-type structure show an incommensurate sinusoidal magnetic ordering at the La concentration x higher than the critical concentration of $x_c = 0.075$ with the wave vector along the a-axis and the magnetic moment along the c-axis as shown by neutron diffraction studies[1,2)]. An additional squaring of the sinusoidal ordering is seen for $x \ge 0.2$ below $T_L \sim 2$ K lower than the Néel temperature T_N[3)]. Resistivity measurements perpendicular to the c-axis[4,5)] have revealed different anomalies at T_N depending on the concentration as a gap-type upturn for $0.1 \le x < 0.2$ and a maximum for $0.3 \le x \le 0.9$. These facts suggest La-concentration dependent two magnetic phases: one is a partial gap-type heavy-fermion SDW state for $x < 0.2$ and the other is an antiferromagnetic (AF) ground state without gap on the Fermi surface for $x > 0.2$. We could expect existence of another critical concentration of $x_{c2} \sim 0.2$ in the magnetic ground state. There is so far no detailed experimental information on the effect of the chemical La substitution to the magnetic interaction and to the Fermi surface along the c- and a-axes at around the critical concentrations. In order to study detailed magnetic properties on these compounds, we have measured the lattice constant a and c by powdered X-ray diffraction experiment, the susceptibility χ and the resistivity ρ of single-crystalline $Ce_{1-x}La_xRu_2Si_2$ in directions parallel (//) and perpendicular (⊥) to the c-axis.

2. Sample Preparations and Experiment

Single-crystalline samples of $Ce_{1-x}La_xRu_2Si_2$ were grown by the Czochralski method with a stoichiometric amount of Ce(3N), La(3N), Ru(3.5N) and Si(5N) in a tri-arc furnace with Ar atmosphere and were annealed at 800 °C in vacuumed quartz tube for one week. The X-ray diffraction measurements by Laue method confirm each sample as a single phase with the B.C.T. structure. The lattice constants were also measured by the powdered X-ray diffraction. The samples were cut by a spark cutter into a typical size of about $0.5 \times 0.5 \times 2$ mm^3. The susceptibility was measured in field of 0.05 T parallel and perpendicular to the c-axis in the temperature range $1.9\ \text{K} \le T \le 300$ K by using a commercial SQUID magnetometer. The resistivity was measured with current parallel and perpendicular to the *c*-axis in the temperature range 0.4 K $\leq T \leq$ 300 K by using a standard AC four-probe technique.

3. Results and Discussion

The lattice constant *a* and *c* obtained from the X-ray diffraction measurements and the calculated unit volume *V* increase almost in proportion to *x* following Vegard's law. Therefore, the La substitution in the present compounds exactly introduces successive negative chemical pressure. We will soon present the data elsewhere. Fig. 1 shows the temperature dependence of the susceptibility χ parallel (//)

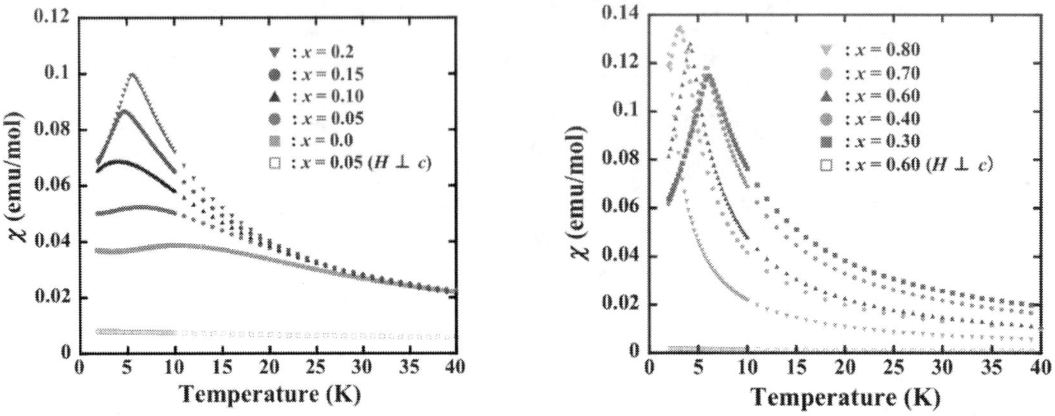

Fig. 1 Temperature dependence of the susceptibility χ of $Ce_{1-x}La_xRu_2Si_2$ along the *c*-axis at $x \leq 0.2$ on the left side and $x \geq 0.3$ on the right. We define T_N by the temperature of the maximum for $x > 0.1$.

and perpendicular ($\perp$) to the *c*-axis of $Ce_{1-x}La_xRu_2Si_2$ in field of 0.05 T at each La concentration *x*. The χ// shows almost a Curie-Weiss form at $T > 20$ K for $0 \leq x \leq 0.8$ with nearly the same paramagnetic effective moment of $\mu_{eff} \sim 3\ \mu_B$/Ce. Below 10 K we observe the saturated temperature dependence of χ// for $0 \leq x \leq 0.05$ as shown in typical heavy-fermion systems. We observe a maximum in χ// at x = 0.1 and observe a sharp maximum for $0.15 \leq x \leq 0.7$ in agreement with the AF transition by the neutron diffraction[1-3] and specific heat measurements[6]. Therefore, we define T_N as the temperature of the maximum in χ//. The T_N increases with increasing *x* and shows a maximum value of $T_N = 6$ K at around $x = 0.3$ and decreases almost linearly for $x > 0.4$ extended to $T_N = 0$ at $x = 1$ as shown in Fig. 2. However, the Weiss temperature Θ is -12 K at $x = 0$ and its magnitude $|\Theta|$ rapidly decreases with increasing *x* to reach ~ 1 K at $x = 0.4$ as shown in this figure, in contrast to the T_N. We have found a significant crossover between T_N and $|\Theta|$ near the expected second critical concentration of $x_{c2} = 0.2$. In Kondo alloys a Curie-Weiss form with the negative Weiss temperature Θ is often observed because of the negative *c-f* exchange interaction due to the hybridization between conduction electrons and the magnetic *f*-electron, and the magnitude $|\Theta|$ roughly corresponds to the spin-fluctuation temperature T_{sf}, or the Kondo temperature T_K[7]. In the present case it is necessary to take account of the effect of the RKKY interaction between the local *f*-electrons and crystal-field

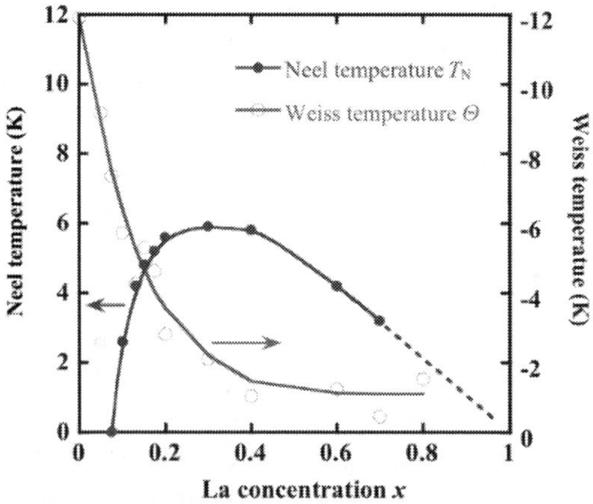

Fig. 2 La concentration *x* dependence of T_N and Θ of $Ce_{1-x}La_xRu_2Si_2$

effect in order to evaluate the T_K. However, the rapid decrease of the $|\Theta|$ suggests the decrease of the *c*-*f* hybridization by a small amount of the La substitution. Then, the crossover at about $x = 0.2$ in Fig. 2 would suggest the quantum phase transition or quantum crossover from the itinerant SDW state in the heavy 4*f*-band due to the Fermi-surface nesting to another AF state with local 4*f* character. Thus, the squared sinusoidal ordering observed by the neutron diffraction[3] would be induced below T_L for $x \geq 0.2$.

Temperature dependence of the resistivity $\rho(T)$ in directions parallel (//) and perpendicular ($\perp$) to the *c*-axis of $Ce_{1-x}La_xRu_2Si_2$ has been measured in the concentration region of $0.05 \leq x \leq 0.6$ for ρ// and $0.05 \leq x \leq 0.9$ for $\rho\perp$. Both ρ// and $\rho\perp$ for $x = 0.05$ show T^2 dependence below 1.5 K with a large

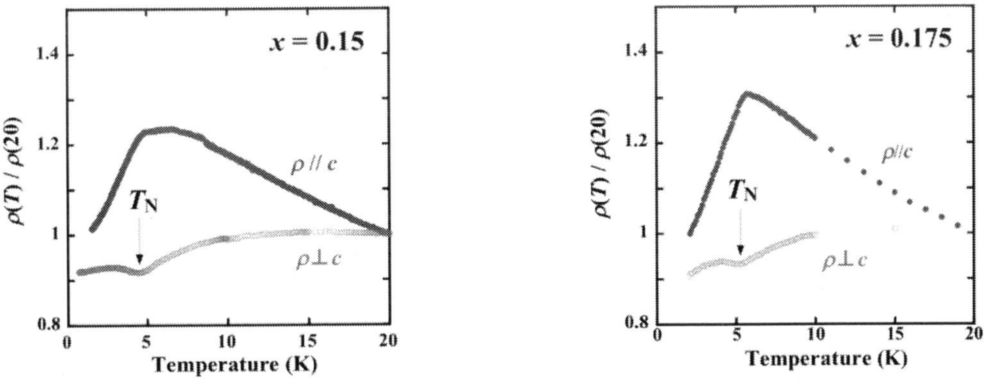

Fig. 3 Temperature dependence of the resistivity ratio $\rho(T)/\rho(20)$ parallel and perpendicular to the *c*-axis of $Ce_{1-x}La_xRu_2Si_2$ at $x = 0.15$ on the left side and $x = 0.175$ on the right

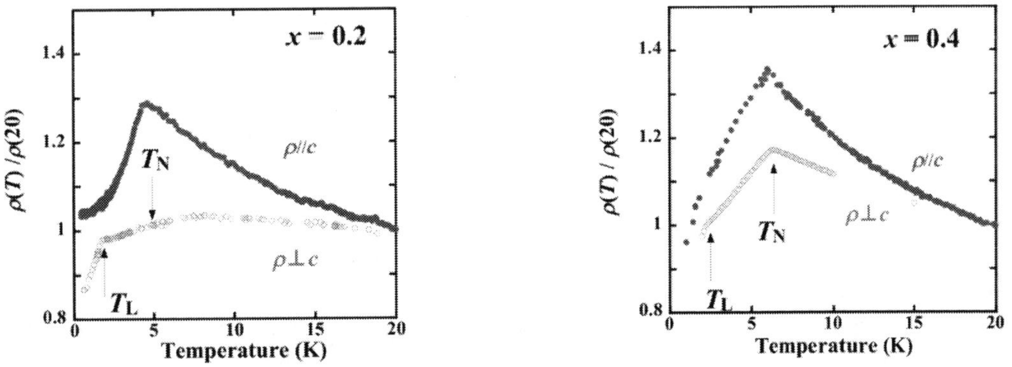

Fig. 4 Temperature dependence of the resistivity ratio $\rho(T)/\rho(20)$ parallel and perpendicular to the *c*-axis of $Ce_{1-x}La_xRu_2Si_2$ at $x = 0.2$ on the left side and $x = 0.4$ on the right

coefficient of $A// = 0.72$ μΩcm and $A\perp = 0.48$ μΩcm, respectively. The resistivity in the Fermi-liquid ground state follows $\rho(0) + AT^2$ in the low-temperature limit due to the electron correlation and the *A* value is known to be very large for typical heavy-fermion compounds. The present results are comparable with published values of $A// = 0.55$ μΩcm and $A\perp = 0.42$ μΩcm in pure $CeRu_2Si_2$[8] and indicate that the compound of $x = 0.05$ is still in the paramagnetic heavy-fermion state. Fig. 3 shows the resistivity ratio $\rho(T)/\rho(20)$ along the both directions for $x = 0.15$ and 0.175. As shown in this figure we have observed a gap-type upturn in $\rho\perp$ at T_N. We have obtained a similar upturn in $\rho\perp$ for $0.1 \leq x \leq 0.185$, though whole data are not shown here (the data for x = 0.13 has recently been published in ref. 9). However, such a gap-type upturn is not observed in ρ// and rather a downward bend is seen at around T_N, indicating the partial gapping due to the nesting of the Fermi-surface only along the *a*-axis. Fig. 4 shows the temperature dependence of the resistivity ratio $\rho(T)/\rho(20)$ parallel (//) and perpendicular (⊥) to the c-axis for $x = 0.2$ and 0.4. For $x = 0.2$, the upturn in $\rho\perp$ at T_N disappears, and a downward kink is exhibited at T_L shown by the neutron diffraction below T_N[3]. This kink has already been seen in the previous resistivity study[4] and is probably due to the squaring of the sinusoidal ordering at T_L. On the other hand, the downward bend in $\rho//$ becomes a sharp maximum at T_N for $x = 0.2$. As shown in the right side of Fig. 4, a sharp peak in $\rho\perp$ takes place at T_N for $x = 0.4$, in addition to the kink due to the squaring at T_L. We have found the sharp peak in $\rho\perp$ at T_N for $0.3 \leq x \leq 0.7$ also as shown in the previous study[4], though our data except for $x = 0.4$ are not presented here. The sharp maximum in $\rho//$ at T_N is still observed for $x = 0.4$ as shown in Fig. 4 up to $x = 0.6$. These resistivity results can be interpreted as the evolution from the itinerant SDW in the heavy-fermion band for $0.1 \leq x < 0.2$ to the AF ordering with the local moment for $x > 0.2$ at least up to $x = 0.7$. The quantum magnetic phase transition or quantum crossover would be realized in the vicinity of $x_{c2} = 0.2$ due to the chemical pressure, which has been suggested by the crossover between T_N and $|\Theta|$ near x_{c2}. From the interpretation of the present results we propose the magnetic x-T phase diagram of the heavy-fermion compounds $Ce_{1-x}La_xRu_2Si_2$ with a critical concentration of $x = 0.2$ as shown in Fig. 5. For $0 \leq x < x_c = 0.075$ paramagnetic Fermi-liquid state remains in the heavy-fermion band. For $0.075 < x < x_{c2} = 0.2$ the SDW state due to the partial nesting of the Fermi surface along the a-axis occurs in the heavy-fermion band. For $x > 0.2$ the sinusoidal ordering without gapping of the Fermi surface takes place in the temperature region of $T_L < T < T_N$ followed by the squaring of the ordered moment at $T < T_L$. Thus, we can conclude occurrence of the local AF ordered moment in the ground state for $0.2 < x < 0.7$.

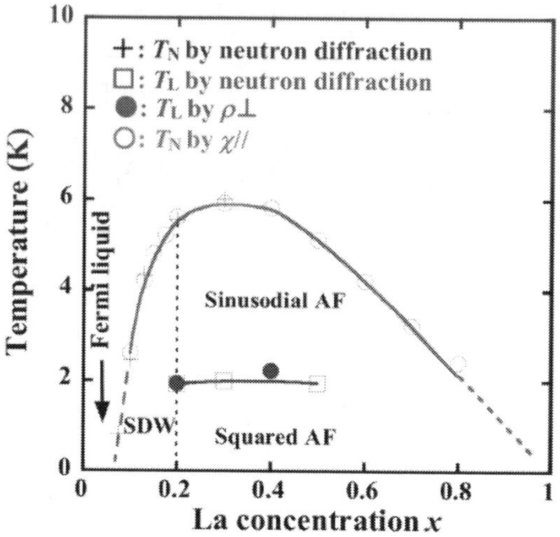

Fig. 5 Magnetic x-T phase diagram of $Ce_{1-x}La_xRu_2Si_2$

This work was carried out by Joint Research in the Institute for Solid State Physics, the University of Tokyo, and supported by a grant for Kidorui (Rare Earth) Program (2011) through Muroran Institute of Technology.

References

[1] Quezel S, Burlet P, Jacoud J-L, Regnault L-P, Rossat-Mignod J, Vettier C, Lejay P and Flouquet J 1988 *J. Magn. & Magn. Mater.* **76 & 77**, 403
[2] Raymond S, Regnault L-P, Kambe S, Mignot J-M, Lejay P. and Flouquet J 1997 *J. Low Temp. Phys.* **109**, 205
[3] Mignot J-M, Jacoud J-L, Regnault L-P, Rossat-Mignod J, Haen P, Lejay P, Boutrouille Ph, Hennion B and Petitgrand D 1990 *Physica B* **163**, 611; Mignot J-M, Jacoud J-L, Regnault L-P, Rossat-Mignod J, Haen P and Lejay P 1991 *Physica B* **171**, 357
[4] Djerbi R, Haen P, Lapierre F, Lehmann P and Kappler J P 1988 *J. Magn. & Magn. Mater.* **76 & 77**, 260
[5] Takeuchi T, Miyako Y, Ocio M, Pari P and Hammann J 1996 *Czech. J. Phys.* **46**, Suppl. S4 2085; Miyako Y, Takeuchi T, Taniguchi T, Kawarazaki S, Marumoto K, Hamada R, Yamamoto Y, Ocio M, Pari P and Hammann J 1996 *J. Phys. Soc. Jpn.* **65**, Suppl. B 12
[6] Besnus M J, Kappler J P, Lehmann P and Meyer A 1985 *Solid State Commun.* **55**, 779
[7] Krishna-murthy H R, Wilson K G and Wilkins J W 1975 *Phys. Rev. Lett.* **35**, 1101
[8] Lapierre F and Haen P 1992 *J. Magn. & Magn. Mater.* **108**, 167
[9] Amakai Y, Harada E, Yokoyama D, Murayama S, Matsumoto K, Takano H, Momono N, Matsubayashi K and Uwatoko Y 2011 *J. Phys. Soc. Jpn.* **80**, SA062

La substitution effect to the heavy-fermion state in structure-disordered Ce-Ru alloys

Y. Amakai[1], D. Yoshii[1], S. Murayama[1], H. Takano[1], N. Momono[1], Y. Obi[2] and K. Takanashi[2]

[1]Graduate School of Engineering, Muroran Institute of Technology, Muroran, Hokkaido, 050-8585, Japan

[2]Institute for Materials Research, Tohoku University, Sendai, 980-8577, Japan

E-mail: a-rain@mmm.muroran-it.ac.jp

Abstract. We have measured the magnetic susceptibility χ, specific heat C_p and resisitivity ρ on sputtered structure-disordered $Ce_{80-y}La_yRu_{20}$ alloys. The χ for $y \leq 70$ follows the Curie-Weiss law in the measurement temperature region. The paramagnetic effective magnetic moment p_{eff} is about 1.8 μ_B/Ce-atom for $y = 0$ and decreases with increasing the La-concentration (0.6 μ_B/Ce-atom for $y = 70$). The C_p follows linear relation of temperature above 7 K for all the samples. The specific heat coefficient γ shows about 200 mJ/molK^2 for $y = 0$ and decreases rapidly with increasing La-concentration. We found the -log T dependence of the ρ in the La-substituted samples. Magnitude and temperature range of the -log T contribution to the whole resistivity increase with increasing the La concentration. On the other hand, we have observed superconducting behaviors of the resistivity $\rho = 0$ and a large diamagnetization at low temperature ($T < 3$ K) for $y = 80$.

1. Introduction

Amorphous alloy system CeRu shows a heavy-fermion like behavior in the Ce-rich side and a superconductivity in the Ru-rich side, which was reported by Homma *et al.*[1] Recently, we have reported the resistivity[2], the specific heat[3] and susceptibility[4] for the structure-disordered (a-)Ce_xRu_{100-x} alloys in the wide Ce-concentration range x. In the Ce-rich side, the electronic specific heat coefficient γ shows a very large value ($\gamma \sim 220$ mJ/Ce-molK^{-2}, at $x = 80$). The resistivity of a-$Ce_{80}Ru_{20}$ follows a T^2 law with a large coefficient A in the low temperature region. However, the γ decreases rapidly with decreasing the Ce-concentration. In the Ru-rich region, the γ becomes less than 10 mJ/molK^{-2} and the resistivity becomes zero (T_c ~3.6 K, at $x = 15$). These facts suggest that a heavy fermion state and a superconducting state appeared, respectively, in the Ce-rich and Ru-rich sides of the structure-disordered system even with no translation symmetry. The susceptibility for $x \geq 39$ shows a Curie-Weiss paramagnetic behavior. The effective magnetic moment p_{eff} is about 1.8 μ_B/Ce-atom at $x = 80$ and decreases rapidly with decreasing the Ce-concentration ($p_{eff} \sim 0.6$ μ_B/Ce-atom, at $x = 39$). The susceptibility for $x \leq 23$ is almost independent of temperature except in the superconducting region. The valence of Ce estimated from recent measurement of x-ray absorption near edge structure is +3.1 in $x = 80$ and +3.45 in $x = 9$.[5] Therefore, it is considered that the 4f-electron of Ce is trivalent configuration in the Ce-rich region of the heavy-fermion state and mixed valence state in the Ru-rich

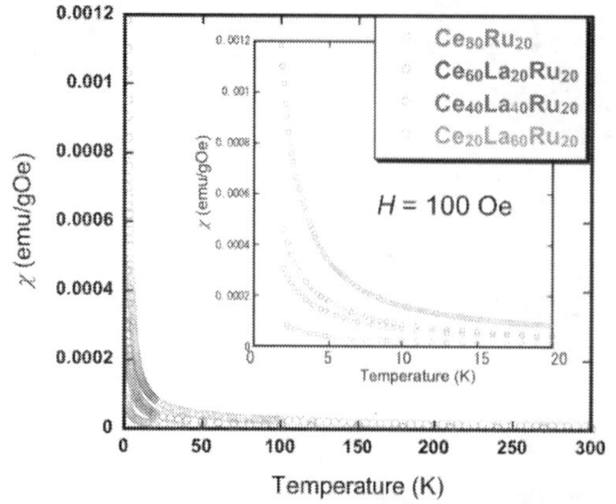

Figure 1. Temperature dependence of susceptibility χ for a-$Ce_{80-y}La_yRu_{20}$ alloys. The inset is lower temperature susceptibility.

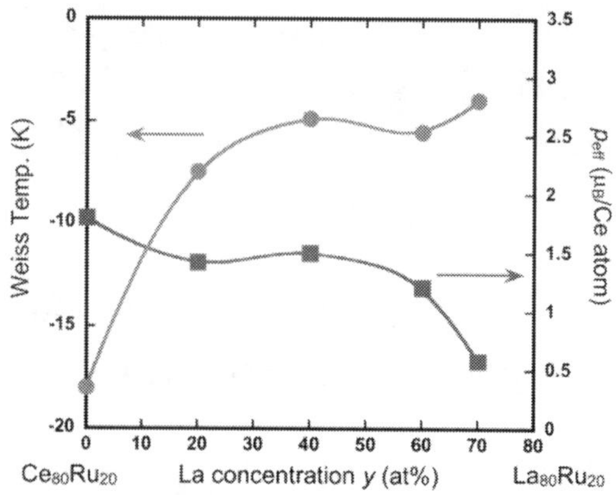

Figure 2. La-concentration dependence of the paramagnetic effective moment p_{eff} and Weiss temperature θ estimated from the Curie-Weiss law.

region of superconducting state. In this study, in order to clarify the effect of 4*f*-electron in *a*-Ce-Ru alloys system, we have measured the susceptibility χ, specific heat C_p and resistivity ρ for the a-$Ce_{80-y}La_yRu_{20}$ alloys (y = 0 - 80) substituted by La.

2. Experimental

Bulk ingots of $Ce_{80-y}La_yRu_{20}$ (y = 0, 20, 40, 60, 70 and 80) were made by arc melting from nominal amount of Ce 99.9 %, La 99.9 % (NIPPON YTTRIUM CO., LTD) and Ru 99.95 % (RARE METALLIC CO., LTD), in argon arc furnace. Amorphous $Ce_{80-y}La_yRu_{20}$ alloys were prepared by a dc sputtering method from the arc-melted ingots onto water-cooled Cu substrate. We confirmed a typical amorphous halo pattern from the x-ray diffraction measurement. The magnetic susceptibility was measured by using a commercial SQUID magnetometer (Quantum Design MPMS) from 2 to 300 K. The specific heat measurement was measured by using a relaxation method (Quantum Design PPMS) from 1.9 to 40 K. The electrical resistivity was measured by using a typical four-terminal method (Quantum Design PPMS) from 1.9 to 300 K.

3. Results and Discussion

Figure 1 shows temperature dependence of the susceptibility χ for a-$Ce_{80-y}La_yRu_{20}$ alloys. The χ for $y \le 60$ increases with decreasing temperature, which follows the Curie-Weiss (CW) law in the measurement temperature region. The value of χ decreases with increasing La-concentration. Figure 2 shows the La-concentration dependence of the paramagnetic effective magnetic moment p_{eff} and Weiss temperature θ estimated from the CW law $N_{Ce}(p_{eff})^2/3k_B(T-\theta)$, where N_{Ce} is the number of Ce-atom per gram. The value of p_{eff} is about 1.8 μ_B/Ce-atom for y = 0, and decreases with increasing the La-concentration. However, the p_{eff} for $20 \le y \le 60$ exhibits almost constant value ($p_{eff} \approx 1.5$ μ_B/Ce-atom). With increasing La-concentration further, the p_{eff} decreases again and becomes about 0.6 μ_B/Ce-atom for y = 70. The value of θ shows about -18 K for y = 0 and the absolute value decreases with increasing the La-concentration. The θ for y >20 is almost constant value ($\theta \approx 5$ K). Figure 3 shows low temperature specific heat C_p over T *vs.* T^2 plots for a-$Ce_{80-y}La_yRu_{20}$ alloys. As shown in this figure, the C_p/T follows almost a T^2-linear relation above 7 K for all the samples. However, the C_p/T for all the samples increases at low temperature with decreasing temperature. We have estimated the γ-value extrapolated from a linear part above 7 K of C_p/T versus T^2 to T = 0. The inset of fig. 3 shows the La-concentration dependence of the specific coefficient γ and value of C_p/T at 2 K for

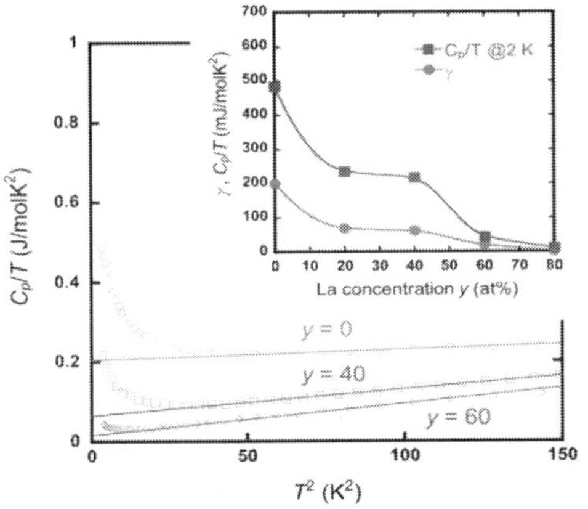

Figure 3. Low temperature specific heat C_p over T *vs.* T^2 plots for a-$Ce_{80-y}La_yRu_{20}$ alloys. The inset is La-concentration dependence of the specific heat coefficient γ and value of C_p/T at 2 K.

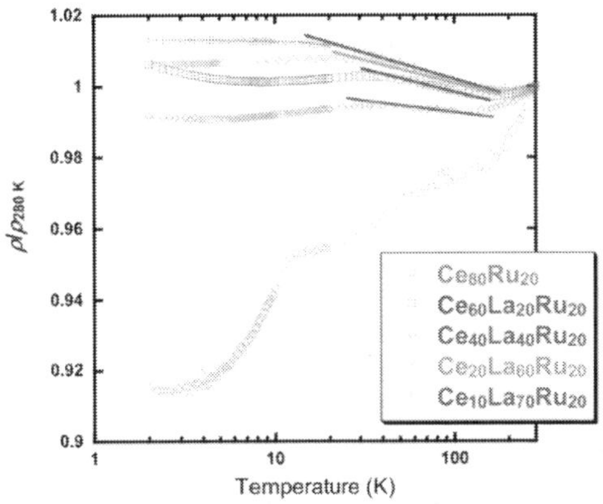

Figure 4. Temperature dependence of the resistivity for a-$Ce_{80-y}La_yRu_{20}$ alloys. The solid lines show a logarithmic guideline for the eyes.

a-$Ce_{80-y}La_yRu_{20}$ alloys. The γ-value shows about 200 mJ/molK^2 for $y = 0$ and decreases rapidly with increasing La-concentration. The γ for $y = 20$ and 40 is almost constant value (~ 50 mJ/molK^2). With increasing the La-concentration further, the γ decreases again and reach about 10 mJ/molK^2 at $y >$ 60. The C_p/T at 2 K exhibits the La-concentration dependence similar to theγ. However, the value of C_p/T at 2 K for $y \leq 40$ is a much larger than γ-value. On the other hand, the C_p/T at 2 K and γ for $y \geq 60$ are almost the same value. Figure 4 shows temperature dependence of the resistivity ρ/ρ_{280K} normalized at 280 K for $y \leq 70$. The ρ for each sample exhibits small temperature change in the whole temperature region and decreases with decreasing the T above 150 K. The ρ for $y = 0$ decreases with decreasing the temperature and a plateau in 10 K $\leq T \leq$ 20 K probably due to the Kondo effect. We found the $-\log T$ dependence of ρ in the higher temperature region by the La substitution for $y \geq 20$. The $-\log T$ contribution to the whole resistivity increases with increasing the La-concentration. This behavior suggests that the Kondo-scattering effect is enhanced by the La substitution. The ρ at $T < 7$ K for $y = 0$ obeys the T^2 dependence with a large coefficient A. Since we have found the large γ value of the specific heat in $y = 0$, the T^2 dependence of the ρ in the low temperature region suggests formation of the heavy-fermion state. However, the T^2 dependence of the ρ was not observed by the La substitution. Therefore, the coherence necessary for the formation of the heavy-fermion state disappeared by the La substitution.

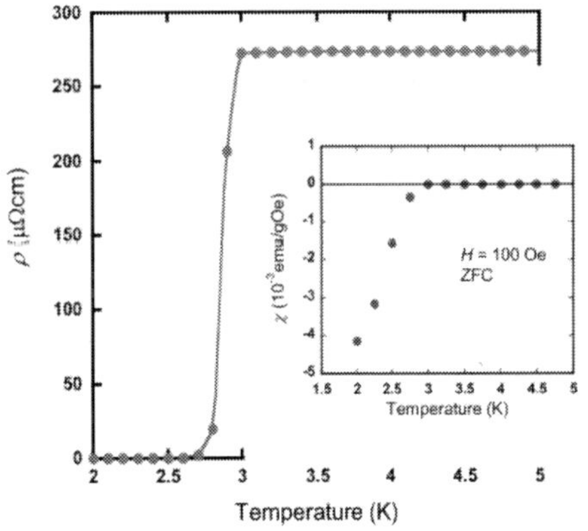

Figure 5. Temperature dependence of the low temperature resistivity ρ and susceptibility χ for $y = 80$.

Figure 5 shows the low temperature resistivity ρ and susceptibility χ for $y = 80$. The ρ at $T > 3$K is almost independent of temperature. We have observed a rapid decrease of ρ in $T < 3$ K and found zero resistivity in $T \sim 2.6$ K. The χ for $y = 80$ shows a small value almost independent of temperature like the Pauli paramagnetic behavior except in the low temperature region. As shown in the inset of Fig. 5, a large diamagnetization was observed at $T < 3$ K. The zero resistivity with a rapid decrease of the ρ and a large diamagnetization of the χ suggest that the superconducting transition have occurred at $T = 2.6$ K.

In the *a*-Ce_xRu_{100-x} alloys case, according as the γ and p_{eff} rapidly decrease with decreasing the Ce-concentration, the superconductivity appeared in the Ru-rich region. The p_{eff} of the present *a*-$Ce_{80-y}La_yRu_{20}$ alloys does not show rapid decrease. Although the p_{eff} decreases down to 0.6 μ_B/Ce-atom in $y = 70$, the superconductivity does not occur up to $y = 80$. The γ for the present alloys decreases with decreasing the Ce-concentration, but does not become small very much in comparison to the case of the *a*-Ce-Ru alloy in the same Ce-concentration. The $\gamma = 61$ and 19 mJ/molK^2 for $y = 40$ and 60, respectively, whereas, $\gamma = 9.6$ and 3.9 mJ/molK^2 for the same Ce-concentration in *a*-Ce-Ru alloys with the superconductivity at $T_c = 1.1$ and 3.0 K, respectively.[3] The $-\log T$ dependence of ρ is observed in the higher temperature region and becomes larger depending on the La substitution. Therefore, the La substitution to the *a*-Ce-Ru alloy would make an increase of the Kondo temperature and make the local Kondo scattering effect clearer in the Ce diluted alloys.

Acknowledgements
This work was performed under the inter-university cooperative research program of the Institute for Materials Research, Tohoku University, and supported by a grant for Kidorui (Rare Earth) Program (2011) through Muroran Institute of Technology.

References
[1] Homma Y, Sumiyama K, Yamaguchi H and Suzuki K 1997 *J. Phys. Soc. Jpn.* **66** 1457
[2] Obi Y, Murayama S, Amakai Y, Okada Y and Asano K 2006 *Physica B* **378-380** 857
[3] Amakai Y, Murayama S, Takano H, Mizutani M, Asano K, Obi Y and Takanashi K 2007 *J. Magn. & Magn. Mater.* **310** 416
[4] Amakai Y, Murayama S, Obi Y, Takano H, Momono N and Takanashi K 2009 *J. Phys.:Conf. Ser.* **150** 042004
[5] Li Y, Onishi N, Nakai I, Amakai Y and Murayama S 2009 *J. Phys. Soc. Jpn.* **78** 094717

Thermodynamic properties of the f-electron-nuclear hyperfine-coupled composite multiplets of Pr ions in filled-skutterudite $PrRu_4P_{12}$

Yuji Aoki[1], **Takahiro Namiki**[1,3], **Shanta R. Saha**[1,4], **Hitoshi Sugawara**[2] **and Hideyuki Sato**[1]

[1] Department of Physics, Tokyo Metropolitan University, Tokyo 192-0397, Japan
[2] Department of Physics, Kobe University, Kobe 657-8501, Japan

E-mail: aoki@tmu.ac.jp

Abstract. It has recently been found that composite multiplets are formed by the hyperfine coupling between the triplet crystalline-electric-field ground state of the $4f$ electrons and the ^{141}Pr nuclear spin in Pr ions of the filled skutterudite $PrRu_4P_{12}$. Entropy analysis of the characteristic Schottky-type peak structure appearing in the low-temperature specific heat demonstrates that it is actually caused by thermal excitations of not only $4f$ electrons but also ^{141}Pr nuclei. The hyperfine coupling constant for Pr ions in nonmetallic $PrRu_4P_{12}$ is determined to be $A = +0.050(2)$ K, which is almost the same as in metallic Pr-based compounds. This fact corroborates that the intra-ion hyperfine coupling of Pr ions is independent of the type of material.

1. Introduction

It has recently been found that composite multiplets are formed by the hyperfine coupling between the crystalline-electric-field (CEF) ground state of the $4f$ electrons and the ^{141}Pr nuclear spin in Pr ions of the filled skutterudite $PrRu_4P_{12}$ [1]. $PrRu_4P_{12}$ is one of the Pr-based filled skutterudites PrT_4X_{12}, which have received intense interest since they display a variety of exotic strongly-correlated-electron behaviors [2, 3, 4]. The charge ordering appearing below $T_{co} = 63$ K [5] has been found to be of unconventional type in the sense that it involves the internal degrees of freedom of the $4f$ electrons through the coupling with the conduction electrons. Below T_{co}, the crystal structure changes into a simple cubic (Pm$\bar{3}$, T_h^1, #200) that contains two crystallographically-inequivalent Pr sites [6] (referred to as Pr1 and Pr2 hereafter) as shown in Fig. 1. The CEF level schemes of the two Pr sites show pronounced T dependence and finally the Pr1 and Pr2 ions have a Γ_1 singlet and a $\Gamma_4^{(2)}$ triplet CEF ground state at $T \ll T_{co}$, respectively [7, 8].

Each Pr ion has a ^{141}Pr nucleus (the natural abundance of 100%) with a nuclear spin $I = 5/2$. Well-established Hamiltonian [9, 10, 11] to deal with the Pr1 and Pr2 single ion is

$$\mathcal{H} = \mathcal{H}_{\rm CEF} + A\boldsymbol{I}\cdot\boldsymbol{J} - (-g_J\mu_{\rm B}\boldsymbol{J} + g_{\rm N}\mu_{\rm N}\boldsymbol{I})\cdot\mu_0\boldsymbol{H}. \quad (1)$$

[3] Present address: Faculty of Engineering, Toyama University, Toyama 930-8555, Japan
[4] Present address: Center for Nano Physics and Advanced Materials, Department of Physics, University of Maryland, College Park, Maryland 20742-4111, USA

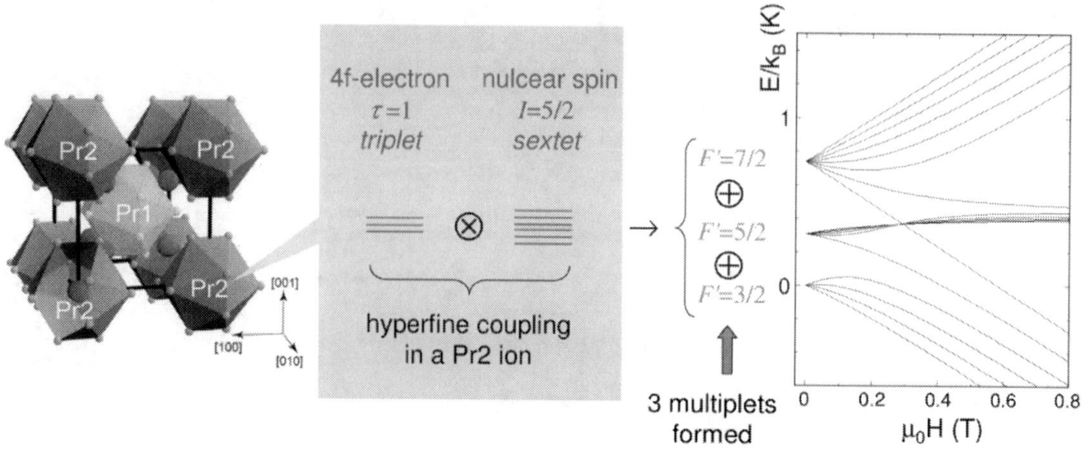

Figure 1. Diagram illustrates the formation of the 4f-electron-nuclear hyperfine-coupled composite multiplets in the Pr2 ion of $PrRu_4P_{12}$. The $4f$-electron triplet CEF state (pseudo-spin $\tau = 1$) and the nuclear spin ($I = 5/2$) are coupled by the hyperfine interaction, transforming into 3 composite multiplets (effective total spin $F' = 7/2, 5/2$ and $3/2$). Calculated energy level scheme in applied magnetic fields is after Ref.[1].

For the CEF Hamiltonian $\mathcal{H}_{\rm CEF}$ [12], we use the CEF parameters determined at 5 K by inelastic neutron scattering measurements [7]. The second term $A\boldsymbol{I}\cdot\boldsymbol{J}$ represents the hyperfine interaction. The coupling constant A was theoretically calculated to be +0.052 K [13, 14] and this value has been confirmed in metallic compounds $PrFe_4P_{12}$ and $PrOs_4Sb_{12}$ by thermodynamical measurements [15, 16, 17]. The third term represents the Zeeman energy of the magnetic moments due to the $4f$-electron $-g_J\mu_{\rm B}\boldsymbol{J}$ and due to the nuclear spin $g_{\rm N}\mu_{\rm N}\boldsymbol{I}$, where $g_J = 4/5$, $\mu_{\rm B}$, $g_{\rm N} = +1.72$, and $\mu_{\rm N}$ are the Landé g-factor, Bohr magneton, the nuclear g-factor, and the nuclear magneton, respectively. Note that there are no fitting parameters in eq. (1).

The term $A\boldsymbol{I}\cdot\boldsymbol{J}$ mixes the $4f$-electron and the nuclear-spin states and can form composite multiplets. The formation of such composite multiplets has been revealed in $PrRu_4P_{12}$ by the observation of a distinct Schottky-type peak structure at $T_{\rm p} = 0.30$ K in specific heat and a remarkable rounding below 1 T in magnetization curve [1]. Note that most Pr-based compounds having a triplet ground state order magnetically at low temperatures, preventing formation of such composite multiplets. In this sense, the formation of *the lattice* of the composite multiplets without showing any magnetic or quadrupolar orderings in $PrRu_4P_{12}$ at least down to 20 mK [18] is an extremely rare case. To the best of our knowledge, this is the first thermodynamical observation of its kind.

In this paper, we analyze the low-temperature specific heat data and discuss the thermodynamic properties of the composite multiplets. We also confirm that the hyperfine coupling constant A for Pr ion is independent of the type of material.

2. Experimental Details

$PrRu_4P_{12}$ single crystals have been grown by the Sn-flux method using high-purity raw materials of 4N(99.99% pure)-Pr, 4N-Ru, 6N-P and 5N-Sn. The single crystalline nature has been checked by X-ray back-reflection Laue technique. No impurity phases were detected in an X-ray powder diffraction experiment for crystals from the same batch. Observed de Haas-van Alphen oscillations in $LaRu_4P_{12}$ single crystals grown by the same manner [19] attest to high quality of the present samples. Specific heat was measured by a quasi-adiabatic heat pulse method using a ^{3}He-^{4}He dilution refrigerator equipped with an 8 T superconducting magnet.

3. Results and discussion

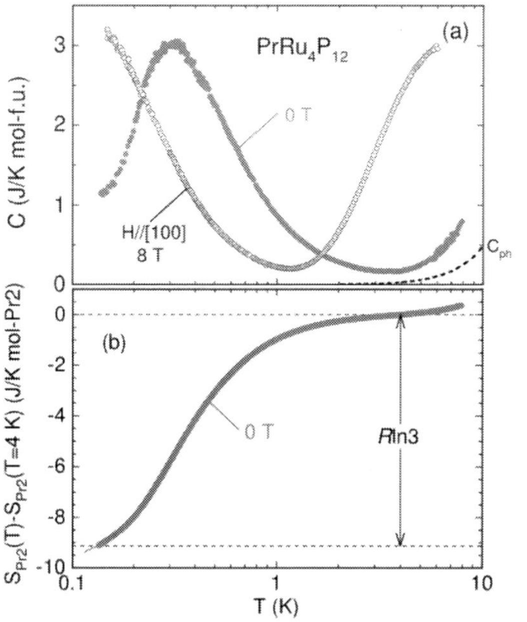

Figure 2. (a) Specific heat $C(T)$ of $PrRu_4P_{12}$ in $\mu_0 H = 0$ and 8 T for $H \parallel [100]$ (after Ref. [1]).(b) Entropy for the Pr2 ion in zero field.

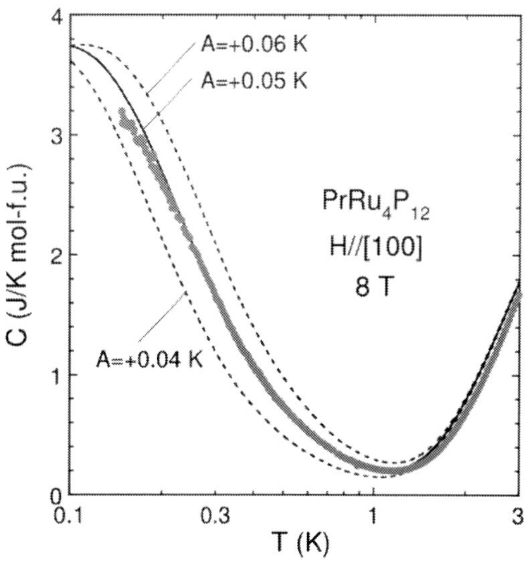

Figure 3. Specific heat $C(T)$ of $PrRu_4P_{12}$ measured in $\mu_0 H = 8$ T for $H \parallel [100]$ (after Ref. [1]) compared with model calculation $C_{Pr2}/2 + C_{Pr1}/2$ for different values of the hyperfine coupling constant A.

Specific heat C shows a Schottky-type peak structure at $T_p = 0.30$ K in zero field as shown in Fig. 2(a) and it shifts to higher temperatures with increasing magnetic field (see Ref. [1] for details). This structure can be accounted for as the thermal excitations in the $4f$-electron-nuclear hyperfine-coupled composite multiplets formed in the Pr2 ions. In this temperature range, the observed C includes contributions from the excitations in the Pr1 ions (C_{Pr1}), the Pr2 ions (C_{Pr2}), phonons (C_{ph}), conduction electrons ($C_{c.e.}$) and the Ru and P nuclei (C_n). Since the CEF ground state of the Pr1 ions is Γ_1 singlet and the 1st excited state $\Gamma_4^{(2)}$ (triplet) is well separated with the excitation energy of 94 K [7], C_{Pr1} has extremely small contribution in the present T range in zero field. Considering the extremely low carrier density in this T range ($T \ll T_{co}$) [20], the conduction-electron contribution $C_{c.e.} = \gamma T$, where γ must be two or three orders of magnitude smaller than 44.4 mJ/K^{2}mol-f.u. of $LaRu_4P_{12}$, should also be negligibly small. For the other contributions, we use $C_{ph} = \beta T^3$ with $\beta = 0.50$ mJ/K^{4}mol-f.u. determined for $LaRu_4P_{12}$ and $C_n = A_N/T^2$ with $A_N = 3.4 \times 10^{-6}$ JK/mol-f.u. estimated by the ^{101}Ru-NQR resonance frequency $\nu_Q = 12.735$ MHz at $T = 4$ K [21]. Subtracting these contributions from the $C(T)$ data, $C_{Pr2}(T)$ can accurately be obtained. The entropy $S_{Pr2}(T)$ calculated using $C_{Pr2}(T)$ is shown in Fig. 2(b). It is clear that the entropy release from the Pr2 ions below 4 K (where C_{Pr2}/T vs T has a minimum, i.e., $S_{Pr2}(T)$ has an inflection point) exceeds $R\ln 3$, which is the maximum value of the entropy release that can be caused by the $\Gamma_4^{(2)}$ triplet CEF ground state of the Pr2 $4f$ electrons. This fact provides unambiguous evidence for the involvement of the nuclear degrees of freedom in the formation of the Schottky-type peak structure along with the

Pr2 $4f$ electrons. In this temperature range, there has been observed no magnetic ordering and no lowering of the Pr site-symmetry from cubic, either of which is necessary for the splitting of the $\Gamma_4^{(2)}$ triplet state and ^{141}Pr nuclear spin state. All of these facts are consistent with the formation of $4f$-electron-nuclear hyperfine-coupled composite multiplets.

The value of A can be accurately determined using high-field $C(T)$ data. In Fig. 3, $C(T)$ data measured in $\mu_0 H = 8$ T for $H \parallel [100]$ are compared with calculated $C_{\text{Pr2}}/2 + C_{\text{Pr1}}/2$ (note that one formula unit of $PrRu_4P_{12}$ contains 0.5×Pr2-ion and 0.5×Pr1-ion) using different values of A. The model calculation is easily performed by diagonalizing the Hamiltonian of eq. (1). Below 1 K, magnetization $M = g_J\langle J\rangle$ is saturated for both Pr1 and Pr2 ions. Therefore, the low-T increase in $C(T)$ is dominated by the Zeeman splitting of the nuclei, which feel the effective hyperfine-enhanced magnetic field of $\mu_0 H - A\langle J\rangle/g_N\mu_N$ for each Pr ion. From the fitting to the $C(T)$ data shown in Fig. 3, $A = +0.050(2)$ K is obtained for Pr ions in extremely-low-carrier $PrRu_4P_{12}$. This value is remarkably close to those in metallic $PrFe_4P_{12}$ [15] and $PrOs_4Sb_{12}$ [16, 17]. This fact corroborates that the intra-ion hyperfine coupling of Pr ions is not affected by the background electronic structure in materials.

Acknowledgments

We thank H. Harima, H. Ishii, K. Iwasa, T. Mito, T. Sakakibara, C. Sekine, H. Shiba, R. Shiina and T. Tayama for valuable discussions. This work was supported by the Grant-in-Aid for Scientific Research on Priority Area "Skutterudite" (15072206) and "Ubiquitous" (20045015), and on Innovative Areas "Heavy Electrons" (20102007) of MEXT and (C: 20540359) and (B: 20340094) of JSPS, Japan.

References

[1] Aoki Y, Namiki T, Saha S R, Tayama T, Sakakibara T, Shiina R, Shiba H, Sugawara H and Sato H 2011 *J. Phys. Soc. Jpn.* **80** 054704
[2] Aoki Y, Sugawara H, Harima H and Sato H 2005 *J. Phys. Soc. Jpn.* **74** 209
[3] Proc. Int. Conf. New Quantum Phenomena in Skutterudite and Related Systems 2008 *J. Phys. Soc. Jpn.* **78** Suppl. A
[4] Sato H, Sugawara H, Aoki H and Harima H 2009 *Handbook of Magnetic Materials* vol 18, ed K H J Buschow (Amsterdam: North-Holland Elsevier) chapter 1 pp 1-110
[5] Sekine C, Uchiumi T, Shirotani I and Yagi T 1997 *Phys. Rev. Lett.* **79** 3218
[6] Lee C H, Matsuhata H, Yamaguchi H, Sekine C, Kihou K, Suzuki T, Noro T and Shirotani I 2004 *Phys. Rev. B* **70** 153105
[7] Iwasa K, Hao L, Kuwahara K, Kohgi M, Saha S R, Sugawara H, Aoki H, Sato H, Tayama T and Sakakibara T 2005 *Phys. Rev. B* **72** 024414
[8] Iwasa K, Hao L, Hasegawa T, Takagi T, Horiuchi K, Mori Y, Murakami Y, Kuwahara K, Kohgi M, Sugawara H, Saha S R, Aoki Y and Sato H 2005 *J. Phys. Soc. Jpn.* **74** 1930
[9] Murao T 1971 *J. Phys. Soc. Jpn.* **31** 683
[10] Ishii H 2004 *J. Low Temp. Phys.* **135** 579
[11] Bleaney B 1973 *Physica* **69** 317
[12] Takegahara K, Harima H and Yanase A 2001 *J. Phys. Soc. Jpn.* **70** 1190
[13] Kondo J 1961 *J. Phys. Soc. Jpn.* **16** 1690
[14] Bleaney B 1963 *J. Appl. Phys.* **34** 1024
[15] Aoki Y, Namiki T, Matsuda T D, Abe K, Sugawara H and Sato H 2002 *Phys. Rev. B* **65** 064446
[16] Aoki Y, Namiki T, Ohsaki S, Saha S R, Sugawara H and Sato H 2002 *J. Phys. Soc. Jpn.* **71** 2098
[17] Sakakibara T, Yano K, Sato H, Tayama T, Custers J, Sugawara H, Aoki Y and Sato H 2008 *J. Phys. Soc. Jpn.* **77** Suppl. A 180
[18] Saha S R, Higemoto W, Koda A, Ohishi K, Kadono R, Aoki Y, Sugawara H and Sato H 2005 *Physica B* **359-361** 850
[19] Saha S R, Sugawara H, Aoki Y, Sato H, Inada Y, Shishido H, Settai R, Ōnuki Y and Harima H 2005 *Phys. Rev. B* **71** 132502
[20] Saha S R, Sugawara H, Namiki T, Aoki Y and Sato H 2009 *Phys. Rev. B* **80** 014433
[21] Mito T: private communication

International Conference on Strongly Correlated Electron Systems (SCES 2011) IOP Publishing
Journal of Physics: Conference Series **391** (2012) 012004 doi:10.1088/1742-6596/391/1/012004

Antiferromagnetic order in vortex states in paramagnetic and d-wave superconductors

Kazushi Aoyama[1,2] **and Ryusuke Ikeda**[2]

[1] Young Researcher Development Center, Kyoto University, Kyoto 606-8302, Japan
[2] Department of Physics, Kyoto University, Kyoto 606-8502, Japan

E-mail: bonn@scphys.kyoto-u.ac.jp

Abstract. We theoretically investigate effects of a vortex lattice structure on antiferromagnetic (AFM) ordering induced inside a *d*-wave superconducting (SC) phase due to a strong Pauli-paramagnetic pair-breaking (PPB) effect. It is shown that a PPB-induced AFM *order* is spatially modulated due to the vortex lattice structure and that, in contrast to the familiar competitive nature of AFM and SC orders, the modulated AFM order is not localized in vortex cores but coexistent with the SC order. Discussion on effects of PPB-induced AFM *fluctuation* on the spin magnetization is also given.

1. Introduction

The heavy-fermion material $CeCoIn_5$ is a spin singlet and *d*-wave superconductor with a strong Pauli-paramagnetic pair-breaking (PPB) effect. In the case with a magnetic field parallel to the basal plane of this material ($H \parallel ab$), there exists a distinct high field and low temperature (HFLT) superconducting (SC) phase in which a Fulde-Ferrell-Larkin-Ovchinnikov (FFLO) vortex lattice is considered to be realized as a result of the strong PPB effect [1]. Recently, it has been clarified through neutron scattering experiments that, in $H \parallel ab$, an antiferromagnetic (AFM) order with its moment vector oriented in the *c*-direction exists inside the HFLT SC phase in spite of the absence of AFM order in the normal state [2]. Considering that, conventionally, AFM and SC orders are competing with each other in zero field, it is surprising that AFM ordering is rather enhanced inside the SC state. In relation to this, it is natural to expect the AFM critical phenomena observed around $H_{c2}(0)$ in a perpendicular field ($H \parallel c$) to originate from the same mechanism as that of the AFM order in $H \parallel ab$ case.

Previously, we have theoretically shown that the PPB effect enhanced strongly by increasing field and decreasing temperature induces an AFM ordering in a *d*-wave SC state [3]. Then, the problem is how an AFM order resulting from this PPB effect is stabilized in the FFLO vortex lattice, i.e., whether it is localized in the normal-state region such as vortex cores and FFLO nodal planes or not. In this report, we examine the stability of the PPB-induced AFM order in the SC vortex lattice, for brevity, without taking into account the FFLO spatial modulation. In addition, as an example of physical quantities reflecting the PPB-induced AFM fluctuation, we discuss the field dependence of the spin magnetization with the AFM fluctuation included.

2. Model

We start from the following electronic Hamiltonian

$$\begin{aligned}\mathcal{H} &= \sum_{\sigma}\int_{\mathbf{r}} \varphi^{\dagger}_{\sigma}(\mathbf{r})\Big[\varepsilon(-i\nabla+|e|\mathbf{A})-\sigma\mu_{\rm B}\,gH\Big]\varphi_{\sigma}(\mathbf{r}),\\ &-\sum_{\mathbf{q},\mathbf{p},\alpha,\beta}\Big[\Delta^{*}(\mathbf{q})\,(-i\,\sigma_y)_{\alpha,\beta}\,w_{\mathbf{p}}\,\hat{c}_{-\mathbf{p}+\frac{\mathbf{q}}{2},\alpha}\,\hat{c}_{\mathbf{p}+\frac{\mathbf{q}}{2},\beta}/2+\mathbf{m}^{*}(\mathbf{q})\cdot\boldsymbol{\sigma}_{\alpha,\beta}\,\hat{c}^{\dagger}_{\mathbf{p}-\mathbf{q},\alpha}\,\hat{c}_{\mathbf{p}+\mathbf{Q}_0,\beta}+\text{H.c.}\Big], \quad (1)\end{aligned}$$

where $\hat{c}_{\mathbf{p},\alpha}$ is the annihilation operator for a quasiparticle with momentum $\mathbf{p}$ and spin projection α, $\varphi_{\sigma}(\mathbf{r}) = \frac{1}{\sqrt{V}}\sum_{\mathbf{p}}\hat{c}_{\mathbf{p},\sigma}\,e^{i\mathbf{p}\cdot\mathbf{r}}$, and $\sigma_i\,(i = x, y, z)$ are the Pauli matrices. In the first line in eq. (1), $\varepsilon(\mathbf{p})$ is a kinetic energy measured from the Fermi level and satisfies the relation $\varepsilon(\mathbf{p}+\mathbf{Q}_0) = -\varepsilon(\mathbf{p}) + T_c\delta_{\rm IC}$ with the commensurate nesting vector $\mathbf{Q}_0 = (\pi, \pi, \pi)$ and the $\mathbf{p}$-independent parameter measuring the deviation from the perfect nesting condition $\delta_{\rm IC}$, and $\mu_{\rm B}\,gH$ is the Zeeman energy. On the second line in eq. (1), the SC gap function $\Delta(\mathbf{q})$ and the AFM moment vector $\mathbf{m}(\mathbf{q})$ play roles of the SC and AFM order parameters, respectively. Throughout this report, the SC transition temperature in zero field is denoted by T_c and the Neel temperature in the nonsuperconducting state with $\delta_{\rm IC} = 0$ is denoted by $T_{\rm N}$. Since the pairing symmetry $w_{\mathbf{p}}$ is of $d_{x^2-y^2}$-type in $CeCoIn_5$ of our interest, the identity $w_{\mathbf{p}+\mathbf{Q}_0} = -w_{\mathbf{p}}$ is satisfied.

The free energy functional of this system can be formally written as

$$\mathcal{F}(\Delta, m) = \mathcal{F}(\Delta, m = 0) + \sum_{\mathbf{q},\mathbf{q}'} X^{-1}(\mathbf{q},\mathbf{q}';\Delta)m(\mathbf{q})m^{*}(\mathbf{q}') + O(|m|^4), \quad (2)$$

where the concrete expressions of $\mathcal{F}(\Delta, m = 0)$ and $X^{-1}(\mathbf{q},\mathbf{q}';\Delta)$ are derived from the microscopic Hamiltonian (1). First, we determine the $H_{c2}(T)$ curve and the amplitude of the spatially averaged SC gap function Δ_0 from the functional $\mathcal{F}(\Delta, m = 0)$. Next, we examine an AFM instability inside the SC phase by evaluating the second order term in $|m|$, $\sum_{\mathbf{q},\mathbf{q}'} X^{-1}(\mathbf{q},\mathbf{q}';\Delta)m(\mathbf{q})m^{*}(\mathbf{q}')$. This procedure in which Δ and m are separately considered is justified at least near the second order AFM transition. Hereafter, we will restrict our discussion to the case with the configuration $\mathbf{m}\perp\mathbf{H}$ which corresponds to the experimentally observed situation in $\mathbf{H}\parallel ab$. In the expressions which follow, t and h are the temperature and magnetic field which are normalized by T_c and $H_{c2}(0)$, respectively.

3. AFM order in the SC vortex lattice

In the vortex lattice in the parallel field $\mathbf{H}\parallel ab$, the SC gap function can be expressed as $|\Delta(X,Y)|^2 = |\Delta_0|^2\sum_{\mathbf{K}_\perp}T^{(2)}_{\mathbf{K}_\perp}\exp\left[i\,\mathbf{K}_\perp\cdot(X,Y)\right]$ with a reciprocal lattice vector of the SC vortex lattice $\mathbf{K}_\perp$, where the magnetic field is applied in the y-direction and the coordinate (X,Y) denotes $(\gamma^{1/2}z, \gamma^{-1/2}x)$ with the ratio of the SC coherence length in the ab plane to that in the c-direction γ. Then, an AFM order induced in the SC vortex lattice would be expressed with the vector $\mathbf{K}_\perp$ as

$$m(X,Y) = N\left(1 + \alpha\sum_{\mathbf{K}_\perp\neq 0}T^{(2)}_{\mathbf{K}_\perp}\exp\left[i\mathbf{K}_\perp\cdot(X,Y)\right]\right), \quad (3)$$

where N is a normalization factor which is determined so that the spatial average of $|m(X,Y)|^2$ is equal to 1 and α is a variational parameter which will be determined by minimizing the second order term in $|m|$. Note that positive (negative) values of α indicate AFM orders suppressed (enhanced) at vortex cores. In the evaluation of $\sum_{\mathbf{q},\mathbf{q}'} X^{-1}(\mathbf{q},\mathbf{q}';\Delta)m(\mathbf{q})m^{*}(\mathbf{q}')$, the expansion with respect to $|\Delta|$ is performed and the parameters $\gamma = 2.85$ and $\alpha_{\rm M,ab} = \sqrt{2}H^{(ab)}_{\rm orb}(0)/(1.25T_c/\mu_{\rm B}\,g) = 7.5$ are used, where $H^{(ab)}_{\rm orb}(0)$ is the orbital limiting field in $\mathbf{H}\parallel ab$

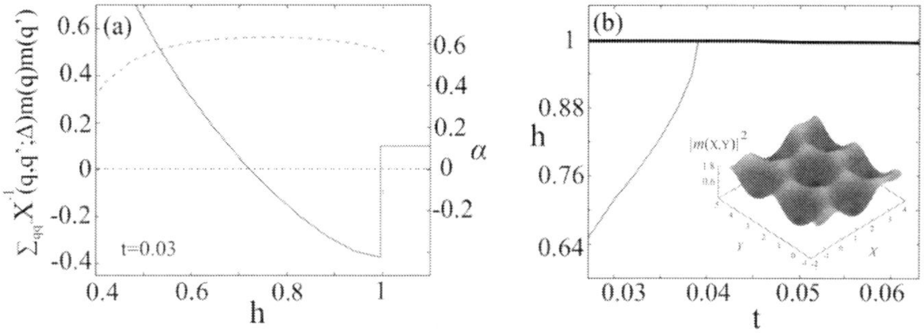

Figure 1. (Color online) (a) Field dependences of the $|m|^2$-term $\sum_{\mathbf{q},\mathbf{q}'} X^{-1}(\mathbf{q},\mathbf{q}';\Delta)m(\mathbf{q})m^*(\mathbf{q}')$ (a solid red curve measured by the left vertical axis) and the variational parameter α (a dashed blue one measured by the right axis) at $t = 0.03$ and (b) the obtained phase diagram, where the triangular lattice is assumed and the parameters $\delta_{\rm IC} = 0.001$ and $T_{\rm N}/T_c = 0.027$ are used. In (a), the $|m|^2$-term becomes negative in the high field region with positive values of α, suggesting the occurrence of an AFM order with a spatial modulation commensurate with the vortex lattice structure. An example of the spatial distribution of this AFM order $|m(X,Y)|^2$ is shown in the inset of (b), where dark thick colored regions correspond to vortex cores. As one can see in the main panel of (b), the AFM order is stabilized in the HFLT corner of the SC phase, where a solid red curve and a thick solid one denote the second order transition curve to the AFM order and the first order H_{c2} transition one, respectively.

case. Details of the calculation can be seen elsewhere [4]. Figure 1 (a) shows the field dependences of the $|m|^2$-term (solid red curve) and α (dashed bule one) at $t = 0.03$. In the low field region where the value of the $|m|^2$-term is positive and larger than that in the normal state, an AFM order never appears in the SC phase due to the competitive nature of these two orders. However, as the magnetic field is increased, the $|m|^2$-term becomes negative and decreases toward the H_{c2} transition, which suggests the occurrence of an AFM order inside the SC phase. The jump seen at the H_{c2} transition of the sign of the $|m|^2$-term is the reflection of that of the amplitude $|\Delta|$ of the SC order parameter at the first order H_{c2} transition. The obtained phase diagram is shown in Fig. 1 (b), where a solid red curve denotes the second order transition to the state with an AFM order. As the positive values of α in Fig. 1 (a) indicate, the AFM order induced in the high field and low temperature corner of the SC phase is not localized in vortex cores but coexistent with the SC order. An example of the spatial distribution of the AFM order is shown in the inset of Fig. 1 (b).

Our result obtained here for the PPB-induced AFM order in the SC *vortex lattice* is consistent with recent NMR data suggesting that the AFM order in the HFLT phase of $CeCoIn_5$ is spatially extended without being localized in the normal-state region [7]. Although the FFLO modulation is ignored in the above calculation, the theoretical study taking account of the longitudinal FFLO modulation along **H** without in-plane vortex-lattice structures included shows that, at least in the high-field side of the FFLO state, the AFM order is not localized in the FFLO nodal planes but coexistent with the SC order [6]. These results obtained in the two approaches, which are complementary to each other, suggest that the PPB-induced AFM order coexistent with the SC order may be stabilized at least in the high-field side of the longitudinal FFLO vortex lattice state.

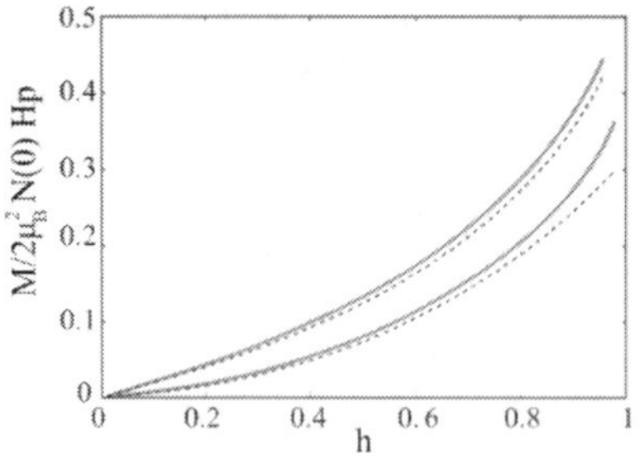

Figure 2. (Color online) Field dependence of the spin magnetization at $t = 0.1$ (lower curves) and $t = 0.3$ (upper ones) in the case with (red solid curves) and without (black dashed ones) the AFM fluctuation. The parameters $T_N/T_{c0} = 0.25975$ and $\delta_{IC} = 0.5$ are used so that the AFM critical point is located at $H_{c2}(0)$. The slope of the magnetization curve becomes steeper as the AFM instability is approached.

4. Effects of AFM fluctuation on the spin magnetization

In this section, effects of AFM fluctuation induced by the PPB effect are investigated. We have pointed out that the AFM fluctuation is strongly reflected in the vortex lattice form factor which is a measure of the internal magnetic field [3]. Here, we examine effects of the AFM fluctuation on the magnetization inside the SC phase. Since the magnetization due to the orbital effect is expected to be small compared with the spin magnetization in the case with strong PPB [5], we will ignore the orbital pair-breaking effect, i.e., the vortex lattice for brevity. The spin magnetization can be expressed as

$$M = 2(\mu_B g)^2 N(0)H - \frac{\partial}{\partial H}\mathcal{F}(\Delta, m = 0) - \frac{T}{2}\frac{\partial}{\partial H}\sum_{\mathbf{q}} \ln X^{-1}(\mathbf{q}, \mathbf{q}; \Delta), \tag{4}$$

where $N(0)$ is the density of state per spin on the Fermi surface and the third term in the right-hand side corresponds to the contribution from the PPB-induced AFM fluctuation. In the Pauli limiting case without the orbital effect included, the SC gap function will be treated as uniform and the evaluation $X^{-1}(\mathbf{q}, \mathbf{q}; \Delta) \simeq X^{-1}(0, 0; \Delta)$ will be used.

The field dependence of the spin magnetization is shown in Fig. 2, where the AFM instability is located at $H_{c2}(0)$. The contribution from the AFM fluctuation is positive and becomes larger as temperature is lowered and field is increased, i.e., as the AFM instability is approached. As a result, the slope of the spin magnetization curve becomes steeper on approaching $H_{c2}(0)$. The obtained result is comparable with the experimental data on the magnetization [8], which supports the scenario that the field-induced AFM critical phenomenon observed below $H_{c2}(0)$ and in $\mathbf{H} \parallel c$ case is a consequence of the PPB-induced AFM instability. Although we have assumed the configuration $\mathbf{m} \perp \mathbf{H}$ in this example calculation, we believe that the result on the PPB-induced AFM fluctuation is not qualitatively changed even for the configuration $\mathbf{m} \parallel \mathbf{H}$.

References

[1] Ikeda R 2007 *Phys. Rev.* B **76**, 134504.
[2] Kenzelmann M et al. 2008 *Science* **321**, 1652; Kenzelmann M et al. 2010 *Phys. Rev. Lett.* **104**, 127001.
[3] Ikeda R, Hatakeyama Y, and Aoyama K 2010 *Phys. Rev.* B **82**, 060510(R).
[4] Aoyama K and Ikeda R 2011 arXiv:1107.5577.
[5] Adachi H, Ichioka M, and Machida K 2005 *J. Phys. Soc. Jpn.* **74**, 2181.
[6] Hatakeyama Y and Ikeda R 2011 *Phys. Rev.* B **83**, 224518.
[7] Kumagai K, Shishido H, Shibauchi T, Matsuda Y 2011 *Phys. Rev. Lett.* **106**, 137004.
[8] Paulsen C, Aoki D, Knebel G, and Flouquet J 2011 *J. Phys. Soc. Jpn.* **80**, 053701.

Hall effect in single crystal $CeCu_2Si_2$ under high pressure

S Araki[1], Y Shiroyama[1], T Shinohara[1], Y Ito[1], Y Ikeda[1], T C Kobayashi[1], S Seiro[2], C Geibel[2] and F Steglich[2]

[1]Graduate School of Natural Science and Technology, Okayama University, Okayama, 700-8530, Japan

[2]Max Planck Institute for Chemical Physics of Solids, Nothnitzer Straße 40, 01187 Dresden, Germany

E-mail: araki@science.okayama-u.ac.jp

Abstract. Hall effect measurements were carried out under high pressures up to 4.9 GPa in a single crystal of $CeCu_2Si_2$. The temperature dependence of the Hall coefficient is interpreted to be composed of two peaks below room temperature. The high-temperature peak shows strong pressure dependence, where the peak shifts from 20 K at ambient pressure to 125 K at 4.9 GPa. This peak is a consequence of the anomalous Hall effect due to skew scattering. The low-temperature peak shifts from 5 K at ambient pressure to 15 K at 4.9 GPa and the magnitude of the peak shows the maximum around 4.1 GPa at which the superconducting transition temperature also reaches the maximum.

1. Introduction

The heavy-fermion compound $CeCu_2Si_2$ is a superconductor with $T_c \simeq 0.6$ K at ambient pressure [1]. This unconventional superconductivity is considered to be mediated by antiferromagnetic (AFM) spin fluctuations around an AFM quantum critical point such as in $CePd_2Si_2$ and $CeIn_3$ [2]. With increasing pressure, T_c is enhanced above 3 GPa and becomes maximum around $P_c \sim 4.5$ GPa, although the AFM spin fluctuations are drastically suppressed under pressure [3]. The enhanced superconductivity soon disappears with further application of pressure above P_c. In the chemical substitution system $CeCu_2(Si_{1-x}Ge_x)_2$, the superconducting phase in the P-T phase diagram splits into two domes, which suggests a different origin for superconductivity at ambient pressure and high pressures around P_c [4].

Recently it was suggested theoretically that the possible origin of the enhancement of T_c around P_c is critical valence fluctuations [5, 6]. The T-linear dependence of resistivity and the enhancement of the residual resistivity around P_c were explained by the valence fluctuation scenario [5]. The NQR study at 4.2 K revealed the downward deviation of $^{63}\nu_Q$ of Cu-nuclei above 3.5 GPa from the linear pressure dependence which is predicted from the volume compression [3]. This result may reflect the change of valence. The pressure dependence of the Ce valence at 14 K was recently measured by XAS [7]. The valence of Ce smoothly decreases with increasing pressure and no sharp valence transition at P_c was observed. In summary, although a number of low-temperature properties suggest the presence of a valence transition, no direct evidence of such a transition has been obtained so far.

The Hall coefficient reflects the transport properties of carriers and is expected to be sensitive to the valence instability. We have measured the Hall coefficient in polycrystalline $CeCu_2Si_2$ under pressure [8]. The temperature dependence of the Hall coefficient has a peak at around 10 K. The magnitude of the Hall coefficient at the peak temperature shows the decreasing trend with increasing pressure and is enhanced from 3.1 to 4.2 GPa at which T_c is enhanced. In this study, the Hall effect was measured under high pressure in single crystal of $CeCu_2Si_2$.

2. Experimental

The single crystal of S-type $CeCu_2Si_2$ was grown by a self-flux method combined with a Bridgman cooling technique [9]. The pressure up to 4.9 GPa was applied in an indenter cell [10] with Daphne oil 7474 [11] as a pressure transmitting medium. The pressure was determined from the superconducting transition temperature of Pb [12]. The Hall voltage V_H was measured down to 1.4 K with AC resistance bridge in the field-sweep measurement. External magnetic field was applied along the [100] direction, which is the hard axis of magnetization. The Hall resistivity is given by $\rho_H(H) = \{V_H(H) - V_H(-H)\}a/2I$, where a and I are the thickness of the sample and the electric current, respectively. The Hall coefficient R_H was obtained from the slope of $\rho_H(H)$. The electrical resistivity was measured up to 4.5 GPa for the sample cut from the same grain to check the T_c using a dilution refrigerator. The pressure dependence of T_c in the present sample reproduces the previous result [5].

3. Results and Discussion

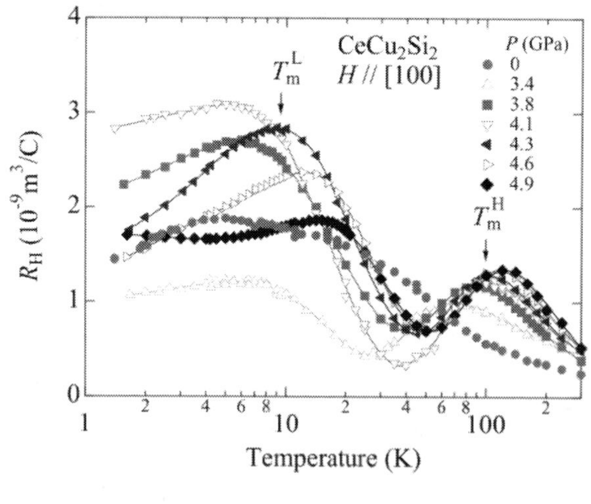

Figure 1. Temperature dependence of the Hall coefficient R_H in $CeCu_2Si_2$ at various pressures.

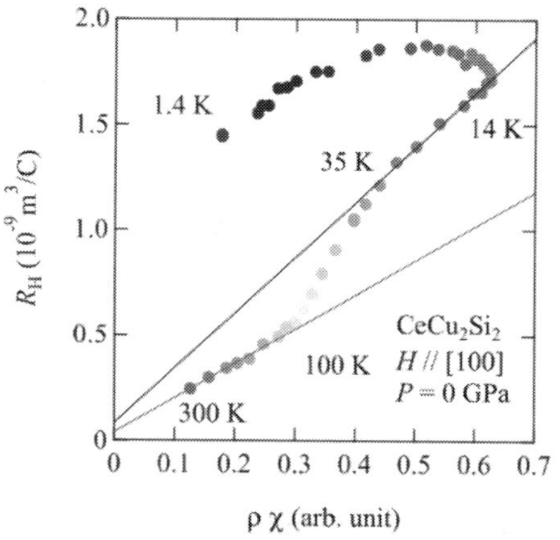

Figure 2. Hall coefficient R_H as a function of $\rho\chi$ in $CeCu_2Si_2$ at ambient pressure.

Figure 1 shows the temperature dependence of the Hall coefficient R_H in $CeCu_2Si_2$ at various pressures. The Hall resistivity increases linearly with increasing field up to 5 T in the present experimental range of temperature and pressure. The R_H at ambient pressure increases with decreasing temperature and reaches the maximum at ~ 5 K. This behavior is almost the same as our previous result in the polycrystalline sample [8]. In heavy fermion systems, the anomalous Hall effect R_H^a due to skew scattering causes this kind of strong temperature dependence. The R_H^a is theoretically proportional to the magnetic susceptibility χ [13] or the product of χ and the electrical resistivity ρ [14] at temperatures higher than the coherence temperature of the Kondo effect. The R_H^a decreases with decreasing temperature below the coherence temperature.

The peak temperature of $R_{\rm H}$ roughly corresponds to the coherence temperature. The $R_{\rm H}$ in $CeCu_2Si_2$ at ambient pressure is plotted against $\rho\chi$ in Fig. 2. The linear relation between $R_{\rm H}$ and $\rho\chi$ holds above 100 K.

The $R_{\rm H}$ below 100 K deviates from the linear dependence at higher temperatures. The $R_{\rm H}$ between 35 and 14 K is proportional to $\rho\chi$ with different slope above 100 K. Note that 14 and 100 K correspond to the characteristic temperatures $T_1^{\rm max}$ and $T_2^{\rm max}$ determined from the peaks in the resistivity, respectively. The $R_{\rm H}$ deviates from the linear relationship below 14 K. The $\rho\chi$ in this temperature region is expected to decrease with decreasing temperature because $T_1^{\rm max}$ corresponds to the coherence temperature. Therefore, the maximum of $R_{\rm H}$ at 5 K cannot be assigned to the contribution of $R_{\rm H}^{\rm a}$ due to skew scattering.

The $R_{\rm H}$ above 3.4 GPa clearly has two peaks, which are observed, for example, at $T_{\rm m}^{\rm L} = 9$ K and $T_{\rm m}^{\rm H} = 105$ K at 4.3 GPa, as shown in Fig. 1. The low-temperature and high-temperature peaks are denoted as $T_{\rm m}^{\rm L}$ and $T_{\rm m}^{\rm H}$, respectively. Note that the high-temperature peak at $T_{\rm m}^{\rm H}$ was not clearly seen up to 4.8 GPa in our previous report for polycrystalline sample [8]. This difference may be caused by the anisotropic Hall coefficient. The pressure dependences of $T_{\rm m}^{\rm L}$ and $T_{\rm m}^{\rm H}$ are summarized in Fig. 3 (a). The $T_{\rm m}^{\rm L}$ and $T_{\rm m}^{\rm H}$ below 2.6 GPa, at which two peaks are not clearly separated, are determined from the peak and the inflection point of $R_{\rm H}$, respectively. The pressure variation of $T_{\rm m}^{\rm H}$ almost follows that of $T_1^{\rm max}$. Therefore, the high-temperature peak of $R_{\rm H}$ at $T_{\rm m}^{\rm H}$ is likely due to the anomalous Hall effect.

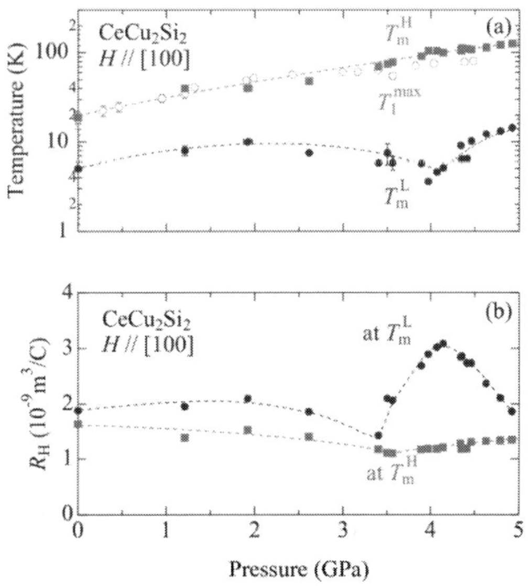

Figure 3. Pressure dependence of (a) $T_{\rm m}^{\rm H}$, $T_{\rm m}^{\rm L}$ in $R_{\rm H}$ and $T_1^{\rm max}$ in ρ (see text), and (b) the magnitude of $R_{\rm H}$ at $T_{\rm m}^{\rm H}$ and $T_{\rm m}^{\rm L}$. The dotted curves are guide to eye.

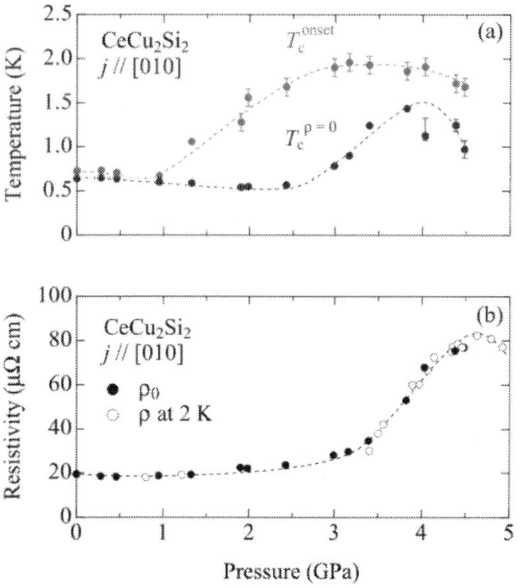

Figure 4. Pressure dependence of (a) T_c and (b) residual resistivity ρ_0 (solid circle) and resistivity at 2 K (open circle). The dotted curves are guide to eye.

The magnitude of $R_{\rm H}$ at $T_{\rm m}^{\rm L}$, shown in Fig. 3 (b), is almost unchanged up to 1.9 GPa, and decreases with increasing pressure up to 3.4 GPa. Then, it increases rapidly and exhibits the maximum at 4.1 GPa. The T_c of our sample reaches its maximum value at around 4.0 GPa, as shown in Fig. 4 (a). Therefore, we conclude that the enhancement of both the $R_{\rm H}$ at $T_{\rm m}^{\rm L}$ and the T_c under pressure occurs simultaneously. The temperature $T_{\rm m}^{\rm L}$ shows a minimum at 4.1 GPa and increases above 4.1 GPa, as shown in Fig. 3 (a). These results imply the existence of fluctuations around 4.1 GPa. The residual resistivity of this sample reaches the maximum at 4.6 GPa, as shown in Fig. 4 (b). This implies the existence of P_c = 4.6 GPa within the valence fluctuations scenario. The valence fluctuations may also influence the R_H. However, the pressure at which R_H is enhanced is slightly lower than the pressure at which ρ_0 reaches the maximum.

The enhancement of the R_H due to AF fluctuations was reported for $CeRhIn_5$ [15]. The R_H of $CeRhIn_5$ is critically enhanced around 2 GPa, at which the AF metallic and superconducting states are separated. The enhancement of the R_H decreases with increasing the magnetic field due to the suppression of AF fluctuations. In the case of $CeCu_2Si_2$, we did not observe a field dependent R_H. Therefore, the possibility of the AF fluctuations for the origin of the enhancement of the R_H at 4.1 GPa is ruled out.

4. Conclusion

At ambient pressure, the temperature dependence of the Hall coefficient R_H along [100] has a linear relation to $\rho\chi$ above 100 K, which is explained by the sum of the ordinary Hall coefficient and the anomalous Hall coefficient due to skew scattering. The maximum in R_H at 5 K is not assigned to the anomalous Hall effect. The R_H under pressure above 3.4 GPa exhibits two peaks. The pressure dependence of the high-temperature peak roughly follows that of T_1^{max} in the resistivity. The origin of high-temperature peak is assigned to the anomalous Hall effect due to skew scattering. The magnitude of R_H at the low-temperature peak increases with increasing pressure above 3.4 GPa and reaches the maximum at 4.1 GPa, at which T_c also reaches its maximum. This result indicates the close relationship between the enhancement of R_H at low temperatures and the increase of T_c.

Acknowledgments

This work was partially supported by a Grant-in-Aid for Scientific Research on Innovative Areas "Heavy Electrons" (No. 20102008) from the Ministry of Education, Culture, Sports, Science and Technology. This work was also supported by Scientific Research B (20340090) from the Japan Society for the Promotion of Science.

References

[1] Steglich F, Aarts J, Bredl C D, Lieke W, Meschede D, Franz W and Schäfer H 1979 *Phys. Rev. Lett.* **43** 1892.
[2] Mathur N D, Grosche F M, Julian S R, Walker I R, Freye D M, Haselwimmer R K W and Lonzarich G G 1998 *Nature* **394** 39.
[3] Fujiwara K, Hata Y, Kobayashi K, Miyoshi K, Takeuchi J, Shimaoka Y, Kotegawa H, Kobayashi T C, Geibel C and Steglich F 2008 *J. Phys. Soc. Jpn.* **77** 123711.
[4] Yuan H Q, Grosche F M, Deppe M, Geibel C, Sparn G and Steglich F 2003 *Science* **302** 2104.
[5] Holmes A T, Jaccard D and Miyake K 2004 *Phys. Rev. B* **69** 024508, and references therein.
[6] Miyake K 2007 *J. Phys.: Condens. Matter* **19** 125201.
[7] Rueff J -P, Raymond S, Taguchi M, Sikora M, Itié J -P, Baudelet F, Braithwaite D, Knebel G and Jaccard D 2011 *Phys. Rev. Lett.* **106** 186405.
[8] Araki S, Shiroyama Y, Ikeda Y, Kobayashi T C, Seiro S, Geibel C and Steglich F 2011 *J. Phys. Soc. Jpn.* **80** SA061.
[9] Seiro S, Deppe M, Jeevan H, Burkhardt U and Geibel C 2010 *Phys. Stat. Solid. B* **247** 614.
[10] Kobayashi T C, Hidaka H, Kotegawa H, Fujiwara K and Eremets M I 2007 *Rev. Sci. Instrum.* **78** 023909.
[11] Murata K *et al* 2008 *Rev. Sci. Instrum.* **79** 085101.
[12] Eiling A and Schilling J S 1981 *J. Phys. F* **11** 623.
[13] Fert A and Levy P M 1987 *Phys. Rev. B* **36** 1907.
[14] Kontani H and Yamada K 1994 *J. Phys. Soc. Jpn.* **63** 2627.
[15] Nakajima Y *et al* 2007 *J. Phys. Soc. Jpn.* **76** 024703.

International Conference on Strongly Correlated Electron Systems (SCES 2011) IOP Publishing
Journal of Physics: Conference Series **391** (2012) 012006 doi:10.1088/1742-6596/391/1/012006

Quantum criticality in $CePt_{1-x}Ni_xSi_2$

R. E. Baumbach, X. Lu, F. Ronning, J. D. Thompson, and E. D. Bauer

MPA-CMMS Los Alamos National Laboratory

E-mail: rbaumbach@lanl.gov

Abstract. We report measurements of the specific heat, electrical resistivity, and magnetic susceptibility for $CePt_{1-x}Ni_xSi_2$ from which we develop a $T-x$ phase diagram that includes a quantum critical point near $x_{cr} \approx 0.125$ and accompanying non-Fermi-liquid behavior in a "v"-shaped region. This phase diagram is strikingly similar to that of $CePtSi_2$ under applied pressure P, suggesting that $CePt_{1-x}Ni_xSi_2$ provides a model system in which a $T-P-x$ phase diagram can be smoothly generated, thereby allowing a systematic study of the influence of disorder on quantum criticality.

1. Introduction

In strongly correlated electron systems, it is often possible to tune antiferromagnetic (AFM) order to zero temperature T by an external parameter: e.g., pressure P or chemical substitution x. For many such systems, the resulting quantum critical point (QCP) is protected by a dome of unconventional superconductivity (SC), above which normal state non-Fermi liquid (NFL) behavior is seen in a "v"-shaped region. As the tuning parameter is increased, a heavy fermion (HF) ground state often emerges [1]. While many QC materials fall into this category, a handful of others exhibit even richer phase diagrams. For instance, two domes of SC are observed for $CeCu_2Si_{2-x}Ge_x$, one of which is associated with an AFM QCP while the other is associated with a valence QCP [2]. It is also well known that P and x, while often similar, do not have identical effects: e.g., most systems which exhibit P induced SC near a QCP do not show SC when tuned by x. As a result, the connection between $T-x$ and $T-P$ phase diagrams is unclear: i.e., the influence of disorder on QC behavior remains an open question.

We were thus motivated to consider the isoelectronic substitution series $CePt_{1-x}Ni_xSi_2$. The compound $CePtSi_2$ is a HF AFM, for which P suppresses the Néel temperature T_N and a QCP is accessed near 1.5 GPa. The QCP is surrounded by a dome of SC that is followed by a "v"-shaped NFL region [3, 4]. In contrast, $CeNiSi_2$ exhibits intermediate valence (IV) behavior [5]. For $CePt_{1-x}Ni_xSi_2$, we find pronounced lattice compression with increasing x and commensurate suppression of T_N towards a putative QCP near $x_{cr} = 0.125$, where NFL behavior is observed in a "v" shaped region. In the vicinity of the QCP the characteristic temperatures T_1 and T_2, which are associated with an interplay between Kondo physics and crystalline electric field effects, respectively, collapse to a single value. This result is similar to what is observed for $CePtSi_2$ when it is tuned toward its QCP by P. We also find that near the QCP, the magnetic susceptibility $\chi(T)$ is anomalously enhanced, suggesting ferromagnetic (FM) fluctuations. Finally, with increasing x above the QCP, a FL region is eventually recovered. This substitution series is of great interest, as it shares a striking resemblance to the $T-P$ phase diagram of $CePtSi_2$ and, therefore, provides a model QC system in which a $T-P-x$ phase diagram can be smoothly constructed.

2. Experimental Details

Polycrystalline specimens of $CePt_{1-x}Ni_xSi_2$ were synthesized by arc melting stoichiometric quantities of Ce (99.99%), Pt (99.95%), Ni (99.99%), and Si (99.9999%) under an Ar atmosphere. The resulting boules were flipped and remelted several times in order to encourage homogeneity and subsequently annealed in evacuated quartz tubes at 900 ^oC for two weeks. X-ray diffraction (XRD) measurements were performed on powdered specimens. The resulting patterns were fitted by means of Rietveld analysis [6] implemented in the program suite GSAS [7]. We investigated the specific heat $C(T)$ for $T = 0.4$ K - 50 K in zero applied magnetic field H and the electrical resistivity $\rho(T)$ for $T = 2$ K - 300 K in $H = 0$ using a Quantum Design Physical Properties Measurement System. Magnetic susceptibility $\chi(T)$ measurements were carried out for $T = 2$ K - 300 K and $H = 1000$ Oe using a Quantum Design Magnetic Properties Measurement System.

3. Results

Rietveld analysis of the XRD data reveals that samples across the entire chemical substitution series crystallize in the $CeNiGe_2$-type orthorhombic structure (space group Cmcm). This structure is nearly tetragonal, consists of stacked Ce and Pt-Si layers along the b-axis, and is closely related to several other prototypical structures which have received intense scrutiny in the context of strongly correlated electron physics and unconventional SC (e.g., $ThCr_2Si_2$-type). The XRD patterns also include several peaks which are attributable to impurity phases, which makes a quantitative analysis of the low T transport and thermodynamic data challenging. Subsequent attempts to reduce or remove the impurity phases by further annealing were unsuccessful. In order to identify the doping region where the Néel temperature might be suppressed to $T = 0$, we estimated the effective "chemical pressure" using the formula $P(x) = B_0/B_0'[(V(x)/V_0)^{-B_0'} - 1]$, where $B_0 = 119$ GPa and $B_0' = 3$ are the $P = 0$ values of the isothermal bulk modulus and its pressure derivative for $CePtSi_2$ [8], V_0 is the unit cell volume for $CePtSi_2$, and $V(x)$ is the unit cell volume for $CePt_{1-x}Ni_xSi_2$ as obtained from XRD measurements (Fig. 1a). From this calculation, we estimate that between $x = 0.1$ - 0.2 the lattice contraction might produce a chemical pressure that is similar to the critical pressure $P_{cr} \approx 1.5$ GPa in $CePtSi_2$ (Fig. 1b).

The $C(T)/T$ data are shown in Fig. 1c. For $x = 0$, C/T increases with decreasing T until a peak is observed near $T_N = 1.9$ K, marking the onset of AFM ordering. We also find a large value for C/T at low T, consistent with the previous conclusion that this system exhibits strong electronic correlations due to hybridization between the f- and conduction electron states. This point of view is further supported by the observation that the Kondo effect removes a significant fraction of the magnetic entropy S_{mag}, giving $S_{mag} \approx 0.4$Rln2 at T_N. As Ni is substituted for Pt, T_N is suppressed to lower T and is no longer observable near the critical concentration x_{cr} = 0.125, where we suggest that a QCP is found (Fig 2f). This result is in agreement with our estimate for x_{cr} from the bulk modulus of $CePtSi_2$, showing that x and P are nearly equivalent tuning parameters. In the vicinity of the QCP, C/T exhibits approximately logarithmic behavior at low T, revealing the onset of NFL behavior. We point out that for several of these curves, broad humps are seen between 1-2 K which may be related to magnetic impurity phases: e.g., $CePt_2$ which exhibits antferromagnetism near $T_N = 1.6$ K. If an impurity $CePt_2$ phase is present, then we estimate that it is no more that 10% of the total sample, by comparing to the jump in C/T at T_N for pure $CePt_2$ [9]. For $x_{cr} < x$, $C/T = \gamma$ recovers FL behavior at low T and γ is gradually reduced with increasing x, revealing that the strong electronic correlations that result in an enhanced effective mass m* for low x are weakened with increasing x. Finally, in order to estimate the single ion Kondo scale T_K, we consider that for a spin 1/2 system, approximately

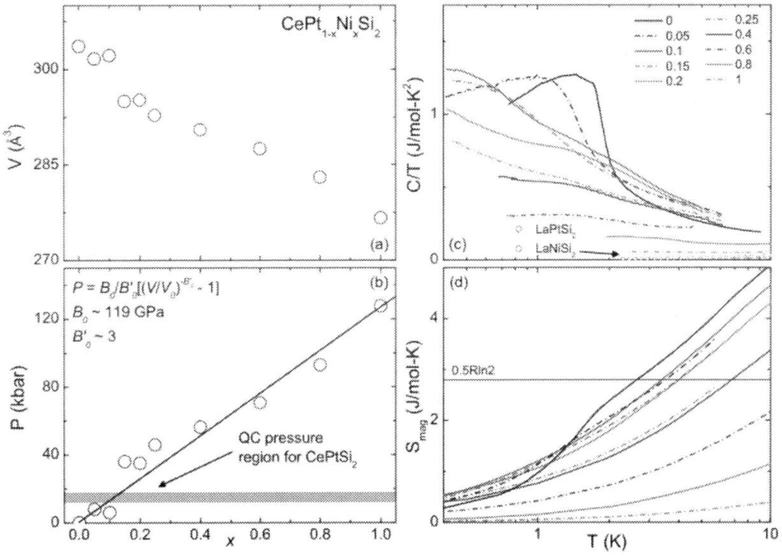

Figure 1. (a) Unit cell volume V versus concentration x from powder x-ray diffraction measurements. (b) Chemical pressure P calculated from the bulk modulus B of pure $CePtSi_2$ (see text) versus x. (c) Specific heat divided by temperature C/T versus T. (d) Magnetic entropy S_{mag} versus T.

1/2Rln2 develops at roughly $0.8T_K$ [10]. From this estimate, we see that T_K smoothly increases with increasing x, as is expected with a gradual crossover from HF to IV behavior (Fig. 2e).

The $\rho(T)$ data are shown in Figs. 2(a,b). The $x = 0$ T-dependence is similar to previous results, where $\rho(T)$ initially increases with decreasing T due to spin fluctuation scattering and subsequently evolves through two maxima near $T_1 \approx 6$ K and $T_2 \approx 35$ K, which are associated with Kondo lattice behavior and crystalline electric field effects [3, 4]. With increasing x, T_1 increases slightly, while the hump associated with T_2 moves to higher T and broadens significantly. Between $x = 0.1$ - 0.2, T_1 and T_2 merge into a single peak which then moves to higher T with increasing x (Fig. 2e). We point out that this behavior is in good agreement with what is observed for $CePtSi_2$ under applied pressure. In order to search for SC in the vicinity of the QCP, we measured $\rho(T)$ for $x = 0.1$, 0.15, 0.2, and 0.25 down to 100 mK. These measurements revealed no evidence for SC.

The $\chi(T)$ data are shown in Fig. 2c where, up to $x = 0.4$, Curie-Weiss (CW) behavior given by the expression $\chi(T) = C/(T-\Theta)$ is observed, resulting in effective magnetic moments $\mu_{eff} \approx 2.5(4)$ μ_B for $x \leq 0.4$, as expected for localized Ce^{3+} ions. The CW temperatures Θ are plotted in Fig. 2d, where we find that they are strongly peaked in the vicinity of the QCP, reflecting an anomalous enhancement of $\chi(T)$ which may be related to the emergence of FM fluctuations. It is interesting to note that for several other systems (e.g., $YbRh_2Si_2$ [11] and β-$YbAlB_4$ [12]) FM may also play a role near the QCP.

Taken together, these measurements reveal a striking similarity between the $T-P$ phase diagram for $CePtSi_2$ and the $T-x$ phase diagram for $CePt_{1-x}Ni_xSi_2$, as summarized in Fig. 2(d,e,f). Thus, the $CePt_{1-x}Ni_xSi_2$ substitution series may provide a platform on which the influence of P and x can be smoothly connected in a $T-P-x$ phase diagram. Such studies are of interest because they will allow for greater understanding of the influence of disorder on QC and associated phenomena (e.g., unconventional SC).

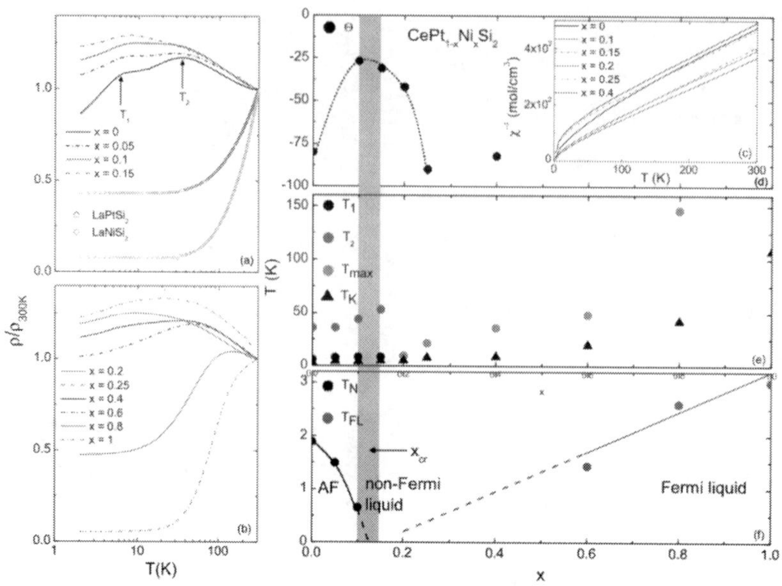

Figure 2. (a,b) Normalized electrical resistivity ρ/ρ_0 versus temperature T for $CePt_{1-x}Ni_xSi_2$ and the nonmagnetic analogue compounds $LaPtSi_2$ and $LaNiSi_2$. (c) Inverse magnetic susceptibility χ^{-1} versus T measured in an applied magnetic field H = 1000 Oe.(d) Curie-Weiss temperature Θ obtained from fits to $\chi(T)$ (see text) versus x. The dotted line is a guide to the eye. (e) Temperatures T_1, T_2, and T_{max} where the electrical resistivity exhibits either a maximum or a hump and Kondo temperature T_K (see text). (f) Phase diagram of $CePt_{1-x}Ni_xSi_2$ where T_N is the Néel temperature obtained from specific heat measurements and T_{FL} is the temperature where C/T becomes constant at low T, consistent with the onset of Fermi liquid behavior. The dashed lines are extrapolations for T_N and T_{FL}. The solid grey bar near $x_{cr} \approx 0.125$ denotes the putative quantum critical region.

4. Acknowledgements
Work at Los Alamos National Laboratory was performed under the auspices of the U.S. Department of Energy, Office of Basic Energy Sciences, Division of Materials Sciences and Engineering and funded in part by the Los Alamos Laboratory Directed Research and Development program.

References
[1] P. Gegenwart, Q. Si, and F. Steglich 2008 *Nature Physics* **4** 186
[2] H. Q. Yuan, F. M. Grosche, M. Deppe, G. Sparn, C. Geibel, and F. Steglich 2006 *Phys. Rev. Lett.* **96** 047008
[3] W. H. Lee, K. S. Kwan, P. Klavins, and R. N. Shelton 1990 *Phys. Rev. B* **42** 6542
[4] T. Nakano, M. Ohashi, G. Oomi, K. Matsubayashi, and Y. Uwatoko 2009 *Phys. Rev. B* **79** 172507
[5] E. Mun, Y. S. Kwon, and M. H. Jung 2003 *Phys. Rev. B* **67** 033103
[6] H. M. Rietveld 1969 *J. Appl. Chryst.* **2** 65
[7] A. Larson and R. V. Dreele 2000 *Los Alamos National Laboratory Report LAUR 86-748*
[8] G. Oomi, T. Kagayama, Y. Uwatoko, H. Takahashi, and N. Mori 1994 *J. Alloys and Compd.* **207/208** 278
[9] R. R. Joseph, K. A. Gschneidener Jr., and R. E. Hungsberg 1972 *Phys. Rev. B* **5** 1878
[10] P. D. Sacramento and P Schlottmann 1989 *Phys. Rev. B* **40** 431
[11] P. Gegenwart, J. Custers, Y. Tokiwa, C. Geibel, and F. Steglich 2005 *Phys. Rev. Lett.* **94** 076402
[12] S. Nakatsuji, K. Kuga, Y. Machida, T. Tayama, T. Sakakibara, Y. Karaki, H. Ishimoto, S. Yonezawa, Y. Maeno, E. Pearson, G. G. Lonzarich, L. Balicas, H. Lee, and Z. Fisk 2008 *Nat. Phys.* **4** 603

International Conference on Strongly Correlated Electron Systems (SCES 2011) IOP Publishing
Journal of Physics: Conference Series **391** (2012) 012007 doi:10.1088/1742-6596/391/1/012007

Cu-NMR spectra in UCu_4Ni uncover site disorder

O O Bernal[1], D A Rose[1], Hsin-Ju Wu[1], M Chiang[1], D E MacLaughlin[2], G R Stewart[3] and J S Kim[3]

[1]Department of Physics and Astronomy, California State University, Los Angeles, CA 90032
[2]Department of Physics and Astronomy, University of California, Riverside, CA 92125
[3]Department of Physics, University of Florida, Gainesville, FL 32611

E-mail: obernal@calstatela.edu

Abstract. Cu-NMR measurements in a random powder of UCu_4Ni reveal two types of spectral lines for each of the two isotopes of naturally abundant Cu in the material. These lines, which we label L_1 and L_2, point to the existence of two inequivalent Cu sites in the sample. We present a study of the NMR line shape in UCu_4Ni at three different frequencies (in the range from 40-70 MHz) and two temperature values (10 K and 150 K), that allow us to assign the lines to particular Cu sites. L_1 is strongly broadened as the frequency decreases, but changes less with increasing temperature. In contrast, the width of L_2 grows in proportion to frequency and decreases noticeably with increasing temperature. This behavior indicates that the crystallographic site corresponding to L_1 is exposed to electric field gradients and has lower point symmetry than the site corresponding to L_2, which displays some anisotropy but no discernible quadrupole effects. By comparison with the Cu-NMR spectra in UCu_4Pd, where only one type of Cu-NMR line has been observed clearly, we can associate L_1 with $Cu^{(16e)}$ nuclei: Cu nuclei sitting at the 16*e* site (Wyckoff notation) in the $AuBe_5$ structure of the parent compound UCu_5. This leaves L_2 as originating from $Cu^{(4c)}$ nuclei; i.e., those sitting at the 4*c* site of the same structure. Unlike in UCu_4Pd, the appearance of signal from $Cu^{(4c)}$ nuclei in the Ni compound is clear evidence of site disorder in UCu_4Ni.

1. Introduction

UCu_4Ni is a non-Fermi liquid, quantum critical system, derived from dilution of Cu with Ni in antiferromagnetic (AFM) UCu_5 [1]. Introduction of Ni in the lattice structure of the parent compound suppresses the AFM transition temperature from 16.5 Kelvin down to zero at $x = 1$. This is very similar to the behavior of $UCu_{5-x}Pd_x$ upon dilution of Cu with Pd [2]. The Pd system has been the subject of extensive study (see e.g., references [2, 3, 4] and references therein). At first and for some time, the lattice structure of UCu_4Pd was considered to be ordered; Pd ions would occupy the 4*c* site of the $AuBe_5$ structure of UCu_5, and Cu would, in turn, occupy the 16*e* site (Wyckoff notation). However, the possibility of site disorder came after Cu-NMR measurements were performed in $UCu_{5-x}Pd_x$ ($x = 1, 1.5$) and interpreted as indicating the existence of inhomogeneity in the magnetic susceptibility [5]. Such inhomogeneity, which could also explain other thermodynamic and transport properties of the system [5, 6], can only come from Cu/Pd site interchange. Nevertheless, NMR only detected an unambiguous signal from one Cu site, which still leaves room for controversy regarding the amount of disorder (and its importance for the underlying physics) when derived from other measurements [7, 8]. Therefore, the existence of site disorder in UCu_4Pd could not be established directly, but only indirectly, by NMR alone.

The case of UCu_4Ni is somewhat different. The material was developed in part to understand the effects of disorder on the physics of quantum critical systems [1]. Because Ni ions are very similar in size to Cu ones (perhaps only slightly smaller), Ni can be expected to distribute itself randomly between the two available crystal sites creating strong site disorder. This expectation is consistent with resistivity and x-ray measurements, which display large values and broad spectra respectively [1]. Disorder in this material is also inferred in the work by López de La Torre et al [9], who have modeled the magnetic susceptibility of UCu_4Ni within the Kondo-Disorder [5] and Griffith Phase [6] pictures. Furthermore, it has been established by μSR measurements that the Ni system behaves similarly to UCu_4Pd down to the lowest temperatures [10]. Nevertheless, to date there has been no direct evidence for site disorder in UCu_4Ni.

In this paper, we present Cu-NMR spectra in a random powder of UCu_4Ni which clearly show the existence of Ni/Cu interchange between $4c$ and $16e$ sites. We find two unambiguous types of Cu-NMR lines with the necessary properties to demonstrate the phenomenon. By comparison with UCu_4Pd, we can directly identify one of the lines as coming from the nuclei sitting at site $16e$, or $Cu^{(16e)}$. The "new" line in UCu_4Ni is therefore associated with nuclei at the $4c$ site, or $Cu^{(4c)}$. Furthermore, consistent with the point symmetry of these sites (trigonal for $16e$ and tetrahedral for $4c$), we find that $Cu^{(16e)}$ nuclei are strongly affected by electric field gradients, whereas $Cu^{(4c)}$ nuclei experience little if any quadrupole effects. On the other hand, $Cu^{(4c)}$ nuclei seem to be strongly affected by the U magnetism. This makes the $4c$ site a particularly suitable Cu-NMR probe (not available in the Pd material) to study magnetic disorder effects in non-Fermi liquid, quantum critical, $UCu_{5-x}Ni_x$.

2. NMR Spectra

2.1. Frequency Dependence at Low Temperature

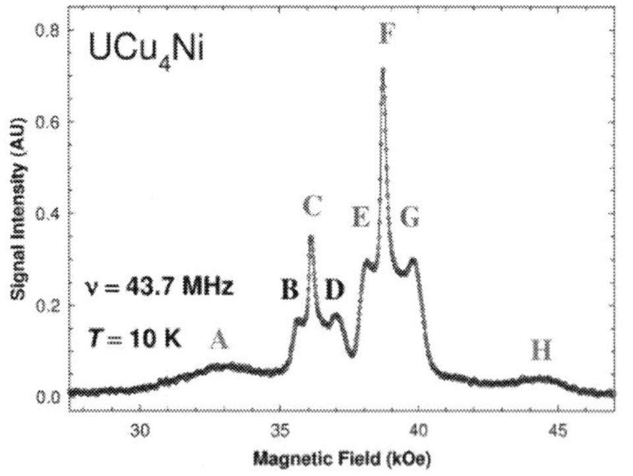

Figure 1. Cu-NMR spectrum in UCu_4Ni at low frequency ν and low temperature T as indicated. The spectral features labeled from A to H are described in the text.

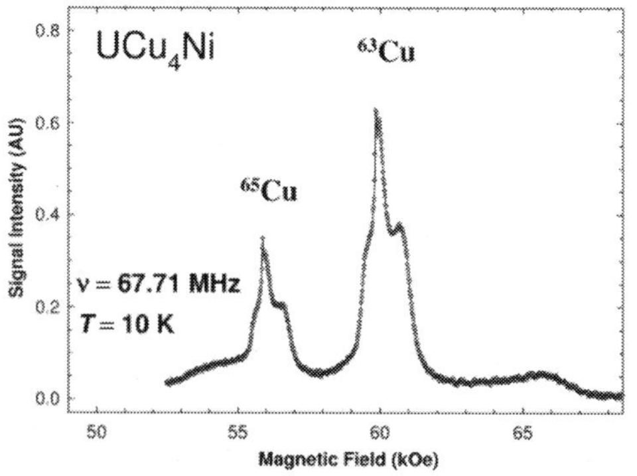

Figure 2. Spectrum at higher value of ν but for the same value of T as in Figure 1. Isotopic assignments are indicated to aid in identifying the lines described in the text.

Figure 1 presents the NMR spectra in UCu_4Ni at $T = 10$ K and $\nu = 43.7$ MHz (the lowest frequency at which spectra were obtained). As can be seen, there is substantial structure in the spectrum. The main and prominent (repeating) patterns correspond to signals from the two isotopes in naturally abundant Cu (^{65}Cu and ^{63}Cu; expected abundance/intensity 30.9% and 69.1% respectively). The low intensity broad peaks on each side of the spectrum (labeled A and H) are made up mostly by signal from quadrupolar-satellite transitions of ^{63}Cu. The corresponding satellites of ^{65}Cu are not resolved. These satellites, together with features (B,D) and (E,G) constitute one type of NMR line which we refer to as L_1. The sets of features

(B,D) and (E,G) of L_1 are second-order quadrupole-split, central transitions of ^{65}Cu and ^{63}Cu respectively. L_1 has therefore two isotopic components, which we designate as $^{65}L_1$ and $^{63}L_1$ for ^{65}Cu and ^{63}Cu respectively. Features C and F constitute the second type of Cu-NMR line (L_2) found in this material—with corresponding isotopic components $^{65}L_2$ and $^{63}L_2$. The assignments become more clear after analyzing the frequency and temperature dependences of the spectral features, and matching them with the corresponding spectra in UCu_4Pd, as described below.

We compare the spectrum in Figure 1 with that displayed in Figure 2. The latter corresponds to the same low temperature of 10 K, but different frequency; i.e., $\nu = 67.71$ MHz (the highest frequency at which spectra were obtained). Using the notation introduced above, we observe first that the distance between A and H in field units remains approximately the same at higher frequencies. This yields a value of the quadrupolar frequency $\nu_Q = 12.6(5)$ MHz (ν_Q is proportional to the z-component of the electric field gradient tensor at the site where $^{63}L_1$ originates). In contrast, we can clearly see that lines from ^{65}Cu and ^{63}Cu are now separated by a larger amount. This follows directly from the fact that the main resonance frequencies are proportional to the applied field. More interestingly, the distance between B and D (of $^{65}L_1$), on the one hand, and E and G (of $^{63}L_1$), on the other, have decreased linearly with $1/\nu^2$, whereas the widths of C (of $^{65}L_2$) and F (of $^{63}L_2$) have increased in proportion to ν. This behavior represents predominantly quadrupolar/electric-field-gradient effects for L_1 and spin/magnetic ones for L_2. There is no discernible sign that L_2 is affected by any quadrupole splitting.

2.2. NMR Spectra at Intermediate Frequency: Temperature Dependence

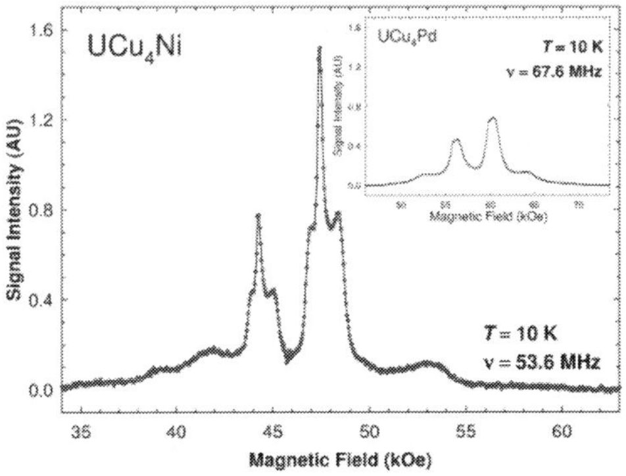

Figure 3. Main Panel: Cu-NMR spectrum in UCu_4Ni at 53.60 MHz and $T = 10$ K as indicated. Inset: low-temperature spectrum in UCu_4Pd (from reference [11]) plotted on a similar scale as that of the main panel. A line of type L_2 is missing in UCu_4Pd.

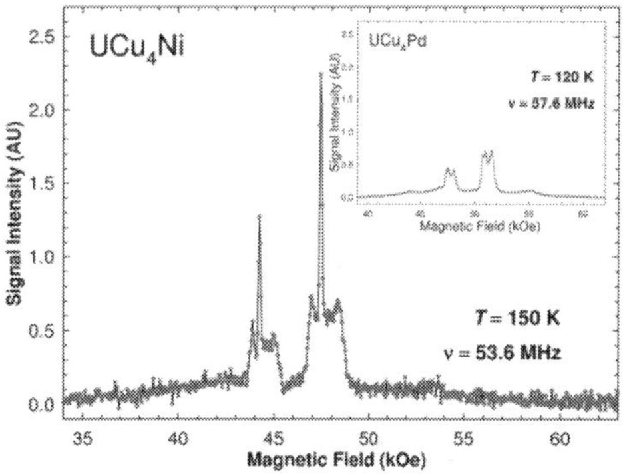

Figure 4. Main Panel: Cu-NMR spectrum in UCu_4Ni at 53.60 MHz and $T = 150$ K. Inset: high-temperature spectrum of UCu_4Pd plotted on a similar scale as that of the main panel. Again, a line of type L_2 is missing in UCu_4Pd.

The spectrum at $\nu = 53.60$ MHz and $T = 10$ K is presented in the main panel of Figure 3. At the same frequency, but for $T = 150$ K, the spectrum looks as depicted in the main panel of Figure 4. It is seen that the distances between B and D, and E and G, do not change appreciably as functions of temperature. This is consistent with the typical lack of T dependence for quadrupole effects. On the other hand, the linewidths of C and F decrease noticeably as they should if affected by the sample paramagnetism. In effect, both types of lines (L_1 and L_2) experience line narrowing, in the sense that they become considerably sharper at higher temperatures. Since the quadrupole-split features remain at the same distance from each other, the narrowing of L_1 should therefore be magnetic in nature also.

3. Discussion

The behavior of the spectra with frequency and temperature variations is consistent with two types of Cu sites in UCu_4Ni. Nuclei at the site represented by L_1 are strongly affected by quadrupolar effects, whereas nuclei at the site represented by L_2 are not. In fact L_1's behavior closely resembles that of the only line clearly resolved in UCu_4Pd (Insets in Figures 3 and 4), which represents the majority Cu site in that material; i.e., $16e$. We therefore associate L_1 with Cu nuclei sitting at the $16e$ site of the crystal structure, or $Cu^{(16e)}$. Line L_2 is absent in the Pd material. Here, in the Ni case, this line could in principle be ascribed to some impurity phase, such as pure Cu or *Cu*Ni. In that case, however, one would expect very narrow and temperature independent line(s) near the field at which $2\pi\nu = \gamma H$ (where $\gamma/2\pi = 11.285$ MHz/T, and H is the applied external field) [12]. However, it is seen that L_2 is strongly affected by frequency and temperature variations. The relaxation time of this signal is also seen to be shorter than that of pure Cu or *Cu*Ni phases whose intensity would be suppressed by the fast pulsing required in the current experiments (50 ms repetition times). This leaves L_2 as originating from $Cu^{(4c)}$ nuclei: those nuclei sitting at site $4c$ of the crystal structure. Unlike in UCu_4Pd, the appearance of a line from $Cu^{(4c)}$ nuclei in the Ni compound is clear evidence of site disorder in UCu_4Ni.

Our preliminary analysis of these data thus yields some quantitative measure of the different effects dominating the appearance of the spectra. In particular the value of ν_Q at $Cu^{(16e)}$, 12.6(5) MHz, should be compared with 10(1) MHz for the Pd compound [5]. This indicates stronger electric field gradients at the trigonal-symmetry $16e$ sites in the Ni sample. On the other hand, $Cu^{(4c)}$ nuclei are subjected to little if any electric field gradients, whereas they are strongly affected by the U paramagnetism. This would be consistent with the tetrahedral point symmetry of the $4c$ site, which would not allow and electric field gradient. The $Cu^{(4c)}$ nuclei are therefore very suitable to study magnetic disorder effects in $UCu_{5-x}Ni_x$ using NMR. That study will be reported elsewhere in the near future.

Acknowledgments

Work supported by NSF/DMR-0604015/1105380 (Los Angeles) and 0801407 (Riverside). Work at the University of Florida was performed under the auspices of the US DOE, contract No. DE-FG02-86ER45268.

References

[1] Mixson D J, Kim J S, Swick M, Jones T, Stewart G R, Scheidt E-W, Scherer W, Murphy T and Palm E C 2006 *Phys. Rev. B* **73** 125106
[2] Andraka B and Stewart G R 1993 *Phys. Rev. B* **47** 3208
[3] Stewart G R 2001 *Rev. Mod. Phys.* **73** 797.
[4] MacLaughlin D E, Heffner R H, Bernal O O, Ishida K, Sonier J E, Nieuwenhuys G J, Maple M B and Stewart G R 2004 *J. Phys.: Condens. Matter* **16** S4479
[5] Bernal O O, MacLaughlin D E, Lukefahr H G and Andraka B 1995 *Phys. Rev. Lett.* **75** 2023
[6] de Andrade M C *et al.* 1998 *Phys. Rev. Lett.* **81** 5620
[7] Booth C H, MacLaughlin D E, Heffner R H, Chau R, Maple M B and Kwei G H 1998 *Phys. Rev. Lett.* **81** 3960
[8] Bauer E D, Booth C H, Kwei G H, Chau R and Maple M B 2002 *Physica B* **65** 245114
[9] López de la Torre M A, González J A, Izquierdo A, Vieira S, Ellerby M and McEwen K A 2000 *J. App. Phys.* **87** 5126
[10] Bernal O O, MacLaughlin D E, Lei Shu, Gavilano J L, Stewart G R, Burnette D J, and Baines C 2010 *J. Phys.: Conference Series* **200** 012012
[11] Bernal O O 2000 *Phys. Stat. Sol. B* **220** 335
[12] Carter G C, Bennett L H and Kahan D J 1977 *Prog. Mater. Sci.* **20** 1181

Crystal structure and physical properties of the novel stannide $Yb_3Pd_2Sn_2$

I. Čurlik[1], M. Reiffers[1,2], M. Giovannini[3], P. Solokha[3], A. Saccone[3], S. Il'kovič[2], D.H. Ryan[4]

[1]Institute of Experimental Physics, Slovak Academy of Sciences, Watsonova 47, 043 53 Košice, Slovakia

[2]Faculty of Sciences, University of Prešov, 17.novembra, SK-080 78 Presov, Slovakia

[3]Department of Chemistry, University of Genova, Via Dodecaneso 31, I-16146 Genova, Italy

[4]Centre for the Physics of Materials and Physics Department, McGill University, 3600 University St., Montreal, QC, H3A 2T8, Canada

E-mail: reiffers@saske.sk

Abstract. The crystal structure and the physical properties at low temperatures of the novel stannide $Yb_3Pd_2Sn_2$ have been studied. The compound crystallizes in a new structure type, space group *Pbcm*, with Yb occupying four different positions in the lattice. Measurements of DC susceptibility and ^{170}Yb Mössbauer suggest a close to divalent Yb-ions behaviour, ruling out a scenario of heterogeneous mixed valence for the Yb-ions .

A large number of unusual low temperature features are found in Yb and Ce intermetallic compounds due to hybridization of *f*-electrons and conduction (*f-c*) electrons. A recent investigation on the Yb-Pd-Sn system was published [1], motivated by the intriguing physical properties of some of the ternary compounds in the system. As an example, in Yb_2Pd_2Sn magnetic order appears and then vanishes by applying increasing values of pressure, suggesting the existence of two quantum critical points within the same compound [2]. Similar behaviour occurs by doping Sn sites by In [3], Yb_2Pd_2In being isotypic to Yb_2Pd_2Sn [4]. Other interesting examples in this system are $YbPd_2Sn$, which is one of the few Yb-compound where superconductivity coexists with antiferromagnetism [5], and αYbPdSn where coexistence of localized and delocalized f-electrons seems to occur [6].

In the investigation of the Yb-Pd-Sn system, a new compound with the formula $Yb_3Pd_2Sn_2$ and an orthorhombic cell was found [1]. Here we present results on the study of the structure and the low temperature physical properties of this compound. The metals used were ytterbium (pieces, 99.993 % Yb/TREM purity, Smart Elements GmbH, Vienna, Austria), palladium (foil, 99.95 mass % purity, Chimet, Arezzo, Italy) and tin (bar, 99.999 mass% purity). The sample was prepared by weighing stoichiometric amounts of the elements. The elements were enclosed in a small tantalum crucible sealed by arc welding under pure argon, and melted in an induction furnace under a stream of pure argon. The alloy was then annealed in a resistance furnace at 600 °C for three weeks and finally quenched in cold water. After quenching, the sample was characterized by scanning electron microscopy (SEM) and electron probe micro-analysis (EPMA) based on energy dispersive X-ray spectroscopy. Since all the attempts to solve the structure of the compound from single crystal data have been unsuccessful, the crystal structure has been determined *ab initio* from powder X-ray diffraction (XRD) data. The structure solution was found in the direct space by global optimization of the parameters using the program FOX [7]. The trial model was tested and refined with the Rietveld method using the program FULLPROF [8]. (for details of crystal structure determination see [9]).

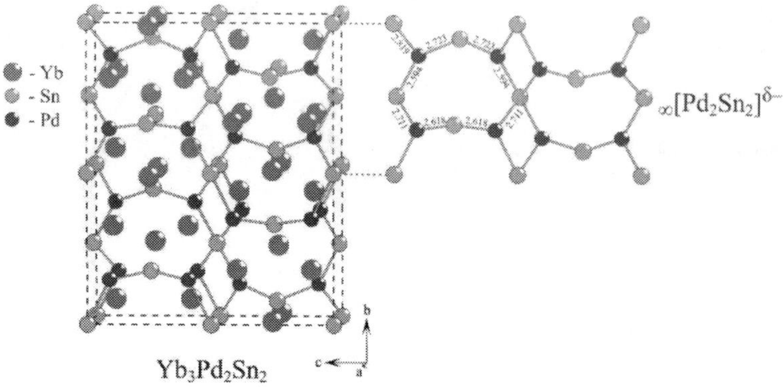

Figure 1. Crystal structure of $Yb_3Pd_2Sn_2$ with outlined polyanionic networks of $_\infty[Pd_2Sn_2]^{\delta-}$ composition. In the top right corner the distances of heterogeneous Pd–Sn contacts within $_\infty[Pd_2Sn_2]^{\delta-}$ network are indicated.

Atom	Wyckoff position	*x/a*	*y/b*	*z/c*	CN
Pd1	8*e*	0.019(2)	0.3517(5)	0.0903(10)	11
Pd2	8*e*	0.281(2)	0.1146(6)	0.0889(9)	10
Yb1	8*e*	0.196(1)	0.6778(4)	0.0981(7)	17
Yb2	8*e*	0.505(2)	0.4296(4)	0.0973(6)	17
Yb3	4*d*	0.004(3)	0.5108(6)	¼	17
Yb4	4*d*	0.295(2)	0.2437(7)	¼	17
Sn1	4*d*	0.503(3)	0.0650(8)	¼	12
Sn2	4*d*	0.209(2)	0.8275(8)	¼	12
Sn3	4*c*	0.308(2)	¼	0	16
Sn4	4*a*	0	0	0	14

Table 1. Crystallographic data of the $Yb_3Pd_2Sn_2$ structure refined in the space group *Pbcm*. R_F = 6.35%, R_B = 8.47%, R_{wp} = 15.2%

The compound crystallizes as a new structure type in the orthorhombic space group *Pbcm*. The refined lattice parameters are a = 0.58262(2) nm, b = 1.68393(8) nm and c = 1.38735(7) nm. $Yb_3Pd_2Sn_2$ is composed of a complex $_\infty[Pd_2Sn_2]^{\delta-}$ polyanionic network in which Yb ions are embedded (Fig.1). In the crystal structure of $Yb_3Pd_2Sn_2$ Yb ions occupy four inequivalent crystallographic sites. Furthermore, all these positions are characterized by similar coordination environments and interatomic distances within the first coordination sphere are comparables (Table 1). The DC magnetization was measured using a Quantum Design MPMS commercial device down to 2 K and up to 5 T. Measurements till 1000 K were performed in VSM in VERSALAB device. ^{170}Yb

Mössbauer spectrum was obtained with both the source and the sample located in a helium flow cryostat at the same temperature, whereas ^{119}Sn Mössbauer spectrum was taken at room temperature. Fig. 2(a) shows the magnetic field dependence of magnetization *M(B)* of $Yb_3Pd_2Sn_2$ for different temperatures. The dependencies are characteristic of paramagnetic behaviour without a tendency to saturation at lower temperatures. The DC susceptibility as a function of temperature (Fig. 2(b)) shows a Pauli paramagnetic contribution suggesting a close to divalent behaviour of Yb ions. The small increase of susceptibility shown at low temperatures is likely due to paramagnetic impurities, whereas a tiny broad bump above 200 K suggests some degree of *c-f* hybridization. Extended measurements till 1000 K confirmed this behaviour.

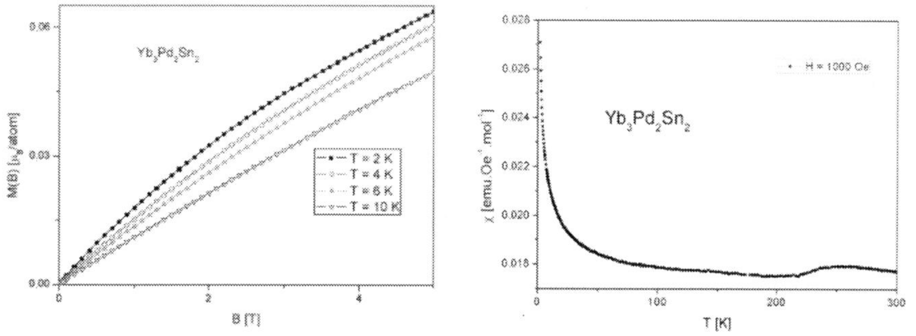

(a) (b)

Figures 2 (a) Field dependence of magnetization of $Yb_3Pd_2Sn_2$ at different temperatures

(b) DC magnetic susceptibility of $Yb_3Pd_2Sn_2$ at 1000 Oe.

The scenario of close to divalent Yb-ions is corroborated by the ^{170}Yb Mössbauer measurements (Fig. 3a). Only a broadened single feature centred on zero velocity appears, suggesting the existence of divalent ytterbium (Yb^{2+}) and ruling out the presence of trivalent ytterbium (Yb^{3+}) between 5 K and 40 K. A detailed fitting of the spectum is precluded by the multiple Yb-site occupation. However, a hightly constrained fit (shown in Fig.3(a)) yields quite reasonable values for linewidth and average magnitude for eQV_{zz}. The ^{119}Sn Mössbauer spectrum is shown in Fig.3(b). It fits well with a

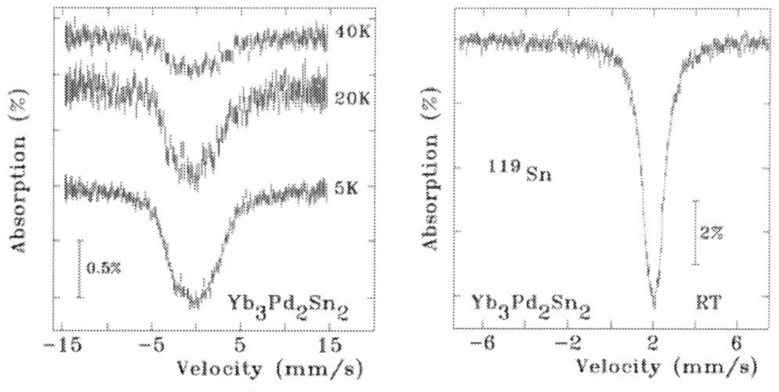

(a) (b)

Figures 3 (a) ^{170}Yb Mössbauer spectrum of $Yb_3Pd_2Sn_2$ at 5, 20 and 40 K. (b) ^{119}Sn Mössbauer spectrum of $Yb_3Pd_2Sn_2$ at room temperature. The solid line is a fit described in the text.

single component, implying that the environments of the four tin sites in the structure are too similar to be resolved. The parameters are typical of tin in a metallic environment.

In summary, structure, magnetism and Mössbauer spectra (^{170}Yb and ^{119}Sn) of a novel polyanionic stannide have been studied. From the structural features, magnetic properties and ^{170}Yb Mössbauer spectra, a close to Yb^{2+} behavior arise for this compound, and a scenario of heterogeneous mixed-valence system, where Yb^{2+} and Yb^{3+} coexist, can be ruled out. Nevertheless, the high-T feature of magnetic susceptibility suggest a certain degree of hybridization of 4f- and conduction electrons. This is corroborated also by electron structure calculation and heat capacity measurements done for this compound [9].

The work has been partly supported by the Slovak-Italian Scientific-Technological Exchange Program for fellowships in Genova and Kosice; partly by the Slovak grant agency VEGA 2/0007/09; the CEX Nanofluid as the Centre of Excelence SAS. The liquid nitrogen for the experiment has been sponsored by the U.S. Steel Kosice, s.r.o.. Financial support was provided by the Natural Sciences and Engineering Research Council of Canada and Fonds Québécois de la Recherche sur la Nature et les Technologies.

References

[1] M. Giovannini, R. Pasero, A. Saccone, Intermetalics 18 (2010) 429.

[2] T. Muramatsu, T. Kanemasa, T. Kagayama, K. Shimizu, Y. Aoki, H. Sato, M. Giovannini, P. Bonville, V. Zlatic, I. Aviani, R. Khasanov, C. Rusu, A. Amato, K. Mydeen, M. Nicklas, H. Michor, E. Bauer, Phys. Rev. B 18 (2011) 180404(R).

[3] E. Bauer, G. Hilscher, H. Michor, C. Paul, Y. Aoki, H. sato, DT Adroja, JG Park, P. Bonville, C. Godart, J. Sereni, M. Giovannini, A. Saccone, J. Phys. Cond. Mat. 17 (2005) S999.

[4] M. Giovannini, E. Bauer, H. Michor, G. Hilscher, A. Galatanu, A. Saccone, P. Rogl, Intermetallics 9 (2001) 481.

[5] A. Amato, B. Roessli, P. Fischer, N. Bernhoeft, A. Stunault, C. Baines et al., Physica B 326 (2003) 369.

[6] T. Görlach, S. Putselyk, A. Hamann, T. Tomanic, M. Uhlarz, Phys, Rev. B 76 (2007) 205112.

[7] V. Favre-Nicolin, R.Černý, 'FOX, free objects for crystallography': a modular approach to ab initio structure determination from powder diffraction", J. Appl. Crystallogr. 35 (2002) 734.

[8] J.Rodriguez-Cavajal, "Recent developments in the program FullProf", Newsletter 26 (2001) 12.

[9] P. Solokha, I. Čurlik, M. Giovannini, N.R. Lee-Hone, M. Reiffers, D.H. Ryan, A. Saccone, J. Solid State Chem. (2011), doi:10.1016/j.jssc.2011.07.031

Crystal Growth and Magnetic properties of RMg_3 (R = La, Ce and Nd)

Pranab Kumar Das, Neeraj Kumar, R. Kulkarni and A. Thamizhavel

Department of Condensed Matter Physics and Materials Science, Tata Institute of Fundamental Research, Homi Bhabha Road, Colaba, Mumbai 400 005, India.

E-mail: `pkd@tifr.res.in`

Abstract. We have successfully grown the single crystals of RMg_3 (R = La, Ce and Nd) compounds by Bridgman method, in a sealed molybdenum crucible. These compounds crystallize in the cubic BiF_3-type structure with the space group $Fm\bar{3}m$. The powder x-ray diffraction pattern clearly revealed that the samples were single phase without any impurity. The crystals were oriented by means of Laue back reflection method for the magnetic and transport measurements. It has been found that $CeMg_3$ orders antiferromagnetically at T_N = 2.6 K, while $NdMg_3$ shows two magnetic orderings with T_{N1} = 6.8 K and T_{N2} = 2.8 K respectively. The magnetic part of the electrical resistivity of $CeMg_3$ exhibited a double peak structure and the magnetization measurement revealed a reduced magnetic moment of 0.5 μ_B/Ce. The heat capacity measurement of $CeMg_3$ measured down to 0.5 K exhibited an enlarged Sommerfeld coefficient of 370 mJ/K^2 mol. From the Schottky heat capacity we have estimated the crystal electric field split energy levels of $CeMg_3$ and $NdMg_3$.

1. Introduction

The magnetism due to the localized $4f$ electrons in the rare-earth based intermetallic systems is always interesting due to the wide range of physical properties exhibited by these compounds. The binary rare-earth intermetallic systems have been studied over several decades. In particular, the Ce based binary compounds are the most extensively studied as they exhibit many novel magnetic behavior owing to the close proximity of the $4f$ level to the Fermi level. In this paper we report on the mangetic properties of a simple binary system, RMg_3 (R = La, Ce, and Nd), which crystallize in the cubic BiF_3- type structure. In one of the early reports on RMg_3 system, Buschow has studied the magnetic properties of $CeMg_3$ and $NdMg_3$ and reported no magnetic ordering down to 4.2 K in these two compounds [1]. However, Galera *et al* and Pierre *et al* have found that $CeMg_3$ orders antiferromagnetically at 3.4 K [2, 3], while $NdMg_3$ orders at 6 K [3]. From the previous neutron scattering experiments on polycrystalline sample [4, 5], it has been found that $NdMg_3$ undergoes an antiferromagnetic ordering at T_N = 6 K with the propagation vector $\mathbf{k}$ = (0.5, 0.5, 0.5). In this paper we report on the magnetic properties of single crystalline RMg_3 (R = La, Ce, and Nd) by studying the magnetic susceptibility, electrical transport and heat capacity measurements.

2. Experiment

From the binary phase diagrams of R-Mg (R = La, Ce, and Nd) by Nayeb-hasehmi and Clark [6], it was found that these compounds melt congruently at temperature close to 800 °C. Hence these compounds can be grown directly from the melt. Because of the high vapor pressure of Mg, the single crystals were grown by Bridgman method by sealing it in a Molybdenum crucible. The molybdenum crucible was then subsequently sealed in a quartz ampoule and loaded into a box type furnace. The furnace was then raised to 850 °C, well above the melting point of these compounds and held at this temperature for 24 hours, then slowly cooled down to 780 °C over a period of 3 days. Bulk shiny single crystals were obtained by gently taping on the alumina crucible. The crystals were then subjected to powder X-ray diffraction to check the phase purity. The orientation of the crystals were done by back reflection Laue method. The dc magnetic susceptibility and the magnetization measurements were performed in the temperature range 1.8-300 K using a superconducting quantum interference device (SQUID) and vibrating sample magnetometer (VSM). The heat capacity measurements were done in a Quantum Design physical property measurement system (PPMS).

3. Results

3.1. X-ray diffraction

Small pieces of the single crystals were crushed into fine powder and subjected to powder x-ray diffraction, using a PANalytical x-ray diffractometer with Cu-K_α radiation, to check the phase purity and to estimate the lattice constant values. No traces of any secondary phases were seen in the x-ray diffractogram indicating the samples are phase pure. A Rietveld analysis was performed and the representative powder x-ray diffraction pattern of $LaMg_3$ is shown in Fig. 1. A reasonably good fit of the experimental pattern confirms the space group $Fm\bar{3}m$ (#225) and the estimated lattice constant was 7.468 Å for $LaMg_3$, 7.422 Å for $CeMg_3$, and 7.363 Å for NdM_3. The crystals were then oriented along the principal crystallographic direction namely [100], by means of Laue back reflection. The four fold symmetry in the Laue pattern confirmed the cubic structure of these

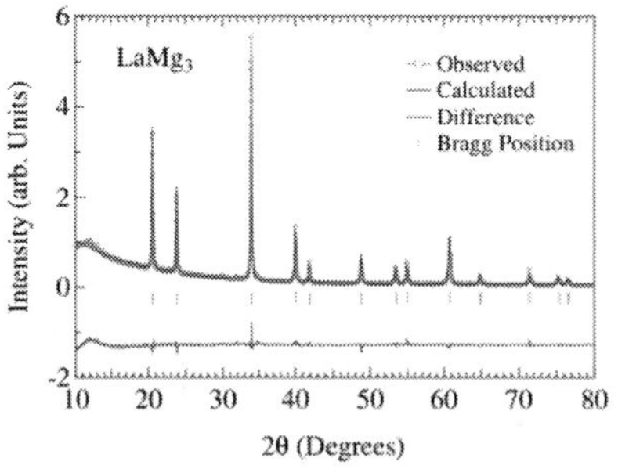

Figure 1. A representative powder X-ray diffraction pattern of fine powders of $LaMg_3$ single crystal. The solid line through the experimental plots is the Rietveld refinement. The Bragg positions and the difference between the calculated and observed pattern is also shown.

compounds. The crystals were then cut along [100] direction by means of a spark erosion cutting machine for the magnetic and transport property measurements.

3.2. Magnetization

The temperature dependence of magnetic susceptibility of $CeMg_3$ and $NdMg_3$ are measured from 1.8 to 300 K in an applied magnetic field of 1 kOe for H parallel to [100] direction. Fig. 2(a) shows the χ vs. T plots, the inset shows the low temperature part of the magnetic susceptibility where the susceptibility exhibit a cusp at 2.6 K for $CeMg_3$ and drops with the decrease in the

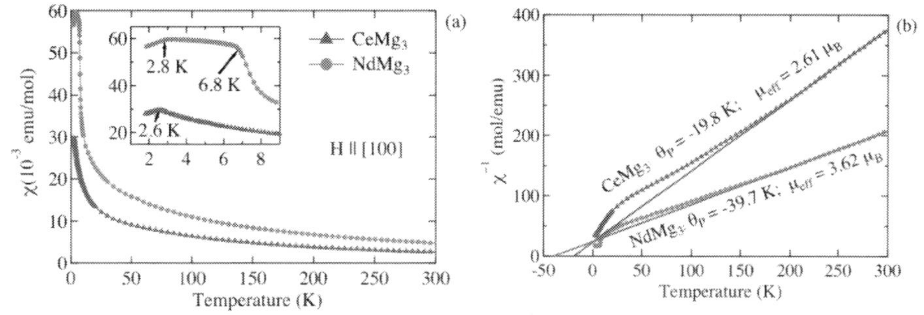

Figure 2. (a) Temperature dependence of magnetic susceptibility of $CeMg_3$ and $NdMg_3$ for field parallel to $H \parallel [100]$. The inset shows the low temperature part of the magnetic susceptibility. (b) Inverse magnetic susceptibility of $CeMg_3$ and $NdMg_3$, the solid lines are fit to Curie-Weiss law.

temperature, while $NdMg_3$ shows two transitions at 6.8 K and 2.8 K. At respective temperatures the compounds order antiferromagnetically. The heat capacity measurement to be discussed later confirms the bulk magnetic ordering at these temperatures. Figure 2(b) shows the inverse magnetic susceptibility. At high temperature the inverse magnetic susceptibility is linear and follows Curie-Weiss law for temperature greater than 200 K. The solid line in Fig. 2(b) is a fit to the Curie-Weiss law. From the fitting, effective magnetic moment $\mu_{\rm eff}$ and the paramagnetic Curie temperature $\theta_{\rm p}$ were found to be 2.61 $\mu_{\rm B}$/Ce and -19.8 K for $CeMg_3$; and 3.62 $\mu_{\rm B}$/Ce and -39.7 K for $NdMg_3$ respectively. The estimated effective moment is close to the free ion values of Ce and Nd in their trivalent state. The negative sign of the $\theta_{\rm p}$ and the linear increase in the magnetization at the lowest measured temperature indicates the antiferromagnetic nature of the magnetic ordering.

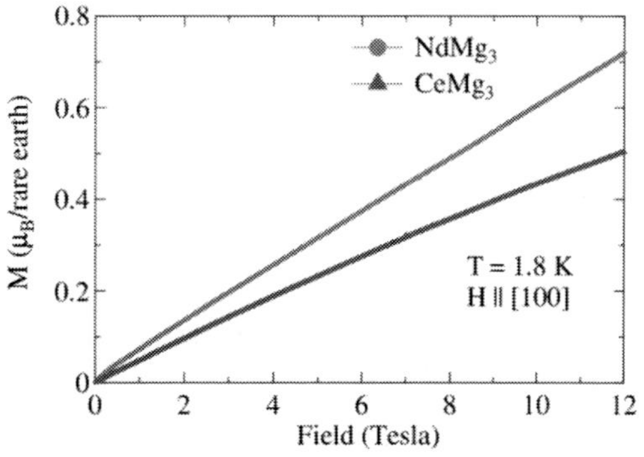

Figure 3. Isothermal magnetization of $CeMg_3$ and $NdMg_3$ for $H \parallel [100]$ at T = 1.8 K.

The field dependence of magnetization $M(H)$ at a constant temperature T = 1.8 K is shown in Fig. 3. The magnetization is almost linear up to a field of 12 T, thus confirming the antiferromagnetic ordering of the Ce and Nd moments. The magnetization does not show any signature of saturation up to a field of 12 T and attains a value of only 0.5 $\mu_{\rm B}$/Ce and 0.7 $\mu_{\rm B}$/Nd, which are much less than the free ion moment values of respective atoms. The reduced value of the moment in $CeMg_3$ is attributed to the Kondo effect.

3.3. Heat Capacity

The temperature dependence of the specific heat capacity of single crystalline $LaMg_3$, $CeMg_3$ and $NdMg_3$ are shown in Fig. 4. The heat capacity of $LaMg_3$ does not show any anomaly, and its temperature dependence is typical for a non-magnetic reference compound. The low temperature part of the heat capacity of $CeMg_3$ is shown in the top inset of Fig. 4, where a clear jump is seen at $T_N = 2.6$ K, confirming the bulk magnetic ordering in this compound. The bottom inset shows the low temperature part of $NdMg_3$ heat capacity, and we see $NdMg_3$ shows two bulk ordering at 6.8 K and 2.8 K. The magnitude of the Sommerfeld coefficient is obtained by fitting the expression $C/T = \gamma + \beta T^2$. The γ, for $CeMg_3$ the value thus obtained is 370 mJ/mol K^2, indicating the heavy fermion nature of $CeMg_3$. Also shown in Fig. 4 is the Schottky heat capacity. From the analysis of Schottky heat capacity, we have estimated the crystal electric field splitting of the 5 fold and 10 fold degnerate levels of $CeMg_3$ and $NdMg_3$ The ground state of $CeMg_3$ is a doublet with an excited quartet state while for $NdMg_3$ the ground state is again a doublet with two excited quartet states. The corresponding energy level sechems are shown in the figure. We did not attempt to estimate the γ value for $NdMg_3$ as the measurement was done only down to 2 K.

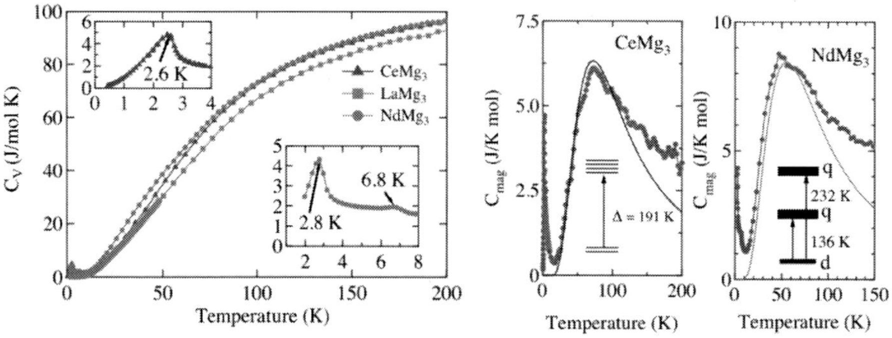

Figure 4. Temperature dependence of the specific heat capacity in $LaMg_3$, $CeMg_3$ and $NdMg_3$. Insets show the low temperature parts. The right panels show the Schottky heat capacity and the estimated energy levels. The 'q' and 'd' in the Schottky heat capacity of $NdMg_3$ represent the quartet and doublet levels respectively.

4. Conclusion

We have successfully grown the single crystal and studied the magnetic properties of RMg_3 (R = La, Ce and Nd). From the magnetic measurements it is evident that $CeMg_3$ orders antiferromagnetically at 2.6 K with a reduced moment and large Sommerfeld coefficeint γ thus indicating that $CeMg_3$ is a heavy fermion compound. On the otherhand, $NdMg_3$ exhibit two antiferromagnetic ordering at $T_{N1} = 6.8$ K and $T_{N2} = 2.8$ K which is confirmed by magnetic susceptibility, magnetization and heat capacity measurements.

References

[1] K.H.J. Buschow, J. Less Comm. Metals **44**, 301 (1976).
[2] J. Pierre, A. P. Murani and R.M. Galera, J. Phys. F: Met. Phys. **11**, 679 (1981).
[3] R. M. Galera, J. Pierre and J. Pannetier, J. Phys. F: Met. Phys., **12**, 993 (1982).
[4] R. M. Galera, A. P. Murani and J. Pierre, J. Magn. Magn. Materials, **23** 317, (1981).
[5] T. Chatterji, G. J. Schneider and R. M. Galera, Phys. Rev. B, **78**, 012411 (2008).
[6] B. Nayeb-hashemi and J. Clark, Bull. Alloy Phase Diagrams, **9**, 916 (1988).

Magnetic phase diagram and Néel critical field in $CeCu_2Ge_2$

Takao Ebihara[1], Yusuke Sugiyama[1], Yasuo Narumi[2], Yoshiki Sawai[3], Koichi Kindo[3], Eoin O'Farrell[4], Michael Sutherland[4] and Neil Harrison[5]

[1] Department of Physics, Shizuoka University, 836 Oya, Shizuoka 422-8529, Japan
[2] Institute for Materials Research, Tohoku University, 2-1-1 Katahira, Sendai, Miyagi 980-8577, Japan
[3] Institute of Solid State Physics, University of Tokyo, 5-1-5 Kashiwanoha, Chiba 277-8581, Japan
[4] Cavendish Laboratory, University of Cambridge, J. J. Thomson Avenue, Cambridge CB3 0HE, UK
[5]National High Magnetic Field Laboratory, Los Alamos, NM 85926, USA

E-mail: `sptebih@ipc.shizuoka.ac.jp`

Abstract. We present magnetic phase diagrams to determine Neel critical field on the basis of electrical resistivity, specific heat and magnetic susceptibility measurements at magnetic fields. The magnetic phase diagrams indicate that Neel temperatures were suppressed completely at 35T for [100] and 31T for [001], respectively. These results indicate Quantum Critical Point exists in a range from 30 to 40T. This phase diagrams should be powerful guides for investigating electronic states near QCP.

1. Introduction

The heavy-fermion antiferromagnet $CeCu_2Ge_2$ is a good reference compound to the prototype heavy fermion superconductor $CeCu_2Si_2$ because both compounds crystallize into the same $ThCr_2Si_2$-type tetragonal structure. $CeCu_2Ge_2$ also shows superconductivity above 7.5GPa when antiferromagnetic order is suppressed by high pressure. $CeCu_2Ge_2$ orders at 4K to incommensurate magnetic structure with a propagation vector of $\mathbf{Q}$=(0.28, 0.28, 0.54). The magnetic moment is $0.74\mu_B$ at T=1.5 K, which is considerably smaller than the moment $1.54\mu_B$ estimated from the Γ_7-ground state split by crystalline electrical eld (CEF) [1].

A quantum critical point (QCP) occurs when an ordered state of matter becomes unstable at the absolute zero temperature. When antiferromagnetic order parameter vanishes, quantum uctuations emerge to mediate exotic superconductivity sometimes. In $CeCu_2X_2$(X=Si, Ge), many high pressure (p) versus temperature (T) and alloying (Si-Ge substitution) versus T studies have been performed to draw magnetic phase diagram [2, 3, 4]. Contrasting to various high pressure studies among heavy fermion systems including $CeCu_2Ge_2$, H-T phase diagram is not well investigated, although magnetic eld is an alternative powerful external parameter for tuning electronic states of condensed matter. In this paper, we present the magnetic phase diagrams and critical elds for a- and c-axes in $CeCu_2Ge_2$.

2. Experiment

Single crystals were grown by the self- ux method [5]. Typical dimension of single crystals are $3\times3\times1$ mm^3. The crystals are con rmed to be in single phase of $CeCu_2Ge_2$ by X-ray powder di raction. Those crystals are characterized by back Laue photography. Electrical resistivity was measured by using standard four-probe method in Physical Property Measurement System (PPMS). Speci c heat was measured by quasi-adiabatic method in PPMS. Magnetic susceptibility is measured by SQUID magnetometer of Magnetic Property Measurement System (MPMS).

3. Results and discussions

3.1. Electrical resistivity, Specific Heats and Magnetic Susceptibility measurements

3.1.1. Electrical resistivity Figure 1 shows temperature dependences of electrical resistivity ratio (/ $_{300K}$) at magnetic elds where (a) H ||[100] (a-axis) and (b) H ||[001] (c-axis). Electric current is along [100] axis. We see double peaks around 100K and 6K in $CeCu_2Ge_2$. This behavior is typical to heavy fermion system. Black solid arrows point shoulders seen near 4K, which relate to the Néel temperature in both panels. These kinks shift to lower temperatures when magnetic eld increases in a- and c-axes. We summarized eld dependence of Néel temperature from electrical resistivity to Fig.4

3.1.2. Specific Heat Figure 2 shows temperature dependence speci c heat (C) below 8K at magnetic elds where (a) $H\|$a-axis and (b) $H\|$c-axis. In these gures, we observed two anomalies in a-axis and one anomaly in c-axis. We indicate anomalies near 4K in both axes by black solid arrows. Black arrowed anomalies in Fig. 2 (a) and (b) are due to antiferromagnetic ordering. The Néel temperature is suppressed with increasing magnetic eld. Although no anomaly is seen without antiferromagnetic ordering in c-axis, one small anomaly is observed around 2.5K at magnetic elds ranging from 11T to 14T for a-axis. This anomaly, indicated by blue arrow in a-axis, corresponds to small jump of magnetization for a-axis around 11T reported by Sugiyama *et.al.* [7]. The small jump in magnetization may be arisen by spin- ip in the a-a plane. With respect to **Q**-vector with (0.28, 0.28, 0.54), the magnitude of magnetic moment is $0.28\times2^{1/2}$ $0.2\mu_B$ for a-axis before spin- ip. If magnetic eld increases along a-axis, minor spin- ip may occur. The magnitude of aligned moment is $0.28\mu_B$ in a-axis. The di erence of moment before and after spin- ip, if there is, is calculated to be about $0.08\mu_B$. In magnetization measurement for $H\|$a-axis, magnitude of small jump in a range from 11T to 14T is read approximately $0.06\mu_B$ from Ref. [7]. Estimated and read values are comparable to each other. Moreover Sugiyama *et.al.* reported that the small jump merges to saturation eld when temperature increases. Thus, the additional small anomalies in speci c heat are from the same magnetic origin as the small jump of magnetization. We summarize eld dependence of Néel temperature and temperature at small anomaly in Fig. 4.

3.1.3. Magnetic Susceptibility Figure 3 shows temperature dependence of magnetic susceptibilities at 10mT (red circles and lines), 5T (green circles and lines), 3.5T (blue circles and lines) and 7T (black circles and lines) for $H\|$a- and c-axes. In a-axis, sharp peaks near 4K were observed. As is seen in hard axis of standard antiferromagnet, magnetic susceptibility saturates in c-axis below 4K. The shoulders are caused from antiferromagnetic ordering. We summarize eld dependence of Néel temperature in Fig. 4.

3.2. Magnetic Phase Diagrams

Figure 4 shows T_N versus H for (a) a- and (b) c-axes. Green, black and blue circles show T_Ns determined by electrical resistivity, speci c heat and magnetic susceptibility measurements.

Gray open circles show an additional anomaly observed in speci c heat in a range from 11T to 14T in a-axis. When we estimate Néel critical eld for $H||a$ axis, we used black and blue data for tting. Because it is rare that H-T phase diagram is drawn in heavy fermion system, the H-T power low scaling is not well established. In high eld study in heavy fermion antiferromagnet $CeIn_3$, the power law scaling of T_N is determined to be expressed pretty well by T_N^* [1-$(H/H_c)^2$] where T_N^* is estimated Néel temperature and H_c is Néel critical eld [6]. We adopt this power law to estimate the critical eld of $CeCu_2Ge_2$.

In Fig. 4, black lines are tting curve as a function of $T_N^*=T_{N,0}[1-(H/H_c)^2]$. The $T_{N,0}$, which indicates T_N at 0T, is estimated to be 4.2K for both axes. The Néel critical elds are estimated to be 35T for a-axis and 31T for c-axis. Critical eld for a-axis is about 30% overestimated than saturation eld (27T) reported in Ref. [7]. In c-axis, the estimated eld is about twice as large as saturation eld of magnetization in the paper. While the a-a plane in $CeCu_2Ge_2$ shows the same symmetry as that in $CeIn_3$, a-c plane in $CeCu_2Ge_2$ is di erent. Quadratic tting is more appropriate to be used for a-a plane of $CeCu_2Ge_2$. Moreover, precise measurements at static high magnetic elds and low temperature could help to estimate critical elds more precisely. However, it is obvious that electronic states are in paramagnetic region above estimated critical elds in this work. We could present phase lines for a- and c- axes in $CeCu_2Ge_2$.

Summary

We revealed H-T phase diagrams in single crystal of $CeCu_2Ge_2$. The antiferromagnetic order was gradually suppressed as magnetic eld increased. By using quadratic tting, Néel critical elds were estimated to be 35T for a-axis and 31T for c-axis. At those points, it is suggested that eld induced quantum critical points emerge. These phase diagrams are powerful guide to investigate electronic states near quantum critical points.

Acknowledgments

We express many thanks to Dr. Watanabe and Professor Miyake for their fruitful discussions. Works by T. E. were performed under the auspices of "100 Tesla spin science" and of basic research (C) by MEXT. T. E. is also acknowledges to Asahi Glass Foundation, CASIO Science Promotion Foundation and Suzuki Foundation.

References

[1] A. Krimmel, A. Loidl, H. Schober, P. C. Canfeld: Physical Review B **55** (1997) 6416.
[2] H. Q. Yuan, F. M. Grosche, M. Deppe, C. Geibel, G. Sparn and F. Steglich: Science **302** (2003) 2104.
[3] H. Q. Yuan, F. M. Grosche, M. Deppe, G. Sparn, C. Geibel and F. Steglich: Physical Review Letters **96** (2006) 047008.
[4] D. Jaccard and A. T. Holmes: Physica B **359-361** (2005) 333.
[5] P. C. Canfield and Z. Fisk: Philosophical Magazine B **65** (1992) 1117.
[6] T. Ebihara , N. Harrison, M. Jaime, S. Uji, and J. C. Lashley: Physical Review Letters **93** (2004) 246401.
[7] K. Sugiyama, T. Miyauchi, Y. Ota, T. Yamada, Y. Oduchi, N.D. Dung, Y. Haga, T.D. Matsuda, M. Hagiwara, K. Kindo, T. Takeuchi, R. Settaia and Y. Ōnuki: Physica B **403** (2008) 769.

Journal of Physics: Conference Series **391** (2012) 012010 doi:10.1088/1742-6596/391/1/012010

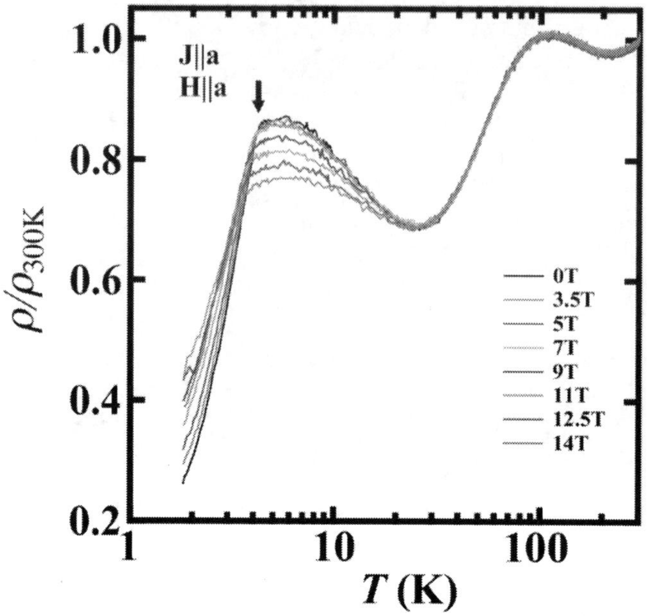

(a)

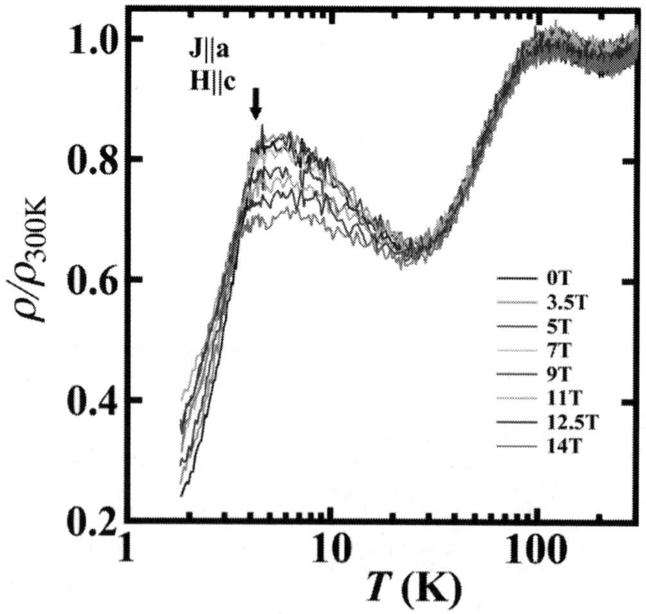

(b)

Figure 1. Electrical resistivity at various magnetic elds for (a) $H\|$ a- and (b) c-axes where $J\|$ a-axis. Antiferromagnetic ordering causes shoulders around 4K indicated by arrows. The shoulder shifts to lower temperature when magnetic eld increases.

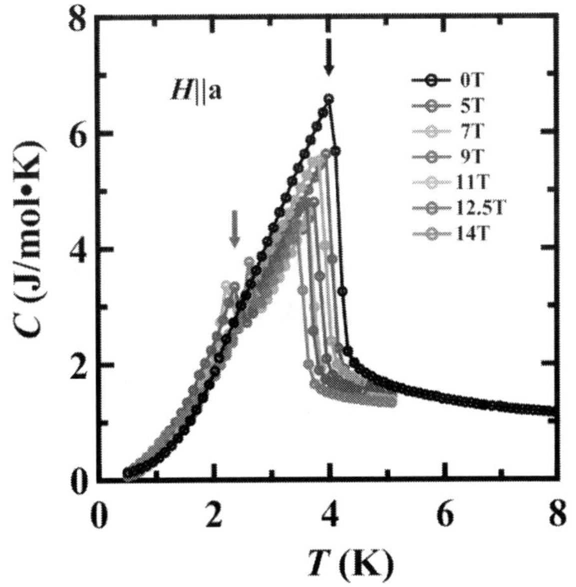

(a)

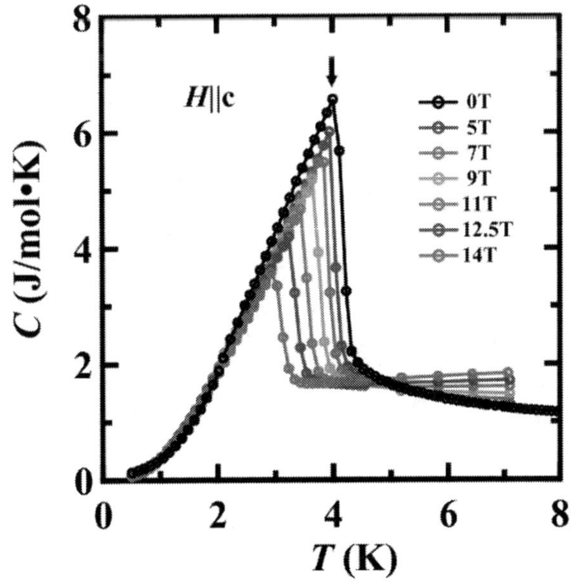

(b)

Figure 2. Low temperature speci c heat for (a) $H\|$ a- and (b) c-axes. Black solid arrows indicate anomalies due to antiferromagnetic transitions. Blue arrow points an additional anomaly that emerges when magnetic eld applied above 11T.

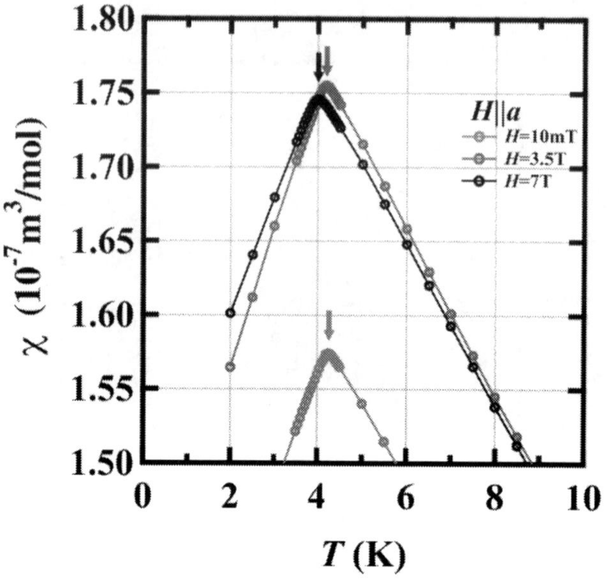

(a)

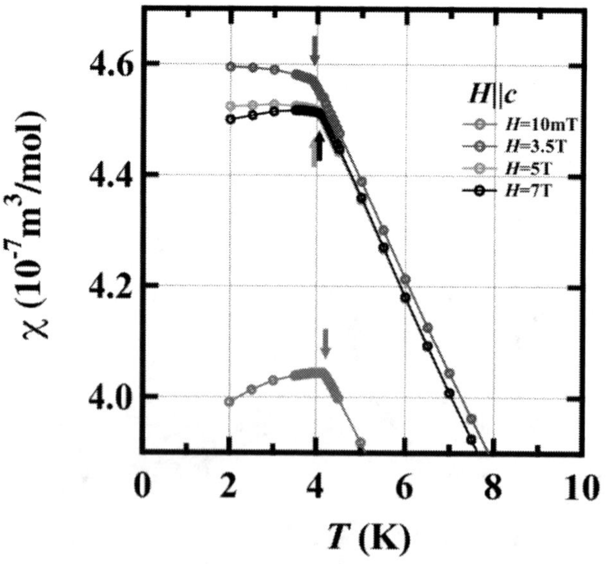

(b)

Figure 3. Magnetic susceptibility for (a) $H\|$ a- and (b) c-axes. The Néel orders are indicated by arrows.

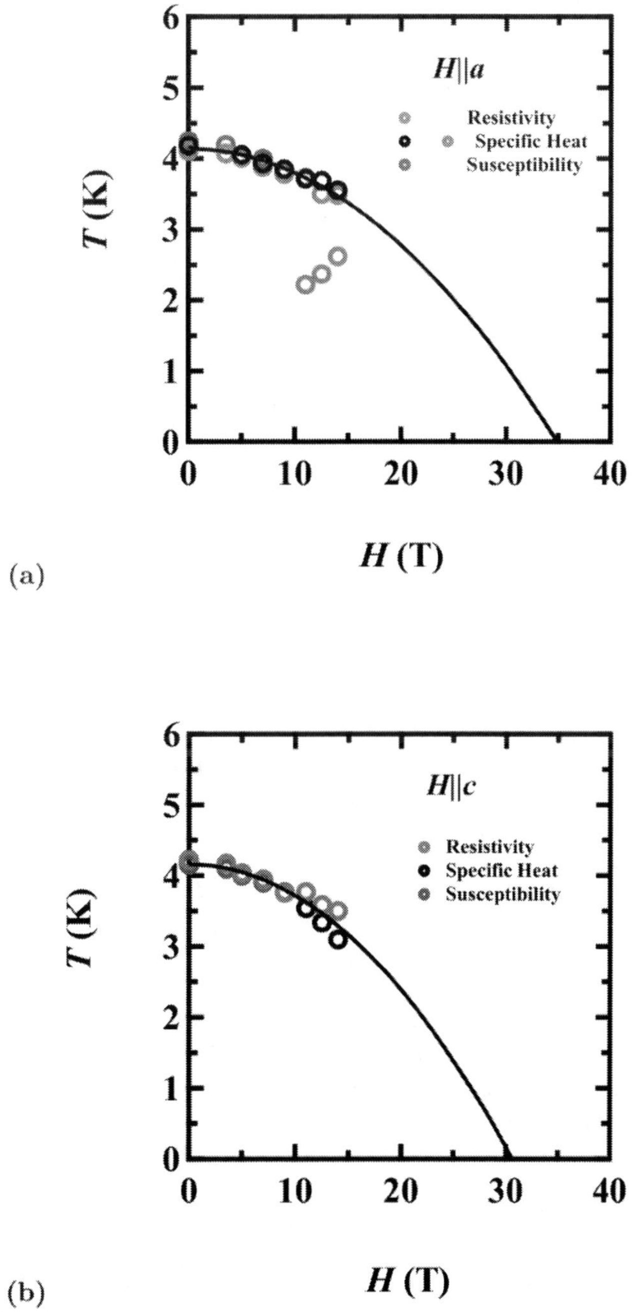

Figure 4. Magnetic phase diagrams for (a) $H\|$ a- and (b) c-axes. Green, black and blue circles represent Néel temperatures determined from electrical resistivity, speci c heat and magnetic susceptibility, respectively. Gray circles indicate anomalies seen in speci c heat.

Shubnikov-de Haas measurements on $LuRh_2Si_2$

S Friedemann[1], S K Goh[1], F M Grosche[1], Z Fisk[2], M Sutherland[1]
[1]Cavendish Laboratory, University of Cambridge, JJ Thomson Avenue, CB3 0HE Cambridge, United Kingdom [2]*Department of Physics and Astronomy, University of California, Irvine, CA 92697-4575, USA*

E-mail: sf425@cam.ac.uk

Abstract. We present Shubnikov-de Haas measurements on $LuRh_2Si_2$, the non-magnetic reference compound to the prototypical heavy-fermion system $YbRh_2Si_2$. We find an extensive set of orbits with clear angular dependences. Surprisingly, the agreement with non-correlated band structure calculations is limited. This may be related to an uncertainty in the calculations arising from a lack of knowledge about the exact Si atom position in the unit cell. The data on $LuRh_2Si_2$ provide an extensive basis for the interpretation of measurements on $YbRh_2Si_2$ indicative of discrepancies between the high-field Fermi surface of $YbRh_2Si_2$ and the "small" Fermi surface configuration.

1. Introduction

Interest in the intermetallic compound $LuRh_2Si_2$ is related to its relevance as a reference compound for the heavy-fermion system $YbRh_2Si_2$ [1, 2]. In addition, $LuRh_2Si_2$ turned out to feature peculiar anomalies in the Hall effect [3]. For both purposes a good understanding of the electronic structure of $LuRh_2Si_2$ is advantageous.

$YbRh_2Si_2$ features a quantum critical point (QCP), i.e., a continuous phase transition at zero temperature. This QCP is accessed at small magnetic fields (60 mT for fields perpendicular to the crystallographic c direction). The QCP in $YbRh_2Si_2$ appears to be a rare case of an unconventional scenario [4, 5, 6]. The basic characteristic of such an unconventional QCP is that the entire Fermi surface becomes critical as part of the electrons become localized when tuning through the QCP. In heavy-fermion systems this partial localization is associated with the breakdown of the Kondo effect, with the composite quasiparticles formed through the Kondo effect disintegrating into their constituents, f- and conduction electrons. Evidence for the change of the Fermi surface is seen in measurements of the Hall effect across the QCP [7, 2]. The Hall coefficient passes through a crossover at finite temperatures which sharpens to a discontinuity in the extrapolation to zero temperature with the values on the low and high field side of the QCP reflecting the "small" and "large", conduction electron and composite quasiparticle Fermi surface, respectively.

The picture of the Fermi-surface reconstruction has been backed by de Haas-van Alphen (dHvA) studies on $YbRh_2Si_2$ [1]. Although the observed angular dependence of the dHvA frequencies at fields above 14 T are best described by the "small" Fermi surface picture the increase of the dHvA frequencies below 10 T is interpreted in terms of the large Fermi surface. The transition at 10 T is speculated to be of Lifshitz type [8]. A large Fermi surface configuration as concluded from the dHvA measurements below 10 T is consistent with the Hall effect measurements indicative for a large Fermi surface configuration above 0.06 T. In later dHvA studies a higher frequency of $F = 14\,\mathrm{kT}$ was observed at fields above 13 T [9]. This frequency was associated with the hole-like Fermi surface sheet of the "small" Fermi surface configuration consistent with the other frequencies. However, it remains an open question why no other orbit of this sheet could be observed. In addition, the agreement of the observed angular dependences with calculated orbits of either the "small" or "large" Fermi surface configuration are limited. Consequently, a better understanding of the "small" Fermi surface may help to interpret the dHvA measurements on $YbRh_2Si_2$.

Here, we present measurements of the Shubnikov-de Haas (SdH) effect on $LuRh_2Si_2$, thereby providing an experimental reference for the "small" Fermi surface of $YbRh_2Si_2$. The heavy-fermion behaviour of $YbRh_2Si_2$ is based on the magnetic moment provided by an incompletely filled $4f$ shell. In $LuRh_2Si_2$ the $4f$ shell is completely filled and thus $LuRh_2Si_2$ may serve as a non-magnetic reference to $YbRh_2Si_2$. In fact, the comparison can be extended to a quantitative level: $LuRh_2Si_2$ features identical lattice parameters within experimental resolution which means that not only the topology but also the absolute dimensions of the Fermi surface of $LuRh_2Si_2$ should resemble the "small" Fermi surface of $YbRh_2Si_2$. Consequently, $LuRh_2Si_2$ is a very powerful reference system to $YbRh_2Si_2$.

2. Experimental setup

Single crystal samples were grown in indium flux as described earlier [10]. They grow preferentially in form of thin ($\lesssim 30\,\mu\mathrm{m}$) platelets which makes them ideal for SdH measurements. SdH oscillations were measured using a standard four probe ac resistivity technique. Contacts were provided by 25 µm gold wires spot welded to the sample and stabilized with silver epoxy The sample size was approximately $0.02 \times 0.1 \times 2\,\mathrm{mm}^3$. The current was applied within the basal plane at an angle of $\approx 10°$ from the (100) axis. Measurements were performed in a ^{3}He/^{4}He-dilution refrigerator in fields up to 16 T. The oscillatory part was deduced by subtraction of a 4th order polynomial fit from the raw data. Oscillation frequencies were determined after Fourier transformation. In order to deduce the angular dependence of the oscillation frequencies the magnetic field was rotated within the crystallographic basal plane.

3. Experimental Results and Discussion

Figure 1 shows a representative trace of the oscillatory part of the resistivity taken at 95 mK for fields between 6 T and 16 T with the field applied along the (110) direction. The power spectrum exhibits numerous oscillation frequencies. Some of the frequencies have a signal to noise ratio of more than 100. For better visibility of the high frequency peaks, data in the Fourier spectrum above 13 kT are multiplied by a factor of 10. 19 frequencies are detected for this field orientation of which some are identified as harmonics and some arise from mixing of fundamental frequencies.

The dominant frequencies are δ (3.7 kT), ϵ' (5.1 kT), and η (10.78 kT). A group of two fundamental frequencies γ and γ' is found at 2.54 kT and 2.4 kT, respectively. The beating seen in panel (a) is associated with the two frequencies η and η' (10.81 kT). In fact, the group of four frequencies around 11 kT comprises two fundamental frequencies very close to each other (η and η') whereas the others arise from mixing with once and twice the low frequency α (0.38 kT). In addition, all frequencies above 11 kT were identified to arise from mixing of η with δ and α in various combinations.

The angular dependence for field rotated in the basal plane is shown in Fig. 2 (a). Note that the trace depicted in Fig. 1 corresponds to an angle of 45° in Fig. 2(a). The main frequencies observed in Fig. 1 obey distinct angular dependences. α is shifted to lower values and vanishes for angles of more than 20° away from (110). Here, a new frequency ι of similar size is seen all the way to field orientation parallel to (100) with small angular dependence. β is not seen for field orientations away from the (110) direction. γ is shifted to lower values and reaches

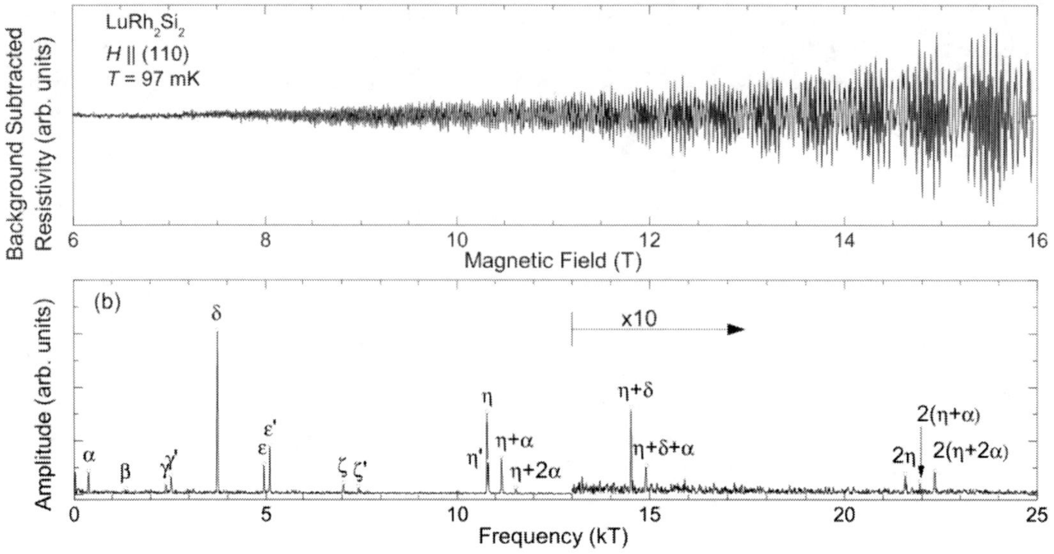

Figure 1. Shubnikov-de Haas oscillations. (a) Oscillatory part of the resistivity between 6 T and 16 T for field along the (110) direction measured at 100 mK. (b) Fourier transform indicating oscillation frequencies. Data above 13 kT are 10 times magnified for better view of the high frequency peaks.

1.6 kT for field along (100). δ extends only a few degree with virtually no change of its value. ϵ obeys non-monotonic behaviour extending over the whole angular range. A new frequency θ emerges for intermediate angles, whereas ζ vanishes for more than 5° away from (110). η shows a strong increase from 10.78 kT to 12.36 kT as the field is rotated towards (100). The splitting into two frequencies, i.e., η and η' was not observed away from (110). The mixing of frequencies discussed for the single trace in Fig. 1 is confirmed in the rotation study. In summary, fundamental frequencies are observed in the range 0.08 kT $< F <$ 12.5 kT.

These fundamental frequencies can be compared to the calculated quantum oscillation frequencies of $YbRh_2Si_2$. Calculations were performed by P.M.C. Rourke et al. [1, 8] using the local density approximation including spin orbit interactions. This method is known to give good descriptions of the Fermi surface topology for non-correlated compounds (like $LuRh_2Si_2$) as well as for Ce-based heavy fermion systems. We note that for Yb-based heavy fermion systems however, a renormalized band method might be necessary [11]. The comparison of calculated and observed quantum oscillation frequencies is done in Fig. 2 (b)-(d). First, Fig. 2 (b) depicts the predicted orbits for $LuRh_2Si_2$ which was used to approximate the "small" Fermi surface configuration of $YbRh_2Si_2$. We note that these calculations employed the lattice parameters of $YbRh_2Si_2$ (a = 4.010 and c = 9.841) which differ only marginally from those of $LuRh_2Si_2$ (a = 4.007 and c = 9.839). Due to these almost identical lattice parameters the comparison of our SdH measurements with the calculated orbits should hold to a quantitative level. However, there appears to be rather limited agreement between measured (Fig. 2 (a)) and calculated frequencies (Fig. 2 (b)). Best agreement is seen for the γ orbit with both satisfactory agreement of the absolute value and the angular dependence to the predicted values of an orbit related to the so-called "doughnut" Fermi surface sheet. Rough agreement of ϵ with a second orbit of the "doughnut" sheet is seen although with significantly different angular dependence. For η the

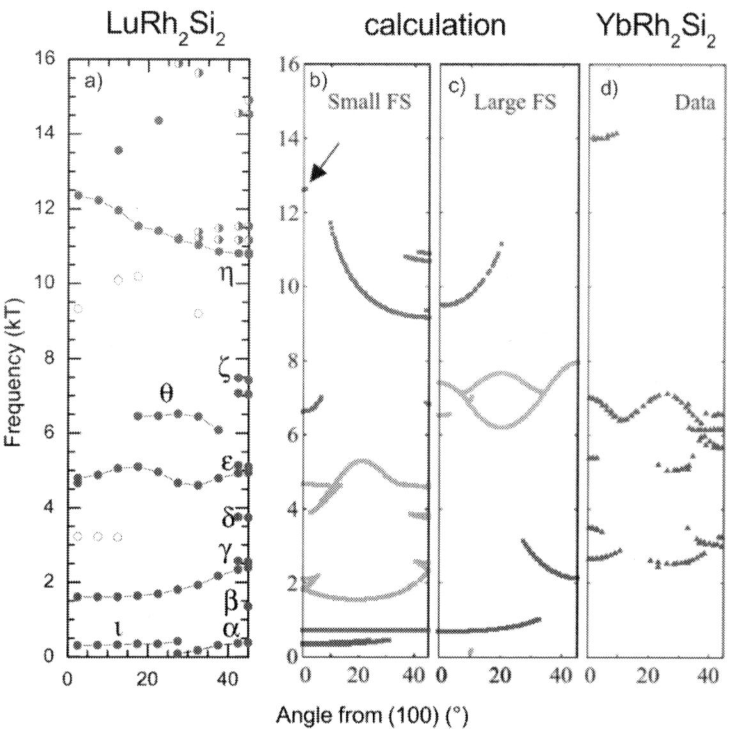

Figure 2. Angular dependence of quantum oscillation frequencies in $LuRh_2Si_2$ and $YbRh_2Si_2$. (a) Angular dependence of the SdH oscillation frequencies of our study on $LuRh_2Si_2$. Fundamental frequencies correspond to filled symbols whereas open symbols mark frequencies which have been identified as harmonics whereas half-filled symbols denote frequencies which were identified to arise from mixing of fundamental frequencies. Lines are guides to the eye, (b) and (c) reproduce the calculated angular dependences of quantum oscillation frequencies for $LuRh_2Si_2$ and $YbRh_2Si_2$, respectively from Sutton et al. [9]. Green, red, and purple represent the "doughnut", "jungle gym", and "pillbox" Fermi surface sheet, respectively. (d) Shows the dHvA results on $YbRh_2Si_2$ of Sutton et al. [9]. All panels have the same scales and show data for rotation of the magnetic field in the basal plane from (100) to (110) with the angle measured against (100).

agreement becomes unsatisfactory. It may be related to an orbit of the "jungle gym" although this should have a stronger angular dependence and is not predicted to extend to (100). ι might be mapped to another orbit of the "jungle gym" sheet whereas no agreement can be found for the α orbit with its strong angular dependence. The orbit predicted for the "pillbox" sheet expected at 1.4 kT is not observed.

An agreement with the predictions for the "large" Fermi surface calculation (Fig. 2 (a)) is not expected as these were done for electronic configuration of $YbRh_2Si_2$. The comparison with the dHvA results on $YbRh_2Si_2$ shows only remote similarities. The ϵ and θ orbits lie in a range where dHvA oscillations were observed but their angular dependence seems to be very different. No SdH oscillations were seen resembling the 14 kT orbit detected in $YbRh_2Si_2$, which was assigned to the "jungle gym" sheet of the "small" Fermi surface configuration. Although it is difficult to draw conclusions regarding $YbRh_2Si_2$ from the absence of this orbit in the SdH measurements on $LuRh_2Si_2$, this seems to be significant. With the η, ι, and probably the ζ orbit we detected key orbits assigned to the "jungle gym" surface. Consequently, the absence of an orbit similar to the 14 kT orbit of $YbRh_2Si_2$ indicates that the assignment of this orbit to the "small" Fermi surface configuration is questionable. In addition, the band structure calculations predict the 14 kT orbit to be limited to field orientations virtually parallel to the (100) direction, only. More detailed band structure calculations such as a renormalized treatment may be required for quantitative comparisons to the measured Fermi surface orbits of $YbRh_2Si_2$ [9].

4. Conclusion

In summary, a large set of SdH orbits has been observed for $LuRh_2Si_2$. The limited agreement with band structure calculations calls for further experimental and theoretical studies. We suggest to improve the band structure calculations by refining the underlying crystallographic parameters. Although the lattice parameters are known to high accuracy, the z-parameter specifying the position of the Si atoms has not yet been measured. In fact, preliminary results indicate a significant dependence of the Fermi surface structure on the z-parameter. Better understanding of the electronic structure of $LuRh_2Si_2$ may help to improve the band structure calculations for $YbRh_2Si_2$ which, in the case of the renormalized treatment, are based on the non-f band structure. In addition, these calculations may provide a better understanding of the temperature and field dependence of the Hall effect in $LuRh_2Si_2$.

Acknowledgments

We acknowledge fruitful discussions with P.M.C. Rourke and G. Zwicknagl. This work was partially supported by the Alexander von Humboldt foundation, the Royal Society, and the EPSRC.

References

[1] Rourke P M C *et al.* 2008 *Phys. Rev. Lett.* **101** 237205
[2] Friedemann S *et al.* 2010 *PNAS* **107** 14547–14551
[3] Friedemann S *et al.* 2010 *Phys. Rev. B* **82** 035103
[4] Si Q, Rabello M S, Ingersent K and Smith J L 2001 *Nature* **413** 804–808
[5] Coleman P, Pépin C, Si Q and Ramazashvili R 2001 *J. Phys.: Condens. Matter* **13** R723–R738
[6] Senthil T, Sachdev S and Vojta M 2003 *Phys. Rev. Lett.* **90** 216403
[7] Paschen S *et al.* 2004 *Nature* **432** 881–885
[8] Rourke P M C *et al.* 2009 *Journal of Physics: Conference Series* **150** 042165–
[9] Sutton A B *et al.* 2010 *Phys. Status Solidi B* **247** 549–552
[10] Maquilon S 2007 Ph.D. thesis UC Davis
[11] Zwicknagl G 2011 *Journal of Physics: Condensed Matter* **23** 094215

Cu-NQR of $CeCu_2Si_2$ under high pressure

K. Fujiwara[1], M. Iwata[1], Y. Okazaki[1], Y. Ikeda[2], S. Araki[2], T. C. Kobayashi[2], K. Murata[3], C.Geibel[4] and F. Steglich[4]

[1] Department of Materials Science, Shimane University, Matsue 690-8504, Japan
[2] Graduate School of Natural Science and Technology, Okayama University, Okayama 700-8530, Japan
[3] Division of Molecular Materials Science, Graduate School of Science, Osaka City University, Osaka 558-8585, Japan
[4] Max Planck Institute for Chemical Physics of Solids, Nothnitzer Strase 40, 01187 Dresden, Germany

E-mail: `fujiwara@riko.shimane-u.ac.jp`

Abstract. We have measured ^{63}Cu-NQR spectral lines and nuclear spin-lattice relaxation time T_1 under pressure up to 5.4 GPa. Daphne oil 7474 is used as the pressure-transmitting medium to obtain good hydrostaticity. NQR frequency $^{63}\nu_Q$ above 4 GPa suddenly decreases from the linear pressure dependence in the low pressure range ($P \leq 3.5$ GPa). The observed sudden downward deviation of $^{63}\nu_Q$ is associated with an increase of Ce valence. Above 4.5 GPa, the linear pressure dependence of $^{63}\nu_Q$ is observed again, most likely related to the sharp crossover to the high valence state.

1. Introduction

Heavy fermion compound $CeCu_2Si_2$ shows an anisotropic superconductivity with a line-node superconducting (SC) gap, mediated by antiferromagnetic (AFM) spin fluctuations. T_c increases drastically around 4 GPa under high pressure. The SC phase diagram has been precisely determined from resistivity and ac specific heat measurements under pressure. [1] The residual resistivity is enhanced around the critical pressure $P_v = 4.5$ GPa, while the coefficient A of T^2-low of resistivity decreases drastically at ~ 5 GPa. A similar increase in T_c has been observed in the isostructural compound $CeCu_2Ge_2$ above 8 GPa. [2] In addition, resistivity measurement on $CeCu_2(Si_{0.9}Ge_{0.1})_2$ have revealed two distinct SC phases under pressure, [3] suggestive of two distinct SC pairing mechanisms. It has been proposed that these anomalies are associated with an abrupt change of the Ce valence. [1, 4] They pointed out that, in $CeCu_2Si_2$, if the critical end temperature of the first-order valence transition is close to 0 K, the ensuing critical valence excitations can mediate superconductivity without being overwhelmed by thermal fluctuations.

We reported NQR study in $CeCu_2Si_2$ under pressure with using the pressure-transmitting medium Daphne oil 7373 in order to elucidate Ce spin and valence state. [5] Nuclear spin-lattice relaxation rate $1/T_1$ is drastically suppressed by applying pressure, suggesting that AFM spin fluctuations are also largely suppressed. On the other hand, T_c goes up from 0.7 K to 1.64 K above 3 GPa. Hence it is difficult to apply the spin fluctuation scenario to the high pressure phase. $1/T_1$ shows T^3-law in the superconducting state and this behavior is ascribed to anisotropic superconductivity with a line node gap. Bulk superconductivity suddenly disappears at 4.8 GPa accompanies with a rapid decrease in the density of state at the Fermi level.

NQR frequency $^{63}\nu_Q$ for ^{63}Cu nuclei linearly increases with pressure up to 3 GPa. $^{63}\nu_Q$ deviates from the linear relation with applying more pressure. The observed deviation is associated with change of the Electric Field Gradient at Cu sites because of valence change of Ce. However, line-width of NQR spectrum becomes broad above 4 GPa owing to the solidification of Daphne oil 7373. Therefore, we measure Cu-NQR under hydrostatic pressure with using the new medium (Daphne oil 7474) to clear pressure dependence of valence state above 4 GPa.

2. Experimental

Nearly hydrostatic pressure was applied by an indenter cell [6] below 4.1 GPa, and by a opposed-anvil cell [7] up to 5.4 GPa. We improved the opposed-anvil cell with a conical stopper, as seen in Fig. 1 (b). The anvil cell is capable of generating pressure up to $P \sim 7$ GPa at low temperature. Daphne oil 7474 is used as the pressure-transmitting medium. The pressure values are determined by measuring T_c of a lead (Pb). A fine grain of Pb is inserted into the NQR coil with the sample powder. The T_c of Pb is determined by ac susceptibility measurement with an outer coil wound around the cell and pickup NQR coil to detect the signal. NQR measurement is performed by standard spin-echo pulse techniques. The NQR spectrum is obtained by Fourier transform of the spin-echo signal.

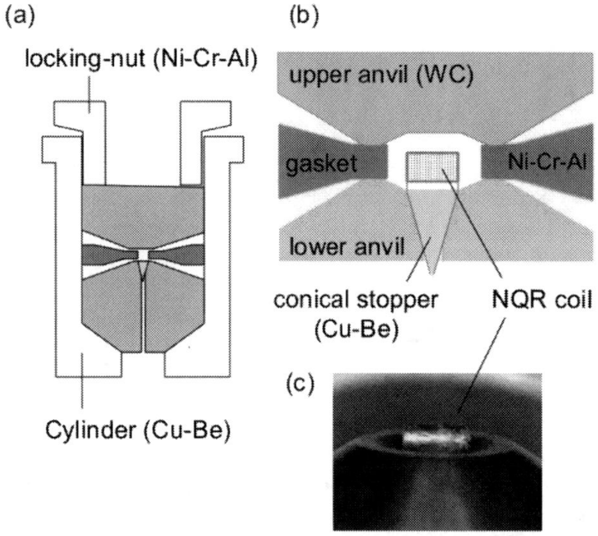

Figure 1. (a) Cross-sectional view of the opposed-anvil cell. (b) The enlarged view around sample space. The upper and lower anvils are made of nonmagnetic tungsten carbide and the gasket is made of Ni-Cr-Al alloy. The typical size of sample space is 2.5 mmϕ in diameter and 1.7 mm in depth at ambient pressure. (c) Arrangement for NQR measurement in the cell. NQR coil is 1 mm in diameter and 1.8 mm in length.

3. Results and Discussion

A solidification pressure of Daphne oil 7474 is 3.7 GPa at room temperature. [8] Therefore, NQR spectrum becomes drastically broad due to pressure inhomogeneity above 3.7 GPa. Since the value of solidifying pressure increases almost linearly with temperature up to 6.7 GPa at 100 °C, [8] NQR spectrum recovers sharp line-width after melting 7474 at high temperature. The full width at half maximum (FWHM) of the ^{63}Cu-NQR spectrum is 40 kHz for 7474 around 5 GPa, while FWHM is 64 kHz for 7373 at 4.8 GPa, [5] suggesting that 7474 has high-quality hydrostaticity above 4 GPa.

Figure 2 shows the P-dependence of $^{63}\nu_Q$ with Daphne oil 7373 (open circles, previous data [5]) and 7474 (solid circles, present data). Here $^{63}\nu_Q$'s with 7373 and 7474 were measured at 4.2 and 3.1 K, respectively. A linear increase in $^{63}\nu_Q$ was observed below 3.5 GPa, which is in agreement with a previous study below 2.6 GPa ($d\nu_Q/dp$ = 0.0952 MHz/GPa). [9] However, for $P \geq 4$ GPa, $^{63}\nu_Q$ deviates rapidly from the solid line with the slope of 0.0952 MHz/GPa. It should be noteworthy that the linear relation of $^{63}\nu_Q$, with the same slope in the low-P region, is observed again above the critical pressure $P_v = 4.5$ GPa, that is , $d\nu_Q/dP$ changes only near the critical pressure P_v.

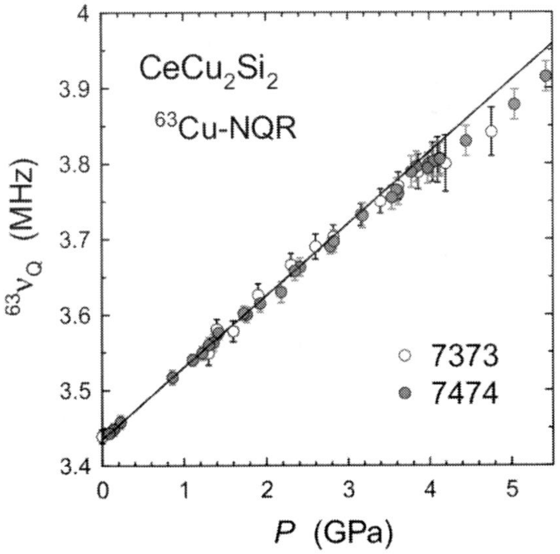

Figure 2. Pressure dependence of the NQR frequency $^{63}\nu_Q$ with Daphne oil 7373 and 7474 at 4.2 K and 3.1 K, respectively. The data with 7373 was previously obtained by the indenter cell. [5] The linear increase in $^{63}\nu_Q$ ($d\nu_Q/dp$ = 0.0952 MHz/GPa) below $P = 2.58$ GPa, [9] is shown by the solid line.

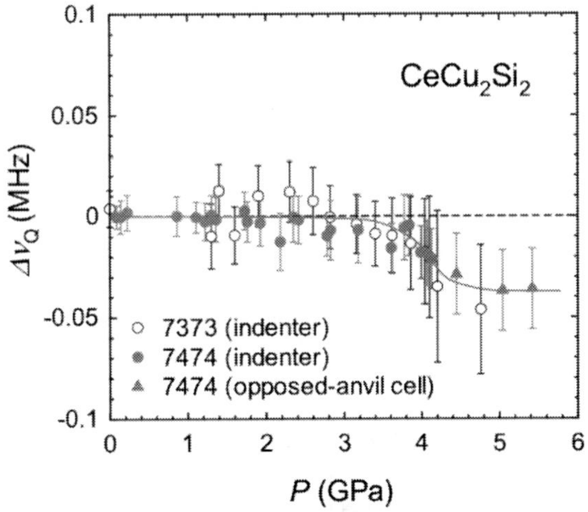

Figure 3. The deviations from the linear P-dependence of NQR frequency $\Delta\nu_Q = {}^{63}\nu_Q - 3.435 - 0.0952\times P$ against pressure. The solid circles ($P \leq 4.1$ GPa) and triangles ($P \geq 4.4$ GPa) indicate data measured with using the indenter cell and the opposed anvil cell, respectively.

Figure 3 shows the deviations from the linear P-dependence of NQR frequency $\Delta\nu_Q = {}^{63}\nu_Q - 3.435 - 0.0952\times P$. The pressure dependence of $^{63}\nu_Q$ in the case of localized/itinerant 4f electrons for Ce ions is obtained by the band calculation in $LaCu_2Si_2/CeCu_2Si_2$ using the lattice parameters determined by the XRD measurement. [10] The results shows that the $^{63}\nu_Q$ for itinerant state (Ce^{4+}) lies below that for localized state (Ce^{3+}) with the same slope as the experimental result. Since $\Delta\nu_Q$ definitely decreases in the narrow pressure region, suggesting that Ce valence suddenly increase near P_v. However, very recently, from X-ray spectroscopy at Ce L_3 edge, the valence state of $CeCu_2Si_2$ have been investigated at $T = 14$ K. [11] Although a clear increase of the Ce valence is found, the valence changes continuously with applying pressure. Thus pressure dependence of NQR frequency seems to be inconsistent with X-ray results. The disagreement may be related to the temperature difference between NQR ($T = 3$ K) and X-ray absorption ($T = 14$ K) experiment.

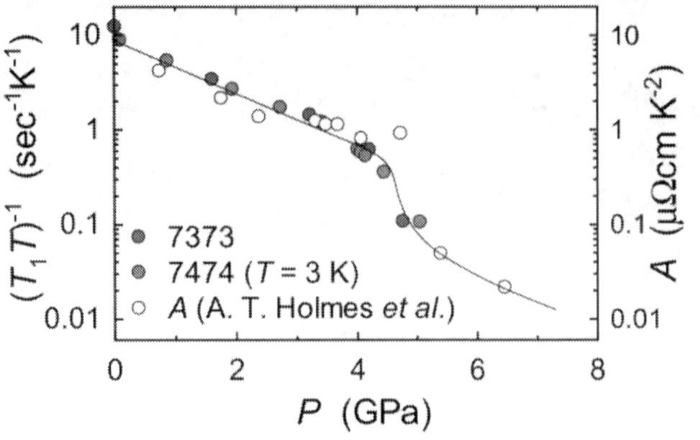

Figure 4. P-dependence of $(T_1T)^{-1}$ and coefficient A of the T^2 resistivity law. The solid and open circles indicate $(T_1T)^{-1}$ and A (taken from ref. 1), respectively. $(T_1T)^{-1}$'s for Daphne 7474 are obtained at $T \sim 3$ K. The solid line is a guide for the eye.

Figure 4 shows that the P-dependence of $(T_1T)^{-1}$ is in good agreement with that of the coefficient A of the T^2 resistivity law. Here $(T_1T)^{-1}$ for 7474 is the value at $T \sim 3$ K. These results are in accordance with a Fermi liquid picture, as both $(T_1T)^{-1}$ and the coefficient A are proportional to the square of the density of states at the Fermi level. It is noted that the effective electron mass clearly drops above P_v.

In conclusion, we succeeded in measuring pressure dependence of NQR frequency $^{63}\nu_Q$ for ^{63}Cu nuclei up to 5.4 GPa, using the opposed-anvil high pressure cell. It was confirmed that $^{63}\nu_Q$ deviates rapidly from the linear P-dependence near the critical pressure $P_v = 4.5$ GPa, associated with increase of Ce valence. The linear relation of $^{63}\nu_Q$ is observed again above the critical pressure P_v, suggesting crossover to high valence state.

4. Acknowledgment

This work was partially supported by a Grant-in-Aid for Scientific Research C (No.21540338) and B (No.20340090) from Japan Society for the Promotion of Science.

References

[1] A. T. Holmes, D. Jaccard, and K. Miyake: Phys. Rev. B **69** (2004) 024508.
[2] E. Vargoz and D. Jaccard: J. Magn. Magn. Mater. **177-181** (1998) 294.
[3] H. Q. Yuan, F. M. Grosche, M. Deppe, C. Geibel, G. Sparn, and F. Steglich: Science **302** (2003) 2104, H. Q. Yuan, F. M. Grosche, M. Deppe, G. Sparn, C. Geibel, and F. Steglich: Phys. Rev. Lett. **96** (2006) 047008.
[4] S. Watanabe and K. Miyake: J. Phys.: Condens. Matter **23** (2011) 094217.
[5] K. Fujiwara, Y. Hata, K. Kobayashi, K. Miyoshi, J. Takeuchi, Y. Shimaoka, H. Kotegawa, T. C. Kobayashi, C. Geibel, and F. Steglich: J. Phys. Soc. Jpn. **77** (2008) 123711.
[6] T. C. Kobayashi, H. Hidaka, H. Kotegawa, K. Fujiwara, and M. I. Eremets: Rev. Sci. Instruments **78** (2) (2007) 023909.
[7] K. Kitagawa, H. Gotou, T. Yagi, A. Yamada, T. Matsumoto, Y. Uwatoko, and M. Takigawa: J. Phys. Soc. Jpn. 79 (2010) 024001.
[8] K. Murata, K. Yokogawa, H. Yoshino, S. Klotz, P. Munsch, A. Irizawa, M. Nishiyama, K. Iizuka, T. Nanba, T. Okada, Y. Shiraga, and S. K. Aoyama: Rev. Sci. Instrument **79** (2008) 085101.
[9] Y. Kawasaki, K. Ishida, T. Mito, C. Thessieu, G.-q. Zheng, Y. Kitaoka, C. Geibel, and F. Steglich: Phys. Rev. B **63** (2001) 140501.
[10] H. Harima: unpublished.
[11] J.-P. Rueff, M. Taguchi, M. Sikora, J.-P. Itié, F. Baudelet, D. Braithwaite, G. Knebel, and D. Jaccard: Phys. Rev. Lett. **106** (2011) 186405.

Inelastic neutron scattering study of Ni-substituted $Ce_{0.5}Fe_4Sb_{12}$ skutterudite compounds

L. Girard[a,b], D. T. Adroja[c], L. Chapon[c], J. W. Taylor[c], R. Viennois[a,d], D. Ravot[a], S. Paschen[e]

[a] Institut Charles Gerhardt Montpellier, Université Montpellier 2, 34095 Montpellier, France

[b] Institut de Chimie Séparative de Marcoule, Université Montpellier 2, 34095 Montpellier, France

[c] ISIS Facility, Rutherford Appleton Laboratory, CCLRC, Chilton, Didcot, Oxon, OX11, 0QX, United Kingdom

[d] Max Planck Institut für Chemische Physik fester Stoffe, Nöthnitzer Str. 40, 01187 Dresden, Germany

[e] Institut of Solid State Physics, Vienna University of Technology, Wiedner Hauptstr. 8-10, 1040, Austria

* Corresponding author

Abstract. An inelastic neutron scattering study of the filled and partially-filled skutterudite compounds RFe_4Sb_{12} and $R_{0.5}Fe_{2.75}Ni_{1.25}Sb_{12}$ (where R = Ce and La) was carried out to understand the nature of the spin dynamics. Strong magnetic scattering was observed in $Ce_{0.5}Fe_{2.75}Ni_{1.25}Sb_{12}$ at ~ 5 meV. The integrated intensity of this peak does not follow the Ce^{3+} form factor, but exhibits a maximum at a momentum transfer ($|Q|$) of 2 $Å^{-1}$. We attribute this feature to a Ce^{3+} crystal field excitation in the presence of magnetic exchange interactions. This picture is supported by thermodynamic and magnetic properties. Finally, we confirm the presence of a spin gap in $CeFe_4Sb_{12}$ suggested by our previous work.

Skutterudites of the MT_4X_{12} type (M = rare earth or actinide, T = transition metal, X = Sb, P) which crystallize in a body centred cubic structure have attracted much attention because of their high thermoelectric efficiency. This is generally attributed to the reduction of the lattice thermal conductivity by the interaction between acoustic phonons of the framework and low-energy vibration modes of the rare-earth ions that are weakly bonded in oversized atomic cages [1,2]. Filled skutterudites are also of considerable interest in fundamental physics because of their various ground states: superconductivity, heavy fermion behaviour, magnetic ordering, metal-insulator transitions, hybridization gap semiconducting or non-Fermi liquid behaviour. These different ground states result from a competition between the intersite RKKY and the onsite Kondo interaction. When the Kondo interaction prevails, the hybridization between the 4f and conduction electrons may lead to the formation of a hybridization gap at the Fermi level (E_F) [3]. In $CeFe_4Sb_{12}$, both the Kondo effect and a hybridization gap have been observed [4,5]. The relatively low specific heat, magnetic Grüneisen parameter and Kondo temperature and the presence of trivalent cerium [4,6-9] suggest that $CeFe_4Sb_{12}$ is situated between the limits of heavy fermion behaviour (as in $CeAl_3$) and intermediate valence behaviour (as in $CeSn_3$) [3,5]. In the partially-filled $Ce_yFe_{4-x}Ni_xSb_{12}$ skutterudites, however, no clear evidence for incoherent Kondo scattering was observed. This could be due to the much smaller charge carrier concentration in this latter system [7-9].

Therefore, in order to further understand the spin dynamics of this class of compounds, we have investigated $Ce_{0.5}Fe_{2.75}Ni_{1.25}Sb_{12}$ using inelastic neutron scattering (INS), heat capacity and magnetic measurements. Our study clearly shows strong magnetic scattering at 5 meV in $Ce_{0.5}Fe_{2.75}Ni_{1.25}Sb_{12}$, in strong contrast to the signal observed in $CeFe_4Sb_{12}$ at about 40 meV [5].

The synthesis details of the single phase polycrystalline samples were reported earlier [4-9]. INS measurements were carried out using the time-of-flight chopper spectrometer MARI at the ISIS Facility (UK). Samples were enclosed in aluminium foil and positioned with a hollow geometry in a thin wall Al sample can. An incident neutron energy of E_i = 20 meV was used for all the measurements. This gives an energy resolution of about 0.65 meV (FWMH) at the elastic scattering position. For details of the experimental set-up, see Ref. 10. All our results are normalized with a vanadium standard. The heat capacity was measured between 2 and 300 K using a microcalorimeter in a commercial PPMS apparatus from Quantum Design. The magnetic susceptibility was measured between 2 and 300 K under zero field cooling conditions with 0.5 T applied magnetic field in a vibrating sample magnetometer from Oxford.

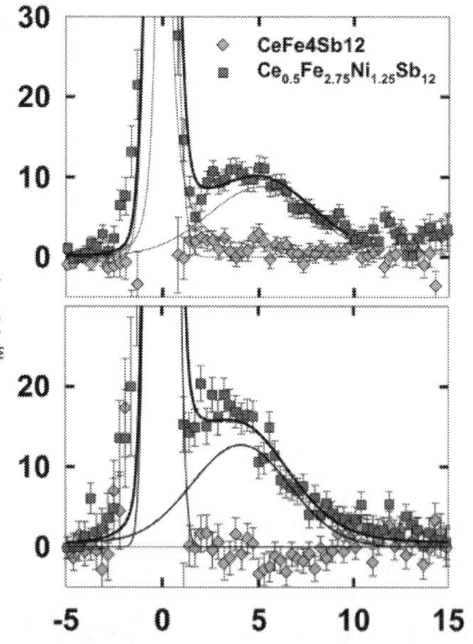

(a)

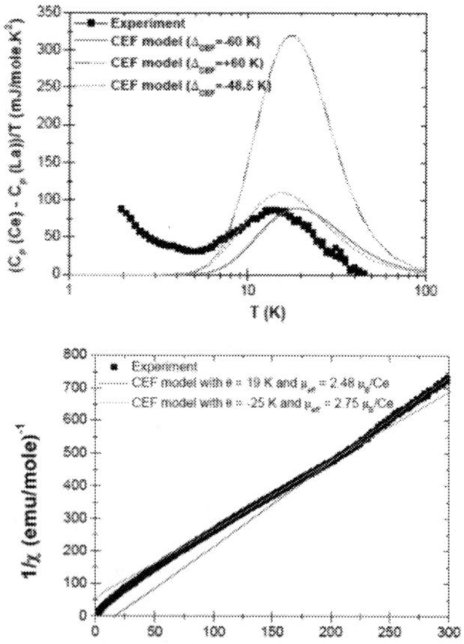

(b)

Figure 1 Magnetic response, obtained as explained in the text, for $CeFe_4Sb_{12}$ and $Ce_{0.5}Fe_{2.75}Ni_{1.25}Sb_{12}$ at 15 K and at low angle, ϕ_{avg} (wave vectors Q), (a) 8.1° (0.44 Å^{-1}) and (b) 21.5° (1.16 Å^{-1}). The thick solid lines represent the fit based on our CEF model, the thin lines its different components (see text).

Figure 2 (a) Cerium contribution of the heat capacity of $Ce_{0.5}Fe_{2.75}Ni_{1.25}Sb_{12}$ and calculated heat capacity using the model described in the text. (b) Magnetic susceptibility of $Ce_{0.5}Fe_{2.75}Ni_{1.25}Sb_{12}$ with the calculated magnetic susceptibility using the model described in the text.

In Fig. 1 we show the magnetic part of the inelastic response at two different average scattering angles (or wave vectors Q), ϕ_{avg}= 8.1° (0.44 Å^{-1}) and 21.5° (1.16 Å^{-1}) for both $CeFe_4Sb_{12}$ and $Ce_{0.5}Fe_{2.75}Ni_{1.25}Sb_{12}$. At low Q, we see a strong inelastic response in $Ce_{0.5}Fe_{2.75}Ni_{1.25}Sb_{12}$ at about 5 meV that is absent in both $CeFe_4Sb_{12}$ and in the non-magnetic reference compound $La_{0.5}Fe_{2.75}Ni_{1.25}Sb_{12}$. The magnetic scattering was determined by subtracting the signal of $LaFe_4Sb_{12}$ and $La_{0.5}Fe_{2.75}Ni_{1.25}Sb_{12}$ from the signal of the respective Ce compounds. In addition, we accounted for the difference in total scattering cross-sections, σ(Ce compound) –h.σ(La compound), where h is 0.94 for $CeFe_4Sb_{12}$ and 0.97 for $Ce_{0.5}Fe_{2.75}Ni_{1.25}Sb_{12}$ (see Refs. 5 and 10 for details). When Q is increased above 2 $Å^{-1}$ (data not shown here), the difference between $Ce_{0.5}Fe_{2.75}Ni_{1.25}Sb_{12}$ and all other compounds decreases and becomes very small above 4 $Å^{-1}$(see also discussion of Fig. 3). Hence, the observed Q dependence clearly indicates the presence of a strong magnetic contribution to the scattering in $Ce_{0.5}Fe_{2.75}Ni_{1.25}Sb_{12}$, because its intensity is maximal at low Q. The absence of a magnetic response for $CeFe_4Sb_{12}$ at low Q agrees with the previous observation of a spin gap at low energy for $CeFe_4Sb_{12}$ [5]. Because $Ce_{0.5}Fe_{2.75}Ni_{1.25}Sb_{12}$ has, like $La_yFe_{4-x}Ni_xSb_{12}$ [8,9,11] a low charge carrier concentration, we suggest that the magnetic peak at 5 meV corresponds to a crystalline electric field (CEF) excitation of the J = 5/2 ground state multiplet of the Ce^{3+} ion, as previously observed in the cubic system $CeSbNi_x$ [12]. In skutterudites, we have the following CEF Hamiltonian for the Ce^{3+} ion in T_h point group symmetry:

$$H_{CF} = B_4^0 (O_4^0 + 5O_4^4)$$

where B_4^0 is the CEF parameter and O_m^n are the Stevens operators [13]. Under the cubic (T_h) CEF potential the J = 5/2 multiplet of the Ce^{3+} ions splits into a doublet (Γ_7) and a quartet (Γ_8), with energy eigenvalues of -240 B_4^0 and 120 B_4^0, respectively and therefore we have Δ_{CEF}=360 B_4^0 [13].
From the peak position alone, the sign of B_4^0 cannot be determined and thus it cannot be decided whether Γ_7 or Γ_8 is the ground state. However, this assignment is possible when also the intensities are taken into account. By fitting the peaks to the above CEF model, we obtain B_4^0= -14.3(6)μeV and -11.6(7) μeV for the best fit of the peaks for Q = 0.44 $Å^{-1}$ and 1.16 $Å^{-1}$, respectively; thus Γ_8 is the ground state according to this analysis. The slightly different value found for B_4^0 at the two Q positions is due to the unusual Q dependence of the intensity observed in the magnetic scattering..

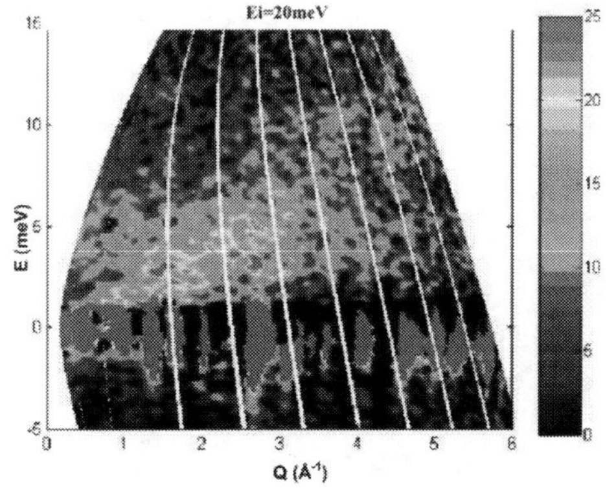

Figure 3 Contour plot of the momentum (Q) and energy (E) transfer of magnetic scattering $S_M(Q,E)$ from $Ce_{0.5}Fe_{2.75}Ni_{1.25}Sb_{12}$ at 15 K.

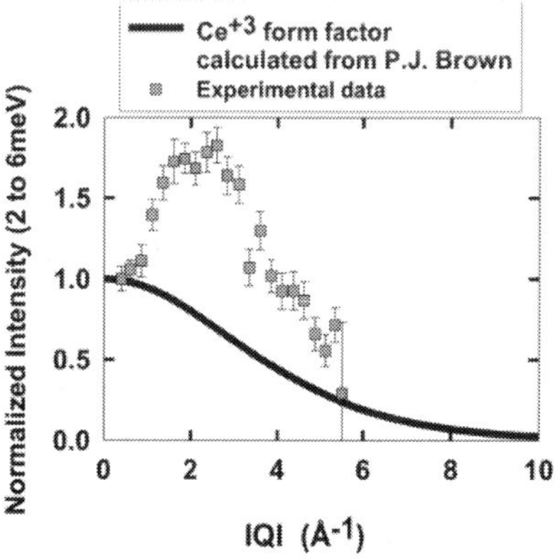

Figure 4 The Q-dependence of the inelastic peak intensity integrated over energies from 2 to 6 meV (symbols).

To test the above model, we have measured the heat capacity and the magnetic susceptibility of $Ce_{0.5}Fe_{2.75}Ni_{1.25}Sb_{12}$ (see Fig. 2). The Ce contribution of the heat capacity of $Ce_{0.5}Fe_{2.75}Ni_{1.25}Sb_{12}$, obtained by subtracting the phonon part estimated from the measured heat capacity of the La reference, is shown in Fig. 2-a. The agreement between the data and the theoretical curve calculated with the above B_4^0 values is much better than in the case where the ground state is Γ_7. Note that the electronic heat capacity of the non-magnetic reference $La_{0.5}Fe_{2.75}Ni_{1.25}Sb_{12}$ is quite low (about 7 mJ/mole.K^2). For the magnetic susceptibility, the above CEF model is insufficient to reproduce the experimental data at low temperature quantitatively whatever the ground state used in the calculation. The discrepancy points to the limit of the model and might be due to additional magnetic interactions. The absence of a peak in the magnetic susceptibility shows the absence of a hybridization gap.

In Fig. 3 and 4, we show a contour plot of the magnetic scattering at 20 K as a function of energy transfer and |Q|, and the integrated intensity of the magnetic scattering (integrated between 2 and 6 meV energy transfer) as a function of |Q|, respectively. A peak appears at Q ~ 2 $Å^{-1}$, which is a very unusual behaviour that has been observed so far only in two Ce-based compounds ($CeFe_4Sb_{12}$ [5] and $Ce_3Pd_{20}Si_6$ [14]). For magnetic excitations of either crystal field or spin gap type, one expects that the Q dependence of the peak intensity follows the square of the magnetic form factor of a free Ce^{3+} ion (see solid line in Fig. 4, where we use the $F(Q)^2$ obtained by P. J. Brown's method [15]). "Unusual" magnetic scattering, with departure from the Ce^{3+} free-ion form factor only at low Q, was observed in the intermediate-valence compound $CeSn_3$ [16]. This was interpreted as a contribution from the polarization of the Ce 5d electrons. It may be that the same mechanism acts in $Ce_{0.5}Fe_{2.75}Ni_{1.25}Sb_{12}$. In the case of $Ce_3Pd_{20}Si_6$ [14], this was interpreted as due to quantum critical fluctuations.

In conclusion we observe magnetic scattering in $Ce_{0.5}Fe_{2.75}Ni_{1.25}Sb_{12}$ and attribute it to a CEF excitation. The fact that such excitation is absent in $CeFe_4Sb_{12}$ is believed to be due to its larger charge carrier concentration and stronger 4f-conduction electron hybridization that broaden the 4f levels and thus collapses the CEF excitations [5]. A further conclusion can also be drawn from our present results: the CEF splitting is relatively small in Ce based skutterudites as proposed earlier [4-8].

We would like to thank K. A. McEwen, S. M. Bennington, J. Lorenzana, J.-G. Park and Dr E.A. Goremychkin, for fruitful discussions. S. P. acknowledges financial support from the Austrian Science Fund (project P19458-N16).

References

[1] B.C. Sales, D. Mandrus and R.K. Williams, Science 272, 1325 (1996).
[2] M. M. Koza, M. R. Johnson, R. Viennois, H. Mutka, L. Girard, D. Ravot, Nat. Mater. 7, 805 (2008).
[3] For areview, see P. Riseborough, Adv. Phys. 49, 257 (2000).
[4] R. Viennois, S. Charar, D. Ravot, A. Mauger, P. Haen, J. C. Tedenac, J. Phys. : Cond. Mat. 18, 5371 (2006).
[5] R. Viennois, L. Girard, L. C. Chapon, D. T. Adroja, R. I. Bewley, D. Ravot, P. S. Riseborough, S. Paschen, Phys. Rev. B 76, 174438 (2007) and Refs. therein.
[6] R. Viennois, L. Girard, D. Ravot, S. Paschen, S. Charar, A. Mauger, P. Haen, J. C. Tedenac, Phys. Rev. B 80, 155109 (2009).
[7] L. Chapon, D. Ravot, J. C. Tedenac, J. Alloys Compds 299, 68 (2000).
[8] R. Viennois, D. Ravot, J. C. Tedenac, S. Charar, A. Mauger, Mat. Sci. Eng. B 119, 1 (2005).
[9] R. Viennois, L. Girard, D. Ravot, F. Terki, S. Charar, J. C. Tedenac, A. Bentien, S. Paschen, F. Steglich, Proc. of the Int. Conf. of Thermoel. ICT03, p. 60 (2004).
[10] J. G. Park, K. A. McEwen, M. J. Bull, Phys. Rev. B 66, 94502 (2002).
[11] R. Viennois, L. Girard, D. Ravot, S. Charar, S. Paschen, Proc. of the 12th Int. Conf. on Narrow-gap Semiconductors, Inst. Phys. Conf. Series 187, p. 229, (2006), edited by J. Kono and J. Léotin, Taylor and Francis (2006).
[12] D. T. Adroja, J. G. M. Armitage, P. C. Riedi, M. H. Jung, Z. Tun, T. Takabatake, Phys. Rev. B 62, 12181 (2000).
[13] K.R. Lea, M.J.M. Leask, W.P. Wolf, J. Phys. Chem. Solids, 23, 1381 (1962) and Refs.therein.
[14] P. P. Deen, A. M. Strydom, S. Paschen, D. T. Adroja, W. Kockelmann, and S. Rols, Phys. Rev. B 81, 064427 (2010) and Refs. therein.
[15] P.J. Brown, International tables for crystallography; Mathematical, Physical and Chemical tables, edited by A.J.C. Wilson (Kluwer Academic Publisher, Amsterdam, 1999).
[16] J. X. Boucherle, G. Fillion, J. Flouquet, F. Givord, P. Lejay, J. Schweizer, J. Mag. Mag. Mat. 54-57, 421 (1986) and Refs. therein.

Detection of the Influence of the Magnetic Field Direction on a Paramagnetic Curie Temperature of $EuB_{6-x}C_x$

Yu V Goryunov[1], **A V Levchenko** [2]

[1] Kazan Physical-Technical Institute of the RAS, 420029, Kazan, Russia
[2] Institute of Material Science Problems of the NASU, 03680, Kiev, Ukraine

e-mail: gorjunov@kfti.knc.ru

Abstract. For the first time found that the Curie-Weiss law depends on the crystallographic direction along which the measuring field is directed. The ESR measurements of the $EuB_{6-x}C_x$ (x<0.1) single crystal indicated that the sample with x = 0.02 behaves in fields along the [111] - direction as a ferromagnetic, while for field along the [100] - the anti-ferromagnetic behavior.

1. Introduction
Complex forms of magnetic ordering, which are realized in the compounds of rare earth metals may be due to effects such as Jahn-Teller's effect occurring in the anionic complexes of these compounds under the influence of the magnetic field. There is suspicion of such behavior in respect europium hexaboride. Recently we have observed [1] in single crystal of europium hexaboride the anisotropy of magnetic susceptibility as anisotropy of induced magnetization $\boldsymbol{M}$. Such behavior of magnetic susceptibility in paramagnetic state is observed by other researchers [2, 3]. In general, the magnetic susceptibility χ is a tensor and components of magnetization can have not a proportional relation with the components of the external magnetic field and the Curie-Weiss's law can be different for different directions. Curie-Weiss's temperature can be anisotropic in case of anisotropic exchange interaction, it can depend on magnetic field direction and its value, it can indicate on magnetic quantum transition.

2. Results and Discussion

2.1. Experimental details
The ESR measurements were performed on frequency 9.25 GHz in TE_{102} rectangular cavity at the temperature 10÷300 K. We used for measurements the EuB_6 single crystal samples of identical form and size but different orientation of crystal axis, to estimate the temperature dependences of the ESR spectra parameters, namely: resonance field, linewidth, intensity, lineshape. The crystals were grown by zone melting as [1,4]. Curie-Weiss's temperatures were estimated by measuring the temperature dependences of inverse intensity of ESR signal. The induced by magnetic field magnetization of the sample was measured by measuring the shift of the resonance line under the influence of the demagnetizing field at perpendicular orientation of the sample plane to external magnetic field [1, 4].

2.2. Influence of Magnetic field direction on Curie-Weiss's temperature.
Observed ESR line had lorentzian shape with dysonian distortion in temperature range 150 -300 K. Also as before (see [4]), at T = 10 – 30 K was observed the magnetic phase separation in samples with

small impurities of carbon. For sample with more content of carbon the splitting of single ESR line was observed at higher temperature and the dependences of resonance field and linewidth are quite different. Figure 1 shows the temperature dependences of resonance fields and linewidth of the ESR for the two orientation of the plates [100] and [111] $EuB_{5.98}C_{0.02}$ single crystal sample. Dependence in figure 1a is a classic example of the temperature dependences for the strong paramagnet. Dependence in figure 1b is a typical temperature dependences for case the strong paramagnet with strong magnetic fluctuation.

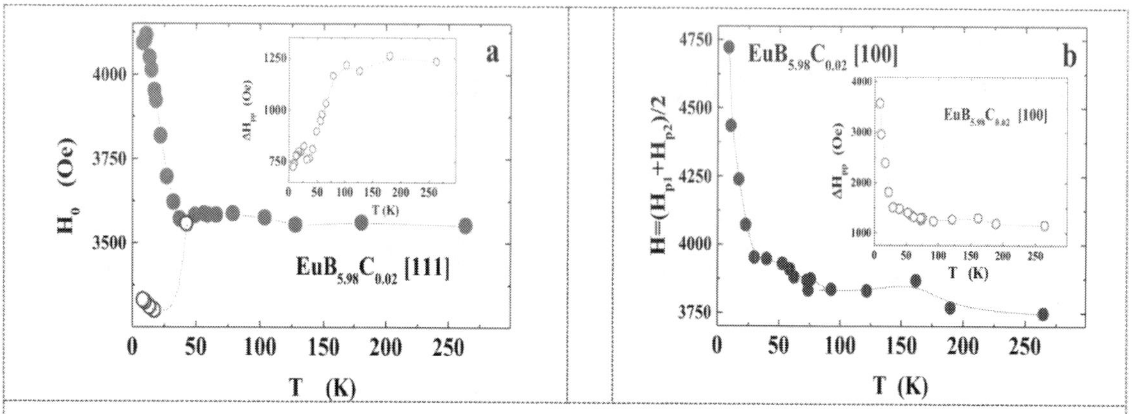

Figure 1. The temperature dependences on resonance fields and linewidth of the ESR in the $EuB_{5.98}C_{0.02}$ single crystal sample..

Recently in [1], we found a strong anisotropy of the induced magnetization of pure europium hexaboride. For pure hexaboride this anisotropy was in considerable amplification of the magnetization along [111] direction. Also, there is an amplification of the ferromagnetic properties along this direction and for anti-ferromagnetic samples with impurity carbon. This is most clearly evident for the samples of the $EuB_{5.98}C_{0.02}$. We found (see Fig. 2) that the single-crystal europium hexaboride with substituting 2% of boron by carbon behaves in fields around 4 kOe along the [111] - direction as a ferromagnetic, while for field along the [100] - direction its behavior resembles the

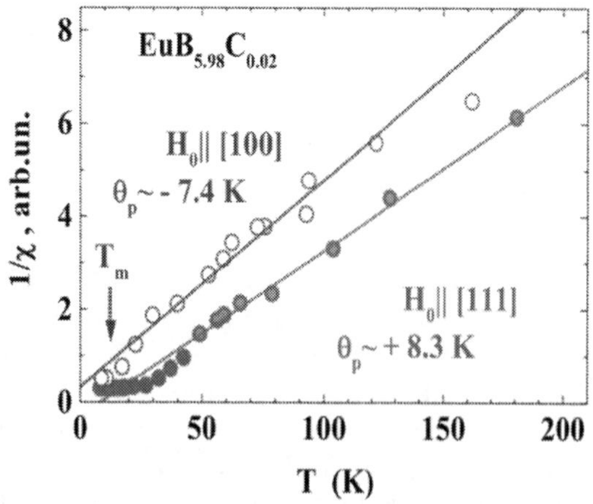

Figure 2. Temperature dependences of the inverse intensity of the ESR spectra for two orientations of the EuB_6 single crystal in sweep magnetic field.

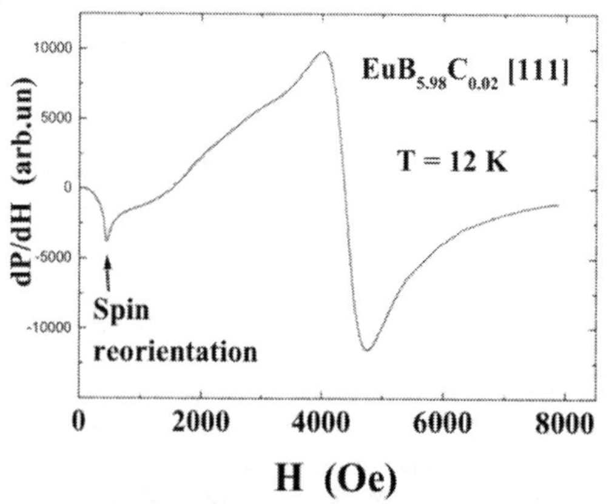

Figure 3. The ESR spectrum for the $EuB_{5.98}C_{0.02}$ [111] single crystal sample for perpendicular orientation of the sweep magnetic field to the sample plate. There is The low field features is not for [100] sample.

behavior of an anti-ferromagnetic. Namely, paramagnetic Curie-Weiss's temperature was about +8 K in case an external magnetic field parallel to [111] axis and it is negative temperature about -7 K for the magnetic field along the [100] axis. It should be noted that for the orientation [100] the deviation of inverse susceptibility from the linear dependence (see Fig. 2) is similar to sample with content of carbon x = 0.07, but point T_m finds at lower temperatures. Difference in the behaviour of the EuB_6 samples for orientations of external magnetic field along the [111] and [100] at low temperatures is evident not only in the behavior of the linewidth, but also in views of the ESR spectra. Anomaly in dependence of the first derivation of absorption dP/dH on H, which is visible in the spectrum on Fig. 3 for the [111] orientation is absent for the same spectra in the orientation [100]. We attribute this spectral feature with the reorientation of the spins of europium ions along the [111] direction.

2.3. Influence of Carbon on Magnetic order and Curie-Weiss's temperature.
A direct consequence the results of section 2.2 is a distinction between the dependences of the paramagnetic Curie temperature on carbon concentration for different crystallographic directions. Figure 4 shows the Curie – Weiss's temperature dependences on the carbon impurities concentration obtained earlier [5] and by us in this work. The figure clearly shows that dependence for the [100] - direction is very different previously known results [5], obtained without taking into account the effects of anisotropy, and it is almost identical with them to the [111]. Indirect exchange interaction is more ferromagnetically in the [111] direction, whereas for other crystal direction it is significantly weakened or even can change sign. Thus, upon application of the measuring field along the [111]-direction with a total antiferromagnetic order arises ferromagnetic component, which manifests itself as a reorientational transition in low fields region of spectra (see Fig.3).

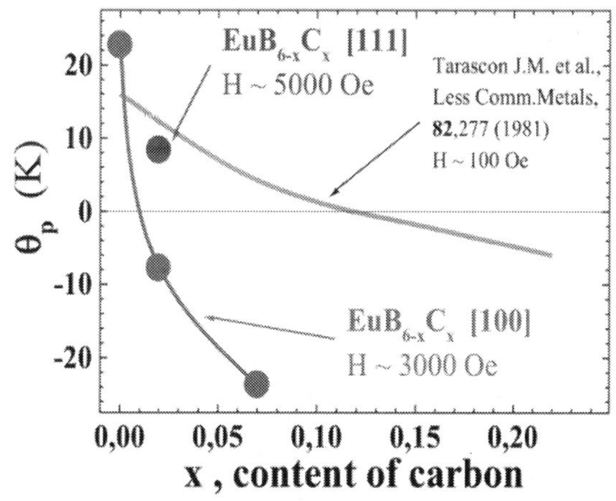

Figure 4. Dependence of paramagnetic Curie-Weiss temperature on content of carbon and orientation of the $EuB_{6-x}C_x$ crystal axes relative sweep magnetic field.

2.4. Structure of EuB_6 and the role of boron in the indirect exchange interaction between Eu ions.
As well known, the structure of europium hexaboride represents two inserted into each other the simple cubic lattices of europium ions Eu^{2+} and B_6 octahedra with lattice constant 4.18 Å. (CaB_6-type structure, space group *Pm*3*m*). In general, the boron octahedra are charged negatively and they have a deficiency of valence electrons. Moving an electron from one octahedron to another should cause to expansion of one and compression another octahedron due to the strengthening of the covalent bonds.

Assuming equidistant of the B-B bonds we obtain that Eu^{2+} ion is surrounded by 24 boron atoms, placed on a sphere of radius 3.08 Å. The spheres have a common boron atoms in the directions {100} and {110}, and in the direction of {111} there are not a common boron atoms. In other words, the planes (111) of Eu^{2+} ions are separated by two planes of boron atoms (Eu-B-B-Eu, FM-super-exchange), the planes (100) of Eu^{2+} ions are separated by a single layer of boron atoms and the planes (110) contain a boron atoms (Eu-B-Eu AFM-superexchange). Direct exchange interaction between Eu^{2+} magnetic moments is practically excluded. Conventional RKKY interaction can not be effective due to the small number of electrons in the conduction band. From effective mechanisms of the indirect exchange interaction remain the mechanisms of indirect exchange via the valence electrons, namely superexchange Kramers-Anderson [6,7] and the modified RKKY-interaction [8] Bloembergen-Rowland [9]. These mechanisms are practically indistinguishable for such alternation of layers.

Indirect exchange interaction (superexchange Kramers-Anderson) via p-orbitals has an anisotropic character, it can be both antiferromagnetic and ferromagnetic, thus leading to the formation of helical magnetic structures [10, 11]. In our case, assuming a character of indirect exchange interaction via sp-orbitals of boron is antiferromagnetic, we automatically obtain the same for directions {100} and {110}, and the ferromagnetic exchange for the directions {111}. It should be also noted that the boron-boron bonds have the directions only {100} and {110}.

Because the sp-orbitals boron of the octahedra impede to the direct moving of the valence $4f^7$ electron from one europium ion to other ion, the localized 4f- electrons impede the free movement of sp - electrons, forming with them the bound states.

The appearance of free electrons leads to filling of sp hybrid orbitals of the octahedra B_6, which should lead to their local compression, as by strengthening the real covalent bonds, and due to their compression by positive ions europium which are attracted to the excess negative charge of the B_6 octahedra. In fact, this is the formation of lattice polarons with the ferromagnetic axis along the {111} direction and symmetry breaking. In this case, the behavior of the B_6 octahedron is somewhat similar the behavior of the Jahn-Teller ion. The decrease in lattice energy due to compression promotes self-localization of free electrons.

3. Conclusion

Thus, summing up the discussion, we conclude that we have detected the influence of magnetic field direction on view of the Curie-Weiss's temperature dependence on the impurity concentration. The external magnetic field changes the distribution of spin density in electron shells by means polarization of the localized spins and these shells, thereby changing the character of the exchange interaction depending on the direction of the external magnetic field. Changing the exchange integrals leads to a change in the average energy of interaction between the localized moments, to the change of the magnetic structure in the ground state and, consequently, a change in the Curie-Weiss's temperature. This means that not such simple behavior of the Curie-Weiss's temperature can point on the magnetic quantum transition associated with the restructuring of the electronic structure of compounds under the influence of a magnetic field and impurities.

4. Acknowledgments

This work was supported grants RFBR and Presidium of the RAS

References

[1] Altshuler T S, Goryunov Y V, Shitsevalova N Y et al, 2010 J. Phys: Conf. Series, **200** 032019
[2] Nateprov A et al., 2009 Abstracts book of inter. confer. ICM-2009 (26-31.07.09, Karlsruhe) 342
[3] Weber F, Cosceev A, Drobnik S, Faisst A, Grube K et al., 2006 Phys. Rev. **B 73** 014427
[4] Altshuler T S, Goryunov Y V, Shitsevalova N Y, Dukhnenko A, 2008 JETP Lett. **88**, 779
[5] Tarascon J M et al., 1981 Less. Comm. Metals. **81**, 277
[6] Kramers H A, 1934 Physica 1, 182
[7] Anderson P W, 1950 Phys. Rev. 79, 350
[8] Ruderman MA, Kittel C, 1954 Phys. Rev. **96**, 99
[9] Bloembergen N, Rowland T J, 1955 Phys. Rev. **97**, 1679
[10] Vonsovskii S V, Magnetism: magnetic propertis dia-, para-, ferro-, antiferro- and ferrimagnets, 1971 (Science, Moskow) 102-410
[11] Nagaev E L, Physics of magnetic semiconductors, 1979 (Science, Moskow) 252-344

The ESR Study of Eu Ternary Pnictides $EuCd_2Sb_2$, $EuZn_2As_2$

Yu Goryunov [1], **V Fritsch** [2], **H v Löhneysen** [2], **A Nateprov** [3]

[1] Kazan Physical-Technical Institute of the RAS, 420029, Kazan, Russia
[2] Karlsruhe Institute of Technology, Phys.Institut, 76131 Karlsruhe, Germany
[3] Institute of Applied Physics of the ASM, 2028, Kishinev, Moldova

gorjunov@kfti.knc.ru

Abstract. The magnetic properties of the Zintl compounds $EuZn_2As_2$ and $EuCd_2Sb_2$ have been studied. The measurements of resistance, magnetic susceptibility and the ESR (X band) was performed in the range 4 - 300 K. The symmetric Eu^{2+} ESR lines had the g-factors 2.00 ($EuZn_2As_2$), 2.03 ($EuCd_2Sb_2$) and a Lorentzian shape above 120 K. The change of shape to Gaussian indicates on the magnetic fluctuations. $EuZn_2As_2$ and $EuCd_2Sb_2$ order antiferromagnetically with the $\mathbf{T}_N$ 16.5 and 7.4 K, respectively; but with positive Curie-Weiss temperatures. The study revealed metallic conductivity for $EuCd_2Sb_2$ and semiconductor type for $EuZn_2As_2$.

1. Introduction

Recently some ternary Eu and Yb pnictides with $CaAl_2Si_2$-type structures have become the subject of particular interest, because they exhibit very promising thermoelectric properties [1]. The $CaAl_2Si_2$ structure is a trigonal form of the AB_2X_2 family, where, similar as in the better known tetragonal $ThCr_2Si_2$ structure the cations on the A-site form layers interspersed with B_2X_2-layers, where the B-ions sit in the center of an X_4 tetrahedron. The way of packing these layers distinguishes between the $ThCr_2Si_2$ and structure [2]. This indicates a possible structural instability B_2X_2-layers. Europium has a half-filled 4f - shell. The electronic configuration of the f-shell in Eu - compounds is $4f^6$ (nonmagnetic) or $4f^7$ (magnetic) depending on whether the valence of the europium is either 3+ or 2+. The energetic proximity of these configurations may lead to valence fluctuations. The magnetic properties and the type of magnetic ordering in $EuCd_2Sb_2$ and $EuZn_2As_2$ are still unclear.

2. Results and discussion

2.1. Experiment

$EuCd_2Sb_2$ and $EuZn_2As_2$ were prepared by direct reaction of the stoichiometric amounts of Zn, Sb, As (99,999 %) and Eu (99,99 %) in an Al_2O_3 crucible enclosed in an evacuated quartz ampoule. The trigonal $CaAl_2Si_2$-type crystal structure (space group $P3\bar{m}1$) was confirmed by X-ray powder difraction. The ESR (electron spin resonance) was measured on the powders (size ~ 4 μm) in X-band (9.3 GHz) in TE_{102} rectangular cavity of spectrometer "Bruker" at 4.2 - 300 K.The dc susceptibility was measured at 2 - 400 K with a commercial Vibrating Sample Magnetometer (Oxford Instruments).

2.2. Electron spin resonance

2.2.1. Resonance fields, g-factor, linewidth, and shape of the ESR lines

The observed single resonance lines of the Eu^{2+} ions were symmetric and they are described very good by a Lorentzian lineshape with linewidths 530 Oe ($EuZn_2As_2$) and 1050 Oe ($EuCd_2Sb_2$) above 120 K.

Position and width of the resonance lines were not changed and the g-factors were 2.00 and 2.03, respectively, in this temperature range. Bivalent europium has the electron spin 7/2 and two stable isotopes with a nuclear spin of 5/2. The absence of the fine and hyperfine structures is associated with the averaging due to strong exchange interactions. At the temperature decreasing well before the magnetic ordering temperature we was observed an increasing of linewidth and a decreasing of the resonance field, indicating the magnification of antiferromagnetic (AFM) fluctuations and the effective molecular Weis's field. A small asymmetry is described by a small (few percent) admixture of the dispersion associated with the Dyson's distortions caused by the presence of the skin layer. At further decrease in temperature occurs the deviation from the Lorentzian lineshape and resonance line is mixture of Lorentzian and Gaussian shapes (Foight's lineshape). The most dramatic change in the lineshape occur in the temperature range 10 – 23 K (see Fig. 1), where it was required of additions of the 40% of the admixture of Gaussian shape line for an acceptable description of the shape. In this temperature range, a sharp increase in the asymmetry of the spectral shape and a decrease in its intensity due to narrowing of the resonance lines. Thus we observed a change lineshape from symmetric Lorentzian to asymmetric Gaussian lineshape. The change of shape from Lorentzian to Gaussian indicates on the magnification of antiferromagnetic fluctuations in paramagnetic range. The increase in asymmetry of the spectral shape may be caused by a decrease of the skin layer and the changing nature of the spins motion at the transition in the AFM- state. The ESR line has pure Dyson's shape at T = 4.2 K in deep AFM-state.
Large deviation of the g-factor ($\Delta g \sim 0.03$) from 2.0 (g-factor of free electron, see also [3]) indicates on the strong hybridization of the europium f-electron states with the p- s- states of the band electrons.

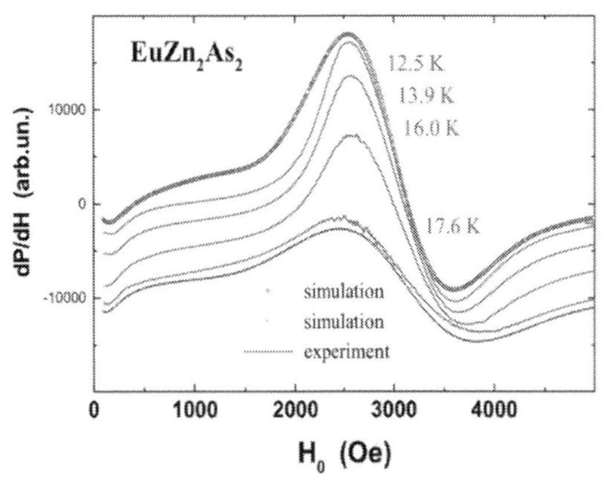

Figure 1. The ESR spectrum at low temperatures. The upper and lower spectra are the result of calculation. These calculation spectra consist from three lines. The first line (initial part of the spectrum) arising at temperatures near to $\mathbf{T}_N$ = 16.5 K is interpreted as non-resonant absorption due to the spin-flop transition. The second and third indistinguishable resonance lines may be due to the phase separation of AFM in a magnetic field.

Since ground state of Eu^{2+} ion is pure spin state ${}_8S^{7/2}$, the resultant orbital angular momentum equal to zero. In a crystalline field of cubic symmetry, the averaged energy of the spin-spin interaction between electrons of a single paramagnetic ion does not depend on their orientation with respect to each other and the ground state ion is completely degenerate in spin. Under the influence of crystal field of tetragonal or trigonal symmetry of the ion electronic cloud deforms slightly and averaged over the electron cloud energy of the spin-spin interaction depends on the relative orientation of electron spins of magnetic ions - the spin degeneracy is lifted. Symmetry and isotropy of ESR line for the powder samples allows us to conclude that we are dealing with a real resonance line, but not with the convolution of the line with anisotropic g - factor. We associate significant increase in the g factor and the ESR line width, the reducing $\mathbf{T}_N$ and $\boldsymbol{\theta}_{CW}$ on going from $EuZn_2As_2$ to $EuCd_2Sb_2$ with the change of the conductivity from the semiconductor-like to the metal.

2.2.2. *Intensity of the ESR line.*

Intensity of the ESR spectrum is proportional to magnetic susceptibility of a spin systems if there is no influence of the skin layer. Peculiarities in the behavior of the spectrum baseline ESR is usually associated features of the dependence of microwave surface impedance of the sample on the magnetic field. Such causes are not in our case (the skin layer is much larger than the size of the powder particles). The most suitable explanation for our case of the low field ESR features of the spectrum is the restructuring of the spin system $EuZn_2As_2$ in relatively small fields. If we turn to the previously investigated similar compounds $EuZn_2Sb_2$ [4], we see that in fields less than 400 Oe is the spin-flop transition in the magnetic sub-lattices of the antiferromagnet. In our case, the region of features of the baseline, which we associate with the spin-flop transition, begins with a temperature slightly higher than the Neel temperature, determined by the maximum intensity of the ESR. This may be due to the induced external magnetic field ferromagnetic order on the background of strong AFM fluctuations. As the Fig.2 shows the changing of intensity of the ESR line with temperature corresponds to a typical temperature dependence of the susceptibility of an antiferromagnet. Also the temperature dependence of inverse intensity is typical. However, it should be emphasized that, unlike the typical for antiferromagnet negative paramagnetic Curie-Weiss temperature we obtained positive $\boldsymbol{\theta}_{ESR}$. This fact of positiveness paramagnetic Curie-Weiss tempereature for the antiferromagnet is confirmed by direct measurements of the dc susceptibility. The obtained values of paramagnetic temperatures were quite close. $\boldsymbol{\theta}_{ESR}$ = +20.4 K and $\boldsymbol{\theta}_{CW}$ = +21.2 K. Respectively for $EuCd_2Sb_2$ $\boldsymbol{\theta}_{ESR}$ = +5.4 K and $\boldsymbol{\theta}_{CW}$ = +3 K.

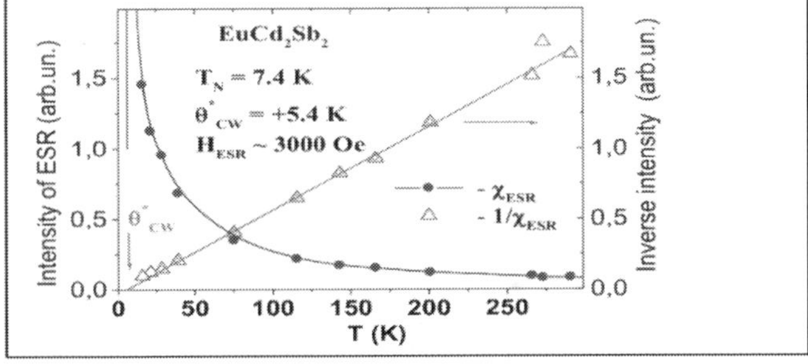

Figure 2. Magnetic susceptibility χ_{ESR} of the powder $EuCd_2Sb_2$ from the date on intensity of the ESR. The temperature dependence of inverse susceptibility indicates the positive Curie-Weiss temperature for this antiferromagnet. DC susceptibility gives the temperature of AFM-ordering $\mathbf{T}_N$ = 7.4 K. .

2.3. DC – susceptibility

The magnetic properties of $EuZn_2As_2$ were studied on polycrystalline samples. In the temperature range 10 - 300 K the susceptibility M/B follows a Curie-Weiss law. The effective paramagnetic moment $\boldsymbol{\mu}_{eff}$ = 7.1 μ_B/Eu and antiferromagnetic order with $\mathbf{T}_N$ = 16.5 K. As the ESR date the inverse susceptibility shows a positive paramagnetic Curie-Weiss temperature $\boldsymbol{\theta}_{cw}$ = 21.2 K. The saturation moment of the magnetization is 5.7 μ_B/Eu. The reduced values point out the presence of some Eu^{3+} ions in $EuZn_2As_2$. The magnetic properties of $EuCd_2Sb_2$ were studied on single-crystalline samples. In the temperature range 2 - 300 K the susceptibility M/B follows a Curie-Weiss law with a Neel temperature $\mathbf{T}_N$ = 7.4 K (see Fig.3). The effective paramagnetic moment $\boldsymbol{\mu}_{eff}$ = 7.83 μ_B/Eu is very close to the S = 7/2 value $\boldsymbol{\mu}_{eff}$ = 7.94μ_B. This indicates that all Eu ions are in the divalent state, as expected

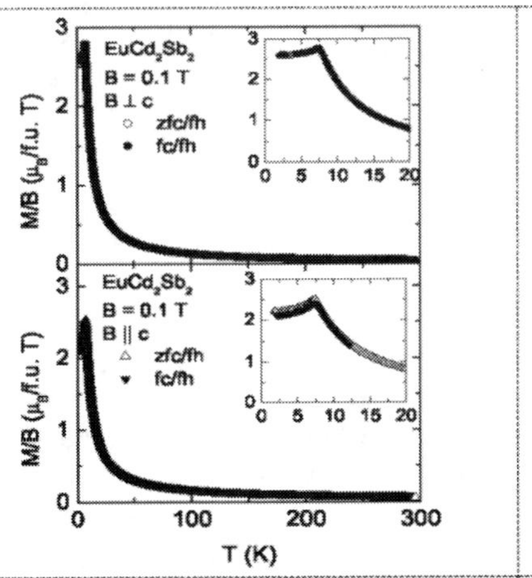

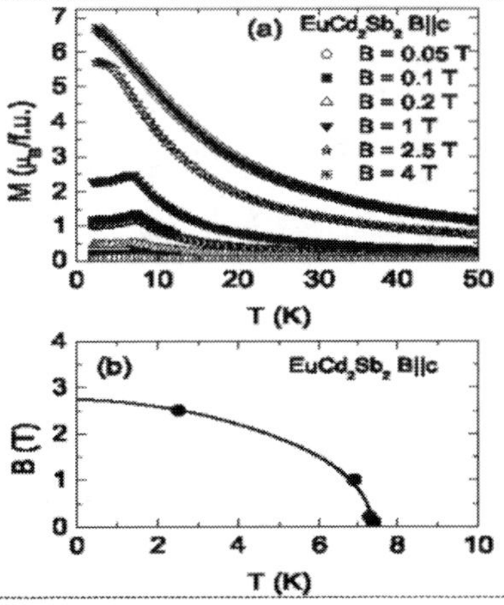

Figure 3. Magnetic susceptibility of $EuCd_2Sb_2$. Inset: magnification of the low-temperature range up to 20 K.

Figure 4. (a) Magnetization of $EuCd_2Sb_2$ in different magnetic fields with the magnetic field aligned parallel to the c-axis. (b) Resulting B-T phase diagram.

from the electron count, when describing this compound within the Zintl concept. With application of an external magnetic field $\mathbf{T_N}$ is shifted downwards and finally suppressed (see Fig. 4). Saturation magnetization at ~ 3 T and 2.5 K gives an absolute magnetic moments μ_S = 6.6 μ_B/Eu. For divalent europium μ_S = 7 μ_B/Eu is expected. We found no evidence for ferromagnetic order [5], however the positive Weiss temperature θ_{CW} = +3 K points to the existence of FM correlations in this material or quantum phase transition in magnetic field induced by Jan-Teller effect in anionic complex.

2.4. Electrical resistance

The electrical resistivity ρ of $EuCd_2Sb_2$ is typical for a metal or semimetal with ρ = 0.25 Ωcm at room temperature (see Fig. 5). This suggests scattering of charge carriers by magnetic fluctuations near $\mathbf{T_N}$ and predominant electron-magnon scattering at lower temperatures. This behavior of resistivity is typically for the 4f compounds with well-localized magnetic moments. The electrical resistivity ρ(T) from 10 to 300 K for $EuZn_2As_2$ (Fig. 5) suggests a narrow gap semiconductor. The activation energy is estimated to be $E_g \approx 0.1$ eV. The carrier concentrations at room temperature are $7.2{\cdot}10^{18}$ cm^{-3} for $EuCd_2Sb_2$ and $5.6{\cdot}10^{18}$ cm^{-3} for $EuZn_2As_2$.

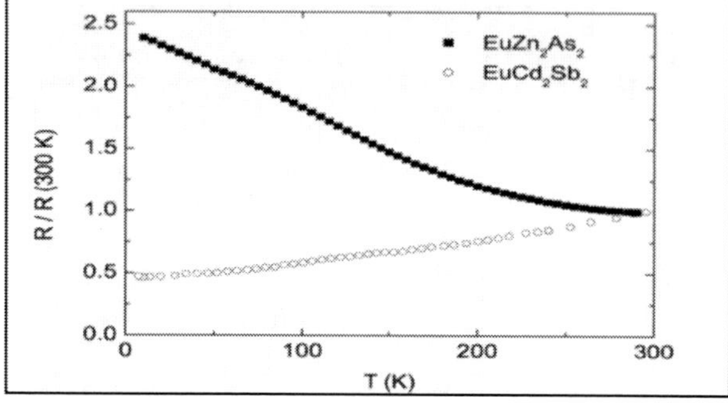

Figure 5. Temperature dependence of the electrical resistance of the polycrystalline $EuZn_2As_2$ and $EuCd_2Sb_2$.

3. Conclusion.

The ternary Eu based compounds $EuCd_2Sb_2$, and $EuZn_2As_2$ with $CaAl_2Si_2$-type crystal structures have been synthesized and their magnetic and electric properties were studied. We can conclude on the base of our date that these compounds are antiferromagnets, in which there are strong ferromagnetic correlations in the fields above 1000 Oe. Taking into account that the density of electrons in the conduction band is very low, we believe that g-actor shift is due to hybridization and the formation of bound states of 4f-electrons Eu^{2+} ion with the electrons of the valence band. As a consequence the indirect exchange interaction between the divalent ions of europium through the electrons of the valence band, the so-called modified-RKKY Bloembergen-Rowland's [6, 7], is responsibly for anti-ferromagnetism of these compounds. Fluctuations in the density of valence band electrons may well cause fluctuations in the valence state of europium between the magnetic state of Eu^{2+} and nonmagnetic Eu^{3+}. This leads to AFM fluctuation above $\mathbf{T_N}$ and reduce the saturation magnetization.

4. Acknowledgments

This work was supported in part by grants of DFG, ASM, RFBR and Presidium of the RAS.

References

[1] Yu C, Zhu T J, Zhang S N, Zhao X B, He J, Su Z, Tritt T M, 2008 *J. Appl. Phys.* **104** 013705
[2] Zheng C and Hoffmann R 1988 Journal of Solid State Chemistry **72** 58
[3] Altshuler S A, Kozyrev B M, EPR of the intermediate group compounds.1972 (M.Science), 478
[4] Weber F, Cosceev A, Drobnik S, Faisst A, et al.., 2006 Phys. Rev. B **73** 014427
[5] Artmann A, Mewis A, Roepke M and Michels G 1996 Z. anorg. allg. Chem. 622 679
[6] Ruderman M A, Kittel C, 1954 Phys. Rev. **96** 99
[7] Bloembergen N, Rowland T J, 1955 Phys. Rev. **97** 1679

First-principles calculations of lattice vibrations on LaT_2Zn_{20} (T = Ru and Ir)

Takumi Hasegawa[1], **Norio Ogita**[1] **and Masayuki Udagawa**[1,2]

[1] Graduate School of Integrated Arts & Sciences, Hiroshima University, Higashi-Hiroshima 739-8521, Japan

[2] Institute for Advanced Material Research, Hiroshima University, Higashi-Hiroshima 739-8530, Japan.

E-mail: hase@hiroshima-u.ac.jp

Abstract. Lattice vibrations of $LaRu_2Zn_{20}$, and $LaIr_2Zn_{20}$ have been investigated by the local density approximation. Although these compounds have a cage structure in which a La atom is surrounded by Zn, the calculated results show that La vibrations have large dispersion and relatively high frequencies. In both compounds, it is found that the vibrations of Zn at 16c site in the plane perpendicular to the three-fold axis have very low frequencies and can be unstable. This result suggests that the transitions in $LaRu_2Zn_{20}$ at 150 K and in $LaIr_2Zn_{20}$ at 200 K are caused by these unstable modes.

1. Introduction

We have calculated lattice dynamical properties for $LaRu_2Zn_{20}$ and $LaIr_2Zn_{20}$ with the $CeCr_2Zn_{20}$ structure[1]. In this structure, the La atoms are well separated from the nearest La and are surrounded by Zn. The cage structural feature has also been found in skutterudite compounds, where large amplitude vibrations of rare-earth ions have attracted much attention, because such vibrations could participate in exotic phenomena[2, 3, 4]. Therefore, it is worth to examine the possibility of such large amplitude vibrations of La in these compounds, and such modes might be one of the possible keys to understand the anomalous behavior observed in $PrIr_2Zn_{20}$[1, 5].

In general, frequencies of atomic vibrations depend on wavevector, but all frequencies become low when atoms vibrate with large amplitude. This can be judged by a partial density of states (partial DOS) of lattice vibrations with a narrow peak at low frequency. Thus, we have calculated the partial DOS in these compounds. On the other hand, $LaRu_2Zn_{20}$ and $LaIr_2Zn_{20}$ show a transition at 150 K and 200 K, respectively, which have been suggested as a structural transition[1]. Therefore, it is important to understand the lattice dynamical properties in these compounds.

2. Procedures

All calculations were made using ABINIT package[6, 7, 8], which is able to obtain lattice dynamical properties of crystals under the density functional theory with plane-wave basis set. We performed firstly the ground-state calculation for each compounds and obtained their optimized structures, minimizing pressures and forces on atoms until stresses are less than

0.01 GPa and forces are less than 3 meV/Å, respectively. Then, inter-atomic force constants (IFCs) were calculated in the framework of the density functional perturbation theory[9] and the Fourier interpolation method[10]. Using the IFCs, phonon dispersion curve were obtained. DOS of phonon modes were calculated by the tetrahedron method with 8×8×8 grid.

For all calculations, local density approximation (LDA) was applied with the exchange-correlation energy parameterized by Perdew and Wang[11]. The norm-conserved pseudo-potential (PP) method was used and the potentials were generated by FHI98PP[12] due to the Troullier-Martins method[13]. The plane-wave cut-off energy was 56 Ry, where 1 Ry = 13.6 eV. The integrations in the reciprocal space were replaced by summation on the 2 × 2 × 2 Monkhost-Pack grid[14]. Since these compounds are metallic, smeared occupation[15] is useful to decrease the grid in the reciprocal space. The smearing parameter, which works like temperature, was set to 0.02 Ry. The IFCs have been determined by the Fourier interpolation method based on 1 × 1 × 1 grid (LaT_2Zn_{20}) of the cubic cell in the reciprocal space. All these parameters were determined until the accuracy of phonon frequencies to be less than 10 cm^{-1}, except for low-energy modes below 40 cm^{-1}. The convergence of the phonon frequencies at $\mathbf{q} = \mathbf{0}$ was examined up to 60 Ry for plane-wave cut-off, down to 0.01 Ry for the smearing parameter, and up to 4 × 4 × 4 Monkhost-Pack grid.

3. Result and discussion

The crystal structure of LaT_2Zn_{20} has the space group $Fd\bar{3}m$. La locates at 8a position (0, 0, 0), Ru at 16d (0.625, 0.625, 0.625), Zn(1) at 16c (0.125, 0.125, 0.125), Zn(2) at 48f (x, 0.25, 0.25), and Zn(3) at 96g (y, y, z). In the optimized structure, a = 14.159 Å, x = 0.6134, y = 0.1839, and z = 0.4514 for $LaRu_2Zn_{20}$ and a = 14.143 Å, x = 0.6110, y = 0.1847, and z = 0.4490 for $LaIr_2Zn_{20}$, where a is a lattice constant.

The phonon frequencies of $LaRu_2Zn_{20}$ are sensitive to the lattice constants. The optimized lattice constant a is 14.159 Å, while the experimental value[1] is 14.426 Å. We also calculated phonon frequencies using the experimental a, where atomic positions were also optimized. The calculated phonon frequencies are quite different between these two lattice constants. The maximum value of the differences is more than 30 cm^{-1}, while the maximum frequency of phonon is about 300 cm^{-1}. The result with the experimental lattice shows better agreement with the Raman scattering result[16]. From now on, we discuss using the experimental one. The calculated parameters of the atomic positions become x = 0.6139, y = 0.1836, and z = 0.4524 for $LaRu_2Zn_{20}$ with a = 14.426 Å, and x = 0.6111, y = 0.1845, and z = 0.4494 for $LaIr_2Zn_{20}$ with a = 14.354 Å[1].

The calculated partial DOS in $LaRu_2Zn_{20}$ is shown in Figs. 1. Some phonon modes have imaginary frequencies, i.e. $\omega^2 < 0$. These modes are plotted at the negative frequency with the same absolute value. The imaginary frequency mode is unstable, that is, this unstable mode decreases the electronic energy. Therefore, some structural transition is expected. As shown in Fig. 1, the low-frequency modes are vibrations of Zn(1). Zn(1) site has two nearest-neighbor La atoms on a straight line (see Fig. 2). The La-Zn(1)-La line coincides with one of the three-fold axes. The modes above 30 cm^{-1} are vibrations of Zn(1) along this three-fold axis, while the modes below 30 cm^{-1} are vibrations of Zn(1) perpendicular to the axis. Therefore, the unstable modes are vibrations of Zn(1) perpendicular to the axis.

The La vibrations are distributed in wide frequency range, which is almost the same as the distribution range of the vibrations of Zn(1) along the three-fold axis. Namely, the La-Zn(1) network is tight and the La vibrations show large dispersion in $LaRu_2Zn_{20}$. Therefore, La atoms in $LaRu_2Zn_{20}$ are unlikely to vibrate with large amplitude.

The calculated frequencies of the Raman active phonons and their experimental result[16] are listed in Table 1. The agreement is not good, since the maximum difference is 18 cm^{-1}. As described above, the small change of the lattice constant by about 2 % makes large change

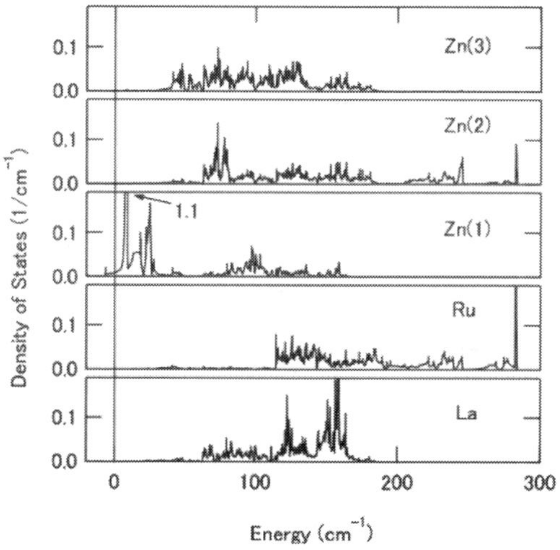

Figure 1. Partial density of states of lattice vibrations in $LaRu_2Zn_{20}$. Some modes have imaginary frequencies and those modes are shown at negative frequencies in Zn(1) DOS.

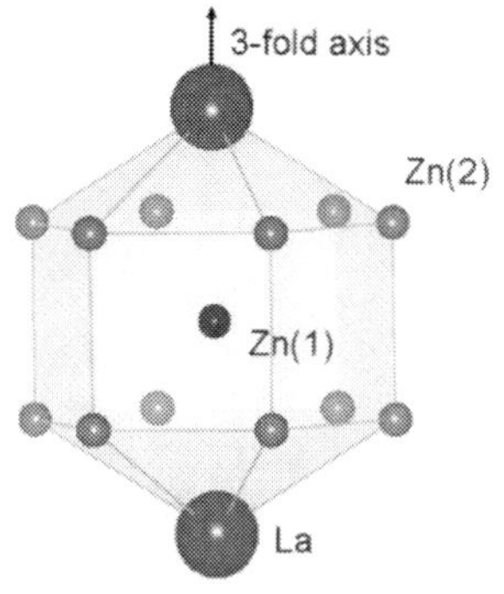

Figure 2. Zn(1) atom and its ligand atoms. Zn(1) atom is tightly bonded with two La atoms. Zn(1) atom can easily move toward Zn(2), since the distance to Zn(2) atom is very long.

Table 1. Calculated frequencies of the Raman active phonons and their experimental values at room temperature[16] in $LaRu_2Zn_{20}$.

A_{1g} (exp.)	-	159.1	201.6						
A_{1g} (calc.)	138.7	152.1	183.9						
E_g (exp.)	63.3	81.3	141.6	165.5					
E_g (calc.)	71.5	81.3	124.2	149.5					
T_{2g} (exp.)	-	81.3	-	113.1	-	-	-	171.2	241.1
T_{2g} (calc.)	71.4	81.8	90.5	109.7	133.8	141.7	157.1	161.4	238.0

of phonon frequencies by about 10 %. Since the calculated lattice constants in LDA usually shows smaller values by a few % than experimental ones, this large difference might be a systematic difference in LDA. As another possibility, the structure at room temperature may not be described by the averaged structure with the space group $Fd\bar{3}m$, because the structure will be deformed at low temperatures.

We discuss the vibration modes of Zn(1) perpendicular to the three-fold axis. If we consider the arrangement of atoms around Zn(1), as shown in Fig. 2, it is found that Zn(1) is surrounded by two La atoms and 12 Zn(2) atoms. Along the three-fold axis, Zn(1) is tightly bonded to two La atoms. However, the atomic distance between Zn(1) and Zn(2) is very long compared with the other nearest-neighbor Zn-Zn distances. This means that there is a free space for Zn(1) toward perpendicular to the axis. As the result, the vibrations of Zn(1) perpendicular to the axis have low frequencies. In this case, this atomic displacements of Zn(1) may decrease the electronic energy. Therefore, some modes become unstable. The most unstable mode is triply degenerated T_{2u} mode at $\mathbf{q} = \mathbf{0}$ in the present result. If the structural transition at 150 K is driven by this mode, Landau theory predicts that the space group of deformed structures is

$I\bar{4}3d$ or R32. However, the calculated frequencies of these low-frequency modes are sensitive to the lattice constant and parameters of the calculation, because their frequencies are too low. We cannot conclude which mode is the most unstable at this stage. The calculated result of $LaIr_2Zn_{20}$ also shows low-frequency modes around 20 cm^{-1} for Zn(1) vibrations, but they are not the unstable modes, that is, they have the positive energies. The origin of the transition at 200 K in $LaIr_2Zn_{20}$ is likely to be the Zn(1) vibrations, since the other optical modes have much higher frequencies.

4. Conclusion

The phonon frequencies of $LaRu_2Zn_{20}$ and $LaIr_2Zn_{20}$ have been calculated. Both compounds have cage-like structures, where a La atom is surrounded by Zn. In such structures, the guest mode due to La might have small dispersion and low frequencies. However, the calculated results show that La vibrations have large dispersion and high frequencies. Therefore, large amplitude vibrations of La atoms will not be found. On the other hand, it is found that the vibrations of Zn(1) perpendicular to the three-fold axis have very low frequencies and that some of them are unstable. Therefore, the structural transition reported in $LaRu_2Zn_{20}$ and $LaIr_2Zn_{20}$ can be caused by these modes, and the Zn(1) modes have large amplitude in the plane perpendicular to the axis above the transition temperature.

Acknowledgments

This work is supported by Grant-in-Aids for Scientific Research on Innovative Areas "Heavy Electrons" (No. 20102005) of the Ministry of Education, Culture, Sports, Science, and Technology, Japan, and for Young Scientists (B) (No. 21740257) of Japan Society for the Promotion of Science, and supported by N-BARD and IAMR of Hiroshima University.

References

[1] Onimaru T, Matsumoto K T, Inoue Y F, Umeo K, Saiga Y, Matsushita Y, Tamura R, Nishimoto K, Ishii I, Suzuki T and Takabatake T 2010 *J. Phys. Soc. Jpn.* **79** 033704
[2] Sanada S, Aoki Y, Aoki H, Tsuchiya a, Kikuchi D, Sugawara H and Sato H 2005 *J. Phys. Soc. Jpn.* **74** 246
[3] Hattori K, Hirayama Y and Miyake K 2005 *J. Phys. Soc. Jpn* **74** 3306
[4] Hotta T 2009 *J. Phys. Soc. Jpn.* **78** 073707
[5] Onimaru T, Matsumoto K T, Inoue Y F, Umeo K, Sakakibara T, Karaki Y, Kubota M and Takabatake T 2011 *Phys. Rev. Lett.* **106** 177001
[6] The ABINIT code is a common project of the Université Catholique de Louvain, Corning Incorporated, and other contributors (URL http://www.abinit.org).
[7] Gonze X, Amadon B, Anglade P -M, Beuken J -M, Bottin F, Boulanger P, Bruneval F, Caliste D, Caracas R, Cote M, Deutsch T, Genovese L, Ghosez Ph, Giantomassi M, Goedecker S, Hamann D R, Hermet P, Jollet F, Jomard G, Leroux S, Mancini M, Mazevet S, Oliveira M J T, Onida G, Pouillon Y, Rangel T, Rignanese G -M, Sangalli D, Shaltaf R, Torrent M, Verstraete M J, Zerah G and Zwanziger J W 2009 *Computer Phys. Commun.* **180** 2582
[8] Gonze X, Rignanese G -M, Verstraete M, Beuken J -M, Pouillon Y, Caracas R, Jollet F, Torrent M, Zarah G, Mikami M, Ghosez Ph, Veithen M, Raty J -Y, Olevano V, Bruneval F, Reining L, Godby R, Onida G, Hamann D R and Allan D C 2005 *Zeit. Kristallogr.* **220** 558
[9] Goze X 1997 *Phys. Rev.* B **55** 10337
[10] Gonze X and Lee C 1997 *Phys. Rev.* B **55** 10355
[11] Perdew J P and Wang Y 1992 *Phys. Rev.* B **45** 13244
[12] Fuchs M and Scheffler M 1999 *Computer Phys. Commun.* **119** 67
[13] Troullier N and Martins J L 1992 *Phys. Rev.* B **43** 1993
[14] Monkhorst H J and Pack J D 1976 *Phys. Rev.* B **13** 5188
[15] Cold smearing of N. Marzari with $a = -0.5634$ was used. See Marzari N 1996 *PhD dissertation* Univerity of Cambridge, and Marzani N, Vanderbilt D, Vita A D and Payne M C 1999 *Phys. Rev. Lett.* **82** 3296. See also *occopt* input variable in the ABINIT web page.
[16] Preliminary Raman scattering data in our research group.

Anisotropic Magnetic Fluctuations and Superconductivity in Ferromagnetic Superconductor UCoGe

T Hattori[†,1], Y Ihara[1], K Karube[1], Y Nakai[1,2], K Ishida[1,2], K Deguchi[3], N K Sato[3] and I Satoh[4]

[1]Department of Physics, Graduate School of Science, Kyoto University, Kyoto 606-8502, Japan. [2]Transformative Research Project on Iron Pnictides (TRIP), Japan Science and Technology Agency (JST), Tokyo 102-0075, Japan. [3]Department of Physics, Graduate School of Science, Nagoya University, Nagoya 464-8602, Japan. [4]Institute for Materials Research, Tohoku University, Sendai 980-8577 Japan.

E-mail: [†]`t.hattori@scphys.kyoto-u.ac.jp`

Abstract. We have performed ^{59}Co NQR / NMR measurements on the single-crystalline UCoGe, in order to investigate the relationship between ferromagnetism and superconductivity. The measurements of Knight-shift and nuclear spin-lattice relaxation rate provide clear evidence that both static and dynamic susceptibilities are ferromagnetic with Ising anisotropy. In addition, H_{c2} also shows extremely large anisotropy which can not be explained by the GL model with anisotropy of effective mass. These data suggest intimate relationship between Ising magnetization and anisotropic superconductivity in UCoGe.

1. Introduction

After the discovery of superconductivity in the ferromagnetic UGe_2 under pressure, the relationship between ferromagnetism and superconductivity has attracted much interest. In 2007, Huy *et al.* reported superconductivity with $T_{SC} \sim 0.6$ K in the U-based compound UCoGe at ambient pressure, where ferromagnetic(FM) ordering occurs at $T_{Curie} \sim 3.0$ K, and ferromagnetism and superconductivity are considered to coexist[1]. From our NQR / NMR measurements, we have shown that superconductivity coexists with ferromagnetism microscopically[2]. For further study on the relationship between ferromagnetism and superconductivity, we investigated the anisotropic properties of the superconductivity via the angle resolved Meissner-signal measurements, as well as of magnetic fluctuations via the direction dependence of nuclear spin-lattice relaxation rate $1/T_1$ measurements[3].

2. Experiment

We used a 55 mg single crystal sample with $1.65 \times 1.65 \times 1.89$ mm^3, grown by the Czochralski method in a tetra-arc furnace. T_{Curie} of the single-crystal sample was evaluated to be 2.55 ± 0.1 K from the Arrot plots[2]. Figure 1 represents the resistivity measured along each direction, and the temperature dependence below 2 K is approximately expressed $\rho = \rho_0 + AT^2$ as shown in the inset. From the anisotropy of A coefficient, the anisotropy of effective mass is estimated as $m_c^*/m_b^* \sim 1.65$ using a relation of $A \propto \sqrt{m^*}$. In contrast to the small anisotropy of m^*,

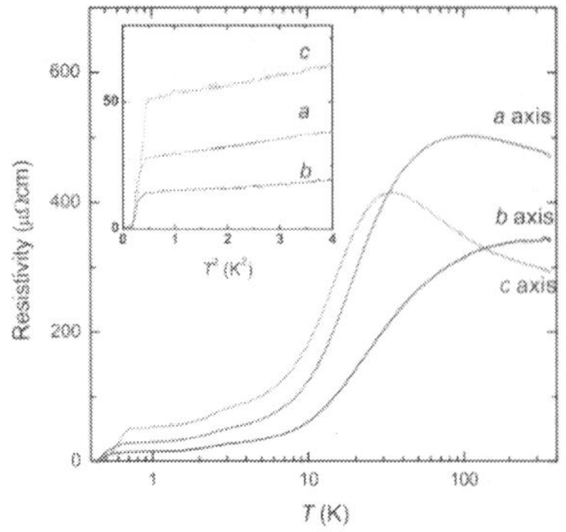

Figure 1. Temperature dependence of resistivity along each axis in the single-crystal UCoGe. The inset shows the plot of the resistivity against T^2. From the anisotropy of A coefficient, the effective mass anisotropy is estimated as $m_c^*/m_b^* \sim 1.65$ using relations of $A \propto \sqrt{m^*}$, $\rho = \rho_0 + AT^2$.

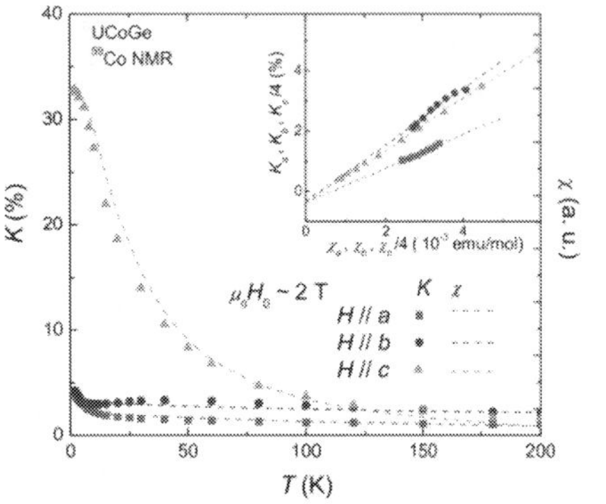

Figure 2. Temperature dependence of ^{59}Co NMR Knight shift for each crystal axis. K-χ plot is shown in the inset.

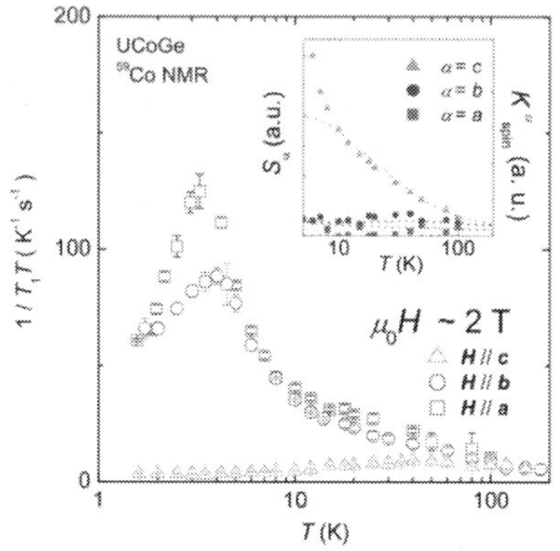

Figure 3. Temperature dependence of $1/T_1T$ with each direction. The dynamic susceptibility S_α and K against T is in the inset.

magnetization has large Ising-type anisotropy with the easy axis along the c axis. The FM and superconducting(SC) properties of our single crystal are in good agreement with previous reports by Huy *et al.*[4] and Aoki *et al.*[5]. Large Residual-Resistivity-Ratio ($RRR \parallel b$ axis $\sim$ 30) and the clear specific-heat jump at $T_{\rm Curie}$ and $T_{\rm SC}$ ensure high quality of the sample[6].

We measured the direction dependences of the Knight-shift (K) and $1/T_1$ with field orientations controlled carefully *in situ* using a single-axis rotator. The nuclear quadrupole splitting of the ^{59}Co NMR spectra allowed us to determine the field directions precisely[7].

3. Result and Discussion

Figure 2 shows the temperature dependence of K. Large enhancement of K_c and small $K_{a,b}$ ensures the Ising anisotropy from microscopic view. K_α is related to the static susceptibility $\chi(\boldsymbol{q}=0)$ as $K_\alpha = A_{\rm hf}^\alpha \chi^\alpha(0) + K_{\rm orb}^\alpha$, where $A_{\rm hf}^\alpha$ is a hyperfine coupling constant along the α direction, and $K_{\rm orb}$ is an orbital contribution, and temperature independent. K versus χ plot, presented in Fig. 2 inset, indicates that a hyperfine coupling constant $A_{\rm hf}^\alpha$ is nearly isotropic. This indicates that the U-5f spins themselves possess the anisotropy due to the spin-orbit interaction, since $A_{\rm hf}^\alpha$ is positive and isotropic when the U-5f electrons are transferred to the Co-4s orbital and interact with the Co nuclei directly.

Dynamic susceptibility which is detected with the measurement of $1/T_1$ has also Ising anisotropy. Figure 3 presents the temperature dependence of $1/T_1T$, which is proportional to the magnetic fluctuations perpendicular to the magnetic field;

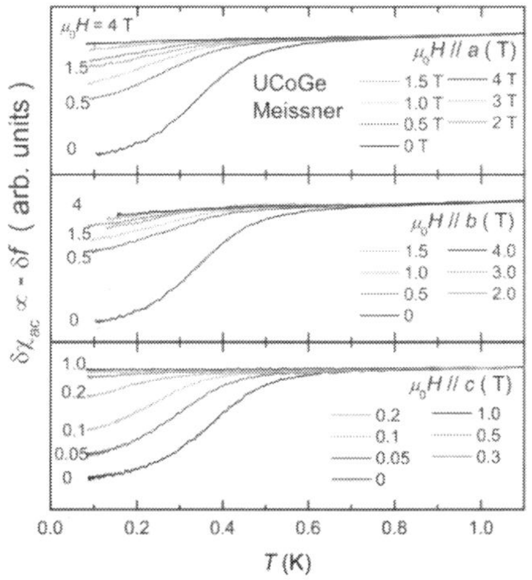

Figure 4. Temperature dependence of ac susceptibility with some fields along each axis.

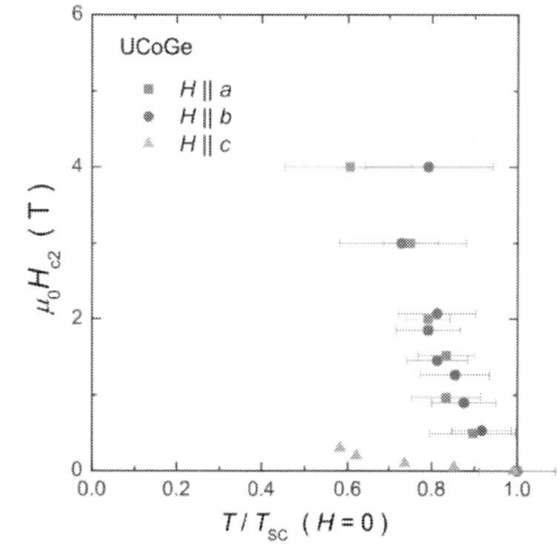

Figure 5. H_{c2} - T Phase diagram of UCoGe estimated by the Meissner signal.

$\left(\frac{1}{T_1 T}\right)_\alpha = \frac{\gamma_n^2 k_B}{(\gamma_e \hbar)^2} \sum_{\boldsymbol{q}} \left[|A_{\rm hf}^\beta|^2 \frac{\chi_\beta''(\boldsymbol{q},\omega_0)}{\omega_0} + |A_{\rm hf}^\gamma|^2 \frac{\chi_\gamma''(\boldsymbol{q},\omega_0)}{\omega_0} \right]$. Therefore large $(1/T_1T)_{a,b}$, and small $(1/T_1T)_c$ means that magnetic fluctuation along c axis is dominant in UCoGe. With this eqn. and the $1/T_1T$ measured in three different field directions, each term represented by $\sum_{\boldsymbol{q}} [|A_{\rm hf}^\alpha|^2 \chi_\alpha''(\boldsymbol{q},\omega_0)/\omega_0] \equiv S_\alpha$ is separated out, and the results are shown in Fig. 3 inset. It is clear that dynamic susceptibility S_α has also the Ising anisotropy along the c axis. The identical angle dependency in both uniform and dynamic susceptibility reveals that the dominant magnetic fluctuations are along the magnetic easy axis (longitudinal mode). In addition, it is noteworthy that S_c scales linearly to $K_{\rm spin}^c$ above 8 K, indicative of the predominance of the FM fluctuations, since this scaling is anticipated for three-dimensional FM fluctuations on the basis of the self-consistent renormalization (SCR) theory[8]. As for the relationship between magnetic fluctuations and superconductivity, P. Monthoux and G. G. Lonzarich[9] and Fujimoto[10] have pointed out that the Ising FM fluctuations are favorable for spin-triplet superconductivity since the Ising FM fluctuations with only the longitudinal mode can minimize the pair breaking caused by the transverse spin fluctuations.

In order to investigate the anisotropic properties of superconductivity, the temperature dependence of ac susceptibility $\chi_{\rm ac}$ along each direction was measured (Fig. 4). From the onset of the Meissner signal, the temperature dependence of H_{c2} was plotted in Fig. 5. The anisotropy of H_{c2} is in good agreement with the previous reports [4, 5]. The smallest H_{c2} along c axis is one of the mysterious behaviors of UCoGe, since spin-triplet superconductivity seems to be stable when field is applied along the spin easy axis (c axis).

To further examine, we measured the angular dependence of H_{c2} in the ac plane. H_{c2} is determined from the angular dependence of $\chi_{\rm ac}$ in various applied fields at 85 mK, and the onset of the Meissner signal is plotted in Fig. 6. H_{c2} is drastically suppressed by the magnetic field along c axis. This angular dependence of H_{c2} is qualitatively the same as previous report[5], but the values are different with a factor 3 with each other, as shown in Fig. 7. It should be noted that this steep angular dependence of H_{c2} cannot be explained at all with the anisotropic GL formula even if field dependence of m^* is taken into account, since $m^*(H^c = 1\text{T})/\ m^*(H = 0) \simeq 0.75$ was reported around H_{c2} along c axis[5].

Therefore interpretation of the extraordinary anisotropy of H_{c2} is crucial for understanding the SC pairing mechanism in UCoGe, and it is considered that another mechanism might work for suppression of the superconductivity when the magnetic field is applied to c axis.

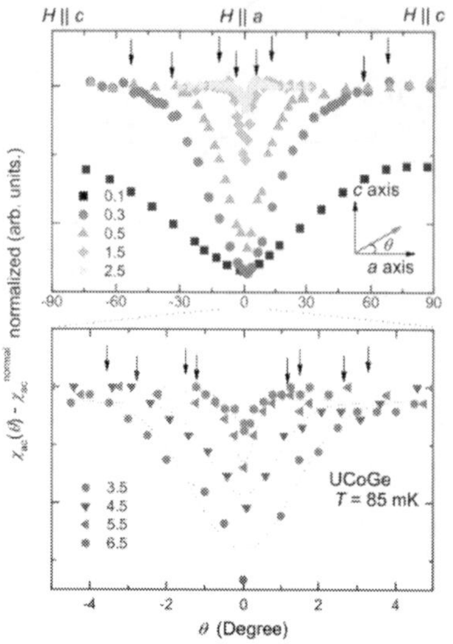

Figure 6. Angular dependence of the Meissner signals detected by the ac susceptibility in various magnetic fields in *ac* plane at $T = 85$ mK.

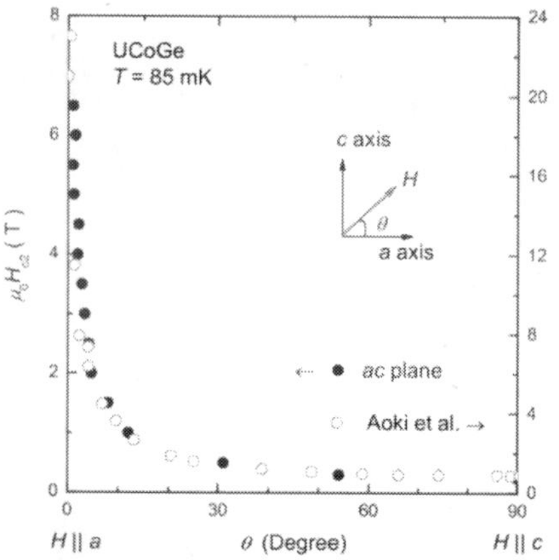

Figure 7. Angle dependence of H_{c2} in *ac* plane at $T = 85$ mK, which is determined by the onset of Meissner signal. H_{c2} data reported by D. Aoki *et al.* are also plotted, which are determined by resistive measurements[5].

4. Summary

Direction-dependent ^{59}Co NMR was performed, and the K and $1/T_1$ measurements provide microscopic evidence that both static and dynamic susceptibilities are ferromagnetic with strong Ising anisotropy along c axis. In addition, we measured the angular dependence of H_{c2} from observation of the Meissner signal probed by ac susceptibility. H_{c2} also shows extremely large anisotropy, which cannot be explained by anisotropic GL model. Interpretation of these behaviors is crucial for understanding the SC pairing mechanism in UCoGe.

The authors thank S. Yonezawa, and Y. Maeno for experimental support and valuable discussions, and D. Aoki, J. Flouquet, A. de Visser, Y. Tada, S. Fujimoto, H. Harima, and H. Ikeda for valuable discussions. This work was partially supported by Kyoto Univ. LTM centre, Yukawa Institute, the "Heavy Electrons" Grant-in-Aid for Scientific Research on Innovative Areas (No. 20102006) from MEXT of Japan, a Grant-in-Aid for the Global COE Program "The Next Generation of Physics, Spun from Universality and Emergence" from MEXT of Japan, a grant-in-aid for Scientific Research from JSPS, KAKENHI (S) (No. 20224015).

References

[1] Huy N T, Gasparini A, de Nijs D E, Huang Y, Klaasse J C P, Gortenmulder T, de Visser A, Hamann A, Görlach T and v Löhneysen H 2007 *Phys. Rev. Lett.* **99** 067006
[2] Ohta T, Hattori T, Ishida K, Nakai Y, Osaki E, Deguchi K, Sato N K and Satoh I 2010 *J. Phys. Soc. Jpn.* **79** 023707
[3] Ihara Y, Hattori T, Ishida K, Nakai Y, Osaki E, Deguchi K, Sato N K and Satoh I 2010 *Phys. Rev. Lett.* **105** 206403
[4] Huy N T, de Nijs D E, Huang Y K and de Visser A 2008 *Phys. Rev. Lett.* **100** 077002
[5] Aoki D, Matsuda T D, Taufour V, Hassinger E, Knebel G and Flouquet J 2009 *J. Phys. Soc. Jpn.* **78** 113709
[6] Deguchi K, Osaki E, Ban S, Tamura N, Simura Y, Sakakibara T, Satoh I and Sato N K 2010 *J. Phys. Soc. Jpn.* **79** 083708
[7] Hattori T, Ihara Y, Ishida K, Nakai Y, Osaki E, Deguchi K, Sato N K and Satoh I 2011 *J. Phys. Soc. Jpn. Suppl.* **80** SA077
[8] Moriya T 1991 *J. Magn. Magn. Matter.* **100** 261
[9] Monthoux P and Lonzarich G G 1999 *Phys. Rev. B* **59** 14598
[10] Fujimoto S 2004 *J. Phys. Soc. Jpn.* **73** 2061

Journal of Physics: Conference Series **391** (2012) 012018 doi:10.1088/1742-6596/391/1/012018

Cluster formation in quantum critical systems

T. Heitmann[1], **J. Gaddy**[2], **J. Lamsal**[1,2] **and W. Montfrooij**[1,2]
[1]Missouri Research Reactor, University of Missouri, Columbia, MO 65211 USA
[2]Department of Physics and Astronomy, University of Missouri, Columbia, MO 65211 USA

E-mail: heitmannt@missouri.edu

Abstract. The presence of magnetic clusters has been verified in both antiferromagnetic and ferromagnetic quantum critical systems. We review some of the strongest evidence for strongly doped quantum critical systems ($Ce(Ru_{0.24}Fe_{0.76})_2Ge_2$) and we discuss the implications for the response of the system when cluster formation is combined with finite size effects. In particular, we discuss the change of universality class that is observed close to the order-disorder transition. We detail the conditions under which clustering effects will play a significant role also in the response of stoichiometric systems and their experimental signature.

Heavy fermion systems that have been prepared via chemical doping to be at a quantum critical point (QCP) at zero Kelvin display a distribution in Kondo shielding temperatures T_K. In heavily doped systems, such as $UCu_{5-x}Pd_x$ [1] and $Ce(Ru_{0.24}Fe_{0.76})_2Ge_2$ [2], the distribution of shielding temperatures can easily span decades of temperature. Upon cooling of such heavily doped systems, the magnetic lattice will be diluted; some magnetic moments will be shielded before others with the result that percolative behavior will ensue. As a consequence, isolated clusters of surviving magnetic ions can and will emerge. Such clusters have been observed in neutron scattering experiments on single crystal $Ce(Ru_{0.24}Fe_{0.76})_2Ge_2$. Following this direct observation of clusters in this antiferromagnetic (AF) quantum critical compound, the presence of magnetic clusters has since been inferred from uniform susceptibility measurements on various ferromagnetic quantum critical systems[3, 4]. In fact, the presence of clusters has even been suggested for almost stoichiometric compounds[5].

In here, we first discuss the importance of finite size effects for isolated magnetic clusters, then we briefly review the evidence for cluster formation in AF $Ce(Ru_{0.24}Fe_{0.76})_2Ge_2$ and the evidence for finite size effects specifically in this material. Next, we scrutinize finite size effects from both a theoretical and experimental point of view, and we finish by speculating how cluster formation might even shape the response in stoichiometric quantum critical systems.

1. Finite size effects in isolated magnetic clusters

The formation of clusters in systems subject to percolation is no surprise, and in heavily doped compounds it is an unavoidable consequence of the dependence of T_K on the immediate surroundings of a magnetic ion. What is, perhaps, surprising is that finite size effects (FSE) force all the moments within the isolated cluster to order at relatively modest temperatures. We detail this in the following where we use the dispersion for AF $RbMnF_3$ as an example (Fig. 1).

Finite size effects come into play when the ordered entity is limited in size, such as is the case for a cluster of radius R. The minimum thermal energy required to excite a magnon in a cluster is $k_BT = E(q = 2\pi/2R)$. For instance, the minimum magnon energy for a spherical

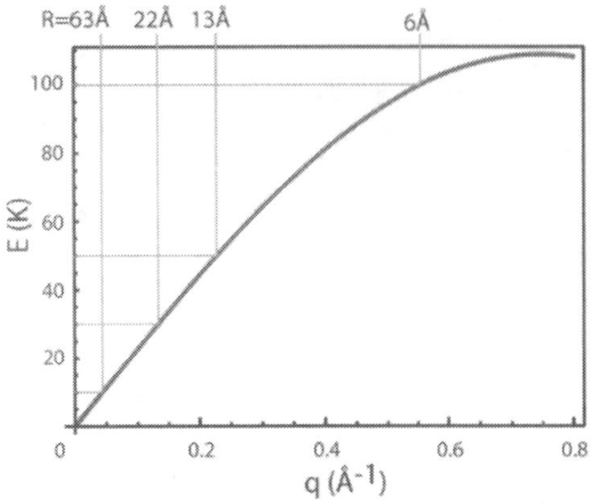

Figure 1. A typical magnon dispersion for an AF system, and the minimum excitation energy induced by finite size effects. The ordering temperatures of some clusters (with radius R listed at the top of the figure) are shown by the solid lines. For example, at $T=$ 10 K clusters of radius 63$\AA$ and smaller would order.

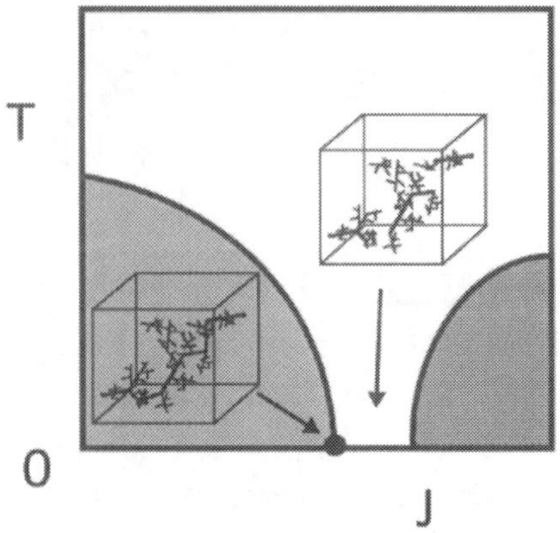

Figure 2. The phase diagram for a chemically doped quantum critical system. Clusters are present both in the ordered phase (left dome), and in the disordered phase. The right dome is the region where all moments are shielded. The QCP (black dot) is the point where a lattice spanning cluster survives down to $T=$ 0 K. (After Ref. [2])

cluster of radius 10Å(which typically has some 100 members) for $RbMnF_3$ is about 60 K (Fig. 1). Therefore, at low T, we should not expect to find any magnons present in such a cluster and, therefore, the cluster will be ordered. Smaller clusters would undergo this forced ordering at even higher temperatures, clusters of 10,000 members will start to order around $\sim$ 10 K.

This implies that once an isolated cluster materializes, FSE automatically order the moments within such a cluster. Even when the spin-spin interaction within a cluster has been weakened due to missing (or partially shielded) moments, we can most definitely expect FSE to dominate in the sub-Kelvin range where quantum critical behavior has been studied.

2. Cluster formation in $Ce(Ru_{0.24}Fe_{0.76})_2Ge_2$

The formation of clusters in quantum critical $Ce(Ru_{0.24}Fe_{0.76})_2Ge_2$ has been observed by means of neutron scattering experiments. Montfrooij et al.[2] noted the emergence of short range magnetic order upon cooling in experiments on a piece of single crystal that displayed the tell-tale quantum critical response in uniform susceptibility, specific heat and resistivity measurements. The short range order was found to span equal number of moments at all temperatures, independent of crystallographic direction. Since the magnetic body centered Ce-ions are separated by much larger distances along the c-axis than along the a-axis in this tetragonal compound (c/a= 2.5), the only viable explanation for the scattering was that magnetic clusters had formed, and that the moments within these clusters had aligned themselves with their neighbors. This lead the authors to propose a phase diagram that included surviving magnetic moments at $T=$ 0 K in the heavy fermion phase (Fig. 2).

3. Percolation theory combined with finite size effects

The Kondo shielding mechanism effectively removes individual moments from the magnetic lattice at a temperature determined by the local surroundings of the magnetic ion. Invariably

Table 1. The $3d$ critical exponents [6] for standard and restricted percolation. The last column shows the exact scaling relationships; primed values reflect the restricted percolation case.

	standard	restricted	relationship
α	-0.62	1	analytic
β	0.41	0.291	$\beta' = \beta/(\beta+1)$
γ	1.80	0.709	$\gamma' = 1 - \beta$
ν	0.88	1/3	$\nu' = 1/d$

this magnetic dilution will lead to the formation of magnetic clusters. Finite size effects, however, do alter the critical behavior in this percolation problem.

The Kondo shielding mechanism is in direct competition with the ordering tendency of the magnetic moments. Once moments have ordered, Kondo shielding will be inhibited by a local molecular field as the shielding mechanism involves spin-flip scattering of the conduction electrons. Therefore, once a cluster becomes isolated from the rest of the magnetic lattice, FSE will force its moments to order (Fig. 1), rendering these clusters impervious to Kondo screening. From a percolation point of view this implies that only moments from the infinite cluster can be removed, isolated clusters cannot be altered.

This restriction on the percolation problem increases the percolation threshold p_c (p is the fraction of occupied sites); every moment removal serves to diminish the strength $P(p)$ of the infinite cluster. It also changes the critical exponents describing the demise of the infinite cluster. This has been calculated in Ref.[6], but it can easily be reasoned out without resorting to calculations as follows. In standard percolation, moments are removed randomly from any site, including those within finite sized clusters. The closer to the percolation threshold, *increasingly* more moments are removed from finite clusters before the infinite clusters loses another moment. On average it will take p/P steps before the infinite cluster is hit again. In contrast, in our restricted scenario it will take only 1 step to remove the next moment. While the infinite clusters end up with an identical morphology in both cases, the difference in occupation probability p at which it occurs *increases* by $p/P-1$ at every p. Because of the p/P factor, the critical exponents will change. We list the modified critical exponents in Table 1.

Given that a distribution of T_K must necessarily lead to a restricted percolation scenario describing the dilution of the magnetic lattice near a QCP, we expect to see the critical exponents listed in Table 1 to be reflected in the response of the system. The exact reflection will depend on the distribution of T_K: the exponents in Table 1 are given as a function of p, whereas in experiments they are measured as a function of T. The Kondo distribution determines $p(T)$, and the low temperature part of $p(T)$-well below the average Kondo shielding temperature-will determine the critical exponents as a function of temperature[7]. The details have not been resolved yet, however, the overall mechanism of how the critical behavior near the QCP is determined by the emergence of magnetic clusters has now been qualitatively established.

4. Clusters in stoichiometric systems

We argue that it is much more likely than not that magnetic clusters also shape the response in (nearly) stoichiometric quantum critical systems. Close to any phase transition, fluctuations of the order parameter will result in ordered regions. Close to a QCP, we can expect fluctuations in the tendency of individual moments to become shielded. These fluctuations can happen because of the bosonic degrees of freedom (zero-point-motion changing the interatomic separation with a concomitant change in T_K) or because of the fermionic degrees of freedom (associated with the shielding conduction electrons themselves). Because of such fluctuations, regions materialize that have magnetic moments isolated from the rest of the lattice by shielded moments; close to the phase transition, such regions (clusters) can contain large numbers of moments.

Once large clusters form, finite size effects dictate that the moments within these clusters order. The main difference with heavily doped systems is that these clusters have a limited lifetime; this lifetime determines whether these clusters will shape the overall response of the system. Typically, zero-point-motion induced atomic displacements happen on a much slower time scale than the electronic motion; in this case, clusters will shape the electronic response of the system. For the case of clusters finding their origin in the fermionic degrees of freedom we cannot make such a statement. However, it is clear that clusters should be expected to emerge, what is unclear is their exact influence on the response of the system.

Emerging clusters in stoichiometric systems can be stabilized with the aid of an external magnetic field. When isolated clusters form, they will likely have dangling moments, bestowing the entire cluster with a net magnetic moment even in the case of AF-coupling. An external field lowers the energy of such clusters to the point that they can last long enough to dominate the response of the system. A good candidate for such a field stabilized system would be $YbRh_2Si_2$[8]. We finish with a list of experimental signatures that would be observed should clusters indeed form in stoichiometric systems.

First, neutron scattering experiments would observe the emergence of short range order characterized by equal numbers of magnetic moments coupled along any crystallographic direction, independent of actual lattice spacings. Second, once clusters form, we should see a sharp drop in resistivity since conduction electrons will no longer be scattered through the Kondo mechanism in these clusters. Third, we expect to see the formation of the clusters reflected in the specific heat given the loss of entropy once the moments within clusters order. Fourth, the uniform susceptibility will reflect the presence of field stabilized magnetic clusters, and a strong signal will be expected given the net moment of most clusters. The response should also show a ferromagnetic component, once again reflecting dangling moments of fully aligned clusters. All measurements should reflect the underlying percolation critical exponents.

Standard tests for the presence of magnetic clusters might not work in quantum critical systems. The clusters do not live forever, and hence, standard hysteresis measurements might only see clusters appear, disappear and appear again without any hysteresis. Resistivity measurements are expected to show a sharp drop anyways once the coherence temperature is reached. The clusters are not expected to provide a measurable residual resistivity as there is no real disorder in the system; rather, the clusters provide low resistance pathways for the conduction electrons to go through, or even to diffract around. Neutron scattering experiments will provide unambiguous evidence, or lack thereof, for the formation of clusters.

This research is supported by the U.S. Department of Energy, Basic Energy Sciences, and the Division of Materials Sciences and Engineering under Grant No.DE-FG02-07ER46381.

References

[1] Aronson M *et al.* 1995 Phys. Rev. Lett. **75**, 725
[2] Montfrooij W *et al.* 2007 Phys. Rev. B **76**, 052404
[3] Westerkamp T *et al.* 2009 Phys. Rev. Lett. **102**, 206404
[4] Garcia Soldervilla J *et al.* 2000 Phys. Rev. B **61**, 6821
[5] Lausberg S *et al.*, in these proceedings
[6] Gaddy J *et al.*, arXiv:1101.6035v1
[7] Stauffer D 1985 *Introduction to Percolation Theory* (Philadelphia: Taylor and Francis)
[8] Custers J *et al.* 2010 Phys. Rev. Lett. **104**, 186402

Effects of Pressure on Cage-Structural Compound $Ce_3Pd_{20}Ge_6$

H. Hidaka[1,2], S.M. Ramos[2], E.N. Hering[2], M.B. Fontes[2], E.B. Saitovitch[2], S. Otani[1], T. Wakabayashi[1], Y. Shimizu[1], T. Yanagisawa[1,3] and H. Amitsuka[1]

[1]Department of Physics, Hokkaido University, Sapporo 060-0810, Japan,
[2]Centro Brasileiro de Pesquisas Físicas, Rio de Janeiro 22290-180, Brazil,
[3]Creative Research Institution, Hokkaido University, Sapporo 001-0021, Japan

E-mail: `hidaka@phys.sci.hokudai.ac.jp`

Abstract.
We have measured the electrical resistivity of clathrate $Ce_3Pd_{20}Ge_6$ under high pressure up to 3.86 GPa in order to investigate pressure effects on two successive phase transitions: O_2^0-type ferroquadrupole order on Ce-8c site at $T_Q = 1.2$ K and antiferromagnetic order on Ce-4a site at $T_N = 0.75$ K. With applying pressure, T_Q stays nearly constant at low pressures and gradually decreases above 2 GPa, while T_N first increases and then decreases above 2.5 GPa. From these results, critical pressures, where T_N and T_Q become 0, are roughly estimated to be around 5 - 6 GPa. In addition, we found that the magnetic resistivity, which is obtained by subtracting the resistivity of $La_3Pd_{20}Ge_6$ from that of $Ce_3Pd_{20}Ge_6$, exhibits $-\ln T$ dependence in two different temperature ranges, suggesting presence of two different Kondo temperatures originated in the two crystallographic Ce-sites.

Cage-structural intermetallic compounds, such as filled skutterudite, pyroclore and clathrate, have attracted much interest because of their appealing physical properties: novel heavy fermion state, unconventional superconductivity, higher multipole ordering, high-thermoelectric figure of merit, and anharmonic oscillations of the guest ion, so called rattling [1, 2]. These exotic phenomena have been proposed to originate from the characteristic cage structure in these systems, and several studies have been made vigorously to make clear their relationship [3, 4]. Among them, the clathrate compounds $R_3Pd_{20}Ge_6$ (R = La-Nd) crystallize in a $Cr_{23}C_6$-type cubic structure where R atoms are located in two different crystallographic sites (4a and 8c sites) [5]. It is known that $Ce_3Pd_{20}Ge_6$ shows two successive 2nd-order phase transitions; O_2^0-type ferroquadrupole (FQ) order on the 8c site at $T_Q = 1.2$ K and antiferromagnetic (AFM) order with a propagation vector $\boldsymbol{q} = (0, 0, 1)$ on the 4a site at $T_N = 0.75$ K [6, 7, 8]. Temperature dependence of electrical resistivity exhibits $-\ln T$ behavior between 2 K and 10 K, suggesting conventional (magnetic) Kondo behavior [6]. To our knowledge, hydrostatic-pressure studies have not been reported in this system so far, although high pressure is a powerful tool for both investigating the role of the electronic correlation in f electron systems and probing an exotic electronic behavior emerging due to the competition of various interactions.

In the present study, we have performed electrical resistivity measurements using polycrystalline $Ce_3Pd_{20}Ge_6$ under hydrostatic pressure up to 3.86 GPa in order to investigate pressure effects on both phase transitions and the Kondo behavior.

The polycrystalline $Ce_3Pd_{20}Ge_6$ and $La_3Pd_{20}Ge_6$ were prepared by the arc-melting method at Hokkaido university, Japan. The ac-electrical resistivity was measured by a conventional four-probe method in the temperature range 0.1 K – 300 K at Centro Brasileiro de Pesquisas Físicas (CBPF), Brazil. Pressure was applied by an indenter-type high-pressure cell with glycerol as the pressure transmitting medium [9]. The pressure value was determined by lead superconducting transition temperature. In the high-pressure study of $Ce_3Pd_{20}Ge_6$, we have measured the resistivity on two different samples from the same batch.

Figure 1(a) shows the temperature dependence of the electrical resistivity $\rho(T)$ on $Ce_3Pd_{20}Ge_6$ at several pressures and $La_3Pd_{20}Ge_6$ at ambient pressure. At the lowest pressure of $Ce_3Pd_{20}Ge_6$ (P = 0.28 GPa), $\rho(T)$ shows $-\ln T$ behavior below 10 K and a maximum around 2.5 K. At low temperatures, the resistivity decreases almost linearly with $\ln T$ below $T_Q \sim$ 1.3 K and then shows a kink at $T_N \sim$ 0.57 K. This behavior is similar to the previous works at ambient pressure [6, 10], showing slightly lower T_N. The difference of T_N may be due to the off-stoichiometry composition, since it has been known that T_N is highly sensitive to Pd concentration [11].

By applying pressure, T_N increases in low pressure region and decreases above $\sim$ 2.5 GPa, while T_Q is almost constant up to $\sim$ 2 GPa decreasing after that, although the feature that determines T_Q becomes vague at high pressures. The obtained pressure-temperature phase diagram of $Ce_3Pd_{20}Ge_6$ is shown in Fig. 2(a). The critical pressures, where T_N and T_Q become 0, could be roughly estimated to be around 5 - 6 GPa. Higher-pressure experiments are now in progress to determine the critical pressures and possible induced critical phenomena.

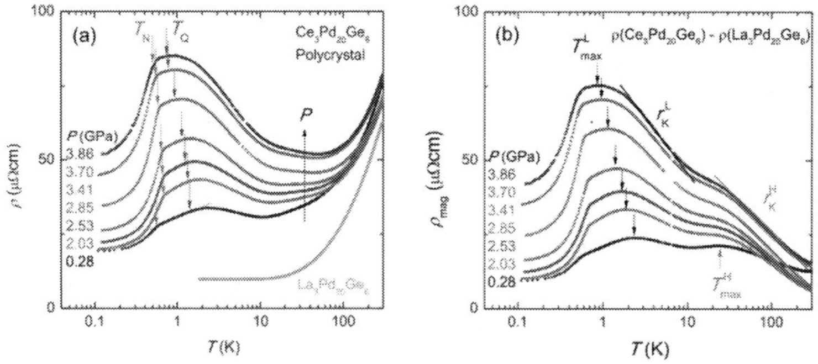

Figure 1. (a) Temperature dependence of the electrical resistivity of $Ce_3Pd_{20}Ge_6$ for high pressures up to 3.86 GPa and $La_3Pd_{20}Ge_6$ at ambient pressure. Red and blue downwards arrows represent T_N and T_Q, respectively. (b) Temperature dependence of the magnetic resistivity on $Ce_3Pd_{20}Ge_6$ at several pressures. The red-upwards and black-downwards arrows indicate the temperature where $\rho_{mag}(T)$ exhibits the maximum in the high and low temperature ranges, T^{H}_{max} and T^{L}_{max}, respectively. The lines with indices r^{H}_{K} and r^{L}_{K} denote examples of the $-\ln T$ fit for the magnetic resistivity in high and low temperature ranges.

In Fig. 1(b), we plot the magnetic contribution to the electrical resistivity, ρ_{mag}, which was obtained by subtracting $\rho(T)$ of $La_3Pd_{20}Ge_6$ for ambient pressure from the experimental data of $Ce_3Pd_{20}Ge_6$ taken under high pressures. We have checked that $\rho(T)$ of $La_3Pd_{20}Ge_6$ is almost unchanged at least up to 2.4 GPa (not shown). It has been revealed that there are two temperature ranges where ρ_{mag} shows $-\ln T$ dependence. Such behavior in $\rho_{mag}(T)$ has been observed in some Kondo-lattice compounds, such as $CeAl_2$ and $CeCu_2Si_2$, and understood as a consequence of the interplay between the crystalline electric field (CEF) and the Kondo effects [12, 13]. Within the model developed by Cornut and Coqblin [14], a magnetic-resistivity slope r_K

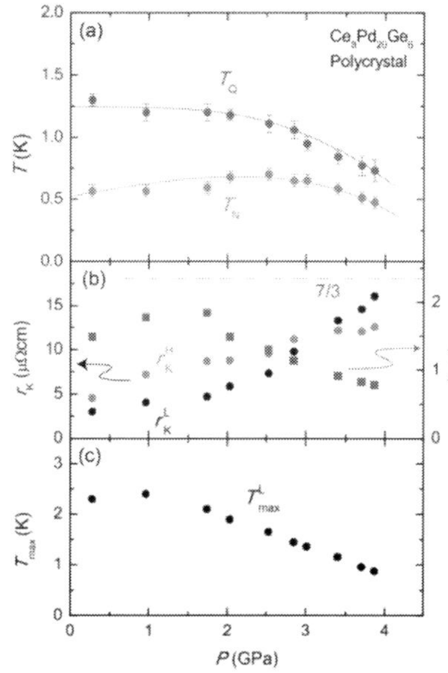

Figure 2. Pressure dependence of (a) T_N and T_Q of $Ce_3Pd_{20}Ge_6$, (b) the magnetic-resistivity slope r_K^H and r_K^L (left-hand scale), the ratio of the slope between the high and low temperature ranges r_K^H/r_K^L (right-hand scale), and (c) the lower resistivity-maximum temperature T_{max}^L in $\rho_{mag}(T)$. The lines are guides to the eye.

($= -d\rho_{mag}(T)/d\ln T$) is proportional to $2J_{cf}^3 D(E_F)^2(\lambda_n^2 - 1)/(2j+1)$, where J_{cf} is the exchange integral between $4f$ electrons and conduction electrons, $D(E_F)$ the density of state at the Fermi energy, and λ_n the degeneracy of the $4f$ level. In the case of $Ce_3Pd_{20}Ge_6$, it has been reported that both the Ce sites have the Γ_8 quartet ground state, and the CEF level splitting between the Γ_8 ground state and the Γ_7 excited state ($\equiv \Delta$) is 60 K at the 4a site and 46 K at the 8c site [6, 15]. According to this model, the ratio, r_K^H/r_K^L, of the $-\ln T$ slope of $\rho_{mag}(T)$ between the high and low temperature ranges can be rewritten as $(\lambda_H^2 - 1)/(\lambda_L^2 - 1)$, and calculated to be 7/3 (~ 2.33). However, the ratio determined experimentally is significantly smaller than this value, varying in the range between 0.8 and 1.8 with pressure, as shown in Fig. 2(b). In addition, the higher Kondo temperature T_K^H and the lower one T_K^L should have the following theoretical relation: $\Delta(T_K^L)^2 = (T_K^H)^3$ [16]. If T_K^L is assumed to be 2 K, the T_K^H values can be estimated to be 6.2 K for the 8c site and 5.7 K for the 4a site, which are considerably smaller than $T_{max}^H \sim 25$ K, the temperature where $\rho_{mag}(T)$ exhibits the maximum in the high temperature range. These facts strongly suggest that the observed double $-\ln T$ behavior is not due to the effects of simple combination of the CEF splitting and the Kondo scattering. Taking the crystal structure of the system into consideration, it should be noted that there might be two different Kondo temperatures, which are originated in the two crystallographic Ce sites. In fact, the similar double $-\ln T$ behavior has been reported in a high-pressure study on $Ce_3Pd_{20}Si_6$, where much different T_K's have been suggested for the two crystallographic Ce sites [17]. From this similarity, we suggest that the double $-\ln T$ behavior on $Ce_3Pd_{20}Ge_6$ should also be attributed to the difference in T_K for the two crystallographic Ce sites.

Interestingly, the lower $\rho_{mag}(T)$-maximum temperature T_{max}^L decreases by applying pressure as shown in Figs. 1(b) and 2(c), which is opposite to the behavior observed in the other Ce compounds. Generally, in the Ce-based compounds T_{max} is understood as the onset of the coherent scattering of conduction electrons on Kondo centers arranged periodically ($= T_{coh}$), and expected to increase by applying pressure, since the pressure leads to an increase of T_{K} as a result of an enhancement of the $c-f$ hybridization. In $Ce_3Pd_{20}Ge_6$, the $c-f$ hybridization is also considered to be enhanced by pressure, because both $r_{\mathrm{K}}^{\mathrm{H}}$ and $r_{\mathrm{K}}^{\mathrm{L}}$, which depend on $|J_{\mathrm{cf}}^3 D(E_{\mathrm{F}})^2|$ [14], increase with increasing pressure (see Fig. 2(b)). This discrepancy between the decreasing of $T_{\mathrm{max}}^{\mathrm{L}}$ and the enhancement of J_{cf} implies that $T_{\mathrm{max}}^{\mathrm{L}}$ can not be simply identified with T_{coh} in the present case. One possible reason for the decrease in $T_{\mathrm{max}}^{\mathrm{L}}$ would be a change of Δ by applying pressure. However, it may be unreasonable to consider that the change of the Δ's affects the resistivity maximum around 2 K, since the Δ's are estimated to be 60 K and 46 K for the 4a and the 8c sites, respectively [15]. Another possible explanation comes from the FQ ordering on the 8c site. The specific-heat and thermal-expansion measurements have detected a broad anomaly around T_{Q}, and suggest the presence of short-range order (SRO) of the FQ ordering and/or of a local Jahn-Teller distortion (LJTD) that set in above T_{Q}. Both the effects may induce the splitting of the Γ_8-quartet ground state, and suppress of the Kondo effect originating from the Γ_8 state. Therefore, if the SRO and/or the LJTD occurs above T_{coh}, the resistivity maximum could be observed at a higher temperature naturally. The similar pressure dependence of T_{Q} and $T_{\mathrm{max}}^{\mathrm{L}}$ might support this latter scenario regarding the SRO and/or LJTD. At the present stage, the reason for the unusual pressure dependence of $T_{\mathrm{max}}^{\mathrm{L}}$ is not clear, and the study using other high-pressure experiments, such as ultrasonic and inelastic-neutron-scattering measurements, will be needed.

In summary, we have studied the pressure dependence of T_{N} and T_{Q} of $Ce_3Pd_{20}Ge_6$ up to 3.86 GPa, and revealed that both the transitions are suppressed and show a tendency to disappear at pressures higher than 5 GPa. The magnetic resistivity exhibits a double $-\ln T$ dependence in different temperature ranges, which will probably be due to the difference in the magnetic scattering at the two crystallographic Ce sites. In addition, we found that $T_{\mathrm{max}}^{\mathrm{L}}$ exhibits the negative pressure dependence in spite of a possible enhancement of the $c-f$ hybridization by pressure.

This work was supported by Grant-in-Aid for Scientific Research B (19340086) and S (20224015) from the Ministry of Education, Culture, Sports, Science and Technology (MEXT), Japan, FAPERJ-PRONEX, FAPERJ-CNE, and CNPq, Brazil. H. Hidaka, S.M. Ramos, and E.N. Hering acknowledge the fellowship given by PCI-CBPF, PAPD-FAPERJ-CAPES, and PNPD-CAPES, respectively.

References

[1] Sato H, Kikuchi D, Tanaka K, Aoki H, Kuwahara K, Aoki Y, Kohgi M, Sugawara H and Iwasa K 2007 *J. Magn. Magn. Matter.* **310** 188-194
[2] Nagao Y, Yamaura J, Ogusu H, Okamoto Y and Hiroi Z 2009 *J. Phys. Soc. Jpn.* **78** 064702
[3] Hattori K, Hirayama Y and Miyake K 2005 *J. Phys. Soc. Jpn.* **74** 3306
[4] Takimoto T 2006 *J. Phys. Soc. Jpn.* **75** 034714
[5] Grivanov A V, Seropegin Yu D and Bodak O I 1994 *J. Alloys Comp.* **204** L9
[6] Kitagawa J, Takeda N and Ishikawa M 1996 *Phys. Rev. B* **53** 5101
[7] Nemoto Y, Yamaguchi T, Horino T, Akatsu M, Yanagisawa T, Goto T, Suzuki O, Dönni A and Komatsubara T 2003 *Phys. Rev. B* **68** 184109
[8] Dönni A, T Herrmannsdörfer T, Fischer P, Keller L, Fauth F, McEwen K A, Goto T and Komatsubara T 2000 *J. Phys:Condens. Matter* **12** 9441
[9] Kobayashi T C, Hidaka H, Kotegawa H, Fujiwara K and Eremets M I 2007 *Rev. Sci. Instrum.* **78** 023909
[10] Yamamizu T, Nakayama M, Kimura N, Komatsubara T and Aoki H 2002 *Physica B* **312–313** 495
[11] Kitagawa J, Takeda N, Sakai F and Ishikawa M. 1999 *J. Phys. Soc. Jpn.* **68** 3413
[12] Ōnuki Y, Furukawa Y and Komatsubara T 1984 *J. Phys. Soc. Jpn.* **53** 2734
[13] Horn S, Edwards M A, Thompson J D and Parks R D 1985 *J. Magn. Magn. Matter.* **52** 385
[14] Cornut B and Coqblin B 1972 *Phys. Rev. B* **5** 4541
[15] Keller L, Dönni A, Zolliker A M and Komatsubara T 1999 *Physica B* **259–261** 336–337
[16] Ueda K and Ōnuki Y 1998 *Physics of Heavy Electron Syetems* (*Shokabo, Tokyo*)
[17] Hashiguchi T, Takeda N, Kitagawa J, Wada N, Takayanagi S, Ishikawa M and Mōri N 2000 *J. Phys. Soc. Jpn.* **69** 667

Journal of Physics: Conference Series **391** (2012) 012020 doi:10.1088/1742-6596/391/1/012020

Pressure-induced Suppression of the Antiferromagnetic Transition in $YbNi_3Al_9$ Single Crystal

T. Hirayama[1], K. Matsubayashi[1], T. Yamashita[2], S. Ohara[2], K. Munakata[3], Y. Uwatoko[1]

[1]Institute for Solid State Physics, The university of Tokyo, Kashiwanoha, Kashiwa, Chiba 277-8581, Japan

[2]Department of Engineering Physics, Electronics and Mechanics, Graduate School of Engineering, Nagoya Institute of Technology, Nagoya 466-8555, Japan

[3]Research Center for Neutron Science and Technology, Comprehensive Research Organisation for Science and Society (CROSS), 162-1 Shirakata, Tokai, Ibaraki 319-1106, Japan

E-mail: `hirayama3@issp.u-tokyo.ac.jp`

Abstract. We have measured the electrical resistivity $\rho(T)$ of heavy fermion antiferromagnet $YbNi_3Al_9$ under hydrostatic pressure up to 8 GPa. We found that the antiferromagnetic ordering temperature T_N is suppressed with increasing pressure, whereas the localized character of the trivalent Yb state becomes stronger inferred from the reduction of the Kondo scattering. We will discuss the possible origin of the suppression of T_N under pressure.

1. Introduction

In Ce- and Yb-based heavy fermion systems, tuning the ground state by pressure and/or magnetic field from a non-magnetic state to a magnetic state or vice versa has attracted attention because the anomalous behavior, such as unconventional superconductivity or non-Fermi liquid state, appears in the vicinity of quantum critical point (QCP). Due to the hole-electron analogy, it is known that the application of pressure in Yb-based compounds has an opposite effect compared to Ce-based compounds, i.e., applying pressure induces magnetically ordered state from nonmagentic state. Once the system enters the magnetically ordered state away from the QCP, the magnetic order is stabilized and thus the magnetic transition temperature increases with increasing pressure. However, here we introduce the suppression of the antiferromagnetic (AF) ordered state by the application of pressure in heavy fermion compound $YbNi_3Al_9$.

Recently, Ohara *et.al* succeeded in growing $YbNi_3Al_9$ single crystal and reported low temperature physical properties [1]. $YbNi_3Al_9$ crystallizes in the $ErNi_3Al_9$ type crystal structure. There are three kinds of layers stacked along the *c*-axis: Yb_2Ni_3-layers, Al_3-layers and Ni_3-layers. Note that Yb ions are arranged in a two-dimensional honeycomb lattice in the Yb_2Ni_3-layers. The temperature dependence of electrical resistivity shows typical behavior of heavy fermion antiferromagnet with a Néel temperature of $T_N \sim 3.4$ K. Residual resistivity is less than $1\mu\Omega$ cm and RRR$\sim$30, indicating high quality of single crystal. In order to investigate the effect of pressure on Yb-based heavy fermion antiferromagnet, we have measured the electrical resistivity of high quality $YbNi_3Al_9$ single crystal up to 8 GPa.

2. Experimental

High pressure was generated by using two types of pressure cells; a hybrid piston cylinder cell (PCC) and cubic anvil cell (CAC) [2]. CAC is known to generate hydrostatic pressure owing to the multiple anvil geometry with a maximum pressure of ~8 GPa. The pressure at low temperature in PCC was determined by the pressure dependence of a superconducting transition temperature of Pb. The applied pressure in CAC is calibrated by the measurement of the resistivity changes of Bi and Te associated with their structural phase transitions at room temperature. Pressure transmitting medium for PCC and CAC was a glycerin and mixture of Fluorinert (FC70 and FC77), respectively. The electrical resistivity was measured by the four-terminal method with an excitation current of 500-1000 μA. The electrical current flows along the a-axis.

3. Result and Discussion

Figure 1 shows temperature dependence of electrical resistivity $\rho(T)$ of $YbNi_3Al_9$ at ambient pressure. $\rho(T)$ exhibits the broad maximum centered at T_{max}~40 K, which is related to incoherent Kondo-scattering processes on the ground state and the excited crystalline electronic field (CEF) level. According to specific heat measurement, the Schottky anomaly is observed with an energy splitting of $\Delta \sim 40$ K [3], consistent with the resistivity peak at T_{max}. At lower temperatures, a clear kink at $T_N \sim 3.2$ K is associated with the antiferromagnetic transition of Yb moments. We define T_N as the minimum in the second derivative $d^2\rho(T)/dT^2$. The insets of Fig. 1(a) and (b) show $\rho(T)$ at selected pressure and the pressure dependence of T_N, respectively. Interestingly, T_N decreases with increasing pressure in the initial ratio of -0.23 K/GPa. This observation is unexpected because the magnetic order of Yb compounds is stabilized with increasing pressure, as mentioned introduction. In order to confirm the suppression of T_N under pressure, we have performed resistivity measurement using CAC at higher pressure region.

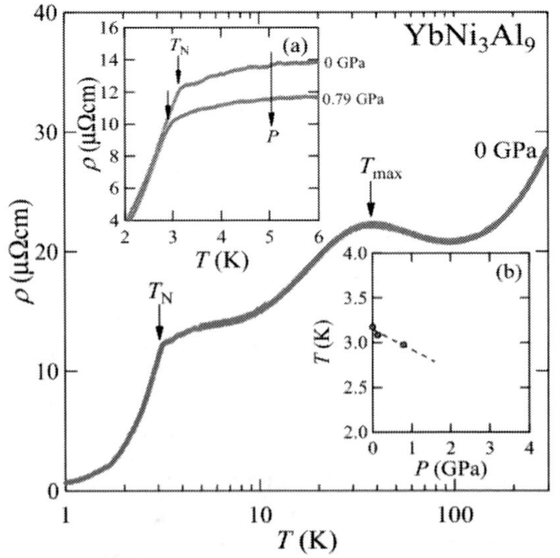

Figure 1. Temperature dependence of the electrical resistivity for $YbNi_3Al_9$ at ambient pressure. The electrical current flows along the a-axis. The inset (a) shows the temperature dependence of the electrical resistivity under pressure using PCC. The arrows indicate the AF transition temperature. The inset (b) shows the pressure dependence of T_N for $YbNi_3Al_9$. Dotted line is a guide to the eye.

Figure 2 shows the temperature dependence of electrical resistivity up to 8 GPa. Applying pressure does not change the position of T_{max} (see the inset of Fig. 2), however, the magnitude of the resistivity at T_{max} significantly decreases with increasing pressure. Thus, the energy splitting of the CEF effect is almost pressure independent and the Kondo scattering is suppressed under pressure. This result indicates a strong localized character of Yb ions under pressure, namely the localized f electrons are effectively decoupled from the conduction electrons. Next, we focus on the low temperature properties. Above 2 GPa, the resistivity anomaly at T_N shifts to lower temperatures and no longer observable in the temperature range down to ~2.4 K. This is consistent with the results obtained at lower pressure region using PCC.

Here we briefly discuss the possible origin of the unusual pressure dependence of T_N in $YbNi_3Al_9$. Recently similar anomalous suppression of T_N under pressure is also reported in some Yb-based compounds, such as Yb_2Pd_2Sn and YbAgGe [4, 5]. In these systems, it is suggested that T_N is affected by a complex interplay of RKKY interaction, Kondo effect and the geometrical frustration. First, we consider the competition between Kondo and RKKY interaction. In Yb-based heavy fermion compounds, starting from the non-magnetic intermediate valence regime, the Kondo interaction is suppressed with increasing pressure and a long range magnetically ordered state appears. Thus, it is expected that the magnetic transition temperature increases with increasing pressure if the system is located close to a magnetic QCP. On the other hand, once the system enters the strongly localized regime, recent theoretical model predicts the enhancement of the Kondo interaction under sufficiently high pressures, resulting in the reduction of the magnetic interaction [6]. In practice, the Kondo scattering in $YbNi_3Al_9$ is monotonically suppressed with increasing pressure, although the valence state of Yb ions is close to trivalent state confirmed by X-ray absorption spectroscopy [7]. Hence, above theory is not applicable to the present system.

From the structural point of view, the Yb ions of $YbNi_3Al_9$ indeed lie on a honeycomb lattice with the potential for magnetic frustration. The magnetic frustration in a honeycomb lattice

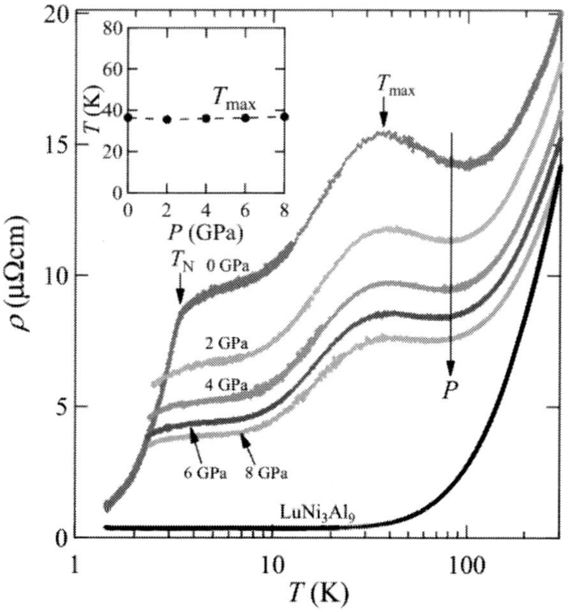

Figure 2. Temperature dependence of the electrical resistivity of $YbNi_3Al_9$ under pressure using CAC, together with the that of $LuNi_3Al_9$. The inset shows pressure dependence of T_{max}.

is only expected if both nearest and next-nearest neighbor interaction is antiferromagnetic, but the magnetization data of $YbNi_3Al_9$ suggest that Yb moments align ferromagnetically in the honey-comb layers while the each plane is coupled antiferromagnetically along the c-axis [1, 7]. Therefore, the scenario of the magnetic frustration is ruled out assuming that the aforementioned magnetic structure persists under pressure. At this stage, the origin of suppression T_N is an open question. Further investigations, such as magnetic susceptibility and neutron diffraction measurements under pressure, are necessary to clarify this issue.

4. Summary

We present the effect of pressure on $YbNi_3Al_9$ single crystal under pressure. The pressure variation of the resistivity indicates the localized character of the trivalent Yb state under pressure. In contrast, we found the pressure-induced suppression of T_N probably due to the complex interplay of the low energy interactions.

5. Acknowledgments

This work was financially supported by Grants-in-Aid for Scientific Research (No. 21340092) and for Scientific Research on Innovative Areas ‘‘Heavy Electrons’’ (No. 20102007). KM was supported by a Grant-in-Aid for Young Scientists B (No. 22740217) from the Ministry of Education, Culture, Sports, Science and Technology.

6. References

[1] Ohara S, Yamashita T, Mori Y and Sakamoto I 2011 *Journal of Physics: Conference Series* **273** 012048
[2] Uwatoko Y, Todo S, Ueda K, A, Uchida, Kosaka M, Mori N and Matsumoto T 2002 *Journal of Physics: Condensed Matter* **14** 11291
[3] Ohara S *et al* private communication
[4] Muramatsu T, Kanemasa T, Kagayama T, Shimizu K, Aoki Y, Sato H, Giovannini M, Bonville P, Zlatic V, Aviani I, Khasanov R, Rusu C, Amato A, Mydeen K, Nicklas M, Michor H and Bauer E 2011 *Phys. Rev. B* **83** 180404
[5] Sengupta K Forthaus M K, Kubo H, Katoh K, Umeo K, Takabatake T and Abd-Elmeguid M M 2010 *Phys. Rev. B* **81** 125129
[6] Goltsev A V and Abd-Elmeguid M M 2005 *Journal of Physics: Condensed Matter* **17** S813
[7] Tobash P H, Jiang Y, Ronning F, Booth C H, Thompson J D, Scott B L and Bauer E D 2011 *Journal of Physics: Condensed Matter* **23** 086002

Metamagnetic Behavior in Heavy Fermion Compounds UCo_2Zn_{20} and UIr_2Zn_{20}

Y Hirose[1], K Enoki[1], S Yoshiuchi[1], T Takeuchi[2], F Honda[1], K Sugiyama[1], E Yamamoto[3], Y Haga[3], M Hagiwara[4], K Kindo[5], R Settai[1], and Y Ōnuki[1,3]

[1]Graduate School of Science, Osaka University, Toyonaka Osaka 560-0043, Japan
[2]Low Temperature Center, Osaka University, Toyonaka, Osaka 560-0043, Japan
[3]Advanced Science Research Center, Japan Atomic Energy Agency, Tokai, Ibaraki 319-1195, Japan
[4]KYOKUGEN, Osaka University, Toyonaka, Osaka 560-8531, Japan
[5]ISSP, University of Tokyo, Kashiwanoha, 5-1-5, Kashiwa, Chiba 277-8581, Japan

E-mail: yhirose@crystal.phys.sci.osaka-u.ac.jp

Abstract. The metamagnetic behavior, which is typical in heavy fermion compounds, is observed in UCo_2Zn_{20} and UIr_2Zn_{20} at H_m= 80 and 20 kOe, respectively. Reflecting the metamagnetic behavior, the magnetoresistance is suppressed in magnetic fields larger than H_m, and the corresponding A-value of a Fermi liquid relation in the electrical resistivity $\rho = \rho_0 + AT^2$ indicates a peak at around H_m. Pressure suppresses this behavior in UIr_2Zn_{20}.

1. Introduction

UT_2Zn_{20}(T : Co, Ir) crystallizes in the cubic structure with a large lattice constant $a \simeq 14$ Å and the U-U distance $d \simeq 6$Å. UCo_2Zn_{20} was reported to be a non-magnetic heavy fermion compound[1]. The magnetic susceptibility follows a Curie-Weiss law at high temperatures and possesses a maximum at $T_{\chi_{\rm max}} = 7$ K. The corresponding specific heat coefficient C/T also possesses a broad maximum at 5 K. On the other hand, UIr_2Zn_{20} was a ferromagnet with a Curie temperature $T_C = 2.1$ K[1]. Both compounds, however, possess large electronic specific heat coefficients $\gamma = 230$ and 450 mJ/(K^2·mol), respectively. Another experimental results were reported very recently, claiming that UCo_2Zn_{20} and UIr_2Zn_{20} are non-magnetic heavy fermion compounds with $T_{\chi_{\rm max}} = 8.2$ and 4.5 K, and the electronic specific heat coefficient $\gamma = 350$ and 400 mJ/(K^2·mol), respectively[2].

Recently, a non-linear increase of magnetization, namely a metamagnetic behavior was observed at the magnetic field H_m in the similar non-magnetic heavy fermion compounds YbT_2Zn_{20} (T : Co, Rh, and Ir)[3]. A simple relation between $T_{\chi_{\rm max}}$ and H_m in Ce-, U-, and Yb-based heavy fermion compounds including YbT_2Zn_{20} was obtained as $15T_{\chi_{\rm max}}({\rm K}) = H_m({\rm kOe})$, namely $k_B T_{\chi_{\rm max}} = \mu_B H_m$. Here, $T_{\chi_{\rm max}}$ roughly corresponds to the Kondo temperature, and the corresponding metamagnetic behavior is observed below $T_{\chi_{\rm max}}$. The almost localized f-electron nature at high temperatures is changed into a heavy fermion state at temperatures lower than $T_{\chi_{\rm max}}$. The heavy fermion state can be, however, changed into the almost localized f-electron nature at magnetic fields larger than H_m.

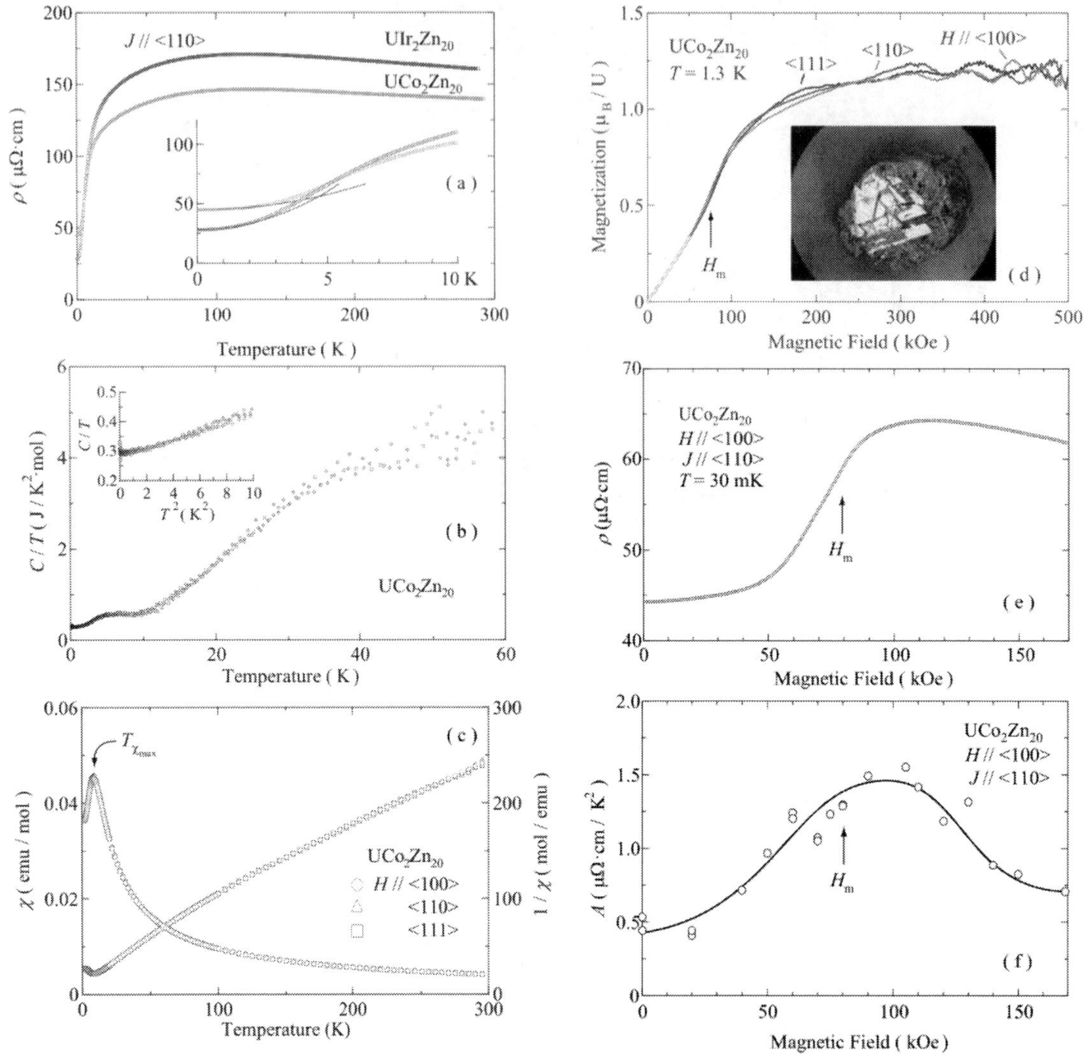

Figure 1. (a) Electrical resistivity of UCo_2Zn_{20} and UIr_2Zn_{20}, (b) specific heat in the form of C/T(inset:C/T vs T^2), (c) magnetic susceptibility and the corresponding inverse susceptibility, (d) high-field magnetization and (inset:single crystal ingots), (d) transverse magnetoresistance, and (e) magnetic field dependence of the A value in UCo_2Zn_{20}.

In the present study, we grew single crystals of UCo_2Zn_{20} and UIr_2Zn_{20}, and measured the electrical resistivity, specific heat, magnetic susceptibility and magnetization to investigate the metamagnetic behavior.

2. Experimental results and analyses

2.1. UCo_2Zn_{20}

First, we show in Fig. 1(a) the temperature dependence of the electrical resistivity ρ in UCo_2Zn_{20}, together with that in UIr_2Zn_{20}. The resistivity increases slightly with decreasing temperature, has a broad maximum around 100 K, and decreases steeply at lower temperatures. The resistivity follows a Fermi liquid relation of $\rho = \rho_0 + AT^2$ below 3 K, as shown by a thin solid line in inset

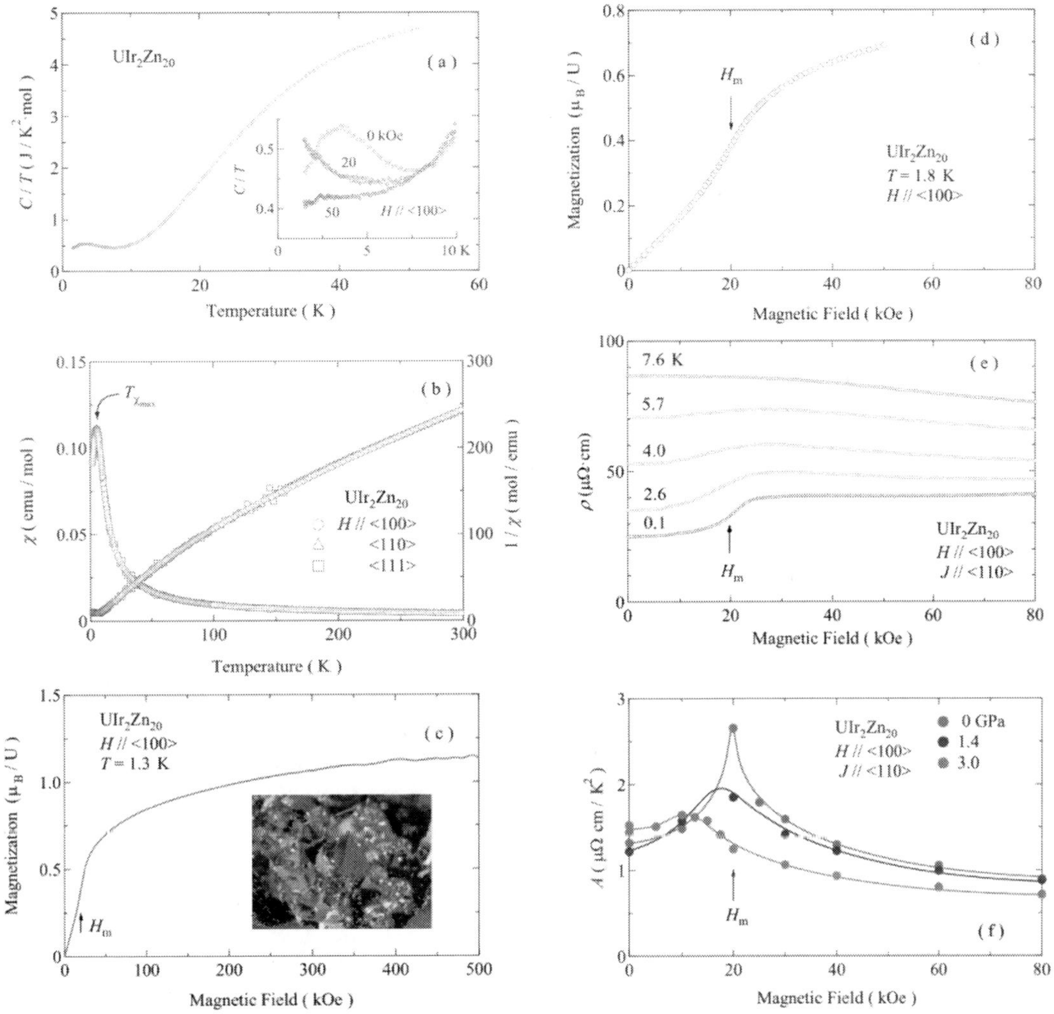

Figure 2. (a) Specific heat in the form of C/T(inset:C/T under 0, 20, and 50 kOe), (b) magnetic susceptibility and the corresponding inverse susceptibility, (c) high-field magnetization (inset:single crystal ingots), (d) magnetization, (e) transverse magnetoresistance at several temperatures, and (e) field dependence of the A value under pressures in UIr_2Zn_{20}.

of Fig. 1(a). The A value is 0.49 $\mu\Omega$·cm/K^2, which corresponds to $\gamma \simeq 220$ mJ/(K^2·mol) from a Kadowaki-Woods relation $A/\gamma^2 = 1.0\times10^{-5}$ $\mu\Omega$·cmK^{-2}/(mJK^{-2}mol^{-1})$^{-2}$[4]. We note that $A = 1.3$ $\mu\Omega$·cm/K^2 and $\gamma = 360$ mJ/(K^2·mol) are obtained for UIr_2Zn_{20}. The low-temperature specific heat coefficient C/T possesses a broad peak at 6 K, as shown Fig. 1(b), and the γ value is 300 mJ/(K^2·mol), which are approximately consistent with the previous results[1, 2], together with $\gamma = 220$ mJ/(K^2·mol) estimated from the A value, mentioned above. The magnetic susceptibility χ increases with decreasing temperature, and possesses a peak at $T_{\chi_{\max}} = 8.5$ K. No anisotropy is observed below and above $T_{\chi_{\max}}$ for H // $\langle 100\rangle$, $\langle 110\rangle$, and $\langle 111\rangle$. The present peak is therefore not due to a magnetic ordering, but is based on a characteristic heavy fermion behavior, as mentioned in Introduction. The effective magnetic moment and the paramagnetic Curie temperature are obtained as $\mu_{\rm eff} = 3.44$ $\mu_{\rm B}$/U and $\theta_{\rm P} = $ -63 K, respectively. The present magnetic moment is close to 3.58 μ_B/U of $5f^2$ and 3.62 μ_B/U of $5f^3$ configuration.

A metamagnetic behavior, which is characteristic in heavy fermion compounds, is clearly observed at $H_m \simeq$ 80 kOe for H // ⟨100⟩, ⟨110⟩, and ⟨111⟩, as shown in Fig. 1(d). The magnetic moment at 500 kOe is small, 1.1 μ_B/U. This means that the $5f$ electrons are not fully localized even at high magnetic fields and still contribute to the conduction electrons. The present metamagnetic behavior is also reflected in the transverse magnetoresistance, as shown in Fig. 1(e). We also obtained the A value from the temperature dependence of the resistivity under constant magnetic fields. The magnetic field dependence of the A value is shown in Fig. 1(f), revealing a broad maximum at around H_m.

2.2. UIr_2Zn_{20}

A broad maximum in C/T is observed at 3.5 K in UIr_2Zn_{20}, as shown in Fig. 2(a), and the γ value is roughly estimated as 450 mJ/(K^2·mol), which is also consistent with the previous results[2] and γ = 370 mJ/(K^2·mol) estimated from the A value. The present γ value is reduced to 400 mJ/(K^2·mol) at 50 kOe, as shown inset of Fig. 2(a). Note that C/T at 20 kOe, which is a metamagnetic field, shown below, increases with decreasing temperature. No anisotropy of the magnetic susceptibility is also observed for H // ⟨100⟩, ⟨110⟩, and ⟨111⟩, as shown in Fig. 2(b). The μ_{eff} and θ_P are 3.55 μ_B/U and -85.8 K, respectively. $T_{\chi_{max}}$ is obtained at 5.3 K.

The metamagnetic behavior is observed at H_m = 20 kOe, as shown in Figs. 2(c) and 2(d). An increase of transverse magnetoresistance below 4.0 K is suppressed above H_m, as shown in Fig. 2(e). It is noted that 4.0 K is below $T_{\chi_{max}}$ = 5.2 K. The A value indicates a sharp peak at H_m = 20 kOe. This peak is, however, reduced with increasing pressure, and most likely diminishes at high pressures of 4 GPa. It is also noted that the A value at H = 0 kOe is approximately unchanged with respect to pressure. This behavior is in contract with that in $YbIr_2Zn_{20}$, where the peak structure is shifted to lower magnetic fields with increasing pressure, and finally H_m becomes zero at a critical pressure P_c = 5.2 GPa, reaching quantum critical point. The A value is increased from A = 0.29 $\mu\Omega$·cm/K^2 at ambient pressure to A = 380 $\mu\Omega$·cm/K^2 at P_c[5].

3. Conclusion

We observed a metamagnetic behavior at H_m = 80 kOe for $T_{\chi_{max}}$ = 8.5 K in UCo_2Zn_{20} and at H_m = 20 kOe for $T_{\chi_{max}}$ = 5.2 K in UIr_2Zn_{20}. The present metamagnetic behavior is typical in heavy fermion Ce, Yb, and U compounds, and a simple relation of 15 $T_{\chi_{max}}$(K) = H_m(kOe) is roughly applied to UCo_2Zn_{20}. UIr_2Zn_{20} is found to deviate from the relation. This is needed to study in more detail.

Acknowledgments

This work was supported by a Grant-in-Aid for Scientific Research on Innovative Areas "Heavy Electrons" (No. 20102002), Specially Promoted Research (No. 20001004), Young Scientists (B), (No. 18740219) and Osaka University Global COE Program " Core Research and Engineering of Advanced Materials-Interdisciplinary Education Center for Materials Science " (G10) from The Ministry of Education, Culture, Sports, Science, and Technology, Japan. This work was also supported by the " Exciting Leading-Edge Research Project " at Osaka University.

References

[1] E. D. Bauer, J. D. Thompson, J. L. Sarrao, and M. F. Hundley : J. Magn. Magn. Mater. **310** (2007) 449.
[2] P. Swatek and D. Kaczorowski : J. Phys. Soc. Jpn. **80** (2010) SA106.
[3] Y. Hirose, M. Toda, S. Yoshiuchi, S. Yasui, K. Sugiyama, F. Honda, M. Hagiwara, K. Kindo, R. Settai, and Y. Ōnuki : J. Phys. : Conf. Ser. **273** (2011) 012003.
[4] K. Kadowaki and S. B. Woods : Solid State Commun. **58** (1986) 507.
[5] F. Honda, S. Yasui, S. Yoshiuchi, T. Takeuchi, R. Settai, and Y. Ōnuki : J. Phys. Soc. Jpn. **79** (2010) 083709.

Spin and Magnetic Field Dependences of Quasiparticle Mass in Ferromagnetic State of Heavy Fermions

Olga Howczak[1*] **and Jozef Spałek**[1,2†]

[1] Marian Smoluchowski Institute of Physics, Jagiellonian University, 30-059 Krakow, Poland,
[2] Faculty of Physics and Applied Computer Science, AGH University of Science and Technology, 30-059 Kraków, Poland.

E-mail: *olga.howczak@uj.edu.pl, †ufspalek@if.uj.edu.pl

Abstract. We investigate the mechanism underlying the suppression of heavy-fermion mass enhancement in the presence of a magnetic field. In the framework of statistically consistent Gutzwiller method (SGA) we study the periodic Anderson model in the strong correlation limit. The finite-U corrections are included systematically allowing to describe the coexistence of Kondo compensation effect and ferromagnetic ordering, as well as weak delocalization of the f-electrons. In particular, we observe that the resulting mass enhancement factor of spin-up electrons and that of spin-down are not equal in ferromagnetic phases and depend strongly on the applied field and the f-level occupancy. We predict that mass enhancement for the spin-up quasiparticles is larger then that of spin-down and both of them decrease in the applied magnetic field. We argue that above features, as well as a nonmonotonic variation of the quasiparticle effective masses observed in our model are in good agreement with earlier experimental measurements for $Ce_xLa_{1-x}B_6$.

1. Introduction

Heavy-fermion (HF) metals are customarily regarded as lattice of strongly correlated f-electrons and coupled to them conduction (band) electrons via an antiferromagnetic Kondo exchange interaction. The f-electrons are associated with the rare-earth or actinides ions. The f-f on-site Coulomb repulsion is strong and the f-electrons behave often in a solid as localized (or almost localized) magnetic moments, even in the presence of the interband hybridization.

One of characteristic properties of the HF compounds is the *metamagnetic behavior*, i.e. a sudden magnetization increase at a critical value of the applied magnetic field, and observed in HF compounds such as $CeRu_2Si_2$ [1], $CeCoGe_3$ [2], UPd_2Al_3, URu_2Si_2, and UPt_3 [3]. Moreover, de Haasvan Alphen (dHvA) oscillation measurements [4, 5, 6] show that the spin-direction dependent effective masses and their field dependences are observed in a number of HF compounds. These features of HF electron systems makes the applied magnetic field a useful experimental probe of observing the strong correlation behavior. In particular, it is crucial to understand the field dependence of the mass enhancement.

We present a scenario in which the heavy itinerant f-electron quasiparticles tend to localize at metamagnetic transition (MMT), with a concomitant suppression of the HF state at that critical field. This can be visualized as a gradual decrease of effective mass with the increasing applied field.

To account for the interplay between the localized- and the itinerant-type behavior of f-electrons we start from the periodic Anderson model (PAM) in an external magnetic field included via the Zeeman term. We also use the large, but finite U version of PAM formulated before [7, 8], in which both residual hybridization and the Kondo-type coupling are accounted for in a systematic manner. Additionally, we utilize the so-called *statistically-consistent* extension of the Gutzwiller-type approach (SGA) [9, 10]. Such formulation allows for a consistent mean-field discussion of correlated states.

2. Model and method

The starting point of our investigation is periodic Anderson model subjected to canonical perturbation expansion [11, 8] in the strong-correlation and mixed-valence limits. We use this transformation in order to include large (but finite) U effects within the Anderson lattice model, with both the effective hybridization and the Kondo interaction effects incorporated in a systematic manner. Explicitly, we start from effective two-orbital Hamiltonian of the form

$$\begin{aligned}\mathcal{H} &= \hat{\mathcal{P}}\left\{\sum_{mn\sigma} t_{mn}c^{\dagger}_{m\sigma}c_{n\sigma} + \sum_{imn\sigma}\frac{V^{*}_{im}V_{in}}{4(U+\epsilon_f)}\hat{\nu}_{if\bar{\sigma}}c^{\dagger}_{m\sigma}c_{n\sigma} + \sum_{im\sigma}(1-\hat{N}_{i\bar{\sigma}})(V_{im}f^{\dagger}_{i\sigma}c_{m\sigma}+h.c.)\right\}\hat{\mathcal{P}}\\ &+ \hat{\mathcal{P}}\left\{\sum_{im\sigma}\frac{2|V_{im}|^2}{U+\epsilon_f}\left(\hat{\mathbf{S}}_i\cdot\hat{\mathbf{s}}_m - \frac{\hat{\nu}_{if}c^{\dagger}_{m\sigma}c_{m\sigma}}{4}\right) + \sum_{i\sigma}\epsilon_f\hat{\nu}_{if\sigma}\right\}\hat{\mathcal{P}}\\ &- \hat{\mathcal{P}}\left\{\frac{1}{2}g_f\mu_B H\sum_{i\sigma}\sigma\hat{N}_{i\sigma} - \frac{1}{2}g_c\mu_B H\sum_{m\sigma}\sigma\hat{n}_{m\sigma}\right\}\hat{\mathcal{P}}, \qquad (1)\end{aligned}$$

where energies of conduction and f-electrons are referenced against the chemical potential. Also, in this Hamiltonian we have projected out completely (with the help of projector $\hat{\mathcal{P}}$) the double occupancies of f states, as well as have defined the following projected quantities for f electrons $\hat{\nu}_{if\sigma} \equiv \hat{N}_{i\sigma}(1-\hat{N}_{i\bar{\sigma}})$, $\hat{\nu}_{if} \equiv \sum_{\sigma}\hat{\nu}_{if\sigma}$, $\hat{\mathbf{S}}_i \equiv (\hat{S}^{\sigma}_i, \hat{S}^{z}_i) = [f^{\dagger}_{i\sigma}f_{i\bar{\sigma}}, 1/2(\hat{\nu}_{if\uparrow}-\hat{\nu}_{if\downarrow})]$, and the corresponding (non-projected) quantities for the conducting band electrons are $\hat{n}_{m\sigma} = c^{\dagger}_{m\sigma}c_{m\sigma}$, $\hat{n}_m = \sum_{\sigma}\hat{n}_{m\sigma}$ and $\hat{\mathbf{s}}_m \equiv (\hat{s}^{\sigma}_m, \hat{s}^{z}_i) = [c^{\dagger}_{m\sigma}c_{m\bar{\sigma}}, 1/2(\hat{n}_{m\uparrow}-\hat{n}_{m\downarrow})]$. Additionally, as we assume that $g_c = g_f = g = 2$, the applied field is defined as $h \equiv \frac{1}{2}g\mu_B H$.

We construct a mean-field Hamiltonian using the Gutzwiller type approach. Proper ground variation state can be represented in the form $|\Psi\rangle = \hat{\mathcal{P}}|\Psi_0\rangle$, where $|\Psi\rangle$ indicates the correlated state and $|\Psi_0\rangle$ is an eigenstate of the effective single-particle Hamiltonian. Then the expectation value of any operator can be in principle calculated as $\langle\hat{\mathcal{O}}\rangle \equiv \frac{\langle\Psi|\hat{\mathcal{O}}|\Psi\rangle}{\langle\Psi|\Psi\rangle} = \frac{\langle\Psi_0|\hat{\mathcal{P}}\hat{\mathcal{O}}\hat{\mathcal{P}}|\Psi_0\rangle}{\langle\Psi_0|\hat{\mathcal{P}}^2|\Psi_0\rangle} \equiv \frac{\langle\hat{\mathcal{P}}\hat{\mathcal{O}}\hat{\mathcal{P}}\rangle_0}{\langle\hat{\mathcal{P}}^2\rangle_0}$.

In the subsequent analysis, we use projector in more general form $\hat{\mathcal{P}} = \prod_i \lambda_{i\uparrow}^{\hat{N}_{i\uparrow}/2}\lambda_{i\downarrow}^{\hat{N}_{i\downarrow}/2}(1-\hat{N}_{i\downarrow}\hat{N}_{i\uparrow})$, which eliminates the double occupancies of f-electrons in real space and conserves average number of f-electrons at each site before and after projection; the latter is accomplished by fixing the fugacity factors $\lambda_{i\sigma} = \frac{1-n_{if\sigma}}{1-n_{if}}$[12].

In effect, this kind of Gutzwiller approximation gives us possibility to calculate the mean value $\langle\mathcal{H}\rangle$ of the Hamiltonian. We have additionally assumed, that all the mean fields appearing in our model represent spatially homogeneous quantities, namely, $n_{if} \equiv \sum_{\sigma=\uparrow,\downarrow}\langle\hat{N}_{i\sigma}\rangle \equiv n_f$, $n_{mc} \equiv \sum_{\sigma=\uparrow,\downarrow}\langle\hat{n}_{m\sigma}\rangle \equiv n_c$, $m_f \equiv \frac{1}{2}\sum_{\sigma=\uparrow,\downarrow}\langle\sigma\hat{N}_{i\sigma}\rangle \equiv m_f$, $\gamma_{im\sigma} \equiv \langle f^{\dagger}_{i\sigma}c_{m\sigma}\rangle_0 \equiv \gamma_\sigma$, $m_{mc} \equiv \frac{1}{2}\sum_{\sigma=\uparrow,\downarrow}\langle\sigma\hat{n}_{m\sigma}\rangle \equiv m_c$, $\xi_{mn\sigma} \equiv \langle c^{\dagger}_{m\sigma}c_{n\sigma}\rangle_0 \equiv \xi$.

In order to provide a *statistically-consistent mean-field approximation* we supply the expectation value of Hamiltonian $\langle\mathcal{H}\rangle$ with a number of constrains containing appropriate

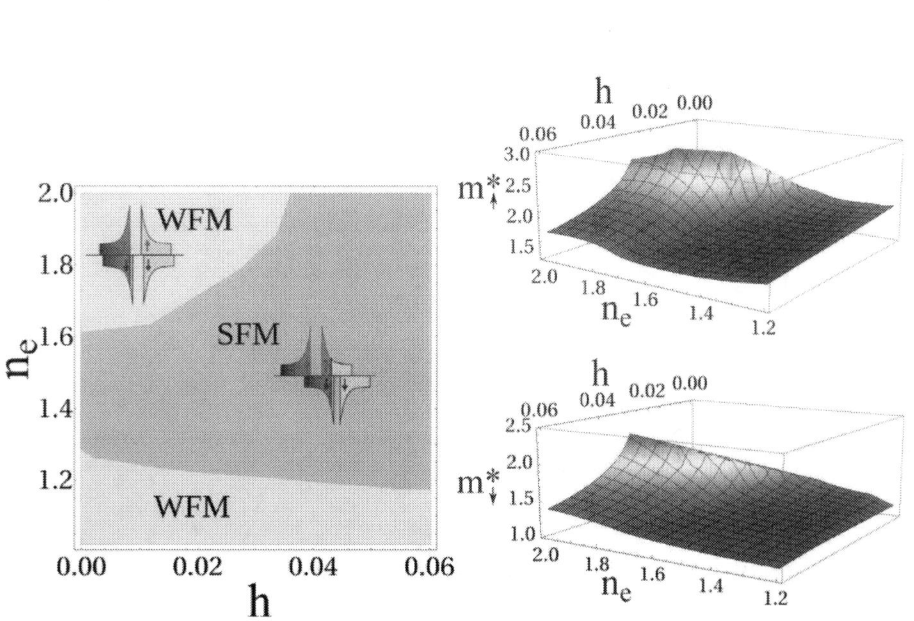

(a) Phase diagram (b) Spin-resolved effective masses

Figure 1: (Color online) (a): Phase diagram on $n_e - h$ plane displaying transition from WFM (weakly ferromagnetic) phase to SFM (strongly ferromagnetic) phase. The phase diagram is drawn for $U = 4.5$, $\epsilon_f = -0.75$, $|V| = 0.3$ and $t = -1/8$, all in units of the conduction electron bandwidth $W = 1$. The insets visualize the density-of-states for both WFM and SFM phases. (b): The *spin-resolved* quasiparticle effective masses on the $n_e - h$ plane. Also, $h \equiv (1/2)g\mu_B H$. Note the relatively low values of the effective mass enhancements, $m^*_\sigma \equiv m^*_\sigma/m$.

Lagrange multipliers regarded as effective mean fields induced by the correlations [9, 13]. Finally, the effective MF Hamiltonian takes the form

$$\begin{aligned}\mathcal{H}_{MF} &= -\sum_{<mn>\sigma}\Big(\eta(c^\dagger_{m\sigma}c_{n\sigma}-\xi)+H.c.\Big)-\lambda_{cm}\sum_{m\sigma}\Big(\sigma c^\dagger_{m\sigma}c_{m\sigma}-m_c\Big)-\mu\sum_{i\sigma}\Big(c^\dagger_{i\sigma}c_{i\sigma}+f^\dagger_{i\sigma}f_{i\sigma}\Big)\\ &- \sum_{i\sigma}\Big(\tau_\sigma(f^\dagger_{i\sigma}c_{i\sigma}-\gamma_\sigma)+H.c.\Big)-\lambda_f\sum_{i\sigma}\Big(f^\dagger_{i\sigma}f_{i\sigma}-n_f\Big)-\lambda_{fm}\sum_{i\sigma}\Big(\sigma f^\dagger_{i\sigma}f_{i\sigma}-m_f\Big)\\ &- \lambda\sum_{i\sigma}\Big(f^\dagger_{i\sigma}f_{i\sigma}+c^\dagger_{i\sigma}c_{i\sigma}-n_e\Big)+\langle\mathcal{H}\rangle, \end{aligned} \tag{2}$$

This Hamiltonian can be brought to a particularly simple 2×2 matrix form in reciprocal space, namely:

$$\mathcal{H}_{MF} = \sum_{\mathbf{k}\sigma}\begin{pmatrix} c^\dagger_{\mathbf{k}\sigma} & f^\dagger_{\mathbf{k}\sigma}\end{pmatrix}\begin{pmatrix}\epsilon^c_{\mathbf{k}\sigma} & -\tau_\sigma \\ -\tau_\sigma & \epsilon^f_\sigma\end{pmatrix}\begin{pmatrix}c_{\mathbf{k}\sigma}\\ f_{\mathbf{k}\sigma}\end{pmatrix} + C, \tag{3}$$

where $n_e = n_f + n_c$ is the total number of electrons per site and the constant C includes all the wavevector independent quantities: $C \equiv \langle\mathcal{H}\rangle + \Lambda(16\xi\eta + \lambda_{cm}m_c + 2\sum_\sigma \tau_\sigma\gamma_\sigma + \lambda_f n_f + \lambda_{fm}m_f + \lambda n)$. For example, the band energy of c-electrons in a two-dimensional case is represented explicitly by $\epsilon^c_{\mathbf{k}\sigma} \equiv -4\eta(\cos k_x + \cos k_y) - \mu - \lambda - \lambda_{cm}\sigma$ and the energy of the localized f-electrons equals $\epsilon^f_\sigma \equiv -\mu - \lambda - \lambda_f - \lambda_{fm}\sigma$. It is easy to see that the eigenvalues of the Hamiltonian (3) are then $E^\alpha_{k\sigma} = \frac{1}{2}\left(\epsilon^c_{k\sigma} + \epsilon^f_\sigma + \alpha\sqrt{(\epsilon^c_{k\sigma} - \epsilon^f_\sigma)^2 + 4\tau^2_\sigma}\right)$, where $\alpha = \pm 1$ labels the quasiparticle band (antibonding and bonding components, respectively). The quasiparticles form two bands with renormalized characteristics.

The mean-field values can be determined in terms of minimizing the free-energy functional $\mathcal{F} = \mathcal{F}_0 - \frac{1}{\Lambda\beta}\sum_{\mathbf{k}\sigma n} \ln\left(1 + e^{-\beta E^n_{\mathbf{k}\sigma}}\right) + \mu n_e$ with respect to all expectation values and all mean fields.

3. Results

Here we focus on the spin-dependent effective-mass behavior in the applied field; a number of strongly correlated effects such as metamagnetism, or tendency towards localization, are discussed elsewhere [14]. The effective mass enhancement of quasiparticles is defined in a standard manner

$$m^*_\sigma \equiv \frac{m^*_\sigma}{m} = \left(\frac{\partial E^\alpha_{k\sigma}}{\partial \epsilon^c_{k\sigma}}\right)^{-1}\Bigg|_\mu . \tag{4}$$

Applied magnetic field affects strongly the low-energy physics of the system, as shown in Figure 1(a). Depending on the number of electrons $n_e = n_f + n_c$ and the magnetic field, the system can be found in two different spin-polarized phases. In the weakly ferromagnetic (WFM) phase as well as in strongly ferromagnetic (SFM) phase we observe a compensation of localized moments by those coming from the conduction band. This is a clear evidence of the antiferromagnetic Kondo interaction [14]. In the SFM phase, in which the system is close to f-electron localization, both the effective spin-dependent masses (with $\sigma = \pm 1$) decrease in the stronger applied field (cf. Figure 1(b)). The effective mass of quasiparticles depends also on the total number of electrons n_e. We believe that obtained behavior can account for the field dependence of the quasiparticle effective mass of $Ce_xLa_{1-x}B_6$ [6, 15] in strong fields. Our claims are based on the fact that our approach reproduce: (i) the field suppression of effective masses in high field, (ii) the non-monotonic $x \equiv n_f$ dependence of m^* for the dominant spin component and its rather small value for this heavy-fermion compound.

Acknowledgments

The number of discussions with Jakub Jędrak and Jan Kaczmarczyk are acknowledged. The work was supported by Ministry of Science and Higher Education, Grants Nos. N N202 128736 and N N202 489839, as well as by Project *TEAM* awarded by the Foundation for Polish Science.

References

[1] Haen P, Flouquet J, Lapierre F, Lejay P and Remenyi G 1987 *Journal of Low Temperature Physics* **67**(5) 391–419
[2] Pecharsky V K, Hyun O B and Gschneidner K A 1993 *Phys. Rev. B* **47** 11839–11847
[3] Sugiyama K, Nakashima M, Futoh M, Ohkuni H, Inoue T, Kindo K, Kimura N, Yamamoto E, Haga Y, Honma T, Settai R and Onuki Y 2000 *Physica B* **281-282** 244 – 246
[4] Aoki H, Uji S, Albessard A K and Ōnuki Y 1993 *Phys. Rev. Lett.* **71** 2110–2113
[5] McCollam A, Daou R, Julian S, Bergemann C, Flouquet J and Aoki D 2005 *Physica B* **359-361** 1 – 8
[6] Goodrich R G, Harrison N, Teklu A, Young D and Fisk Z 1999 *Phys. Rev. Lett.* **82** 3669–3672
[7] Spalek J and Gopalan P 1989 *J. Phys. France* **50** 2869 – 2893
[8] Spałek J 1988 *Phys. Rev. B* **38** 208–212
[9] Jedrak J and Spałek J 2010 *Phys. Rev. B* **81** 073108
[10] Jędrak J, Kaczmarczyk J and Spałek J unpublished (*Preprint* `arXiv:1008.0021`)
[11] Spalek J and Gopalan P 1990 *Physica B* **163** 151 – 154
[12] Fukushima N 2008 *Phys. Rev. B* **78** 115105
[13] Jedrak J and Spałek J 2011 *Phys. Rev. B* **83** 104512
[14] Howczak O and Spałek J unpublished (*Preprint* `arXiv:1110.6336`)
[15] Harrison N, Goodrich R G, Teklu A, Balicas L, Brooks J S, Young D, Fisk Z, Cooley J C and Smith J L 2001 *Physica B* **294-295** 234 – 239

Ultrasound velocity measurements in the vortex-state of YNi_2B_2C

Shin-ichi Ishikawa and Tadataka Watanabe

Department of Physics, College of Science and Technology (CST), Nihon University, Chiyoda-ku, Tokyo 101-8308 Japan

E-mail: tadataka@phys.cst.nihon-u.ac.jp
shinishi0205@gmail.com

Abstract. We performed ultrasound velocity measurements in single-crystalline YNi_2B_2C (T_c = 15.6 K) in the vortex state. The elasticity of the flux line lattice was probed in temperature and magnetic-field dependence of the ultrasound velocities. Magnetic-field dependence of the ultrasound velocities revealed elastic anomalies due to a first-order 45° reorientation transition of the hexagonal flux line lattice, and a second-order hexagonal-to-square lattice transition of the flux line lattice.

1. Introduction

The borocarbide RNi_2B_2C (R = Y, Lu) has attracted much interest due to the occurrence of unconventional superconductivity with the nodal gap function despite the *s*-wave symmetry. A number of experiments provided the compelling evidence for the anisotropic *s*-wave gap function with point nodes along [100] and [010] directions of the tetragonal crystal [1-10]. Another interesting feature of this compound is that the flux line lattice (FLL) undergoes magnetic-field-induced structural transitions. With the magnetic field applied parallel to the c axis, $\boldsymbol{H}\|\boldsymbol{c}$, the small-angle neutron scattering (SANS) experiments revealed the occurrence of a first-order 45° reorientation transition of the hexagonal FLL, and a second-order hexagonal-to-square FLL transition [11-14]. It is considered that these FLL transitions are relevant to the underlying anisotropy of the superconducting gap as well as the Fermi surface. Thus it is crucial to study the FLL structure for understanding the electronic structure in the vortex state in RNi_2B_2C.

Ultrasound velocity measurement is a unique tool for the study of the FLL structure. In the vortex state, this tool can examine the FLL elasticity in addition to the conventional crystal-lattice elasticity, since the sound wave with the propagation $\boldsymbol{k}$ and the polarization $\boldsymbol{u}$ can couple to the FLL via the Lorentz force in the configuration of $\boldsymbol{u}\perp\boldsymbol{H}$ [15]. The FLL elasticity provides important information of the electronic structure in the vortex state.

In this study, we report results of ultrasound velocity measurements in single-crystalline YNi_2B_2C (T_c = 15.6 K) in the vortex state with $\boldsymbol{H}\|\boldsymbol{c}$. We investigated the FLL elasticity focusing on the magnetic-field induced FLL structural transitions in YNi_2B_2C.

2. Experimental

A large single crystal of YNi_2B_2C was grown by the floating zone method. For the ultrasound velocity measurements, the single crystal was cut into cuboid shape with dimensions of $3.5\times3.6\times3.1$ mm^3.

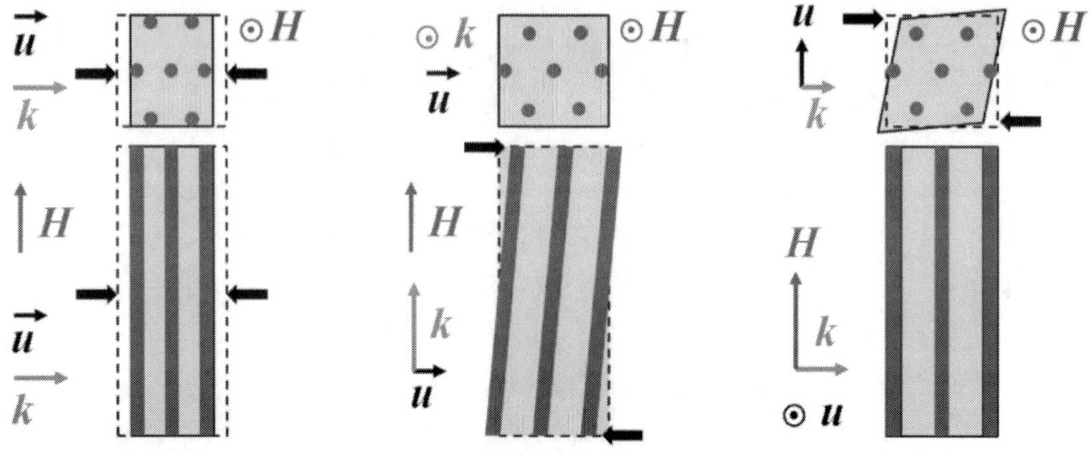

(a) C_{11}^{FLL} (b) C_{44}^{FLL} (c) C_{66}^{FLL}

Figure 1 Schematic picture of FLL deformation by sound wave with propagation *k* and polarization *u* in (a) compression modulus C_{11}^{FLL} , (b) tilt modulus C_{44}^{FLL}, and (c) shear modulus C_{66}^{FLL} .

Table 1 Elastic modulus of crystal lattice and flux line lattice with *H*||*c* in YNi_2B_2C, and the corresponding configuration of sound wave with propagation *k* and polarization *u*.

Elastic modulus		Magnetic field ***H***	propagation ***k***	polarization ***u***
(a) Compression modulus	C_{11}^{FLL+CL}	[001]	[100]	[100]
(b) Tilt modulus	C_{44}^{FLL+CL}	[001]	[001]	[100]
(c) Shear modulus	C_{66}^{FLL+CL}	[001]	[100]	[010]

Mirror surfaces of the sample were prepared by careful polishing using the 1 μm diamond slurry because the ultrasound measurements are quite sensitive to the roughness of the sample surface. The ultrasound velocity measurements were performed using the phase comparison technique which can measure the relative change in the ultrasound velocities with high resolution of about 1 ppm. The ultrasounds were generated and detected by $LiNbO_3$ transducers glued on the parallel mirror surfaces of the sample by the epoxy adhesive. The ultrasound velocities were measured in temperatures (*T*) from 2 K to 100 K with magnetic field (*H*) up to 7 T. For the FLL elasticity, there are three symmetrically independent elastic modes: compression modulus C_{11}^{FLL}, tilt modulus C_{44}^{FLL}, and shear modulus C_{66}^{FLL}, as illustrated in Figs. 1 (a), (b), and (c), respectively. We performed ultrasound velocity measurements in all of these elastic modes in the vortex state with ***H***||***c***. The configuration of propagation ***k*** and polarization ***u*** of the sound wave for each elastic mode is summarized in Table 1.

3. Results and discussion

Figures 2(a), (b), and (c) show temperature dependence of the ultrasound velocities in C_{11}, C_{44}, and C_{66}, respectively. In the normal state, $C_{44}(T)$ and $C_{66}(T)$ exhibit softening with decreasing temperature, which is most likely attributed to the Fermi surface nesting [16]. $C_{11}(T)$ in the normal state, on the other hand, exhibits ordinal hardening with decreasing temperature. In the superconducting state with $H = 0$, $C_{44}(T)$ and $C_{66}(T)$ exhibit a discontinuous change of slope at T_c and harden below T_c with decreasing temperature, while $C_{11}(T)$ exhibits no superconductivity-driven anomaly. The elastic properties mentioned here imply that the nesting-sensitive elastic modes, C_{44} and C_{66}, couple to the superconducting order parameter much stronger than the nesting-insensitive mode C_{11}.

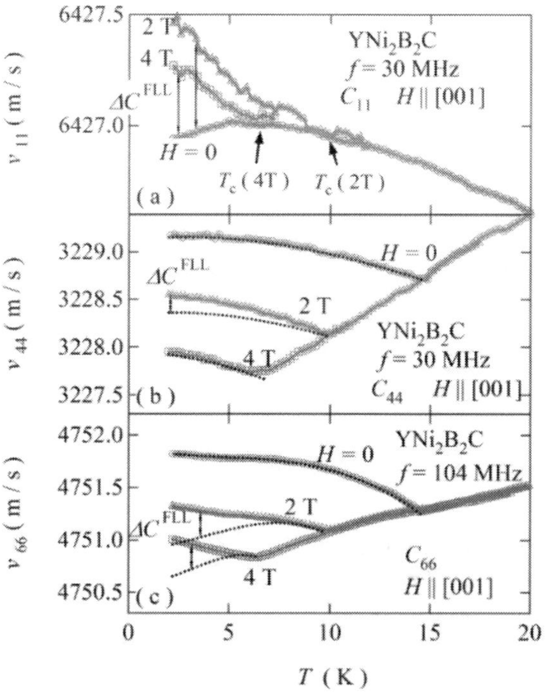

Figure 2 Temperature dependence of the ultrasound velocities with $H \parallel [001]$ in (a) C_{11}, (b) C_{44}, and (c) C_{66}.

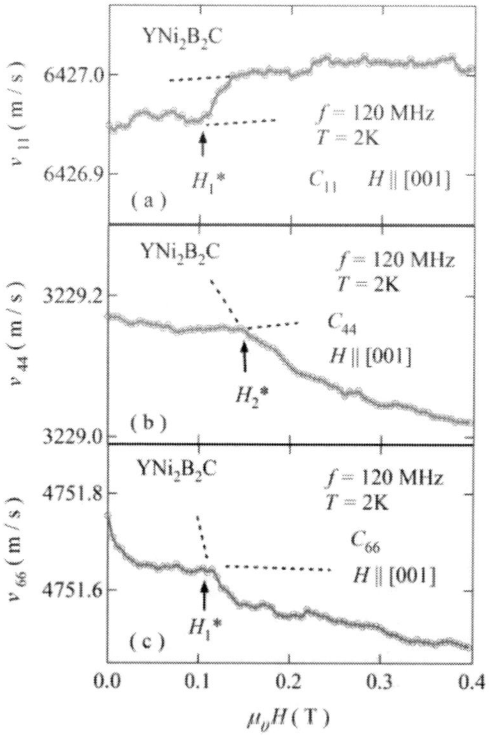

Figure 3 Magnetic-field dependence of the ultrasound velocities at 2 K with $H \parallel [001]$ in (a) C_{11}, (b) C_{44}, and (c) C_{66}.

In the vortex state, the measured elasticity C_{ij} should consist of the crystal elasticity in the normal state (the background) C_{ij}^{0}, the superconductivity-driven change of the crystal elasticity ΔC_{ij}^{CL}, and the FLL elasticity ΔC_{ij}^{FLL}, $C_{ij} = C_{ij}^{0} + \Delta C_{ij}^{CL} + \Delta C_{ij}^{FLL}$. As shown in Fig. 2 (a), $C_{11}(T)$ at 2 T and 4 T exhibits gradual hardening below T_c with decreasing temperature. Since $C_{11}(T)$ with $H = 0$ manifests $\Delta C_{11}^{CL} \cong 0$, this hardening below T_c only in the vortex state should be due to the hardening of the FLL elasticity ΔC_{11}^{FLL}.

In contrast to $C_{11}(T)$, the hardening below T_c in $C_{44}(T)$ and $C_{66}(T)$ with $H = 0$, as shown in Figs. 2 (b) and (c), manifests the presence of the ΔC_{ij}^{CL} component in these elastic modes. Thus the hardening below T_c in $C_{44}(T)$ and $C_{66}(T)$ at 2 T and 4 T should in part arise from the hardening in ΔC_{ij}^{CL}. However, the measured $C_{44}(T)$ and $C_{66}(T)$ at 2 T and 4 T are harder than the components of $C_{ij}^{0}(T) + \Delta C_{ij}^{CL}(T)$ in C_{44} and C_{66} which are estimated in assumption of $\Delta C^{CL}(T = 0) \propto T_c$, as shown in Figs.2 (b) and (c). Thus the measured elasticity in C_{44} and C_{66} in the vortex state should contain the component of the FLL elasticity ΔC_{ij}^{FLL}. And we conclude that all elastic modes measured in the present study probe the FLL elasticity ΔC_{ij}^{FLL}.

Figures 3(a), (b), and (c) show the low-field ultrasound velocities at 2 K in C_{11}, C_{44}, and C_{66} as a function of magnetic field, respectively. Elastic-mode dependent anomalies are observed at $H_1^* \sim 0.1$ T and $H_2^* \sim 0.15$ T: a sudden jump at H_1^* in $C_{11}(H)$, a cusp at H_2^* in $C_{44}(H)$, and a cusp at H_1^* in $C_{66}(H)$. It is noteworthy that H_1^* and H_2^* respectively correspond to the first-order 45° reorientation transition of hexagonal FLL, and the second-order hexagonal-to-square FLL structural transition which were observed in SANS experiments [14]. Thus the observed elastic anomalies at H_1^* and H_2^*

should be attributed to these FLL structural transitions. Taking into account that the measured elasticity C_{ij} contains the contribution of the FLL elasticity $\Delta C_{ij}^{\mathrm{FLL}}$ in all the elastic modes, as mentioned above in conjunction with Figs. 2, the elastic anomalies at H_1* and H_2* most likely arise in the FLL elasticity $\Delta C_{ij}^{\mathrm{FLL}}$. Thus the present results suggest that the first-order 45° reorientation transition of the hexagonal FLL gives rise to the elastic anomalies in the in-plane elastic modes of the compression modulus C_{11}^{FLL} and the share modulus C_{66}^{FLL}, while the hexagonal-to-square FLL structural transition gives rise to the elastic anomaly in the out-of-plane elastic mode of the tilt modulus C_{44}^{FLL}.

4. Summary

In summary, ultrasound velocity measurements in the vortex state in YNi_2B_2C probed the FLL elasticity and discovered the elastic anomalies due to the low-field FLL structural transitions. Further detailed ultrasound velocity measurements, analysis, and discussion could explain the correlation between the FLL elasticity and the electronic state in the vortex state in YNi_2B_2C.

5. Acknowledgement

This work was partly supported by Grants-in-Aid for Young Scientists (B) (21740266) and Scientific Research on Priority Areas (22014001) from MEXT of Japan.

References

[1] Nohara M, Isshiki M, Sakai F, and Takagi H 1999 J. Phs. Soc. Jpn. **68**, 1078.
[2] Nohara M, Suzuki H, Mangkorntong N, and Takagi H 2000 Physica C **341-348**, 2177.
[3] Izawa K, Shibata A, Matsuda Y, Kato Y, Takeya H, Hirata K, Van der Beek C J, and Konczykowski M 2001 Phys. Rev. Lett. **86**, 1327.
[4] Yang I S, Klein M V, Cooper S L, Canfield P C, Cho B K and Lee S I 2000 Phys. Rev. B **62**, 1291.
[5] Yokoya T, Kiss T, Watanabe T, Shin S, Nohara M, Takagi H, and Oguchi T 2000 Phys. Rev. Lett. **85**, 4952.
[6] Boaknin E, Hill R W, Proust C, Lupein C, Taillefer L, and Canfield P C 2001 Phys. Rev. Lett. **87**, 237001.
[7] Izawa K, Kamata K, Nakajima Y, Matsuda Y, Watanabe T, Nohara M, Takagi H, Thalmeier P, and Maki K 2002 Phys. Rev. Lett. **89**, 137006 .
[8] Park T, Salamon M B, Choi E M, Kim H J, and Lee S I 2003 Phys. Rev. Lett. **90**, 177001.
[9] Watanabe T, Nohara M, Hanaguri T, and Takagi H 2004 Phys. Rev. Lett. **92**, 147002.
[10] Baba T, Yokoya T, Tsuda S, Watanabe T, Nohara M, Takagi H, Oguchi T, and Shin S 2010 Phys. Rev. B **81**, 180509(R).
[11] Paul McK D, Tomy C V, Aegerter C M, Cubitt R, Lloyd S H, Forgan E M, Lee S L, and Yethiraj M 1998 Phys. Rev. Lett. **80**, 1517.
[12] Eskildsen M R, Abrahamsen A B, Kogan V G, Gammel P L, Mortensen K, Andersen N H, and Canfield P C 2001 Phys. Rev. Lett **86**, 5148.
[13] Levet S J, Dewhurst C D, and Paul McK D 2002 Phys. Rev. B **66**, 014515.
[14] Dewhurst C D, Levet S J, and Paul McK D 2005 Phys. Rev. B **72**, 014542.
[15] Watanabe T, Kasahara Y, Izawa K, Sakakibara T, Matsuda Y, Van der Beek C J, Hanaguri T, Shishido H, Settai R, and Onuki Y 2004 Phys. Rev. B **70**, 020506(R).
[16] Isida S, Matsushita A, Takeya H, and Suzuki M 2001 Physica C **349**, 150.

Spin dynamics of the new phosphides $YbRh_6P_4$ and $CeIr_2P_2$ as studied by electron spin resonance

V.A. Ivanshin[1], **E.M. Gataullin,**
Kazan (Volga region) Federal University, Kremlevskaya str. 18, 420008 Kazan, Russia
A.A. Sukhanov
Zavoisky Physical-Technical Institute, Sibirsky trakt 10/7, 420029 Kazan, Russia
U. Pfannenschmidt, R. Pöttgen
Institut für Anorganische und Analytische Chemie, Universität Münster, Corrensstrasse 30, D-48149 Münster, Germany

E-mail: Vladimir.Ivanshin@ksu.ru

Abstract. We report the electron spin resonance (ESR) measurements in the new phosphides $YbRh_6P_4$ and $CeIr_2P_2$. Details of preparation of both materials which proceeds by different techniques are given. Electronic spin-lattice relaxation (SLR) processes with an involvement of the first excited Stark sublevel of the Yb^{3+} ion with an energy $\Delta \approx 82.1$ K describe well a drastic broadening and vanishing of the ESR signal in $YbRh_6P_4$ above 10 K. The observed ESR behavior provides evidence for the relatively weak *f-p* hybridization effects in both phosphides in contrast to silicides $YbRh_2Si_2$ and $YbIr_2Si_2$.

1. Introduction

Hybridization between local *f* states and conduction electrons (CE) can lead to a dichotomy of localized/itinerant characters of electrons in some strongly correlated electron systems (SCES). ESR probes microscopically both the local moment (LM) spins and CE and could be observed in some dense SCES even without any kind of paramagnetic doping [1-3]. According to several different theories which have been very recently reviewed in [4], the key origin of the unexpected narrow ESR signals observed in some undoped intermetallics can be related to a strong hybridization between localized *f* or *d* electrons and conduction bands in conjunction with ferromagnetic (FM) correlations. In the present contribution, we report the temperature-dependent ESR spectra in two ternary phosphides $YbRh_6P_4$ and $CeIr_2P_2$.

2. Sample preparation and experimental

Single crystals of the metal rich phosphide $YbRh_6P_4$ [5] were prepared by the bismuth flux technique. Starting materials were pieces of ytterbium (Smart Elements, > 99.9 %), rhodium powder (Heraeus, > 99.9 %), red phosphorus (Hoechst, Knapsack, ultrapure) and bismuth shots (ABCR GmbH, 99.99 %). Pieces of ytterbium, rhodium powder, red phosphorus and bismuth shots were mixed in a molar ratio of 1:2:2:60 (Yb:Rh:P:Bi) and sealed in an evacuated silica tube. The latter was placed in a muffle furnace and first heated to 770 K at a rate of 20 K/h and kept at that temperature for 24 h. Subsequently the temperature was raised to 1320 K at a rate of 50 K/h, kept at that temperature for 100

[1] To whom any correspondence should be addressed.

h, followed by slow cooling to 970 K at a rate of 2 K/h and further to 570 K at a rate of 4 K/h. The sample was quenched on air and the excess bismuth flux was slowly dissolved with a 1:1 molar mixture of glacial acetic acid (VWR International) and H_2O_2 (ACROS 35%). The resulting sample was washed with demineralized water. The reaction product consisted of pillar-shaped crystals with metallic lustre. The $YbRh_6P_4$ crystals were mechanically separated for the ESR measurements. The route of flux-free synthesis has been used for preparation of $CeIr_2P_2$. Starting materials were pieces of cerium (Smart Elements, >99.9 %), iridium powder (Heraeus, >99.9 %) and red phosphorus (Hoechst, Knapsack, ultrapure). For the preparation of the precursor $CeIr_2$, the elements were weighed in a ratio of 1.1:2 (Ce:Ir). The mixture was pressed to a pellet and placed in a tantalum tube which was sealed in an evacuated silica tube. In a muffle furnace the sample was heated to 770 K at a rate of 20 K/h and pre-reacted at that temperature for 24 h. Subsequently, the temperature was raised to 1470 K at a rate of 50 K/h and kept for another 24 h. After expiration of the reaction time the sample has been cooled to room temperature by turning the power off. The sample was ground, pressed to a pellet, placed in a tantalum tube and sealed in a silica tube under vacuum again. Then it was heated to 1370 K at a rate of 50 K/h and kept at that temperature for ten days. This annealing process was repeated three times. The polycrystalline sample of $CeIr_2P_2$ was analyzed through a Guinier powder pattern (CuKα_1 radiation, α-quartz: a = 491.30, c =540.46 pm as internal standard). The Guinier camera was equipped with an imaging plate technique (Fuji-film, BAS-READER 1800). The tetragonal lattice parameters were refined by a least-squares routine. Proper indexing was confirmed by a comparison of the experimental pattern with a calculated one. The Guinier pattern revealed a small amount of IrP_2 as by-product. ESR experiments were carried out on an continuous wave X-band EMX/plus spectrometer (Bruker BioSpin GmbH, Germany) equipped with Oxford cryogenics ESR900 cryostat. Measurements were performed on the powder samples in the temperature range between 4.5 and 25 K with the temperature stability better than ±0.1 K.

3. Results and discussion

A well-defined ESR line could be detected in $CeIr_2P_2$ at the lowest temperature of measurements T = 4.5 K only (Figure 1(a)). This signal at the resonance field $H_{res} \sim 2570$ Oe (effective ESR g-factor g ≈ 2.68) with intensity lower than that of the cavity background can be related to the Ce^{3+} ESR which is rarely observed in the Ce-based undoped compounds [6-8]. Contrary, Figure 1(b) shows the temperature evolution of the ESR spectrum in $YbRh_6P_4$ which could be followed up to 20 K. Compared with $YbRh_2Si_2$ [9] and $YbIr_2Si_2$ [10] ESR signal intensity in $YbRh_6P_4$ is more than two times smaller than in both these silicides as it was also recently found for $YbRh_2Pb$ [11]. No angular investigations of the ESR spectra have been done because of a twinned crystal structure of the $YbRh_6P_4$ crystals. The nature of the relaxation mechanisms of the ESR-active spins is reflected by the temperature dependences of the ESR g-factor and peak-to-peak linewidth ΔH which are presented in Figure 2. The temperature evolution of the effective ESR g-value in $YbRh_6P_4$ at T > 10 K (Figure 2(a)) is typical for the random transitions of the $4f$ ions from the ground state to the first excited Kramers doublet with activation energy Δ [12] and can be fitted by the solid line according to the equation [9, 11]:

$$g(T) = g_0 + \Delta g_0 \exp(-\Delta/T),\ \Delta g_0 = g_{exc} - g_0, \quad (1),$$

where Δ = 82.1 K, g_0 = 3.34, and g_{exc} = - 25.66. The value of g_{exc} is significantly larger than usual g-values for Yb^{3+} ion in different crystal symmetries and cannot be explained in the framework of the used model which has been proposed for a nonmetallic lanthanide compounds. The dotted line in Figure 2(b) has been related to the common LM relaxation in a metallic environment which contains a linear Korringa-like contribution and an exponential Orbach-type relaxation term with the involvement of the same first excited Kramers doublet of the Yb^{3+} ion [9-11, 13]:

$$\Delta H = A + BT + C\exp(-\Delta/T), \quad (2),$$

where $A = 348.7$ Oe, $B = 5$ Oe/K, $C = 5.6$ kOe.

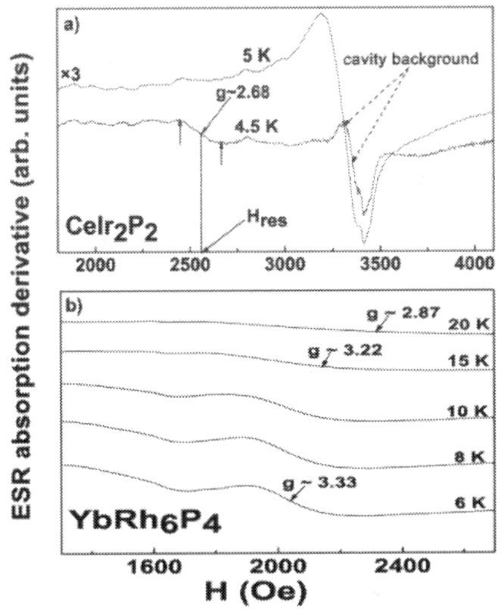

Figure 1. X-band ESR spectra at various temperatures for $CeIr_2P_2$ (panel (a)) and $YbRh_6P_4$ (panel (b)). Solid arrows indicate the positions of the ESR signals at g = 2.68 (H_{res} ~ 2600 Oe) in panel (a) and g = 3.33, 3.22, and 2.87 in panel (b) which can be attributed with the Ce^{3+}- or Yb^{3+}-ions, correspondingly. Dotted arrows mark background signals from ESR cavity ((H_{res} ~ 3300 Oe).

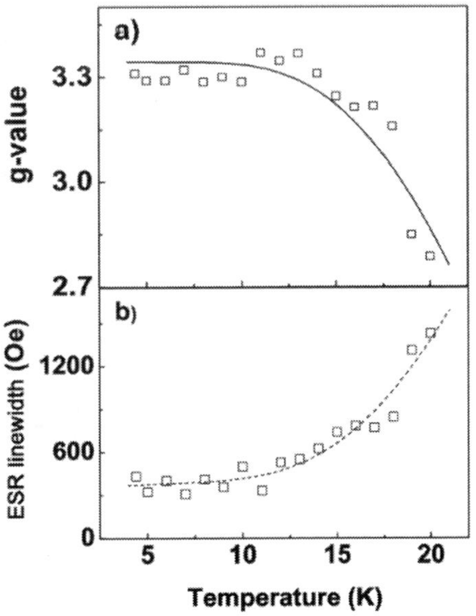

Figure 2. Temperature dependence of the ESR (a) g-value, (b) linewidth in $YbRh_6P_4$. The solid and dotted lines represent the data as explained in the text.

An analysis of these data and the ESR behavior of some other dense Yb-based systems investigated by ESR method [1-3, 9-11, 13] shows that Si-containing compounds $YbRh_2Si_2$ and $YbIr_2Si_2$ exhibit very strong hybridization effects and intense ESR absorption. Thus, the replacement of Si ([Ne]$3s^23p^2$) by Pb ([Hg]$6p^2$) or by P ([Ne]$3s^23p^3$) does not cause any significant qualitative change of spin dynamics but dramatically reduces the *f-p* hybridization and the intensity of ESR signals in $YbRh_6P_4$ and $YbRh_2Pb$. On the other hand, the *f-d* hybridization is mainly responsible for contribution of CE to the ESR relaxation of the collective spin mode consisting of the *4f* LM and CE [10, 13] and shows no visible influence on the intensity of the observed ESR signals. Moreover, recent ARPES investigations of both silicides $YbRh_2Si_2$ and $YbIr_2Si_2$ [14] showed that substituting iridium for rhodium results in a shift of the energy of the valence bands, where the valence and the *f* states are interacting and thus to a change in the strength of the hybridization. In contrast to some Co-containing *4f* intermetallics $YbCo_2Zn_{20}$ [3], $Ho(Co_{1-x}Si_x)_2$ [15], CeCoPO [16], and CeCoAsO [17] which exhibit a possible strong influence of the Co *3d* electrons on their magnetic properties, main features of ESR in $YbRh_6P_4$ and $CeIr_2P_2$ (e.g., observability of ESR at low temperatures only and characteristic temperature dependences of ESR parameters) can be understood from the point of view of the localized *4f* moments. Further studies are needed in order to clarify the microscopic nature and formation process of the *f-d-p* hybridization states in the intermetallic compounds with strong correlation of the 4*f* electrons.

4. Conclusion

We detected and studied the ESR of the ternary phosphides $CeIr_2P_2$ and $YbRh_6P_4$. A relatively weak ESR signals in both compounds reflect the extremely weak *f-p* hybridization effects in contrast to the case of silicides $YbRh_2Si_2$ and $YbIr_2Si_2$. Our results in $YbRh_6P_4$ reveal temperature dependences of the ESR g-factor and peak-to-peak linewidth for an Orbach-like ESR relaxation of the Yb^{3+} LM which can be related to the first excited crystal-field state at 82.1 K.

References

[1] Krellner C, Förster T, Jeevan H, Geibel C, Sichelschmidt J 2008 *Phys. Rev. Lett.* **100** 066401
[2] Holanda LM, Vargas JM, Iwamoto W, Rettori C, Nakatsuji S, Kuga K, Fisk Z, Oseroff SB, Pagliuso PG 2011 *Phys. Rev. Lett.* **107** 026402
[3] Ivanshin VA, Litvinova TO, Sukhanov AA 2011 *J. Phys: Conf. Ser.* **273** 012035
[4] Abrahams E, Wölfle P 2011 *Ann. Phys. (Berlin)* **523** 591, and references therein
[5] Pfannenschmidt U, Rodewald UC, Pöttgen R 2011 *Monatsh. Chem.* **142** 219
[6] Huang CY, Sugawara K and Cooper BR 1976 *Proc. Int. Conf. Cryst. Field Effects Met. Alloys* pp 51–60
[7] Demishev SV, Semeno AV, Bogach AV, Samarin NA, Ishchenko TV, Filipov VB, Shitsevalova NY, Sluchanko NE 2009 *Phys. Rev.* B **80** 245106
[8] Förster T, Sichelschmidt J, Krellner C, Geibel C, Steglich F 2010 *J. Phys.: Condens. Matter* **22** 435603
[9] Ivanshin VA, Aminov LK, Kurkin IN, Sichelschmidt J, Stockert O, Ferstl J, Geibel C 2003 *JETP Lett.* **77** 526
[10] Sichelschmidt J, Wykhoff J, Krug von Nidda H-A, Fazlishanov II, Hossain Z, Krellner C, Geibel C, Steglich F 2007 *J. Phys.:Condens. Matter* **19** 016211
[11] Ivanshin VA, Litvinova TO, Sukhanov AA, Sokolov DA, Aronson MC 2009 *JETP Lett.* **90** 116
[12] Aminov LK, Malkin BZ, Teplov M 1996 *Handbook on the Physics and Chemistry of Rare Earths* vol 22, ed K A Gschneidner Jr and L Erwin (Amsterdam: Elsevier) p 378
[13] Sichelschmidt J, Ivanshin VA, Ferstl J, Geibel C, Steglich F 2003 *Phys. Rev. Lett.* **91** 156401
[14] Vyalikh DV, Danzenbächer S, Kucherenko Yu, Krellner C, Geibel C,. Laubschat C, Shi M, Patthey L, Follath R, Molodtsov SL 2009 *Phys. Rev. Lett.* **103** 137601
[15] Vasylyev D, Prokleška J, Sechovský V 2008 *Acta Physica Polonica A* **113** 347
[16] Krellner C, Burkhardt U, Geibel C 2009 *Physica B* **404** 3206
[17] Sarkar R, Jesche A, Krellner C, Baenitz M, Geibel C, Mazumdar C, Poddar A 2010 *Phys. Rev. B* **82** 054423¶

Neutron scattering study on magnetic ordering in a partially rare-earth filled skutterudite $Pr_xFe_4Sb_{12}$

K Iwasa[1], T Orihara[1], K Saito[1], K Tomiyasu[1], Y Murakami[2], H Sugawara[3], K Kuwahara[4], H Kimura[5], R Kiyanagi[5], Y Ishikawa[5], Y Noda[5], Y Aoki[6], H Sato[6] and M Kohgi[6]

[1] Department of Physics, Tohoku University, Sendai 980-8578, Japan
[2] CMRC, IMSS, High Energy Accelerator Research Organization, Ibaraki 305-0801, Japan
[3] Department of Physics, Kobe University, Kobe 657-8501, Japan
[4] Institute of Applied Beam Science, Ibaraki University, Ibaraki 310-8512, Japan
[5] IMRAM, Tohoku University, Sendai 980-8577, Japan
[6] Department of Physics, Tokyo Metropolitan University, Hachioji, Tokyo 192-0397, Japan

E-mail: iwasa@m.tohoku.ac.jp

Abstract. Neutron scattering studies reveal that a crystal-field-splitting ground state of Pr^{3+} $4f^2$ electrons in a partially filled $Pr_xFe_4Sb_{12}$ ($x < 1$) is a triplet. A magnetically ordered structure below 4 K consists of both magnetic moments at the Pr and Fe sites, in contrast to the $x = 1$ system without any magnetic ordering due to the $4f$-electron singlet ground state.

1. Introduction

Among rare-earth-filled skutterudite compounds, magnetic ordering sensitive to the Pr-ion filling fraction x for $Pr_xFe_4Sb_{12}$ has been studied [1, 2, 3]. No magnetic ordering takes place in a fully filled system ($x = 1$), in which a singlet ground state for Pr $4f$-electron crystal-field splitting (CF) was exemplified by inelastic neutron scattering [4]. On the other hand, partially filled compounds ($x = 0.7 \sim 0.9$) undergo magnetically ordering transitions with ferromagnetic character below 4 K [1, 2]. The earlier inelastic neutron scattering study for $x = 0.73$ proposed that a CF ground state is a magnetic triplet [5]. The magnetic moments of the Fe $3d$ electrons are suggested to contribute to the paramagnetic susceptibility. Thus, the magnetic behaviors in $Pr_xFe_4Sb_{12}$ are supposed to arise from an interplay between the Pr and Fe ions. In the present study, we revisit a CF level scheme in the partially filled system by measuring temperature evolution of absolute neutron-scattering cross sections. We also carried out neutron diffraction experiments to shed further insight into the magnetic structure that has not fully been solved yet.

2. Experimental procedure

Samples were synthesized by the Sb self-flux method [2]. Magnetic properties were measured by a SQUID magnetometer, after confirming crystal structures by a powder X-ray diffraction. Neutron scattering studies were performed at JRR-3, JAEA (Tokai, Japan). Structure of the single crystalline sample at 6 K was determined at the four-circle diffractometer FONDER. Magnetic excitations and magnetic structures were investigated at the triple-axis spectrometer TOPAN. For the latter, both polarized and unpolarized neutrons were used.

3. Results and analysis

Inelastic neutron scattering data for the polycrystalline sample are shown by symbols in Fig. 1. The result were obtained by subtracting background intensity evaluated from data for $LaFe_4Sb_{12}$, and by evaluating the absolute scattering function $S(E)$ using intensity from a standard vanadium sample. The samples by the same synthesis procedure show hysteresis behaviors of magnetization measured at 2 K (data from the single-crystal sample will be shown later). It is consistent with the previously report on the partially filled system. The filling factor of the present sample could not be determined precisely by the powder diffraction techniques. Lines represent results of a least-squares fitting analysis based on the CF Hamiltonian for T_h space group for the Pr sites [6]. This analysis succeeds in reproducing quantitatively the temperature variation of the spectra with the set of parameters $W = 0.13$ meV, $x = 0.479$, and $y = 0.285$. The level scheme is $\Gamma_4^{(2)}$ (0 meV) - Γ_1 (2.4 meV) - $\Gamma_4^{(1)}$ (10.3 meV) - Γ_{23} (10.7 meV). The scale factor for a ratio of calculated $S(E)$ to the experimental result is around 1.3, which is close to the ideal

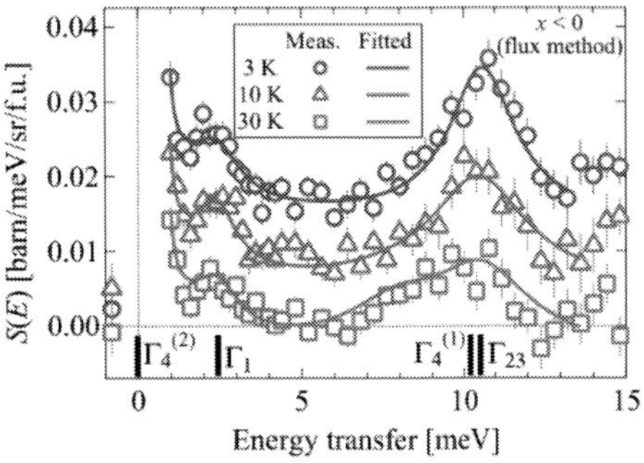

Figure 1. Symbols represent the measured spectra. Lines are the fitting analysis. Origin for each temperature set is shifted by 0.01.

$\lambda = 1.24$ Å, Reliable Factor $R = 0.0528$			
	Pr	Fe	Sb
occupation	0.895(82)	1	1
coordinate x_c	0	0.25	0
coordinate y_c	0	0.25	0.3362(2)
coordinate z_c	0	0.25	0.1600(2)
B [Å^2]	1.05(32)	0.295(47)	0.382(50)

Table 1. Structural parameters determined by the single-crystal diffraction using FONDER. The parameters with error magnitudes in the parentheses were adjusted in the analysis, and the others were fixed.

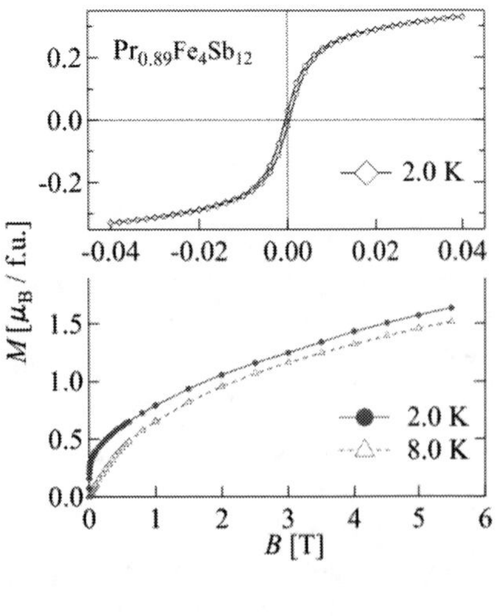

Figure 2. Measured magnetization curves of the sample used for the neutron diffraction studies.

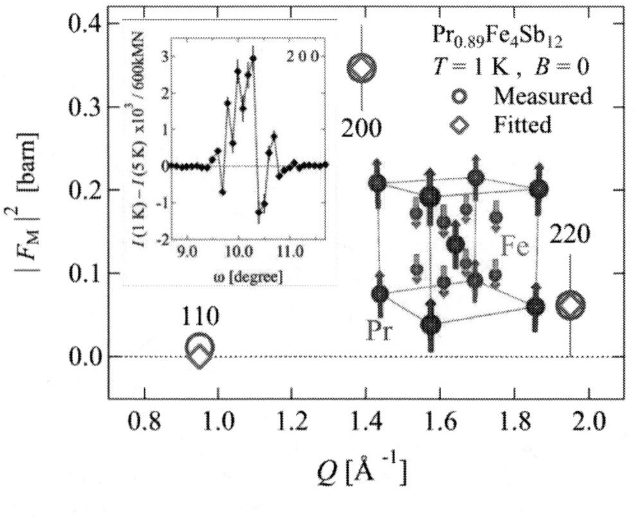

Figure 3. Blue circles are the measured result of magnetic structure factors. Red squares are the fitted result for the depicted model.

value of unity. On the other hand, the CF scheme for the fully filled compound is Γ_1 (0 meV) - $\Gamma_4^{(2)}$ (2.3 meV) - $\Gamma_4^{(1)}$ (11.4 meV) - Γ_{23} (18 meV), parameterized with $W = 0.230$ meV, $x = 0.398$ and $y = 0.101$ [4]. The low-lying CF-level sequence in the present partially filled compound is reverse from the fully filled compound, and the Γ_{23} level is located at the energy lower than the fully filled one.

In Table 1, we summarize the structural parameters for the synthesized single-crystal sample determined from the data by FONDER. The Pr-ion fraction was determined to be $x = 0.89$. Magnetization up to 5.5 T of this sample, shown in Fig. 2, is the same as in the previously reported data [2]. Small hysteresis in the low magnetic-field region, shown in an upper part, is also consistent with the reported ferromagnetic behavior. Therefore, neutron diffraction was performed for investigating a magnetic structure of this sample. We observed very tiny magnetic reflection peaks from this sample below around 4 K, superimposed on the nuclear diffractions. A rocking-curve profile at the reciprocal-lattice point 200 by subtracting intensities at 5 K from those at 1 K is shown in an inset of Fig. 3. The determined magnetic structure factors for the three reflections at 1 K are plotted in Fig. 3. The dependence on the scattering vector magnitude Q deviates from that expected for a uniform ferromagnetic structure obeying a monotonic curve of the magnetic form factor. As discussed before, both of the magnetic moments at Pr and Fe are anticipated to play a role in the magnetic ordering. We took into account a simple model based on anti-parallel arrangement of the magnetic moments at the Pr and Fe sites, as shown in the structure picture of Fig. 3. The present structure factors results are well reproduced by the magnetic moments $0.29\mu_B$ at the Pr site and $0.19\mu_B$ at the Fe site. Polarized neutron diffractions were measured at a magnetic field 1 T applied along the [001] axis and 1.8 K. The observed flipping ratios $I_{\rm off}/I_{\rm on}$ at the selected reflection points are plotted against Q in Fig. 4, where $I_{\rm off}$ or $I_{\rm on}$ represent the peak intensities for the incident neutron spins parallel or anti-parallel to the magnetic field applied to the sample, respectively. The data are well reproduced by the Pr magnetic moments of $\sim 0.79\mu_B$ being parallel to the magnetic field, as shown in the inset. The Fe magnetic moments do not contribute to the observed flipping ratios. They might flop toward the direction perpendicular to the magnetic field, and have zero magnetic structure factors for the measured reflections. The net magnetization value $\sim 0.79\mu_B$ per chemical formula unit for 1 T is close to the magnetization. Because no additional magnetic reflection appears, the magnetic unit cell is identical with that of the crystal structure.

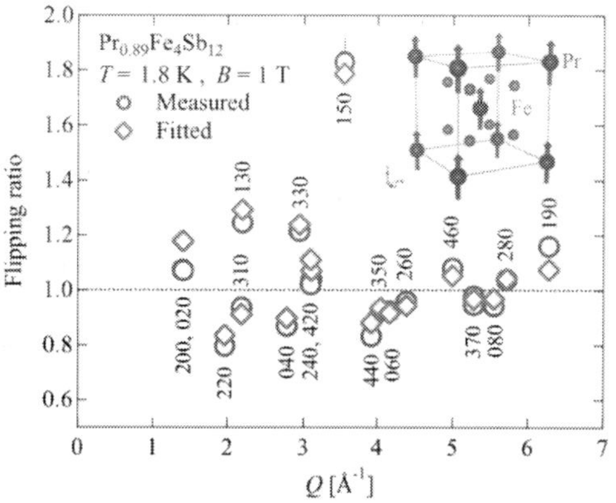

Figure 4. Blue circles are the polarized-neutron flipping ratios. Red circles are the calculated result for the model in an inset.

4. Discussion

As demonstrated by the simple collinear model for the zero magnetic field condition, the spontaneous magnetization is $0.47\mu_B$/f.u, dominated by the Fe-site magnetic moments (the number of the Fe sites is four times as large as the Pr one). The ordered magnetic moment $0.29\mu_B$ at the Pr site is smaller than the eigenvalue of $1.52\mu_B$ for the $\Gamma_4^{(2)}$ ground state. The suppression of the magnetic moment is thought to be due to fluctuation even at 1 K induced by the Pr-site randomness, since the neutron-diffraction intensity does not show a saturation behavior even at 1 K. On the other hand, under the magnetic field of 1 T, the magnetization of $\sim 0.79\mu_B$ is explained only by the Pr-site magnetic moments aligning along the applied magnetic field without distinct contribution of the Fe sites. Recently, the magnetically ordered structure below 18 K for an isomorphic material $PrFe_4As_{12}$ was reported [8]. For 0.5 T applied along the [001] axis at 2 K, a magnetic moment of $1.59\mu_B$ for the Pr ion directs along the magnetic field, and that of $0.56\mu_B$ for the Fe ion tilting from the [001] axis by 79 degrees. This magnetic structure composed of the almost perpendicular moments is very similar to the present case in $Pr_{0.89}Fe_4Sb_{12}$ at 1 T and 1.8 K.

The CF ground state for Pr $4f$ electron switches from the singlet in the fully filled materials to the triplet in the partially filled materials. The concentration of Pr-site vacancy by 11% in the present sample means almost one vacancy in the five unit cells. It is not sufficient for changing uniformly point-charge potential at the Pr sites. It was reported that the CF scheme is sensitive to the hybridization of $4f$ with conduction electrons. $PrRu_4P_{12}$ shows the $4f$ electron multipole ordering on the metal-nonmetal transition [9], and $PrOs_4P_{12}$ shows the CF level shift with temperature [10]. The shift of the CF levels with temperature in these materials is 1 - 3 meV, which is close to the ground state switch from $x = 1$ to $x = 0.89$ for $Pr_xFe_4Sb_{12}$. Thus, the lack of the Pr sites in $Pr_xFe_4Sb_{12}$ may induce the uniform switch in the CF ground state, with the assistant of a spatially extent effect thorough the hybridized conduction band.

The magnetic moments of the Fe $3d$ electrons are commonly present in the skutterudite materials composed of Fe and Sb. Almost the same magnitudes of the effective magnetic moment per Fe site were evaluated from the high-temperature magnetic susceptibilities: $\mu_{\rm eff}^{\rm Fe} = 0.68\mu_B$ for $x = 1$ [3], $0.60\mu_B$ for our $x = 0.89$ sample, $0.69\mu_B$ for $x = 0.87$ [2], and $0.59\mu_B$ for $LaFe_4Sb_{12}$ [7]. On the other hand, the fully filled $x = 1$ system for $Pr_xFe_4Sb_{12}$ does not undergo any ordering phase transition only by the Fe magnetic moments [3]. Thus, the magnetic ordering in the skutterudite composed of Fe and Sb is given by the interplay between the magnetic moments at Pr and Fe sites. Recently, a microscopic theory for the inter-site interaction between the transition-metal and rare-earth sites in skutterudite was proposed on the basis of the d-f hybridization effect between these atoms [11]. It concludes that a ferrimagnetic ordering is enhanced for the triplet CF ground state, which is consistent with the observation $Pr_xFe_4Sb_{12}$.

5. Summary

The combination of the magnetic moments of Fe and those due to the relevant threefold CF ground state of Pr is necessary for the ferrimagnetic ordering below 4 K in $Pr_xFe_4Sb_{12}$.

Acknowledgments

Mr. M. Onodera and Mr. M. Ohkawara are acknowledged for their supports for the experiments. This study is supported by the Grants-in-Aid for Scientific Research from MEXT, Japan (No. 20102005, 21224048, 23244068, and 23-3397 (JSPS Fellows)). This work was carried out under the Joint-use Research Program for Neutron Scattering, ISSP, the University of Tokyo, at the Research Reactor JRR-3, JAEA (Proposal No. 10402, 10560, and 10774).

References

[1] Bauer E *et al.* 2002 *Phys. Rev. B* **66** 214421
[2] Butch N P *et al.* 2005 *Phys. Rev. B* **71** 214417
[3] Tanaka K *et al.* 2007 *J. Phys. Soc. Jpn.* **76** 103704
[4] Kuwahara K *et al.* 2010 *J. Phys.: Conf. Series* **200** 012101
[5] Bauer E *et al.* 2007 *J. Magn. Magn. Mater.* **310** 286
[6] Takegahara K *et al.* 2001 *J. Phys. Soc. Jpn.* **70** 1190, 2001 **70** 3468, 2002 **71** 372
[7] Viennois R *et al.* 2005 *Eur. Phys. J. B* **46** 257
[8] Wiśniewski P *et al.* 2011 *J. Phys. Soc. Jpn.* **80** Supplement A012
[9] Iwasa K *et al.* 2005 *Phys. Rev. B* **72** 024414
[10] Iwasa K *et al.* 2009 *Phys. Rev. B* **79** 235113
[11] Shiina R 2012 *J. Phys. Soc. Jpn.* **81** 024706

Effect of magnetic field in heavy-fermion compound $YbCo_2Zn_{20}$

Koji Kaneko[1], Shingo Yoshiuchi[2], Tetsuya Takeuchi[3], Fuminori Honda[2], Rikio Settai[2] and Yoshichika Ōnuki[2]

[1] Quantum Beam Science Directorate, Japan Atomic Energy Agency, Tokai, Naka Ibaraki 319-1195, Japan
[2] Graduate School of Science, Osaka University, Toyonaka 560-0043, Japan
[3] Low Temperature Center, Osaka University, Toyonaka 560-0043, Japan

E-mail: kaneko.koji@gmail.com

Abstract. Inelastic neutron scattering experiments on poly crystalline sample of heavy-fermion compound $YbCo_2Zn_{20}$were carried out in order to obtain microscopic insights on the ground state and its magnetic field response. At zero field at 300 mK, inelastic response consists of two features: quasielastic scattering and a sharp peak at 0.6 meV. With increasing temperature, a broad peak comes up around 2.1 meV, whereas quasielastic response gets broader and the peak at 0.6 meV becomes unclear. By applying magnetic field, the quasielastic response exhibits significant broadening above 1 T, and the peak at 0.6 meV is obscure under fields. The peaks in inelastic spectra and its temperature variation can be ascribed to the suggested crystal-field model of Γ_6 - Γ_8 - Γ_7 with the overall splitting of less than 3 meV. The observed quasielastic response and its rapid broadening with magnetic field indicates that the heavy-electron state arises from the ground state doublets, and are strongly suppressed by external field in $YbCo_2Zn_{20}$.

1. Introduction

The new family of ternary compounds RT_2X_{20} (R=rare earth or actinide, T = transition metal, X = Al, Zn) with the cubic $CeCr_2Al_{20}$ structure get surge of interest due to its wide variety of physical properties including superconductivity[1, 2]. The characteristic structure with an R ion incorporated in the nearly spherical cage formed by 16 X ions, gives two essential aspects on physical properties; one is strong hybridisation between f and conduction electrons due to a large coordination number around the R ion, and the other is highly quasi-degenerated crystal electric field levels. Moreover, a large distance between R ions of ~6 Å leads to weaker interactions, and allows various attractive ground states to be realised in this system. Among this family, it should be noted that all of ytterbium members YbT_2Zn_{20} (T= Fe, Co, Ru, Rh, Os, Ir) exhibit heavy-fermion behaviour with the large electronic specific coefficient of $\gamma > 400$ mJ/mol K^2[1-9]. In particular, $YbCo_2Zn_{20}$ shows extremely large value of $\gamma \sim 8$ J/mol K^2, one of the heaviest fermion compounds so far. The A coefficient of electrical resistivity is correspondingly large, and the Kondo temperature is estimated to be less than 3 K in this compound[1, 8, 9]. Based on these observation, $YbCo_2Zn_{20}$ is considered to lie at the proximity of a quantum critical point (QCP).

In fact, recent study revealed the appearance of ordered phases by applying small pressure and/or external magnetic field in $YbCo_2Zn_{20}$[3, 9, 10]. In the case of Yb based compounds, the application of hydrostatic pressure tunes the system from a non-magnetic ground state to a magnetic ordered one through the quantum critical point. Saiga *et al.* revealed pressure induced antiferromagnetic order in $YbCo_2Zn_{20}$ at around ~1 GPa, where deviation from the Fermi liquid behaviour was observed[3, 11]. The application of a magnetic field leads to drastic effects. Metamagnetic transitions were found at low field of B_M~0.6 T, where non-Fermi liquid behaviour was observed as well[8, 9]. Further increasing the field, the A coefficient of electrical resistivity in $YbCo_2Zn_{20}$ exhibits rapid decrease, which suggests that heavy quasiparticles are strongly suppressed by external magnetic field[3, 4, 8, 9]. Besides, a field-induced ordered phase, of which origin is still under investigation, occurs when a field of more than 6 T is applied along the [1 1 1] direction[9, 10]. These facts suggest that $YbCo_2Zn_{20}$ is located very close to a QCP, and can be one of the model compounds in Yb-system to study dynamical properties around QCP. In order to get microscopic insights on f-electron state in $YbCo_2Zn_{20}$,inelastic neutron scattering experiments down to low temperature and under magnetic fields were carried out.

2. Experiment

Polycrystalline samples of $YbCo_2Zn_{20}$ were grown by the Zn self-flux method. Details of the sample preparation is published elsewhere[5, 6, 8]. Neutron inelastic scattering experiments were performed on the cold triple-axis spectrometer LTAS installed at the C2-1 port in the guide hall of the research reactor JRR-3 in Tokai, Japan. A combination of vertical and horizontal focusing PG monochrometer and analyser, respectively were employed. The collimation of guide-80'-120'(radial)-open with k_f=1.3 Å gives the energy resolution of about 140 μeV. The sample was attached to the ^{3}He-^{4}He dilution refrigerator, and inserted into the 5 T vertical field superconducting magnet.

3. Results and Discussion

Figure 1 shows inelastic scattering spectra of $YbCo_2Zn_{20}$ at $\boldsymbol{Q}$=2 Å^{-1} up to 3 meV measured at (a) 300 mK and (b) 10 K. At 300 mK, strong quasielastic response which extends above present measured range was observed in $YbCo_2Zn_{20}$. Besides, a distinct inelastic peak was superposed onto the quasielastic response at around 0.6 meV. With increasing temperature up to 10 K, inelastic spectra of $YbCo_2Zn_{20}$ undergoes substantial change. It should be noted that an additional excitation appears around 2.1 meV. On the other hand, the excitation at 0.6 meV becomes unclear, and the quasielastic part gets broader.

Figure 2 displays magnetic field dependence of inelastic scattering spectra of $YbCo_2Zn_{20}$ measured at 50 mK under several fields up to 5 T. Note that inelastic scattering spectra of $YbCo_2Zn_{20}$ exhibit strong magnetic-field variation. At low magnetic field up to 1 T, the identical spectra to that at zero field were obtained, in other words, no marked difference was observed in the present spectra below and above the metamagnetic transition at B_M. In contrast, with further increasing field above 1 T, the quasielastic component shows remarkable broadening up to 5 T. The peak at 0.6 meV is buried in the broadened quasielastic response with the field. No additional peak was found under magnetic field.

Hereafter, we focus on the quasielastic response observed in $YbCo_2Zn_{20}$. The observed spectra can be reproduced by a sum of quasielastic Lorentzian and a Gaussian peak at 0.6 meV or 2.1 meV. The result was shown as solid lines in Fig.1. By fitting the spectra at 300 mK at zero field, the quasielastic line width was derived to be 0.3 meV. This value coincides with the reported Kondo temperature of a few Kelvin, suggesting that the observed response arise from Kondo-type fluctuation.

This quasielastic component exhibits significant broadening with increasing field. In YbT_2Zn_{20} system, strong reduction of heavy electron state was suggested in $YbCo_2Zn_{20}$

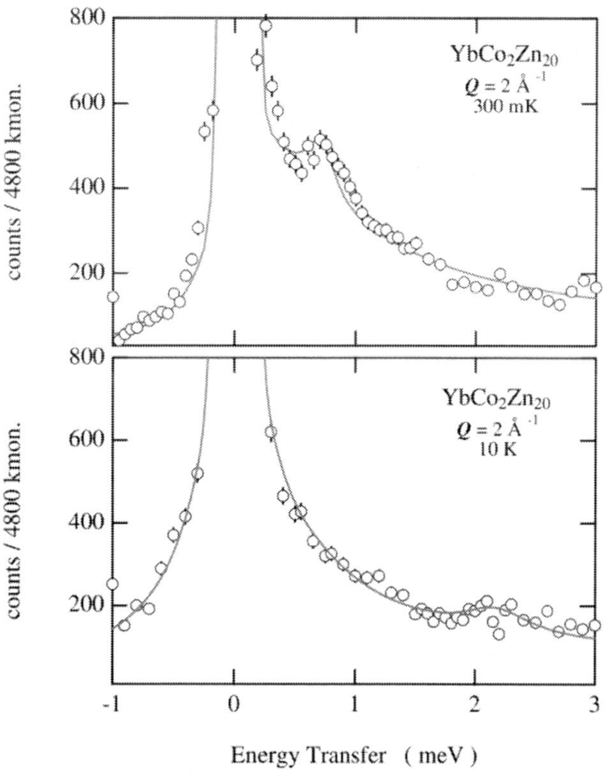

Figure 1. Inelastic spectra measured on polycrystalline sample of $YbCo_2Zn_{20}$ at (a) 300 mK and (b) 10 K with k_f=1.3 Å^{-1}.

as well as $YbIr_2Zn_{20}$ based on specific heat, electrical resistivity and de Haas-van Alphen measurements[6, 7, 8, 9]. The $\sqrt{A}$-coefficient of the electrical resistivity in $YbCo_2Zn_{20}$ decreases two orders of magnitude decreases from zero field to 14 T for the field along the <1 0 0>, which is supported by the effective mass derived from the dHvA measurements; the mass shows rapid decrease with increasing field. The broadening of the quasielastic response with field in $YbCo_2Zn_{20}$ supports the reduction of the heavy-electron state from the microscopic point of view. Since significant anisotropic magnetic field response was found in $YbCo_2Zn_{20}$, an experiment using single crystal is indispensable for further quantitative analysis.

Together with the quasielastic scattering, the distinct peaks were clearly observed, especially at zero field. The observed peak can be ascribed to crystal-field excitation. Takeuchi *et al.* proposed a crystal-field level scheme for $YbCo_2Zn_{20}$ as Γ_6(0 K) - Γ_8(9.1 K) - Γ_7(28.5 K) based on the results of specific heat, magnetic susceptibility and magnetization measurements[9] The peak at 0.6 meV observed in the present study is consistent with to the excitation from the ground state to the first excited state. The second peak at 2.1 meV and its evolution at high temperature of 10 K agrees to that expected for the excitation from Γ_8 to Γ_7. While there is small discrepancy in the energy of the first excited state, the overall splitting and its temperature dependence are consistent with the suggested model.

The present result clearly reveal the small overall splitting of the crystal field in $YbCo_2Zn_{20}$. On the other hand, the difference between the ground state and the first excited state is distinctly larger than the energy scale of Kondo temperature. This fact indicates that the large entropy of ~8 J/mol K^2 in $YbCo_2Zn_{20}$ does not come from degenerated crystal-field ground state, but

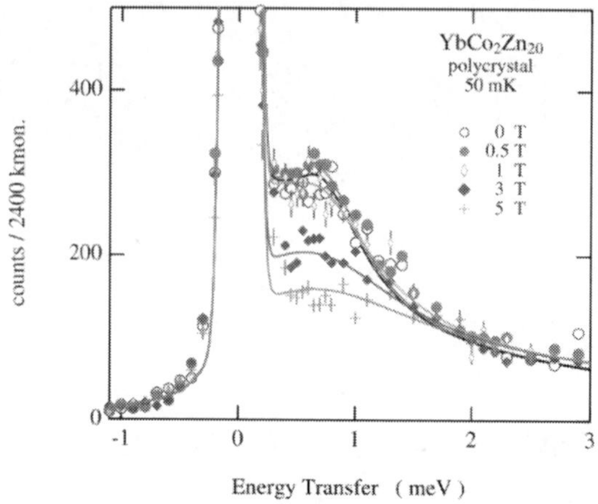

Figure 2. Magnetic field variation of inelastic scattering spectra of $YbCo_2Zn_{20}$ at 50 mK under several magnetic fields up to 5 T.

mainly originate from the doublet ground state.

For more detailed discussion for anisotropy, magnetic correlation effects, and an origin of the field induced ordered phase, neutron scattering experiments using single crystals are in progress.

4. Acknowledgments

The authors would like to thank H. Yamauchi and Y. Shimojo for their assistance in the neutron scattering experiment. This work was supported by a Grant-in-Aid for Scientific Research on Innovative Areas "Heavy Electrons" (#20102002, #21102524, #23102715), on Priority Areas of New Material Science Using Regulated Nano Spaces (#22013009, #22013022), and Young Scientists B (#23740247) of The Ministry of Education, Culture, Sports, Science, and Technology, Japan.

References

[1] Torikachvili M S, Jia S, Mun E D, Hannahs S T, Black R C, Neils W K, Martien D, Bud'ko S L and Canfield P C 2007 Proc. Natl. Acad. Sci. **104** 9960.
[2] Onimaru T, Matsumoto K T, Inoue Y F, Umeo K, Saiga Y, Matsushita Y, Tamura R, Nishimoto K, Ishii I, Suzuki T, Takabatake T 2009 *J. Phys. Soc. Jpn.* **79** 033704.
[3] Saiga Y, Matsubayashi K, Fujiwara T, Kosaka M, Katano S, Hedo M, Matsumoto T and Uwatoko Y 2008 *J. Phys. Soc. Jpn.* **77** 053710.
[4] Matsubayashi K, Saiga Y, Matsumoto T and Uwatoko Y 2009 *JPhys.: Conf. Ser.* **150** 042117.
[5] Yoshiuchi S, Toda M, Matsushita M, Yasui S, Hirose Y, Ohya M, Katayama K, Honda F, Sugiyama K, Hagiwara M, Kindo K, Takeuchi T, Yamamoto E, Haga Y, Settai R and Ōnuki Y 2009 *J. Phys. Soc. Jpn.* **78** 123711.
[6] Takeuchi T, Yoda M, Matsushita M, Yoshiuchi S, Ohya M, Katayama K, Hirose Y, Yoshitani N, Honda F, Sugiyama K, Hagiwara M, Kindo K, Yamamoto E, Haga Y, Tanaka T, Kubo Y, Settai R and Ōnuki Y 2010 *J. Phys. Soc. Jpn.* **79** 064609.
[7] Honda F, Yashui S, Yoshiuchi S, Takeuchi T, Settai R and Ōnuki Y 2010 *J. Phys. Soc. Jpn.* **79** 083709.
[8] Ohya M, Matsushita M, Yoshiuchi S, Takeuchi T, Honda F, Settai R, Tanaka T, Kubo Y and Ōnuki Y 2009 *J. Phys. Soc. Jpn.* **79** 083601.
[9] Takeuchi T, Yoshiuchi S, Ohya M, Taga Y, Hirose Y, Sugiyama K, Honda F, Hagiwara M, Kindo K, Settai R, Ōnuki Y 2011 *J. Phys. Soc. Jpn.* **80** 114703.
[10] Shimura Y, Sakakibara T, Yoshiuchi S, Honda F, Settai R and Ōnuki Y 2011 *J. Phys. Soc. Jpn.* **80**, 073707.
[11] Matsubyashi K, unpublished.

Thermoelectric power investigation of SmS under pressure

S. Kanematsu, K. Imura[A], H. S. Suzuki[B], K. Deguchi, N. K. Sato

Department of Physics, Nagoya University, Nagoya 464-8602, Japan
[A]UVSOR Facility, Institute for Molecular Science, Okazaki, 444-8585, Japan
[B]Quantum Beam Center, National Institute for Materials Science, Tsukuba, Ibaraki 305-0047, Japan

E-mail: `kanematsu.shinkichi@c.mbox.nagoya-u.ac.jp`

Abstract. SmS undergoes a pressure-induced phase transition from a semiconducting to a mixed-valent state. Although the mixed valence state has been extensively studied, the semiconducting state with a divalent ionic state is not well understood yet. Here we report transport properties of the low-pressure semiconducting phase. Combining the thermoelectric power with the electrical resistivity, we observe that the mobility of electrons excited onto the conduction band is much greater than that of holes left behind in the valence band, the electron effective mass is nearly equal to the hole effective mass, and the relaxation time of the electron is much larger than that of the hole.

1. Introduction

SmS undergoes successive phase transitions under pressure (P). As P increases, the system first undergoes a valence transition at a critical pressure ($\sim$ 7 kbar at room temperature) from divalence (Sm^{2+}) to mixed valence (Sm^{2+} and Sm^{3+}) accompanied by a color change from black to golden-yellow. With further increasing P within the high pressure phase (named golden SmS), a magnetic transition occurs at a second critical pressure ($\sim$ 18 kbar) [1, 2]. According to a magnetic susceptibility measurement, the ground state at $P > P_c$ is antiferromagnetic [3]. Golden SmS has attracted much interest because of its Kondo insulator behavior [4], but little is known about the low pressure phase (called black SmS). Here we focus on black SmS.

Black SmS is a narrow-gap semiconductor with an energy gap E_g of the order of 0.1 eV at ambient pressure. E_g is decreased by the application of P as well as magnetic field H, and collapses to zero at an effective pressure $P_{\text{eff}} \sim 3$ - 4 kbar, where $P_{\text{eff}} = P + bH$ with $b = 0.14$ kbar/T [5]. Above this pressure, the system becomes metallic [6]. Of particular interest is to know what happens in the vicinity of the insulator to metal transition. In order to elucidate the black phase more in detail, we measure the thermoelectric power (TEP) S and the electrical resistivity ρ at the same pressure.

2. Experimental details

Single crystals of SmS were grown by the Bridgman method using a high-frequency induction furnace. We prepared two ingots with different starting compositions; stoichiometric sample #8, and S-excess sample #9. Samples with a typical dimension of $1 \times 1 \times 4$ mm^3 were cleaved for the present study. Electrical resistivity was measured by means of the conventional DC four-terminal method at ambient and high pressures. TEP was measured by means of the differential DC method at ambient and high pressures [7], using a thermocouple consisting of Au0.07at%Fe (abbreviated as AuFe hereafter) and Chromel wires of diameter $\phi = 76$ μm. A strain gauge was used for a heater to generate a temperature gradient (ΔT) between both ends of the sample. To obtain the Seebeck coefficient S, we need to measure the voltage ratio, $r = V_A/V_B$. Here, V_A and V_C is a voltage generated in the AuFe-sample-AuFe and the Chromel-sample-Chromel closed circuit, respectively. Then, S is evaluated using the following equation

$$S = \frac{S_A - rS_C}{1-r}, \tag{1}$$

where S_A and S_C indicate an absolute Seebeck coefficient of AuFe and Chromel, respectively. An external pressure was applied using a NiCrAl-BeCu hybrid piston-cylinder with a WC piston. Daphne oil 7373 was used as a pressure transmitting medium. Pressure was determined using a superconducting transition of indium [8].

3. Results

Figure 1 shows the temperature dependence of S under pressure. At ambient pressure (left panel), we observe the $1/T$ dependence characteristic of a semiconductor. The absolute value of S is as large as about 300 μV/K at room temperature, and exceeds 1.4 mV/K at 40 K. The negative sign of S indicates that the electric current is carried dominantly by electrons.

At $P = 0.78$ kbar, the overall T dependence of S is same as the above, but its absolute value $|S|$ decreases. This decrease in $|S|$ arises from the reduction in the energy gap $E_{\rm g}$, as shown below. At $P = 1.36$ kbar, the $1/T$ dependence is still observed, but another feature appears at around 130 K, where a broad maximum emerges. At $P = 1.81$ kbar (right panel), in addition to the broad maximum, there emerges a shallow minimum at around 20 K. This is understood as follows: In the semiconducting state, $|S|$ increases with lowering temperature, and then decreases toward zero for $T \to 0$ according to the thermodynamics. This gives rise to the minimum. With further increasing P, this minimum becomes shallow and shifts toward high temperatures.

The inset of the right panel displays S in a log T vs log $|S|$ plot at $P = 2.04$ kbar. We observe the nearly T linear dependence at low temperatures.

At $P = 2.49$ kbar (left panel), the system transforms into the golden phase. At room temperature, the sign of S is positive, in accordance with a previous result [7]. We also observe that S changes sign at a low temperature and shows a maximum below 10 K. Detailed results of the golden phase are given elsewhere [9].

Figure 2 compares the temperature dependences of S and the electrical resistivity ρ. In the left panel, the result at $P = 0.74$ kbar is displayed. Corresponding to the semiconducting $1/T$ dependence of S, we observe the resistivity to increase exponentially with lowering temperature. In the right panel, the result at $P = 2.53$ kbar is shown. With decreasing temperature, S initially increases, and then decreases below approximately 80 K before increasing again below about 30 K. As mentioned, the shallow minimum is ascribed to the gap formation. In the same temperature region where the semiconducting feature is observed in S, we observe the resistivity to increase with lowering temperature. The gap energy obtained from these two methods will be compared below.

Figure 3 shows the temperature dependences of S and ρ together with the pressure dependence of $E_{\rm g}$. In the left panel, S is plotted against $1/T$ at pressures indicated in the figure. We evaluate some parameters using the relation [10]

$$S = \frac{k_{\rm B}}{e}\frac{b-1}{b+1}\frac{E_{\rm g}}{2k_{\rm B}T} + \frac{3}{4}\ln\frac{m_{\rm n}}{m_{\rm p}}, \tag{2}$$

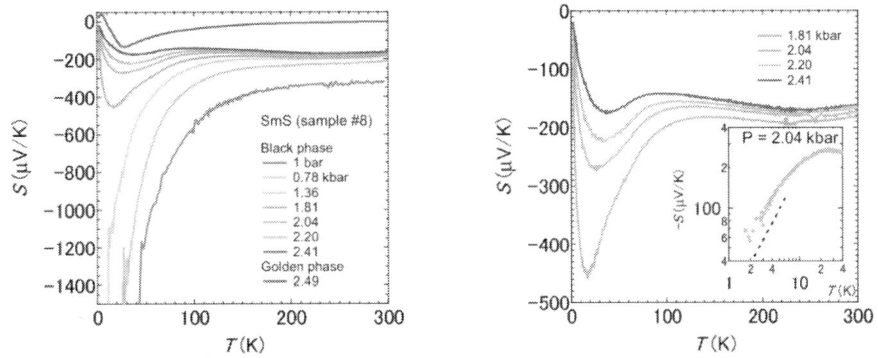

Figure 1. Temperature dependence of the Seebeck coefficient S at pressures indicated in the figure. Some data of the left panel are plotted in the right panel on an expanded scale. The inset of the right panel shows the T dependence of S at low temperatures in logarithmic plots.

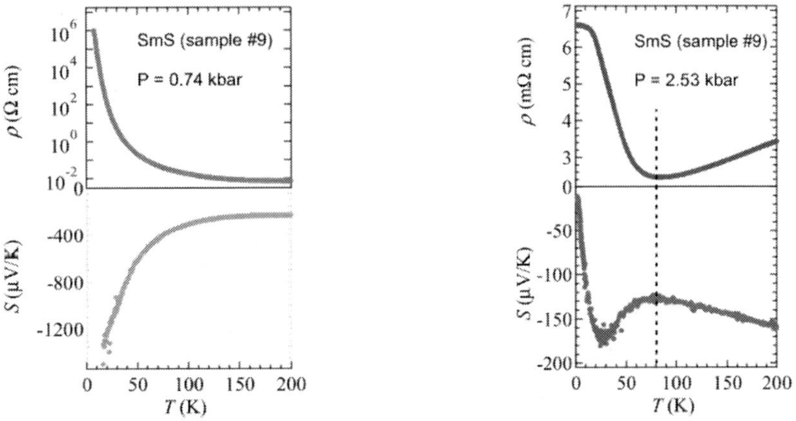

Figure 2. Temperature dependences of S and ρ at $P = 0.74$ kbar (left panel) and at $P = 2.53$ kbar (right panel). Note that the $1/T$ dependence of S correlates with the exponential like T dependence of ρ.

where b is the mobility ratio, $b = \mu_n/\mu_p$ (μ_n and μ_p are electron and hole mobility, respectively), m_n and m_p are the electron and hole effective mass, respectively, k_B is the Boltzmann constant, and e is the electron charge. From the slope of the broken straight line, we evaluate the "gap energy" E_g^S defined as

$$E_g^S = \frac{b-1}{b+1} E_g. \tag{3}$$

The gap E_g in the right-hand side is deduced from the slope of the broken straight line in the Arrhenius plot (middle panel). Thus obtained gap energies are summarized in the right panel. We observe the overall similar P dependence for both the gap energies. This leads to that the prefactor of E_g in eq. 3 is close to 1, leading to that $b \gg 1$. (Strictly speaking, since the prefactor is smaller than 1, E_g^S should be smaller than E_g, contrary to the experiment. The

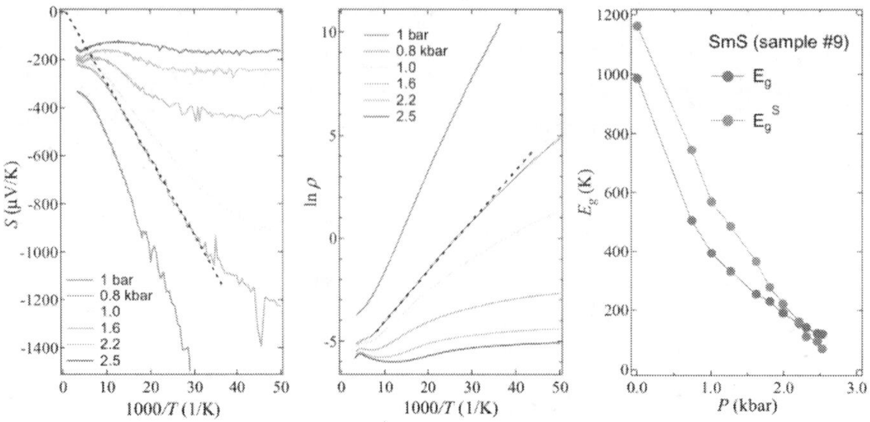

Figure 3. Pressure evolution of the gap energy. The slope of the straight line in the left and middle panel yields the gap energy E_{g}^{S} and E_{g}, respectively. Pressure dependences of E_{g}^{S} and E_{g} are plotted in the right panel. Note that the overall feature is similar.

origin of this discrepancy is not clear.) Therefore, we have the relation, $\mu_{\mathrm{n}} \gg \mu_{\mathrm{p}}$, indicating that conduction electrons with 5d character are much mobile than holes in the valence band having the 4f character dominantly.

According to eq. 2, the quantity of $\ln m_{\mathrm{n}}/m_{\mathrm{p}}$ is evaluated from an interception of the straight line with the vertical axis. Then, the experimental results shown in the left panel of Fig. 3 yields the relation $\ln m_{\mathrm{n}}/m_{\mathrm{p}} \sim 0$, resulting in the relation $m_{\mathrm{n}} \sim m_{\mathrm{p}}$. Combining this with the mobility ratio obtained above, we have the relation $\tau_n \gg \tau_p$.

4. Summary

SmS is a semiconductor at ambient pressure. The energy gap E_{g} deduced from both the electrical resistivity and the thermoelectric power decreases with increasing pressure, and collapses to zero at a pressure, which is roughly estimated as 3 kbar. In the vicinity of this pressure, the Seebeck coefficient shows the complicated behavior. In particular, we observed the emergence of a broad peak whose origin is unknown. Furthermore, using a combination of the resistivity and the thermopower, we obtained the following relations; $\mu_{\mathrm{n}} \gg \mu_{\mathrm{p}}$, $m_{\mathrm{n}} \sim m_{\mathrm{p}}$, and $\tau_{\mathrm{n}} \gg \tau_{\mathrm{p}}$.

References

[1] A. Barla, J. P. Sanchez, Y. Haga, G. Lapertot, B. P. Doyle, O. Leupold, R. Rüffer, M. M. Abd-Elmeguid, R. Lengsdorf and J. Flouquet 2004 *Phys. Rev. Lett.* **92** 066401
[2] Y. Haga, J. Derr, A. Barla, B. Salce, G. Lapertot, I. Sheikin, K. Matsubayashi, N. K. Sato and J. Flouquet 2004 *Phys. Rev. B* **70** 220406
[3] K. Matsubayashi, K. Imura, H. S. Suzuki, G. Chen, N. Mori, T. Nishioka, K. Deguchi and N. K. Sato 2007 *J. Magn. Magn. Mater.* **310** 408
[4] G. Aeppli and Z. Fisk 1992 *Comments Cond. Mat. Phys.* **16** 155
[5] K. Imura, K. Matsubayashi, H. S. Suzuki, K. Deguchi and N. K. Sato 2008 *J. Phys: Conf. Series.* **121** 032014
[6] N. K. Sato *et al.* in preparation
[7] K. Imura, S. Kanematsu, K. Deguchi, H. S. Suzuki, K. Matsubayashi and N. K. Sato 2011 *J. Phys. Soc. Jpn. Suppl.* **80** SA077
[8] L. D. Jennings and C. A. Swenson 1958 *Phys. Rev.* **112** 31
[9] K. Imura, S. Kanematsu, K. Matsubayashi, H. S. Suzuki, K. Deguchi and N. K. sato 2011 *J. Phys. Soc. Jpn.* **80** 113704
[10] N. E. Sluchanko *et al.* 2000 *Phys. Rev. B* **61** 9906

Hall effect of $Ce(Ru_{1-x}Fe_x)_2Al_{10}$ single crystal

Yukihiro Kawamura[1,2,3], Daishi Hirai[1], Takashi Nishioka[1], Kazuyuki Matsubayashi[2], Yoshiya Uwatoko[2], Hideki Yoshizawa[2], Chihiro Sekine[3]

[1]Graduate School of Integrated Arts and Sciences, Kochi University, Kochi 780-8520, Japan
[2]Institute for Solid State Physics, University of Tokyo, Kashiwa 305-0047, Japan
[3]Muroran Institute of Technology, Muroran 050-8383, Japan

E-mail: y_kawamura@mmm.muroran-it.ac.jp

Abstract. The Hall resistivity ρ_H has been measured between 2 K and 300 K for pseudo-ternary compounds $Ce(Ru_{1-x}Fe_x)_2Al_{10}$ (x=0.375, 0.625, 1) single crystals up to 7 T. The ρ_H shows linear dependence against magnetic field except low temperatures. We calculated Hall coefficient R_H by the gradient of ρ_H under the field between 5 T and 7 T, where the anomalous Hall effect expected to be small. While R_H for $CeRu_2Al_{10}$ drastically increases below mysterious antiferromagnetic ordering temperature ($T_0 \sim 27$ K), R_H decreases by replacing Ru with Fe. The R_H becomes negative at lowest temperatures for x = 0.375, x = 0.625 and x = 1. The reversal of the sign suggests that $Ce(Ru_{1-x}Fe_x)_2Al_{10}$ is in a multicarrier system.

1. Introduction

The ternary compounds CeT_2Al_{10} (T = Fe, Ru) crystallize in the orthorhombic $YbFe_2Al_{10}$-type crystal structure (space group No. 63 $Cmcm$)[1, 2]. Both compounds have unique physical properties. $CeRu_2Al_{10}$ has an antiferromagnetic transition at exceptionally high temperature ($T_0 \sim 27$ K) in spite of the content of Ce ions being very low[3], where Ce is the origin of magnetic moment[4]. On the other hand, $CeFe_2Al_{10}$ is reported as Kondo semiconductor with the c-f hybridization gap ($\Delta T \sim 15$ K[5]). In general, Kondo semiconductor is in the itinerant electron system, while antiferromagnetic ordering occurs in the localized electron system. Kondo semiconductor and antiferromagnetic order might coexist on $CeRu_2Al_{10}$, because electrical resistivity on $CeRu_2Al_{10}$ increases with decreasing temperature even above T_0. The relation between Kondo semiconducting behavior and this phase transition on this system has not clarified yet. We have studied $Ce(Ru_{1-x}Fe_x)_2Al_{10}$ ($0 \leq x \leq 1$) system because this system might connect these two characteristic behavior. The substitution effect of Fe for Ru in $CeRu_2Al_{10}$ indicates the effect of applying chemical pressure, because the lattice constants of $CeFe_2Al_{10}$ are smaller than those of $CeRu_2Al_{10}$. Magnetization, electrical resistivity and specific heat measurement on $Ce(Ru_{1-x}Fe_x)_2Al_{10}$ ($0 \leq x \leq 1$) and those on CeT_2Al_{10}(T=Fe, Ru) under pressure indicate that the effect of chemical pressure is similar to the effect of physical pressure[6, 7]. The purpose of this study is to investigate this phase transition and Kondo semiconducting behavior from a viewpoint of carriers. Thus, we measured Hall resistivity of $Ce(Ru_{1-x}Fe_x)_2Al_{10}$ ($0 \leq x \leq 1$) single crystals.

2. Experimental

Single crystals of $Ce(Ru_{1-x}Fe_x)_2Al_{10}$ (x=0.375, 0.625, 1) were grown by using Al self-flux method. The starting composition of $Ce(Ru_{1-x}Fe_x)_2Al_{70}$ was placed in an Al_2O_3 crucible, sealed in a quartz ampule in vacuum. It heated up to 1050 °C and then gradually cooled down to 680 °C at a rate of 5 °C/h. The Al flux was removed by centrifugation and by sinking 5 mol/L NaOH solution for 24 hours. The x-ray analysis confirmed that each sample is in a single phase. The crystallographic axes are identified by X-ray analysis. Hall resistivity was measured with ordinary four probes method by using the PPMS system (Quantum Design Co., Ltd.) in a temperature range between 2 K and 300 K. The magnetic field is applied along b-axis because b-axis is the hard magnetization axis and anomalous Hall coefficient is smallest along b-axis. The current is applied along a-axis and the voltage of the c-axis is read. Hall resistivity ρ_H is given by $\rho_H = V_H ld/Iw$, and Hall coefficient is given by $R_H = \partial V_H/\partial B \cdot d/I$, where V_H is the Hall voltage, l is the length along current direction, d is the thickness, w is the width along the voltage direction, I is the current, B is the magnetic flux density. V_H is given by $V_H = 1/2\{V(+B)-V(-B)\}$ in order to eliminate the contribution of magnetic resistance.

3. Results and Discussion

Fig. 1, 2 and 3 shows ρ_H vs B on $Ce(Ru_{1-x}Fe_x)_2Al_{10}$, where x = 0.375, 0.625 and 1, respectively. While $\rho_H(B)$ is almost linear from 300 K to 10 K, it gradually deviates from linear by decreasing temperature. Under non-interacting circumstances, Hall coefficient R_H is expressed as $R_H = R_0B + R_S(M)$, where R_0 are the ordinary Hall coefficient and R_S are the anomalous Hall coefficient. We assumed ordinary Hall coefficient is dominant in $\rho_H(B)$ between 5 T and 7 T and calculated R_H of $Ce(Ru_{1-x}Fe_x)_2Al_{10}$ from the gradient of Hall resistivity under these magnetic field. The reason of this is as follows.

$CeRu_2Al_{10}$ has a large magnetic anisotropy. The magnetic susceptibility χ for B // a-axis has a strong temperature dependence and shows Curie-Weiss behavior at high temperature. The maximum value of χ for B // a-axis is in the order of 10^{-2} emu/mol. On the other hand, χ for B // b-axis has weak temperature dependence. It almost unchanged even around T_0 and the value of χ for B // b-axis is as low as 10^{-3} emu/mol in all temperature region, except below 10 K. Below 10 K, $\chi(T)$ shows an upturn by decreasing temperature[8]. The magnetization of this upturn is saturated by applying field. Over the field of 5 T, χ for B // b-axis remains unchanged even below 10 K. These behaviors are observed not only in $CeRu_2Al_{10}$ but also in $Ce(Ru_{1-x}Fe_x)_2Al_{10}$ system. The non-linear behavior of $\rho_H(B)$ below 10 K is most probably attributed to the anomalous Hall effect of the upturn of $\chi(T)$. Thus, an anomalous Hall effect of $Ce(Ru_{1-x}Fe_x)_2Al_{10}$ should be small when the field is applied over 5 T.

Fig. 4 shows the temperature dependence of Hall coefficient $R_H(T)$. The R_H of x = 0 is strong enhanced below T_0 as shown in the previous report[3, 9]. The enhancement at lowest temperatures suppressed by substituting Fe. The R_H of x = 0.375 and x = 0.625 are enhanced below T_0 and the value of them becomes negative at low temperatures. These behaviors are also observed in related compound $CeOs_2Al_{10}$[10]. The R_H of x = 1 increases with decreasing temperature from 300 K and shows a maximum at ~ 80 K, followed by the negative value below 30 K. This seems different from the previous report measured by polycrystalline sample[5]. However, the Hall coefficient of $CeFe_2Al_{10}$ in the previous report was measured at 2 T. The positive Hall coefficient is obtained if we calculate the R_H from the data of ρ_H under 2 T, which can be seen from Fig. 3. Thus, this result is consistent with the previous report. The maximum at ~ 80 K in R_H of x = 1 is consistent with the value of the pseudo-gap ($\Delta/k_B \sim 70$ K) observed in ^{27}Al-NQR measurement[11] The temperature, where the R_H becomes negative, increases with x. The reversal of the sign suggests that $Ce(Ru_{1-x}Fe_x)_2Al_{10}$ is in a multi-carrier system. Our results indicate that the gap open at T_0 in $CeRu_2Al_{10}$ or the pseudo-gap observed in ^{27}Al-NQR measurement in $CeFe_2Al_{10}$ exhibit as positive value of R_H and the c-f hybridization gap proposed

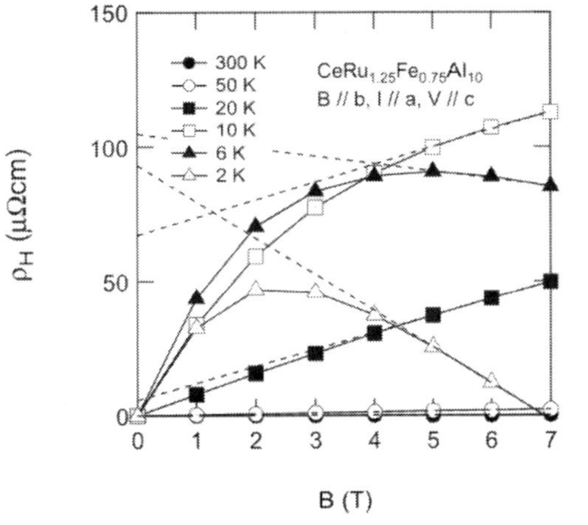

Figure 1. ρ_H vs B on $CeRu_{1.25}Fe_{0.75}Al_{10}$ (x=0.375 in $Ce(Ru_{1-x}Fe_x)_2Al_{10}$) on the temperature of 300 K(filled circle), 50 K(open circle), 20 K(filled square) , 10 K(open square), 6 K(closed triangle) and 2 K(open triangle) for H // b-axis. The dotted line indicates the fitting line of ρ_H(B) between 5 T and 7 T.

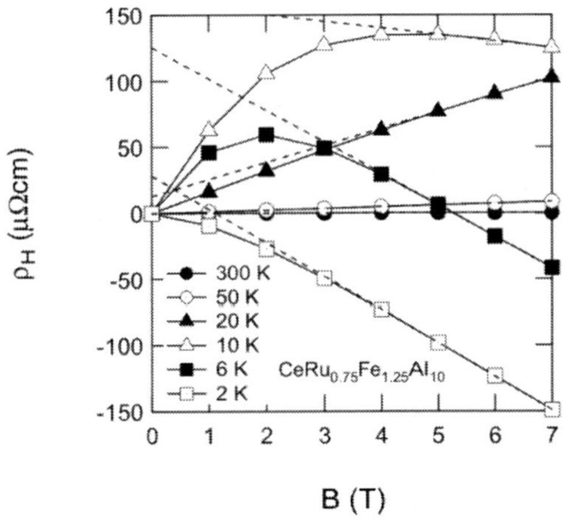

Figure 2. ρ_H vs B on $CeRu_{0.75}Fe_{1.25}Al_{10}$ (x=0.625 in $Ce(Ru_{1-x}Fe_x)_2Al_{10}$).

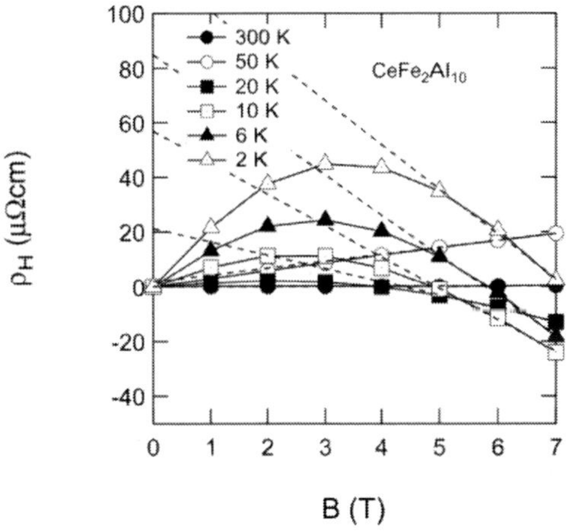

Figure 3. ρ_H vs B on $CeFe_2Al_{10}$ (x=1 in $Ce(Ru_{1-x}Fe_x)_2Al_{10}$).

in $CeFe_2Al_{10}$[5] exhibit as negative value of R_H

As we have shown in our previous report, the R_H of $CeRu_2Al_{10}$ at lowest temperatures are suppressed by pressure[12]. From this paper, the R_H of $Ce(Ru_{1-x}Fe_x)_2Al_{10}$ are also suppressed by Fe substitution. The consistency of chemical pressure and physical pressure is observed in the Hall effect as well as in the electrical resistivity, the magnetization and the specific heat. However, the corresponding pressure is somewhat different in Hall coefficient from in the electrical resistivity, the magnetization and the specific heat[6]. The value of R_H shows positive at lowest temperatures even at 2 GPa in the pressure measurements. This is probably because $CeRu_2Al_{10}$ has sample dependence. T_0 of one sample of $CeRu_2Al_{10}$ disappears at 3 GPa by applying pressure but T_0 of another sample of $CeRu_2Al_{10}$ disappears at 4 GPa in our measurements with using the cubic anvil pressure cell, which has the high quality of hydrostatic

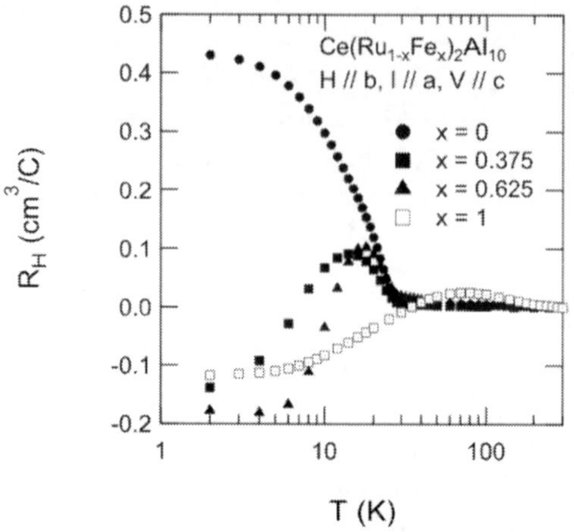

Figure 4. Temperature dependence of Hall coefficient R_H of $Ce(Ru_{1-x}Fe_x)_2Al_{10}$ for x = 0 (closed circle), x = 0.375 (closed square), x = 0.625 (closed triangle) and x=1 (open square).

pressure. It seems that pressure dependence of $CeRu_2Al_{10}$ depends on samples.

4. Summary

We have studied the Hall effect of the seudo-ternary compounds $Ce(Ru_{1-x}Fe_x)_2Al_{10}$ (x=0.375, 0.625, 1). Magnetic field dependence of Hall resistivity shows almost linear behavior except below 10 K. Hall coefficient below T_0 strongly enhanced for x = 0. This enhancement is suppressed by increasing x and Hall coefficient becomes negative at lowest temperatures for x = 0.375, x = 0.625 and x = 1. The reversal of the sign suggests that $Ce(Ru_{1-x}Fe_x)_2Al_{10}$ is a multicarrier system.

Acknowledgment

This work was partially supported by a Grant-in-Aid for for Science Research (C) of the Ministry of Education, Culture, Sports, Science and Technology.

References

[1] A. I. Tursina, S. N. Nesterenko, E. V. Murashova, I. V. Chernyshev, H. Noel and Y. D. Seropegin: Acta Cryst. E **61** (2005) i12.
[2] V. M. T. Thiede, T. Ebel, W. Jeitschoko: J. Mater. Chem. **8** (1998) 125.
[3] A. M. Strydom: Physica B **404** (2009) 2981.
[4] H. Tanida, D. Tanaka, M. Sera, C. Moriyoshi, Y. Kuroiwa, T. Takesaka, T. Nishioka, H. Kato and M. Matsumura: J. Phys. Soc. Jpn. **79** (2010) 043708
[5] Y. Muro, K. Motoya, Y. Saiga, and T. Takabatake: J. Phys. Soc. Jpn. **78** (2009) 083707.
[6] T. Nishioka, D. Hirai, Y. Kawamura, H. Kato, M. Matsumura, H. Tanida, M. Sera, K. Matsubayashi and Y. Uwatoko: J. Phys. Conf. Seri. **273** (2011) 012046
[7] T. Nishioka, Y. Kawamura, T. Takesaka, R. Kobayashi, H. Kato, M. Matsumura, K. Kodama, K. Matsubayashi and Y. Uwatoko: J. Phys. Soc. Jpn. **78** (2009) 123705.
[8] H. Tanida, D. Tanaka, M. Sera, C. Moriyoshi, Y. Kuroiwa, T. Takesaka, T. Nishioka, H. Kato, M. Matsumura: J. Phys. Soc. Jpn. **79** (2010) 083701
[9] H. Tanida, D. Tanaka, M. Sera, C. Moriyoshi, Y. Kuroiwa, T. Takesaka, T. Nishioka, H. Kato, M. Matsumura: J. Phys. Soc. Jpn. **79** (2010) 063709
[10] K. Umeo, T. Ohsuka, Y. Muro, J. Kajino and T. Takabatake: J. Phys. Soc. Jpn. **80** (2011) 064709.
[11] Y. Kawamura, S. Edamoto, T. Takesaka, T. Nishioka, H. Kato, M. Matsumura, Y. Tokunaga, S. Kambe and H. Yasuoka: J. Phys. Soc. Jpn. **79** (2010) 103701
[12] Y. Kawamura, Y. Ogane, T. Nishioka, H. Kato, M. Matsumura, D. Tanaka, H. Tanida, M. Sera, A. Kondo K. Matsubayashi and Y. Uwatoko: J. Phys. Soc. Jpn **80** (2011) SA046

Journal of Physics: Conference Series **391** (2012) 012029 doi:10.1088/1742-6596/391/1/012029

Magnetic phase diagram and crystalline electric field of $NdRu_2Al_{10}$ single crystal

Yukihiro Kawamura[1,2,3], Sakiyo Tanimoto[1], Takashi Nishioka[1], Hiroshi Tanida[4], Masafumi Sera[4], Kazuyuki Matsubayashi[2], Yoshiya Uwatoko[2], Akihiro Kondo[2], Koichi Kindo[2], Chihiro Sekine[3]

[1]Graduate School of Integrated Arts and Sciences, Kochi University, Kochi 780-8520, Japan
[2]Institute for Solid State Physics, University of Tokyo, Kashiwa 305-0047, Japan
[3]Muroran Institute of Technology, Muroran, Hokkaido 050-8585, Japan
[4]Department of Quantum Matter, AdSM, Hiroshima University, Higashihiroshima, Hiroshima 739-8530, Japan

E-mail: y_kawamura@mmm.muroran-it.ac.jp

Abstract.
We measured specific heat of $NdRu_2Al_{10}$ single crystal under magnetic field along a-axis in order to investigate basic physical properties of this compound. We found three anomalies at $T_1 \sim 2.4$ K, $T_2 \sim 1.2$ K and $T_3 \sim 0.9$ K in the temperature dependence of specific heat. These anomalies are attributed to antiferromagnetic order. In addition, we observed a Schottky-type anomaly, which reflects splitting of crystalline electric field (CEF). The entropy reaches Rln2 just above T_1 and almost reaches Rln10 at room temperature. These results indicate that these three phase transitions occur in a doublet ground state and that all CEF levels of J = 9/2 are below 300K. The results of specific heat are consistent with the CEF level scheme proposed by magnetization measurement.

1. Introduction

$NdRu_2Al_{10}$ is a new ternary intermetallic compound, which crystallizes in the orthorhombic $YbFe_2Al_{10}$ ($Cmcm$) crystal structure[1]. $NdRu_2Al_{10}$ is studied as a reference compound of $CeRu_2Al_{10}$. $CeRu_2Al_{10}$ has been extensively studied recently because of its exceptionally high phase transition temperature ($T_0 \sim 27$ K)[2-16]. Taking into account the de Gennes factor for isostructural compound $GdRu_2Al_{10}$ which shows the antiferromagnetic order ($T_N \sim 17$ K), T_0 is too high for Ce compound[2]. Since the neutron-scattering study of $CeRu_2Al_{10}$ revealed that long-range magnetic order exists below T_0[3, 4], the phase transition at T_0 on $CeRu_2Al_{10}$ is considered to be antiferromagnetic order. However, the cause of the exceptionally high transition temperature has not beeen clarified yet and another mechanism besides antiferromagnetic order seems to occur at T_0. Another feature of $CeRu_2Al_{10}$ that grabs our attention is its crystalline electric field (CEF) levels. The calculated first excited doublet of CEF on $CeRu_2Al_{10}$ is as high as 500 K[5]. This temperature seems too high for 4f electrons to be localized. One of the methods to clarify these curious behaviors of $CeRu_2Al_{10}$ is comparing with other isostructural compounds. That is why we have studied $NdRu_2Al_{10}$. The ultrasonic measurements of $NdRu_2Al_{10}$ revealed a hardening at $T_1 \sim 2.4$ K followed by a softening at $T_2 \sim 1.3$ K and $T_3 \sim 1.0$ K [17]. These anomalies are thought to be antiferromagnetic transitions. Unlike T_0 of $CeRu_2Al_{10}$, T_N of

$NdRu_2Al_{10}$ follows the de Gennes factor by taking into account the T_N of $GdRu_2Al_{10}$. In the magnetization measurements of $NdRu_2Al_{10}$, an obvious magnetic anisotropy in the paramagnetic phase attributed to CEF is observed[18]. In order to investigate $CeRu_2Al_{10}$, the origin of T_1, T_2 and T_3 as well as CEF level scheme on $NdRu_2Al_{10}$ must be clarified. In this paper, we report the results of specific heat on $NdRu_2Al_{10}$ under magnetic field.

2. Experimental

Single crystal of $NdRu_2Al_{10}$ was grown by using Al self-flux method. The starting composition of $NdRu_2Al_{60}$ was placed in an Al_2O_3 crucible, sealed in a quartz ampule in vacuum. We heat it up to 1050 °C and then cooled it down to 900 °C at a rate of 1 °C/h, followed by 5 °C/h down to 680 °C. The Al flux was removed by centrifugation. The remaining flux was removed by submerging the sample in 5 mol/L NaOH solution for 24 hours. The resulting single crystals were confirmed as single phase through powder X-ray diffraction analysis. Specific heat under magnetic field is measured with relaxation method by using PPMS systems (Quantum Design Co., Ltd.). The magnetic field is applied along a-axis at the range between 0 T and 7 T.

3. Results and Discussion

Figure 1 (a) shows the temperature dependence of specific heat devided by temperature C/T (T) under the magnetic field of 0 T $\sim$ 2 T. At 0 T, a distinct peak is observed at $T_1 \sim 2.4$ K, followed by anomalies at $T_2 \sim 1.2$ K and $T_3 \sim 0.9$ K. The temperature of the peak at T_1 decreases with the increase of the applied magnetic field, while the anomalies at T_2 and T_3 are almost independent to the magnetic field. Although the anomaly at T_2 is obscure at 0 T, it enhances and becomes distinct at 1 T, which suggests that magnetic structure is aligned at T_2. Figure 1 (b) shows C/T (T) under the field of 3 T $\sim$ 7 T. The peaks at T_1 and T_2 merged at 3 T and only one distinct peak was observed, while anomaly at T_3 was too obscure to be observed. At 5 T and 7 T, no distinct peak was observed, indicating the disappearance of the antiferromagnetic transition. A broad hump is observed at 5 T and 7 T and the temperature of the peak increases along with the increase of magnetic field. Thus, the entropy used by phase transition is thought to be used at high temperatures after disappearing phase transition.

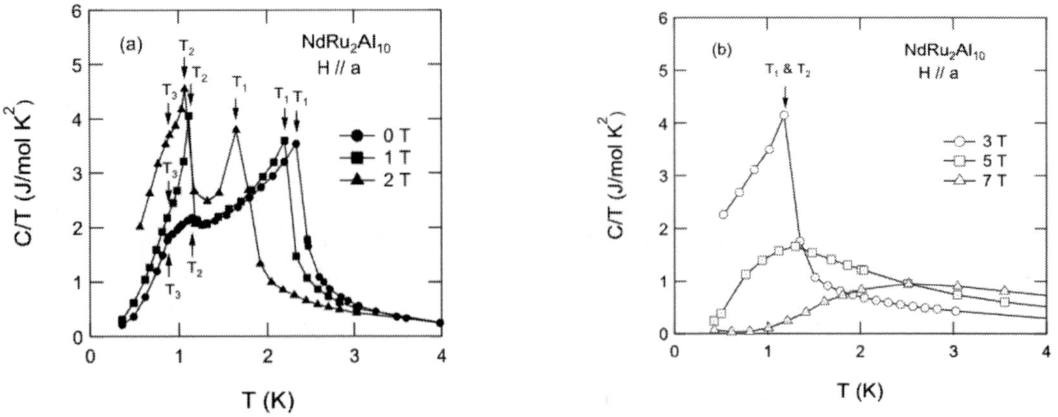

Figure 1. (a) C/T on $NdRu_2Al_{10}$ under the field of 0 T(closed circle), 1 T(closed square), 2 T(closed triangle) (b) C/T on $NdRu_2Al_{10}$ under the field of 3 T(open circle), 5 T(open square) and 7 T(open triangle).

Figure 2 (a) shows the temperature dependence of C_{4f}/T on $NdRu_2Al_{10}$ at 0 T, where C_{4f} indicates the contribution of 4f electrons to the specific heat on $NdRu_2Al_{10}$. C_{4f} is obtained from assuming $LaRu_2Al_{10}$ to the phonon contribution of $NdRu_2Al_{10}$. Schottky anomaly, which is attributed to CEF splitting, was observed around 35 K. The inset of the figure 2 (a) shows the schematic view of the CEF splitting, which can be predicted by magnetization measurement[18]. The schottky anomaly is consistent with this proposed level scheme. Unlike CEF levels of $CeRu_2Al_{10}$, CEF levels of $NdRu_2Al_{10}$ are not so high that it can be regarded as ordinary localized electron system.

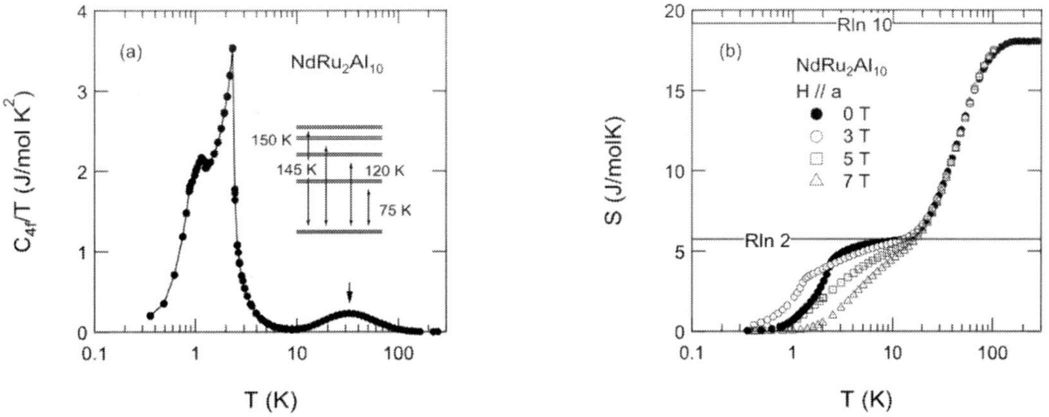

Figure 2. (a)Temperature dependence of the 4f contribution to the Specific heat C_{4f} on $NdRu_2Al_{10}$ (see text in detail). The inset shows CEF level scheme predicted by magnetization measurements. (b)Temperature dependence of entropy at the field of 0 T (closed circle), 3 T (open circle), 5 T (open square), 7 T (open triangle).

Figure 2 (b) shows the temperature dependence of entropy (S). At 0 T, the entropy reaches the value of Rln2 just above T_1 ($\sim$ 2.4 K) and reaches almost Rln10 at room temperature. The results indicate that these three phase transitions occur in a doublet ground state and all CEF level of J = 9/2 are below 300K. The value at room temperature barely reached Rln10. The reason is probably due to the difference of the phonon contribution between $NdRu_2Al_{10}$ and $LaRu_2Al_{10}$. The phonon contribution of the specific heat of $NdRu_2Al_{10}$ is not exactly the same as that of $LaRu_2Al_{10}$, even though we take $LaRu_2Al_{10}$ as the phonon contribution of $NdRu_2Al_{10}$. The value of entropy above 30 K is independent to the magnetic field. This result indicates that the entropy is conserved even at the magnetic field where the antiferromagnetic transition disappears.

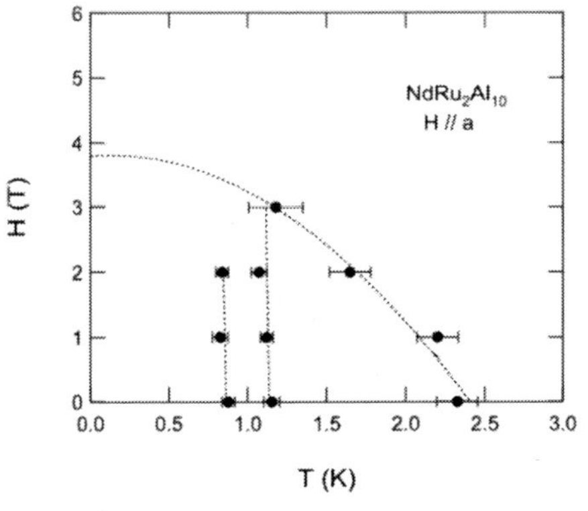

Figure 3. H-T phase diagram of $NdRu_2Al_{10}$ for H//a.

Figure 3 shows the magnetic phase diagram of $NdRu_2Al_{10}$. Up to 2 T, T_1 decreases with the increase of magnetic field, while T_2 and T_3 are almost independent to the magnetic field. At 3 T, only one phase transition is observed. T_1 and T_2 seems to be merged and an anomaly at T_3 is invisible. All phase transition disappears at 5 T. This phase diagram is consistent with the ultrasonic measurements[17].

4. Summary

We measured the specific heat of $NdRu_2Al_{10}$ single crystal along a-axis under magnetic field. In the temperature dependence of specific heat, a distinct anomaly is observed at $T_1 \sim 2.4$ K, followed by anomalies at $T_2 \sim 1.2$ K and $T_3 \sim 0.9$ K corresponding to the antiferromagnetic transition. In addition, a Schottky-type anomaly is observed around 45 K in the temperature dependence of 4f contribution of the specific heat, which is reflected through the splitting of crystalline electrical field. The entropy reaches Rln2 just above T_1 and almost reaches Rln10 at room temperature. These results indicate that these three phase transitions occur in a doublet ground state and all CEF level of $J = 9/2$ are below 300K. The result of specific heat is consistent with the CEF level scheme proposed by magnetization measurement. We obtained the magnetic phase diagram of $NdRu_2Al_{10}$ for H // a.

Acknowledgment

The authors gratefully acknowledge the support from Muroran Institute of Technology (a grant for the Kidorui (Rare Earth) Program).

References

[1] V. M. T. Thiede, T. Ebel, W. Jeitschoko: J. Mater. Chem. **8** (1998) 125.
[2] T. Nishioka, Y. Kawamura, T. Takesaka, R. Kobayashi, H. Kato, M. Matsumura, K. Kodama, K. Matsubayashi and Y. Uwatoko: J. Phys. Soc. Jpn. **78** (2009) 123705.
[3] D.D. Khalyavin, A. D. Hillier, D. T. Adroja, A. M. Strydom, P. Manuel, L. C. Chapon, P. Peratheepan, K. Knight, P. Deen, C. Ritter, Y. Muro and T. Takabatake: Phys. Rev. B **82** (2010) 100405
[4] J. Robert, J. M. Mignot, G Andre, T. Nishioka, R. Kobayashi, M. Matsumura, H. Tanida, D. Tanaka and M. Sera: Phys. Rev. B **82** (2010) 100404
[5] K. Hanzawa: J. Phys. Soc. Jpn **80** (2011) 023707
[6] A. M. Strydom: Physica B **404** (2009) 2981.
[7] Y. Muro, K. Motoya, Y. Saiga and T. Takabatake: J. Phys. : Conf. Seri. **200** (2010) 012136
[8] M. Matsumura, Y. Kawamura, S. Edamoto, T. Takesaka, H. Kato, T. Nishioka, Y. Tokunaga, S. Kambe and H. Yasuoka: J. Phys. Soc. Jpn. **78** (2009) 123713.
[9] H. Tanida, D. Tanaka, M. Sera, C. Moriyoshi, Y. Kuroiwa, T. Takesaka, T. Nishioka, H. Kato and M. Matsumura: J. Phys. Soc. Jpn. **79** (2010) 043708.
[10] K. Hanzawa: J. Phys. Soc. Jpn. **79** (2010) 043710.
[11] I. Ishii, Y. Suetomi, T. K. Fujita, T. Takesaka, T. Nishioka and T. Suzuki: J. Phys. Soc. Jpn. **79** (2010) 053602.
[12] S. Kambe et al.,: J. Phys. Soc. Jpn. **79** (2010) 053708.
[13] H. Tanida, D. Tanaka, M. Sera, C. Moriyoshi, Y. Kuroiwa, T. Takesaka, T. Nishioka, H. Kato and M. Matsumura: J. Phys. Soc. Jpn. **79** (2010) 063709
[14] A. Kondo, J. Wang, K. Kindo, T. Takesaka, Y. Kawamura, T. Nishioka, D. Tanaka, H. Tanida and M. Sera: J. Phys. Soc. Jpn. **79** (2010) 073709
[15] H. Tanida, D. Tanaka, M. Sera, C. Moriyoshi, Y. Kuroiwa, T. Takesaka, T. Nishioka, H. Kato and M. Matsumura: J. Phys. Soc. Jpn. **79** (2010) 083701
[16] K. Hanzawa: J. Phys. Soc. Jpn. **79** (2010) 084704
[17] T. Suzuki, I. Ishii, Y. Suetomi, H. Muneshige, T. K. Fujita, S. Tanimoto and T. Nishioka: SCES2011 Poster/155/Fr
[18] H. Tanida, D. Tanaka, M. Sera, S. Tanimoto, T. Nishioka, M. Matsumura, M. Ogawa, C. Moriyoshi, Y. Kuroiwa, J. E. Kim, N. Tsuji, and M. Takata: Phys. Rev. B **84** (2011) 115128

Journal of Physics: Conference Series **391** (2012) 012030 doi:10.1088/1742-6596/391/1/012030

Anisotropic gap formation in CeM_2Al$_{10}$ (M = Ru, Os)

Shin-ichi Kimura[1,2], Takuya Iizuka[2], Yuji Muro[3], Junpei Kajino[3], and Toshiro Takabatake[3,4]

[1] UVSOR Facility, Institute for Molecular Science, Okazaki 444-8585, Japan
[2] School of Physical Sciences, The Graduate University for Advanced Studies (SOKENDAI), Okazaki 444-8585, Japan
[3] Department of Quantum Matter, ADSM, Hiroshima University, Higashi-Hiroshima, Hiroshima 739-8530, Japan
[4] Institute for Advanced Materials Research, Hiroshima University, Higashi-Hiroshima, Hiroshima 739-8530, Japan

E-mail: kimura@ims.ac.jp

Abstract. We report an anomalous energy gap formation in CeM_2Al$_{10}$ (M = Ru, Os) by using a polarized optical conductivity [$\sigma(\omega)$] measurement. Near 28 K (= T_0), both compounds undergo an antiferromagnetic transition order, the origin of which is not yet clear. We found that the $\sigma(\omega)$ spectra along the orthorhombic b-axis have characteristic peak structure at the energy gap edge, which implies the existence of the band nesting due to a charge instability. The peak structure does not originate from the conventional Kondo semiconducting property but can be explained by the charge density wave (CDW) fluctuation based on the mean-field theory. This suggests that the opening of energy gap along the b-axis originates from the CDW or spin-density wave formation.

1. Introduction

Cerium-based compounds CeM_2Al$_{10}$ (M = Fe, Ru, Os) with the orthorhombic YbFe$_2$Al$_{10}$-type crystal structure [1] are classified into Kondo semiconductors/insulators (KI) because of the anisotropic electronic structure [2]. CeRu$_2$Al$_{10}$ and CeOs$_2$Al$_{10}$ show anomalous antiferromagnetic transition associated with insulator-to-metal and insulator-to-insulator transitions, respectively, at T_0 = 28 K [3, 4, 5], while CeFe$_2$Al$_{10}$ has no magnetic transition. Because T_0 is higher than the Neel temperature of Nd- and Gd-counterparts, the phase transition is considered to be driven by other mechanism than the Ruderman-Kittel-Kasuya-Yoshida (RKKY) interaction, which gives rise to magnetic transitions in conventional rare-earth compounds [4, 6]. To clarify the origin of the anomalous magnetic transition at T_0, we have investigated the electronic structure by using temperature-dependent polarized optical conductivity [$\sigma(\omega)$] spectra of CeM_2Al$_{10}$ along the three principal axes. As a result, we observed the strong anisotropy in the electronic structure, in spite of the rather isotropic Kondo semiconducting energy gap due to the hybridization between the conduction band and the nearly localized Ce $4f$ state, namely c-f hybridization [2]. In CeRu$_2$Al$_{10}$ and CeOs$_2$Al$_{10}$, there is another gap structure with a peak at the gap edge below the hybridization gap energy along the b-axis [7]. The gap structure and its temperature dependence are inconsistent with those of conventional Kondo semiconductors previously reported but can be explained by the charge-density wave (CDW) fluctuation based on the mean-field theory by Lee, Rice, and Anderson [8].

In this paper, we show the evidences that the antiferromagnetic ordering is induced by the charge instability with charge or spin density wave formation along the *b*-axis.

2. Experimental

Near-normal incident polarized optical reflectivity [$R(\omega)$] spectra were acquired in a very wide photon-energy region of 2 meV – 30 eV to ensure an accurate Kramers-Kronig analysis (KKA). We used two FTIRs in the far-infrared to near-infrared range ($\hbar\omega$ = 2 meV 1.5 eV) and a grating monochromator in the visible to vacuum-ultraviolet range (1.5 30 eV) with synchrotron radiation. In order to obtain $\sigma(\omega)$ via KKA of $R(\omega)$, the spectra were extrapolated below 2 meV with a Hagen-Rubens function, and above 30 eV with a free-electron approximation $R(\omega) \propto \omega^{-4}$ [9]. Single-crystalline samples of CeM_2Al_{10} (M =Fe, Ru, Os) were synthesized by the Al-flux method [5] and was well-polished using 0.3 μm grain-size Al_2O_3 lapping film sheets for the $R(\omega)$ measurements.

3. Result and discussion

Obtained polarized $\sigma(\omega)$ spectra of the three compounds along all principal axes at the temperature of 10 K are shown in Fig. 1. Sharp peaks in the photon energy range of 15 50 meV originate from optical absorptions by TO-phonons, which were discussed elsewhere [10]. Besides these peaks, energy gaps (cross-hatchings) are observed at around 40-55 meV in all materials. In $CeFe_2Al_{10}$, the identical gap sizes of about 55 meV along all principal axes suggest the isotropic c-f hybridization energy, which have already been reported [2]. In $CeRu_2Al_{10}$ and $CeOs_2Al_{10}$, there are shoulder and peak structures at about 40 and 45 meV, respectively, along all axes suggesting the isotropic c-f hybridization gaps.

In $CeFe_2Al_{10}$, there is no peak structure below the energy of the hybridization gap except for sharp peaks of TO-phonons. In both $CeRu_2Al_{10}$ and $CeOs_2Al_{10}$, however, there is a broad peak at around 20 meV (namely "20-meV peak" hereafter) only along the *b*-axis. Such a peak

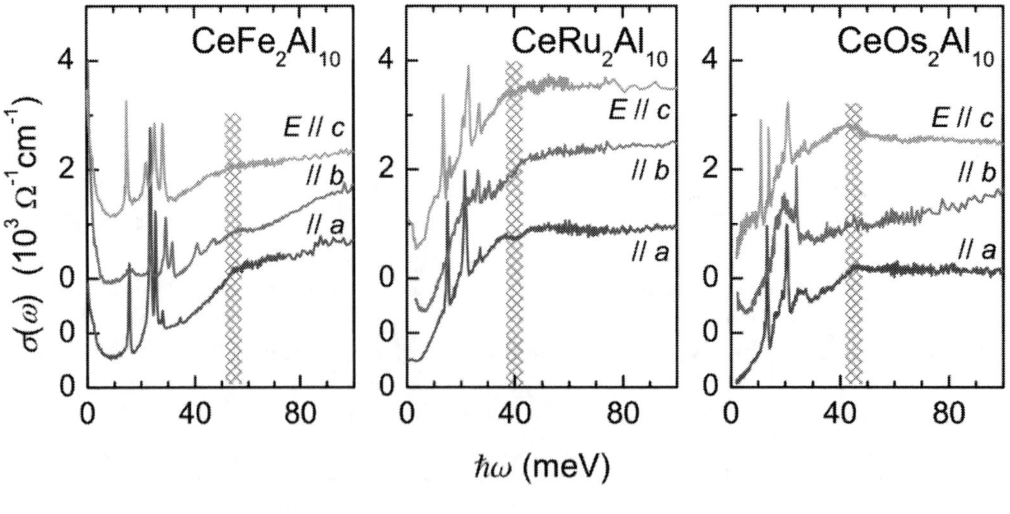

Figure 1. Polarized optical conductivity [$\sigma(\omega)$] spectra of CeM_2Al_{10} (M = Fe, Ru, Os) at 10 K. Sharp peaks in the photon energy range of 10 40 meV are due to TO-phonons. The cross-hatchings indicate the hybridization gap energy between the conduction band and the Ce 4f states. The gap energies are about 55 meV in $CeFe_2Al_{10}$, 40 meV in $CeRu_2Al_{10}$, and 45 meV in $CeOs_2Al_{10}$. Each of the lines is shifted by 1×10^3 Ω^{-1}cm^{-1} for clarity.

has never been observed within the hybridization gap of conventional KIs.

To investigate the evolution of the 20-meV peak along the b-axis, we measured the temperature dependence by a small temperature step. The obtained $\sigma(\omega)$ spectra of $CeRu_2Al_{10}$ and $CeOs_2Al_{10}$ are shown in Fig 2. Some sharp peaks at around 25 30 meV originate from TO-phonons described above. A shoulder at 40 meV in $CeRu_2Al_{10}$ and a small peak at 44 meV in $CeOs_2Al_{10}$ originate from the optical transition across the c-f hybridization gap. The c-f hybridization gaps in both materials still appear at 34 K in $CeRu_2Al_{10}$ and 40 K in $CeOs_2Al_{10}$. In the figures, the tails of the c-f hybridization gap, are drawn by dashed lines. The 20-meV peaks become visible below the temperatures (namely T^*) of 31 ± 1 K in $CeRu_2Al_{10}$ and 36 ± 2 K in $CeOs_2Al_{10}$. This suggests that the 20-meV peak develops after the c-f hybridization gap opened. The shape of the 20-meV peak is akin to that due to the band nesting by a CDW or spin-density wave (SDW) formation in $(TMTSF)_2PF_6$ and so on [11]. T^*s are slightly higher than T_0s (27.3 K in $CeRu_2Al_{10}$ and 28.7 K in $CeOs_2Al_{10}$), and are close to the peak temperatures in the magnetic susceptibility data [4].

These spectral features and the temperature dependence are consistent with the fluctuation of the CDW formation. The CDW fluctuation has been discussed to explain the decrease in the magnetic susceptibility data of other quasi one-dimensional CDW compounds [12, 13] using the mean-field theory [8]. Thereby, the CDW fluctuation occurs at the temperature T_C^{MF} higher than the CDW transition temperature T_C. At T_C^{MF}, the magnetic spin susceptibility starts

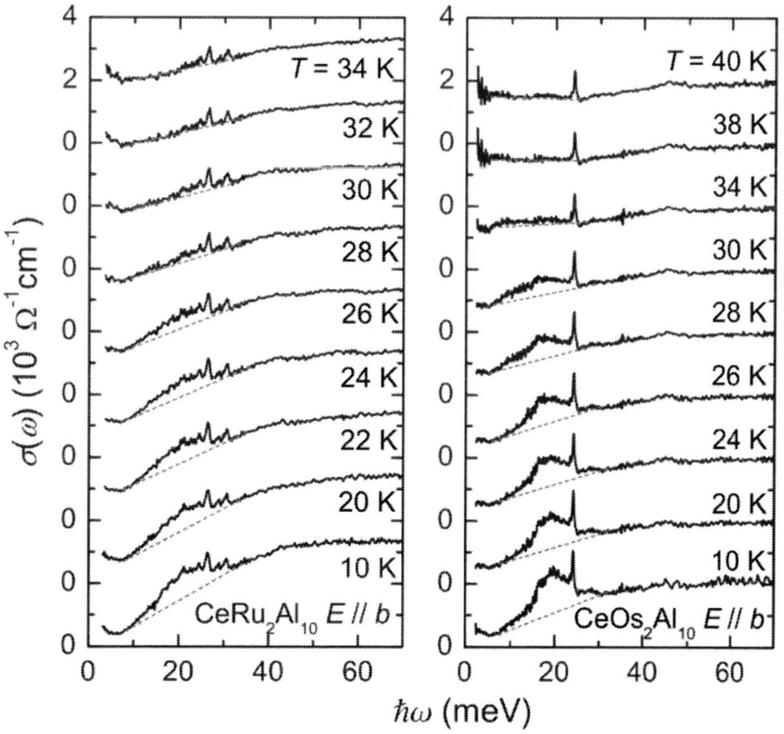

Figure 2. Temperature dependence of optical conductivity $[\sigma(\omega)]$ spectra of $CeRu_2Al_{10}$ and $CeOs_2Al_{10}$ along the b-axis in the photon energy range below 70 meV. The backgrounds due to the c-f hybridization gaps below 40 meV in $CeRu_2Al_{10}$ and 45 meV in $CeOs_2Al_{10}$ are depicted by dashed lines. Sharp peaks at around 25 30 meV are due to TO-phonons. Each of the lines is offset by $2 \times 10^3\ \Omega\ ^1$cm ^{1} for clarity.

to decrease from a constant Pauli spin susceptibility at higher temperature, and the density of states at the Fermi level also decreases indicating the energy gap opening. The relation between T_C^{MF} and T_C can be attributed to a SDW case. If T^* and T_0 are regarded as T_C^{MF} and T_C, respectively, the temperature dependences of both $\sigma(\omega)$ and magnetic susceptibility of $CeRu_2Al_{10}$ and $CeOs_2Al_{10}$ can be explained by the CDW fluctuation theory. This relationship strongly suggests that the energy gaps along the b-axis of $CeRu_2Al_{10}$ and $CeOs_2Al_{10}$ originate from the CDW/SDW formation.

4. Conclusion

To summarize, the mechanism of the energy gap formation of $CeRu_2Al_{10}$ and $CeOs_2Al_{10}$ along the b-axis was studied by using the optical conductivity [$\sigma(\omega)$] spectra. The temperature dependence of the gap formation in $\sigma(\omega)$ spectra and the magnetic susceptibility data are consistent with the behavior of the CDW fluctuation based on the mean-field theory. This result strongly suggests that the energy gap along the b-axis originates from the CDW/SDW formation.

Acknowledgments

This work was partly supported by the Use-of-of-UVSOR Facility Program (BL7B, 2009) of the Institute for Molecular Science. The work was partly supported by a Grant-in-Aid for Scientific Research from JSPS and MEXT of Japan (Grant No. 22340107, 20102004).

References

[1] Thiede V M T, Ebel T and Jeitschko W 1998 *J. Mater. Chem.* **8** 125.
[2] Kimura S, Muro Y and Takabatake T 2011 *J. Phys. Soc. Jpn.* **80** 033702.
[3] Strydom A M 2009 *Physica B* **404** 2981.
[4] Nishioka T, Kawamura Y, Takesaka T, Kobayashi R, Kato H, Matsumura M, Kodama K, Matsubayashi K and Uwatoko Y 2009 *J. Phys. Soc. Jpn.* **78** 123705.
[5] Muro Y, Kajino J, Umeo K, Nishimoto K, Tamura R and Takabatake T 2010 *Phys. Rev. B* **81** 214401.
[6] Muro Y, Kajino J, Onimaru T and Takabatake T 2011 *J. Phys. Soc. Jpn.* **80** SA021.
[7] Kimura S, Iizuka T, Miyazaki H, Irizawa A, Muro Y and Takabatake T 2011 *Phys. Rev. Lett.* **106** 056404.
[8] Lee P A, Rice T M and Anderson P W 1973 *Phys. Rev. Lett.* **31** 462.
[9] Dressel M and Grüner G 2002 *Electrodynamics of Solids* (Cambridge University Press, Cambridge, UK).
[10] Kimura S *et al.* unpublished.
[11] Dressel M, Degiorgi L, Brickmann J, Schwartz A and Grüner G 1997 *Physica B* **230-232** 1008.
[12] Johnston D C, Maki M and Grüner G 1985 *Solid State Commun.* **53** 5.
[13] Grüner G 1988 *Rev. Mod. Phys.* **60** 1129.

Searching for the in-plane anisotropy of the specific heat of UPt_3 in rotating fields

S. Kittaka[a], K. An[a], T. Sakakibara[a], Y. Haga[b], E. Yamamoto[b], N. Kimura[c], Y. Ōnuki[b,d], K. Machida[e]

[a]Institute for Solid State Physics, University of Tokyo, Kashiwa 277-8581, Japan
[b]Advanced Science Research Center, Japan Atomic Energy Agency, Tokai, Ibaraki 319-1195, Japan
[c]Department of Physics, Tohoku University, Sendai 980-8578, Japan
[d]Department of Physics, Osaka University, Toyonaka 560-0043, Japan
[e]Department of Physics, Okayama University, Okayama 700-8530, Japan

E-mail: `kittaka@issp.u-tokyo.ac.jp`

Abstract. In order to identify the gap structure of a spin-triplet superconductor UPt_3, we performed field-angle-resolved specific heat (C_ϕ) measurements in a magnetic field rotated around the [0001] axis. Although various theories predict fourfold symmetry of the gap in the superconducting C phase, we observed no angular variation in C_ϕ in a temperature range down to 88 mK (~ $0.17T_c$). This absence of the variation in C_ϕ implies that the field-angle-dependent quasiparticle excitations, reflecting the gap anisotropy, are averaged out in the bulk of UPt_3.

1. Introduction

The heavy-fermion superconductor UPt_3 has attracted great interest because of its unusual superconducting (SC) properties such as the multiple SC phases (Fig. 1(a)) [1]. NMR Knight shift measurements suggested that the spin state of the Cooper pair is triplet in UPt_3 [2] and various experiments indicated the presence of line nodes in the SC gap [1]. On theoretical grounds, various multi-component order parameters have been proposed to explain the multiple SC phases. [1, 3, 4]. For instance, in the E_{2u} (E_{1g}) scenario with the spin-triplet (spin-singlet) pairing, the three distinct SC phases can be explained by introducing a symmetry-breaking field which lifts the degeneracy of the SC order parameters $k_z(k_x^2 - k_y^2)$ and $2k_xk_yk_z$ (k_xk_z and k_yk_z). Accordingly, for the the E_{2u} (E_{1g}) scenario, the gap function has fourfold (twofold) symmetry in the A and C phases with vertical line nodes, which are absent in the B phase whose gap function becomes $|k_z|(k_x^2 + k_y^2)$ ($|k_z|(k_x^2 + k_y^2)^{1/2}$) with axial symmetry. Schematic gap structures in the B and C phases for the E_{2u} model are depicted in Figs. 1(b) and 1(c), respectively. Nevertheless, the origin of the symmetry-breaking field is still controversial. It could possibly be related to short-range antiferromagnetic fluctuations developing below 5 K [5] and/or distortions in the crystal structure [6].

On experimental grounds, however, the SC gap structures of UPt_3 have not been well resolved yet. A recent junction experiment has reported that vertical nodes exist in the A phase at 45 degrees from the *a* axis in the *ab* plane whereas they are absent in the B phase [7]. The gap structure in the C phase has not yet been examined by the directional probe. In the present study, we have performed field-angle-resolved specific heat measurements on UPt_3 in order to explore the gap structure in the low-temperature B and C phases. This technique is a powerful tool to determine nodal directions of

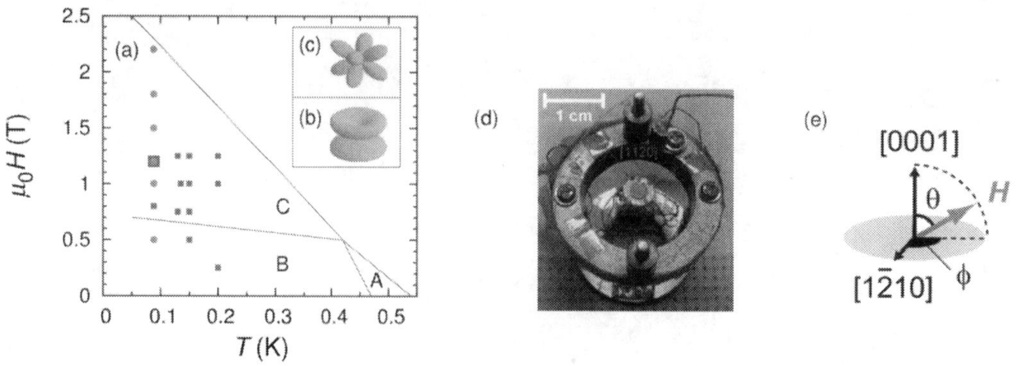

Figure 1. (a) Field-temperature phase diagram of UPt_3 for $H \parallel ab$. Squares (circles) represent the field and temperature at which the ϕ dependence of the specific heat were investigated at $\theta = 90$ (60) deg. Inset: gap structures of (b) $|k_z|(k_x^2 + k_y^2)$ and (c) $|k_x k_y k_z|$ for the B and the C phases, respectively, expected in the E_{2u} model. (d) Picture of the sample on the specific-heat cell used in the present study. (e) Definitions of the azimuthal angle ϕ and the polar angle θ.

bulk superconductors [8], and has recently been applied to $CeCoIn_5$ to conclude the $d_{x^2-y^2}$ symmetry; in a magnetic field H rotated around the [0001] axis, a fourfold oscillation of the specific heat has been observed with maxima occurring along the [100] directions [9] as predicted in the low-T and low-H "Doppler shift" regime [10].

2. Experimental

The single crystal of UPt_3 used in the present study (54.2 mg weight) was grown by the Czochralski pulling method. The sample was shaped into a disk with the flat surfaces parallel to the (0001) plane, as shown in Fig. 1(d). The specific heat was measured by a standard quasi-adiabatic heat-pulse technique as well as a relaxation method in a dilution refrigerator (Oxford Kelvinox AST Minisorb). Magnetic fields were generated in the x-z plane using a vector magnet composed of a vertical solenoid (3 T) and horizontal split-pair coils (5 T). The vector magnetic field was then rotated around the z axis of the specimen by rotating the refrigerator with a stepper motor mounted at the top of a magnet Dewar. For each field angle, the specific-heat value was determined by the average of ten successive measurements. In all the data presented, a nuclear Schottky contribution (C_n) by the Zeeman splitting of ^{195}Pt nuclei ($I = 1/2$) was subtracted. It was confirmed that the addenda contribution is always less than 4% of the sample specific heat and is invariant in rotating fields.

3. Results and Discussion

Figure 2 compares the temperature dependence of C/T measured by the quasi-adiabatic (open circles) and the relaxation (dots) methods. The C/T data measured by the two methods well coincide each other, and are also in agreement with the results of the previous reports [11]. These facts guarantee the high quality of our sample ($T_c = 0.52$ K) and ensure the reliability of our experimental setups. As can be seen in the figure, a rapid increase of C/T was observed below about 0.1 K even in zero field. This Schottky-like behavior was previously reported to originate from an antiferromagnetic long-range ordering at 20 mK [12, 13].

Figure 3 shows the field-strength dependence of C/T at 88 mK for $H \parallel ab$. As shown in the inset of Fig. 3, C/T is linear to the square root of H in the wide field range from 0 T to nearly H_{c2}. This $\sqrt{H}$ behavior is much clearer than the previously reported one obtained at 150 mK [14], and strongly suggests that Doppler-shifted quasi-particle excitations around the line nodes are predominant in the field variation of C/T at 88 mK.

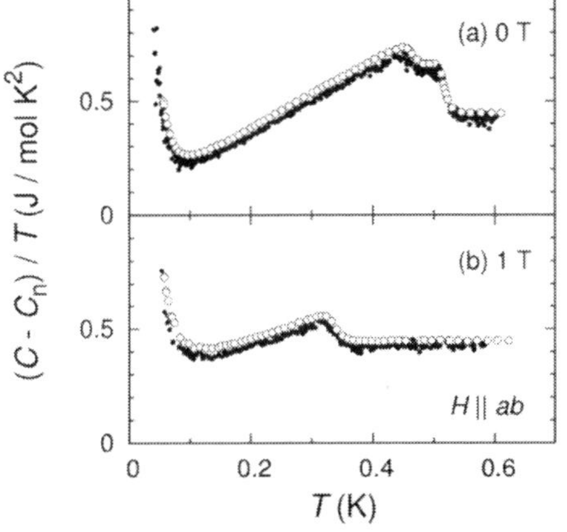

Figure 2. Temperature dependence of the nuclear-subtracted C/T at (a) 0 and (b) 1 T for $H \parallel ab$. Open (closed) symbols are the data measured using the quasi-adiabatic (relaxation) method.

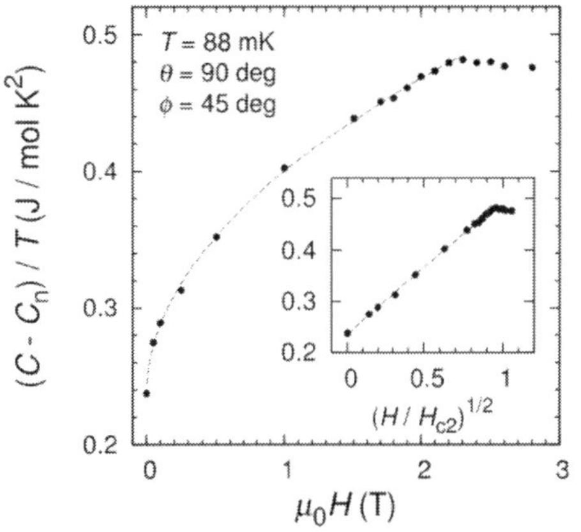

Figure 3. Field dependence of the nuclear-subtracted C/T at 88 mK for $H \parallel ab$. Inset shows the $\sqrt{H}$-dependence. The dashed lines represent fits of the data using the function $a\sqrt{H} + b$.

In Fig. 4, we present the results of the specific heat measurements in magnetic fields rotated around the [0001] axis. The field direction is defined by the angles θ and ϕ, where θ is the polar angle of H with respect to the crystalline [0001] axis and ϕ denotes the azimuthal angle within the (0001) plane measured from the crystalline $[1\bar{2}10]$ axis, as depicted in Fig. 1(e). We investigated the specific heat as a function of ϕ, C_ϕ, under various conditions (T, H, and θ) as summarized in Fig. 1(a) by symbols. It is very important to examine the θ dependence of C_ϕ because the gap functions are considered to possess a horizontal line node. If the order parameter of UPt_3 in the C phase is $k_xk_yk_z$ or k_xk_z, the anisotropy of C_ϕ is expected to become largest at $\theta \sim 60°$ rather than $\theta = 90°$ because the magnetic field is always in the nodal direction for $\theta = 90°$ in the presence of the horizontal line node. Therefore, we investigated C_ϕ with $\theta = 60°$ (conical rotation of H) as well as $\theta = 90°$. Nevertheless, in a wide temperature range between 88 and 200 mK, no angular variation was detected in C_ϕ for both $\theta = 60°$ and $\theta = 90°$ within the limit of our experimental resolution (0.1% of the electronic specific heat) in any field strengths below H_{c2}. This is in striking contrast to the observation of a fourfold oscillation of C_ϕ with an amplitude of larger than 1% in a $d_{x^2-y^2}$-wave superconductor $CeCoIn_5$ [9].

To evaluate the θ dependence of the fourfold oscillation amplitude, we compared $C(\phi = 0)$ with $C(\phi = 45)$ at various θ and H at 88 mK. The difference $\Delta C(\theta) = C(\theta, \phi = 0) - C(\theta, \phi = 45)$ normalized by $C(\theta = 90, \phi = 45)$ is plotted in Fig. 5. For all field strengths studied, $\Delta C(\theta)$ stays nearly zero, indicating that the fourfold oscillation in C_ϕ is absent at any θ.

Considering the results of a number of studies on the gap structure of UPt_3, it is not reasonable to conclude from the present results that the gap of UPt_3 is isotropic around the [0001] axis in the C phase as well as the B phase. As a possible origin of the absence of the anisotropy in C_ϕ, we suggest that the superconducting order parameter forms a domain structure. The order parameters with the fourfold or twofold symmetry around the [0001] axis would have three domain states in the hexagonal crystal structure. If these domains are equally populated, then the in-plane anisotropy of the specific heat might be averaged out in the bulk. It is also possible that the lowest measurement temperature of 88 mK is not sufficiently low enough to detect the gap anisotropy through C_ϕ.

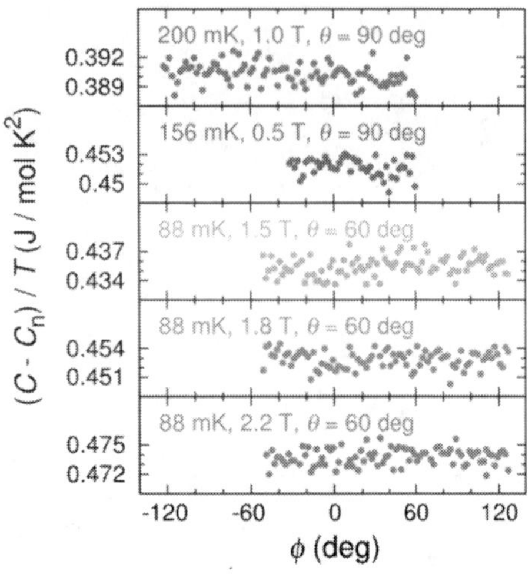

Figure 4. Azimuthal-angle ϕ dependence of the nuclear-subtracted C/T investigated in various conditions.

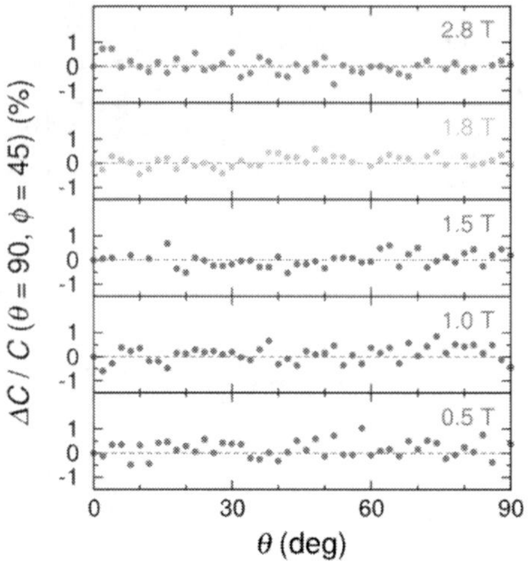

Figure 5. Polar-angle θ dependence of $\Delta C = C(\phi = 0) - C(\phi = 45)$ at 88 mK normalized by $C(\theta = 90, \phi = 45)$ in various magnetic fields.

4. Conclusion

We measured the specific heat of UPt_3 in rotating magnetic fields around the [0001] axis. It was found that the specific heat is invariant in the rotating fields of any strength below H_{c2} at temperatures between 88 and 200 mK. This result indicates that the change in the density of states as a function of in-plane field angle is tiny in the bulk. Although the anisotropy of the gap structure has not been detected, we proposed a possibility that the gap anisotropy is smeared out in the bulk by the presence of domain structures of the symmetry-breaking fields or by thermal excitations of quasi-particles in the present measurement.

Acknowledgement

We thank I. Vekhter, Y. Machida, and K. Izawa for valuable discussions. This work has been partially supported by Grant-in Aid for Scientific Research on Innovative Areas (20102002, 20102007, 23102705) by Grants-in-Aid from the Ministry of Education, Culture, Sports, Science and Technology of Japan, and for Research Activity Start-up (22840013) by Japan Society for the Promotion of Science (JSPS).

References

[1] Joynt R and Taillefer L 2002 *Rev. Mod. Phys.* **74** 235
[2] Tou H, Kitaoka Y, Ishida K, Asayama K, Kimura N, Onuki Y, Yamamoto E, Haga Y and Maezawa K 1998 *Phys. Rev. Lett.* **80** 3129
[3] Sauls J A 1994 *J. Low Temp. Phys.* **95** 153
[4] Machida K, Nishira T and Ohmi T 1999 *J. Phys. Soc. Jpn.* **68** 3364
[5] Aeppli G, Bishop D, Broholm C, Bucher E, Siemensmeyer K, Steiner M and Stüsser N 1989 *Phys. Rev. Lett.* **63** 676
[6] Walko D A, Hong J I, Rao T V C, Wawrzak Z, Seidman D N, Halperin W P and Bedzyk M J 2001 *Phys. Rev. B* **63** 054522
[7] Strand J D, Bahr D J, Harlingen D J V, Davis J P, Gannon W J and Halperin W P 2010 *Science* **328** 1368
[8] Sakakibara T, Yamada A, Custers J, Yano K, Tayama T, Aoki H and Machida K 2007 *J. Phys. Soc. Jpn.* **76** 051004
[9] An K, Sakakibara T, Settai R, Onuki Y, Hiragi M, Ichioka M and Machida K 2010 *Phys. Rev. Lett.* **104** 037002
[10] Vorontsov A and Vekhter I 2006 *Phys. Rev. Lett.* **96** 237001
[11] Brison J P, Keller N, Lejay P, Tholence J L, Huxley A, Bernhoeft N, Buzdin A I, Fak B, Flouquet J, Schmidt L, Stepanov A, Fisher R A, Phillips N and Vettier C 1994 *J. Low Temp. Phys.* **95** 145
[12] Schuberth E A, Strickler B and Andres K 1992 *Phys. Rev. Lett.* **68** 117
[13] Koike Y, Metoki N, Kimura N, Yamamoto E, Haga Y, Onuki Y and Maezawa K 1998 *J. Phys. Soc. Jpn.* **67** 1142
[14] Ramirez A P, Stücheli N and Bucher E 1995 *Phys. Rev. Lett.* **74** 1218

Magnetic anisotropy of $YbNi_4P_2$

Cornelius Krellner[1,2,*] **and Christoph Geibel**[1]

[1] Max Planck Institute for Chemical Physics of Solids, D-01187 Dresden, Germany
[2] Cavendish Laboratory, University of Cambridge, Cambridge CB3 0HE, United Kingdom.

E-mail: *`krellner@cpfs.mpg.de`

Abstract. We report on transport and magnetic measurements between 1.8 and 400 K on single crystalline $YbNi_4P_2$, which was recently reported to be a heavy fermion system with a low lying ferromagnetic transition at $T_C = 0.17$ K, based on data from polycrystals. The tetragonal crystal structure of $YbNi_4P_2$ presents quasi-one-dimensional Yb chains along the c direction. Here we show that at high temperatures, the magnetic anisotropy of $YbNi_4P_2$ is dominated by the crystal electrical field effect with an Ising-type behaviour, which gets more pronounced towards lower temperatures. The electrical resistivity also reflects the strong anisotropy of the crystal structure and favours transport along c, the direction of the Yb chains.

1. Introduction

Within the field of strongly correlated electron systems, the discovery of remarkable phases and transitions is often tightly coupled to the growth and characterization of novel materials. Studying these emergent phenomena necessitate the preparation of high-quality single crystals to investigate the physical properties as a function of crystal orientation and momentum. $YbNi_4P_2$ is such a new stoichiometric compound where we recently observed a very low lying ferromagnetic (FM) transition at $T_C = 0.17$ K [1]. Above this transition distinct deviations from the predictions of Landau-Fermi-liquid theory were observed, indicating the presence of a nearby FM quantum critical point. Among $4f$-based systems this is the clearest experimental example of FM quantum criticality [2], although further investigations are necessary to understand this material in more detail.

A further intriguing property of $YbNi_4P_2$ is the quasi-one-dimensionality, with Yb and Ni-tetrahedra chains along the c direction of the tetragonal crystal structure [3]. These chains are well separated in the ab plane, which is due to the small $c/a \approx 0.5$ ratio. In Fig. 1a the structure is presented using a stereoscopic view along c with the Yb chains located in the channels between chains of edge-connected Ni tetrahedra. Turning the structure by 90° one clearly sees the Yb chains running from bottom to the top in Fig. 1b. The quasi-one-dimensional anisotropy was theoretically confirmed by uncorrelated band-structure calculations, which reveal that the three main Fermi surfaces all have a predominantly one-dimensional character [1].

So far, the anisotropy of the physical properties of $YbNi_4P_2$ have not been investigated, as the previous studies were performed on polycrystalline samples [1, 4]. In this contribution, we report on magnetic and transport measurements on single crystals between 1.8 and 400 K, which reveal a pronounced single-ion anisotropy originating from crystal electric field (CEF) effects.

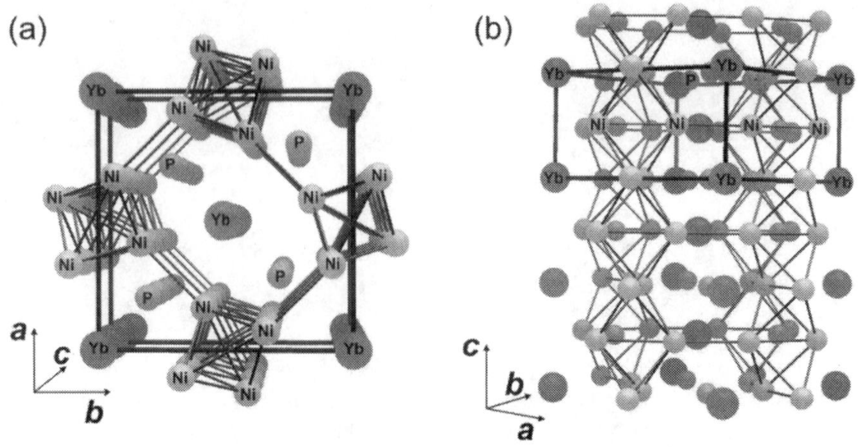

Figure 1. Tetragonal crystal structure of $YbNi_4P_2$. (a) Stereoscopic view along c with the Yb chains located in the channels between chains of edge-connected Ni tetrahedra. (b) Rotated view, with the c direction along the vertical axis, the Yb chains are well separated from each other in the ab plane.

2. Experimental

The single crystals of $YbNi_4P_2$ were grown in a Ni-rich self-flux at 1400°C in a closed Ta container. The excess flux was removed by centrifuging at high temperatures followed by etching in diluted nitric acid. The crystals grow without preferred growth direction and were oriented and subsequently cut using back-scattering Laue and a conventional diamond saw, respectively. Several powder X-ray diffraction patterns of crushed single crystals confirmed the formation of $YbNi_4P_2$ with the $ZrFe_4Si_2$ structure type (space group: $P4_2/mnm$; $Z = 2$). The lattice and structure parameters were found to be in good agreement with the reported structure data [5] and are given in detail in Ref. [1]. In addition, energy dispersive X-ray spectra performed on a scanning electron microscope (Philips XL30) proved the formation of single phase $YbNi_4P_2$. However, the exact stoichiometry could not be refined with the latter method, as the L-line of Yb and the K-line of Ni nearly completely overlap in the energy resolved spectra. The dc-susceptibility and magnetization data were recorded using a Quantum Design (QD) VSM SQUID equipped with a 7 T Magnet. Resistivity measurements were done using standard ac four-probe geometry in a QD PPMS.

3. Results and Discussion

The electrical resistivity $\rho(T)$ shown in Fig. 2 evidences clearly the anisotropy expected from the crystal and electronic structure. Above 100 K, the resistivity along c, the direction of the Yb chains, is a factor of 2 smaller than perpendicular to the chains ($j\|a$). In this temperature range the electrical transport is dominated by phonons, reflected in a linear-in-T dependence for both directions. At 1.8 K, the anisotropy ρ_a/ρ_c amounts already to a factor of 5. Furthermore, the temperature dependences below 50 K are strongly dependent on the direction of current flow. For the easy transport direction ($j\|c$), a pronounced drop is observed below $T = 30$ K, characteristic of coherent Kondo scattering. For $j\|a$, $\rho(T)$ increases below 50 K, presents a well defined maximum at $T = 20$ K and then also strongly decreases towards lower temperatures. Such distinct anisotropic behaviour in this T range is typical for Kondo-lattice systems and results from the single-ion anisotropy induced by the CEF [6].

The anisotropy of the susceptibility, $\chi(T)$, above 1.8 K is similarly dominated by the single-

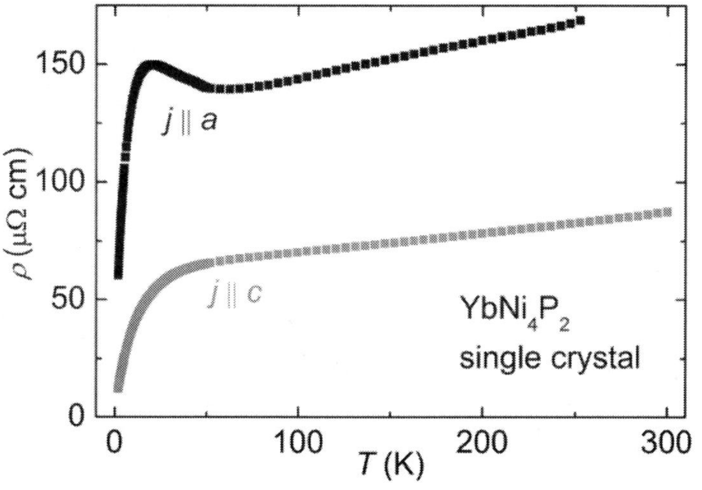

Figure 2. Temperature dependence of the electrical resistivity with the current along the a (black) and the c (red) direction. The resistivity along the Yb chains ($j\|c$) is distinctevely smaller than along a in the whole invetigated temperature range. Both directions display a pronounced drop towards lower temperatures due to the onset of coherent Kondo scattering. A pronounced Kondo maximum at T = 20 K is visible only for $j\|a$.

ion effects, although less pronounced at high temperatures. The inverse susceptibility presents a perfect Curie-Weiss behaviour between 200 and 400 K (blue lines), shown in the inset of Fig. 3a. The effective moments for magnetic field along c, $\mu^c_{eff} = 4.63\,\mu_B$, and perpendicular to c, $\mu^{ab}_{eff} = 4.54\,\mu_B$, match that of the free Yb^{3+} moment ($4.54\,\mu_B$). The anisotropy at high temperatures is reflected in different Weiss temperatures, $\Theta^c_W = -7$ K and $\Theta^{ab}_W = -33$ K, for the applied field parallel to c and perpendicular to c, respectively. Below 50 K, the anisotropy of $\chi(T)$ gets more pronounced with a much stronger increase for $B\|c$ towards lower T and a tendency to saturation for $B \perp c$. The absolute values at $T = 1.8$ K and $B = 0.1$ T are a factor of 5 larger for $B\|c$ than for $B \perp c$, i.e., $\chi^c/\chi^{ab}(0.1\,\mathrm{T}) \approx 5$. From the polycrystalline data it is known, that $\chi(T)$ further increases towards lower temperatures and nearly diverges at T_C [1]. Presently, we cannot state anything about the anisotropy of the inter-ionic magnetic exchange interaction, which most likely will dominate the susceptibility data in small fields towards T_C. Therefore, the anisotropy of $\chi(T)$ needs to be studied on these single crystals down to the mK range in the future.

The magnetization, $M(B)$, as function of applied magnetic field at 1.8 K is presented for both directions in Fig. 3b. For $B\|c$, we observe a strongly curved $M(B)$, in agreement with the strong field dependence of the susceptibility measured on the same sample. At 7 T, $M(B)$ still increases towards higher fields and the polarized moment amounts to $1.33\,\mu_B$. For $B \perp c$, the $M(B)$ curve is nearly linear with a much smaller polarized moment of $0.56\,\mu_B$ at 7 T.

These observations of the magnetic properties above 1.8 K can be well understood in terms of the CEF, which leads to a pronounced single-ion anisotropy with Ising-type behaviour, i.e., the magnetic easy direction is the c axis of the tetragonal unit cell. On the basis of the present results, a complete determination of the CEF parameters cannot be given, which would need further experimental input. However, some constraints of the CEF scheme can be already drawn from the existing data. Specific heat measurements on the polycrystal down to 40 mK revealed the entropy in zero magnetic field [1], from which we can conclude that the CEF ground state of $YbNi_4P_2$ is a doublet, as the entropy at 4 K amounts to $0.5R\ln 2$. Furthermore, we can estimate by means of the minimum in the thermopower at 35 K, the maximum at 20 K in $\rho(T)$ along the a direction and the pronounced deviation from the Curie-Weiss behaviour in the susceptibility below 50 K, that the first excited CEF doublet lies roughly between 10 to 30 K above the CEF ground state. Finally, the fact that we recover the full effective moment in the Curie-Weiss law between 200 and 400 K gives some evidence that the complete splitting of the 4 Kramers

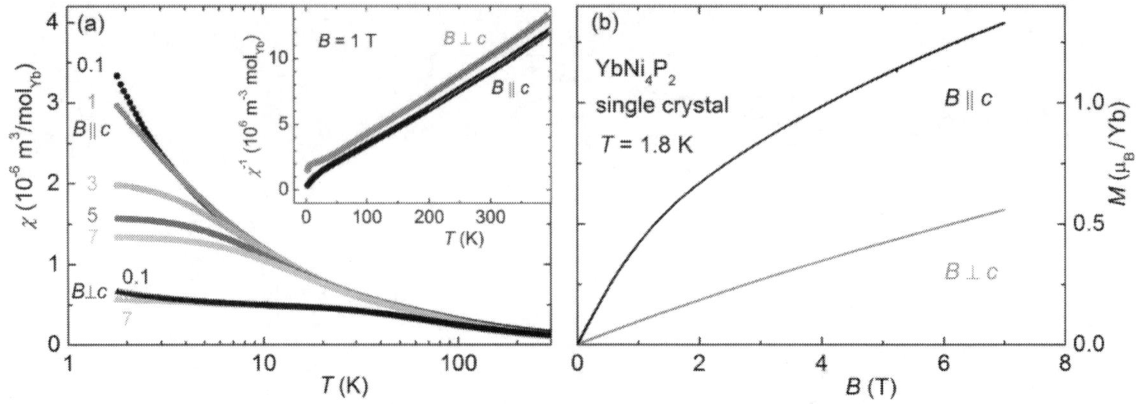

Figure 3. Magnetic anisotropy of $YbNi_4P_2$. (a) $\chi(T)$ for fields parallel (circles) and perpendicular (triangles) to the c direction. In the inset, $\chi^{-1}(T)$ is shown for both directions at $B = 1\,T$. (b) Isothermal magnetization as a function of applied magnetic field at $T = 1.8\,K$.

doublets of the Yb^{3+} is not much larger than 400 K. However, inelastic neutron measurements are underway to determine the CEF scheme more accurately.

4. Conclusion

In summary, we succeeded in growing single crystals of $YbNi_4P_2$ large enough to determine the anisotropy by means of magnetic susceptibility and resistivity measurements. We showed that the anisotropy above 1.8 K is dominated by crystalline electric field effects with an Ising-type single-ion ground state. This is reflected in a much higher polarized moment at 1.8 K for magnetic field along the c direction compared to the in-plane moment. The resistivity presents large anisotropy in the whole investigated temperature range and reveals more efficient transport along the crystallographic c direction, the direction of the quasi-one-dimensional Yb chains. Further investigations to lower temperatures on these single crystals are needed to reveal the anisotropy of the magnetic exchange and the direction of the moment within the ferromagnetic ordered state.

Acknowledgments

The authors thank U. Burkhardt and P. Scheppan for chemical analysis of the samples, as well as C. Bergmann, N. Caroca-Canales and R. Weise for technical assistance in sample preparation. This work is partially supported through DFG Research Unit 960 and the SPP 1458.

References

[1] C Krellner, S Lausberg, A Steppke, M Brando, L Pedrero, H Pfau, S Tencé, H Rosner, F Steglich, and C Geibel. Ferromagnetic Quantum Criticality in the Quasi-One-Dimensional Heavy Fermion Metal $YbNi_4P_2$. *New J. Phys.*, in press, 2011. arXiv:1108.4274.

[2] GR Stewart. Non-Fermi-liquid behavior in *d*- and *f*-electron metals. *Rev. Mod. Phys.*, 73:797–855, 2001.

[3] SI Chikhrij, SV Orishchin, and YB Kuz'ma. New phosphides of rare-earth-metals with the $ZrFe_4Si_2$-type structure. *Dopov. Akad. Nauk Ukr. SSR Ser. A*, 7:81–83, 1986.

[4] S Deputier, O Pena, T Le Bihan, J Y Pivan, and R Guerin. Magnetic properties of the lanthanoid nickel arsenides and phosphides $LnNi_4X_2$ with $ZrFe_4Si_2$-type structure. *Physica B*, 233:26–36, 1997.

[5] YB Kuz'ma, SI Chykhrij, and SL Budnyk. Yb-Ni-P system. *J. Alloy. Compd.*, 298:190–194, 2000.

[6] S Kashiba, S Maekawa, S Takahashi, and M Tachiki. Effect of Crystal Field on Kondo Resistivity in Ce Compounds. *J. Phys. Soc. Jpn.*, 55:1341–1349, 1986.

Journal of Physics: Conference Series **391** (2012) 012033 doi:10.1088/1742-6596/391/1/012033

Strongly correlated electron behavior in single crystalline $U_2Os_3Al_9$

Neeraj Kumar*, **Pranab Kumar Das, Ruta Kulkarni, A. Thamizhavel, S. K. Dhar**

Department of Condensed Matter Physics and Materials Science, Tata Institute of Fundamental Research, Homi Bhabha Road, Colaba, Mumbai 400 005, India.

E-mail: *goyal@tifr.res.in

Abstract. We report the magnetic properties of a single crystal of a new compound $U_2Os_3Al_9$ which crystallizes in the well known $Y_2Co_3Ga_9$ type orthorhombic structure with space group *Cmcm*. The susceptibility of $U_2Os_3Al_9$ shows a peak at 7 K typical of antiferromagnetic ordering. The susceptibility in the paramagnetic state is anisotropic, the easy axis of magnetization lying in the ab-plane of the orthorhombic crystal lattice. The magnetization at 2 K, measured up to a maximum field of 160 kOe, shows a metamagnetic transition near 118 kOe when the field is aligned along [010] in addition to a small metamagnetic transition near 25 kOe. The bulk antiferromagnetic ordering of the uranium ions at T_N = 7 K is confirmed by a peak in the heat capacity with ΔC nearly 7 J/U.mol K. An extrapolation of the heat capacity data from the paramagnetic regime to T = 0 gives an enhanced electronic specific heat coefficient of 120 mJ/U.mol K^2. The electrical resistivity of $U_2Os_3Al_9$ shows a negative temperature coefficient between 300 and T_N which is a signature of spin fluctuations in a narrow band or a Kondo type of interaction. The data thus suggest the presence of strong electron correlations in this compound.

1. Introduction

The study of ternary compounds of the type $R_2T_3X_9$ crystallizing in the orthorhombic $Y_2Co_3Ga_9$-type structure has revealed that Yb, Ce and U based analogues show strongly correlated electron behaviour due to the hybridization of 4f/5f electrons with the itinerant conduction electrons [1, 2, 3, 4]. A host of ground states ranging from Pauli paramagnetic heavy fermion, dense Kondo lattices to magnetically ordered state has been reported for T = Co, Ru, Rh and Ir and X = Al and Ga. Recently we reported magnetic properties of $R_2Ru_3Ga_9$ (R = Ce, U) compounds [5]. $U_2Ru_3Ga_9$ shows a negative temperature coefficient of resistivity at high temperatures below 300 K and orders antiferromagnetically at 16 K. Ce_2Ru_3Ga exhibits valence fluctuating behaviour with a paramagnetic ground state. An extension of our work has led to the discovery of a new iso-structural compound $U_2Os_3Al_9$. In the present work we report the anisotropic magnetic and transport properties of single crystalline $U_2Os_3Al_9$ in the temperature range 1.8-300 K and in applied magnetic fields upto 160 kOe.

2. Experiment

A single crystal of $U_2Os_3Al_9$ was grown in a tetra arc furnace using Czochraski pulling process. About 10 grams of total charge, with at least 3N pure metals, was taken and initially melted a

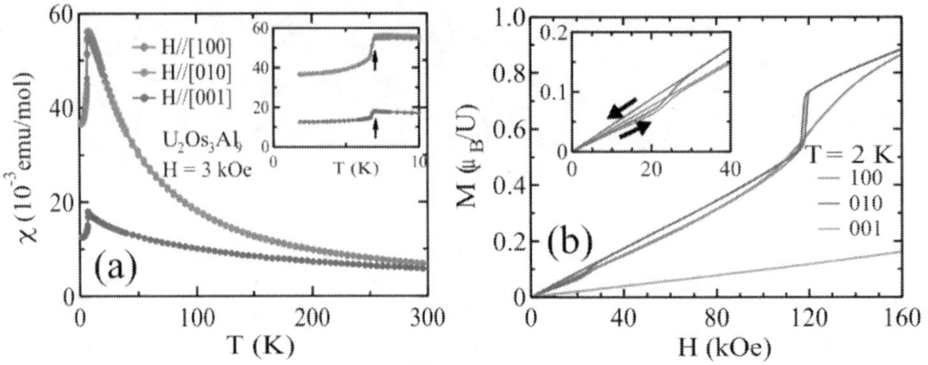

Figure 1. (Color online) (a) Magnetic susceptibility versus temperature from 1.8-300 K is shown in top panel; inset shows the susceptibility below 10 K with a clear transition at 7 K. (b) Magnetization versus field at 2 K up to 160 kOe for field applied along [100], [010], and [001] with field ramped up and down. Inset shows the transition near 25 kOe along [100] and [010] axes

few times to ensure proper homogenization. A crystal was then pulled at a rate of 10 mm/hour. An oriented single crystal was cut from the pulled rod with spark erosion and oriented using Laue diffraction. A powder x-ray diffraction taken on a Phillips Pan-Analytical x-ray diffractometer confirmed the phase purity of the material. The actual crystal composition was confirmed using a CAMECA SX100 Electron Probe Microanalyser (EPMA). The DC magnetic susceptibility and the in-field magnetization measurements were performed in the temperature range 1.8--300 K using a Quantum Design(QD) superconducting quantum interference device (SQUID) magnetometer. The temperature dependence of electrical resistivity in the range 1.8-300 K was measured using a home made DC electrical resistivity set up. The heat capacity was performed using a QD physical properties measurement system (PPMS).

3. Results and Discussion

3.1. Crystal Structure

Powder x-ray diffraction pattern revealed the sample to be single phase with $Y_2Co_3Ga_9$ type crystal symmetry. Lattice parameters obtained from Rietveld analysis using FullProf software [6] package are a = 13.2375 Å, b = 7.6450 Åand c = 9.3992 Å.

3.2. Magnetization

Fig. 1(a) shows the magnetic susceptibility of $U_2Os_3Al_9$ in temperature range 1.8-300 K in a field of 3 kOe along the three principal axes respectively. Susceptibility is nearly isotropic along a and b axes but anisotropy is observed relative to [001] axis. χ_{100}/χ_{001} varies from 1.2 at 300 K to 3.2 at 10 K. Susceptibility at high temperatures varies according to Curie Weiss law $\chi = C/(T - \theta_p)$ and a least squares fitting furnishes μ_{eff} for H parallel to [100], [010] and [001] axes as 2.82 μ_B, 2.96 μ_B and 3.02 μ_B, respectively. The corresponding values of θ_p are -13.6, -19.9 and -148 K. Values of effective magnetic moment are lesser than both U^{4+} (3.58 μ_B) and U^{3+} (3.62 μ_B) moment implying some degree of hybridization between the $5f$ and ligand electronic orbitals. Negative values of θ_p indicate an antiferromagnetic exchange interaction between the uranium ions. Indeed, the susceptibility along each of the three axis shows a sharp peak at 7 K associated with the antiferromagnetic ordering of the uranium ions.

Iso-thermal magnetization versus the applied magnetic field is shown in Fig. 1(b) for field applied along three principal directions. An overall nearly linear magnetization up to about

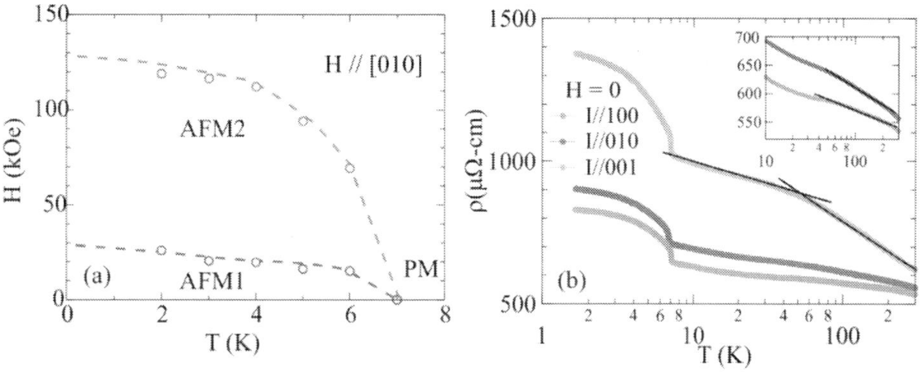

Figure 2. (Color online) (a) Field vs Temperature phase diagram for $H \parallel [010]$ constructed using magnetization data. (b) The variation of resistivity with temperature along the principal axes. Solid lines show linear variation of resistivity with -ln(T).

100 kOe is consistent with the AFM ordering of uranium moments. The magnetization does not saturate in field as high as 160 kOe. The moment at 160 kOe for $H \parallel$ [100],[010], and [001] is 0.77, 0.72, and 0.16 μ_B respectively. As seen in the susceptibility the moment values for $H \parallel$ [100] and [010] axes are higher than for $H \parallel$ [001]. For $H \parallel$ [010] an irreversible jump in the magnetization is observed near 25 kOe in addition to a metamagnetic transition around 118 kOe. Similar behaviour is observed for $H \parallel$ [100] but the transition at 118 kOe is broader. However for $H \parallel$ [001] no metamagnetic transition is observed upto the highest field of 160 kOe, which implies that it is the hard axis of magnetisation. For $H \parallel$ [010] the magnetization was measured at various temperatures (not shown). Metamagnetic transition shifts to lower fields at higher temperatures. A magnetic phase diagram is constructed for $H \parallel$ [010] as shown in Fig. 2(a).

3.3. Electrical Resistivity

Fig. 2(b) shows the temperature dependence of electrical resistivity with current applied along [100], [010] and [001], respectively, from room temperature down to 1.8 K in zero external magnetic field. Some anomalous features of resistivity are noticeable. The magnitude of resistivity at 300 K is high. Although not a common phenomenon, comparable values of resistivity have been previously seen in some uranium systems [7]. The resistivity exhibits a negative temperature coefficient, which is often seen in strongly electron correlated uranium systems. Solid lines in Fig. 2(b) show how the resistivity varies linearly with -ln(T). For $I \parallel$ [001] two such linear regions are observed. Below 7 K the resistivity sharply rises in magnitude which we attribute to the onset of superzone boundaries on the Fermi surface due to the antiferromagnetic ordering at 7 K. The rise in the resistivity below 7 K continues down to the lowest temperature of 1.8 K in our measurement. The negative temperature coefficient of resistivity at high temperatures can tentatively be attributed to the presence of spin fluctuations in a narrow f-band system or to a Kondo-like interaction between the U-5f moments and the itinerant charge carriers.

3.4. Heat Capacity

Fig. 3(a) shows the heat capacity of $U_2Os_3Al_9$ in the range 1.8-30 K and adjusted heat capacity of the reference polycrystalline compound $Lu_2Os_3Al_9$ which is used as a background for the lattice and conduction electron contributions. Bulk nature of the magnetic transition is confirmed by the occurrence of a peak in the heat capacity near 7 K with a peak height ΔC of 7 J/U.mol K.

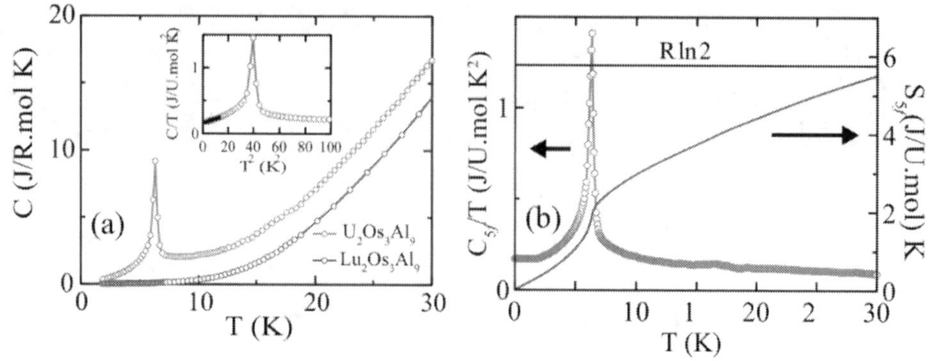

Figure 3. (Color online) (a) Heat capacity vs temperature of $R_2Os_3Al_9$ (R = U, Lu) from 1.8 K to 30 K, Lu heat capacity is adjusted for molecular weight difference. Inset shows C/T vs T^2 of $U_2Os_3Al_9$. (b) $C_{5f/T}$ and $S_{5f/T}$ vs T for $U_2Os_3Al_9$.

The jump is reduced compared to the mean field value of 12.5 J/mol-K for spin S=$\frac{1}{2}$. The $5f$ contribution to the heat capacity, obtained by the standard procedure after adjustment is made for the difference in the atomic weights of U and Lu, is plotted in the Fig. 3(b); the corresponding entropy S_{5f} is also plotted. The entropy at T_N is nearly 0.37Rln2 and attains the value Rln2 (for spin $\frac{1}{2}$ level) at an elevated temperature of nearly 30 K. These two observations indicate a low U-moment state and support the reduced value of μ_{eff} derived from susceptibility. The variation of entropy with temperature nearly 30 K suggest possible Shottaky contribution to the heat capacity from excited crystal field split levels. Data above 30 K are required for verification.

$C_{5f/T}$ vs T is increasing with decreasing temperature seen clearly in Fig. 3(b). Assuming the increase is solely due to an increasing density of states associated with a narrow band, an extrapolation of $C_{5f/T}$ vs. T^2 data, between 30 and 15 K, to 0 K furnishes a γ (C/T as T$\rightarrow$0) of 120 mJ/U.mol.K^2, which suggests a substantial enhancement of the effective mass due to electronic correlations.

4. Conclusion

We have successfully grown a single crystal of orthorhombic $U_2Os_3Al_9$ by Czochraski pulling process. The uranium ions in this compound order antiferromagnetically at $T_N = 7$ K at which a gap opens up at the Fermi surface. The bulk nature of magnetic ordering is confirmed by heat capacity. A negative temperature coefficient of resistivity, magnetic entropy lower than Rln2 up to T_N and a relatively high γ indicate the presence of strong electron correlations in U_2Os_3Al.

References

[1] Trovarelli, C. Geibel, B. Buschinger, R. Borth, S. Mederle, M. Grosche, G. Sparn, and F. Steglich 1999 Phys. Rev. B **60**, 1136
[2] B. Buschinger, C. Geibel, M. Weiden, C. Dietrich, G. Cordier, G. Olesch, J. Khler, and F. Steglich 1997 J. Alloys Compd. **260**, 44
[3] S. K. Dhar, C. Mitra, P. Bonville, M. Rams, K. Królas, C. Godart, E. Alleno, N. Suzuki, K. Miyake, N. Watanabe, Y. Onuki, P. Manfrinetti and A. Palenzona 2001 Phys. Rev. B **64**, 094423
[4] R. Tróc, O. Tougait and H. Noël 2007 Intermetallics **15**, 1091
[5] Neeraj Kumar, K. V. Shah, R. Nagalakshmi and S. K. Dhar 2010 J. Appl. Phys. **107**, 09E113
[6] Rodriguez-Carvajal 1993 Physica B **192**, 55
[7] T. T. M. Palstra, G. J. Nieuwenhuys, R. F. M. Vlastuin, J. van den Berg, J. A. Mydosh and K. H. J. Buschow 1987 J. Mag. Mag. Mat. **67**, 331
[8] H. R. Ott, H. Rudigier, P. Delsing and Z. Fisk 1984 Phys. Rev. Lett. **52**, 15511554

High pressure phase diagram of $CeCoGe_{2.2}Si_{0.8}$

J. Larrea J[1], J. Teyssier[2], H. Ronnow[3], M. Müller [1] and S . Paschen[1]

[1]Institute of Solid State Physics, Vienna University of Technology, Wiedner Haupst. 8 - 10, 1040 Vienna, Austria

[2]Département de Physique de la Matière Condensée, Université de Genève, Quai Ernest-Ansermet 24, 1211 Genève 4, Switzerland

[3]Laboratory for Quantum Magnetism, École Polytechnique Fédérale de Lausanne, 1015 Lausanne, Switzerland

E-mail: `larrea@ifp.tuwien.ac.at`

Abstract. We have investigated the temperature-pressure phase diagram of the heavy fermion compound $CeCoGe_{2.2}Si_{0.8}$ by DC magnetic susceptibility measurements, $\chi_{DC}(T)$, under high pressure. The Néel temperature of $T_N = 4$ K in zero pressure is reduced by pressure up to 3 kbar. At higher pressures antiferromagnetic order appears to gradually transform into a spin glass like-state. Magnetic field decreases both T_N and the spin glass freezing temperature T_f. At 3 T and 6.5 kbar a divergence of $\chi_{DC}(T)$ is observed with a power law that is consistent with a disorder-dominated quantum criticality.

1. Introduction

A challenging problem in the physics of strongly correlated electron systems is to understand the role of disorder, when matter approaches a quantum phase transition. The non-Fermi liquid (NFL) behavior and non-canonical phase diagrams around the quantum critical point (QCP) of a number of U and Ce-based heavy fermion (HF) systems [1, 2, 3] appear to be well described by the *quantum Griffiths* model [4, 5]. It describes the coexistence of a Fermi liquid bath, in which the Kondo interactions dominates and rare magnetically ordered regions, in which the RKKY interaction dominates. To date there are, however, few systematic investigations on the role of disorder on the quantum critical properties at pressure or magnetic field tuned QCPs.

The pseudo-ternary HF compound $CeCoGe_{3-x}Si_x$ [6] with tetragonal $BaNiSn_3$-type structure and antiferromagnetic (AF) order for $x = 0.75$ ($T_N = 6$ K) and 0.9 ($T_N = 3.9$ K) is one of the few examples of heavy fermion compounds that can be driven to a QCP under moderate high pressures (< 10 kbar) [7, 8]. Recent electrical resistivity measurements under pressure (P) [8] for $x = 0.9$ have suggested that the QCP ($P_C = 6.2$ kbar) might be governed by the coexistence or competition between a Griffiths phase and critical spin fluctuations. Here, we report DC magnetic susceptibility measurements under high pressures for $CeCoGe_{2.2}Si_{0.8}$, a material with a Si content similar to that recently investigated by electrical resistivity measurements under pressure [8].

2. Experimental

Polycrystalline $CeCoGe_{2.2}Si_{0.8}$ samples were prepared in two steps as described in Ref.6. Precursors were first melted in an arc furnace and, in the second step, annealed at 900 °C for two weeks in a sealed quartz tube under low pressure of argon. Powder X-ray diffraction, EDX and magnetic susceptibility measurements did not reveal any secondary phases. DC magnetic susceptibility measurements were carried out in a S700X SQUID magnetometer down to 2 K. Hydrostatic pressures up to 10 kbar were applied on a sample with dimension 5.5 mm×1.55 mm×1.35 mm using a CuBe pressure cell (easy lab M10) and Sn as in-situ manometer. The absolute value of the sample magnetic susceptibility was obtained by subtracting the pressure cell contribution at different pressures and magnetic fields from the measured data.

3. Result and discussion

Figure 1(a) shows the temperature variation of the magnetic susceptibility, $\chi_{DC}(T)$, at all investigated pressures. $\chi_{DC}(T)$ at $P = 0$ kbar shows the typical profile of a material that undergoes long range antiferromagnetic ordering at the Néel temperature T_N, defined as the temperature where $\chi_{DC}(T)$ has a maximum. The extracted value of $T_N = 4$ K is in good agreement with values recently derived from specific heat and electrical resistivity measurements on samples of the same batch (not shown). The residual resistivity $\rho_0 = 24$ $\mu\Omega$ cm is comparable to values reported by Eom *et al.* [6], but three times smaller than the value reported for a polycrystalline sample with x=0.9 [8]. This indicates that our sample is of good quality. We would also like to note that ρ_0 of our $CeCoGe_{2.2}Si_{0.8}$ samples is comparable to ρ_0 values of a number of other heavy fermion systems that are not considered to be disorder dominated [3, 9].

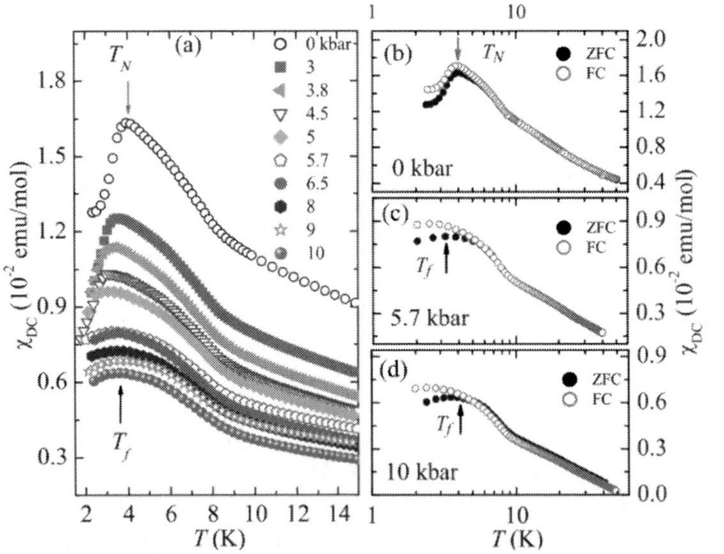

Figure 1. (a) Temperature dependence of the magnetic susceptibility of $CeCoGe_{2.2}Si_{0.8}$ at different selected pressures, measured at zero field cooling (ZFC) conditions with $\mu_0 H = 0.1$ T. We obtained the magnetic susceptibility as $M(H)/H$ where M is the magnetization. (b)-(d) $\chi_{DC}(T)$ at different pressures measured at ZFC and field cooling (FC) conditions with $\mu_0 H = 0.1$ T. The arrows at T_N and T_f are placed at the temperature where a maximum in $\chi_{DC}(T)$ is observed below and above 5 kbar, respectively.

Pressure reduces the magnitude of χ_{DC} and shifts T_N continuously to lower temperatures, at least up to 5 kbar (Fig.1a). Beyond that pressure, the reduction of the magnitude of χ_{DC} is less pronounced and a different profile of $\chi_{DC}(T)$ develops, with a rounded hump centered at T_f. With increasing pressure T_f first increases slightly up to 8 kbar and then becomes constant. This indicates a change in the type of order at a pressure of about 5 kbar. Further evidence for this conjecture comes from zero field cooling (ZFC) and field cooling (FC) $\chi_{DC}(T)$ measurements in an applied field of 0.1 T. While for pressures below 5 kbar a clear maximum is present for both the ZFC and the FC $\chi_{DC}(T)$ curves (Fig. 1 b), for pressures above 5 kbar only the ZFC curve displays a very broad maximum (Fig. 1 c, d). The hysteresis in the former case (Fig. 1 b) is characteristic of antiferromagnets with strong magnetic anisotropy and pronounced domain effects. The hysteresis in the latter case (Fig. 1 c, d) is reminiscent of a spin glass-type behavior

(SG), probably governed by an interplay of AF and ferromagnetic (FM) interactions. This is summarized in the temperature-pressure phase diagram shown in Fig. 2.

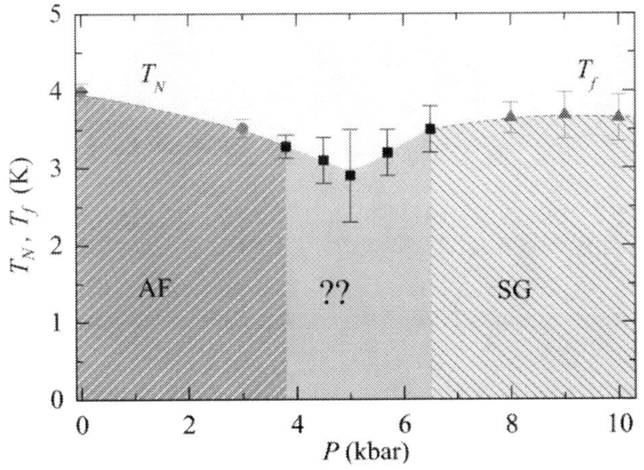

Figure 2. Temperature - pressure phase diagram of $CeCoGe_{2.2}Si_{0.8}$ obtained by magnetic susceptibility experiments (Fig.1). Dashed colored lines separate the antiferromagnetic (AF) and the spin glass-like behavior (SG) regimes. The fully colored area under the square symbols separates the AF and SG regimes and represents an intermediate regime where the magnetic susceptibility show a non-monotonic change with pressure.

A pressure-induced change of the AF ground state may also be inferred from our $\chi_{DC}(T)$ measurements in different magnetic fields (Fig. 3). At zero pressure, $\chi_{DC}(T_N)$ is continuously reduced by the magnetic field (Fig. 3 a) as is T_N (inset of Fig. 3 a). For 6.5 kbar, on the other hand, $\chi_{DC}(T_f)$ at first increases with increasing field (Fig. 3 b). Unfortunately, T_f quickly moves out of the investigated temperature window, calling for measurements below 2 K. At 3 T, $\chi_{DC}(T)$ increases strongly with decreasing temperature and may be approximated by $\chi_{DC} \propto T^{-0.14(2)}$. A power law divergence $\chi \propto T^{-\eta}$ with $0.1 < \eta \leq 0.3$ is expected for a disorder dominated QCP [5].

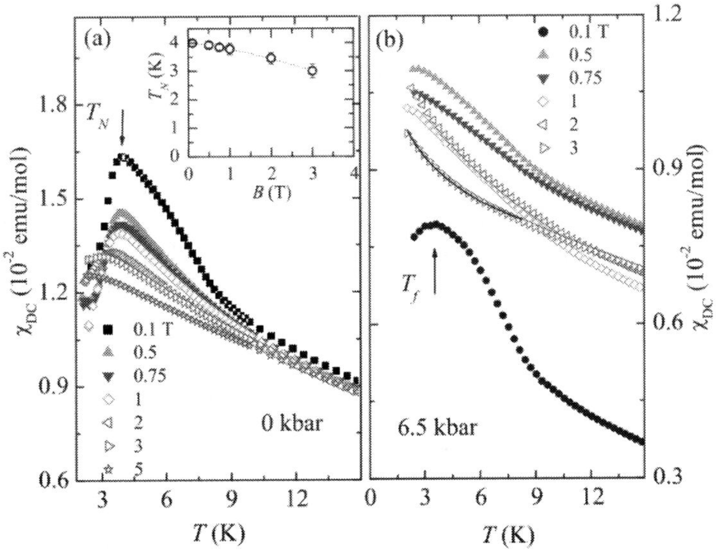

Figure 3. Temperature dependence of the magnetic susceptibility of $CeCoGe_{2.2}Si_{0.8}$ at different magnetic fields for (a) 0 kbar and (b) 6.5 kbar, measured at ZFC conditions with $B = 0.1$ T. The arrow at T_N marks the onset of antiferromagnetic order while the arrow at T_f is the freezing temperature that marks the onset of spin glass-like behavior. The inset shows the magnetic field variation of T_N. The black full line on the $\mu_0 H = 3$ T data represents a fit $\chi_{DC} \propto T^{-0.14(2)}$ below 9 K.

We speculate that the putative SG state induced by pressured above 5 kbar consists of FM domains coupled via short range AF correlations to each other. This scenario is compatible with the formation of a Griffith phase: random substitution of Ge by Si leads to the formation of rare magnetic regions which become critical and play a relevant role when the system approaches the QCP [10].

In conclusion, we have investigated the temperature-pressure phase diagram of the heavy fermion material $CeCoGe_{2.2}Si_{0.8}$ using magnetic susceptibility measurements down to 2 K, at pressures up to 10 kbar. The data suggest that an antiferromagnetically ordered ground state characterized by strong magnetic anisotropy and domain effects is successfully suppressed by pressure and transformed into a state reminiscent of a spin glass. It is noteworthy that for $CeCoGe_{2.1}Si_{0.9}$ a faster suppression of T_N with pressure was observed [8]. From that behavior we would have expected to reach a quantum critical point in $CeCoGe_{2.2}Si_{0.8}$ at a pressure of about 7 kbar. While we could not confirm this, we can also not exclude it since the putative spin glass phase appearing at pressures above 5 kbar might cover an underlying quantum critical point.

Application of magnetic fields suppresses both the antiferromagnetic and the putative spin glass order. Measurements below 2 K are required to further explore field-tuned quantum critical points at different pressures.

4. Acknowledgments

Financial support by the European Research Council (ERC Advanced Grant No 227378) is gratefully acknowledged.

References

[1] Andraka B., Tsvelik A. M. 1991 *Phys. Rev. Lett.* **67** 2886
[2] García Soldevilla J., Gómez Sal C., Blanco J. A., Espeso J. I., Rodríguez Fernández J. 2001 *Phys. Rev.* B **61** 6821
[3] For a review see G. Stewart 2001 *Rev. Mod. Phys.* **73** 797
[4] For a review see T. Vojta 2006 *J. Phys. A: Math. Gen.* **39** R143.
[5] Castro Neto A. H., Castilla G., Jones B. A. 1998 *Phys. Rev. Lett.* **81** 3531.
[6] Eom D., Ishikawa M., Kitagawa J., Takeda N. 1998 *J. Phys. Soc. Jpn.* **67** 2495
[7] Continentino M., Medeiros S. N., Orlando M. T. D., Fontes M. B., Baggio-Saitovitch E. M. 2006 *Phys. Rev.* B **44** 012414
[8] Alzamora M., Fontes M. B., Larrea J., Borges H. A., Baggio-Saitovitch E. M., Medeiros S. N. 2007 *Phys. Rev.* B **76** 125106
[9] Larrea J., Fontes M. B., Baggio-Saitovitch E. M., Eichler A., Abd-Elmeguid M. M., Geibel C., Continentino M. 2007 *J. Phys. Soc. Jpn., Suppl. A* **76** 156
[10] Krishnamurthy V.V., Nagamine K., Watanabe I., Nishiyama K., Ohira S., Ishikawa M., Eom D. H., Ishikawa T., Briere T. M. 2002 *Phys. Rev. Lett.* **88** 046402

Journal of Physics: Conference Series **391** (2012) 012035 doi:10.1088/1742-6596/391/1/012035

Pressure and field tuning in the heavy fermion ferromagnet $CeAgSb_2$

Peter Logg[1], Zhuo Feng[1], Takao Ebihara[2], Swee K. Goh[1], Patricia Alireza[1], F. Malte Grosche[1]

[1] Cavendish Laboratory, University of Cambridge, UK
[2] Department of Physics, Shizuoka University, Shizuoka 422-8529, Japan

E-mail: pwl25@cam.ac.uk

Abstract.
The intermetallic compound $CeAgSb_2$ is an unusual example of a ferromagnetically ordered heavy fermion system. Ferromagnetism sets in below the Curie temperature T_c=9.6 K at ambient pressure. We have investigated the magnetisation of $CeAgSb_2$ under applied hydrostatic pressure of up to 45 kbar. T_c is suppressed rapidly, and at pressures > 35 kbar it is replaced by an unidentified ordered phase, possibly antiferromagnetism. The ordered magnetic moment in $CeAgSb_2$ is aligned along the c-axis. We investigate the effect of transverse field tuning on T_c, and show that magnetic order at low temperature is suppressed by in-plane fields exceeding about 3 T.

1. Introduction

Condensed systems which couple localised and itinerant electronic states have been found to display a range of intriguing ordered states. Typically, magnetically ordered Ce compounds have a tendency to antiferromagnetism at low temperatures. However, due to the oscillatory nature of the RKKY interaction governing the indirect coupling between spins, this is not necessarily the case. The relative scarcity of ferromagnetically ordering heavy fermion compounds make $CeAgSb_2$ an intriguing system to investigate. It crystallises in the $ZrCuSi_2$ type structure, with space group $P4/\mathrm{nmm}$ and two formula units per cell [1]. Above the ordering temperature, $CeAgSb_2$ is a Kondo-lattice compound [2], with inelastic neutron scattering suggesting a Kondo temperature T_K of around 60-80 K [3], agreeing well with estimates made from a crystalline electric field (CEF) analysis [4]. Neutron diffraction [3], magnetisation and electronic transport [4] measurements have unequivocally confirmed the ferromagnetic ground state. The magnetism is strongly anisotropic, with a small ordered moment of around 0.4 μ_B uniaxially aligned along the c-axis. At high temperatures, the Curie-Weiss dependence of the susceptibility would be consistent with a localised character of the $4f^1$ state on the Ce site [4]. A logarithmic increase in the resistivity at around 60 K marks the onset of Kondo scattering [4], followed by a maximum and then sharp downturn, corresponding to the sudden coherence of the scattering and then transition to ferromagnetism. The application of hydrostatic pressure to $CeAgSb_2$ continuously changes the strength of the coupling between the $4f^1$ and conduction electrons. This has the effect of suppressing the transition to long-range magnetic order. At the critical pressure $P_c \sim 35$ kbar, the ferromagnetic ordering temperature extrapolates to zero temperature [4]. The residual resistivity ρ_0 increases suddenly near P_c [5]. There has been no report of superconductivity anywhere on the phase diagram. The exact nature of the high pressure phase remains unclear. Sidorov et al [6] found evidence in high pressure resistivity and ac-calorimetry data that the system is driven antiferromagnetic above 35 kbar, and that the antiferromagnetic phase in turn is fully suppressed at a second quantum critical point by around 50 kbar. Antiferromagnetism is proposed owing to the lattice parameters of $CeAgSb_2$ being similar to the antiferromagnet $PrAgSb_2$ at around this pressure. This has not been independently confirmed.

Field tuning of the ferromagnetic phase via a transverse magnetic field has also been demonstrated through a number of magnetisation and transport measurements [4, 7, 8]. A critical field of around 3 T is sufficient to destroy the magnetic phase, although the character of critical region of the phase diagram has yet to be fully determined.

Here, we report the first detailed high pressure magnetisation measurements on $CeAgSb_2$. Our data confirm that uniaxial ferromagnetism disappears continuously at a critical pressure near 35 kbar and suggest that transverse field tuning offers an interesting alternative route towards a quantum critical point in this material.

2. Experimental Details

Single crystals of $CeAgSb_2$ were grown via a Sb self-flux method, using high purity Ce chips (99.9%), Ag powder (99.99%) and Sb shots (99.99%). Magnetisation measurements were made with a Cryogenics Ltd S700X susceptometer. High pressure measurements were made using a miniature moissanite-anvil SQUID cell, with a sample space of around 0.01 mm^3 [9]. The pressure is determined from the superconducting transition temperature of Sn at low T and from the shift of the ruby fluorescence lines at room T.

3. Results

3.1. Pressure Tuning

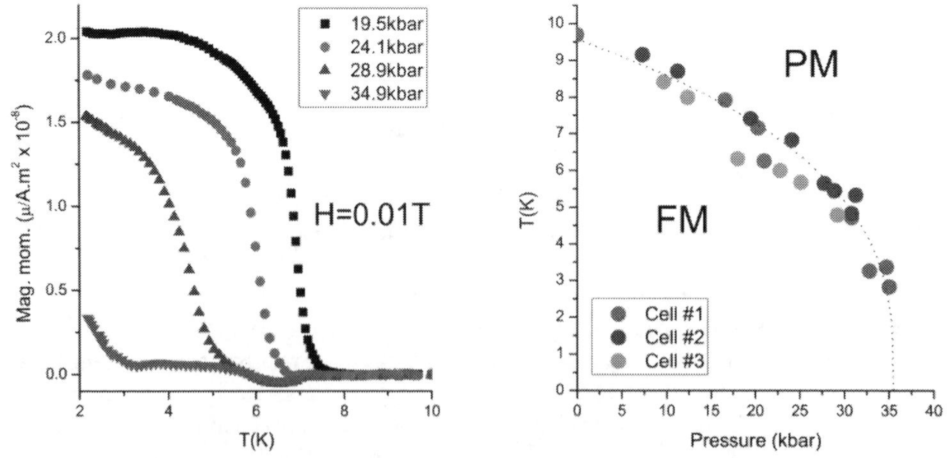

(a) $M(T)$ curves showing the suppression of the ferromagnetic phase with the application of hydrostatic pressure.

(b) P-T phase diagram with pressure points taken from three individual cell runs. Ferromagnetism disappears beyond ~35 kbar.

Figure 1. Pressure tuning data

Figure 1(a) shows the measured temperature dependence of the magnetisation in a small applied field of 0.01 T at a range of pressures. The transition to ferromagnetism is identified as a step-like feature in $M(T)$; at ambient pressure and zero fields, the onset occurs at around 9.6

K. The application of pressure reduces the ordering temperature, as well as the magnitude of the saturated moment. We can resolve the transition down to temperatures of about 3 K. This corresponds to the region of the $P-T$ phase diagram where the FM phase rapidly disappears, on the approach to the critical pressure $P_c \sim 35$ kbar. The $M(T)$ data shown have all been offset and had a linear slope subtracted, fitted to the paramagnetic region in order to eliminate some of the influence from the cell. The feature appearing at around 7 K was due to a contaminant, and disappeared after the cell was cleaned thoroughly.
Three sets of pressure points have been taken, using the same pressure cell and parent sample. Pressure was determined for the second and third cell runs via the ruby fluorescence technique, whereas the superconducting transition of Sn was used for the first. Fig. 1 (b) shows the $P-T$ phase plot, which agrees well with reports elsewhere [5, 6]. For this study, the Curie temperature T_c was extracted from temperature sweeps at a nominally zero field. The dotted line shown in the phase plot is a fit of the Curie temperature to the power law - $T_c \propto |P - P_c|^\beta$ - giving the critical pressure $P_c = 35.5 \pm 0.4$ kbar, and $\beta = 0.33 \pm 0.04$. This exponent is less than the mean-field value $\beta = 0.5$.
At the critical pressure the ferromagnetic order is destroyed. The high pressure, low temperature phase remains ambiguous - the critical pressure could bring about the total screening of the local moment, driving the system to a heavy Fermi-liquid, or it could mark the onset of a different type of magnetic order. High-pressure resistivity measurements by Sidorov et al [6] found that Fermi-liquid ($\rho(T) \sim T^2$) behaviour becomes prominent as the pressure is increased beyond 35 kbar.
Our measurements did not produce any evidence for antiferromagnetic order at $P > P_c$. Antiferromagnetism would be expected to appear as a kink in $M(T)$. However, our high pressure magnetisation measurements are not likely to be accurate enough to resolve the signature of an antiferromagnetic transition.

3.2. *Field Tuning*

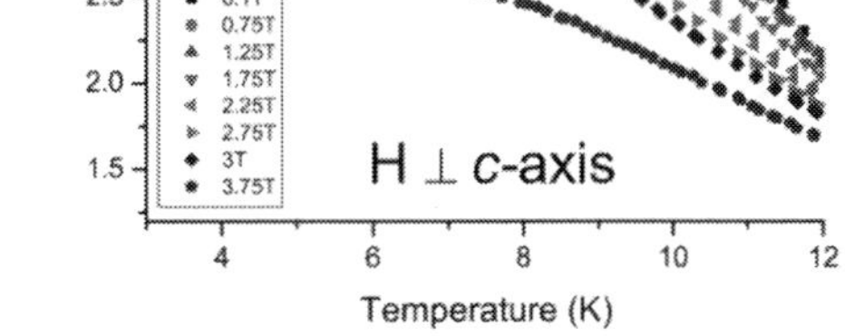

(a) Susceptibility plot showing the suppression of the ferromagnetic phase with a magnetic in-plane tuning field.

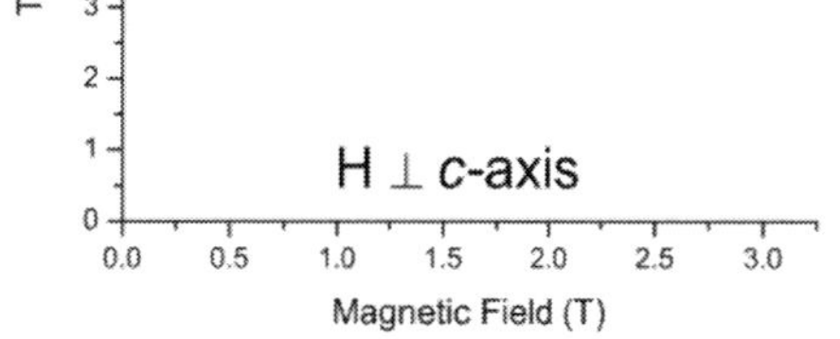

(b) H-T phase diagram with points taken from the location of the kink in $M(T)$. Ferromagnetism is fully suppressed above $\sim 3T$.

Figure 2. Field tuning data

Here, the ferromagnetic state is suppressed via the application of a transverse magnetic field. The evolution of the ordering temperature is tracked through measurement of the in-plane DC susceptibility $\chi^{||ab}$ at various fields. We measure the in-plane $\chi^{||ab}$ response rather than χ^c due to the arrangement of the pick-up system in the SQUID. T_c is identified as a distinct kink-like anomaly in $\chi^{||ab}$. The tuning field has to be perpendicular to the easy axis of magnetisation to allow spontaneous symmetry breaking at T_c. Any components of the tuning field along the c-axis will encourage alignment of the moments in a certain direction, softening the transition point. The sharpness of the kink reflects how transverse our tuning field is.

Figure 2 shows a selection of susceptibility data at various fields, as well as the corresponding $H-T$ phase plot. The effect of field is to reduce the tendency to ferromagnetism. This is seen as a shift in the kink in fig. 2(a) to lower temperatures as the field is increased. The susceptibility anomaly broadens with increasing field, and could no longer be resolved for $\mu_0 H > 2.75$ T.

The observed transverse field tuning phase diagram is consistent with results from pulsed field experiments [7] and resistivity measurements [4]. Our data suggests that ferromagnetic order can be suppressed by transverse fields exceeding the critical field $H_c \sim 3$ T. More detailed studies of the pressure dependence of H_c and of the low temperature magnetic, thermodynamic and transport properties of $CeAgSb_2$ near the field and pressure induced ferromagnetic quantum phase transition are under way.

References

[1] Sologub O, Nol H, Leithe-Jasper A, Rogl P and Bodak O I 1995 *J. Sol. St. Chem.* **115** 441 – 446
[2] Houshiar M, Adroja D T and Rainford B D 1995 *J. Mag. Mag. Mat.* **140-144** 1231 – 1232
[3] Araki S, Metoki N, Thamizhavel A and Onuki Y 2003 *J. Phys. Cond. Mat.* **15** S2179
[4] Jobiliong E, Brooks J S, Choi E S, Lee H and Fisk Z 2005 *Phys. Rev. B* **72** 104428
[5] Nakashima M *et al.* 2003 *J. Phys. Cond. Mat.* **15** L111–L117
[6] Sidorov V A, Bauer E D, Frederick N A, Jeffries J R, Nakatsuji S, Moreno N O, Thompson J D, Maple M B and Fisk Z 2003 *Phys. Rev. B* **67** 224419
[7] Takeuchi T *et al.* 2003 *Phys. Rev. B* **67**(6) 064403
[8] Strydom A, Mhlungu B and Thamizhavel A 2008 *J. Mag. Mag. Mat.* **320** e453 – e456
[9] Alireza P and Lonzarich G G 2009 *Rev. Sci. Instr.* **80** 023906

Low-temperature electrical resistivity of antiferromagnetic $Ce_4Pt_{12}Sn_{25}$

K-A Lorenzer, M Inamdar, L Shafeek, A Prokofiev, and S Paschen

Institute of Solid State Physics, Vienna University of Technology, Wiedner Hauptstrasse 8-10, 1040 Vienna, Austria

E-mail: lorenzer@ifp.tuwien.ac.at

Abstract. The compound $Ce_4Pt_{12}Sn_{25}$ has a rather complex cubic crystal structure in which the Ce atoms are encapsulated in cages of Pt and Sn resulting in a large Ce-Ce distance. Recently, specific heat and magnetic susceptibility measurements revealed a phase transition from a paramagnetic to an antiferromagnetic (AFM) ground state with a low Néel temperature of $T_N = 0.19\,\mathrm{K}$. Application of a magnetic field was shown to weaken the AFM phase and ultimately suppress it at a critical field of 0.7 T. Here we investigate the observed antiferromagnetic phase transition by electrical resistivity measurements in applied magnetic fields in the vicinity of this field-tuned quantum critical point.

1. Introduction

Landau Fermi liquid (LFL) theory successfully describes the low-temperature behaviour of many metals, even of strongly correlated heavy fermion metals. However, at quantum critical points the LFL description has been shown to break down and non-Fermi liquid (NFL) behaviour arises [1].

$Ce_4Pt_{12}Sn_{25}$ represents an interesting example of a strongly correlated cage compound. In the simple cubic crystal structure Ce atoms occupy a unique position inside a cage formed by Sn and Pt atoms (see upper inset of Fig. 1) [2, 3]. Despite the large Ce-Ce next-neighbour distance of 6.14 Å, that might be expected to favour the Kondo interaction over the RKKY interaction, AFM ordering is observed in $Ce_4Pt_{12}Sn_{25}$ with a low Néel temperature $T_N = 0.19\,\mathrm{K}$ [4, 5, 6]. Here we study the electrical resistivity of the material, with focus on the evolution of the AFM transition in magnetic fields.

2. Experimental

Single crystals of $Ce_4Pt_{12}Sn_{25}$ were grown by the self-flux method. Polycrystalline samples of $Ce_4Pt_{12}Sn_{25}$ and $PtSn_4$, produced by repeated melting of elemental Ce (99.95 %), Pt (99.95 %), and Sn (99.9999 %) in a cold copper boat under argon atmosphere, were placed in a high-quality alumina crucible, which was subsequently sealed in a quartz ampoule. After short heat treatment at 900 °C the ampoule was cooled down to 650 °C. At this temperature, the ampoule was centrifuged to separate the single crystals of $Ce_4Pt_{12}Sn_{25}$ from the $PtSn_4$ flux. The phase and stoichiometry of the sample were confirmed by powder X-ray diffraction on a few ground single crystals and by SEM/EDAX analysis on bigger crystals. For the electrical resistivity measurements single crystals were aligned along the three principal crystallographic directions with the help of Laue back reflection patterns and cut/polished into small platelets of typical size 1.5 mm×1.0 mm×0.40 mm. Four gold wires were spot welded on the surface of each sample. One such sample was glued with GE Varnish to the sample platform of an Oxford MX400 dilution refrigerator, close to a field-calibrated ruthenium oxide temperature sensor. To improve the resolution of the measurement, the voltage signal was amplified using low-temperature transformers before detecting it with a Linear Research LR700 resistance bridge. Magnetic field is provided by a 15 T superconducting magnet. For temperatures above 2 K this sample was measured with the AC transport option in a PPMS from Quantum Design. For all resistivity measurements the current j was choosen to be parallel to the [1 0 0] direction. For the magnetoresistance measurements the sample was thinned down to 0.18 μm to gain resolution. The magnetic field H was applied perpendicular to the current $j \perp H$.

3. Zero-field electrical resistivity

Figure 1 shows the temperature dependence of the electrical resistivity, $\rho(T)$, of $Ce_4Pt_{12}Sn_{25}$ (0.04-295 K) in zero field. Metallic behaviour is observed, with a broad hump around 150 K. The hump might be due to two closely spaced crystal electric field (CEF) excitations that were reveiled by specific heat at energies of about 200 K and 250 K [4] and/or due to Kondo scattering. Indeed the non-magnetic reference compound $La_4Pt_{12}Sn_{25}$ does not show this feature [4]. The lack of four-fold point symmetry of the surrounding cage at the Ce site gives rise to a splitting of the six-fold degeneracy into three Kramers doublets with a Γ_7 doublet as the ground state [6].

Between 1.9 K and 4.5 K, ρ is essentially temperature independent. This indicates that both the phonon contribution ρ_{ph} and the resistivity due to spin-disorder and spin-flip scattering ρ_{mag} are negligably small or independent of temperature. Below 1.9 K, a decrease of ρ sets in, with increasing slope $\partial\rho/\partial T$ as T is lowered. At $T_N = 0.19$ K, a sharp kink is seen in $\rho(T)$, signaling the transition into the AFM state (see lower insert of Fig. 1). Below 90 mK, LFL behaviour $\rho(T) = \rho_0 + AT^2$ is observed (Fig. 2, insert). In order to explain the data up to the Néel temperature, we add a term to account for the scattering of electrons on AFM magnons [7]

$$\rho_{SW}(T) = b\Delta^2\sqrt{\frac{k_BT}{\Delta}}e^{-\Delta/k_BT}\left[1 + \frac{2}{3}\left(\frac{k_BT}{\Delta}\right) + \frac{2}{15}\left(\frac{k_BT}{\Delta}\right)^2\right]. \qquad (1)$$

The coefficient b is a material constant that depends on the spin wave stiffness D. The energy gap Δ of the spin wave dispersion is a result of anistropy, either due to an intrinsic material anisotropy or caused by the presence of a symmetry-lowering external influence like an applied magnetic field [7]. The high symmetry of cubic $Ce_4Pt_{12}Sn_{25}$ in zero-field would a priori suggest a vanishing energy gap. However, Lee et al. [4] observed a LFL contribution with $A = 11\,\mu\Omega\text{cmK}^{-2}$ plus a T^5 term. The large exponent of the latter clearly indicates the presence of spin waves with a moderate energy gap of the order of k_BT_N. Our fit parameters are listed in Tab. 1. The A coefficient we find is slightly larger than the value given in Ref. [4], which is explained by a differently choosen spin wave term. Above T_N the LFL contribution following a T^2 dependence is no longer observed. This behaviour is common for magnetically ordered heavy fermion systems near a quantum critical point, see e.g. Ref. [8]. The larger value of Δ compared to T_N can be interpreted as the onset of spin gap opening at higher temperatures.

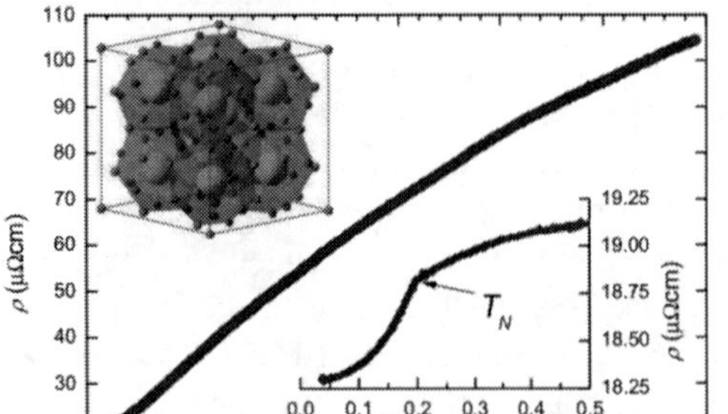

Figure 1. Temperature dependence of the electrical resistivity $\rho(T)$ of $Ce_4Pt_{12}Sn_{25}$. The lower insert displays a close-up of the low-temperature data with the AFM transition indicated by an arrow. The upper insert shows the atomic arrangement in the unit cell of $Ce_4Pt_{12}Sn_{25}$.

Table 1. Fit parameters for the zero-field resistivity below T_N using the expression $\rho(T) = \rho_0 + AT^2 + \rho_{SW}(T)$ with the spin wave contribution according to Eqn. 1.

ρ_0 ($\mu\Omega$cm)	A ($\mu\Omega$cmK^{-2})	b (K^{-2})	Δ (K)
12.79(1)	14.4(5)	35.7(9)	0.54(2)

This gives a natural explanation for the observed long tail in the specific heat [6] as well as the decrease of the resistivity below $T = 1.9$ K. Alternatively, magnetic frustration and Kondo singlet formation were under discussion as possible origins for the latter observation [4].

4. Field tuning of the AFM transition

Application of magnetic fields along the [0 0 1] direction reduces T_N continuously and finally suppresses it altogether at the critical field of $\mu_0 H_N \approx 0.9$ T (Fig. 2). This defines a field-induced quantum critical point (QCP). The electrical resistivity at this field has the non-Fermi liquid form $\Delta\rho \sim T$. Besides the effect of suppression of the AFM state, a positive magntoresistance is observed, seen in Fig. 2 as an upwards shift of the $\rho(T)$ curves with increasing field.

The field dependence of T_N is best tracked by $\partial\rho/\partial T$, where T_N manifests as a maximum (Fig. 3, left panel). With increasing fields the peak position shifts to lower temperatures and broadens. For fields above H_N a much broader peak appears in $\partial\rho/\partial T$. We fitted the data with

$$\rho_{CEF}(T) = \frac{3\pi N m}{\hbar e^2 \epsilon_F} J^2 (g_j - 1)^2 \sum_{i,f} \left\langle m_{s_i}, \Gamma_i \left| \vec{s}\vec{J} \right| m_{s_f}, \Gamma_f \right\rangle^2 \frac{2}{1 + e^{-\frac{E_i - E_f}{k_B T}}} \frac{e^{-\frac{E_i}{k_B T}}}{\sum_j e^{-\frac{E_j}{k_B T}}}, \quad (2)$$

a model that describes the influence of the crystal electric field (CEF) on the spin-disorder scattering of conduction electrons [9]. N is the number of scattering centers, m the effective mass of the charge carriers, ϵ_F the Fermi energy, e the elementary charge, $J^2(g_j - 1)^2$ the de Gennes factor, $|\Gamma\rangle$ and E the crystal field state and its eigenvalue, $|m_s\rangle$ the spin state of the conduction electrons and k_B the Boltzmann constant. An external field lifts the ground state degeneracy by splitting the Γ_7 doublet into two singlets by the Zeeman energy $\pm g_J \mu_B \mu_0 \langle\Gamma_7 | J_z | \Gamma_7\rangle H$. Since the first excited CEF level is about 200 K we can neglect contributions from excited states. The fits give a fairly good description of the

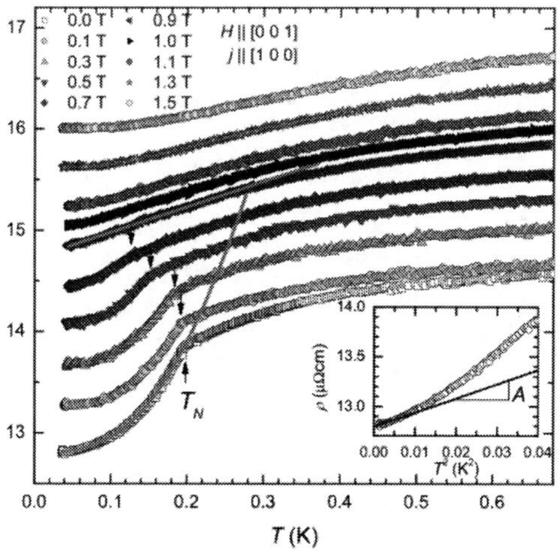

Figure 2. Low-temperature electrical resistivity $\rho(T)$. In zero field the resistivity shows a sharp drop at $T = 0.19$ K which we associate with the Néel temperature T_N. The grey solid line is a fit to the model described in the text. The insert shows the zero-field resistivity plotted vs T^2 and the LFL contribution (dotted line) derived from the model fit. Magnetic fields continuously suppress T_N to zero (black arrows). At the critical field $\mu_0 H_N \approx 0.9$ T non-Fermi liquid behaviour, $\Delta\rho \sim T$ (red solid line) is observed. The upwards shift of the $\rho(T)$ curves with increasing field is due to a sizable positive magnetoresistance.

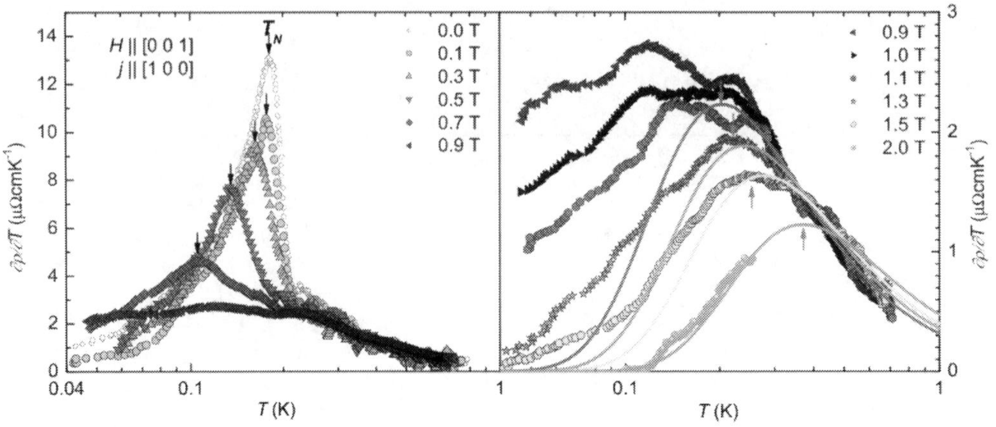

Figure 3. Temperature derivative of the resistivity, $\partial\rho/\partial T$, for the data shown in Fig. 2. At T_N, $\partial\rho/\partial T$ shows a pronounced peak (black arrows). The peak's position is shifted to lower temperatures with increasing field H. Above the critical field H_N a second broad feature can be discerned which is supposed to be due to a Zeeman-split CEF doublet. Solid lines are fits to a CEF model discussed in the text.

data (Fig. 3, right panel). For the lowest temperatures and for fields close to H_N, the CEF model does not reproduce $\partial\rho/\partial T$ accurately. This is not suprising since the model does not include effects due to critical fluctuations nor to a spin gap.

5. Conclusion

We have performed high-resolution low-temperature electrical resisitvity measurements on $Ce_4Pt_{12}Sn_{25}$ in various magnetic fields to elucidate the evolution of its antiferromagnetic state with field. The zero-field resistivity reveals Landau Fermi liquid behaviour and a gapped spin wave contribution with $\Delta = 0.54\,\mathrm{K}$. Above T_N a sublinear temperature dependence is observed up to 1.9 K which might be explained by short-range order fluctuations and/or the opening of the observed spin gap Δ.

Field tuning continuously suppresses the transition temperature T_N to zero at the critical field $\mu_0 H_N \approx 0.9\,\mathrm{T}$ which defines a quantum critical point. At this critical field the electrical resistivity shows non-Fermi liquid behaviour with $\Delta\rho \sim T$. The resistivity data in fields above the critical field are described with a crystal electric field model with a Zeeman split ground state doublet. The agreement between data and model is good for magnetic fields sufficiently away from the quantum critical point.

Acknowledgments

We gratefully acknowledge financial support from the European Research Council/ERC Advanced Grant No 227378 and the TU Vienna Doctoral Program Functional Matter.

References

[1] Löhneysen H et al. 2007 *Rev. Mod. Phys.* **79** 1015
[2] Chafik El Idrissi B et al. 1990 *Mat. Res. Bull.* **25** 807
[3] Shafeek L et al. 2010 *J. Phys.: Conf. Ser.* **200** 012182
[4] Lee H-O et al. 2010 *J. Phys.: Condens. Matter* **22** 065601
[5] Movshovich R et al. 2011 *J. Phys.: Conf. Ser.* **273** 012045
[6] Kurita N et al. 2010 *Phys. Rev.* B **82** 174426
[7] Fontes M B et al. 1999 *Phys. Rev.* B **60** 6781
[8] Gegenwart P et al. 2002 *Phys. Rev. Lett.* **89** 056402
[9] Rao V U S et al. 1970 *Phys. Rev.* B **2** 4613

Electronic properties of Plutonium compounds with the cubic Laves phase structure

Takahiro Maehira[1] and Yasutomi Tatetsu[2]

[1] Faculty of Science, University of the Ryukyus, Nishihara, Okinawa 903-0213

[2] Graduate school of Engineering and Science, University of the Ryukyus, Nishihara, Okinawa 903-0213

E-mail: `maehira@sci.u-ryukyu.ac.jp`

Abstract. Energy band structures of PuX_2 (X=Ru, Rh, and Ir) are investigated by a relativistic linear augmented-plane-wave method with the exchange-correlation potential in a local density approximation. It is found in common that the energy bands in the vicinity of the Fermi level are mainly due to the hybridization between Pu $5f$ and X d electrons.

1. Introduction

It has been one of the important issues in the research field of condensed matter physics to unveil the electronic structure of actinide compounds of which magnetism is one of the fascinating phenomena. Among such compounds, PuX_2 (X=Ru, Rh, and Ir) have been studied intensively both theoretically and experimentally. To understand the electronic properties of these materials, the features of the ground state should be first clarified. The traunsuranium compounds PuX_2 with the $MgCu_2$-type cubic Laves phase structure, where X means Ru, Rh, and Ir in the periodic table, show an interesting trend of $5f$-electron characters. The lattice constant ($PuRu_2$: 7.4740 Å, $PuRh_2$: 7.4880 Å, $PuIr_2$: 7.5310 Å) increases with increasing the atomic number of X (Ru=44, Rh=45, and Ir=77)[1]. The moderately large electron specific heat coefficient $\gamma_{\rm exp}$ is 145 mJ/K^2·mol in $PuRh_2$ [2]. In this paper, we research what the key issues are to construct the energy band structures around the Fermi energy for PuX_2 we attempt to unveil $5f$ electron properties purely originating from actinide atoms.

2. Method of Band Calculation

The calculations for the energy band structures are carried out by using the relativistic linear augmented-plane-wave (RLAPW) method with the exchange and correlation potential in the local density approximation (LDA). All $5f$ electrons are assumed to be itinerant and the calculations are performed in the paramagnetic phase. Note here that relativity should be taken into account, because of large atomic numbers of the constituent atoms. The self-consistent calculation is carried out for the experimental lattice constant.

PuX_2 (X=Ru, Rh, and Ir) materials are categorized into the $MgCu_2$-type structure, which is characterized by the space group Fd3m. The $5f$ electrons in PuX_2 are assumed to be itinerant. The iteration process for solving the Dirac one-electron equation starts with the crystal charge density that is constructed by superposing the relativistic atomic charge densities for neutral atoms Pu([Rn]$5f^5 6d^1 7s^2$), Pu([Kr]$4d^7 5s^1$), Rh([Kr]$4d^8 5s^1$), and Ir([Xe]$5d^7 6s^2$), where [Rn], [Kr],

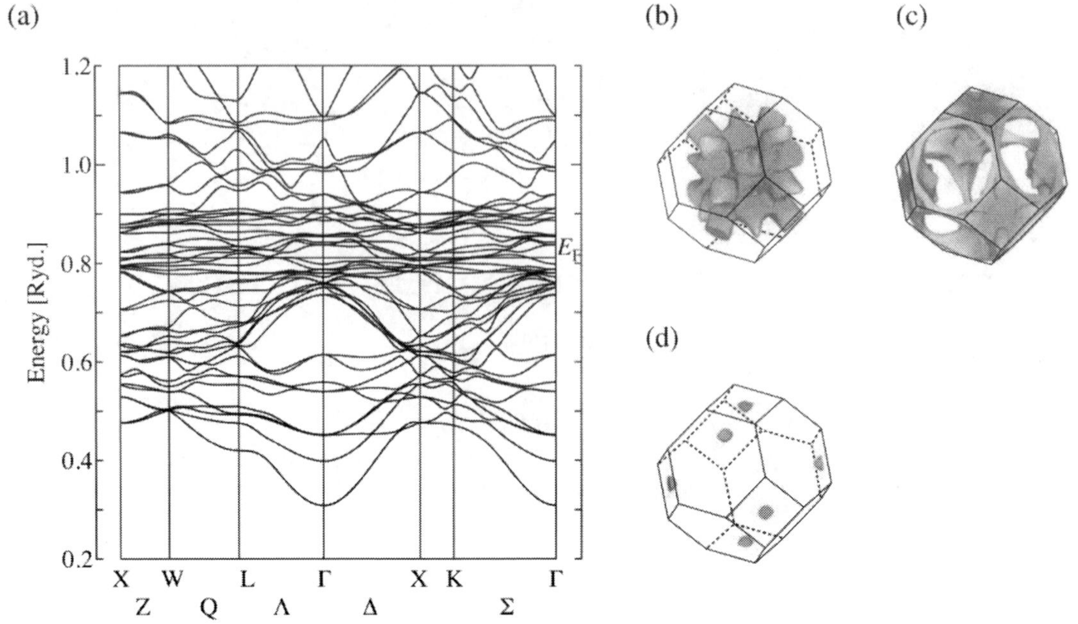

Figure 1. The energy band structure for (a) $PuRu_2$ calculated by using the self-consistent RLAPW method. E_F indicates the position of the Fermi level. (b) The hole-Fermi surfaces in the 24th, (c) electron sheet in the 25th, and (d) the 26th bands for $PuRu_2$. The center point in each Brillouin zone is the Γ point.

and [Xe] symbolically indicate the closed electronic configuration for radon, krypton, and xenon, respectively. In the relativistic atomic calculation, the spin-orbit splitting in Pu $5f$, and Ru $4d$, Rh $4d$, and Ir $5d$ states are found to be 80 mRyd., 24 mRyd., 29 mRyd., and 90 mRyd., respectively. Note here that mRyd. denotes milli-Rydberg and 1 Ryd.=13.6 eV.

3. Results of Band Calculation

First let us discuss the calculated result for $PuRu_2$ as shown in Figure 1(a), in which we depict the energy band structure along the symmetry axes in the Brillouin zone in the energy region from 0.2 Ryd. to 1.2 Ryd.. Note here that the six Pu $6p$ bands in the energy range between −1.2 Ryd. and −0.3 Ryd. are not shown in Figure 1(a), since those bands are irrelevant to the present discussion. The Fermi levels are found to be 0.811 Ryd. for $PuRu_2$. From Figure 1(a), we can observe that fourteen flat bands lie in the narrow energy range just above E_F. These flat bands are identified as the f bands, since they consist dominantly of the Pu $5f$ states. The f-bands apparently separate into two subbands which correspond to the total angular momentum j=5/2 (lower bands) and 7/2 (upper bands). On the other hand, the Bloch states which have the Ru $4d$ components dominantly lie in the wide energy range. A strong hybridization between Pu $5f$ and Ru $4d$ states occurs in the vicinity of E_F, and consequently the Bloch states on the Fermi surface sheet consist mainly of these states. In Figure 1(a), the lowest twenty-three bands are fully occupied. The next three band is partially occupied, while higher bands are empty. Namely, 24th, 25th and 26th bands crossing the Fermi level construct the Fermi surface, as shown in Figure 1(b)-(d). $PuRu_2$ has the even number of electrons per primitive cell, indicating that it becomes an compensated metal where the carrier number of electrons and holes is equal. In fact, the number of carriers contained in these Fermi-surface sheets are 0.322 holes/cell, 0.320

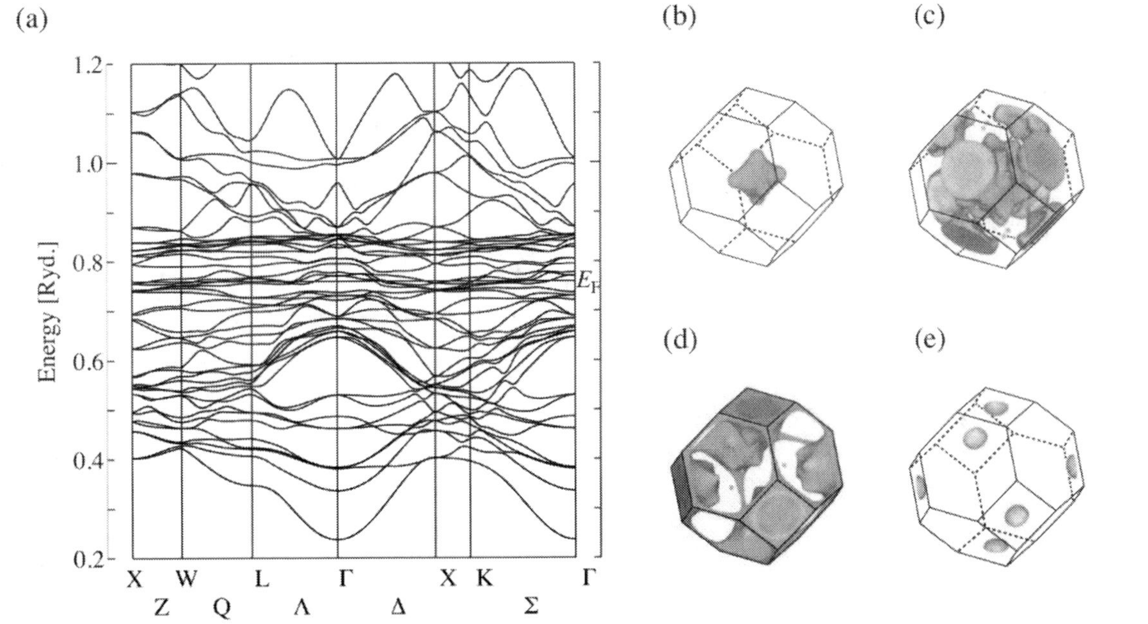

Figure 2. The energy band structure for (a) $PuRh_2$ calculated by using the self-consistent RLAPW method. E_F indicates the position of the Fermi level. (b) The hole-Fermi surfaces in the 25th, (c) the 26th, (d) electron sheet in the 27th, and (e) the 28th bands for $PuRh_2$. The center point in each Brillouin zone is the Γ point.

electrons/cell, and 0.002 electrons/cell in the 24th, 25th, and 26th bands, respectively. The 24th band constructs a hole sheet which is essentially a hollow sphere centered at the Γ point and multiply connected with eight arms along the ⟨111⟩ direction. This hole sheet is considerably rugged, and has one window along the ⟨100⟩ direction, as shown in Figure 1(b). The 25th band constructs a set of three inequivalent electron sheets which are centered at the X points in the Brillouin zone, which exhibits a complex network consisting of "arms" along the edges of the Brillouin zone, as observed in Figure 1(c). The 26th band constructs a set of three small electron sheets centered at the X points and its shape is almost spherical, as shown in Figure 1(d). It is easily recognized from Figure 1(b)-(d) that the Bloch states on all sheets of the Fermi surface consist dominantly of the Pu 5f and Ru 4d components.

Now let us discuss the Fermi surface structure of $PuRh_2$. In Figure 2(a), we show the energy band structure along the symmetry axes in the Brillouin zone in the energy region from 0.2 Ryd. to 1.2 Ryd. Note that Pu 6p bands are omitted, since they are located well below 0.2 Ryd. The Fermi level E_F, indicated by a solid line, is located at 0.760 Ryd. for $PuRh_2$. Concerning the energy band structure in the vicinity of E_F, we emphasize that there always occur a hybridization between Plutonium 5f and Rh 4d states. The 25th, 26th, 27th, and 28th bands construct the Fermi surface, as shown in Figure 2(b)-(e). The 25th band constructs the small hole sheet which consist of a six projections cube centered at the Γ point, as is shown in Figure 2(b). The 26th band constructs the hole sheet which consist of a roughness sphere centered at the Γ point and four cushions centered at the L points, as is shown in Figure 2(c). In Figure 2(d) and 2(e), the 27th and 28th bands construct the Fermi surface electron-sheets, which are very similar to 25th and 26th electron sheets of $PuRu_2$. The numbers of carriers contained in these Fermi surface sheets are 0.034 holes/cell, 0.424 holes/cell, 0.452 electrons/cell, and 0.006 electrons/cell in the 25th, 26th, 27th, and 28th bands, respectively. The total number of holes is equal to that of

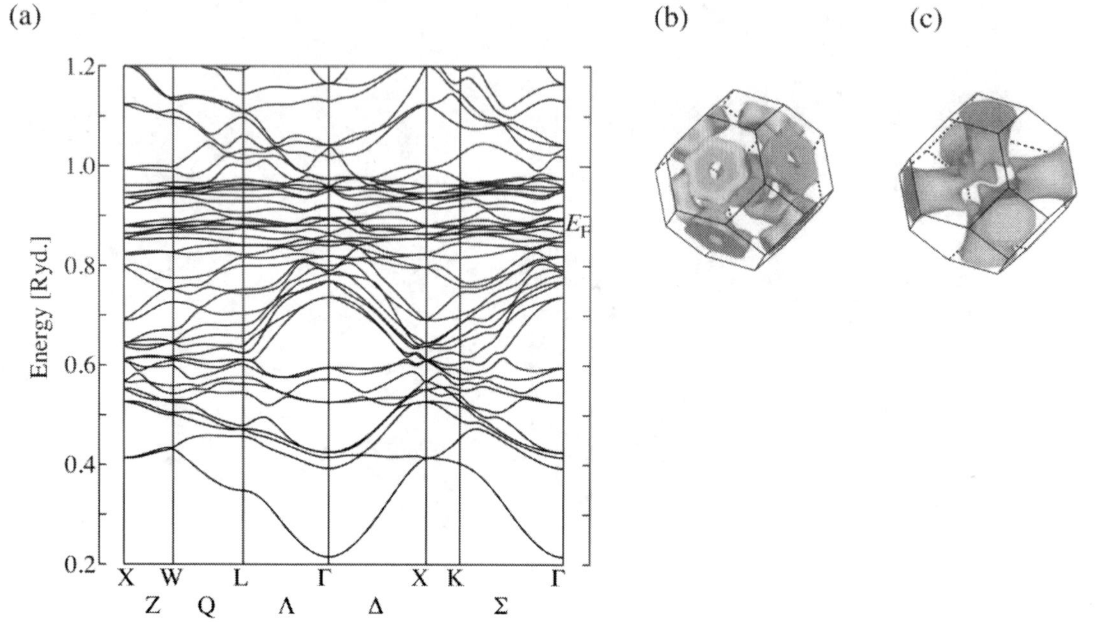

Figure 3. The energy band structure for (a) $PuIr_2$ calculated by using the self-consistent RLAPW method. E_F indicates the position of the Fermi level. (b) The hole-Fermi surfaces in the 26th, and (c) electron sheet in the 27th bands for $PuIr_2$. The center point in each Brillouin zone is the Γ point.

electrons, which represents that $PuRh_2$ is a compensated metal.

In Figure 3(a), two bands cross the E_F so that $PuIr_3$ has two Fermi surfaces, one hole sheet in the 26th band and one electron sheet in the 27th band. Figure 3 shows (b) the hole sheet in the 26th, and (c) electron sheet in the 27th band for $PuIr_2$. The Fermi level E_F, indicated by a solid line, is located at 0.881 Ryd. for $PuIr_2$. Concerning the energy band structure in the vicinity of E_F, we emphasize that there always occur a hybridization between Plutonium 5f and Ir 5d states. In Figure 3(b) and 3(c), the 26th and 27th bands construct the Fermi surface electron-sheets, which are very similar to 26th and 27th electron sheets of $PuRh_2$. The numbers of carriers contained in these Fermi surface sheets are 0.419 holes/cell, 0.419 electrons/cell in the 26th, and 27th bands, respectively. The total number of holes is equal to that of electrons, which represents that $PuIr_2$ is a compensated metal.

4. Discussion and Conclusion

We have applied the RLAPW method to the self-consistent calculation of the electronic structure for PuX_2(X=Ru, Rh, and Ir) on the basis of the itinerant 5f electron picture by assuming the non-magnetic phase. We found that a strong hybridization between Pu 5f and X d states occurs in the vicinity of E_F, and this makes these Fermi surfaces complicated. The theoretical specific heat coefficient γ_{band} are 20.8, 27.4, and 21.0 mJ/K^2·mol for $PuRu_2$, $PuRh_2$, and $PuIr_2$, respectively. The de Haas-van Alphen frequencies of PuX_2 will be calculated and published elsewhere.

References

[1] A R Harvey, M B Brodsky, and W J Nellis 1973 *Phys. Rev.* B **7** 4137
[2] G R Stewart, B Andraka and R G Haire 1994 *Journal of Alloys and Compounds* **213/214** 111

A theoretical study of the spin glass-Kondo-magnetic disordered alloys in the presence of a random field.

S.G. Magalhaes[1*], **F.M. Zimmer**[2] **and B Coqblin**[3]

[1] Instituto de Física, Universidade Federal Fluminense, 24210-346 Niteroi,RJ, Brazil
[2] Departamento de Física, Universidade Federal de Santa Maria, 97105-900 Santa Maria, RS, Brazil
[3] Laboratoire de Physique des Solides, CNRS, Université Paris-Sud, 91405 Orsay, France

E-mail: *sgmagal@gmail.com

Abstract.
We study here the influence of a random applied magnetic field on the competition between the Kondo effect, the spin glass phase and a ferromagnetic order in disordered cerium systems such as $CeNi_{1-x}Cu_x$. The model used here takes an intrasite Kondo coupling and an intersite random coupling; both the intersite random coupling and the random magnetic field are described within the Sherrington-Kirkpatrick model and the one-step replica symmetry breaking procedure is also used here. We present phase diagrams giving Temperature versus the Kondo exchange parameter and the random magnetic field makes decrease particularly the importance of the spin glass and ferromagnetic phases.

1. Introduction

The competition between the Kondo effect, a magnetic order (ferromagnetic or antiferromagnetic) and the spin glass state has been extensively studied from both an experimental and a theoretical point of view. Such a competition has been observed in $CeNi_{1-x}Cu_x$ [1] or $CePd_{1-x}Rh_x$ [2] alloys. A transition from a cluster spin glass state to an inhomogeneous ferromagnetic phase has been observed at very low temperatures in $CeNi_{1-x}Cu_x$ alloys [3]. On the other hand, a phase diagram giving Kondo, ferromagnetic and spin glass phases has been deduced from a theoretical point of view where the random intersite exchange interaction is described by the Sherrington-Kirkpatrick model [4] or recently by the van-Hemmen one [5].

However, some cerium systems, like $CeNi_{1-x}Cu_x$ alloys, exhibit hysteresis cycles with sharp macroscopic jumps in the magnetization at very low temperatures [6]. The effect is attributed to the formation of clusters in which there is competition between a random anisotropy and the intercluster exchange. Recently, it has been suggested that this result could be modelled by the presence of random magnetic fields applied in such clusters [6]. Therefore, besides the presence of disorder related with cluster spin glass formation, it would be possible to have a new form of disorder in the problem given by the random field which would not increase the frustration in the problem. As a conclusion, $CeNi_{1-x}Cu_x$ would be an example of competition among two types of disorder and Kondo effect.

The purpose of the present paper is to study the effect of a random magnetic field on the competition between the Kondo effect, the spin glass state and the ferromagnetic order [7]. We take here a random distribution within the Sherrington-Kirkpatrick model for both the intersite exchange interaction and the magnetic field. We determine, as usual, the phase diagrams Temperature versus the Kondo coupling constant J_K showing the effect of the random magnetic field.

2. The model

The Hamiltonian describes here a Kondo lattice where we consider a random coupling J_{ij} between localized magnetic moments and a random field h_i, in addition to the classical Kondo term for a spin $S = 1/2$ and an exchange constant J_K. We add here a random field with respect to the classical Hamiltonian that we have used previously [4].

Thus the starting hamiltonian is given by :

$$H = \sum_{ij,s} t_{ij} d^\dagger_{js} d_{is} + \epsilon_0 \sum_{i,s} \hat{n}^f_{is} + J_K \sum_i [\hat{S}^+_{fi}\hat{s}^-_{di} + \hat{S}^-_{fi}\hat{s}^+_{di}] - \sum_{i,j} J_{ij}\hat{S}^z_{fj}\hat{S}^z_{fi} - \sum_i h_i \hat{S}^z_{fi} \tag{1}$$

with $J_K > 0$. In Eq. (1), $\hat{S}^z_{fi} = \frac{1}{2}[\hat{n}^f_{i\uparrow} - \hat{n}^f_{i\downarrow}]$, $\hat{S}^+_{fi} = f^\dagger_{i\uparrow} f_{i\downarrow}$, $\hat{S}^-_{fi} = (\hat{S}^+_{fi})^\dagger$, $\hat{s}^+_{di} = d^\dagger_{i\uparrow} d_{i\downarrow}$, $\hat{s}^-_{di} = (\hat{s}^+_{si})^\dagger$, $\hat{n}^f_{is} = f^\dagger_{is} f_{is}$, where $f^\dagger_{is}$ (f_{is}) and $d^\dagger_{is}$ (d_{is}) are fermionic creation (annihilation) operators of f and d electrons, respectively. The spin projections are indicated by $s = \uparrow$ or $\downarrow$. The coupling J_{ij} and the field h_i are random variables following Gaussian independent probability distributions :

$$P(J_{ij}) = \left[\frac{N}{2\pi J^2}\right]^{1/2} \exp\left[-\frac{N}{2J^2}\left(J_{ij} - \frac{J_0}{N}\right)^2\right], \tag{2}$$

$$P(h_i) = \left[\frac{1}{2\pi\Delta^2}\right]^{1/2} \exp\left[-\frac{1}{2\Delta^2}(h_i - h_0)^2\right]. \tag{3}$$

The free energy averaged over the disorder present in the model is obtained as below [7]

$$<< F(\{J_{ij}\},\{h_i\}) >>_{J,h} = \int \prod_{(i,j)} [dJ_{ij} P(J_{ij})] \times \prod_i [dh_i P(h_i)]\, f(\{J_{ij}\},\{h_i\}). \tag{4}$$

The classical replica method is used to perform the average over J_{ij} and h_i in Eq.(4) and the free energy is, therefore, given by :

$$-\beta F = \lim_{N\to\infty} \lim_{n\to 0} \frac{1}{Nn} (<< Z^n >>_{J,h} - 1), \tag{5}$$

where Z^n is the replicated partition function.

From now on, we follow closely the procedure introduced in Refs. [8, 9]. The partition function is obtained within the path integral formalism using two types of Grassmann variables associated to localized electrons $(\psi(\omega))$ and conduction ones $(\varphi(\omega))$. Therefore, we can write :

$$Z = \int D(\psi^*\psi) D(\varphi^*\varphi) e^{[A_0 + A_K + A_{SG} + A_h]}. \tag{6}$$

The different actions introduced in equation (6) are presented in Refs. [8, 9], except A_h which is introduced here. The detailed calculations will be published elsewhere [11]. The static approximation [10] has been used in the Kondo action A_K and in the spin glass one A_{SG}.

The procedure follows closely Refs. [7, 8, 9, 11] and we determine firstly the Kondo and spin glass order parameters. The Kondo order parameters $\lambda_\sigma \approx \lambda = \frac{1}{N}\sum_{j,\omega} \left\langle \psi^*_{j\sigma}(\omega)\varphi_{i\sigma}(\omega)\right\rangle$ and $\lambda^\dagger_\sigma \approx \lambda^\dagger$ are introduced in the partition function Z. Then the Kondo and spin glass order parameters are used here to determine the phase diagrams (T, J_K).

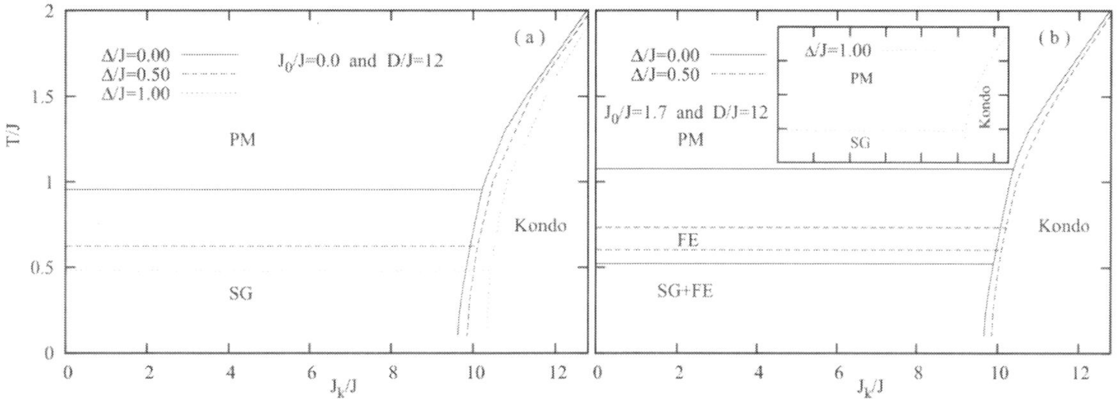

Figure 1. Calculated phase diagrams T/J versus J_K/J for several values of Δ for (a) $J_0/J = 0$ and (b) $J_0/J = 1.7$.

3. Results and Conclusions

Thus, we have derived the phase diagrams giving the different phases versus temperature and the Kondo parameterJ_K for different parameters, in order to study the influence of a random field. We have taken here a gaussian distribution for the intersite spin glass interaction within the Sherrington-Kirkpatrick model, as we have previously done in refs. [4, 8, 9]. In fact, it is a first model, because we have improved in previous papers the description of the Kondo-spin glass cerium systems by taking for the intersite interaction first a generalization of the Mattis model [12] and then the van Hemmen model [5]; in fact, these two recent models give a ferromagnetic phase below the spin glass one in good agreement with experiment in $CeNi_{1-x}Cu_x$ [1], while the Sherrington-Kirkpatrick model gives the opposite behavior. Any way, our purpose is firstly to study the influence of the magnetic random field and the phase diagrams presented here will be already interesting from that point of view.

Figure 1 shows the phase diagrams T/J versus J_K/J for several values of the deviation of the random field, i.e. Δ varying from 0 to a large value. The figure 1(a) describes the case $J_0 = 0$ without a ferromagnetic phase, while the figure 1(b) describes the case with a large J_0 value yielding a ferromagnetic phase (FE), a spin glass one (SG), a Kondo phase and a mixed spin glass-ferromagnetic (SG+FE) phase. For instance, in Fig. 1(a) the temperature of the PM/SG phase transition is decreased as Δ enhances at the same time as the Kondo state appears at higher values of J_K. For $J_0/J = 1.7$, the PM, FE and SG+FE are found as temperature decreases for low values of J_K. When Δ enhances, the FE and SG+FE phases are gradually destroyed. However, the FE order is more affected than the SG one by Δ. For example, inset of Fig. 1(b) ($J_0/J = 1.7$ and $\Delta/J = 1.0$) exhibits only the SG phase at low temperatures with Kondo state at larger J_K.

In conclusion, we see clearly that the disorder introduced by the random magnetic field tends to finally destroy not only the FE phase but also the SG one. Interestingly, in a smaller extend the Kondo state is also affected in the presented case. In other words, our results also indicate that this new form of disorder, i. e. the random magnetic field also disrupts the singlet formation. In our paper, we have obtained a SG phase below the FE one, in contrast with the experimental behaviour observed in $CeNi_{1-x}Cu_x$ alloys; but this result is due to the use of the Sherrington-Kirkpatrick model as previously observed [4] and already corrected by taking a generalization of the Mattis model [12] or the van Hemmen model [5]. Any way, the important result of our paper is to show the role of the random magnetic field which tends to destroy the SG and FE phases. Further works are necessary to study in more detail the effect of the random magnetic field when the problem is formulated in terms of clusters instead of canonical spins [11].

Acknowledgments
This work was partially supported by the Brazilian agency CNPq. FMZ also acknowledges the Brazilian agency FAPERGS.

4. References

[1] N. Marcano, J.C. Gomez Sal, J.I. Espeso, L. Fernandez Barquin and C. Paulsen, Phys.Rev. B **76**, 224419 (2007)
[2] T. Westerkamp, M. Deppe, R. Kuchler, M. Brando, C. Geibel, P. Gegenwart, A.P. Pikul and F. Steglich, Phys. Rev. Lett. **102**, 206404 (2009)
[3] N. Marcano, J.C. Gomez Sal, J.I. Espeso,J.M. de Teresa, P.A. Algarabel, C. Paulsen and J.R. Iglesias, Phys. Rev. Lett. **98**, 166406 (2007)
[4] S.G. Magalhaes, A.A. Schmidt, A. Theumann and B. Coqblin, Eur. Phys. B **30**, 419 (2002)
[5] S.G. Magalhaes, F.M. Zimmer and B. Coqblin, Phys.Rev. B **81**, 094424 (2010)
[6] J.R. Iglesias, J.I. Espeso, N. Marcano and J.C. Gomez Sal, Phys.Rev. B **79**, 195128 (2009)
[7] S.G. Magalhaes, C.A. Morais and F.D. Nobre, J. Stat. Mech **2011**, 07014 (2011)
[8] A. Theumann, B. Coqblin, S.G. Magalhaes and A.A. Schmidt, Phys.Rev. B **63**, 54409 (2001)
[9] A. Theumann and B. Coqblin, Phys.Rev. B **69**,214418 (2004)
[10] A.J. Bray and M.A. Moore, J. Phys.C **13**,L655 (1980)
[11] S.G. Magalhaes, F.M. Zimmer, B. Coqblin, to be published.
[12] S.G. Magalhaes, F.M. Zimmer, P.R. Krebs and B. Coqblin, Phys.Rev. B **74**, 014427 (2006)

International Conference on Strongly Correlated Electron Systems (SCES 2011) IOP Publishing
Journal of Physics: Conference Series **391** (2012) 012039 doi:10.1088/1742-6596/391/1/012039

^{31}P NMR Study of Filled Skutterudite $CeOs_4P_{12}$

K. Magishi[1], H. Sugawara[2], T. Saito[1] and K. Koyama[1]

[1] Institute of Socio-Arts and Sciences, The University of Tokushima, Tokushima 770-8502, Japan

[2] Graduate School of Science, Kobe University, Kobe 675-8501, Japan

E-mail: magishi@ias.tokushima-u.ac.jp

Abstract. We report the results of the electrical resistivity $\rho(T)$ and the ^{31}P-NMR measurements on filled skutterudite $CeOs_4P_{12}$ in order to investigate the magnetic properties at low temperatures from a microscopic point of view. For the polycrystalline sample synthesized under high pressure (HP), $\rho(T)$ increases with decreasing temperature. On the other hand, for the single crystal (SC), $\rho(T)$ reveals a positive temperature dependence between room temperature and 200 K, and increases with decreasing temperature below 200 K. Also, $1/T_1$ obeys the activated temperature dependence $1/T_1 \propto \exp(-\Delta/k_B T)$ above 160 K with an energy gap $\Delta/k_B \sim 500$ K and 540 K for the HP and the SC samples, which are slightly larger than that of a previous report.

1. Introduction

The family of filled skutterudites with the general formula RM_4X_{12} (R = rare earth, M = Fe, Ru, Os, X = P, As, Sb) has attracted a great deal of interest. This is not only due to their promising thermoelectric properties, but also because these compounds show a wide variety of transport and magnetic properties at low temperatures. Among them, while most filled skutterudite compounds show metallic behavior, a series of CeM_4P_{12} (M = Fe, Ru, Os) compounds shows semiconducting one with a hybridization gap of 400−1500 K[1, 2, 3], and there is a tendency that the magnitude of the gap decreases with increasing cell volume[4]. Also, CeM_4P_{12} exhibit an anomalously small lattice constant than that expected from trivalent lanthanides contraction[5]. The reduced lattice constants of CeM_4P_{12} is qualitatively understood by the change of hybridizaton between the Ce $4f$ and the conduction electrons depending on the cell volume.

$CeOs_4P_{12}$ synthesized under high pressure shows a semiconcucting transport property, and its temperature dependence obeys an activation conduction form $\rho(T) = \rho_0 \exp(-\Delta/k_B T)$ with $\Delta/k_B \sim 400$ K[3], which is smaller than those for $CeFe_4P_{12}$[1] and $CeRu_4P_{12}$[2]. Recently, we have succeeded in synthesizing single crystalline sample of $CeOs_4P_{12}$ by the Sn-flux method.

In this paper, we focus on the magnetic properties of filled skutterudite $CeOs_4P_{12}$ at low temperatures via the microscopic probes of ^{31}P-NMR. In order to understand various properties of $CeOs_4P_{12}$ with $4f$ electrons, it is important to clarify the characteristic electronic structure of $LaOs_4P_{12}$ without $4f$ electrons as a reference compound. $LaOs_4P_{12}$ is known to as a superconductor with $T_C = 1.8$ K[2].

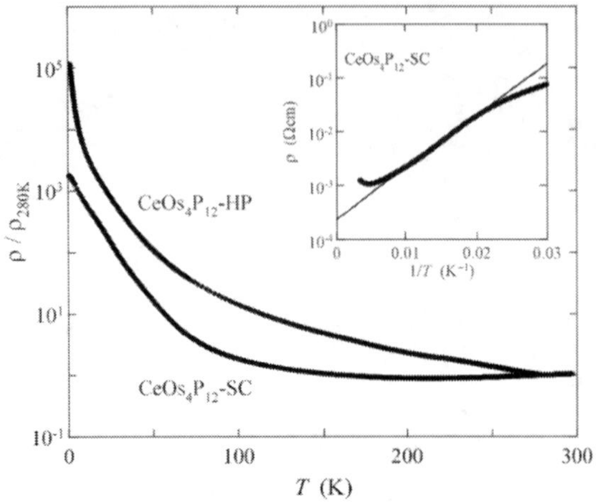

Figure 1. Temperature dependencies of the electrical resistivity $\rho(T)$ for $CeOs_4P_{12}$ normalized at 280 K.

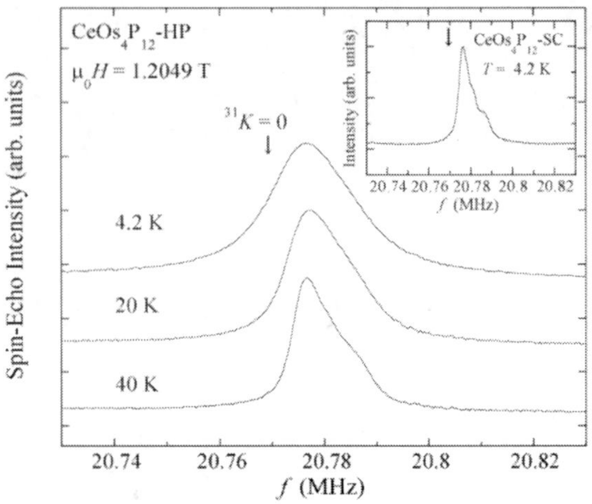

Figure 2. ^{31}P-NMR spectra for the HP sample of $CeOs_4P_{12}$. Inset shows the spectrum for the SC sample at 4.2 K.

2. Experimental

For the NMR measurements, we use two samples, which were crushed into powder. Polycrystalline samples were prepared at high temperature and high pressure using a wedge-type cubic-anvil high pressure apparatus (HP)[6]. Singlecrystals (SC) were synthesized using the Sn-flux method. Samples were characterized by powder X-ray diffraction at room temperature. Electrical resistivity was measured using a standard four-probe DC method in a temperature range of 0.4–300 K using a ^{3}He refrigerator. Magnetic susceptibility was measured using a SQUID magnetometer in a temperature range of 2–300 K. NMR measurements were performed in a temperature range of 1.5–300 K using a conventional phase-coherent pulsed spectrometer. NMR spectrum was obtained by Fast-Fourier-Transformation (FFT) of the FID signal. The nuclear spin-lattice relaxation time T_1 was measured by the saturation recovery method.

3. Results and Discussion

Figure 1 shows the temperature dependencies of the electrical resistivity $\rho(T)$ for $CeOs_4P_{12}$ normalized at 280 K. For the HP sample, $\rho(T)$ increases with decreasing temperature. The temperature dependence of $\rho(T)$ can be fit over the temperature range from 160 to 300 K to an activation conduction form $\rho(T) = \rho_0 \exp(-\Delta/k_B T)$, where Δ is the activation energy and k_B is Boltzmann's constant. The value of Δ/k_B is about 500 K for $CeOs_4P_{12}$, which is slightly larger than the results for the previous results[3]. On the other hand, for the SC sample, $\rho(T)$ reveals a positive temperature dependence like metal between room temperature and 200 K, and shows a minimum around 200 K. Then, it increases with decreasing temperature. As shown in the inset, the value of the activation energy Δ/k_B is estimated about 220 K over the limited temperature range 50 K$\leq T \leq$100 K, which is quite different from the value for the HP sample.

Figure 2 displays the evolution of the ^{31}P-NMR spectra for the HP sample of $CeOs_4P_{12}$. The asymmetrically distributed spectrum, that is, the uniaxially symmetric powder pattern is seen, which may arise from the anisotropic Knight shift. The line width of the spectrum was estimated to be $\sim$ 10 kHz. The spectrum is very narrow and is almost independent of temperature, except for below 10 K, where it is slightly broadened by the effect of the paramagnetic impurity. Inset shows the spectrum for the SC sample at 4.2 K. The spectrum is sharp compared with that for the HP sample, indicating that the quality of the SC sample is high.

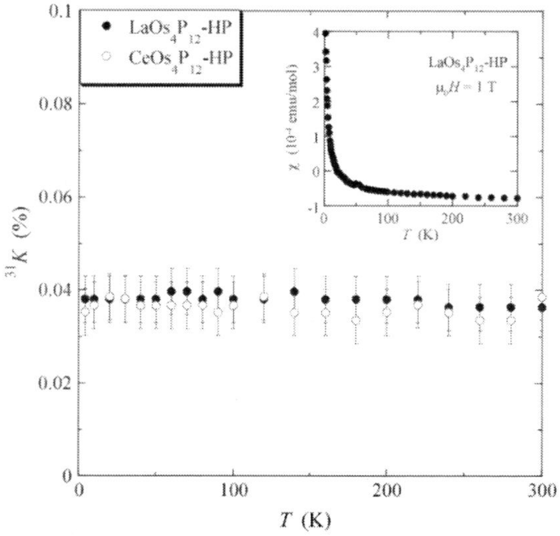

Figure 3. Temperature dependencies of the ^{31}P Knight shift for the HP samples of $CeOs_4P_{12}$(○) and $LaOs_4P_{12}$(●).

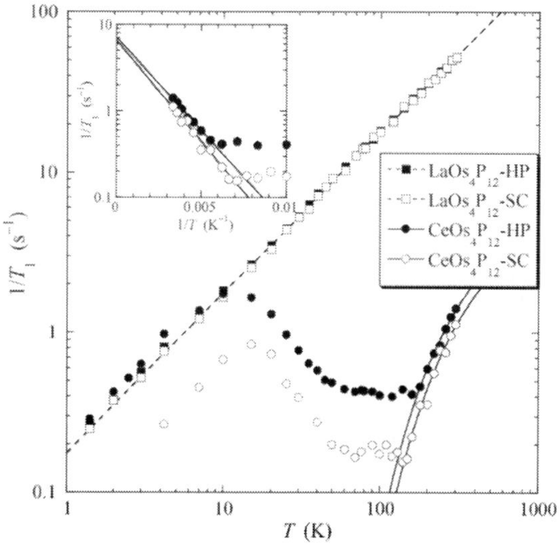

Figure 4. Temperature dependencies of the nuclear spin-lattice relaxation rates $1/T_1$. Inset shows the Ahrrenius plot of $1/T_1$ vs $1/T$.

Figure 3 shows the temperature dependencies of the ^{31}P Knight shifts for the HP sample of $CeOs_4P_{12}$(○), together with that of $LaOs_4P_{12}$(●). The Knight shift was derived from the peak maximum of the NMR spectrum. For both compounds, Knight shifts are very small ($\sim 0.04\%$) and are almost independent of temperature. Results for the SC samples are quite similar (not shown), and seems to be intrinsic and different from the Curie-like upturn at low temperatures in $\chi(T)$ possibly originating from the paramagnetic impurity, as seen in the inset of Fig.3.

Figure 4 indicates the temperature dependencies of the nuclear spin-lattice relaxation rates $1/T_1$ for ^{31}P nuclei in $CeOs_4P_{12}$, together with the results for $LaOs_4P_{12}$[7]. In case for both the HP and the SC samples of $LaOs_4P_{12}$, $1/T_1$ is almost proportional to the temperature between 1.5 K and 300 K, that is, $1/T_1$ obeys the Korringa relation T_1T = constant as expected in the metallic state, indicated by the dotted line in the figure.

On the other hand, for the HP sample of $CeOs_4P_{12}$, $1/T_1$ decreases about one order of magnitude, and follows the activated-type temperature dependence above 160 K as $1/T_1 \propto \exp(-\Delta/k_BT)$ with $\Delta/k_B \sim 500$ K, as indicated by the solid line in the figure. It is thought to originate from the existence of the gap of the magnetic excitation due to the hybridization. The magnitude of the energy gap is slightly larger than the results for the previous results[3] and is smaller than those for $CeFe_4P_{12}$[8] and $CeRu_4P_{12}$[8, 9]. It indicates that the degree of the hybridization between the Ce $4f$ and the conduction electrons is reduced in $CeOs_4P_{12}$ compared with the cases of $CeFe_4P_{12}$ and $CeRu_4P_{12}$. For the SC sample, $1/T_1$ shows the similar behavior as $1/T_1 \propto \exp(-\Delta/k_BT)$ with $\Delta/k_B \sim 540$ K, where the value of Δ/k_B is slightly larger than that of the HP sample, as seen in the inset. Also, the value is larger than that estimated from the electrical resistivity measurements.

Below 160 K, $1/T_1$ deviates from the activated behavior, and is almost independent of the temperature, which suggests that other relaxation effect takes place at low temperatures. It may be attributed to the relaxation by the paramagnetic local moments. Furthermore, at low temperatures below 50 K, $1/T_1$ distributes and increases gradually upon cooling. It has a maximum around 10 K and show a rapid decrease below 10 K, which is a similar behavior in $CeFe_4P_{12}$[8] and $CeRu_4P_{12}$[8, 9]. At this stage, we have no idea to explain these behaviors at low temperatures. But the relaxation rate for the SC sample is smaller than that for the HP sample, which indicates that the effect by impurity is small and the quality of the sample is high, consistent with a sharp spectrum.

4. Summary

We have carried out the electrical resistivity and the ^{31}P-NMR experiments in $CeOs_4P_{12}$ in order to understand the magnetic properties from the microscopic viewpoints. While $\rho(T)$ for the HP sample increases with decreasing temperature, $\rho(T)$ for the SC sample reveals a positive temperature dependence like metal between room temperature and 200 K, and increases with decreasing temperature below 200 K. $1/T_1$ obeys the activated temperature dependence $1/T_1 \propto \exp(-\Delta/k_BT)$ above 160 K with an energy gap $\Delta/k_B \sim 500$ K and 540 K for the HP and the SC samples, which are slightly larger than that of a previous report. Also, the value of the energy gap is smaller than those for $CeFe_4P_{12}$ and $CeRu_4P_{12}$, indicating that the degree of the hybridization between the Ce $4f$ and the conduction electrons is reduced in $CeOs_4P_{12}$.

Acknowledgements

This work was supported by the Grants-in-Aid for Scientific Research (No.21540339) from Japan Society for Promotion of Science, and the Ministry of Education, Culture, Sports, Science and Technology of Japan.

References

[1] G.P. Meisner, M.S. Torikachvili, K.N. Yang, M.B. Maple and R.P. Guertin: J. Appl. Phys **57** (1985) 3073.
[2] I. Shirotani, T. Adachi, K. Tachi, S. Todo, K. Nozawa, T. Yagi and M. Kinoshita: J. Phys. Chem. Solids **57** (1996) 211.
[3] I. Shirotani, T. Uchiumi, C. Sekine, M. Hori, S. Kimura and M. Hamaya: J. Solid State Chem. **142** (1999) 146.
[4] H. Sugawara, S. Osaki, M. Kobayashi, T. Namiki, S.R. Saha, Y. Aoki and H. Sato: Phys. Rev. B **71** (2005) 125127.
[5] C.B.H. Evers, W. Jeitschko, L. Boonk, D.J. Braun, T. Ebel and U.D. Scholz: J. Alloys Comp. **224** (1995) 184.
[6] D. Kikuchi *et al.*, unpublished.
[7] K. Magishi, Y. Iwahashi, H. Sugawara, T. Saito, K. Koyama, K. Tanaka, D. Kikuchi and H. Sato: J. Phys. Soc. Jpn. **77** (2008) SA22.
[8] K. Magishi, H. Sugawara, T. Saito, K. Koyama and H. Sato: Physica B **378-380** (2006) 175.
[9] K. Fujiwara, C. Sekine and I. Shirotani: Physica B **281&282** (2000) 296.

Low temperature properties of a low carrier heavy fermion YbPtSb

K. Matsubayashi[1], R. Yamanaka[1], H. S. Suzuki[2], K. Munakata[3], Y. Uwatoko[1]

[1]Institute for Solid State Physics, The University of Tokyo, Kashiwanoha, Kashiwa, Chiba 277-8581, Japan
[2]National Institute for Materials Science, Tsukuba, Ibaraki 305-0047, Japan
[3]Research Center for Neutron Science and Technology, Comprehensive Research Organization for Science and Society, Bldg. IQBRC, Tokai, Ibaraki 319-1106, Japan.

E-mail: kazuyuki@issp.u-tokyo.ac.jp

Abstract. We have grown single crystals of a low carrier heavy fermion YbPtSb and studied low temperature properties by means of the electrical resistivity (ρ), dc and ac magnetic susceptibility (χ), and specific heat (C). The clear evidence of the antiferromagnetic order at $T_N \sim 0.34$ K was observed by bulk measurements. At lower temperatures below T_N, extrapolating C/T to T=0 yields $\gamma \sim 9$ J/mole K^2, and $\rho(T)$ shows T^2 temperature dependence with a large A coe cient, as observed in YbPtBi. We will discuss the Fermi liquid state realized in the low carrier systems.

1. Introduction

In heavy fermion compounds, unusual low temperature properties have been observed by tuning the system toward a magnetic quantum critical point (QCP) using a control parameter such as magnetic eld and pressure. Among the ternary Yb intermetallic compounds with a cubic structure, YbPtBi and $YbCo_2Zn_{20}$ have attracted as outstanding example because of the huge Sommerfeld coe cient γ exceeding 7 J/mole K^2 [1, 2]. In these systems, the QCP can be tuned by relatively small magnetic eld and pressure: the antiferromagnetic order of YbPtBi can be suppressed by the magnetic eld of 0.3 T, and $YbCo_2Zn_{20}$ exhibits pressure induced antiferromagnetic order above P_C 1 GPa [3]. Since YbPtBi is a low carrier concentration semi-metal, which is in contrast to most of metallic compounds exhibiting the quantum critical behavior, the large γ-value in the vicinity of the QCP is of particular interest. Despite the intensive studies, the origin of the massive electronic state in YbPtBi remain unclear. For better understanding of the heavy fermion state in low carrier systems, we have studied low temperature properties of YbPtSb, which is identical structure to that of YbPtBi. According to the previous report, a value of C/T reaches 1.5 J/mole K^2 at 2 K [4], however, physical properties below 2 K have not been reported so far. Recent ultrasonic measurements reveal the strong renormalization of quasi-particles in YbPtSb, probably due to non- Fermi liquid characteristics close to the QCP [5]. In order to investigate the low temperature properties of YbPtSb, we have grown YbPtSb single samples, and measured electrical resistivity, magnetic susceptibility and speci c heat down to 100 mK.

2. Experimental methods

Single crystals were grown by the Bridgman method in a Mo crucible. X-ray di raction measurements indicate that obtained samples crystallize in the half-Heusler MgAgAs structure and the lattice constants are in agreement with previous report (a = 6.47 Å) [4]. Obtained single crystal was rst aligned with a X-ray Laue and cut with a spark cutter. The electrical resistivity and the Hall resistivity were measured by a standard four-probe dc technique with current ow along [100] direction. The dc magnetization was measured using a commercial (MPMS) magnetometer. An external magnetic eld was applied along [100] direction. The ac magnetic susceptibility was measured at a xed frequency of 317 Hz with a modulation eld of 0.1 mT. Heat capacity measurements were carried out by the adiabatic heat pulse method.

3. Results and discussion

Figure 1 (a) shows the temperature dependence of the electrical resistivity and the Hall coe cient R_H of YbPtSb. (T) monotonically decreases with decreasing temperature and has a rather high value, re ecting the low carrier nature. $R_H(T)$ exhibits weak temperature dependence and is positive in the temperature range studied here, indicating that the dominant carrier type is hole like. Assuming a single carrier, de ned as n_{H} = $-1/eR_{\mathrm{H}}$, the carrier concentration is approximately n_{H} 1×10^{21} cm ^{3}. These overall features of the transport properties agree with the previous report [4]. The magnetic susceptibility (T) of YbPtSb is plotted in Fig. 1 (b), together with the inverse magnetic susceptibility ^{1}. (T) follows the

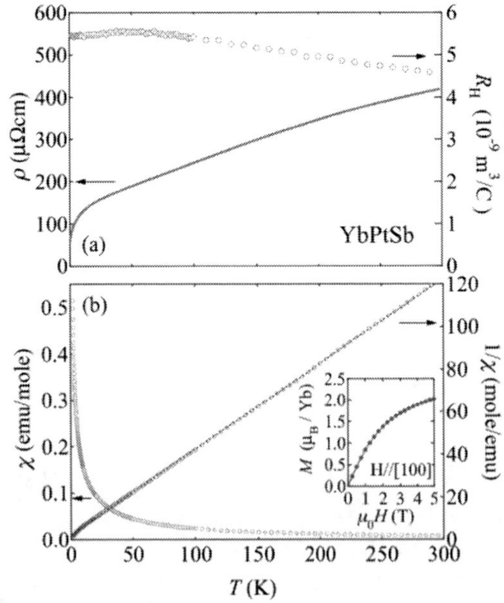

Figure 1. (a) Temperature dependence of electrical resistivity (left axis) and Hall coe - cient (right axis). (b) Magnetic susceptibility (right axis) and inverse magnetic susceptibility (right axis) with $\mu_0 H$ 0.1 T. The inset of Fig. 1 (b) shows magnetization curves at 2 K in magnetic elds of [100] direction.

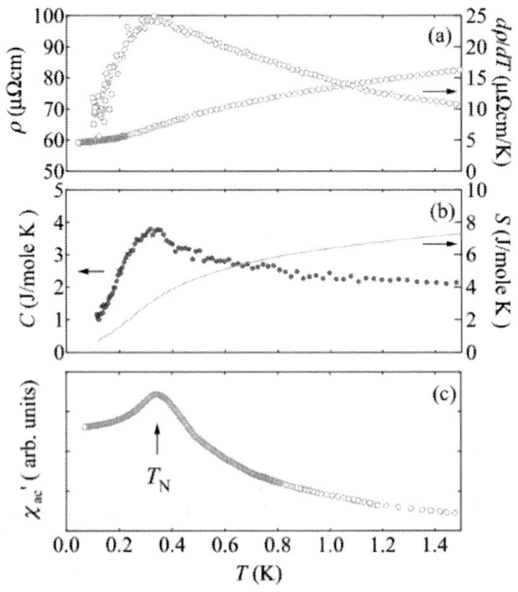

Figure 2. (a) Low-temperature electrical resistivity (left axis) and its temperature derivative (right axis). (b) Speci c heat (left axis) and entropy (right axis). (c) Ac magnetic susceptibility as a function of temperature. Arrow indicates the antiferromagnetic transition temperature T_{N}

Curie-Weiss law above 20 K with the e ective magnetic moment $\mu_{\rm eff} = 4.46\ \mu_{\rm B}$, which is very close to the value of a free Yb^{3+} ion ($\mu_{\rm eff} = 4.53\ \mu_{\rm B}$). The paramagnetic Curie temperature $\theta_{\rm P}$ is -5.7 K, suggesting the presence of weak antiferromagnetic exchange interaction. The localized character of Yb ions at high temperature is consistent with NMR Knight shift [6]. The deviation from the Curie-Weiss law below 20 K is attributed to the crystalline electric eld (CEF) e ect. Since YbPtSb crystallizes in the cubic structure, the ground state multiplet of Yb ions will split into a quartet (Γ_8) and two doublet (Γ_7 and Γ_6) states. As shown in the inset of Fig. 1 (b), the magnetization tends to saturate at $2\mu_{\rm B}$/Yb, which is consistent with the reduction of the magnetic moment of Yb ion re ecting the CEF e ect.

We next focus on the low temperature properties of YbPtSb. As shown in Fig. 2 (a), there is a slight change in slope of the (T) as seen more clearly in the curve of $d\ /dT$ versus temperature, implying the occurrence of the magnetic order. This is con rmed by speci c heat and ac magnetic susceptibility that the anomaly due to the antiferromagnetically ordered state is detected at the temperature $T_{\rm N}$ 0.34 K (Fig. 2 (b) and (c)). Above $T_{\rm N}$, there is the long tail in the speci c heat, implying the presence of short range order. As shown in Fig. 2(b) (right axis), the value of the magnetic entropy S at $T_{\rm N}$ is only $0.6R\ln 2$ and reaches $R\ln 2$ near 0.8 K, suggesting the doublet ground state (Γ_7 or Γ_6).

With further decreasing temperature below $T_{\rm N}$, C/T tends to become constant as shown in Fig. 3. By extrapolating C/T to T=0, we obtained γ 9±1 J/mole K^2, which is almost the same as that of YbPtBi. In accordance with the speci c heat, (T) manifests clear T^2 temperature dependence below $T_{\rm FL}$ 0.2 K, namely Fermi Liquid (FL) behavior (see the inset of Fig. 3). A t to the data of $= {}_0 + AT^2$ yields $A = 43\pm5\ \mu\Omega$cm/K^2. It is worth noting that the FL state with the large value of γ and A coe cient is observed in a low carrier heavy fermion YbPtSb. In the FL regime, most of heavy fermion systems follow a universal relation, so called Kadowaki-Woods (K-W) ratio: A/γ^2 $1.0\times10^{\ 5}$ $\mu\Omega$cm(K/mJmole)^2 [7]. For YbPtSb in the preset study, we obtained the value of A/γ^2=(0.5±0.2)×10$^{\ 6}$ $\mu\Omega$cm(K/mJmole)^2, which is smaller than the typical value of A/γ^2. According to the experimental and theoretical work, this deviation from the K-W scaling is caused by N-fold degeneracy of quasiparticles [8]. Indeed, the A/γ^2 obtained

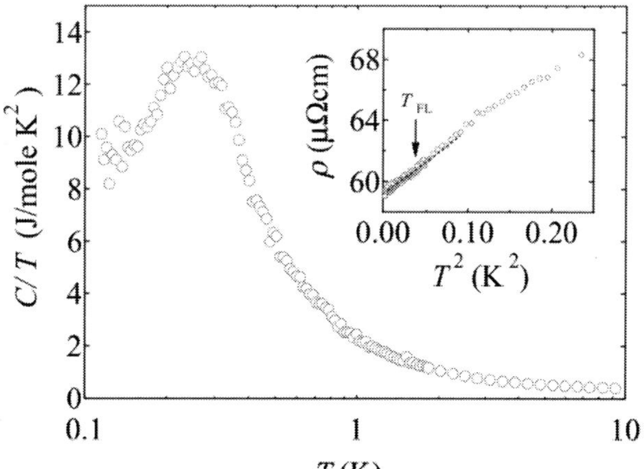

Figure 3. Low temperature speci c heat divided by temperature C/T as a function of logarithmic temperature. Inset shows as a function of T^2. The dashed line represents the tting of the vs T^2 curve.

from our experiment coincides with that with N=6 degeneracy, however, it is not expected from the CEF splitting in the cubic structure for YbPtSb. Furthermore, A/γ^2 also depends on the carrier concentration. Considering the carrier density estimated aforementioned Hall e ect, we found the further deviation of A/γ^2 from the K-W ratio. Therefore, the conventional K-W scaling is not applicable to YbPtSb, probably due to multiband nature. From the view point of QCP, it is expected that the antiferromagnetic order can be suppressed by applying a small magnetic eld as inferred from the large γ value. Further measurements are in progress to investigate the eld-induced QCP of YbPtSb.

4. Summary

We have investigated the low temperature properties of YbPtSb single crystal by means of resistivity, hall e ect, magnetic susceptibility and speci c heat. We present a convincing evidence for the occurrence of an antiferromagnetically ordered state at T_{N} 0.34 K. Below T_{N}, the system enters the FL state with the large γ and A coe cient, however, we found the deviation of A/γ^2 ratio from the K-W scaling due to the multiband nature of YbPtSb.

Acknowledgments

The authors thank T. Koyama, T. Mito, and Y. Nakanishi for valuable discussions. This work was nancially supported by Grants-in-Aid for Scienti c Research (No. 21340092) and for Scienti c Research on Innovative Areas ''Heavy Electrons'' (No. 20102007). KM was supported by a Grant-in-Aid for Young Scientists B (No. 22740217) from the Ministry of Education, Culture, Sports, Science and Technology.

References

[1] Fisk Z, Canfield P C, Beyermann W P, Thompson J D, Hundley M F, Ott H R, Felder E, Maple M B, Lopez de la Torre M A, Visani P and Seaman C L 1991 *Phys. Rev. Lett* **67** 3310.
[2] Torikachvili M S, Jia S, Mun E D, Hannahs S T, Black R C, Neils W K, Martien D, Bud'ko S L, and Canfield P C 2007 *PNAS* **104** 9960
[3] Saiga Y, Matsubayashi K, Fujiwara T, Kosaka M, Katano S, Hedo M, Matsumoto T and Uwatoko Y 2008 *J. Phys. Soc. Jpn.* **77** 053710
[4] Kasaya M, Suzuki H, Yamaguchi T and Katoh K 1992 *J. Phys. Soc. Jpn.* **61** 4187
[5] Nakanishi Y, Ito K, Kamiyama T, Kashiwazaki R, Deto K, Shichinomiya F, Nakamura M, Yhoshizawa M and Suzuki H S 2010 *Phys. Status Solidi B* **247** 754.
[6] Koyama T, Abe M, Mito T, Ueda K, Kohara T and Suzuki H S 2011*J. Phys.: Conf. Ser.* **273** 012040
[7] Kadowaki K and Woods S B 1986 *Solid State Commun.* **58** 507
[8] Tsujii N, Kontani H, and Yoshimura K 2005 *Phys. Rev. Lett.* **94** 057201

T/B scaling of magnetization in the mixed valent compound β-$YbAlB_4$

Yosuke Matsumoto[1], S. Nakatsuji[1], K. Kuga[1], Y. Karaki[1], Y. Shimura[1], T. Sakakibara[1], A. H. Nevidomskyy[2,3], P. Coleman[2,4]

[1] Institute for Solid State Physics, University of Tokyo, Kashiwa 277-8581, Japan
[2] Center for Materials Theory, Department of Physics and Astronomy, Rutgers University, Piscataway, N.J. 08854, USA
[3] Department of Physics and Astronomy, Rice University, Houston, Texas 77005, USA
[4] Department of Physics, Royal Holloway, University of London, Egham, Surrey TW20 0EX, UK

E-mail: matsumoto@issp.u-tokyo.ac.jp, satoru@isso.u-tokyo.ac.jp

Abstract. Here we provide the first clear evidence of Fermi-liquid breakdown in an intermediate valence system. We employ high precision magnetization measurements of the valence fluctuating superconductor β-$YbAlB_4$ to probe the quantum critical free energy down to temperatures far below the characteristic energy scale of the valence fluctuations. The observed T/B scaling in the magnetization over three decades not only indicates unconventional quantum criticality, but places an upper bound on the critical magnetic field $|B_c| < 0.2$ mT, a value comparable with the Earth's magnetic field and six orders of magnitude smaller than the valence fluctuation scale. This tiny value of the upper bound on B_c, well inside the superconducting dome, raises the fascinating possibility that valence fluctuating β-$YbAlB_4$ is intrinsically quantum critical, without tuning the magnetic field, pressure, or composition: the first known example of such a phenomenon in a metal.

1. Introduction

The breakdown of Fermi liquid behavior in metals observed near a magnetic quantum critical point, challenges our current understanding of strongly correlated electrons. Archetypical quantum critical (QC) materials can be found in the heavy-fermion intermetallics, which are usually described as a Kondo lattice with a competition between local moment magnetism and conduction electron screening of the local moments (the Kondo e ect) [1, 2, 3]. To date, all QC materials of this kind are known to have an almost integral valence which stabilizes the local moments considered essential for the criticality. By contrast, departures from integral valence associated with valence uctuations are thought to promote screening of local moments, suppressing critical phenomena.

On the other hand, recent studies by our group revealed the rst example of an Yb-based heavy fermion superconductivity in the new compound β-$YbAlB_4$ ($T_c = 80$ mK) [4, 5]. Pronounced non-Fermi liquid behavior above T_c and its magnetic eld dependence indicate that the system is a rare example of a pure metal that displays quantum criticality at ambient pressure and close to zero magnetic eld [4]. Furthermore, recent hard x-ray photoemission spectroscopy measurements revealed intermediate valence of $Yb^{+2.75}$ providing the rst unique example of

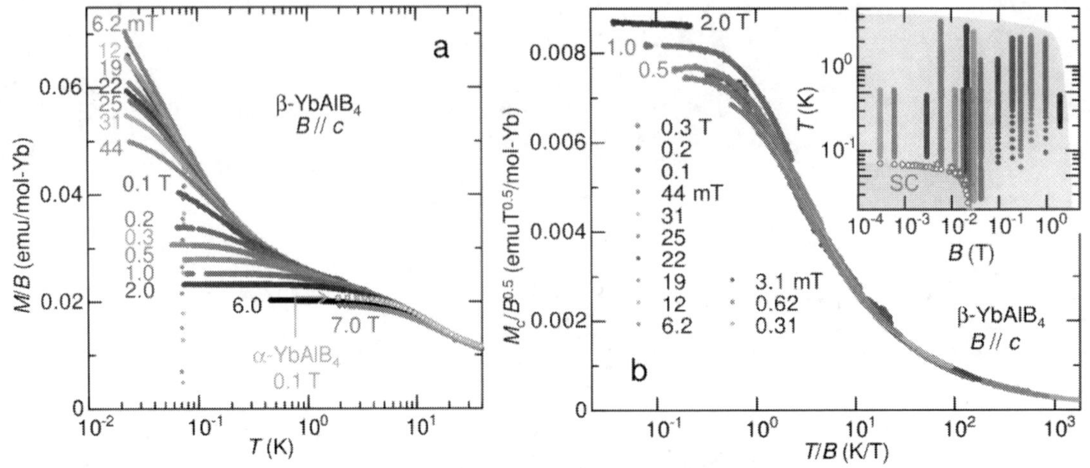

Figure 1. (a) Temperature dependence of the magnetic susceptibility M/B of both β- (solid circles) and α-YbAlB$_4$ (open squares) under the eld $B \parallel c$. (b) Scaling plot of $M_c/B^{0.5}$ vs. T/B for β-YbAlB$_4$. M_c is the critical contribution to the magnetization: $M_c = M \quad M_0$(see text). Inset shows the B-T range where the scaling applies (solid circles in the blue shaded region) and the superconducting (SC) phase boundary (open circles and triangles [5, 8]) of β-YbAlB$_4$.

quantum criticality in a mixed valent system [6]. Interestingly, the system is governed by two distinct energy scales: a high-energy scale, $T_0 \quad 200$ K characterizing valence uctuations, and a low-energy scale, $T^* \quad 8$ K characterizing the emergence of the Kondo lattice behavior [7].

Here we present clear experimental evidence that indicates the zero- eld quantum criticality in β-YbAlB$_4$ [7]. We employed high resolution magnetization measurements and found a scaling equation $dM/dT = B^{\ 1/2}f(T/B)$, which holds over wide ranges of magnetic elds (0.3 mT $< B <$ 2 T) and temperatures (20 mK $< T <$ 3 K). This T/B scaling strongly suggests $B_c = 0$ within the experimental resolution of 0.2 mT, which is comparable to the Earth ' s magnetic eld. This indicates a possibility of quantum critical phase that extends over a nite pressure range. The experimental details of this work have been discussed already in [7, 8].

2. Results and discussion

The divergent magnetic susceptibility at $T \to 0$ is one of the most remarkable feature which characterize the QC in β-YbAlB$_4$. Figure 1 shows the T dependence of the c-axis susceptibility $= M/B$ in the wide range of T and B spanning four orders of magnitude. Below $T \quad 3$ K, the results show a systematic evolution from a non-Fermi liquid (NFL) metal with divergent susceptibility at zero eld to a Fermi liquid (FL) with nite susceptibility in a eld.

Divergent behavior in NFL region is well tted by a sum of the QC contribution $M_c/B \propto T^{\ 0.5}$ and T-independent constant term $_0(= M_0/B) = 0.017$ (emu/mol). The latter constant term, which is close to the zero-T susceptibility of the non-critical α-YbAlB$_4$, may originate from constant Van Vleck contribution to susceptibility and/or from Pauli susceptibility of the non-critical parts of the Fermi surfaces. Figure 1b shows $M_c/B^{0.5}$ vs. T/B for β-YbAlB$_4$ in the region shown in the inset ($T \lesssim 3$ K and $B \lesssim 2$ T). Interestingly, the data collapse on a single curve at $T > B$, indicating a scaling relation $M_c = M \quad M_0 = B^{0.5}\psi(T/B)$. Under elds $B > 0.5$ T, M_c does not follow a single scaling curve. This may be due to a small eld-nonlinear contribution to M_0, which is neglected in the above analysis. In fact, the scaling relation free

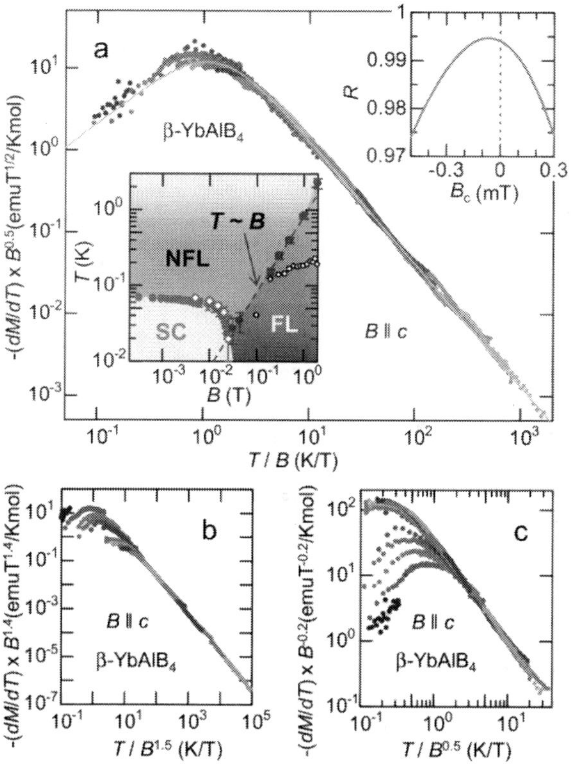

Figure 2. (a) Scaling observed for the c-axis magnetization at $T \lesssim 3$ K and $B \lesssim 2$ T. The light blue line represents a fit to the empirical Eq. 1 with scaling function $\phi(x) = \Lambda x(A + x^2)^{-n}$ (see text). The right inset shows Pearson's correlation coefficient R for the fit with finite B_c (see text). The best fit is obtained with $\alpha = 1.50 \pm 0.02$ and $B_c = -0.1 \pm 0.1$ mT (see text). The left inset shows the B-T phase diagram of β-YbAlB$_4$ in the low T and B region. The filled circles are the peak temperatures of $-dM/dT$ below which the FL ground state is stabilized. The open circles are the temperature scale $T_{\rm FL}$ below which the T^2 dependence of the resistivity is observed [4]. The upper critical fields of the superconducting (SC) phase is shown by open circles and triangles [5, 8]. (b)(c) Scaling plots with $(\delta, \eta) = (1.5, 1.4), (0.5, -0.2)$ for the same data set in Fig. 2a, respectively (see texts).

from this contribution which is obtained after T-derivative of the both parts:

$$\frac{\mathrm{d}M}{\mathrm{d}T} = B^{\alpha-2}\phi\left(\frac{T}{B}\right), \tag{1}$$

with $\alpha = 3/2$ works much better as shown in Figure 2a. Here we can observe a clear power law behavior of $-dM/dT \propto T^{-1.5}$ at $T > B$ (NFL region), which indicates $M \propto T^{-0.5} + const.$, and T-linear behavior of $-dM/dT$ at $T < B$ as expected for a FL. This empirical scaling property implies that close to the QCP, below $T \approx 3$ K and $B \approx 2$ T, β-YbAlB$_4$ has no intrinsic energy scale, and that furthermore, temperature and field are interchangeable variables, whose ratio T/B determines the physical properties. The peak of the scaling curve is located at $T/B \sim 1$ K/T and indicates that the thermodynamic boundary between the FL and NFL regions is on the $T \sim B$ line, as shown in the phase diagram of the inset of Figure 2a. The QC free energy with the form $F_{\rm QC} = B^{\alpha} f(T/B)$ is obtained by integrating both parts of Eq. (1), where f is a scaling function of the ratio T/B with the limiting behavior: $f(x) \propto x^{\alpha}$ in the NFL regime ($x \gg 1$) and $f(x) \propto$ const $+ x^2$ in the FL phase ($x \ll 1$). Indeed, $f(x) \propto (A + x^2)^{\alpha/2}$ fits the observed scaling behavior of dM/dT in Eq. (1) with $\phi(x) = \Lambda x(A + x^2)^{\frac{\alpha}{2}-2}$, as shown by the fit in Figure 2a, achieved with $\alpha = 3/2$.

The T/B scaling implies that the critical field B_c of the quantum phase transition is located just at zero field. A finite value of B_c requires the scaling function $f(x)$ and $\phi(x)$ with a ratio $x = T/|B - B_c|$ rather than $x = T/B$. A bound for B_c can be determined by substitution of $x = T/|B - B_c|$ into Eq. (1) while searching the value of B_c that gives the best fit to the experimental data. The inset of Figure 2a shows the Pearson's correlation coefficient R obtained from this fit, indicating that B_c is optimal at -0.1 ± 0.1 mT. The error-bar is only a few times larger than Earth's magnetic field (~ 0.05 mT), more significantly it is two orders of magnitude smaller than $\mu_0 H_{c2} = 30$ mT. This result is particularly stunning given the large valence uctuation scale T_0 200 K. Thus β-$YbAlB_4$ provides a unique example of essentially *zero-field quantum criticality.*

A possibility of other scaling, such as T/B^{δ} scaling ($dM/dT = B$ $^{\eta}\phi(T/B^{\delta})$) with $\delta \neq 1$, is checked by scaling plot, $(dM/dT)B^{\eta}$ versus T/B^{δ}. Figures 2b and 2c show the trial scaling plots with $(\delta, \eta) = (1.5, 1.4), (0.5, \text{-}0.2)$ for the same data set in Fig. 2a, respectively. Here, η was adjusted against xed δ so that a scaling was realized at high T parts where the NFL power-law behavior was observed. As is clearly seen from the gures, the data at smaller T/B^{δ} do not satisfy the scaling by changing δ away from 1. It turned out that the data in the widest T and B range are scaled with $\delta = 1.0 \pm 0.1$. The detail will be discussed elsewhere [9].

The observation of zero eld quantum criticality in valence uctuating β-$YbAlB_4$ has a number of signi cant implications. Conservatively, one might conclude that β-$YbAlB_4$ is a system in which a fortuitous combination of structure and chemistry ne-tune the critical eld B_c to zero. In such a scenario, the QCP would be characterized by gapless soft modes: spin uctuations in the case of an antiferromagnetic quantum criticality, or critical charge uctuations associated with mixed valence. In the former case, the system would be literally at the edge of an antiferromagnetic instability, while the latter case would involve a critical end-point of valence instability line [10, 11]. In both cases, the application of pressure is expected to immediately restore the Fermi liquid ground state.

Set against these two possibilities is our observation of a tiny upper bound (0.2 mT) on the critical magnetic eld B_c. This critical state is so fragile, that a eld not much larger than the Earth's magnetic eld drives β-$YbAlB_4$ back into a Fermi liquid. This suggests the intriguing possibility that β-$YbAlB_4$ is a *quantum critical phase* of matter, to which in nitesimal magnetic eld is a relevant perturbation converting it into a Fermi liquid. In this case, non-Fermi liquid properties will be una ected by pressure. An example of such a state is a gapless spin liquid, in which the presence of an extensive manifold of gapless spin excitations would lead to a critical phase characterized by power-law correlation functions [12, 13]. It is left for future studies, such as pressure experiments, to discriminate between these two alternative scenarios.

Acknowledgments
We thank H. Ishimoto, D.E. MacLaughlin, K. Miyake, T. Senthil, Q. Si, T. Tomita, K. Ueda, and S. Watanabe for useful discussions. This work is partially supported by Grants-in-Aid (No. 21684019) from JSPS, by Grants-in-Aids for Scienti c Research on Priority Areas (No. 19052003) and on Innovative Areas (No. 20102006, No. 20102007) from MEXT, Japan, by Global COE Program "the Physical Sciences Frontier", MEXT, Japan, by Toray Science and Technology Grant, and by a grant from the National Science Foundation DMR-NSF-0907179 (P. C. and A. H. N.).

References
[1] Mathur N D *et al.* 1998 *Nature* **394** 39–43
[2] v Lohneysen H, Rosch A, Vojta M and Wolfle P 2007 *Rev. Mod. Phys.* **79** 1015–1075
[3] Gegenwart P, Si Q and Steglich F 2008 *Nature Phys.* **4** 186–197
[4] Nakatsuji S *et al.* 2008 *Nature Phys.* **4** 603–607
[5] Kuga K, Karaki Y, Matsumoto Y, Machida Y and Nakatsuji S 2008 *Phys. Rev. Lett.* **101** 137004
[6] Okawa M *et al.* 2010 *Phys. Rev. Lett.* **104** 247201
[7] Matsumoto Y *et al.* 2011 *Science* **331** 316–319
[8] Matsumoto Y, Kuga K, Karaki Y, Tomita T and Nakatsuji S 2010 *Phys. Status Solidi B* **247** 720–722
[9] Matsumoto Y *et al.* 2011 *preprint*
[10] Yuan H Q, Grosche F M, Deppe M, Geibel C, Sparn G and Steglich F 2003 *Science* **302** 2104–2107
[11] Holmes A T, Jaccard D and Miyake K 2004 *Phys. Rev. B* **69** 024508
[12] Senthil T, Sachdev S and Vojta M 2003 *Phys. Rev. Lett.* **90** 216403
[13] Paul I, Pepin C and Norman M R 2007 *Phys. Rev. Lett.* **98** 026402

Journal of Physics: Conference Series **391** (2012) 012042 doi:10.1088/1742-6596/391/1/012042

Delocalization of the f electron in $Ce_xLa_{1-x}Ru_2Si_2$ - the de Haas-van Alphen effect measurement

Y Matsumoto[1,2], N Kimura[2], T Komatsubara[2], H Aoki[2], N Kurita[3], T Terashima[3] and S Uji[3]

[1]Advanced Science Research Center, Japan Atomic Energy Agency, Tokai, Ibaraki 319-1195, Japan

[2]Graduate School of Science and Center for Low Temperature Science, Tohoku University, Sendai 980-8578, Japan

[3]National Institute for Materials Science, Tsukuba, Ibaraki 305-0003, Japan

E-mail: matsumoto.yuji19@jaea.go.jp

Abstract.
We report the first observation of the continuous Fermi surface (FS) variation from La compound ($LaRu_2Si_2$) with no f electron to Ce compound ($CeRu_2Si_2$) with itinerant f electron via the de Haas-van Alphen (dHvA) effect. The dHvA frequency smoothly varies with Ce concentration and there is no discontinuous change with Ce concentration. It is found that the effective mass and signal amplitude with Ce concentration depends strongly on the Fermi surface sheet, and the effective mass is enhanced toward x_c (= 0.91) and the signal amplitude reduces around x_c or somewhere between x_c and $x \sim 0.8$.

1. Introduction

Many anomalous physical properties like non Fermi liquid and heavy fermion (HF) have been observed and studied in the highly correlated f electron systems. How the f electron nature changes from impurity state to lattice system is the important issue to resolve the anomalous properties. The f electron nature has two different states, that is the itinerant and localized state. If the f electron is itinerant, the f electron contributes to FS. While, if the f electron is localized, the f electron does not contribute to FS. The dHvA measurement is a powerful tool to measure the FS, so the dHvA measurement have been performed to resolve the issue. When the f electron can be assumed to be localized in the Ce compound like $Ce_xLa_{1-x}B_6$ [1], the FS properties can be followed via the dHvA effect measurements from the La compound to the Ce compound. These studies indicate that the f electron is likely to be localized in the impurity state. In the case of f electron is itinerant in the Ce compounds, the dHvA signal can be observed in a small concentration range [2], so how the f electron nature changes from impurity state to lattice system is not obvious.

$CeRu_2Si_2$ is a typical paramagnetic heavy fermion compound with the electronic specific heat coefficient γ of 350 mJ/mole K^2 [3]. The f electron of $CeRu_2Si_2$ is itinerant [4]. When a magnetic field is applied parallel to the [001] direction its the tetragonal $ThCr_2Si_2$ - type structure, a metamagnetic like transition takes place at B_m (= 7.7 T) and Fermi surface (FS) drastically changes from the FS of itinernat f electron to $LaRu_2Si_2$ like FS [4]. By replacing Ce with La, $Ce_xLa_{1-x}Ru_2Si_2$ changes to antiferromagnetic at x_c = 0.91 and finally normal metal of $LaRu_2Si_2$ with no f electron [6]. When a magnetic field is applied in the (001) plane, the magnetization increases gradually with field and no metamagnetic behavior is observed at least up to 18 T. We have extensively studied Fermi surface properties in dilute Ce concentration samples for magnetic field parallel to the [100] direction [5]. However we have not observed the obvious dHvA signal in high Ce concentration samples. We have observed the continuous Fermi surface variation from $x = 0.0$ ($LaRu_2Si_2$) with no f electron to 1.0 ($CeRu_2Si_2$) via the dHvA effect for a magnetic field parallel to the [110] direction and we report the variation of dHvA frequency, effective mass and signal amplitude with Ce concentration.

2. Experimentals

Single crystals of $Ce_xLa_{1-x}Ru_2Si_2$ were grown by using the Czochralski pulling method in a tetra-arc furnace under Ar gas atmosphere and were annealed under vacuum for a week at 900 °C. The lattice parameter was determined by a powder X-ray diffraction analysis. Unit cell volume changes almost linearly with Ce concentration x from $x = 0.0$ to 1.0. Several samples ranging from $x = 0.02$ to 0.5 were analyzed by the inductively coupled plasma method. The measured content x agrees with the nominal content within 10 %. We use the nominal concentration x to label the sample.

The dHvA effect measurements were performed by using the conventional field modulation method in dilution refrigerators under magnetic fields up to 18 T. The signal amplitude is an average obtained from the Fourier analysis between 10 and 16 T at about 0.1 K, and the signal amplitude of $LaRu_2Si_2$ is normalized to 1.0 for all of oscillation [6], finally the amplitude make adjustment of the influence of the dHvA frequency and effective mass.

3. Result and discussion

We have successfully observed dHvA oscillation of $Ce_xLa_{1-x}Ru_2Si_2$ from $x = 0.0$ to 1.0 for a magnetic field along the [110] direction. Figure 1 shows the dHvA signal and its Fourier spectrum for $x = 0.02$, 0.51, 0.90 and 0.97. The left panels (a1), (b1), (c1) and (d1) of Fig. 1 show the dHvA signals obtained in the field range between 11 and 17.5 T. The right panels (a2), (b2), (c2) and (d2) of Fig. 1 show their corresponding Fourier spectra in the frequency range between 0 and 2000 T. Fermi surfaces of $LaRu_2Si_2$ (a) and $CeRu_2Si_2$ (b) show in Fig. 2 [7, 8]. The figures indicate the names of the dHvA oscillations and the corresponding orbits when the magnetic field is applied parallel to the [110] direction. We temporally denote the symbol of the dHvA oscillation corresponding to those in $LaRu_2Si_2$ from $x = 0.0$ to x_c, while the symbol of the dHvA oscillations corresponding to those in $CeRu_2Si_2$ from x_c to $x = 1.0$. All the dHvA oscillations corresponding to those in $LaRu_2Si_2$ i.e. the ω, γ', β' and R oscillation can be observed for $x = 0.04$. The β' and γ' oscillations arise from the ellipsoidal small hole surfaces, the R oscillation arises from the ring electron surface and the ω oscillation arises from the large hole surface. In antiferromagnetic state of $x = 0.51$, the two clear oscillations are observed, while the γ' oscillation is only observed for $x = 0.90$, indicating that there is no maximum in the amplitude reduction at $x = 0.5$. In paramagnetic state of $x = 0.97$, the β and γ oscillations from small ellipsoidal hole surfaces are observed.

Figure 3 shows the dHvA frequency, effective mass and signal amplitude plotted against Ce concentration for magnetic field parallel to the [110] direction. We plot the dHvA frequencies of β'-β, γ'-γ and R oscillations as a function of Ce concentration x in the top panel (a). We have observed the γ'-γ oscillation from $x = 0.0$ to 1.0. The frequency of γ'-γ oscillation continuously varies from $x = 0.0$ to 1.0. Especially there is not discontinuous change at x_c.

We also plot the effective masses of observed oscillations as a function of Ce concentration x in the middle panel (b). In low Ce concentration, the effective mass of the R oscillation increases with x, while the effective of the β' and γ' oscillation moderately increase from $x = 0.0$ to about

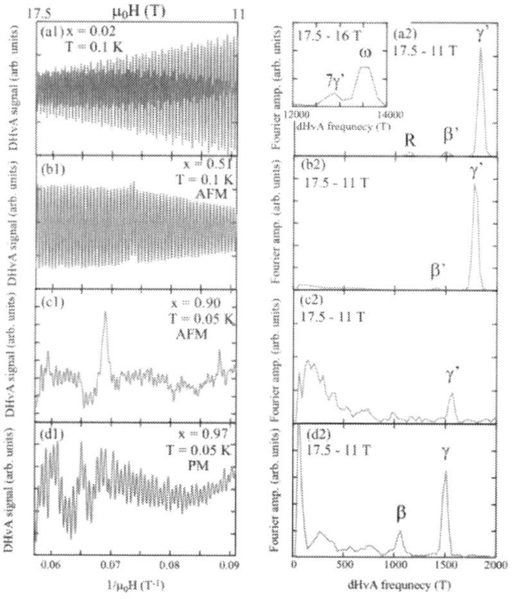

Figure 1. DHvA signals and Fourier spectra for $Ce_xLa_{1-x}Ru_2Si_2$ when a magnetic field is applied parallel to the [110] direction. DHvA signals for x = 0.02 of impurity Kondo state (a1), $x = 0.51$ of antiferromagnetic state (b1), x = 0.90 near x_c (c1) and $x = 0.97$ of paramagnetic state (d1) as a function of inverse magnetic filed and the corresponding Fourier spectra for x = 0.02 (a2), $x = 0.51$ (b2), $x = 0.90$ (c2) and x = 0.97 (d2). The inset of (a2) shows the Fourier spectrum for the ω oscillation.

0.6. The effective mass of γ' oscillation increases toward x_c, indicating that the effective mass of γ-γ' oscillation is enhanced around x_c.

We also plot the signal amplitude of the β'-β, γ'-γ and R oscillations as a function of Ce concentration x in the bottom panel (c). The signal amplitudes of the R and ω oscillation rapidly decreases with increasing x, while the amplitude reduction of the β' and γ' oscillations are much smaller. The signal amplitude of γ-γ' oscillation largely decreases around x_c, and the amplitude is about 10^{-4} times as small as that of $LaRu_2Si_2$.

We discuss the dHvA effect results. We could observe the dHvA oscillation from the system with no f electron system to the itinerant f electron system for the first time. The frequency of the γ'-γ oscillation smoothly varies from $x = 0.0$ to 1.0. Our previous study is indicated that the f electron is itinerant in impurity state. If the f electron is itinerant from the impurity state to $CeRu_2Si_2$, it is expected that all of the dHvA frequencies smoothly varies from $x = 0.0$ to 1.0. Therefore, the result is consistent with our previous study. However we could observe only the γ'-γ oscillation arisen from the small Fermi surface with poor f electron content.

The signal amplitude of the γ'-γ oscillation seems to be minimum around x_c and the effective mass is also enhanced around x_c. Since the signal amplitude is a function of effective mass as well as electron scattering, it is not clearwhether or not the minimum of the signal amplitude is at x_c. In the sister system $CeRu_2(Si_{1-x}Ge_x)_2$, it is found that the residual resistivity becomes maximum in the antiferromagnetic state near x_c not at x_c [9]. The maximum of the residual resistivity is discussed by Hattroi and Miyake in terms of enhanced charge fluctuation due to the competition between the antiferromagnetic interaction and the Kondo effect, and resultant enhancement of the impurity scattering. They predict that the enhancement takes place when $T_N \sim T_K$ [10]. The Ge concentration for the maximum of residual resistivity seems to takes place when $T_N \sim T_K$ in $CeRu_2(Si_{1-x}Ge_x)_2$. On the other hand, T_N is equal to T_K around x = 0.85 in $Ce_xLa_{1-x}Ru_2Si_2$. Then, considering the correspondence between the two alloy systems, the signal amplitude is likely to be minimum around $x = 0.85$ rather than at x_c.

Acknowledgments

We thank M. Suzuki, M. Kikuchi and H. Moriyama for technical support. This work was supported by Grants-in-Aid for Scientific Research (A) 19204034 and for Scientific Research

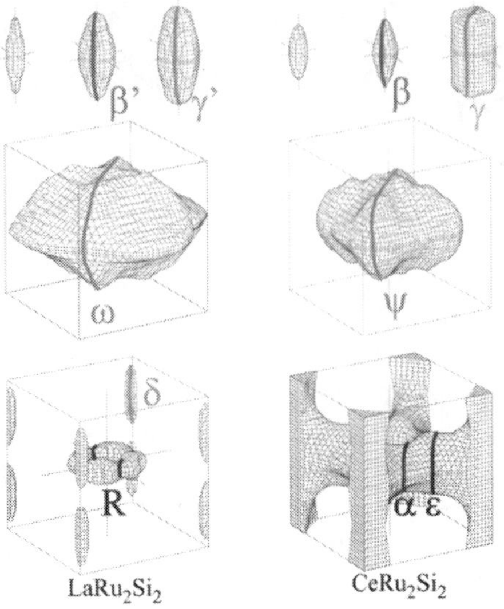

Figure 2. Fermi surfaces of $LaRu_2Si_2$ (a) and $CeRu_2Si_2$ (b) [7, 8]. In both the figures, the Fermi surfaces at the top and middle are hole surfaces and those at the bottom are electron surfaces. The figures indicate the names of the dHvA oscillations and the corresponding orbits when the magnetic field is applied parallel to the [110] direction.

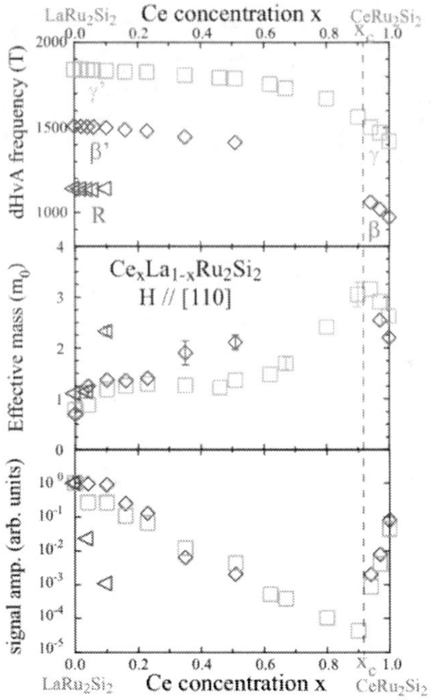

Figure 3. DHvA frequencies (a), effective masses (b) and signal amplitude (c) plotted as a function of Ce concentration x in $Ce_xLa_{1-x}Ru_2Si_2$ for a magnetic field along the [110] direction. The left axis of (c) is logarithmic scale. The open squares denote the frequencies of the γ' and γ oscillations. The open triangles denote the frequencies of the R oscillation. The open diamonds denote the frequencies of the β' and β oscillations.

on Innovative Areas "Emergence of Heavy Electrons and Their Ordering" from MEXT Japan. YM acknowledges financial supports from Global COE program "Weaving Science Web beyond Particle-matter Hierarchy" and from Research fellowship for Young Scientists JSPS.

References

[1] Endo E, Nakamura S, Isshiki T, Kimura N, Nojima T, Aoki H, Harima H and Kunii S 2006 *J. Phys. Soc. Jpn.* **75** 114704.
[2] Harrison N, Alver U, Goodrich R G, Vekhter I, Sarrao J L, Pagliuso P G, Noreno N O, Balicas L and Fisk Z 2004 *Phys. Rev. Lett.* **93** 186405.
[3] Haen P, Flouquet J, Lapierre F, Lejay P and Remenyu G 1987 *J. Low. Temp. Phys.* **67** 391.
[4] Aoki H, Uji S, Albessard A K, Onuki Y 1993 *Phys. Rev. Lett.* **71** 2110.
[5] Matsumoto Y, Kimura N, Komatsubara T, Kimata M, Terashima T and Uji S 2010 *J. Phys. Soc. Jpn.* **79** 083706.
[6] Matsumoto Y, Sugi M, Kimura N, Komotsubara T, Aoki H, Satoh I, Terashima T and Uji S 2008 *J. Phys. Soc. Jpn.* **77** 053703.
[7] Yamagami H and Hasegawa A 1992 *J. Phys. Soc. Jpn.* **61** 2388.
[8] Yamagami H and Hasegawa A 1993 *J. Phys. Soc. Jpn.* **62** 592.
[9] Matsumoto Y, Aoki K, Shimizu Y, Kimura N, Komatsubara T, Kimata M, Terashima T and Uji S 2011 *J. Phys. Soc. Jpn.* **80** 074716.
[10] Hattori K and Miyake K 2010 *J. Phys. Soc. Jpn.* **79** 073702.

Field-induced collective spin-exciton condensation in a quasi-2D $d_{x^2-y^2}$-wave heavy electron superconductor

Vincent P. Michal
Commissariat à l'Energie Atomique, INAC/SPSMS, 38054 Grenoble, France
E-mail: vincent.michal@cea.fr

Abstract. The origin of the spin resonance observed in $CeCoIn_5$ with Inelastic Neutron Scattering is subject to debate. It has been shown recently that in this heavy electron compound at low temperature an instability to a ground state with coexisting $d_{x^2-y^2}$-wave superconductivity and Spin Density Wave (SDW) order in a magnetic field is a corollary of the consideration of a collective spin excitation mode in a quasi-2D $d_{x^2-y^2}$-wave Pauli-limited superconductor. This provides a natural scenario for the occurence of the puzzling high-field-low-temperature phase highlighted in $CeCoIn_5$. We present perspectives on this ground state transition and propose directions for future experiment.

1. Background information and model considerations in the spin-fermion approach

Modelling and understanding collective phenomena in correlated electron systems in general and new superconductors in particular is an interesting and challenging issue. The two-side observation of spin resonances in high-temperature and heavy-fermion superconductors has raised the question of whether the phenomena in these very different systems have a common mechanism. Interpreting Inelastic Neutron Scattering data [1] for $CeCoIn_5$ has recently given rise to interesting debate [2]. We have proposed [3] that the ground state transition in magnetic field observed [4] in superconducting $CeCoIn_5$ (superconducting critical temperature $T_c \simeq 2.3$K) is a consequence of the existence of the superconductor collective spin excitation mode [1] whose condensation[1] explains naturally the confinement of magnetic ordering within the superconductor.

Historically measurement in $CeCoIn_5$ phase diagram of a line of discontinuity in the specific heat indicated a second-order transition to a high-field-low-temperature phase (10T $< H <$ 11.5T, $T < 0.4$K) which was initially thought to be a realisation of the Fulde-Ferrel-Larkin-Ovchinnikov (FFLO) state. Later observation of magnetic Bragg peaks [4] at incommensurate wavevector $\mathbf{q} = (1/2 - \delta, 1/2 - \delta, 1/2)$ (in units of the tetragonal crystal Brillouin zone dimensions) with $\delta \simeq 0.05$ whose appearence coincided with this very transition line gave evidence that the ground state had a magnetic character with magnetic coherence length $\sim 3000\AA$ [4] extanding well beyond the vortex cores and ordered magnetic moments of 15% the Bohr magneton [4]. Nuclear Magnetic resonance (NMR) experiment [5] supported characterise the magnetic phase.

When not superconducting $CeCoIn_5$ is a semi-metal with non-Fermi liquid resistivity exponent and a Kondo temperature $T_K \sim 7$K while the Kondo lattice coherence temperature

[1] Here the term *condensation* is used in the sense of a transition of a collective excitation mode to static ordering.

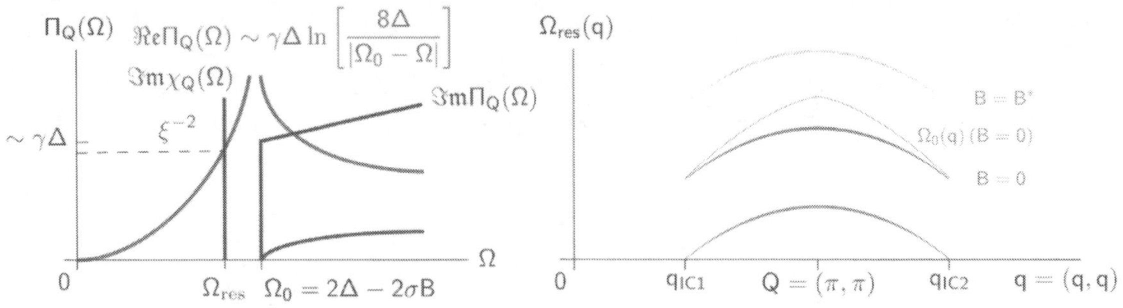

Figure 1. Left: Behaviour of the uniaxial dynamic spin susceptibility (see text for discussion). The effect of the magnetic field is to split the threshold energy Ω_0 into a two-step discontinuity. Right: Cartoon of the predicted resonant mode dispersion and threshold energy at zero field [10], evolution with transverse field in the uniaxial case and criticality at incommensurate wave-vector.

$T_{\rm coh} \sim 50$K [6] yielding low temperature hybridized bands with large effective masses and a large specific heat discontinuity at the transition to the superconductor. In magnetic field the metal to superconductor transition is observed to be first order at temperature $T \lesssim 1$K, a consequence of a large Maki parameter $\alpha_{M0} \sim H_{c20}^{\rm orb}/H_{c20}^{\rm P}$ ($H_{c20}^{\rm orb}$ [$H_{c20}^{\rm P}$] is the orbital [paramagnetic] upper critical field at zero temperature) which itself follows from a large electron effective mass. We consider a constant interaction magnetic instability from the superconducting, non-magnetic side. The model starts with a quasi-2D (α-band in $CeCoIn_5$ [7]) itinerant heavy electron band close to a commensurate SDW ground-state instability [8] with antiferromagnetic hot-spots (points in wave-vector space that belong to the Fermi line and are connected by the diagonal antiferromagnetic ordering wave-vector) which below T_c becomes a superconductor with $d_{x^2-y^2}$-wave symmetry gap [9] and include a magnetic field B. In the spin-fermion approach [2, 10] spin density excitations and fermion excitations couple through a constant g and yield the dynamic spin susceptibility

$$\chi^{-1}(\mathbf{q},\Omega) = \chi_0^{-1}\Big[\xi^{-2} + |\mathbf{q}-\mathbf{Q}|^2 - \Pi^{xx}(\mathbf{q},\Omega)\Big], \tag{1}$$

where ξ is the magnetic correlation length.To account for the magnetic anisotropy of the physical system, spin-density excitations are considered along a single axis (c-axis of the tetragonal crystal) whose direction is set prependicular to the magnetic field. The dimensionless spin excitation self-energy due to fermions (sometimes termed spin polarization function) $\Pi(\mathbf{q},\Omega)$ thus writes[2]

$$\begin{aligned}\Pi^{xx}(\mathbf{q},\Omega) &= -g^2\chi_0 T\sum_{\mathbf{k},m,\sigma}\Big[G_\sigma(\mathbf{k},i\omega_m)G_{-\sigma}(\mathbf{k}+\mathbf{q},i\omega_m+\Omega+i0^+)\\ &+ F_\sigma(\mathbf{k},i\omega_m)F^+_{-\sigma}(\mathbf{k}+\mathbf{q},i\omega_m+\Omega+i0^+)\Big],\end{aligned} \tag{2}$$

where G, F and F^+ are Green's functions in a superconductor with Zeeman field [3] characterized by the spectrum of fermion excitations $E^s_{k\sigma} = s[\epsilon_k^2 + \Delta_k^2]^{1/2} + \sigma B$ ($s, \sigma = \pm 1$ are the particle-hole and spin indices respectively, ϵ_k is the zero-field spectrum in the metal [3], and the $d_{x^2-y^2}$-wave gap in general reads $\Delta_k = (\Delta_0/2)\sum_n a_n[\cos(nk_a) - \cos(nk_b)]$). The properties of the

[2] Eq. 2 includes magnetic field effect in the Pauli limit. The influence of the electron orbital motions on the spin susceptibility can be evaluated [11] in the linearized Doppler shift approximation and gives correction of the order $B/(H^{\rm P}_{c20}\alpha_{M0})$ which is small in a Pauli-limited superconductor as expected.

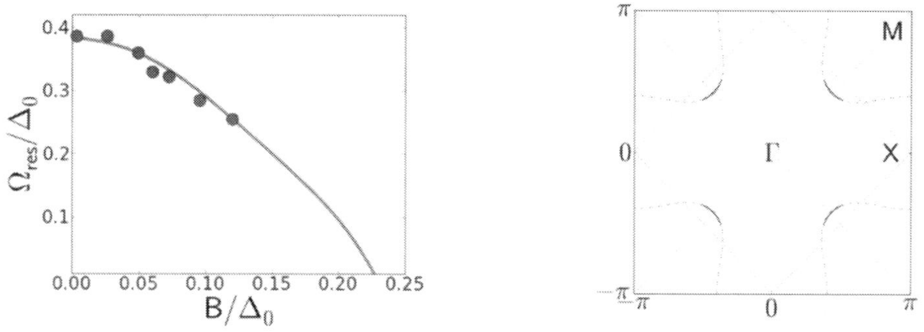

Figure 2. Left: Collective mode energy variation with field at incommensurate wavevector $\mathbf{q} = (0.45, 0.45)$, and comparison with experiment [12]. Right: 2D Brillouin zone including in dotted grey the line where the superconducting gap changes sign and the antiferromagnetic reduced Brillouin zone, and in blue the Fermi pockets near the gap nodes at the condensation field $B^* = 0.23\Delta_0$ [3].

2D function Π at one loop level is known for zero field [2] and is considered here in Zeeman magnetic field (c. f. Fig. 1). Here we consider that the dynamics of the mode is determined by the superconducting gap such that Eq. 1 yields the condition $\xi^{-2} + |\mathbf{q} - \mathbf{Q}|^2 - \Re\mathfrak{e}\Pi(\mathbf{q}, \Omega) = 0$ ($\Omega > 0$) defining the $\mathrm{d}_{x^2-y^2}$-wave superconductor collective spin-excitation mode referred to as spin-exciton. The spectral function (the imaginary part of the dynamic spin susceptibility) of the mode looks like a Dirac delta in absence of Landau damping by particle-hole pair creation, which translates into the condition $\Im\mathfrak{m}\Pi(\mathbf{q}, \Omega)/(\gamma\Delta) \ll 1$ with $\gamma = Ng^2\chi_0/(2\pi|\mathbf{v}_{\mathrm{khs}} \times \mathbf{v}_{\mathrm{khs+q}}|)$, N the number of hot-spots and $\mathbf{v}_{\mathrm{khs}}$ the Fermi velocity at hot-spot.

2. From collective spin-excitation mode in a quasi-2D $\mathrm{d}_{x^2-y^2}$-wave superconductor to ground state instability in a transverse magnetic field

The fluctuation-dissipation theorem relates the dynamic structure factor measured in Inelastic Neutron Scattering experiment [1, 12] with the spectral function of the model (Eq. 1) $\mathcal{S}(\mathbf{q}, \Omega) = 2\Im\mathfrak{m}\chi(\mathbf{q}, \Omega + i0^+)/(1 - e^{-\Omega/T})$, the latter being the analytic continuation of the imaginary time spin correlation function $\chi(\mathbf{q}, i\Omega_n) = \int_0^{1/T} d\tau \int d^2\mathbf{r}\, e^{-i\mathbf{q}\cdot\mathbf{r} + i\Omega_n\tau}\langle S^x(\mathbf{r}, \tau)S^x(0, 0)\rangle$.

The collective spin-excitation mode we discuss is a property of a quasi-2D $\mathrm{d}_{x^2-y^2}$-wave superconductor on the border of a commensurate SDW continuous ground state transition (an s-wave gap yields a negative Π with no divergence and the logarithmic divergence (Fig. 1) associated with the particle-hole excitation threshold in the d-wave case is due to two-dimensionality [2]). This d-wave effect is the one brought about by coherence factors in the BCS formalism.

The mode dispersion is schematically represented in the right-hand side of Fig. 1 at zero field and at the condensation field. The mode spectral weight is maximum at commensurate wave-vector because there the threshold energy is also maximum. Away from this point the downward dispersion [10] follows from the wave-vector dependence of the d-wave gap in the Brillouin zone. A transverse magnetic field splits the mode into a branch which goes down in energy and an upper part which becomes damped by the continuum of particle-hole excitations. The consideration of magnetic isotropy gives a splitting between three undamped modes [13]. Ongoing Inelastic Neutron Scattering experiment [12] is providing information on the actual magnetic anisotropy of the system and is showing deviation from strict uniaxiality.

The left hand side of Fig. 2 shows the evolution of the resonance energy with field at fixed incommensurate wave-vector $\mathbf{q} = (0.45, 0.45)$ [3] where we have taken account of the variation of the gap magnitude with temperature and magnetic field by considering the superconductor gap equation (we found the zero temperature first order critical field $H^p_{c20} = 0.36\Delta_0$ corresponding to the band structure considered in [3]). The right-hand side of Fig. 2 shows Fermi pockets induced by magnetic field approaching the antiferromagnetic hot-spot locations. Before the pockets reach these points the ground state transition occurs together with reconstruction of the Fermi surface. This must be visible with Nuclear Magnetic Resonance.

3. Discussion, perspectives and conclusion

We now turn to some perspectives on this instability of the superconducting ground state. First there is a possibility of a double-q structure which follows from the mode dispersion (Fig. 1) and is consistent with the prediction of Y. Kato et al. [14], there seen as a consequence of the incommensurate ordering wave-vector connecting nested pockets. As was emphasized in [14], this degeneracy is however expected to be lifted by small coupling between the electron orbital motion and the magnetic field. We suspect magnetostriction effects also play a part. It has been also conjectured [15] the existence of a phase where SDW lives in a Larkin-Ovchinnikov (LO) superconducting state (where the order parameter spatially varies along the magnetic field direction). Because the two orders couple, this should give rise to additional length scale for space dependence of the SDW order parameter with experimental hallmark as Bragg peaks measurable with elastic neutron scattering. Such a prediction signalling coexistence between LO and SDW is waiting for experimental verification (this is also true for inhomogeneous superconductivity with a different spin state such as Pair Density Wave state).

We saw that the specificities of the heavy electron superconductor $CeCoIn_5$ brings all conditions (first order transition [4] resulting from Pauli limiting, tendency towards two-dimensionality [7], pairing symmetry [9], border of antiferromagnetism [8]) for realizing a ground state instability which is the natural result of the condensation under magnetic field of a preexisting spin collective excitation mode in the $d_{x^2-y^2}$-wave superconductor. As noted in [2], collective phenomena in heavy electron compounds involve a physics that stems from a lattice of localized f-electrons coupling with conduction electrons. Understanding the nature of the heavy delocalized electrons and their relation with superconductivity remains a significant challenge.

4. Acknowledgments

It is a pleasure to thank V. P. Mineev for support and also D. Agterberg, J. Flouquet, S. Gerber, P. B. Littlewood, S. Raymond, and I. Vekhter for many stimulating discussions.

[1] C. Stock, C. Broholm, J. Hudis, H. J. Kanga and C. Petrovic, Phys. Rev. Lett. **100**, 087001 (2008).
[2] A. V. Chubukov and L. P. Gor'kov, Phys. Rev. Lett. **101**, 147004 (2008).
[3] V. P. Michal and V. P. Mineev, Phys. Rev. B **84**, 052508 (2011).
[4] M. Kenzelmann *et al.*, Science **321**, 1652 (2008). Phys. Rev. Lett. **104**, 127001 (2010).
[5] G. Koutroulakis *et al.*, Phys. Rev. Lett. **104**, 087001 (2010).
[6] Y.-F. Yang, Z. Fisk, H.-O. Lee, J. D. Thompson and D. Pines, Nature **454**, 611 (2008).
[7] R. Settai *et al.*, J. Phys.: Condens. Matter **13** L627 (2001). D. Hall *et al.*, Phys. Rev. B **64**, 212508 (2001).
[8] Pagliuso *et al.*, Physica B **129**, 312 (2002).
[9] A. Vorontsov and I. Vekhter, Phys. Rev. Lett. **96**, 237001 (2006).
[10] A. V. Chubukov, B. Jankó and O. Tchernyshyov, Phys. Rev. B, **63**, 180507(R) (2001) and references therein.
[11] M. Eschrig, M. R. Norman, and B. Jankó, Phys. Rev. B **64**, 134509 (2001).
[12] J. Panarin *et al.*, J. Phys. Soc. Jpn. **78**, 113706 (2009). C. Stock et al., arXiv:1203.2189v1. S. Raymond.
[13] J.-P. Ismer, I. Eremin, E. Rossi, and D. K. Morr, Phys. Rev. Lett. **99**, 047005 (2007).
[14] Y. Kato, C. D. Batista, and I. Vekhter, Phys. Rev. Lett. **107**, 096401 (2011).
[15] Y. Hatakeyama and R. Ikeda, Phys. Rev. B **83**, 224518 (2011)

NMR and NQR investigations of local symmetry in the hidden order phase of URu_2Si_2

T Mito[1], M Hattori[1], G Motoyama[1], Y Sakai[1], T Koyama[1], K Ueda[1], T Kohara[1], M Yokoyama[2], and H Amitsuka[3]

[1] Graduate School of Material Science, University of Hyogo, Hyogo 678-1297, Japan
[2] Faculty of Science, Ibaraki University, Mito 310-8512, Japan
[3] Graduate School of Science, Hokkaido University, Sapporo 060-0810, Japan

E-mail: `mito@sci.u-hyogo.ac.jp`

Abstract. In order to study local symmetries at the 4*e*(Si)- and 4*d*(Ru)-sites above and below hidden order transition temperature $T_{\rm O} = 17.5$ K in URu_2Si_2, we have investigated the nuclear quadrupole interaction by ^{73}Ge-nuclear magnetic resonance (NMR) measurement on a 10% ^{73}Ge-substituted sample $URu_2(Si_{0.9}Ge_{0.1})_2$, and by ^{101}Ru-nuclear quadrupole resonance (NQR) measurement on a pure single crystal sample. The present ^{73}Ge-NMR measurements do not give any evidence for change in the local symmetry at 4*e*-site within experimental accuracy. On the other hand, the precise measurement of ^{101}Ru-NQR frequency has detected an anomaly just below $T_{\rm O}$ as sensitively as the thermal expansion measurement dose. Its temperature dependence shows a linear relation with that of the in-plane lattice parameter, which may be a clue to clarify a modification in microscopic charge distribution at $T_{\rm O}$.

1. Introduction

The mysterious order at $T_{\rm O} = 17.5$ K in URu_2Si_2, so called the "hidden order", has been a topic of interest and controversy for more than two decades. In order to identify how symmetry of the system changes at $T_{\rm O}$, information about local symmetry at each site has recently attracted considerable attention. In this context, investigation of the nuclear quadrupole interaction, which can be probed by nuclear magnetic resonance (NMR) and nuclear quadrupole resonance (NQR) measurements, is useful, because it detects the modification of local charge distribution as follows; the nuclear spin I (restricted to $I \geq 1$) interacts with the electric-field gradient (EFG) caused by surrounding charges though $\mathcal{H}_Q = (h\nu_{\rm Q}/6)\left[3I_z^2 - I^2 + \eta\left(I_x^2 - I_y^2\right)\right]$, where $\nu_{\rm Q}$ is the NQR frequency of the $\pm 1/2 \leftrightarrow \pm 3/2$ transition, Q is the quadrupolar moment of the nucleus, the asymmetry parameter $\eta = (V_{xx} - V_{yy})/V_{zz}$, and $V_{\alpha\alpha}$ is EFG at the nuclear position along the α direction, following $V_{zz} \geq V_{xx} \geq V_{yy}$. In this paper, we report the results of the measurements of the nuclear quadrupole interactions at the 4*e*(Si)- and 4*d*(Ru)-sites. For the 4*e*-site, because ^{29}Si nuclear spin ($I = 1/2$) does not couple to the EFG, we performed ^{73}Ge-NMR measurement on a partially ^{73}Ge-substituted sample, $URu_2(Si_{1-x}Ge_x)_2$.

2. Experimental details

We prepared two kinds of samples; a single crystalline sample of URu_2Si_2 and a 10% Ge-substituted polycrystalline sample, $URu_2(Si_{0.9}Ge_{0.1})_2$. The latter was prepared by arc-melting constituent elements together in an argon atmosphere. Enriched ^{73}Ge (98%) with minimum purity of 99.9% was used for the substitution, which enebles us to measure the nuclear quadrupole interaction at the 4e-site due to $^{73}I = 9/2$. With this sample, we have also measured the susceptibility χ by a commercial SQUID magnetometer at a field $B = 1$ T, and the specific heat C using the adiabatic heat pulse method. As for the single crystal, which was grown along the tetragonal a axis by the Czochralski technique in a tri-arc furnace, we have measured ^{101}Ru-NQR as well as the thermal expansion. The thermal expansion was measured using the strain gauge technique with a copper block as a dummy sample. All the NMR and NQR experiments were carried out by using the spin-echo technique with a phase-coherent pulsed spectrometer.

3. Results and Discussion

Figure 1 shows the T-dependences of χ and C/T for the $URu_2(Si_{0.9}Ge_{0.1})_2$ sample. A second-order transition is clearly observed, which indicates that the transition is still a bulk property. This is consistent with a previous report [1]. Both maxima of C/T and $d\chi/dT$ give $T_O \sim 16.5$ K. Because T_O increases with pressure [2], the slightly lower transition temperature than in pure URu_2Si_2 is explained in terms of a negative pressure effect due to the partial substitution of Si with larger Ge.

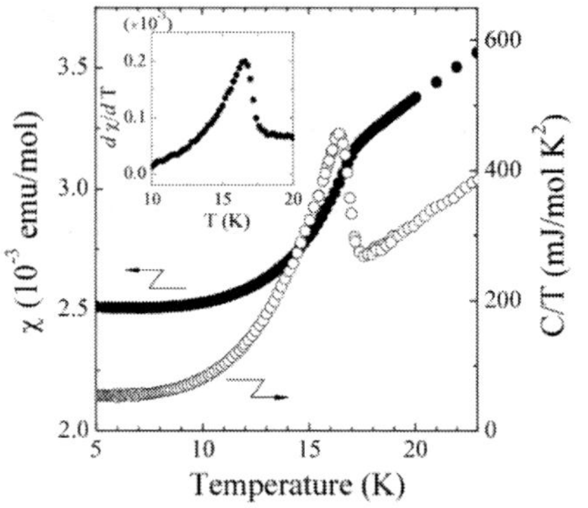

Figure 1. T-dependences of χ (solid circles) and C/T (open circles) [3] for the 10% Ge-substituted sample $URu_2(Si_{0.9}Ge_{0.1})_2$. Inset shows T-derivative of χ, $d\chi/dT$, as a function of T.

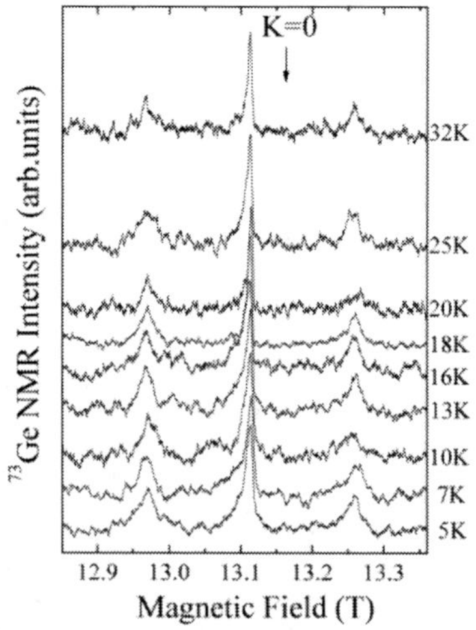

Figure 2. ^{73}Ge-NMR lines of $URu_2(Si_{0.9}Ge_{0.1})_2$ for $B \perp c$, which was measured at a frequency of 19.55 MHz.

Figure 2 shows ^{73}Ge-NMR lines for $B \perp c$ measured above and below T_O. Note we used an oriented powder sample along the c-axis by using strongly anisotropic magnetization. This alignment method leaves a random distribution of a-b basal-plane orientations. Although the data are somewhat scattered because of very weak signal intensity, we observed the definite effect of the nuclear quadupole interaction which is characterized by $\nu_Q = 0.43$ MHz [3]. If the local symmetry at the 4e-site is lowered below T_O, one expect a change in lineshape in the order state. For example, a breakdown of 4-fold symmetry should give rise to a non-zero η value at this site, resulting in a line broadening and/or splitting of quadrupole-split satellite peaks. As a result of more intensive measurements than in our previous report [3], no systematic change in the lineshape was observed. The present experimental accuracy yields $\eta < 0.1$ at the $4e$-site below T_O [3], which may not be enough accurate to conclude the absence of the breakdown of the 4-fold symmetry in the hidden order phase. We therefore need more precise measurements for this issue.

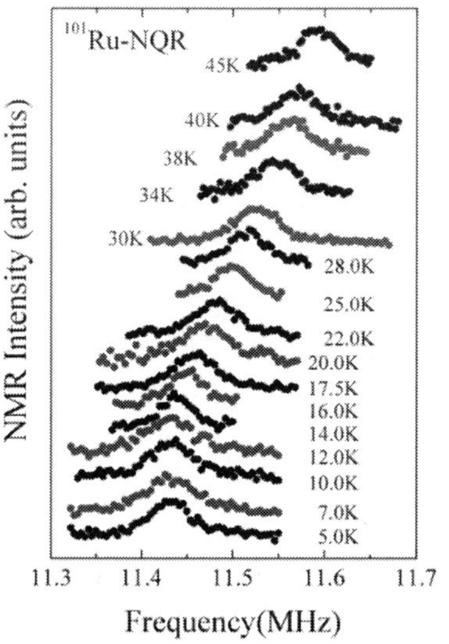

Figure 3. T-dependence of ^{101}Ru-NQR line measured at $\pm 3/2 \leftrightarrow \pm 5/2$ transition.

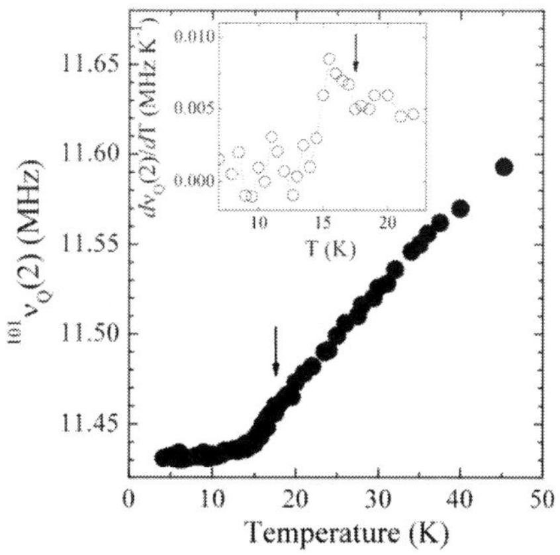

Figure 4. T-dependence of $\nu_Q(2)$, which was obtained by a Lorentzian fit to the spectrum data shown in Fig. 3. Inset: Plot of the T-derivative of $\nu_Q(2)$, $d\nu_Q(2)/dT$, as a function of T. The arrows show $T_O = 17.5$ K.

Figure 3 shows the T-dependence of ^{101}Ru-NQR line measured near the frequency of the $\pm 3/2 \leftrightarrow \pm 5/2$ transition, $\nu_Q(2)$. While the spectrum is almost independent of T below T_O, it gradually shifts to higher frequency with T above T_O. No significant broadening was observed in the whole T-range of the experiment, consistent with previous reports [4, 5, 6]. $\nu_Q(2)$ obtained by a Lorentzian fit to the spectrum data is plotted as a function of T in Fig. 4. The previous works [5, 6] observed no anomaly in $\nu_Q(T)$ around T_O. Our careful measurement however indicates a small deviation downward just below T_O from a roughly linear T-dependence observed above T_O. This is also manifested by a maximum in the plot of $d\nu_Q(2)/dT$ vs T just below T_O, as shown in the inset of Fig. 4. Structural anomaly at T_O in URu_2Si_2 has been so far detected only by few experiments, including the thermal expansion measurement. The present result demonstrates that the NQR experiment probes the anomaly associated with the onset of the hidden order as sensitively as the thermal expansion measurement dose, as described later.

It is interesting to compare the T-dependence of $\nu_Q(2)$ with those of lattice parameters, since generally $\nu_Q(T)$ is supposed to be a function of a unit cell volume change $\Delta V(T)/V$. In Fig. 5(a) and (b), we show the T-dependences of the relative length changes along the a and c axes, $\Delta l_a/l_a$ and $\Delta l_c/l_c$, respectively. The result indicates that $\nu_Q(2)(T)$ cannot be expressed as a function of $\Delta l_c/l_c$, nor consequently $\Delta V(T)/V$, because $\Delta l_c/l_c$ shows a minimum around 33 K while $\nu_Q(2)(T)$ monotonically changes around this temperature. Instead, we have found that the T-dependence of $\nu_Q(2)$ is very similar to that of $\Delta l_a/l_a$. Actually the plot of the relative frequency change $\Delta\nu_Q(2)(T)/\nu_Q(2)$ vs $\Delta l_a(T)/l_a$ with T as an implicit variable demonstrates a linear relation between them for $4 < T < 45$ K (see Fig. 6). The previous paper [5] indicated that $\nu_Q(2)(T)$ is not described by the $\Delta V(T)$ term, but by the singlet ground-state models for

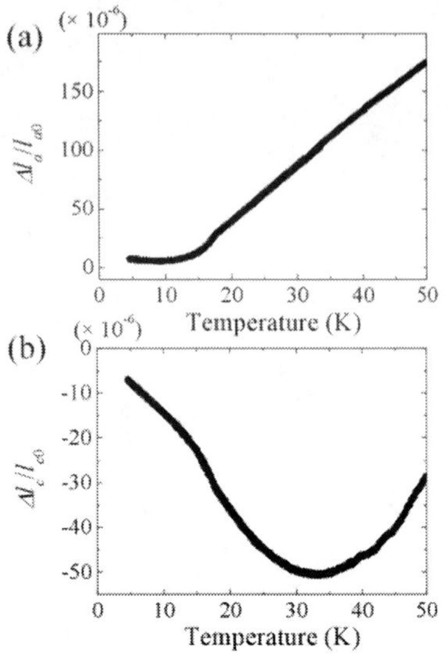

Figure 5. T-dependences of (a) $\Delta l_a/l_a$ and (b) $\Delta l_c/l_c$.

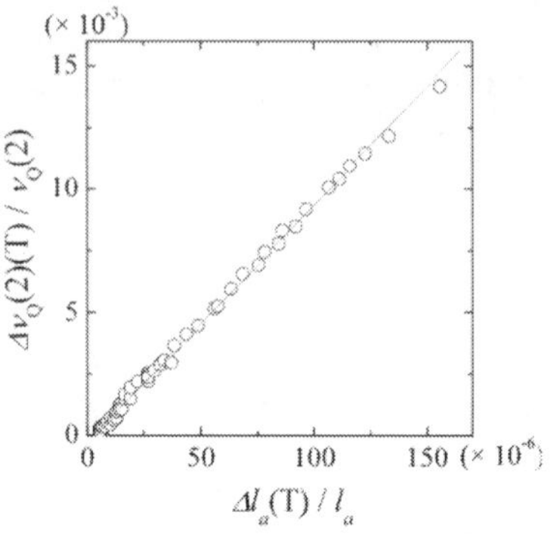

Figure 6. Plot of $\Delta\nu_{\rm Q}(2)/\nu_{\rm Q}(2)$ vs $\Delta l_a/l_a$ with T as an implicit variable. The solid line shows a linear fit to the data.

the $(5f)^2$ configuration of U for $T > T_{\rm O}$. On the other hand, our result reveals that the T-dependence of V_{zz} at the 4d site simply follows $l_a(T)$ in the whole T-range of the measurement. This may be a clue to identify T-induced modification in the microscopic charge distribution.

4. Conclusion

We have investigated local symmetries at the 4d- and 4e-sites above and below the hidden order temperature in URu_2Si_2. The ^{73}Ge-NMR measurement on a $URu_2(Si_{0.9}Ge_{0.1})_2$ sample has not given any evidence for change in lineshape, namely in local symmetry at the 4e-site. We however may need to improve the experimental accuracy to obtain a conclusive evidence for this issue. On the other hand, the T-dependence of ^{101}Ru-NQR frequency which was precisely measured on a pure single crystal has detected an anomaly associated with the hidden order. Also $\nu_{\rm Q}(2)(T)$ turned out to follow the T-dependence of the in-plane lattice parameter at least in the T-range between 4 and 45 K. An analysis based on the microscopic charge distribution will be desired.

Acknowledgments

We thank Prof. H. Harima for valuable discussions. This work was partially supported by Grant-in-Aid for Scientific Research (Grant No. 21540342) and Scientific Research on Innovative Areas "Heavy Electrons" (Grant No. 21102522) from the Ministry of Education, Culture, Sports, Science and Technology of Japan.

References

[1] Park J-G and . Coles B R 1994 *J. Phys.: Condens. Matter* **6** 1425.
[2] Motoyama G, Yokoyama N, Sumiyama A and Oda Y 2008 *J. Phys. Soc. Jpn.* **77** 123710, and references therein.
[3] Mito T, Hattori M, Motoyama G, Sakai H, Koyama T, Ueda K, Kohara T, Yokoyama M and Amitsuka H 2011 *J. Phys. Soc. Jpn.* **80** SA016.
[4] Matsuda K, Kohori Y and Kohara T 1996 *J. Phys. Soc. Jpn.* **65** 679.
[5] Saitoh S, Takagi S, Yokoyama M and Amitsuka H 2005 *J. Phys. Soc. Jpn.* **74** 2209.
[6] Bernal O O, Moroz M E, Ishida K, Murakawa H, Reyes A P, Kuhns P L, MacLaughlin D E, Mydosh J A and Gortenmulder T J 2006 *Physica B* **378-380** 574.

The Novel Phase Diagram of YbPd

Atsushi Miyake[1], Tomoko Kagayama[1], Katsuya Shimizu[1], Takao Ebihara[2]

[1] KYOKUGEN, Osaka University, Toyonaka 560-8531, Japan,
[2] Department of Physics, Shizuoka University, Shizuoka 422-8529, Japan

E-mail: amiyake@cqst.osaka-u.ac.jp

Abstract. Precise electrical resistivity and ac calorimetry measurements under pressures have revealed the (P, T) phase diagram of a mixed-valent compound YbPd. The diagram is rather complicated, re ecting the valence and magnetic instability. In addition to the reported phase transitions at $T_N = 1.9$ K and $T_{MH} = 0.5$ K, the lower temperature transition at $T_{ML} = 0.3$ K has been con rmed thermodynamically. T_N decreases monotonically up to 1.7 GPa and disappears discontinuously at P_c 1.9 GPa, accompanying the drastic enhancement of the effective mass and the residual resistivity, although the ordered phase below T_{ML} still remains. At P_c, the both of magnetic and valence uctuations seems to develop.

1. Introduction

YbPd shows puzzling four phase transitions at 125, 110, 1.9 and 0.5 K, although it crystalizes in a simple CsCl-structure at ambient conditions and has temperature-insensitive intermediate valence of ~2.8 [1]. The two higher temperature transitions may couple to the structural distortions, which are detected as symmetric sharp peaks in temperature dependence of thermal expansion coefficient and specific heat, and a clear thermal hysteresis in the electrical resistivity, implying first-order phase transitions [1, 2]. At $T_N \sim 1.9$ K, an antiferromagnetic phase transition have been confirmed via the magnetic susceptibility and Mössbauer measurements [1, 3]. The susceptibility and the thermal expansion show clear thermal hysteresis at T_{MH} ~0.5 K, while the origin of the phase transition is unclear [1, 4]. Interestingly, it is indicated that the non-magnetic and magnetic Yb-sites are present in almost equal proportions by Mössbauer study [3]. In fact, the magnetization and magnetic entropy were well reproduced by assuming the one half of Yb-atoms are magnetic [2, 4]. Therefore, it is proposed that the one half of magnetic Yb sites form a heavy fermion antiferromagnet [4]. The structural distortion was confirmed by Raman scattering experiments; the crystal symmetry of YbPd at 4 K loses four-fold symmetry, and the distortion may occur below T_1=125 K [5]. Thermal expansion and the electrical resistivity measurement under pressure have revealed that T_1, T_2 and T_N are suppressed with increasing pressure [6]. The magnetic transitions may relate to the high-T structural transitions, but the details at low temperature are not clear yet, i.e. how these transitions are suppressed to 0 K. In this paper, we have revealed the novel pressure temperature phase diagram of YbPd coupling to the magnetic and valence instability through the simultaneous measurements of ac-calorimetry and electrical resistivity at low temperature.

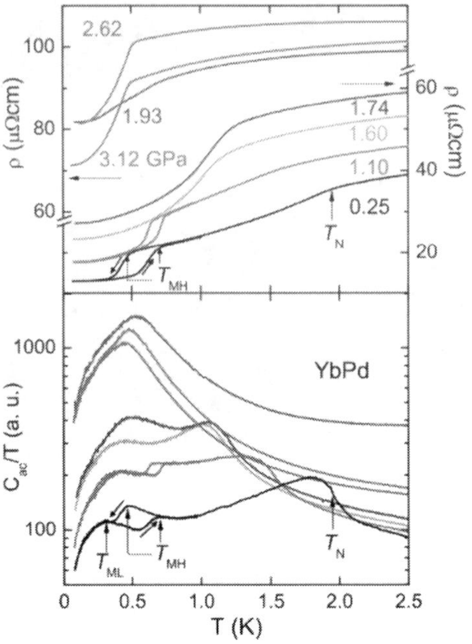

Figure 1. Temperature dependence of (a)ρ and (b)C_{ac}/T of YbPd under several pressures. For (a), it is noted that the right and left scales are for below and above 1.8 GPa, respectively.

2. Experimental

Single crystals were grown by the self flux method [8]. Typical dimensions of the grown crystals were 1 1 1 mm^3. The electrical resistivity ρ and ac calorimetry C_{ac} have been measured simultaneously at low temperature down to 0.06 K and under pressures up to 3.2 GPa. The small sample used here was cut and polished to roughly 230 150 30 m^3. Two Au-AuFe (0.07%) thermocouples (TC) are welded to the both ends of sample, and two Au wires are also attached between TCs for ρ measurement. For C_{ac} measurement, a modulating Joule heat was applied through the sample or TC, and the oscillating temperature was measured by another TC. The relative pressure change of C_{ac} can be estimated directly at a fixed temperature, same modulation current and frequency, though this includes the contribution of the addenda from pressure medium, diamond, etc, which are assumed to be negligibly smaller than that of YbPd and does not vary much as a function of pressure. Pressure was applied at room temperature by a diamond anvil pressure cell and estimated by a ruby fluorescence method at low temperature. Glycerin was used as a pressure medium, which solidifies around 4 GPa and at room temperature [7], indicating that the pressure was applied homogeneously in this experimental pressure region.

3. Result and Discussion

Fig. 1 represents temperature dependence of ρ and C_{ac}/T under several pressures and below 2.5 K. At the lowest measured pressure of 0.25 GPa, $\rho(T)$ shows clear anomalies corresponding to the phase transitions at T_N ~1.9 K and T_{MH} ~0.5 K, being in good agreement with previous reports [1, 4]. A clear thermal hysteresis is observed around T_{MH}. λ-type anomaly at T_N and a peak at T_{MH} also appear in $C_{ac}(T)/T$. In addition, a phase transition at T_{ML} ~0.3 K has been confirmed thermodynamically, which is also seen in the susceptibility at ambient pressure

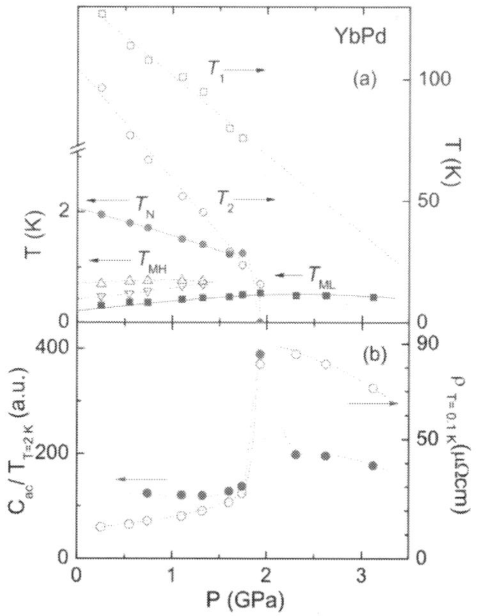

Figure 2. (a)Pressure-temperature phase diagram of YbPd. The symbols △ and ▽ represent T_{MH} on warming and cooling, respectively. (b)Pressure dependence of the C_{ac}/T at 2 K and the ρ at 0.1 K. The lines are guides.

[1], indicating that five phase transitions take place below room temperature. With increasing pressure, T_{N} decreases, T_{MH} increases with suppressing thermal hysteresis, and T_{ML} weakly increases. At 1.6 GPa, T_{MH} disappears as like a critical end point. The C_{ac} anomaly at T_{ML} is more pronounced with higher pressure. Interestingly, T_{N} is suppressed discontinuously at P_{c} ~1.9 GPa. At P_{c}, both of $\rho(T)$ and $C_{\mathrm{ac}}(T)/T$ are strongly enhanced, and a clear anomaly at T_{ML} is observed in only $C_{\mathrm{ac}}(T)/T$. On further compression, a clear kink in $\rho(T)$ appears at T_{ML}.

As shown in Fig.2(a), the obtained phase diagram of YbPd is complicated. T_1 and T_2 defined as temperatures of peak in $\mathrm{d}\rho/\mathrm{d}T$ (not shown) decreases monotonically as a function of pressure, which is in agreement with the previous reports [6]. Due to broadening and suppressing of the $\mathrm{d}\rho/\mathrm{d}T$ peak, T_1 is not unambiguously determined above 2 GPa. T_2 and T_{N} disappear discontinuously at a same pressure P_{c} ~1.9 GPa, where some characteristics are also observed as discussed later. T_{MH} slightly increases accompanied by suppression of the hysteresis and disappears at a critical end point $(P_{\mathrm{cr}}, T_{\mathrm{cr}})$ ~(1.45 GPa, 0.72 K). Because the transition at T_{MH} is of first order in the antiferromagnetic phase, it may correspond to change the magnetic structure. T_{ML} shows broad maximum centered close to P_{c}, but the mechanism of the transition and the relation to the antiferromagnetic order is unclear at present. It is also important to make clear whether the ground states between below and above P_{c} are same origins.

In order to discuss the change in electronic state across P_{c}, we plot pressure dependence of the electronic specific heat $\propto C_{\mathrm{ac}}/T$ in paramagnetic phase at T=2 K, and ρ_0 defined as the ρ at 0.1 K, as shown in Fig.2(b). Although P_{c} is first order critical point rather than second order, both of C_{ac}/T and ρ_0 show a strong enhancement at P_{c}, indicating the development of some fluctuations. The structural transitions at T_1 and/or T_2 seems to be the valence order, i.e. the Yb-sites split into magnetic and non-magnetic ones in equal proportions. The reason why antiferromagnetic ordering disappears discontinuously is due to the disappearance of the magnetic sites at P_c, where valence order temperature $T_2 \to 0$. The increase of across P_c indicates that the magnetic ions form the heavy electron states through the hybridization to the conduction electrons. The sharp peak of at P_c also indicates that the development of the magnetic fluctuation even with the first order critical point. The strong enhancement of ρ_0 may come from the development of the valence fluctuation. Similar phenomena was reported, e.g. in the provable valence crossover of $CeCu_2(Si, Ge)_2$ [9]. If the ordered phase below T_{ML} is magnetic, and the transition at T_1 survives above P_c, not only at T_2 but at T_1 valence order, splitting the magnetic and non-magnetic Yb-sites, may occur. The novel phase diagram of YbPd arises from the interplay between magnetic and valence instability.

4. Summary

We have revealed the pressure temperature phase diagram of YbPd thermodynamically. The antiferromagnetic ordered phase disappears discontinuously, where the transition at T_2 also vanishes at $P_c \sim 1.9$ GPa. It supports the scenario that the transition at T_1 and/or T_2 is splitting the Yb-sites into non-magnetic and magnetic. The strong enhancement of the residual resistivity seems to arise from the development of the valence fluctuation. The novel phase diagram of YbPd is due to the complicated relations between valence and magnetic instability.

Acknowedgment

This work was supported by Osaka University Global COE program "Core Research and Engineering of Advanced Materials-Interdisciplinary Education Center for Materials Science" and a Grant-in-Aid for Young Scientists (B) (No. 23740271) of JSPS, and also by the Exciting Leading-Edge Research Project project "Heavy Electrons and Superconductivity produced by High Quality Single Crystals and High Pressure Techniques" at Osaka University.

References

[1] Pott R, Boksch W, Leson G, Politt B, Schmidt H, Freimuth A, Keulerz K, Langen J, Neumann G, Oster F, Röhler J, Walter U, Weidner P and Wohlleben D 1985 *Phys. Rev. Lett.* **54** 481
[2] Mitsuda A, Yamada K, Sugishima M and Wada H 2009 *Physica B***404** 3002.
[3] Bonville P, Hammann J, Hodges J A, Imbert P and Jéhanno G J 1986 *Phys. Rev. Lett.* **57** 2733
[4] Tokiwa Y, Grünheit S, Jeevan H S, Stingl C, Gegenwart P 2011 *J. Phys.: Conf. Ser.* **273** 012062
[5] Hasegawa T, Ogita N, Sugishima M, Mitsuda A, Wada H and Udagawa M 2011 *J. Phys.: Conf. Ser.* **273** 012030
[6] Sugishima M, Yamada K, Mitsuda A, Wada H, Matsubayashi K, Uwatoko Y, Suga K and Kindo K 2010 *J. Phys.: Condens. Matter* **22** 375601
[7] Cook R L, H E King J, Herbst C A and Herschbach D R 1994 *The Journal of Chemical Physics* **100** 5178
[8] Can eld P C, Fisk Z 1992 *Phil. Mag. B***65** 1117
[9] Holmes A T, Jaccard D, Miyake K 2004 *Phys. Rev. B* **69** 024508

Ru substitution effect on the peak effect in superconducting $PrOs_4Sb_{12}$

Ryoichi Miyazaki[1], Ryuji Higashinaka[1], Yuji Aoki[1], Hitoshi Sugawara[2] and Hideyuki Sato[1]

[1]Department of Physics, Tokyo Metropolitan University, Hachioji, Tokyo 192-0397, Japan
[2]Department of Physics, Kobe University, Kobe 657-8501, Japan

E-mail: `ryoichi-miyazaki@ed.tmu.ac.jp`, `aoki@tmu.ac.jp`

Abstract. Ac susceptibility (χ_{ac}) measurements on $Pr(Os_{1-x}Ru_x)_4Sb_{12}$ single crystals for $x = 0.05$ and 0.1 were performed by mutual inductance method. A peak structure caused by anomalously enhanced flux pinning force, which is so-called the peak effect, appears in the H dependence of χ_{ac}. The peak structure shifts to lower fields as T increases and disappears in $T > 1$ K in both samples, while in $PrOs_4Sb_{12}$ it is observable up to near T_c. This fact indicates that the Ru substitution suppresses the peak effect. We demonstrate that the observed T dependent behavior of the peak structure can be explained roughly by the synchronization model although there remains deviation from the model curve, suggesting some modification may be needed in the model to be applied to $Pr(Os_{1-x}Ru_x)_4Sb_{12}$.

1. Introduction

Since the filled skutterudite compound $PrOs_4Sb_{12}$ was reported to be the first heavy fermion (HF) superconductor among Pr-based compounds, $PrOs_4Sb_{12}$ has been attracting much attention [1]. The superconducting transition temperature $T_c = 1.85$ K being higher than $T_c = 0.74$ K of no-4f-electron reference $LaOs_4Sb_{12}$ [1,2] demonstrates clearly the involvement of Pr 4f electrons in the superconductivity. Actually, several measurements performed so far suggest that it is of unconventional type. The Sb NQR spin-lattice relaxation rate indicates strong-coupling isotropic superconducting gap [3]. The existence of multiple superconducting phases with point nodes is indicated by angle-resolved thermal conductivity measurements [4], although angle-resolved specific heat suggests fully opened gap [5]. The appearance of spontaneous internal fields is indicated by zero field μSR measurements [6]. Considering singlet-triplet crystalline electric field (CEF) level scheme of Pr 4f electrons[7], pairing mediated by quadrupole excitons has been discussed [8].

Another feature that characterizes the unconventional superconducting state in $PrOs_4Sb_{12}$ is the so-called "peak effect", i.e., anomalously enhanced flux pinning force in the mixed state below the upper critical field H_{c2}. An anomalous irreversible behavior is observed in the magnetization curve in the mixed state [9], a considerable peak structure in the field dependence of the surface impedance [10], and minima in the H dependence of the flux-flow resistivity [11]. Similar anomalies have been observed in many of the HF superconductors [12–14] and therefore it is expected that the observed peak effect in $PrOs_4Sb_{12}$ reflects the involvement of the heavy quasiparticles in the HF superconductivity. In this paper, we report the Ru substitution effect on the peak effect of $PrOs_4Sb_{12}$ by ac susceptibility measurements. Since the unconventional HF and conventional superconducting states appear at the opposite ends in the

$Pr(Os_{1-x}Ru_x)_4Sb_{12}$ series with continuous T_c variation in between [15–17], $Pr(Os_{1-x}Ru_x)_4Sb_{12}$ is an appropriate system to study how the anomalous flux pinning depends on the type of superconductivity.

2. Experimental

Ac susceptibility has been measured by mutual inductance method with ac field $H_{ac} \sim 0.3$ Oe down to 0.1 K using a dilution refrigerator equipped with a 8-T superconducting magnet. Single crystals of $Pr(Os_{1-x}Ru_x)_4Sb_{12}$ with $x = 0.05$ and 0.1 were grown by Sb self flux method. Obtained single crystals with the size of about $1 \times 1 \times 2$ mm^3 exhibit sharp superconducting transitions in T dependence measurements under zero field, indicating good sample quality.

3. Results and discussion

3.1. Temperature dependence

Figure 1 shows the T dependence of χ_{ac} ($= \chi' + i\chi''$) in zero field, where χ' and χ'' correspond to the real and the imaginary parts of χ_{ac}, respectively. χ' represents the equilibrium differential susceptibility $\chi' = \mathrm{d}M/\mathrm{d}H$ while χ'' represents the energy dissipation $W = \pi\chi'' H_{ac}^2$. In both samples, χ' decreases below the superconducting transition temperature $T_c \sim 1.81$ K ($x = 0.05$) and 1.76 K ($x = 0.1$), respectively. Broad peaks also appear in the T dependence of χ'' just below T_c in both samples.

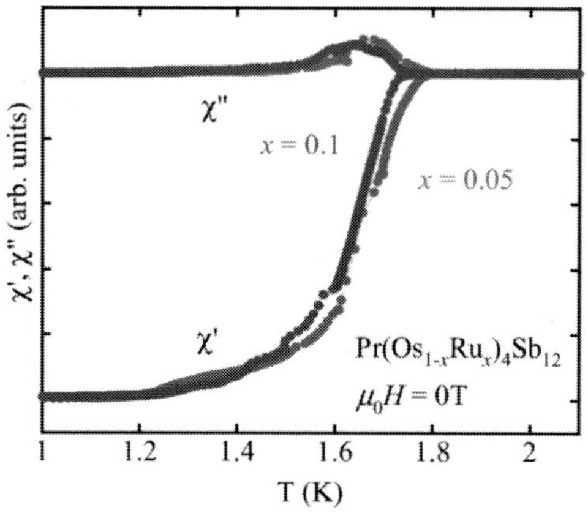

Figure 1. Temperature dependences of χ_{ac} of the single crystals of $Pr(Os_{1-x}Ru_x)_4Sb_{12}$ with $x = 0.05$ and 0.1 measured in zero field. χ' and χ'' represent the equilibrium differential susceptibility and the energy dissipation, respectively.

3.2. Field dependence

The H dependences of χ_{ac} at several temperatures are shown in fig.2. Specific peak structures indicating the peak effect are observed in $\chi'(H)$ just below H_{c2} in both samples. With increasing T, the peak structures shift to the lower fields and becomes suppressed gradually. The H dependence of χ'' also shows a broad peak at the field where χ' shows the peak structure. These behaviors of χ_{ac} in the Ru doped samples are similar to those of $PrOs_4Sb_{12}$ [18]. However, while the peak structure remains up to $T \sim T_c$ in $PrOs_4Sb_{12}$, the peak structure is not visible in $\chi'(H)$ and $\chi''(H)$ for $T > 1$ K in $x = 0.05$ and 0.1. To analyze the feature of the peak structures quantitatively, H_{c2}^*, H_{min} and H_{pk} representing the kink structure just below H_{c2}, the local minimum of $\chi'(H)$ and the local peak in $\chi'(H)$ are determined respectively by using field derivative $\mathrm{d}\chi'/\mathrm{d}H$.

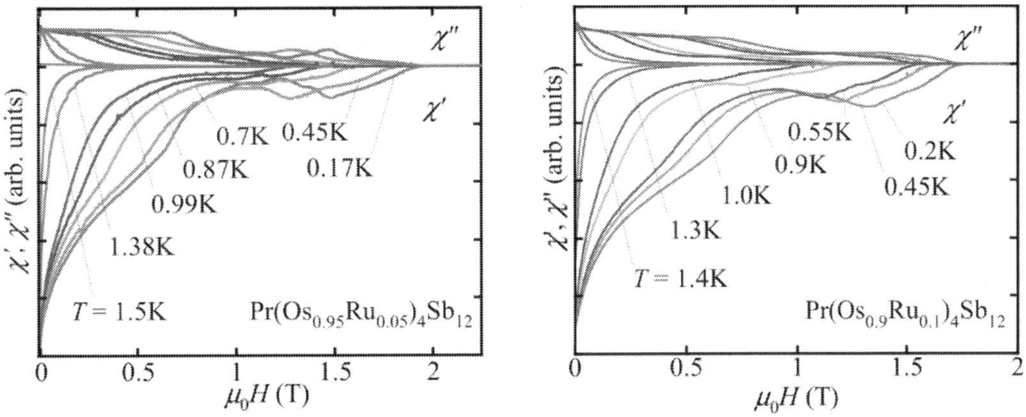

Figure 2. The field dependence of χ_{ac} at several temperatures measured with decreasing field from 2.5 T to 0 T at a sweep rate 0.02 T/min. At $T \sim 0.2$ K, anomalous peak structures and kinks appear just below H_{c2} and in lower field, respectively. With increasing temperature, these structures are suppressed.

3.3. H-T phase diagram

Obtained H-T phase diagrams are shown in fig.3. The overall T dependence of H_{c2} is reproduced by the WHH (Werthamer-Helfand-Hohenberg) curve in the clean limit above 0.2 T [19], below which $H_{c2}(T)$ exhibits slight deviation with a positive curvature possibly caused by the multiband effect as discussed in $PrOs_4Sb_{12}$ [20].

The considerable T dependent shift of the H region of the peak structure cannot be explained by the "matching effect", in which the pinning becomes effective in a certain field where the average pinning site spacing matches the flux line spacing. In the present case, so-called "synchronization effect" remains as a possible origin of the peak effect [21]. With increasing H, the shear modulus C_{66} of the flux-line lattice (FLL) is expected to decrease as $\propto (1 - H/H_{c2})^2$ [22]. Therefore, in high fields, FLL can be deformed so as to be effectively pinned by randomly distributed pinning centers, leading to an increase of the pinning force density. However, when H is close to H_{c2}, each flux line begins to shear due to the further decrease of C_{66} and the pinning force density starts to decrease. This is the scenario of the formation of a peak structure in the synchronization model. Considering that the pinning force density has a maximum at $H_{\rm min}$, we have compared the H-T data with the model.

In fig.3, shown by the dotted line is the synchronization model curve, $H_{\rm min} = H_{c2}(T) - \zeta H_{c2}(T)^{3/2}$, where ζ is a fitting parameter corresponding to the density of the pinning centers and depends on neither T nor H [23]. The best fitting values of ζ are 2.3×10^{-1} $T^{-1/2}$ and 2.4×10^{-1} $T^{-1/2}$ for $x = 0.05$ and 0.1, respectively; these values are comparable to that of $CeRu_2$ ($\sim 5.8 \times 10^{-2}$ $T^{-1/2}$) [23], whose peak effect is nicely explained by the synchronization model. The model curve roughly reproduces the T dependence of $H_{\rm min}$. Quantitatively, however, the T dependence of $H_{\rm min}$ is faster than the model curve.

As demonstrated by the disappearance of the peak effect above 1 K in $Pr(Os_{1-x}Ru_x)_4Sb_{12}$ ($x = 0.05$ and 0.1), the anomalous pinning force enhancement is suppressed by the Ru substitution. However, it is expected naively that the Ru substitution may introduce crystalline defects that could behave as pinning centers leading to the peak effect. In this sense, the suppression of the peak effect by the Ru substitution is anomalous. In $CeRu_2$ [24], it was reported that the peak effect is more pronounced for samples with the smaller RRR values, i.e., including larger amount of conduction electron scattering centers, which may work as flux pinning centers. In $Pr(Os_{1-x}Ru_x)_4Sb_{12}$, it has been found that the singlet-triplet CEF excitation energy increases considerably by the Ru substitution; $E/k_B \sim 8$ K for $x = 0$ and 80 K for $x = 1$ [17, 25]. Because of this, the 4f-electron contribution to the dc magnetic susceptibility χ_{dc} at

$T \sim T_c$ decreases with increasing the Ru substitution. The resulting smaller χ_{dc}, which is less effective for the flux pinning, can be one of the possible reasons for the peak effect suppression for $x \neq 0$. However, the decrease in χ_{dc} for $x = 0.1$ is only 20% and it is questionable if such a small decrease in χ_{dc} can account for the observed suppression of the peak effect. Other changes in the electronic properties might be necessary to be taken into account; e.g., with the Ru substitution, the 5d conduction electrons from Os ions are gradually replaced by the 4d conduction electrons from Ru ions [26, 27]. We expect on-going measurements in higher Ru concentration region will be helpful to clarify the source of the anomalous flux pinning in $Pr(Os_{1-x}Ru_x)_4Sb_{12}$.

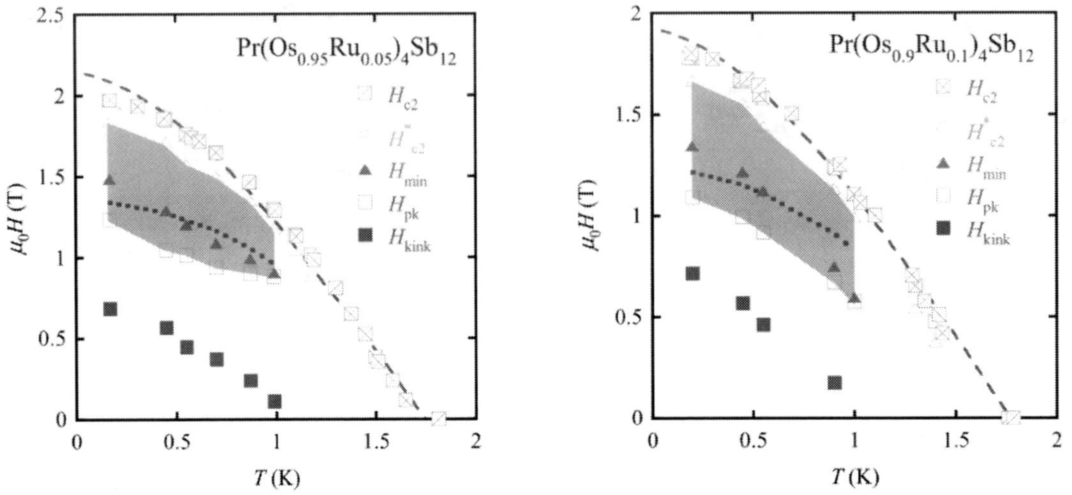

Figure 3. H-T phase diagrams with the characteristic fields, H^*_{c2}, H_{min}, H_{pk} and H_{kink}. The dashed line represents the clean-limit WHH model curve for $H_{c2}(T)$. The shaded area shows the region where the peak effect appears noticeably. The green doted line represents the synchronization model curve. H_{kink} representing the kink structure in low fields decreases almost linearly with increasing T and disappears in $T > 1$K.

Acknowledgments

This work was supported by Grant-in-Aids for Scientific Research on Innovative Areas "Heavy Electrons" (No. 20102007) and on Priority Areas "Physics of Superclean Materials" (No. 20029018) and "Skutterudite" (no. 15072206) and (C)(No.20540359) from MEXT, Japan.

References

[1] Bauer E D, Frederick N A, Ho P C, Zapf V S and Maple M B 2002 *Phys. Rev. B* **65** 100506(R)

[2] Sugwara H, Osaki S, Saha S R, Aoki Y, Sato H, Inada Y, Shishido H, Settai R, Ōnuki Y, Harima H and Oikawa K 2002 *Phys. Rev. B* **66** 220504(R)

[3] Kotegawa H, Yogi M, Imamura Y, Kawasaki Y, q Zheng G, Kitaoka Y, Ohsaki S, Sugawara H, Aoki Y and Sato H 2003 *Phys. Rev. Lett.* **90** 027001

[4] Izawa K, Nakajima Y, Goryo J, Matsuda Y, Osaki S, Sugawara H, Sato H, Thalmeier P and Maki K 2003 *Phys. Rev. Lett.* **90** 117001

[5] Custers J, Namai Y, Tayama T, Sakakibara T, Sugawara H, Aoki Y and Sato H 2006 *Physica B* **378** 179

[6] Aoki Y, Tsuchiya A, Kanayama T, Saha S R, Sugawara H, Sato H, Higemoto W, Koda A, Ohishi K, Nishiyama K and Kadono R 2003 *Phys. Rev. Lett.* **91** 067003

[7] Kohgi M, Iwasa K, Nakajima M, Metoki N, Araki S, Bernhoeft N, Mignot J M, Gukasov A, Sato H, Aoki Y and Sugawara H 2003 *J. Phys. Soc. Jpn.* **72** 1002

[8] Koga M, Matsumoto M and Shiba H 2006 *J. Phys. Soc. Jpn.* **75** 014709
[9] Tayama T, Sakakibara T, Sugawara H, Aoki Y and Sato H 2003 *J. Phys. Soc. Jpn.* **72** 1516
[10] Tou H, Nakai Y, Doi M, Sera M, Sugawara H and Sato H 2006 *Physica B* **378** 209
[11] Kobayashi M, Sato H, Sugawara H, Fujiwara H and Aoki Y 2005 *J. Phys. Soc. Jpn.* **74** 1690
[12] Yagasaki K, Hedo M and Nakama T 1993 *J. Phys. Soc. Jpn.* **62** 3825
[13] Haga Y, Yamamoto E, Inada Y, Aoki D, Tenya K, Ikeda M, Sakakibara T and Ōnuki Y 1996 *J. Phys. Soc. Jpn.* **65** 3646
[14] Tenya K, Ikeda M, Tayama T, Mitamura H, Amitsuka H, Sakakibara T, Maezawa K, Kimura N, Settai R and Ōnuki Y 1995 *J. Phys. Soc. Jpn.* **64** 1063
[15] Frederick N A, Do T D, Ho P C, Butch N P, Zapf V S and Maple M B 2004 *Phys. Rev. B* **69** 024523
[16] Ho P C, Butch N P, Zapf V S, Yanagisawa T, Frederick N A, Kim S K, Yuhasz W M, Maple M B, Betts J B and Lacerda A H 2008 *J. Phys.: Cond. Matt.* **20** 215226
[17] Miyazaki R, Aoki Y, Kikuchi D, Sugawara H and Sato H 2010 *J. Phsy.: Conf. Ser.* **200** 012125
[18] Grube K, Drobnik S, Pfleiderer C, v Löhneysen H, Bauer E D and Maple M B 2006 *Phys. Rev. B* **73** 104503
[19] Werthamer N R, Helfand E and Hohenberg P C 1966 *Phys. Rev.* **147** 295
[20] Measson M A, Braithwaite D, Flouquet J, Seyfarth G, Brison J P, Lhotel E, Paulsen C, Sugawara H and Sato H 2004 *Phys. Rev. B* **70** 064516
[21] Kramer E J 1973 *J. Appl. Phys.* **44** 1360
[22] Campbell A M and Evetts J E 1972 *Adv. Phys.* **21** 199
[23] Tenya K, Yasunami S, Tayama T, Amitsuka H, Sakakibara T, Hedo M, Inada Y, Yamamoto E, Haga Y and Ōnuki Y 1999 *J. Phys. Soc. Jpn.* **68** 224
[24] Hedo M, Yobayashi Y, Inada Y, Yamamoto E, Haga Y, Suzuki J, Metoki N, Ōnuki Y, Sugawara H, Sato H, Tenya K, Tayama T, Amitsuka H and Sakakibara T 1998 *J. Phys. Soc. Jpn.* **67** 3561
[25] Akita H, Yoshino G and Ochiai A 2006 *Physica B* **378** 197
[26] Harima H and Takegahara K 2003 *J. Phys.:Cond. Matt.* **15** S2081
[27] Aoki Y, Sugawara H, Harima H and Sato H 2005 *J. Phys. Soc. Jpn.* **74** 209

Fabrication of single crystalline EuO thin film with SrO buffer layer on $SrTiO_3$ substrate

H Miyazaki[1, 2], H Momiyama[3], T Hajiri[2, 4], T Ito[4], K Imura[2], M Matsunami[2, 5] and S Kimura[2, 5]

[1]Center for Fostering Young and Innovative Researchers, Nagoya Institute of Technology, Nagoya, Aichi 466-8555, Japan
[2]UVSOR Facility, Institute for Molecular Science, Okazaki, Aichi 444-8585, Japan
[3]School of Engineering, Nagoya Institute of Technology, Nagoya, Aichi 466-8555, Japan
[4]Graduate School of Engineering, Nagoya University, Nagoya 464-8603, Japan
[5]School of Physical Sciences, The Graduate University for Advanced Studies (SOKENDAI), Okazaki, Aichi 444-8585, Japan

E-mail: miyazaki@nitech.ac.jp, kimura@ims.ac.jp

Abstract. We report the new fabrication method of single crystalline EuO thin film along the (100) plane and its crystal structure and magnetic property. EuO thin film was fabricated on SrO buffer layer created on $SrTiO_3$ substrate. The obtained EuO thin film was formed a single phase of the rock-salt structure with the lattice constant of 0.516 nm and showed ferromagnetism below the Curie temperature of 71 K. The examined physical properties are consistent with the bulk materials.

1. Introduction

Europium monoxide (EuO), a ferromagnetic semiconductor, with a Curie temperature (T_C) of 69 K, has been attracting attention due to its anomalous magneto-optical and transport properties [1-6]. In the case of electron doping with excess Eu or the substitution of Eu^{2+} ions with Gd^{3+} or La^{3+} ions, T_C increases to as high as 200 K and a metal-insulator transition (MIT) appears in which the electrical resistivity drops by more than 12 orders of magnitude with decreasing temperature [3,7-10]. When an external magnetic field is applied, the MIT temperature is shifted to the higher temperature side and the electrical resistivity decreases due to the colossal magnetoresistive (CMR) effect [9]. Since the CMR effect originates from localized Eu^{2+} $4f$ electrons, the spin polarization of carriers is predicted to be fully 100 % below T_C [11-14]. Therefore EuO is one of candidate compounds for next-generation spintronics applications such as spin valves and spin switches.

To apply EuO for spintronics devices, it is important to grow a few atomic layers single crystalline EuO on the substrate. So far we report the fabrication and physical properties of high quality single crystal thin films of EuO with the thickness of 50 nm [15]. Well defined EuO thin films cannot be grown just on the $SrTiO_3$ substrate, because the other phase of Eu_2O_3 inevitably appears at the interface between the $SrTiO_3$ and EuO film due to oxygen diffusion. To solve this problem, we adopted BaO as a buffer layer between EuO thin films and $SrTiO_3$ substrate [16]. In this case, due to the large lattice mismatch between the BaO buffer layer and EuO (6.3 %), EuO thin films with a few atomic layers cannot be grown epitaxially on the BaO thin film [15-16]. To satisfy small lattice mismatch between EuO and buffer layer, we propose that SrO is used for buffer layer in between EuO thin film and $SrTiO_3$ substrate, as shown in Fig. 1. The relationship of the epitaxial growth is EuO[110] // SrO[110] // $SrTiO_3$[110] and EuO(001) // SrO(001) // $SrTiO_3$(001). SrO forms a rock-salt crystal structure with lattice constant of 0.5144 nm, which is the same crystal structure and similar lattice constant as EuO (The lattice mismatch between SrO buffer layer and EuO thin film is slightly -0.04 %) [17]. Thanks to this quite small lattice mismatch, EuO thin films with a few atomic layers are expected to be grown epitaxially on the $SrTiO_3$ substrate.

In this paper, we report the growth of single crystalline EuO thin film with SrO buffer layer on $SrTiO_3$, its crystal structure and magnetization property. The obtained EuO film with the thickness of 50 nm forms a single phase rock-salt structure with a lattice constant of 0.516 nm and shows a ferromagnetic transition at T_C = 71 K.

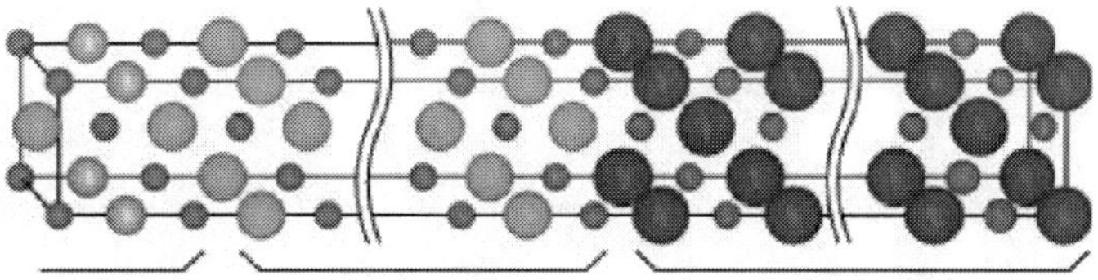

Figure 1. (Color online) Atomic structure of EuO (001) thin film with SrO buffer layer onto $SrTiO_3$ substrate. Lattice constant of $SrTiO_3$, SrO and EuO and lattice mismatch between $SrTiO_3$–SrO and SrO–EuO are shown in the figure.

2. Experimental Procedure

Single-crystalline EuO thin film was grown using a molecular beam epitaxy (MBE) method [16]. The thin films were grown on 0.05 wt% Nb-doped $SrTiO_3$ (001) single-crystalline substrates. First, the substrates were heated at 600 °C for 1 hour under an ultra-high vacuum (2.0×10^{-7} Pa) to obtain clean surface. Then, strontium metal with a purity of 99.9 % was evaporated using an effusion cell under an oxygen pressure of 1.0×10^{-4} Pa to create a buffer layer of SrO with a thickness of 5.0 nm. Finally, the EuO thin film was grown by the evaporation of Eu metal under an oxygen pressure of 8.0×10^{-6} Pa on the SrO buffer layer at a substrate temperature of 400 °C. The crystal structure was observed using an X-ray diffractometer (XRD; RINT-Ultima III, Rigaku Ltd.) with a Cu $K\alpha$ radiation. The magnetization was measured with a superconducting quantum interference device magnetometer (SQUID; MPMS-7, Quantum Design Ltd.).

3. Results and Discussion

Figure 2 shows the XRD patterns of 50 nm EuO thin films with and without SrO buffer layer on a $SrTiO_3$ substrate at room temperature, respectively. The Eu_2O_3 (310) peak at 2θ = 31.5 deg. was observed in the XRD pattern of the case without SrO buffer layer, which is due to an oxygen diffusion from $SrTiO_3$ to EuO. From the photoemission spectroscopy, only Eu^{2+} $4f$ and O^{2-} $2p$ state exist in EuO / SrO / $SrTiO_3$ and no other peak appear [18]. These results suggest that the SrO buffer layer protects the oxygen diffusion. The obtained lattice constant of EuO thin films with SrO buffer layer along the parallel to the surface normal is 0.516 ± 0.001 nm that is in good agreement with a previous report obtained EuO thin films with the case of BaO buffer layer (0.515 ± 0.001 nm) [15]. Therefore the obtained EuO thin film is confirmed to be rock-salt crystal structure as same as bulk materials [16]. The lattice mismatch between SrO buffer layer and EuO thin films is -0.3 %. Therefore due to the quite small lattice mismatch, the atomic flatness of EuO ultra-thin films is expected when SrO buffer layer is used. Note that peaks marked by open circles are unknown peak but are not originated from europium compounds.

Figure 3 shows the temperature-dependent magnetization of EuO thin film with SrO buffer layer, which is measured by applying a magnetic field of 500 Oe along the EuO [010] direction. Since the magnetization curve can be reproduced by the Brillouin function with spin moment $S = 7/2$ and $T_C = 71$ K, the EuO thin film is explained by simple Heisenberg ferromagnetism. T_C is slightly higher than that of the bulk material (69 K). T_C can be controlled by the stoichiometry, the lattice constant and the lattice distortion. In the present case, the obtained EuO thin film is confirmed to be formed a single-crystalline rock-salt structure and the same lattice constant as the bulk materials. Therefore, the inconsistency in the value of T_C is considered to be due to the slight excess of Eu atoms. Our previous magneto-optical infrared imaging experiment on the EuO thin film with BaO buffer layer also suggested the excess of Eu ion leading to a magnetic polaron states [16, 19].

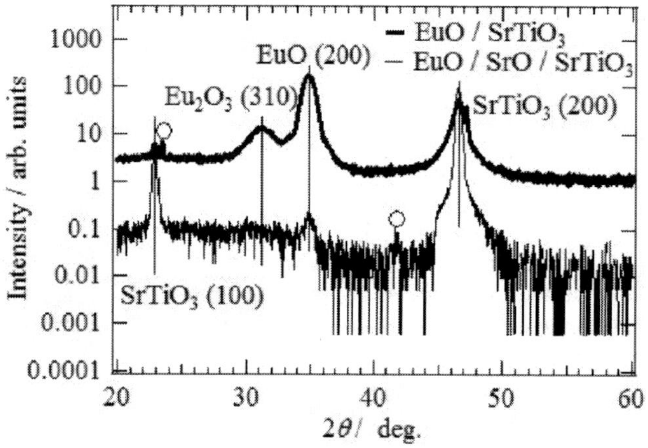

Figure 2. XRD patterns of 50 nm EuO thin films with (thin line) and without (thick line) SrO buffer layer on a $SrTiO_3$ substrate at room temperature, respectively. The origin of the peaks marked by open circles is unknown and must not originate from europium compounds.

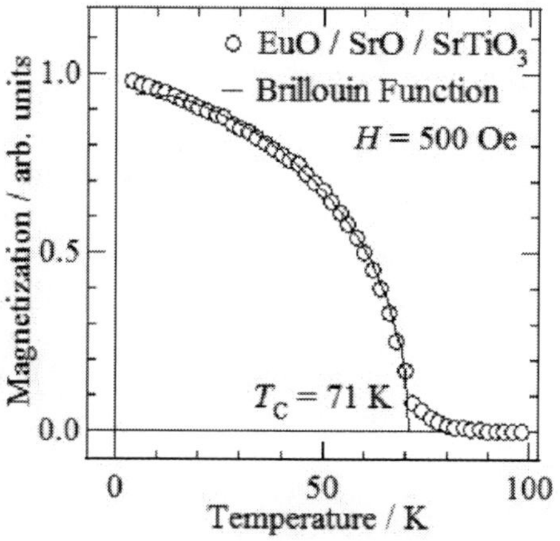

Figure 3. Temperature-dependent magnetization curve (open circle) of EuO thin film (thickness = 50 nm) and the fitting curve (solid line) of Brillouin function with $S = 7/2$ and $T_C = 71$ K. The magnetic field of 500 Oe was applied along the EuO [010] direction.

4. Conclusion

In conclusion, we successfully fabricated single-crystalline EuO thin film with SrO buffer layer on $SrTiO_3$ substrate using a molecular beam epitaxy method. The obtained EuO thin film was found to be a single crystalline ferromagnetic with $T_C = 71$ K. Due to the quite small lattice mismatch at the interface, we expect that single crystalline EuO thin films with a few atomic layers can be fabricated. The technique is useful for next-generation spintronics devices such as a spin filter.

5. Acknowledgements

This work was a joint studies program of the Institute for Molecular Science (2011), and was partially supported by a Center for Fostering Young and Innovative Researchers and Grant-in-Aid for

Scientific Research B (No. 22340107) from the Ministry of Education, Culture, Sports, Science and Technology (MEXT) of Japan.

References

[1] Matthias B, Bozorth R and Vleck J V 1961 *Phys. Rev. Lett.* **7** 160
[2] McGuire T and Shafer M 1964 *J. Appl. Phys.* **35** 984
[3] Mauger A and Godart C 1986 *Phys. Rep.* **141** 51
[4] Wang H, Schoenes J and Kaldis E 1986 *Helv. Phys. Acta* **59** 102
[5] Tu K N Ahn K Y and Suits J C 1972 *IEEE Trans. Magn.* **8** 651
[6] Uspenskii Y A and Harmon B N 2000 *Phys. Rev. B* **61** R10571
[7] Shapira Y, Foner S and Reed T 1973 *Phys. Rev. B* **8** 2299
[8] Shafer M W, Dimmock J B and Penny T 1972 *J. Phys. Chem. Solids* **33** 2251
[9] Oliver M R, Dimmock J O, McWhorter and Reed T B 1972 *Phys. Rev. B* **5** 1078
[10] Miyazaki H, Im H J, Terashima K, Yagi S, Kato M, Soda K, Ito T and Kimura S 2010 *Appl. Phys. Lett.* **96** 232503.
[11] Steeneken P G, Tjeng L H, Elfimov I, Sawatzky G A, Ghiringhehhi, Brookes N B and Huang D –J 2002 *Phys. Rev. B* **88** 047201
[12] Schmehl A, Vaithyanathan V, Herrnberger A, Thiel S, Richter C, Liberati M, Heeg T, Rockerath M, Kourkoutis L F, S. Muhbauer, Boni P, Muller D A, Barash Y, Schubert J, Idzerda Y, Mannhart J and Schlom D G 2007 *Nat. Mater.* **6** 882
[13] Santos T S, Moodera J S, Raman K V, Negusse E, Holroyd J, Dvorak J, Liberati M, Idzerda Y U and Arenholz 2008 *Phys. Rev. Lett.* **101** 147201
[14] Miyazaki H, Ito T, Im H J, Yagi S, Kato M, Soda K and Kimura S 2009 *Phys. Rev. Lett.* **102** 227203
[15] Miyazaki H, Ito T, Im H J, Terashima K, Yagi S, Kato M, Soda K and Kimura S 2009 *Jpn. J. Appl. Phys.* **48** 055504
[16] Iwata N, Pindoria G, Morishita T and Kohn K 2000 *J. Phys. Soc. Jpn.* **69** 230
[17] Wagman D D, Evans W H, Parker V B, Schumm R H, Halow I, Bailey S M, Churney K L and Nuttal R I 1982 *J. Phys. Chem. Ref. Data Suppl.* **11** 2
[18] Miyazaki H *unpublised data*
[19] Kimura S, Ito T, Miyazaki H, Mizuno T, Iizuka T and Takahashi T 2008 *Phys. Rev. B* **78** 052409

NMR spin-lattice relaxation rate of heavy fermion superconductor UBe_{13}

Kyohei Morita[1], Keisuke Kuroda[1], Yudai Hara[1], Hisashi Kotegawa[1], Hideki Tou[1,†], Etsuji Yamamoto[2], Yoshinori Haga[2], and Yoshichika Ōnuki[3]

[1]Department of Physics, Graduate School of Science, Kobe University, Kobe 657-8501, Japan
[2]Advanced Science Research Center, Japan Atomic Energy Agency, Tokai 319-1195, Japan
[3]Department of Physics, Graduate School of Science, Osaka University, Toyonaka 560-0043, Japan

E-mail: † tou@crsytal.kobe-u.ac.jp

Abstract. ^{9}Be NMR measurements have been carried out for a single crystal UBe_{13} with T_c =0.85 K, in order to clarify unusual properties in the normal state. For the applied field parallel to [111] direction, the quadrupole split Be(II) lines gather around the central line. ^{9}Be nuclear spin-lattice relaxation rate was measured for Be(II) sites for $H \| [111]$ at 0.85, 7, 15 T. $1/T_1$ does not depend on applied fields above $T^*_{NMR} \approx 5 - 9$ K and weakly depends on temperature. $1/T_1$ at 0.85 and 7 T is proportional to T^n with $n = 0.5 - 0.6$ down to $T = 2$ K, suggesting that antiferromagnetic spin fluctuations exist. On the other hand, $1/T_1$ is suppressed by applied magnetic field. The present field dependence of $1/T_1$ at low temperatures would give important information about the formation of the heavy quasiparticles in UBe_{13}.

1. Introduction

Since the discovery of heavy Fermion superconductivity in UBe_{13} having the extremely large electronic specific heat coefficient of $\gamma_e = 1100$ mJ/(mol K^2),[1, 2] extensive experimental and theoretical works have been carried out in order to clarify the nature of the superconducting state of UBe_{13}. For instance, specific heat [1, 2], neutron diffraction [3, 4], μSR [5, 6], and NMR measurements [7, 8, 9, 10], etc., have probed the unconventional nature of both normal and superconducting states.

The superconducting transition occurs at around $T_c \approx 0.85$ K with large specific heat jump, $\Delta C/\gamma_e T_c \approx 1$ [1, 2], indicating that the heavy-quasiparticles are responsible for the superconductivity. In the superconducting state, the power law temperature dependence of the specific heat [2], the NMR spin-lattice relaxation rate [8], etc., suggest that the superconducting energy gap vanishes at points and/or lines on the Fermi surface. These features have been interpreted as evidence for an anisotropic pairing state, most likely a *p*-wave triplet state, in UBe_{13}.

In the normal state, the temperature dependence of the electric resistivity shows $\rho(T) \propto -\ln T$ above 40 K [1, 11, 12], associated with the Kondo scattering. On the other hand, T^2 of resistivity and Pauli paramagnetic susceptibility expected for the heavy Fermi liquid state are not observed down to 2 K; $\rho(T)$ is almost temperature independent in the temperature range of 40 K − 2 K, and decreases gradually with decreasing temperature below $T^* = 2$ K down to T_c. $\rho(T_c) \approx 139$ $\mu\Omega$cm is quite large in comparison with that in the other heavy fermion superconductors [12]. Temperature dependence of magnetic susceptibility $\chi(T)$ shows logarithmic divergence down to 2 K. These anomalies have attracted much attention as a non-Fermi liquid behavior. In order to explain the non-Fermi liquid behavior in UBe_{13}, a two-channel Kondo effect is proposed theoretically [13].

In order to investigate the normal state in UBe_{13}, we report magnetic field dependence of ^{9}Be NMR relaxation rate $1/T_1$ in the single crystal UBe_{13} at various magnetic fields.

2. Experimental

Single crystals of UBe_{13} were grown by the Al-flux method. Details of sample preparation techniques for single crystals of UBe_{13} was reported elsewhere [11]. Samples were characterized to be a single phase by a Laue photograph as well as the X-ray diffraction. UBe_{13} crystallizes in the cubic $NaZn_{13}$-type structure [space group (Fm3c)] with the lattice constant of $a = 10.257$ Å, where the U atoms are in the position of 8(a) site of (0, 0, 0), ±(1/2, 1/2, 1/2), and Be atoms have two crystallographically inequivalent sites. Be(I) atoms are in the position of 8(b) of ±(1/4, 1/4, 1/4), and Be(II) atoms are in the position of 96(i) sites. The superconducting phase transition was confirmed to occur at $T_c \approx 0.85$ K by means of DC electrical resistivity and surface impedance measurements [13, 14].

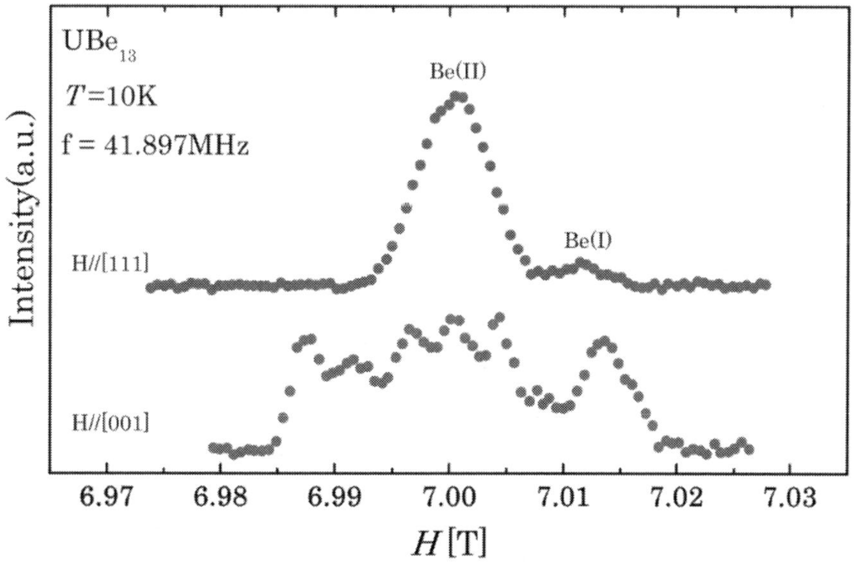

Figure 1. ^{9}Be NMR spectra measured at $H = 7$ T and for $H \| [001]$ and $H \| [111]$.

^{9}Be-NMR measurements were carried out by using a conventional pulsed NMR spectrometer in the temperature range of 1.5-100 K. A field angle was determined from the angular dependence of the ^{9}Be-NMR peak position. The field calibration was carried out with ^{27}Al resonance ($^{27}K \approx 0.161\%$ at 4.2 K) of a reference sample [12]. We confirmed that the linewidth of ^{9}Be-NMR spectrum measured at $H = 0.85$ T and for $H \|$ [001] is as quite narrow as ≈10 G. Furthermore no extra signals associated with impurity phases were observed [12], which guarantee the high-quality of the single crystal from a microscopic level. In order to investigate the normal state in UBe_{13}, we report temperature dependence of the ^{9}Be NMR relaxation rate $1/T_1$ in the single crystal UBe_{13} at $H = 0.85$, 7 and 15 T and for $H \| [111]$.

3. Results and Discussions

Figure 1 shows ^{9}Be NMR spectra measured at $H = 7$ T and for $H \| [001]$ and $H \| [111]$. When the magnetic field is applied to [111] direction, quadrupole split Be(II) lines merge into the central resonance line around 7 T because the effect of the electric field gradient (EFG) is effectively canceled out. The peak observed around 7.012 T corresponds to the signal from Be(I) atoms in 8(b) site [12, 15].

Typical examples of ^{9}Be-NMR relaxation curves for the Be(II) site are shown in Fig. 2(a) and (b). Since the effect of the EFG is canceled out for $H \| [111]$, the relaxation curve shows a single exponential relaxation curve, which can be fitted by $(M_0 - M(t))/M(t) = \exp(-t/T_1)$ by two digits. Therefore $1/T_1$ can be determined uniquely. The solid lines are the results of the fitting. Temperature dependence of $1/T_1$ was measured at the main peak of the Be(II) line.

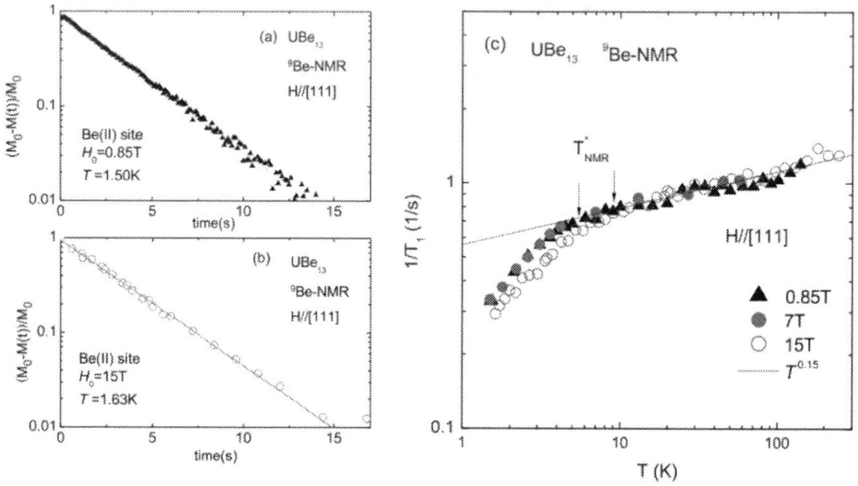

Figure 2. (a) ^{9}Be NMR spin-lattice relaxation curve for $H = 0.85$ T and $T = 1.5$ K. (b) ^{9}Be NMR spin-lattice relaxation curve for $H = 15$ T and $T = 1.6$ K. (c)Temperature dependences of $1/T_1$ for $H \| [111]$ and at $H = 0.85$ (open triangles), 7 (closed circles), and 15 T (open circles). Solid lines in (a) and (b) are fits to relaxation data (see text). Solid line in (c) represents $1/T_1 \propto T^{0.15}$ as a guide for eyes.

Figure 2(c) shows the temperature dependences of $1/T_1$ in the paramagnetic normal state measured at $H = 0.85, 7, 15$ T. Overall temperature dependences of $1/T_1$ for $H = 0.85$ T and 7 T are almost consistent with the temperature dependence of $1/T_1$ reported by MacLaughlin et al.[7]. $1/T_1$ does not depend on applied fields above 10 K and exhibits weak temperature dependence. The solid line in Fig. 2(c) is a guide for the eyes for $1/T_1 \approx T^{0.15}$. With decreasing temperature, $1/T_1$ deviates from the line at T^*_{NMR}. T^*_{NMR} is obtained as $T^*_{NMR} \approx 5$ K for $H = 0.85$ T and 7 T, wheras ≈ 9 K for $H = 15$ T. Below T^*_{NMR}, the Korringa relation of $1/T_1T = const.$, which is connected to the heavy Fermi liquid state, cannot be observed, but rather it obeys $1/T_1 \propto T^n$ with $n = 0.5 - 0.6$ down to $T = 2$ K, suggesting a non Fermi liquid (NFL) state by spin fluctuations near the quantum critical point.

Blow T^*_{NMR}, $1/T_1$ at $H = 15$ T is suppressed compared to that at $H = 0.85$ T and 7 T. Such a suppression of $1/T_1$ by applying magnetic fields is analogous with the breakdown of the heavy Fermi liquid state. Then, if one assumes that the heavy-quasiparticles are formed through Kondo mechanism, the characteristic critical field, H^*, corresponding to the breakdown of the Kondo coherency is estimated to be 7 T $< H^* <$ 15T. This field range is comparable to

$T^*_{NMR} \approx 5-9$ K, suggesting that the Kondo mechanism works for the formation of the heavy quasiparticles in UBe_{13}.

However, one can find that the NFL like behavior observed below T^*_{NMR} is robust against the applied magnetic field. These features are contrasted with the typical relaxation behavior of the heavy Fermion system, in which $1/T_1$ undergoes a moderate crossover from the $1/T_1 =$ *const.* behavior at higher temperatures than a characteristic temperature, $T^* \approx T_K$, to the $T_1T =$ *const.* behavior at low temperatures. Actually, the temperature independent Pauli susceptibility, which is expected for the heavy Fermi liquid state, is not observed down to ≈ 2 K [12]. No observation of the Korringa relation, i.e., the NFL like behavior, implies the competition between the Kondo effect and other effect, e.g. crystal electric field states and/or multipoles, which would disturb the formation of the Kondo singlet[16].

4. Sammary

The NMR results in this study provide insight into how the magnetic response is modified by applied fields. $1/T_1$ does not depend on applied fields above $T^*_{NMR} \approx 5-9$ K and weakly depends on temperature. Below T^*_{NMR}, $1/T_1$ starts to decrease which is reminiscent of the development of the Kondo coherency. However $1/T_1$ does not follow the Korringa relation. Especially for $H = 15$ T, the Korringa relation cannot be observed though $1/T_1$ is suppressed compared with that at low fields. The NFL like behavior implies the competition between the Kondo effect and other effect, e.g. crystal electric field states and/or multipoles, which would disturb the formation of the Kondo singlet.

Acknowledgments

We gratefully acknowledge K. Miyake for informative discussions. This work was supported by a Grant-in-Aid for Scientific Research on Innovative Areas "Heavy Electrons" (20102005) of the Ministry of Education, Culture, Sports, Science and Technology, Japan and Grants-in-Aid for Scientific Research (B:22340102, 16037211) from the Japan Society for the Promotion of Science.

References

[1] Ott H R, Rudigier H, Fisk Z, and Smith J L 1983, *Phys. Rev. Lett.* **50** 1595.
[2] Ott H R, Rudigier H, Rice T M, Ueda K, Fisk Z, and Smith J L 1984, it Phys. Rev. Lett. **52** 1915.
[3] Goldman A I, Shapiro S M, Cox D E, Smith J L, and Fisk Z 1985, *Phys. Rev. B* **32** 6042.
[4] Goldman A I, Shapiro S M, Shirane S. Smith J L and Fisk Z, 1986, *Phys. Rev. B* **33** 1627.
[5] Heffner R H, Cooke D W, Giorgi A L, Hutson R L, Schillaci M E, Rempp H D, Smith J L, Willis J O, MacLaughlin D E, Boekema C, Lichti R L, Oostens J, Denison A B, 1989, *Phys. Rev. B* **39** 11345.
[6] Sonier J E, Heffner R H, MacLaughlin D E, Nieuwenhuys G J, Bernal O, Movshovich R, Pagliuso P G, Cooley J, Smith J L, and Thompson J D, 2000, *Phys. Rev. Lett.* **85** 2821.
[7] MacLaughlin D E, Tien C, Clark W G, Lan M D, Fisk Z, Smith J L and Ott H R 1984, *Phys. Rev. Lett.* **53** 1833.
[8] Tien C and Jiang I M, 1989 *Phys. Rev. B* **40** 229.
[9] Ahrens E T, Heffner R H, Hammel P C, Reyes A P, Thompson J D, Smith J L, and Clark W G, 1999, *Phys. Rev. B* **59** 1432.
[10] Clark W G, Lan M C, Kalkeren G v, Wong W H, Tien C, MacLaughlin D E, Smith J L, Fisk Z, and Ott H R, 1987, *J. Magn. Magn. Mater.* **63-64** 396.
[11] Haga Y, Yamamoto E, Honma T, Nakamura A, Hedo M, Araki S, Ohkuni H, and O nuki Y 1999, *Physica B* **259-261** 627.
[12] Tou H, Tsugawa N, Sera M, Harima H, Haga Y, Onuki Y, 2007, *J. Phys. Soc. Jpn.* **76** 024705.
[13] Cox D L, 1997, *Phys. Rev. Lett.* **59** 1240.
[14] Tou H, Tsugawa N, Doi M, Nakai Y and Sera M, 2006, *J. Phys. Soc. Jpn.* **75** Suppl. 201.
[15] Morita K, Hara Y, Sakano Z, Kotegawa H, Tou H, Haga Y, and Onuki Y, 2011, *J. Phys. Soc. Jpn.* **80** SA099.
[16] Nishiyama S, Matsuura H, Miyake K, 2010, *J. Phys. Soc. Jpn.* **79** 104711.

Journal of Physics: Conference Series **391** (2012) 012049 doi:10.1088/1742-6596/391/1/012049

Fe substitution effect on the phase transition and hybridization gap in $CeOs_2Al_{10}$

Yuji Muro[1], Keisuke Yutani[1], Jumpei Kajino[1], Toshiro Takabatake[1,2]

[1]AdSM and [2]IAMR, Hiroshima University, Higashi-Hiroshima 739-8530, Japan

E-mail: ymuro@pu-toyama.ac.jp

Abstract. We report on the magnetic susceptibility, electrical resistivity and specific heat measurements of the alloys $Ce(Os_{1-x}Fe_x)_2Al_{10}$ ($0 \leq x \leq 1$) in order to understand the anomalous phase transition in $CeOs_2Al_{10}$ at T_0 = 28.5 K. With increasing x, the unit cell volume decreases linearly and the c-f hybridization is enhanced as indicated by the systematic change in the magnetic susceptibility. On the other hand, T_0 decreases slowly to 25 K at x = 0.3 then sharply to 20 K at x = 0.5, and eventually disappears at x = 0.6. However, the jump of C_p at T_0, which is proportional to the entropy released by the phase transition, decreases quickly from 0.22 J/Kmol at x = 0 to 0.08 J/Kmol at x = 0.3, and then linearly to 0 at x = 0.6. These behaviours of T_0 resemble the pressure dependence of T_0 for $CeOs_2Al_{10}$ as well as the x dependence of T_0 in $Ce(Ru_{1-x}Fe_x)_2Al_{10}$ for $0.25 \leq x$. This result suggests that the critical concentration where T_0 disappears depends strongly on the c-f hybridization strength.

1. Introduction

Kondo semiconductors have been one of the central issues in the physics of strongly correlated f-electron system [1]. Kondo semiconductors display an activation-type behaviour in the transport properties at low temperatures because of the formation of a small energy gap at the Fermi level owing to the strong hybridization between the local f state and conduction bands, so-called c-f hybridization. This strong c-f hybridization hinders magnetic order in the Ce-based Kondo semiconductors except $CeOs_4Sb_{12}$, which undergoes a spin-density-wave transition at 0.9 K [2].

Ternary Ce compounds CeT_2Al_{10} (T=Fe, Ru and Os), which crystallize in the orthorhombic $YbFe_2Al_{10}$-type structure [3, 4], are a new series of Kondo semiconductors [5-9]. Especially, $CeRu_2Al_{10}$ and $CeOs_2Al_{10}$ display an unusual phase transition at T_0 = 27 K and 28.5 K, respectively, while $CeFe_2A_{10}$ is a Kondo semiconductor without magnetic phase transitions akin to CeNiSn and CeRhSb [10]. The formations of hybridization gap below 80 K in CeT_2Al_{10} (T=Ru and Os) are recognized in an activation-type electrical resistivity and the reduction of optical conductivity below 45 meV [11]. Below T_0, the resistivity shows a further increase which indicates opening of another energy gap on a part of Fermi surface by the phase transition. Moreover, the muon spin relaxation and neutron diffraction measurements revealed an antiferromagnetic order of ~0.3 μ_B of Ce 4f moments aligned along the orthorhombic c-axis [12-16]. Despite such a small value of Ce moment, T_0's for CeT_2Al_{10} are higher than the Néel temperatures of isostructural counterparts of Nd (T_N=2.4 K for T=Ru and T_N=2.2 K for T=Os) and Gd (T_N=17.5 K for T=Ru and T_N=18 K for T=Os) [7, 17]. In addition to anomalously high T_0, pressure dependence of T_0 for $CeOs_2Al_{10}$ shows a sudden decrease in the range of 0.26 GPa above the critical pressure P_c=2.54 GPa [18]. A similar behaviour is observed in

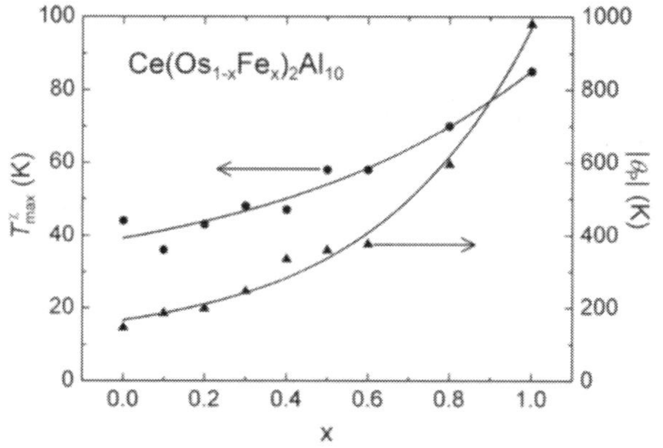

Figure 1. Variations of T^{χ}_{max} and $|\theta_p|$ versus Fe concentration x derived from the magnetic susceptibility. Solid lines are guide to the eye.

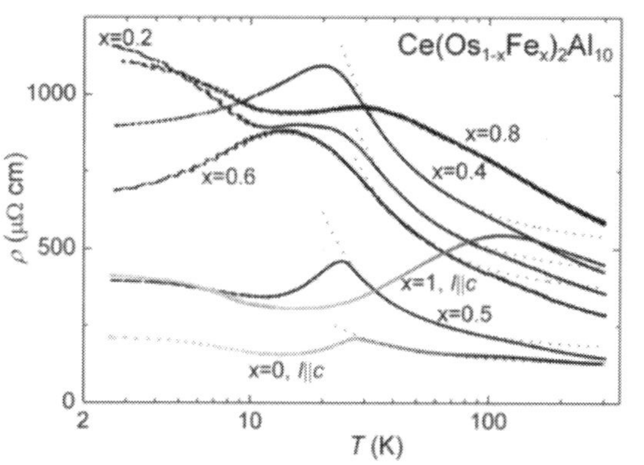

Figure 2. Temperature dependence of electrical resistivity for $Ce(Os_{1-x}Fe_x)_2Al_{10}$. Dashed lines for $0 \leq x \leq 0.6$ represent the thermally activating behaviour.

$Ce(Ru_{1-x}Fe_x)_2Al_{10}$ at $0.75 < x < 0.825$ [19]. These behaviours are quite different from a gradual decrease of T_N observed in conventional Ce-based Kondo antiferromagnets by application of pressure or alloying. In this paper, we report the Fe substitution effect on T_0 of $CeOs_2Al_{10}$ by measuring the magnetic susceptibility, electrical resistivity and specific heat. We compare our results with the effects of pressure on $CeOs_2Al_{10}$ and alloying on $Ce(Ru_{1-x}Fe_x)_2Al_{10}$.

2. Experimental

Polycrystalline samples of $Ce(Os_{1-x}Fe_x)_2Al_{10}$ were prepared by arc-melting the constituent amount of pure elements of Ce, Os, Fe and Al. The obtained samples were annealed at 800 °C for one week. The least-square fitting of powder x-ray diffraction (XRD) patterns showed that the lattice parameters a, c, and unit-cell volume V of $Ce(Os_{1-x}Fe_x)_2Al_{10}$ decrease almost linearly with increasing x, while the parameter b shows significantly non-linear behaviour. A slight deviation from the liner extrapolation at $x = 1$ should result from the intermediate valency in $CeFe_2Al_{10}$. The electron-probe microanalyses (EPMA) and powder XRD indicate that Fe concentrations in the obtained samples agree well with the nominal compositions. The bottom of the ingot with $x = 0.2$ contained an impurity phase of $CeOs_3Al_{12}$, which was eliminated from the sample used for the physical measurements.

The magnetic susceptibility M/B was measured at 1 T by using a commercial SQUID magnetometer (Quantum Design MPMS) in the temperature range from 2 K to 300 K. The electrical resistivity ρ was measured by the ac four probe method at 2.5 K < T < 300 K. The specific heat C_p was measured by the relaxation technique at 2 K < T < 300 K in the Quantum Design PPMS.

3. Results and Discussion

The inverse magnetic susceptibility B/M for $Ce(Os_{1-x}Fe_x)_2Al_{10}$ in the whole range $0 \leq x \leq 1$ was found to obey a Curie-Weiss law from 300 K down to 100 K. With increasing x, the slope of B/M above 100 K decreases owing to the enhancement of c-f hybridization. This enhancement is also recognized by the increase of T^{χ}_{max}, which is the temperature at a maximum of M/B vs T. The x dependence of T^{χ}_{max} and the paramagnetic Curie temperature θ_p are shown in Fig. 1. Not only T^{χ}_{max} but also $|\theta_p|$ increase exponentially with increasing x. This behaviour agrees with that observed in the pressure study for $CeOs_2Al_{10}$ [18]. Therefore, the contraction of unit cell volume by Fe substitution strengthens the c-f hybridization.

Figure 2 displays the variations of $\rho(T)$ for $Ce(Os_{1-x}Fe_x)_2Al_{10}$. For $0 < x < 0.6$, a thermally activation behaviour is observed between 70 K and 30 K as shown by the dashed lines. The activation energy Δ is estimated as 30~50 K by the fitting of $\rho(T)$ to the equation $\rho(T) = \rho_0 \exp(\Delta/2k_BT)$. The values of Δ agree with that obtained for a single-crystal $CeOs_2Al_{10}$ [9]. Thus the activation behaviour

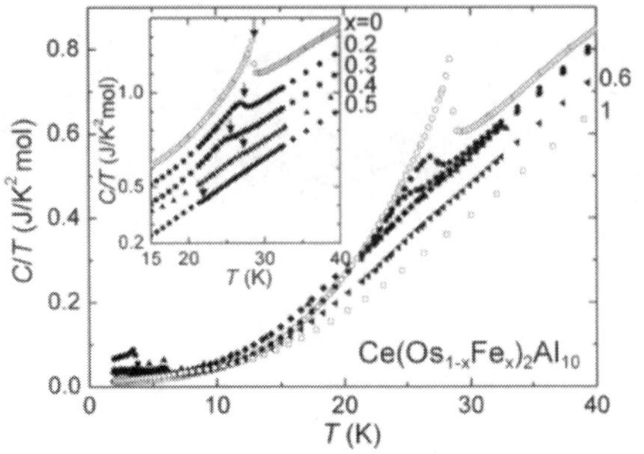

Figure 3. Temperature dependence of C/T for $Ce(Os_{1-x}Fe_x)_2Al_{10}$. The inset shows C/T vs T between 15 K and 40 K. The curves except x=0.5 are shifted by 0.1 J/Kmol for clarity. The arrows indicate the C/T anomaly due to the phase transition.

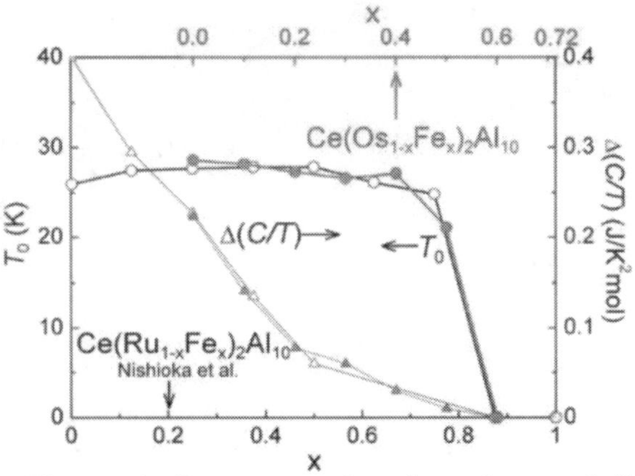

Figure 4. Fe-concentration dependence of T_0 and the value of jump at T_0 in C/T for $Ce(Os_{1-x}Fe_x)_2Al_{10}$. The open symbols represent the results for $Ce(Ru_{1-x}Fe_x)_2Al_{10}$ deduced from ref. 19.

in $Ce(Os_{1-x}Fe_x)_2Al_{10}$ originates from the formation of hybridization gap observed in the optical conductivity for $CeOs_2Al_{10}$ [11]. The constant behaviour of Δ in the range $0 < x < 0.6$ suggests that the gap size is independent of x. This result is inconsistent with the pressure dependence of Δ_h for $CeOs_2Al_{10}$ [18]. Because of the slow decrease of lattice parameter b, the anisotropic hybridization, especially that along the b-axis, should play an important role for the hybridization-gap formation. Below 30 K, $\rho(T)$ shows a broad maximum and then increases with decreasing T which is also observed in $CeOs_2Al_{10}$ [9]. The low-temperature upturn disappears at x = 0.4 and 0.6, but revives at x = 0.8 and 1. The reason of disappearance of upturn in x = 0.4 and 0.6 is unclear, but it may be related to the disappearance of phase transition because the low-temperature upturn disappears above P_c for $CeOs_2Al_{10}$ [18]. The $\rho(T)$ for x = 0.8 resembles that of $CeFe_2Al_{10}$ for $I//a$. Thus the phase transition should be absent at x = 0.8.

Figure 3 shows the specific heat divided by temperature C/T versus T for $Ce(Os_{1-x}Fe_x)_2Al_{10}$. A jump of C/T by the phase transition is observed clearly for x up to 0.3. For x = 0.4 and 0.5, a small but significant anomaly is recognized at 27 K and 21 K, respectively, as shown by an arrow in the inset of Fig. 3. Additional anomalies are observed at temperatures below 7 K for x = 0.2, 0.4 and 0.6. They can be attributed to the magnetic transition of $CeOs_3Al_{12}$ impurity phase because no anomalies are observed in $\rho(T)$ below 10 K as shown in Fig. 2. At x = 0.6, the values of C/T below 40 K decrease with respect to those for x = 0 ~ 0.5. This decrease results from the suppression of the magnetic contribution of C_p at x = 0.6 where the phase transition disappears and the c-f hybridization is enhanced.

The x dependences of T_0 and the jump in C/T at T_0, $\Delta(C/T)$, are plotted in Fig. 4. On going from x = 0 to 0.5, T_0 decreases gradually from 28.5 K to 21 K, then drops sharply to 0 K. On the other hand, $\Delta(C/T)$, which represents the contribution of phase transition for the entropy, decreases rapidly from x = 0 then vanishes at x = 0.6. These behaviours are very similar to the pressure dependences of T_0 and $\Delta(C/T)$ for $CeOs_2Al_{10}$ [18]. Moreover, as shown in Fig. 4, the dependences of T_0 and $\Delta(C/T)$, respectively, on x for $Ce(Os_{1-x}Fe_x)_2Al_{10}$ coincides with those for $Ce(Ru_{1-x}Fe_x)_2Al_{10}$ [19] if the value of x is shifted by 0.25. This feature indicates that the phase transition in CeT_2Al_{10} results from the same origin. Moreover, this also suggests that the critical Fe concentration x_c where the phase transition disappears depends not on the unit cell volume V but on the c-f hybridization strength because the hybridization strength for T=Os is stronger than that for T=Ru despite V of T=Os is only 0.3% larger than that of T=Ru. The importance of c-f hybridization for the phase transition is also suggested by the comparison between the result of $Ce(Os_{1-x}Fe_x)_2Al_{10}$ and pressure effects on $CeOs_2Al_{10}$; T_0 disappears when T^{χ}_{max} is increased to 60 K in both cases. However, the discrepancy between x dependence of T_0 and $\Delta(C/T)$, especially the sudden disappearance of T_0, cannot be explained solely by the enhancement of *c-f* hybridization. Therefore, we have to consider following magnetic interactions in addition to the *c-f* hybridization: a superexchange interaction proposed by the pressure study [18], enhancement of on-site 4*f*-5*d* mixing due to the lack of inversion symmetry at Ce site pointed out by the calculation of crystalline electric field effect [20], antiferromagnetic order driven by the formation of charge-density-wave along the *b*-axis proposed by the study of optical conductivity [11]. In order to clarify the origin of anomalous response of T_0 to pressure and Fe-substitution, we are doing neutron scattering experiment for $Ce(T_{1-x}Fe_x)_2Al_{10}$.

4. Summary

We studied Fe substitution effect on the magnetic transition in the Kondo semiconductor $CeOs_2Al_{10}$ by the measurements of magnetic susceptibility, electrical resistivity and specific heat for polycrystalline $Ce(Os_{1-x}Fe_x)_2Al_{10}$. Exponential increases of paramagnetic Curie temperature and temperature at the maximum in the susceptibility indicate that the *c-f* hybridization strength in $Ce(Os_{1-x}Fe_x)_2Al_{10}$ increases with increasing x. The formation of hybridization gap in the range $0 < x < 0.6$ is manifested by the thermally activation behaviour in the resistivity between 70 K and 30 K. The weak variations of activation energy and lattice parameter for *b*-axis in the range $0 < x < 0.6$ suggest the importance of anisotropic *c-f* hybridization for the gap formation. The specific-heat anomaly due to the phase transition is observed in the range $0 < x < 0.5$. The variations of transition temperature and the magnitude of specific-heat anomaly in $Ce(Os_{1-x}Fe_x)_2Al_{10}$ agree with the pressure dependence of those for $CeOs_2Al_{10}$ and x dependence in $Ce(Ru_{1-x}Fe_x)_2Al_{10}$ at $x > 0.25$. This agreement suggests strongly that the critical point where the phase transition disappears is determined by the strength of *c-f* hybridization.

Acknowledgement
We thank T. Onimaru and K. Umeo for valuable discussions. This work was supported by a Grant-in-Aid for Scientific Research on Innovative Areas "Heavy Electrons" (Grant No. 20102004) of MEXT, Japan.

References

[1] Misra P 2008 *Handook of Metal Physics—Heavy Fermion Systems* (Amsterdam: Elsevier)
[2] Yogi M *et al.* 2005 *J. Phys. Soc. Jpn.* **74** 1950
[3] Thiede V M, Ebel T and Jeitschko W 1998 *J. Mater. Chem.* **8** 125
[4] Tursina A I *et al.* 2005 *Acta Crystallogr. Sect. E: Struct. Rep. Online* **61** i12
[5] Strydom A M 2009 *Physica B* **404** 2981
[6] Muro Y, Motoya K, Saiga Y and Takabatake T 2009 *J. Phys. Soc. Jpn.* **78** 123705
[7] Nishioka T *et al.* Y 2009 *J. Phys. Soc. Jpn.* **78** 123705
[8] Muro Y, Motoya K, Saiga Y and Takabatake T 2010 *J. Phys.: Conf. Ser.* **200** 012136
[9] Muro Y *et al.* 2010 *Phys. Rev. B* **81** 214401
[10] Takabatake T *et al.*, 1998 *J. Mag. Mag. Mat.* **177-181**, 277
[11] Kimura S *et al.* 2011 *Phys. Rev. Lett.* **106** 056404
[12] Kambe S *et al.* Y 2010 *J. Phys. Soc. Jpn.* **79** 053708
[13] Robert J *et al.* 2010 *Phys. Rev. B* **82** 100404 R
[14] Khalyavin D D *et al.* 2010 *Phys. Rev. B* **82** 100405 R
[15] Adroja D T *et al.* 2010 *Phys. Rev. B* **82** 104055
[16] Kato H *et al.* 2009 *J. Phys. Soc. Jpn.* **78** 123705
[17] Muro Y, Kajino J, Onimaru T and Takabatake T 2011 *J. Phys. Soc. Jpn.* **80** SA021
[18] Umeo K, Ohsuka T, Muro Y, Kajino J and Takabatake T 2011 *J. Phys. Soc. Jpn.* **80** 064709
[19] Nishioka T *et al.* 2011 *J. Phys.: Conf. Ser.* **273** 012046
[20] Hanzawa K 2011 *J. Phys. Soc. Jpn.* **80** 023707

Raman scattering spectra of $LaRu_2Al_{10}$ and $CeRu_2Al_{10}$

K Nagano [1], T Hasegawa [1], N Ogita [1], M Udagawa [1, 2], H Tanida [3], D Tanaka [3], M Sera [3], T Nishioka [4] and M Matsumura [4]

[1] Graduate School of Integrated Arts and Sciences, Hiroshima University, Higashi-Hiroshima 739-8521, Japan
[2] Institute for Advanced Material Research, Hiroshima University, Higashi-Hiroshima 739-8530, Japan
[3] Graduate School of Advanced Sciences of Matter, Hiroshima University, Higashi-Hiroshima 739-8530, Japan
[4] Graduate School of Integrated Arts and Sciences, Kouchi University, Kochi 780-8520, Japan

E-mail: m114128@hiroshima-u.ac.jp

Abstract. Phonon spectra of $LaRu_2Al_{10}$ and $CeRu_2Al_{10}$ have been studied by Raman scattering from 3.4K to 300K. Polarization dependence at room temperature gives the precise mode assignment for the sake of first principles calculation results. The measured Raman spectra are well explained by the space group Cmcm and there is no remarkable difference between $LaRu_2Al_{10}$ and $CeRu_2Al_{10}$. Since we observed no anomaly at $T_0 = 27K$ in $CeRu_2Al_{10}$, the lattice distortion due to the structural transition has not been detected. From the temperature dependence of the vibrational frequency of Ce and La, their vibrational amplitude is not large in $CeRu_2Al_{10}$ and $LaRu_2Al_{10}$, that is, close to harmonic case.

1. Introduction

$CeRu_2Al_{10}$ has the orthorhombic $YbFe_2Al_{10}$-type structure with space group Cmcm [1]. $CeRu_2Al_{10}$ shows the transition at $T_0 = 27K$ [2-4], while $LaRu_2Al_{10}$ shows no anomaly. It is reported that this transition is the antiferromagnetic order [5, 6], but the Ce moments appear along c-axis determined by neutron diffraction [5], in spite of the magnetic easy axis along the a-axis. This discrepancy and the origin of this magnetic order are not well understood, and the transition temperature is too high for the RKKY interaction. On the other hand, spin-Peierls transition was suggested to explain the peculiar temperature dependence of resistivity [7]. In this case, the lattice distortion should follow [8].

To clarify whether the lattice distortion exists or not below T_0, the phonon Raman scattering spectra have been measured for $LaRu_2Al_{10}$ and $CeRu_2Al_{10}$. Since the lattice distortion accompanies with new peaks in Raman scattering spectra, the additional peaks judge the existence of the distortion. In addition, $CeRu_2Al_{10}$ was suggested as a cage compound [9]. In the large cage with a small guest atom, the vibration of the guest atom becomes anharmonic and this anharmonicity is experimentally determined by the energy decrease of the vibrational frequency with decreasing temperature [10, 11]. To clarify the characteristic feature of $CeRu_2Al_{10}$, $LaRu_2Al_{10}$ was employed as the reference, where $LaRu_2Al_{10}$ does not show the phase transition.

2. Experimental

We employed single crystals of $LaRu_2Al_{10}$ and $CeRu_2Al_{10}$. They were grown by the Al self-flux method. Three different surfaces with (100), (010), and (001) planes were polished by alumina powder with 0.05 μ m in diameter to reduce the diffuse scattering of the incident light. Raman scattering spectra were analyzed by the triple-monochromator (JASCO NR-1800) and detected by the liquid-N_2 cooled charge coupled device (CCD) detector (Princeton Instruments Inc. LN/CCD-1100PB). For excitation light, the 514.5nm line of the Ar^+ laser (Spetra Physics Stabilite 2017 Ion Laser) was employed and the natural emission lines of the Ar^+ laser were excluded by the premonochromator. For the measurements at low temperatures, the cryostat cooled by the GM cryocooler (SHI SRDK-205) was used. The energy of the measured spectra was calibrated using natural emission lines of a He-Ne laser.

$YbFe_2Al_{10}$ type structure gives the Raman active modes as $10A_g + 9B_{1g} + 6B_{2g} + 8B_{3g}$. These modes can be separately observed in the six polarization geometries. The polarization geometry is denoted by (I,S), where I and S are the polarization directions of the incident and scattered light, respectively. In this study x, y, z were [1,0,0], [0,1,0], and [0,0,1], respectively. In (x,x), (y,y), and (z,z) polarization geometries A_g modes can be observed, and B_{1g}, B_{2g}, and B_{3g} mode are observed in (x,y), (z,x), and (y,z), respectively. Since there are two rare-earth atoms in the primitive cell, their degree of freedom is decomposed as $A_g + B_{1g} + B_{3g} + B_{1u} + B_{2u} + B_{3u}$ and the Raman active modes are A_g, B_{1g} and B_{3g} among them. Other Raman active modes are the vibrations of Al.

3. Results and discussion

Polarization dependence at room temperature is shown in Fig.1, where solid triangles denote the phonon peaks in each symmetry and cross marks denote the natural emission peaks of Ar^+ laser. The open square marks observed in (y,y) polarization are B_{3g} modes, because the polarization direction of the incident laser was slightly inclined from the b-axis. The observed lattice vibrations are $10A_g + 8B_{1g} + 6B_{2g} + 8B_{3g}$, which are close to $10A_g + 9B_{1g} + 6B_{2g} + 8B_{3g}$ predicted by group theory.

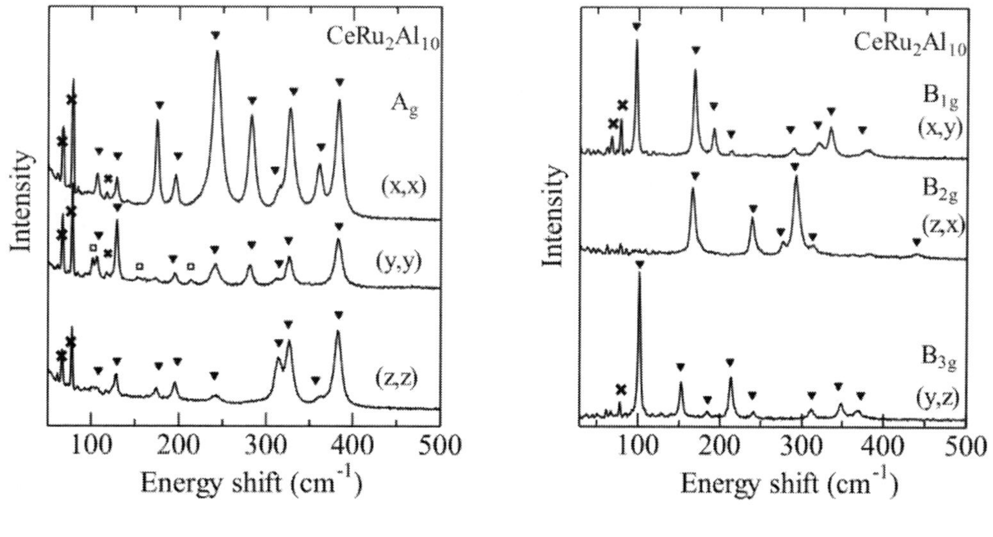

Fig.1: Polarization dependence of Raman spectra of $LaRu_2Al_{10}$ and $CeRu_2Al_{10}$ at room temperature. The triangle marks denote the assigned modes and cross marks show the natural emission peaks of the incident laser. The open square marks observed in (y,y) polarization are B_{3g} modes.

Table1: Phonon energies of $LaRu_2Al_{10}$ and $CeRu_2Al_{10}$ and their assignment. Energy unit is cm^{-1}.

No.		1	2	3	4	5	6	7	8	9	10
A_g	La	107.4	133.5	180.1	197.8	246.5	281.4	320.4	336.3	363.2	388.6
	Ce	106.1	129.8	174.7	197.3	245.3	285.9	315.9	330.6	364.8	386.8
B_{1g}	La	99.4	171.0	193.5	217.4	–	288.5	319.2	334.9	378.2	
	Ce	96.6	166.7	192.6	215.9	–	288.2	322.6	336.5	384.9	
B_{2g}	La	165.5	234.9	273.6	288.2	311.7	435.0				
	Ce	165.7	241.0	278.7	294.4	314.6	443.5				
B_{3g}	La	106.1	156.8	187.5	210.1	242.8	313.4	349.6	376.6		
	Ce	102.2	154.7	189.3	216.4	244.6	315.3	354.0	369.3		

Thus, the present result is consistent with the crystal structure with space group Cmcm, though one B_{1g} mode is missing due to its weak intensity. Remarkable difference between $LaRu_2Al_{10}$ and $CeRu_2Al_{10}$ has not been found.

The observed frequencies and their assignments are summarized in Table1. The vibrational frequencies of the rare-earth ions are about $105cm^{-1}$ and $130cm^{-1}$ in A_g modes, $100cm^{-1}$ in B_{1g} mode, and $105cm^{-1}$ in B_{3g}. These were determined by the comparison with the first principle calculation result [12]. The rare-earth vibration in A_g mode becomes two vibrations due to the hybridization between the rare-earth and Al modes.

We discuss the phonon frequency difference between La and Ce from the structural point of view. The lattice parameter of $LaRu_2Al_{10}$ is larger by a = 0.041 Å, b = 0.028 Å, and c = 0.031 Å than those of $CeRu_2Al_{10}$ [1]. The ionic diameter of La^{3+} is larger by 0.054 Å than that of Ce^{3+} [13]. Therefore, the change of ionic diameter is larger than that of the lattice parameters. Thus, the force constants between La and Al should be stronger than those between Ce and Al. While the force constants between Al and Al for $LaRu_2Al_{10}$ are weaker than those for $CeRu_2Al_{10}$, because of the lattice parameter increase. Therefore, the frequency of the Ce vibration will be lower than that of La, and the frequencies of the Al vibrations in $CeRu_2Al_{10}$ may be higher than those of $LaRu_2Al_{10}$. The La/Ce vibration satisfies this tendency. However, some modes of the Al vibrations violate this tendency. In table1, the corresponding modes with larger energy-decrease than $4cm^{-1}$ are shown by square. This anomaly can be explained by a difference of charge distribution between $LaRu_2Al_{10}$ and $CeRu_2Al_{10}$.

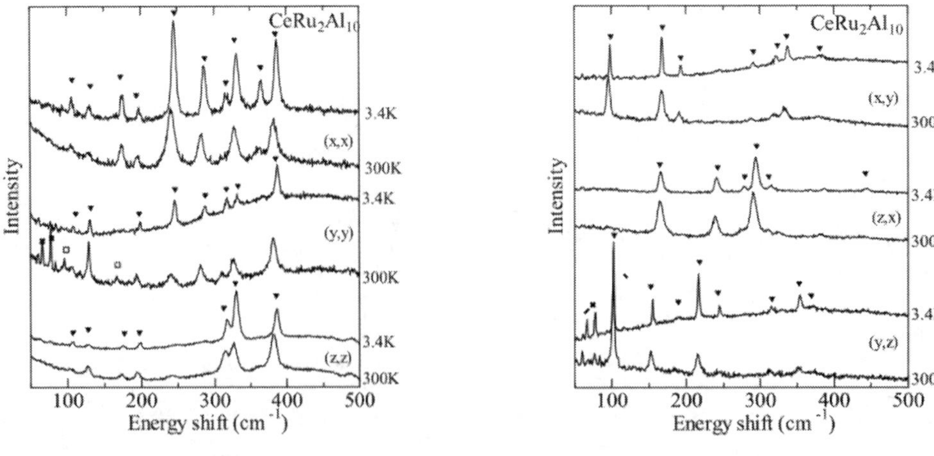

Fig.2: Raman spectra of $CeRu_2Al_{10}$ at 3.4K and 300K. The triangle marks denote the assigned modes in table1. The cross marks are the natural emissions of the incident laser. The square marks observed in (y,y) polarization are B_{1g} modes, because the polarization direction of the incident laser was slightly inclined from the b-axis.

Temperature dependence of Raman spectra has been measured from 3.4K to 300K for $LaRu_2Al_{10}$ and $CeRu_2Al_{10}$. Raman spectra of $CeRu_2Al_{10}$ at 3.4K and 300K are shown in Fig.2. If the lattice deforms, new peaks should appear in Raman spectra. However we have not found any new peak between 3.4K and 300K in $LaRu_2Al_{10}$ and $CeRu_2Al_{10}$ in the present experimental accuracy. Therefore, the lattice distortion does not occur below T_0 =27K in $CeRu_2Al_{10}$. The vibration of La/Ce is harmonic, since the remarkable energy decrease has not been observed.

4. Conclusions

We have investigate the phonon Raman spectra of single crystalline $LaRu_2Al_{10}$ and $CeRu_2Al_{10}$. We have successfully assigned the observed 32 peaks as $10A_g + 8B_{1g} + 6B_{2g} + 8B_{3g}$ by the polarization dependence. The observed Raman spectra are well explained by the space group Cmcm. The lattice distortion has not been observed below T_0 = 27K in $CeRu_2Al_{10}$, judging from no additional peaks at 3.4K. We have found the possibility of the electronic state difference by the phonon energy difference between $LaRu_2Al_{10}$ and $CeRu_2Al_{10}$.

5. Acknowledgments

This work is supported by Grant-in-Aids for Scientific Research on Innovative Areas "Heavy Electrons" (No. 20102005) of The Ministry of Education, Culture, Sports, Science, and Technology, Japan, and also supported by N-BARD and IAMR of Hiroshima University.

References

[1] Thieda V M T, Ebel T, and Jeitxchko W 1998 *J. Mater. Chem.* **8** 125.
[2] Strydom A M 2009 *Physica* B **404** 2981
[3] Nishioka T, Kawamura Y, Takesaka T, Kobayashi R, Kato H, Matsumura M, Kodama K, Matsubayashi K, and Uwatoko Y 2009 *J. Phys. Soc. Jpn.* **78** 123705
[4] Muro Y, Motoya K, Saiga Y, and Takabatake T 2010 *J. Phys.: Conf. Series* **200** 012136
[5] Khayavin D D, Adroja D T, Strydom A M, Manuel P, Chapon L C, Peratheepan P, Knight K, Deen P, Ritter C, Muro Y, and Takabatake T 2010 *Phys. Rev.* B **82** 100405(R)
[6] Robert J, Mignot J M, Andre G, Nishioka T, Kobayashi R, Matsumura M, Tanida H, Tanaka D, and Sera M 2010 *Phys. Rev.* B **82** 100404(R)
[7] Hanzawa K 2010 *J. Phys. Soc. Jpn.* **79** 043710
[8] Hanzawa K 2010 *J. Phys. Soc. Jpn.* **79** 084704
[9] Tanida H, Tanaka D, Sera M, Tanimoto S, Nishioka T, Matsumura M, Ogawa M, Moriyoshi C, Kuroiwa Y, Kim J E, Tsuji N and Takata M 2011 *Phsy. Rev.* B **84** 115128
[10] Udagawa M, Hasegawa T, Takasu Y, Ogita N, Suekuni K, Avila M A, Takavatake T, Ishikawa Y, Takeda N, Nemoto Y, Goto T, Sugawara H, Kikuchi D, Sato H, Sekine C, Shirotani I, Yamaura J, Nagao Y and Hiroi Z 2008 *J. Phys. Soc. Jpn.* **77** Suppl. A 142
[11] Takasu Y, Hasegawa T, Ogita N, Udagawa M, Avila M A, Suekuni K, Ishii I, Suzuki T and Takabatake T 2006 *Phys. Rev.* B **74** 174303
[12] Hasegawa unpublished.
[13] Templeton D H, and Dauben C H 1961 *The rare earth*, ed F I Spedding and A H Daane (New York, John Wiley) p 20

International Conference on Strongly Correlated Electron Systems (SCES 2011) IOP Publishing
Journal of Physics: Conference Series **391** (2012) 012051 doi:10.1088/1742-6596/391/1/012051

Nonmagnetic Γ_3 doublet ground state in a caged compound $PrRh_2Zn_{20}$

N. Nagasawa[1], T. Onimaru[1], K. T. Matsumoto[1], K. Umeo[2] and T. Takabatake[1,3]

[1]AdSM, Hiroshima University, Higashi-Hiroshima 739-8530, Japan
[2]N-BARD, Hiroshima University, Higashi-Hiroshima 739-8526, Japan
[3]IAMR, Hiroshima University, Higashi-Hiroshima 739-8530, Japan

E-mail: n-nagasawa@hiroshima-u.ac.jp

Abstract.
We have prepared single crystals of $PrRh_2Zn_{20}$, which is isostructural and isoelectronic to $PrIr_2Zn_{20}$ where coexistence of superconductivity and antiferroquadrupolar order has been recently found. The electrical resistivity ρ, magnetic susceptibility χ, isothermal magnetization M, and specific heat C have been measured. Hysteretic behavior appears in $\rho(T)$ for 170 K $< T <$ 470 K, suggesting a structural phase transition. On cooling below 10 K, χ approaches a constant value, indicating a non-magnetic ground state. Anisotropic behavior in $M(B)$ at 1.8 K for $B > 2$ T and a broad peak in $C(T)$ at around 10 K are reproduced by the two crystal-field levels split by 32 K; a nonmagnetic Γ_3 doublet ground state and a magnetic Γ_4 triplet excited state.

1. Introduction

Praseodymium-filled skutterudites show a variety of phenomena owing to the $4f^2$ electronic configuration of the Pr^{3+} ions, such as heavy-fermion superconductivity in $PrOs_4Sb_{12}$ [1], scalar-type multipole ordering in $PrFe_4P_{12}$ [2, 3, 4], and a metal-insulator transition in $PrRu_4P_{12}$ [5]. Recently, another class of Pr encaged compounds PrT_2X_{20} (T: transition metal, X: Al, Zn) have been synthesized, which crystallize in the cubic $CeCr_2Al_{20}$-type structure with the space group of $Fd\bar{3}m$ [6]. The Pr atom is encapsulated in a highly symmetric Frank-Kasper cage formed by 16 zinc atoms. This leads crystalline electric field (CEF) ground states to be degenerated so that multipolar degrees of freedom remain active and the $4f$ electrons can be hybridized with conduction electrons. Therefore, phase transitions and/or Kondo effect of the multipolar degrees of freedom are expected to occur. Recently, $PrIr_2Zn_{20}$ with a nonmagnetic Γ_3 doublet ground state has been found to undergo a superconducting transition at T_c=0.05 K in the presence of antiferroquadrupolar (AFQ) order below T_Q=0.11 K [7, 8, 9]. It is suggested that the quadrupolar fluctuations play an important role in the formation of the Cooper pairs. PrT_2Al_{20} (T=Ti, V, Nb) also display anomalous temperature dependences of the electrical resistivity and the specific heat which is possibly related to quadrupole Kondo effect [10, 11]. These facts suggest that strong correlation between the $4f^2$ electronic state and the conduction electrons plays a crucial role in forming the unusual ground states of the PrT_2X_{20} family.

In the present work, we focus on $PrRh_2Zn_{20}$ which is isoelectronic to $PrIr_2Zn_{20}$. Because the CEF level scheme is expected to be similar to that for $PrIr_2Zn_{20}$, the ground state would be the Γ_3 doublet. To study how the ground state changes on going from $PrIr_2Zn_{20}$ to $PrRh_2Zn_{20}$, we have grown single-crystalline samples of $PrRh_2Zn_{20}$ and studied the magnetic and transport properties.

2. Experimental

Single-crystalline samples of $PrRh_2Zn_{20}$ were grown by the melt-growth method with high-purity elements of Pr (4N), Rh (3N) and Zn (6N). The $CeCr_2Al_{20}$-type structure was confirmed by powder x-ray diffraction method. The lattice parameter at room temperature was refined to be 14.294(8) Å, whose value agrees with the reported one [6]. The crystal compositions were determined by electron-probe microanalysis (EPMA) to be 1 : 1.98(1) : 20.02(5). The EPMA detected binary alloys $RhZn_{6.4}$ and $Pr_{14}Zn$ whose volume fractions are less than 5 % in view of back-scattering electron images. The single-crystalline sample was oriented by the back reflection Laue method using an imaging plate camera, IPXC/B (TRY-SE).

The electrical resistivity ρ was measured by a standard four-probe AC method in a laboratory-built system with a Gifford-McMahon-type refrigerator between 30 and 600 K. The specific heat C was measured by a relaxation method using a commercial calorimeter (Quantum Design PPMS) at temperatures between 0.4 and 300 K. The magnetization M was measured using a SQUID magnetometer (Quantum Design MPMS) between 1.8 and 300 K and in magnetic fields up to 5 T.

3. Results and Discussion

Figure 1 shows the result of $\rho(T)$ for $PrRh_2Zn_{20}$. In the temperature range between 170 K and 470 K, hysteretic behavior appears, indicating a first-order phase transition. Hysteretic behaviors were also observed in $LaRu_2Zn_{20}$ and $PrRu_2Zn_{20}$ at T_S=150 and 138, respectively [7]. Since superlattice reflections were observed in $PrRu_2Zn_{20}$ below T_S, the hysteretic behavior was attributed to a structural modulation. If the local symmetry at the Pr site was lowered in $PrRu_2Zn_{20}$, quadrupolar degrees of freedom would vanish below the transition temperature. In $PrIr_2Zn_{20}$, on the other hand, no evidence for structural transition was found in $\rho(T)$ from 530 K to 30 K, as shown in Fig. 1. Therefore, the quadrupolar degree of freedom remains active in the Γ_3 doublet. The inset of Fig. 2 shows $\chi(T)$ of $PrRh_2Zn_{20}$ in a magnetic field of

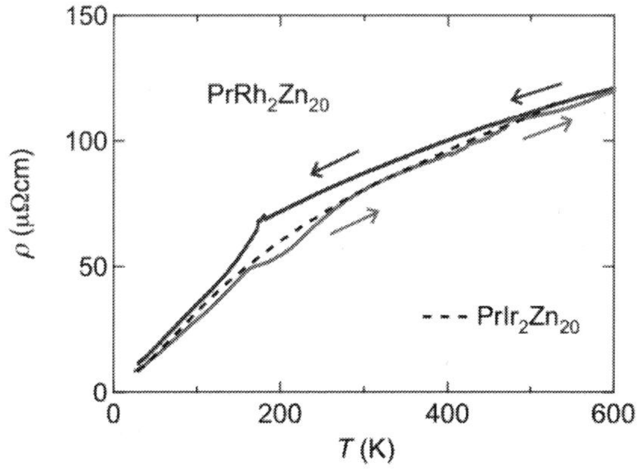

Figure 1. Temperature dependence of the electrical resistivity ρ of $PrRh_2Zn_{20}$ (solid lines) and $PrIr_2Zn_{20}$ (broken lines).

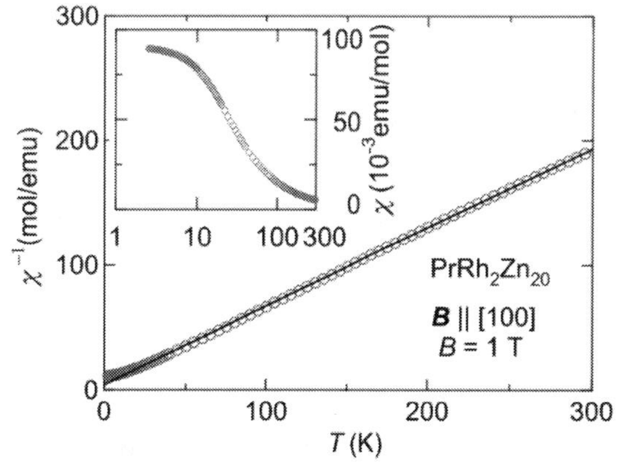

Figure 2. Temperature dependence of the inverse of magnetic susceptibility $\chi(T)$ for $PrRh_2Zn_{20}$. The inset shows the Van-Vleck behavior in $\chi(T)$.

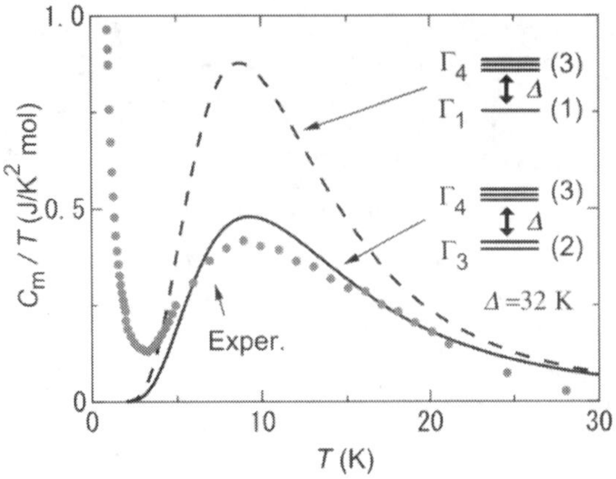

Figure 3. Temperature dependence of the magnetic part of the specific heat divided by temperature, C_m/T. The solid and broken lines are calculated by the models with Δ=32 K as shown in the inset.

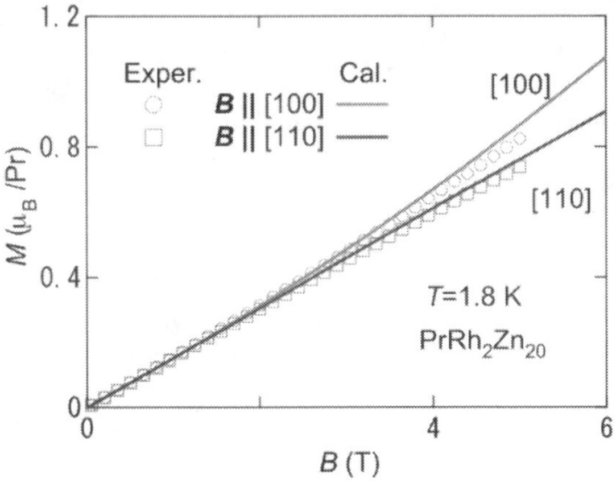

Figure 4. Isothermal magnetization $M(B)$ for $B \parallel$ [100] (○) and B $\parallel$ [110] (□) at 1.8 K. The solid lines were calculated using the Γ_3 - Γ_4 model as shown in the inset of the Fig. 3.

B=1 T. On cooling below 10 K, $\chi(T)$ gradually approaches a constant value, indicating a Van-Vleck paramagnet. This behavior suggests that the CEF ground state is nonmagnetic, either Γ_1 singlet or Γ_3 doublet. The main panel of Fig. 2 shows $\chi^{-1}(T)$ measured at B=1 T. Above 30 K, χ^{-1} follows the Curie-Weiss law with the effective magnetic moment of 3.55 μ_B/f.u., which is in agreement with the value of the trivalent Pr free ion. The T-linear behavior of $\chi^{-1}(T)$ in the wide temperature range 30 K $< T <$ 300 K suggests the weak CEF effect on the Pr $4f$ electrons. The paramagnetic Curie temperature θ_p of $-$ 4.8 K indicates that the intersite interaction between the magnetic moments of Pr ions is antiferromagnetic but rather weak.

Figure 3 shows the temperature dependence of the magnetic part of the specific heat divided by temperature, C_m/T. For the lattice part of C in $PrRh_2Zn_{20}$, we used the data of $LaIr_2Zn_{20}$ because the more relevant compound $LaRh_2Zn_{20}$ does not form. A broad peak manifests itself in C_m/T at around 10 K. To reproduce this Schottky-like anomaly, we assumed the cubic point group T_d of the Pr sites. Then, two models as depicted in Fig. 3 are considered for two levels split by Δ=32 K; one is the Γ_3 doublet-Γ_4 triplet model and the other is the Γ_1 doublet-Γ_4 triplet model. Here, we neglect the higher excited levels. The solid and broken lines represent the calculations, respectively. The peak at around 10 K is better reproduced by the former than the latter. This fact is favorable to the Γ_3 doublet ground state. The Γ_3 doublet ground state is also supported by the anisotropic field dependence of $M(B)$. Figure 4 shows the isothermal magnetization $M(B)$ for $\boldsymbol{B} \parallel$ [100] and $\boldsymbol{B} \parallel$ [110] at 1.8 K. The two data are almost the same up to 2 T, above which two curves gradually diverge. This anisotropic behavior can be well reproduced by the calculation of $M(B)$ using the Γ_3-Γ_4 model, as shown with the solid lines in Fig. 4.

On cooling below 3 K, C/T continuously increases and reaches 4 J/K^2 mol at 0.4 K as was observed in $PrIr_2Zn_{20}$ [7]. This increase of C/T possibly results from the gradual release of the entropy of the Γ_3 doublet. Since the magnetic properties of $PrRh_2Zn_{20}$ at temperatures down to 0.4 K resemble those of $PrIr_2Zn_{20}$, we expect that an AFQ order and/or a superconducting transition occur at temperatures below 0.4 K.

It remains as an important question why the Γ_3 doublet ground state is stable in $PrRh_2Zn_{20}$ in spite of the possible structural transition above 170 K. Because the Γ_3 doublet is realized only in the cubic point groups, the low temperature phase needs to belong to another cubic point group.

We recall that the metal-insulator transition in $PrRu_4P_{12}$ at T_{MI}=63 K is accompanied with a structural transition from a body-centered cubic structure in the high-temperature metallic phase to a simple cubic one in the low-temperature insulator phase [13]. In order to determine the structure of the low-temperature phase of $PrRh_2Zn_{20}$, x-ray diffraction analysis on a single-crystalline sample is on progress.

4. Conclusion

We performed measurements of ρ, χ, M, and C on single-crystalline samples of $PrRh_2Zn_{20}$. The hysteretic behavior in $\rho(T)$ above 170 K indicates a structural transition. A Van-Vleck behavior in $\chi(T)$ below 10 K, the Schottky anomaly in $C(T)$ at around 10 K and the anisotropic behavior of $M(B)$ above 2 T are consistent with the nonmagnetic Γ_3 doublet ground state. However, no phase transition attributed to the Γ_3 doublet has been observed down to 0.4 K. At lower temperatures, an AFQ order and/or a superconducting transition would be observed.

Acknowledgments

This work was supported by a Grant-in-Aid for Scientific Research on Innovation Areas "Heavy Fermions" (Grant Nos. 20102004, 21102513, and 23102718) of MEXT, Japan.

References

[1] Bauer E D, Frederick N A, Ho P C, Zapf V S and Maple M B 2002 *Phys. Rev.* B **65** 100506(R)
[2] Aoki Y, Namiki T, Matsuda, Abe K, Sugawara H and Sato H 2002 *Phys. Rev.* B **65** 064446
[3] Hoshino S, Otsuki J and Kuramoto Y 2010 *J. Phys. Soc. Jpn.* **79** 074720
[4] Kikuchi J, Takigawa M, Sugawara H and Sato H 2007 *J. Phys. Soc. Jpn.* **76** 043705
[5] Sekine C, Uchiumi T, Shirotani I and Yagi T 1997 *Phys. Rev. Lett.* **79** 3218
[6] Nasch T, Jeitschko W and Rodewald U C 1997 *Z Naturforsch* B **52** 1023
[7] Onimaru T, Matsumoto K T, Inoue Y F, Umeo K, Saiga Y, Matsushita Y, Tamura R, Nishimoto K, Ishii I, Suzuki T and Takabatake T 2010 *J. Phys. Soc. Jpn.* **79** 033704
[8] Onimaru T, Matsumoto K T, Inoue Y F, Umeo K, Sakakibara T, Karaki Y, Kubota M and Takabatake T 2011 *Phys. Rev. Lett.* **106** 177001
[9] Ishii I, Muneshige H, Suetomi Y, Fujita T K, Onimaru T, Matsumoto K T, Takabatake T, Araki K, Akatsu M, Nemoto Y, Goto T and Suzuki T 2011 *J. Phys. Soc. Jpn.* **80** 093601
[10] Sakai A and Nakatsuji S 2011 *J. Phys. Soc. Jpn.* **80** 063701
[11] Higashinaka R, Maruyama A, Ando M, Watanabe M, Aoki Y and Sato H 2011 *J. Phys. Soc. Jpn.* **80** SA048
[12] Izumi F and Ikeda T 2000 *Matter. Sci. Forum* 321-324 198
[13] Lee C H, Matsuhata H, Yamamoto A, Ohta T, Takazawa H, Umeo K, Sekine C, Shirotani I and Hirayama T 2001 *J. Phys.:Condens. Matter* **13** L45

Pressure effect of transport properties of hexagonal $Yb_2Ni_{12}P_7$

Tomohito Nakano[1], Ryohei Satoh[2], Katsuhiko Tsuchiya[2], Naoya Takeda[1], Kazuyuki Matsubayashi[3], Yoshiya Uwatoko[3]
[1]Faculty of Engineering, Niigata University, Ikarashi 2, Niigata, 950-2181, Japan
[2]Graduate School of Science and Technology, Niigata University, Ikarashi 2, Niigata 950-2181, Japan
[3]Institute for Solid State Physics, Tokyo University, Kashiwa, 277-8581, Japan

E-mail:t-nakano@eng.niigata-u.ac.jp

Abstract. We prepared single crystals of $Yb_2Ni_{12}P_7$ using a Sn-flux method and measured their resistivity under hydrostatic pressure up to 8 GPa. At ambient pressure, the magnetic part of resistivity $\rho_{mag}(T)$ shows a peak around T_{max} ~ 100 K and T^2 dependence below the characteristic temperature of the Fermi liquid, T_{FL} = 3 K. When pressure increases up to 2.06 GPa, the value of $\rho_{mag}(T)$ above T_{max} decreases slightly but ρ_{mag} below T_{max} increases slightly. T_{FL} decreases with increasing pressure, and is expected to become zero at 2.5 GPa. Around this pressure, the coefficient of T^2 in ρ_{mag} significantly increases, suggesting the existence of a quantum critical point. Above 2.5 GPa, overall ρ_{mag} gradually decreases with increasing pressure up to 6 GPa, but it below 15.5 K increases above 6 GPa.

1. Introduction

In Yb-based intermetallic compounds, it is well known that magnetic/valence instability is induced by hybridization between 4f- and conduction-electrons. Hybridization is controlled by pressure, and nonmagnetic-magnetic and/or valence quantum phase transitions are expected. $Yb_2Ni_{12}P_7$ is one of the candidates in which a nonmagnetic-magnetic quantum critical point (QCP) is expected by applying pressure.

The crystal structure of $Yb_2Ni_{12}P_7$ is a hexagonal $Zr_2Fe_{12}P_7$-type structure (Space group $P\bar{6}$)[1]. There are 1c and 1f crystallographic sites with hexagonal point symmetry for each Yb ion which occupies voids of the $Cr_{12}P_7$-type structure. Recently, rare earth intermetallic compounds which crystallize in a cage structure (such as $Yb_2Ni_{12}P_7$) have attracted much attention because of their various properties, such as magnetism, superconductivity, heavy fermion, multipole order, and rattling. The susceptibility of $Yb_2Ni_{12}P_7$ follows Curie-Weiss law down to 50 K from room temperature with an effective Bohr magnetron 4.5 μ_B, indicating the trivalent state of Yb-ions[2]. This compound shows no magnetic transition down to 50 mK and heavy fermion like behaviour with the electric specific heat coefficient γ = 200 mJ/K^{2}mol-Yb[2, 3]. In order to explore the QCP, we prepared high quality single crystals of $Yb_2Ni_{12}P_7$ using a Sn-flux method, and measured their electrical resistivity under hydrostatic pressure.

2. Experimental detail

The single-crystal samples of $Yb_2Ni_{12}P_7$ were prepared by a molten-metal flux method using Sn as the flux. Detail of sample synthesis is described in [4]. Hexagonal prism crystals with a typical length of 0.5 mm were obtained after eliminating the flux using hydrochloric acid. The X-ray powder diffraction patterns confirmed the $Zr_2Fe_{12}P_7$-type structure. The electrical resistivity ρ was measured using a DC four-probe method with excitation current parallel to the cylindrical axis of the hexagonal prism crystal in a ^{3}He cryostat down to 280 mK. Hydrostatic pressure was generated using a NiCrAl-CuBe hybrid-piston-cylinder cell up to 2.06 GPa[5] and a cubic-type anvil cell up to 8 GPa. Daphne oil 7373 and mixture of Flourinert FC 70 & 77 were used as pressure transmitting media for the piston cylinder cell and the cubic anvil cell, respectively.

3. Results and discussion

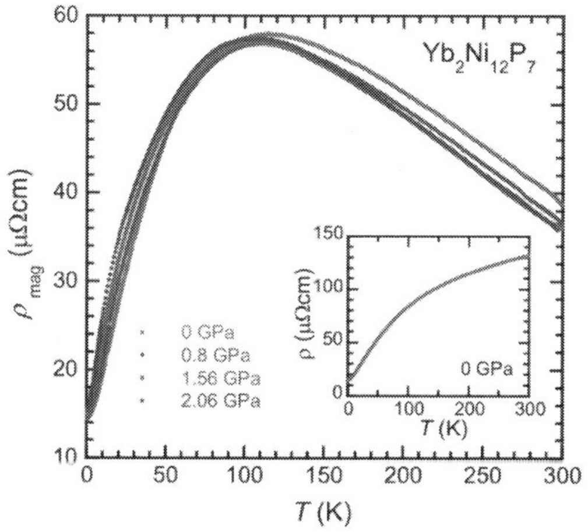

Figure 1. Magnetic part of resistivity ρ_{mag} of $Yb_2Ni_{12}P_7$ as a function of temperature under hydrostatic pressure up to 2.06 GPa. Inset: resistivity ρ of $Yb_2Ni_{12}P_7$ at ambient pressure.

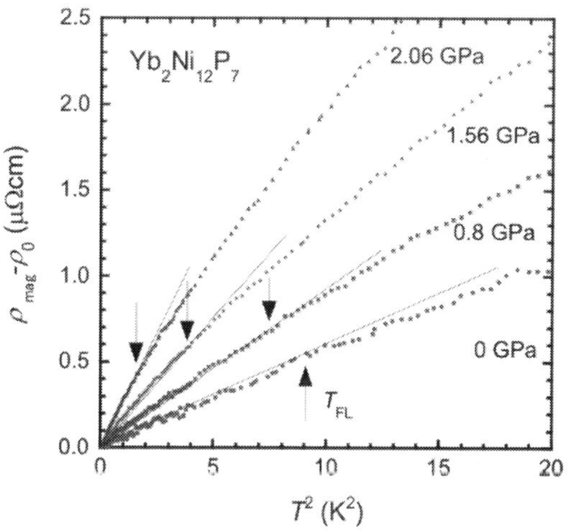

Figure 2 ρ_{mag} - ρ_0 of $Yb_2Ni_{12}P_7$ under selected pressures as a function of T^2.

The temperature dependence of ρ at ambient pressure is metallic, and shows a shoulder around 100 K as shown in the inset of Fig. 1[2]. The magnetic part of ρ, ρ_{mag}, was evaluated by subtracting phonon contribution using the ρ-ρ_0 of non-magnetic $Lu_2Ni_{12}P_7$, where ρ_0 is a temperature independent residual resistivity. Figure 1 shows $\rho_{mag}(T)$ under selected pressures. $\rho_{mag}(T)$ roughly shows -log T dependence from room temperature down to 150 K, and shows a broad maximum around T_{max} ~ 100 K. These behaviours are characteristic of the Kondo effect. $\rho_{mag}(T)$ smoothly decreases below T_{max} and shows T^2 dependence below T_{FL} = 3 K as shown in Fig 2, indicating that the electronic state of $Yb_2Ni_{12}P_7$ at ambient pressure is the Fermi liquid state. Here, T_{FL} is defined as the upper limit of T^2 dependence in $\rho_{mag}(T)$.

The overall $\rho_{mag}(T)$ does not change drastically by applying pressure, as shown in Fig. 1. With increasing pressure, the value of $\rho_{mag}(T)$ above T_{max} decreases slightly, but it below T_{max} increases slightly. In contrast with the overall $\rho_{mag}(T)$, T_{FL} clearly decreases with increasing pressure with initial slope of -0.55 K/GPa as shown in Fig. 3(a). T_{FL} abruptly decreases above 1.5 GPa, and hence the disappearance of T_{FL} and the existence of QCP are expected around 2.5 GPa. The temperature

Journal of Physics: Conference Series **391** (2012) 012052 doi:10.1088/1742-6596/391/1/012052

dependence of ρ_{mag} for Fermi liquid in the low-temperature region is described by the well-known formula,

$\rho_{mag} = \rho_0 + AT^2$ (1),

where, the first term is the residual resistivity from scattering by impurities. The second term is the resistivity from scattering between conduction electrons. ρ_0 and A are displayed in Fig. 3(b) and (c), respectively. ρ_0 and A increase with increasing pressure up to 2.06 GPa. These behaviors of ρ_0 and A are observed around QCP with superconducting phase as $CeCu_2Ge_2$[6].

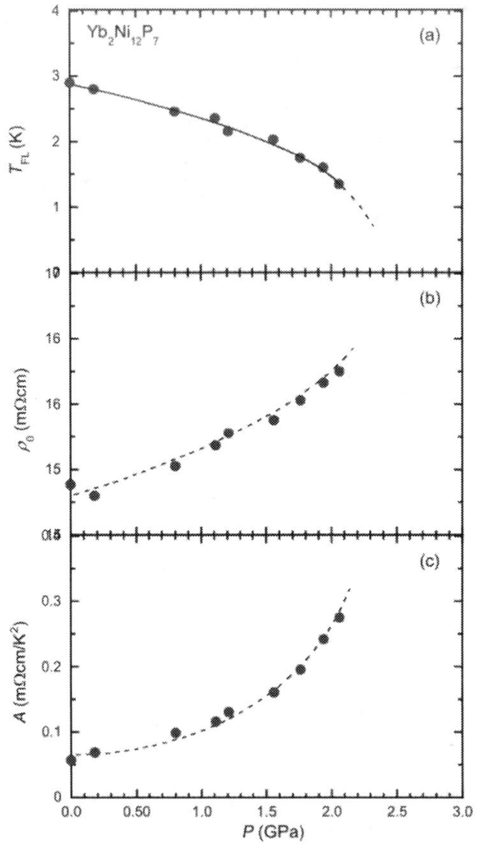

Figure 3. . (a) T_{FL}, (b) ρ_0 and (c) A of $Yb_2Ni_{12}P_7$ as a function of pressure

In order to explore the QCP, we measured resistivity under high pressures from 1.5 to 8 GPa using a cubic-type anvil cell and the results are shown in Fig. 4(a). The overall ρ_{mag} gradually decreases above 2.5 GPa, however, the value of ρ_{mag} below 15.5 K at 8 GPa becomes larger than that at 6 GPa as shown in Fig. 4(b). This behavior would be consistent with that an ordered phase is induced by applying pressure as observed in the Yb-based compound $YbCo_2Zn_{20}$ [7].

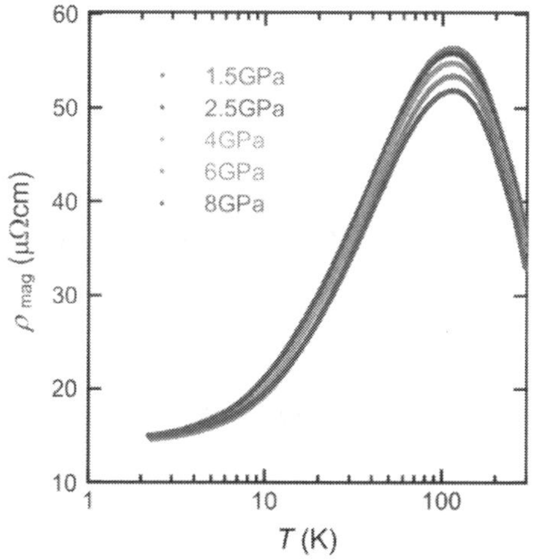

Figure 4(a) Temperature dependence of electrical resistivity in $Yb_2Ni_{12}P_7$ under pressure up to 8 GPa.

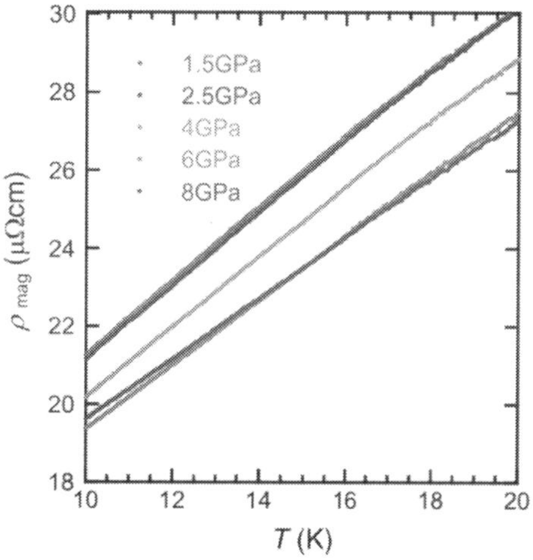

Figure 4(b) ρ_{mag} around 15 K of $Yb_2Ni_{12}P_7$ under pressure up to 8 GPa.

4. Summary

We have measured ρ of $Yb_2Ni_{12}P_7$ single crystals under hydrostatic pressure up to 8 GPa. At ambient pressure, $\rho_{mag}(T)$ shows a peak around T_{max} ~ 100 K and Fermi liquid like T^2 dependence below T_{FL} = 3 K. With increasing pressure, overall $\rho_{mag}(T)$ does not show drastic change but T_{FL} decreases with divergence of A and the significant increase of ρ_0. It is expected that T_{FL} becomes zero and QCP exists at 2.5 GPa. Above 2.5 GPa, ρ_{mag} gradually decreases up to 6 GPa but the value of ρ_{mag} below 15.5 K increases with increasing pressure. This behavior would be consistent with an ordered phase induced by applying pressure, as observed in some Yb-based compounds.

Acknowledgment

This work was supported by a Grant-Aid for Scientific Research on Innovative Area "Heavy Electron" (No. 20102004) of the Ministry of Education, Culture, Sports, Science, and Technology, Japan.

References

[1] Zeppenfeld K and Jeitschko W 1993 *J. Phys. Chem. Solids.* **54** 1527
[2] Tsuchiya K 2011 *Master thesis, Niigata University*
[3] Cho B K, DiSalvo F J, Kim J S, Stewart G R and Bud'ko S L 1998 *Physica B* **253** 40
[4] Tsuchiya K, Ohashi A, Nashimoto T, Ono H, Tatematsu K, Nakano T and Takeda N 2011 *J. Phys. Soc. Jpn.* **80** SA081
[5] Uwatoko Y, Hedo M, Kurita N, Koeda M, Abliz M and Matsumoto T 2003 *Physica B* **329** 1658
[6] Jaccard D, Wilhelm H, Alami-Yadri K and Vargoz E 1999 *Physica B* **259-261** 1
[7] Saiga Y, Matsubayashi K, Fujwara T, Kosaka M, Katano S, Hedo M, Matsumoto T and Uwatoko Y 2008 *J. Phys. Soc. Jpn.* **77** 053710

Shubnikov-de Haas oscillations in the heavy fermion α-$YbAlB_4$

E.C.T. O'Farrell[1,2], D.A. Tompsett[1], N. Horie[2], S. Nakatsuji[2] and M.L. Sutherland[1]

[1] Cavendish Laboratory, University of Cambridge, J.J. Thomson Ave, Cambridge CB3 0HE, UK
[2] Institute for Solid State Physics, University of Tokyo, Kashiwa, Japan 277-8581

E-mail: eoin@issp.u-tokyo.ac.jp

Abstract.
Shubnikov-de Haas oscillations were measured in α-$YbAlB_4$ at fields up to $B = 16$ T. Quantum oscillations were used to directly extract the quasiparticle mean free path by fitting to the Dingle decay yielding a mean free path $\sim$ 550 Å. We also describe a novel fitting procedure used to extract a low frequency oscillation from the non-oscillatory background magneto-resistance.

1. Introduction:

When crossing a quantum phase transition the correlation structure of the ground state changes. For this reason quantum critical points (QCP) are a central topic in the study of strongly correlated electron systems. Tuning to the vicinity of a QCP has become a reliable experimental heuristic for discovering new electronic states of matter [1].

The $YbAlB_4$ [2, 3, 4, 5] systems α-$YbAlB_4$ and β-$YbAlB_4$ (see Figure 1A inset) have received considerable interest in this context. β-$YbAlB_4$ is the first example of a Yb-based heavy fermion superconductor [3] around a QCP that occurs at exactly zero magnetic field [4]. Superconductivity in β-$YbAlB_4$ is suggested to be unconventional on the basis that superconductivity is only found in the clean limit and pairing involves highly renormalized quasiparticles [5]. In contrast, α-$YbAlB_4$ has been found to be a Fermi liquid at zero magnetic field in measurements to $T = 0.01$ K [6].

However, in several respects α-$YbAlB_4$ closely resembles β-$YbAlB_4$. At temperatures greater than the putative Kondo lattice coherence temperature ($T^* = 8$K [4, 6]) both the specific heat and magnetization of α and β are almost identical suggesting almost identical local moment properties. This is consistent with the local isomorphism of the Yb sub-lattice whereby up to nearest neighbour the Yb-Yb bond distances and angles are almost the same. Therefore, a detailed comparison of the low energy excitations in α and β polymorphs should yield information on the relevant parameter to realize quantum criticality in β-$YbAlB_4$.

By comparing the electronic structure of α and β polymorphs we can extract differences in the strength and/or symmetry of the hybridization between f-moments and conduction electrons (c) that will have important implications for a microscopic approach to this problem. Such differences have been considered theoretically by Tompsett et al. [7] and it was found that the boron heptagon that surrounds and strongly hybridizes with the Yb ion (in both α and β

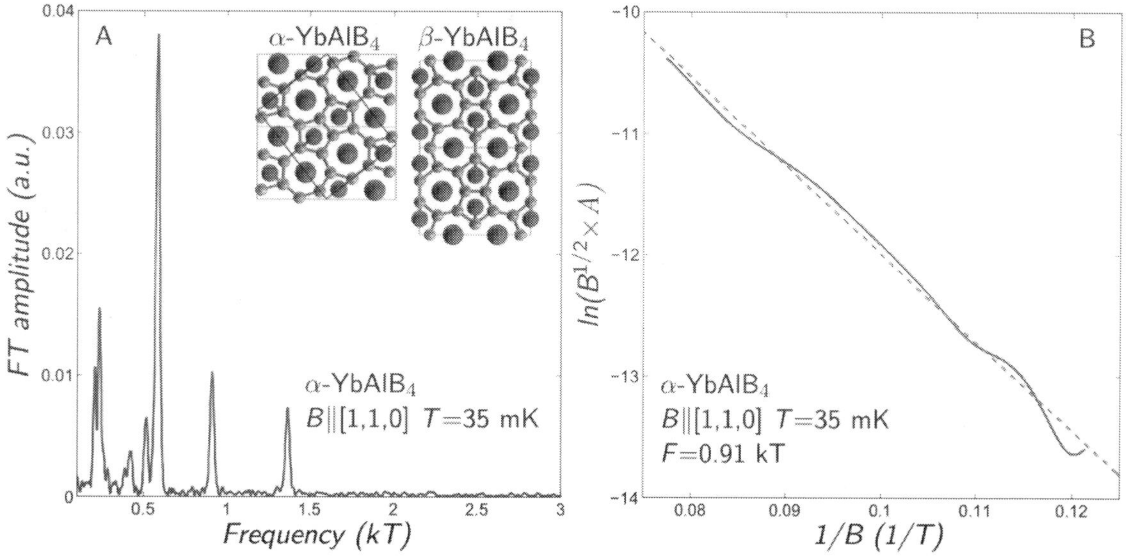

Figure 1. A: Quantum oscillation spectrum for magnetic field $B \parallel [1,1,0]$ *inset*: Crystal structures of β-YbAlB$_4$ and α-YbAlB$_4$ viewed along the c-axis from Tompsett et al. [7], illustrating the distortion in lattice of α-YbAlB$_4$; Yb (red), Al (pink), B (grey). B: Dingle plot for the $F = 0.91$ kT frequency at 35 mK, where A is the amplitude of the oscillation; assuming a circular cross section for the orbit we extract the mean free path as ~550 Å as described in the text.

polymorphs) is distorted in the structure of the α polymorph; this leads to a more significant anisotropy of the $c-f$ hybridization within the ab crystal plane.

An experimental investigation of electronic structure has been carried out for β-YbAlB$_4$ by quantum oscillation measurement of the Fermi surface [8]. In this paper we present a partial result of a Shubnikov-de Haas study of α-YbAlB$_4$. Here we use the Dingle relation for the decay of quantum oscillations to determine the quasiparticle mean free path and verify the high quality of the crystals used. We then describe the analysis of a particularly low frequency oscillation and show how the oscillatory component was extracted from the non-oscillatory background magneto-resistance.

2. Methods

Single crystals of α-YbAlB$_4$ were synthesized according to the method of Macaluso *et al.* [2]. Crystals of α-YbAlB$_4$ grow as long needles along the c-axis. Therefore, electrical contacts were made to apply current (I) parallel to the c-axis ([0,0,1]) and resistivity measurements were made using a current of approximately 50 μA, having checked that Joule heating effects were absent. In this article we present measurements made with fields applied parallel to the [1,1,0] and [0,0,1] crystal axes.

3. Results

Figure 1A shows the quantum oscillation spectrum measured for the magnetic field $B \parallel [1,1,0]$ crystal axis. The spectrum shows several frequencies and we analyze the Dingle decay to characterize the quality of our samples. To fit the Dingle decay we need to extract amplitude of some frequency; for this we choose the peak at $F = 0.91$ kT as it is well separated from other frequencies so can be easily filtered in the Fourier domain.

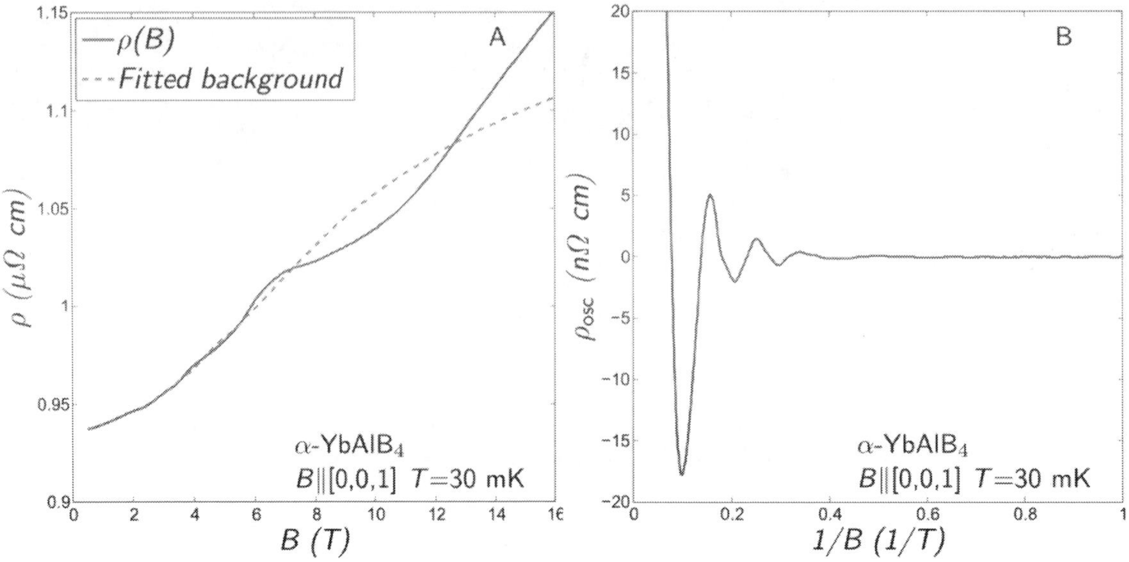

Figure 2. A: Resistivity $\rho(B \parallel c \parallel I)$ measured at constant temperature T =30 mK, together with the fit to the non-oscillatory background $\rho_{\rm MR}(B)$. B: Oscillatory component of the resistivity $\rho_{\rm osc}(B) = \rho(B) - \rho_{\rm MR}(B)$

Figure 1B shows the result for the amplitude of $F = 0.91$ kT, plotted on the appropriate Dingle scale for Shubnikov-de Haas oscillations. The approximately linear decay on this scale is consistent with the expected decay of a single oscillatory component without interference. The gradient of this line is $-\pi B_D/l$, where $B_D = (2\hbar F/e)^{\frac{1}{2}}$, l is the mean free path and we have assumed the orbit has a circular cross section to convert from the real space area to circumference.

From this fitting we obtain quasiparticle mean free path 550±20 Å and because of the assumption of a circular orbit this is a lower bound for the sample measured. This is comparable to the mean free path (900 Å) measured for β-$YbAlB_4$ [8], which shows that α-$YbAlB_4$ can be synthesized at a similar level of purity to β-$YbAlB_4$ and furthermore suggests that the measured differences with respect to superconductivity and quantum criticality [4] should reflect the intrinsic behaviour of the system rather than the effects of disorder in either polymorph.

Figure 2A shows resistivity measured with $I \parallel B \parallel c$ at a constant temperature of 30 mK. From the raw data it is clear that the resistivity has an oscillatory component. We now consider how to extract this from the non-oscillatory background magneto-resistance.

The total resistivity is assumed to be of the form $\rho(B) = \rho_{\rm MR}(B) + \rho_{\rm osc}(B)$, where $\rho_{\rm MR}(B)$ is the background magneto-resistance, $\rho_{\rm osc}(B)$ is the quantum oscillatory component and is assumed to be essentially periodic in $1/B$. $\rho_{\rm MR}(B)$ is expected to have a polynomial dependence on B, it is then usually adequate to fit the entire resistivity to some polynomial and then subtract this fitted polynomial. If the polynomial order is sufficiently low and the oscillation frequency is sufficiently high then the projection of the oscillation onto the polynomial fit is essentially zero and as such the oscillatory component is extracted intact. For the data in Figure 2A the number of periods observed is clearly small (< 5) over the entire B range so the usual criteria for making a simple polynomial fit to the background does not apply; even a small polynomial order will interfere with the oscillation.

A particular difficulty is that the oscillation is periodic in $1/B$ while the background is expected to be polynomial in B, therefore a fit in B does not weight all parts of the oscillation equally. To overcome this problem we used a robust local regression technique well known in the field of applied statistics (LOWESS) [9] using the 'locfit' implementation by Loader [10].

We briefly describe the procedure as we are unaware any previous application of this technique in the context of quantum oscillations, we apply the fit in $1/B$ so define $x = 1/B$; the data is first weighted around the fit location (x_0) using a weighting function $W(x)$ which has the property that it is zero beyond some cut-off distance (x_c) from the fit location *i.e.* $W(|x - x_0| > x_c) = 0$, the locally weighted data is then regressed to a polynomial [1] in the usual least squares way. The weighting function we used is the common tri-cube model $W(x) = (1 - |x/x_c|^3)^3 \quad |x/x_c| < 1,\ W(x) = 0 \quad x/x_c \geq 1$ [9]. The advantage of this procedure is that it acts as a high pass filter in $1/B$ without requiring a model for $\rho_0(B)$. In our application of this procedure we use a cutoff that corresponds to an oscillation frequency ($1/x_c = F_c < 2$T).

Figure 2A shows the background produced in this way, the apparent tendency to saturate at higher fields is likely an artifact appearing at the end of the fit range (in $1/B$). Figure 2B shows the oscillatory component extracted in this way. ρ_{osc} is periodic in $1/B$ with frequency $F = 10.4$ T and decays at low field as expected. The oscillation has significant anharmonicities, given the difficulty in removing the background we do not ascribe significance to these.

4. Conclusion

We measured quantum oscillations in α-$YbAlB_4$ at fields up to $B = 16$ T. For $B \parallel [1,1,0]$ we used the Dingle decay relation to extract the quasiparticle mean free path as 550Å. From this we conclude that high purity single crystals of α-$YbAlB_4$ can be prepared similarly to β-$YbAlB_4$. We also measured resistivity for $B \parallel [0,0,1]$, in this configuration we found a low oscillation frequency $F = 10.4$T and we described the technique used to remove the non-oscillatory background.

Acknowledgments

We acknowledge useful discussions with K. Kuga and Y. Matsumoto. This work is partially supported by the Royal Society, U.K., the Engineering and Physical Science Research Council (EPSRC), U.K., Grants-in-Aid (No.21684019) from the Japan Society for the Promotion of Science (JSPS), Grants-in-Aid for Scientific Research on Innovative Areas "Heavy Electrons" of MEXT, Japan , Toray Science and Technology Grant, and by an Overseas Postdoctoral Fellowship of the JSPS (E.O'F).

References

[1] Löhneysen H, Rosch A, Vojta M and Wölfle P 2007 *Reviews of Modern Physics* **79** 1015–1075
[2] Macaluso R, Nakatsuji S, Kuga K, Thomas E, Machida Y, Maeno Y, Fisk Z and Chan J 2007 *Chem. Mater* **19** 1918–1922
[3] Nakatsuji S, Kuga K, Machida Y, Tayama T, Sakakibara T, Karaki Y, Ishimoto H, Yonezawa S, Maeno Y, Pearson E, Lonzarich G, Balicas L, Lee H and Fisk Z 2008 *Nature Physics* **4** 603–607
[4] Matsumoto Y, Nakatsuji S, Kuga K, Karaki Y, Horie N, Shimura Y, Sakakibara T, Nevidomskyy A and Coleman P 2011 *Science* **331** 316
[5] Kuga K, Karaki Y, Matsumoto Y, Machida Y and Nakatsuji S 2008 *Phys. Rev. Lett.* **101** 137004
[6] Matsumoto Y, Kuga K, Tomita T, Karaki Y and Nakatsuji S 2011 *Phys. Rev. B* **84**(12) 125126
[7] Tompsett D A, Yin Z P, Lonzarich G G and Pickett W E 2010 *Phys. Rev. B* **82** 235101
[8] O'Farrell E C T, Tompsett D A, Sebastian S E, Harrison N, Capan C, Balicas L, Kuga K, Matsuo A, Kindo K, Tokunaga M, Nakatsuji S, Csanyi G, Fisk Z and Sutherland M L 2009 *Phys. Rev. Lett.* **102**
[9] Cleveland W and Devlin S 1988 *Journal of the American Statistical Association* **83** 596–610
[10] Loader C 1996 *Annals of Statistics* **24** 1602–1618

[1] we used a quadratic which does not have oscillatory character (*c.f.* order$\geq$ 3) as the second derivative is constant

Specific heat of $Ce_3Pd_{20}Si_6$ single crystal in magnetic fields

H Ono[1] , T Nakano[2] , N Takeda[2] , G Ano[1] , M Akatsu[1] , Y Nemoto[1] , T Goto[1] and H Kitazawa[3]

[1] Graduate School of Science and Technology, Niigata University, Niigata, Japan

[2] Faculty of Engineering, Niigata University, Niigata, Japan

[3] National Institute for Materials Science, Tsukuba, Japan

E-mail: f09k001f@mail.cc.niigata-u.ac.jp

Abstract. We report the specific heat measurement of single crystalline $Ce_3Pd_{20}Si_6$ and the resultant magnetic phase diagram for the direction of B // [100]. In zero field, we confirmed two successive phase transitions at T = 0.24 K and 0.4 K. The magnetic phase diagram clearly shows a phase boundary of paramagnetic phase I, quadrupole ordered phase II and II'.

1. Introduction

The crystal structure of $Ce_3Pd_{20}Si_6$ is a $Cr_{23}C_6$-type cubic with a space group $Fm\bar{3}m$ (No. 225) [1]. Ce atoms occupy two different crystallographic sites; Ce atom at 4a-site is inside a cage of 12 Pd atoms and 6 X atoms, and forms a face-centered cubic lattice, the other Ce atom at 8c-site is surrounded by a cage of 16 Pd atoms, which forms a simple cubic lattice.

The magnetic properties of $Ce_3Pd_{20}Si_6$ at low temperatures are controversial. In an earlier investigation, $Ce_3Pd_{20}Si_6$ is a spin glass below T = 0.15 K [2], afterward, it was found that $Ce_3Pd_{20}Si_6$ shows two successive phase transitions at T_L = 0.31 K and T_U = 0.5 K [3]. The effect of T_L and T_U on magnetic field is different. T_L shifts to lower temperature with increase magnetic field and vanishes above 1 T, whereas T_U shifts to higher temperature and becomes unclear above 8 T. Very recently, Goto *et al.* measured the elastic constant on single crystal for three principal axes and constructed the magnetic phase diagram [4]. They found that $Ce_3Pd_{20}Si_6$ has three ordered phases II, II', III and paramagnetic phase I (refer to Fig. 4 which will be described later). The phase III is antiferromagnetic and the phase II and II' are identified as antiferro-quadrupole ordered phases. The phase boundary between phase-I and phase-II is strongly anisotropic in the magnetic field. In zero field, however, they found only one transition at 0.17 K. Mitamura *et al.* carried out magnetization measurements for principal axes using specimen cut from the same single crystal used in ultrasonic measurements [5]. The magnetic phase diagram obtained from magnetization measurements has two phase transitions at 0.23 K and 0.45 K. Slightly lower transition temperature than T_L and T_U is due to a strong sample dependence which is thought to be due to sample homogeneity and annealing conditions [6]

As a mentioned above, the low temperature physical properties of $Ce_3Pd_{20}Si_6$ is still controversial even in the same single crystal. The purpose of the present study is to establish the magnetic phase diagram with B // [100] by specific heat measurement.

2. Experimental

Starting materials were 5N-Ce, 3N5-Pd and 10N-Si. They were mixed with an atomic ratio Ce : Pd : Si = 3 : 20 : 6, and melted into a button in an argon atmosphere using a mono-arc furnace. A single crystal of $Ce_3Pd_{20}Si_6$ was grown by a floating zone furnace equipped with double ellipsoidal mirrors. The specimen used in the present study was cut from the same single crystal used in ultrasound and magnetization measurements and annealed for 4 weeks at 1173 K. The specific heat was measured by an adiabatic heat pulse method with a dilution refrigerator.

3. Results and Discussions

Figure 1 shows the temperature dependence of the specific heat, $C(T)$, of $Ce_3Pd_{20}Si_6$ for $B \leq$ 1.0 T. In zero field, we observed a sharp peak at $T = 0.24$ K and a shoulder at $T = 0.4$ K, which indicates antiferromagnetic and antiferro-quadrupole transition. Hereafter, we use a notation of $T_{\text{II-III}}$ and $T_{\text{I-II}}$ etc. as a transition temperature. The sharp peak at $T_{\text{II-III}}$ shifts to lower temperature with increasing magnetic field and vanishes at $B = 1$ T. On the other hand, $T_{\text{I-II}}$ shifts to higher temperature with increasing B. These behaviors are consistent with the magnetization measurement.

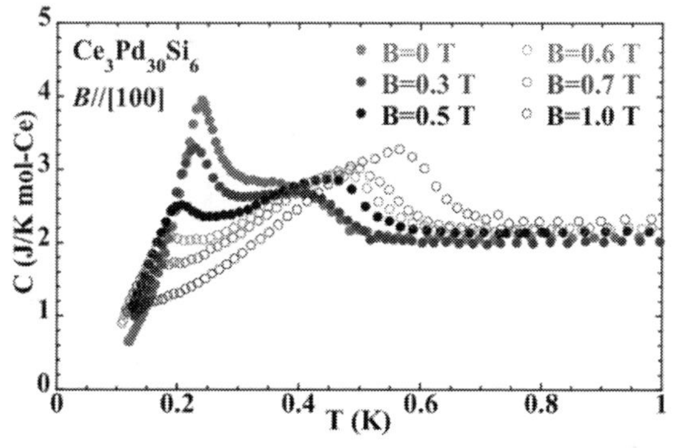

Figure 1. (Color online) The temperature dependence of specific heat of $Ce_3Pd_{20}Si_6$ for $B \leq 1.0$ T. The lower peak and higher peak indicates antiferromagnetic and quadrupole transition, respectively.

Figure 2 shows $C(T)$ for 1.0 T $\leq B \leq$ 2.5 T. It has to be noted that simple extrapolation of $C(T)$ at $B = 1.0$ T to $T = 0$ K does not go to zero. Since $C(T)$ goes to zero at $T = 0$ K, $C(T)$ at $B = 1.0$ T should decrease rapidly below 0.1 K. Although it is difficult to see a sign of decrease in $B = 1.0$ T data, a faint peak appears around 0.15 K above 1.2 T. The faint peak for $B \leq 1.8$ T is too tiny to conclude as a phase transition, but it becomes clear above 2.0 T and moves to higher temperature and merges with a peak at $T_{I\text{-}II}$. This peak above B = 2.0 T just corresponds to the transition of phase II to II'.

Figure 3 exhibits $C(T)$ above 2.5 T. The peak indicates the transition between phase II' and I. $T_{I\text{-}II'}$ decreases monotonically and vanishes at 4 T.

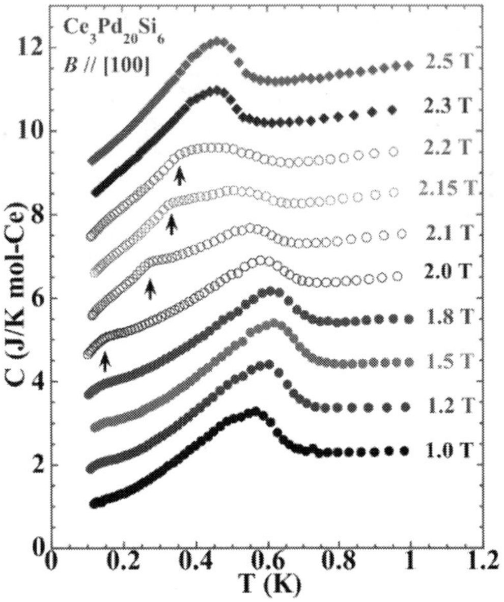

Figure 2. (Color online) The temperature dependence of specific heat. The date from bottom to top show B = 1.0, 1.2, 1.5, 1.8, 2.0, 2.1, 2.15, 2.2, 2.3 and 2.5 T, which are shifted by 1 J / K mol-Ce, respectively. The arrow indicates the transition between phase II and II'.

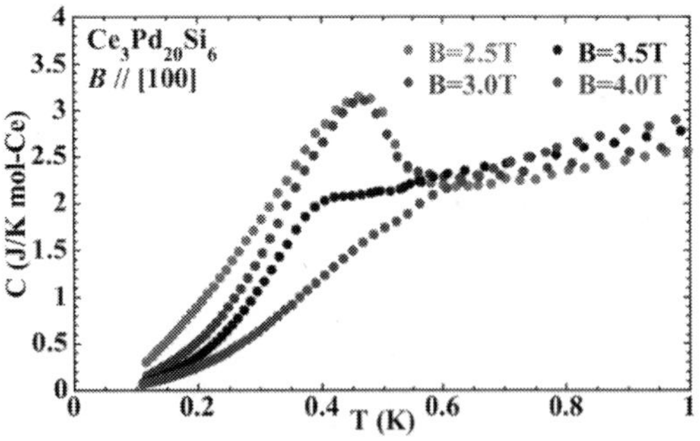

Figure 3. (Color online) The temperature dependence of specific heat of $Ce_3Pd_{20}Si_6$ for $B \geq 2.5$ T. This peak indicates the transition between phase I and II'.

We show in Figure 4 the magnetic phase diagram of $Ce_3Pd_{20}Si_6$ for the B // [100] obtained from the present measurements. We plot a peak temperature in C as a phase transition temperature. The transition temperature $T_{I\text{-}II}$ significantly increases with B at low fields. This is a common feature of quadrupole transition. The phase boundary between paramagnetic phase I and quadrupole ordered phase II or II' around B = 2 T clearly shows a cusp structure, which has not been observed in the magnetization and elastic constant measurements. This cusp structure would suggest that the order parameter of two quadrupole ordered phase have a competitive effect each other.

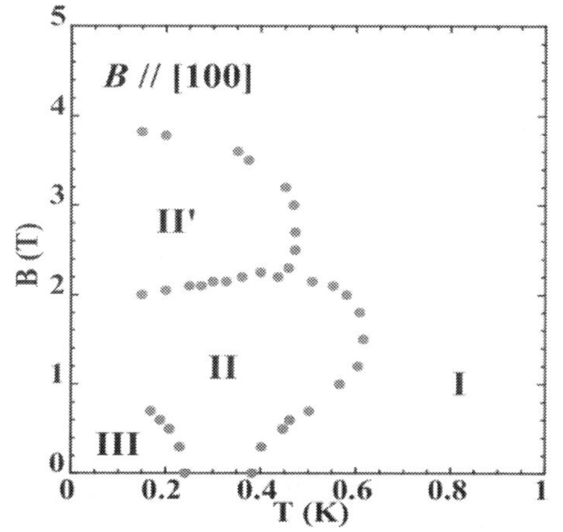

Figure 4. (Color online) The magnetic phase diagram of $Ce_3Pd_{20}Si_6$ for B // [100], determined from the present specific heat measurements. The phase I is paramagnetic phase. The phase II and II' are antiferro-quadrupole ordered phase. The phase III is antiferromagnetic ordered phase.

4. Conclusion

We measured the specific heat of $Ce_3Pd_{20}Si_6$ and constructed the magnetic phase diagram with B // [100]. We confirmed two successive transition at $B = 0$ T. The phase diagram below 1T is in good accordance with the magnetization measurement.

We established clear phase boundary between I, II and II' around $B = 2$ T for the first time. The faint peak around 0.15 K for 1.2 T $\leq B \leq$ 1.8 T is still ambiguous.

Acknowledgment

This work was supported by a Grand-Aid of Scientific Research on Innovative Anea "Heavy Fermion" (No.20102004) of Ministry of Education, Culture, Sports, Science and Technology, Japan.

4. References

[1] Grivanov A V, Seropegin Yu D and Bodak O I, (1994) *J. Alloys. Comp.* **204** L9
[2] Takeda N, Kitagawa J and Ishikawa M, (1995) *J. Phys. Soc. Japan* **64** 387
[3] Strydom A M, Pikul A, Steglich F and Paschen S, (2006) *J. Phys.: Conf. Ser.* **51** 239
[4] Goto T, Watanabe T, Tsuduku S, Kobayashi H, Nemoto Y, Yanagisawa T, Akatsu M, Ano G, Suzuki O, Takeda N, Dönni A and Kitazawa H, (2009) *J. Phys. Soc. Japan* **78** 024716
[5] Mitamura H, Tayama T, Sakakibara T, Tuduku S, Ano G, Ishii I, Akatsu M, Nemoto Y, Goto T, Kikkawa A and Kitazawa H, (2010) *J. Phys. Soc. Japan* **79** 0747102
[6] Shiobara D, Master Thesis (2009) Niigata University

Magnetic order in non-centrosymmetric $CePt_3B$

D Rauch[1], S Süllow[1], M Bleckmann[2], A Buchsteiner[3], N Stüßer[3], H-H Klauss [4], H Luetkens[5] and E Bauer[6]

[1] Institute of Physics of Condensed Matter, Technical University of Braunschweig, D-38106 Braunschweig, Germany
[2] Wehrwissenschaftliches Institut für Werk- und Betriebsstoffe WIWeB, D-85435 Erding, Germany
[3] Berlin Neutron Scattering Center, Helmholtz Zentrum Berlin, D-14109 Berlin, Germany
[4] Institute of Solid State Physics, University of Technology Dresden, D-01069 Dresden, Germany
[5] Laboratory for Muon-Spin Spectroscopy, Paul Scherrer Institute, CH-5232 Villigen, Switzerland
[6] Institute of Solid State Physics, Vienna University of Technology, A-1090 Vienna, Austria

E-mail: d.rauch@tu-bs.de

Abstract.
$CePt_3B$ exhibits two magnetic phases, an antiferromagnetic and a weak ferromagnetic one. To determine these magnetic structures of $CePt_3B$, neutron diffraction and μSR experiments have been carried out. Neutron diffraction experiments provided no evidence of either antiferromagnetic or ferromagnetic order. In contrast, in zero field muon-spin relaxation (μSR) experiments, magnetic transition temperatures $T_N = 7.8$ K and $T_C \sim 6$ K are observed from the onset of spontaneous muon precession.

In recent years, the observation of heavy fermion superconductivity in non-centro symmetric intermetallics such as $CePt_3Si$ [1] has initiated a multitude of experimental and theoretical studies. Especially, in terms of superconductivity the nature of the pairing states has been of interest [2]. Conversely, the lack of inversion symmetry gives rise to an additional magnetic exchange term, the Dzyaloshinsky-Moriya (DM) interaction [3, 4]. This term tends to align spins perpendicular to each other, which combined with the common (for instance Heisenberg like) magnetic exchange effectively gives rise to spin canting or complex magnetic structures.

Already for $CePt_3Si$, which crystallizes in a tetragonal lattice with the space group $P4mm$ [1], the DM interaction could be a relevant issue for the magnetic behaviour, since the system orders antiferromagnetically below $T_N = 2.2$ K. However, it has been shown that the antiferromagnetically ordered state is collinear ($Q = (0, 0, 1/2)$), with small magnetic moments $\mu_{ord} = 0.16\ \mu_B$ [5]. In contrast, $CePt_3B$, which is isostructural and related in electron count to $CePt_3Si$ [6, 7, 8, 9], exhibits a complex magnetically ordered state at low temperatures, with an antiferromagnetic phase below $T_N = 7.8$ K and a second weakly ferromagnetic transition below $T_C \sim 6$ K. While the weak ferromagnetism might point to the relevance of the DM interaction, the microscopic nature of these magnetic phases has not been revealed so far. Here, we present a first study of these magnetic phases by means of neutron diffraction and μSR experiments.

The polycrystalline sample $CePt_3B$, using ^{11}B in the production, was prepared by melting high-purity elements in a water-cooled copper crucible using a high frequency generator under argon atmosphere. Subsequently the sample was annealed at 880°C for 14 days. The magnetization and susceptibility was measured employing a commercial SQUID magnetometer. Neutron diffraction experiments were performed using the E6 spectrometer of the Berlin Neutron Scattering Center (BENSC) of the Helmholtz Zentrum Berlin, with a neutron wave length of $\lambda = 2.444$ Å in a standard Orange cryostat ($T \geq 1.6$ K). The μSR experiments were carried out at the Swiss Muon Source of the Paul Scherrer Institute. Measurements were performed on the GPS instrument using a He-flow cryostat ($T \geq 1.7$ K).

For sample characterization, magnetization measurements on $CePt_3B$ at low temperatures $T = 1.8 - 8$ K have been carried out (Fig. 1). A ferromagnetic hysteresis below $T = 5$ K is observable, and which has vanished for 8 K. In the limit of zero temperature, an extrapolated remanent magnetization of $\mu \approx 0.09\mu_B$/Ce atom is obtained, in agreement with Ref. [6] (inset of Fig. 1). Further, at low temperatures the inverse susceptibility χ^{-1} indicates two magnetic phase transitions into long range ordered states, an antiferromagnetic and a weak ferromagnetic one (Fig. 2). The antiferromagnetic T_N and ferromagnetic temperatures T_C for $CePt_3B$ are determined as $T_N = 7.8$ K and $T_C \sim 6$ K from χ^{-1}, in good agreement with the Refs. [6, 8].

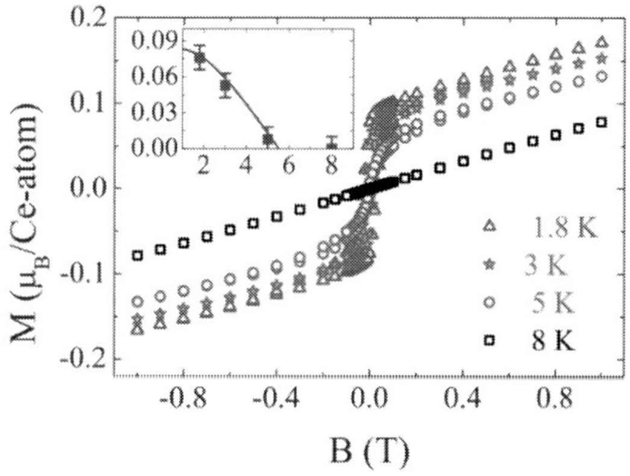

Figure 1. The magnetization of $CePt_3B$, as function of temperature. Inset: The remanent ferromagnetic moment as function of temperature of $CePt_3B$; solid line as guide to the eye.

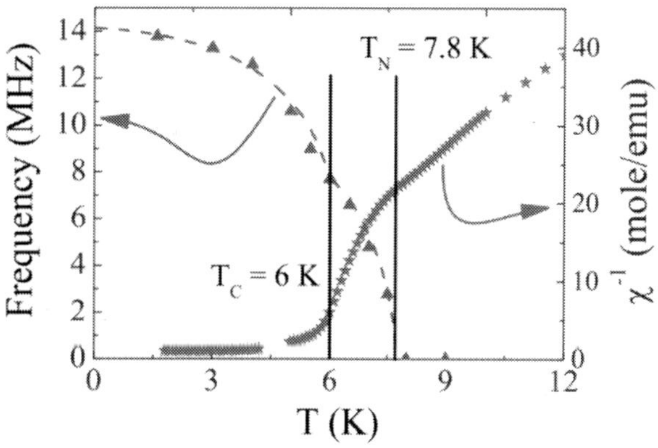

Figure 2. Oscillation frequencies of the muon signal (red), together with the inverse susceptibility χ^{-1} (green) of $CePt_3B$; lines are guides to the eye, for details see text.

Neutron diffraction patterns of $CePt_3B$, taken at 1.6, 6.6 and 15.3 K are depicted in Fig. 3. The data at 15.3 K are successfully fitted with the tetragonal structure model (space group *P4mm*) of Sologub *et al.* [7] (not shown in the plot for clarity). In these fits, a few minor peaks could not be associated to this structure, indicating the presence of a second phase (volume fraction ~ 10 %). In the refinement, by excluding the secondary phase peaks a value $R_{Bragg} = 6.2$ % is obtained. In Tab. 1 we summarize the result of our refinement. All parameter are in a good agreement with the values obtained of Sologub *et al.* [7]. The smaller lattice parameters are reflecting a shrinking of the lattice due to the lower temperature used in our experiments, as compared to the room temperature experiments carried out by Sologub *et al.* [7].

A comparison of the spectra at 1.6, 6.6 and 15.3 K and the difference spectra (lower panel) at 1.6 K/15.3 K and 6.6 K/15.3 K are also shown in Fig. 3. All the data are normalized to a secondary phase peak, which should not change intensity upon transition into the magnetically ordered state. Surprisingly, within experimental solution no additional diffraction intensity has been found for any of our examined Bragg peaks, nor have additional peaks been observed.

To estimate the magnetic moment that ought to be detectable in our experiment, magnetic

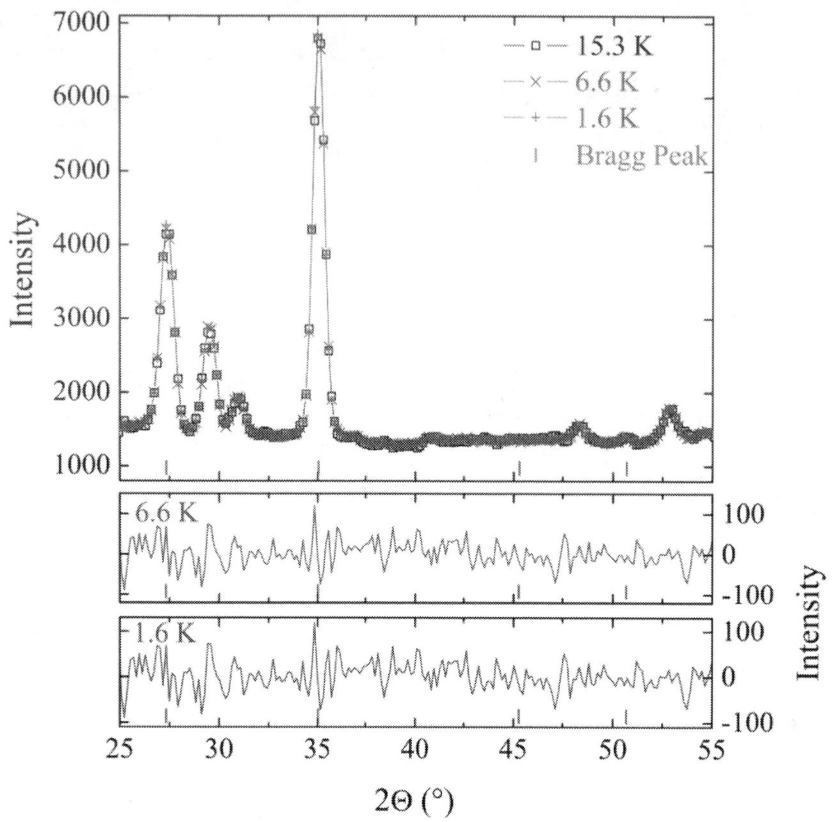

Figure 3. Neutron diffraction pattern of $CePt_3B$, taken at 1.6 (red), 6.6 (green) and 15.3 K (black). In the lower panels, the difference between the data at 1.6/6.6 K and the 15.3 K data are shown.

Table 1. Results of a refinement of powder neutron diffraction data on $CePt_3B$ at $T = 15$ K, in comparison to x-ray powder diffraction at room temperature carried out by Sologub *et al.* [7]. Here, the lattice parameter a, b and c and the positional parameter x, y, z are summarized.

	15.3 K			**Sologub** *et al.* **[7]**		
a (Å)	3.9943(4)			4.0031(3)		
b (Å)	3.9943(4)			4.0031(3)		
c (Å)	5.0620(1)			5.0736(4)		
R_{Bragg} (%)	6.2			4.8		
Ce	0	0	0	0	0	0
Pt(1)	0	0.5	0.522(1)	0	0.5	0.5132(13)
Pt(2)	0.5	0.5	0.131(3)	0.5	0.5	0.1174(11)
B	0.5	0.5	0.713(8)	0.5	0.5	0.688 (17)

structure simulations with magnetic moments of different size and arrangement have been performed. In particular, a ferromagnetic arrangement of the magnetic moments and an antiferromagnetic one with a doubling of the unit cell in a direction have been assumed. These simulations indicate that a magnetic moment of 0.5 μ_B in the ferromagnetic and 0.2 μ_B in the antiferromagnetic case should be observable.

Since the neutron diffraction experiments did not deliver information about the magnetic phases, zero field μSR measurements were carried out on the same sample. In these experiments a distinctive μSR oscillatory signal below the antiferromagnetic transition temperature T_N is observed, clearly indicating the presence of static magnetic fields from long range magnetic order in the bulk of the sample (Fig. 4). In a next step, we determined the temperature dependence of the muon oscillation frequencies.

These zero field spectra have been analyzed using a sum of three precession signals associated with different muon sites in the lattice. Plotted is the highest frequency as a function of temperature, the other frequencies are proportional to it. The results are depicted in Fig. 2, together with the inverse susceptibility χ^{-1}.

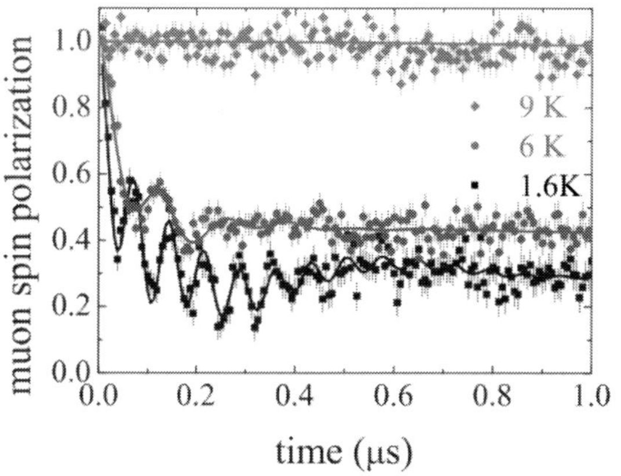

Figure 4. Time dependence of the μ^+ spin polarization of $CePt_3B$ for temperatures T = 1.6 K (black), 6 K (red) and 9 K (green).

Both transition temperatures T_N and T_C can also be identified in the μSR measurements. From Fig. 2 it is seen that T_N denotes the temperature of onset of spontaneous muon precession, while at T_C there is a change of slope of the temperature dependence of the muon oscillation frequency. The weak ferromagnetic behavior likely is formed by canting of magnetic moments of the antiferromagnetic structure of $CePt_3B$.

In conclusion, the structural properties, investigated by neutron diffraction experiments, are in good agreement with those reported in Sologub *et al.* [7]. Further, we have unsuccessfully attempted to identify scattering intensity in our neutron study upon transition into the magnetically ordered phases. However, our μSR experiments clearly show the existence of long range magnetic order in the bulk of our sample. Taken together, neutron diffraction and μSR experiments suggest that the magnetically ordered phases in $CePt_3B$ carry magnetic moments of medium size, probably with an ordering vector different from the one in $CePt_3Si$.

Altogether, $CePt_3B$ exhibits an antiferromagnetic and a weak ferromagnetic phase, with T_N = 7.8 K and $T_C \sim$ 6 K. In view of the data presented here, a possible explanation for the weak ferromagnetic behavior below T_C could be the canting of antiferromagnetic ordered spins with decreasing temperature. The DM interaction might be, most likely, the cause for this canting of magnetic moments in non-centro symmetric $CePt_3B$.

References

[1] Bauer E et al. 2004 *Phys. Rev. Lett.* **92** 2
[2] Fujimoto S 2007 *J. Phys. Soc. Jpn.* **76** 051008
[3] Dzyaloshinsky I 1958 *J. Phys. Chem. Sol.* **4** 241
[4] Moriya T 1960 *Phys. Rev.* **120** 91
[5] Metoki N et al. 2004 *J. Phys.: Condens. Matter* **16** L207-212
[6] Süllow S et al. 1994 *Physica B* **199** & **200** 644
[7] Sologub O et al. 2002 *J. Alloys Comp.* **337** 10
[8] Lackner R et al 2005 *J. Phys.: Condens. Matter* **17** S 905
[9] Vejpravov J, Prokleš J, Sechovsk V 2006 *J. Alloys Comp.* **408** 54

Temperature-dependent hybridization gaps: A cause of phonon anomalies in f electron systems?

Peter S. Riseborough

Department of Physics, Temple University, Philadelphia, Pa 19122, USA

E-mail: `prisebor@temple.edu`

Abstract. There is evidence that a number of heavy-fermion/mixed-valent materials show strongly renormalized hybridization gaps either at the Fermi-energy or close to the Fermi-energy. In the former case, a heavy-fermion semiconducting state ensues and in the later case, the system remains metallic at low temperatures. Due to the temperature-dependence of the electronic correlations, the magnitudes of the hybridization gaps decrease with increasing temperatures. The existence of a temperature-dependent low-energy electronic energy scale opens up the possibility that the Born-Oppenheimer approximation may fail and that there may be a resonant coupling between the phonons and the electronic excitations. It is argued that such a mechanism may be the cause of the anomalous phonon mode observed in α-uranium at high temperatures.

1. Introduction

The low temperature state of heavy fermion semiconductors or Kondo insulators are commonly described by a hybridization gap model in which the hybridization gap is strongly reduced due to the presence of strong electronic correlations[1]. Archetypes of this class of materials are given by $Ce_3Bi_4Pt_3$[2], SmB_6[3] and YbB_{12}[4]. The gap has been observed in $Ce_3Bi_4Pt_3$ by inelastic neutron scattering experiments[5, 6], optical conductivity[7, 8] and photoemission measurements[9, 10]. Substitutional doping rapidly closes the gap[11, 12] which illustrates that the gap has a coherent character as is expected for a hybridization gap[13]. Furthermore, measurements of the gap in $Ce_3Bi_4Pt_3$ show that it is strongly temperature dependent[14] which is in agreement with the theory. In these semiconducting materials, the gap opens up at the Fermi-energy. Low-lying hybridization gaps or pseudogaps have been observed in materials which are always metallic such as $YbFe_4Sb_{12}$[15], $CeRu_4Sb_{12}$[15] and UPd_2Al_3[16]. The existence of either pseudo-gaps or gaps in the electronic spectrum with magnitudes in the range 12 to 50 meV raises the possibility of the existence of excitations which through resonant interactions have mixed electronic and phonon character. In particular, this mixing could result in the softening of the optic phonon spectrum and the formation of more than three modes for each lattice degree of freedom[17, 18].

The phonon spectrum of α-uranium shows a marked harmonic softening at high temperatures[19, 20] and also shows the formation of a new mode[21, 22] at temperatures around 450 K. However, those authors argue that the anomalies are due to the formation of intrinsically localized modes. Intrinsically Localized Modes (ILMs) are persistent oscillatory and spatially localized excitations of homogeneous lattices that are stabilized by strong anharmonicity[23].

Such modes have been extensively studied in classical one-dimensional lattices via numerical simulations[24, 25]. In the classical limit, these excitations should give rise to a broad continuous spectral density which is in direct contrast to the discrete nature of the observed spectrum[21]. This discrepancy can be resolved by quantizing the internal degree of freedom of the ILM, in which case, the observed excitation could correspond to the lowest non-trivial ILM that has the quantum number $n = 2$. However, an investigation of quantized ILMs[26] shows that the ILM should exist for all momentum transfers, and is not restricted to a small range of momentum transfers as found in the experiments[21, 22]. Therefore, due to the harmonic nature of the lattice[20] and due to the localized nature of the excitation in momentum space, the experimental measurements are not fully in accord with the interpretation in terms of ILMs. Hence, in this manuscript, we shall indicate how the existence of a small temperature-dependent hybridization gap can result in the softening of the phonon modes and how it may lead to the formation of a new breathing mode.

2. Calculation and Results

The system is described by the Anderson Lattice Hamiltonian which includes an f-electron phonon interaction. The on-site Coulomb interaction is treated within the mean-field slave boson approximation[6], which at low temperatures produces a coherent renormalized band. The coherent quasiparticle band is characterized by a renormalized temperature-dependent hybridization matrix element $V(T)$ and an f-level energy $E_f(T)$ which is shifted close to the Fermi-energy. The resulting system of renormalized quasiparticles and phonons is described by

$$\begin{aligned} \hat{H} &= \sum_{i,\sigma} E_f(T)\ f^\dagger_{i\sigma} f_{i\sigma} + \sum_{\underline{k},\sigma} \epsilon_{\underline{k}}\ d^\dagger_{\underline{k}\sigma} d_{\underline{k}\sigma} + \sum_{\underline{k},\sigma} \Big(V(T)\ f^\dagger_{\underline{k}\sigma} d_{\underline{k}\sigma} + V(T)^*\ d^\dagger_{\underline{k}\sigma} f_{\underline{k}\sigma} \Big) \\ &+ \sum_{\underline{q},\gamma} \hbar\omega_{\underline{q}\gamma}\ a^\dagger_{\underline{q}\gamma} a_{\underline{q}\gamma} + \sum_{\underline{k},\sigma,\underline{q},\gamma} \lambda_{\underline{q}\gamma} \Big(a^\dagger_{\underline{q}\gamma} + a_{-\underline{q}\gamma} \Big) f^\dagger_{\underline{k}+\underline{q}\sigma} f_{\underline{k}\sigma} \end{aligned} \tag{1}$$

where $f^\dagger_{i,\sigma}$ creates an electron of spin σ on the i-th lattice site and $d^\dagger_{\underline{k},\sigma}$ creates an electron of spin σ in the conduction band state labeled Bloch wave vector $\underline{k}$. The operator $a^\dagger_{\underline{q},\gamma}$ creates a

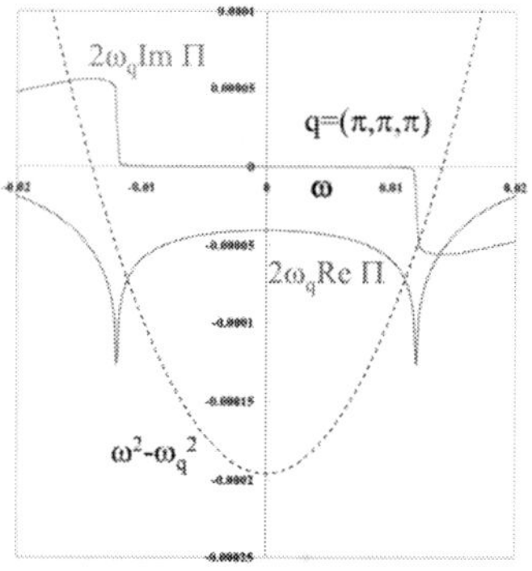

Figure 1. The real (blue line) and the imaginary parts (red line) of the polarization part $2\omega_{\underline{q}}\Pi_{\underline{q}}(\omega)$ in units of $(\mathrm{eV})^2$ versus frequency (in units of eV), for $\underline{q} = (\pi, \pi, \pi)$. The renormalized phonon frequency is given by the intersection of the line $2\omega_q\ \Pi_q(\omega)$ with the parabola $\omega^2\ - \omega_q^2$ (dashed line). The plot shows that the renormalized phonon frequency exhibits a significant softening when the bare frequency ω_q is comparable to the indirect gap.

phonon with wave vector $\underline{q}$ and polarization γ. The phonon propagator is calculated from the Dyson equation

$$D_{\underline{q}}(\omega) \;=\; D^{0}_{\underline{q}}(\omega) \;+\; D^{0}_{\underline{q}}(\omega)\;\Pi_{\underline{q}}(\omega)\;D_{\underline{q}}(\omega) \tag{2}$$

where the unperturbed phonon propagator is given by

$$D^{0}_{\underline{q}\gamma}(\omega) \;=\; \frac{2\;\hbar\omega_{\underline{q}\gamma}}{\hbar^2\omega^2 \;-\; \hbar^2\omega^2_{\underline{q}\gamma}} \tag{3}$$

and the polarization part $\Pi_{\underline{q}}(\omega)$ is calculated to order $\lambda^2_{\underline{q}}$. The polarization part shows a dramatic low frequency variation for wave vectors q near the boundary of the Brillouin zone. The frequency dependence of the real and the imaginary parts of the polarization part are shown in fig.(1). The characteristic energy scale of the peaks in the real part and the threshold of the imaginary part is given by the indirect hybridization gap[6]. The renormalized phonon frequencies are given by the intersection of the curve $2\hbar\omega_{\underline{q},\gamma}\Re e\;\Pi_{\underline{q}}(\omega)$ with the parabola $\hbar^2(\omega^2 \;-\omega^2_{\underline{q},\gamma})$. The lifetime of the excitation is governed by $2\hbar\omega_{\underline{q},\gamma}\Im m\;\Pi_{\underline{q}}(\omega)$. It is seen that as the temperature is increased and the hybridization gap decreases towards the optic phonon frequency, the electron-phonon coupling can result in a softening of the phonon spectrum. For phonon frequencies smaller than the indirect gap, the curves may have more than one intersection. In this case, resonant mixing occurs between the electronic excitations and the optic phonon modes which leads to the formation of a new broad mode at high energies.

In summary we have argued that, in an f-electron system with a sufficiently small hybridization gap or pseudo-gap, the electron-phonon interactions may couple the electronic excitations with the phonon modes. If, as the temperature is decreased, the hybridization gap is lowered towards the optic phonon frequency, then the optic modes are expected to soften with increasing temperature. As the hybridization gap is further reduced below the optic phonon frequency, one expects that the electronic excitations will resonantly mix with the phonons resulting in a splitting of the optic modes into two branches. The predictions of this mechanism are consistent with experimental results on α-uranium[19, 21], even though at the present time there is no direct experimental evidence for the existence of a hybridization pseudo-gap. The proposed mechanism could be active in the large class of materials which are known to exhibit small hybridization gaps.

Acknowledgments
This work was supported by the U.S. Department of Energy, Office of Basic Energy Sciences, Materials Science through the award DEFG02-84ER45872. The author would also like to thank T. Durakiewicz, M.E. Manley and A.J. Sievers for stimulating conversations.

3. References

[1] Riseborough PS, 2000 *Adv. in Phys.* **49** 257-320
[2] Hundley MF, Canfield PC, Thompson JD, Fisk Z and Lawrence JM, 1990 *Phys. Rev.* B, **42** 6842-6845
[3] Kasuya T, Takegahara K, Fujita T and Banai E, 1979 *J. Phys. Colloq.* **5** 308
[4] Kasaya M, Iga F, Negishi K, Nakai S and Kasuya T, 1983 *J. Mag. Mag. Mat.* **31-34** 437
[5] Severing A, Thompson JD, Canfield PC, Fisk Z and Riseborough PS, 1991 *Phys. Rev.* B, **44** 6832-6837
[6] Riseborough PS, 1992 *Phys. Rev.* B, **45** 13984-13995
[7] Bucher B, Schlesinger Z, Canfield PC and Fisk Z, 1994 *Phys. Rev. Lett.* **72** 522-525
[8] Riseborough PS, 1994 *Physica* B, **199&200** 466
[9] Breuer K, Messerli S, Purdie D, Garnier M, Hengsberger M, Panaccione G, Baer Y, Takahashi T, Yoshii S, Kasaya M, Katoh K, *et al.*, 1998 *Europhys. Lett.* **41** 565-570
[10] Riseborough PS, 1988 *Phys. Rev.* B, **58** 15534-15547

[11] Iga F, Kasaya M and Kasuya T, 1985 *J. Mag. Mag. Mat.* **52** 279
[12] Hundley MF, Canfield PC, Thompson JD and Fisk Z, 1994 *Phys. Rev.* B, **50** 18142
[13] Riseborough PS, 2003 *Phys. Rev.* B, **69** 235213
[14] Hundley MF, Thompson JD, Canfield PC and Fisk Z, 1994 *Physica* B, **199** 443-444
[15] Dordevic SV, Basov DN, Dilley NR, Bauer ED and Maple MB, 2001 *Phys. Rev. Lett.* **86** 684-687
[16] Yang X-D, Riseborough PS and Durakiewicz T, 2011 *J. Phys.* CM. **23** 094211
[17] Riseborough PS and Yang X-D, 2007 *J. Mag. Mag. Mat.* **310** 938-940
[18] Riseborough PS and Yang X-D, 2010 *Phys. Rev.* B, **82** 094303
[19] Manley ME, Fultz B, McQueeney RJ, Brown CM, Hults WL, Smith JL, Thoma DJ, Osborn R and Robertson JL, 2001 *Phys. Rev. Lett.* **86** 3076
[20] Manley ME, Lander GH, Sinn H, Alatas A, Hults WL, McQueeney RJ, Smith JL and Willit J, 2003 *Phys. Rev.* B, **67** 052302
[21] Manley ME, Yethiraj M, Sinn H, Volz HM, Alatas A, Lashley JC, Hults WL, Lander GH and Smith JL, 2006 *Phys. Rev. Lett.* **96** 125501
[22] Manley ME, Lynn JW, Chen Y and Lander GH, 2008 *Phys. Rev. B*, **77** 052301
[23] Sievers AJ and Takeno S, 1988 *Phys. Rev. Lett.* **61** 970
[24] Flach S and Willis CR, 1998 *Phys. Rep.* **295** 181-264
[25] Flach S and Gorbach AV, 2008 *Phys. Rep.* **467**, 1-116
[26] Basu S and Riseborough PS, 2011 *Philos. Mag.* http://dx.doi.org/10.1080/14786435.2011.607141

NMR/NQR Study of pressure-induced superconductor $CePt_2In_7$

Hironori Sakai[1], Yo Tokunaga[1], Shinsaku Kambe[1], Hanoh Lee[2], Vladimir A. Sidorov[2], Paul H. Tobash[2], Filip Ronning[2], Eric D. Bauer[2] and Joe D. Thompson[2]

[1] Advanced Science Research Center, Japan Atomic Energy Agency, Tokai, Ibaraki 319-1195, Japan

[2] Los Alamos National Laboratory, Los Alamos, New Mexico 87545, USA

E-mail: sakai.hironori@jaea.go.jp

Abstract. Zero- eld ^{115}In-nuclear magnetic resonance (NMR) has been used to study single crystals of $CePt_2In_7$, which is a heavy-fermion antiferromagnet with a Néel temperature (T_N) of 5.2 K at ambient pressure. The NMR spectra under zero eld near the $3\nu_Q$ line of the orthorhombic In(3) sites are taken at 1.6 K under hydrostatic conditions at ambient pressure and 2.4 GPa. These data reveal the coexistence of commensurate and incommensurate antiferromagnetic (AFM) orders at ambient pressure and that the commensurate ordering is stabilized by increasing pressures. The nuclear spin-lattice relaxation rates ($1/T_1$) for In(3) sites indicate the localized nature of f electrons far above T_N. The values of $1/T_1$ in the paramagnetic state decrease by applying pressure. In contrast, the residual values of $1/T_1$ much below T_N increase by pressure.

1. Introduction

Recently, an intermetallic compound $CePt_2In_7$ [1] has been added to the so-called Ce115 family. The Ce115 family members with a chemical formula of $CeTIn_5$ (T=Co, Rh, Ir) have attracted much attention since each is a heavy-fermion superconductor either at atmospheric pressure ($CeCoIn_5$ [2] and $CeIrIn_5$ [3]) or when antiferromagnetic (AFM) order is suppressed by applied pressure ($CeRhIn_5$ [4]). While the Ce115 structure (space group $P4/mmm$) can be regarded as an alternate stacking of $CeIn_3$ and TIn_2 layers along the c axis, the structure of $CePt_2In_7$ ($I4/mmm$) is comprised of one $CeIn_3$ plane followed by two $PtIn_2$ layers stacked along the c axis to form the tetragonal body-centered unit cell. Initial specific heat and electrical resistivity measurements on polycrystalline $CePt_2In_7$ showed that it is a heavy-fermion antiferromagnet with T_N =5.2 K and that applying pressure induces a broad dome of superconductivityabout the critical pressure $P_c \approx$ 3 GPa where $T_N(P_c)$ extrapolated to $T = 0$ [5]. Very recently, single crystals of $CePt_2In_7$ have become available, and like their polycrystal analogs, they order antiferromagnetically at T_N = 5.2 K out of a heavy-fermion paramagnetic state [6]. Resistivity and *ac*-specific heat studies on these crystals as a function of pressure find a much narrower dome of pressure-induced superconductivity, even though the maximum T_c and critical pressure P_c are the same as in polycrystalline samples [7]. Compared to the Ce115, the dimensionality of $CePt_2In_7$ is reduced relative to the Ce115, as indicated by its crystal structure and band structure calculations [5] that are confirmed by quantum oscillation studies [8].

With its ability to study small single crystals, extreme sensitivity to lattice distortions and compatibility with high pressure environments, NQR/NMR is a powerful probe of the microscopic nature and pressure-induced evolution of magnetism in $CePt_2In_7$. Recently, we have reported the pressure and temperature dependences of ^{115}In-NMR spectra under zero field for this material [9]. This experiment revealed the coexistence of commensurate (C-) and incommensurate (IC-) AFM orderings at ambient pressure and the stabilization of C-AFM state as the superconducting phase is approached with increasing pressure. In this brief paper, we report the characteristic changes of spectral shape and relaxation rates between ambient pressure and the hydrostatic pressure of 2.4 GPa.

2. Experimental

Single crystals for $CePt_2In_7$ were grown by a self-flux method. NQR and NMR measurements under zero field were performed by using a conventional phase-coherent pulsed NMR spectrometer in the temperature range 1.5-200 K. A dozen single crystals were mounted into an rf-coil in order to avoid decomposition and/or surface strain caused by crushing the crystals. It is noted that powdering or crushing crystals of $CePt_2In_7$ influences the physical properties, especially the volume fraction of IC-/C- AFM states [9]. Hydrostatic pressures up to 2.4 GPa were applied with a standard hybrid piston-cylinder-type clamp cell using Daphne7373 oil as a pressure medium.

3. Results and discussions

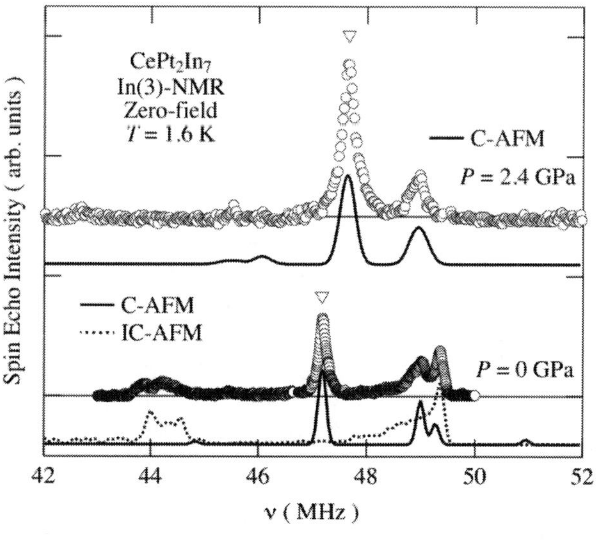

Figure 1. Zero field NMR spectra under the ambient pressure and the hydrostatic pressure of 2.4 GPa at 1.6 K for In(3) sites in $CePt_2In_7$. The solid and dotted curves represent the simulated curves on the assumption of commensurate and incommensurate magnetic structures, respectively. The open triangles (▽) indicate the spectral positions for no internal field.

All the nuclear quadrupolar resonance (NQR) lines for ^{115}In were measured at ambient pressure in the paramagnetic state. These NQR lines can be ascribed to three crystallographically inequivalent In sites, that is, In(1) ($4/mmm$ in Hermann-Mauguin notation), In(2) ($4m2$), and In(3) ($2mm$). Because the overlap with other lines is minimal, the detailed spectral structure around $3\nu_Q$ of the In(3) sites is tracked with temperature and pressure [9]. Figure 1 shows the zero-field NMR spectra under ambient pressure and 2.4 GPa at 1.6 K for In(3) sites in $CePt_2In_7$. The spectrum at ambient pressure can be interpreted as coexistence of IC-AFM and C-AFM orders. The IC-AFM ordered portion provides a wing-shape spectrum (dotted curve in Fig. 1), and the C-AFM portion yields the remaining Gaussian-like peaks (solid curve in Fig. 1), including the peak at the same position as the NQR line determined above

T_N (indicated by ∇ in Fig. 1). With applying pressures, the wing-shape spectrum gradually weakens until 2.4 GPa where the spectrum shows the Gaussian-like peaks. The simulations of these spectra are specified in Ref. [9]. The pressure dependence of the NMR spectra in zero field below T_N reveals the stabilization of C-AFM order as the superconducting phase is approached. It appears, therefore, that superconductivity in $CePt_2In_7$ emerges as the C-AFM phase boundary is suppressed toward $T = 0$.

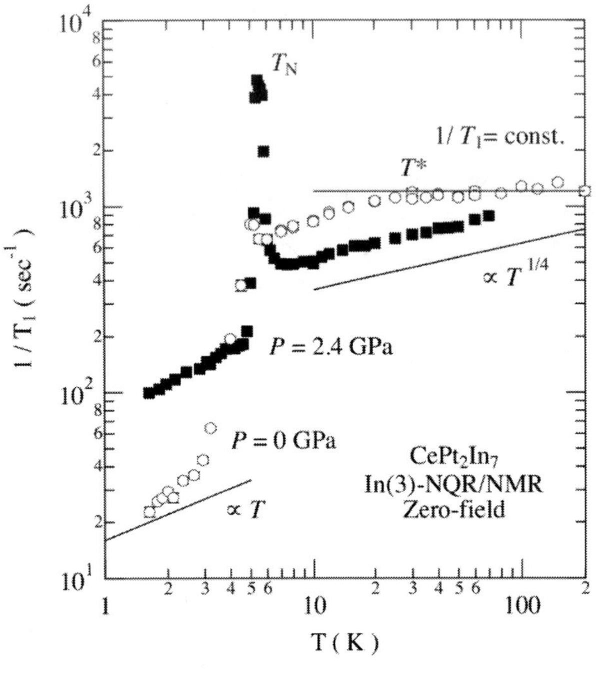

Figure 2. Temperature dependence of $1/T_1$ under the ambient pressure (○) and the hydrostatic pressure of 2.4 GPa (■) for In(3) sites in $CePt_2In_7$. The solid lines are guides to the eye.

As noted above, the partly lingering NQR line implies the existence of sites partially canceling internal fields by the C-AFM arrangement. Indeed, as shown in Fig. 2, a sharp drop of nuclear spin-lattice relaxation rate $1/T_1$ at T_N on this lingering line is observed, which reflects the emergence of an AFM energy gap. At the ambient pressure, the temperature dependence of $1/T_1$ in the paramagnetic state above 25 K shows a constant behavior, which indicates the localized character of $4f$ electrons. Below T $\sim$ 20 K, $1/T_1$ starts to decrease gradually from the constant value. It suggests that an onset of c-f hybridization may occur around T . As seen in Fig. 2, the critical slowing down just above T_N is observed over a quite narrow temperature window. Thus, T_N can be precisely determined by the $1/T_1$ measurement.

The temperature dependence of $1/T_1$ at 2.4 GPa is also shown in Fig. 2. At 2.4 GPa, T_N is determined to be 5.4 K from the peak of $1/T_1$. This confirms microscopically that the AFM state is robust against pressures up to 2.4 GPa, which is consistent with the macroscopic measurements of resistivity and ac-specific heat [7]. In the paramagnetic state at 2.4 GPa, the values of $1/T_1$ decrease from those at ambient pressure, and it does not show a constant value even around 100 K. By applying pressure, T seems to increase above 100 K, that is, $4f$ electrons procure itinerancy via c-f hybridization far above T_N. Simultaneously, $1/T_1$ shows a power-law temperature dependence of $T^{1/4}$, which can be seen near an AFM instability. A similar temperature dependence of $1/T_1$ has been observed in the well-known heavy fermion superconductor $CeCoIn_5$ [10, 11]. On the other hand, the residual $1/T_1$ after opening an AFM gap is found to increase with pressure. It may suggest an increase of residual density of states by mismatched nesting condition of Fermi surfaces, and/or persistent magnetic fluctuations even below T_N. The origin of the residual $1/T_1$ under pressure may relate to the pressure-induced superconductivity in this material.

4. Conclusion

We have presented zero-field NMR spectra of the antiferromagnt $CePt_2In_7$, which indicates the coexistence of IC/C- AFM orderings at ambient pressure. The pressure dependence of the NMR spectra reveals the stabilization of C-AFM order as the superconducting phase is approached. The $1/T_1$ in the paramagnetic state under 2.4 GPa show a characteristic power-law temperature dependence near an antiferromagnetic instability. In order to understand the interplay among magnetism, quantum criticality, and superconductivity in $CePt_2In_7$, further NMR experiments under pressures are currently in progress.

Acknowledgments

Work in Japan was supported by the JSPS KAKENHI for Young Scientists (B) (No. 21750067), the MEXT KAKENHI for Scientific Research on Innovative Areas "Heavy Electrons" (No. 20102006, 20102007), and the Reimei Research Program of JAEA. Work at Los Alamos National Laboratory was performed under the auspices of the U.S. DOE, Office of Basic Energy Sciences, Division of Materials Sciences and Engineering.

References

[1] Kurenbaeva Z M, Murashova E V, Seropegin Y D, Noël H and Tursina A I 2008 *Intermetallics* **16** 979
[2] Petrovic C, Pagliuso P G, Hundley M F, Movshovich R, Sarrao J L, Thompson J D, Fisk Z and Monthoux P 2001 *J. Phys. :Condens. Matter* **13** L337
[3] Petrovic C, Movshovich R, Jaime M, Pagliuso P G, Hundley M F, Sarrao J L, Fisk Z and Thompson J D 2001 *Europhys. Lett* **53** 354
[4] Hegger H, Petrovic C, Moshopoulou E G, Hundley M F, Sarrao J L, Fisk Z and Thompson J D 2000 *Phys. Rev. Lett.* **84** 4986
[5] Bauer E D, Lee H O, Sidorov V A, Kurita N, Gofryk K, Zhu J X, Ronning F, Movshovich R, Thompson J D and Park T 2010 *Phys. Rev. B* **81** 180507(R)
[6] Tobash P H et al, unpublished.
[7] Sidorov V A et al, unpublished.
[8] Altarawneh M M, Harrison N, McDonald R D, Balakirev F F, Mielke C H, Tobash P H, Zhu J X, Thompson J D, Ronning F and Bauer E D 2011 *Phys. Rev. B* **83** 081103(R)
[9] Sakai H, Tokunaga Y, Kambe S, Lee H O, Sidorov V A, Tobash P H, Ronning F, Bauer E D and Thompson J D 2011 *Phys. Rev. B* **83** 140408(R)
[10] Kohori Y, Yamato Y, Iwamoto Y, Kohara T, Bauer E D, Maple M B and Sarrao J L 2001 *Phys. Rev. B* **64** 134526
[11] Kawasaki Y, Kawasaki S, Yashima M, Mito T, q Zheng G, Kitaoka Y, Shishido H, Settai R, Haga Y and Onuki Y 2003 *J. Phys. Soc. Jpn.* **72** 2308

Thermal properties of the nonmagnetic cubic Γ_3 Kondo lattice systems PrTr_2Al$_{20}$ (Tr = Ti, V)

Akito Sakai and Satoru Nakatsuji

Institute for Solid State Physics, University of Tokyo, Kashiwa 277-8581, Japan.

Abstract. We report the results of the specific heat measurements for PrTr_2Al$_{20}$ (Tr = Ti and V), both of which have the cubic nonmagnetic Γ_3 ground state of the crystal electric field scheme. Sharp peak anomalies observed at $T = 2.0$ K for $PrTi_2Al_{20}$ and 0.6 K for PrV_2Al_{20} indicate the second order phase transition due to the multipole degree of freedom. The application of the magnetic field along [111] does not change the anomalies observed in the specific heat of either $PrTi_2Al_{20}$ or PrV_2Al_{20}, suggesting the quadrupole nature of the phase transition. A large Sommerfeld coefficient γ found above the ordering temperature in $PrTi_2Al_{20}$ suggests heavy quasi particle formation due to the scattering process of conduction electrons by quadrupole moments.

1. Introduction

A various interesting phenomena associated with the Kondo effect have been discovered in condensed matter physics such as heavy Fermions, Kondo insulators, quantum criticality and unconventional superconductivity. The Kondo effect in these systems is normally regarded as the "magnetic" Kondo effect since conduction (c) electrons screen the f electrons' magnetic dipole degree of freedom. Therefore, an important question is whether a "nonmagnetic" analog of the Kondo effect is possible by using multipole degree of freedom. Theoretically, Cox predicted that quadrupole moments, which may have two channels for hybridization, can be over-compensated by the c electrons when the f^2 ion has a cubic symmetry and crystal electric field (CEF) ground state is cubic nonmagnetic Γ_3 [1]. According to this theory, the ground state is no longer a Fermi liquid and anomalous metallic behaviors appear as a result of the quadrupolar Kondo effects [2].

Experimentally, there have been a few studies on a Pr-based cubic compound with the nonmagnetic Γ_3 ground CEF state such as $PrInAg_2$, $PrMg_3$ and $PrPb_3$ [3, 4, 5]. These report a large enhancement of the Sommerfeld coefficient and discuss the possibility of a heavy fermion state due to the quadrupolar Kondo effects. However, the origin of the enhanced specific heat is still controversial partly because the degeneracy of the non-Kramers Γ_3 doublet can be easily lifted by structural disorder or distortion. On the other hand, if the system with the Γ_3 CEF ground doublet exhibits a quadrupole long-range order, it ensures that the degeneracy remains at least down to the transition temperature. Therefore, if we find a system that shows quadrupolar ordering and strong hybridization effect, it might allow us to observe Kondo effects based on the quadrupolar degree of freedom above the transition temperature.

PrTr_2Al$_{20}$ (Tr = Ti, V) are such candidate systems and have the cubic $CeCr_2Al_{20}$-type crystal structure, as shown in Fig. 1 with the space group $Fd\bar{3}m$. The symmetry of R site is T_d and cubic [6]. Strong hybridization between $4f$ and c electrons is expected because R

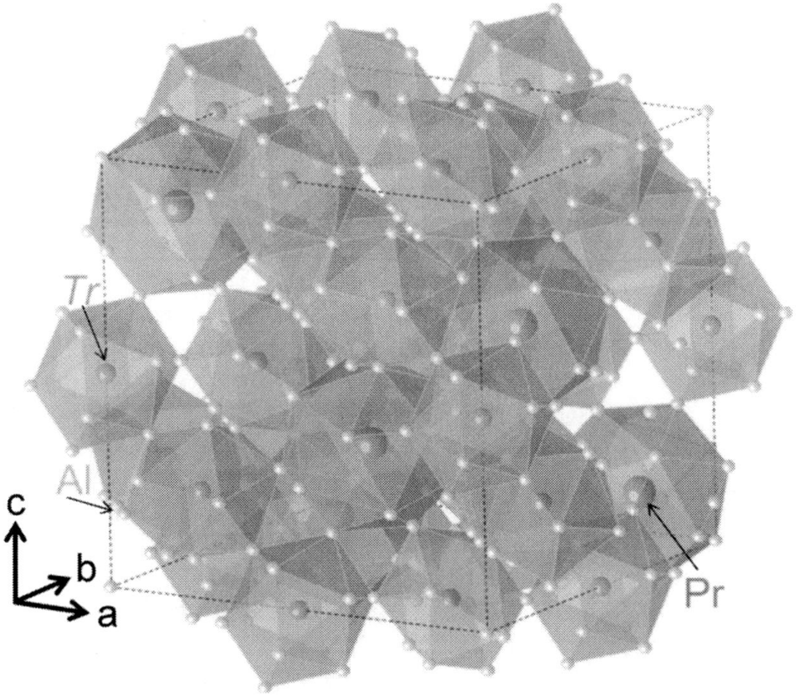

Figure 1. Crystal structure of PrTr_2Al$_{20}$. The yellow, pink and aqua bolls represent Pr, transitional metal and Al respectively.

ions are surrounded by sixteen Al ions, which is the largest coordination number for tetrahedral groupings of spheres[7], indicating a large number of channels of hybridization .

Our previous report of PrTr_2Al$_{20}$ revealed that [8]: (1) The CEF ground state of both PrTi$_2$Al$_{20}$ and PrV$_2$Al$_{20}$ is the nonmagnetic Γ_3 ground doublet and well isolated from the first excited state by the gap of about 40 to 50 K. (2) Both materials exhibit quadrupole order at $T_{\rm O} = 2.0$ K (Ti) and 0.6 K (V). (3) Both materials show strong hybridization effects including the conventional Kondo effects. The $-\ln T$ increase of the resistivity and large Weiss temperature of the susceptibility indicate the Kondo effect, which is found stronger in PrV$_2$Al$_{20}$ than the Ti counterpart. (4) PrV$_2$Al$_{20}$ exhibits anomalous metallic behavior such as divergent $C_P/T \sim T^{-3/2}$, $-T^{1/2}$ dependent Van Vleck susceptibility, and $T^{1/2}$ dependent resistivity over a decade of temperature between 2 and 20 K above $T_{\rm O} = 0.6$ K. (5) All the above observations suggest that the increase in Kondo coupling by changing the Tr site from Ti to V suppresses the quadrupolar ordering, and instead leads to stronger hybridization between the quadrupoles of ground state and conduction electrons, and thus the quadrupolar Kondo effect.

Here, we report the results of the specific heat measurements of PrTr_2Al$_{20}$ performed under $\mu_0 H = 0$ and 5 T // [111]. A large Sommerfeld coefficient $\gamma \sim 100$ (mJ/mole K^2) is observed above $T_{\rm O}$ for PrTi$_2$Al$_{20}$, which suggests the formation of heavy fermions.

2. Experimental

Single crystals of RTr_2Al$_{20}$ (R = Pr and La, Tr =Ti and V) were grown by an Al self-flux method under vacuum, using 4N(99.99%)-Pr, 3N-La, 3N-Ti,V and 5N-Al. The crystal structure was verified by the X-ray powder and single crystal diffraction measurements at room temperature and is found to have the lattice parameters a = 14.723(7) Å (Ti) and 14.591(2) Å (V). The

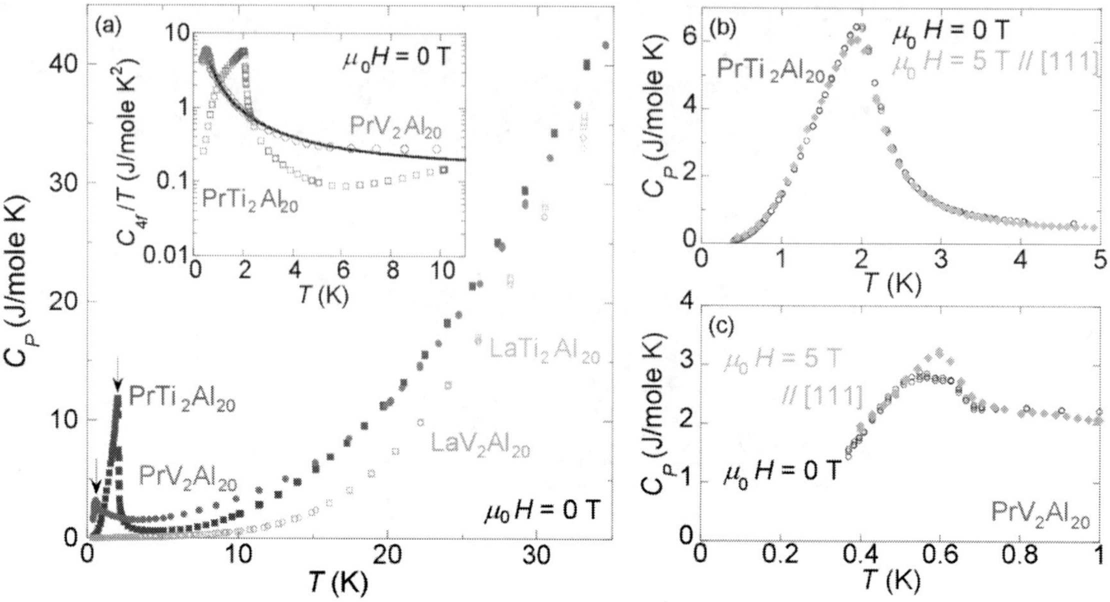

Figure 2. (a) Temperature dependence of the specific heat C_P of $PrTi_2Al_{20}$ (filled square), PrV_2Al_{20} (filled circle), $LaTi_2Al_{20}$ (open square) and LaV_2Al_{20} (open circle). Arrows indicate the transition temperatures defined at the peak of C_P. Inset: Temperature dependence of the $4f$ electron contribution to the specific heat divided by temperature C_{4f}/T. Solid line indicates the fitting to a power law. (b) C_P of $PrTi_2Al_{20}$ under zero field (open circle) and $\mu_0H = 5$ T // [111] (filled diamond) at low temperatures. (c) C_P of PrV_2Al_{20} under zero field (open circle) and $\mu_0H = 5$ T // [111] (filled diamond) at low temperatures

specific heat C_P above 0.4 K were measured by a standard relaxation method.

3. Results and Discussion

Figure 2(a) shows the temperature dependence of the specific heat $C_P(T)$ of $PrTi_2Al_{20}$ (filled square), PrV_2Al_{20} (filled circle), $LaTi_2Al_{20}$ (open square) and LaV_2Al_{20} (open circle) under zero field. A peak indicated by an arrow is observed at $T_O = 2.0$ K for $PrTi_2Al_{20}$ and 0.6 K for PrV_2Al_{20}, indicating the second order phase transition. As we discussed in [8], the CEF ground state of both $PrTi_2Al_{20}$ and PrV_2Al_{20} is the nonmagnetic Γ_3 doublet. Besides, the first excited CEF state is located at $\Delta = \sim$55 K (Ti) and ~40 K (V), which is one order of magnitude larger than T_O. Therefore, these peaks do not come from a magnetic order but from a multipole order.

The inset of Fig. 2(a) shows the temperature dependence of the $4f$ electron contribution to the specific heat divided by temperature C_{4f}/T of $PrTi_2Al_{20}$ (open square) and PrV_2Al_{20} (open circle). This is determined by subtracting C_P of the La analogs. For $PrTi_2Al_{20}$, C_{4f}/T remains $\sim$ 100 (mJ/mole K^2) above T_O up to $\sim$ 10 K. Note that in $PrTi_2Al_{20}$ the Fermi liquid behavior is observed in the temperature dependence of the resistivity in the same T range above T_O up to 20 K, namely, $\rho(T) \propto T^2$ [8]. The T^2 coefficient is estimated to be $A \sim 3.3$ ($\mu\Omega$cm/K^2). Therefore, the constant $C_{4f}/T \sim 100$ (mJ/mole K^2) should provide the estimate for the Sommerfeld coefficient, and as the result, the Kadowaki-Woods ratio is given by $A/\gamma^2 \sim 3.3 \times 10^{-6}$ ($\mu\Omega$cmK^{-2}/(mJK^{-2}mole^{-1})^2). Interestingly, this is close to the universal value known for heavy fermion systems [9]. Because the temperature range is much lower than the CEF gap scale, the large Sommerfeld coefficient should come from the mass enhancement due to the scattering process of conduction electrons by quadrupole moments, not by magnetic moments. In contrast, C_{4f}/T of PrV_2Al_{20} above T_O is found to be strongly T dependent and divergent down to T_O. The solid line in Fig. 2(a) inset shows the fit to the power law, namely $C_{4f}/T \propto T^{-\alpha}$ with $\alpha = 1.5$. As we discussed in detail in [8], this dramatic increase of C_{4f}/T is attributable to the quadrupolar Kondo effect.

Figures 2(b) and (c) show C_P under zero field (open circle) and $\mu_0 H = 5$ T // [111] (filled diamond) at low temperatures of $PrTi_2Al_{20}$ and PrV_2Al_{20}, respectively. Unlike the C_P under $\mu_0 H$ // [100] (Fig. 3(a) and (b) in [8]), the peak temperature and the shape of C_P under $\mu_0 H$ // [111] for both $PrTi_2Al_{20}$ and PrV_2Al_{20} show almost no change from the zero field results. This anisotropy and field insensitiveness are the signatures of multipole ordering.

4. Conclusions

We have measured the specific heat of $PrTr_2Al_{20}$ and $LaTr_2Al_{20}$ (Tr = Ti, V) under $\mu_0 H = 0$ and 5 T // [111]. Sharp peaks are observed owing to the quadrupole ordering at $T_O = 2.0$ K (Ti) and 0.6 K (V). In $PrTi_2Al_{20}$, a large $\gamma \sim 100$ (mJ/mole K^2) and large A coefficient observed above T_O indicate the formation of heavy fermions due to the scattering process of conduction electrons by quadrupole moments. Field insensitive peak shape of C_P for both $PrTi_2Al_{20}$ and PrV_2Al_{20} supports the quadrupolar origin of the phase transition.

5. Acknowledgements

We thank T. Sakakibara, T. Tomita, E.T.C. O'Farrell, T. J. Sato, J. Y. Chan, M. Kangas, D. C. Schmitt, K. Ueda and P. Coleman for useful discussions. This work is partially supported by Grants-in-Aid (No.21684019) from the Japan Society for the Promotion of Science(JSPS), by Grants-in-Aids for Scientific Research on Innovative Areas "Heavy Electrons" of MEXT, Japan, by Grant-in-Aid for JSPS Fellows and by Toray Science and Technology Grant.

6. References

[1] Cox D L 1987 *Phys. Rev. Lett.* **59** 1240–1243
[2] Cox D L and Zawadowski A 1998 *Adv. Phys.* **47** 599–942
[3] Yatskar A, Beyermann W P, Movshovich R and Canfield P C 1996 *Phys. Rev. Lett.* **77** 3637–3640
[4] Tanida H, Suzuki H S, Takagi S, Onodera H and Tanigaki K 2006 *J. Phys. Soc. Jpn.* **75** 073705
[5] Morin P, Schmitt D and du Tremolet de Lacheisserie E 1982 *J. Magn. Magn. Mater.* **30** 257 – 264
[6] Niemann S and Jeitschko W 1995 *J. Solid State Chem.* **114** 337–341
[7] Frank F C and Kasper J S 1958 *Acta Cryst.* **11** 184–190
[8] Sakai A and Nakatsuji S 2011 *J. Phys. Soc. Jpn.* **80** 063701
[9] Kadowaki K and Woods S 1986 *Solid State Commun.* **58** 507 – 509

Elastic Constants of $DyRhIn_5$

Naoyuki Sanada, Ryuta Watanuki, and Kazuya Suzuki

Department of Advanced Materials Chemistry, Yokohama National University, Yokohama 240-8501, Japan

Abstract. The components-separated magnetic transition in $DyRhIn_5$ was investigated by measuring the magnetic susceptibility and elastic constants. The magnetic susceptibility along the [001] direction indicates the antiferromagnetic ordering of *c*-component of the magnetic moments at $T_N = 28.1$ K whereas the susceptibility along those of [100] and [110] direction show the paramagnetic behavior. The results of the elastic constant measurements suggest that the degeneracy of quadrupolar degrees of freedom exists in spite of the formation of magnetic order because the elastic softenings are observed below T_N. The quadrupolar effect in $DyRhIn_5$ is discussed in terms of the symmetry classification.

1. Introduction

"Successive component-separated magnetic transitions", which is unique successive transitions characterized by the independent magnetic ordering of the *c*- and *ab*-components of the magnetic moments at different temperatures, is known as one of the interesting magnetic behavior caused by the existence of a competing interaction in an *f*-electron system. DyB_4 [1] and $TbCoGa_5$ [2] are well known examples in rare-earth compounds which show such successive transitions. More recently, this type of successive transitions is reported in $HoRh_2Si_2$ [3].

In our study of successive components-separated magnetic transitions, we focused on $DyRhIn_5$ that crystallizes into the tetragonal $HoCoGa_5$-type structure ($D^1_{4h} - P4/mmm$). N. V. Hieu *et al.* reported that the magnetic susceptibility of $DyRhIn_5$ for $\mathbf{B}||[001]$ shows a clear peak as $T_N = 28.1$ K, whereas that for $\mathbf{B}||[100]$ shows no anomaly at T_N and increases even below T_N [4, 5]. This magnetically ordered phase in $DyRhIn_5$ is quite similar to the intermediate phase of DyB_4 or $TbCoGa_5$. In the case of DyB_4 and $TbCoGa_5$, we suggested that the presence of the degeneracy of quadrupolar degrees of freedom is an important key to the occurrence of the successive components-separated magnetic transitions. Therefore, the research on quadrupolar degrees of freedom is essential to clarify the magnetic transition in $DyRhIn_5$. In order to investigate the magnetic transition in $DyRhIn_5$, we have performed measurements of the magnetic susceptibility and elastic constants on the single crystals. The results and discussion are reported below.

2. Experimental

Single crystals of $DyRhIn_5$ were grown by the self-flux method using In as flux. The starting materials were inserted in a quartz tube in the ratio of Dy:Rh:In=1:1:30 and sealed under high vacuum. This tube was heated up to 1000°C and cooled down to room temperature at a rate of −10°C/hr. The direction of the single crystals was determined by the X-ray back Laue method. The magnetic susceptibility of $DyRhIn_5$ was measured using a SQUID magnetometer in the temperature range from 1.8 to 300 K under a magnetic field of 0.1 T. The elastic constants were examined by means of the ultrasonic measurement in the temperature range from 1.8 K to 50 K using the single crystalline sample. The measurement of the elastic constants is an effective technique to investigate the quadrupolar effect in an f-electron system with an orbitally degenerate ground state, because the quadrupole moment of the f-electron couples to the elastic strain. A relative change in the elastic constant, $\Delta C_\Gamma/C_\Gamma$, of the sample was obtained using $\Delta C_\Gamma/C_\Gamma = 2\Delta v/v + (\Delta v/v)^2$, which is derived from $C = \rho v^2$, where v and ρ represent the sound velocity and the density of the crystal, respectively. $\Delta v/v$ was measured with a phase comparator using double-balanced mixers.

3. Experimental results and discussion

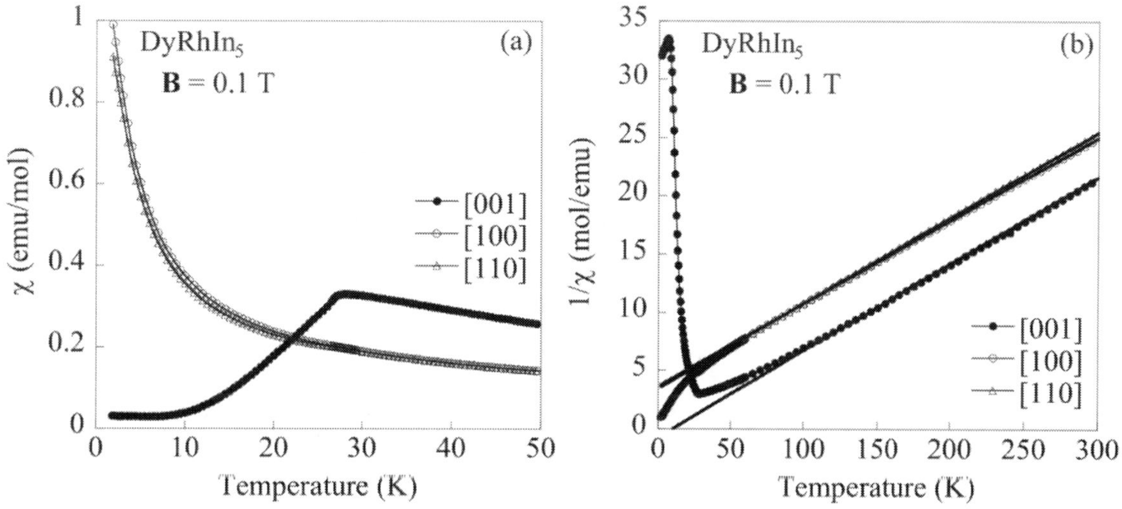

Figure 1. (a) Temperature dependence of the magnetic susceptibility χ measured for magnetic field $\mathbf{B} = 0.1$ T applied parallel to the main symmetry directions in $DyRhIn_5$ single crystal. (b) Temperature dependence of inverse magnetic susceptibility $1/\chi$ along the main symmetry directions in $DyRhIn_5$ single crystal. The solid lines show the Curie-Weiss fit to the data.

Figure 1 shows the temperature dependence of the magnetic susceptibility χ for the single crystal of $DyRhIn_5$. No difference between the field cooling (FC) and zero-field cooling (ZFC) susceptibility in any direction is observed throughout the whole temperature range. Therefore, only ZFC data are shown for clarification. The coincidence of the ZFC and FC data suggests that no spontaneous magnetization is present in $DyRhIn_5$. The susceptibility for $\mathbf{B}\|[001]$ shows a clear maximum at $T_N = 28.1$ K. In contrast, those for $\mathbf{B}\|[100]$ and $\mathbf{B}\|[110]$ show no clear anomaly around T_N and increase even below T_N. These results are mostly consistent with Ref. [4], but it is slightly different in that the susceptibility for $\mathbf{B}\|[100]$ and $\mathbf{B}\|[110]$ continue to increase down to 1.8 K whereas the one for $\mathbf{B}\|[100]$ saturated toward the lowest temperature in previous report [4].

The magnetic susceptibility above 150 K follows the Curie-Weiss law (shown in Fig. 1(b)) with the effective paramagnetic moments μ_{eff} and the Weiss temperatures θ_p: $\mu_{\text{eff}} = 10.39\mu_B$ and $\theta_p = 9.6$ K for $\mathbf{B}\|[001]$, $\mu_{\text{eff}} = 10.59\mu_B$ and $\theta_p = -51.2$ K for $\mathbf{B}\|[100]$ and $\mu_{\text{eff}} = 10.44\mu_B$ and $\theta_p = -47.1$ K for $\mathbf{B}\|[110]$, respectively. The effective paramagnetic moments are close to $10.63\mu_B$ of the free ion value of Dy^{3+}.

Figure 2 shows the temperature dependence of the relative change in the elastic constants $\Delta C_\Gamma/C_\Gamma$ of $DyRhIn_5$. The transverse mode $\Delta C_{44}/C_{44}$ related with elastic strain ε_{xy} or ε_{zx} shows

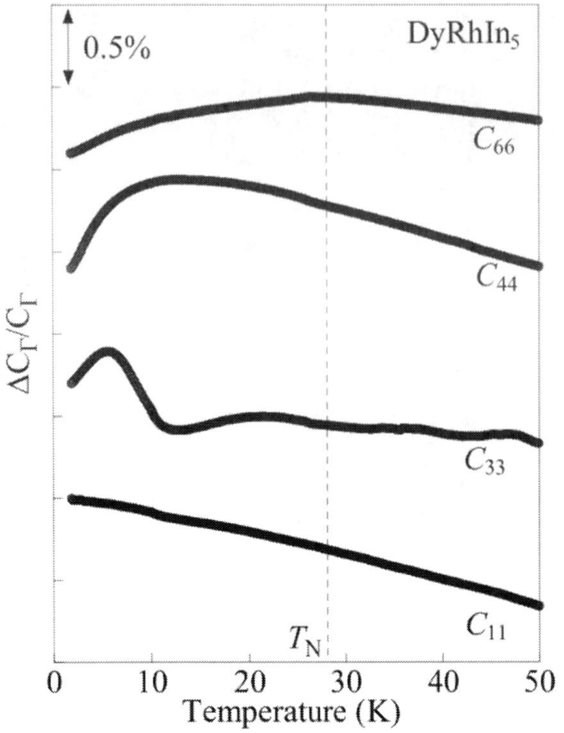

Figure 2. Temperature dependence of the relative change in the elastic constants $\Delta C_{11}/C_{11}$, $\Delta C_{33}/C_{33}$, $\Delta C_{44}/C_{44}$ and $\Delta C_{66}/C_{66}$ of $DyRhIn_5$. The vertical broken line indicate the magnetic transition temperature T_N.

a softening of 0.5% from ~13 K down to 1.8 K and the slight anomaly at approximately T_N. Meanwhile the longitudinal mode $\Delta C_{33}/C_{33}$ and the transverse mode $\Delta C_{66}/C_{66}$ show a smaller softening and the longitudinal mode $\Delta C_{11}/C_{11}$ shows the monotonic hardening with decreasing temperature down to 1.8 K. $\Delta C_{33}/C_{33}$ shows a nonmonotonic temperature dependence below T_N. However, a reason of the nonmonotonic behavior of $\Delta C_{33}/C_{33}$ is still unknown. The elastic softenings below T_N suggest that the degeneracy of the quadrupolar degrees of freedom still exist in magnetically ordered phase. Moreover the quadrupole moments related with C_{44} are hardly influenced by the magnetic ordering in the [001] direction.

Table 1. The symmetry of quadrupole moments, elastic constants and dipole moments in a tetragonal system with D_{4h}

Symmetry	Quadrupole Moments	Elastic constants	Dipole moments
Γ_1	$O_2^0 = \frac{1}{2}(2J_z^2 - J_x^2 - J_y^2)$	$C_u = \frac{1}{6}(C_{11} + C_{12} - 4C_{13} + 2C_{33})$	
Γ_2			J_z
Γ_3	$O_2^2 = \frac{\sqrt{3}}{2}(J_x^2 - J_y^2)$	$\frac{1}{2}(C_{11} - C_{12})$	
Γ_4	$O_{xy} = \frac{\sqrt{3}}{2}(J_xJ_y + J_yJ_x)$	C_{66}	
Γ_5	$O_{yz} = \frac{\sqrt{3}}{2}(J_yJ_z + J_zJ_y)$ $O_{zx} = \frac{\sqrt{3}}{2}(J_zJ_x + J_xJ_z)$	C_{44}	J_x, J_y

Here, we show the symmetry classification of the quadrupole moments, elastic constants and dipole moments in a tetragonal system with D_{4h} of $HoCoGa_5$-type structure in Table 1. The behaviors of the magnetic susceptibility and the elastic constant C_{44} of $DyRhIn_5$ are quite similar to those in the intermediate phase of $TbCoGa_5$ [2]. The *ab*-component of the magnetic moments seems to be influenced by the quadrupole moment system, but not the *c*-component in $DyRhIn_5$ and $TbCoGa_5$. The largest softening of C_{44} in the measured C_Γ modes indicates that the dominant components of the quadrupole moments are $O_{yz} = J_yJ_z + J_zJ_y$ and $O_{zx} = J_zJ_x + J_xJ_z$ with Γ_5 symmetry. These results strongly suggest the relation between the magnetic ordering of the *c*-component of the magnetic moments with the paramagnetic behavior along the *ab*-plane and the largest softening of the elastic constant C_{44}.

It is known that rare-earth compounds may have degenerate crystalline electric field (CEF) ground state where orbital degrees of freedom remain. In general, if quadrupolar degrees of freedom are added in a magnetic system, the magnetic transition temperature may be suppressed from the original one due to the increase in entropy change during the magnetic transition. In $DyRhIn_5$ and $TbCoGa_5$, the dominant components of the quadrupole moments O_{yz} and O_{zx} related with the elastic constant C_{44} are coupled with the *ab*-component of the magnetic dipole moment (J_x, J_y) in Γ_5 symmetry. Therefore, the transition temperature with the magnetic ordering of the *ab*-component can be strongly suppressed, but the ordering of the *c*-component (J_z) is not affected. Consequently, the magnetic ordering only the *c*-component occurs whereas the *ab*-components are paramagnetic.

Acknowledgments

We are very grateful to Dr. M Akatsu for technical assistance. This work was supported by a Grant-in-Aid for Scientific Research (C) (Grant No. 21540355) from Japan Society for the Promotion Science (JSPS). One of the authors (N. S.) acknowledges the support from the Research Fellowship of JSPS for Young Scientists.

References

[1] Watanuki R, Sato G, Suzuki K, Ishihara M, Yanagisawa T, Nemoto Y and Goto T 2005 *J. Phys. Soc. Jpn.* **74** 2169
[2] Sanada N, Watanuki R, Suzuki K, Akatsu M and Sakakibara T 2009 *J. Phys. Soc. Jpn.* **78** 073709
[3] Shigeoka T, Fujiwara T, Munakata K, Matsubayashi K and Uwatoko Y 2011 *J. Phys.: Conf. Series* **273** 012127
[4] Hieu N V, Shishido H, Takeuchi T, Thamizhavel A, Nakashima H, Sugiyama K, Settai R, Matsuda T D, Haga Y, Hagiwara M, Kindo K, and Ōnuki Y 2006 *J. Phys. Soc. Jpn.* **75** 074708
[5] Hieu N V, Takeuchi T, Shishido H, Tonohiro C, Yamada T, Nakashima H, Sugiyama K, Settai R, Matsuda T D, Haga Y, Hagiwara M, Kindo K, Araki S, Nozue Y, and Ōnuki Y 2007 *J. Phys. Soc. Jpn.* **76** 064702

T-H Phase Diagram of $PrPb_3$ in [001] and [110] Magnetic Field Directions

Y. Sato, S. Makiyama, H. Morodomi, Y. Inagaki, T. Kawae
Department of Applied Quantum Physics, Kyushu University, Fukuoka. 819-0385, Japan
E-mail: y.sato.483@s.kyushu-u.ac.jp

H. S. Suzuki
National Institute for Material Science, Tsukuba, 305-0047, Japan

T. Onimaru
Department of Quantum Matter, Hiroshima University, Higashi-Hiroshima. 739-8530, Japan

Abstract. We have measured the magnetic field dependence of the specific heat in $PrPb_3$ for $H \parallel [001]$ and [110] up to $H = 8\,\mathrm{T}$. At $H = 0\,\mathrm{T}$, a sharp λ-type peak due to antiferro-quadrupolar(AFQ) ordering and a small anomaly appears at $T_Q = 0.43\,\mathrm{K}$ and $T^* = 0.25\,\mathrm{K}$, respectively. The field dependence of T_Q shows a reentrant behavior and a high-field phase appear above $H \sim 6\,\mathrm{T}$ in the both field directions. T^* shows a large difference between $H \parallel [001]$ and [110], suggesting that the anomaly is related to the anisotropic feature of the quadrupolar moment in magnetic fields. From these results, we describe the T-H phase diagram $PrPb_3$ in $H \parallel [001]$ and [110].

1. Introduction

In $4f$-electron systems, the spin-orbital interaction is larger than the crystalline-electric-field (CEF) effect so that the magnetic properties of $4f$-electrons are characterized by the total angular momentum $\boldsymbol{J}(= \boldsymbol{L} + \boldsymbol{S})$. In this case, the orbital degrees of freedom can be represented by quadrupolar moments. Five independent quadrupolar moments with two Γ_3-type and three Γ_5-type are defined under cubic(O_h) symmetry. $PrPb_3$ has a $AuCu_3$-type cubic structure and a trivalent Pr ion forms $4f^2$ configuration in a crystal. The ninefold degenerate $\boldsymbol{J}$ multiplet $^3H_4$ of the trivalent Pr ion splits into four levels: a singlet Γ_1, a doublet Γ_3, a triplet Γ_4 and a triplet Γ_5 by the CEF effect. The CEF ground state of $PrPb_3$ has been studied by various experiments, such as nuclear inelastic scattering [1], specific heat [2] and magnetization measurements [3]. It is established that the CEF ground state is a non-Kramers doublet Γ_3, which carries two quadrupolar moments, $O_2^0(= [3J_z^2 - \boldsymbol{J}^2]/2)$ and $O_2^2(= \sqrt{3}[J_x^2 - J_y^2]/2)$, and the first exited one is a triplet Γ_4 with an energy gap $\sim 19\,\mathrm{K}$ [2]. As the temperature is decreased, $PrPb_3$ exhibits antiferro-quadrupolar (AFQ) ordering at $T_Q = 0.4\,\mathrm{K}$ [4].

The T-H phase diagram of AFQ ordering in $PrPb_3$ was studied by magnetization measurements, which revealed that the field dependence of T_Q is very anisotropic for $H \parallel [001]$, [110] and [111] and the high-field phase appears only in $H \parallel [110]$ [3] [5]. Moreover, the order parameter of the low-field phase is determined to be O_2^0 by inelastic neutron diffraction measurements [6].

In this paper, we report the T-H phase diagram of $PrPb_3$ in $H \parallel [001]$ and [110] measured by the specific heat up to $H = 8\,T$. At $H = 0\,T$, we observe a sharp peak due to the AFQ ordering at $T_Q = 0.43\,K$ and a small anomaly at $T^* = 0.25\,K$. The field dependence of T_Q shows a reentrant behavior and a high-field phase appears above $H \sim 6\,T$ in the both field directions. T^* shows a large difference between $H \parallel [001]$ and [110], suggesting that the anomaly is related to the anisotropic feature of the quadrupolar moment.

2. Experimental

A single crystal of $PrPb_3$ were prepared by the Bridgman method. The stoichiometric amounts of Pr and Pb metals were heated in a molybdenum crucible sealed by electric beam welding in a vacuum. A high quality single crystal suitable for neutron diffraction experiments was obtained by this method. The specific heat was measured by a pseudo-adiabatic heat pulse method with a ^{3}He-^{4}He dilution refrigerator from $T = 0.1\,K$ to $T = 1\,K$ and a superconducting magnet up to $H = 8\,T$.

3. Results and Discussion

3.1. $H \parallel [001]$

In Fig. 1(a), we show the temperature dependence of the specific heat at low fields for $H \parallel [001]$. At $H = 0\,T$, a sharp λ-type peak due to AFQ transition appears at $T_Q = 0.43\,K$, while AFQ transition peak is observed at $T_Q = 0.4\,K$ in the previous studies. We reported that the AFQ transition temperature is suppressed to $T_Q = 0.28\,K$ only by 1% nonmagnetic La substitution of Pr, which demonstrates that T_Q is very sensitive to sample quality, such as randomness and the presence of grain boundaries. In our previous sample with $T_Q = 0.4\,K$, the tail of the neutron reflection peak was broad compared with that of the present sample, implying that the sample included some grain boundaries. From these results, we consider that the increase in T_Q comes from the high quality of the present sample.

In addition, a small anomaly appears at $T^* = 0.25\,K$ at $H = 0\,T$. Although the clear anomaly was not observed in the previous specific heat measurements, the slope in C/T changes at around $T = 0.25\,K$. This indicates that the anomaly reflects the essential feature of $PrPb_3$. This anomaly was also observed with ultra sound measurements [7]. It seems that the T^* anomaly

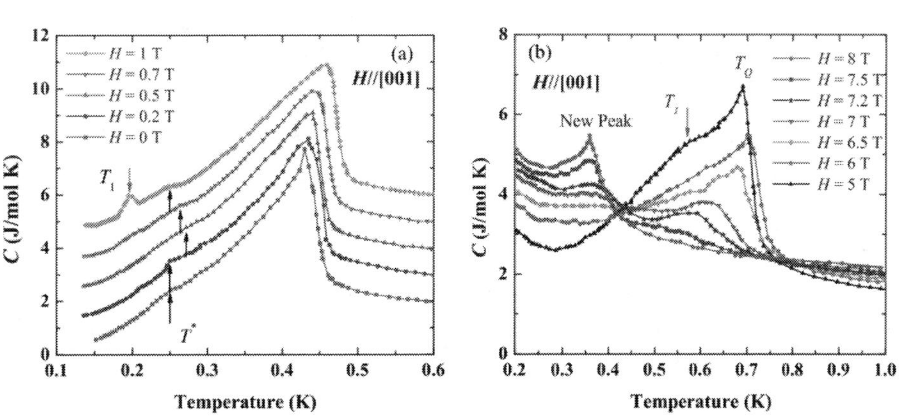

Figure 1. (a) Magnetic field dependence of the specific heat in $H \parallel [001]$ up to $H = 1\,T$, where the data are shifted by 1 J/(mol K), respectively. (b) Magnetic field dependence of the specific heat in $H \parallel [001]$ between $H = 5$ and $8\,T$.

stays approximately $T^* = 0.25\,\mathrm{K}$ below $H = 1\,\mathrm{T}$. Above $H = 1\,\mathrm{T}$, the large peak appears at T_1, which is due to the first-order transition between the commensurate and incommensurate structures [8]. T_1 shows a strong hysteresis in the temperature scan as follows. The anomaly at T_1 cannot be observed in the specific heat with decreasing temperature, while we can observe the sharp anomaly at T_1 after decreasing the temperature to a value much lower than T_1. With increasing magnetic field, T_1 shifts to higher temperatures.

In Fig. 1(b), we show the magnetic field dependence of the specific heat from $H = 5\,\mathrm{T}$ to $H = 8\,\mathrm{T}$. As the magnetic field is increased, a sharp peak at T_Q broadens, while a new peak becomes obvious at temperatures lower than T_Q. At $H = 7.2\,\mathrm{T}$, for instance, two peaks are clearly observed at $T \sim 0.35\,\mathrm{K}$ and $T_Q \sim 0.6\,\mathrm{K}$. At $H = 8\,\mathrm{T}$, the new peak develops into a λ-shaped anomaly, while the peak due to the AFQ ordering disappears. From these results, it is reasonable to consider that a new phase emerges in the high-field region in $H \parallel [001]$. It is noted that the new peak does not show the hysteresis between increasing and decreasing temperatures, implying a second-order transition between the paramagnetic and high-field phases. We have already reported the detail of these results in Ref. [9].

3.2. $H \parallel [110]$

Fig. 2(a) shows the temperature dependence of the specific heat at low fields for $H \parallel [110]$. As shown in the figure, T_Q increases gradually to higher temperatures with increasing magnetic field. However, the increase of T_Q in $H \parallel [110]$ is much lower than in $H \parallel [001]$. At $H = 2\,\mathrm{T}$, for instance, T_Q is at $\sim 0.45\,\mathrm{K}$ for $H \parallel [110]$, while it reaches $\sim 0.52\,\mathrm{K}$ for $H \parallel [001]$.

The application of magnetic field induces staggered magnetic dipolar and octupolar moments depending on the field direction with respect to the crystal axis [10]. The induced moments stabilize the AFQ phase, leading to the increase of T_Q. Therefore, the field dependence of the increase of T_Q is explained by the difference of the induced moments. The order parameter of the low-field phase is O_2^0 for both directions so that the increase of T_Q for $H \parallel [110]$ is lower than that for $H \parallel [001]$.

The anomaly at T^* shifts to a higher temperature with increasing the magnetic field, and is observed up to $H = 2.5\,\mathrm{T}$. We consider that the anomaly is related to changing the structure of the quadrupolar moments. The partially-disordered structure is observed down to T=0.1 K in the neutron experiments [8]. However it is plausible at low temperatures that some of the disordered moments change into an ordered structure at T^* to reduce the entropy and the rest

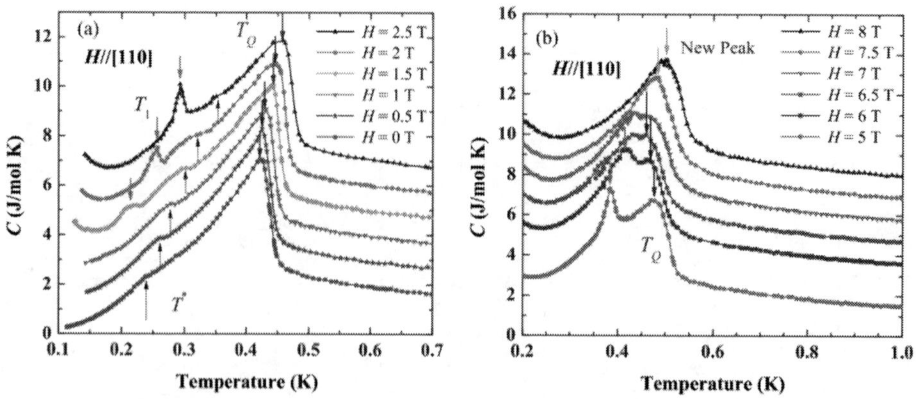

Figure 2. (a) Magnetic field dependence of the specific heat in $H \parallel [110]$ up to $H = 2.5\,\mathrm{T}$, where the data are shifted by 1 J/(mol K), respectively. (b) Magnetic field dependence of the specific heat in $H \parallel [110]$ between $H = 5$ and 8 T, which are shifted by 1 J/(mol K), respectively.

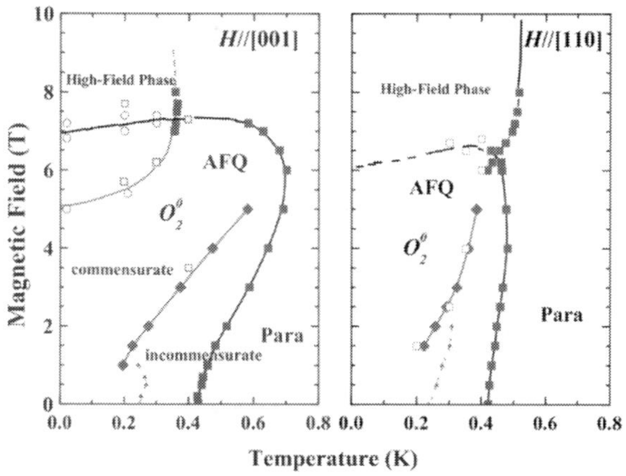

Figure 3. The T-H phase diagram of $PrPb_3$ in $H \parallel [001]$ and $H \parallel [110]$. Filled and open squares indicate the results obtained by the specific heat measurements at fixed H and by the C/T plot at fixed T, respectively. Filled triangles plot T^*. Open circles are given by the magnetization measurements at fixed T [3]. The solid line draws the phase diagram determined by the experiments, while the dotted line serves as a visual guide.

are frozen by the quadrupole Kondo effect as shown in its diluted system [11]. A large peak due to the first-order transition appears above $H = 1.5\,\mathrm{T}$ as shown by T_1 in Fig. 2(a). The field dependence of T^* and T_1 are clearly different from those for $H \parallel [001]$. The difference is likely caused by the anisotropic feature of the quadrupolar moments in magnetic fields.

We show the temperature dependence of the specific heat at high fields for $H \parallel [110]$ in Fig. 2(b). It is seen that T_Q starts to decrease above $H = 5\,\mathrm{T}$, while a new peak appears at high temperatures above $H = 7\,\mathrm{T}$, suggesting the emergence of a high field phase.

We describe the T-H phase diagrams for $H \parallel [001]$ and $H \parallel [110]$ in Fig. 3 based on the present results. The AFQ ordering temperatures at T_Q and the first-order transition temperatures at T_1 are in good agreement with those in the previous experiments [3] [8]. We note that the high-field phase appears not only in $H \parallel [110]$ but also in $H \parallel [001]$.

4. Conclusion

We have measured the magnetic field dependence of the specific heat in $PrPb_3$ for $H \parallel [001]$ and [110] up to $H = 8\,\mathrm{T}$. At $H = 0\,\mathrm{T}$, two anomalies are observed at $T_Q = 0.43\,\mathrm{K}$ and $T^* = 0.25\,\mathrm{K}$, respectively. The field dependence of T_Q shows a reentrant behavior and a high-field phase appears above $H \sim 6\,\mathrm{T}$ in the both field directions. T^* shows a large difference between $H \parallel [001]$ and [110]. From these results, we describe the T-H phase diagram $PrPb_3$ in $H \parallel [001]$ and [110].

References

[1] Gross W, Knorr K, Murani A P and Buschow K H J, 1980 *Z. Phys. B*, **37** 123.
[2] Aoki D, Katayama Y, Settai R, Inada Y, Onuki Y, Harima H, Kletowski Z, 1997 *J. Phys. Soc. Jpn.*, **66** 3988.
[3] Tayama T, Sakakibara T, Kitami K, Yokoyama M, Tenya K, Amitsuka H, Aoki D, Onuki Y and Kletowski Z, 2001 *J. Phys. Soc. Jpn.*, **70** 248.
[4] Bucher E, Andres K, Gossard A C and Maita J P, 1972 *J. Low. Temp.*, **2** 322.
[5] Onimaru T, Sakakibara T, Harita A, Tayama T, Aoki D and Onuki Y, 2004 *Phys. Rev. lett.*, **73** 2377.
[6] Onimaru T, Sakakibara T, Tayama T, Aso N, Yoshizawa H, Aoki D, Onuki Y, Kawae T, Kitai T and Takeuchi T, 2005 *Physica. B*, **359-361** 935.
[7] Jumonji S, 2005 *Master Thesis, Niigata University.*
[8] Onimaru T, Sakakibara T, Aso N, Yoshizawa H, Suzuki H S and Takeuchi T, 2005 *Phys. Rev. Lett.*, **94** 197201.
[9] Sato Y, Morodomi H, Ienaga K, Inagaki Y, Kawae T, Suzuki H S and Onimaru T, 2010 *J. Phys. Soc. Jpn.*, **79** 093708.
[10] Shiina R, Shiba H and Sakai O, 1999 *J. Phys. Soc. Jpn.*, **68** 2390.
[11] Kawae T, Kinoshita K, Nakaie Y, Tateiwa N, Takeda K, Suzuki H S and Kitai T, 2006 *Phys. Rev. Lett.*, **96** 027210.

Magnetic and electrical properties of filled skutterudite compound $(\mathrm{Sm}_{1-x}R_x)\mathrm{Ru}_4\mathrm{P}_{12}$ (R=La, Y)

C. Sekine [1], **Y. Kiyota** [1], **Y. Kawamura** [1] **and T. Yagi** [2]

[1] Graduate School of Engineering, Muroran Institute of Technology, Muroran, Hokkaido 050-8585, Japan
[2] Institute for Solid State Physics, University of Tokyo, Kashiwa, Chiba 277-8581, Japan

E-mail: `sekine@mmm.muroran-it.ac.jp`

Abstract. Sm-site substitution e ect on filled skutterudite compound $\mathrm{SmRu_4P_{12}}$, which shows an intriguing metal–insulator (MI) transition, has been studied by electrical resistivity, magnetic susceptibility and specific heat measurements. Single-phase polycrystalline $(\mathrm{Sm}_{1-x}\mathrm{La}_x)\mathrm{Ru_4P_{12}}$ (x=0.1, 0.2, 0.3) and $(\mathrm{Sm}_{1-x}\mathrm{Y}_x)\mathrm{Ru_4P_{12}}$ (x=0.1, 0.2, 0.3) have been prepared at high temperature and high pressure. Both La and Y substitutions cause a large suppression of the semiconductor-like behavior at low temperature. The MI transition temperature (T_{MI}) shifts to lower temperature with increasing La concentration. This change is qualitatively consistent with external pressure experiments. It could be due to a negative chemical pressure e ect induced by La substitution. However, T_{MI} also shifts to lower temperature with increasing Y concentration, which is expected for a positive chemical pressure e ect. The results suggest that introducing a disorder into the periodicity of Sm a ects sensitively the MI transition in this system.

1. Introduction

The lled skutterudite compound $\mathrm{SmRu_4P_{12}}$ exhibits a metal-insulator (MI) transition at T_{MI}=16.5K [1]. The MI transition is accompanied by large anomalies in thermodynamic quantities such as speci c heat, magnetization [2, 3, 4], thermal expansion [5] and elastic constant [6, 7]. T_{MI} shifts toward higher temperatures with increasing magnetic eld [1, 2], as observed for the antiferro-quadrupoler (AFQ) ordering in $\mathrm{CeB_6}$. The magnetic entropy of Rln4 is released below T_{MI} [2]. This indicates that the crystalline electric eld (CEF) ground state is a Γ_{67} quartet with dipolar, quadrupolar, and octupolar degree of freedom in T_h symmetry. In addition to the MI transition, another anomaly gradually appears at a lower temperature T^* in thermodynamic quantities which becomes sharper with increasing magnetic eld [2-7]. This complex H-T phase diagram suggests that multipolar degrees of freedom are involved in the ordering. The phases for $T>T_{\mathrm{MI}}$, $T^*<T<T_{\mathrm{MI}}$ and $T<T^*$ are referred as phase I, II and III, respectively, in the literature. Phase I is a paramagnetic metal. The nature of phase II and III is not clear so far. NMR and μSR measurements detect internal magnetic eld below T_{MI} [8, 9]. The ordered dipole moment is found below T_{MI} by ^{149}Sm nuclear resonant forward scattering in zero eld [10]. From nuclear speci c heat and recent neutron di raction experiments, the ordered dipolar moment was estimated to be $0.29\mu_{\mathrm{B}}$ [11] and $0.35\mu_{\mathrm{B}}$ [12], respectively, suppressed in comparison with the value for usual dipolar ordering. Furthermore, a (111)-type distortion of 4f wave function was observed below T_{MI} by ^{101}Ru-NQR measurements [13]. These results and group

theoretical considerations give a new scenario for the mechanism of MI transition, which is due to the coupling between the dipole moment and the multipole moment [7, 11, 14]. However, the nature of the ordered state and the mechanism of the MI transition are still enigmas. In order to investigate the origin of the MI transition, we have studied Sm-site substitution e ect on $SmRu_4P_{12}$.

2. Experimental

Polycrystalline samples of $(Sm_{1-x}La_x)Ru_4P_{12}$ (x=0.1, 0.2, 0.3) and $(Sm_{1-x}Y_x)Ru_4P_{12}$ (x=0.1, 0.2, 0.3) were prepared at high temperatures and high pressures using a wedge-type cubic anvil high-pressure apparatus [15]. x represents the stoichiometric ratio of La or Y to Sm in starting materials. The samples were prepared by reacting stoichiometric amounts of 3N (99.9% pure)-Y, La, Sm chips, 4N-Ru and 6N-P powders at 4.5 GPa. The reaction temperatures were between 1000 and 1100 °C. The prepared samples were characterized by powder x-ray di raction using $CoK_{\alpha1}$ radiation and silicon as a standard. $(Sm_{1-x}R_x)Ru_4P_{12}$ (R=La, Y) maintained the cubic- lled skutterudite-type structure for $0 \leq x \leq 0.3$. $(Sm_{1-x}R_x)Ru_4P_{12}$ (R=La, Y) solid solution exists for all the value of x ($\leq$0.3), and crystallizes in the cubic skutterudite structure. The lattice parameter of alloys increases (decreases) linearly with increasing La (Y) concentration (g. 1). Resistivity was measured by a standard dc four-probe method. Magnetization and dc magnetic susceptibility were measured with a Quantum Design MPMS superconducting quantum interference device magnetometer. The speci c heat was measured by thermal relaxation method (PPMS; Quantum Design).

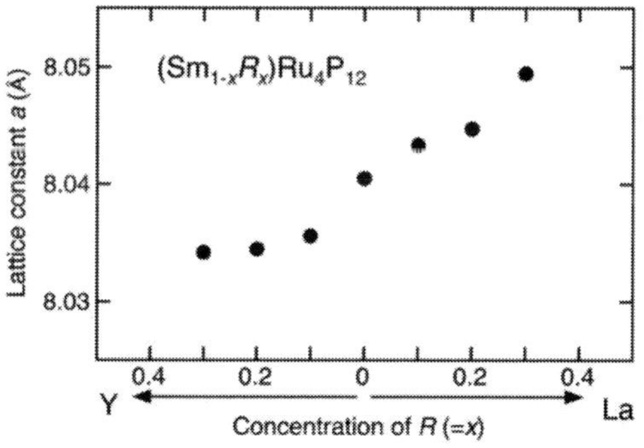

Figure 1. Lattice constants of $(Sm_{1-x}La_x)Ru_4P_{12}$ and $(Sm_{1-x}Y_x)Ru_4P_{12}$.

3. Results and discussion

Fig. 2 shows the temperature dependence of electrical resistivity (T) for (a) $(Sm_{1-x}La_x)Ru_4P_{12}$ (x=0.1, 0.2, 0.3) and (b) $(Sm_{1-x}Y_x)Ru_4P_{12}$ (x=0.1, 0.2, 0.3). Both La and Y substitutions cause a suppression of the semiconductor-like behavior at low temperatures. The MI transition temperature (T_{MI}) shifts to lower temperature with increasing La concentration. This change is qualitatively consistent with external pressure experiments [16]. It could be due to negative chemical pressure induced by La substitution. However, T_{MI} also shifts to lower temperature with increasing Y concentration, which is expected for positive chemical pressure e ect. The results suggest that introducing a disorder into the periodicity of Sm a ects sensitively the MI transition in this system. A resistivity minimum, which may be due to the Kondo e ect, was observed around 50K (T_{min}) for all alloys and the minimum shifts with x.

The temperature dependence of magnetic susceptibility (T) is shown in g. 3 for (a) $(Sm_{1-x}La_x)Ru_4P_{12}$ (x=0.1, 0.2, 0.3) and (b) $(Sm_{1-x}Y_x)Ru_4P_{12}$ (x=0.1, 0.2, 0.3). T_{MI} indicates the MI transition temperature deduced from (T) data of $SmRu_4P_{12}$. The cusp anomaly due to

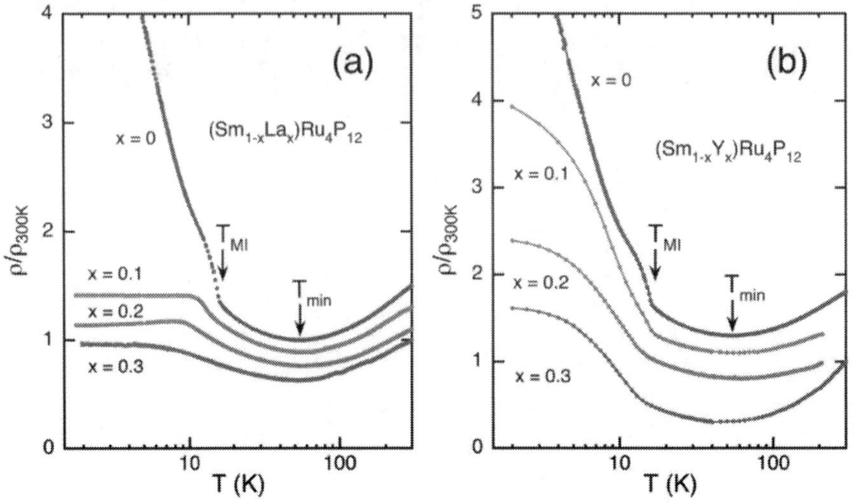

Figure 2. Temperature dependence of resistivity for (a) (Sm1 $_x$La$_x$)Ru$_4$P$_{12}$ (x=0, 0.1, 0.2, 0.3) and (b) (Sm1 $_x$Y$_x$)Ru$_4$P$_{12}$ (x=0, 0.1, 0.2, 0.3). They are normalized to room-temperature resistivity value.

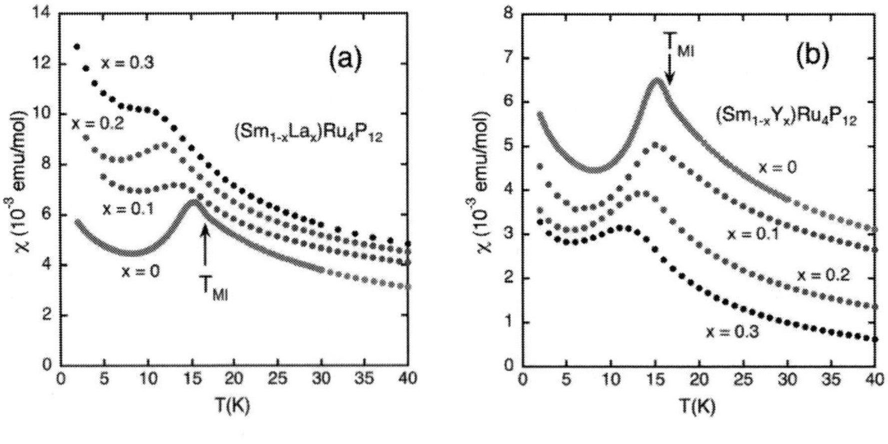

Figure 3. Temperature dependence of magnetic susceptibility for (a) (Sm1 $_x$La$_x$)Ru$_4$P$_{12}$ (x=0, 0.1, 0.2, 0.3) and (b) (Sm1 $_x$Y$_x$)Ru$_4$P$_{12}$ (x=0, 0.1, 0.2, 0.3).

the MI transition shifts toward lower temperature with increasing both La and Y concentration, which is consistent with the result of (T).

Fig. 4 shows the temperature dependence of speci c heat $C(T)$ for (Sm$_{0.9}$La$_{0.1}$)Ru$_4$P$_{12}$ at 0T and 7T. For SmRu$_4$P$_{12}$, the single peak in $C(T)$ at zero eld changes into a double peak in elds and the high-temperature anomaly slightly shifts to higher temperature with increasing eld [2]. However, $C(T)$ for (Sm$_{0.9}$La$_{0.1}$)Ru$_4$P$_{12}$ reveals only a single round peak at zero eld and even in 7T. Furthermore, the anomaly slightly shifts to lower temperature. Thus, a small amount of La substitution seems to wipe out Phase II in H-T phase diagram for SmRu$_4$P$_{12}$.

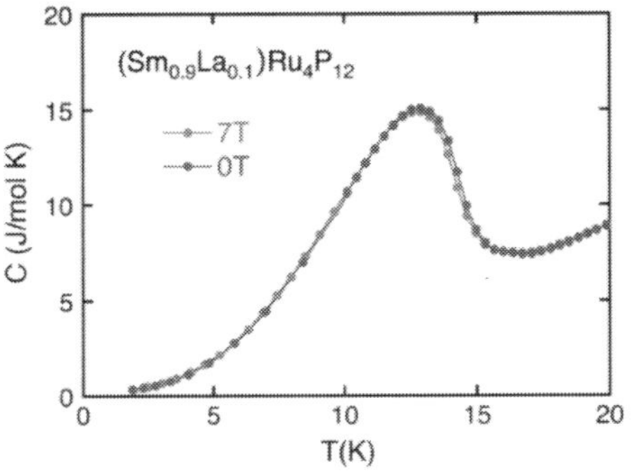

Figure 4. Temperature dependence of speci c heat for $(Sm_{0.9}La_{0.1})Ru_4P_{12}$ at 0T and 7T.

4. Summary

We have synthesized the lled skutterudite compounds $(Sm_{1-x}R_x)Ru_4P_{12}$ (R=La, Y) at high pressure and performed the measurements of electrical resistivity, magnetic susceptibility and speci c heat. Sm-site substitution e ects on $SmRu_4P_{12}$ has been studied. The results suggest that introducing a disorder into the periodicity of Sm a ects sensitively the MI transition in this system. Especially, phase II is quite fragile for doping, which suggests that the formation of phase II is strongly related to multipole interactions.

Acknowledgments

This work was supported by a Grant-in-Aid for Scienti c Research on Innovative Areas "Heavy Electrons" (No. 20102004) of The Ministry of Education, Culture, Sports, Science, and Technology, Japan and a Grant-in-Aid for Scienti c Research (B) (No. 23340092) of the Japan Society for the Promotion of Science. The authors gratefully acknowledge the support from Muroran Institute of Technology (a grant for the Kidorui (Rare Earth) Program).

References

[1] Sekine C, Shirotani I, Matsuhira K, Haen P, Brion S D, Chouteau G, Suzuki H and Kitazawa H 2003 *Acta Phys. Pol. B* **34** 983
[2] Matsuhira K, Hinatsu Y, Sekine C, Togashi T, Maki H, Shirotani I, Kitazawa H, Takamasu T, and Kido G 2002 *J. Phys. Soc. Jpn.* **71** Suppl. 237
[3] Matsuhira K, Doi Y, Wakeshima M, Amitsuka H, Shimaya Y, Giri R, Sekine C and I. Shirotani I 2005 *J. Phys. Soc. Jpn.* **74** 1030
[4] Kikuchi D, Tanaka K, Aoki H, Kuwahara K, Aoki Y, Sugawara H and Sato H 2007 *J. Magn. Magn. Mater.* **310** e225
[5] Sekine C, Shimaya Y, Shirotani I and Haen P 2005 *J. Phys. Soc. Jpn* **74** 3395
[6] Yoshizawa M, Nakanishi Y, Kumagai T, Oikawa M, Sekine C and Shirotani I 2004 *J. Phys. Soc. Jpn.* **73** 315
[7] Yoshizawa M, Nakanishi Y, Oikawa M, Sekine C, Shirotani I, Saha S R, Sugawara H and Sato H 2005 *J. Phys. Soc. Jpn.* **74** 2141
[8] Hachitani K, Amanuma H, Fukazawa H, Kohori Y, Koyama K, Kumagai K.-i, Sekine C and I. Shirotani I 2006 *J. Phys. Soc. Jpn.* **75** 124712
[9] Hachitani K, Fukazawa H, Kohori Y, Watanabe I, Sekine C and Shirotani I 2006 *Phys. Rev. B* **73** 052408
[10] Tsutsui S, Kobayashi Y, Okada T, Haba H, Onodera H, Yoda Y, Mizumaki M, Tanida H, Uruga T, Sekine C, Shirotani I, Kikuchi D, Sugawara H and Sato H 2006 *J. Phys. Soc. Jpn.* **75** 093703
[11] Aoki Y, Sanada S, Kikuchi D, Sugawara H and Sato H 2007 *J. Phys. Soc. Jpn.* **76** 113703
[12] Lee C H, Tsutsui S, Kihou K, Sugawara H and Yoshizawa H 2011 *J. Phys. Soc. Jpn.* submitted
[13] Masaki S, Mito T, Takemura M, Wada S, Harima H, Kikuchi D, Sato H, Sugawara H, Takeda N, Zheng G 2007 *J. Phys. Soc. Jpn.* **76** 043714
[14] Kiss A and Kuramoto Y 2009 *J. Phys. Soc. Jpn.* **78** 124702
[15] Kihou K, Shirotani I, Shimaya Y, Sekine C and Yagi T 2004 *Mater. Res. Bull.* **39** 317
[16] Miyake A, Ando I, Kagayama T, Shimizu K, Sekine C, Kihou K, and Shirotani I 2006 *J. Alloys and Compounds* **408-412** 238

Searching for a Quantum Critical Point in Rh doped ferromagnetic $Ce_{2.15}Pd_{1.95}In_{0.9}$

J.G. Sereni[1], **M. Giovannini**[2], **M. Gómez Berisso**[1], **A. Saccone**[2]

[1] Div. Bajas Temperaturas, CAB - CNEA and Conicet, 8400 S.C. de Bariloche, Argentina
[2] Dipartimento di Chimica e Chimica Industriale, Università di Genova, I-16146 Genova, Italy

E-mail: `jsereni@cab.cnea.gov.ar`

Abstract.
Low temperature magnetic and thermal (C_m) properties of the ferromagnetic (FM) alloys $Ce_{2.15}(Pd_{1-x}Rh_x)_{1.95}In_{0.9}$ were investigated in order to explore the possibility for tuning towards a quantum critical point (QCP) by doping Pd with Rh. As expected, the magnetic transition decreases from $T = 4.1$ K at $x = 0$ with increasing Rh concentration. However, the phase boundary splits into two transitions, the upper being antiferromagnetic (AF) whereas the lower FM. The AF phase boundary extrapolates to $T_N = 0$ for $x_{cr} \approx 0.65$ whereas the first order FM transition vanishes at $x \approx 0.3$. The quantum critical character of the $T_N \to 0$ point is inferred from the divergent T dependence of the tail of C_m/T observed in the $x = 0.5$ and 0.55 alloys, and the tendency to saturation of the maximum of $C_m(T_N)/T$ currently observed in exemplary Ce compounds when $T_N \to 0$. Beyond the critical concentration the unit cell volume deviates from the Vegard's law in coincidence with a strong increase of the Kondo temperature.

1. Introduction

The $Ce_{2\pm t}Pd_{2\mp u}In_{1-v}$ family of alloys shows an extended range of solid solution [1] with a peculiar magnetic behavior since the Ce-rich branch ($t > 0 > u$) behaves ferromagnetic (FM) whereas the Pd-rich ($t > 0 > u$) is antiferromagnetic (AF). Such a difference of magnetic structure under a slight variation of the alloy composition indicates that the energies of both phases are very similar. Another evidence for the competition between magnetic structures in this type of compounds is given by Ce_2Pd_2Sn [2], which shows a competition between ferro- and antiferromagnetic phases with a finite temperature critical point tuned by $25\% Ni$ doping. Consequently this family of alloys are appropriate materials for testing the stability of exotic order parameters.

For this work, we have exploited the FM character of the $Ce_{2.15}Pd_{1.95}In_{0.9}$ composition to search for a quantum critical point (QCP) [3] by tuning the chemical potential through the doping of the Pd lattice with Rh, like in the previously studied pseudo-binary compound $CePd_{1-x}Rh_x$ [4]. For such a purpose, we have investigated the low temperature properties through magnetic (M) and specific heat (C_P) measurements performed on the Rh doped $Ce_{2.15}(Pd_{1-x}Rh_x)_{1.95}In_{0.9}$.

2. Experimental and results

This alloyed system forms continuously all along the complete Pd/Rh concentration range with the Mo_2B_2Fe type crystalline structure. The samples were prepared using a standard arc melting procedure under an argon atmosphere, and remelted several times to ensure good homogeneity. The range of concentration investigated covers up to 50% of the total *Rh* by *Pd* substitution and sample characterization by EPMA and X-ray diffraction indicates that a small percentage of $Ce(Pd_{1-x}Rh_x)$ with x composition slightly higher than the bulk is present.

As reported in Ref. [1] the Ce-plane, at $z = 0.5$, contains Ce atoms placed in the crystallographic site $4h$ whereas Ce atoms exceeding stoichiometric concentration (i.e. $u = 0.15$) replace In atoms at the $2a$ site at the $z = 0$ and 1 planes. Concerning structural properties, one observes that the volume of the unit cell decreases with *Rh* content following Vegard's law up to $x = 0.3$. The main structural variation as a function of concentration is observed along the c-axis.

The low temperature behavior was investigated through thermal and magnetic parameters. Concerning specific heat measurements, in Fig. 1 we present the thermal dependence of the magnetic contribution divided by temperature (C_m/T), after phonon subtraction extracted from the non-magnetic La isotypic reference compound.

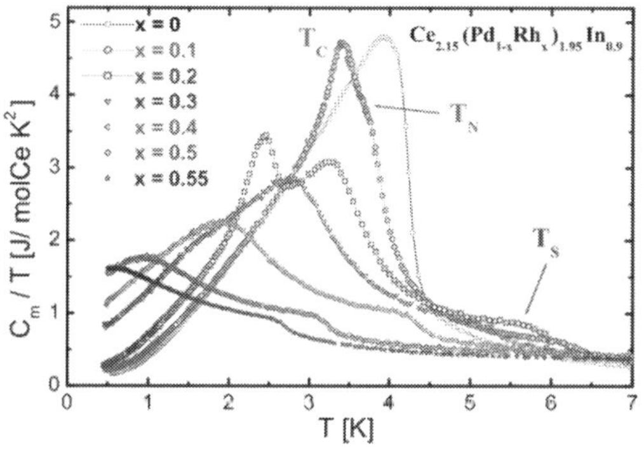

Figure 1. (Color on line) Thermal and concentration dependence of the magnetic contribution to specific heat divided temperature. T_C and T_N represent respective FM and AF transitions whereas T_S identifies a concentration dependent satellite transition discussed in the text.

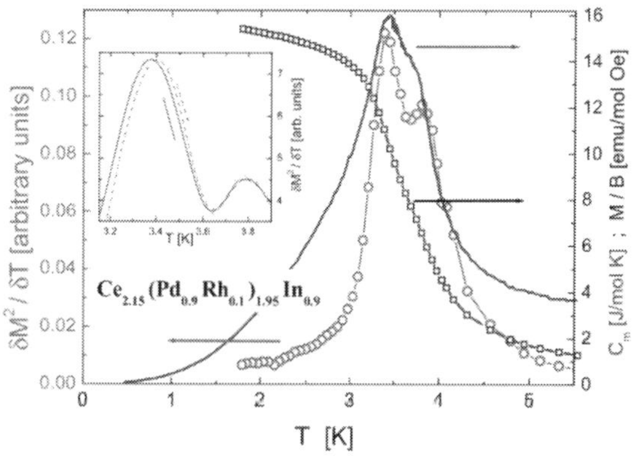

Figure 2. (Color on line) Comparison between thermal derivative of $M^2(T)$ (left axis) with $C_m(T)$ (right axis) showing the split between FM and AF transitions. The original $M(T)$ dependence is also included (right axis). Inset: hysteresis of $\partial M^2/\partial T$ between heating and cooling procedures.

As it can be seen in Fig. 1, the magnetic transition in the parent compound ($x = 0$) shows the characteristic C_m/T jump of second order character. However, a slight broadening at the maximum of C_m/T can be observed. This broadening was formerly interpreted as an intrinsic experimental effect [1]. The present results indicate that the sample at x = 0 is actually close to a bi-critical point at $T_{cr} = 4.1$ K since the phase boundary splits into two transitions by *Rh* doping. Similar feature was observed in the intensively studied $CeRu_2(Ge_{1-x}Si_x)_2$ system [5]. Also in Fig. 1 a *Rh* concentration dependent satellite anomaly ($T_S(x)$) is observed. The intrinsic character of this satellite transition is discussed below.

Magnetic $M(T)$ measurements on the $x = 0$ sample do not improve the identification of the bi-critical point because the FM signal largely exceeds the underlying AF cusp. The presence of this bi-critical point is not surprising if we consider the mentioned study of $Ce_{2\pm t}Pd_{2\mp u}In_{1-v}$ solid solutions [6]. From that study, one concludes that $T_{cr} = 4.1$ K, lies in the extrapolated values of $T_N(u)$ from the Pd-rich AF samples (i.e. $Pd \geq 2 + u$) to the Ce-rich FM-branch (i.e. $Pd \geq 2-u$). The same coincidence is observed for the paramagnetic temperature $\theta_P(u)$ extrapolated to $Pd = 1.95$ from the $Pd \geq 2$ side.

In the following we will analyze the evolution of the low temperature properties under further increase of Rh content in $Ce_{2.15}(Pd_{1-x}Rh_x)_{1.95}In_{0.9}$. The splitting between $T_N(x)$ and $T_C(x)$ becomes more pronounced in the thermal and magnetic properties of sample $x = 0.1$. That feature is observed as an incipient structure in the cusp of C_m/T shown in Fig. 1 and Fig. 2. A further evidence for the splitting of both phase boundaries is obtained from $M(T)$ measurements. Taking into account that from thermodynamic properties the internal magnetic energy U_m of a FM phase is related to the spontaneous magnetization, i.e. $U_m \propto M^2$, its temperature derivative $\partial M^2/\partial T$ is proportional to $C_m(T)$ [7]. In Fig. 2 we compare both parameters (left axis for $\partial M^2/\partial T$ and right axis for $C_m(T)$) showing that the mentioned structure at the transition is better defined by a $\partial M^2/\partial T$ versus T dependence. The same features are observed in the $x = 0.2$ sample since the transitions increase their thermal difference ($T_N = 3.3$ K and $T_C = 2.5$ K) as observed in Fig. 1 from C_m/T results. The first order character of the (lower) T_C transition is evidenced by an hysteresis in $\partial M^2/\partial T$ at the lower transition which shows a shift between heating or cooling procedures as depicted in the inset of Fig. 2. The first order character of the lower transition is in agreement with the fact that the AF order parameter developing from T_N suddenly changes to an FM structure at T_C.

Also from Fig. 1, a change of the nature of both transitions can be appreciated for $x \geq 0.3$. While $T_C(x)$ tends to vanish becoming a weak shoulder at $x = 0.3$, the $C_m(T_N)$ jump transforms into a cusp at that concentration and then into a broad maximum for $x \geq 0.4$ centered at $T_{max} =$ 1.8 K ($x = 0.4$), 0.9 K ($x = 0.5$) and 0.5 K ($x = 0.55$) respectively. Those maxima are followed at higher temperature by a large tail denouncing non-Fermi-liquid behavior [3]. Coincidentally, the value of the C_m/T maxima decrease, extrapolating to ≈ 1.4 J/molK^2 for $T_{max} \to 0$ (see Fig. 1) as currently observed in the proximity of a QCP [8]. Particularly, sample $x = 0.55$ follows a modified power law divergency $C_m/T \propto 1/(T^{1.25}+1)$ [9] once the T_S anomaly contribution is subtracted.

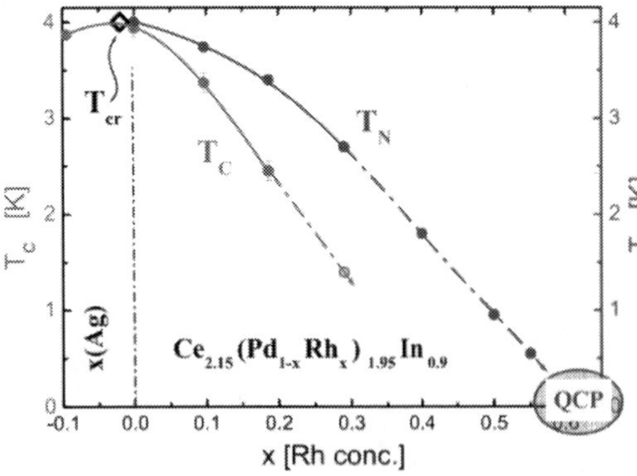

Figure 3. (Color on line) Magnetic phase diagram showing $T_N(x)$ and $T_C(x)$ phase boundaries decrease. Hollow circle in $T_C(x{=}0.3)$ corresponds to the kink in $C_m(T)/T$ of Fig. 1. Negative x values correspond to Ag doping and hollow symbol at $Ag(x{=}-0.03)$ to the bi-critical point.

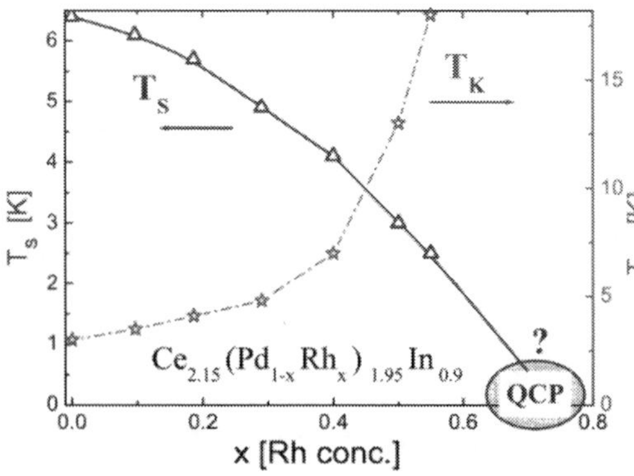

Figure 4. (Color on line) Rh concentration dependence of the satellite anomaly T_S and the $T_K(x)$ increase (right axis).

Concerning the satellite transition at $T_S(x)$, it can be mapped out to the T_C transition of the pseudo-binary compound $CePd_{1-x}Rh_x$ [4] at slightly higher ($\approx$ 10%) *Rh* concentration. The relevant features observed in this anomaly are the following: i) the height of the $C_m(T_S)$ contribution and its associated entropy corresponds to about 7% of a $CePd_{1-x}Rh_x$ phase; ii) in samples with $x \geq 0.4$, $C_m(x, T = T_S)$ nicely coincide once represented as a function of a normalized temperature $t = T/T_S$. These characteristics do not coincide with random contributions expected from spurious phases. Preliminary magnetic measurements indicate that: iii) the T_S anomaly is suppressed by the application of a moderate magnetic field ($B \approx 0.1$ T). Such a rapid suppression is in contrast with the magnetic behavior of pure $CePd_{1-x}Rh_x$ because that compound shows a robust FM phase [4]. Coincidentally, iv) the FM signal observed around $M(T_S)$ is significantly weaker that the expected from the binary compound. All these features raise the question whether the origin of the T_S anomaly is due to a spurious $CePd_{1-x}Rh_x$ phase or to the intrinsic excess of Ce atoms placed at the '2a' crystalline sites [1], with an atomic environment mimicking that of $CePd_{1-x}Rh_x$.

3. Magnetic Phase Diagrams

In Fig. 3 we present the phase diagram showing that the upper (T_N) and the lower (T_C) transitions converge into a bi-critical point at $x \to 0$. Both boundaries are well defined for $x \leq 0.2$, but they broaden and smear respectively for $x \geq 0.3$. Since Rh doping effect is expected to introduce *holes* into the conduction band, in order to confirm the existence of that critical point we have tested the possibility to move on the other direction, i.e. introducing more *electrons* into the band. For such a purpose we investigated the Ag doped alloy $Ce_{2.15}(Pd_{0.90}Ag_{0.10})_{1.95}In_{0.9}$, whose preliminary thermal and magnetic properties indicate a clear FM behavior with the consequent disappearance on the AF phase at the bi-critical point, estimated at Ag$\approx$ 0.03 and $T \approx 4$ K as indicated in Fig. 3. The FM character of the $T_C(x)$ phase boundary is proved by $M(T)$ measurements on respective samples.

Fig. 4 presents the concentration dependence of the satellite $T_S(x)$ anomaly with its extrapolation to zero at $x \approx 0.75$, and the Kondo temperature $T_K(x)$ evaluated using the Desgranges-Schotte criterion [10]. According to this criterion, T_K can be computed as the temperature at which the entropy reaches the value $S(T_K) = 2/3RLn2$. It is worth noting that the rapid increase of $T_K(x)$ coincides with the extrapolation of $T_N(x)$ to the quantum critical point and the deviation of $V(x)$ from the Vegard's law. On the contrary, the degrees of freedom involved in the $T_S(x)$ anomaly seems not to be affected by Rh increase at least up to $x = 0.55$, see Fig. 1.

Further studies at higher Rh concentration are in progress to better determine $T_S(x)$ at lower temperature and its dependence on magnetic field in elucidate the origin of that anomaly. Simultaneously, other $Ce_{2.15}(Pd_{1-x}Ag_x)_{1.95}In_{0.9}$ alloys are being investigated to better tune the electronic concentration on the bi-critical point.

References

[1] Giovannini M., Michor H., Bauer E., Hilscher G., Rogl P., Bonelli T., Fauth F., Fischer P., Herrmannsdorfer T., Keller L., Sikora W., Saccone A., Ferro R., 2000, *Phys. Rev.* B **61** 4044.
[2] Sereni J. G., Schmerber G., Gomez Berisso M., Chevalier B., Kappler J. P., 2012, *Phys. Rev.* B **85** 134404.
[3] Loehneysen H. v., Rosch A., Vojta M., Wölfle P., 2007, *Rev. Mod. Phys.* **79** 1015.
[4] Sereni J.G., Westerkamp T., Gegenward P., Canales N., Geibel C., 2007, *Phys. Rev.* B **75** 024432.
[5] Haen P., Biuod H., T. Fukuhara T., 1999, *Physica* B **259-261** 85.
[6] Sereni J.G., Giovannini M., Gomez Berisso M., Saccone A., 2011, *Phys. Rev.* B **83** 064491.
[7] See for example: Belov K.P., in *Magnetic Transitions*, Consultants Bureau Enteprisses Inc., N.Y., 1961.
[8] J.G. Sereni; arXiv 1202.1724v1 cond-matter 8 Feb.2012.
[9] Sereni J.G., 2007, *J. Low Temp. Phys.* **147** 179.
[10] Desgranges H.-U. and Schotte K.D., 1982, *Physics Letters* **91A** 240.

Magnetization processes and phase diagram of $HoRh_2Si_2$ single crystal having a component-separated magnetic transition

T Shigeoka[1], T Fujiwara[1], K Matsubayashi[2], Y Uwatoko[2]

[1]Graduate School of Science and Engineering, Yamaguchi University, Yamaguchi 753-8512, Japan.
[2]Institute for Solid State Physics, University of Tokyo, Kashiwa-shi Chiba 277-8581, Japan.

shigeoka@yamaguchi-u.ac.jp

Abstract. Thermal variations of magnetization under various magnetic fields and magnetization processes at various temperatures have been studied on a $HoRh_2Si_2$ single crystal compound. Three anomalies at T_{N1}=29.1 K, T_t=27.3 K and T_{N2}=12.0 K exist at low field in the thermal variation. The thermal dependence of M/B and the magnetization change drastically with magnetic field. Step-like metamagnetic processes appear along the all directions below T_{N2}; two-step ones appear along the [001] and [100] directions, and a one-step one appears along the [110] direction. For $T_{N2}<T<T_t$, a two-step metamagnetic one persists along the [001] direction whereas the processes along directions in the basal plane become a paramagnetic like one. The B_{001}-T magnetic phase diagram is constructed, in which there are at least five ordered phases. Two field-induced magnetic phases appear for low and high temperatures.

1. Introduction

The ternary compound $HoRh_2Si_2$, which crystallizes in the tetragonal $ThCr_2Si_2$-type structure, has an antiferromagnetic ordering below T_N=27 K and an additional transition at T_N'=11 K [1, 2]. The neutron diffraction study has reported that the antiferromagnetic structure is a simple one; the propagation vector $\boldsymbol{k}$=(0,0,1) with magnetic moments tilted from the c-axis by 28° at low temperatures and has given no information on the transition at T_N'. Takano et al. have proposed that the transition at T_N' is corresponding to the temperature, at which magnetic moments become parallel to the c-axis with increasing temperature [3].

A recent magnetic study on a single crystal compound has reported that $HoRh_2Si_2$ shows so called "a successive component-separated magnetic transition"; the magnetic moments of *c*- and *ab*-components independently order at the different temperatures, T_{N1}=29,1 K and T_{N2}=12.1 K [4]. Such a transition often appears in frustration systems. The geometrical frustration of exchange interactions between magnetic ions is responsible for most of this-type transitions [5, 6]. The similar type transition has been reported in the tetragonal compound $TbCoGa_5$ where a new mechanism, no geometrical frustration, has been proposed [7]. $HoRh_2Si_2$ shows similar behavior to one of $TbCoGa_5$. In this compound, besides ordering temperatures T_{N1} and T_{N2}, there is another transition at T_t=27.3 K whose origin is not yet known [4]. So, the detail magnetic behavior is very interesting. In this work, temperature dependences of magnetization under various magnetic fields and magnetization processes at various temperatures have been studied on the $HoRh_2Si_2$ single crystal compound and the B_{001}-T magnetic phase diagram is constructed.

2. Experimental

The single crystal $HoRh_2Si_2$ was grown by the tetra-arc Czochralski method. The single phase nature was confirmed by X-ray powder diffraction. The good quality of the single crystal was confirmed and crystallographic orientations were determined by back Laue method. The crystal was fixed on a thin plastic plate so that the desired direction is perpendicular to the plate within experimental angle accuracy of one degree. Then the PPMS and MPMS, SQUID magnetometer (Quantum Design) were used to study the magnetic property of this crystal. The measurements using MPMS were performed in the Institute for Solid State Physics, University of Tokyo.

3. Results and discussion

Figure 1 shows the thermal variations of magnetization divided by magnetic field M/B under various magnetic fields along the easy magnetization direction of [001] around the ordering temperature. Under B=0.01 T, two anomalies, a peak at T_{N1}=29.1 K and a cusp at T_t=27.3 K, can be seen, and another small anomaly also appears at T_{N2}=12.1 K as reported previously [4] though it is not shown here. There are three ordered phases under very low magnetic field. We call hereafter these the phase I ($T_{N1}>T>T_t$), the phase II ($T_t>T>T_{N2}$) and the phase III ($T_{N2}>T$). The temperature dependency changes drastically with magnetic field. On applying magnetic field any anomaly of T_{N2} was not detected while a clear cusp appears along the directions in the basal plane and a λ-type anomaly is observed for the specific heat measurement [4]. Around B=1 T with decreasing temperature, the M/B changes almost linearly above about 27 K at which a rapid decrease appears; only one anomaly is seen. Then at higher field, a change in M/B becomes small around this temperature range and a broad peak and two knees appear. To see the change clearer, the temperature dependences of M/B and its field derivative dM/dB at B=2.5 T are shown in Fig. 2. The derivative shows two clear peaks and a change of sign which correspond to two knees and a peak in the M/B curve, respectively. Above B=3 T, a peak and a rapid decrease appear; the knee at lower temperature disappears. This drastic change indicates an existence of field-induced magnetic phase near the ordering temperature.

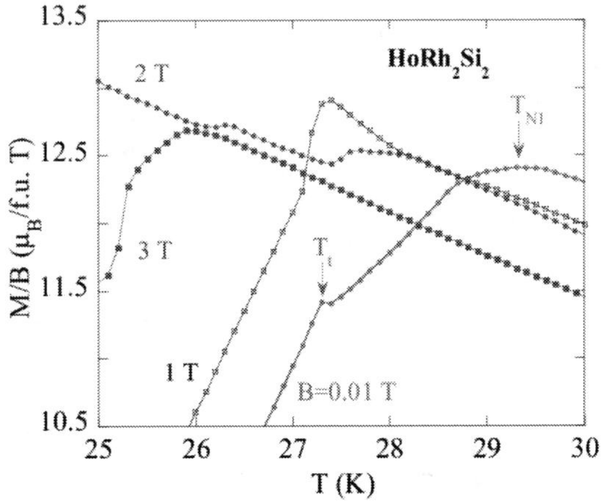

Figure 1 The thermal variations of magnetization divided by magnetic field M/B under various magnetic fields along the [001] direction around the ordering temperature on the $HoRh_2Si_2$ single crystal.

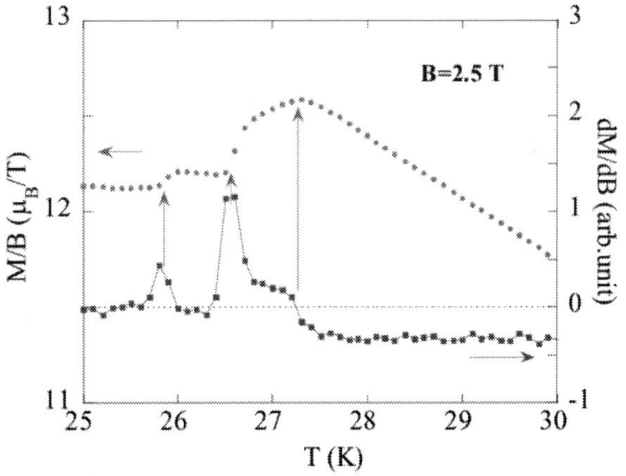

Figure 2 Temperature dependence of M/B and its field derivative dM/dB at 2.5 T along the [001] direction.

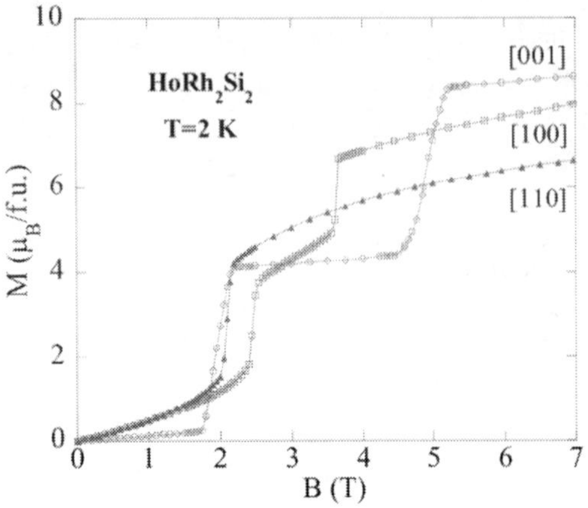

Figure 3 Magnetization curves at 1.5 K along the main symmetry axes on the $HoRh_2Si_2$ single crystal.

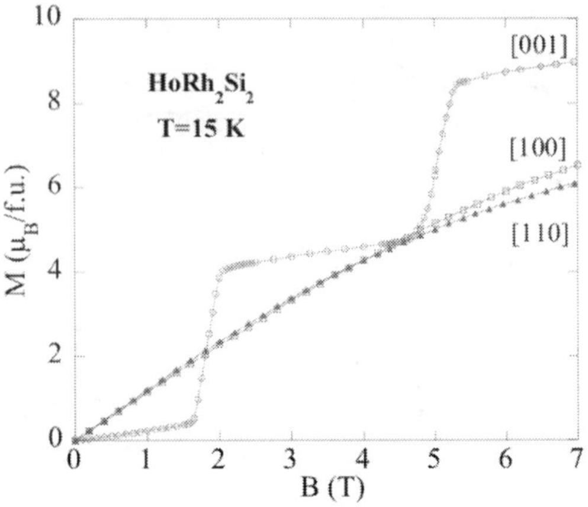

Figure 4 Magnetization curves at 15 K.

The magnetization processes at various temperatures have been investigated to obtain information on the magnetic phases. As an example, the processes at 2 K: $T< T_{N2}$ and 15 K: $T_{N2}<T< T_t$ are shown in Figs. 3 and 4, respectively. Sharp step-like metamagnetic transitions appear along the all directions below T_{N2}: for the phase III, which indicates that magnetic moments tilt from the main symmetry axes. The [001] magnetization process is a two-step metamagnetic one; the magnetization increases rapidly at critical fields B_{c1}=1.9 T and B_{c2}=4.9 T. The magnetization curve is almost linear with a slight slope except around transition fields. The magnetization just above B_{c1} and B_{c2} is M_1=4.1 μ_B/f.u. and M_2=8.3 μ_B/f.u., respectively. The M_2 is almost two times as large as M_1. If a Ho atom in this compound has the theoretical Ho^{3+} free ion moment of 10 μ_B, the tilt angle of magnetic moments from the c-axis can be deduced from M_2 to be 34° which is nearly equal to one estimated by neutron study [1]. The magnetization process can be speculated as follow; the simple antiferromagnetic structure: +-+- below B_{c1} changes a ferrimagnetic one: +++- above B_{c1} and a ferromagnetic one: ++++ above B_{c2} keeping the tilt angle of magnetic moments. The [100] magnetization process is also a two-step one with critical magnetic fields at B_{a1}=2.5 T and B_{a2}=3.6 T. The magnetization jumps of 1.9 μ_B/f.u. and 1.8 μ_B/f.u. occur at B_{a1} and B_{a2}, respectively. The magnetization just above B_{a2} is 6.7 μ_B, deducing the tilt angle of 48° from the [100] direction. The slope of the magnetization curve for $B_{c1}<B< B_{c2}$ is larger than one of the [001] direction, indicating this field-induced phase is less stable than one of [001]. A one-step metamagnetic process appears along the [110] directions; the critical field is B_{b1}=2.1 T and magnetization above the critical field is 4.2 μ_B/f.u. The magnetization curves along the directions in the basal plane have a downward curvature for low fields and an upward curvature for high fields, which is different from one of the [001] direction. For $T_{N2}<T<T_t$: the phase II, the two-step metamagnetic process along the c-axis remains whereas metamagnetic ones disappear along the basal plane directions as shown in Fig.4. The magnetization just above B_{c2} is 8.5 μ_B/f.u. which is larger than one at 2K in spite of higher temperature. This indicates the tilt angle of magnetic moments from the c-axis become smaller; the moments become parallel to the c-axis at T_{N2}. This is also supported by our preliminary neutron diffraction experiment where antiferromagnetic reflections (00*L*) (*L*=odd numbers) disappear above T_{N2}, indicating that magnetic moments are along the *c*-axis. The details will be published soon elsewhere. The processes along the basal plane directions become paramagnetic-like with a considerably large gradient. It should be suggested that the phase II is a partial order of magnetic component; the magnetic moments of *c*-component order while *ab*-components disorder [4]. This behaviour is not inconsistent with the suggestion and may come from a frustration. The metamagnetic transition in the [001] process smears with increasing temperature and above T_t was not

observed. Any anomaly was hardly detected along the all directions; the processes become paramagnetic-like ones. Above T_{N1}, the processes along all directions become a paramagnetic one.

The B_{001}-T magnetic phase diagram was constructed from temperature and magnetic field dependences of magnetization and is shown in Fig. 5. There are at least five phases in the ordered state.

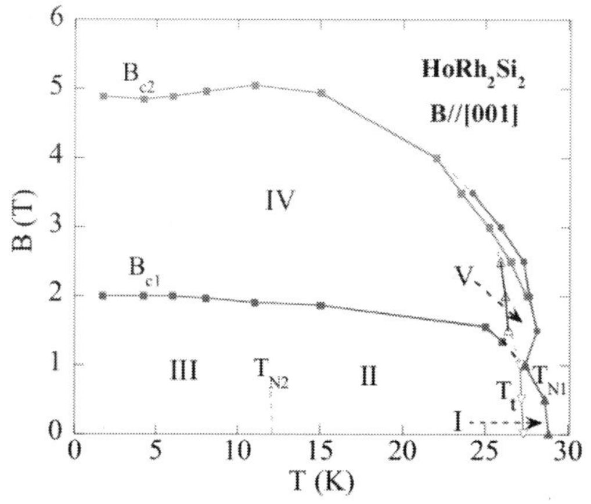

Figure 5 B_{001}-T magnetic phase diagram of $HoRh_2Si_2$. The marks■ below 20 K and ■are determined from the field dependence of magnetization: critical fields of metamagnetic transition B_{C1} and B_{c2}.
The others are from the temperature dependence of M/B and dM/dB. ●: peak temperatures, △ and ■: temperatures of rapid change; that is, peak temperatures of dM/dB. ▲: peak temperatures for lower magnetic fields: T_{N1}. ▽: temperatures of cusps: T_t.

Two field-induced phases are referred to as the phase IV and V for low and high temperatures, respectively. Peak temperatures in the M/B vs. T curves as shown in Fig. 2 are plotted by closed circles together with temperatures of rapid decrease by squares though it is ambiguous whether there are one or two transitions. The field variation of transition temperature shown by closed circles, boundary between the phase V and the paramagnetic phase, shows a re-entrant behaviour. It is strange behaviours, indicating competitions of some interactions such as quadrupolar interaction. The phase boundary between the phase II and III which should be connected to T_{N2} is not determined because any anomaly was hardly detected around T_{N2} by magnetic measurements for the [001] directions. This shows that a significant change of ordering at T_{N2} occurs only in the basal plane, which is consistent with a partial order state of the phase II; The magnetic moments of *ab*-component order below T_{N2} while the *c*-component ones have already ordered below T_{N1}, suggesting a quadrupole interaction play a important role and a frustration occurs for $T_{N2}<T<T_{N1}$. No geometrical frustration is considered from an arrangement of Ho atoms in this compound. So we believe that $HoRh_2Si_2$ is a candidate for a new type frustration. Further studies, such as elastic constant and neutron measurements are now underway to map out these interesting behaviours

References

[1] Felner I, Nowik I 1984 *J. Phys. Chem. Solids* **45** 419.
[2] Sekizawa K, Takano Y, Takigami H, Takahashi Y 1987 *J. Less-Common Metals* **127** 99.
[3] Takano Y, Ohhata K, Sekizawa K 1987 *J. Magn. Magn. Mater.* **66** 187.
[4] Shigeoka T, Fujiwara T, Munakata K, Matsubayashi K. Uwatoko Y 2011 *J. Phys. Conf. Ser.* **273** 012127.
[5] Achiwa A 1969 *J. Phys. Soc. Jpn.* **27** 561. .
[6] Watanuki R, Sato G, Suzuki K, Ishihara M, Yanagisawa T, Nemoto Y, Goto T 2005 *J. Phys. Soc. Jpn.***74** 2169.
[7] Sanada N, Watanuki R, Suzuki K, Akatsu M, Sakakibara T 2009 *J. Phys. Soc. Jpn.* **78** 073709.

Anisotropic Form Factors of Neutron Scattering by Magnetic Octupole in CeB_6

Ryousuke Shiina

Department of Materials Science and Technology, Niigata University, Niigata 950-2181, JAPAN

E-mail: `shiina@eng.niigata-u.ac.jp`

Abstract. The form factor of elastic neutron scattering is calculated for the quadrupole order phase of CeB_6 in magnetic fields. It is shown that the dependence of the form factor on the direction of neutron momentum transfer is very small for the Bragg reflections, whereas the scattering due to field-induced octupoles gives rise to a significant anisotropy of the form factor for the super-lattice reflections. These results are discussed in quantitative comparison with recent exprimental results. The similarity and dissimilarity with the results for phase IV in $Ce_{1-x}La_xB_6$ is also discussed.

1. Introduction

Recently, the multipole effects in f electron systems have been widely studied in relation with their anomalous ordered phases. A well-known example having competing interactions of high-rank multipoles is CeB_6, which exhibits an antiferro-quadrupole (AFQ) order phase at low temperatures. It is established that some unusual properties of the AFQ phase in magnetic fields originate from large influences of field-induced magnetic octupoles.[1, 2] A number of proposals on other materials after clarifying the AFQ problem in CeB_6 have been found in recent review articles such as [3, 4]. Among them, it is quite remarkable that the realizations of pure octupole phases at zero magnetic field are promising for $Ce_{1-x}La_xB_6$ ($x < 0.8$) and NpO_2.[5, 6, 7]

In general, however, the direct observation of high-rank multipoles and the quantitative estimation of their magnitudes remain to be a difficult problem. In this paper, we report theoretical results on the magnetic form factor of elastic neutron scattering for the AFQ phase of CeB_6 in magnetic fields. It is shown that the form factor as a function of neutron momentum transfer becomes very anisotropic for the super-lattice reflections, under an influence of the field-induced octupoles. The comparison with the recent experimental results is discussed.

2. General Aspects of Magnetic Form Factor

In neutron scattering by a periodic array of magnetic ions, the differential cross section for the momentum transfer $\boldsymbol{\kappa} = \boldsymbol{k}' - \boldsymbol{k}$ of neutrons is expressed as [8]

$$\frac{d^2\sigma}{d\Omega dE'} \propto \sum_{n,n'} P_n |\langle n'|\mathbf{Q}_{\rm t}^{\perp}(\boldsymbol{\kappa})|n\rangle|^2 \delta(\hbar\omega + E_n - E_{n'}). \quad (1)$$

Here, $n(n')$ denotes an initial (final) state of electrons with energies $E_{n(n')}$, and the probability for the initial state is represented by P_n. In this paper, we study the elastic scattering with $n' = n$ for the ground state of magnetic ions. It is assumed that the ground state consists of fixed-J multiplets with J being the total angular momentum of f electrons.

The full informations on the interaction between neutrons and magnetic ions are involved in the total scattering operator $\mathbf{Q}_t^{\perp}$ which corresponds to a lattice sum of local operators $\mathbf{Q}_i^{\perp}$ with the site index i. In order to clarify the physical meaning, however, it is more useful to consider local unprojected operators $\mathbf{Q}_i$ defined in $\mathbf{Q}_i^{\perp} = \boldsymbol{\kappa} \times (\mathbf{Q}_i \times \boldsymbol{\kappa})$. In particular, it is well known that $\mathbf{Q}_i \sim \mathbf{Q}_i^{\mathrm{D}} = \frac{g}{2}F(\kappa)\mathbf{J}_i$ in the long wave-length region $\boldsymbol{\kappa} \sim 0$, where $\mathbf{J}$ represents the dipole operator of f electrons with g being the Landé g factor. Since F is a decreasing function of $\kappa = |\boldsymbol{\kappa}|$, the dominant contribution usually comes from scattering by the dipoles. In order to study multipole orders, however, one should go beyond such a dipole approximation. Since all the operators in the J basis must be reduced to a set of tensor operators, the correction to the dipole approximation is generally given by a linear combination of odd-rank multipoles. Therefore, omitting the site index i, the m component of the unprojected scattering operator $\mathbf{Q}$ can be expressed as

$$Q_m(\boldsymbol{\kappa}) = Q_m^{\mathrm{D}}(\boldsymbol{\kappa}) + \sum_n C_{mn}(\boldsymbol{\kappa})X_n, \quad (2)$$

where X_n denotes a component of higher-rank magnetic multipole operators and $C_{mn}(\boldsymbol{\kappa})$ are the expansion coefficients.

As we have shown in ref. [9], this mixture of multipoles can be predicted in terms of the symmetry of the scattering process. A basis of the symmetry analysis is the fact that the cross section must be invariant under transformations that do not change either crystal and momentum transfer. In the case of cubic CeB_6, the group theory results in $\mathbf{Q} = 0$ at the superlattice spots for the following octupole orders: T_z^{β} and T_{xyz} for $\boldsymbol{\kappa}\,||(0,0,1)$, $T_x^{\beta}+T_y^{\beta}+T_z^{\beta}$ for $\boldsymbol{\kappa}\,||(1,1,1)$ and T_z^{β} for $\boldsymbol{\kappa}\,||(1,1,0)$. Note that the symbols for the multipole operators classified by the symmetry are defined in terms of the angular momentum operator of $J = 5/2$ in ref. [1]. In addition, one can find that $\mathbf{Q}$ should be in parallel to $\boldsymbol{\kappa}$ in some cases, such as T_z^{α} for $\boldsymbol{\kappa}\,||(0,0,1)$. Since the relevant quantity in the scattering is the component of $\mathbf{Q}$ projected onto the $\boldsymbol{\kappa}$ plane, the scattering is forbidden in all of the above cases.

3. Calculated Results

The component of the AFQ order parameter of CeB_6 is believed to be O_{xy} for $H||(001)$, $O_{yz}+O_{zx}+O_{xy}$ for $H||(111)$, $O_{yz}+O_{zx}$ for $H||(110)$, all of which induces strongly an octupole T_{xyz}. [2, 3, 4, 10] We consider these orders in the framework of the quartet Γ_8 as the crystal field ground state of $4f$ electrons. The ordering wave vector is $\boldsymbol{k} = (1/2, 1/2, 1/2)$ in the simple-cubic lattice of the Ce ions. Note that the AFQ ground state inevitably leaves a Kramers degeneracy, which should be lifted by an infinitesimal field strength in the model analysis. We have used for the present calculation such a state with lifting the degeneracy, which can be comparable to the experimental results for 5~10T at around 1K. See ref. [2] for the details of multipolar interactions and the mean-field theory.

First, the numerical results for the Bragg reflections in the AFQ phase in $H||(001)$ are presented as continuous functions of the momentum transfer of neutrons in Fig. 1, which are compared with the results for the normal phase in the same figure. It is shown that the momentum dependence of the form factor in the AFQ phase is of a quite conventional form with little anisotropy, which is almost equivalent to that in the normal phase. In fact, the observation of the form factor has been reported in the literatures and is consistent with this calculation.[11, 12] Note that since the available experiments were performed at higher temperatures, the thermal fluctuation should be taken into account for the quantitative comparison.

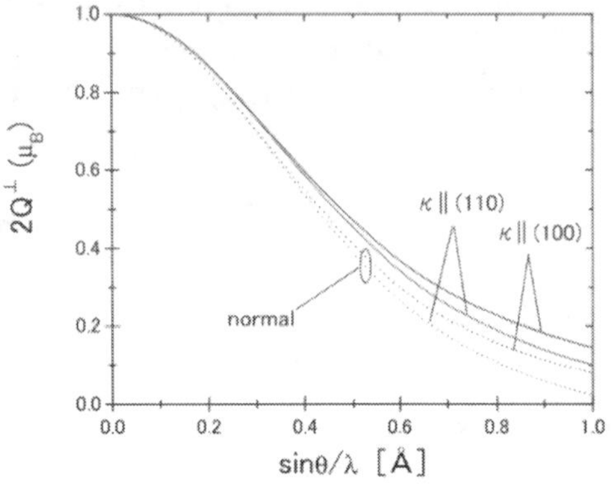

Figure 1. Magnetic form factor for the Bragg spots as a function of neutron momentum transfer $\kappa/4\pi = \sin\theta/\lambda$. The blue and red solid lines are the results for the AFQ phase in $\boldsymbol{H}\,||(001)$. The dashed lines are the corresponding results for the normal phase which is scaled by the saturation value of the magnetization $M_z = 11/7\mu_{\rm B}$.

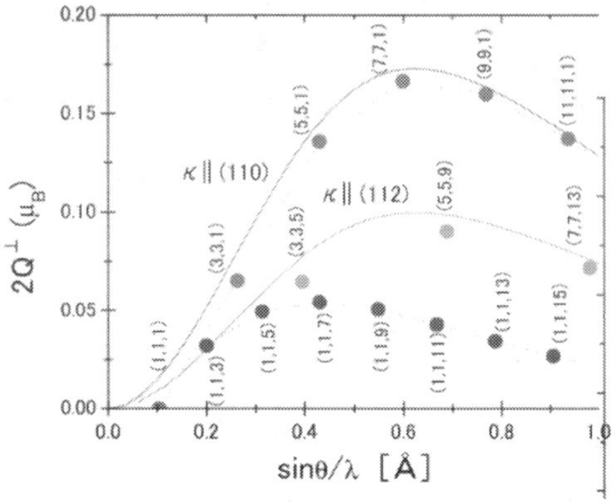

Figure 2. Magnetic form factor for the super-lattice reflections as a function of neutron momentum transfer. The dots are the results at the super-lattice spots for the AFQ phase in $\boldsymbol{H}\,||(001)$. The indices (l, m, n) for the dots represent $\boldsymbol{\kappa} = (l/2, m/2, n/2)$. The solid lines are the results for the limit of high symmetry directions.

The calculated results for the superlattice spots in the AFQ phase in $H||(001)$ are plotted in Fig. 2. It is apparent that the form factor increases with κ in the region of small momentum transfer and takes a maximum around $\sin\theta/\lambda \sim 0.6\AA$, with the intensity $2Q^{\perp} \sim 0.17\mu_B$. The qualitative features are consistent with previous calculations.[13, 14] We should also remark the anisotropy for the direction of momentum transfer. It is clear that $Q^{\perp}$ for a series of $\boldsymbol{\kappa} \propto (m, m, 1)$ have large values compared with others. In particular, we confirmed that the form factor for (m, m, m) vanishes, irrespective of m. This is nothing but the symmetry restriction discussed in §2, which characterizes the scattering by T_{xyz}. We have also calculated the form factor in other field directions and have found characteristic momentum dependences and anisotropies reflecting dipole and octupole ordering induced by the fields, details of which will be published elsewhere.

Quite recently, a neutron scattering experiment for phase II in $H||(001)$ has been carried out by Kuwahara *et al.*.[15] It shows that the form factor is enhanced as κ increases. The maximum appears around $\sin\theta/\lambda \sim 0.5\AA$ with the value slightly less than 0.2 μ_B. In addition, the distinct anisotropy is observed, which indicates that intensities are large for $\boldsymbol{\kappa}$ close to the direction $(1, 1, 0)$ and almost vanishing for the direction (1, 1, 1). These features are even quantitatively consistent with the theoretical results. The agreements clearly show that the octupole T_{xyz} has an almost saturation value in a weak magnetic field ($\sim 5T$), giving a strong support to the multipole RKKY model in ref. [2].

Finally, it is useful to compare the above results with the form factor in phase IV, which appears in La doped CeB_6. As shown in ref. [9], momentum dependence of the form factor and the absolute value of the intensities calculated for phase IV at zero field are quite similar to the present results for phase II. Note, however, that the anisotropy for the momentum transfer is completely different. It is shown that the largest intensity in phase IV appears for $\boldsymbol{\kappa}\,||(1, 1, 1)$, whereas in phase II, high intensities are obtained for $\boldsymbol{\kappa}$ close to the direction (1, 1, 0). Such a difference in the anisotropy of the form factor in the same material clearly reflects the realizations of different octupole orders in phases II and IV.

4. Summary

The magnetic form factor for elastic neutron scattering due to field-induced octupolar order in the AFQ phase of CeB_6 was studied theoretically. First, we reviewed the result of the symmetry properties on the basis of the general framework of neutron scattering due to octupoles. Then, we presented the numerical results of the form factor whose dependence on the neutron momentum transfers exhibits a significant anisotropy for the super-lattice reflections. It was reported that the intensities and the anisotropy of the calculated results account well for the recent experimental observations by Kuwahara *et al.*. The similarity and difference in the form factors for phases II and IV was also discussed. The present study provides a typical example of an identification of the symmetry and the magnitude of hidden octupole moments by neutron scattering. This method is in principle applicable to any other materials showing the multipole ordering.

5. Acknowledgments

The author would like to acknowledge K. Kuwahara for valuable discussion on the experimental studies and H. Shiba and O. Sakai for collaboration on the initial stage of this study. The author is supported financially by a Grant-in-Aid for Scientific Research on Innovative Areas "Heavy Electrons" (No. 21102520) from MEXT and a Grant-in-Aid for Scientific Research C (No. 21540368) from JSPS.

References

[1] R. Shiina, H. Shiba and P. Thalmeier: J. Phys. Soc. Jpn. **66** (1997) 1741.
[2] R. Shiina, O. Sakai, H. Shiba and P. Thalmeier: J. Phys. Soc. Jpn. **67** (1998) 941.
[3] Y. Kuramoto, H. Kusunose, and A. Kiss: J. Phys. Soc. Jpn. **78** (2009) 072001.
[4] P. Santini, S. Carretta, G. Amoretti, R. Caciuffo, N. Magnani, and G. H. Lander: Rev. Mod. Phys. 81 (2009) 807.
[5] D. Mannix, Y. Tanaka, D. Carbone, N. Bernhoeft, and S. Kunii: Phys. Rev. Lett. **95** (2005) 117206.
[6] K. Kuwahara, K. Iwasa, M. Kohgi, N. Aso, M. Sera, and F. Iga: J. Phys. Soc. Jpn. **76** (2007) 093702.
[7] J. A. Paixão, C. Detlefs, M. J. Longfield, R. Caciuffo, P. Santini, N. Bernhoeft, J. Rebizant, and G. H. Lander: Phys. Rev. Lett. **89** (2002) 187202.
[8] See for example: S. W. Lovesey: *Theory of Neutron Scattering from Condensed Matter* (Oxford University Press, 1984).
[9] R. Shiina, O. Sakai, and H. Shiba: J. Phys. Soc. Jpn. **76** (2007) 094702.
[10] T. Matsumura, T. Yonemura, K. Kunimori, M. Sera, and F. Iga: Phys, Rev. Lett. **103** (2009) 017203.
[11] P. Burlet, J. X. Boucherle, J. Rossat-Mignod, J. W. Cable, W. C. Koehler, S. Kunii and T. Kasuya: J. Phys. Paris **C 7** (1982) 273.
[12] F. Givord, J.-X. Boucherle, P. Burlet, B. Gillon and S. Kunii: J. Phys. Condens. Matter **15** (2003) 3095.
[13] K. Hanzawa: J. Phys. Soc. Jpn. **70** (2001) 1900.
[14] P. Santini and G. Amoretti: J. Phys. Soc. Jpn. **71** (2002) Suppl. 11.
[15] K. Kuwahara *et al.*: in preparation.

Anomalous Upper Critical Field in the Heavy-Fermion Superconductor UBe_{13} Studied by DC Magnetization Measurements

Yusei SHIMIZU[1,†], **Yoshinori HAGA**[2], **Kenichi TENYA**[3], **Tatsuya YANAGISAWA**[4], **Hiroyuki HIDAKA**[1], **and Hiroshi AMITSUKA**[1]

[1]Graduate School of Science, Hokkaido University, Sapporo 060-0810, Japan
[2]Advanced Science Research Center, Japan Atomic Energy Agency, Tokai 319-1195, Japan
[3]Faculty of Education, Shinshu University, Nagano 390-8621, Japan
[4]Creative Research Institution, Hokkaido University, Sapporo 001-0021, Japan

E-mail: †yusei@phys.sci.hokudai.ac.jp

Abstract. We have studied anomalous upper critical field $B_{c2}(T)$ of a single crystal of UBe_{13} by DC magnetization measurements down to 0.14 K and up to 8 T. We have observed a broad anomaly and a peak effect below B_{c2} in magnetization curves. Although a possibility of the Fulde-Ferrell-Larkin-Ovchinnikov (FFLO) state has been proposed as an origin of the unusual upper critical field by previous works, our results have shown that thermal equilibrium magnetization curves reveal no step-like behavior, which could be observed at 1st-order phase transition from a superconducting mixed state to the FFLO state. We have also investigated anisotropy of B_{c2} for $B \parallel$ <001> and $B \parallel$ <110>, and found that B_{c2} for $B \parallel$ <001> is larger than B_{c2} for $B \parallel$ <110> below $\sim$ 0.5 K. These results might indicate that the anomalous upward curvature of B_{c2} in UBe_{13} is related to anisotropy of B_{c2} rather than the FFLO state.

1. Introduction

The heavy-fermion system UBe_{13} is one of the most exotic superconducting materials in uranium compounds. After the discovery of this unconventional superconductivity by Ott *et al.* [1], it has been shown that this material has an anisotropic superconducting gap structure by specific heat $C(T)$ [2], nuclear magnetic resonance (NMR) spin-relaxation rate $1/T_1$ [3], and ultrasonic attenuation $\alpha(T)$ [4], and magnetic field penetration depth $\lambda(T)$ [5]. However, in fact, no consensus has been obtained about the superconducting gap structure of UBe_{13}. As for the parity of Cooper pairing, it has been said that Cooper pairing state of UBe_{13} is triplet because Knight shift does not decrease below T_c [6]. However, a recent muon spin rotation (μ^+SR) study for $B \parallel < 001 >$ has shown that Knight shift decreases below T_c, and recently Tou *et al.* have reported that NMR Knight shift exhibits an anisotropic behavior below T_A, which is lower than T_c. Therefore, the parity of superconductivity in UBe_{13} remains unclear even now.

UBe_{13} reveals an anomalous upper critical field curvature B_{c2} which can not be explained in a framework of the Bardeen-Cooper-Schrieffer (BCS) theory. First, the B_{c2} of UBe_{13} is much larger than the Pauli limit $\mu_0 H_P(T=0)$ of 1.84 T_c T ($\sim$ 1.66 T for $T_c = 0.9$ K) [9]. On the other hand, the value of B_{c2} for UBe_{13} estimated experimentally by extrapolation to $T = 0$ is 8-13 T. Second,

$B_{c2}(T)$ of UBe_{13} shows an upturn-like behavior at around $T_c/2$. So far, several explanations for this unusual B_{c2} curve have been proposed, including two-coexisting superconducting order parameters [10], a recovering of coherent Fermi-liquid state in the normal state [11], and the Fulde-Ferrell-Larkin-Ovchinnikov (FFLO) state [12]. In particular, the FFLO state, in which a Cooper pair has a non-zero momentum $\boldsymbol{q}$, is generally possible in a singlet superconductor with a strong paramagnetic effect [13]. Therefore, it is noteworthy to clarify whether the unusual upward curvature of UBe_{13} is due to the appearance of the FFLO state or not. In addition, anisotropy of B_{c2} in UBe_{13} is also important to clarify the superconducting symmetry and still in discussion [14, 15, 16].

In the present work, we have performed low temperature DC magnetization measurements for a single crystal of UBe_{13} in order to elucidate the origin of anomalous upper critical field B_{c2} of UBe_{13}. We have also investigated anisotropy of H_{c2} for $B \parallel < 001 >$ and $B \parallel < 110 >$.

2. Experimental Procedures

A single crystal of UBe_{13} was grown by an Al-flux method reported in [17]. Previous low-temperature DC magnetization [18] and specific-heat measurements [16] were measured for the same single crystal. The superconducting transition temperature of the sample is 0.81 K from specific-heat measurement in zero field obtained by entropy conservation analysis [16]. The static DC magnetization measurements were carried out by using a capacitive Faraday-force magnetometer installed in a ^{3}He-^{4}He dilution refrigerator [19] down to $\sim$ 0.14 K. The measurements have been done in a field gradient of 5-9 T/m applied in addition to the uniform magnetic field up to $\sim$ 8 T. Since the dimension of the used sample is $1.8\times1.9\times0.8$ mm^3, distribution of the magnetic field due to the field gradient is estimated to be less than $\sim$ 18 mT in a field gradient of 9 T/m.

3. Results and Discussion

Figure 1 shows the magnetization curves of UBe_{13} in the superconducting state at 0.18 K for $B \parallel < 001 >$. Up and down arrows denote increasing and decreasing processes, respectively. An irreversibility observed in low-field region below $\sim$ 2 T is due to a vortex flux pinning, which is typically seen in type-II superconductors. $B_{c2}(= \mu_0 H_{c2})$ can be defined as a field where the hysteresis of magnetization vanishes completely. Almost linear component of above B_{c2} is the normal state magnetization; there is a small contribution of non-linear susceptibility, $(1/3!)\chi_3 B^3$, as well as that of linear susceptibility, $\chi_1 B$. Assuming that the normal-state susceptibility in the superconducting mixed state due to vortices is the same as that in the normal state above B_{c2}, we obtain the contribution of superconducting diamagnetic magnetization by subtracting the normal-state magnetization. Figure 2 shows a diamagnetic magnetization curves in the mixed state at 0.18 K for $B \parallel <001>$. Since the hysteresis of magnetization is small in our sample, we can obtain a thermal equilibrium magnetization M_{eq}^{SC} (Red Marker) by averaging increasing and decreasing processes. We find that the thermal equilibrium magnetization process reveals a minimum in the mixed state. In other words, the absolute value of superconducting diamagnetic magnetization exhibits a maximum in the mixed state. In this work, we define B^*_{Mag} as the field where the diamagnetic response shows the maximum. The temperature dependence of B^*_{Mag} is similar to that of $B^*(T)$, which is observed in a magnetic field dependence of the specific heat and temperature dependence of thermal expansion [20]. This magnetization curve is extremely anomalous because an absolute value of magnetization in a type-II superconductor exhibit a maximum at lower critical field B_{c1} and then decreases with increasing magnetic field monotonically in general. We have observed this maximum of diamagnetic response also for $B \parallel < 110 >$, and B^*_{Mag} for $B \parallel < 110 >$ is larger than that for $B \parallel < 001 >$. Further study for this broad anomaly will be reported in another paper.

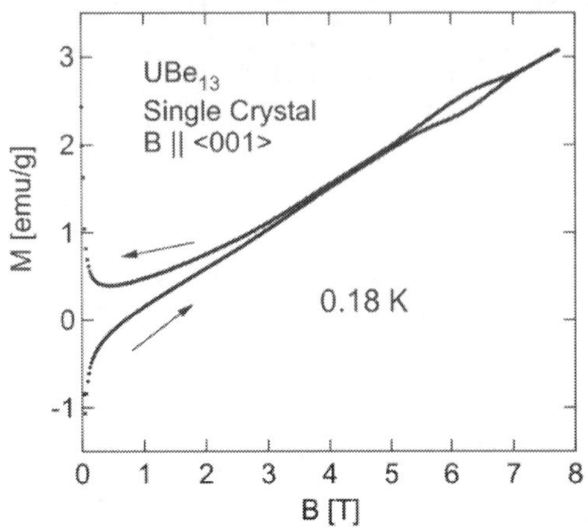

Figure 1. Magnetization curves of single crystal of UBe_{13} at 0.18 K for B || <001>. Up and down arrows denote increasing and decreasing process, respectively.

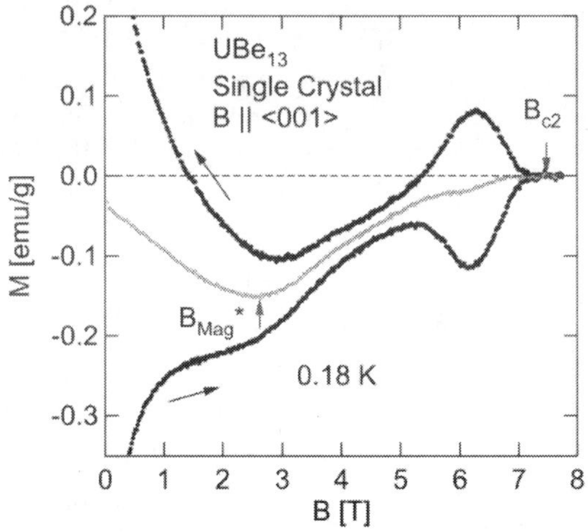

Figure 2. Superconducting diamagnetic magnetization curves at 0.18 K for B || <001> subtracted the normal state magnetization. Red marker indicates the thermal equilibrium magnetization.

Next, we remark about a possibility of the FFLO state in UBe_{13}. In a type-II singlet superconductor with a strong paramagnetic effect, the FFLO state could occur in high-field region of superconducting phase. In this case, its superconducting diamagnetic curve should exhibit a step-like behavior due to a 1st-order phase transition from the mixed state to the FFLO state [21]. However, we could not observe such a step-like behavior in the magnetization curves. Therefore, the results of our magnetization measurements do not support a possibility of the FFLO state in UBe_{13}. The FFLO state due to a strong paramagnetic effect occurs usually for even-parity Cooper pairs which are formed by anti-parallel spins. Therefore, our results are consistent with the possibility of an odd-parity (triplet) superconductor in which Cooper pairs have parallel-spin parings.

Another key to solve the unusual superconductivity in UBe_{13} is anisotropy of B_{c2}. Figure 3 (a) and (b) shows magnetization curves at 0.50 K and 0.18 K for B || <001> and B || <110> in high-field regions, respectively. The B_{c2} is almost isotropic within the experimental error at $\sim$ 0.5 K. This is consistent with the results of our previous specific-heat measurements [15, 16]. However, at 0.18 K, we have found an anisotropic B_{c2} in UBe_{13} between B_{c2} for $B \parallel < 001 >$ and that for $B \parallel < 110 >$. This anisotropy begins to appear below 0.4-0.5 K, and becomes larger with decreasing temperature. Figure 4 shows the temperature dependence of the difference between $B_{c2}^{<001>}$ and $B_{c2}^{<110>}$. Here, we define ΔB_{c2} as $B_{c2}^{<001>} - B_{c2}^{<110>}$. This anisotropy of B_{c2} is sufficiently larger than the error of the magnetic field due to distribution of gradient field. This anisotropy might be associated with the upturn of the B_{c2} in UBe_{13}, which is observed at around $T_c/2$. It is needed to understand why the upper critical field of UBe_{13} is anisotropic only at low-temperature in order to elucidate the origin of anomalous upward upper critical field and its superconducting symmetry.

4. Conclusion

We have performed static DC magnetization measurements for a single crystal of UBe_{13} and observed the broad anomaly in the superconducting mixed state and the peak effect just below

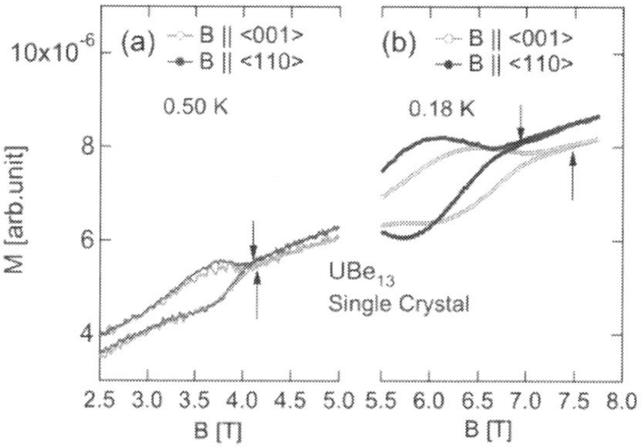

Figure 3. Magnetization curves at (a) 0.18 K, and (b) 0.50 K for B || <001>. Up and down arrows denote the B_{c2} for B || <001> and B || <110>, respectively.

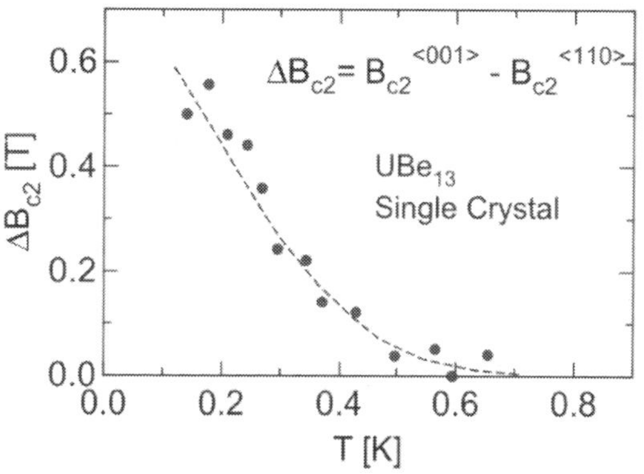

Figure 4. Difference between the upper critical field $B_{c2}^{<001>}$ and $B_{c2}^{<110>}$. We define ΔB_{c2} as $B_{c2}^{<001>} - B_{c2}^{<110>}$. The dashed line is guide to the eye.

B_{c2}. The thermal equilibrium magnetization reveals no step-like behavior in the superconducting phase. On the other hand, the anisotropy of B_{c2} for between B || <001> and B || <110> has been observed only below ~ 0.5 K, where $B_{c2}{}^{<001>}$ is larger than $B_{c2}{}^{<110>}$. These facts indicate that the upward curvature of $B_{c2}(T)$ in UBe_{13} is probably originated from the low-temperature anisotropy of B_{c2} rather than the occurrence of the FFLO state.

Acknowledgments

One of authors (YS) would like to thank H. Tou for valuable discussions. The present work was supported by Grant-in-Aid for Scientific Research on Innovative Areas 'Heavy Electrons' (20102002) and Scientific Research B (19340086), and S (20224015) from the Ministry of Education, Culture, Sports, Science and Technology, Japan.

References

[1] Ott H R *et al.*,1983 *Phys. Rev. Lett.* **50** 1595
[2] Ott H R *et al.*, 1984 *Phys. Rev. Lett.* **52** 1915
[3] MacLaughlin D E *et al.*, 1984 *Phys. Rev. Lett.* **53** 1833
[4] Golding B *et al.*, 1985 *Phys. Rev. Lett.* **55** 2479
[5] Einzel D *et al.*, 1986 *Phys. Rev. Lett.* **56** 2513
[6] Tien C, and Jiang I M 1989 *Phys. Rev. B* **40** 229
[7] Sonier J E *et al.*, 2003 *Physica B* **326** 414
[8] Tou *et al.* private communications.
[9] Maple M B *et al.*, 1985 *Phys. Rev. Lett.* **54** 477
[10] Rauchschwalbe U *et al.*, 1987 *Europhys. Lett.* **3** 757
[11] Tachiki M *et al.*, 1985 *Physica* **135B** 57
[12] Thomas F *et al.*, 1996 *J. Low. Temp. Phys.* **102** 117
[13] Fulde P, and Ferrell A 1964 *Phys. Rev.* **135** 550
[14] Aliev F G *et al.*, 1991 *J. Low. Temp. Phys.* **85** 359
[15] Signore P J C *et al.*, 1995 *Phys. Rev. B* **52** 10315
[16] Shimizu Y *et al.*, 2011 *J. Phys. Soc. Jpn.* **80** 093701
[17] Haga Y *et al.*, 1999 *Physica B* **259** 627
[18] Shimizu Y *et al.*, 2011 *J. Phys. C.*: **273** 012084
[19] Sakakibara T *et al.*, 1994 *Jpn. J. Appl. Phys.* **33** 5067
[20] Kromer F *et al.*, 1998 *Phys. Rev. Lett.* **81** 4476
[21] Gruenberg L W, and Gunther L 1966 *Phys. Rev. Lett.* **1966** 996

International Conference on Strongly Correlated Electron Systems (SCES 2011) IOP Publishing
Journal of Physics: Conference Series **391** (2012) 012066 doi:10.1088/1742-6596/391/1/012066

Field-Induced Ordering in the Heavy Fermion Compound $YbCo_2Zn_{20}$

Yasuyuki Shimura[1], Toshiro Sakakibara[1], Shingo Yoshiuchi[2], Fuminori Honda[2], Rikio Settai[2], and Yoshichika Ōnuki[2]

[1] Institute for Solid State Physics, University of Tokyo, Kashiwa 277-8581, Japan
[2]Graduate School of Science, Osaka University, Toyonaka, Osaka 560-0043

E-mail: simu@issp.u-tokyo.ac.jp

Abstract. DC Magnetization of the heavy-fermion compound $YbCo_2Zn_{20}$ was examined at very-low temperatures down to 0.07 K in the field up to 14.5 T. In addition to the nearly-isotropic metamagnetic behavior at 0.6 T, a new metamagnetic behavior is observed in the field variation of magnetization $M(H)$ at 6 T only for $H \parallel [111]$. For the same field direction, a clear kink appears in the temperature dependence of the magnetization $M(T)$ for fields above 6 T. These results strongly suggest the existence of a new ordered phase induced by magnetic fields. The $H-T$ phase diagram as well as the anisotropy in $M(H)$ above $\sim$ 2 T are best explained by a level crossing of the low-lying crystalline-electric-field states by magnetic fields.

1. Introduction

The majority of heavy fermion systems studied so far are Ce- or U-based compounds. Ytterbium with $4f^{13}$ electronic configuration, which is considered the hole counterpart of the $4f^1$ configuration of Ce, also possesses the same potential to exhibit the heavy fermion behavior. While properties of Yb-based heavy fermions are in many respects similar to those of Ce-based systems owing to the electron-hole analogy, there appears to be some seeming differences between Yb and Ce based heavy fermions regarding the quantum critical behavior, as observed in β-$YbAlB_4$ [1] and $YbRh_2Si_2$ [2]. In this regards, it is of fundamental interest to study the Yb-based heavy fermion compounds which are in proximity to the quantum critical point.

Recently, a class of heavy fermion systems YbT_2Zn_{20}, (T = Fe, Co, Ru, Rh, Os and Ir), with a cubic $CeCr_2Al_{20}$ type structure have been synthesized [3, 4]. All these compounds show a large electronic specific heat coefficient γ in excess of 500 mJ/mol·K^2. The electrical resistivity $\rho(T)$ also shows a strong T^2 dependence at low temperatures with the coefficient A exceeding 5×10^{-2} $\mu\Omega$cm/K^2. Among the YbT_2Zn_{20} compounds, $YbCo_2Zn_{20}$ is of particular interest because of its gigantic γ value as large as 7900 mJ/mol·K^2. This compound exhibits a metamagnetic peak structure in the field dependence of the AC susceptibility $\chi_{ac}(H)$ at a very low field of $\mu_0 H_m =$ 0.57 T applied along the [100] direction [5]. In particular, a magnetic phase has been reported to appear under a relatively low pressure of 1 GPa [6]. It is therefore considered that $YbCo_2Zn_{20}$ is in close proximity to a quantum critical point. In this paper, we report on the low temperature magnetizations of a single crystal of $YbCo_2Zn_{20}$ along the three principal field directions. We found that a distinct magnetic anisotropy develops in fields above 2 T with [111] being the

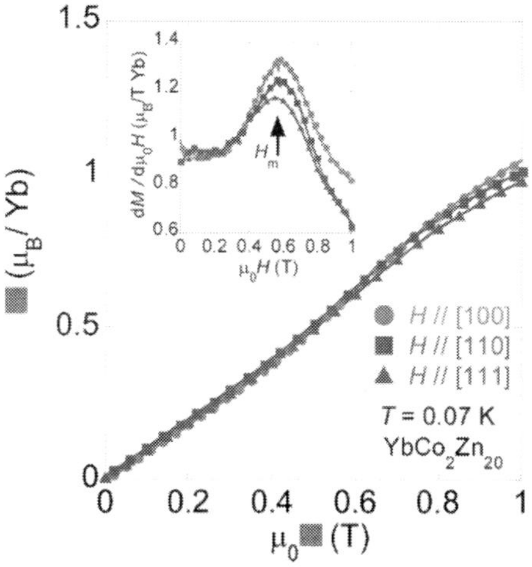

Figure 1. Low field part of the magnetization $M(H)$ of $YbCo_2Zn_{20}$ below 1 T along the three principle field directions [100] (circles), [110] (squares) and [111] (triangles) obtained at 0.07 K. The inset shows the field variation of the differential susceptibility $\mathrm{d}M(H)/\mathrm{d}H$.

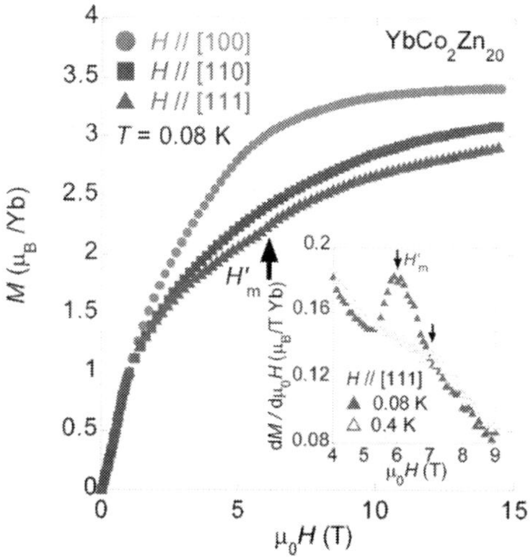

Figure 2. $M(H)$ of $YbCo_2Zn_{20}$ up to 14.5 T for $H \parallel$ [100] (circles), [110] (squares) and [111] (triangles) measured at 0.08 K. The inset displays $\mathrm{d}M(H)/\mathrm{d}H$ for $H \parallel$ [111] obtained at 0.08 K (solid triangles) and 0.4 K (open triangles).

hard axis. More importantly, we obtained evidence of a field-induced ordering along the [111] direction at low temperatures below 0.6 K [7].

2. Experimental Results and Discussion

Single-crystal samples of $YbCo_2Zn_{20}$ were grown by the Zn-self-flux method. DC magnetization measurements down to 0.07 K in magnetic fields up to 14.5 T were performed using a capacitive Faraday magnetometer with a dilution refrigerator [8]. All the measurements were done in a gradient field of 5 T/m.

Figure 1 displays the field variation of the magnetization $M(H)$ of $YbCo_2Zn_{20}$ below 1 T for $H \parallel$ [100] (solid circles), [110] (solid squares) and [111] (solid triangles) obtained at the base temperature of 0.07 K. A metamagnetic upturn is observed in $M(H)$ at around 0.6 T for all three field directions. This metamagnetic behavior can be seen more clearly as a peak structure in $\mathrm{d}M(H)/\mathrm{d}H$ as shown in the inset, which well reproduces the previous $\chi_{\mathrm{ac}}(H)$ data in a [100] magnetic field. As can be seen from the figure, the peak position is nearly independent of the field directions implying that the anisotropy of this low field metamagnetism is very weak. This result is in contrast to the metamagnetic behavior seen in other non-magnetic heavy fermion systems. For instance, metamagnetism in $CeRu_2Si_2$ is driven only by the field component parallel to the c axis [9]. Even in the cubic system $YbIr_2Zn_{20}$, the metamagnetic crossover field H_{m} is anisotropic, with $\mu_0 H_{\mathrm{m}} = 9.7$ and 12 T for $H \parallel$ [100] and [110], respectively [10]. The nearly isotropic metamagnetism in $YbCo_2Zn_{20}$ is probably due to the small critical field (0.6 T) in addition to the cubic symmetry.

Figure 2 shows $M(H)$ of $YbCo_2Zn_{20}$ measured for the three field directions at $T = 0.08$ K. As can be seen from the figure, a distinct anisotropy develops above ~ 2 T: $M_{[100]} > M_{[110]} > M_{[111]}$. Remarkably, $M(H)$ for the hard axis [111] exhibits metamagnetic behavior at $\mu_0 H'_{\mathrm{m}} = 6$ T, which is more clearly seen in the $\mathrm{d}M(H)/\mathrm{d}H$ plot as shown in the inset. The peak in $\mathrm{d}M(H)/\mathrm{d}H$,

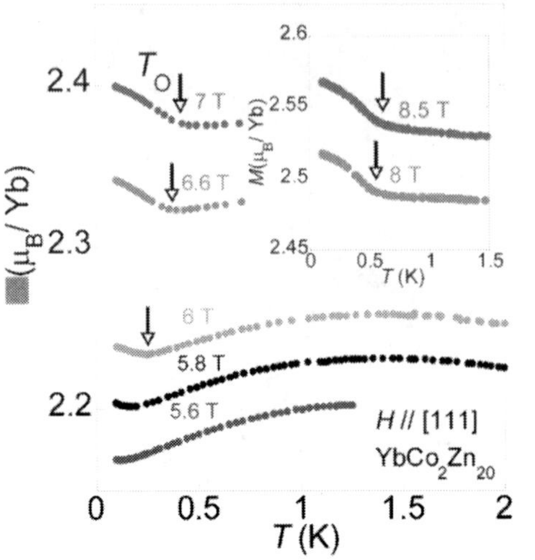

Figure 3. Temperature variation of the magnetization $M(T)$ in the magnetic fields above 5.6 T for $H \parallel [111]$. The arrow indicates T_o where a kink appears.

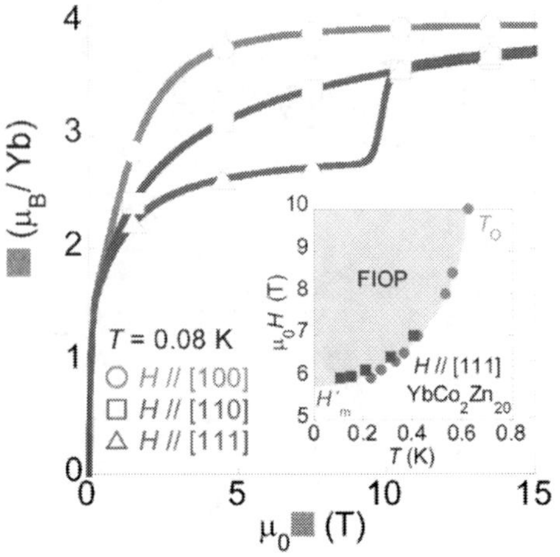

Figure 4. The magnetization process calculated on the basis of CEF model (see text). The inset indicates $H - T$ phase diagram of the field-induced ordered phase in $YbCo_2Zn_{20}$. Solid circle and solid square shows T_o and H'_m, respectively.

which marks the metamagnetism, moves to the high field side with increasing T. The amplitude of the peak rapidly diminishes at $T = 0.4$ K where only a small anomaly remains around 7 T.

In order to further examine this high-field feature in the magnetization, we measured the temperature variation $M(T)$ in [111] magnetic fields and the results are shown in Fig. 3. As can be seen from the figure, we observed a clear kink (an upward bending) in $M(T)$ for fields above 6 T. The position of the kink in $M(T)$, defined as $T_o(H)$, moves to the higher temperature side with increasing H. In the inset of Fig. 4, we plotted both $H'_m(T)$ and $T_o(H)$ in a $H - T$ diagram. The results strongly suggest the presence of a field-induced ordered phase (FIOP) in fields parallel to [111]. It should be noticed that the positive slope of the phase boundary ($dT_o/dH > 0$) and the upward bending in $M(T)$ at T_o are qualitatively in agreement with the Ehrenfest relation for a second-order phase transition [7].

The FIOP in Fig. 4 is reminiscent of a field-induced ordering in the Pr-based skutterudites $PrOs_4Sb_{12}$ and $PrFe_4P_{12}$ [11, 12, 13], in which a crossing among the Zeeman split low-lying crystalline-electric-field (CEF) states drives FIOP. We discuss this possibility in $YbCo_2Zn_{20}$, on the basis of a localized f electron picture. The $J = 7/2$ multiplet of Yb^{3+} splits into two doublets (Γ_6 and Γ_7) and a Γ_8 quartet in the cubic field. Magnetic moment values ($H \parallel [100]$) of these levers are 1.3, 1.7 and 2.1 μ_B/Yb for Γ_6, Γ_7 and Γ_8 states, respectively. The observed magnetization in excess of 3 μ_B/Yb thus implies that at least two CEF states are involved in the low-lying state. Actually, the specific heat divided by temperature, C/T, measured by Takeuchi *et al.* [14] revealed a shoulder structure around 3 K, which can be attributed to a Schottky peak due to a CEF excited state with the energy gap Δ/k_B below ~10 K. In order to reproduce the experimental results, we adopt the CEF scheme Γ_6 (0 K) - Γ_8 (6 K) - Γ_7 (27 K). The calculated field variation of the magnetization at 0.08 K is shown in Fig. 4, which well explains the observed magnetic anisotropy $M_{[100]} > M_{[110]} > M_{[111]}$. The CEF level scheme yields a crossing field of 10 T existing only for $H \parallel [111]$, in agreement with the FIOP whose transition temperature T_O becomes highest at this field. Moreover, the broad peak in $M(T)$ appearing around 1.5 K in Fig. 3 can also be qualitatively understood by the CEF model [15]. From these results, we consider that the FIOP is due to a CEF level crossing in magnetic fields.

In general, Kondo spin fluctuations smear the CEF levels in heavy fermion systems. It is thus highly remarkable that a CEF level crossing drives the field-induced ordering in $YbCo_2Zn_{20}$. In this compound, we consider that Yb ions become close to a trivalent state under high magnetic fields. This would also be a consequence of the fact that $YbCo_2Zn_{20}$ is in proximity to QCP.

3. Summary

We measured DC magnetization of $YbCo_2Zn_{20}$ along the three principal field directions $H \parallel$ [100], [110] and [111]. In addition to the nearly isotropic metamagnetic behavior at 0.6 T, we obtained evidence of a field-induced ordered phase above 6 T for $H \parallel$ [111]. The observed anisotropic phase diagram as well as the magnetic anisotropy strongly suggest that the field-induced ordered phase is driven by a crossing of low-lying CEF levels in a [111] magnetic field.

Acknowledgments

This work has been supported by a Grant-in-Aid for Scientific Research on Innovative Areas "Heavy Electrons" (No. 20102007) of the Ministry of Education, Culture, Sports and Technology, Japan and a Grant-in-Aid for Scientific Research (A) (No. 20244053) from the Japan Society for the Promotion of Science.

References

[1] Matsumoto Y, Nakatsuji S, Kuga K, Karaki Y, Horie N, Shimura Y, Sakakibara T, Nevidomskyy A H and Coleman P 2011 *Science* **331** 316
[2] Custers J, Gegenwart P, Wilhelm H, Neumaier K, Tokiwa Y, Trovarelli O, Geibel C, Steglich F, Pepin C and Coleman P 2003 *Nature* **424** 524
[3] Nasch T, Jeitschko W and Rodewald U C 1997 *Z. Naturforsch. B* **52** 1023
[4] Torikachvili M S, Jia S, Mun E D, Hannahs S T, Black R C, Neils W K, Martien D, Bud'ko S L and Canfield P C 2007 *Proc. Natl. Acad. Sci. U.S.A.* **104** 9960
[5] Ohya M, Matsushita M, Yoshiuchi S, Takeuchi T, Honda F, Settai R, Tanaka T, Kubo Y and Ōnuki Y 2010 *J. Phys. Soc. Jpn.* **79** 083601
[6] Saiga Y, Matsubayashi K, Fujiwara T, Kosaka M, Katano S, Hedo M, Matsumoto T and Uwatoko Y 2008 *J. Phys. Soc. Jpn.* **77** 053710
[7] Shimura Y, Sakakibara T, Yoshiuchi S, Honda F, Settai R and Ōnuki Y 2011 *J. Phys. Soc. Jpn.* **80** 073707
[8] Sakakibara T, Mitamura H, Tayama T and Amitsuka H 1994 *Jpn. J. Appl. Phys.* **33** 5067
[9] Haen P, Flouquet J, Lapierre F, Lejay P and Remenyi G 1987 *J. Low Temp. Phys.* **67**(5) 391
[10] Takeuchi T, Yasui S, Toda M, Matsushita M, Yoshiuchi S, Ohya M, Katayama K, Hirose Y, Yoshitani N, Honda F, Sugiyama K, Hagiwara M, Kindo K, Yamamoto E, Haga Y, Tanaka T, Kubo Y, Settai R and Ōnuki Y 2010 *J. Phys. Soc. Jpn.* **79** 064609
[11] Aoki Y, Namiki T, Ohsaki S, Saha S R, Sugawara H and Sato H 2002 *J. Phys. Soc. Jpn.* **71** 2098
[12] Tayama T, Sakakibara T, Sugawara H, Aoki Y and Sato H 2003 *J. Phys. Soc. Jpn.* **72** 1516
[13] Tayama T, Custers J, Sato H, Sakakibara T, Sugawara H and Sato H 2004 *J. Phys. Soc. Jpn.* **73** 3258
[14] Takeuchi T, Ohya M, Yoshiuchi S, Matsushita M, Honda F, Settai R and Ōnuki Y 2011 *J. Phys: Conf. Ser.* **273** 012059
[15] Takeuchi T, Yoshiuchi S, Ohya M, Taga Y, Hirose Y, Sugiyama K, Honda F, Hagiwara M, Kindo K, Settai R and Ōnuki Y 2011 *J. Phys. Soc. Jpn.* **80** 114703

Evolution from antiferromagnetic to paramagnetic Kondo insulator with increasing hybridization; XPS studies

A. Ślebarski and J. Goraus
Institute of Physics, University of Silesia, Uniwersytecka 4, 40-007 Katowice, Poland

E-mail: andrzej.slebarski@us.edu.pl, jerzy.goraus@us.edu.pl

Abstract. We present the Ce 3*d* x-ray photoemission (XPS) spectra for CeM_2Al_{10} (M=Ru, Os, Fe) from which we determined the on-site hybridization between the f and conduction electron states, Δ_{cf}, and the 4f-level occupancy, n_f. Those parameters have been obtained using the Gunnarsson-Schönhammer approach. We found Δ_{cf} stronger for the Kondo insulator $CeFe_2Al_{10}$ than for the remaining compounds with Ru and Os. We discuss the type of behaviour of CeM_2Al_{10} on the base of the earlier theoretical phase diagram obtained within the Anderson-lattice model.

1. Introduction

Kondo insulators (KI) are characterized as a class of *nonmagnetic* narrow-gap Δ semiconductors and semimetals, which exhibits a metallic heavy-fermion (HF) state at temperatures $T > \Delta$[1]. Within the framework of the periodic Anderson model [2], the small energy gap in Kondo insulators arises from the strong hybridization V between the conduction band and the f-electron states. The insulating gap Δ diminishes, however, with increasing temperature [3], which indicates that its origin is not the same as in the conventional band semiconductors, instead it is a result of strong many-body correlations [4]. Most of the known Kondo insulators; e.g., CeRhSb [5] or $Ce_3Bi_4Pt_3$ [6] are paramagnetic because the magnetic moments of Ce are quenched due to the Kondo singlet state via strong hybridization between itinerant (s,p) and localized (f) states. However, Kondo semiconductors $CeRu_2Al_{10}$ [7] and $CeOs_2Al_{10}$ [8] that show anomalous magnetic phase transition at $T^* \sim 28$ K below the Kondo temperature T_K ~ 100 K have been found, in contrast to isostructural paramagnetic Kondo insulator $CeFe_2Al_{10}$ [9] and remaining well known KIs. The T^* phase transition was first attributed to an antiferromagnetic order of the Ce sublattice with strongly reduced magnetic moment of Ce atoms [10], while the magnetic ordering was not confirmed by NMR [11]. Very recently Kimura *et al.* suggested that the antiferromagnetic ordering at T^* in $CeOs_2Al_{10}$ [12] and $CeRu_2Al_{10}$ [13] can result from the hybridization effect between the conduction and f-electrons which causes charge-density wave (CDW) instability which can induce magnetic ordering. The nature of the T^*-phase transition has been, however, still controversial since the Ce-Ce distance in the unit cell is larger than 5 Å and Gd-based counterparts have a lower Néel temperature T_N. We try to interpret this new *magnetic* KI state on the base of the ground-state phase diagram for periodic Anderson model [2] on the V-n_e plane (n_e is the total number of electrons per site). This approach allowed to determine in the mean-field approximation the phase boundary between the antiferromagnetic Kondo-insulating state (AKI) with almost compensated magnetic moments and the paramagnetic Kondo insulator phase (for details, see [2]).Hybridization energy $\Delta_{cf} \sim$ V and the occupation number of the f shell, n_f, have been determined experimentally from the XPS

Published under licence by IOP Publishing Ltd

spectra using the Gunnarsson-Schönhammer (GS) approach [14]. We measured the energy Δ_{cf} for $CeRu_2Al_{10}$ and $CeOs_2Al_{10}$ significantly smaller than Δ_{sf} of $CeFe_2Al_{10}$, which inside the Anderson-lattice model [2] suggests the antiferromagnetic Kondo insulator ground state (or antiferromagnetic phase AFM2, for details see Ref. [2]) for the both *magnetic* materials and paramagnetic KI low-T state for $CeFe_2Al_{10}$. Our experimental results agree with very recent local-density approximation (LDA) band structure calculations [13], which confirmed that the hybridization energy V of $CeRu_2Al_{10}$ and $CeOs_2Al_{10}$ is slightly weaker than that of $CeFe_2Al_{10}$.

2. XPS spectra; results, analysis and discussion

Polycrystalline samples CeM_2Al_{10}; M = Ru, Os and Fe, were prepared by arc-melting stoichiometric amounts of the elemental metals in an ultra-high-purity argon atmosphere and annealing for 2 weeks at 800°C. The XPS spectra were obtained with monochromatized Al Kα radiation using a PHI 5700 ESCA spectrometer. We also measured the electrical resistivity ρ and magnetic ac susceptibility χ_{ac} using a Quantum Design PPMS platform, the results are very consistent with those, recently obtained for the polycrystalline samples.

Figure 1 compares the valence band (VB) XPS spectra obtained for the series of CeM_2Al_{10} compounds. The main peak in these spectra originates mainly from the M-element d states hybridized with other valence band electrons. The results shown in figure 1 indicate the different electronic structure of $CeRu_2Al_{10}$ and $CeOs_2Al_{10}$ in respect to $CeFe_2Al_{10}$. The Fe 3d states of $CeFe_2Al_{10}$ are located near the Fermi level, whereas the Ru 4d states of $CeRu_2Al_{10}$ and Os 5d states of $CeOs_2Al_{10}$ are widely distributed in the valence bands. Our results well agree with very recent local-density approximation (LDA) band structure calculations [13], and suggest that hybridization between the Fe 3d and Ce 4f states is stronger than that in the case of Ru 4d and Os 5d.

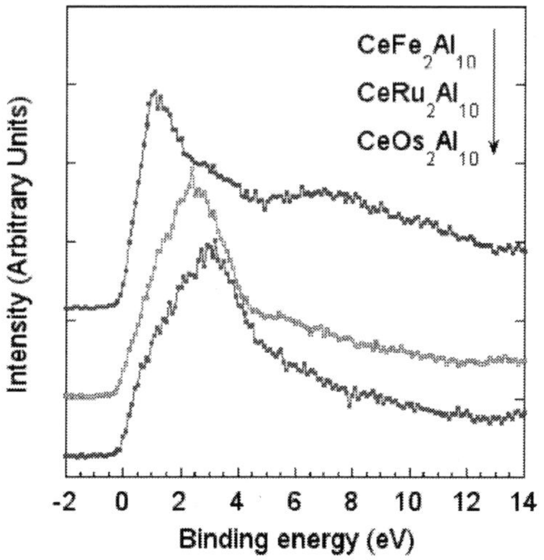

Figure 1. Valence band XPS spectra for CeM_2Al_{10}, M = Fe, Ru, and Os.

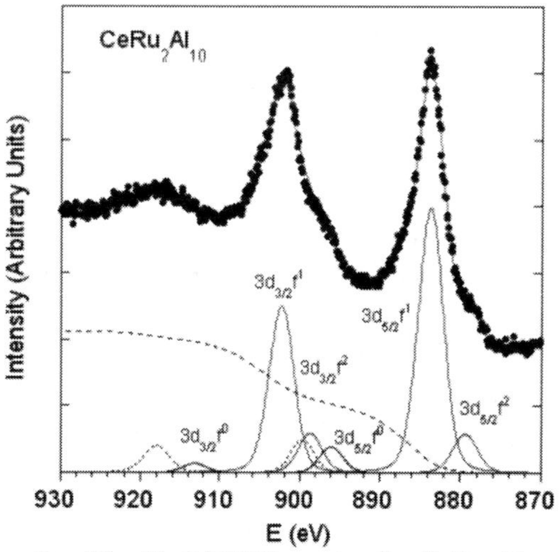

Figure 2. The Ce 3d XPS spectra for $CeRu_2Al_{10}$. The f^n (n = 0, 1, 2) components are separated on the basis of the Doniach-Šunjić theory [16]. The dotted lines indicate the plasmon excitations, the second dotted line shows the background. Similar spectra were measured for CeM_2Al_{10}, M = Os, Fe.

The XPS spectra of the 3d core levels provide detailed information about the 4f shell configuration and the f-conduction-electron hybridization. Due to many-body interactions, the Ce 3d XPS spectra show different final states depending on the occupation of the f shell; f^0, f^1 and f^2 (detailes in [14]). Figure 2 displays the Ce 3d XPS spectra for the series of CeM_2Al_{10} compounds. All the final state-contributions f^n are observed in the spectra. The quantitative analysis of the Ce 3d XPS spectra was performed on the basis of the Gunnarsson-Schönhammer [14] model. The hybridization width $\Delta_{cf} = \pi V^2 N(E_F)$, where $N(E_F)$ is the density of states (DOS) at the Fermi level E_F, was estimated from the ratio of the proper intensities $I(f^2)/[I(f^1) + I(f^2)]$. The analysis suggests that the hybridization Δ_{cf} is ~56 meV for $CeRu_2Al_{10}$ and ~58 meV for $CeOs_2Al_{10}$, while for $CeFe_2Al_{10}$ $\Delta_{cf} \approx 76$ meV is distinctly increased. Although the deconvolution of the XPS spectra as well as the model are subject to error (less than 20%, [15]) the trend in increasing of the hybridization effect in the order of Ru ≈ Os → Fe is clear. The f^0 components in the 3d XPS spectra of CeM_2Al_{10} also shown in the figure 2, marks the fractional valence character of Ce atoms. From the GS method the fractional valence of Ce is roughly equal to the intensity ratio $v = I(f^0)/[I(f^0) + I(f^1) + I(f^2)]$, which is ~0.04 for $CeRu_2Al_{10}$ and $CeOs_2Al_{10}$ and $v \approx 0.03$ for $CeFe_2Al_{10}$. Such a small value of v is characteristic of the correlated and almost localized f-electron systems.

The stability of paramagnetic vs. magnetic ground state in the Kondo-lattice limit is strongly dependent on the site hybridization magnitude V and the number n_e of electrons per site. The Doradziński-Spałek (DS) phase diagram [2] on the V-n_e can be used for qualitative interpretation of the ground state of the CeM_2Al_{10} series on the base of the XPS experimental data. Considering, that the number of the conduction electrons $n_e = n_f + n_c \approx 2$, the large hybridization energy $V \sim [\Delta_{cf}/N(E_F)]^{1/2}$ decides about the KI state formation. For the series CeM_2Al_{10}, energy Δ_{sf} is about 20 meV larger for $CeFe_2Al_{10}$ in respect to Δ_{cf} of $CeRu_2Al_{10}$ and $CeOs_2Al_{10}$. One expects the small DOS at E_F for the Kondo-gap state in $CeFe_2Al_{10}$ and large value of hybridization energy V. Therefore, in DS diagram $CeFe_2Al_{10}$ would be located in the Kondo-insulator region. The transition from KI region to a metallic region can be possible vs. the change of hybridization energy V, this would be a case in the series of CeM_2Al_{10} compounds. The most probable state for $CeRu_2Al_{10}$ and $CeOs_2Al_{10}$, predicted from the DS phase diagram is the antiferromagnetic Kondo insulator, or semimetallic antiferromagnet with a pseudogap at the Fermi energy.

3. $CeFe_2Al_{10}$, universal scaling $\chi\rho$ = *const*

Recently, it was reported [17,18] that the formation of the Kondo-insulator gap is due to the presence of collective spin-singlet Kondo state, which is singled out by magnetic susceptibility $\chi(T) \rightarrow 0$ with decreasing temperature, and activated behaviour of the resistivity $\rho(T)$. In consequence, the universal scaling law $\chi(T)\rho(T) = const$ completes the definition of a Kondo semiconductor. Figure 3 shows that the $\chi(T)\rho(T) = const$ behaviour is also observed for $CeFe_2Al_{10}$, which is a typical paramagnetic Kondo insulator.

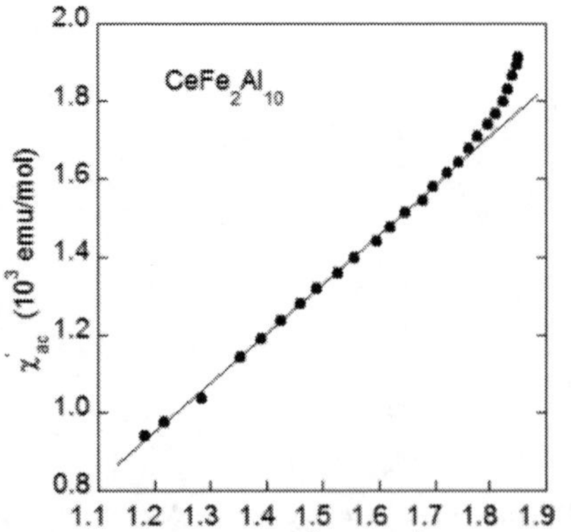

Figure 3. The ac susceptibility χ'_{ac} as a function of the inverse resistivity ρ^{-1} for $CeFe_2Al_{10}$. The $\chi(T)\rho(T) = const$ behaviour is observed between ~ 6 K and 20 K. χ'_{ac} is subtracted by impurity effect (yC/T; $y = 0.008$ and C is Curie constant for paramagnetic Ce).

The main conclusions from our magnetic and XPS data are the following: $CeFe_2Al_{10}$ obeys the features characteristic of the known paramagnetic Kondo insulators (see Ref. [19]). The analysis of the Ce 3d XPS spectra suggests that the hybridization energy between the f and conduction electron states is the largest for $CeFe_2Al_{10}$ inside the series of CeM_2Al_{10} compounds, where M = Ru, Os, and Fe. The magnetic/paramagnetic ground state properties of CeM_2Al_{10} can be grasped by the periodic Anderson model. This model suggests the paramagnetic KI state of $CeFe_2Al_{10}$, whereas for the remaining compounds (Ru, Os) the most probable seems to be an antiferromagnetic KI state. The model is, however, too simple to explain the nature of the phase transition below T^*.

Acknowledgments
The work was supported by the Ministry of Science and Higher Education within the research project No. N N202 032137.

References

[1] For review, see Aeppli G and Fisk Z 1992 *Comments Condens. Matter. Phys.* **16**, 155
[2] Doradziński R and Spałek J 1998 *Phys. Rev.* B **58** 3293
[3] Ekino T, Takabatake T, Tanaka H, and Fujii H 1995 *Phys. Rev. Lett.* **75** 4262
[4] Sanchez-Castro C, Bedell K S, and Cooper B R 1993 *Phys. Rev.* B **47** 6879
[5] Malik S K and Adroja D T 1991 *Phys. Rev.* B **43** 6277
[6] Hundley M F, Canfield P C, Thompson J D, and Fisk Z 1990 *Phys. Rev.* B **42** 6842
[7] Strydom A M 2009 *Physica* B **404** 2981
[8] Nishioka T, Kawamura Y, Takesaka T, Kobayashi R, Kato H, Matsumura M, Kodama K, Matsubayashi K, and Uwatoko Y 2009 *J. Phys. Soc. Jpn.* **78** 123705
[9] Muro Y, Motoya K, Saiga Y, and Takabatake T 2010 *J. Phys.: Conf. Ser.* **200** 12136
[10] Khalyavin D D, Hillier A D, Adroja D T, Strydom A M, Manuel P, Chapon L C, Peratheepan P, Knight K, Deen P, Ritter C, Muro Y, and Takabatake T 2010 *Phys. Rev.* B **82** 100405
[11] Matsumura M, Kawamura Y, Edamoto S, Takesaka T, Kato H, Nishioka T, Tokunaga V, Kambe S, and Yasuoka H 2009, *J. Phys. Soc. Jpn.* **78** 123713
[12] Kimura S, Iizuka T, Miyazaki H, Irizawa A, Muro Y, and Takabatake T 2011 *Phys. Rev. Lett.* **106** 56404
[13] Kimura S, Iizuka T, Miyazaki H, Hajri T, Matsunami M, Mori T, Irizawa A, Muro Y, Kajino J, and Takabatake T 2011 *Phys. Rev.* B **84** 165125
[14] Gunnarsson O, Schönhammer K 1983 *Phys. Rev.* B **28** 4315; Fuggle J C, Hillebrecht F U, Zolnierek Z, Lässer R, Freiburg Ch, Gunnarsson O, and Schönhammer K 1983 *Phys. Rev.* B **27** 7330
[15] Ślebarski A, Zawada T, Spałek J, and Jezierski A 2004 *Phys. Rev.* B **70** 235112
[16] Doniach S and Šunjić M 1970 *J. Phys.* C **3** 286
[17] Ślebarski A and Spałek J 2005 *Phys. Rev. Lett.* **95** 46402
[18] Spałek J, Ślebarski A, Goraus J, Spałek L, Tomala K, Zarzycki A, and Hackerem A 2005 *Phys. Rev.* B **72** 155112
[19] Spałek J and Ślebarski A 2011 *J. Phys.:Conf. Ser.* **273** 012055

Is $CeCoSi_3$ a superconductor?

M Smidman[1], R P Singh[1], M R Lees[1], D Mck Paul[1], D T Adroja[2] and G Balakrishnan[1]

[1] Department of Physics, University of Warwick, Coventry CV4 7AL, UK
[2] ISIS Facility, STFC, Rutherford Appleton Laboratory, Didcot, OX11 0QX, UK

Email: m.smidman@warwick.ac.uk

Abstract. We report low temperature magnetization measurements on several samples of $CeCoSi_3$ synthesized by induction melting and the Czochralski method. We aim to investigate whether the noncentrosymmetric material $CeCoSi_3$ is a superconductor as there are conflicting reports in the literature. Induction melted samples were generally found to display superconducting signals with up to a 12% superconducting volume fraction. However most samples grown by the Czochralski method produced no superconducting signal although the boules were generally inhomogeneous. Superconducting impurity phases could not be identified from powder x-ray diffraction studies. Further work is necessary to determine whether $CeCoSi_3$ may be classified as another noncentrosymmetric superconductor in the same series as $CeRhSi_3$, $CeIrSi_3$ and $CeCoGe_3$ and the first at ambient pressure.

1. Introduction

There has been recent interest in noncentrosymmetric superconductors. These are good candidates for unconventional gap symmetries since the lack of an inversion centre in the crystal structure may lead to the mixing of singlet and triplet pairing states. In particular $CeRhSi_3$, $CeCoGe_3$ and $CeIrSi_3$ become heavy fermion (HF) superconductors under pressure whilst displaying antiferromagnetism (AFM) in ambient conditions [1,2,3]. The requirement of pressure for the emergence of superconductivity in the rest of the series limits experimental investigations, whilst the existence of AFM order further complicates the analysis of data.

$CeCoSi_3$ crystallizes in the same noncentrosymmetric $BaNiSn_3$ structure as the aforementioned HF superconductors. There are conflicting reports about whether $CeCoSi_3$ becomes superconducting at ambient pressure. Superconductivity was first reported in $CeCoSi_3$ in ac magnetic susceptibility and resistivity measurements in polycrystalline samples with a T_c of 1.2 - 1.4 K [4]. A subsequent NMR study confirmed superconductivity in the material and further ac susceptibility measurements indicated a T_c of 1.1 - 1.2 K in as cast samples and 0.7 - 0.8 K in annealed ones [5]. However, other more recent studies do not report superconducting transitions in the resistivity down to 0.5 K in both polycrystalline [6,7] and single crystal samples [8]. The confirmation of superconductivity at ambient pressure in this system would provide a ready means of studying the nature of the superconducting states in the $CeTX_3$ (T = Rh, Co, Ir, X = Si, Ge) series. Due to the conflicting reports in the literature, the emphasis of this study is on producing a larger number of samples and checking the reproducibility of results.

2. Experimental

Polycrystalline samples of $CeCoSi_3$ were made using stoichiometric quantities of elemental Ce, Co and Si. These were melted together using rf induction heating under an argon atmosphere. Attempts at growing single crystals were made by the Czochralski method using a Cyberstar tetra-arc furnace. Prior to these growths, the constituent materials were melted in an rf induction furnace. For the majority of growths a stoichiometric Ce : Co : Si ratio of 1 : 1 : 3 was used. Following recent reports [9] some growths were made using a ratio of 1 : 1 : 3.5. However, we were unable to isolate large enough single crystals for measurements with either of these starting ratios. Samples were wrapped in Ta foil, sealed under vacuum in a quartz tube and annealed at 900°C for between five days and two weeks. High resolution powder x-ray diffraction (XRD) measurements were made using CuKα radiation on a Panalytical X-Pert Pro MPD. AC resistivity measurements were collected down to 1.8 K using a Quantum Design Physical Property Measurement System (PPMS) using the 4 probe technique. Magnetization measurements were made down to 1.8 K using a Quantum Design SQUID magnetometer. Magnetization measurements down to 0.48 K were made using an iQuantum ^{3}He insert.

3. Results and Discussion

A number of samples were synthesized by the methods described. We have chosen to present the results for two of the samples made by induction melting and these will be referred to as RF-A and RF-B. X-ray diffraction measurements made on the sample RF-A showed that, as well as peaks corresponding to the main phase, a small number of low intensity impurity peaks are present which we were unable to match to any known superconducting phases. The resistivity measurements on the samples indicate a residual resistivity ratio (RRR) of 29 for RF-A and 31 for RF-B after annealing for 170 hours at 900°C.

The magnetization (M) against applied field (H) is shown for two induction melted samples in figure 1. The negative gradient at low fields indicates approximately a 12% superconducting fraction for RF-A and a 2% fraction for RF-B. These values were calculated assuming the superconducting fraction had the same density as the reported values for $CeCoSi_3$. Figure 2 shows M vs H for an arc melted ingot from which a growth by the Czochralski method was made. This sample has approximately a 35% superconducting volume fraction.

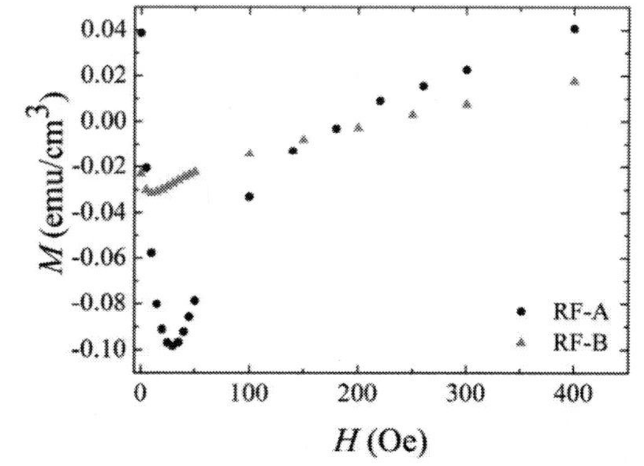

Figure 1. M vs H of $CeCoSi_3$ at $T = 0.5$ K for two samples made by induction melting.

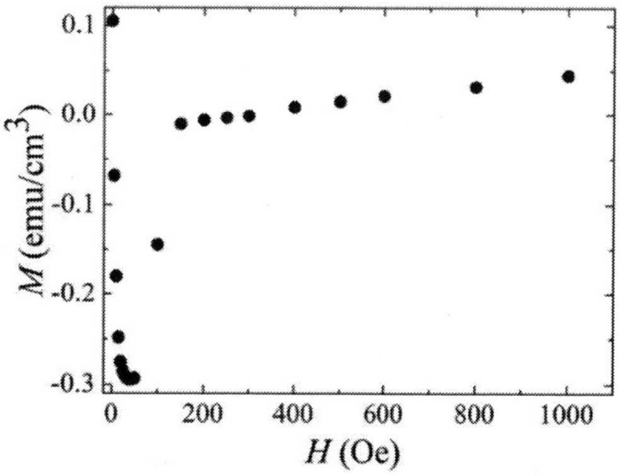

Figure 2. M vs H of $CeCoSi_3$ at $T = 0.48$ K for an arc melted ingot from which a crystal growth was attempted.

As sizeable crystals could not be isolated from the samples pulled by the Czochralski method, small sections of the grown boule were used for magnetization measurements. Figure 3 shows M vs H for samples grown by the Czochralski method with a stoichiometric starting composition. All samples, apart from sample C display no evidence of superconductivity at $T = 0.48$ K. Sample C displays superconductivity with approximately a 3% volume fraction at 0.48 K. No superconductivity is measured in sample B despite being cut from the same growth as sample C, around 1 cm further down the boule. This boule was pulled from the arc melted ingot for which M vs H data are shown in figure 2 with a superconducting volume fraction of around 35%. Figure 4 shows M vs H for a growth with a starting composition Ce : Co : Si of 1 : 1 : 3.5. No superconducting signal is observed in either the as cast or annealed samples. The data in the figures have not been offset and are shown as measured. Further measurements are required to establish the origin of the finite magnetization in zero applied field.

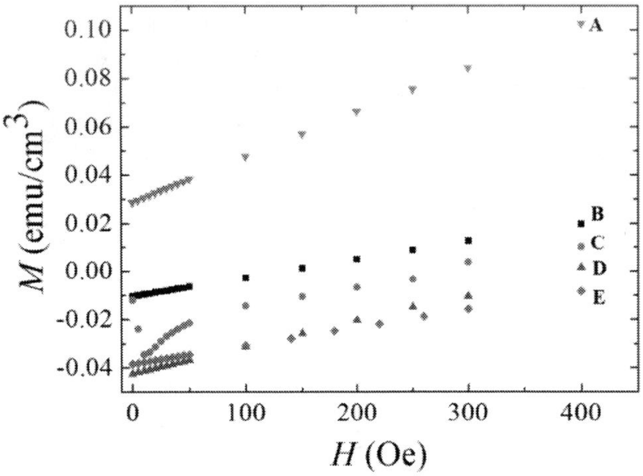

Figure 3. M vs H of $CeCoSi_3$ at $T = 0.48$ K for various samples grown by the Czochralski method with a stoichiometric starting composition. Sample A was annealed for two weeks, E for one week whilst D is as cast. Samples B and C are from the same growth, annealed for five days.

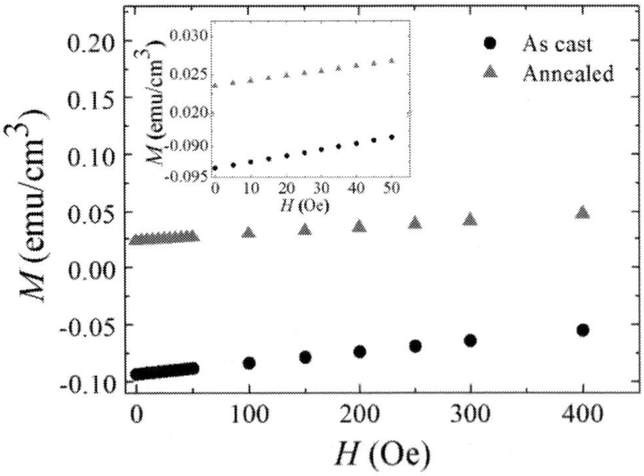

Figure 4. M vs H of $CeCoSi_3$ at $T = 0.48$ K for a growth with an off stoichiometric starting composition (Ce:Co:Si ratio of 1 : 1 : 3.5). The sample was annealed for five days at 900°C.

It can be seen that the induction melted samples generally had an observable superconducting fraction of varying magnitude, whilst most of the samples pulled using the Czochralski method were entirely paramagnetic. However the one growth where a small superconducting fraction was observed also had sections further along the boule where superconductivity was not detectable. It is therefore possible that samples which showed no superconductivity when tested may contain superconducting regions which were not measured. Annealing did not appear to have an effect on the occurrence of superconductivity, in line with previous measurements [9]. The maximum superconducting volume fraction that we have measured amongst the induction melted samples was 12% for Sample RF-A. If the superconductivity in this sample was due to the presence of an impurity phase, this could in principle be detected by high resolution powder x-ray diffraction. As stated earlier, we were unable to categorically assign any of the observed low intensity extraneous peaks to any superconductor.

4. Conclusion

Several samples of $CeCoSi_3$ were synthesized by rf induction melting and the Czochralski method. Superconducting fractions of approximately 12% and 2% were detected in induction melted samples, but were not detectable in those made by the Czochralski method. However, a superconducting fraction of around 35% was measured in an arc melted ingot used for one of these growths. Further studies are necessary to resolve whether $CeCoSi_3$ is a superconductor and this work is currently in progress. In particular powder diffraction studies and compositional analysis of a large number of samples will be useful methods of determining whether an impurity superconducting phase is present. Large single crystals would also be desirable to check the reproducibility of previous reported results where they do not show superconductivity [8].

Acknowledgements

The authors would like to thank the EPSRC for financial support. The equipment used in this research was obtained through the Science City Advanced Materials: Creating and Characterising Next Generation Advanced Materials Project, with support from Advantage West Midlands (AWM) and part funded by the European Regional Development Fund (ERDF).

References

[1] Pfleiderer C 2009 *Rev. Mod. Phys.* **81** 1551-1624
[2] Kimura N, Ito K, Saitoh K, Umeda Y, Aoki H and Terashima T 2005 *Phys. Rev. Lett.* **95** 247004
[3] Sugitani I *et al.* 2006 *J. Phys. Soc. Jpn.* **75** 043703
[4] Haen P, Lejay P, Chevalier B, Lloret B, Etourneau J and Sera M 1985 *J. Less-Common Met.* **110** 321-5
[5] Iwamoto Y, Ueda K, Kohara T and Yamada Y 1995 *Physica* B **206 & 207** 276-8
[6] Eom D, Ishikawa M, Kitagawa J and Takeda N 1998 *J. Phys. Soc. Jpn.* **67** 2495-500
[7] Okuda Y *et al.* 2007 *J. Phys. Soc. Jpn.* **76** 044708
[8] Shimoda T *et al.* 2007 *J. Magn. Magn. Mater.* **310** 308-9
[9] Iida H, Kadota Y, Kogure M, Sugawara T, Aoki H and Kimura N 2011 *J. Phys. Soc. Jpn.* **80** 083701

Elastic anomalies at successive phase transitions in $NdRu_2Al_{10}$

T Suzuki[1,2,3], I Ishii[1], Y Suetomi[1], H Muneshige[1], T K Fujita[1], S Tanimoto[4] and T Nishioka[4]
[1] Department of Quantum Matter, AdSM, Hiroshima University, Higashi-Hiroshima 739-8530, Japan
[2] Institute for Advanced Materials Research, Hiroshima University, Higashi-Hiroshima 739-8530, Japan
[3] Cryogenics and Instrumental Analysis Division, N-BARD, Hiroshima University, Higashi-Hiroshima 739-8526, Japan
[4] Graduate School of Integrated Arts and Sciences, Kochi University, Kochi 780-8520, Japan

E-mail: tsuzuki@hiroshima-u.ac.jp

Abstract. We carried out ultrasonic measurements on $NdRu_2Al_{10}$ single crystals in order to clarify what kind of phase transition occurs in this compound. The elastic moduli were measured as a function of temperature from 0.5 K to 150 K using the phase comparison-type pulse echo method. As the temperature is decreased from 150 K, longitudinal elastic modulus C_{11} and transverse elastic modulus C_{44} show a monotonic enhancement. With further decreasing temperature, both elastic moduli show abrupt hardening at 2.4 K and then show two-step softening at 1.3 K and 1.0 K. We found obvious hysteresis around both transitions at 1.3 K and 1.0 K in the temperature sweep, whereas no hysteresis was observed around the transition at 2.4 K within our experimental accuracy. The abrupt hardening at 2.4 K is very similar to that found in $CeRu_2Al_{10}$ and $CeOs_2Al_{10}$, suggesting that the transition at 2.4 K originates from the same mechanism of the transitions in $CeRu_2Al_{10}$ and $CeOs_2Al_{10}$. This is the first report that $NdRu_2Al_{10}$ undergoes the successive phase transitions.

1. Introduction

The cage compounds with the composition formula RT_2Al_{10} (R: rare earth; T = Fe, Ru and Os), which have the $YbFe_2Al_{10}$-type orthorhombic structure, have attracted much attention because $CeRu_2Al_{10}$ and $CeOs_2Al_{10}$ undergo a novel phase transition at T_0 = 27.3 K and 28.7 K, respectively [1-3]. There have been many proposals about a mechanism of ordering. Nishioka *et al.* suggested a charge-density-wave (CDW)-like transition at T_0 in $CeRu_2Al_{10}$ [2]. Meanwhile, Tanida *et al.* proposed a magnetic-type phase transition in which the singlet spin pair between spins on neighboring Ce ions is formed below T_0 from the measurements on the thermal and transport properties of La-diluted $Ce_{1-x}La_xRu_2Al_{10}$ [4]. Recently, an antiferromagnetic structure along the c-axis below T_0 is reported in inelastic neutron scattering and muon spin relaxation experiments [5].

We already measured temperature dependences of elastic moduli of $CeRu_2Al_{10}$ and reported that abrupt hardening occurs at T_0 in all elastic moduli without hysteresis, suggesting that the transition is of the second order and there is a $Q\varepsilon^2$ -type coupling term in an effective Hamiltonian where Q and ε represent an order parameter and an elastic strain, respectively [6]. This elastic behaviour is almost identical with that of the typical CDW-transition-compounds TaS_3 and $NbSe_3$ [7,8]. Therefore we pointed out that the plausible origin for the transition at T_0 in $CeRu_2Al_{10}$ is the CDW ordering [6]. From the optical conductivity measurements, Kimura *et al.* proposed that the phase transition in the same $CeRu_2Al_{10}$ originates from the CDW formation which induces antiferromagnetic ordering [9], supporting our conclusion for $CeRu_2Al_{10}$.

Very recently, successive phase transitions were found in $NdOs_2Al_{10}$ [10]. In order to investigate what kind of phase transition occurs in $NdRu_2Al_{10}$, we synthesized single crystals of $NdRu_2Al_{10}$ and measured elastic moduli using an ultrasonic technique. This is the first report on the successive phase transitions in $NdRu_2Al_{10}$.

2. Experimental

Single crystals of $NdRu_2Al_{10}$ were grown by the Al self-flux method. The lattice constants of $NdRu_2Al_{10}$ are referred from ref. 11, such as $a = 9.115$ Å, $b = 10.261$ Å and $c = 9.152$ Å. The elastic moduli C_{11} and C_{44} were measured as a function of temperature from 0.5 to 150 K by the phase comparison-type pulse echo method using a ^{3}He cryostat with a superconducting magnet. The modulus C_{11} is the longitudinal mode propagating along the *a*-axis, and C_{44} is the transverse mode propagating along the *b*-axis with the polarization direction along the *c*-axis. The elastic modulus C was calculated using $C = \rho v^2$ with a room-temperature mass density $\rho = 4.782$ g/cm^3, where v is the sound velocity in a sample.

3. Results and discussion

Figure 1 shows temperature dependence of the longitudinal elastic modulus C_{11} in $NdRu_2Al_{10}$, which was measured with the ultrasonic frequency of 99 MHz. The inset shows C_{11} in an expanded scale below 3.5 K. The modulus C_{11} is the linear response to the ε_{xx} strain. The modulus C_{11} increases monotonically with decreasing temperature at high temperatures. With further decreasing temperature, abrupt hardening occurs at $T_1 = 2.4$ K and then two-step softening appears around $T_2 = 1.3$ K and $T_3 = 1.0$ K, respectively.

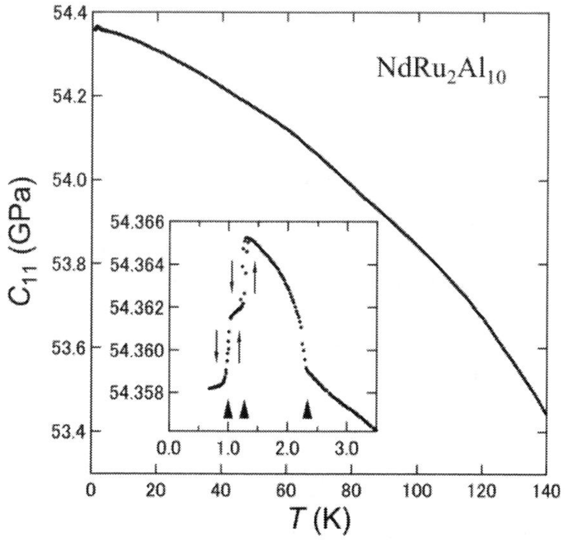

Figure 1. Temperature dependence of C_{11}

When we repeated the temperature sweep around three transitions, we found hysteresis around T_2 and T_3 in contrast to no hysteresis around T_1 in the temperature sweep. Here we report, for the first time, that $NdRu_2Al_{10}$ undergoes three phase transitions: the second order transition at T_1 and the first order transitions at T_2 and T_3, respectively.

Figure 2 shows temperature dependence of the transverse elastic modulus C_{44} in $NdRu_2Al_{10}$, which was measured with the ultrasonic frequency of 102 MHz. The modulus C_{44} is the linear response to the ε_{yz} strain. This transverse elastic modulus shows almost the same temperature dependence of C_{11}, such as the monotonic increase at high temperatures, the abrupt hardening at T_1 and the two-step softening around T_2 and T_3. However, the hysteresis around T_2 and T_3 is more obvious than that in the C_{11} mode. This may be related to the symmetry of ordering structure corresponding to the symmetry of the ε_{yz} strain.

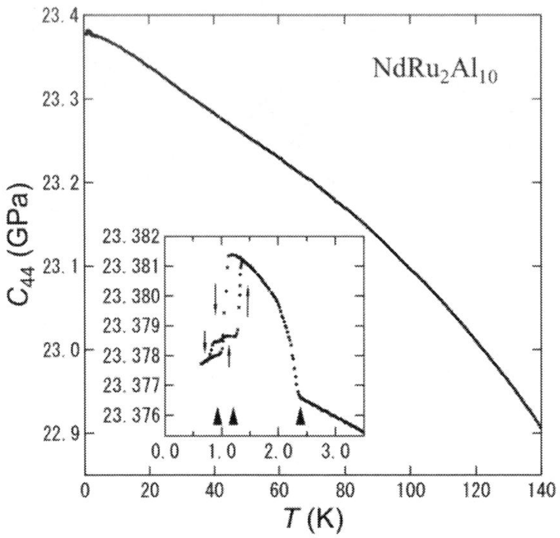

Figure 2. Temperature dependence of C_{44}

At the second order transition at T_1, both of longitudinal and transverse moduli abruptly enhance without hysteresis, suggesting that there is a $Q\varepsilon^2$-type coupling term in the effective Hamiltonian [12]. This elastic behaviour is almost identical with that in $CeRu_2Al_{10}$ [6]. Therefore, we estimate that this transition also originates from the CDW transition accompanied by antiferromagnetic ordering which was found in $CeRu_2Al_{10}$. The isomorphic compounds RT_2Al_{10} may have similar instability commonly.

As for the first order transitions at T_2 and T_3, the origin is unclear at present. However, in our preliminary elastic measurements under magnetic fields, we found that both transition temperatures of T_2 and T_3 decrease with increasing the magnetic field. This may suggest that both transitions originate from an antiferromagnetic ordering. In order to clarify the origin of phase transitions, we need to perform experiments on magnetic properties of $NdRu_2Al_{10}$.

4. Conclusion

We carried out ultrasonic experiments on $NdRu_2Al_{10}$ single crystals to measure temperature dependence of longitudinal and transverse elastic moduli in the temperature range between 0.5 and 150 K. We found for the first time that $NdRu_2Al_{10}$ undergoes the second order transition at T_1 = 2.4 K and the first order transitions at T_2 = 1.3 K and T_3 = 1.0 K, respectively. We estimate that a CDW transition accompanied by antiferromagnetic ordering occurs at T_1.

5. References

[1] Strydom A M 2009 *Physica B* **404** 2981
[2] Nishioka T, Kawamura Y, Takesaka T, Kobayashi R, Kato H, Matsumura M, Kodama K, Matsubayashi K and Uwatoko Y 2009 *J. Phys. Soc. Jpn.* **78** 123705
[3] Muro Y, Kajino J, Umeo K, Nishimoto K, Tamura R and Takabatake T 2010 *Phys. Rev. B* **81** 214401
[4] Tanida H, Tanaka D, Sera M, Moriyoshi C, Kuroiwa Y, Takesaka T, Nishioka T, Kato H and Matsumura M 2010 *J. Phys Soc. Jpn.* **79** 043708
[5] Khalyavin D D, Hillier A D, Adroja D T, Strydom A M, Manuel P, Chapon L C, Peratheepan P, Knight K, Deen P, Ritter C, Muro Y and Takabatake T 2010 *Phys. Rev. B* **82** 100405(R)
[6] Ishii I, Suetomi Y, Fujita T K, Takesaka T, Nishioka T and Suzuki T 2010 *J.Phys. Soc. Jpn.* **79** 053602
[7] Xiang X -D and Brill J W 1989 *Phys. Rev. Lett.* **63** 1853
[8] Xiang X -D and Brill J W 1989 *Phys. Rev. B* **39** 1290
[9] Kimura S, Iizuka T, Miyazaki H, Hajiri T, Matsunami M, Mori T, Irizawa A, Muro Y, Kajino J and Takabatake T 2011 *Phys. Rev. B* **84** 165125
[10] Muro Y and Takabatake T 2011 *Private communication*
[11] Thiede V M T, Ebel T and Jeitschko W 1998 *J. Mater. Chem.* **8** 125
[12] Rehwald W 1973 *Adv. Phys.* **22** 721

Acknowledgments

This work was supported by a Grant-in-Aid (No. 20340093) and a Grant-in-Aid for Scientific Research on Innovative Areas "Heavy Electrons" (No. 20102005) from the Ministry of Education, Culture, Sports, Science and Technology of Japan.

Crystal-field effect on anisotropic magnetic properties of CeT_2Al_{10} (T = Ru and Os)

K. Yutani[1]**, Y. Muro**[1]**, J. Kajino**[1]**, T. J. Sato**[2] **and T. Takabatake**[1,3]

[1] Department of Quantum Matter, ADSM, and [3]IAMR, Hiroshima University, Higashi-Hiroshima 739-8530, Japan
[2]Institute for Solid State Physics, University of Tokyo, Kashiwa 277-8581, Japan

E-mail: takaba@hiroshima-u.ac.jp

Abstract. The magnetic susceptibility of single crystalline samples of CeT_2Al_{10} (T=Ru, Os) showing unusual phase transitions at T_0 = 27 K and 29 K, respectively, has been measured up to 700 K along the three principal axes of the orthorhombic structure. The crystal field analysis of the data led to the crystal field scheme with excited levels at 326 K and 530 K for T = Ru and 433 K and 704 K for T = Os. For the ground state, the butterfly-like wave function elongates along the *b*-axis in the distorted hexagon composed of Al(3) and Al(4) atoms in the *b*-*c* plane. This charge distribution allows the 4*f* electron state to hybridise with the 3*p* states of the Al(3) and Al(4) atoms. We propose that this hybridisation plays an important role in both the formation of the hybridisation gap and the enhancement of the antiferromagnetic ordering temperature.

1. Introduction

A family of cerium-based intermetallic compounds CeT_2Al_{10} (T = Fe, Ru, Os) attracts much attention because $CeRu_2Al_{10}$ and $CeOs_2Al_{10}$ undergo unusual transitions at rather high temperatures T_0 = 27 K and 28.5 K, respectively, while $CeFe_2Al_{10}$ shows behaviour of a paramagnetic Kondo semiconductor [1-4]. At the initial stage of the studies of $CeRu_2Al_{10}$, a structural transition at T_0 was proposed by the ^{27}Al NQR measurement [5]. However, neutron diffraction and μSR experiments indicate that the magnetic moments on the Ce sites of approximately 0.3 μ_B order antiferromagnetically below T_0 [6-8]. In the ordered phase, the magnetic moments orient along the orthorhombic *c* axis alternately. [7] This finding of magnetic order raised a question why T_0's are much higher than the Néel points of the Nd- and Gd-counterparts [2, 9].

The magnetic susceptibility $\chi(T)$ for CeT_2Al_{10} (T = Fe, Ru, Os) exhibits strong anisotropy, $\chi_a > \chi_c > \chi_b$ in the paramagnetic state [2, 3]. The dominant origin of the magnetic anisotropy may be the crystalline electric field (CEF) effect on the Ce ion which is surrounded by four T atoms and 16 Al atoms in the $YbFe_2Al_{10}$-type structure. The CEF of the *m*2*m* point symmetry splits the $^2F_{5/2}$ multiplet of the Ce^{3+} ion into three Kramers doublets. However, a rather weak CEF effect on the Pr and Nd ions in the isostructural compounds ROs_2Al_{10} was inferred from the small paramagnetic Curie temperatures less than 10 K [9]. This suggests that the hybridisation effect in the Ce compounds is stronger than the CEF effect due to the point charges of the ligand atoms. It is important to determine the crystal field level scheme for the Ce compounds to understand the mechanism of the enhancement of T_0. Therefore, we have measured χ_a, χ_c, and χ_b for single crystals of $CeRu_2Al_{10}$ and $CeOs_2Al_{10}$ up to 700 K. Our

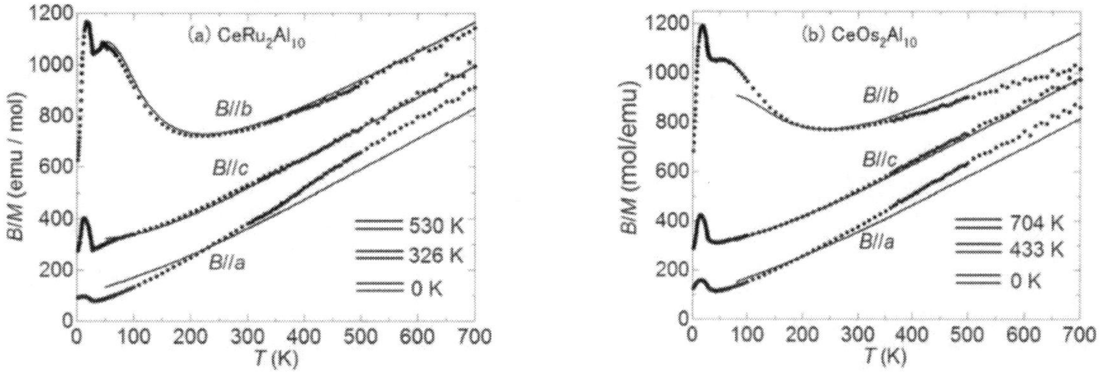

Figure 1. Inverse magnetic susceptibilities of (a) $CeRu_2Al_{10}$ and (b) $CeOs_2Al_{10}$ along the three principal axes as a function of temperature. The solid curves represent results of the CEF analysis. The energy levels of 4*f* electron states are shown in the insets.

results of the CEF level scheme will be compared with those obtained by neutron inelastic scattering experiments and the recent calculation [10, 11].

2. Experimental

Single crystalline samples of CeT_2Al_{10} (T = Ru, Os) were grown using an Al self-flux method, which was reported previously [3]. Binary alloys of $CeRu_2$ and $CeOs_2$ were prepared by arc melting of pure elements. The crushed ingots of the alloy together with an excess amount of Al in the ratio 1:2:20 were loaded into an alumina crucible, which was sealed in a quartz tube under an Ar atmosphere of 1/3 atm. The crucible was heated to 1150°C and then cooled slowly to 720°C at which point the molten Al flux was separated by centrifuging. The obtained single crystals have typical size of 4 mm and 2 mm in diameter for T = Ru and T = Os, respectively. No deviation from the ideal stoichiometry of 1:2:10 was detected by electron-probe microanalysis. Powder x-ray patterns confirmed the orthorhombic $YbFe_2Al_{10}$-type structure. For the measurement of $\chi(T)$ along the principal axes, the crystal was oriented by the back reflection Laue method using an imaging plate camera.

The measurement of $\chi(T)$ was performed at an external field $B = 1$ T in two temperature ranges using a superconducting quantum interference device magnetometer, MPMS. The lower range is from 2 K to 350 K and the higher one is from 300 K to 700 K. For the latter, the crystal was wrapped in Al foil and placed in an oven installed in the MPMS. The two sets of data were smoothly connected each other in the overlapped range between 300 K and 350 K.

3. Results and Discussions

Figure 1 shows the inverse of χ_a, χ_c, and χ_b for single crystals of $CeRu_2Al_{10}$ and $CeOs_2Al_{10}$ up to 700 K. At temperatures above 300 K, the data of χ_a and χ_c follow the Curie-Weiss law with an effective magnetic moment of 2.5-2.8 μ_B/Ce, being close to the trivalent free-ion value 2.54 μ_B/Ce. However, the effective magnetic moments for χ_b are 3.0 and 3.7 μ_B/Ce for T = Ru and T = Os, respectively. The extrapolation of the fitting to the inverse of $\chi_b(T)$ yields large negative values for the paramagnetic Curie temperature; −550 K and −820 K for T = Ru and T = Os, respectively. Such a large negative value can be attributed to the Kondo effect, which manifests in $\chi(T)$ as a negative molecular field in the CEF analysis [12].

In CeT_2Al_{10}, the Ce ion with total angular momentum $J = 5/2$ occupies the site with the *m*2*m* (C_{2v}) point symmetry. Hence, the Stevens operators O_n^m for n = 6 are zero, and therefore the CEF Hamiltonian is described by five terms as

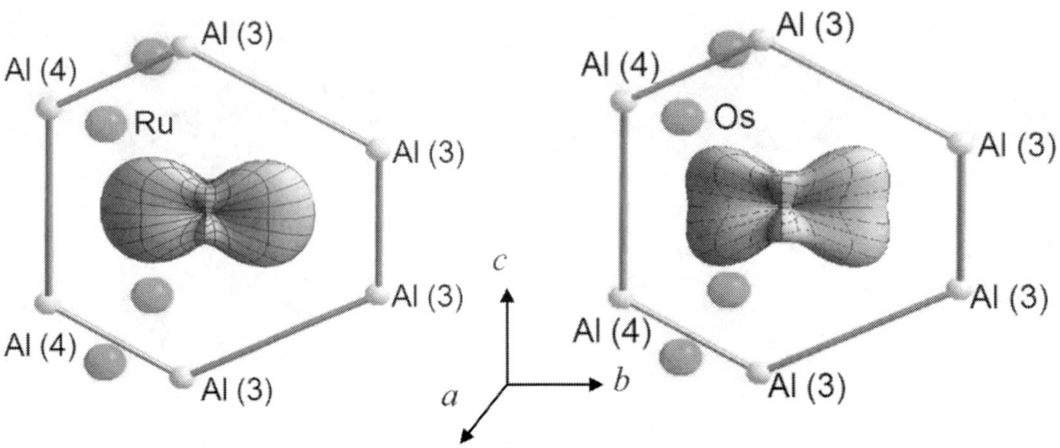

Figure 2. Spatial charge distribution of the ground-state wave function of the Ce 4*f* electron in (a) $CeRu_2Al_{10}$ and (b) $CeOs_2Al_{10}$. The atomic coordination of Al(3), Al(4), Ru and Os is drawn.

$$\mathcal{H}_{\mathrm{CEF}} = B_2^0 O_2^0 + B_2^2 O_2^2 + B_4^0 O_4^0 + B_4^2 O_4^2 + B_4^4 O_4^4 \quad (1)$$

where B_n^m stand for CEF parameters describing the strength of the interaction [13]. In the mean-field approximation, $\chi_i(T)$ (i = *a*, *b*, *c*) is expressed as $\chi_i(T) = \chi_{i\mathrm{CEF}}(T)/[1 - \lambda\chi_{i\mathrm{CEF}}(T)]$, where λ is the mean-field parameter. We took the *c* axis as the quantization axis although *b* axis was chosen by Hanzawa [11]. The expression for $\chi_{i\mathrm{CEF}}(T)$ based on the CEF model is given by

$$\chi_{i\mathrm{CEF}} = \frac{(g_J\mu_B)^2}{\sum_m e^{-E_m/k_BT}}\left(\frac{\sum_m |\langle m|J_i|m\rangle|^2 e^{-E_m/k_BT}}{k_BT} + \sum_m \sum_{n(\neq m)} |\langle n|J_i|m\rangle|^2 \frac{e^{-E_m/k_BT} - e^{-E_n/k_BT}}{E_n - E_m}\right) \quad (2)$$

where the notations of the symbols are given in our previous report [14]. The solid lines in Figs. 1(a) and 1(b) are the calculated curves using the parameters obtained by a least-squares fitting to the data above 80 K. The obtained CEF parameters and 4*f* level schemes are listed in Table 1. We confirmed that our CEF parameters for $CeRu_2Al_{10}$ are comparable with those obtained by Hanzawa [11] when the quantization axis is converted from the *c* axis to *b* axis. The dominant term is B_2^2 whose value is −39 K for both compounds. The rather large molecular field parameters of −90 and −60 mol/emu for T = Ru and T = Os, respectively, can be attributed to the strong Kondo effect in addition to the inter-site antiferromagnetic interaction. The energies of the excited doublets Δ_1 and Δ_2 are, respectively, 326 K and 530 K for T = Ru and 433 K and 704 K for T = Os, whose values agree with those obtained by inelastic neutron scattering experiments [10]. It is worth noting that the overall splitting is rather large compared to other Ce-based intermetallic compounds. Because Δ_1 exceeds 300 K, we consider that only the ground state doublet takes part in the magnetic order below T_0 as well as the unusual magnetic excitations at 8 and 11 meV for T = Ru and T = Os, respectively, which develop below T_0 [10, 15].

The crystal field parameters allow us to draw the spatial charge distribution of the ground-state wave function. As shown in Fig. 2, the butterfly-like wave function lies in the *b*-*c* plane. This distribution is consistent with the easy magnetization axis along the *a* axis in the paramagnetic state. The wave function elongates along the *b*-axis in the distorted hexagon made of Al(3) and Al(4) atoms in the *b*-*c* plane. This charge distribution suggests that the 4*f* electron state hybridises strongly with the

Table 1. Crystal field parameters, molecular field coefficient, energy levels, and wave functions of the Ce $4f$ electron in $CeRu_2Al_{10}$ (left) and $CeOs_2Al_{10}$ (right).

CEF para- meters (K)	B_2^0	B_2^2	B_4^0	B_4^2	B_4^4	λ(mol/emu)
$CeRu_2Al_{10}$	1.58	−38.9	0.714	−1.95	−4.0	−90
$CeOs_2Al_{10}$	1.6	−39	1.4	−2.5	−6.0	−60

Energy levels	$CeRu_2Al_{10}$ Wave functions					
E(K)	|+5/2>	|+3/2>	|+1/2>	|−1/2>	|−3/2>	|−5/2>
530	0	−0.049	0	0.736	0	−0.676
530	+0.676	0	−0.736	0	0.049	0
326	0	0.684	0	−0.468	0	−0.559
326	0.559	0	0.468	0	−0.684	0
0	0	0.728	0	0.490	0	0.480
0	−0.480	0	−0.490	0	−0.728	0

Energy levels	$CeOs_2Al_{10}$ Wave functions					
E(K)	|+5/2>	|+3/2>	|+1/2>	|−1/2>	|−3/2>	|−5/2>
704	0	−0.043	0	−0.740	0	0.671
704	−0.671	0	0.740	0	0.043	0
433	0	0.532	0	−0.585	0	−0.612
433	0.612	0	0.585	0	−0.532	0
0	0	0.846	0	0.331	0	0.419
0	−0.419	0	−0.331	0	−0.846	0

$3p$ states of the Al(3) and Al(4) atoms. This hybridisation should play an important role in both the enhancement of the ordering temperature and the formation of the hybridisation gap. Hanzawa, on the other hand, pointed out that onsite $4f$-$5d$ mixing may occur due to the lack of inversion symmetry at the Ce site [11]. It was suggested that the $4f$-$5d$ mixing may enhance the Ruderman-Kittel-Kasuya-Yoshida interaction which causes the magnetic ordering of Ce magnetic moments. The onsite $4f$-$5d$ and inter-site $4f$-$3p$ hybridisations would be observed by the undergoing photoemission and linear-polarised soft-x-ray absorption experiments. Remaining problems to be clarified are the mechanisms for the enhancement of T_0 and for the change in the direction of the magnetic moments from the a axis in the paramagnetic state to the c axis in the ordered state. To answer these problems, further studies are necessary which take into account the anisotropic hybridisation based on the CEF scheme obtained by this work.

Acknowledgement
We thank T. Onimaru, K. Umeo and T. Matsumura for valuable discussions, and D.T. Adroja for providing his unpublished CEF parameters estimated from neuron data. This work was supported by a Grant-in-Aid for Scientific Research on Innovative Areas "Heavy Electrons" (Grant No. 20102004) of MEXT, Japan.

References

[1] Strydom A M 2009 *Physica B* **404** 2981
[2] Nishioka T *et al.* 2009 *J. Phys. Soc. Jpn.* **78** 123705
[3] Muro Y *et al.* 2010 *Phys. Rev. B* **81** 214401
[4] Muro Y *et al.* 2009 *J. Phys. Soc. Jpn.* **78** 083707
[5] Matsumura M *et al.* 2009 *J. Phys. Soc. Jpn.* **78** 123705
[6] Kambe S *et al.* 2010 *J. Phys. Soc. Jpn.* **79** 053708
[7] Khalyavin D D *et al.* 2010 *Phys. Rev. B* **82** 100405 R
[8] Adroja D T *et al.* 2010 *Phys. Rev. B* **82** 104055
[9] Muro Y *et al.* 2011 *J. Phys. Soc. Jpn.* **80** SA021
[10] Deen P P *et al.* unpublished.
[11] Hanzawa K 2011 *J. Phys. Soc. Jpn.* **80** 023707
[12] Maekawa S *et al.* 1985 *J. Phys. Soc. Jpn.* **54** 1955
[13] Hutchings M T 1965 *Solid State Phys.* **16** 227
[14] Onimaru T *et al.* 2008 *J. Phys. Soc. Jpn.* **77** 074708
[15] Robert J *et al.* 2010 *Phys. Rev. B* **82** 100404 R

Magnetic properties of cubic compound $Ce_6Ni_6P_{17}$ with geometric frustrations

N Takeda[1], K Izumi[1], H Ono[2], S Yodono[2] and T Nakano[1]

[1]Faculty of Engineering Niigata University, Niigata 950-2181, Japan
[2]Graduate School of Science and Technology, Niigata University, Niigata 950-2181, Japan

E-mail: ntakeda@eng.niigata-u.ac.jp

Abstract. We report the magnetic susceptibility and the low-temperature specific heat of $Ce_6Ni_6P_{17}$ with geometric frustrations and $La_6Ni_6P_{17}$ as a nonmagnetic counterpart. The magnetic susceptibility of $Ce_6Ni_6P_{17}$ decreases monotonically with decreasing temperature and the specific heat shows a broad peak around 1.4K. The evaluated magnetic entropy is about a half of Rln2 at 5.0 K. This result suggests that the frustrations persist down to very low temperatures and the residual entropy originate from frustration is present.

1. Introduction

Recently, a cage structure has attracted much attentions. Ion in a cage frequently shows local or unharmonic vibration, which is conjectured to induce glass-like transport properties[1]. On the course of researching new compounds with a cage structure, we succeeded to prepare $R_6Ni_6P_{17}$ (R=La,Ce). $R_6Ni_6P_{17}$ is reported to be a cubic compound with a space group $I\bar{4}3m$ but no physical properties is reported so far[2]. The crystal structure is very unique and shown in Fig.1. On contrary to lots of rare-earth compounds with a cage structure, where rare-earth ion is inside a cage, R-ions in $R_6Ni_6P_{17}$ form an octahedron cage encapsulating phosphorous. The octahedra form bcc lattice[2]. If the interaction between Ce-ions in octahedron is antiferromagnetic, an octahedron induces geometric frustrations as schematically shown in Fig.2. The frustration in octahedron is not well known compared with triangular, pyrochlore and kagomé lattices. As far as we know, Mo_3Sb_7, which is a superconductor with $T_c = 2.3$ K, has a bcc lattice which composed of octahedron[3]. The geometric coordination is very similar to the present case of $Ce_6Ni_6P_{17}$. The structural phase transition, however, releases the frustration below 50 K, and hence the really frustrated ground state is not realized. Theoretically, the antiferromagnetic Ising model on a simple cubic lattice which is composed of corner-sharing octahedron is investigated by Monte Carlo simulations[4]. The simulations revealed that the system with nearest-neighbor (NN) interactions alone does not exhibit any phase transition, leading to a degenerate ground state with large residual entropy. This result does not apply simply to the $Ce_6Ni_6P_{17}$ with bcc lattice, but large residual entropy is expected in $Ce_6Ni_6P_{17}$. Aside from a geometric frustration, a bcc lattice composed of octahedron has another interesting problem as pointed out in Mo_3Sb_7[3]. The NN-bond is edge of octahedron and the next-nearest-neighbor (NNN) -bond is a Ce-Ce dumbbell along cubic edge in $Ce_6Ni_6P_{17}$. If antiferromagnetic interaction of NNN-bond is much stronger than that of NN-bond, dimerization of Ce-Ce dumbbell would be expected. The NN- and NNN-distance for $Ce_6Ni_6P_{17}$ are very close, 4.139 Å and 4.262 Å, respectively, so there would be a competition of frustration and dimmerization. In this paper, we present the magnetic

Published under licence by IOP Publishing Ltd

susceptibility and the specific heat of $Ce_6Ni_6P_{17}$ with geometric flustrations and $La_6Ni_6P_{17}$.

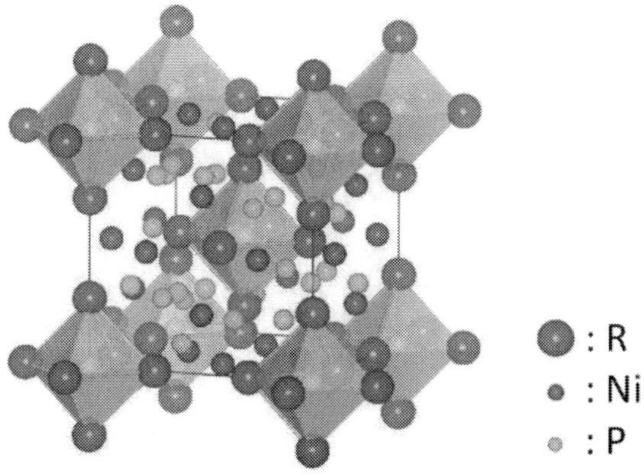

Figure 1. The crystal structure of $R_6Ni_6P_{17}$. Octahedron composed of R-ions forms a bcc lattice. The edge of octahedron is the NN-bond and the R-R dumbbell along cubic edge is the NNN-bond.

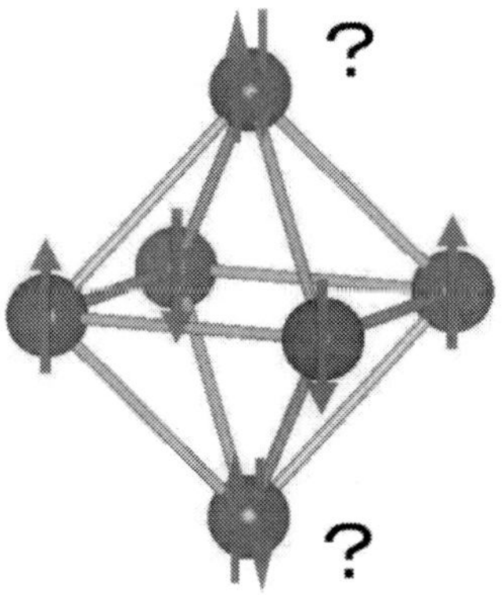

Figure 2. Schematic representation of frustration in an octahedron. If the interaction is antiferromagnetic, two spins frustrate.

2. Experiment

The samples used in the present experiments were prepared by molten Sn-flux method with compositions varied between R:Ni:P:Sn=1:1:8:30 and 1:1:8:50[2]. The constituents in an evacuated quartz were heated up to 900 °C and cooled slowly down to 400 °C for 168 hours followed by natural cooling in a furnace. Very small single crystals were obtained after eliminating Sn-flux by hydrochloric acid. The X-ray diffraction examination exhibits a small amount of NiP_3 phase which is a normal metal[5]. The lattice parameter is 10.169 Å for $La_6Ni_6P_{17}$ and 10.114 Å for $Ce_6Ni_6P_{17}$, respectively. These values are coincide with the reported values[2]. Single crystals less than 50μm were grounded and pressed to pellet. The magnetic susceptibility, $\chi(T)$, was measured by a SQUID magnetometer. The specific heat, $C(T)$, was measured by an adiabatic method with a ^{3}He-cryostat.

3. Results and discussion

Figure 3 shows $\chi(T)$ of $La_6Ni_6P_{17}$ and $Ce_6Ni_6P_{17}$. $\chi(T)$ of $La_6Ni_6P_{17}$ is positive and exhibits weak temperature dependence, indicating Pauli paramagnetism. χ(300K) is 1.34×10^{-4}

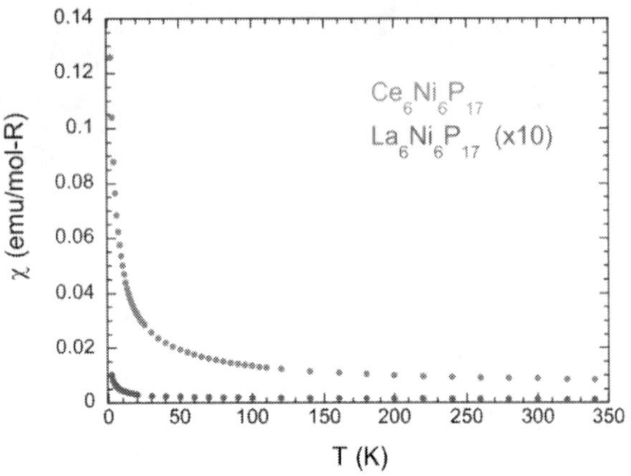

Figure 3. The temperature dependence of the magnetic susceptibility of $R_6Ni_6P_{17}$. $\chi(T)$ of $La_6Ni_6P_{17}$ is multiplied by 10. $La_6Ni_6P_{17}$ is a Pauli paramagnet. $\chi(T)$ of $Ce_6Ni_6P_{17}$ does not show a simple Curie-Weiss law.

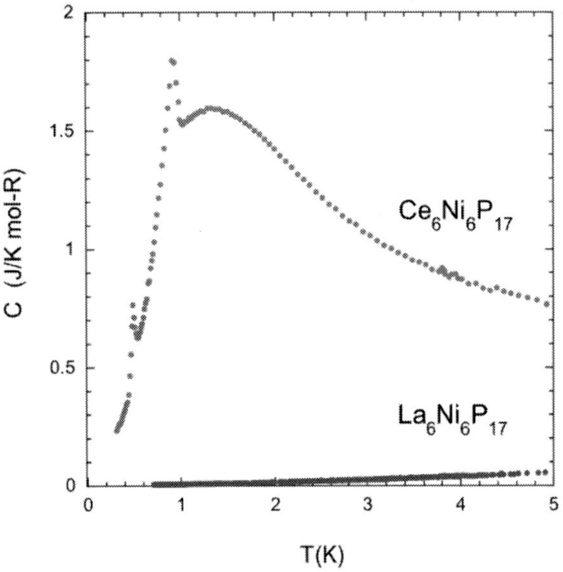

Figure 4. The specific heat of $Ce_6Ni_6P_{17}$ and $La_6Ni_6P_{17}$. $La_6Ni_6P_{17}$ is normal metal with the electronic specific-heat coefficient $\gamma = 45$ mJ/mol·K^2 and the Debye temperature of 380 K. $C(T)$ of $Ce_6Ni_6P_{17}$ shows a Schottky-like broad peak around 1.4 K implying the presence of frustrations. The magnetic phase transitions are evident from two small peaks at 0.5 K and 0.9 K.

emu/mol-La. $\chi(T)$ of $Ce_6Ni_6P_{17}$ increases monotonically with decreasing temperature down to 2.0 K. The monotonous increase down to low temperatures suggests that the presence of frustrating spins as calculated in the antiferromagnetic triangular Ising lattice[6]. The dimmerization as expected apparently is not seen in the low-temperature behavior. We analyzed the susceptibility by $\chi(T) = \chi_0 + C/(T - \theta)$. The temperature-independent constant term $\chi_0 \sim 0.01$ emu/mol-Ce is unknown origin. The effective Bohr magneton μ_{eff} is evaluated to be $2.0 \sim 2.3\mu_B$, which depends on fitting range and is slightly smaller than 2.54 μ_B of Ce^{3+}. The Weiss temperature θ is nearly equal to 0 K, which indicates that the magnetic interaction is very weak or that ferromagnetic interaction and antiferromagnetic one coexist. No magnetic phase transition was found down to 2.0 K.

The low-temperature specific-heat, $C(T)$, of $Ce_6Ni_6P_{17}$ and $La_6Ni_6P_{17}$ is displayed in Fig.4. $La_6Ni_6P_{17}$ is a normal metal with the electronic specific heat coefficient $\gamma = 45$ mJ/mol·K^2 and the Debye temperature of 380 K. $C(T)$ of $Ce_6Ni_6P_{17}$ exhibits a Shottky-like anomaly with a broad peak around 1.4 K. A Schottky-like specific heat is also calculated in typical triangular lattice[6] and previously mentioned octahedron simple cubic lattice[4]. The octahedron simple cubic model predicts that the strength of magnetic interaction J is related to a peak position T_p by $J \sim 0.5T_p$. The small value of T_p corresponds to the small Weiss temperature. In addition to this broad peak, two sharp but small peaks were also observed, which suggests magnetic phase transitions. The nature of these magnetic transitions is not clear at present. $J = 5/2$ multiplet of Ce^{3+} splits into three doublets under low point symmetry and the ground state has the entropy of Rln2, where R is gas constant. We evaluated the magnetic entropy S_m and the result is shown in Fig.5 as a ratio to Rln2. The evaluated entropy at 5 K is much smaller than Rln2. This result implies that $Ce_6Ni_6P_{17}$ is really frustrate system down to very low temperatures and the residual entropy is present.

In many Ce-based compounds, it is well known that the Kondo effect plays very important roles. The Kondo effect reduces the magnetic entropy and leads to heavy Fermi-liquid ground state which has a nearly $T-$independent large C/T-value at low temperatures. In many heavy-electron systems with an antiferromagnetic ground state, an reduced entropy is observed as in the present case. The lowest temperature in the present study is not low enough, and it is difficult to estimate $C/T-$value at $T = 0$ K. The shape of $C(T)$ and the strong $T-$dependent C/T, however, convince us to believe that the Kondo effect does not play a dominant role. This assertion is also evident from that fact of no tendency to saturate in $\chi(T)$ at low temperatures.

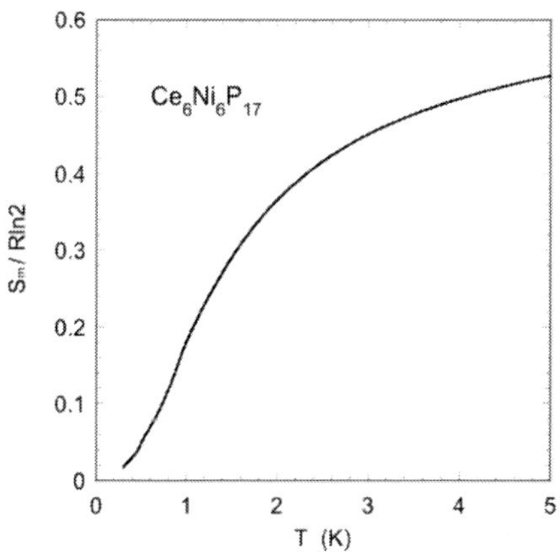

Figure 5. The magnetic entropy per Ce-ion of $Ce_6Ni_6P_{17}$. The entropy is about a half of Rln2 at 5 K.

4. Conclusions

We presented the magnetic susceptibility and specific heat measurements of $R_6Ni_6P_{17}$ with a unique crystal structure. We showed that $La_6Ni_6P_{17}$ is a normal metal and $Ce_6Ni_6P_{17}$ is a geometrically frustrated system originated from octahedrally aligned Ce-ions. We also showed the presence of possible residual entropy of about a half of Rln2.

Acknowledgments

This work was supported by a Grant-in-Aid for Scientific Research on Innovative Area "Heavy Electron" (No.20102004) of The Ministry of Education, Culture, Sports, Science, and Technology, Japan.

References

[1] Suekuni K, Avila M A, Umeo K and Takabatake T 2007 *Phys. Rev.* **B75** 195210
[2] Braun D J and Jeitschko W 1978 *Acta. Cryst.* B**34** 2069
[3] Koyama T, Yamashita H, Takahashi Y, Kohara T, Watanabe I, Tabata Y and Nakamura H 2008 *Phys. Rev. Letters* **101** 126404
[4] Tahara D, Motoe Y, Imada M 2007 *J. Phys. Soc. Jpn.* **76** 013708
[5] Takeda N, Nakajo T, Ono H, Tatematsu K and Nakano T 2011 *J. Phys. Soc. Jpn.* **80** SA035
[6] Wada K and Ishikawa T 1982 *J. Phys. Soc. Jpn.* **52** 1774

Characteristic magnetic phase diagram in heavy-fermion compound $YbCo_2Zn_{20}$

T Takeuchi[1,3], S Yoshiuchi[2], M Ohya[2], M Matsushita[2], S Yasui[2], Y Hirose[2], K Sugiyama[2,3], F Honda[2], R Settai[2] and Y Ōnuki[2]

[1] Low Temperature Center, Osaka University, Toyonaka, Osaka 560-0043, Japan
[2] Graduate School of Science, Osaka University, Toyonaka, Osaka 560-0043, Japan
[3] Center for Quantum Science and Technology under Extreme Conditions (KYOKUGEN), Osaka University, Toyonaka, Osaka 560-8531, Japan

E-mail: takeuchi@cqst.osaka-u.ac.jp

Abstract. We studied the magnetic field versus temperature phase diagram of the heavy-fermion compound $YbCo_2Zn_{20}$ with the cubic $CeCr_2Al_{20}$-type structure by measuring the specific heat, ac-susceptibility, and electrical resistivity under magnetic fields. In a low magnetic field region, a metamagnetic behavior was observed at $H_m = 6$ kOe for three principal field directions below $T_{\chi_{max}} = 0.32$ K, at which the temperature dependence of ac-susceptibility shows a local broad maximum. Two crossover lines were found below and above H_m. At high magnetic fields, the heavy-fermion state was rapidly suppressed with increasing magnetic field, and the field-induced ordered phase appeared in a very limited field direction around $H \parallel \langle 111 \rangle$ above the transition field $H_Q = 60$ kOe in the temperature range below $T_Q = 0.6$ K.

1. Introduction

Yb-based compounds have been extensively investigated since the Yb^{3+} has one $4f$-hole and are considered as a countersystem to one-$4f$-electron Ce-based compounds. The electron-hole analogy between $Yb^{3+}(4f^{13})$ and $Ce^{3+}(4f^1)$ electronic configurations tends to induce the localization of $4f$ electrons, enhancing a magnetically ordered state when pressure is applied to nonmagnetic Yb-based compounds. Namely, the pressure reduces the unit cell volume, which leads to the trivalent magnetic $Yb^{3+}(4f^{13})$ state rather than to the divalent nonmagnetic $Yb^{2+}(4f^{14})$ state. Another interesting feature of Yb-based compounds is valence transition or fluctuation due to the small energy gap for $Yb^{3+}(4f^{13})$– $Yb^{2+}(4f^{14})$ charge instabilities. While the number of cerium compounds showing such anomalous behavior is fairly large in number, there are relatively few Yb-based compounds showing typical properties of valence fluctuation or heavy-fermion behavior. This may be due to a relatively smaller spatial extension of the $4f$ shell in Yb^{3+} ion as compared to that in the Ce^{3+} ion. Namely, the smaller extension of the $4f$ shell in Yb-based compounds leads to the localization of $4f$ electrons and hence smaller hybridization between $4f$ electrons and conduction electrons. It is therefore needed to identify new Yb-based compounds and study their low-temperature properties.

Recently, a new family of Yb-based heavy-fermion compounds YbT_2Zn_{20} (T = Fe, Co, Ru, Rh, Os, and Ir) with the cubic $CeCr_2Al_{20}$-type ($Fd\bar{3}m$ space group) structure has been reported [1]. The lattice constant of these compounds is very large, i.e., $a \sim 14$ Å, and hence the distance between Yb-Yb atoms is large, more than 6 Å. In YbT_2Zn_{20}, the Yb atom forms the diamond structure and is encapsulated in the Frank-Kasper cage formed by 16 zinc atoms, while the transition atom T has the icosahedral zinc coordination, forming caged structures. These compounds show no long-range magnetic ordering at temperatures down to 0.02 K, and the electronic specific heat coefficient γ at low temperatures was found to be very large, ranging between 500 and 800 mJ/(K^2·mol) [1]. For $YbCo_2Zn_{20}$, $YbRh_2Zn_{20}$, and $YbIr_2Zn_{20}$, we succeeded in growing single crystals and found an abrupt nonlinear increase of magnetization or the metamagnetic behavior in the nonmagnetic heavy-fermion state [2, 3, 4]. The metamagnetic anomaly was observed at $H_m = 6$, 63, and 97 kOe below $T_{\chi_{max}} = 0.32$, 5.8, and 7.4 K for $YbCo_2Zn_{20}$, $YbRh_2Zn_{20}$, and $YbIr_2Zn_{20}$, respectively. Here, $T_{\chi_{max}}$ is the temperature at which the temperature dependence of magnetic susceptibility χ shows a local broad maximum.

Since the magnetic susceptibility of $YbCo_2Zn_{20}$ increases continuously with decreasing temperature in the temperature range from room temperature down to 2 K, following the Curie-Weiss law, the Kondo temperature in $YbCo_2Zn_{20}$ is expected to be much lower than that in $YbIr_2Zn_{20}$ with $T_{\chi_{max}} = 7.4$ K ($T_K = 21$ K) [1]. The corresponding electronic specific heat coefficient of $YbCo_2Zn_{20}$ is extremely large: $\gamma \simeq 8$ J/(K^2·mol) at low temperatures [1]. The cyclotron effective mass at zero magnetic field was estimated to be 100 - 500m_0 from the magnetic field dependences of the coefficient $\sqrt{A}$ in the Fermi liquid relation of the electrical resistivity, $\rho = \rho_0 + AT^2$, and cyclotron effective mass determined from the dHvA experiment [4]. The extremely large $A = 160$ $\mu\Omega$·cm/K^2 at $H = 0$ kOe was found to be reduced largely with increasing magnetic field to $A = 0.021$ $\mu\Omega$·cm/K^2 at 150 kOe. Here, we note that the coefficient $\sqrt{A}$ correlates with the γ value in the Fermi liquid relation.

In this paper, we present the characteristic H-T phase diagrams of the heavy-fermion compound $YbCo_2Zn_{20}$ obtained from specific heat, ac-susceptibility, and electrical resistivity measurements under various magnetic fields.

2. Experimental Procedure

Single crystal samples of $YbCo_2Zn_{20}$ were grown by the same Zn-self flux method as that used for $YbIr_2Zn_{20}$ [2, 3]. The specific heat was measured by the quasi-adiabatic heat pulse method under magnetic fields up to 90 kOe. The electrical resistivity and magnetoresistance were measured by an ordinary four-probe dc and/or ac method under magnetic fields of up to 150 kOe. The ac-susceptibility was measured by the mutual inductance method. A dilution refrigerator was utilized for the measurement below 1 K.

3. Experimental results and discussion

In a low magnetic field region, the magnetic field dependence of χ_{ac} shows a broad peak at $H_m = 6$ kOe at temperatures below $T_{\chi_{max}} = 0.32$ K, as shown in figure 1(b), which corresponds to the metamagnetic behavior of magnetization. As shown in figure 1(c), the temperature dependence of χ_{ac} at $H = 0$ kOe shows a broad peak at $T_{\chi_{max}} = 0.32$ K and decreases slightly at lower temperatures. $T_{\chi_{max}}$ shifts to lower temperatures when the magnetic field is increased and disappears around $H_m = 6$ kOe. For further increasing the magnetic field, a broad peak in χ_{ac} appears again above H_m and shifts to higher temperatures, as indicated by open circles in figure 1(a). Interestingly, the magnetic field dependence of the specific heat in the form of C/T at 0.095 K shows distinct two peaks at about 4.0 and 7.5 kOe, as indicated by arrows in figure 1(d), and exhibits a local minimum at H_m. With increasing temperature, the peak at a higher (lower) magnetic field shifts to higher (lower) fields, as indicated by arrows for the data at 0.2 K. These points are also plotted in figure 1(a) by open triangles.

Evolution of the temperature dependence of electrical resistivity under magnetic fields is also shown in figure 1(a) as the contour plot of the power law exponent n for the temperature dependent resistivity in the form of T^n. This plot was constructed by interpolating the results of temperature sweeps at sixteen values of applied magnetic fields: a 1 kOe interval from 0

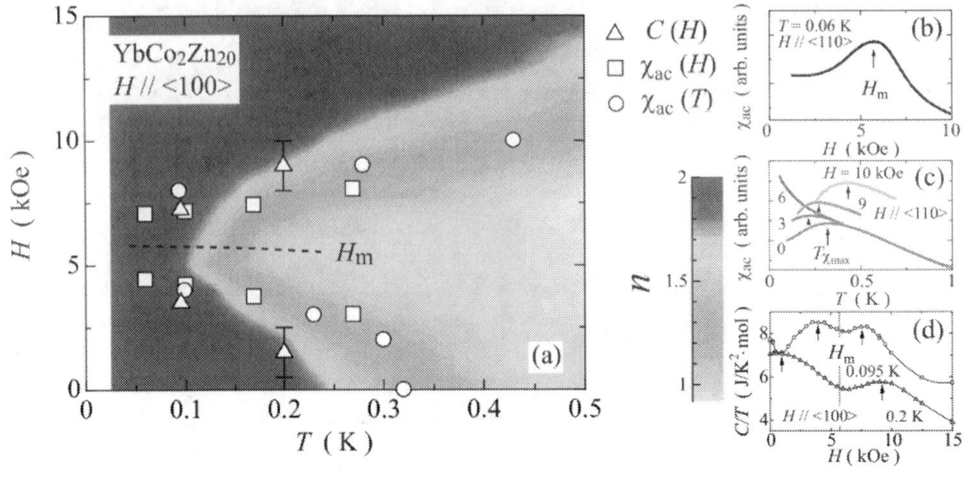

Figure 1. (a) Contour plot of the power law exponent n for the temperature dependent resistivity in the form of T^n. The plot was constructed by interpolating the results of temperature sweeps at magnetic fields from 0 to 15 kOe with a 1 kOe interval. The open triangles and circles were determined from the magnetic field dependence of C/T and the temperature dependence of χ_{ac} under constant magnetic fields, respectively. (b) Magnetic field and (c) temperature dependences of χ_{ac} for $H \parallel \langle 110 \rangle$, and (d) magnetic field dependences of C/T at 0.095 and 0.2 K for $H \parallel \langle 100 \rangle$ of $YbCo_2Zn_{20}$.

to 15 kOe. From this plot, it is found that there are two crossover lines below and above the metamagnetic anomaly at $H_m = 6$ kOe. The Fermi-liquid state develops at lower temperatures than these crossover lines. Note that these characteristic features at low magnetic fields, including the metamagnetic behavior at H_m, are almost the same for the three principal field directions, $H \parallel \langle 100 \rangle$, $\langle 110 \rangle$, and $\langle 111 \rangle$.

In contrast to the low field behavior, the field-induced ordered phase (FIOP) was observed in a limited angular range around the $\langle 111 \rangle$ direction above the transition field $H_Q = 60$ kOe below $T_Q = 0.6$ K, as shown in figure 2(a). At the phase boundary of the FIOP, the electrical resistivity and specific heat show a clear anomaly at H_Q or T_Q, as shown in figures 2(b) and 2(c). At $T = 0.08$ K, the magnetoresistance shows peaks at $H_m = 6$ kOe and $H_Q = 60$ kOe, which correspond to the metamagnetic anomalies. With increasing temperature, the peak at H_m rapidly smears out and becomes obscure at a temperature between 0.30 and 0.45 K, which is similar to the results for $H \parallel \langle 100 \rangle$ [4]. The peak at H_Q, on the other hand, slightly shifts to higher fields and becomes broader with increasing temperature.

The specific heat at 50 kOe, which is below $H_Q = 60$ kOe, decreases linearly below 1 K without any anomalous behavior. When we apply the magnetic field of 70 kOe, a λ-shaped peak was observed at 0.4 K, indicating the evolution of FIOP. The λ-shaped peak shifts to about 0.6 K and the peak becomes larger at the magnetic field of 90 kOe. These results strongly support the existence of the *field-induced long-range-ordered phase* for $H \parallel \langle 111 \rangle$. Very recently, the FIOP of $YbCo_2Zn_{20}$ for $H \parallel \langle 111 \rangle$ has been confirmed by the capacitance magnetization measurement at low temperatures [5].

The characteristic features of the FIOP in $YbCo_2Zn_{20}$ are very similar to those in $PrFe_4P_{12}$, where the FIOP was observed in a limited angular range around the $\langle 111 \rangle$ direction above the other ordered phase [6]. The FIOP in $PrFe_4P_{12}$ has been explained by considering a two-

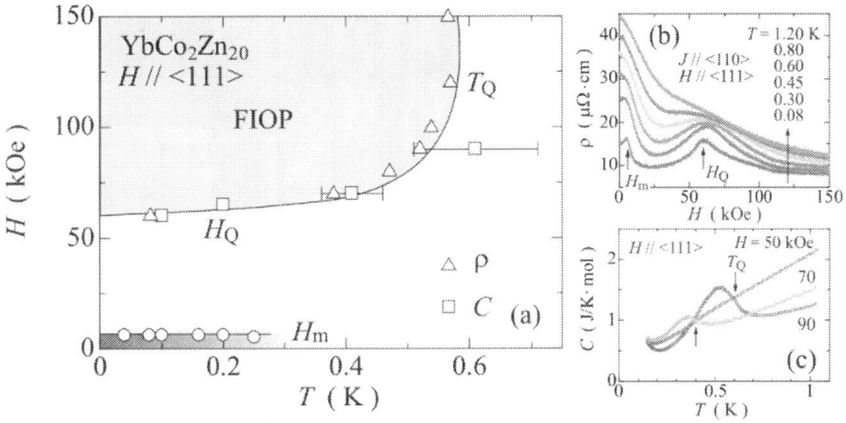

Figure 2. (a) H-T phase diagram for $H \parallel \langle 111 \rangle$ in $YbCo_2Zn_{20}$. Squares are determined by the specific heat measurement, and triangles are obtained by the resistivity measurement. (b) Magnetoresistances with the current $J \parallel \langle 110 \rangle$ for $H \parallel \langle 111 \rangle$ at several selected temperatures, and (c) temperature dependences of the specific heat C under several magnetic fields along the $\langle 111 \rangle$ direction.

sublattice mean-field model with quadrupole interactions for O_2^0 and O_2^2. In this case, the level crossing of the two lowest levels occurs for $H \parallel \langle 111 \rangle$ in the proposed CEF scheme. The numerical analyses for $YbCo_2Zn_{20}$ based on the CEF model suggest that the FIOP for $H \parallel \langle 111 \rangle$ is the antiferro-quadrupolar ordered phase with the quadrupole moment of O_2^0 or O_2^2. In order to clarify the order parameter of the FIOP, a further experimental study is necessary.

4. Conclusion

H-T phase diagrams of the heavy-fermion compound $YbCo_2Zn_{20}$ were studied precisely using single crystal samples. The characteristic crossover phase diagram was obtained for the low field region below about 15 kOe, which was found to be independent of the direction of magnetic field. On the other hand, the FIOP was observed in a limited angular range around the $\langle 111 \rangle$ direction above the transition field $H_Q = 60$ kOe in the temperature range below $T_Q = 0.6$ K.

Acknowledgments

This work was supported by a Grant-in-Aid for Scientific Research on Innovative Areas "Heavy Electrons" (No. 20102002), Specially Promoted Research (No. 20001004), Young Scientists (B), (No. 18740219) and Osaka University Global COE Program "Core Research and Engineering of Advanced Materials-Interdisciplinary Education Center for Materials Science" (G10) from the Ministry of Education, Culture, Sports, Science, and Technology, Japan.

References

[1] Torikachvili M S, Jia S, Mun E D, Hannahs S T, Black R C, Neils W K, Martien D, Bud'ko S L, and Canfield P C 2007 *Proc. Natl. Acad. Sci. U.S.A.* **104** 9960
[2] Yoshiuchi S *et al* 2009 *J. Phys. Soc. Jpn.* **78** 123711
[3] Takeuchi T *et al* 2010 *J. Phys. Soc. Jpn.* **79** 064609
[4] Ohya M, Matsushita M, Yoshiuchi S, Takeuchi T, Honda F, Settai R, Tanaka T, Kubo Y and Ōnuki Y 2010 *J. Phys. Soc. Jpn.* **79** 083601
[5] Shimura Y, Sakakibara T, Yoshiuchi S, Honda F, Settai R and Ōnuki Y 2011 *J. Phys. Soc. Jpn.* **80** 073707
[6] Tayama T, Custers J, Sato H, Sakakibara T, Sugawara H and Sato H 2004 *J. Phys. Soc. Jpn.* **73** 3258

Journal of Physics: Conference Series **391** (2012) 012073 doi:10.1088/1742-6596/391/1/012073

First-principle calculations for electronic properties of PuX_3 (X=Rh, Pd, Pt)

Yasutomi Tatetsu[1] and Takahiro Maehira[2]

[1] Graduate school of Engineering and Science, University of the Ryukyus, Nishihara, Okinawa 903-0213

[2] Faculty of Science, University of the Ryukyus, Nishihara, Okinawa 903-0213

E-mail: giappoliano@phys.u-ryukyu.ac.jp

Abstract. Energy band structures of PuX_3 (X=Rh, Pd, and Pt) are investigated by a relativistic linear augmented-plane-wave method with the exchange-correlation potential in a local density approximation. It is found in common that the energy bands in the vicinity of the Fermi level are mainly due to the hybridization between Pu $5f$ and X d electrons.

1. Introduction

The traunsuranium compounds PuX_3 with the $AuCu_3$-type cubic crystal structure, where X means Rh, Pd, and Pt in the periodic table, show an interesting trend of $5f$-electron characters. In the cerium and uranium compounds with this cubic type structure have studied a lot in experimental and theoretical ways. On the other hand, there are only few theoretical studies for these compounds.

These three compounds order antiferromagnetically with Neel temperature T_N of 6.6, 2.4, and 40 K, respectively[1]. From the other magnetic susceptibility and heat capacity measurements on $PuPd_3$ indicate an antiferromagnetic transition at $T_N = 24.6$ 0.4 K reported by M. D. Le *et al.*[2]. Photoelectron spectroscopy measurements indicate that the Pu ions in $PuPd_3$ are well localized[3]. The lattice constant ($PuRh_3$: 4.008 A, $PuPd_3$: 4.102A, $PuPt_3$: 4.103A) increases with increasing the atomic number of X (Rh=45, Pd=46, and Pt=78)[1].

In this paper, we research what the key issues are to construct the energy band structures around the Fermi energy for PuX_3 we attempt to unveil $5f$ electron properties purely originating from actinide atoms.

2. Method of Band Calculation

The calculations for the energy band structures are carried out by using the relativistic linear augmented-plane-wave (RLAPW) method[4]. All $5f$ electrons are assumed to be itinerant and the calculations are performed in the paramagnetic phase. To explain the electronic structure of $PuPd_3$, a LDA+U calculation might be needed because photoelectron spectroscopy measurements indicate that the Pu f-states are localized. We will discuss a LDA+U method applying to PuX_3 in a future paper. Note here that relativity should be taken into account, because of large atomic numbers of the constituent atoms. The spatial shape of the one-electron

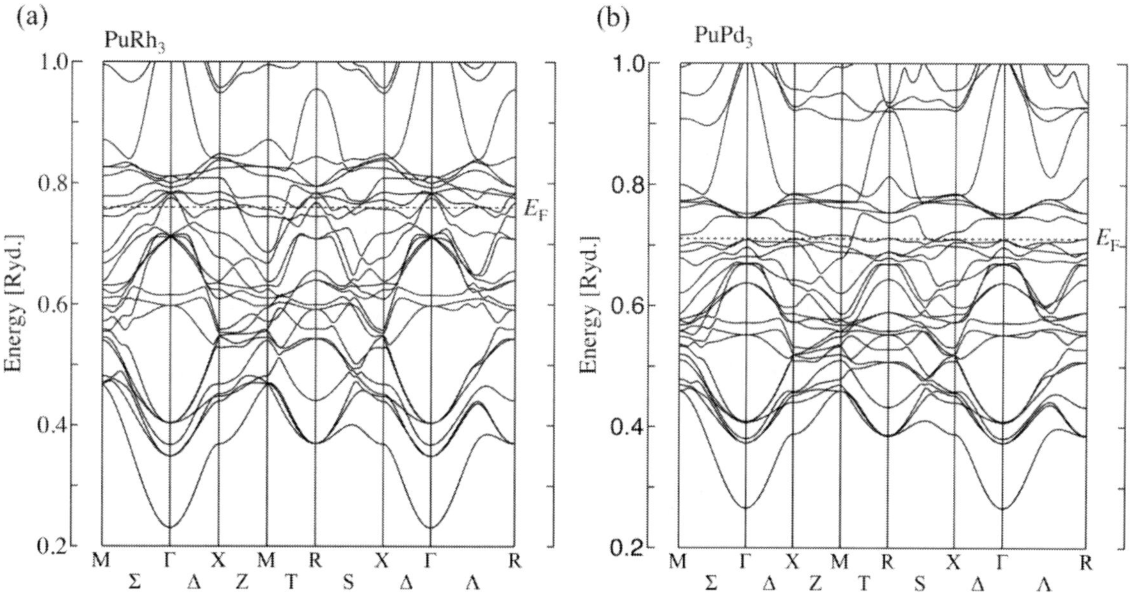

Figure 1. The energy band structure for (a) $PuRh_3$ and (b) $PuPd_3$ calculated by using the self-consistent RLAPW method. E_F indicates the position of the Fermi level and is shown by the dashed line.

potential is determined in the muffin-tin approximation, and we use the exchange and correlation potential in the local density approximation (LDA). The self-consistent calculation is carried out for the experimental lattice constant.

PuX_3 (X=Rh, Pd, and Pt) materials are categorized into the $AuCu_3$-type structure, which is characterized by the space group Pm3m. The $5f$ electrons in PuX_3 are assumed to be itinerant. The iteration process for solving the Dirac one-electron equation starts with the crystal charge density that is constructed by superposing the relativistic atomic charge densities for neutral atoms Pu([Rn]$5f^5 6d^1 7s^2$), Rh([Kr]$4d^8 5s^1$), Pd([Kr]$4d^{10}$), and Pt([Xe]$5d^9 6s^1$), where [Rn], [Xe], and [Kr] symbolically indicate the closed electronic configuration for radon, xenon, and krypton, respectively. In the relativistic atomic calculation, the spin-orbit splitting in Pu $5f$, and Rh $4d$, Pd $4d$, and Pt $5d$ states are found to be 80 mRyd., 29 mRyd., 35 mRyd., and 98 mRyd., respectively. Note here that mRyd. denotes milli-Rydberg and 1 Ryd.=13.6 eV.

3. Results of Band Calculation

First let us discuss the calculated results for $PuRh_3$ and $PuPd_3$ as shown in Figure 1(a) and (b), in which we depict the energy band structure along the symmetry axes in the Brillouin zone in the energy region from 0.2 Ryd. to 1.0 Ryd..

The Fermi levels are found to be 0.760 Ryd., 0.711 Ryd., and 0.854 Ryd. for $PuRh_3$, $PuPd_3$, and $PuPt_3$, respectively. Concerning the energy band structure in the vicinity of E_F, we emphasize that there always occurs a hybridization between actinide $5f$ and X d states among three materials. Above E_F near the point in $PuPd_3$, the flat $5f$ bands split into two groups, corresponding to the total angular momentum j=5/2 (lower bands) and 7/2 (upper bands).

The calculated total density of states (DOS) for (a) $PuPd_3$ and (b) $PuPt_3$ are shown in Figure 2. The bold line, dashed line, and thin line indicate the total and partial DOS for d and f components, respectively. The $5f$ bands are split into two subbands by the spin-orbit

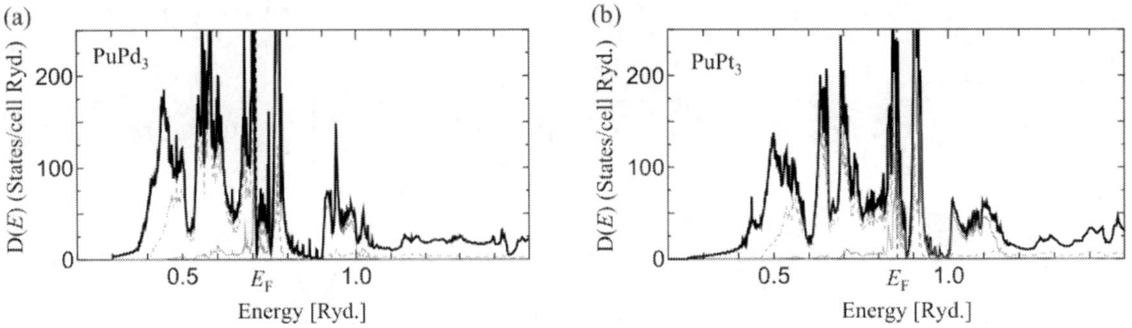

Figure 2. Density of states for (a) $PuPd_3$ and (b) $PuPt_3$. The total density of states, d components in X atoms, and f components in Pu atom are shown by the bold line, dashed line, and thin line, respectively. E_F indicates the Fermi energy.

interaction and due to the hybridization between Pu $5f$ and X d electrons, finite DOS always appear at the Fermi level. Since the Fermi level E_F lies between the 5f subbands, the density of states has sharp peaks in the vicinity of E_F. The total density of states at E_F are evaluated as $D(E_F) = 131.7$ states/Ryd.cell, 4.5 states/Ryd.cell, and 151.9 states/Ryd.cell in $PuRh_3$, $PuPd_3$, and $PuPt_3$, respectively. By using these values, the theoretical specific heat coefficient $_{\text{band}}$ are 22.8, 0.79, and 26.3 mJ/K^2·mol for $PuRh_3$, $PuPd_3$, and $PuPt_3$, respectively. The experimental value for $PuPd_3$ is 71.1 0.1 mJ/K^2·mol [2]. The d components in Rh, Pd, and Pt are distributed around the E_F.

In figure 1(a), four bands cross the E_F so that $PuRh_3$ has four Fermi surfaces, three hole sheets in the 16th-18th bands and one electron sheet in the 19th band. Figure 3 shows (a) the hole sheets in the 17th, (b) 18th band, and (c) electron sheet in the 19th band for $PuRh_3$. The Fermi surface in the 16th band not shown here is a small cube-like hole sheet which is centered at the point. The 17th band has a rugged hole sheet centered at the point and a spherical small centered at the R point. Also, there are eight hole pockets lying on the axis and small pockets on general points in the Brillouin zone. The 18th band is composed of a small spherical hole sheet centered at the R point and a large hole sheet centered at the point, which

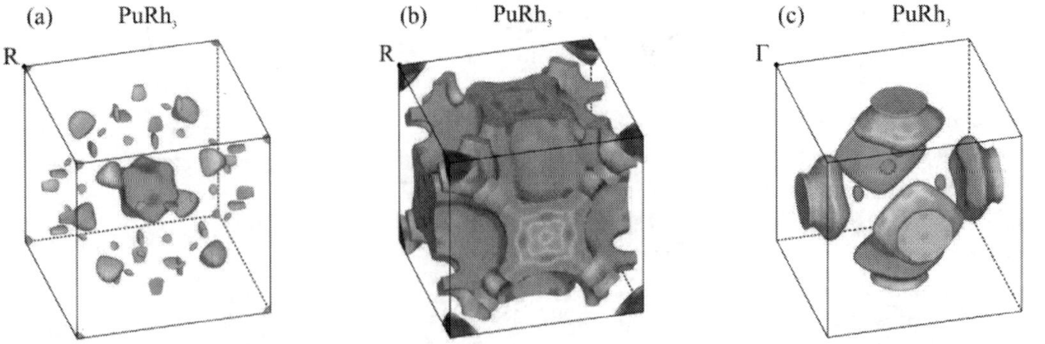

Figure 3. (a) The hole-Fermi surfaces in the 17th, (b) 18th bands, and (c) electron sheet in the 19th band for $PuRh_3$. The center point in each Brillouin zone is the in (a) and (b), and R point in (c).

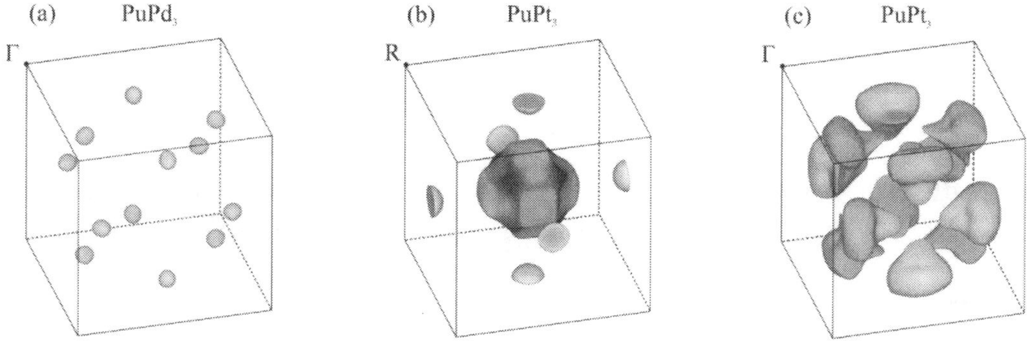

Figure 4. (a) The electron Fermi surface in the 20th band for $PuPd_3$ centered at the R point. (b) The hole- and (c) electron Fermi surfaces in the 19th and 20th bands for $PuPt_3$ centered at the and R points, respectively.

is multiply connected from zone to zone on each cubic plane. The 19th band has three dog's bone-like electron sheets centered at the each M point and eight small electron pockets lying on the T axis.

The Fermi surfaces of $PuPd_3$ are composed of two kinds of electron sheets. The 19th band has a small spherical electron sheet centered at the R point and six tablet-like electron sheets lying on the T axis. The 20th band has twelve small pockets, as shown in Figure 4 (a). The total number of holes equal to that of electrons in these Fermi surface sheets, which represents that $PuPd_3$ is a compensated metal.

$PuPt_3$ has three Fermi surfaces. The Fermi surfaces in the 18th and 19th bands of $PuPt_3$ look like same shapes but the 18th bands has a smaller Fermi surface than that of 19th band shown in Figure 4 (b), which is constructed by a octahedron-like hole sheet centered at the point and three small spherical hole sheets centered at the each X point. Figure 4 (c) shows the electron sheet in the 20th band of $PuPt_3$ centered at the R point. This Fermi surface is composed of twelve complicated electron sheets. The number of carries contained in these Fermi surface sheets are 0.016 holes, 0.114 holes, and 0.130 electrons in the 18th, 19th, and 20th bands, respectively. The total number of holes is equal to that of electrons, which represents that $PuPt_3$ is a compensated metal.

4. Discussion and Conclusion

We have applied the RLAPW method to the self-consistent calculation of the electronic structure for $PuRh_3$, $PuPd_3$, and $PuPt_3$ on the basis of the itinerant $5f$ electron picture by assuming the non-magnetic phase. We found that a strong hybridization between Pu $5f$ and X d states occurs in the vicinity of E_F, and this makes these Fermi surfaces complicated. To clarify the magnetism in PuX_3, the study in a hybridization between the $5f$ and d characters (f-d model) is now in progress.

References

[1] Harvey A R , Brodsky M B , and Nellis W J 1973 *Phys. Rev.* B **7** 4137
[2] Le M D , McEwen K A, Wastin F, Colineau E, Jardin R, and Rebizant J 2008 *Physica* B **403** 1035-1037
[3] Le M D, Walker H C, McEwen K A, Gouder T, Huber F and Wastin F 2008 *J. Phys.: Condens. Matter* **20** 275220
[4] Higuchi M and Hasegawa A 1995 *J. Phys. Soc. Jpn.* **64** 830
[5] Gunnarsson O and Lundqvist B I 1976 *Phys. Rev.* B **13** 4274

Inelastic X-ray Scattering of Heavy Fermion Compounds ROs$_4$Sb$_{12}$ (R: Light Rare-Earth)

S. Tsutsui [1], H. Uchiyama [1], J. P. Sutter [2], A. Q. R. Baron [1,3], J. Yamaura [4], H. Sugawara [5], and H. Sato [6]

[1] Japan Synchrotron Radiation Research Institute, SPring-8, Sayo, Hyogo 679-5198, Japan

[2] Diamond Light Source, Didcot, Oxfordshire OX11 0DE, United Kingdom

[3] RIKEN, SPring-8, Sayo, Hyogo 679-5148, Japan

[4] Institute for Solid State Physics, University of Tokyo, Kashiwa, Chiba 277-8581, Japan

[5] Graduate School of Science, Kobe University, Kobe, Hyogo 657-8501, Japan

[6] Department of Physics, Tokyo Metropolitan University, Hachioji, Tokyo 192-0397, Japan

E-mail: satoshi@spring8.or.jp

Abstract. Inelastic X-ray scattering experiments were carried out in the series of filled skutterudites ROs$_4$Sb$_{12}$ (R: light rare-earth). Phonon energy of the rare-earth guest modes in the longitudinal (1 0 0) direction decreases with increase of atomic number of rare-earth elements from La to Sm inserted into a Sb cage. This fact demonstrates that the guest mode energy is affected by the ionic radius of the rare-earth atom. In addition, comparison between inelastic X-ray scattering and X-ray diffraction experiments reveals the correlation of the guest mode energy in the longitudinal (1 0 0) direction and the Einstein temperature obtained by the temperature dependence of the atomic displacement factor.

1. Introduction

Low-lying optical modes in cage-structured compounds, which are called a guest mode, have been an interesting problem in strongly correlated electron systems for recent years. Correlation of the ground states such as superconductivity and heavy fermion behavior with the guest modes has been discussed in these compounds, filled skutterudites, β-pyrochlore oxides and so on, for example [1, 2]. In some of filled skutterudites, the presence of low-lying rare-earth guest mode was reported previously by inelastic neutron and X-ray scattering and second-order Raman scattering [3-5]. Since the importance of the precise estimation of the guest mode energy was pointed out to interpret the results of specific heat measurements [6], rare-earth dependence of the guest mode is crucial to discuss the physical properties at least in filled skutterudites. We reported rare-earth dependence of the guest mode energy in some of $R\mathrm{Os_4Sb_{12}}$ compounds [7].

We measured inelastic X-ray scattering (IXS) spectra of $\mathrm{CeOs_4Sb_{12}}$ and $\mathrm{EuOs_4Sb_{12}}$ at 300 K. The spectra of these compounds are similar to the other $R\mathrm{Os_4Sb_{12}}$ compounds measured previously. The fact is clarified that the lowest mode at the zone boundaries is the motion of the rare-earth atoms inserted into the Sb icosahedral cage in $R\mathrm{Os_4Sb_{12}}$ compounds. In other words, the lowest energy mode is the guest mode in the $R\mathrm{Os_4Sb_{12}}$ compounds. Evident rare-earth dependence of the phonon energy in the lowest mode of every $R\mathrm{Os_4Sb_{12}}$ was found in the present work. However, the results obtained are to understand by change of the rare-earth mass. Here, we discuss the rare-earth dependence of the guest modes and comparison of the present results with the recent results of X-ray diffraction experiments using single crystalline samples.

2. Experimental Setups

IXS experiments were carried out at BL35XU in SPring-8 [8, 9]. The Si(11 11 11) backscattering setup whose energy resolution is about 1.5 meV was chosen in the present work. The IXS spectra of $R\mathrm{Os_4Sb_{12}}$ (R: light rare-earth) were measured at room temperature. The measured samples were single crystals prepared by the Sb self-flux method with the typical size less than 1 $\mathrm{mm^3}$. The rocking curve width of each sample is about 0.02 degrees, very close to the resolution of the spectrometer.

3. Experimental Results and Discussion

Figure 1 shows IXS spectra of filled skutterudites $R\mathrm{Os_4Sb_{12}}$ (R = La, Ce, Sm and Eu) around (9 0 0), a zone boundary in the longitudinal (1 0 0) direction, at room temperature. Phonon excitation around 5 meV shows strong R-dependence, whereas that around 12 meV is independent of R atom. Comparison of the phonon dispersion between $\mathrm{IrSb_3}$ and $\mathrm{LaOs_4Sb_{12}}$ demonstrates that the excitation around 5 meV comes from the rare-earth atomic motion in the icosahedral Sb cage [10]. The R-dependence of the excitation energy is difficult to interpret the mass of rare-earth elements. If the

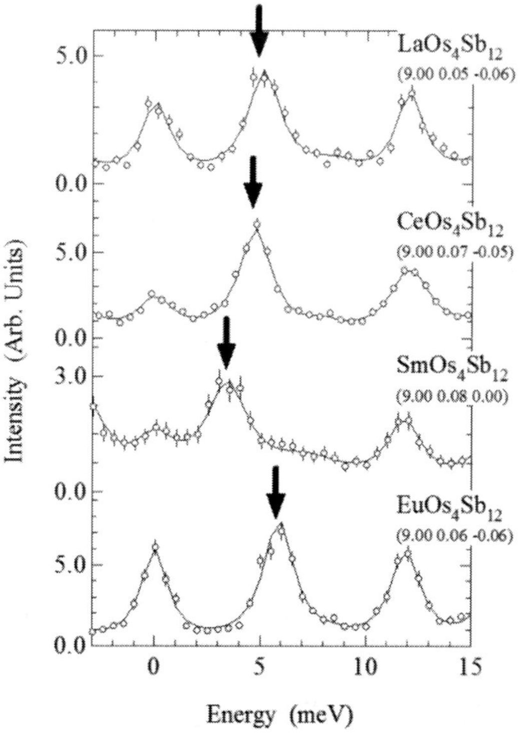

Fig 1. IXS spectra of $R\mathrm{Os_4Sb_{12}}$ (R = La, Ce, Sm and Eu) at room temperature. Arrowed excitations show the phonon modes associated with R atoms' motion in the icosahedral Sb cage.

excitation energy correlates with the rare-earth mass, the *R*-dependence energy decreases with increase of the atomic number. The result of $EuOs_4Sb_{12}$ rules out the contribution of the rare-earth mass as an important factor to determine the guest mode energy in ROs_4Sb_{12} compounds.

Figure 2 shows the plot of the guest mode energy at the zone boundary in the longitudinal (1 0 0) direction determined by the previous IXS experiments [7] and the present ones and Einstein temperature determined by X-ray diffraction experiments [11] which are associated with the rare-earth motion in the icosahedral Sb cage. Although the Einstein temperature in $EuOs_4Sb_{12}$ has not reported yet, the Einstein temperature reasonably agrees with the guest mode energy in the longitudinal (1 0 0) direction. In $SmOs_4Sb_{12}$, the Einstein temperature obtained by X-ray diffraction agrees very well with the guest mode energy measured by IXS. As was reported in Ref. 7, the phonon energy in the transverse (1 1 0) direction is by about 1 meV lower than that in the longitudinal (1 0 0) direction. Judging from the phonon dispersion in $SmOs_4Sb_{12}$ [12], the guest mode energy in the longitudinal (1 0 0) direction is the highest and that in the transverse (1 1 0) direction is the lowest. The results of IXS and X-ray diffraction experiments imply that the Einstein temperatures obtained by X-ray diffraction provides the highest guest mode energy.

What is important factor in the guest mode energy? This was already discussed experimentally in filled skutterudites [6, 13]. The previous systematic studies using specific heat and EXAFS measurements suggest that one of the important factors is guest free space, free space at the rare-earth site in filled skutterudites. From the aspect of the guest free space in ROs_4Sb_{12} compounds, the present result infers that *R*-dependence of the guest mode energy is difficult to explain under the assumption of the trivalent rare-earth guest ions in icosahedral Sb cage. In $SmOs_4Sb_{12}$ and $EuOs_4Sb_{12}$, the previous works suggest that both Sm and Eu valence are an intermediate state between divalent and trivalent ones [14, 15]. In addition, the ferromagnetic ground states of both $SmOs_4Sb_{12}$ ($EuOs_4Sb_{12}$) suggests Sm (Eu) ion occupies the valence state close to a trivalent (divalent) state [15, 16]. Since divalent rare-earth ions are larger than trivalent one, the result shown in Fig. 2 depicts the correlation between guest mode energy and guest free space.

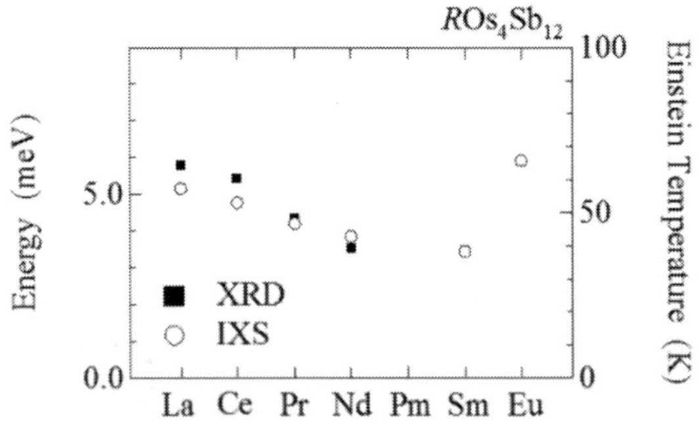

Fig. 2. Rare-earth dependence of the phonon excitation energies of the rare-earth atomic motion around (9 0 0) determined by IXS (labelled as IXS) and the Einstein temperatures at the rare-earth site obtained by X-ray diffraction (labelled as XRD) in ROs_4Sb_{12} compounds.

4. Summary

We have succeeded in investigation of *R*-dependence of the guest modes in ROs_4Sb_{12} and comparison between the guest mode energy determined by IXS and Einstein temperature by X-ray diffraction. Since the guest mode energy is close to the Kondo temperature in typical HF compounds, the present results imply that the guest mode may be contributed to large electronic specific heat coefficient in the series of ROs_4Sb_{12}, which is predicted by theories based on the HF mechanism induced by electron-phonon coupling [17-21]. The agreement between the guest mode energy and Einstein temperature indicates that the Einstein model is reasonable approximation to estimate the guest mode energy in $RO s_4 Sb_{12}$.

Acknowledgments

The author ST thanks M. Udagawa, N. Ogita, T. Hasegawa, H. Kusunose, C. Lee and K. Kaneko for their fruitful discussion and K. Miyake and P. Thalmeier for their interest. The IXS experiments were carried out under the approval of JASRI (Proposal Nos. 2005A0369-ND3d-np, 2006A1039, 2007A1301, 2007B2077, 2008A1058 and 2009B1126). The present work is partially supported by the Grant-in-Aid for Young Scientists (No. 20740181) and for Scientific Research (C) (22540345) from the Japan Society of Promotion of Science, and the Grant-in-Aids for Scientific Research Priority Area "Skutterudite" (No. 18027017) and for Scientific Research on Innovative Area "Heavy Electrons" (No. 20102005) of the Ministry of Education, Culture, Science and Technology, Japan.

References

[1] Sato H, Kikuchi D, Tanaka K, Ueda M, Aoki H, Ikeno T, Tatsuoka S, Kuwahara K, Aoki Y, Kohgi M, Sugawara H, Iwasa K and Harima H 2008 *J. Phys. Soc. Jpn.* **77** *Suppl.A* 1 and references therein.
[2] Nagao Y, Yamaura J, Ogusu H, Okamoto Y and Hiroi Z, 2009 *J. Phys. Soc. Jpn.* **78** 064702.
[3] Ogita N, Kondo T, Hasegawa T, Takasu Y, Udagawa M, Takeda N, Ishikawa K, Sugawara H, Kikuchi D, Sato H, Sekine C and Shirotani I, 2006 *Physica B* **383** 128.
[4] Lee C H, Hase I, Sugawara H, Yoshizawa H and Sato H, 2006 *J. Phys. Soc. Jpn.* **75** 123602.
[5] Tsutsui S, Kobayashi H, Ishikawa D, Sutter J P, Baron A Q R, Hasegwa T, Ogita N, Udagawa M, Yoda Y, Onodera H, Kikuchi D, Sugawara H, Sekine C, Shirotani I and Sato H, 2008 *J. Phys. Soc. Jpn.* **77** 033601.
[6] Matsuhira K, Wakeshima M, Hinatsu Y, Sekine C, Shirotani I, Kikuchi D, Sugawara H and Sato H, 2007 *J. Magn. Magn. Mater.* **310** 226.
[7] Tsutsui S, Uchiyama H, Sutter J P, Baron A Q R, Sugawara H, Yamaura J, Hiroi Z, Ochiai A and Sato H, 2010 *J. Phys.: Conf. Ser.* **200** 012213.
[8] Baron A Q R, Tanaka Y, Goto S, Takeshita K, Matsushita T and Ishikawa T, 2000 *J. Phys. Chem. Solids* **61** 461.
[9] Baron A Q R, Sutter J P, Tsutsui S, Uchiyama H, Masui T, Tajima S, Heid R and Bohnen K -P, 2008 *J. Phys. Chem. Solids* **69** 3100.
[10] Tsutsui S, Kobayashi H, Sutter J P, Uchiyama H, Baron A Q R, Yoda Y, Kikuchi D, Sugawara H, Sekine C, Shirotani I, Ochiai A and Sato H, 2008 *J. Phys. Soc. Jpn.* **77** *Suppl. A* 257.
[11] Yamaura J and Hiroi Z, 2011 *J. Phys. Soc. Jpn.* **80**, 054601.
[12] Tsutsui S, Mizumaki M, Tsubota M, Uchiyama H, Baron A Q R, Uruga T, Kikuchi D, Sugawara H and Sato H, 2011 *J. Phys. Soc. Jpn.* **80** *Suppl. A* SA036.
[13] Mizumaki M, Tsutsui S, Uruga T, Tanida H, Kikuchi D, Sugawara H and Sato H, 2011 *J.Phys. Soc. Jpn.* **80** 074603.
[14] Mizumaki M, Tsutsui S, Tanida H, Uruga T, Kikuchi D, Sugawara H and Sato H, 2007 *J. Phys. Soc. Jpn* **76** 053706.
[15] Bauer E D, Ślebarski A, Frederick N A, Yuhasz W M, Maple M B, Cao D, Bridges F, Giester G, and Rogl P 2004 *J. Phys.: Condens. Matter* **16** 5095.
[16] Sanada S, Aoki Y, Aoki H, Kikuchi D, Sugawara H and Sato H, 2005 *J. Phys. Soc. Jpn.* **74** 246.
[17] Yu C and Anderson P W, 1984 *Phys. Rev. B* **29** 6165.
[18] Kusunose H and Miyake K, 1996 *J. Phys. Soc. Jpn.* **65** 3032
[19] Mitsumoto K and Ōno Y, 2005 *Physica C* **426-431** 330.
[20] Hattori K and Miyake K, 2007 *J. Phys. Soc. Jpn.* **76** 094603.
[21] Hotta T, 2008 *J. Phys. Soc. Jpn* **77** 103711.

International Conference on Strongly Correlated Electron Systems (SCES 2011) IOP Publishing
Journal of Physics: Conference Series **391** (2012) 012075 doi:10.1088/1742-6596/391/1/012075

Pressure effect on the ferromagnetism of an off-center rattling system $Eu_8Ga_{16}Ge_{30}$

K. Umeo[1,2], H. Yamane[2], M. A. Avila[3], T. Onimaru[2], T. Takabatake[2,4]

[1]N-BARD, Hiroshima University, Higashi-Hiroshima, 739-8526, Japan
[2]Department of Quantum Matter, ADSM, and [4]IAMR, Hiroshima University, Higashi-Hiroshima 739-8530, Japan
[3]Federal University of ABC, Santo Andre, SP, 09210-170, Brazil

E-mail: kumeo@sci.hiroshima-u.ac.jp

Abstract. We report the pressure effect on the ferromagnetism of type-I clathrate $Eu_8Ga_{16}Ge_{30}$ with T_C = 36 K by means of the resistivity ρ and Hall coefficient R_H measurements. With increasing pressure up to 11.4 GPa, T_C increases up to 45 K. The concomitant decrease of ρ is attributed to the increase of the carrier concentration because the normal part of R_H decreases with the ratio of -3.5%/GPa. At ambient pressure, $\rho(T)$ has broad peak at T^*=23 K. Upon pressurizing, the peak is weakened but remain visible at T^* = 27 K even at P = 11.4 GPa. Our findings suggest that the anomaly at T^* is not so sensitive to the cage size up to 11 GPa, where the cage volume is compressed to 87% of the value at ambient pressure.

1. Introduction

Intermetallic clathrates showing low thermal conductivity and high thermopower have attracted much attention because of their potential in thermoelectric applications [1-3]. Among a large number of clathrate compounds, type-I clathrates $X_8Ga_{16}Ge_{30}$ (X = Ba, Sr and Eu) have been most extensively investigated. The unit cell consists of two dodecahedra and six tetrakaidecahedra composed of Ga and Ge atoms, which incorporate X1 atoms (2a site) and X2 atoms (6d site), respectively [4]. Neutron and x-ray diffraction studies revealed that the 6d site for X = Sr and Eu is split into four-fold sites displaced from the cage center, while the 6d site for X = Ba is not split [2, 5] . The reduced themal conductivity for $Sr_8Ga_{16}Ge_{30}$ and $Eu_8Ga_{16}Ge_{30}$ was attributed to highly anhamonic and localized vibrations (so called rattling) of X2 atoms in the tetrakaidecahedral cage [2].

$Eu_8Ga_{16}Ge_{30}$ is the only clathrate compound where magnetic ions fully occupy the guest positions in the cages. Furthermore, it is dimorphic, having type-I and type-VIII structures. The compounds of the two structures exhibit ferromagnetic transitions at T_C = 36 K and 10.5 K, respectively [1, 2, 6]. The divalent state of the Eu ions in the two compounds manifests itself in the Curie-Weiss-type temperature dependence of the magnetic susceptibility with the effective moment of 7.9 μ_B/Eu and the saturated magnetization M_s of 7 μ_B/Eu. The magnetic and transport measurements for $Eu_8Ga_{16-x}Ge_{30+x}$ with both type structures revealed that the thermal and transport properties depend on the carrier concentration, whereas T_C and M_s do not [1, 7, 8]. Pacheco *et al.* reported that the higher T_C in type-I than in type-VIII is due to the enhanced effective mass of the conduction electrons in type-I [7]. For type-I $Eu_8Ga_{16}Ge_{30}$, both the electrical resistivity ρ and the temperature derivative of magnetization M, dM/dT exhibit peaks at T^* = 23 K in addition to the anomalies at T_C [8, 9]. The specific heat jump at

T_C is 30 % of the value expected from the mean-field model for a three dimensional ferromagnet of the Heisenberg type [9]. Furthermore, ^{151}Eu Mössbauer spectroscopy revealed that only 25 % of Eu moments ordered ferromagnetically below T_C [10]. This ratio is in agreement with that for the 2a site with respect to the 8 Eu atoms in the unit cell. These observations led to a model that a ferromagnetic state with modulated moments appears at T_C, and it gradually changes to a uniform one on cooling through T^*[9]. The anomaly at T^* is suppressed by either the change of the composition x in $Eu_8Ga_{16-x}Ge_{30+x}$ or the partial substitution of Si for Ge [8, 11]. When the cage is compressed, the rattling of the guest Eu ions would be suppressed, and then T^* would be depressed. To study the pressure dependence of T^* and T_C, we have measured ρ and Hall coefficient R_H under pressures up to 11 GPa.

2. Experimental

A single-crystalline sample of $Eu_8Ga_{16}Ge_{30}$ was grown using a Ga self flux method, as described previously [9]. The elemental composition Eu : Ga : Ge was determined by the electron-probe microanalysis to be 8 : 15 : 29.5. The ρ and R_H were measured for $P < 3$ GPa by an ac four-terminal method using a piston-cylinder pressure cell. The R_H was measured between 2 and 50 K under magnetic fields up to 6 T. The ρ for $P \leq 11.4$ GPa was measured by a dc four-terminal method using a Bridgman anvil cell [12]. Daphne oil was used as a pressure medium in each pressure cell. The pressure was estimated by the pressure dependence of the superconducting transition temperature of tin.

3. Results and Discussions

Figure 1 shows the temperature dependence of ρ under pressures. The $\rho(T)$ over the measured temperature range decreases upon pressurizing, probably as the result of the increase of the carrier concentration. The peak of ρ at T_C shifts to higher temperatures with pressure. The maximum of ρ at T^* shifts to higher temperatures and becomes broader, but remains visible even at P=11.4 GPa as shown in the inset of Fig. 1.

In order to examine the pressure dependence of the carrier concentration, we have measured the magnetic field dependence of the Hall resistivity ρ_{xy} at 2 K far below T_C. The data at P = 0.1, 1.32, and 2.81 GPa are depicted in Fig. 2. The $\rho_{xy}(B)$ increases with B up to 0.5 T but decreases linearly with further increasing B. Empirically, $\rho_{xy}(B)$ of a ferromagnet is described by the formula, $\rho_{xy}(B) = R_oB + R_e\mu_0M(B)$, where R_o and R_e are ordinary and extraordinary Hall coefficients, respectively [7]. The R_o is inversely proportional to the carrier concentration, n, within the one carrier model. The initial increase of $\rho_{xy}(B<0.5$ T) can be explained by the increase of M in the ferromagnetic state. For $B > 0.5$ T, M is saturated and therefore the first term of R_oB gives rise to the linear dependence of $\rho_{xy}(B)$. Thus, R_o can be determined from the slope of the linear fit to $\rho_{xy}(B)$ for $B > 0.5$ T. As shown in the inset, the absolute value of R_o decreases with the ratio of -3.5%/GPa. This finding means the linear increase of the carrier concentration with pressure.

We summarize the pressure dependence of T_C and T^* of $Eu_8Ga_{15}Ge_{29.5}$ in Fig. 3. The T_C increases linearly with an initial ratio of 0.94 K/GPa, and reaches 45 K at 11.4 GPa. The pressure dependence of T_C allows us to determine the Grüneisen parameter for T_C, $\Gamma = -d\ln T_C/d\ln V$. Assuming the bulk modulus of 100 GPa, we obtain $\Gamma = 2.6$, which is same order of magnitude as those for Eu-filled skutterudites $Eu_{0.83}Fe_4Sb_{12}$ ($\Gamma = 6$), and $Eu_{0.44}Co_4Sb_{12}$ ($\Gamma = 9.6$) [13]. The T^* increases to 26 K at 3 GPa, and stays at a constant value for $P > 3$ GPa.

At first, we discuss the reason for the increase of T_C with pressure. The ferromagnetic ordering of $Eu_8Ga_{16}Ge_{30}$ arises by the Ruderman-Kittel-Kasuya-Yosida (RKKY) interaction of Eu^{2+} magnetic moments. Because T_C is in proportion to the RKKY coupling constant, J_{RKKY}, $|\Gamma| = d\ln T_C/d\ln V$ is proportional to $d\ln J_{RKKY}/d\ln V$. Within the free-electron model, the J_{RKKY} is described as $J_{RKKY} \propto F(k_FR_{ij})=[2k_FR_{ij}\cos(2k_FR_{ij})-\sin(2k_FR_{ij})]/(2k_FR_{ij})$, where k_F and R_{ij} are the Fermi wave number and

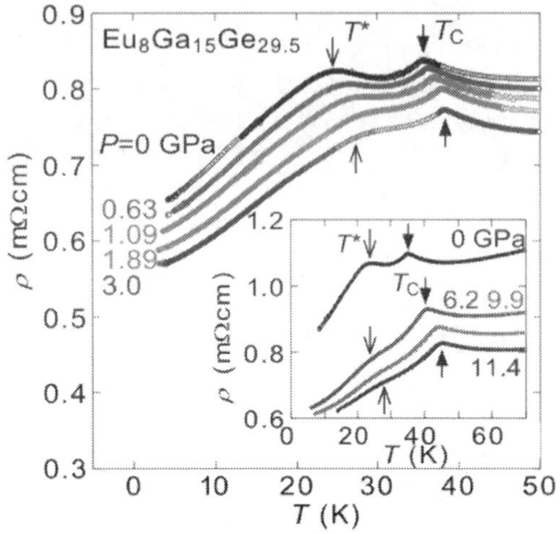

Figure 1. Temperature dependence of electrical resistivity ρ of $Eu_8Ga_{15}Ge_{29.5}$ under pressures. The inset shows the data up to 11.4 GPa applied using a Bridgman anvil cell.

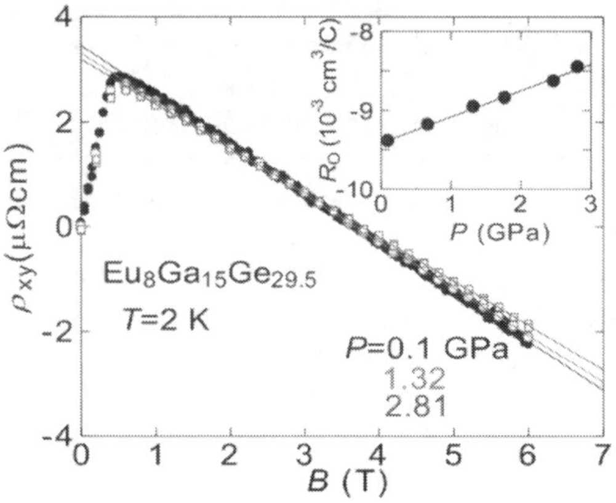

Figure 2. Magnetic field dependence of Hall resistivity under pressures. The inset shows the pressure dependence of ordinary Hall coefficient R_o.

the distance between the i-th and j-th Eu ions, respectively. It is noteworthy that $F(k_F R_{ij})$ is independent of volume V, since $k_F \propto V^{-1/3}$ times $R_{ij} \propto V^{1/3}$ becomes constant. Then, the J_{RKKY} remains constant against the change in volume, or pressure within this model. Experimentally, T_C of $Eu_8Ga_{15}Ge_{29.5}$ is enhanced by applying pressure, whereas it is suppressed in a similar way with increasing x in $Eu_8Ga_{16-x}Ge_{30+x}$ and y in $Eu_8Ga_{16}Ge_{30-y}Si_y$ [8, 11]. The unit cell volume V_{cell} is almost constant with increasing x in $Eu_8Ga_{16-x}Ge_{30+x}$, whereas V_{cell} decreases with increasing y in $Eu_8Ga_{16}Ge_{30-y}Si_y$. These relations suggest that T_C has no correlation with V_{cell}. Usually, the pressure dependence of T_C of rare-earth compounds is caused by the change of the band structure near the Fermi level. Indeed, it was pointed out that the higher T_C in type-I (36 K) than in type-VIII (10.5 K) results from the heavier effective mass of carriers in type-I, i. e., m^*(type-I)$\cong 3m^*$ (type-VIII) [7].

Next, we discuss the relation between T^* and the cage size in $Eu_8Ga_{16}Ge_{30}$. The high-pressure x-ray diffraction study showed that the unit cell volume at P =11 GPa is compressed to 87% of the value at ambient pressure [14]. The lattice parameter, accordingly, is shrunk to 95.5%. From these data, we will estimate the cage size at P =11 GPa. A universal relation was found between the guest free space in the tetrakaidecahedron and the magnitude of thermal conductivity at 150 K in type-I clathrates [3]. The free space for guest vibration is written as $R_{free}=R_{cage}-r_{cage}-r_{guest}$, where R_{cage} is the cage radius defined as the distance between the 6d (guest) site and the 24k (cage) site. At ambient pressure, R_{free} for $X_8Ga_{16}Ge_{30}$ (X = Ba, Sr and Eu) are 1.34, 1.50 and 1.58 Å, respectively [3]. Assuming that R_{cage} is compressed by applying pressure, we obtain R_{free}=1.39 Å for X = Eu at P =11 GPa. Because this value is close to that for R_{free}(X=Ba) of the on-center counterpart, Eu atoms in the tetrakaidecahedral cage would return to center. This conjecture contradicts with the fact that the anomaly at T^* remains even at 11 GPa. On the other hand, the anomaly at T^* vanishes by changing the composition; increasing either x up to 1.01 from 0.53 in $Eu_8Ga_{16-x}Ge_{30+x}$ or y up to 7.3 in $Eu_8Ga_{16}Ge_{30-y}Si_y$ [8, 11], although the V_{cell} hardly decreases [8, 11]. These findings suggest that T^* has no correlation with the cage size. Instead, the carrier concentration increases in the three cases; with increasing x, y and pressure in $Eu_8Ga_{16-x}Ge_{30+x}$, $Eu_8Ga_{16}Ge_{30-y}Si_y$, and $Eu_8Ga_{15}Ge_{29.5}$, respectively. Therefore, we conclude that the increase of the carrier concentration leads to the suppression of T^*. In order to reveal the origin of T^*,

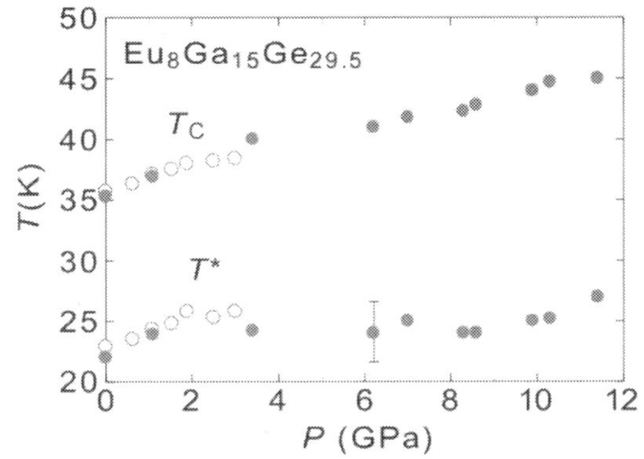

Figure 3. Pressure dependence of T_c and T^* for $Eu_8Ga_{15}Ge_{29.5}$.

more detailed structural and magnetic studies of $Eu_8Ga_{16-x}Ge_{30+x}$ with various values of x are necessary.

In summary, we performed the resistivity ρ and Hall coefficient R_H measurements of type-I clathrate $Eu_8Ga_{15}Ge_{29.5}$ with T_C = 36 K and T^* = 23 K under high pressures up to 11.4 GPa. With increasing pressure, the T_C increases up to 45 K at P =11.4 GPa and broad peak at T^* in $\rho(T)$ is weakened. The anomaly remains at T^* even at P = 11.4 GPa. Contrary to our expectation, T^* is found to be not so sensitive to the cage size up to 11 GPa, where the cage volume is compressed to 87% of the value at ambient pressure.

Acknowledgements

We thank F. Iga, and K. Suekuni for fruitful discussions. The measurement of Hall coefficient under pressures was performed at N-BARD, Hiroshima University. This work was supported by Grants-in-Aid for the Science Research form MEXT of Japan, Grant Nos. 20102004, 21540363 and 23740275.

References

[1] Paschen S, Carrillo-Cabrera W, Bentien A, Tran V H, Baenitz M, Grin Y and Steglich F 2001 *Phys. Rev.* B **64** 214404
[2] Sales B C, Chakoumakos B C, Jin R, Thompson J R and Mandrus D 2001 *Phys. Rev.* B **63** 245113
[3] Suekuni K, Avila M A, Umeo K, Fukuoka H, Yamanaka S, Nakagawa T and Takabatake T 2008 *Phys. Rev.* B **77** 235119
[4] Eisenmann B, Schäfer H and Zagler R 1986 *J. Less-Common Met.* **118** 43
[5] Chakoumakos B C, Sales B C and Mandrus D G 2001 *J. Alloys Compd.* **322** 127
[6] Nolas G S, Weakley T J R, Cohn J L and Sharma R 2000 *Phys. Rev.* B **61** 3845
[7] Pacheco V, Bentien A, Carrillo-Cabrera W, Paschen S, Steglich F and Grin Y 2005 *Phys. Rev.* B **71** 165205
[8] Bentien A, Pacheco V, Paschen S, Grin Yu and Steglich F 2005 *Phys. Rev.* B **71** 165206
[9] Onimaru T, Yamamoto S, Avila Marcos A, Suekuni T and Takabatake T 2010 *J. Phys. Conf. Ser.* **200** 022044
[10] Hermann R P, Keppens V, Bonville P, Nolas G S, Grandjean F, Long G J, Christen H M, Chakoumakos B C, Sales B C and Mandrus D 2006 *Phys. Rev. Lett.* **97** 017401
[11] Onimaru T, unpublished
[12] Ohashi M and Oomi G 2009 *Jpn. J. Appl. Phys.* **48** 070221
[13] Grytsiv A, Rogl P, Berger St, Paul Ch, Bauer E, Godart C, Ni B, Abd-Elmeguid M M, Saccone A, Ferro R and Kaczorowski D 2002 *Phys. Rev.* B**66** 094411
[14] Kume T, unpublished

Successive phase transitions induced by magnetic fields in a cubic system, $NdPd_3S_4$

M Watahiki[1], E Matsuoka[2], K Iwasa[1], Y Matsumoto[1], S Nakamura[3], T Nojima[3], K Ohoyama[3], K Tanigaki[1], H Aoki[1] and H Onodera[1]

[1] Department of Physics, Tohoku University, Sendai 980-8578, Japan
[2] Department of Physics, Kobe University, Kobe 657-8501, Japan
[3] Institute for Materials Research, Tohoku University, Sendai 980-8577, Japan

E-mail: watahiki@s.tohoku.ac.jp

Abstract. To investigate the role of multipoles in the phase transitions of $NdPd_3S_4$, measurements of specific heat, magnetization, and electrical resistivity were performed for single-crystalline samples. We also carried out powder neutron diffraction experiments to identify a magnetically ordered structure. At zero magnetic field, an antiferromagnetic (AFM) transition occurs at $T_N = 1.8$ K. When a magnetic field is applied, the AFM ordered phase, characterized by a propagation vector $\boldsymbol{k} = (1, 0, 0)$, switches into field-induced phases. We propose that the field-induced phases are associated with quadrupolar ordering.

1. Introduction

The interplay between various electronic degrees of freedom results in a complex phase diagram in terms of the magnetic field (B) and temperature (T) in correlated electron systems. The phase transitions associated with electron multipoles have been reported in several rare-earth compounds, such as CeB_6 [1] and TbB_2C_2 [2].

Recently, complex phase diagrams were also discovered for the RPd_3S_4 (R: rare-earth ion) family, crystallizing into a $NaPt_3O_4$-type cubic structure. Trivalent rare-earth ions form a body-centered cubic sublattice and are subjected to a cubic crystalline electric field (CEF) with a $\mathbf{T_h}$ point group. For example, field-induced antiferroquadrupolar (AFQ) ordering in $TbPd_3S_4$ [3] and AFQ ordering in $DyPd_3S_4$ [4] have been reported previously.

Here, we focus on $NdPd_3S_4$. A previous study on a polycrystalline sample revealed that $NdPd_3S_4$ shows an antiferromagnetic (AFM) transition at 2.0 K and that the CEF ground state is a quartet [5]. Since the quartet ground state has multipole degrees of freedom, in addition to a magnetic dipole, we expect complex phase diagrams. Multipolar ordering in Nd compounds is rare as far as we know. The purpose of this work is to clarify the contribution of multipole degrees of freedom in the ordered phases. For this purpose, we studied anisotropic behavior under magnetic fields using single-crystalline samples of $NdPd_3S_4$.

In this paper, we report phase transitions in $NdPd_3S_4$, investigated by measurements of specific heat, magnetization, and electrical resistivity. B-T phase diagrams are determined from the results of these measurements. In addition, powder neutron diffraction experiments were performed to identify a magnetically ordered structure in the AFM phase of $NdPd_3S_4$.

2. Experiments

Single-crystalline samples of $NdPd_3S_4$ were synthesized by a chemical transport method with the transport medium KI, as described in Ref. [6].

The specific heat was measured by the relaxation method over the temperature range of 0.5 to 50 K, under applied magnetic fields up to 8 T. Magnetization measurements down to 0.5 K were performed by the capacitive Faraday method in magnetic fields up to 8 T. The electrical resistivity was measured by the DC four-probe method between 0.5 and 300 K and up to 10 T.

Powder neutron diffraction experiments were carried out using the diffractometer HERMES of the Institute of Materials Research, Tohoku University, installed at the cold-neutron guide of JRR-3 at the Japan Atomic Energy Agency, Tokai, Japan. We used neutrons with a wavelength of 1.84772(9) Å provided by the 331 reflection of the single-crystalline Ge monochromator.

3. Results

At first, neutron diffraction data taken at zero magnetic field is shown in order to reveal magnetic ordering. We observed Bragg peaks appearing below 3 K, in addition to the nuclear scattering from the $NaPt_3O_4$-type structure. Figure 1 shows the evolution of the 100 Bragg peak intensity as a function of temperature, which is evaluated by integration of a scan profile as shown in the inset. This peak disappears at 3 K, and thus it is clear that the 100 peak indicates the magnetic ordering. We observed other peaks growing at low temperature, corresponding to 111, 300, 311, 410, 621, and so on, which are not shown. These peaks can be characterized by a propagation vector $\boldsymbol{k}$ = (1, 0, 0) for an AFM structure. No magnetic peak indexed by any other propagation vector was observed. This observation means that the direction of the magnetic moments of the corner Nd^{3+} ions is opposite to those of the body-centered sites. The same AFM ordered structure was also observed in $TbPd_3S_4$ [7]. The integrated intensity of the 100 peak starts increasing below 3 K and rapidly increases below 2.1 K, which suggests that fluctuation of magnetic dipoles exists below 3 K.

Figure 2 shows the temperature dependence of the specific heat $C(T)$ under various magnetic fields applied along the [110] direction. At B = 0 T, $C(T)$ shows a λ-type peak at T_N = 1.8 K. The peak at T_N shifts to the lower temperature side with increasing magnetic field, and it splits into two anomalies at T_N' and T_{Q1} above 1 T. The lower temperature anomaly at T_N' also shifts with the applied magnetic field. This anomaly vanishes above 2 T, while that at T_{Q1} remains robust. The T_{Q1} anomaly separates into two anomalies at T_{Q2} and T_{Q3} above 2 T. The anomaly at T_{Q3} shifts to the high temperature side when applying magnetic fields up to 8 T, and at T_{Q2}, it rapidly shifts to the lower temperature side, vanishing above 3.5 T.

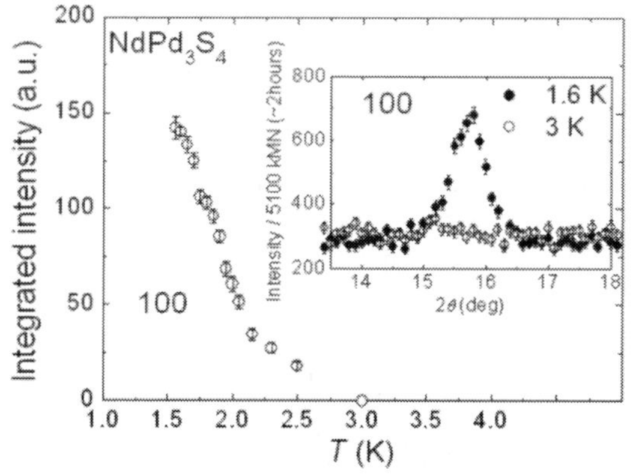

Figure 1. Temperature dependence of the integrated intensity of the 100 peak. The inset shows profiles through the 100 position measured at 1.6 K and 3 K.

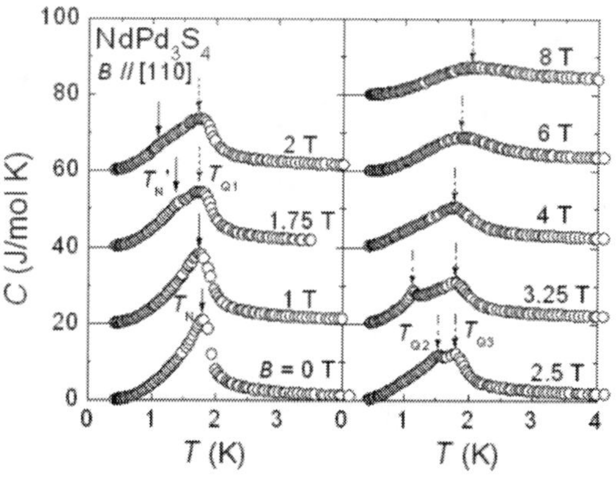

Figure 2. Temperature dependence of the specific heat of $NdPd_3S_4$ in several magnetic fields applied along the [110] direction.

Journal of Physics: Conference Series **391** (2012) 012076 doi:10.1088/1742-6596/391/1/012076

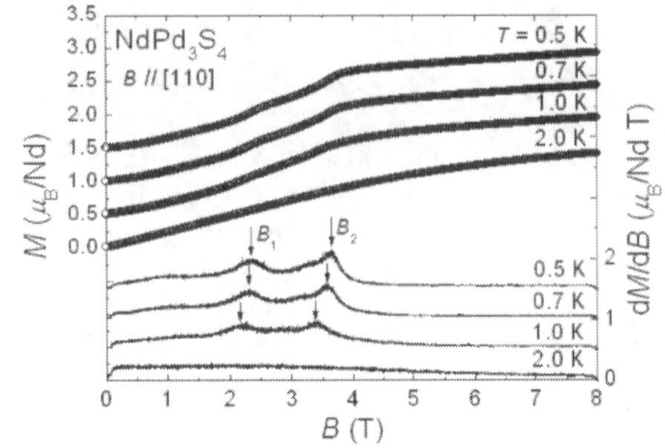

Figure 3. Magnetization curves (top) with their derivative (bottom) at several temperatures. Magnetic fields were applied along the [110] direction.

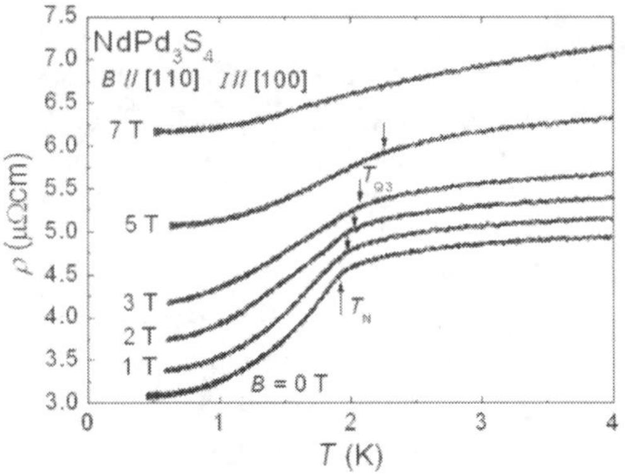

Figure 4. Temperature dependence of electrical resistivity in several magnetic fields along applied the [110] direction

The magnetization curves $M(B)$ and their derivatives dM/dB are shown in the upper and lower parts of Fig. 3, respectively. At 0.5 K, two anomalies in dM/dB at $B_1 = 2.2$ T and $B_2 = 3.6$ T were observed. As the temperature increases to T_N, B_1, and B_2 move toward lower fields, vanishing above T_N.

At $B = 0$ T, the electrical resistivity $\rho(T)$ shows a sharp drop at 1.9 K near T_N as shown in Fig. 4. As the magnetic field increases, this drop becomes weaker with the shift of the anomaly temperature.

4. Discussion

We measured $C(T, B)$, $M(T, B)$, and $\rho(T, B)$ under magnetic fields along the [100] and [111] directions. Based on the temperatures and magnetic fields of the observed anomalies, we determined the phase diagrams, as summarized in Fig. 5. Para and AFM correspond to the paramagnetic phase and the AFM ordered phase with the propagation vector (1, 0, 0), respectively. When magnetic fields are applied, the AFM ordered phase switches into field-induced phases ($I_{[100]}$, $I_{[110]}$, $II_{[110]}$, $I_{[111]}$, $II_{[111]}$, III).

We will discuss the phase diagrams for $NdPd_3S_4$ in comparison with those of the isostructural compound $TbPd_3S_4$ [3, 7]. The Tb-based system also undergoes the AFM ordering transition in the lower magnetic field region below 2.55 K. This AFM ordered structure includes collinear alignment of the magnetic moment at the Tb sites, which is the same magnetic ordered structure found for $NdPd_3S_4$ in the present study. The AFM phase in $TbPd_3S_4$ switches into the canted magnetic structure phases under finite magnetic fields above 0.5 T. The AFQ order is suggested to coexist with the canted magnetic structure in the field-induced phases for $TbPd_3S_4$. Because $NdPd_3S_4$ exhibits the same AFM ordered structure and because the AFM phase switches into the similar field-induced phases $I_{[100]}$, $I_{[110]}$, and $I_{[111]}$ under finite magnetic fields, we expect that quadrupolar ordering occurs in these field-induced phases.

As shown in the phase diagrams in Fig. 5, the transition temperatures from Para to the field-induced phases $I_{[110]}$, $I_{[111]}$, $II_{[110]}$, $II_{[111]}$ and III slightly increase with the applied magnetic field, similar to the behavior of the transition temperatures from the paramagnetic phase to the field-induced phases with the AFQ order in $TbPd_3S_4$. Furthermore, an increase in transition temperatures is commonly observed for quadrupolar ordered phases in other typical quadrupolar ordering systems. We therefore propose that these field-induced phases ($I_{[110]}$, $II_{[110]}$, $I_{[111]}$, $II_{[111]}$, III) of $NdPd_3S_4$ are due to the AFQ ordered phases with the canted magnetic structure.

Next, we will discuss the order parameter in the field-induced phase $I_{[100]}$, which shows different features from those in the ordered phases discussed above. The anomaly corresponding to the phase transition from $I_{[100]}$ to Para was not clearly observed in the magnetization curves (not shown), and the transition temperature from Para to $I_{[100]}$ does not increase by applying magnetic fields (Fig. 5). These results suggest that the magnetic moment in $I_{[100]}$ is close to a saturated moment. In other words, the canted angle is considerably smaller than 90°. We consider that $I_{[100]}$ is also associated with the AFQ ordered phase with the canted magnetic structure which is smaller than 90°. We need to carry out further studies in order to elucidate the low-lying CEF states responsible for the successive phase transitions with strongly anisotropic behaviors and the actual order parameter in $NdPd_3S_4$, for example, by using the neutron diffraction technique.

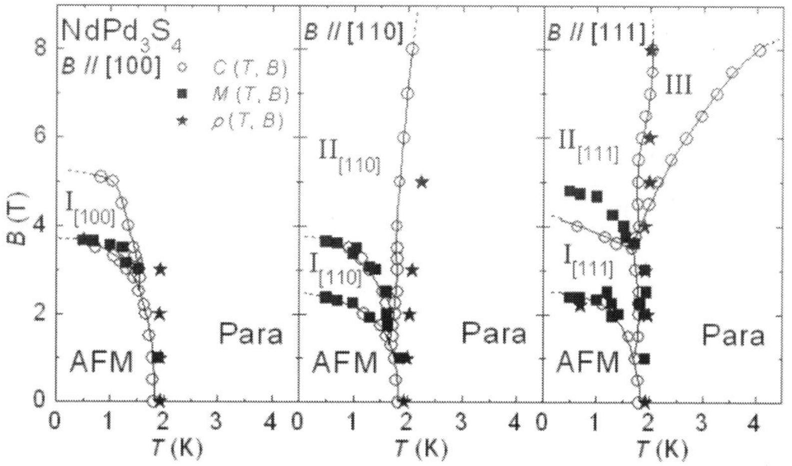

Figure 5. *B-T* phase diagrams for fields applied along the [100], [110], and [111] directions.

5. Conclusion

We synthesized single-crystalline samples of $NdPd_3S_4$; to the best of our knowledge, this is the first report of such a synthesis. From the results of the specific heat, magnetization, electrical resistivity, and powder neutron diffraction measurements, we constructed the *B-T* phase diagrams. The AFM ordered phase was confirmed by neutron diffraction. The properties of the phase diagrams as well as the behaviors suggested by the measurements indicate that the AFM phase switches into quadrupolar ordered phases under finite magnetic fields.

Acknowledgments

We thank Mr. M. Ohkawara and Mr. K. Nemoto for their support at JAEA and Mr. H. Shida for his support at Tohoku University. The neutron experiments were performed under User Programs conducted by the ISSP, University of Tokyo. This work was supported by MEXT, Japan (Global COE Program, and Grants-in-Aid for Scientific Research (S) (No.21224048), Young Scientists (B) (21740238) and Innovation Areas ''Heavy Electrons'' (20102005 and 21102503)).

References

[1] Effantin J M, Rossat-Mignod J, Burlet P, Bartholin H, Kunii S and Kasuya T 1985 *J. Magn. Magn. Mater.* **47&48** 145
[2] Kaneko K, Onodera H, Yamauchi H, Sakon T, Motokawa M and Yamaguchi Y 2003 *Phys. Rev. B* **68** 012401
[3] Matsuoka E, Shida H, Matsumura T, Ohoyama K and Onodera H 2011 *J. Phys. Soc. Jpn.* **80** SA085
[4] Matsuoka E, Tayama T, Sakakibara T, Hiroi Z, Shirakawa N, Takeda N and Ishikawa M 2007 *J. Phys. Soc. Jpn.* **76** 084717
[5] Abe K, Kitagawa J, Takeda N and Ishikawa M 1999 *Phys. Rev. Lett.* **83** 5366
[6] Matsuoka E, Usui D, Tanida H, Nakamura S, Nojima T and Onodera H 2007 *J. Phys. Soc. Jpn.* **76** 073707
[7] Matsuoka E, Takahashi F, Kitagawa J, Ohoyama K, Yoshizawa H and Ishikawa M 2001 *J. Magn. Magn. Mater.* **231** L23

Anisotropic electrical resistivity of the Kondo insulator $CeRu_4Sn_6$

H Winkler, K-A Lorenzer, A Prokofiev and S Paschen

Institute of Solid State Physics, Vienna University of Technology, Wiedner Hautptstr. 8-10, 1040 Vienna, Austria

E-mail: `winkler@ifp.tuwien.ac.at`

Abstract. The intermetallic compound $CeRu_4Sn_6$ has tentatively been classified as anisotropic Kondo insulator. Here we present electrical resistivity measurements on $CeRu_4Sn_6$ single crystals, both along the main directions a and c of the tetragonal crystal structure, and along the diagonal c' of the a-a plane. This direction was selected because $c' = \sqrt{2}a$ differs from c by only 0.2 % , suggesting that the coumpound might alternatively be regarded as quasi-cubic. Amazingly, strong anisotropy is observed not only between a and c but also between c and c'. We analyse the temperature dependent resistivities both with a simple semiconductor model and within a Kondo insulator picture.

1. Introduction

In Kondo insulators, frequently also referred to as heavy fermion semiconductors, a narrow gap (or pseudo gap) opens at low temperatures in the electronic density of states (DOS) at the Fermi level [1]. This is thought to be caused by the hybridization between localized (f-shell) and conduction (p, d-shell) electrons [2].

Most Kondo insulators investigated so far exhibit a cubic crystal structure (e.g., YbB_{12}, SmB_6, FeSi, $Ce_3Bi_4Pt_3$). Only a few materials (e.g., CeNiSn, CeRhSb) are orthorhombic. The anisotropic behaviour of physical properties observed in the orthorhombic compounds was discussed in terms of nodes of the energy gap along certain directions in k-space [3, 4].

The intermetallic compound $CeRu_4Sn_6$ was first synthesized by Das and Sampathkumaran [5]. It crystallizes in a tetragonal crystal structure (space group: $I\bar{4}2m$) with the lattice constants a=6.8810 Å and c=9.7520 Å. Since the diagonal of the tetragonal plane $c'=\sqrt{2}a$ differs from c by less than 0.2 %, the structure may alternatively be approximated by a quasi-cubic cell.

2. Experimental

The polycrystalline starting material for crystal growth was synthesized by melting first Ce and Ru, and subsequently the resulting Ce/Ru alloy with Sn in a horizontal water cooled copper boat using high-frequency heating. The purity of the starting materials was 99.99 % for Ce (Ames Lab) and Ru, and 99.9999 % for Sn. The single crystal growth was performed by the floating zone melting technique using optical heating in a four-mirror furnace of Crystal Corporation. All steps were performed under Ar 99.9999 % protective atmosphere after several purgings with Ar. Since $CeRu_4Sn_6$ melts incongruently we used a self-flux technique [6]. X-ray diffraction and SEM/EDX investigations revealed that the crystals are single phased and have the correct stoichiometric composition.

Electrical resistivity measurements were done by a standard 4-point technique, with alternating dc current in a physical property measurement system (PPMS) from Quantum Design between 350 K and 2 K, and with low-frequency ac current in a ^{3}He/^{4}He dilution refrigerator from Oxford Instruments at temperatures between 2 K and 50 mK.

3. Electrical resistivity

Figure 1 shows the electrical resistivity ρ along the principal directions of the tetragonal cell (a, c) and along the diagonal of the tetragonal plane (c') from 350 K down to 50 mK on a semi-logarithmic scale. Each curve shows a remarkable increase of ρ with decreasing temperature down to approximately 11 K (9.5 K) for the directions a and c' (c). With further decreasing temperature ρ decreases slightly until another increase sets in, at about 2.5 K (1.9 K) for the a- and c'- (c-)direction. The c-direction curve shows additionally a small hump at about 150 K, which is more clearly seen in Fig. 2.

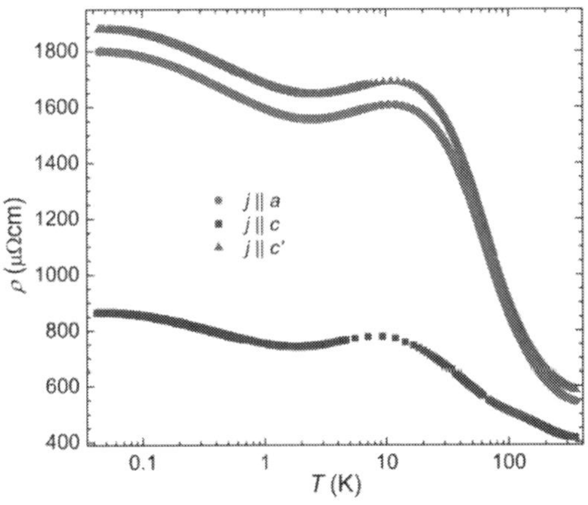

Figure 1. Temperature dependence of the electrical resistivity, $\rho(T)$, along the three directions a, c and c'. Pronounced anisotropy between the tetragonal plane directions a and c', and the c-direction is seen over the whole temperature range. The two different in-plane directions (a, c') show similar behaviour.

At first we try to model the behaviour of $\rho(T)$ by a two-gap structure in the DOS which is frequently found in simple semiconductors. In this picture, the high-temperature behaviour can be explained with an intrinsic gap. The second upturn at low temperatures is caused by a smaller extrinsic gap between impurity states and the intrinsic electron bands of the DOS. We fitted $\rho(T)$ at high temperatures with $\rho = \rho_1\ e^{(\Delta/2k_BT)}$ where ρ_1 is a constant, Δ is the energy gap, and k_B the Boltzmann constant. For each direction fits were made in two different temperature ranges. The resulting curves are shown in Fig. 2 and the fit parameters are listed in Tab. 1. A good description of the data is only possible in relatively narrow temperature ranges. A similar fit at low temperatures (between about 0.5 K and 2 K) yields gap values that are much smaller than the temperatures of the fitted range. Thus these fits are considered unphysical and are not shown.

An alternative explanation for the measured $\rho(T)$ curves is based on the interpretation of $CeRu_4Sn_6$ as Kondo insulator. In these materials a temperature-dependent hybridization gap is expected to develop below a characteristic temperature. In the absence of quantitative predictions for this behaviour fits to the data cannot be made. However, models for Kondo metals can be of some use. An important contribution to the temperature dependence of the resistivity comes from Kondo scattering. In the regime of incoherent scattering this results, for a metal, in the well-known -ln T behaviour. To analyse the data with this relation we have to first subtract the phonon contribution to the electrical resistivity, $\rho_{ph}(T)$, which we approximate by the electrical resistivity of the non-f reference compound $LaRu_4Sn_6$ [7]. The

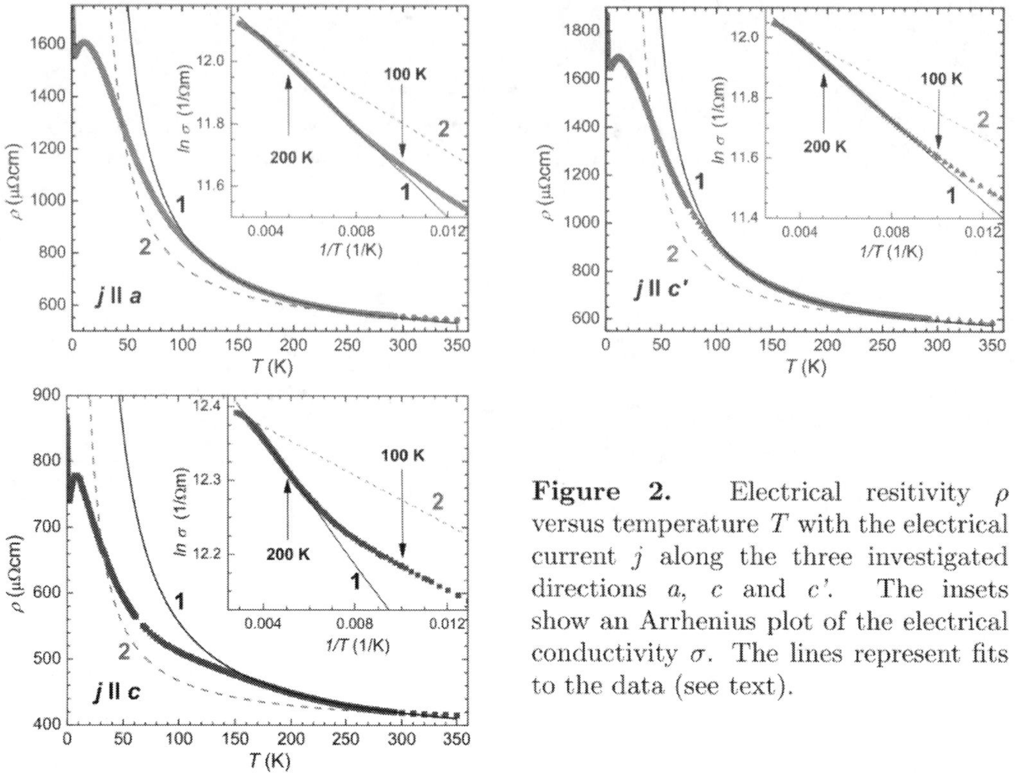

Figure 2. Electrical resitivity ρ versus temperature T with the electrical current j along the three investigated directions a, c and c'. The insets show an Arrhenius plot of the electrical conductivity σ. The lines represent fits to the data (see text).

Table 1. Parameters obtained by fitting $\rho = \rho_1\ e^{(\Delta/2k_BT)}$ to the $\rho(T)$ data for the indicated current directions and the indicated temperature ranges (see Fig. 2).

j along	fitnumber	fitted temperature range (K)	ρ_1 ($\mu\Omega$cm)	Δ/k_B (K)
a	1; 2	120 - 220; 300 - 350	434.7; 477.6	140.8; 89.9
c'	1; 2	120 - 220; 300 - 350	476.3; 519.7	132.6; 83.1
c	1; 2	190 - 270; 300 - 350	362.7; 395.9	85.0; 33.1

resulting $\Delta\rho = \rho - \rho_{ph}$ data are shown in Fig. 3. For each investigated direction three temperature regions can be identified where $\Delta\rho \sim$ -ln T is observed. Fits in these three ranges (1 - 3) are shown as straight lines in Fig. 3. A noteable observation is that the slopes of these straight lines decrease from range 1 to range 3 for the c-direction. For the directions within the tetragonal plane (a, c'), however, the slope in range 2 is larger than the one in range 1.

Some hints to understand this behaviour come from the Coqblin-Schrieffer model [8, 9]. While this is a single impurity model it has successfully explained some of the electrical resistivity features of Kondo lattice systems such as $CeAl_3$ or $CeAl_2$. One result of the Coqblin-Schrieffer model is that, for a constant DOS, different regions of -ln T behaviour can be resolved if the crystalline electric field-split multiplets of the magnetic impurity are energetically well enough separated. Since the prefactor of the -ln T-term decrease as the excited levels get thermally depopulated, the absolute value of the slope is expected to decrease, as qualitatively observed for the c-direction $\Delta\rho(T)$ of $CeRu_4Sn_6$. The three linear regions would then correspond to the Kondo effect in the full $J = 5/2$ multiplet (range 1), the two lowest-lying doublets (range 2),

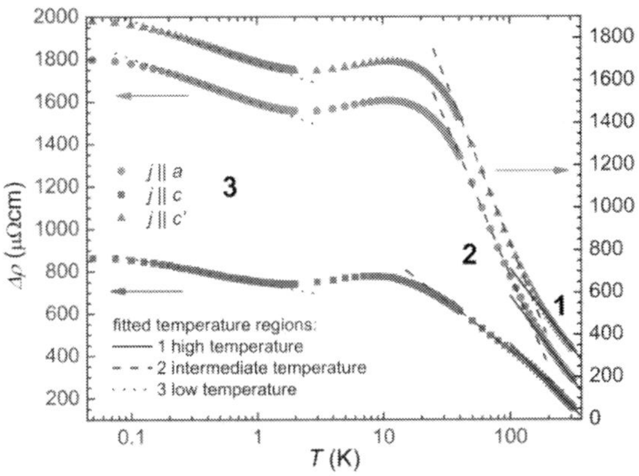

Figure 3. Magnetic contribution to the electrical resitivity $\Delta\rho$ versus temperature T, with the electrical current j along the three investigated directions a, c and c'. The lines indicate -ln T behaviour in three different temperature ranges (1 - 3).

and the ground-state doublet (range 3).

However, another effect should not be ignored. This is the change of the slope with the DOS at the Fermi level. At otherwise identical model parameters the absolute value of the slope of the -ln T-term decreases with decreasing DOS. Thus, the $\Delta\rho(T)$ behaviour for the c-direction could alternatively be interpreted as being due to incoherent Kondo scattering in a metal in which the DOS at the Fermi level decreases with temperature. The fact that within the tetragonal plane no such decrease of the absolute value of the slope is observed between range 1 and range 2 may then indicate that the gap opening is less relevant for the tetragonal plane, possibly due to the formation of nodes in the hybridization. Clearly, more realistic calculations are required to further sharpen our understanding of the electrical resistivity in this and other Kondo insulators.

4. Conclusion

Our electrical resistivity measurements on single crystalline $CeRu_4Sn_6$ revealed a remarkable anisotropy between the c-direction and directions within the tetragonal plane (a and c' directions) perpendicular to it. This confirms our previous observations from specific heat and magnetization measurements [6]. We have analysed the data qualitatively in terms of two different models, a simple semiconductor model with a two-gap structure and the Coqblin-Schrieffer model with temperature dependent electronic density of states. While these analyses will certainly be a guide for future experiments a more realistic theoretical description, possibly involving treatments with band structure calculations combined with dynamical mean field theory, is needed for a more complete understanding.

Acknowledgments

We gratefully acknowledge financial support from the European Research Council under the European Community's Seventh Framework Programme (FP7/2007-2013)/ERC grant agreement No. 227378 and from the Austrian Science Fund (Project P19458-N16).

References

[1] Aeppli G and Fisk Z 1992 *Comments Condens. Matter Phys.* **16** 155
[2] Riseborough P S 2000 *Adv. Phys.* **49** 257
[3] Ikeda H and Miyake K 1996 *J. Phys. Soc. Jpn.* **65** 1769
[4] Moreno J and Coleman P 2000 *Phys. Rev. Lett.* **84** 342
[5] Das I and Sampathkumaran E V 1992 *Phys. Rev.* B **46** 4250
[6] Paschen S et al. 2010 *J. Phys.: Conf. Series* **200** 012156
[7] Guo Z 2005 *PhD thesis* University of Johannesburg
[8] Cornut B and Coqblin B 1972 *Phys. Rev.* B **5** 4541
[9] Hewson A C 1993 The Kondo Problem to Heavy Fermions *Cambridge University Press*

Heat Capacity Measurement of Heavy Fermion $YbCo_2Zn_{20}$ under Magnetic Field

R. Yamanaka[1], K. Matsubayashi[1], Y. Saiga[2], T. Kawae[3], Y. Uwatoko[1]

[1]The Institute for Solid State Physics, The University of Tokyo, Kashiwanoha, Kashiwa, Chiba 277-8581, Japan

[2]Institute for Advanced Materials Research, Hiroshima University, Higashihiroshima, Hiroshima 739-8530, Japan

[3]Department of Applied Quantum Physics, Faculty of Engineering Kyushu University, Fukuoka, Fukuoka 819-0395, Japan

E-mail: rina_y@issp.u-tokyo.ac.jp

Abstract. We have measured the specific heat of heavy fermion $YbCo_2Zn_{20}$ down to 0.1 K under magnetic field up to 3 T. At zero-field, the electronic contribution to the specific heat C_e/T shows logarithmic increase with decreasing temperature and tends to saturate at lower temperature, reflecting the Fermi-liquid state. With increasing magnetic field, C_e/T exhibits a power-law divergence at $H_m \sim 0.6$ T, which corresponds to the critical field of the metamagnetic crossover. The field dependence of C_e/T shows the two peak structure and the local minimum point of this structure coincides with H_m as observed in prototypical metamagnetic compound $CeRu_2Si_2$. We will compare the characteristic energy scale relevant to the metamagnetic behavior of $YbCo_2Zn_{20}$ and $CeRu_2Si_2$.

1. Introduction

In the heavy fermion, Ce and Yb systems, the hybridization between $4f$ and conduction electrons induces instability of magnetic moments and charge configurations. Recently, one of the most interesting topics has been the ground state properties of heavy fermion compounds located close to a magnetic Quantum Critical Point (QCP). Applying pressure or magnetic field controls the electronic configuration, and thus anomalous behaviors appear at around magnetic QCP, such as an unconventional superconductivity and non Fermi-liquid state. In the case of Yb based heavy fermion compounds, it is known that $YbRh_2Si_2$ is a prototypical compound of a field-tuned QCP from an antiferromagnetic to a paramagnetic state [1]. Recently, YbT_2Zn_{20} (T = Co, Rh, Ir) family of compounds have attracted much attention because of their heavy fermion state and proximity to the pressure-induced QCP [2]. These compounds have a cubic $CeCr_2Al_{20}$ type structure, in which a rare earth ion is located in 16 Zn ions, forming a Frank-Kasper polyhedron. The high temperature magnetic susceptibility shows the Curie-Weiss law with the effective moment close to the value of Yb^{3+} ($\mu_{\rm eff} = 4.5\ \mu_B$), although there is no indication of magnetic order down to 20 mK. In the series of YbT_2Zn_{20} family, $YbCo_2Zn_{20}$ exhibits outstanding features as follows. The low temperature electrical resistivity and specific heat follows $\rho = \rho_0 + AT^2$ and $C/T = const.$ which indicates the Fermi-liquid behavior: the values of A and γ are 165 $\mu\Omega$cm/K^2 and 7900 mJ/moleK^2, respectively. These large values of A and γ are due to the small Kondo temperature, suggesting the closeness to the pressure induced magnetic order. According to the resistivity measurements under pressure, magnetic ordered phase appears around critical pressure $P_c \sim 1$ GPa [3]. In general, applying pressure stabilizes the magnetic Yb^{3+} configuration and resulting in the appearance of a magnetically ordered state in $YbCo_2Zn_{20}$. It is worth noting that the critical pressure of $YbCo_2Zn_{20}$ is lower than that of other Yb based compounds such as $YbNi_2Ge_2$ ($P_c \sim 5$ GPa) [4] and $YbCu_2Si_2$ ($P_c \sim 8$ GPa) [5]. Therefore $YbCo_2Zn_{20}$ has an advantage of investigating the behavior of Yb-based compounds near QCP. More interestingly, it is reported that a metamagnetic crossover occurs at $H_m \sim 0.57$ T [6]. The origin of this metamagnetic behavior is not fully understood, however relevance to the valence fluctuation is considered [7]. To shed light on the nature of the metamagnetic behavior, we have measured specific heat of $YbCo_2Zn_{20}$ single crystal under magnetic field.

2. Experimental methods

Single crystal of $YbCo_2Zn_{20}$ was grown by a Zn self-flux method as described previously [3]. The sample used in this experiment was cut into square pole whose size is $1.4 \times 1.4\times 3.5$ mm^3 and weight is 40.2 mg. The specific heat was measured by the adiabatic method in temperature range 0.1 K $< T <$ 3 K with the dilution refrigerator. An external magnetic field was applied parallel to [110] direction.

3. Results and discussion

Figure 1 shows the temperature dependence of the electronic contribution of the specific heat C_e/T under magnetic field on a logarithmic temperature scale. Here, the electronic contribution to the specific heat C_e of $YbCo_2Zn_{20}$ was obtained by subtracting that of $LuCo_2Zn_{20}$. At zero-

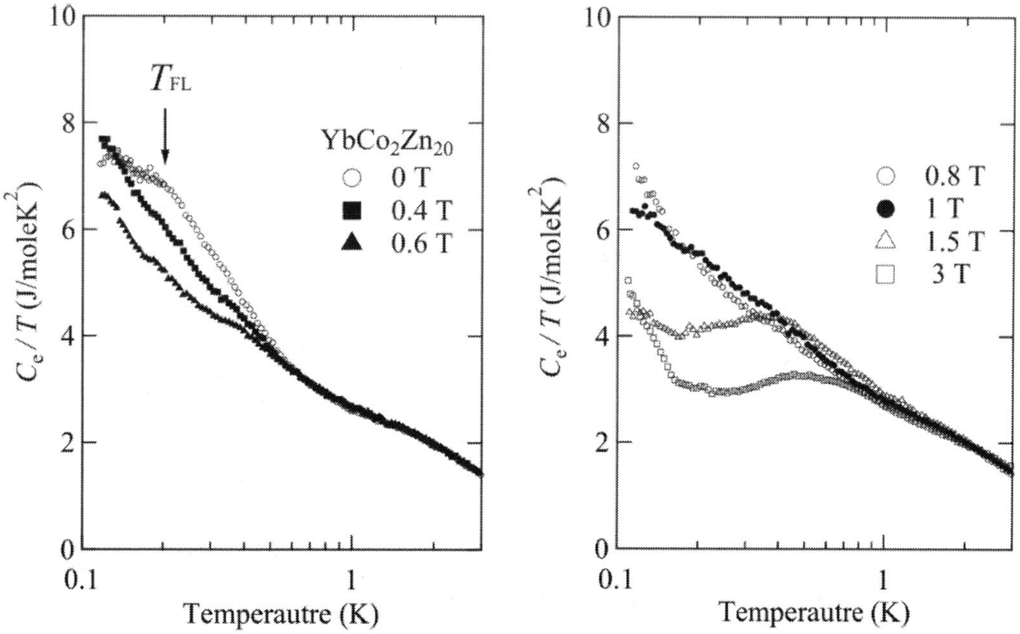

Figure 1. Temperature dependence of the electronic contribution of the specific heat for $YbCo_2Zn_{20}$ at selected magnetic field. Arrow indicates the onset temperature of the Fermi-liquid behavior.

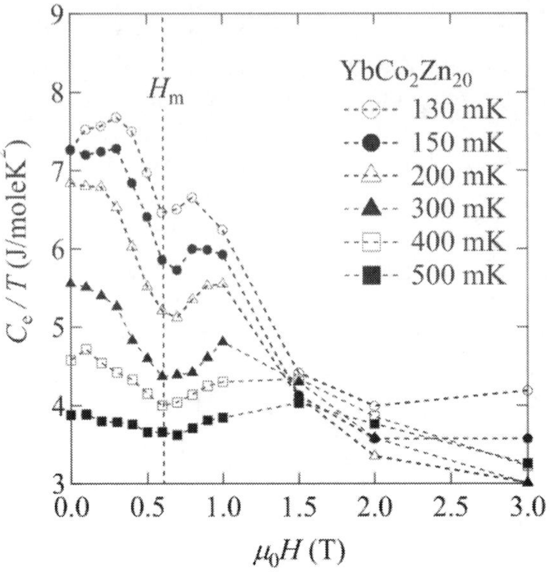

Figure 2. Field dependence of C_e/T at selected temperature. Two peak structure, which has local minimum point at H_m, was observed at around 0.6 T.

field, C_e/T exhibits a broad shoulder structure at around $\sim$ 2 K due to the combined effect of the Kondo effect and the Schottky anomaly associated with the crystalline electronic field (CEF) splitting. At lower temperature, C_e/T increases logarithmically and tends to saturate below $T_{FL} \sim 0.2$ K, coinciding with the T^2 dependence in the resistivity, namely Fermi-liquid behavior as observed in the previous report [2]. With increasing magnetic field, the T_{FL} is gradually suppressed to a lower temperature. It is worth noting that C_e/T diverges more rapidly than a logarithmic dependence at $H_m \sim 0.6$ T, which corresponds to the metamagnetic field. With further increasing magnetic field, the temperature dependence changes to a peak structure as typical behavior of Kondo systems under magnetic field.

To see the field variation of C_e/T, we plotted C_e/T as a function of magnetic field at selected temperature in Fig. 2. The most intriguing feature obtained here is the two peak structure which has local minimum point at $H_m \sim 0.6$ T, coinciding again with the critical field of the metamagnetic crossover. Although the double peak structure becomes weaker with increasing temperature, the minimum point is temperature independent. Similar feature is also observed in the prototypical heavy fermion compound $CeRu_2Si_2$ which shows metamagnetic transition at $H_m \sim 7.7$ T [8]. In the case of $CeRu_2Si_2$, the field dependence of C/T exhibits the double peak structure above 1 K, and thus that changes to a single peak structure at lower temperatures [9]. Here we compare the characteristic energy scales of $YbCo_2Zn_{20}$ to those of $CeRu_2Si_2$ summarized in Table 1. It is obvious that the values of H_m and $T_{\chi max}$ of $YbCo_2Zn_{20}$ are ten times smaller than that of $CeRu_2Si_2$, i.e. the characteristic energy scale of $YbCo_2Zn_{20}$ is much smaller than that of $CeRu_2Si_2$. Considering the difference of the energy scale, it is expected that the double peak structure in $YbCo_2Zn_{20}$ may evolve into a single peak structure at lower temperature, suggesting the development of the critical fluctuation. Further experimental and theoretical studies are needed to clarify the origin of the metamagnetic behavior and the relationship with the pressure-induced magnetically ordered state of $YbCo_2Zn_{20}$. The next experimental challenge is to measure the specific heat under pressure and magnetic field.

	$H_{\rm m}$	$T_{\chi{\rm max}}$	γ (J/moleK^2)	Refs.
$CeRu_2Si_2$ [H//c]	7.7	10	0.37	[8]
$YbCo_2Zn_{20}$ [H// [110]]	0.6	0.35	7.8	[10]

Table 1. The values of $H_{\rm m}$, $T_{\chi{\rm max}}$ and γ of $CeRu_2Si_2$ and $YbCo_2Zn_{20}$.

4. Summary

We have measured the specific heat of $YbCo_2Zn_{20}$ (H // [110]) down to 0.1 K under magnetic field up to 3 T. At zero-field, the temperature dependence of the specific heat tends to saturate at low temperature which indicates the Fermi-liquid description. At $H_{\rm m} \sim 0.6$ T corresponding to the critical field of metamagnetic crossover, $C_{\rm e}/T$ exhibits a power-law divergence, and double peak structure appears in the field dependence of $C_{\rm e}/T$ at lowest temperature.

Acknowledgments

This work was financially supported by Grants-in-Aid for Scientific Research (No. 21340092) and for Scientific Research on Innovative Areas ''Heavy Electrons'' (No. 20102007). KM was supported by a Grant-in-Aid for Young Scientists B (No. 22740217) from the Ministry of Education, Culture, Sports, Science and Technology.

References

[1] Philipp G, Qimiao S and Frank S *Nature Physics* 2008 **4** 186
[2] Torikachibili M S, Jia S, Mun E D, Hannahs S T, Black R C, Neils W K, Martein D, Bud' ko S L and Canfield P C *Proc. Natl. Acad. Sci. U.S.A.* 2007 **104** 9960
[3] Saiga Y, Matsubayashi K, Fujiwara T, Kosaka M, Katano S, Hedo M, Matsumoto T and Uwatoko Y *J. Phys. Soc. Jpn* 2008 **77** 053710
[4] Knebel G, Braithwaite D, Lapertot T, Canfield T and Flouquet J *J.Phys. : Condense.Matter* 2001 **14** 10935
[5] Colombier E, Braithwaite G, Lapertot G, Salce B and Knebel G *Phys. Rev. B* 2009 **79** 245113
[6] Ohya M, Matsushita M, Yoshiuchi S, Takeuchi T, Honda F, Settai R, Tanaka T, Kubo Y and Onuki Y *J. Phys. Soc. Jpn* 2010 **79** 083601
[7] Watanabe S and Miyake K *J. Phys. : Condense. Matter* 2011 **23** 094217
[8] Flouquet J, Kambe S, Regnault L P, Haen P, Brison J P, Lapierre F and Lejay P *Physica B* 1995 **215** 77
[9] Aoki Y, Matsuda T D, Sugawara H, Sato H, Ohkuni H, Settai R, Onuki Y, Yamamoto E, Haga Y, Andreev A V, Sechovsky V, Havela L,, Ikeda H and Miyake K *J. Magn. Magn. Mater.* 1998 **177** 271
[10] Matsubayashi K *Private Communication*

A Study of Elastic Properties of URu_2Si_2 in Comparison with the Non-5*f* Contribution of $ThRu_2Si_2$

T. Yanagisawa[1,2], H. Saito[2], Y. Watanabe[2], Y. Shimizu[2], H. Hidaka[2], and H. Amitsuka[2]

[1] Creative Research Institution, Hokkaido University, Sapporo 001-0021, Japan
[2] Graduate School of Science, Hokkaido University, Sapporo 060-0810, Japan

E-mail: tatsuya@phys.sci.hokudai.ac.jp

Abstract. Ultrasonic investigations of the elastic constants of non-5*f* compound $ThRu_2Si_2$ were performed in order to highlight possible quadrupolar response in the elastic constants of heavy fermion compound URu_2Si_2, by using the $ThRu_2Si_2$ for backgrounds to subtract the phonon contributions in the elastic constants of URu_2Si_2. We confirmed that the all-elastic constants of $ThRu_2Si_2$ show no elastic anomaly, *i.e.*, increasing monotonically and saturating toward low temperatures, which can be understood regular phonon contributions for the solid states. The background-subtracted $(C_{11}-C_{12})/2$ of URu_2Si_2 keeps decreasing with decreasing temperatures from 300 K, while other share modes C_{44} and C_{66} show little change. These contrasts suggest that the Γ_3-type quadrupole moment O_2^2 will be still dominant and active in URu_2Si_2 at low temperatures.

1. Introduction

Regarding a mysterious phase transition in URu_2Si_2 at $T_O = 17.5$ K, so-called 'Hidden Order (HO)', there are several experimental and theoretical approaches for the unknown order parameter [1, 2, 3]. In particular, non-magnetic multipolar orderings, including O_{xy}-type antiferro-quadrupole (AFQ) order [4], have recently been emerged as a candidate for the HO, where any intrinsic magnetic dipole moment and symmetry breaking lattice distortion should not be detected under ambient pressure in zero magnetic fields [5]. A direct verification experiment for the AFQ order has been performed by using resonant X-ray scattering (RXS), however, no significant scattering, which indicate quadrupole O_{xy} and O_2^2 order, have been detected thus far [6]. Since another theoretical predictions for the HO, such as antiferro-electric-hexadecapole order [7] or electronic-nematic phase [8, 9], have been proposed more recently, many experimental efforts such as RXS and neutron scattering under magnetic fields are ongoing in order to verify them microscopically.

On the other hand, ultrasonic measurement is one of the useful macroscopic methods to investigate the quadrupole response in the solid state. Generally in the localized 4*f*-electron system, a temperature dependence of the elastic constant, observed by ultrasound, can be understood as quadrupole susceptibility. When the system possesses quadrupole degree of freedom, a Curie-type decreasing (softening) appears on the elastic constant, corresponds to volume conservative and symmetry breaking ultrasonic modes [10]. Mostly in the case of U-based compound, such elastic softening in the shear mode is rarely found due to enhancement of its itinerant character of the *f*-electrons. However, if the hidden order parameter breaks the local symmetry of the crystal, some kind of elastic response via electron-phonon interaction would still be expected. In the present work, we have performed ultrasonic measurement both on URu_2Si_2 and on non-5*f* system $ThRu_2Si_2$ single crystals in order to search possible contributions of electric quadrupoles onto the elastic constants of URu_2Si_2.

Table 1. Absolute values of elastic constants at 80 K for URu_2Si_2 and $ThRu_2Si_2$ in the present study.

Elastic Constant	Quadrupole (Symmetry)	**URu_2Si_2** (10^{10} J m^{-3})	**$ThRu_2Si_2$** (10^{10} J m^{-3})
C_{11}	O_2^0 (Γ_1)	26.2	26.8
$(C_{11}-C_{12})/2$	O_2^2 (Γ_3)	5.7	6.2
C_{44}	O_{yz}, O_{zx} (Γ_5)	9.6	10.7
C_{66}	O_{xy} (Γ_4)	12.1	10.2

2. Experimental Details

Single crystalline URu_2Si_2 and $ThRu_2Si_2$ were prepared by Czochralski tri-arc technique and no annealing process was performed. A hexagonal-cylinder-shaped sample for URu_2Si_2 and two parallelepiped samples for $ThRu_2Si_2$ were cut from each bulk crystal by means of high-precision spark erosion. URu_2Si_2 sample has dimensions of 5.097 mm for [001], 4.161 mm for [110] and 3.864 mm for [100]. $ThRu_2Si_2$ samples have orientation of [001]-[010]-[001] and [110]-[1-10]-[001] with corresponding dimensions 1.970 × 0.896 × 0.799 mm^3 and 2.229 × 1.259 × 0.697 mm^3, respectively. A phase comparator using double balanced mixer can detect the relative change of sound velocity down to 4.2 K by using ^{4}He refrigerator. Piezoelectric $LiNbO_3$ wafers of 36°Y-cut with 100 μm thickness were used for generating and detecting longitudinal ultrasonic waves. The absolute value of sound velocity was estimated by monitoring ultrasonic echoes on a digital oscilloscope and calculated by sample length L and mass density ρ, which was calculated from the lattice constants.

3. Results and Discussions

The obtained absolute values of the elastic constants on URu_2Si_2 and $ThRu_2Si_2$ are summarized in Table 1 with corresponding quadrupole and symmetry, where measurement frequencies varying between 100-175 MHz were used. Comparing the elastic constants, $ThRu_2Si_2$ represents larger values in C_{11}, $(C_{11}-C_{12})/2$, C_{44} than that of URu_2Si_2. It is simply expected that additional 5*f*-electron contribution result the elastic constant in the U-based compound lower in general. Intriguingly, C_{66} mode, however, exhibits the opposite tendency of magnitude, which implies that the bonding force of the lattice has strong anisotropy due to some 5*f*-electron's contribution. Though it is a matter of speculation and open question.

Figure 1 shows a comparison of elastic constants C_{11}, $(C_{11}-C_{12})/2$, C_{44} and C_{66} of URu_2Si_2 and $ThRu_2Si_2$. The elastic constants C_{11} and $(C_{11}-C_{12})/2$ of URu_2Si_2 exhibit maximum at ~100 K while C_{44} and C_{66} do not show any anomaly in Para state above T_O, which are consistent with previous reports [11, 12]. On the other hand, all elastic constants of $ThRu_2Si_2$ increase monotonically with decreasing temperature and exhibit no anomaly. In the present study, the elastic constants of $ThRu_2Si_2$ were used for the backgrounds to subtract the phonon contributions in the elastic constants of URu_2Si_2.

Figure 2 represents the elastic constants of URu_2Si_2 with those of $ThRu_2Si_2$ subtracted, which are displayed by relative change as,

$$\frac{C_{ij(U-Th)}-C_{ij(U-Th@300K)}}{C_{ij(U-Th@300K)}}=\frac{\{C_{ij(URu_2Si_2)}(T)-C_{ij(ThRu_2Si_2)}(T)\}-\{C_{ij(URu_2Si_2)}(300K)-C_{ij(ThRu_2Si_2)}(300K)\}}{\{C_{ij(URu_2Si_2)}(300K)-C_{ij(ThRu_2Si_2)}(300K)\}}. \quad (1)$$

The softening appears in C_{11} down to 30 K implies that the system has the instabilities of Kondo volume collapse [13]. The background-subtracted $(C_{11}-C_{12})/2$ keeps decreasing almost linearly towards low temperature, while other share modes C_{44} and C_{66}, corresponding to Γ_5-type quadrupoles O_{yz} $(=J_yJ_z+J_zJ_y)$, O_{zx} $(=J_zJ_x+J_xJ_z)$ and Γ_4-type O_{xy} $(=J_xJ_y+J_yJ_x)$, respectively, show little change. These contrasts suggest that the Γ_3-type quadrupole moment will be active in URu_2Si_2, if one can assume that the softening of $(C_{11}-C_{12})/2$ is explained by quadrupole susceptibility.

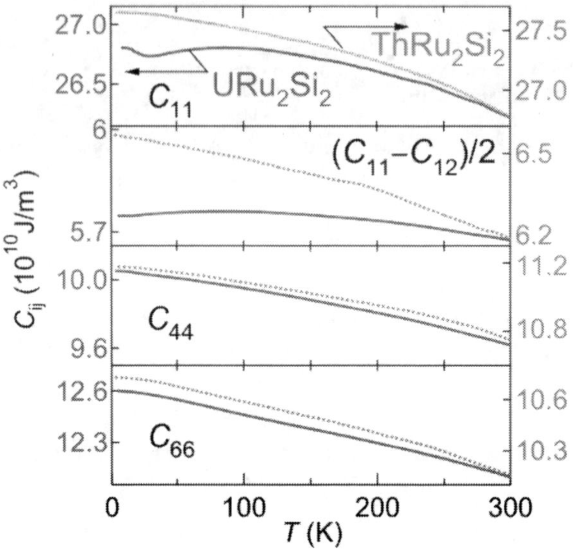

Figure 1. Elastic constants of URu_2Si_2 (Blue-line, left axis) and $ThRu_2Si_2$ (Red-dots, right axis).

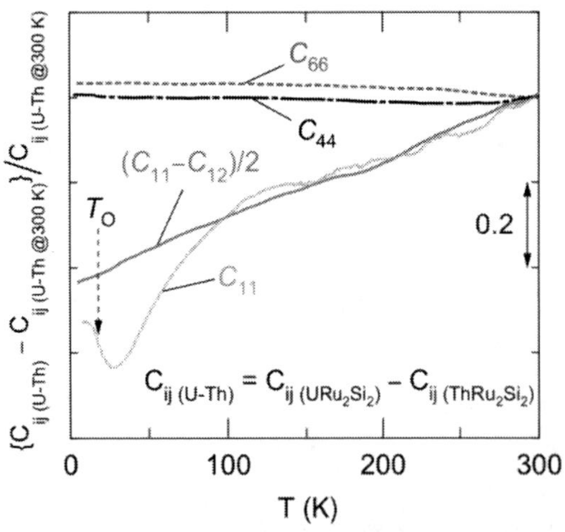

Figure 2. Relative change of the elastic constants as a function of temperature for URu_2Si_2 with the effect of $ThRu_2Si_2$ subtracted.

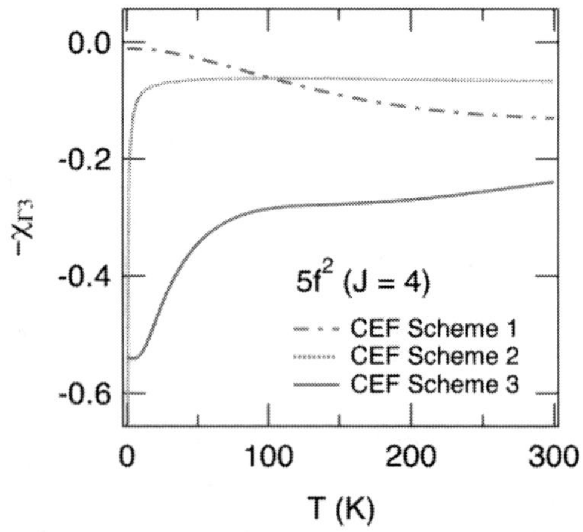

Figure 3. Calculated quadrupole susceptibility $-\chi_{\Gamma3}$ vs. temperature for three proposed CEF level schemes of $5f^2$ ($J = 4$).

The CEF scheme-1 is $\Gamma_1^{(1)}$ (0 K) – Γ_2 (50 K) – $\Gamma_1^{(2)}$ (171 K) – $\Gamma_5^{(2)}$ (500 K) – Γ_4 (1254 K) – Γ_3 (1451 K) – $\Gamma_5^{(1)}$ (1727 K), proposed by Nieuwenhuys. [14] The CEF scheme-2 is $\Gamma_5^{(1)}$ (0 K) – $\Gamma_1^{(1)}$ (404 K) – Γ_2 (1076 K) – $\Gamma_1^{(2)}$ (1119 K) – $\Gamma_5^{(2)}$ (2017 K) – Γ_4 (3253 K) – Γ_3 (3613 K), proposed by Galatanu *et al.* [15] The CEF scheme-3 is Γ_4 (0 K) – $\Gamma_1^{(1)}$ (44 K) – Γ_2 (111 K) – $\Gamma_5^{(2)}$ (485 K) – Γ_3 (534 K) – $\Gamma_5^{(2)}$ (826 K) –$\Gamma_1^{(2)}$ (875 K), proposed by Santini *et al.* [16]

Temperature dependence of the elastic constant $(C_{11}\text{-}C_{12})/2$ can be described by CEF effect of localized *f*-electron's Γ_3 quadrupole susceptibility as,

$$\frac{(C_{11}-C_{12})}{2} = \frac{(C_{11}^0-C_{12}^0)}{2} - \frac{N g_{\Gamma 3}^2 \chi_{\Gamma 3}(T)}{1-g'_{\Gamma 3}\chi_{\Gamma 3}(T)} \quad (2)$$

Here, $(C_{11}^0-C_{12}^0)/2$ is background, *i.e.*, other than the *f*-electron contributions such as phonon contributions on $ThRu_2Si_2$, N is number of the ions in a unit volume, $g_{\Gamma 3}$ and $g'_{\Gamma 3}$ is coupling constant of quadrupole-strain interaction and quadrupole-quadrupole intersite interactions for Γ_3-type quadrupole moment O_2^2 ($= J_x^2-J_y^2$), respectively. Only the calculated quadrupole susceptibility by using the CEF scheme-3 shows decreasing for all temperature ranges, while the CEF scheme-2 gives broad local maximum at around 100 K and the CEF scheme-1 results increase in $-\chi_{\Gamma 3}$ from room temperature as shown in Fig. 3. However, the CEF scheme-3 still cannot be perfectly reproducing the temperature dependence of the background-subtracted $(C_{11}-C_{12})/2$ below 100 K. In the present stage, a strong *c-f* hybridization effect of URu_2Si_2 at low temperature region are not take into account. The possible Kondo effect had been evidenced by several physical properties such as log$T$ increasing of resistivity around room temperature or strong suppression of resistivity and magnetic susceptibility below 60 K. Since the many-body effect might suppress not only the Γ_3-quadrupole susceptibility but also other quadrupole responses, it is rather difficult to understand the temperature dependence of the elastic constant of 5*f*-electron systems only from the localized picture. We could, however, conclude that at least the Γ_3-type quadrupole moment O_2^2 in URu_2Si_2 is still dominant in comparison to O_{yz}, O_{zx} and O_{xy} and/or this compound has some electric instability, which well couples to the strain wave with Γ_3 symmetry.

Acknowledgements
The present work was supported by Grant-in-Aid for Young Scientists (B) (2374025001), Scientific Research for Innovative Areas 'Heavy Electrons' (2374025001), and Scientific Research S (20224015) from the Ministry of Education, Culture, Sports, Science and Technology. One of the authors (T.Y.) was supported by Hokkaido Univ. Leader Development System in the Basic Interdisciplinary Research Areas.

References
[1] Palstra T T M *et al.* 1985 Phys. Rev. Lett. **55** 2727.
[2] Schlabitz W *et al.* 1986 Z. Phys. B **62** 171.
[3] Maple M B *et al.* 1986 Phys. Rev. Lett. **56** 185.
[4] Harima H *et al.* 2010 J. Phys. Soc. Jpn. **79** 033705.
[5] Amitsuka H 2010 *JPSJ Online—News and Comments* [March 10, 2010].
[6] Amitsuka H *et al.* 2010 J. Phys.: Conf. Ser. **200** 012007.
[7] Kusunose H and Harima H 2005 J. Phys. Soc. Jpn. **74** 1582.
[8] Okazaki R *et al.* 2011 Science **331** 439.
[9] Thalmeier P and Takimoto T 2011 Phys. Rev. B **83** 165110.
[10] Thalmeier P and Lüthi B1991 in *Handbook on the Physics and Chemistry of Rare Earths*, edited by K.A. Gshchneider, Jr. and L. Eyring (North-Holland, Amsterdam) Vol. 14 p. 311.
[11] Kuwahara K *et al.* 1997 J. Phys. Soc. Jpn. **66** 3251.
[12] Wolf B *et al.* 1994 J. Low Temp. Phys. **94** 307.
[13] Lüthi B *et al.* 1993 Phys. Lett. A **175** 237.
[14] Nieuwenhuys G J 1987 Phys. Rev. B **35** 5260.
[15] Garatanu A *et al.* 2005 J. Phys. Soc. Jpn. **74** 1582.
[16] Santini P and Amoretti G 1994 Phys. Rev. Lett. **73** 1027.

^{121}Sb-NMR Knight shift study of filled skutterudite $CeOs_4Sb_{12}$

M Yogi[1], H Niki[1], H Mukuda[2], Y Kitaoka[2], H Sugawara[3], and H Sato[4]

[1]Department of Physics and Earth Sciences, Faculty of Science, University of the Ryukyus, Okinawa 903-0213, Japan
[2]Department of Materials Engineering Science, Graduate School of Engineering Science, Osaka University, Osaka 560-8531, Japan
[3]Department of Physics, Graduate School of Science, Kobe University, Kobe 657-8501, Japan
[4]Department of Physics, Tokyo Metropolitan University, Tokyo 192-0397, Japan

E-mail: myogi@sci.u-ryukyu.ac.jp

Abstract. We carried out an ^{121}Sb nuclear magnetic resonance (NMR) measurement on $CeOs_4Sb_{12}$ to investigate its electric state. The precise temperature (T) dependence of the Knight shift was obtained by using an aligned powdered sample and NQR parameters. The T dependence of the Knight shift below 300 K can not be explained by the c-f hybridization gap model with $2D = 3000$ K and $2\Delta = 320$ K obtained by the $1/T_1$. This result indicates that there is q-dependence in the gap. It was revealed that a ferromagnetic correlation develops with decreasing temperature below 50 K from an increase of the Knight shift. Considering this result and previous $1/T_1$ measurements, there are both ferromagnetic and antiferromagnetic fluctuations in $CeOs_4Sb_{12}$.

1. Introduction

Filled skutterudite compound $ReTx_4Pn_{12}$ (Re = rare earth; Tx = Fe, Ru and Os; Pn = P, As and Sb) shows various ground states in many combinations of Re, Tx and Pn atoms. These varieties of intriguing phenomena arise from the characteristic crystal structure of the filled-skutterudite compounds (space group: $Im\bar{3}$, T_h^5 No.204); twelve pnictogen atoms surround a rare-earth ion, which causes strong hybridization between conduction and f-electrons. Among filled-skutterudite compounds, most Ce-based compounds show insulating or semiconducting behavior, on the basis of which they are called hybridization-gap semiconductors. The magnitude of the gap is related to a lattice constant of the crystal, which collapses for Sb-based compounds [1]. However, $CeOs_4Sb_{12}$, which has the largest lattice constant among $CeTx_4Pn_{12}$ compounds, exhibits Kondo insulating behavior in electric resistivity with a very small gap $\Delta/k_B \sim 10$ K [2]. A large specific heat coefficient $\gamma \sim 92$ mJ/K^{2}mol was also reported [3]. We previously reported that $CeOs_4Sb_{12}$ has a hybridization gap of $2\Delta/k_B \sim 320$ K, though a large residual density of states exists within the gap, as determined from a measurement of the nuclear spin-lattice relaxation rate $1/T_1$ [4]. Furthermore, the nuclear spin-lattice relaxation rate $1/T_1$ was observed to show an increase in spin fluctuation on the verge of an antiferromagnetic (AFM) quantum critical point (QCP) below 20 K. Moreover, $CeOs_4Sb_{12}$ was reported to exhibit AFM order at

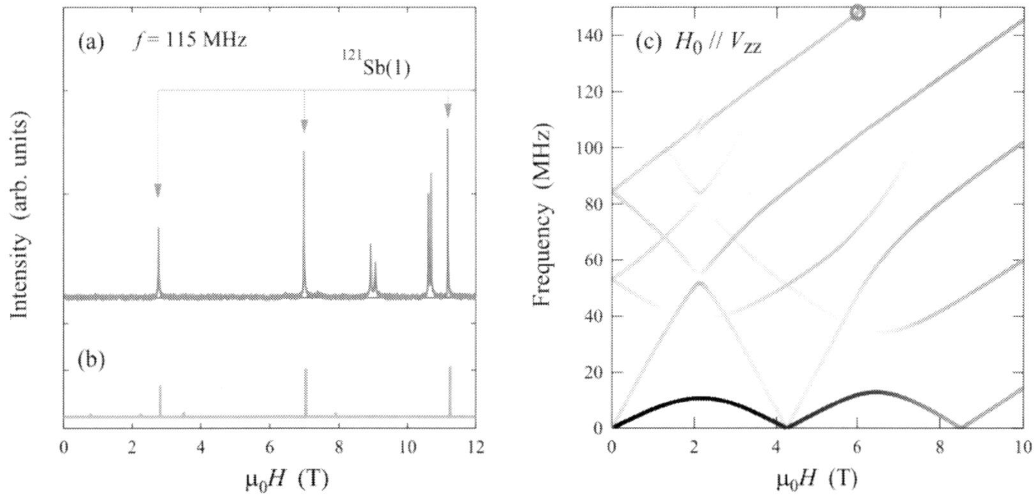

Figure 1. (a) ^{121}Sb NMR spectrum as a function of $\mu_0 H$ at the resonance frequency $f = 115$ MHz and $T = 4.2$ K. (b) Calculated spectrum at ^{121}Sb(1) site for $H \parallel [100]$. (c) Numerical calculation of the resonance frequency for ^{121}Sb(1) as a function of $\mu_0 H$. The color density of the plot indicates the transition probability. The open circle indicates a transition which was used for measurement of the Knight shift.

approximately $T \sim 1$ K and $H = 0$ [5]. A discriminative $H - T$ phase diagram was obtained by resistivity, specific heat and ^{121}Sb-NMR $1/T_1T$ measurements [1, 3, 6].

In order to investigate the electric state of $CeOs_4Sb_{12}$ further, we carried out an Sb-NMR experiment and measured the precise temperature (T) dependence of the Knight shift.

2. Experimental

A single crystal of $CeOs_4Sb_{12}$ was prepared by the Sb-flux method [1]. For the NMR measurements, the sample was powdered to facilitate applied rf-field penetration. Pulse NMR measurements were performed on two Sb isotopes ^{121}Sb ($I = 5/2$, $^{121}\gamma/2\pi = 10.189$MHz/T and $^{121}Q = -0.543 \times 10^{-28}$ m^2) and ^{123}Sb ($I = 7/2$, $^{123}\gamma/2\pi = 5.5176$MHz/T and $^{123}Q = -0.692 \times 10^{-28}$ m^2) using the conventional spin-echo method in the range of $T = 1.5 - 300$ K. Here, I, γ and Q are the nuclear spin, the nuclear gyromagnetic ratio and the nuclear quadrupole moment, respectively. The NMR spectrum was obtained by integrating a spin-echo signal point by point as a function of the magnetic field.

3. Results and Discussion

The electric field gradient at the Sb site is quite large and non-axially symmetric in the filled skutterudite compound; therefore, the nuclear quadrupole frequency increases. For example, $CeOs_4Sb_{12}$ has the nuclear quadrupole frequency $^{121}\nu_Q(^{123}\nu_Q) = 43.847(26.628)$ MHz with asymmetry parameter $\eta = 0.463$ at $T = 4.2$ K [4]. Therefore, a broad complex NMR spectrum would be expected for a powdered sample. However, as shown in Fig. 1(a), we observed many sharp peaks close to the case of the spectrum using a single crystal. This is because $CeOs_4Sb_{12}$ has a small anisotropy of magnetization above ~2 T ($M^{[100]} > M^{[110]} > M^{[111]}$), and the powdered sample is aligned to $H \parallel [100]$. Although Sb has two isotopes, namely, ^{121}Sb and ^{123}Sb, most of the peaks arise from ^{121}Sb because of the differences in nuclear gyromagnetic ratio γ_n and nuclear quadrupole frequency ν_Q. The filled skutterudite structure

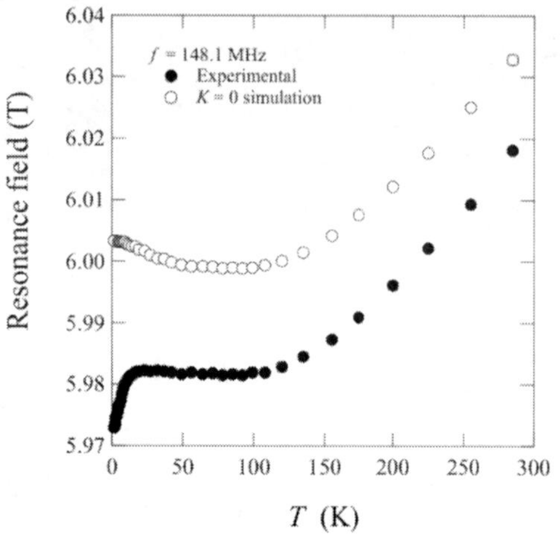

Figure 2. The open circles represent a simulation of resonance fields assuming $K = 0$. The closed circles represent resonance fields of $CeSb_4Sb_{12}$ at $f = 148.1$ MHz.

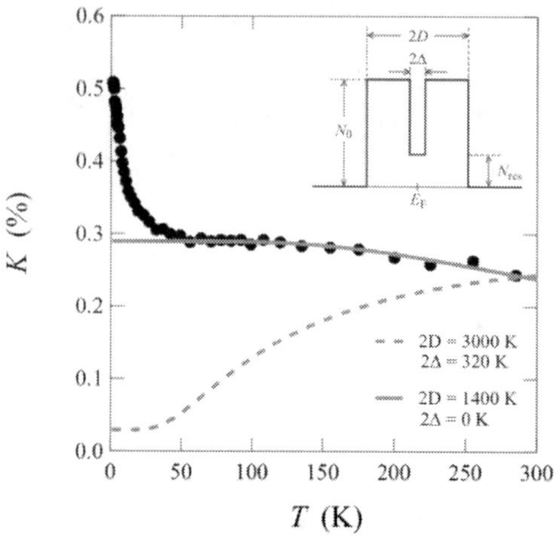

Figure 3. T dependences of Knight shift. The solid and dashed lines are the T dependence of $1/T_1$ calculated by assuming a rectangular shape of density of states.

has one crystallographically equivalent Sb site; however, three ^{121}Sb NMR spectra appear for $H \parallel [100]$ owing to the anisotropy of the Sb 24g site [7]. Among them, we focused on Sb(1) site, which is located in $H \parallel [100] \parallel V_{zz}$, for a measurement of the Knight shift in this study. Here, V_{zz} is the main principal axis of the electric field gradient (EFG) at the Sb position.

The nuclear spin Hamiltonian in the case of $H \parallel V_{zz}$ is given by

$$\mathcal{H} = -\gamma_n \hbar (1+K) H_0 I_z + \frac{h\nu_Q}{6}\left[3I_z^2 - I^2 + \eta(I_x^2 - I_y^2)\right]. \tag{1}$$

The first term of the Hamiltonian represents the Zeeman interaction between the ^{121}Sb nuclear spin $\boldsymbol{I}$ and the applied field $\boldsymbol{H}_0$. K is the Knight shift caused by a hyperfine interaction between the nucleus and electrons, and is related to the spin susceptibility. The second term in the Hamiltonian represents the nuclear quadrupole interaction between the EFG and the ^{121}Sb nuclear quadrupole moment. Here, ν_Q is defined by $\nu_Q \equiv 3eQV_{zz}/2I(2I-1)h$, and η is the asymmetry parameter defined by $\eta = (V_{xx} - V_{yy})/V_{zz}$, which is a measure of the deviation from the axial symmetry of the EFG tensor. The numerically calculated field-swept NMR spectrum for the Sb(1) site is shown in Fig. 1(b) which well explains some of peaks of the experimental result. Here, we assumed $K = 0$. Other peaks are attributed to the Sb(2) and Sb(3) sites, respectively.[6] Figure 1(c) shows the numerical calculation of the resonance frequency at the ^{121}Sb(1) site as a function of $\mu_0 H$. As for the measurement of the Knight shift, we used a transition at $f = 148.1$ MHz and $H \sim 6$ T.

We measured the NMR spectrum around 6 T and determined the resonance field at each T. The closed circles in Fig. 2 represent the T dependence of the resonance field, which indicates strong T dependence. However, this does not directly indicate T dependence of the Knight shift. This is because $^{121}\nu_Q$ and η also show T dependence. To estimate the precise T dependence of the Knight shift, we measured the T dependence of $^{123}\nu_Q$ and η, then we calculated the resonance fields in the case of $K = 0$. From these data, we obtained the Knight shift using the relation $K = (H_0 - H_m)/H_m$. Here, H_0 is the resonance field in the case of $K = 0$ (open circles in Fig. 2), and H_m is the experimental result (closed circles in Fig. 2).

Figure 3(a) shows the T dependence of the Knight shift, which shows a slight increase as T decreases below 300 K. We try to explain this behavior using the $c-f$ hybridization gap model with a rectangular shape of the density of states (DOS) as shown in the inset of Fig. 3. As for $1/T_1$, this model explains well the T dependence with parameters $2D = 3000$ K and $2\Delta = 320$ K [4]. However, these parameters can not explain the variation of the Knight shift. If we choose smaller values of $2D$ and 2Δ, this model can explain the T variation of the Knight shift above 50 K. For example, the solid line in Fig. 3 is a calculation with $2D = 1400$ K and $2\Delta = 0$ K, which fits the data well. This discrepancy of the gap size might be caused by the q-dependence of the gap. The Knight shift shows an apparent increase as T decreases below ~ 50 K, indicating the development of a ferromagnetic correlation because the Knight shift probes spin susceptibility $\chi_s = (0, 0)$. From this result and our previous T_1 experiment, it is indicated that there are both ferromagnetic and antiferromagnetic fluctuations in $CeOs_4Sb_{12}$. The anomalous ordered phase above ~ 1 T is considered to be caused by such complex electron correlations.

4. Summary

In summary, we have performed Sb-NMR measurement on $CeOs_4Sb_{12}$. The powdered sample is aligned to $H \parallel [100]$, which allows us to estimate a precise value of the Knight shift even where the ν_Q and η are large. The T dependence of the Knight shift below 300 K can not be explained by the c-f hybridization gap model with $2D = 3000$ K and $2\Delta = 320$ K obtained by the $1/T_1$. This result indicates that there is q-dependence in the gap. It was revealed from an increase of the Knight shift that a ferromagnetic correlation develops as T decreases below 50 K. Considering this result and previous $1/T_1$ measurements, it can be concluded that there are both ferromagnetic and antiferromagnetic fluctuations in $CeOs_4Sb_{12}$.

Acknowledgments

This work was supported by a Grant-in-Aid for Scientific Research on Priority Area "Skutterudite" (Nos. 15072204 and 15072206) and on Innovative Areas "Heavy Electrons" (No. 20102007) of The Ministry of Education, Culture, Sports, Science, and Technology, Japan; and a Grant-in-Aid for Young Scientists (B: 18740209) from the Japan Society for the Promotion of Science.

References

[1] Sugawara H, Osaki S, Kobayashi M, Namiki T, Saha S R, Aoki Y and Sato H 2005 *Phys. Rev.* B **71** 125127.
[2] Bauer E D, Ślebarski A, Freeman E J, Sirvent C and Maple M B 2001 *J. Phys.: Condens. Matter* **13** 4495.
[3] Namiki T, Aoki Y, Sugawara H, and Sato H 2003 *Acta Phys. Pol.* B **34** 1161.
[4] Yogi M, Kotegawa H, Zheng G-q, Kitaoka Y, Ohsaki S, Sugawara H and Sato H 2005 *J. Phys. Soc. Jpn.* **74** 1950.
[5] Iwasa K, Itobe S, Yang C, Murakami Y, Kohgi M, Kuwahara K, Sugawara H, Sato H, Aso N, Tayama T, and Sakakibara T 2007 *J. Phys. Soc. Jpn.* **77** Suppl. A 318.
[6] Yogi M, Niki H. Yashima M, Mukuda H, Kitaoka Y, Sugawara H, and Sato H 2009 *J. Phys. Soc. Jpn.* **79** 053703.
[7] Tou H, Doi M, Sera M, Yogi M, Kitaoka Y, Kotegawa H, Zheng G.-q, Harima H, Sugawara H, Sato H 2005 *Physica* B **395-361** 892.

International Conference on Strongly Correlated Electron Systems (SCES 2011) IOP Publishing
Journal of Physics: Conference Series **391** (2012) 012081 doi:10.1088/1742-6596/391/1/012081

Low temperature magnetic structure of geometrically frustrated $SrHo_2O_4$

O Young[1], L C Chapon[2,3] and O A Petrenko[1]

[1] Department of Physics, University of Warwick, Coventry CV4 7AL, United Kingdom
[2] ISIS Facility, Rutherford Appleton Laboratory, Chilton, Didcot OX11 0QX, United Kingdom
[3] Institut Laue-Langevin, 6 Jules Horowitz, BP156, 38042 Grenoble Cedex 9, France

E-mail: o.young@warwick.ac.uk

Abstract. $SrHo_2O_4$ is a geometrically frustrated compound because the Ho^{3+} ions are arranged in a network of triangles and hexagons and coupled with antiferromagnetic exchange interactions. Powder neutron diffraction measurements show that long-range magnetic order is established at $T_N = 0.68$ K. The low temperature magnetic structure is made up of two distinct components: a $\mathbf{k} = 0$ long-range structure as well as a shorter-range low-dimensional structure with scattering intensity around the $(0\ 0\ \frac{1}{2})$ position. The diffuse component of the magnetic scattering from $SrHo_2O_4$ remains very intense at temperatures well above T_N.

1. Introduction

Geometric frustration occurs in magnetic systems that cannot satisfy all the competing interactions because of their underlying lattice structure [1]. The Kagome [2], pyrochlore [3] and garnet [4] structures are well known frustrated systems and their magnetic ions are arranged in lattices that are made up of either corner- or edge-sharing triangles or tetrahedra. Interest in these systems comes from the rich variety of magnetic behaviour that is seen due to the influence of small perturbations (such as single ion anisotropy) and fluctuations (thermal or quantum) that will stabilise a ground state out of the macroscopic degeneracy of the classical Hamiltonian. The honeycomb lattice - made up of a planar network of hexagons - has become the subject of much theoretical interest since it has the lowest possible coordination number in two dimensions and thus quantum fluctuations are expected to play a big role in establishing its ground state [5]. Bipartite lattices, such as the honeycomb, are not expected to be frustrated, however they can be if next-nearest-neighbour exchange terms are not negligible.

Recently, a number of compounds based on the honeycomb lattice have been realised experimentally in $S = \frac{3}{2}$ systems $Bi_3Mn_4O_{12}(NO_3)$ [6] and β-$CaCr_2O_4$ [7], as well as the classical spin systems $BaLn_2O_4$ [8] and $SrLn_2O_4$ [9]. Previous bulk property measurements of $SrLn_2O_4$ compounds (which crystallize in the $CaFe_2O_4$-type structure [10], space group *Pnam*) have revealed a discrepancy between the high Curie-Weiss constants [11] and the low Néel ordering temperatures [12, 13]. For one of the members of this family, $SrEr_2O_4$, we have already established the magnetic structure that appears below $T_N = 0.75$ K [12], despite having $\theta_{CW} = -13.5$ K [11]. We have now begun an extended investigation into the effects of frustration as a function of the rare earth element in the $SrLn_2O_4$ systems, firstly by studying $SrHo_2O_4$.

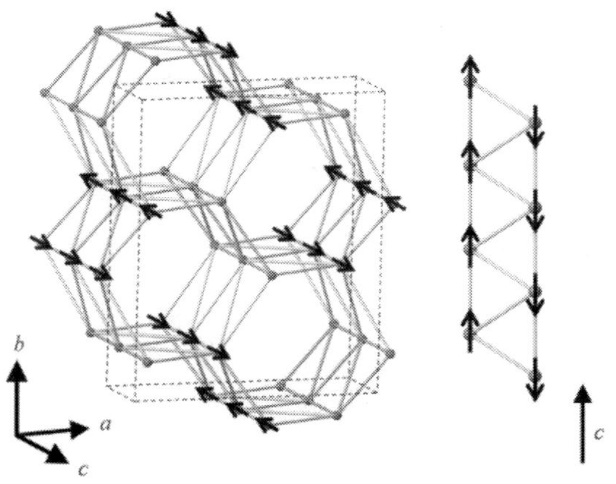

Figure 1. (Colour online) Magnetic structure of $SrHo_2O_4$ at $T = 45$ mK determined from a Rietveld refinement performed using the neutron scattering data collected on GEM (ISIS). When viewed in the a-b plane, honeycombs of the Ho^{3+} ions are visible. Triangular chains running along the c-axis connect the honeycomb layers and give rise to geometric frustration. The two crystallographically inequivalent Ho^{3+} sites are shown in different colours as only one (blue) carries a significant magnetic moment of about 5 μ_B (the second Ho^{3+} has no more than 0.2 μ_B). All of the moments point along the c-axis, with neighbouring chains coupled antiferromagnetically. Swapping the moments between the red and blue sites has a negligible impact on the quality of the refinement.

From powder neutron diffraction data, we have established that the long-range ordered magnetic structure of $SrHo_2O_4$ (shown in figure 1) appears at $T_N = 0.68$ K. The temperature dependence of the magnetic scattering shows that in addition to the $\mathbf{k} = 0$ Néel phase there exists some diffuse scattering, around the $(0\ 0\ \frac{1}{2})$ and symmetry related positions, that persists at temperatures well above T_N. In a related, isostructural compound, $SrEr_2O_4$, the coexistance of two magnetic phases well below T_N [14] is thought to arise because one of the two crystallographically inequivalent Er^{3+} sites orders in a long-range three-dimensional structure, whereas the other Er^{3+} partially orders in a short-range magnetic phase.

2. Experimental Details

The $SrHo_2O_4$ sample was prepared from high purity starting compounds $SrCO_3$ and Ho_2O_3, in an off-stoichiometric ratio 1 : 0.8775. The powders were ground together and heated in air at 1400°C for a total of 48 hours (in an alumina crucible) with one intermediate grinding to ensure homogeneity of the mixture. The purity of the $SrHo_2O_4$ powder was checked by performing a Rietveld refinement using x-ray diffraction data, and this indicated the final purity to be greater than 99 %. Magnetic susceptibility measurements, on powder $SrHo_2O_4$, show a cusp at 0.7 K and this suggests the presence of a phase transition. For a single crystal of $SrHo_2O_4$ we have previously reported $T_N = 0.62$ K [13] from low temperature susceptibility measurements. We believe that the difference in T_N, between the powder and the single crystal $SrHo_2O_4$, may be due to slight non-stochiometry of the samples resulting from the crystal growth process. The GEM instrument at ISIS (Rutherford Appleton Laboratory, UK) was used to collect neutron scattering data in a range of temperatures from 45 mK to 10 K. The FULLPROF software suite was employed to refine the magnetic structure of $SrHo_2O_4$ using the powder data.

3. Results and Discussion

The proposed magnetic structure of $SrHo_2O_4$ that appears at $T_N = 0.68$ K was refined from powder neutron scattering data, and is shown in figure 1. The moments are aligned along the c-axis, since no $(00l)$ reflections were observed. The moments are coupled ferromagnetically in chains running along the c direction, with neighbouring chains in the same ladder coupled antiferromagnetically. Only one of the crystallographically inequivalent sites of the Ho^{3+} ions carries a significant magnetic moment of 5 μ_B, the second has no more than 0.2 μ_B. Swapping the moments between the sites in our model has a negligible impact on the quality of the

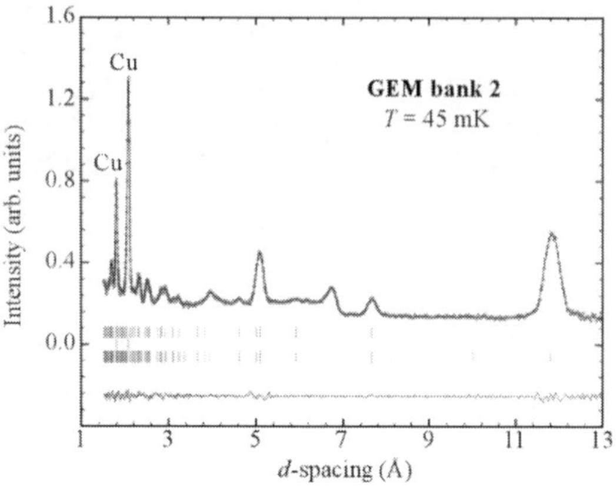

Figure 2. (Colour online) $SrHo_2O_4$ powder neutron diffraction refinement using the data collected in detector bank 2 on the GEM instrument at T = 45 mK. Data points and fit are shown in the upper curve, the difference between the observed and calculated pattern is shown in the lowest curve. Three sets of vertical tick marks (from top to bottom) correspond to the expected positions for nuclear peaks from $SrHo_2O_4$, reflections from the copper sample container and magnetic peaks from $SrHo_2O_4$ respectively. The peaks coming from the copper sample holder are labelled for clarity.

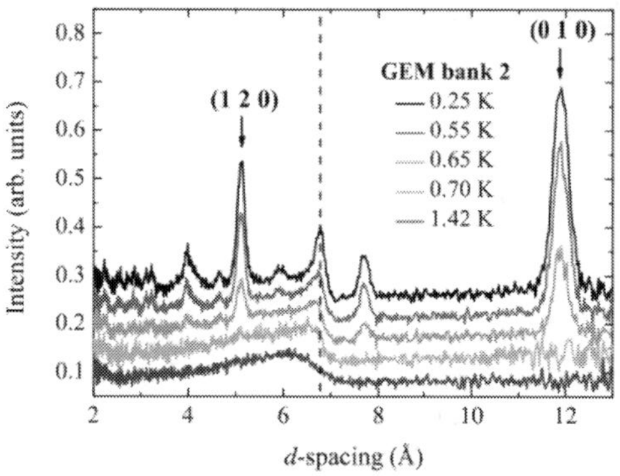

Figure 3. (Colour online) Temperature evolution of the magnetic scattering of $SrHo_2O_4$ (curves have been offset for clarity). Below 0.7 K magnetic Bragg peaks can be seen indicating the presence of long-range order. Above T_N however, short-range correlations around $d \simeq 6.7$ Å (denoted by the dashed line) are also observed. The magnetic peak due to these short-range correlations gets sharper (but does not change position) upon decreasing the temperature. The magnetic diffraction pattern is isolated by subtracting a 10 K background from the low temperature runs.

refinement. The FULLPROF refinement was performed using data collected in detector banks 2 to 5 of the GEM instrument simultaneously. Figure 2 only shows the magnetic structure refinement of the data collected in detector bank 2 at 45 mK, since in this detector bank all of the low Q magnetic reflections from $SrHo_2O_4$ are visible. The diffuse scattering, seen around $d \simeq 6.7$ Å (represented by a dashed line in figure 3) corresponds to the $(0\ 0\ \frac{1}{2})$ position in reciprocal space, was treated as part of the background in the structure refinement as the limited information from powder data does not allow for a complete solution of this component of the magnetic scattering. The agreement factor for the magnetic phase in the detector bank shown in figure 2 is $R_{\mathrm{mag}} = 1.32$ %.

Figure 3 shows five neutron diffraction patterns collected in the temperature range 0.25 K to 1.42 K. Broad, diffuse scattering peaks are seen at temperatures much higher than T_N (for example see the bottom pattern in figure 3 which was taken at 1.42 K). Bragg peaks begin to appear below 0.7 K, indicating the onset of long-range order. The peaks get more intense upon lowering the temperature, but in reciprocal space their positions remain the same. This long-range magnetic structure was found to be commensurate with the lattice and it can be indexed with the propagation vector $\mathbf{k} = 0$. The diffuse scattering that coexists with the long-range order, however, has the characteristic powder profile of a low-dimensional magnetic structure.

Within the refined magnetic structure there is a large difference between the values of the Ho^{3+} moments that sit on crystallographically inequivalent sites. We believe that this is due to the coexistance of two separate magnetic structures in this material. The first is the long-range ordered three-dimensional magnetic structure (discussed above) that appears at $T_N = 0.68$ K, see figure 4a); the second is a short-range ordered low-dimensional structure that develops well

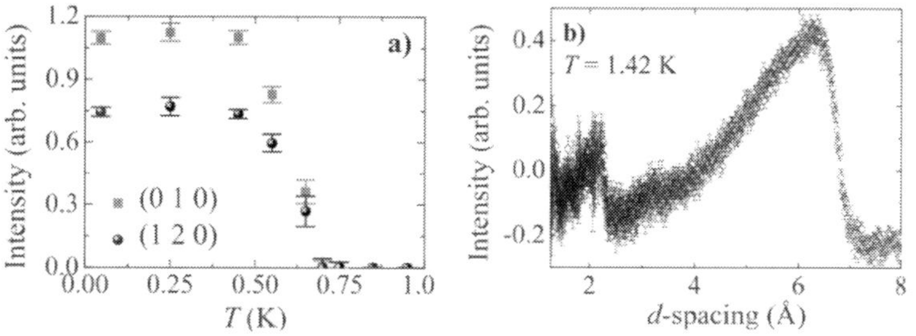

Figure 4. a) Temperature dependence of the integrated magnetic intensity for the two strongest magnetic peaks: (0 1 0) and (1 2 0), and b) magnetic scattering above T_N, at $T = 1.42$ K isolated by subtracting a 10 K background from data collected at 1.42 K.

above T_N, see figure 4b). It is useful to note that the magnetic scattering shown in figure 4b), at $T = 1.42$ K, is practically identical to the magnetic scattering shown in figure 16b) in the paper by Karunadasa *et al.* [11], where the data were obtained by subtracting the 5.05 K background from the data collected at 1.72 K. Further neutron scattering experiments on single crystals of $SrHo_2O_4$, which are already available [15], are now needed to give us greater insight into the nature of the unusual diffuse scattering features seen in the powder data.

4. Conclusions

We have determined the magnetic structure of $SrHo_2O_4$ which forms below $T_N = 0.68$ K. The powder neutron scattering data have revealed the coexistance of two magnetic phases in the same material: a fully ordered commensurate phase with the propagation vector $\mathbf{k} = 0$, and a short-range ordered, low-dimensional phase which persists at temperatures much higher than T_N. Further neutron scattering experiments on single crystals of $SrHo_2O_4$ are planned to fully determine the nature of the diffuse scattering seen at low temperatures.

References

[1] Diep H T (ed) 2005 *Frustrated Spin Systems* (Singapore: World Scientific); Ramirez A P 1994 *Annu. Rev. Mater. Sci.* **24** 453; Collins M F and Petrenko O A 1997 *Can. J. Phys.* **75** 605
[2] Ramirez A P, Espinosa G P and Cooper A S 1990 *Phys. Rev. Lett.* **64** 2070
[3] Bramwell S T and Gingras M J P 2001 *Science* **294** 1495
[4] Schiffer P, Ramirez A P, Huse D A and Valentino A J 1994 *Phys. Rev. Lett.* **73** 2500
[5] Mattsson A, Frojdh P and Einarsson T 1994 *Phys. Rev.* B **49** 3997; Okumura S, Kawamura H, Okubo T and Motome Y 2010 *J. Phys. Soc. Jpn.* **79** 114705
[6] Matsuda M, Azuma M, Tokunaga M, Shimakawa Y and Kumada N 2010 *Phys. Rev. Lett.* **105** 187201
[7] Damay F, Martin C, Hardy V, Maignan A, André G, Knight K, Giblin S R and Chapon L C 2010 *Phys. Rev.* B **81** 214405
[8] Doi Y, Nakamori W and Hinatsu Y 2006 *J. Phys.: Condens. Matter* **18** 333
[9] Lopato L M and Kushchevskii A E 1973 *Ukrainskii Khimicheskii Zhurnal* **39** 7 [in russian]
[10] Decker B F and Kasper J S 1957 *Acta Cryst.* **10** 332
[11] Karunadasa H, Huang Q, Ueland B G, Lynn J W, Schiffer P, Regan K A and Cava R J 2005 *Phys. Rev.* B **71** 144414
[12] Petrenko O A, Balakrishnan G, Wilson N R, de Brion S, Suard E and Chapon L C 2008 *Phys. Rev.* B **78** 184410
[13] Hayes T J, Young O, Balakrishnan G and Petrenko O A 2012 *J. Phys. Soc. Jpn.* **81** 024708
[14] Hayes T J, Balakrishnan G, Deen P P, Manuel P, Chapon L C and Petrenko O A 2011 *Phys. Rev.* B **84** 174435
[15] Balakrishnan G, Hayes T J, Petrenko O A and M^{c}K Paul D 2009 *J. Phys.: Condens. Matter* **21** 012202

Peculiar magnetic phase diagrams of $GdPd_2Si_2$ single crystal

Y Zhang[1], T Fujiwara[1], K Matsubayashi[2], A Kondo[2], K Kindo[2], Y Uwatoko[2] and T Shigeoka[1]

[1]Graduate School of Science and Engineering, Yamaguchi University, Yamaguchi 753-8512, Japan.
[2]Institute for Solid State Physics, University of Tokyo, Kashiwa-shi Chiba 277-8581, Japan.

shigeoka@yamaguchi-u.ac.jp

Abstract. High field magnetization measurement were performed up to 45 T using a long pulse magnet on a $GdPd_2Si_2$ single crystal compound having the tetragonal $ThCr_2Si_2$-type crystal structure and an antiferromagnetic ordering of T_N=16.7 K. The magnetization processes are almost typical antiferromagnetic ones with a weak magnetic anisotropy though small anomalies exist in the processes. The magnetization saturates 7 μ_B/f.u. above B_s=32 T. The B_{001}-T and B_{100}-T magnetic phase diagrams were constructed. There are three phases, the phase I, II and III, under non-magnetic field in the ordered state for both diagrams. The field-induced phase, phase IV, appears below the paramagnetic phase by the [100] magnetic field, B_{100}, which is accompany with a re-entrant behaviour. The B_{100} makes the phase II be stable and the phase I and III be instable. These strange behaviours indicate the direction of the magnetic moment changes in each phase.

1. Introduction

The ternary compound $GdPd_2Si_2$, which crystallizes in the tetragonal $ThCr_2Si_2$-type crystal structure, shows an antiferromagnetic order below 13 K or 16.5 K and a metamagnetic behavior [1, 2]. The previous reports suggest Gd magnetic moments lie in the basal plane. The magnetic moment behavior in the basal plane is interesting, which is less studied in this tetragonal system. The recent study on a single crystal $GdPd_2Si_2$ [3] has reported a strange magnetic behavior; it orders antiferromagnetically below T_N=16.7 K. With decreasing temperature, the [001] susceptibility increases down to T_2=12.3 K via a small kink at T_N, then decreases steeply, goes through a deep valley, increases rapidly and shows a clear kink at T_1=6.8 K followed by almost constant value (see in the Figure 1). There are three anomalies at T_N, T_2 and T_1, indicating phase transitions. The existence of three order phases has been confirmed below T_N from a specific heat measurement. On the other hand, the [100] and [110] susceptibilities in the basal plane are almost identical with each other. They increase up to the maximum value at T_N and decrease with a rapid change around T_1. No clear anomaly is seen around T_2. The appearance of an additional field-induced phase has been also reported. In order to obtain information for a mechanism of this magnetic behavior, high field magnetization processes have been studied under pulsed magnetic fields up to B=45 T. The magnetic phase diagrams were constructed.

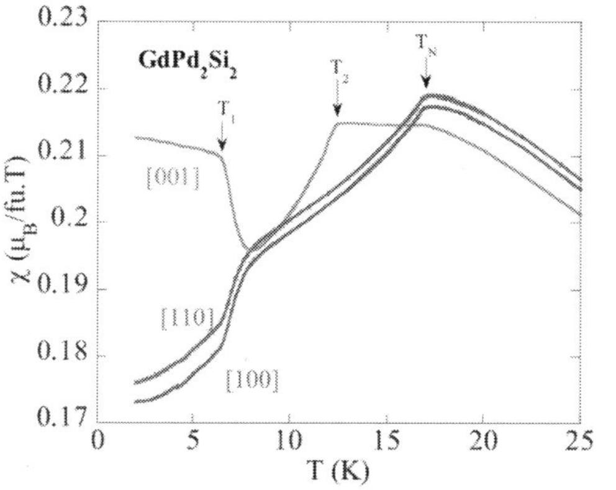

Figure 1 Temperature dependence of magnetic susceptibilities along the main symmetry axes on the $GdPd_2Si_2$ single crystal, after ref.[3].

2. Experimental

The single crystal was grown by the Czochralski method using a tetra-arc furnace. The tetragonal $ThCr_2Si_2$-type structure and a single phase nature were confirmed by X-ray powder diffraction. The good quality of the single crystal was confirmed and crystallographic orientations were determined by back Laue method. The crystal was fixed on a thin plastic plate so that the desired direction is perpendicular to the plate within experimental angle accuracy of one degree. Then it was subjected to magnetic measurements. The high-field magnetization measurements were performed up to 45 T by a pick-up coil method using a non-destructive long-pulse magnet installed at the International MegaGauss Science Laboratory in the Institute for Solid State Physics, the University of Tokyo.

3. Results and discussion

Figure 2 shows high field magnetization processes at 1.3 K along the main symmetry axes in the tetragonal cell of the $GdPd_2Si_2$ single crystal. The processes look like a typical antiferromagnetic one with a weak magnetic anisotropy; that is, the magnetization curves are almost identical for all directions and almost linear below a saturation field. The saturation field B_S is slightly different between the directions in the basal plane and the [001] direction and is about 32 T and about 34 T, respectively. The saturation moment M_S is 7 μ_B/f.u., which is in good agreement with the Gd^{3+} free ion moment. This shows that a field-induced ferromagnetic configuration is realized above B_S and the 4f electrons are well localized in this compound. The saturation field is considerably high among

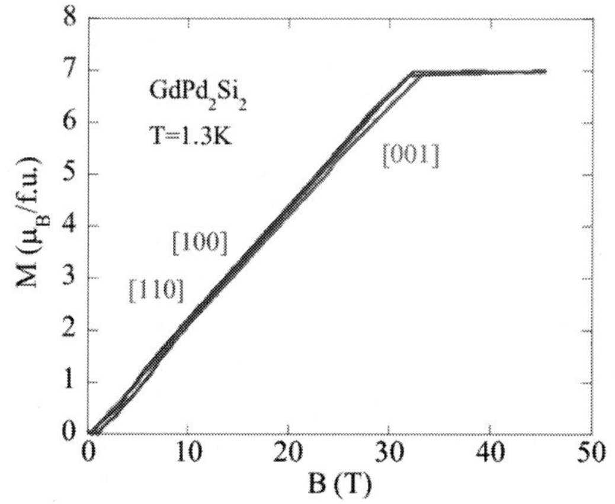

Figure 2 Single crystal magnetization processes along the main symmetry axes at 1.3 K.

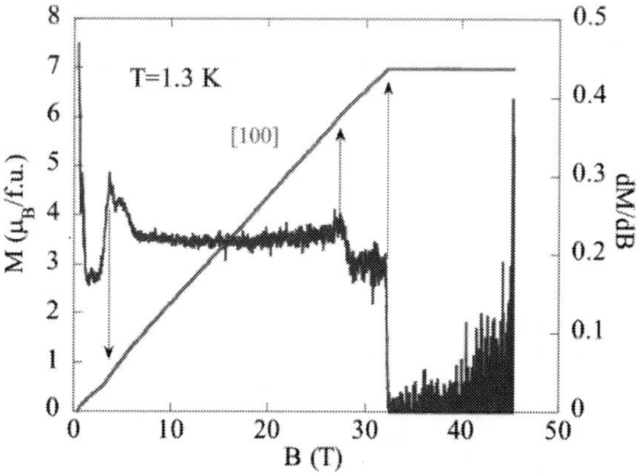

Figure 3 The [100] magnetization and dM/dB vs. B plots at 1.3 K

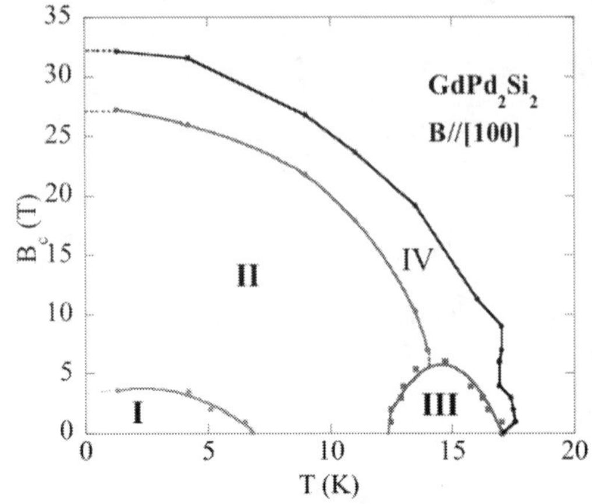

Figure 4 $B_{[100]}$-T magnetic phase diagram

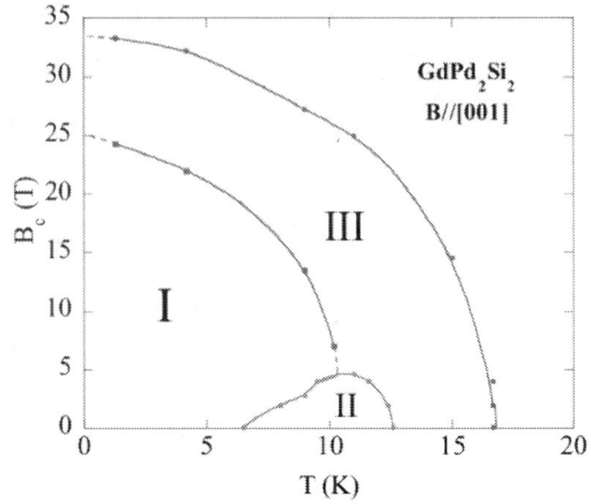

Figure 5 $B_{[001]}$-T magnetic phase diagram

isostructure compounds GdT_2Si_2 (T=transition metals) in spite of relatively low ordering temperature. For example, it is comparable to B_s=37 T of $GdCo_2Si_2$ which has a higher Neel temperature of 45 K and a simple antiferromagnetic structure [4, 5]. This comparison suggests that the antiferromagnetic structure of $GdPd_2Si_2$ is not simple and a non-collinear one, and some effects which prevent a ferromagnetic configuration exist. On observing the magnetization curves in detail, there are some noticeable anomalies in the curves, suggesting changes in magnetic structure. The [100] magnetization curve on the ascending process at 1.3 K together with the dM/dB curve is shown in Fig.3 as an example. There are at least three anomalies which are pointed by arrows in the dM/dB vs. B curve: a peak at B_1=3.5 T and rapid changes in dM/dB at B_2=27.4 T and B_S=32.2 T. The peak shows an appearance of a metamagnetic transition. A small hysteresis loop is observed around B_1, indicating the first order transition. A change in the magnetic structure should be responsible for this transition. The transitions at B_2 and B_s are of the second order. These successive transitions indicate the existence of competing exchange interactions in this compound.

Magnetic phase diagrams $B_{[100]}$-T and $B_{[001]}$-T were constructed from magnetization processes at various temperatures and temperature dependences of M/B at various magnetic fields, and are shown in Figs. 4 and 5, respectively. The $B_{[110]}$-T diagram is similar to the $B_{[100]}$-T one. So the $B_{[100]}$-T diagram is only shown here. Three ordered phases exist under non-magnetic field for both diagrams, which are referred to as the phase I, II and III for $T<T_1$, $T_1<T<T_2$ and $T_1<T<T_N$, respectively. The strange temperature dependence of M/B comes from the appearance of these phases. Both diagrams are peculiar, the $B_{[100]}$-T in particular. An additional field-induced phase, referred to as phase IV, appears just below the paramagnetic phase. The magnetic field dependency of ordering temperature, the boundary between the phases III and IV, is strange for low fields. The ordering temperature firstly increases and above about 3 T decreases with increasing magnetic field. A re-entrant behaviour appears. Such behaviour often appears in a quadrupole ordering where a quadrupolar interaction play an important role. However, the phase IV is not a quadrupole ordering state because a Gd ion is S-state. Some other effects should be required to understand an appearance of the phase IV. In contrast to the $B_{[100]}$-T one, the field dependency of the ordering temperature is ordinary in the $B_{[001]}$-T diagram. The comparison of both diagrams is interesting. The magnetic behaviour is anisotropic nevertheless a Gd compound. A field-induced phase, the phase IV, appears only by magnetic field applied to the [100] direction, leading the re-entrant behaviour. The [100] magnetic field makes the phase II be stable and the phase I and III instable. In contrast, the [001] magnetic field does not induce an additional field-induced phase. The phase I and III become stable and the phase II becomes instable by the [001] magnetic field. These results suggest that the direction of magnetic moment in the phase II is different from those in the phase I and III. This strange behaviour is, however, not explained now. A new

mechanism to determine magnetic moment direction should be required. Before proposing it, the magnetic structures should be determined.

References

[1] Czjzek C, Oestreich V, Schmitt H, Latek K, Tomala K, 1989 *J .Magn. Magn. Mater* **79** 42
[2] Yakinttos J.K, Gamari-Seale H, 1982 *Z. Phys*. B **48** 251
[3] Shigeoka T, Fujiwara T, Kanto K, Zhang Y, Uwatoko Y, 2011 *J. Phys. Conf. Series* **273** 012121.
[4] Shigeoka T, Iwata N, Koiwai S, Uwatoko Y, Hane S, Goto T, 2001 *Physica* B **294-295** 258.
[5] Yung L D, Franse J J M, Buschow K H J, Brommer P E, Thuy N P, 1997 *J. Alloys Comp.* **260** 35.

International Conference on Strongly Correlated Electron Systems (SCES 2011) IOP Publishing
Journal of Physics: Conference Series **391** (2012) 012083 doi:10.1088/1742-6596/391/1/012083

Magnetic and optical properties of multiferroic $GdMnO_3$ film

Mohammed S. Al Qahtani[1], Marzook S. Alshammari[1], H.J Blythe[1], A.M. Fox[1], G.A. Gehring[1], N. Andreev[2], V. Chichkov[2], and Ya. Mukovskii[2]

[1] Department of Physics and Astronomy, University of Sheffield, Sheffield S3 7RH, UK
[2] National University of Science and Technology "MISiS", Leninskii Prospect 4, 119049 Moscow. Russia

Abstract
We report on magneto-optical studies of an epitaxial thin film of $GdMnO_3$ grown on $SrTiO_3$. Significant differences were found between the magneto-optical and absorption spectra, which highlight the importance of magneto-optics in determining the magnetic character of the optical transitions. Two main features were identified in the magnetic circular dichroism (MCD) spectra: the charge transfer transition between Mn *d* states at ~2eV and the band edge transition from the oxygen *p* band to the *d* states at ~3eV. The strength of the MCD at ~2eV correlates well with the Mn spin ordering, indicating the applicability of magneto-optics for understanding the magnetic properties.

1. Introduction

In recent years there has been great interest in the physics of "multiferroics". Discovered in the 1970s, these unusual materials exhibit ferroelectric and magnetic ordering simultaneously. These two order parameters have strong magnetoelectric coupling, allowing the reversal of the ferroelectric polarization by the application of a magnetic field or the control of the magnetic order parameter by an electric field. Together with recent unprecedented progress in epitaxial thin film growth, this unique property makes these "old" materials very good candidates with which to construct multifunctional spintronic devices [1, 2].

In this paper we report magneto-optical measurements on the orthorhombic multiferroic manganite $GdMnO_3$. This material exhibits several magnetic phase transitions as the temperature is lowered. First, magnetic frustration due to competing exchange integrals between successive neighbours stabilizes a spiral magnetic phase below the Néel temperature T_N =47K [3-5]. Next there is a lock-in transition to a canted A-type antiferromagnetic phase at 19K for $GdMnO_3$. At even lower temperatures, the Gd moments order and the oxygen atoms are pushed off the Mn-Mn bond, driving an electric polarization. It is thus clear that the charge, spin, orbital, and lattice subsystems are linked very tightly in this material, making their investigation by magneto-optical techniques particularly interesting

We focus particularly on a spectral feature at ~2 eV. In $LaMnO_3$ this peak is observed clearly in the absorption spectrum and is attributed to a charge transfer excitation between neighbouring Mn ions. However, the peak is suppressed in absorption for heavier rare earth manganites including $GdMnO_3$ [6-9]. This reduction in intensity has been related to the different canting of the Mn *d*-orbitals and also to the directions of Mn spins. In our work we find that the charge transfer peak appears very prominently in the magnetic circular dichroism (MCD) spectra, which allows us to correlate the magneto-optical properties with the magnetization as the temperature is lowered through the magnetic transitions. We find that the strength of the peak correlates with the Mn spin ordering, thus confirming the strong magnetoelectric coupling of $GdMnO_3$.

2. Sample preparation and magnetic characterisation

Epitaxial thin film samples of $GdMnO_3$ were deposited on a [110] $SrTiO_3$ substrate using RF magnetron sputtering. $SrTiO_3$ was chosen as the substrate because it facilitated high quality epitaxial thin film growth. However it did have some disadvantages for the optical studies reported here. The band gap energy at 3.2eV reduced the useful energy range over which absorption measurements could be made, and the structural transition at ~100K gave rise to birefringence, making magneto-optical measurements more difficult [10].

The targets of $GdMnO_3$ were prepared from stoichiometric mixtures of Gd_2O_3 and MnO_2. $GdMnO_3$ films of thickness of 100-200nm were sputtered onto single crystal $SrTiO_3$ [110] substrates in an atmosphere of Ar and O_2 at a pressure of 1–2 mTorr. The substrate temperature was 650°C. This produced an epitaxial $GdMnO_3$ film with the [110]-direction parallel to the plane of the substrate. X-ray data taken at room temperature are shown in Fig. 1(a), confirming the high crystalline quality.

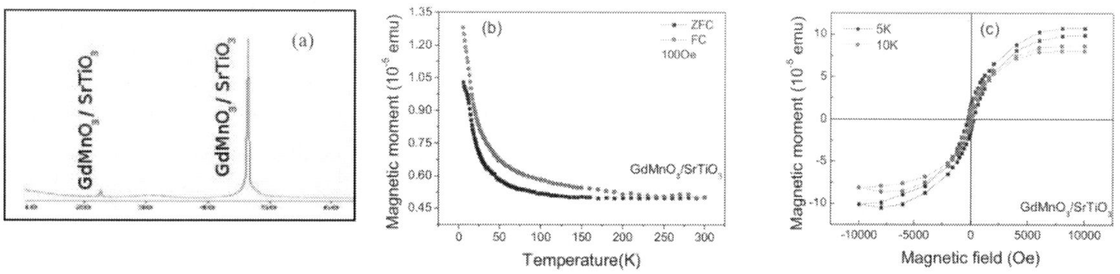

Figure 1. (a) X-ray data showing epitaxial growth, **(b)** Field cooled (red) and zero field cooled (black) magnetisation data of the film (contribution from STO was subtracted) **(c)** Magnetic hysteresis loops of the film at 5K and 10K (linear background subtracted.)

The magnetic properties of the films were investigated by SQUID magnetometry with the magnetic field perpendicular to the samples. This geometry was chosen to coincide with the Faraday geometry used for the magneto-optical studies. Field cooled and zero field cooled magnetisation plots are shown in Fig. 1(b). The data were taken in a field of 100 Oe and the contribution from the diamagnetic substrate has been subtracted. We note there is a positive, temperature-independent susceptibility above ~100 K.

Magnetic hysteresis loops were only observed for temperatures below the transition to the canted phase at 19 K. Typical hysteresis loops at 5 K and 10 K are shown in Fig. 1(c), and coercive fields are 250 Oe and 140 Oe at temperatures 5 K and 10 K are deduced. These hysteresis loops were obtained by subtracting off the diamagnetic signal from the substrate and the paramagnetic signal from the film, both of which are linear in the field. As shown in Fig. 1(b), this paramagnetic signal was large at low temperatures, but could be clearly identified from the linear variation of the magnetization for fields in the range 7–10 kOe.

3. Optical Measurements

The absorption spectrum of the film was measured at room temperature and is shown in Fig. 2(a). Previous work has identified a strong, electric-dipole-allowed optical transition at the band edge around ~3 eV between the oxygen *p* states and unoccupied Mn *d* states. This absorption band is observed in our samples, although it was not possible to make measurements above 3.2 eV due to the absorption of $SrTiO_3$ substrate [11]. (See inset to Fig. 2(a).) Weak absorption at the charge-transfer transition around 2 eV was also observed in the absorption, but without any structure.

Magnetic circular dichroism (MCD) measurements were made using the apparatus described in [12]. The MCD spectrum measures the difference in absorption between left and right circularly polarised light, and thus reflects the magnetic character of the electronic states involved in the optical transition [13]. In our experiments we had to be careful to orient our samples so that the *E* vector of the light was along the optic axis of the substrate, in order to avoid complications due to the birefringence of $SrTiO_3$ below 100 K.

We present two sets of data. In the first, we measured the MCD spectrum in zero magnetic field after having reduced the field to zero from values of ±5 kOe. This gives the spectrum at remanence. In the second, the data were obtained from the difference in the induced ellipticity taken in fields $\pm H$ where H=5 kOe. This is the standard MCD spectrum and corresponds to the MCD at saturation. In the second type of measurements it was necessary to subtract the contribution from the substrate. This was done by measuring the temperature dependence of the MCD from a blank substrate and subtracting it from the spectra of the $GdMnO_3/SiTiO_3$ sample. The $SrTiO_3$ substrate was found to have a large and strongly temperature-dependent circular dichroism just below its band gap, making saturation MCD measurements on the $GdMnO_3$ film difficult in this spectra region. Since the substrate dichroism varied linearly with the magnetic field it did not contribute to the remanent spectrum.

The MCD spectra of the $GdMnO_3$ film at remanence and saturation are compared in Figs 2(b) and 3(a). The remanent spectrum should have the same shape as that taken in field but be scaled by the ratio M_r / M_{sat}, where M_r and M_{sat} are the magnetizations measured at remanence and saturation, respectively. This was observed for the signal at ~2 eV at 10 K, but not for the signal near the band edge, where the large paramagnetic signal from the substrate leads to the possibility of errors. The saturation data shown in Fig. 3(a) are therefore only presented for the energy region below 2.3eV where we were confident that there was no contamination from the substrate.

The remanent spectrum at 10 K shown in Fig. 2(a) appears to shows the three spectral components at 1.9, 2.3 and 2.6 eV that were observed in $LaMnO_3$ but only at high temperature [8]. The remanent signal is strong near 2eV and weakens where the interband transition starts at ~3eV. We attribute this to the fact that the transitions near 2 eV correspond to transitions between two magnetic ions, and consequently have much stronger magneto-optic effects than the transitions from the oxygen states to an unoccupied Mn level which only involve one magnetic ion.

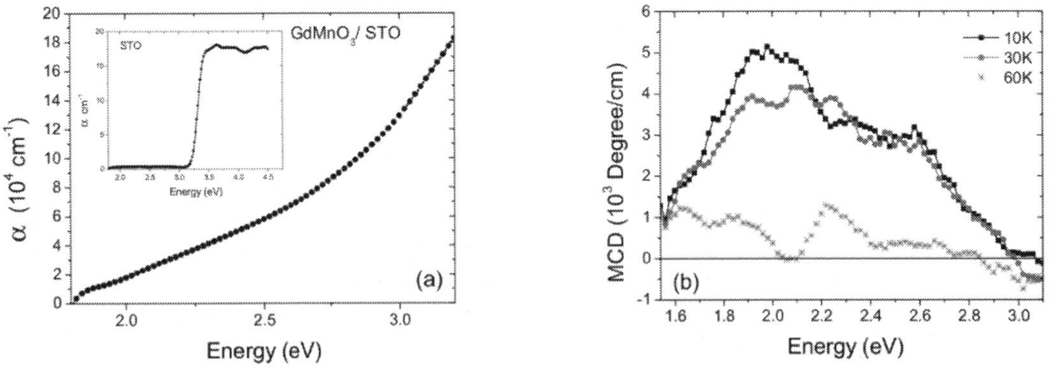

Figure 2. (a) The absorption of the film on the substrate, the absorption of a bare substrate is shown in the inset; **(b)** the MCD signal taken at remanence between 10K and 60K.

The MCD signal is sensitive to magnetic order, making it interesting to study its temperature dependence. Figure 3(b) shows the MCD signal at 2 eV in field over the temperature range 5–300 K. The signal drops off sharply with temperature up to 50 K, and then more slowly to 100 K, following the behaviour of the magnetisation shown in Fig. 1(b). The close correspondence between the MCD at 2 eV and the magnetization gives a clear indication of the magnetic character of the charge transfer transitions between neighbouring Mn ions.

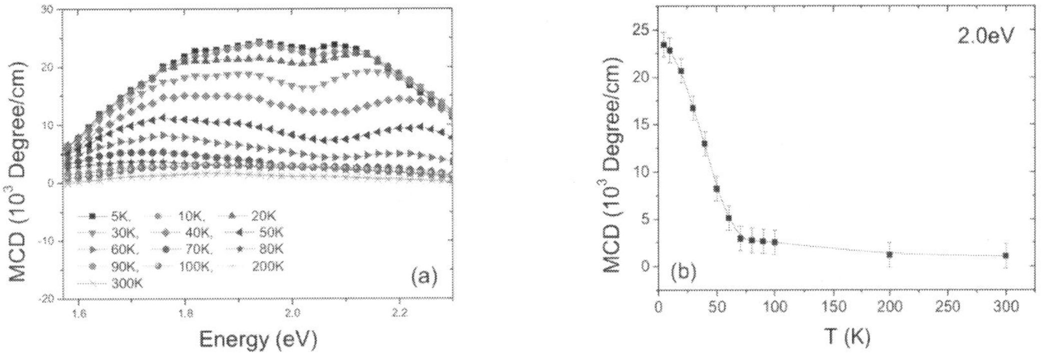

Figure 3(a). The MCD signal taken in 5KOe over a temperature range 5-300K (the MCD signal from the substrate was subtracted for each temperature);(**b**) The intensity of this MCD signal as a function of temperature at 2eV.

4. Conclusions

We have demonstrated the importance of making MCD measurements at remanence for the understanding of the ordering of the Mn spins in $GdMnO_3$. We find that the charge transfer transition at ~2 eV in $GdMnO_3$ gives a strong signal in the MCD spectrum at remanence, even though it was not visible as a distinct peak in the absorption. The temperature dependence of the charge-transfer MCD signal was found to follow the magnetization. These results clarify the importance of magnetoelectric coupling in determining the magneto-optical properties of multiferroic materials such as $GdMnO_3$.

5. Acknowledgments
The work in Sheffield was supported for studentships for M.S. Al Qahtani and M.S. Alshammari from K.A.C.S.T. (Saudi Arabia) and the work in Moscow was supported by Joint RFBR-NSF Grant 09-02-92661 and NA was supported by the Programme of Creation and Development of the NUST "MISiS".

6. References
1. L. Martin et al J Phys Cond Matt 20, 434220 (2008)
2. H. Bea *et al* J. Phys. Cond Matt 20, 434221 (2008))
3. T. Kimura *et al.* Nature 426, 55 (2003)
4. J. Baier *et al* Phys. Rev. B **73**, 100402 (2006)
5. J. Baier *et al* J. Magn. Magn. Mat. **310,** 1165 (2007)
6. X.L Wang *et al* J. Appl. Phys. **107**, 09B510 (2010)
7. M.W. Kim *et al* Phys. Rev. Lett. **96,** 247205 (2006)
8. M.W. Kim *et al* New J. Phys **6**, 156 (2004)
9. M. Bastjan *et al* Phys. Rev. B **77**, 193105 (2008)
10. M.A. Geday and A.M. Glazer J. Phys. Cond. Matt. **16**, 3303 (2004)
11. K. van Benthem *et al* J. Appl. Phys. **90**, 6156 (2001)
12. J.R. Neal, et al, Phys. Rev. Lett. **96**, 197208 (2006)
13. F.J. Kahn *et al* Phys Rev **186**, 891 (1969)

Non-magnetic impurity and defect doping in superconductors Li_2Pd_3B with noncentrosymmetric crystal structures

G Bao[1], Y Inada[1,2,*], S Harada[1] and G-q Zheng[1]
[1]Department of Physics, Okayama University, Okayama 700-8530,
[2]Graduate School of Education, Okayama University. Okayama 700-8530

E-mail: y-inada@okayama-u.ac.jp

Abstract. The spin singlet and spin triplet superconducting states can coexist in noncentrosymmetric crystal structures like Li_2T_3B (T: Pd, Pt), while it is prohibited in ordinary crystal structures. In this study, we focused our research on non-magnetic impurity effect in Li_2T_3B(T: Pd, Pt). The nature of the pair breaking by non-magnetic impurity in the parity mixing superconducting state is still unclear. We investigated the effect of non-magnetic impurity on *s*-wave dominant parity mixing superconductor Li_2Pd_3B by substituting B with Al. In this substitution process, we controlled the quality of the samples and showed their quality deteriorations through the measurement of residual resistivity at T_c. Furthermore, we checked the relations of residual resistivity and T_c as well as its superconducting phase diagram. Our results showed that *s*-wave dominant parity mixing superconductor Li_2Pd_3B exhibits the small T_c suppression attributed by the non-magnetic impurities and defects, while the H_{c2} value clearly increased about 1.5 times in the poor quality samples. Such behaviours are similarly observed in ordinary *s*-wave superconductors.

1. Introduction

The superconductors with noncentrosymmetric crystal structures has been a hot topic in condensed matter physics. In ordinary centrosymmetric crystal structures, Cooper-pair symmetry was classified into either spin singlet or spin triplet state, however, the broken inversion symmetry in the crystal causes finite antisymmetric spin orbit coupling (ASOC) [1-3], which causes the parity mixing. The first discovery of noncentrosymmetric heavy fermion unconventional superconductor $CePt_3Si$ [4] opened the new physics in superconductivity.

Li_2Pd_3B and Li_2Pt_3B crystalize with same noncentrosymmetric crystal structures ($P4_332$ in space group [5]). They have attracted much attention because they exhibit completely different superconducting properties [6-10]. Previous studies of NMR [8] and penetration depth measurements [9] suggested that Li_2Pt_3B is a spin triplet dominant superconductor with the ratio of singlet to triplet order parameter of 0.6, with the inclusion of sufficiently large amount of spin-singlet order parameter components, while Li_2Pd_3B is an *s*-wave spin singlet dominant superconductor with the ratio of 4. We expect they are candidate to study the parity mixing superconductors. Furthermore, no strong electron correlation is observed in Li_2Pd_3B and Li_2Pt_3B, as contrasted with $CePt_3Si$. $CePt_3Si$ has a tetragonal crystal structure with no inversion center due to the lack of the mirror plane perpendicular to the only

* To whom any correspondence should be addressed.

one axis (c-axis), while Li_2T_3B (T: Pd, Pt) is in a cubic structure composed of distorted octahedral units of BPt_6 or BPd_6. There is no mirror plane in all directions and no inversion center. These differences may be important in the discussion of superconductivity in nonentrosymmetric crystal structures with ASOC without strong electron correlation.

It is known that *s*-wave superconductor is not affected by non-magnetic impurity and defects doping in contrast to that of a non *s*-wave superconductor. It is caused by the sign inversion of the order parameter on the Fermi surface, although the nature of the pair breaking by non-magnetic impurity in the parity mixing superconducting state is still unclear. We expect that parity mixing ratio can be controlled by non-magnetic impurity effect. At this point, we investigate that the non-magnetic impurity and defect doping effect in Li_2T_3B (T: Pd, Pt). In this paper, our experimental results are presented in detail on the spin-singlet dominant superconductor of Li_2Pd_3B. We prepared the different quality samples by substituting B with Al, and the sample quality is checked by the residual resistivity.

2. Experimental Details

Polycrystalline samples of $Li_2Pd_3(B_{1-x}Al_x)$ were prepared by two-step arc melting method [6]. In the first step, $Pd_3(B_{1-x}Al_x)$ was synthesized by using Pd (99.95%), B (99.5%), Al (99.999%). In the second step, excess Li (99.9%) by 10% was added to $Pd_3(B_{1-x}Al_x)$ alloys. The powder XRD measurements showed that there was no other phase in all samples. Measurements of electrical resistivity vs. temperature have been done by standard four-probe method under magnetic fields by using the PPMS (Quantum Design). Temperature dependence of magnetization was measured by the MPMS (Quantum Design). All samples showed more than 90% of the perfect shielding effect and the bulk superconductivity.

3. Results and Discussion

Figure 1(a) shows the temperature dependence of the electrical resistivity in the four different quality samples of $Li_2Pd_3(B_{1-x}Al_x)$. Sharp transitions and clear zero resistivity are observed. The aluminum content of this samples are $x = 0$ (#4), 0.01 (#2), 0.05 (#1), 0.1 (#3), respectively. Their quality deteriorations are shown through the measurement of residual resistivity (RR). The #1 with smallest RR is expected to include the least impurities and defects. Contrary to expectations, The #4 (no aluminum content) does not show the smallest RR. In synthesis process of these compounds, sufficient repetitional melting process is difficult to prevent Li loss, so it is difficult to control the homogeneity of samples. It is probably a reason why the #4 does not show the smallest RR, although the sample qualities (non-magnetic impurities and defects) can be estimated by the residual resistivity. The critical temperatures T_c were determined at the mid point between normal and zero resistivity. Figure 1(b) shows the T_c values plotted against the RR. Small T_c suppression was observed by non-magnetic impurities and defects. This is similar behaviour in ordinary *s*-wave superconducting state.

Figure 2(a) shows the temperature dependence of the upper critical field $H_{c2}(T)$ in four samples. The values of T_c are almost same (10% reduction in worse sample #4), while the $H_{c2}(0)$ value clearly increased about 1.5 times in worse sample #4. This is similar behaviour in ordinary *s*-wave superconducting state explained by nonlocal generalization of the London equation introduced by Pippard [11]. The coherence length ξ in the presence of scattering were assumed to be related to that of pure material ξ_0 by

$$\frac{1}{\xi} = \frac{1}{\xi_0} + \frac{1}{\alpha l} \qquad (1)\ [11],$$

where l is the mean free path and α is a numerical constant. From the relation of $H_{c2}(0) = \phi_0/2\pi\xi^2$ ($\phi_0 = h/2e$ is the quantum fluxoid), we can expect $H_{c2}(0)$ to increase when l becomes short.

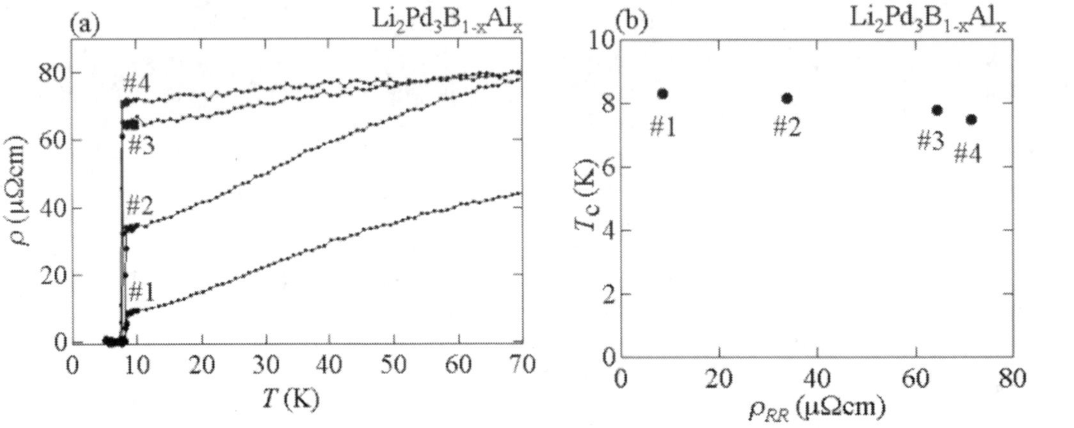

Figure 1. (a) Temperature dependence of resistivity in the four different quality samples, and (b) the critical temperature T_c plotted against the RR. The aluminum content of this samples are $x = 0$ (#4), 0.01 (#2), 0.05 (#1), 0.1 (#3), respectively. It is expected that the #1 will contain the least defects and impurities meanwhile, the #4 will contain the larger defects. Small T_c suppression is observed.

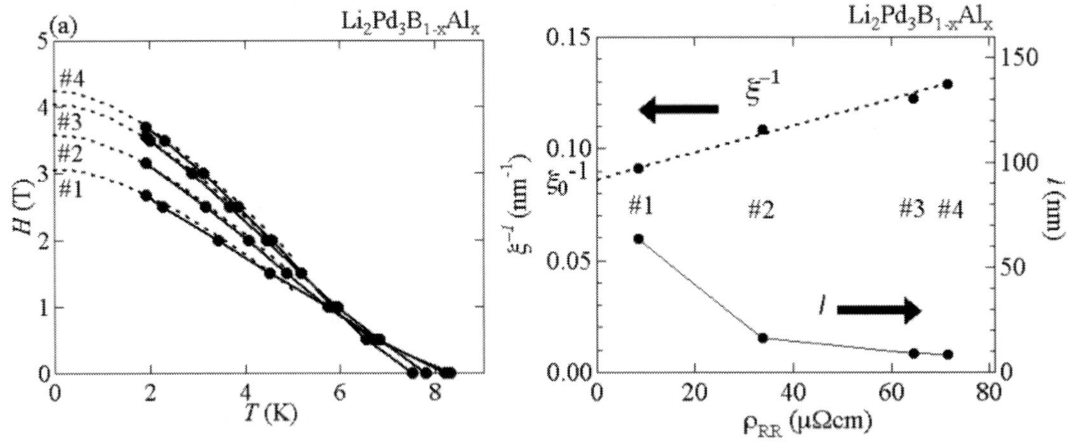

Figure 2. (a) H-T phase diagram of the four different quality samples, and (b) $1/\xi$ and the mean free path l plotted against residual resistivity (RR) in the four different quality samples. The ξ_0 represents the coherence length in pure material, which is estimated from Eq.(1). The mean free path l is obtained from Eq.(2). The $H_{c2}(0)$ value clearly increased about 1.5 times in worse sample #4.

We can estimate the coherence length of ξ_0 in an ideally pure material from Fig.2 (b) under the assumption of $(\alpha l)^{-1} \propto \rho_{RR}$. The value of ξ_0^{-1} is obtained from the y-intercept of the $1/\xi$ vs. ρ_{RR} graph in Fig.2 (b). The estimated value of ξ_0 is 11.6 nm. The mean free path l is obtained from

$$l = 1.00965 \times 10^{-21} (\xi_0 \gamma T_c \rho_{RR})^{-1} \quad (2)\ [12],$$

where ξ_0 is the BCS coherence length of 1.16×10^{-10} m as mentioned above, γ is the electronic specific-heat coefficient of 1.94×10^2 $Jm^{-3}K^{-2}$ in Li_2Pd_3B [10]. The values of T_c(K) and ρ_{RR} (Ωm) are critical temperature and residual resistivity in four different samples, respectively. The values of ξ_0 are plotted in Fig.2(b). The values of ξ are 11.0 nm (#1), 9.2 nm (#2), 8.2 nm (#3) and 7.7 nm (#4), respectively. The values of l are 63.7 nm (#1), 16.2 nm (#2), 8.9 nm (#3) and 8.4 nm (#4), respectively. As the sample quality deteriorates, the mean free path l and ξ shortens. In this system, superconductivity is gradually deviating from the clean limit condition in the poor quality sample (#3 or #4).

The α in Eq.(1) is a numerical constant of order unity in ordinary superconductors. In our results, the value of α is about 0.3, and it is smaller compared to the experimental value of Tin, which is 0.8 [11]. The small value of α brings differences between the values of ξ and ξ_0 (pure material) regardless of the enlargement of the value of l. The small α may be attributed by ASOC.

4. Conclusion

Our results showed that *s*-wave dominant parity mixing superconductor Li_2Pd_3B exhibits the small T_c suppression attributed by the non-magnetic impurities and defects, while the H_{c2} value clearly increased about 1.5 times in the poor quality samples. They are similar behaviour of ordinary *s*-wave superconductor. The ξ_0 (a coherence length in an ideal pure sample) was estimated in Li_2Pd_3B at 11.6 nm.

Li_2Pt_3B is a spin-triplet dominant superconductor with the inclusion of sufficiently large amount of spin-singlet order parameter components. It is interesting how non-magnetic impurity influences superconducting nature of the material. It is expected that non-magnetic impurities have the ability to selectively influence the spin-triplet order parameter components in the parity mixing superconducting state.

This work was supported by the "Topological Quantum Phenomena" (No. 22103004) Grant-in Aid for Scientific Research on Innovative Areas from the Ministry of Education, Culture, Sports, Science and Technology (MEXT) of Japan.

References

[1] Dresselhaus G 1955 *Phys. Rev. Lett.* **100** 580
[2] Rashba E and Tverd Fiz (Leiningrad) Tela, 1959 *Sov. Phys. Solid State* **1** 368
[3] Gorkvand L P and Rashba E I, 2001 *Phys. Rev. Lett.* **87** 037004
[4] Bauer E, Hilscher G, Michor H, Paul Ch, Scheidt E W, Grivanov A, Seropegin Y, Noel H, Sigrist M, and Rogl P, 2004 *Phys. Rev. Lett.* **92** 027003
[5] Eibenstein U and Jung W J 1997 *Solid State Chem.***133** 21
[6] Badica P, Kondo T, Orimo and Togano K 2005 *J. Phys. Soc. Jpn.* **74** 1014-1019
[7] Nshiyama M, Inada Y and Zheng Guo-qing 2005 *Phys.Rev. B* **71**, 220505
[8] Nshiyama M, Inada Y and Zheng Guo-qing 2007 *Phys. Rev. Lett.* **98,** 047002
[9] Yuan H Q, Agterberg D.F, Hayashi N, Badica P, Vadervelde D, Togano K, Sigrist K and Salmon M B 2006 *Phys. Rev. Lett.* **97** 017006
[10] H. Takeya, M. ElMassalami, S. Kasahara and K. Hirata, 2007 *Phys. Rev.* B **76**, 104506.
[11] Pippad A B, 1953 *Proceedings the Royal Society A*. **216** 547-568
[12] e.g., the appendix in T. P. Orlando *et.al.*, 1979 Phys.Rev.B**19** 4545

Spin-Orbital Entanglement Emerging from Frustration in the Kugel-Khomskii Model

W Brzezicki[1] **and A M Oleś**[1,2]

[1]Marian Smoluchowski Institute of Physics, Jagellonian University, Reymonta 4, PL-30059 Kraków, Poland

[2] Max-Planck-Institut FKF, Heisenbergstrasse 1, D-70569 Stuttgart, Germany

E-mail: `w.brzezicki@uj.edu.pl`

Abstract. We investigate the zero-temperature phase diagrams of the bilayer square-lattice Kugel-Khomskii (d^9) model involving entangled and singlet phases using mean-field cluster approach. This diagram includes interlayer singlet phase observed in copper fluoride $K_3Cu_2F_7$ and exotic entangled spin-orbital phases. For a monolayer case, realized in K_2CuF_4, we perform similar calculations at finite temperature and show that the alternating-orbital ferromagnet decays first to an entangled uniform ferromagnet and then to a paramagnet.

It has been shown that quantum fluctuations are enhanced near the orbital degeneracy and could suppress long-range order in the Kugel-Khomskii (KK) model [1], called below the d^9 model. This model was introduced long ago for a perovskite $KCuF_3$ [2], a strongly correlated system with a single hole within degenerate e_g orbitals at each Cu^{2+} ion. Kugel and Khomskii showed that alternating orbital (AO) order can be stabilized by a purely electronic superexchange mechanism. This happens for strongly frustrated orbital superexchange [3], and columnar Ising-type of order is obtained [4] in the two-dimensional quantum compass model. This model exhibits nontrivial symmetry properties which may be employed to perform efficient calculations for square compass clusters [5].

Orbital order occurs in a number of compounds with active orbital degrees of freedom, where strong Coulomb interaction localizes electrons (or holes) and gives rise to spin-orbital superexchange [6]. When spin and orbital pseudospins couple to each other, their order is usually complementary — AO order accompanies ferromagnetic (FM) spin order, and ferro-orbital (FO) order coexists with antiferromagetic (AF) spin order. However, the above Goodenough-Kanamori rules, see also [3], are not satisfied in cases when spin-orbital entanglement (SOE) dominates [7], as for instance in the spin-orbital d^1 model on a triangular lattice [8].

The spin-orbital superexchange KK model for Cu^{2+} (d^9) ions in $KCuF_3$ with $S = 1/2$ spins and e_g orbitals described by $\tau = 1/2$ pseudospin was derived from the degenerate Hubbard Hamiltonian with hopping t, intraorbital Coulomb interaction U and Hund's exchange J_H [9]. It describes the Heisenberg SU(2) spin interactions coupled to the orbital problem by superexchange $J = 4t^2/U$,

$$\mathcal{H} = -\frac{1}{2}J\sum_{\langle ij\rangle||\gamma}\left\{\left(r_1\,\Pi^t_{\langle ij\rangle} + r_2\,\Pi^s_{\langle ij\rangle}\right)\left(\frac{1}{4} - \tau_i^\gamma\tau_j^\gamma\right) + r_3\,\Pi^s_{\langle ij\rangle}\left(\frac{1}{2} - \tau_i^\gamma\right)\left(\frac{1}{2} - \tau_j^\gamma\right)\right\} - E_z\sum_i \tau_i^z\,. \tag{1}$$

where $\{r_1, r_2, r_3\}$ depend on $\eta \equiv J_H/U$ [9], and $\gamma = a, b, c$ is the bond direction. In a bilayer two ab planes are connected by interlayer bonds along the c axis [10] (a monolayer has only bonds within a single ab plane). Here

$$\Pi^s_{\langle ij\rangle} = \frac{1}{4} - \mathbf{S}_i \cdot \mathbf{S}_{i+\gamma}, \qquad \Pi^t_{\langle ij\rangle} = \frac{3}{4} + \mathbf{S}_i \cdot \mathbf{S}_{i+\gamma}, \tag{2}$$

are projection operators on a spin triplet (singlet) configuration on a bond $\langle ij\rangle$, and τ_i^γ are the orbital operators for bond direction $\gamma = a, b, c$. They are defined in terms of Pauli matrices $\{\sigma_i^x, \sigma_i^z\}$ as follows:

$$\tau_i^{a(b)} \equiv \frac{1}{4}(-\sigma_i^z \pm \sqrt{3}\sigma_i^x), \qquad \tau_i^c = \frac{1}{2}\sigma_i^z. \tag{3}$$

Finally, E_z is the crystal-field splitting which favors either $x \equiv x^2 - y^2$ (if $E_z > 0$) or $z \equiv 3z^2 - r^2$ (if $E_z < 0$) orbitals occupied by holes. Thus the model Eq. (1) depends on two parameters: E_z/J and η.

The spin-orbital model Eq. (1) describes also CuO_2 planes in La_2CuO_4, where indeed $U \gg t$ and large $E_z/J_H \simeq 0.27$ favors holes within x orbitals [9]. The superexchange between Cu^{2+} ions ~ 0.127 eV reproduces there the experimental value. In this paper we consider the model Eq. (1) for $K_3Cu_2F_7$ bilayer and K_2CuF_4 monolayer compound where nearly degenerate e_g orbitals are expected. It has been shown that the magnetic state of $K_3Cu_2F_7$ is described by the interlayer valence bond (VB) phase [11] with AO configuration whereas K_2CuF_4 undergoes a pressure driven phase transition (PT) from the FM phase with alternating orbitals at low pressure to the AF phase with with x orbitals uniformly occupied [12].

The simplest approach is a single-site mean field (MF) approximation applied to the model Eq. (1). It excludes any spin fluctuations so the spin projectors $\Pi^{t(s)}_{\langle ij\rangle}$ ($\Pi^s_{\langle ij\rangle}$) can be replaced by their mean values depending on the assumed magnetic order. In the orbital sector we apply then the MF decoupling for the products $\{\tau_i^\gamma \tau_{i\pm\gamma}^\gamma\}$ along the axis γ:

$$\tau_i^\gamma \tau_{i\pm\gamma}^\gamma \simeq \langle\tau_i^\gamma\rangle\tau_{i\pm\gamma}^\gamma + \tau_i^\gamma\langle\tau_{i\pm\gamma}^\gamma\rangle - \langle\tau_i^\gamma\rangle\langle\tau_{i\pm\gamma}^\gamma\rangle. \tag{4}$$

As order parameters we take $t^a \equiv \langle\tau_1^a\rangle$ and $t^c \equiv \langle\tau_1^c\rangle$ for a chosen site $i = 1$ (which is sufficient in the orbital sector as $t^b = -t^a - t^c$) and we assume two orbital sublattices: each neighbor of the site i is rotated by $\pi/2$ in the ab plane meaning that $\langle\tau_{i+\gamma}^{a(b)}\rangle = t^{b(a)}$. The self-consistency equations can be solved analytically and the phase diagram can be obtained by comparing the ground state energies for different points in the $(E_z/J, \eta)$ plane (see Ref. [10]). One finds two classes of solutions: (i) uniform orbital configurations ($t^c = \pm 1/2$, $t^{a(b)} = \mp 1/4$) for global FO order, and (ii) nontrivial AO order with orbitals staggering from site to site in ab planes.

In a better cluster MF (or Bethe-Peierls-Weiss) approach, introduced to capture the effects of quantum fluctuations, one divides the bilayer square lattice into separate cubes containing 8 sites each and treats the bonds inside a cube exactly, while the bonds connecting different cubes are included in MF. This approach has at least three advantages over the single-site MF: (i) spins can fluctuate, (ii) elementary cell can double, and (iii) ordered phases may be characterized by an independent spin-orbital order parameter. The MF leads in a cluster to five order parameters: magnetic $s \equiv \langle S_1^z\rangle$, orbital $t^{a(b)} \equiv \langle\tau_1^{a,b}\rangle$, and spin-orbital $v^{a(b)} \equiv \langle S_1^z\tau_1^{a,b}\rangle$. The self-consistency equations can be solved only numerically by iterative Lanczos diagonalization of a cluster followed by the update of the mean fields (for details see Ref. [10]). To capture the effect of the SOE we first obtain the phase diagram assuming $v^{a(b)} \equiv st^{a,b}$ (factorizable SO mean field)— see Fig. 1(a) and then performed the calculation with true $v^{a(b)}$ to get Fig. 1(b).

Including spin fluctuations in this approach stabilizes three VB phases absent in the single-site approach. These are VBz, VBm and PVB phases, see Fig. 1(a). The VBz replaces

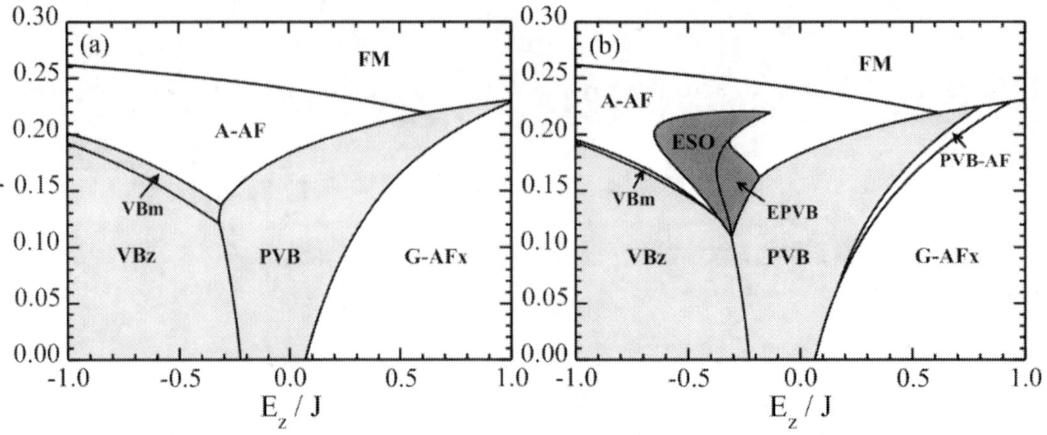

Figure 1. Cluster MF phase diagrams of the d^9 bilayer model (1) obtained with: (a) factorizable SO mean field, and (b) independent, potentially non-factorizable SO mean field. Light shaded (yellow) areas mark singlet phases with spin disorder and dark (orange) shading indicates phases with SO entanglement.

ordinary antiferromagnet in the negative E_z region of the phase diagram and involve interlayer singlets accompanied by FOz configuration which can smoothly evolve for growing η towards AO configuration in the VBm being the phase observed in $K_3Cu_2F_7$ by Manaka *et al.* [11]. Here we explain it for realistic $\eta \simeq 0.14$. In the PVB (plaquette VB) phase spin singlets are pointing uniformly in a or b direction within the cluster and the clusters form a checkerboard pattern so the unit cell is doubled. Other phases, i.e., FM, G-AFx and A-AF, exhibit long range magnetic order and can be obtained in ordinary MF approach (see Ref. [10]).

A different class of phases stems from the non-factorizable SO mean field and involves SOE — these are the ESO, EPVB and PVB-AF phases shown in Fig. 1(b) . The PVB-AF phase connects PVB and G-AF phases by second order PTs and is characterized by fast changes in orbital order and appearance of global magnetization. The ESO phase has no magnetization and weak FO order. When E_z grows, the ESO phase does change continuously into the EPVB configuration, being an entangled precursor of the PVB phase with highly non-uniform AF magnetization. As shown in Ref. [10], both ESO and EPVB are characterized by strong on-site SOE defined as $r^{a,b} \equiv v^{a,b} - st^{a,b}$.

Using the same cluster MF approach as above one can easily study the properties of the KK model for a single layer at zero and finite temperature T. At $T = 0$ one finds the phase diagram of the form shown in the Fig. 2(a). In the low-η region it involves only AF and PVB phases and the VBz is replaced by AFz one in the absence of the second layer. The PVB area is strongly elongated in the horizontal directions and caps AFz down to $E_z < -2J$ where one finds a narrow stripe of a paramagnetic (PM) phase separating PVB from FMz phase (FM phase with FOz order). The presence of paramagnetizm at $T = 0$ shows that a monolayer system is even more frustrated than a bilayer. The E_z-induced FM-AFx PT above $\eta \simeq 0.225$ agrees with the one observed in K_2CuF_4 (see Ref. [12]) with E_z acting as uniaxial pressure.

In the Fig. 2(b) we show the thermal evolution of spin and orbital order parameters and SOE for a point ($E_z = -2.1J, \eta = 0.19$) lying in the FM area of the monolayer phase diagram. This evolution involves passage through the FMz phase and ends in the PM phase. The FM magnetization s exhibits critical behaviour staying close to 1/2 in the FM phase and dropping quickly to zero at the FMz-PM PT. The orbital parameters $\{t^a, t^b\}$ show inverse bifurcation at

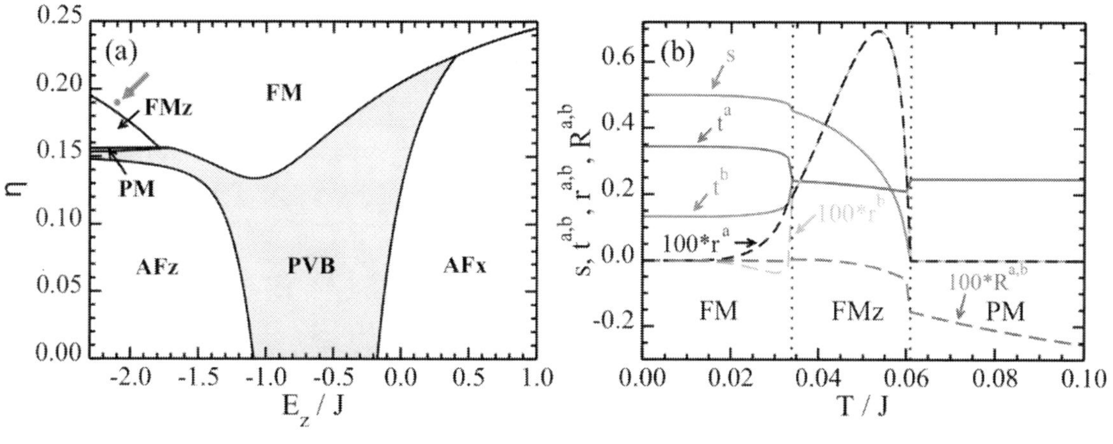

Figure 2. Panel (a)— phase diagram of the d^9 monolayer model at $T = 0$. Shaded (yellow) area marks singlet phase with spin disorder. Panel (b)— spin s and orbital $t^{a,b}$ order parameters, on-site $r^{a,b}$ and bond $R^{a,b}$ SOE parameters (multiplied by 100) for point ($E_z = -2.1J, \eta = 0.19$), marked with thick (red) arrow in (a), as functions of temperature. Thermal evolution of the system involves intermediate FMz phase before reaching PM phase.

the FM-FMz PT being the end of the AO order but remain finite in the paramagnet which is induced by finite E_z. The on-site SOE represented by $r^{a,b}$ are small everywhere but in FMz phase they become relatively large which suggests that the FMz is an entangled phase. The bond SOE defined by $R^{a,b} \equiv \langle \mathbf{S}_1\mathbf{S}_i\tau_1^{a,b}\tau_i^{a,b}\rangle - \langle \mathbf{S}_1\mathbf{S}_i\rangle\langle\tau_1^{a,b}\tau_i^{a,b}\rangle$ take considerable values only in the PM phase and shows rather purely statistical covariance than a true quantum effect.

Summarizing, we have shown exotic types of order with spin-orbital entanglement in a bilayer system at $T = 0$ and entangled intermediate phase between ferromagnet and paramagnet for a monolayer at finite temperature. Within the same cluster MF approach we were able to capture the most important experimental features of both types of the Kugel-Khomskii systems.

Acknowledgments
We acknowledge support by the Foundation for Polish Science (FNP) and by the Polish Ministry of Science and Higher Education under Project No. N202 069639.

References
[1] Feiner L F, Oleś A M and Zaanen J 1997 *Phys. Rev. Lett.* **78** 2799
Feiner L F, Oleś A M and Zaanen J 1998 *J. Phys. Condens. Matter* **10** L555
[2] Kugel K I and Khomskii D I 1982, *Usp. Fiz. Nauk* **136** 621
[3] Oleś A M 2010, *Acta Phys. Polon. A* **115** 36
[4] Cincio L, Dziarmaga J and Oleś A M 2010 *Phys. Rev. B* **82** 104416
Trousselet F, Oleś A M and Horsch P 2010 *Europhys. Lett.* **91** 40005
[5] Brzezicki W and Oleś A M 2010 *Phys.Rev. B* **82** 060401
Brzezicki W 2010 *AIP Conf. Proc.* **1297** p 407
[6] Khaliullin G 2005 *Prog. Theor. Phys. Suppl.* **160** 155
[7] Oleś A M, Horsch P, Feiner L F and Khaliullin G 2006 *Phys. Rev. Lett.* **96** 147205
[8] Normand B and A.M. Oleś A M 2008 *Phys. Rev. B* **78** 094427
Chaloupka J and Oleś A M 2011 *Phys. Rev. B* **83** 094406
[9] Oleś A M, Feiner L F and Zaanen J 2000 *Phys. Rev. B* **61** 6257
[10] Brzezicki W and Oleś A M 2011 *Phys.Rev. B* **83** 214408
[11] Manaka H, Miyashita Y, Watanabe Y and Masuda T 2007 *J. Phys. Soc. Jpn.* **76** 044710
[12] Ishizuka M, Yamada I, Amaya K and Endo S 1996 *Endo S., J. Phys. Soc. Jpn.* **65** 1927

Magnetoresistance studies of homogenously disordered 3-dimensional NbN thin films

Madhavi Chand[1]*, Mintu Mondal[1], Sanjeev Kumar[1], Anand Kamlapure[1], Garima Saraswat[1], S. P. Chockalingam[1,2], John Jesudasan[1], Vivas Bagwe[1], Vikram Tripathi[1], Lara Benfatto[3] and Pratap Raychaudhuri[1]

[1]Tata Institute of Fundamental Research, Homi Bhabha Road, Navy Nagar, Colaba, Mumbai-400005, India
[2]Columbia University, 538 West 120th Street, New York, NY 10027, USA
[3]INFM-CNR Statistical Mechanics and Complexity Center, University of Rome "La Sapienza," P.le A. Moro 5, 00185 Rome, Italy

*email: chand@tifr.res.in

Abstract. We present magnetoresistance studies on a series of homogenously disordered 3-dimensional NbN thin films with disorder ranging from the moderately clean limit ($k_F l$~10.12) to the extremely dirty limit where the superconducting ground state is completely destroyed ($k_F l$~0.42). We find that for samples with $k_F l$ >1, the magnetoresistance is positive up to 12T and as disorder increases it decays more slowly with temperature. On the other hand, for samples with $k_F l$<1, we observe a peak in the magnetoresistance which vanishes at a temperature close to the pseudogap temperature of disordered superconducting samples. These observations are consistent with the idea of the disorder-driven non-superconducting state being comprised of pre-formed Cooper pairs.

1. Introduction

The nature and properties of the disorder-driven superconductor-insulator transition has been a subject of active research for many years. Detailed experimental investigations have been reported on a number of systems like quench condensed films of Bi [1,2], homogenously disordered TiN [3] and NbN [4] and amorphous InO [5,6], to name a few. In these experiments, many interesting phenomena have been observed such as the existence of a pseudogap state at high disorder [4], a large peak in the magnetoresistance [1,5], and the evidence of pairing in films with a non-superconducting ground state in the form of magnetoresistance (MR) oscillations [2]. Numerical studies also predict phenomena like the spontaneous formation of superconducting islands [7,8] and consequently a pseudogap state [7] and are therefore in agreement with many of the experiments. One of the main open questions is the nature of charge carriers at the SIT and in the regime where the system is driven into a non-superconducting ground state. The idea of the insulating state being comprised of Cooper pairs has received a lot of support in recent times [9,10,6] from tunneling and MR studies. Different explanations have been suggested recently for the peaks observed in the MR based on ideas of percolation [11] and localization [12]. In this paper we present MR measurements on a series of 3-dimensional, homogenously disordered NbN thin films. Samples with $k_F l$ >1 have a positive MR up to 12T and as disorder increases it decreases at a slower rate with temperature. On the other hand, for samples with $k_F l$<1, there is a peak in the MR which vanishes at a temperature close to the pseudogap temperature.

2. Experimental Details

NbN thin films were prepared by reactive DC magnetron sputtering of a Nb target in a gas mixture of Ar and N_2 onto single crystal MgO (100) substrates. Stoichiometric NbN is a conventional BCS superconductor with a relatively high superconducting transition temperature (T_c) of 17K. However, the T_c is extremely sensitive to the growth conditions, which enables us to tune the disorder over a wide range. Details of fabrication and characterization can be found in References [13] and [14]. Disorder is characterized by the Ioffe Regel parameter $k_F l$ ($k_F \equiv$ Fermi wave vector and $l \equiv$ mean free path), obtained experimentally from Hall effect and resistivity (ρ) measurements at 285K using free electron theory [13]. MR was measured in standard four-probe geometry, on a sample patterned using a shadow mask, in a conventional He3/He4 cryostat with a lowest temperature of 300mK and a highest field of 12T. Two kinds of measurements were made on the samples, namely resistance (R) *vs.* temperature (T) at different fields (H) and R *vs.* H at different temperatures.

3. Results and Discussion

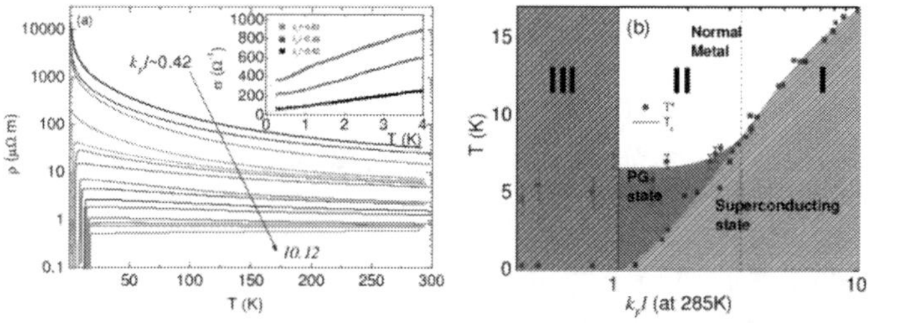

Figure 1: (a) Resistivity vs. Temperature for samples with $k_F l$~10.12, 8.82, 8.13, 8.01, 5.5, 4.98, 3.65, 3.27, 1.95, 1.68, 1.58, 1.23, 0.82, 0.49 and 0.42, on a logarithmic scale and. Inset: σ *vs.* T for the most disordered samples. (b) Phenomenological phase diagram showing superconducting critical temperature, T_c and pseudogap temperature (T^*) vs $k_F l$. The labels I, II and III demarcate the regions conforming to mean field theory, with a pseudogap state and no superconducting ground state respectively.

Figure 1(a) shows ρ *vs.* T curves on a logarithmic scale for the full range of samples. We can see that barring the cleanest sample ($k_F l$~10.12) the rest have a negative temperature coefficient of resistance that gets more and more pronounced as disorder is increased. Simultaneously, the T_c decreases continuously till $k_F l$~1 where it vanishes completely. This has been shown in Figure 1(b) in the form of a phenomenological phase diagram. We have shown previously through scanning tunnelling spectroscopy [4] that in the strongly disordered regime ($k_F l$<4), a gap is observed in the tunnelling density of states at temperatures higher than T_c. This gap, known as the pseudogap, finally vanishes at a higher temperature T^*. This suggests that the superconducting transition is governed by phase fluctuations and Cooper pairs continue to exist at temperatures above which the phase coherence across the sample is lost.

In Figure 2 we present the MR data of samples with a well defined superconducting transition temperature. Figure 2(a) shows R as a function of applied magnetic field at different temperatures (T) above T_c as well as R *vs.* T for different applied fields. Figures 2(a) and 2(b) show the raw data for samples with T_c ~13.6K and 3.2K respectively.

In Figure 3 we show the MR, defined as the relative change in resistance with magnetic field $\frac{R(12T)-R(0)}{R(12T)}$ as a function of T/T_c for a series of films with T_c~16.2, 13.57, 11.5, 8.13, 7K, 4.5K, 3.2K and 1.65K. This plot clearly indicates that as the system gets more disordered, the MR decreases less rapidly and persists up to higher temperatures above T_c.

These MR observations are qualitatively consistent with the phase fluctuations and pseudogap scenario described in Ref [4]. In the pseudogap state the system can be viewed as consisting of

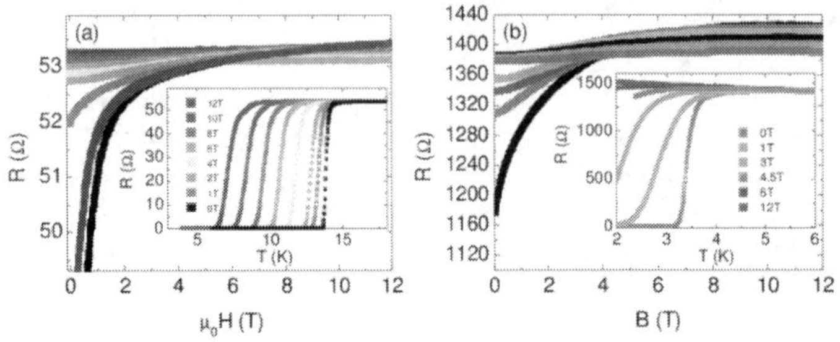

Figure 2: (a) Resistance (R) *vs.* applied magnetic field (H) for a sample with T_c ~13.6K, measured at different temperatures (T): 13.85K, 14K, 14.3K, 14.7K, 15K, 15.5K, 16K, 17K, 18K, 20K, 24K and 30K. The inset shows R *vs.* T for the same sample measured at different H (b) R vs. H for sample with T_c ~3.2K measured at 3.5K, 3.7K, 3.9K, 4.1K, 4.3K, 4.6K, 5K, 5.5K and 6K, again, inset shows R *vs.* T for the same sample measured at different H

superconducting domains. Although the superconducting domains cannot produce a zero resistance state above T_c, because of phase incoherence between them, the superconducting correlations persist and lead to a lower resistance in this regime $T_c < T < T^*$. On the application of a magnetic field, these intra-domain superconducting correlations are also suppressed and the resistance increases to what it would have been in the absence of pairing correlations.

We now look at the samples that do not show a superconducting transition. Figures 4(a), (b) and (c) show ρ *vs.* T for samples with k_Fl~0.82, 0.49 and 0.42. Here we observe a prominent peak in the MR. We label the position of the MR peak in the data measured at the lowest temperature (300mK) as H_{peak}. It is evident from Figures 4(a), (b) and (c) that H_{peak} shifts to lower fields with increasing disorder. In Figure 4(d) we plot these values of H_{peak} along with the experimentally measured upper critical field (H_{c2}) for the superconducting ones. We see that H_{peak} evolves smoothly from H_{c2} across the SIT. Details of measurement of H_{c2} can be found in references [13] and [15]. Figures 3(e-f) show the temperature evolution of the MR peaks. The peaks vanish at ~5K, which is close to T^* in the disordered films.

We now try to understand the origin of the MR peak. We have seen in the phenomenological phase diagram (Figure 1(b)) that T^* saturates at high disorder and remains ~6K even very close to k_Fl~1. This seems to suggest that Cooper pairing continues to exist even when k_Fl<1. Different theoretical frameworks have provided different explanations for this MR peak seen in many disordered superconductors. A recent paper suggests that the non-superconducting state comprises of localized Cooper pairs [12]. Such a scenario can also exist due to the fractal nature of the wavefunction in the insulating state [9]. However, we do not believe that this interpretation is applicable in the case of our films as even the non superconducting films are not deep in the insulating regime. We have found that on plotting the conductivity (σ) *vs.* T and extrapolating to zero temperature (inset of Figure 1(a)), $\sigma(0)\neq0$ which indicates that these samples are dirty metals and not insulators. Therefore the interpretation more suitable here is the one due to Dubi et. al. [11] wherein the systems can be thought of as comprising of superconducting puddles separated by metallic regions of which shrink in size with increasing magnetic field due to suppression of superconducting correlations leading to a positive MR. Beyond a certain field, H_{peak} these puddles shrink so much that it is no longer favorable for the current to pass through them resulting in the fact that the current now passes through only the normal regions, which

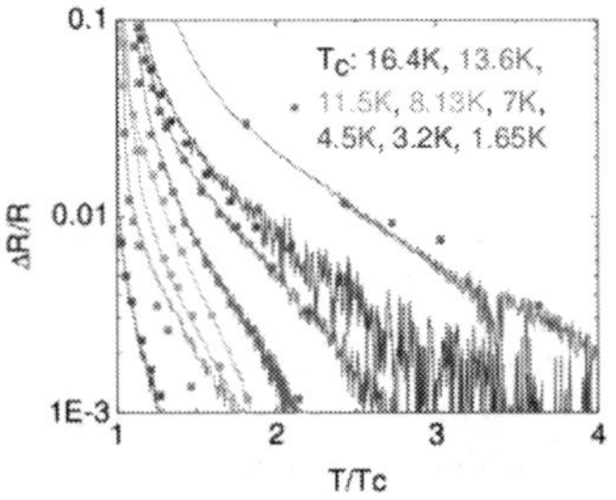

Figure 3: (R(12T)-R(0T))/R(12T) as a function of T/T_c for a series of films with T_c~16.2, 13.57, 11.5, 8.13, 7K, 4.5K, 3.2K and 1.65K. The points are obtained from the MR scans and the solid lines are from the R-T scans at 0T and 12T.

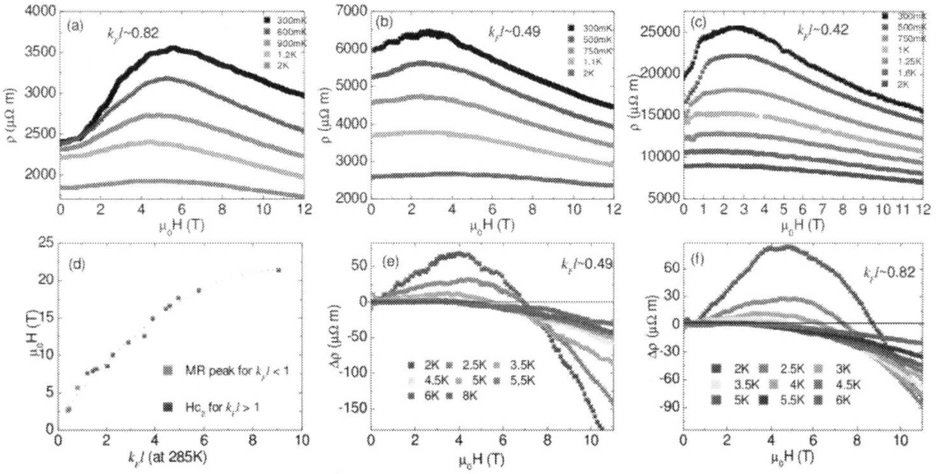

Figure 4: Resistivity as a function of magnetic field at different temperatures for 3 NbN thin films with (a) k_Fl~0.82, (b) k_Fl~0.49 and (c) k_Fl~0.42. (d) Variation of $H_{c2}(0)$ (for k_Fl>1) and H_{peak} (for k_Fl<1) as a function of k_F. (e)-(f) Expanded view of ($\rho(H)$-$\rho(H$=0)) vs. magnetic field close to temperatures where the peak in the MR vanishes.

increase with further increase in H, resulting in negative MR. We therefore interpret H_{peak} as the field beyond which superconducting correlations become insignificant, which is consistent with the fact that H_{peak} evolves from H_{c2}. In addition, the fact that the peak vanishes at a temperature close to T^* further corroborates the existence of Cooper pairs in the non superconducting state [16].

Conclusions

We have measured magnetoresistance for a series of NbN samples over a wide range of disorder and found it to be consistent with the idea of persistence of Cooper pairing above T_c in strongly disordered superconductors where the superconducting state is destroyed by phase fluctuations and also in the regime where the ground state is no longer superconducting.

References

[1] H. Q. Nguyen et. al., Phys. Rev. Lett. 103, 157001 (2009).
[2] M. D. Stewart Jr. et. al., Science **318,** 1273 (2007).
[3] B. Sacepe et. al., Nat Commun. 1, 140 (2010)
[4] M. Mondal et. al., Phys. Rev. Lett. **106**, 047001 (2011)
[5] G. Sambandamurthy, et. al. Phys. Rev. Lett. **94**, 017003 (2005).
[6] B. Sacépé et. al., Nature Physics 7, 239–244 (2011).
[7] A. Ghosal, M. Randeria, and N. Trivedi, Phys. Rev. Lett. **81** 3940 (1998); Phys. Rev. B **65**, 014501 (2001)
[8] Y. Dubi, Y. Meir, and Y. Avishai, Nature London **449** 876 (2007); Phys. Rev. B **78**, 024502 (2008).
[9] M. V. Feigel'man, L. B. Ioffe, V. E. Kravtsov, and E. A.Yuzbashyan, Phys. Rev. Lett. **98** 027001 (2007)
[10] K. Bouadim, Y. L. Loh, M. Randeria and N. Trivedi, *Nature Physics* doi:10.1038/nphys2037
[11] Y. Dubi, Y. Meir, and Y. Avishai, Phys Rev. B **73**, 054509 (2006)
[12] M. Müller, arXiv: 1109.0245v1
[13] S. P. Chockalingam et. al., Phys. Rev. B **77** 214503 (2008)
[14] M. Chand et. al., Phys. Rev. B **80**, 134514 (2009)
[15] M. Mondal et. al., J. Supercond. Nov. Magn. **24,** 341 (2011)
[16] A more detailed discussion can be found in M. Chand et. al. arXiv: 1107.0705

Magnetic properties of correlated one-band metals: the effect of the Roth band shift in the single site approximation

C M Chaves[1], E J Calegari[2], S G Magalhães[3], A Troper[1]

[1] Centro Brasileiro de Pesquisas Físicas, Rio de Janeiro 22290-180, RJ, Brazil
[2] Departamento de Física, Universidade de Santa Maria, 97105-900, Santa Maria, RS,Brazil
[3] Instituto de Física, Universidade Federal Fluminense, Niterói 24210-346, RJ, Brazil

E-mail: cmch@cbpf.br

Abstract. The magnetic properties of one-band metals are discussed within the single site approximation, which is a picture to take into account electron correlations beyond the standard mean-field approach. The metal is described by a Coulomb correlated band. The Roth band shift arising from a two pole approximation is included. Here we present results for the infinite Coulomb correlation limit.

1. Introduction

In his pioneering paper on electron correlation in narrow one-band metals, Hubbard [1], now referred to as Hubbard I, proposed a model that reduces to the exact solution in the zero bandwidth limit. This model however, contrary to what is observed for real metals, found that at low temperatures, for any finite value of the correlation parameter U, the metal would be in a insulating state. In order to correct this, he proposed[2] another approximation, now referred to as Hubbard III. It consists of freezing say, the down spins, and let the up spins move and be scattered by this background (scattering correction). As these ideas were to become later the coherent potential approximation (CPA) , the approximation in Hubbard III became also known as alloy analogy [3].

Following Roth [4] we include scattering correction but allow the dynamics of the down spin electrons through the single site approximation (SSA). In this formulation the full Coulomb correlation exists in only one site, say the origin; in the others there is a sort of 'external field', the site independent but spin and energy dependent self-energy Σ^σ. Σ^σ is self-consistently determined by imposing the vanishing of the scattering T matrix associated to the given site that exhibits the full Coulomb interaction.

The metal is described by one non-degenerate conventional Hubbard-like narrow band with in-site interaction U:

$$\mathcal{H} = \sum_{i,j,\sigma} t_{ij} a^+_{i\sigma} a_{j\sigma} + \sum_i U n_{i\uparrow} n_{i\downarrow}, \tag{1}$$

where $n_{i\sigma} = a^+_{i\sigma} a_{i\sigma}$; σ denotes spin. t_{ij} is the tunneling amplitude between neighboring sites i and j. The single site approximation (SSA) [4] adopts instead the following effective

Hamiltonian:

$$\mathcal{H}_{eff} = \sum_{i,j,\sigma} t_{ij} a^{+}_{i\sigma} a_{j\sigma} + \sum_{i \neq 0,\sigma} n_{i,\sigma} \Sigma^{\sigma} + U n_{0\uparrow} n_{0\downarrow} \tag{2}$$

The method then replaces a translationally invariant problem, as defined by (1), by an impurity problem where only the origin incorporates the Coulomb interaction, the other sites being acted by the field Σ^{σ}. But the effective Hamiltonian (2) is far from being simple due to the presence of the Coulomb intra-atomic term at the origin and we have to resort to some approximation.

We use the two-pole approximation developed by Roth [4]; after some calculation we end up with the following Green's function (in this paper we'll be working at temperature $T = 0$):

$$G_{kk',\sigma}(w) = \frac{\delta_{kk'}}{w - \epsilon_k - \Sigma^{\sigma}(w) - \mathcal{W}_{k,-\sigma}}. \tag{3}$$

In this expression

$$\epsilon_k = \frac{t(cos(k_x a) + \cos(k_y a) + \cos(k_z a))}{A}, \tag{4}$$

is the recursion relation of the bare band . We use $t = 1$ and $A = 3$, in arbitrary energy units. All energy magnitudes are taken in units of t, making them dimensionless. The bare band width is then $W = 2$. On the other hand the band shift $\mathcal{W}_{k,\sigma}$ is

$$\mathcal{W}_{k,\sigma} = \frac{(2n_{k,-\sigma} - 1)\sum_{k'} \epsilon_{k'} n_{k',\sigma}}{n_{\sigma}(1 - n_{\sigma})}, \tag{5}$$

The vanishing of the T-matrix gives a self-consistent equation for the self-energy:

$$\Sigma^{\sigma}(w) = U < n_{0-\sigma} > + \Sigma^{\sigma}(w)(U - \Sigma^{\sigma}(w)) F^{\sigma}(w, \Sigma^{\sigma}(w)), \tag{6}$$

with

$$F^{\sigma} = \sum_k G_{kk,\sigma} \tag{7}$$

The self-consistency is implemented as follows: we start giving initial values for $< n_{0\sigma} >$ and take for Σ^{σ}, the Hartree-Fock(HF) expression $U < n_{0-\sigma} >$ and $\mathcal{W}_{k,\sigma} = 0$. Then F is calculated (eq(7)) and a new Σ^{σ} is found through eq(6). Now it is possible to calculate the density of states (DOS), the Fermi energy and $\mathcal{W}_{k,\sigma}$ and new $< n_{\sigma} >$. This proceeds until a prescribed convergence in Σ^{σ} is reached ; in our case the absolute value of the difference between the input and the output values differs by less than $\epsilon = 5 \times 10^{-3}$. The self-consistency applies for $< n_{\sigma} >$ as well. The total $< n > = < n_{\uparrow} > + < n_{\downarrow >}$ is preserved but the dynamics and the self-consistency redistribute the down and up spins.

2. Numerical results and conclusions

We focus on the very strong coupling limit, namely, $U = \infty$. In fig (1) the total area of the density of states (DOS) is displayed versus $< n >$. It is seen that the total area decreases with $< n >$ *linearly*, a consequence of the elimination of states involving double occupied sites (Hubbard-like contraction [1]). In the present method, as in alloy analogy approaches, there is a strong dependence of DOS -and so of other physical quantities- on $< n_{\sigma} >$. Fig (2) displays the zero temperature magnetization versus $< n >$; magnetization grows with higher electron occupancy, but not monotonically since we are dealing with renormalized CPA DOS; fig(3)compares the energy of the paramagnetic and of the ferromagnetic states.

Ferromagnetism is more favorable for $< n > \geq 0.4$. As fig (4)shows the strong correlation distorts the bands considerably; the initial bare bands were just the conventional d-band DOS of (4).

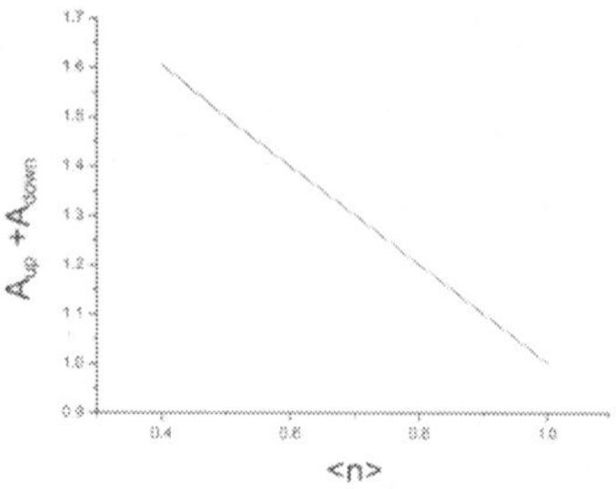

Figure 1. (Color online) Plot of DOS total area $A_{up} + A_{down}$ versus $< n >$. The decrease with $< n >$ is due to elimination of states involving double occupied sites; the number of these states increases with $< n >$.

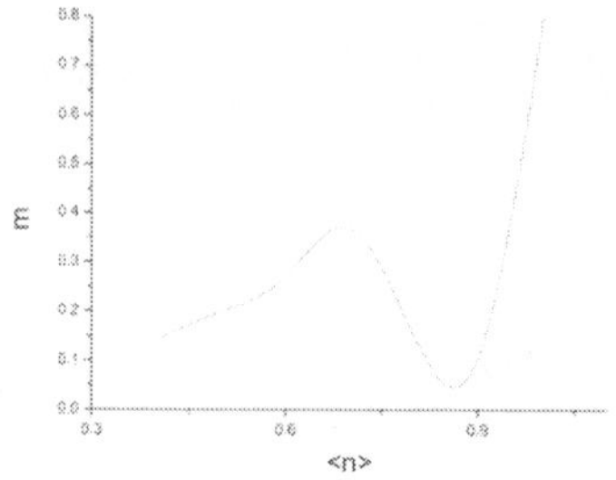

Figure 2. (Color online) Magnetization m versus $< n >$. The non-monotonic behavior is due to the strong dependence of DOS on $< n >$.

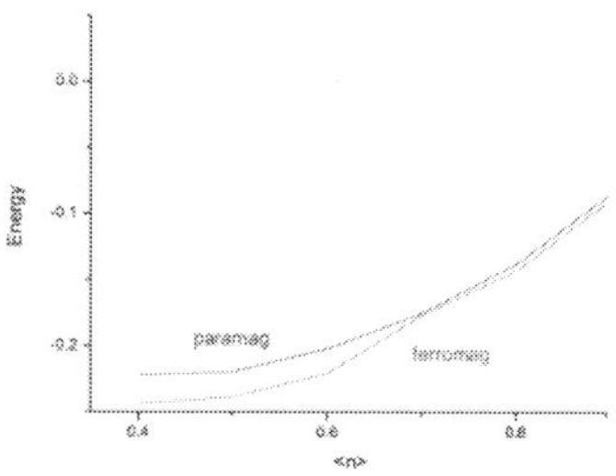

Figure 3. (Color online) Energy of the paramagnetic and of the ferromagnetic states versus $< n >$.

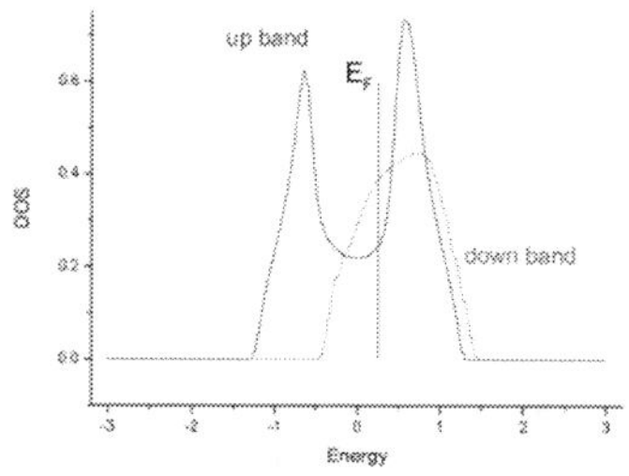

Figure 4. (Color online) Density of states for $< n >= 0.6$. The Fermi level is at $E_F = 0.247$ giving a magnetization $m = 0.276$.

A more complete calculation for finite U in the strong limit regime ($U/W > 1$) as well as including one-electron hybridization in a two band model, suitable to describe transition metals, is in progress.

Acknowledgments

CMC acknowledges the support from the Brazilian agency *FAPERJ* and AT and SGM the support from the Brazilian agency CNP_q and EJC the support from the Brazilian agency *FAPERGS*.

References

[1] Hubbard J 1963 *Proc. Royal Soc.* A **276** 238.
[2] Hubbard J 1964 *Proc. Royal Soc.* A **281** 401.
[3] Janiš V, Mašek J and Vollhardt D 1993 *Z. Phys.* B **91** 325.
[4] Roth L M 1973 *AIP Conference Proceedings on Magnetism and Magnetic Materials* **18** 668.

Magnetism and electronic properties of Mn:Ge(111) interfaces probed by core level photoemission spectroscopy

S Dash[1], M C Mozzati[2], P Galinetto[2], G Drera[1], S Pagliara[1] and L Sangaletti[1]

[1] Dipartimento di Matematica e Fisica, Università Cattolica, via dei Musei 41, 25121 Brescia (Italy)
[2] Dipartimento di Fisica "A. Volta", Università degli Studi di Pavia, via Bassi 6, 27100 Pavia (Italy)

E-mail: sangalet@dmf.unicatt.it

Abstract. The electronic and magnetic properties of Mn:Ge(111) interfaces have been investigated by photoelectron spectroscopy and SQUID magnetometry. An ordered, metallic and ferromagnetic, Mn:Ge(111) interface and a disordered, semiconducting and paramagnetic, Mn_xGe_{1-x} surface alloy have been considered. An analysis of the Mn 2p X-ray photoemission core line shows that the former interface can be described by a single-configuration Mn $3d^6$ initial state, while the latter presents satellite features typical of Mn-based diluted magnetic semiconductors, characterized by relevant ligand-to-metal charge transfer effects.

1. Introduction

Since the discovery of ferromagnetism in the Mn_xGe_{1-x} diluted magnetic semiconductor (DMS) [1], the Mn-Ge system as played a central role for possible applications in the field of spintronics. This role has been recently tested with devices based on $Mn_{0.05}Ge_{0.95}$ quantum dots [2]. The Mn-Ge system includes a large set of compounds, ranging from the Mn_xGe_{1-x} diluted alloy ($x < 0.05$) [1], to the $MnGe_2$ nanocolumns [3], and from Mn:Ge(100) or Mn:Ge(111) interfaces [4,5] to amorphous, ferromagnetic (FM), Mn-Ge thin films [6].

In the present study we compare the electronic properties of an ordered metallic Mn:Ge(111) thin interface with those of a Mn_xGe_{1-x} diluted alloy, that can be regarded as a DMS. Our attention is focused on the DMS, as core level photoemission data on this kind of system are virtually missing, in spite of a large set of X-ray absorption data collected at the Mn L-edge. In many respects, the two compounds represent the end-points of the Mn-Ge system: while the former is FM and metallic, the latter is semiconducting and, in the present case, weakly paramagnetic.

2. Experimental

The samples have been prepared by e-beam evaporation of Mn contained in a Mo crucible in ultra-high-vacuum conditions, at a base pressure better than 1×10^{-9} mbar. The metallic sample is obtained by annealing at 400 °C a Mn layer (5 ML) evaporated onto a c(2x8) reconstructed Ge(111) surface. This annealing induces the formation of an ordered surface layer, which can be regarded as the seed layer for the growth of thicker Mn_5Ge_3 thin films. The surface ordering after annealing (a $(\sqrt{3}\times\sqrt{3})R30°$ reconstruction [5]) has been checked by low-energy electron diffraction. The DMS alloy is obtained by evaporating at room temperature a thin Mn layer on a sputtered Ge(111) surface. The sputtering prevents the formation of an ordered metallic layer, and favors the formation of a diluted alloy through the diffusion, already at room temperature, of Mn atoms into the Ge host lattice. Photoemission spectroscopy data ruled out oxygen contaminations in both samples.

X-ray photoemission (XPS) data have been obtained by exciting the sample with the Al k_α line of a twin-anode X-ray source and by collecting the spectra with a VG-Scienta R3000 electron spectrometer. Static magnetization (M) was measured at 1000 Oe in field cooling regime from 350 K to 2 K with a SQUID Quantum Design Magnetometer. For the metallic sample, magnetization loops were collected at 50 K for magnetic fields ranging between 0 and ±2000 Oe.

3. Results and Discussion

3.1. Electronic properties

Figure 1-a shows the Mn 2p XPS spectrum of the ordered surface alloy, with the expected Mn $2p_{3/2}$ and Mn $2p_{1/2}$ spin-orbit split peaks, with a binding energy (BE) of 638 and 649 eV, respectively. The lineshape is quite similar to that measured on a reference thick (0.2 μm) Mn film (Figure 2-d), obtained by e-beam evaporation on a silicon substrate. Following Ref.7, the Mn 2p XPS lineshape was calculated on the basis of a single configuration Mn $3d^6$ initial state (Figure 1-b). This was done under the assumption that the metallic character of the sample yields a $3d^6$ electron configuration in the outer shell of the Mn ion, i.e. that the overlap between the 3d and 4s/4p bands increases by one unit the count of 3d electrons of atomic Mn. The results show that the lineshape of Mn 2p can be well described by the present single configuration atomic approach, which fully accounts for multiplet effects arising from the coupling of the core hole with the $3d^6$ electron shell, while the $3d^5$ atomic calculations (Figure 1-c) fail in reproducing the main features of the experimental data.

Figure 2 shows the Mn 2p core levels of the Mn_xGe_{1-x} DMS. The spectrum of the as-grown film (Figure 2-a) shows a manifold of features, ascribed to both metallic Mn and Mn diluted in the Ge lattice. In fact, a comparison with the Mn 2p XPS core line from metallic Mn (Figure 2-d) indicates that the M_1 and M_2 features (marked by dashed vertical lines) can be ascribed to metallic Mn. These features are progressively quenched with annealing treatments (Figure 2-b,c), indicating that the annealing induces a diffusion of Mn into the Ge lattice, rather than a clustering of Mn on the Ge(111) surface. After the second annealing treatment (Figure 2-c), the spectrum can be described by two broad peaks (A and C) separated by the spin-orbit interaction. The width of each peak is ascribed to disorder effects, related to both surface sputtering effects and diffusion of Mn into the Ge lattice. On the high BE side of these peaks, two satellites, B and D are also detected. These satellites are often found in Mn-based DMS, such as $Cd_{1-x}Mn_xTe$ (Figure 2-e), $Zn_{1-x}Mn_xS$ and $Ga_{1-x}Mn_xAs$ [8, 9], and are ascribed to charge transfer (CT) effects from the ligand anions (Te, S or As, respectively) to the 3d levels of Mn cations. These effects can be accounted for in the frame of a configuration interaction (CI) model where the electronic states involved in the photoemission process are described by a linear combination of several configurations (see, e.g. Ref. 9 and Refs. therein) such as $3d^n$, $3d^{n+1}\underline{L}$, $3d^{n+2}\underline{L}^2$, where $\underline{L}$ represents a hole in the anion created by the CT to the 3d cation orbitals. The CT energy is defined by $\Delta=E(d^{n+1})-E(d^n)$. The intensity of B and D satellites varies depending on Δ, as well as on the hybridization strength (T) between the p and d orbitals involved in the CT process (here from Ge 4p to Mn 3d). Therefore, the lineshape analysis of the Mn 2p core levels shown in Figure 2-c provides an evidence of Mn-Ge hybridization effects for the Mn_xGe_{1-x} DMS compound. A CI calculation of the Mn 2p spectral weight for a Mn^{2+} impurity in a Ge matrix is shown in Figure 1-d and compared with the DMS alloy spectrum of Figure 2-c. The calculated curve (shaded area) has been obtained by setting $\Delta=1.0$ eV, the on-site Coulomb repulsion $U_{dd} = 3.3$ eV, $T = 1.0$ eV and the Mn 2p-3d Coulomb interaction $Q_{pd} = 4.4$ eV. These findings are consistent with those obtained on Mn-doped DMS [8, 9]. As expected, the CT energy is rather low as compared to other DMS systems, as it roughly scales with the energy gap of the host crystal, that for Ge is smaller than for CdTe and GaAs.

Therefore, for the Mn_xGe_{1-x} DMS, the observed lineshape indicates that CT-related configurations contribute to the XPS spectrum, resulting in the B and D satellites shown in Figure 2. In turn, the metallic surface alloy virtually shows a single configuration behavior. In principle, the ground state of an itinerant transition metal is assumed to be described by a broad distribution of different d^n weights which converge towards a single d count when electron localization is increased by the effects of the U_{dd} correlation energy (see Ref.[10], and Refs. therein). Large electronic correlations are supposed to quench charge fluctuations and limit the number of d^n weights. These effects have been invoked to explain the suppression of valence fluctuations observed for the ground state of Mn in MnSi [11].

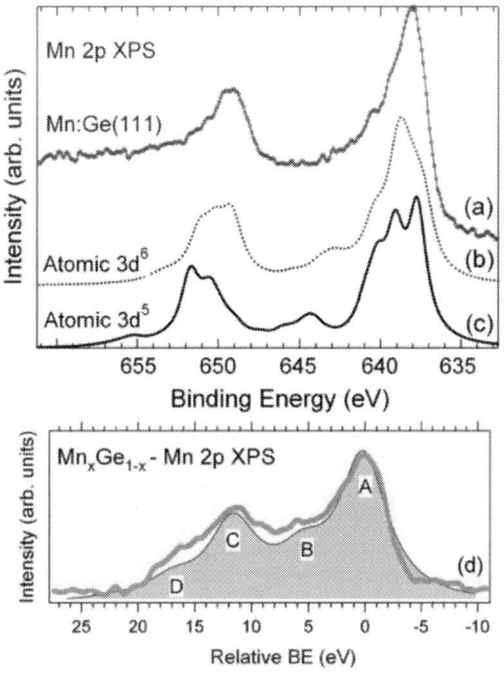

Figure 1. Mn 2p XPS spectrum of the Mn: Ge(111) ordered surface alloy (a). Calculated Mn 2p XPS spectra of single-configuration Mn $3d^6$ (b) and Mn $3d^5$ (c) electron systems. CI calculation of the Mn 2p XPS spectrum for a Mn^{2+} impurity in a Ge matrix (shaded area, d). Calculations are compared to XPS data obtained from Figure 2-c after background subtraction (thick line, d).

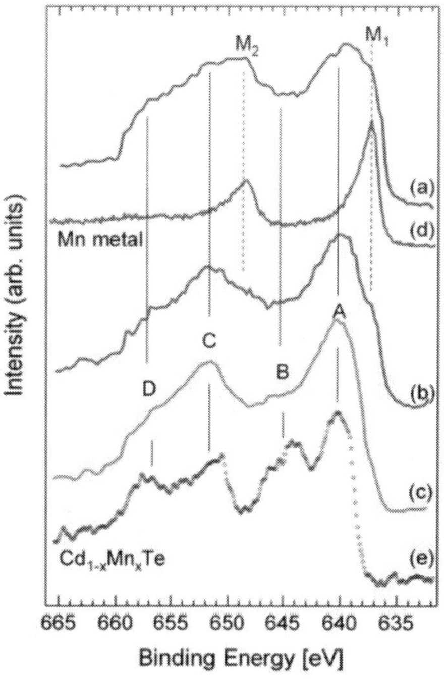

Figure 2. Mn 2p XPS spectra of the Mn_xGe_{1-x} DMS. (a) as grown; (b) after the first annealing at 473 K for 15'; (c) after the second annealing at 473 K for 30'. Mn 2p XPS spectrum of a metallic thick Mn film (d) and of a $Cd_{1-x}Mn_xTe$ DMS [(e), from Ref.8].

3.2 *Magnetic properties*

The magnetization curves of both samples are shown in Figure 4 and Figure 5. The ordered surface alloy shows a dependence of magnetization on temperature (Figure 4) characteristic of a ferromagnetic (FM) ordering established just about 300 K. The inset of Figure 4 shows the magnetization loop measured at 50 K. This behavior is consistent with that observed in similar metallic surface alloys [5]. For the Mn_xGe_{1-x} DMS, the main features of the M vs. T curve (Figure 5) are: (i) an overall Curie-like behavior, revealing the presence of paramagnetic ions (substitutional Mn in the Ge host), (ii) a broad maximum in the 200-250 K range, that could be ascribed to the effects of Mn_5Ge_3 nanoparticle inclusions [12], and (iii) a magnetization increase above 270 K. A similar behavior was observed for $Mn_{0.06}Ge_{0.94}$ thin films grown on a Ge(100) substrate [12]. For a higher Mn content, a FM behavior is expected [12]. The lack of a dominant FM contribution in the present sample indicates that the density of Mn atoms diffused into the Ge bulk lattice is still too low to yield a long-range FM behavior.

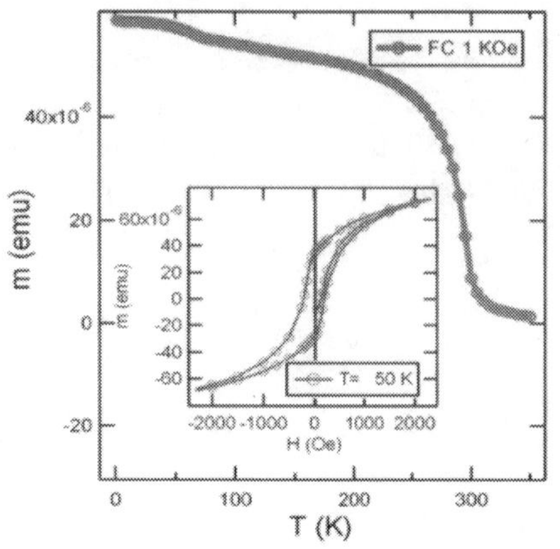

Figure 4. M vs. T curve obtained from the ordered surface alloy. Inset: hysteresis loop measured at 50 K.

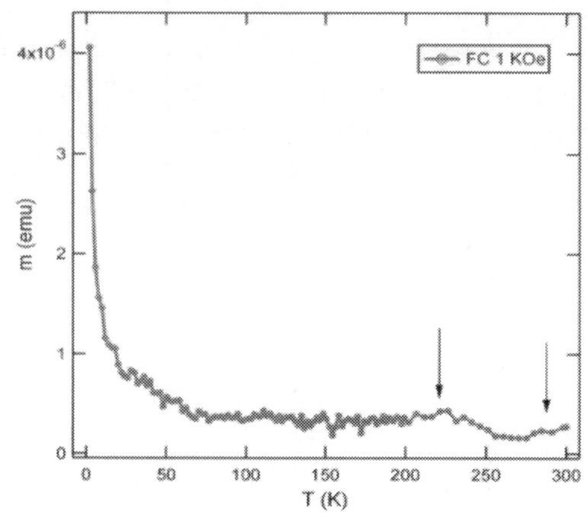

Figure 5. M vs. T curve obtained from the Mn_xGe_{1-x} DMS.

4. Conclusions

Mn evaporation on Ge(111) may yield quite different surface alloys, depending on the surface preparation prior to Mn deposition. The difference in surface ordering significantly affects the magnetic and electronic properties. The Mn 2p photoemission core level data provide evidence of Mn-Ge hybridization effects in the DMS alloy. These effects can be accounted for in the frame of a CI model by assuming CT from Ge to Mn. In turn, the metallic, FM surface alloy is virtually described by a single $3d^6$ configuration, suggesting that electronic correlations among 3d electrons are large enough to suppress the 3d charge fluctuations expected in metallic systems.

References

[1] Park Y D *et al.* 2002 *Science* **295** 651
[2] Xiu F, Wang Y, Kim J, Upadhyaya P, Zhou Y, Kou X, Han W, Kawakami R, Zou J and Wang K, 2010 *ACS Nano* **4** 4948
[3] Jamet M, *et al.* 2006 *Nature Materials* **5** 653
[4] Zeng C, Erwin S C, Feldman L C, Li A P, Jin R, Song Y, Thomson J R and Weitering HH 2003 *Appl. Phys. Lett.* **83** 5002
[5] Sangaletti L, Ghidoni D, Pagliara S, Goldoni A, Morgante A, Floreano L, Cossaro A, Mozzati M C and Azzoni C B 2005 *Phys. Rev. B* **72** 35434
[6] Guchhait S, *et al.* 2011 *Phys. Rev. B* **84** 024432
[7] Sangaletti L, Drera G, Magnano E, Bondino F, Cepek C, Sepe A and Goldoni A 2010, *Phys. Rev. B* **81** 085204
[8] Sangaletti L, Verdini A, Drera G, Pagliara S, Floreano L, Morgante A and Goldoni A, 2010 *Phys. Rev. B* **81** 245320
[9] Okabayashi J, Kimura A, Rader O, Mizokawa T, Fujimori A, Hayashi T and Tanaka M, 1998 *Phys. Rev. B* **58** R4211
[10] van der Laan G, Dhesi S S, and Dudzik E 2000 *Phys. Rev. B* **61** 12277
[11] Carbone F, Zangrando M, Brinkman A, Nicolaou A, Bondino F, Magnano E, Nugroho A A, Parmigiani F, Jarlborg Th and van der Marel D 2006 *Phys. Rev. B* **73** 085114
[12] De Padova P *et al.* 2008 *Phys. Rev. B* **77** 045203

International Conference on Strongly Correlated Electron Systems (SCES 2011) IOP Publishing
Journal of Physics: Conference Series **391** (2012) 012089 doi:10.1088/1742-6596/391/1/012089

Search for Third Harmonic Magnetic Bragg Reflections in Charge-Stripe Ordered $La_{1.725}Sr_{0.275}NiO_4$.

P. G. Freeman[1,2], **C. Niedermayer** [3], **and D. Prabhakaran**[4]

[1]Helmholtz-Zentrun Berlin, Hahn-Meitner-Platz 1, D-14109 Berlin, Germany.
[2]Institut Laue-Langevin, BP 156, 38042 Grenoble Cedex 9, France.
[3]Laboratory for Neutron Scattering, ETHZ and PSI, CH-5232 Villigen PSI, Switzerland.
[4]Department of Physics, Oxford University, Oxford, OX1 3PU, United Kingdom.

E-mail: paul.freeman@helmholtz-berlin.de

Abstract.
The magnetic Bragg reflections of charge-stripe ordered $La_{1.725}Sr_{0.275}NiO_4$ where investigated by neutron diffraction to determine the intensity ratio of first to third order harmonics. The third harmonic Bragg reflections were not observed at temperatures across that magnetically ordered phase, creating an upper bound for their intensity of 0.16 % of the first harmonic intensity.

1. Introduction

The charge stripe interpretation of La-based cuprates has been under investigation since the discovery of charge stripe order in a non-superconducting La-Nd based cuprate[1]. A great advance on the magnetic structure of La-Nd cuprates was achieved by polarized neutron diffraction, confirming a co-linear spin arrangement at 1/8 doping level that is consistent with charge stripe order[2]. The striking hourglass magnetic excitation spectrum of the cuprates can also be understood within a charge stripe ordered picture[3, 4]. Despite this effort, the centring of the charge stripe order remains a undesirable free parameter for modelling the magnetic excitations of cuprates[5].

Recently calculations were made of the relative intensity of first to third harmonic magnetic Bragg reflections for both site (Cu or Ni) and bond (O) centred charge stripes measured by neutron diffraction, and the ratio is found to be significantly larger for site centred charge stripes[6]. Due to a small Cu moment in the magnetically ordered cuprates, resolving the third harmonics will be very challenging in the cuprates. The relative intensity of magnetic third harmonic Bragg reflections in charge stripe ordered $La_2NiO_{4.13}$ was determined to be 1.7 %[7]. The measured value can be compared to the calculated values of 1.1 % for site centring or 11.8 % for bond centring, implying site centred charge stripes[6]. But as the charge-stripe structure of $La_2NiO_{4.13}$ locks into an oxygen ordering supercell, this supercell could be the cause of site centred charge order[7]. It is therefore desirable to investigate the third harmonics in charge-stripe ordered $La_{2-x}Sr_xNiO_4$, where there is no Sr ordering, to further test the calculation and determine the effect of the oxygen supercell on the charge stripe structure of this material.

2. Experimental details

A single crystal of $La_{1.725}Sr_{0.275}NiO_{4+\delta}$ was grown by the floating-zone technique[8]. The sample was cut from the crystal used in our earlier neutron studies of $La_{1.725}Sr_{0.275}NiO_{4+\delta}$ [9] , and is a rod of 8 mm diameter and 15 mm length used in our recent diffraction measurements[10]. Thermo-Gravimetric Analysis (TGA) of an as-grown crystal determined the oxygen excess to be $\delta = 0.02 \pm 0.01$. Data on the bulk magnetization of an as-grown crystal are published elsewhere[11]. As our previous studies showed the excess oxygen has little effect on the properties of this material, we label the sample $x = 0.275$.

Neutron diffraction measurements were performed on the cold Triple-Axis Spectrometer (TAS) RITA II at the P.S.I., and the experimental setup has been described in detail previously[10]. The neutrons initial and final energy were selected from Bragg reflections off a vertically focusing pyrolytic graphite (PG) monochromator, and a flat PG analyzer. The sample was oriented so that (h, h, l) positions in reciprocal space could be accessed. In this work we refer to the body centred tetragonal unit cell of LSNO, which has unit cell parameters $a \approx$ 3.8 Å and $c \approx 12.7$ Å.

3. Results

In figure 1(a) we show the charge stripe ordered model for site/Ni centred charge stripes in $x =$ 0.275, which produces magnetic Bragg reflections indicate in Fig. 1(b) in the $(hk0)$ reciprocal plane, and in the (hhl) reciprocal plane in Fig. 1(c). First order magnetic Bragg reflections are observed at $(h + \frac{1}{2}, k + \frac{1}{2}, l) \pm (\varepsilon/2, \varepsilon/2, 0)$ with $\varepsilon = 0.300 \pm 0.001$ at base temperature, and third harmonic magnetic Bragg reflections should be observed at $(h + \frac{1}{2}, k + \frac{1}{2}, l) \pm (3\varepsilon/2, 3\varepsilon/2, 0)$. The magncetic ordering temperature of the $x = 0.275$ is $T_{SO} = 140 \pm 5$ K[10].

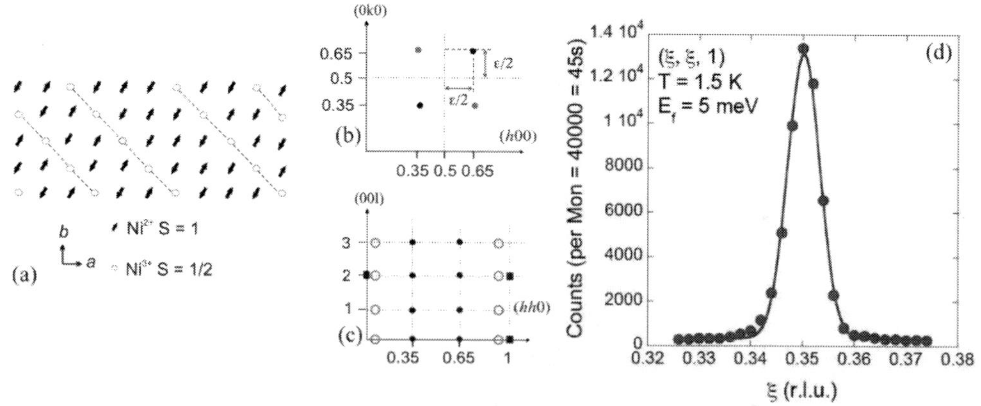

Figure 1. (a) A model of the Ni sites of a single Ni-O layer of LSNO $x = 0.275$ for site centred charge order. (b) The first harmonic magnetic Bragg reflections of $x = 0.275$ in the $(hk0)$ reciprocal plane. There are two sets of reflections creating a quartet of magnetic Bragg reflections, two for the order in (a) and two for the charge stripe twin with stripes rotated 90° to those in (a). (c) The structural Bragg reflections and magnetic Bragg reflections of of the (hhl) reciprocal plane. Solid squares indicate structural Bragg reflections, solid circles are first harmonic magnetic Bragg reflections and hollow circles indicate the position of third harmonic magnetic Bragg reflections. For the purpose of clarity we have omitted the positions of charge order Bragg reflections. (d) An example of a scan prarallel to $(hh0)$ of a first harmonic Bragg reflection in charge stripe ordered $x = 0.275$. The solid line is the result of a fit of a Gaussian lineshape on a flat background

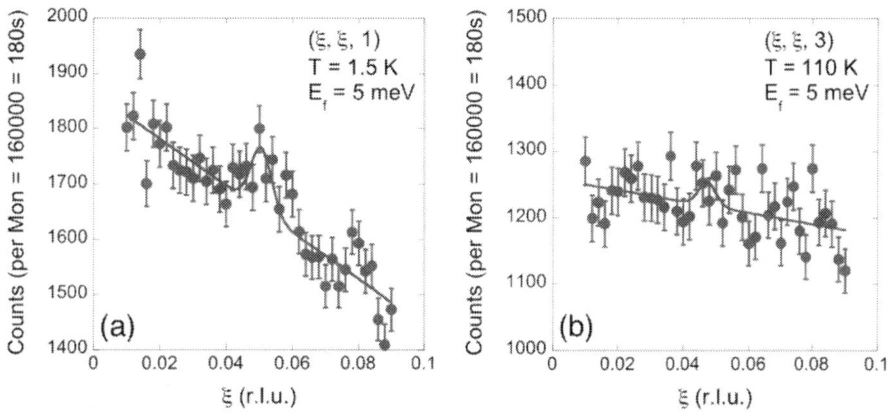

Figure 2. Scans of reciprocal space parallel to $(hh0)$ at the position of third harmonic magnetic Bragg reflections, (a) 1.5 K, $l = 1$, and (b) 110 K, $l = 3$. The solid lines are the result of fits of a Gaussian lineshape on a sloping background, where centre is determined from the position of first harmonic reflections, and the width is fixed to be the same as that of the first harmonic.

Table 1. Fitted amplitude of peaks at the $(0.05, 0.05, l)$ position of third harmonic magnetic Bragg reflections of charge-stripe ordered $La_{1.725}Sr_{0.275}NiO_4$.

l	Temperature (K)	Amplitude Counts of Third Harmonic	Background Noise	Intensity Compared to First Harmonic (%)
1	1.5	113 ± 27	41	0.22 ± 0.05
3	1.5	54 ± 26	37	0.12 ± 0.06
3	40	69 ± 26	36	0.16 ± 0.06
3	110	35 ± 24	35	0.13 ± 0.09

To avoid the high intensity structural Bragg reflections we decided to search for the third order magnetic Bragg reflections around (001) and (003). In Fig. 1(d) we show a scan through the first harmonic reflection centred at $(0.35, 0.35, 1)$ measured at 1.5 K . In figure 2(a) we show a scan through the position of the third harmonic at $(0.05, 0.05, 1)$. At this position the background is tilted with no definite observation of a magnetic Bragg reflection. Using the position of the first order harmonic and the width of this reflection as fitted by a Gaussian lineshape, we fitted the amplitude of the expected reflection in Fig. 2(a). The fit determined an amplitude of 113 ± 27 counts compared to a statistical error of 41 counts of the background. In figure 2(b) we show how we repeated this search for third order harmonics near $(0.35, 0.35, 3)$ at T = 110 K, having also measured the first harmonic at this temperature.

In table 1 we compare the amplitude of the fit of the scans of Fig. 2, the statistical error of background for each point, and relative percentage of the fitted amplitude to the first harmonics intensity. Although the fitted peak amplitude is positive, it is never large than three times the standard deviation of the background. We therefore take these values as upper limits on the relative intensity of the third harmonic Bragg reflection, that is the third harmonic has less than 0.16±0.03 % the intensity of the first harmonic magnetic Bragg reflection of $x = 0.275$.

4. Discussion and Conclusions

We have searched for, and failed to observe, third order harmonic magnetic Bragg reflections in charge stripe ordered $La_{2-x}Sr_xNiO_4$ $x = 0.275$, at temperatures across the magnetically ordered phase. Our results differ from the observation of higher harmonic reflections in charge stripe ordered $La_2NiO_{4.133}$[7].

Analysis of the scans indicates an upper limit on the strength of the third order magnetic Bragg reflections in $x = 0.275$ of 0.16 % of the first harmonic. Our results indicate the model of the spin structure proposed in ref. [6] does not fully encompass the characteristics of the charge ordering in $x = 0.275$. One characteristic the model does not include is the extended nature of the charge ordered site, that is indicated by the energy dependence of resonant soft x-ray diffraction of $La_{1.8}Sr_{0.2}NiO_4$[12]. Alternatively the model may require mixing of bond and site centred charge ordering, as observed by transmission-electron-microscopy (TEM) at the surface of this material[13]. In $La_2NiO_{4.133}$ the magnetic domain walls have been determined to be narrow, consistent with the square wave profiles used in the third harmonic calculation[6], and the observed ratio of third to first harmonic reflections is close to the predicted value for site centred charge order. We ascribe the presence of higher harmonics in $La_2NiO_{4.133}$ as due to the distortion caused by the oxygen supercell ordering, that may play an important roll in causing the anomalous magnetic excitation spectrum of $La_2NiO_{4.11}$[14].

Resonant soft x-ray diffraction and TEM results remain the strongest experimental evidence for predominantly site centred charge order in $La_{2-x}Sr_xNiO_4$[12, 13]. Searching for third harmonic magnetic Bragg reflections in the cuprates may determine the centring of the charge order in these materials, but it is likely that third harmonics are only observed in materials in which structural distortions stabilise charge order, i.e. La-Nd based cuprates near 1/8 doping[1]. However, in systems where structural distortions stabilise the charge order, the order may not represent that of the undistorted systems.

5. Acknowledgements

This work is based on experiments performed at the Swiss spallation neutron source SINQ, Paul Scherrer Institute, Villigen, Switzerland. Crystal growth was supported by the Engineering and Physical Sciences Research Council of Great Britain.

References

[1] Tranquada J. M., Sternleib B. J., Axe J. D., Nakamura Y. and Uchida S., 1995 Nature (London) **375**, 561.
[2] Christensen N. B., Ronnow H. M., Mesot J., Ewings R. A., Momono N., Oda M., Ido M.,Enderle M., McMorrow D. F. , and Boothroyd A. T. 2007 Phys. Rev. Lett. **98**, 197003.
[3] Tranquada J. M., Woo H., Perring T. G., Goka H., Gu G. D., Xu G., Fujita M. and Yamada K., 2004 Nature (London) **429**, 534.
[4] Boothroyd A. T., Babkevich P., Prabhakaran D., and Freeman P. G., 2011 Nature (London) **375**, 561.
[5] Seibold G., and Lorenzana J. , 2005 Phys. Rev. Lett. **94**, 107006.
[6] Phillabaum B., and Carlson E.W., 2003 Phys. Rev. B **77**, 104526.
[7] Wochner P., Tranquada J. M. , Buttery D. J. , and Sachan V. , 1998 Phys. Rev. B **57**, 1066.
[8] Prabhakaran D., Isla P., and Boothroyd A. T., 2002 J. Cryst. Growth **237**, 815.
[9] Freeman P. G. , Boothroyd A. T. , Prabhakaran D., Enderle M. and Niedermayer C., 2004 Phys. Rev. B **70**, 024413.
[10] Freeman P. G., Christensen N. B., Prabhakaran D. and Boothroyd A. T., 2010 J. Phys.: conf. **200**, 012037.
[11] Freeman P. G. , Boothroyd A. T. , Prabhakaran D. and Lorenzana J., 2006 Phys. Rev. B **73**, 014434.
[12] Schussler-Langeheine C., Schlappa J., Tanaka A., Hu Z., Chang C. F., Schierle E., Benomar M., Ott H., Weschke E., Kaindl G., Friedt O., Sawatzky G. A., Lin H. J., Chen C. T., Braden M., and Tjeng L. H., 2005 Phys. Rev. Lett. **95**, 156402.
[13] Li J., Zhu Y., Tranquada J. M. , Yamada K. , and Buttrey D. J. 2003 Phys. Rev. B **67**, 012404.
[14] Freeman P. G., Enderle M., Hayden S. M., Frost C. D., Yao D. X., Carlson E. W., Prabhakaran D. and Boothroyd A. T., 2009 Phys. Rev. B **80**, 144523.

Substitution effect on the superconductivity in $W_{5-x}Ta_xSiB_2$ with the T_2 phase structure

M. Fukuma, K. Kawashima and J. Akimitsu

Department of Physics and Mathematics, Aoyama Gakuin University, Fuchinobe 5-10-1, Chuo-ku, Sagamihara-shi, Kanagawa 252-5258, Japan

fukuma.0816@phys.aoyama.ac.jp

Abstract. We report here the magnetic susceptibility and electrical resistivity measurements of $W_{5-x}Ta_xSiB_2$. $W_{5-x}Ta_xSiB_2$ crystallizes in a tetragonal (the Mo_5SiB_2 (T_2 phase) type) structure. We evaluated the effect of the partial substitution of Ta in W_5SiB_2. The superconductivity was observed below $x = 2$ in $W_{5-x}Ta_xSiB_2$. As substituted Ta for W, the superconducting transition temperature (T_c) clearly enhanced from 5.8 K to 6.5 K. The largest volume fraction exhibits at $x = 0.2$ sample. Thus, this work shows, for the first time, the superconductivity in $W_{5-x}Ta_xSiB_2$. Moreover, we determined that the dominant carrier sign of W_5SiB_2 are the hole from the thermoelectric power measurement and discussed about the enhancement of T_c of $W_{5-x}Ta_xSiB_2$ system.

1. Introduction

One of major phase of the boro-silicide TM_5SiB_2 (TM = transition metal element), which is named as the T_2 phase (Mo_5SiB_2-type structure), has a tetragonal structure with the space group of *I4/mcm* (No. 140) [1]. The TM atoms occupy at the Wyckoff positions of 16*l* and 4*c*, and Si and B atoms are positioned at 4*a* and 8*h* sites, respectively [2]. The TM_5SiB_2 compound show the various interesting phenomena, such as ferromagnetic ordering in Fe_5SiB_2 and Mn_5SiB_2 [3,4]. Moreover, superconductors have been reported in three compounds of Mo_5SiB_2 ($T_c = 5.6$ K), Nb_5SiB_2 ($T_c = 7$ K), and W_5SiB_2 ($T_c = 5.8$ K) [5,6,7].

We focused on the superconductor of W_5SiB_2 with T_c of 5.8 K. The electrical resistivity, magnetization, and specific-heat data reveal that the W_5SiB_2 exhibits a conventional type-II superconductor and can be described by the BCS theory. In addition, W_5SiB_2 has been reported that the *d*-orbital of W plays an important role for the superconductivity from a viewpoint of the calculated electronic density of state [8]. In the framework of BCS theory, the higher density of state (DOS) near Fermi energy (E_F), lighter elements resulting in higher frequency phonon modes, and large electron-phonon interaction strength are needed for enhancing the superconducting state and T_c. Considering the existence of these previous studies in the BCS superconductors [9,10], it is expected that the superconducting transition temperature of W_5SiB_2 is enhanced by the substitution effect which has a increase the DOS near E_F.

In this paper, we successfully synthesized the polycrystalline sample of $W_{5-x}Ta_xSiB_2$ with T_2 phase structure using a conventional solid state reaction, and we reported the results of magnetic susceptibility and electrical resistivity measurements of a boro-silicide superconductor, $W_{5-x}Ta_xSiB_2$, with the maximum transition temperature of $T_c = 6.5$ K with $x = 0.2$. In addition, we clarify that the dominant carriers of W_5SiB_2 at 300 K are whether the holes or electrons by thermoelectric power evaluation.

2. Experimental details

Polycrystalline sample of $W_{5-x}Ta_xSiB_2$ were synthesized by the conventional solid state reactions following synthesis conditions in Ref. 7.

The synthesized samples were analyzed by a powder X-ray diffraction technique using a conventional X-ray spectrometer with a graphite monochromator. Intensity data were collected with CuKα radiation over a 2θ range from 5° to 80° at 0.02° step width. The magnetic susceptibilities were determined by a SQUID magnetometer (MPMS-R2, Quantum Design) at temperatures between 1.8 K and 10 K in and applied magnetic fields of $H = 10$ Oe. The electrical resistivity was measured using a conventional dc four-probe method with the measuring current of $I = 1$ mA in the temperature range between 2.5 K and 300 K, using PPMS (Quantum Design) and a laboratory-built apparatus. Electrical leads were fabricated by spot-welding gold wires (25 μm) onto the polished surface of a specimen. The thermoelectric power measurement was performed by the conventional four-probe method for bar-shape sample.

3. Experimental results and discussion

Fig. 1 displays the powder X-ray diffraction (PXRD) pattern of the synthesized sample of $W_{4.8}Ta_{0.2}SiB_2$. The diffraction peaks in PXRD pattern from the T_2 phase structure was observed as a main contribution. The residual minor peaks were a small amount of impurity phase, which was identified as the WSi_2. The PXRD data of $W_{4.8}Ta_{0.2}SiB_2$ in Fig. 1 is in good agreement with the previous one for the non-doped sample of W_5SiB_2 [5]. The $W_{4.8}Ta_{0.2}SiB_2$ can be indexed as a tetragonal unit cell with the space group *I4/mcm* (No. 140).

The temperature dependence of the magnetic susceptibility of $W_{5-x}Ta_xSiB_2$ ($x = 0$ and 0.2) under an applied magnetic field of $H = 10$ Oe is shown in Fig. 2. The magnetic susceptibility data were measured on the field cooling (F.C.) process. The magnetic susceptibility of W_5SiB_2 ($x = 0$) exhibited a marked drop at around 5.8 K, suggesting the occurrence of superconductivity. On the other hand, a sharp transition at around 6.5 K was observed in the data of $W_{5-x}Ta_xSiB_2$ with $x = 0.2$ in Fig. 2. We confirmed that the solution limit of Ta substitution was $x = 0.2$ because the diffraction peaks of

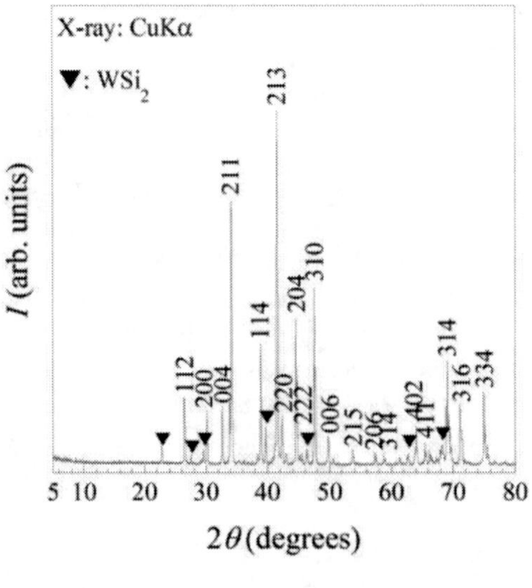

Figure 1. (Color online) Powder X-ray diffraction pattern of $W_{4.8}Ta_{0.2}SiB_2$. Arrows indicate the impurity phase of WSi_2.

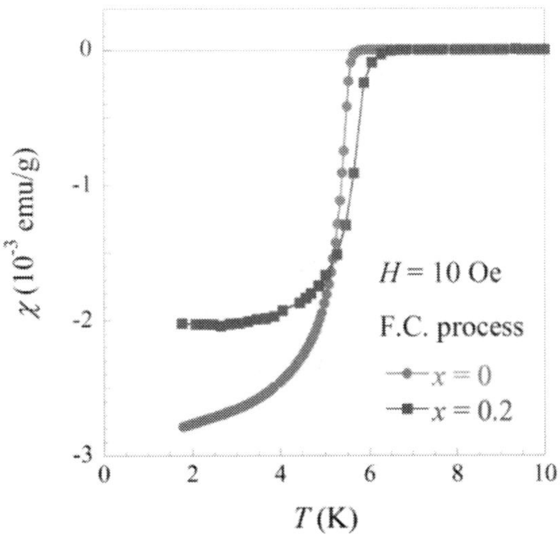

Figure 2 (Color online) Temperature dependence of the magnetic susceptibility of $W_{5-x}Ta_xSiB_2$ with $x = 0$ and 0.2 under an applied magnetic field of $H = 10$ Oe (F.C. process).

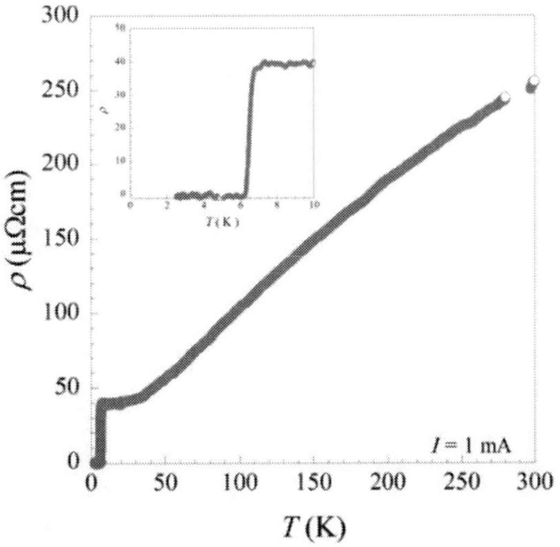

Figure 3 (Color online) Temperature dependence of the electrical resistivity of $W_{4.8}Ta_{0.2}SiB_2$ with measuring current of $I = 1$ mA. The inset shows an enlarged view of the plot in a low-temperature range ($T \leqq 10$ K) of electrical resistivity.

impurity phases in the PXRD patterns increased with x above $x = 0.3$ region in $W_{5-x}Ta_xSiB_2$ system. The largest superconducting volume fraction, $V.F.$ = ~15% defined of χ data of $W_{4.8}Ta_{0.2}SiB_2$ at 1.8 K was confirmed.

Superconductivity can also be inferred from the result of the electrical resistivity measurements, as shown in Fig. 3. The onset transition temperatures (T_c^{onset}) and the zero resistivity temperature (T_c^{zero}) were observed to be 6.6 K and 6.4 K. The transition width is defined to be a temperature interval between 10 and 90% of the transition and was observed to be approximately ~0.15 K. Such a sharp transition is characteristic of a sample with a good quality. The residual resistivity ratio (RRR), ρ_{300K}/ρ_0 (ρ_0 is the residual resistivity at T_c^{onset}), was obtained to be ~6.5, which is a fine value despite the existence of a minor impurity phase in measured sample. The defined T_c in the electrical resistivity measurement is in good agreement with the magnetic susceptibility result. From the results of both the magnetic susceptibility and the electrical resistivity measurements, T_c of $W_{4.8}Ta_{0.2}SiB_2$ was determined to be 6.5 K.

In order to determine a type of dominant carrier of W_5SiB_2, the Sebeek coefficient, S, was determined at 300 K by using the thermoelectric-power measurement. The value of the S of W_5SiB_2 at 300 K was to be about S = ~0.85 μV/K. The positive sign of S indicates that the dominant carriers of W_5SiB_2 are holes. The electrons of *d*-orbital in Ta atom are less one electron than that of *d*-orbital in W atoms. This fact suggests that the enhancement of T_c of W_5SiB_2 was induced by increasing a density of hole carriers by partial substituted Ta for W in $W_{5-x}Ta_xSiB_2$ system.

In summary, we validated the effect of Ta partial substitution on the superconductivity in $W_{5-x}Ta_xSiB_2$ system. The superconducting transitions of $W_{5-x}Ta_xSiB_2$ were observed from the electrical resistivity and magnetic susceptibility measurements. The maximum superconducting volume fraction was confirmed from the data of $W_{4.8}Ta_{0.2}SiB_2$ with T_c = 6.5 K. This enhancement of T_c in $W_{5-x}Ta_xSiB_2$ system suggests the results of increment of the density of state near Fermi energy. Further experiments on the superconducting state are necessary and desirable to clarify superconducting state for $W_{5-x}Ta_xSiB_2$, such as specific heat measurements and theoretical calculation of electronic state.

Acknowledgements

This work was supported by the High-Tech Research Center Project for Private Universities from the Ministry of Education, Culture, Sports, Science and Technology (MEXT), a Giant-in-Aid for Specially Promoted Research and a Grant-in-Aid for Scientific Research. Moreover, this research was supported by Strategic International Collaborative Research Program between EU and Japan (LEMSUPER), Japan Science and Technology Agency. Two of authors (KK and JA) were supported by IMRA MATERIAL CO., LTD.

References

[1] R. Sakidja, J. H. Perepezko, S. Kim, and N. Sekido: Acta Materialia **56** (2008) 5223.
[2] C. J. Rawn, J. H. Schneibel. C. M. Hoffmann, and C. R. Hubbard: Intermetallics 9 (2001) 209.
[3] D. M. Almeida, C. Bormio-Nunes, C. A. Nunes, A. A. Coelho, and G. C. Coelho: J. Magn. Magn. Mater. **321** (2009) 2578.
[4] V. Reghavan: J. Phase Equilibria Diffusin **28** (2007) 380.
[5] P. Peshev, G. Gyurov, R. Stoyanchev, Izv. Khim. **19** (1986) 267.
[6] A. Brauner, C. A. Bortolozo, G. Rodrigues, A. J. S. Machado, Solid State Commun. **149** (2009) 467.
[7] M. Fukuma, K. Kawashima, M. Maruyama, J. Akimitsu, J. Phys. Soc. Jpn. **80** (2011) 024702.
[8] M. Fukuma, K. Kawashima, M. Maruyama, J. Akimitsu, Physica C **471** (2012) 714.
[9] E. E. Havinga, H. Damsma and M. H. van Maarn, J. Phys. Chem. Solids **31** (1970) 2653.
[10] E. E. Havinga, H. Damsma and J. M. Kanis, J. Less-Common Met. **27** (1972) 281.

Broadband microwave study of $SrRuO_3$ and $CaRuO_3$ thin films

Diana Geiger[1], **Marc Scheffler**[1], **Martin Dressel**[1], **Melanie Schneider**[2], **Philipp Gegenwart**[2]

[1]1. Physikalisches Institut, Universität Stuttgart, D-70550 Stuttgart, Germany
[2]I. Physikalisches Institut, Georg-August-Universität, D-37073 Göttingen, Germany

E-mail: `scheffl@pi1.physik.uni-stuttgart.de`

Abstract. We study the microwave properties of $SrRuO_3$ and $CaRuO_3$ thin films in the frequency range 45 MHz to 40 GHz, and at temperatures between 80 K and 300 K, using a broadband Corbino technique. The films are grown on $SrTiO_3$ and MgO substrates, and we discuss how dielectric substrate resonances complicate the study of the film properties, in particular in the case of $SrTiO_3$ with its large dielectric constant. For $SrRuO_3$ on MgO substrate we find signatures of the ferromagnetic transition throughout our spectral range.

1. Introduction

Quantum phase transitions (QPTs) and their role for non-Fermi-liquid (NFL) behavior in electronic systems are a highly investigated topic, both theoretically and experimentally [1, 2]. While antiferromagnetic QPTs are found in a number of different materials (e.g. heavy fermion compounds [3]), one of the established examples of a ferromagnetic QPT [4] is the doped system $Sr_{1-x}Ca_xRuO_3$ [5, 6, 7] at the critical concentration $x \approx 0.8$. In optical measurements the frequency acts as tuning parameter which can be adjusted to probe the relevant energy scales of the system. For the pure compounds of our material system, $SrRuO_3$ and $CaRuO_3$, NFL behavior of the optical conductivity has been reported for infrared and THz frequencies [8, 9, 10], but it is unclear to which extent this is related to the ferromagnetic QPT. To match the characteristic low energy scales of QPTs and strongly correlated electron systems [11], in the present study we address the optical properties of $SrRuO_3$ and $CaRuO_3$ at very low energies, in the microwave frequency range.

2. Experiment

We have grown epitaxial thin films of $SrRuO_3$ and $CaRuO_3$ on $SrTiO_3$ substrates and polycrystalline $SrRuO_3$ films on MgO substrates using metalorganic aerosol deposition [7, 12, 13]. These films are in the range of 50 nm thick, i.e. much thinner than their microwave skin depth. We have measured the microwave properties of these samples in the frequency range 45 MHz to 40 GHz and at temperatures between 80 K and 300 K with a broadband Corbino technique as described in [14], but with a different cryostat that was optimized for intermediate and higher temperatures (similar to the one in [15]). Here the thin film sample under investigation terminates a coaxial cable: a vector network analyzer (NWA) sends the microwave signal through the cable onto the sample, where it is reflected and therefore probes its electromagnetic properties. The reflected signal is detected by the NWA. To allow for a precise data analysis, we perform a thorough three-standard calibration procedure [14]. From the calibrated reflection coefficient we directly calculate the impedance and conductivity of the sample.

3. Results

In Fig. 1 we present reflection coefficient S_{11}, impedance Z, and conductivity $\sigma = \sigma_1 + i\sigma_2$ spectra for $SrRuO_3$ films on $SrTiO_3$ and MgO substrates, measured at 300 K. In this frequency range and at this temperature we expect a nearly constant σ_1 and a vanishing σ_2, consistent with metallic Drude behavior [16]. This is what we find for the film on MgO, see Fig. 1(f): $\sigma_1 \approx 0.23 \cdot 10^4\ \Omega^{-1}\ \mathrm{cm}^{-1}$ is almost constant and σ_2 is smaller. On top of these flat curves, there are two weak resonant features around 7 GHz and 13 GHz; these are not intrinsic to the $SrRuO_3$ film, but they are due to dielectric resonances in the substrate [17]. As these resonances are limited to small frequency regions, we can still analyze the microwave properties of the $SrRuO_3$ film in the broad frequency range outside the substrate resonances.

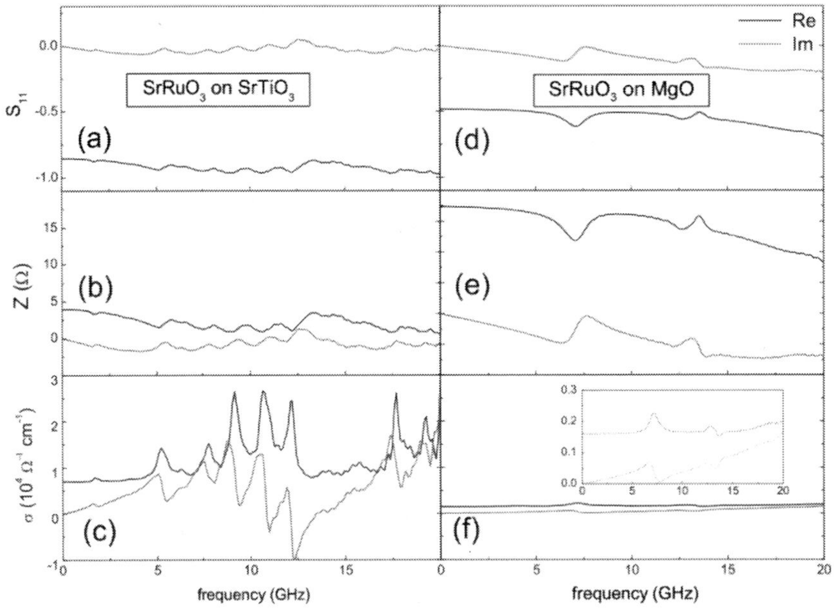

Figure 1. Real (black) and imaginary (red) parts of reflection coefficient S_{11}, impedance Z, and conductivity at 300 K for $SrRuO_3$ on $SrTiO_3$ and $SrRuO_3$ on MgO.

The $SrRuO_3$ on $SrTiO_3$ spectra in Fig. 1(a)-(c) show a rather different behavior: strong resonances cover the entire frequency range. These numerous resonances are a result of the very high dielectric constant of the $SrTiO_3$ substrate [18, 19], which shifts substrate resonances from higher frequencies into the measured microwave frequency range [17]. Here the influence of the substrate overwhelms the contributions of the $SrRuO_3$ film, and we cannot deduce the intrinsic microwave properties of the film from the measured overall response.

With decreasing temperature, the influence of the $SrTiO_3$ substrate gets even stronger, due to its increasing dielectric constant [18, 19]. This is shown in Fig. 2, where we plot $\mathrm{Re}(Z)$ of the measured impedance for $SrRuO_3$ and $CaRuO_3$ on $SrTiO_3$ substrates as a function of frequency and temperature. Upon cooling, substrate resonances move to lower frequencies, thus increasing the total number of resonances in our frequency range. With this strong influence of the substrate at GHz frequencies, we find signatures of the intrinsic response of the metallic films only at frequencies below 1 GHz, with a temperature dependence consistent with dc measurements [7].

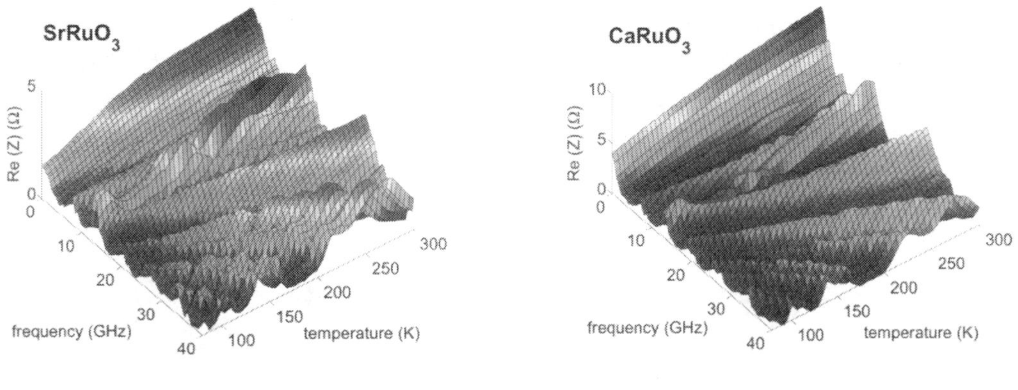

Figure 2. Real part of the impedance for $SrRuO_3$ and $CaRuO_3$ on $SrTiO_3$.

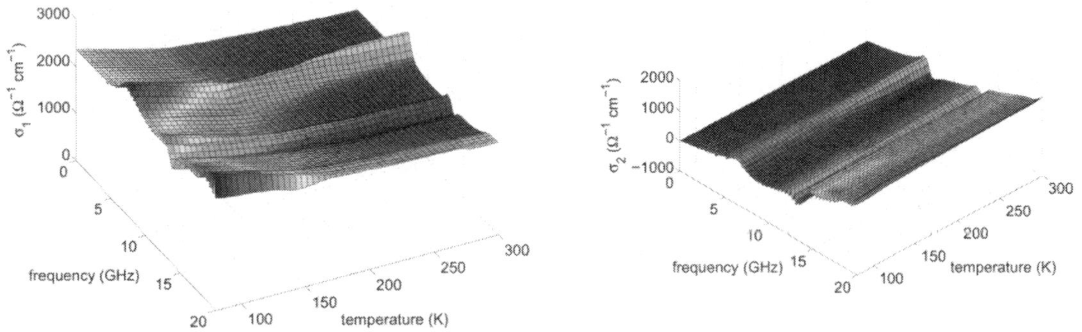

Figure 3. Real and imaginary parts of the conductivity of $SrRuO_3$ on MgO.

The situation is quite different for the $SrRuO_3$ film on MgO, as evident from the frequency and temperature dependence of σ_1 in Fig. 3. The surface of this plot is smooth up to 20 GHz, and the two substrate resonances in this range remain at the same frequency for all temperatures. For this sample, we measure intrinsic properties of the $SrRuO_3$ film. In Fig. 4 we show the temperature dependence of the real part ρ_1 of the resistivity ($\rho = 1/\sigma$) for different frequencies. The kink around 150 K indicates the ferromagnetic transition in the thin film and is known from the dc resistivity of $SrRuO_3$ [6, 7]. With increasing frequency, the kink slightly shifts toward lower temperatures. This is reminiscent of the behavior of a ferromagnetic antiresonance (FMAR) [20] in the sense that at higher probing frequencies the observed onset of ferromagnetism moves to lower temperatures. But contrary to FMAR, here we are not sensitive to the magnetic permeability of the sample and only sense the conductivity or resistivity, respectively. This kink at the ferromagnetic transition can also be identified for the $SrRuO_3$ film on $SrTiO_3$, see Fig. 2 at low frequencies, but the strong temperature dependence of the substrate excludes a detailed analysis. For $CaRuO_3$ no such kink is observed, consistent with dc data and the absence of ferromagnetic ordering [6, 7].

While we can easily determine the microwave properties of the $SrRuO_3$ film on MgO, this choice of substrate is disadvantageous compared to $SrTiO_3$ from a film growth point of view:

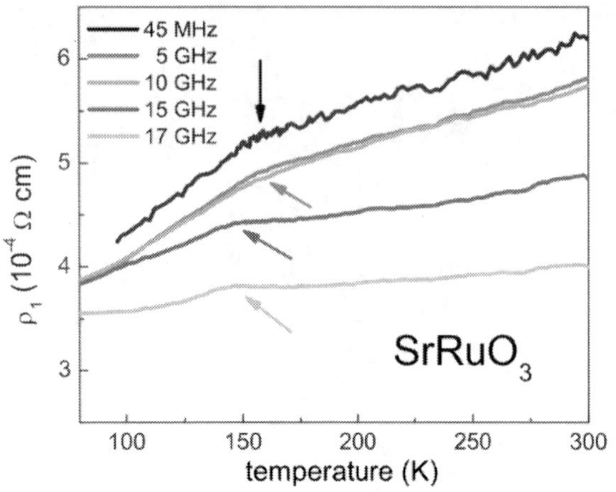

Figure 4. Temperature dependence of the real part of the resistivity of $SrRuO_3$ (on MgO substrate) at different frequencies. Arrows indicate the kink which heralds the ferromagnetic transition.

the lattice mismatch to the $Sr_{1-x}Ca_xRuO_3$ system is larger, and as a result the film quality is reduced [13]. This is evident from our data for the case of $SrRuO_3$, where for MgO substrate compared to $SrTiO_3$ the room temperature conductivity is much lower (Fig. 1); also the resistivity ratio upon cooling is smaller. For $CaRuO_3$ the lattice mismatch to MgO is even too big for successful film growth.

4. Conclusions and Outlook

We have measured the frequency-dependent impedance of $SrRuO_3$ and $CaRuO_3$ thin films on $SrTiO_3$ substrate and a $SrRuO_3$ thin film on MgO substrate at microwave frequencies. For $SrRuO_3$ on MgO, we clearly observe the signature of the ferromagnetic transition throughout our frequency range. While thin films on MgO substrate are well suited for optical measurements at low frequencies, thin films on $SrTiO_3$ show strong substrate contributions, which hamper a proper analysis. Since the high lattice mismatch of MgO and $Sr_{1-x}Ca_xRuO_3$ limits the film quality, an alternative substrate with low and temperature-independent dielectric constant and lower lattice mismatch should be tried for future studies e.g. $NdGaO_3$ [21].

Acknowledgements

We acknowledge financial support by the DFG, including SFB 602.

References

[1] Vojta T 2000 *Ann. Phys.* **9** 403
[2] Löhneysen H v, Rosch A, Vojta M, Wölfle P 2007 *Rev. Mod. Phys.* **79** 1015
[3] Gegenwart P, Si Q, Steglich F 2008 *Nature Physics* **4** 186
[4] Vojta T, Belitz D, Kirkpatrick T, Narayanan R 1999 *Ann. Phys.* **8** 593
[5] Yoshimura K, Imai T, Kiyama T, Thurber K R, Hunt A W, Kosuge K 1999 *Phys. Rev. Lett.* **83** 4397
[6] Khalifah P, Ohkubo I, Christen H M, Mandrus D G 2004 *Phys. Rev. B* **70** 134426
[7] Schneider M, Moshnyaga V, Gegenwart P, 2010 *Phys. Status Solidi B* **247** 577
[8] Kostic P, Okada Y, Collins N C, Schlesinger Z, Reiner J W, Klein L, Kapitulnik A, Geballe T H, Beasley M R 1998 *Phys. Rev. Lett.* **81** 2498
[9] Dodge J S, Weber C P, Corson J, Orenstein J, Schlesinger Z, Reiner J W, Beasley M R 2000 *Phys. Rev. Lett.* **85** 4932
[10] Lee Y S, Yu J, Lee J S, Noh T W, Gimm T H, Choi H Y, Eom C B 2002 *Phys. Rev. B* **66** 041104
[11] Scheffler M, Dressel M, Jourdan M, Adrian H 2005 *Nature* **438** 1135
[12] Moshnyaga V, Khoroshun I, Sidorenko A, Petrenko P, Weidinger A, Zeitler M, Rauschenbach B, Tidecks R, Samwer K 1999 *Appl. Phys. Lett.* **74** 2842
[13] Schneider M, Moshnyaga V, Gegenwart P, 2010 *J. Phys.: Conf. Ser.* **200** 012178
[14] Scheffler M, Dressel M 2005 *Rev. Sci. Instrum.* **76** 074702
[15] Booth J C, Wu D H, Anlage S M 1994 *Rev. Sci. Instrum.* **65** 2082
[16] Dressel M and Grüner G 2002 *Electrodynamics of Solids* (Cambridge: Cambridge University Press)
[17] Kitano H,Ohashi T, Maeda A 2008 *Rev. Sci. Instrum.* **79** 074701
[18] Weaver H E 1959 *J. Phys. Chem. Solids* **11** 274
[19] Krupka J, Geyer R G, Kuhn M, Hinken J H 1994 *IEEE Trans. Microwave Theory Tech.* **42** 1886
[20] Schwartz A, Scheffler M, Anlage S A 2000 *Phys. Rev. B* **61** R870
[21] Ricci F, Bevilacqua M F, Granozio F M, di Uccio U S 2002 *Phys. Rev. B* **65** 155428

International Conference on Strongly Correlated Electron Systems (SCES 2011) IOP Publishing
Journal of Physics: Conference Series **391** (2012) 012092 doi:10.1088/1742-6596/391/1/012092

Structural and Magnetic Properties in Spinel Type $Fe_{1-x}Zn_xV_2O_4$

Hiroki Ishibashi and Yuki Kitadai

Department of Physical Science, Graduate School of Science, Osaka Prefecture University, Sakai, Osaka 599-8531, Japan

E-mail: hiroki@p.s.osakafu-u.ac.jp

Abstract. Structural and magnetic properties in spinel type $Fe_{1-x}Zn_xV_2O_4$ have been investigated for $x \leq 0.4$. Successive structural transitions of cubic-to-tetragonal ($c < a$), tetragonal-to-orthorhombic and orthorhombic-to-tetragonal ($c > a$) are observed for $x \leq 0.2$. On the other hand, the only cubic-to-tetragonal ($c > a$) structural transition is observed for $0.3 \leq x \leq 0.4$. The result of the magnetic measurements indicates that the ferrimagnetic transition is observed for $x \leq 0.4$ and the transition temperature is close to the tetragonal-to-orthorhombic transition temperature for $x \leq 0.2$, and the cubic-to-tetragonal transition one for $0.3 \leq x \leq 0.4$. These results are discussed in terms of Jahn-Teller distortions of FeO_4 tetrahedron and VO_6 octahedron, and spin-orbital interaction at Fe^{2+} ion.

1. Introduction

The orbital degrees of freedom in 3d transition metal compounds play important roles in appearance of their various physical properties. In particular, spinel vanadium oxides AV_2O_4 (A = Mg^{2+}, Zn^{2+}, Mn^{2+}, etc.), in which V^{3+} ion at the octahedral site (B-site) has two 3d electrons in the triply degenerate t_{2g} states, have received considerable attention in the last decade because of their peculiar orbital and magnetic ordering [1-4]. Among such spinel vanadium oxides, FeV_2O_4 is a unique compound in which both Fe^{2+} and V^{3+} ions have orbital degrees of freedom; Fe^{2+} ion at the tetrahedral site (A-site) has three 3d electrons in the doubly degenerate e states. Recently, structural and magnetic properties of the spinel FeV_2O_4 were reported [4]. This compound exhibits successive structural transitions from cubic to compressed tetragonal with the lattice constants of $c < a$ at $T_{s1} \sim 140$ K, from tetragonal to orthorhombic accompanied by a ferrimagnetic transition at $T_{s2} \sim 110$ K, and from orthorhombic to elongated tetragonal with $c > a$ at $T_{s3} \sim 60$ K with decreasing temperature for polycrystalline samples. The structural transition at T_{s1} is due to the cooperative Jahn-Teller distortion of FeO_4 tetrahedron, which is caused by the orbital degree of freedom of the e orbital electrons in Fe^{2+} ion. For the structural transitions below T_{s2}, it was suggested that the possible scenario is the competition and cooperation of Fe^{2+} and V^{3+} orbitals, however, the origin is not completely understood. In this study, we investigate the effect of substitution of Zn^{2+} for Fe^{2+} at A-site in FeV_2O_4, i.e., the structural and magnetic properties of $Fe_{1-x}Zn_xV_2O_4$. Because Zn^{2+} has neither orbital degree of freedom nor spin, the substitution of Zn^{2+} for Fe^{2+} induces voids in terms of orbital degree of freedom and spin at A-site. This substitution may reduce the effect of the cooperative Jahn-Teller distortion of FeO_4 tetrahedron, and it is therefore expected to clarify roles of the e orbital electrons in Fe^{2+} at A-site and t_{2g} ones in V^{3+} at B-site for the successive structural transition in FeV_2O_4.

2. Experimental

Polycrystalline samples of $Fe_{1-x}Zn_xV_2O_4$ with $x = 0.1$, 0.2, 0.3 and 0.4 were prepared by a solid state reaction. The stoichiometric amounts of high purity powders of Fe, Fe_2O_3, ZnO and V_2O_3 were mixed and pressed into pellets and heated at the temperatures between 1273 K and 1373 K in sealed quartz tubes. The samples were identified as single phases with a cubic spinel structure at room temperature by conventional powder diffraction measurement. Temperature dependences of magnetization were measured by SQUID magnetometer. High-resolution X-ray powder diffraction experiments using synchrotron radiation were performed at a beamline of BL-6C at the Photon Factory, KEK in Japan. The wavelength of incident beam used was 1.198 Å. A flat Si(111) crystal analyzer was used in order to obtain the high angular resolution. The temperature dependences of the peak profiles of 111, 400 and 440 reflections of the cubic spinel structure were measured.

3. Results and discussion

Figure 1(a) shows the temperature dependence of peak profiles of 400 reflection corresponding to the cubic spinel structure for $x = 0.1$. The 400 peak of the cubic structure is split into two peaks at ~110 K and three peaks at ~90 K, and then these are merged into two peaks again at ~60 K with decreasing temperature. This result indicates that the successive structural transitions from cubic to tetragonal

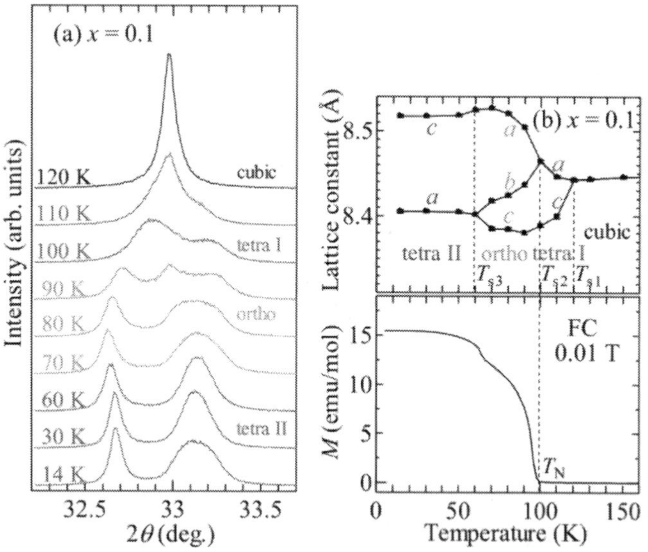

Figure 1. Temperature dependence of (a) X-ray diffraction pattern of 400 reflection, (b) lattice constants and magnetization with an applied field of 0.01 T for $x = 0.1$. tetra I, ortho and tetra II represent tetragonal I, orthorhombic and tetragonal II phases, respectively.

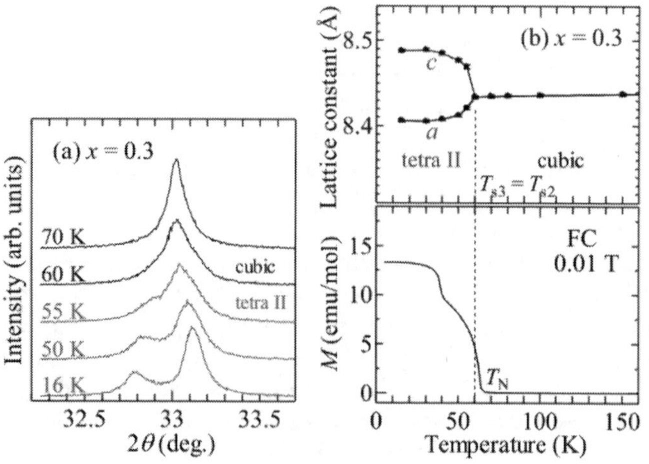

Figure 2. Temperature dependence of (a) X-ray diffraction pattern of 400 reflection, (b) lattice constants and magnetization with an applied field of 0.01 T for $x = 0.3$. tetra II represents the tetragonal II phase.

($c < a$), tetragonal ($c < a$) to orthorhombic and from orthorhombic to tetragonal ($c > a$) occur, which are similar to those observed in FeV_2O_4 though the transition temperatures are different. Hereafter, we call the compressed tetragonal ($c < a$) and the elongated tetragonal ($c > a$) phases to 'tetragonal I' and 'tetragonal II', respectively. The temperature dependence of the lattice constants estimated from the peak position of the Bragg reflections and the magnetization for $x = 0.1$ are shown in figure 1(b). The structural transition temperatures from cubic to tetragonal I, from tetragonal I to orthorhombic and from orthorhombic to tetragonal II for $x = 0.1$ are $T_{s1} \sim 120$ K, $T_{s2} \sim 100$ K and $T_{s3} \sim 70$ K, respectively. The diffraction pattern below ~ 30 K indicates peak broadening as shown in figure 1(a). This is probably due to the tetragonal-to-orthorhombic transition observed in the single crystal of FeV_2O_4 [4], however, we regard this phase as tetragonal in here because of the difficulties of the peak decomposition. The magnetization indicates the sharp increase corresponding to the ferrimagnetic transition at $T_N \sim 100$ K, which is close to T_{s2}.

Figure 2 shows the temperature dependence of peak profiles of 400 reflection, lattice constants and magnetization for $x = 0.3$. A single structural transition from cubic to tetragonal II is observed at $T_{s3} = T_{s2} \sim 60$ K, and both tetragonal I and orthorhombic phases are absent. The magnetization indicates the ferrimagnetic transition at $T_N \sim 55$ K.

The phase diagram of $Fe_{1-x}Zn_xV_2O_4$ for $x \leq 0.4$ is summarized in figure 3. The cubic-to-tetragonal I structural transition temperature T_{s1} decreases with increasing x for $x \leq 0.2$ and the tetragonal I phase disappears for $x \geq 0.3$. The ferrimagnetic transition temperature T_N also decreases with decreasing temperature for $x \leq 0.4$. Furthermore, it is found that T_N is close to the tetragonal I-to-orthorhombic structural transition temperature T_{s2} for $x \leq 0.2$ and the cubic-to-tetragonal II one T_{s3} (= T_{s2}) for $0.3 \leq x \leq 0.4$, which suggests a strong correlation between structure and magnetism.

Here we discuss the origin of the phase diagram on the assumption that the lattice distortion is corresponding to the distortion of FeO_4 tetrahedron and VO_6 octahedron. In FeV_2O_4, the tetragonal I phase is caused by the cooperative Jahn-Teller distortion of FeO_4 octahedron, which is compressed along c-axis because two of three e orbital electrons in Fe^{2+} ion occupy $3z^2 - r^2$ orbital with lower energy [4, 5]. The substitution of Zn^{2+} for Fe^{2+} which induces voids in terms of orbital degree of freedom may reduce the cooperative Jahn-Teller distortion, and consequently T_{s1} decreases with decreasing temperature and the tetragonal I phase is supressed for $x \geq 0.3$.

The structural transition accompanied by the ferrimagnetic transition similar to that observed in the present study was reported in the spinel-type $Fe_{1-x}Mn_xCr_2O_4$ system, in which the e orbital electron in Fe^{2+} ion at A-site is dominant to their phase transitions [5]. It was proposed that the magnetic ordering stabilizes the orbital with the shape of $x^2 - y^2$ type by the second term of spin-orbit coupling. For example, when the magnetization direction is parallel to x axis, $y^2 - z^2$ orbital becomes stable, in consequence FeO_4 octahedron elongates along a-axis. In the present case, the lattice constant a and the

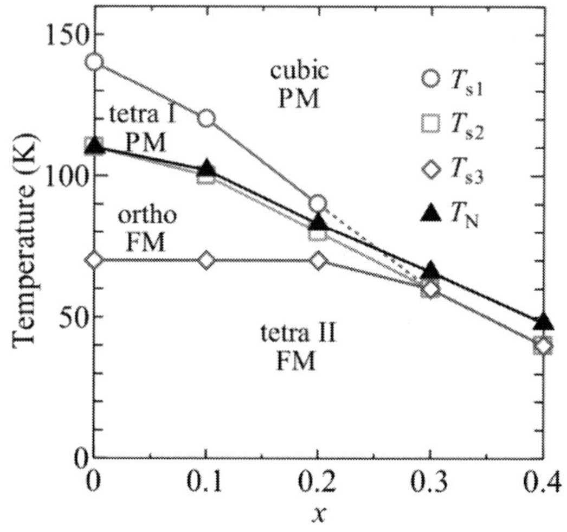

Figure 3. Phase diagram of $Fe_{1-x}Zn_xV_2O_4$ system for $0 \leq x \leq 0.4$. PM and FM represent paramagnetic and ferrimagnetic phases, respectively. Note that $T_{s2} = T_{s3}$ for $x \geq 0.3$ because the orthorhombic phase is absent.

magnetization simultaneously increase with decreasing temperature for the orthorhombic phase as shown in figure 1(b). Therefore it is suggested that the magnetization direction is parallel to *a*-axis, which is consistent with the results of X-ray diffraction and stain experiments using single crystals of FeV_2O_4 in an applied magnetic field [4]. Just below T_{s2}, $3z^2 - r^2$ electron is still dominant and FeO_4 tetrahedron is compressed along *c*-axis, however, increase in the magnetization with decreasing temperature makes $y^2 - z^2$ electron stabilized by spin-orbit coupling, which gives rise to the elongation of FeO_4 tetrahedron along *a*-axis. From the above consideration, it is suggested that the orthorhombic distortion is due to the competition of the $3z^2 - r^2$ and $y^2 - z^2$ orbital electrons at A-site. On the other hand, the reason why the orthorhombic phase is absent for $0.3 \leq x \leq 0.4$ is as follows: the long-range ordering of $3z^2 - r^2$ electron is suppressed, and thus the ferrimagnetic transition stabilizes only $x^2 - y^2$ orbital, inducing the tetragonal II phase.

The structure of the ground state of the present system is different from that of $Fe_{1-x}Mn_xCr_2O_4$ for small x region, i.e., $Fe_{1-x}Mn_xCr_2O_4$ remains orthorhombic [5] though $Fe_{1-x}Zn_xV_2O_4$ becomes elongated tetragonal. It is noted that Cr^{3+} ion at B-site in $Fe_{1-x}Mn_xCr_2O_4$ does not have orbital degree of freedom. This suggests that the orbital state of V^{3+} at octahedral site plays important role to stabilize the tetragonal II phase appearing in $Fe_{1-x}Zn_xV_2O_4$. If the VO_6 octahedron is elongated along *c*-axis, energy level of t_{2g} orbital splits into two; doubly degenerate *zx* and *yz* orbitals with lower energy and *xy* orbital with higher energy. V^{3+} ion in $Fe_{1-x}Zn_xV_2O_4$ has two t_{2g} electrons, and thus elongated VO_6 octahedron becomes stable by occupying the electrons in the *zx* and *yz* orbitals. However, the structural analysis indicated that VO_6 is compressed for the elongated tetragonal phase [4], which is inconsistent with the present results. The further studies are necessary to understand the role of t_{2g} electron V^{3+} for the tetragonal II phase.

4. Summary

We have investigated the structural and magnetic properties of $Fe_{1-x}Zn_xV_2O_4$ system. Successive structural transitions of cubic-to-tetragonal ($c < a$), tetragonal-to-orthorhombic accompanied by ferrimagnetic transition and orthorhombic-to-tetragonal ($c > a$) are observed for $x \leq 0.2$. On the other hand, the only cubic-to-tetragonal ($c > a$) structural transition accompanied by ferrimagnetic transition is observed for $0.3 \leq x \leq 0.4$. The ferrimagnetic transition induces the $x^2 - y^2$ type orbital in Fe^{2+} ion at A-site due to the spin-orbit coupling, causing the orthorhombic phase for $x \leq 0.2$ and tetragonal ($c > a$) one for $0.3 \leq x \leq 0.4$. Further, t_{2g} electrons in V^{3+} ion at B-site may also play an important role to stabilize the tetragonal ($c > a$) phase.

Acknowledgments

This work has been performed under the approval of the Photon Factory Program Advisory Committee (Proposal No. 2008G119).

References

[1] Wheeler M, Lake B, Nazmul Islam A T M, Reehuis M, Steffens P, Guidi T, and Hill A H 2010 *Phys. Rev.* B **82** 140406

[2] Lee S H, Louca D, Ueda H, Park S, Sato T J, Isobe M, Ueda Y, Rosenkranz S, Zschack P, Íñiguez J, Qiu Y and Osborn R 2004 *Phys. Rev. Lett.* **93** 156407

[3] Suzuki T, Katsumasa M, Taniguchi K, Arima T and Katsufuji T 2007 *Phys. Rev. Lett.* **98** 127203

[4] Katsufuji T, Suzuki T, Takei H, Shingu M, Kato K, Osaka K, Takata M, Sagayama H and Arima T 2008 *J. Phys. Soc. Jpn.* **77** 053708

[5] Ohtani S, Watanabe Y, Saito M, Abe N, Taniguchi K, Sagayma H, Arima T, Watanabe M and Noda Y 2010 *J. Phys. Condens. Matter* **22** 176003

International Conference on Strongly Correlated Electron Systems (SCES 2011) IOP Publishing
Journal of Physics: Conference Series **391** (2012) 012093 doi:10.1088/1742-6596/391/1/012093

Substitution effects in the Iron-Silicide Superconductor $Lu_2Fe_3Si_5$

J. Koshio[a], H. Okuyama, K. Takase, Y. Takano, and T. Watanabe[b]

Department of Physics, College of Science and Technology (CST), Nihon University, Chiyoda-ku Tokyo 101-8308, Japan

[a]eternity.ck.2020@gmail.com
[b]tadataka@phys.cst.nihon-u.ac.jp

Abstract. We studied effects of Fe-site substitutions on superconductivity in the iron silicide $Lu_2Fe_3Si_5$ by investigating electric and magnetic properties of the mixed crystal $Lu_2(Fe_{1-x}M_x)_3Si_5$ (M = Mn, Co and Ru). The Mn and Co substitutions induce Curie tails in temperature dependence of magnetic susceptibilities indicating the enhancement of the localized character of 3*d* electrons. The Ru substitution reveals the disorder-sensitive superconductivity, suggesting the sign-reversed superconducting gap function in $Lu_2Fe_3Si_5$.

1. Introduction

Ternary iron silicide $Lu_2Fe_3Si_5$ is a superconductor which exhibits relatively high T_c = 6.1 K despite the inclusion of Fe as an element [1]. This compound crystallizes in the $Sc_2Fe_3Si_5$-type tetragonal structure (space group, *P4/mnc*) consisting of a quasi-one-dimensional iron chain along the *c* axis and quasi-two-dimensional iron squares parallel to the basal plane [2]. Since the isoelectronic $Lu_2Ru_3Si_5$ and $Lu_2Os_3Si_5$ do not exhibit superconductivity, Fe 3*d* electrons in $Lu_2Fe_3Si_5$ should play significant role for the appearance of superconductivity. And it is expected that the electronic correlations of the 3*d* electrons make the pairing mechanism of superconductivity exotic in $Lu_2Fe_3Si_5$. Recent specific heat [3], penetration depth [4], and thermal conductivity measurements [5] provided evidences for the multigap superconductivity with fully opened gaps on the whole Fermi surfaces. Furthermore, it was recently reported that the small amounts of the Lu-site substitution by the nonmagnetic elements Y and Sc rapidly depress T_c in accordance with the increase in the residual resistivity [6,7]. Such a disorder-sensitive superconductivity strongly suggests the sign reversal of the superconducting gap function. Very recently, the detailed analysis of the crystal structure revealed the Fe-Si covalent-bonding network with the fully ionized Lu-site [8]. Thus it is expected that the Fe-site substitution affects on the superconductivity and the electronic structure differently from the Lu-site substitution.

We investigated structural, magnetic, and electric properties of $Lu_2(Fe_{1-x}Mn_x)_3Si_5$, $Lu_2(Fe_{1-x}Co_x)_3Si_5$, and $Lu_2(Fe_{1-x}Ru_x)_3Si_5$ to study the site dependence of the substitution effects in $Lu_2Fe_3Si_5$.

2. Experimental

Polycrystalline samples of $Lu_2(Fe_{1-x}M_x)_3Si_5$ (M = Mn, Co and Ru) were prepared from stoichiometric amounts of high-purity elements (Lu ingot = 99.9%, Fe powder = 99.99%, Si powder = 99.99%, Mn powder = 99.9%, Co powder = 99.9% and Ru powder = 99.9%) by arc melting under Ar atmosphere. The as-grown samples were wrapped in tantalum foil, vacuum-sealed in a quartz tube, and then annealed at 1050℃ for 10 days. The crystal structure of the samples was analyzed by powder X-ray diffraction (XRD) measurements at room temperature using Cu Kα radiation. DC magnetic susceptibilities and electrical resistivities were measured by using the Quantum Design PPMS.

3. Results and Discussion

According to the measured XRD patterns, $Lu_2(Fe_{1-x}Co_x)_3Si_5$ crystallizes in the $Sc_2Fe_3Si_5$-type structure in the Co concentration range of $x < 0.3$, while $Lu_2(Fe_{1-x}Mn_x)_3Si_5$ and $Lu_2(Fe_{1-x}Ru_x)_3Si_5$ belong to the $Sc_2Fe_3Si_5$-type structure in the whole concentration range of $0 \leqq x \leqq 1$. Figure 1 shows the unti-cell volumes of $Lu_2(Fe_{1-x}M_x)_3Si_5$, (M = Mn, Co, and Ru) as functions of the impurity concentration x. The unti-cell volumes of $Lu_2(Fe_{1-x}Mn_x)_3Si_5$ and $Lu_2(Fe_{1-x}Ru_x)_3Si_5$ increase with x, while that of $Lu_2(Fe_{1-x}Co_x)_3Si_5$ decreases with x.

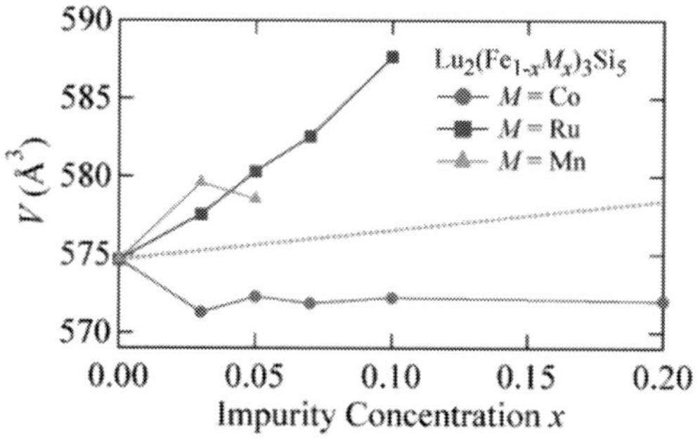

Figure 1. The unti-cell volumes of $Lu_2(Fe_{1-x}M_x)_3Si_5$ (M = Mn, Co, and Ru) as functions of impurity concentration x.

Figures 2 (a) and (b) depict temperature dependence of the magnetic susceptibilities in $Lu_2(Fe_{1-x}M_x)_3Si_5$ (M = Mn, Co, and Ru). The magnetic susceptibilities of $(Lu_{1-x}Dy_x)_2Fe_3Si_5$ are also shown in Fig. 2 (a) for comparison [6]. The Curie tails are induced at low temperatures in $Lu_2(Fe_{1-x}Mn_x)_3Si_5$ and $Lu_2(Fe_{1-x}Co_x)_3Si_5$ by the small-amount substitution of Mn and Co for Fe but are much weaker than $(Lu_{1-x}Dy_x)_2Fe_3Si_5$, which seems to be consistent with the crystal structure composed of the Fe-Si covalent-bonding network with the fully ionized Lu-sites [8]. Nevertheless, as shown in the inset to Fig. 2 (a), $Lu_2Mn_3Si_5$ exhibits a ferrimagnetic transition at 71K [9]. And the Curie tail at low temperatures in $Lu_2(Fe_{1-x}Co_x)_3Si_5$ is enhanced with increasing x as shown in Fig. 2 (b). These results indicate the enhancement of the localized character of 3d electrons by replacing Fe by Mn and Co, and imply that the tuning of the population of 3d electrons in $Lu_2Fe_3Si_5$ causes the competition between itinerant and localized characters of 3d electrons. As shown in Fig. 2, $Lu_2(Fe_{1-x}Ru_x)_3Si_5$ exhibits almost non-magnetic behavior similar to $Lu_2Fe_3Si_5$.

Figure 3 shows temperature dependence of the normalized electrical resistivities $\rho(T)/\rho(300\ K)$ of $Lu_2(Fe_{1-x}M_x)_3Si_5$ (M = Mn, Co, and Ru). It is evident that the small amounts of the Fe-site substitutions rapidly depress T_c with the systematic increase in the residual resistivity. Figure 4 plots the T_c depression in the non-magnetic $Lu_2(Fe_{1-x}Ru_x)_3Si_5$ as a function of the normalized residual resistivity $\rho(T_c)/\rho(300\ K)$. In Fig. 4, the T_c depressions in $(Lu_{1-x}R_x)_2Fe_3Si_5$ (R = Sc and Y) are also shown for comparison [6]. The rapid T_c depression of $Lu_2Fe_3Si_5$ by non-magnetic impurities is, regardless of the doped site, universally scaled with the residual resistivity. Thus the disorder-sensitive superconductivity revealed in the present study strongly suggests the sign-reversed superconducting gap function in $Lu_2Fe_3Si_5$.

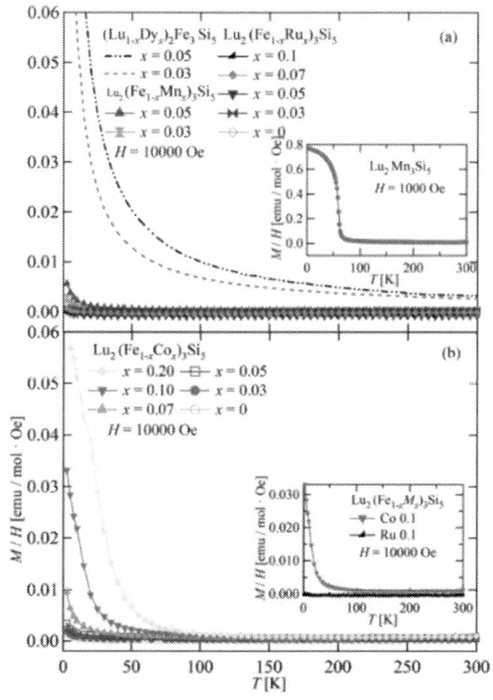

Figure 2. Temperature dependence of the magnetic susceptibilities in $Lu_2(Fe_{1-x}M_x)_3Si_5$ (M = Mn, Co and Ru), and $(Lu_{1-x}Dy_x)_2Fe_3Si_5$ [6]. The inset to (a) shows the magnetic susceptibility of $Lu_2Mn_3Si_5$. The inset to (b) compares the magnetic susceptibilities between $Lu_2(Fe_{1-x}Co_x)_3Si_5$ and $Lu_2(Fe_{1-x}Ru_x)_3Si_5$ at x = 0.1.

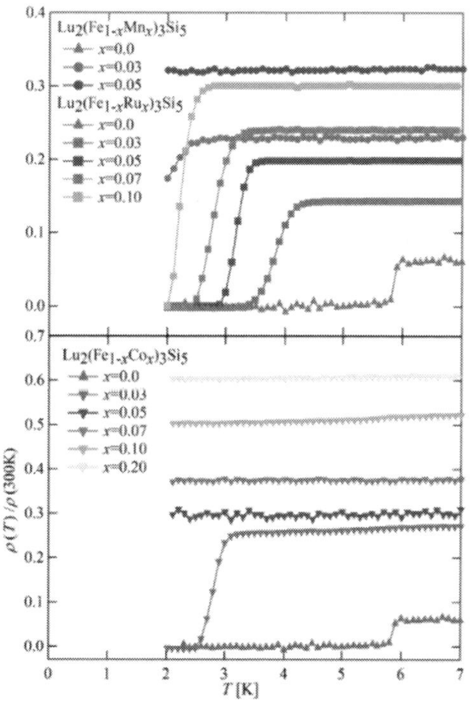

Figure 3. Temperature dependence of the normalized electrical resistivities in $Lu_2(Fe_{1-x}M_x)_3Si_5$ (M = Mn, Co and Ru).

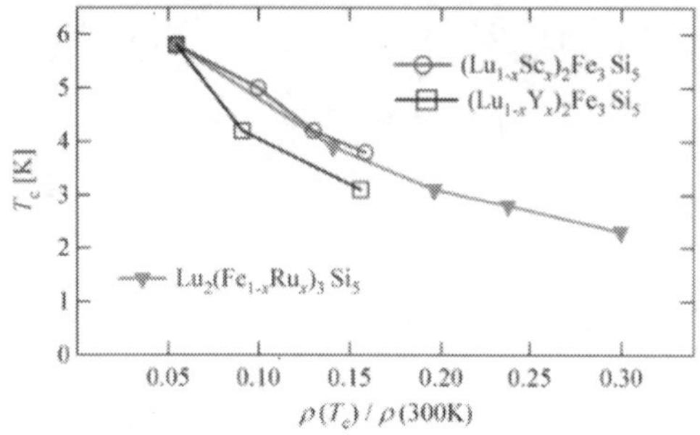

Figure 4. The T_c depression in $Lu_2(Fe_{1-x}Ru_x)_3Si_5$ as a function of the normalized residual resistivity $\rho(T_c)/\rho(300K)$. The T_c depressions in $(Lu_{1-x}R_x)_2Fe_3Si_5$ (R = Sc and Y) are also shown for comparison [6].

4. Acknowledgement

This work was partly supported by a Grant-in-Aid for Young Scientists (B) (21740266) from MEXT of Japan.

References

[1] Braun H F 1980 Phys. Lett. A **75**, 386.
[2] H. F. Braun, in Ternary Superconductors, edited by G. K. Shenoy, B. D. Dunlap, and F. Y. Fradin (North-Holland, Amsterdam, 1981), p. 255.
[3] Nakajima Y, Nakagawa T, Tamegai T, and Harima H 2008 Phys. Rev. Lett. **100**, 157001.
[4] Gordon R T, Vannette M T, Martin C, Nakajima Y, Tamegai T, and Prozorov R 2008 Phys. Rev. B **78**, 024514.
[5] Machida Y, Sakai S, Izawa K, Okuyama H, and Watanabe T 2011 Phys. Rev. Lett. **106**, 107002.
[6] Watanabe T, Sasame H, Okuyama H, Takase K, and Takano Y 2009 Phys. Rev. B **80**, 100502(R).
[7] Hidaka H, Nakajima Y, and Tamegai T, 2009 Physica C **469**, 999.
[8] Okuyama H, Takase H, Takano Y, Watanabe T, Yoshida F, Moriyoshi C, and Kuroiwa Y *unpublished.*
[9] Mèot-Meyer M, Venturini G, Malaman B, Rae E M, and Roques B 1985 Mat. Res. Bull. **20**, 1009.

Field reference technique for ac magnetic susceptibility measurements under pressure

N. Kabeya, K. Deguchi and N. K. Sato

Department of Physics, Graduate School of Science, Nagoya University, Nagoya 464-8602, Japan

E-mail: kabeya.noriyuki@d.mbox.nagoya-u.ac.jp

Abstract. In the present study, we propose a new simple technique based on Faraday low of induction for the ac magnetic susceptibility measurement under high pressure. Although the technique needs a simple additional change from the conventional method, it performs certain separation a real (′) from imaginary (″) variable of magnetic susceptibility and free from the change of the ac eld amplitude. A typically result obtained from this technique and the shielding effect of ac magnetic eld by a pressure cell are shown.

1. Introduction

An ac magnetic susceptibility measurement has been employed to investigate magnetic and superconducting properties of heavy fermion systems for a wide range of temperature and magnetic field [1, 2, 3, 4]. The conventional technique of a high-pressure ac magnetic susceptibility experiment has a difficulty arising from a shielding effect of modulation field: An eddy current induced in a metallic pressure cell will modulate an amplitude and phase of the oscillation field at a sample site. This effect makes it difficult to unambiguously separate a real (χ') from imaginary (χ'') part of the magnetic susceptibility. This motivates us to develop a method for a more precise magnetic susceptibility measurement under high pressure.

2. Experimental method

An external pressure was applied using a CuBe-NiCrAl hybrid piston-cylinder. Daphne oil 7373 was used as a pressure transmitting medium. Pressure was determined using a superconducting transition of indium. In what follows, we give a detailed description of the measuring system of the ac magnetic susceptibility.

Figure 1a shows a schematic illustration of coil setting. A primary coil generates an ac magnetic field $B = B_0 e^{i\omega t}$. Here, B_0 and ω is an amplitude and frequency of the magnetic field, respectively. The primary coil is placed outside the pressure cell. A pair of secondary (signal detection) coils marked (iii) are put in the pressure cell with a sample. (One may notice that there is another coil marked (iv); this is described below.) These coils and the sample are immersed in the pressure transmitting medium. The field felt by the sample is modulated by an eddy current produced in the pressure cell and described as $B' = B|a|e^{i\theta}$. Here, $|a|$ and $e^{i\theta}$ is an amplitude ratio and phase of the modulated field, respectively. These experimental settings are the same as those of the conventional method. The difference lies in the secondary coil system; to monitor the modulated filed B' which the sample feels, we put an additional

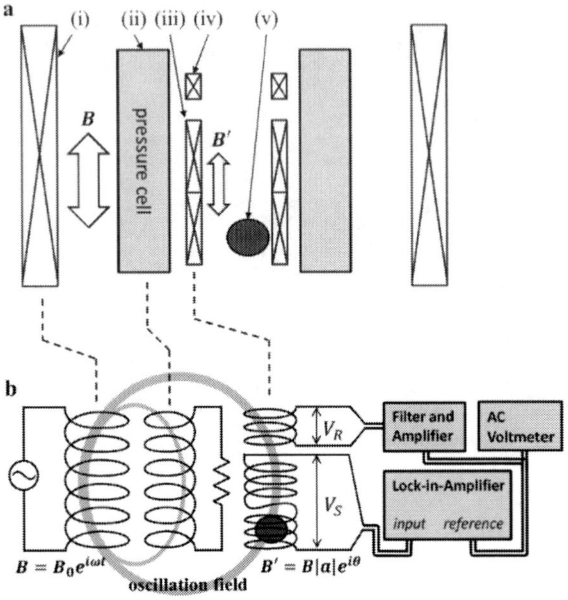

Figure 1. Schematic illustration of the ac measurement system presented. **a.** Schematic illustration of coil setting. (i): primary coil, (ii): pressure cell, (iii): field reference coil, (iv): secondary coil, (v): sample. B and B' represent the oscillated magnetic field outside and inside the pressure cell, respectively. **b.** Schematic illustration of the measurement system. Gray circles represent oscillated magnetic fields. The pressure cell is described as an effective circuit consisting of coil and resistance. $V_{\rm R}$ and $V_{\rm S}$ represent the voltage induced in the field reference coil and the secondary coil, respectively. The field reference coil is connected to the reference input of the lock-in-amplifier via a filter and amplifier.

pick-up coil, referred to as "field reference coil" hereafter, which is often used in a vibrating sample magnetometer (VSM) [5]. This coil is isolated from the secondary coil.

Figure 1b shows a schematic illustration of the measurement system. The reference coil generates an ac voltage $V_{\rm R}$ as output. The voltage is amplified and then input into a lock-in-amplifier as a reference signal. This signal correctly gives a phase of the filed felt by the sample. Therefore, X and Y outputs of the lock-in-amplifier with respect to $V_{\rm S}$ unambiguously yield χ' and $-\chi''$, respectively.

The secondary coil consists of a pair of coils as usual, and each (clockwise and counter-clockwise) coil has 600 turns. The reference coil has about 100 turns. For the frequency f, we obtained good performance between 10 Hz and 1.3 kHz. For $f < 10$ Hz, the voltage is too small to use as a reference signal. This is because the voltage induced by the field is in proportion to frequency. For $f > 2$ kHz, on the other hand, the capacitance component of the coil produces a spurious phase in the voltage $V_{\rm R}$. The results presented below were obtained using the amplitude and frequency of 1.0 Oe and 100 Hz, respectively.

3. Results

Figure 2 shows the temperature dependence of the ac magnetic susceptibility of indium and lead, both of which were put inside the coil together. In the real part of the susceptibility χ', a step like diamagnetic signal is observed at 3.4 K and 7.2 K. The former corresponds to the indium, and the latter to the lead. Although a small peak is observed at 3.4 K, the imaginary part χ'' is almost independent of temperature as expected. This clearly demonstrates that the present technique works properly.

We show the temperature dependence of the ratio B'/B in Fig. 3a. For a reference, we give the result for the case in which no pressure cell was used: The ratio of $B'/B = 1$ means that there is no shielding effect. For the case in which the samples were put into the pressure cell although they were not pressurized, the ratio was decreased down to about 0.95, which means that the

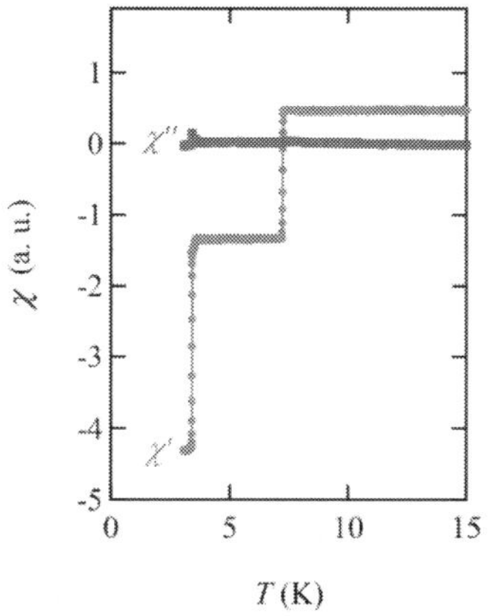

Figure 2. Temperature dependence of susceptibility of indium and lead.

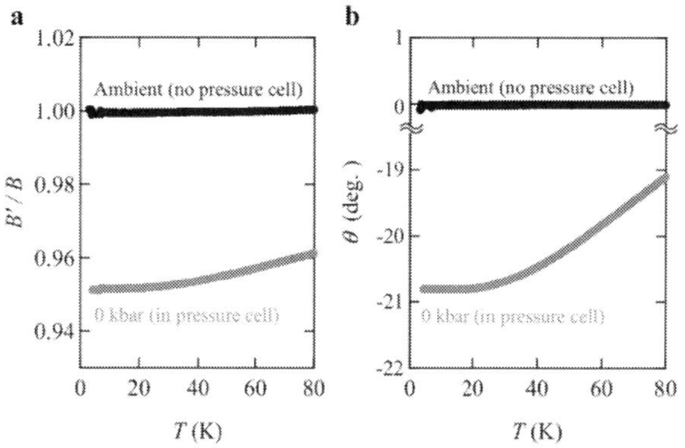

Figure 3. Shielding effect of the pressure cell. **a.** Temperature dependence of field penetration rate B'/B. **b.** Temperature dependence of phase shift between B' and B.

eddy current produced in the pressure cell shields 5 percent of the external magnetic field. This shielding effect decreases with increasing temperature, due to the increase in the electrical resistivity of the pressure cell with temperature. We note that similar phenomena occur in a vacuum can made of copper: more than 50 percent of the oscillation field was shielded (not shown here).

Figure 3b shows the phase difference between B' and B as a function of temperature. At ambient pressure with no pressure cell, we observe $= 0$, indicating that there is no phase shift between B' and B. For the case in which the sample is put in the pressure cell, a phase difference emerges, ~ -21 degree at low temperature.

4. Conclusion

We presented a new, simple technique for the ac magnetic susceptibility measurement under high pressure. This technique only needs a small modification; an additional pick-up coil to monitor the magnetic field felt by the sample is put into the pressure cell together with a pair of pick-up coils to probe the sample signal. Using this technique, we avoid the difficulty arising from the eddy current effect of the pressure cell, which enables us to safely separate the real part χ' from the imaginary part χ'' of the ac magnetic susceptibility.

Acknowledgments

This work was partially supported by Grant-in-Aid for Scientific Research (Nos. 20224015, No. 20102006 and No. 21008593). One of the author (N. K.) was supported by a Grant-in-Aid for JSPS Fellows.

References

[1] Huxley A, Sheikin I and Braithwaite D 2000 *Physica* B **284-288** 1277
[2] Petrova A E, Krasnorussky V, Sarrao J and Stishov S M 2006 *Phys. Rev.* B **73** 052409
[3] Slooten E, Naka T, Gasparini A, Huang Y K and de Visser A 2009 *Phys. Rev. Lett.* **103** 097003
[4] Kabeya N, Iijima R, Osaki E, Ban S, Imura K, Deguchi K, Aso N, Homma Y, Shiokawa Y and Sato N K 2009 *Physica* B **404** 3241
[5] Foner S 1956 *Rev. Sci. Instrum.* **27** 548

International Conference on Strongly Correlated Electron Systems (SCES 2011) IOP Publishing
Journal of Physics: Conference Series **391** (2012) 012095 doi:10.1088/1742-6596/391/1/012095

Structural Analysis of Spinel Compound CuV_2S_4 with Incommensurate Charge-Density Wave

Shogo Kawaguchi[1], Yoshiki Kubota[1], Naruki Tsuji[2], Jungeun Kim[2], Kenichi Kato[3], Masaki Takata[3] and Hiroki Ishibashi[1]

[1] Department of Physical Science, Graduate School of Science, Osaka Prefecture University, Sakai, Osaka 599-8531, Japan
[2] Japan Synchrotron Radiation Institute, SPring-8, 1-1-1 Kouto, Sayo-cho, Sayo-gun, Hyogo 679-5198, Japan
[3] RIKEN SPring-8 Center, 1-1-1 Koute, Sayo-cho, Sayo-gun, Hyogo 679-5148, Japan

E-mail: s_s.kawaguchi@p.s.osakafu-u.ac.jp

Abstract. Spinel Compound CuV_2S_4 exhibits incommensurate charge–density wave (CDW) transition at $T_t \sim 90$ K accompanied by a structural phase transition. We have performed synchrotron powder diffraction experiments in order to determine the crystal structure for low temperature phase. The CDW phase has an orthorhombic symmetry with the superspace group $Imm2(0\beta0)$. It was found that the crystal structure is characterized by the significant change of bond lengths between vanadium atoms along a–axis. We suggested that this change is related to the orbital ordering of t_{2g} orbitals, which is the origin of the formation of the CDW.

1. Introduction

Spinel compounds with chemical formula AB_2X_4 (X is O, S, or Se) have been studied for decades, because of their unique physical properties. The spinel structure consists of face–centered cubic lattice of X anions within which A and B cations occupy the tetrahedral and octahedral interstitial sites, respectively. Moreover, the B site cations have a corner sharing tetrahedral network known as a pyrochlore lattice with three dimensional geometrical frustration. When the B cations are mixed in valence, exciting physics accompanied by orbital ordering often arises. For example, $CuIr_2S_4$ ($Ir^{3.5+}$) exhibits a metal–insulator transition at 230 K accompanied by charge and orbital ordering simultaneously [1]. AlV_2O_4 ($V^{2.5+}$) shows the orbital ordering of vanadium t_{2g} orbitals with the formation of a molecular–like vanadium heptamer below 700 K [2]. Furthermore, CuV_2S_4 ($V^{3.5+}$) displays anomalies in magnetic susceptibility and electrical resistivity accompanied by a structural phase transition at $T_t \sim 90$ K [3]. It has been discussed that these anomalies are related to the formation of the incommensurate charge–density wave (CDW) with wave vector $\mathbf{q} = (1/4 - \delta)$ [110] below T_t. The CDW state of CuV_2S_4 with three dimensional lattice has attracted significant interest over recent decades [3–6] since the CDW formation is well known as general phenomena of low–dimensional systems. However, the origin of CDW formation in CuV_2S_4 is not understood at present. Because only the crystal system for CDW phase is reported [4] and the detailed crystal structure below T_t has not determined so far.

In this paper, we report the details of the crystal structure of CDW phase in CuV_2S_4, which were investigated by X–ray powder diffraction measurements using synchrotron radiation.

2. Experimental details

A polycrystalline sample was synthesized by a solid state reaction. The stoichiometric mixture of high purity Cu, V and S powders was heated in an evacuated quartz tube at 1123 K for a week. Subsequently, the sample was ground and pressed into pellets in a glove box filled with argon gas to avoid oxidation of the sample. The pellet was sintered in an evacuated silica tube at 1123 K for 5 days.

X–ray powder diffraction experiments using synchrotron radiation source were carried out on the large Debye–Scherrer camera installed at the beamline BL02B2 in SPring–8 by using an imaging plate as a detector [7]. The wavelength of an incident X–ray was 0.4999 Å. The diffraction data were collected in the range between 20 K and 150 K using a low temperature helium gas blower. The JANA 2006 program [8] was used for the structural refinements by the Rietveld method from the synchrotron powder diffraction data.

3. Results and Discussion

X–ray powder diffraction patterns at 70 K and 150 K are shown in figure 1 (a). At 150 K, CuV_2S_4 has a cubic normal spinel structure with a lattice parameter $a = 9.8203(1)$ Å, which is obtained by Rietveld refinement. Figure 1 (b) shows that the peak profile of 800 reflection, which splits into two peaks at 70 K. This peak splitting means the structural phase transition from cubic symmetry to tetragonal or orthorhombic one at T_t as previous report [4]. The lattice constants calculated from the peak positions of some fundamental reflections corresponding to the pseudo–cubic axes, were $a_c = b_c \sim 9.82$ Å and $c_c \sim 9.80$ Å, where it was assumed that the crystal symmetry is tetragonal since it is difficult to distinguish between tetragonal and orthorhombic symmetry from the peak splitting.

In addition, we observed weak superlattice reflections whose intensities are less than about 0.2 % of maximum intensity of the fundamental reflections. These reflections can be indexed with four integer *hklm* with respect to the reciprocal basis vectors **a***, **b*** and **c*** of the fundamental structure and modulation vector **q**, according to $\mathbf{H} = h\mathbf{a}^* + k\mathbf{b}^* + l\mathbf{c}^* + m\mathbf{q}$, where **H** is a reciprocal lattice vector.

In order to determine the superspace group, the peak positions of both fundamental and superlattice reflections observed are checked carefully. We should note that the tetragonal symmetry should be excluded because one dimensional modulation perpendicular to c–axis cannot possess a

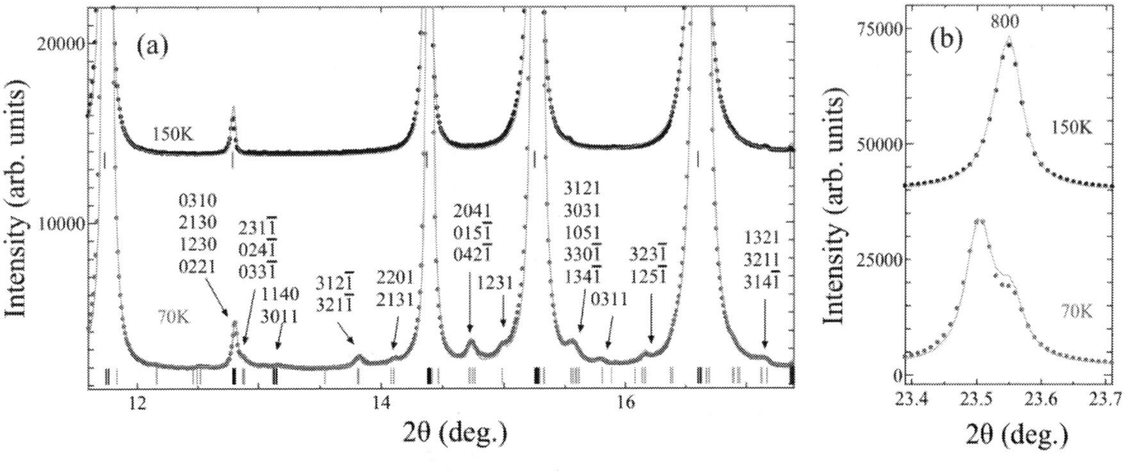

Figure 1. (a) Portion of the Rietveld fitting of the synchrotron powder diffraction patterns at 150 K and 70 K (arrows show superlattice reflections). The dots (black and red) and the solid line (green) exhibit observed pattern and calculated pattern, respectively. (b) X–ray scans in the vicinity of the 800 cubic Bragg peak at 150 K and 70 K.

Table 1. Structural parameters of average structure at 70 K from the Rietveld refinement of synchrotron powder diffraction data. The superspace group is *Imm*2 (0*β*0).

Structural parameters		Atom	Wyck.	x	y	z	$U_{iso}(Å^2)$
a (Å)	6.9477(1)	Cu1	2a	0	0	7/8	0.0071(2)
b (Å)	6.9416(1)	Cu2	2b	0	1/2	0.127(1)	0.0071(2)
c (Å)	9.8011(1)	V1	4c	0.250(4)	0	0.250(3)	0.0027(6)
q	0.7391(5)	V2	4d	0	0.252(3)	0.500(3)	0.0027(6)
R_{wp} (%)	4.52	S1	4c	0.267(3)	0	0.003(3)	0.0007(8)
R_B (%)	1.82	S2	4c	0.235(4)	0	0.494(3)	0.0007(8)
R_{wF} overall (%)	2.71	S3	4d	0	0.238(2)	0.254(3)	0.0007(8)
Main R_{wF} (%)	2.80	S4	4d	0	0.742(3)	0.742(3)	0.0007(8)
First–order R_{wF} (%)	2.66						

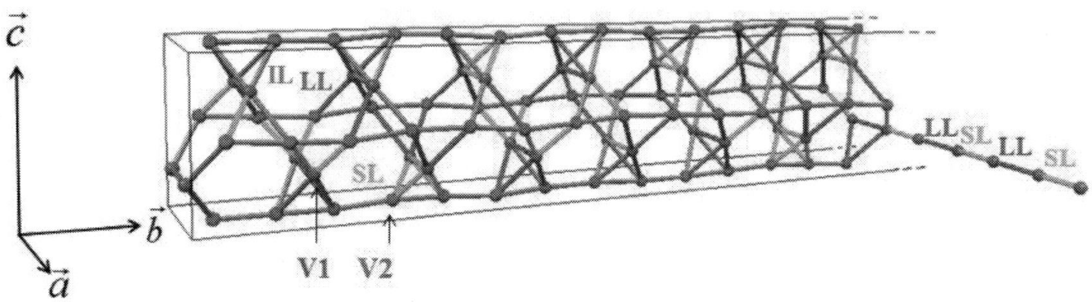

Figure 2. (a) Crystal structure of CuV_2S_4 at 70 K schematically. There are two inequivalent V atoms (V1, V2). SL (green), IL (grey) and LL (red) bonds are three different types of the V–V bond lengths.

four–fold rotation axis normal to c–axis. Hence, we examined the orthorhombic space groups among the group–subgroup relation of $Fd\bar{3}m \rightarrow I4_1/amd \rightarrow Imma \rightarrow Ima2$, $I2_12_12_1$ or $Imm2$ because the phase transition at T_t is the second order one [4]. Here, the orthorhombic space groups *Imma*, *Ima*2 and $I2_12_12_1$ cannot explain the peak positions of the superlattice reflections. As a result, we concluded that the structure for the low–temperature phase has a superspace group *Imm*2 (0*β*0) with a lattice parameter a = 6.94 Å, b = 6.94 Å, c = 9.80 Å and the modulation vector of $\mathbf{q} \sim 0.74\ \mathbf{b}^*$, where all superlattice reflections can be indexed by $m = \pm 1$.

As a next step, we have performed incommensurate structural analysis by Rietveld refinement on the basis of the initial structure model so as to determine the magnitude and the direction of the atomic displacements for each atom. In this analysis, the positional modulation was considered. The overall quality of the fitting was fairly good as shown in figure 1(a), and the details of structural parameters of the average structure and modulation vector are listed in Table 1.

The displacements of V atoms are much larger than those of Cu and S atoms: for instance, the amplitudes of $\mathbf{u}^{V1}$ and $\mathbf{u}^{V2}$, where $\mathbf{u}^{\mu}$ is the displacement from the atomic position of the μ–th site for average structure [9, 10], are 0.47 Å, whereas those of $\mathbf{u}^{Cu1}$ and $\mathbf{u}^{Cu2}$ are less than 0.16 Å. These results mean that the displacements of V atoms play an important role in the phase transition at T_t. Figure 2 shows the crystal structure including modulation along b–axis schematically, where the Cu and S atoms are not shown for clarity. There are two independent V sites. The blue and red atoms are V1 and V2 site, respectively. Note that the V–V bond lengths at 70 K range from 3.0 Å to 4.0 Å, though those at 150 K are 3.49 Å for all V–V bonds. Here, we define three kinds of the V–V bonds in terms of bond lengths at 70 K; Shorter Lengths (SL): 3.0 – 3.3 Å, Longer Lengths (LL): 3.7 – 4.0 Å, and

Intermediate Length (IL): 3.4 − 3.6 Å. Most of the V1−V1 bonds along *a*−axis have SL and LL bonds alternately. On the other hand, all the V2−V2 bonds along *b*−axis have IL bonds.

We suggest that the feature of V1–V1 bonds in CuV_2S_4 is similar to that in AlV_2O_4 on the kagomé net planes in which the electron transfer is dominated by the *d*–*d* direct overlap of the σ bond [2, 11]. In CuV_2S_4, the V1–V1 bonds alternating SL and LL bonds can be formed by orbital ordering of d_{xy} orbitals in vanadium t_{2g} orbitals. In addition, the temperature dependence of magnetic susceptibility and electrical resistivity show abrupt decrease and increase, respectively at T_t [4–6]. These facts indicate that conduction electrons are likely to be localized. Thereby, the orbital ordering transition most likely occurs in V1–V1 bonds, and the V1–V1 bonds are formed by strong electronic bonding such as spin–Peierls transition in $CuIr_2S_4$ [1]. It is conceivable that the orbital ordering of d_{xy} orbitals is related to the formation of incommensurate CDW. However, to understand the origin of the CDW completely, it is necessary to consider not only V1–V1 bonds lengths but also V1–V2 bond lengths. The construction of the novel orbital ordering model is now in progress.

4. Conclusion

We have first determined the crystal structure for the low temperature phase of CuV_2S_4 by synchrotron powder diffraction experiments at 70 K. Considering the superlattice reflections, the crystal symmetry is orthorhombic with the wave vector $\mathbf{q}$ = 0.7391(5) $\mathbf{b}^*$. The difference in bond lengths between V1–V1 and V2–V2 bonds occurs in the structural phase transition at T_t. We suggest that orbital ordering of d_{xy} orbitals should occur in the V1 atoms and is expected to be related to the formation of incommensurate CDW.

Acknowledgements

This study was supported by the Ministry of Education, Culture, Sports, Science and Technology of Japan through a Grant-in Aid for Scientific Research. The synchrotron radiation experiments were performed at BL02B2 of SPring–8 with the approval of the Japan Synchrotron Radiation Research Institute (JASRI) (Proposal No. 2010B0084, 2011A0084). The authors thank Mr. M. Takemoto for his help in the synchrotron powder diffraction experiment.

References

[1] Radaelli P G, Horibe Y, Gutmann M J, Ishibashi H, Chen C H, Ibberson R M, Koyama Y, Hor Y S, Kiryukhin Y and Cheng S W 2002 *Nature* (London) **416** 155-158
[2] Horibe Y, Shigeru M, Kurushima K, Ishibashi H, Ikeda N, Kato K, Motome Y, Furukawa N, Mori S and Katsufuji T 2006 *Phys. Rev. Lett.* **96** 086406
[3] Fleming R M, Disalvo F J, Cava R J and Waszczak J V 1981 *Phys. Rev. B* **24** 2850
[4] Okada H, Koyama K and Watanabe K 2004 *J. Phys. Soc. Jpn.* **73** 3227
[5] Okada H, Koyama K, Nakatake M, Shimada K, Nakatame H, Taniguchi M and Watanabe K 2008 *J. Phys. Soc. Jpn.* **77** 114711
[6] Naik I, Yadav C S and Rastogi A K 2007 *Phys. Rev. B* **75** 115112
[7] Nishibori E, Takata M, KatoK, Sakata M, Kubota Y, Aoyagi S, Kuroiwa Y, Yamakata M and Ikeda N 2001 *Nucl. Instrum. Methods Phys. Res.* A **467** 1045
[8] Petricek V, Dusek M and Palatinus L *Jana2006*. The cryastallographic computing system, Institute of Praha, Czech Republic 2006.
[9] Petricek V and Dusek M 2004 *Z. Kristallgr.* **219** 692
[10] M. Dusek, Petricek V, Wunschel M, Dinnebier R E and van Smaalen S 2001 *J. Appl. Cryst.* **34** 398
[11] Croft M, Kiryukhin V, Horibe Y and Cheng S W 2007 *New J. Phys.* **9** 86

μSR study of A-site ordered perovskite manganite $LaBaMn_2O_6$

Yu Kawasaki, Takeshi Minami, Yutaka Kishimoto, Takashi Ohno, Kohki H. Satoh[A], Akihiro Koda[A,B], Ryosuke Kadono[A,B], Tomohiko Nakajima[C] and Yutaka Ueda[D]

Department of Quantum Materials Science, University of Tokushima, Tokushima 770-8506, Japan
[A]Graduate University for Advanced Studies (SOKENDAI), Tsukuba 305-0801, Japan
[B]Muon Science Laboratory, Institute of Materials and Structure Science, High Energy Accelerator Research Organization (KEK), Ibaraki 319-1106, Japan
[C]AMRI, National Institute of Advanced Industrial Science and Technology, Tsukuba 305-8565, Japan
[C]Institute for Solid State Physics, University of Tokyo, Kashiwa 277-8581, Japan

E-mail: yu@pm.tokushima-u.ac.jp

Abstract. We have investigated magnetic properties of the A-site ordered perovskite manganite $LaBaMn_2O_6$ by muon spin relaxation. In the ferromagnetic metal (FM) phase below $T_C \sim 330$ K, no muon spin oscillation is observed in the time spectra, where the muon spin relaxation significantly deviates from a single exponential form. These results may suggest a large spatial distribution of spin dynamics, *i.e.* the inhomogeneous magnetic nature in the FM phase of $LaBaMn_2O_6$, which has been also observed in the FM phase of A-site disordered perovskite manganites. It is, therefore, suggested that there exists the spatial distribution of spin dynamics in the FM phase of the perovskite manganites irrespective of the order/disorder at the A-site of perovskite structure.

1. Introduction

In these several decades, the perovskite manganites $R_{1-x}A_xMnO_3$ (R = rare earth element, A = Ca, Ba, Sr) have been investigated intensively, because they show a rich variety of fascinating electromagnetic properties, such as a colossal magnetoresistance (CMR), charge and orbital ordering (CO) and metal-insulator transition. Recently, the A-site ordered perovskite manganites $RBaMn_2O_6$ have been synthesized in half-doped manganites ($x = 0.5$). The crystal structure of $RBaMn_2O_6$, shown in the inset of Fig. 1, has a layer-type ordering of R and Ba ions along the c-axis, resulting in the two dimensional structure with the stacking sequence of -RO-MnO_2-BaO-MnO_2- planes. Interestingly, the order of R and Ba elements at the A site of a perovskite structure dramatically changes their physical properties from the A-site disordered manganites, $R_{0.5}Ba_{0.5}MnO_3$, with the same chemical composition [1, 2, 3, 4, 5, 6].

One of the features peculiar to the A-site ordered $RBaMn_2O_6$ is their much higher CO and magnetic ordering temperatures compared with the A-site disordered manganites. For example, the CO transition temperature of $YBaMn_2O_6$ is as high as 500 K, while $Y_{0.5}Ba_{0.5}MnO_3$ has only a spin glass phase below 30 K. The increase in the ordering temperature is considered to

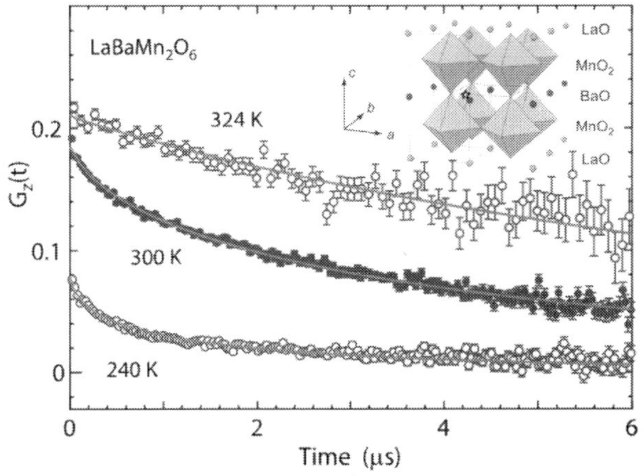

Figure 1. μSR time spectra measured at 324 K, 300 K and 240 K in zero magnetic field. Inset shows the crystal structure of $LaBaMn_2O_6$ with its tetragonal unit cell (dashed lines).

originate mainly from the ordering of R and Ba in the structure, which results in the absence of electrostatic potential disorder. The very high ordering temperature in $R$$BaMn_2O_6$ is considered to be a key feature for the development of new materials that exhibit CMR at room temperature [7, 8].

In this report, we focus on another A-site ordered perovskite manganite, $LaBaMn_2O_6$. This compound shows a transition from a paramagnetic metal (PM) to a ferromagnetic metal (FM) phase around $T_C \sim 330$ K. The transition temperature is enhanced from 280 K in $La_{0.5}Ba_{0.5}MnO_3$. Noticeably, it is found by the neutron diffraction measurements that a part of the FM phase transforms to antiferromagnetic insulating (AFI(CE)) phase below $\sim$ 200 K and that the AFI(CE) phase coexists with the FM phase at the ground state [3]. The NMR study has revealed from the Mn nuclear spin-spin relaxation rate that the magnetic properties of the FM phase in $LaBaMn_2O_6$ is the same type as that of the disordered form $La_{0.5}Ba_{0.5}MnO_3$ [6]. It seems that the magnetic properties of the FM phase are not affected by the absence of disorder at the A site in $LaBaMn_2O_6$, although the transition temperature is enhanced. In order to get further information on magnetic properties of the FM phase, we have performed μSR experiments on $LaBaMn_2O_6$ in zero field.

2. Experimental

Powder sample of $LaBaMn_2O_6$ for the present μSR measurements is the same one used in the previous NMR experiments [6]. This sample was prepared by a solid-state reaction and subsequent sintering process in pure Ar gas, followed by annealing in flowing O_2 gas. The details of the preparation method are described in the literature [2]. The crystal structure and magnetic properties of this sample are confirmed to be consistent with the conclusions based on the previous data [1, 2, 3, 4, 5, 6]. From the Mn-NMR spectra, majority of the sample is confirmed to be A-site ordered, although it contains disordered $La_{0.5}Ba_{0.5}MnO_3$ with several percent [6]. The μSR data presented here have been taken at the M20 muon channel at TRIUMF (Vancouver, Canada) and on πA muon channel at KEK (Tsukuba, Japan). All the μSR data have been measured in zero field.

3. Results and Discussion

We show the μSR time spectra measured at 324 K, 300 K and 240 K in Fig.1. In the FM phase below $T_C \sim 330$ K, no clear muon spin oscillation is observed in the time spetra, where the time spetra significantly deviates from a single exponential form. The time spectra data have been

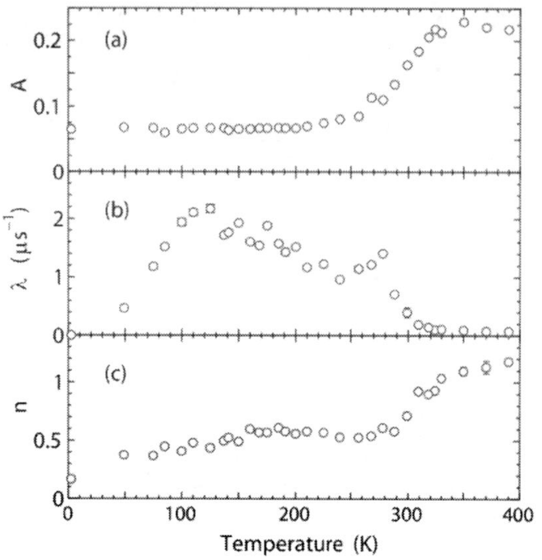

Figure 2. Temperature dependence of (a) A, (b) λ and (c) n, obtained from the stretched exponential model in $LaBaMn_2O_6$.

fitted by using the stretched exponential model expressed as

$$G_z(t) = A\exp(-\lambda t)^n,$$

where A is the initial asymmetry and λ is the longitudinal muon spin relaxation rate and n is the stretched exponent. $n = 1$ means a single exponential relaxation, while smaller values, $n < 1$, suggests a distribution of relaxation rates. The best fits are represented by the red solid lines in Fig. 1.

The parameters A, λ and n obtained at various temperatures are summarized in Figs. 2 (a), (b) and (c), respectively. $A \simeq 0.22$ has very weak temperature dependence in the PM phase above T_C. With decreasing temperature, A starts to decrease at T_C and reaches about 1/3 of the value in the PM phase. The value of 1/3 is expected if all implanted muons are exposed to a static local field in a polycrystalline or powder sample, *i.e.*, the volume fraction of magnetic phase is almost 100%.

In the temperature dependence of λ, two broad peaks are observed at 280 K and 125 K. The relaxation peak at 280 K, close to the inflection point of A, is considered to be due to a critical slowing down of Mn magnetic moments accompanied by the ferromagnetic order. The very broad peak around 125 K, ranging from 50 K to 200 K, suggests a crossover-like change in spin dynamics. It may be due to the change in magnetic structure from the FM one to the AFI(CE) one in the part of sample [3, 6]. Actually, oscillating components with frequency of 69.6 MHz and 31.6 MHz, which may originate from muons in the AFI(CE) phase, appear in the muon time spectra below 160 K (not shown).

The muon relaxation curve changes from an exponential ($n = 1$) in the PM phase to a root exponential ($n = 1/2$) in the FM phase. The small value of n may suggest a spatial distribution of spin dynamics in the FM phase, that has been observed in the A-site disordered ferromagnetic manganites $La_{1-x}Ca_xMnO_3$ ($x = 0.18$, 0.33 and 0.375) and $La_{0.85}Sr_{0.15}MnO_3$ [9, 10, 11, 12]. This result may suggest that there exists spatially inhomogeneous spin dynamics even in the FM phase of $LaBaMn_2O_6$, although the disorder at the A-site is largely diminished in this material. The further decrease in n below 160 K may suggest an increase in distribution of spin dynamics, most probably reflecting the appearance of the AFI(CE) phase inside the FM one.

Finally, we discuss on a possible muon site in $LaBaMn_2O_6$. It is reasonable to assume that the muon site in $LaBaMn_2O_6$ is the position corresponding to that in $YBaMn_2O_6$, *i.e.*, the position

1 Å away from oxygen in the BaO plane [13]. In the tetragonal unit cell of $LaBaMn_2O_6$, there are three oxygen sites; O(1) at (1/2, 1/2, 0) in the LaO plane, O(2) at (1/2, 0, 0.2373(5)) in the MnO_2 plane and O(3) at (1/2, 1/2, 1/2) in the BaO plane at 400 K [3]. One may, therefore, consider the muon site to be at (0.755, 1/2, 1/2) in the BaO plane, indicated by open star in the inset of Fig. 1. We have calculated the magnitudes of dipolar field at the muon site in the FM phase. Here, we used the structural parameters listed in Table I of the literature [3] and the size of Mn magnetic moment 3.0 μ_B estimated from magnetization in 5 T [3]. An easy axis is assumed to be [110], corresponding to the direction of Mn moment in the sublattice of the AFI(CE) phase of $YBaMn_2O_6$. The calculation gives the dipolar field of 0.38 T. The corresponding muon frequency of 51.5 MHz is sufficiently observable in the resolution of our experimental condition, ruling out possibility that muon frequency is too high to be detected it in the FM phase.

4. Conclusions

Magnetic properties of the A-site ordered perovskite manganite $LaBaMn_2O_6$ have been investigated by μSR experiments in zero field. In the FM phase below 330 K, no clear muon spin oscillation has been observed in the time spectra, while the muon relaxation significantly deviates from a single exponential form. These results may suggest a spatial distribution of spin dynamics, which has been observed in the FM phase of A-site disordered ferromagnetic manganites, although the disorder at the A-site is largely diminished in $LaBaMn_2O_6$. With decreasing temperature, oscillating components appear in the muon time spectra below 160 K. The appearance of the oscillating components may be due to the existence of the antiferromagnetic insulating phase inside the FM one.

References

[1] F. Millange, V. Caignaert, B. Domengés, B. Raveau and E. Suard, Chem. Mater. **10**, 1974 (1998).
[2] Y. Ueda and T. Nakajima, J. Phys.:Condens. Matter, **16**, S573 (2004).
[3] T. Nakajima, H. Kageyama, H. Yoshizawa, K. Ohoyama and Y. Ueda, J. Phys. Soc. Jpn., **72**, 3237 (2003).
[4] D. Akahoshi, M. Uchida, Y. Tomioka, T. Arima, Y. Matsui and Y. Tokura, Phys. Rev. Lett., **90**, 177203 (2003).
[5] S.V. Trukhanov, I. O. Troyanchuk, H. Szymczak and K. Bärner, Phys. Rev. B, **66**, 184424 (2002).
[6] Y. Kawasaki, T. Minami, Y. Kishimoto, T. Ohno, K. Zenmyo, H. Kubo, T. Nakajima and Y. Ueda, Phys. Rev. Lett., **96**, 037202 (2006).
[7] T. Nakajima and Y. Ueda, J. Appl. Phys., **98**, 046108 (2005).
[8] S.C. Bhargava, M. Isobe and Y. Ueda, Jpn. J. Appl. Phys., **47**, 7866 (2008).
[9] R.H. Heffner, L.P. Le, M.F. Hundley, J.J. Neumeier, G.M. Luke, K. Kojima, B. Nachumi, Y.J. Uemura, D.E. MacLaughlin and S-W. Cheong, Phys. Rev. Lett., **77**, 1869 (1996).
[10] R.H. Heffner, J.E. Sonier, D.E. MacLaughlin, G.J. Nieuwenhuys, G.M. Luke, Y.J. Uemura, William Ratcliff II, S.-W. Cheong and G. Balakrishnan, Phys. Rev. B, **63**, 094408 (2001).
[11] R.H. Heffner, J.E. Sonier, D.E. MacLaughlin, G.J. Nieuwenhuys, F. Mezei, G. Ehlers, J.F. Mitchell, S.-W. Cheong, Physica B, **326**, 494 (2003).
[12] S.G. Barsov, A.L. Getalov, V.P. Koptev, S.A. Kotov, S.M. Mikirtychyants, G.V. Shcherbakov, A.A. Arsenov, and Ya.M. Mukovskii *et al.*, Physica B, **289-290**, 81 (2000).
[13] Y. Kawasaki, T. Minami, Y. Kishimoto, T. Ohno, A. Koda, K.H. Satoh, R. Kadono, J.L. Gavilano, H. Luetkens, T. Nakajima and Y. Ueda, Physica B, **404**, 781 (2009).

Phase transition in the normal state of the non-centrosymmetric superconductor Mo_3Al_2C

T. Koyama[1], Y. Ozaki[1], K. Ueda[1], T. Mito[1], T. Kohara[1], T. Waki[2], Y. Tabata[2], and H. Nakamura[2]

[1] Graduate School of Material Science, University of Hyogo, Kamigori, Hyogo 678-1297, Japan
[2]Department of Materials Science and Engineering, Kyoto University, Kyoto 606-8501, Japan

E-mail: t-koyama@sci.u-hyogo.ac.jp

Abstract. We found a phase transition in the normal state of superconducting carbide, Mo_3Al_2C by nuclear magnetic resonance (NMR) measurement. At ~130 K the sharp central NMR line broadens and has additional broad components at the foot. In addition, the Knight shift and the spin-lattice relaxation rate exhibit a significant decrease, suggesting the reduction in the conduction electron density. From these results, Mo_3Al_2C undergoes the phase transition which can be understood as the charge-density wave ordering. The decrease in $1/T_1$ confirms the superconductivity appears in the partially quenched Fermi surface.

1. Introduction

The cubic β-Mn type (space group $P4_132$) crystal structure is unique. The Mn-II (12d) sites construct a corner sharing regular triangular network, where antiferromagnetic interaction between moments at 12d site would give rise to geometrical frustration[1]. Then the lack of long-range magnetic ordering in pure β-Mn is interpreted as this effect in the β-Mn lattice. Looking from another point of view, this structure does not posses a center of inversion. In superconducting materials with an inversion center, the Cooper pairs are either in the spin-singlet state or in the spin-triplet state due to the Pauli exclusion principle and parity conservation. However, when the inversion symmetry is broken, the absence of inversion symmetry along with parity-violating antisymmetric spin-orbit coupling allows admixture of spin-singlet and spin-triplet components. The extent of parity mixing depends on the strength of the spin-orbit coupling that is enhanced by the lack of inversion symmetry. In such materials, novel superconducting properties can be expected.

A metallic carbide, Mo_3Al_2C, crystallizing a β-Mn type crystal structure exhibits the superconductivity (SC) at T_c=9 K[2]. Recently, Bauer *et al.* and Karki *et al.* independently investigated the normal and superconducting state properties in poloycrystalline sample of Mo_3Al_2C[3, 4]. It was reported that Mo_3Al_2C shows temperature-independent Pauli paramagnetic behavior in the magnetic susceptibility and no phase transition in the normal state[4]. The Sommerfeld constant is estimated to be $\gamma \approx 18$ mJ/mol K^2. The electronic specific heat jump at the transition point $\Delta/\gamma T_c \approx 2.2$ suggests the strong-coupling SC. The notable features of this compound are power-law behaviors of the specific heat and the nuclear spin-lattice relaxation rate $1/T_1$ in the superconducting state, implying that the order parameter is not of conventional s-wave. In this paper, we reported the first observation of the phase

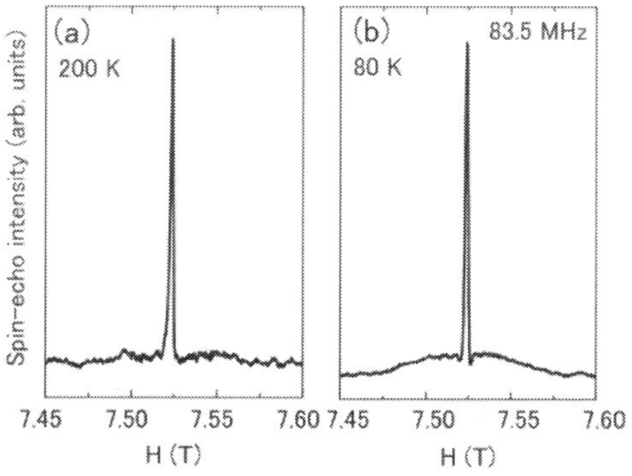

Figure 1. (a), (b) ^{27}Al NMR spectra obtained by sweeping magnetic field at fixed frequency of 83.5 MHz.

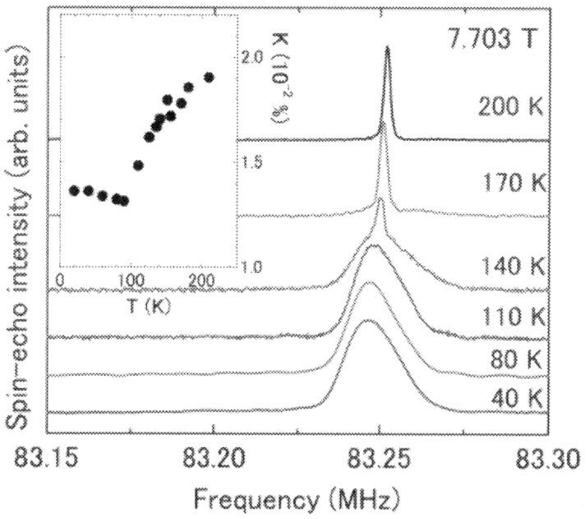

Figure 2. (Color on line) Temperature variation of the central line of ^{27}Al NMR spectra. Data were obtained by fast-Fourier-transformation technique at a fixed magnetic field. Inset shows the temperature dependence of ^{27}Al Knight shift.

transition in the normal state of Mo_3Al_2C investigated by nuclear magnetic resonance (NMR) measurement.

2. Experimantal

Pollycrystalline samples of Mo_3Al_2C were arc-melted under an argon atmosphere as described in detail elsewhere[4]. NMR experiment was carried out using a conventional phase-coherent pulsed spectrometer.

3. Result and discussions

We have performed ^{27}Al NMR experiment. Al atoms in Mo_3Al_2C are located at Mn-I (8*c*) site in the β-Mn type structure. Figure 1 (a) and (b) show field-swept NMR spectra at fixed frequency of 83.5 MHz. The spectrum at 200 K is sharp and symmetric, indicating that quadrupole interaction is negligible. On the other hand, the spectrum at 80 K has broad components at the foot of the intense peak. The width of the broad components is independent of the magnetic field, suggesting that these components originate from the appearance of quadrupole satellite lines due to the enhancement of quadrupole interaction at low temperatures. To see the temperature evolution in the sharp peak, we show, in Fig. 2, the NMR spectra obtained from fast-Fourier transformation (FFT) technique at a constant magnetic field of H=7.503 T. Above 180 K, the spectrum is a symmetric Lorentzian-type line shape. With decreasing temperature, a broad component develops at the foot of the sharp peak. The full width at half maximum of the spectra below 100 K is six times larger than that above 200 K.

The inset of Fig. 2 shows the temperature dependence of the Knight shift K measured at the central peaks of the ^{27}Al NMR spectrum. K takes positive values which is smaller than that of Al metal (0.16 %) in all temperatures. K decreases with temperature followed by almost temperature-independent behavior below 100 K. The change in K between 170 and 100 K

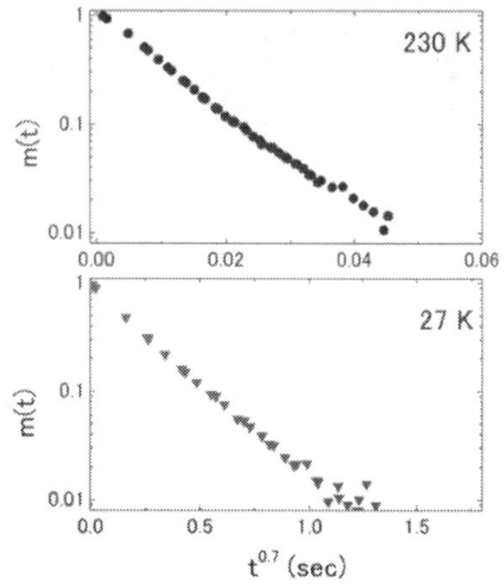

Figure 3. (Color on line) Nuclear magnetization recovery curve $m(t)$ at 27 and 230 K plotted against $t^{0.7}$, where t is the time between the saturation pulse and the first spin-echo pulse.

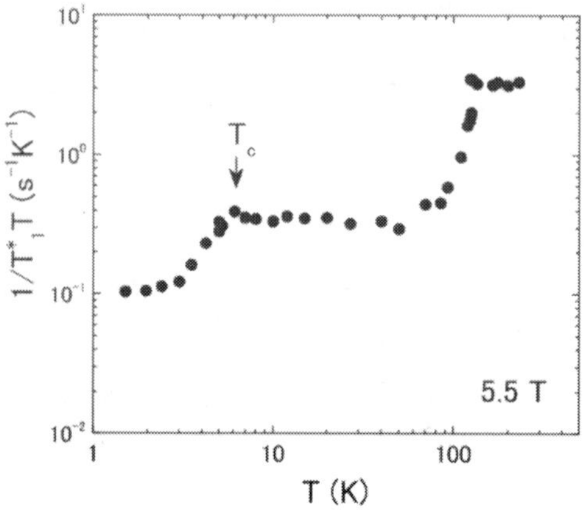

Figure 4. Temperature dependence of $1/T_1^*T$ at 5.5 T.

implies the electronic structure modifications. For the nonmagnetic materials, the total K can be written as,

$$K = K_s + K_{\rm dia} + K_{\rm orb}, \tag{1}$$

where K_s, $K_{\rm dia}$, and $K_{\rm orb}$ are the Fermi-contact, the orbital, and the diamagnetic terms, respectively[5]. Using the density of states at Fermi level $D(E_F)$, the sum of the first two terms can be expressed by

$$K = K_s + K_{\rm dia} = \left\{\frac{H_s}{N\mu_{\rm B}} - \frac{8\pi}{9}\left(\frac{m}{m^*}\right)^2\right\}\chi_{\rm P} \propto D(E_F), \tag{2}$$

where $H_s, N, \mu_{\rm B}, m^*/m$, and $\chi_{\rm P}$ are the Fermi contact hyperfine field, Avogadro's number, the Bohr magneton, the band effective mass ratio, and the Pauli paramagnetic susceptibility, respectively. By considering that $K_{\rm orb}$ is temperature-independent, the drop of K can be interpreted as a decrease in $D(E_F)$.

It is well known that $1/T_1$ probes $D(E_F)$. The nuclear magnetization $M(t)$ was measured using the free-induction-decay (FID) amplitude so as to selectively obtain the T_1 value of the sharp line. The nuclear magnetization recovery $m(t)$, which is defined by $m(t) = [M(\infty) - M(t)]/M(\infty)$, does not follow the expected theoretical function[6]. This is probably due to the distribution of T_1 values stemming from some impurities, such as Mo_3Al_8, MoC, and Mo. In order to extract a characteristic relaxation rate $1/T_1^*$, we fitted the recovery curves to a stretched exponential function,

$$m(t) = \exp\left[-\left(\frac{t}{T_1^*}\right)^{\beta}\right], \tag{3}$$

where β is the stretching exponent with $0 < \beta \leq 1$[7]. As shown in Fig. 3, obtained $\beta \sim 0.7$ is nearly temperature-independent both above and below the transition temperature, indicating that the characteristic of $m(t)$ does not vary with temperature. Figure 4 shows the temperature dependence of $1/(T_1^*T)$. The present $1/(T_1^*T)$ value below 90 K is four times smaller than the earlier report[3]. The discrepancy is ascribed to the different recovery function employed to extract $1/T_1$. The remarkable feature is the rapid decrease of $1/(T_1^*T)$ at $\sim$ 120 K. The values of $1/(T_1^*T)$ at low temperatures are one order smaller than those at high temperatures. Because of the relation of $1/(T_1^*T) \propto D(E_F)^2$, $D(E_F)$ at low temperatures is roughly estimated to be 30% of that at high temperature. This result demonstrates that Mo_3Al_2C exhibits the phase transition accompanying the partially quenching of Fermi surface. The decrease in $D(E_F)$ is also inferred from the K. By further decreasing temperature, $1/(T_1^*T)$ decreases below T_c, indicating SC occurred at the partially quenched Fermi surface.

We speculate that this transition is connected with the formation of the charge- density wave (CDW). The reductions in ^{27}K and $1/(T_1^*T)$ at the transition are ascribed to the appearance of gap at $D(E_F)$ owing to the CDW formation. The broadening and the change in NMR line shape arise via a spatial distribution of the electric-field gradient (EFG) tensor at the observed nuclear site. Since CDW transition is predominantly driven by Fermi surface nesting, such transition require low-dimensional character in crystal structure. The highly anisotropic Fermi surface is responsible for the formation of the CDW. Further band calculation and experiment using single crystalline sample is needed to clarify the transition at $\sim$130 K.

4. Conclusion

We have investigated the non-centrosymmetric Mo_3Al_2C using NMR measurement. The reduction in the conduction electron density at $\sim$130 K is inferred from the decrease in ^{27}K and $1/(T_1^*T)$. In addition ^{27}Al NMR line shape shows a sudden change owing to the distribution of EFG tensor at Al position. We thus conclude that the reduction in $D(E_F)$ is associated with the appearance of a gap at the Fermi surface resulting from a CDW transition in Mo_3Al_2C. The decrease in $1/T_1^*$ confirms the SC appears in the partially quenched Fermi surface.

5. Acknowledgments

This work was partially supported by a Grant-in-Aid for Scientific Research on Priority Areas "Novel States of Matter Induced by Frustration" (19052003) and a Grant-in-Aid for Young Scientists (23740276).

References

[1] Nakamura H, Yoshimoto K, Shiga M, Nishi M and Kakurai K 1997 *J. Phys.: Condens. Matter* **9** 4701
[2] Johnston J, Toth L E, Kennedy K and Parker E. R. 1964 *Solid State Commun.* **2** 123
[3] Bauer E, Rogl G, Chen Xing-Qiu, Khan R. T, Michor H, Hilscher G, Royanian E, Kumagai K, Li D. Z, Li Y. Y, Podloucky R and Rogl P 2010 *Phys. Rev. B* **82** 064511
[4] Karki A. B, Xiong Y. M, Vekhter I, Browne D, Adams P. W, Young D. P, Thomas K. R, Chan J. Y, Kim H, and Prozorov R 2010 *Phys. Rev. B* **82** 064512
[5] Carter G. C, Bennett L. H, and Kahan D. J, *Metallic Shifts in NMR, Part I* (Pergaon Press, Oxford, 1977)
[6] Narath A 1967 *Phys. Rev.* **162** 320.
[7] Johnston D. C 2006 *Phys. Rev. B* **74** 184430

Frustration effects in the spinel oxide $Ge(Fe_{1-x}Co_x)_2O_4$

Takahiro Matsuda[1], Ryotaro Agata[1] Tadataka Watanabe[1], Norio Umeyama[2], Sigeo Hara[3], Madoka Tokumoto[2], Yoshiaki Hata[4], Hiroshi Yasuoka[4], and Shin-Ichi Ikeda[2]

[1] Department of Physics, College of Science and Technology (CST), Nihon University, Chiyoda-ku, Tokyo 101-8308, Japan

[2] ESPRI, AIST, Tsukuba, Ibaraki 305-8568, Japan

[3] Department of Physics, Chuo University, Bunkyou-ku, Tokyo 112-8551, Japan

[4] Department of Applied Physics, National Defense Academy, Yokosuka, Kanagawa 239-8686, Japan

Abstract. We studied magnetic properties of Ge-based spinel oxides $Ge(Fe_{1-x}Co_x)_2O_4$ to verify the frustration effect. We discovered that the spin-glass-like behavior in $Ge(Fe_{1-x}Co_x)_2O_4$ exists in $0 \leq x \leq 0.9$, and coexists with the antiferromagnetic order in $0.1 \leq x \leq 0.9$. Taking into account the coexistence of spin molecular excitations (frustration effect) and antiferromagnetic order in $GeCo_2O_4$ ($x = 1$), the spin-glass-like behavior coexisting with the antiferromagnetic order in $Ge(Fe_{1-x}Co_x)_2O_4$ might be attributed to the freezing of the spin molecules by the substitution of Fe for Co.

1. Introduction

Geometrical frustration is a concept which provides an intriguing playground for condensed matter physics. Cubic spinels AB_2O_4 with magnetic B ions have attracted much interest in light of the geometrical frustration which is inherent in the B-site sublattice of corner sharing tetrahedra. Ge-based spinel oxides GeM_2O_4 (M = Fe and Co) consist of magnetic M^{2+} ions (Fe^{2+} ($3d^6$) and Co^{2+} ($3d^7$)) with threefold degeneracy of t_{2g} orbitals on the octahedral B sites, and non-magnetic Ge^{4+} ions on the tetrahedral A sites. Thus it is expected that these compounds provide a rich field for the orbital physics in the magnetically frustrated system.

$GeCo_2O_4$ exhibits the occurrence of an antiferromagnetic transition at $T_N = 21.0$ K accompanied with cubic-to-tetragonal structural elongation [1-3]. On the other hand, the magnetic susceptibility in the paramagnetic state exhibits Curie-Weiss behavior with the positive Weiss temperature $\Theta_W = 81.0$ K indicating the dominant contribution of the ferromagnetic interactions [4]. For $GeCo_2O_4$, recent ultrasound velocity measurements and in elastic neutron scattering experiments suggested that this compound is a promising candidate for the spin-orbit-coupled frustrated system [5-7]. For $GeFe_2O_4$, although the detailed study of the frustration effect has not been performed so far, the magnetic susceptibility measurements reported the spin-glass like behavior which is probably due to frustration [8]. The difference of the magnetic properties between $GeCo_2O_4$ and $GeFe_2O_4$ indicates that these compounds have different types of exchange interactions and frustration effects. And it is expected that the orbital degrees of freedom play significant roles for the magnetism of $GeCo_2O_4$ and $GeFe_2O_4$ in the different ways. Furthermore, it is expected that, in the mixed crystal $Ge(Fe_{1-x}Co_x)_2O_4$, the tuning of the population of d electrons with sustaining threefold degeneracy of t_{2g} orbitals leads to novel phenomena relevant to the orbital degrees of freedom. In this study,

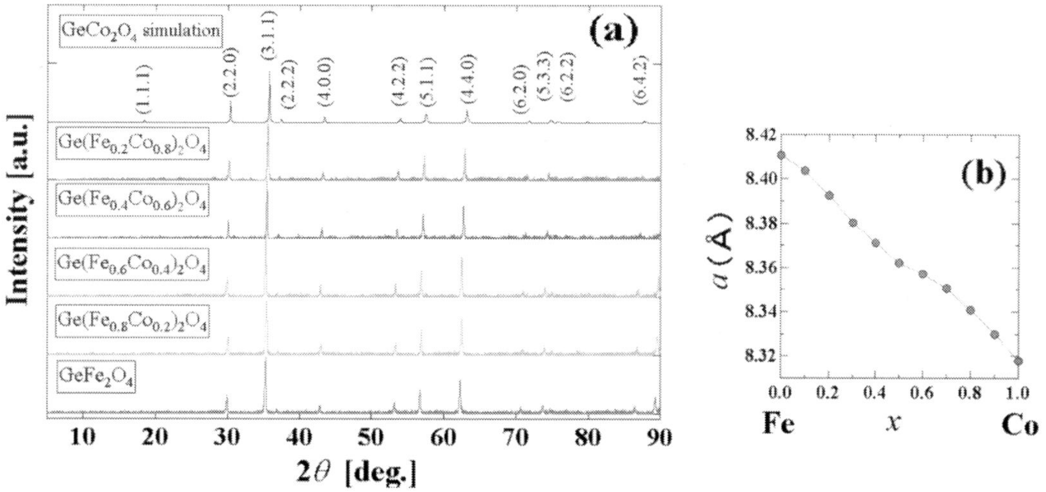

Figure 1 (a) The XRD patterns of Ge(Fe$_{1-x}$Co$_x$)$_2$O$_4$.(b) The *a*-axis length of Ge(Fe$_{1-x}$Co$_x$)$_2$O$_4$ as a function of Co concentration *x*.

we report magnetic properties of the mixed crystal Ge(Fe$_{1-x}$Co$_x$)$_2$O$_4$.

2. Experimental

Polycrystalline samples of Ge(Fe$_{1-x}$Co$_x$)$_2$O$_4$ ($0.0 \leq x \leq 1.0$) were prepared by solid state reaction. Mixtures of stoichiometric amounts of GeO_2, FeO, and CoO powders were sealed into evacuated quartz tube, and heated for 40 hours at 800 ℃ followed by 24 hours at 950 ℃. The crystal structure of the samples was analyzed by powder X-ray diffraction (XRD) measurements at room temperature using Cu Kα radiation. The magnetic susceptibility measurements were carried out using a superconducting quantum interference device (SQUID) magnetometer in temperatures from 3K to 300K with magnetic field of H = 1000 Oe in zero-field-cooled (ZFC) and field-cooled (FC) processes.

3. Results and discussion

Figure 1 (a) shows the XRD patterns of the polycrystalline Ge(Fe$_{1-x}$Co$_x$)$_2$O$_4$ samples. All the samples crystallize in the spinel-type structure without any additional peak. Figure 1 (b) shows the *a*-axis length of Ge(Fe$_{1-x}$Co$_x$)$_2$O$_4$ as a function of Co concentration *x*. It is evident that all the samples obey Vegard's law: the *a*-axis length of Ge(Fe$_{1-x}$Co$_x$)$_2$O$_4$ decreases with increasing Co concentration *x*. These results ensure that the *B*-site Fe atoms are properly replaced by Co atoms.

Temperature dependence of the FC and ZFC magnetic susceptibilities with H = 1000 Oe in Ge(Fe$_{1-x}$Co$_x$)$_2$O$_4$ is shown in Fig. 2(a). GeFe$_2$O$_4$(x = 0) exhibits the spin-glass-like behavior with the evolution of the irreversibility of the magnetic susceptibility below T_g = 56 K. In Ge(Fe$_{0.9}$Co$_{0.1}$)$_2$O$_4$ (x = 0.1), the magnitude of the magnetic susceptibility is enhanced compared to GeFe$_2$O$_4$ (x = 0), and T_g shifts to the higher temperature of T_g = 126 K. With further increase of Co concentration x in $0.1 \leq x \leq 0.9$, the magnitude of the magnetic susceptibility of $Ge(Fe_{1-x}Co_x)_2O_4$ is depressed, but the spin-glass-like behavior persists in this

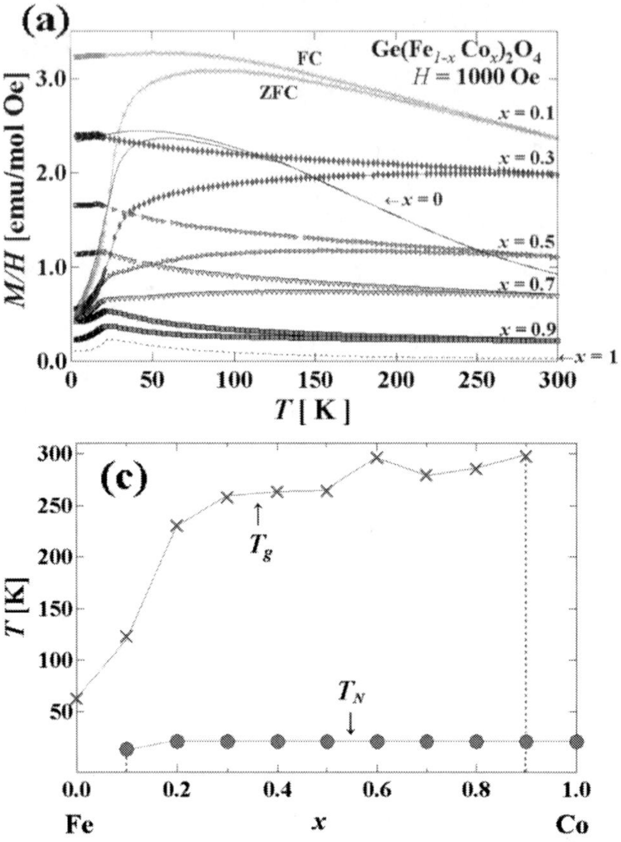

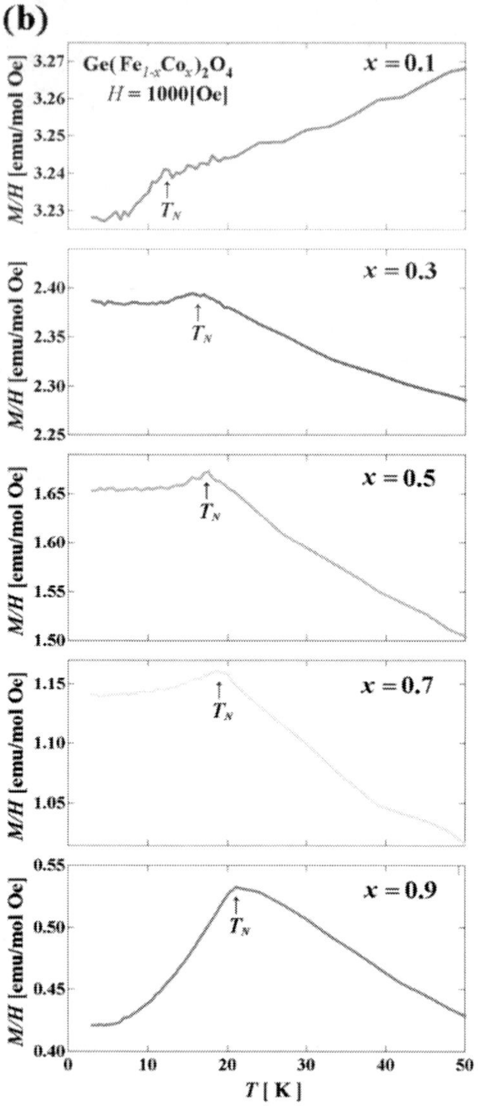

Figure 2 Temperature dependence of (a) the FC and ZFC magnetic susceptibilities, and (b) the low-temperature FC magnetic susceptibilities in $Ge(Fe_{1-x}Co_x)_2O_4$ with H = 1000 Oe.
(c) The plots of spin glass transition temperature T_g (×) and Néel temperature T_N (•) of $Ge(Fe_{1-x}Co_x)_2O_4$ as functions of Co concentration x.

x range. $GeCo_2O_4$ (x = 1) exhibits the occurrence of the antiferromagnetic transition at T_N = 20.8K, and the disappearance of the spin-glass-like irreversible behavior in the magnetic susceptibility. It is noted that, as shown in Fig. 2(b), the cusp in the magnetic susceptibility due to the antiferromagnetic transition exists in $Ge(Fe_{1-x}Co_x)_2O_4$ with $0.1 \leq x \leq 1$, which indicates the coexistence of spin glass and antiferromagnetic order in $Ge(Fe_{1-x}Co_x)_2O_4$ with $0.1 \leq x \leq 0.9$.

On the basis of the magnetic susceptibilities shown in Figs. 2 (a) and (b), Fig. 2 (c) plots spin glass transition temperature T_g (×) and Néel temperature T_N (•) of $Ge(Fe_{1-x}Co_x)_2O_4$ as functions of Co concentration x. Here, T_g is defined as the temperature below which the irreversibility in the magnetic susceptibility evolves.

With increasing the Co concentration x, T_g increases up to ~230 K at $x = 0.2$ and exhibits weak x dependence in the wide x range of $0.2 \leq x \leq 0.9$. Additionally, Fig. 2 (c) tells that the spin-glass-like behavior is rapidly suppressed in $x > 0.9$ and disappears in $GeCo_2O_4$ ($x = 1$). A salient feature indicated in Fig. 2 (c) is the coexistence of the spin-glass-like behavior and the antiferromagnetic order in $Ge(Fe_{1-x}Co_x)_2O_4$ with the wide x range of $0.1 \leq x \leq 0.9$. Very recent ultrasound velocity measurements and inelastic neutron scattering experiments in $GeCo_2O_4$ ($x = 1$) provided the compelling evidence for the presence of spin-molecular excitations due to frustration in not only the paramagnetic phase but also the antiferromagnetic phase [6,7]. These reports suggest the coexistence of spin molecules and antiferromagnetic order in $GeCo_2O_4$ ($x = 1$). Thus the appearance of the spin-glass-like behavior which coexists with the antiferromagnetic order in $Ge(Fe_{1-x}Co_x)_2O_4$ might be due to the freezing of the spin molecules by the substitution of Fe for Co. Further experimental and theoretical works are expected to verify the frustration effect and the possible exotic ground state in $Ge(Fe_{1-x}Co_x)_2O_4$.

4. Acknowledgement

This work was partly supported by Grants-in-Aid for Young Scientists (B) (21740266) and Scientific Research on Priority Areas (22014001) from MEXT of Japan.

References

[1] Hubsch J, and Gavoille G 1987 J. Magn. Magn. Mater. 66,17.

[2] Diaz S, de Brion S, Chouteau G, Canals B, Simonet V, and Strobel P 2006 Phys. Rev. B 74, 092404.

[3] Hoshi T, Aruba-Katori H, Modals M, and Takagi H 2007 J. Magn. Magn. Mater. 310, e448.

[4] Diaz S, de Brion S, Holzapfel M, Chouteau G, and Stepel P 2004 Physical B 346-347, 146.

[5] Watanabe T, Hara S, and Ikeda S. I 2008 Phys. Rev. B 78, 094420.

[6] Watanabe T, Hara S, Ikeda S. I, and Tomiyasu K 2011 Phys. Rev. B 84, 020409(R).

[7] Tomiyasu K, Crawford M K, Adroja D T, Manuel P, Tominaga A, Hara S, Sato H, Watanabe T, Ikeda S. I, Lynn J W, Iwasa K, and Yamada K 2011 Phys. Rev. B 84, 054405.

[8] Aruga-Katori H, Hayashi H, Honda Z, and Takagi H, unpublished.

International Conference on Strongly Correlated Electron Systems (SCES 2011) IOP Publishing
Journal of Physics: Conference Series **391** (2012) 012099 doi:10.1088/1742-6596/391/1/012099

Probing Spin Chirality of the Equilateral Triangular-Lattice Antiferromagnet $RbFe(MoO_4)_2$ through Multiferroicity

H Mitamura[1], R Watanuki[2], N Onozaki[2], Y Shimura[1], S Kittaka[1], T Sakakibara[1] and K Suzuki[2]

[1]Institute for Solid State Physics, University of Tokyo, Kashiwa 277-8581
[2]Faculty of Engineering, Yokohama National University, Yokohama 240-8501

E-mail: mitamura@issp.u-tokyo.ac.jp

Abstract. Ferroelectricity was studied in magnetic fields up to 14 T on the XY-like perfect-triangular-lattice antiferromagnet $RbFe(MoO_4)_2$, which exhibits a magnetically induced ferroelectric polarization along the *c* axis in the low-field 120°structure phases P1 and P2. We have experimentally confirmed that the high-field P4/P5 phases of $RbFe(MoO_4)_2$ for $B \parallel a$ are nonferroelectric. The results favor the scenario that the ferroelectricity in the low-field phases of this system is driven by the in-plane spin chirality.

1. Introduction

Magnetically driven ferroelectricity (MFE) has attracted much interest. Most of the phenomena have successfully been explained by the spin current mechanism in which the local electric polarization is expressed as $\boldsymbol{P}_{ij} \propto \boldsymbol{r}_{i,j} \times \left(\boldsymbol{S}_i \times \boldsymbol{S}_j\right)$ [1]. It has also been pointed out that MFE can be a probe for spin helicity [2]. Recently, ferroelectricity with $\boldsymbol{P} \parallel c$ has been found to emerge concomitantly with an antiferromagnetic transition at $T_N = 3.9$ K in the triangular lattice compound $RbFe(MoO_4)_2$ (hereafter referred to as RFMO) [3], in which Fe^{3+} ions carrying spin $S = 5/2$ form a perfect triangular lattice [4]. According to the neutron diffraction measurements, the zero-field ordering wave vector of this compound is (1/3, 1/3, 0.468), where the in-plane spin configuration is a so-called 120°structure with spins lying in the triangular plane while incommensurability along the *c* axis is present [3, 5]. In this situation, $\boldsymbol{P}_{ij}$ lies in the basal plane and the macroscopic polarization $\boldsymbol{P} = \sum \boldsymbol{P}_{ij}$ vanishes by symmetry. Thus, the spin current mechanism does not work. Recently, Kenzelmann *et al.* [3] discussed on the basis of phenomenological arguments that both in-plane triangular spin chirality as well as the inter-plane spin helicity in RFMO lead to MFE when a mirror plane perpendicular to the *c* axis is absent.

Crystal structure of RFMO at room temperature belongs to the space group $P\bar{3}m1$. Below 190 K (T_S), the symmetry is lowered to $P\bar{3}$ by a lattice distortion which preserves the threefold axis [5]. When a magnetic field *B* is applied along the *a* axis at 1.6 K, five phases P1~P5 appear successively with increasing *B* before reaching a paramagnetic saturation at 19.9 T [7]. All these phases are characterized by the wave vector $(1/3, 1/3, q_z)$, where $q_z = 0.468$ for P1, 1/3 for P3, and 0.4 ~ for P5 [3, 5, 8]. The in-plane spin configurations of these phases can be compared with the calculated results on the ideal Heisenberg spin triangular antiferromagnet done by Kawamura and Miyashita [8, 9]. P1 and P2 correspond to the low-field noncolinear state (phase I). P3 is equivalent to phase II with a 1/3 magnetization plateau, whereas P4 and P5 correspond to the high-field noncollinear state (phase III).

Adopting their calculations, P1 and P2 have finite in-plane spin chirality but P3~P5 do not. Considering the c-axis incommensurability of the magnetic structure of RFMO, on the other hand, inter-plane helicity may exist in all the phases except for P3 (collinear up-up-down state). The experiment by Kenzelmann *et al.* [3] has revealed that P1 and P2 are ferroelectric but P3 is nonferroelectric. Their results are, however, not sufficient to discriminate which chirality/helicity is most relevant for the MFE. If the in-plane spin chirality is responsible for MFE, then the high-field P4 and P5 phases must be nonferroelectric. Conversely, if P4 or P5 phase is ferroelectric, then one can conclude that the in-plane spin chirality is irrelevant for the MFE. We measured the dielectric properties of RFMO in much higher fields up to the P5 phase region and discuss the possible origin of MFE.

2. Experimental

Single crystals of $RbFe(MoO_4)_2$ with thin plate shape were prepared by a flux technique. Specific heat C, magnetization M and dielectricity are measured in fields up to 14 T applied along the a axis in a temperature range down to 0.25 K.

3. Results and Discussion

In order to give an overview of the dielectric and magnetic properties of RFMO, we first show in Fig. 1 the magnetic phase diagram for $B \parallel a$ determined by $C(T)$ (open diamonds), $M(T)$ (open squares), $M(H)$ (open circles), $\epsilon(T)$ (solid squares) and $\epsilon(H)$ (solid circles) measurements. Apart from the P4-P5 boundary, which is not clearly resolved in the present experiment, our phase diagram is in good agreement with the previous results [8].

Figure 2 shows the $\epsilon(T)$ data in various magnetic fields. For $B = 0$, a small but distinct peak of $\epsilon(T)$ is observed at $T_N = 3.9$ K. At this temperature, a remarkable change is observed in the $P - E$ loop as shown in the inset; a clear hysteresis as well as an S-shape feature appear. Thus, we consider that the peak in $\epsilon(T)$ reflects the ferroelectric transition. Increasing B from zero, this peak in $\epsilon(T)$ moves to the low temperature side along the P1 or P2 boundary showing a significant increase in the amplitude. This confirms that P1 and P2 are both ferroelectric. On the other hand, no dielectric anomaly is observed upon the transitions from paramagnetic to P3, P5 phases nor at the P4-P5 transition. This result indicates that not only P3 but also P4/P5 phases are nonferroelectric.

We first discuss why MFE is allowed in the P1/P2 phases of RFMO. The appearance of $P \parallel c$ implies that any symmetry with respect to the c-axis inversion is broken. For a triangular lattice with the highest symmetry of $P6/mmm$, there are six symmetry operations that make the c-axis inversion. These are m_z, i, S_3, S_6, C_{2x} and C_{2y}. How the lattice and the zero-field spin structures in RFMO transform under these symmetry operations is illustrated in Fig. 3. For instance, the lattice structure ($P\bar{3}$) below T_S is invariant under i and S_6 (Table 1). This fact ensures that no ferroelectricity with $P \parallel c$ appears in the paramagnetic state. When the magnetic order sets in, transformations of the spin structures should also be considered. As can be seen from Table 1, the in-plane spin chirality changes sign by i and S_6. Similarly, the helical spin alignment along the c axis also breaks the symmetry with respect to i and S_6 as given in Table 1. These symmetry properties also hold in a magnetic field applied along the c plane, in which case the relevant symmetry operation becomes only the inversion i. Thus, both of the in-plane and the inter-plane spin configurations can drive MFE in the P1/P2 phases in RFMO [3, 10]. For the P3 phase, the spin configuration becomes the up-up-down state without spin chirality/helicity. Then, the inversion symmetry is recovered and MFE vanishes in agreement with the experiment. Thus, the ferroelectric properties in P1~P3 phases can equally be explained by the in-plane spin chirality as well as the spin helicity along the c axis in the anitferromagnetically ordered phase.

Our experiment has shown that MFE is absent both in the P4 and P5 phases. This observation is consistent with the scenario that MFE is caused by the in-plane spin chirality of the triangular lattice, since the in-plane spin chirality is considered to vanish in the P4/P5 phases. If the inter-plane spin helicity still exists in these phases, then we can conclude that the MFE in RFMO is driven exclusively

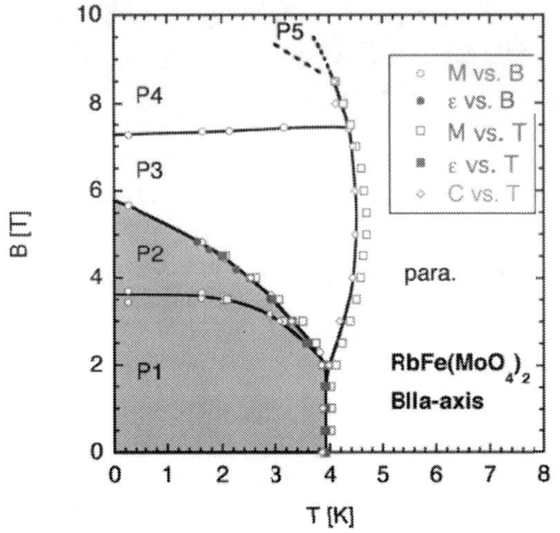

Figure 1. Magnetic phase diagram of $RbFe(MoO_4)_2$ for $B \parallel a$.

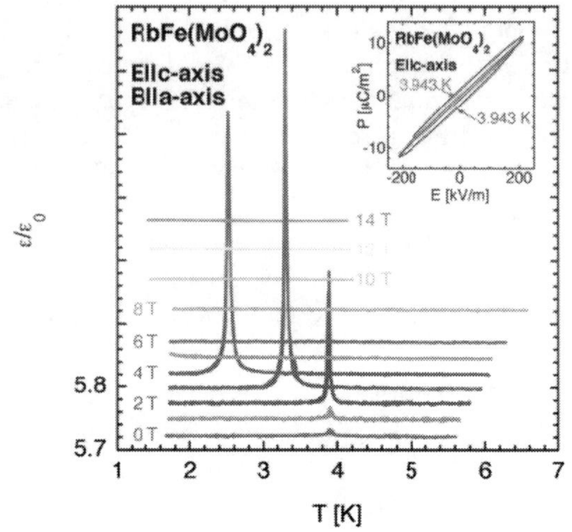

Figure 2. ϵ vs. T plot in various magnetic fields. (inset) P vs. E curves above and below the Néel temperature at $B = 0$.

by the in-plane triangular spin chirality. It is therefore highly desired to determine the spin configuration in the P4/P5 phases.

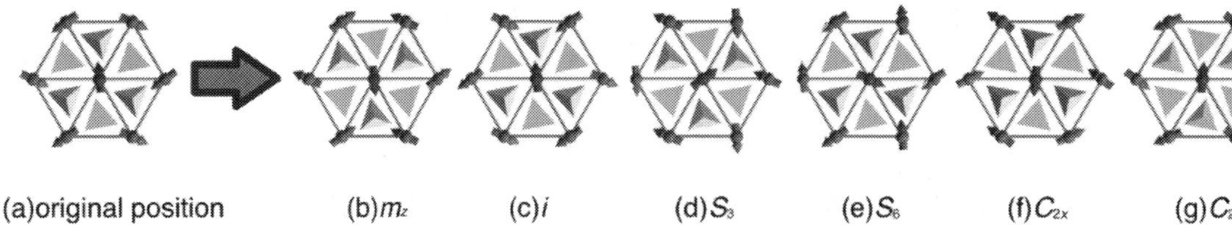

Figure 3. Transformations of the lattice and the triangular spins in the low temperature phase of $RbFe(MoO_4)_2$ at zero magnetic field. Only those involving the c-axis inversion are shown. Original position of the lattice and spins (a) changes into (b) ~ (f) by the symmetry operations m_z, i, S_3, S_6, C_{2x} and C_{2y}, respectively. The lattice (spin chirality) is invariant under i and S_6 (m_z, S_3 and C_{2y}).

Table 1. Symmetry relations that involve the c-axis inversion in the low temperature phase of $RbFe(MoO_4)_2$ at zero field.

	symmetry operations					
	m_z	i	S_3	S_6	C_{2x}	C_{2y}
lattice	×	○	×	○	×	×
in-plane spin chirality	○	×	○	×	×	○
inter-plane spin helicity	×	×	×	×	○	○

4. Summary

Whether or not a perfect triangular lattice XY antiferromagnet with the 120° spin structure exhibits macroscopic ferroelectricity perpendicular to the plane is a question which has not yet been resolved. $RbFe(MoO_4)_2$ with the $P\bar{3}$ structure is one of the candidate compounds to study the magnetically induced ferroelectricity driven by the in-plane spin chirality of the 120°structure. In this compound, however, the ordered spin structure possesses inter-layer spin helicity which also breaks c-axis inversion symmetry. We measured the dielectric polarization of $RbFe(MoO_4)_2$ in the high-field non-collinear P4/P5 phases and confirmed that they are non-ferroelectric. Unless the spin alignment in the P4/P5 phases takes a fan type structure, our results imply that the magnetically induced ferroelectricity in $RbFe(MoO_4)_2$ is driven by the in-plane spin chirality.

5. References

[1] Katsura H, Nagaosa N, and Balatsky A V 2005 *Phys. Rev. Lett.* **95** 057205
[2] H. Sagayama, N. Abe, K. Taniguchi, T. Arima, Y. Yamasaki, D. Okuyama, Y. Tokura, S. Sakai, T. Morita, T. Komesu, H. Ohsumi, and M. Takata: J. Phys. Soc. Jpn. **79** (2010) 043711.
[3] Kenzelmann M, Lawes G, Harris A B, Gasparovic G, Broholm C, Ramirez A P, Jorge G A, Jaime M, Park S, Huang Q, Shapiro A Ya and Demianets L A 2007 *Phys. Rev. Lett.* **98** 267205
[4] Inami T, Ajiro Y and Goto T 1996 *J. Phys. Soc. Jpn.* **65** 2374
[5] Inami T 2007 *J. Solid State Chem.* **180** 2075
[6] Arima T 2007 *J. Phys. Soc. Jpn.* **76** 073702
[7] Smirnov A I, Yashiro H, Kimura S, Hagiwara M, Narumi Y, Kindo K, Kikkawa A, Katsumata K, Shapiro A Ya and Demianets L A 2006 *Phys. Rev. B* **74** 024412
[8] Svistov L E, Smirnov A I, Prozorova L A, Petrenko O A, Micheler A, Büttgen N, Shapiro A Ya and Demianets L A 2006 *Phys. Rev. B* **74** 024412
[9] Kawamura H and Miyashita S 1985 *J. Phys. Soc. Jpn.* **54** 4530
[10] Harris A B 2007 *Phys. Rev. B* **76** 054447

Long-time variation of magnetic structure in multistep metamagnets $Ca_3(Co\text{-}M)_2O_6$

Taketo Moyoshi and Kiyoichiro Motoya

Department of Physics, Tokyo University of Science, 2641 Yamazaki, Noda 278-8510, Japan

E-mail: `moyoshi@ph.noda.tus.ac.jp`

Abstract. A long-time variation of magnetic structure has been observed in a geometrically frustrated magnet $Ca_3Co_2O_6$. In order to clarify that the long-time variation in this material does not originate in randomness due to impurities, magnetization measurements have been made on single crystals of diluted compounds $Ca_3(Co\text{-}M)_2O_6$ (M=Ga, Mg and Al). The long-time variations of the magnetization were detected in these compounds and characteristics of the time variation behavior were not much modi ed by the introduction of disorder produced by the substitution of non-magnetic atoms. These results show that the long-time variation of the magnetic structure in $Ca_3Co_2O_6$ is not caused by randomness or imperfections.

1. Introduction

In spin-glass systems, long-time variations of magnetic properties have been regarded as a characteristic behavior due to the multi-valley structure of the free energy which comes from random magnetic interactions [1]. Therefore, in a system without randomness, we have not expected to observe long-time variation of magnetic properties. However, a long-time variation of magnetic structure has been detected in $CeIr_3Si_2$ which has no randomness or imperfections [2, 3]. This compound shows successive magnetic transitions and multi-step metamagnetic transitions in low magnetic eld, which indicate the presence of frustrating magnetic interactions. We examined several compounds which show those magnetic transitions and found a long-time variation of magnetic structure in $Ca_3Co_2O_6$ and several rare-earth intermetallic compounds [4-6]. We think that the long-time variation of magnetic structure in these materials is caused by the competing magnetic interactions.

$Ca_3Co_2O_6$ crystallizes in a hexagonal structure [7]. It consists of Co chains along the c-axis and these chains are arranged in a triangular lattice in the c-plane. This material exhibits two magnetic transitions at T_{c1}=25 K and T_{c2}=13 K and 5-step metamagnetic transitions below 7 T at 1.9 K [8].

In neutron scattering measurements, we found that the magnetic Bragg peak observed at (1, 0,) immediately after cooled below T_{c1} moved towards =0 with time. We determined that the magnetic moments of Co atoms form a series +++ +++/ /+++ +++ along a Co chain where + and mean the moment direction parallel and antiparallel to the c direction, respectively. Therefore, the shift of the peak position with time corresponds to an increase of the ferromagnetically coupled length of the Co chain.

In order to clarify that the long-time variation of magnetic structure in this material does not originate in randomness due to impurities or defects, we have investigated the time variation of the magnetization of $Ca_3(Co\text{-}M)_2O_6$ (M=Ga, Mg and Al).

2. Experimental procedure

A single crystal of $Ca_3(Co_{0.986}Ga_{0.014})_2O_6$ was prepared by the ux method [9]. The size of the crystal is 0.5 mm 0.5 mm 3 mm with the longest direction parallel to the c-axis. The lattice parameters at room temperature are a=9.094 A and c=10.441 A (hexagonal unit cell). The concentration of Ga was determined by Electron Probe Micro Analyzer.

Magnetization measurements were made using a superconducting quantum interference device (SQUID) magnetometer (Quantum Design model MPMS). A magnetic eld was applied parallel to the c-axis. The temperature variation of magnetization was measured with zero- eld-cooled (ZFC) and eld-cooled (FC) processes at a eld of B=5 mT. Magnetization processes was measured at T=1.9 K with increasing and decreasing magnetic eld of up to B=5.5 T. The time variations of magnetization were measured after the sample was cooled to various target temperatures from T=30 K ($T>T_{c1}$) at B=5 mT. A typical time needed to reach the target temperature was 7 min. Then the magnetization was repeatedly measured as a function of time t. Speci c heat measurements were made using a Quantum Design PPMS. The same measurements were made on single crystals of the Mg-doped and the Al-doped compounds.

3. Results and Discussion

Figure 1 (a) shows the temperature variation of magnetization of $Ca_3(Co_{0.986}Ga_{0.014})_2O_6$ measured with B=5 mT along the c-axis. Solid and open symbols are the data measured in the ZFC and FC processes, respectively. The ZFC curve shows a sudden increase at T_{c1}=23 K with decreasing T and it shows peaks at T_{c2}=13 K and T_{c3}=8 K. In the FC process, an increase in magnetization at T_{c1} was observed, but peaks at T_{c2} and T_{c3} were not observed. The transition temperatures T_{c1} and T_{c2} are close to those of $Ca_3Co_2O_6$. The behaviors at T_{c1} and T_{c2} are also similar to those of $Ca_3Co_2O_6$. However, the magnetization value is about 1/5 of that of $Ca_3Co_2O_6$ and a peak at T_{c3} was not observed in $Ca_3Co_2O_6$.

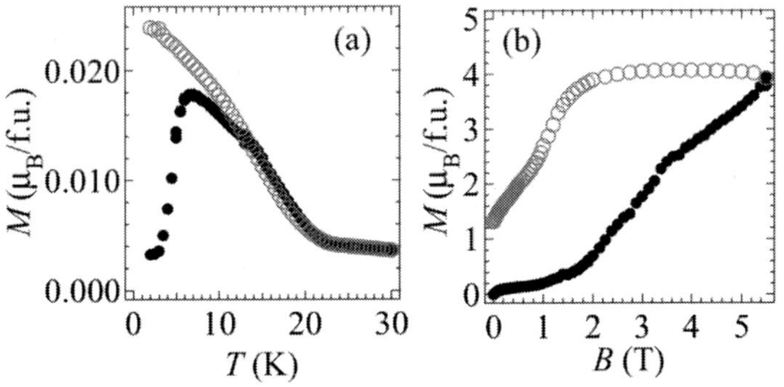

Figure 1. (a) Temperature variation of magnetization of $Ca_3(Co_{0.986}Ga_{0.014})_2O_6$ measured with B=5 mT applied along the c-axis. Solid and open symbols are the data measured in the zero- eld-cooled and eld-cooled processes, respectively. (b) Magnetization processes of $Ca_3(Co_{0.986}Ga_{0.014})_2O_6$ measured at T=1.9 K. Solid and open symbols are the data measured in the increasing and decreasing processes, respectively.

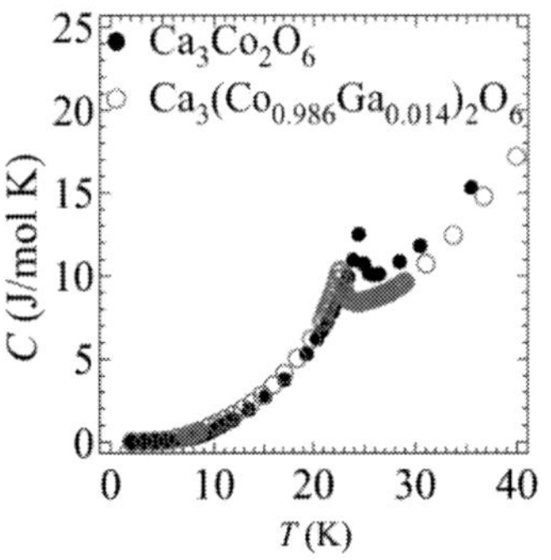

Figure 2. Temperature variation of speci c heat of $Ca_3Co_2O_6$ (closed circle) and $Ca_3(Co_{0.986}Ga_{0.014})_2O_6$ (open circle).

Figure 1(b) shows the magnetic eld variation of the magnetization of $Ca_3(Co_{0.986}Ga_{0.014})_2O_6$ measured at T=1.9 K up to B=5.5 T. Solid and open symbols are the data measured in the increasing and decreasing processes, respectively. In this measurement, the broad metamagnetic transitions were observed. The magnetization values and critical magnetic elds are di erent between two processes. In the increasing process, the magnetization curve shows metamagnetic transitions at B_{c1}=0.1 T, B_{c2}=1.0 T, B_{c3}=1.8 T and B_{c4}=3.6 T. In the decreasing process, metamagnetic transitions were observed at B'_{c1}=0.1 T and B'_{c2}=0.8 T. The metamagnetic transitions are much broader than those observed in $Ca_3Co_2O_6$. The saturation magnetization was 4 B per formula unit, which is smaller than that of $Ca_3Co_2O_6$.

Figure 2 shows the temperature variation of the speci c heat of $Ca_3Co_2O_6$ and $Ca_3(Co_{0.986}Ga_{0.014})_2O_6$. These curves show sharp peaks at 24.4 K and 22.6 K. These temperatures are close to the magnetic transition temperature T_{c1} observed in the temperature variation of magnetization. Moreover, the shapes of the peaks of two curves are quite similar. These results show that Ga atoms are substituted uniformly into the Co positions.

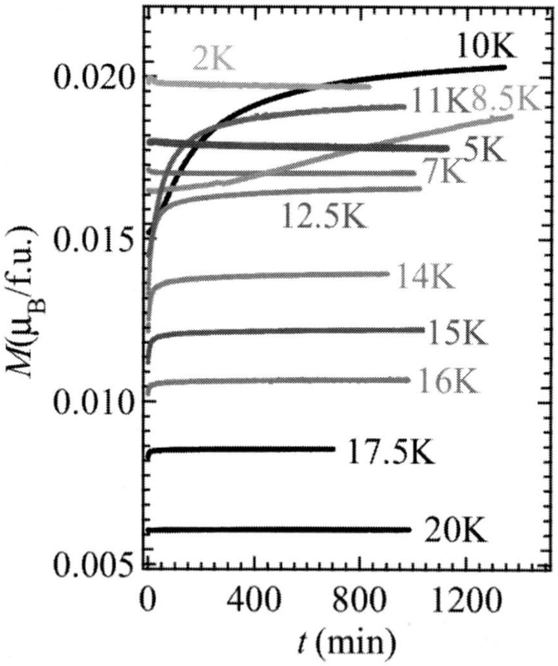

Figure 3. Time variations of magnetization of $Ca_3(Co_{0.986}Ga_{0.014})_2O_6$ after cooling to various temperatures from T=30 K in B=5 mT

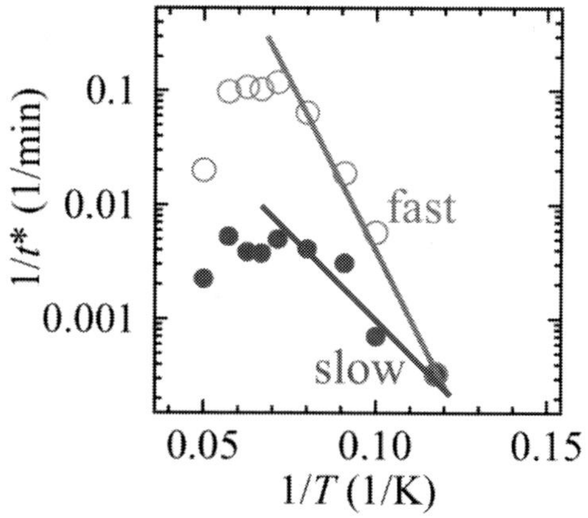

Figure 4. Arrhenius plot of t_{MF} (open circle) and t_{MS} (closed circle).

Figure 3 shows the time variation of magnetization $Ca_3(Co_{0.986}Ga_{0.014})_2O_6$ measured with FC process at B=5 mT. The time variation behavior systematically changes with temperature. At T=2, 5 and 7 K, the magnetization (M) decreases with time (t) up to t 1000 min. At 8.5 K, M increases with t and changes the curvature at 400 min. Above 10 K, M monotonically increases with t. These time variation behavior of magnetization is quite similar to that of $Ca_3Co_2O_6$. In order to investigate the detail of the time variation behavior, for $T \geq 8$ K, we t the data by a sum of two exponential functions as

$$M(t) = M_0 + \quad M_{MF}\exp(\quad t/t_{MF}) + \quad M_{MS}\exp(\quad t/t_{MS}) \quad , \tag{1}$$

Table 1. Activation energies $E_{\mathrm{aMF}}/k_{\mathrm{B}}$ and $E_{\mathrm{aMS}}/k_{\mathrm{B}}$ of $\mathrm{Ca_3Co_2O_6}$ and $\mathrm{Ca_3(Co\text{-}M)_2O_6}$.

	$E_{\mathrm{aMF}}/k_{\mathrm{B}}$ (K)	$E_{\mathrm{aMS}}/k_{\mathrm{B}}$ (K)
$\mathrm{Ca_3Co_2O_6}$	90 8	48 7
$\mathrm{Ca_3(Co_{0.986}Ga_{0.014})_2O_6}$	137 3	76 7
$\mathrm{Ca_3(Co_{0.985}Mg_{0.015})_2O_6}$	120 8	63 5
$\mathrm{Ca_3(Co_{0.996}Al_{0.004})_2O_6}$	90 7	49 5

which is the same function used in the case of $\mathrm{Ca_3Co_2O_6}$ in ref. 4. In this equation, t_{MF} and t_{MS} mean the fast and slow components of the characteristic times for the time variation in M. The Mg-doped and the Al-doped samples also showed similar time variation of M and we tted the data by the same function.

Figure 4 shows the Arrhenius plot ($\log(1/t\ )$ vs $1/T$ plot) of the two characteristic times t_{MF} and t_{MS}. Below T_{c2} (*i.e.*, $1/T>0.08$ $\mathrm{K}^{\ 1}$) two characteristic times follow the Arrhenius law as shown by the straight lines. The value and the temperature dependence of t_{MF} and t_{MS} are nearly same to those of $\mathrm{Ca_3Co_2O_6}$. Table 1 shows the activation energies E_{aMF} and E_{aMS} for the fast and slow components of the time variation of M for all samples observed in the present measurements. The activation energies for $\mathrm{Ca_3(Co\text{-}M)_2O_6}$ are not largely modi ed from those of $\mathrm{Ca_3Co_2O_6}$.

To summarize, we investigated the long-time variation of the magnetization of $\mathrm{Ca_3(Co\text{-}M)_2O_6}$ (M=Ga, Mg and Al). Characteristics of the time variation behavior were not much modi ed by the disorder produced by the substitution of non-magnetic atoms. These results show that the long-time variation of the magnetic structure in $\mathrm{Ca_3Co_2O_6}$ is not caused by randomness or imperfections.

We thank the members of the Neutron Scattering Laboratory, Institute for Solid State Physics, the University of Tokyo for supporting our experiments. This work is partly supported by a Grant-in-Aid for Scienti c Research from the Ministry of Education, Culture, Sports, Science, and Technology.

References

[1] Fischer K H and Hertz J H 1991 Spin Glasses *Cambridge University Press, Cambridge*
[2] Motoya K, Muro Y and Takabatake T 2010 *J. Phys. Conf. Series* **200** 032048
[3] Motoya K, Muro Y and Takabatake T 2010 *J. Phys. Conf. Series* **251** 012019.
[4] Moyoshi T and Motoya K 2011 *J. Phys. Soc. Jpn.* **80** 034701
[5] Moyoshi T, Takahashi R and Motoya K 2011 *J. Phys. Conf. Series* **273** 012125
[6] Motoya K, Moyoshi T and Shigeoka T 2011 *J. Phys. Conf. Series* **273** 012124
[7] Fjellvåg H, Gulbrandsen E, Aasland S, Olsen A and Hauback B C 1996 *J. Solid State Chem.* **124** 190
[8] Kageyama H, Yoshimura K, Kosuge K, Mitamura H and Goto T 1997 *J. Phys. Soc. Jpn.* **66** 1607
[9] Maignan A, Michel C, Masset A C, Martin C and Raveau B 2000 *Eur. Phys. J.* B **15** 657

Journal of Physics: Conference Series **391** (2012) 012101 doi:10.1088/1742-6596/391/1/012101

NMR studies of CDW transition in the spinel-type $Cu(Ir_{1-x}V_x)_2S_4$

H Niki[1], K Higa[1], M Oshiro[1], Y Okada[1], M Yogi[1], S Ebisu[2] and S Nagata[2]

[1] Department of Physics, Faculty of Science, University of the Ryukyus, Nishihara, Okinawa 903-0213, Japan

[2] Muroran Institute of Materials Research, Muroran, Hokkaido 980-8577, Japan

E-mail: niki@sci.u-ryukyu.ac.jp

Abstract. The spinel CuV_2S_4 exhibits two-step anomalies at about 92 K (T_{t1}) and 56 K (T_{t2}), reflecting the CDW state. In order to investigate the CDW state on V rich side in $Cu(Ir_{1-x}V_x)_2S_4$ microscopically, NMR measurements of ^{51}V for $Cu(Ir_{1-x}V_x)_2S_4$ (x = 1.00, 0.98 and 0.95) have been carried out between 4.2 and 300 K. The line width of ^{51}V for x = 1.00 is very narrow above T_{t1} and become wider with decreasing x. For x = 1.00 and 0.98, the line width is suddenly broadened at T_{t1} due to an electrical quadruple interaction between ^{51}V nucleus and CDW charge modulation, but for x = 0.95, the line width is continuously increased around T_{t1}. The Knight shifts of ^{51}V for x = 1.00, 0.98 and 0.95 show almost the same negative shift due to the influence of d-electrons above T_{t1}. At T_{t1}, the values of the Knight shift for x = 1.00 and 0.98 suddenly change to positive values, but for x = 0.95 continuously shifts to a positive one. The $1/T_1T$'s of ^{51}V for x = 1.00, 0.98 and 0.95 obey the Korringa law above T_{t1}. Below T_{t1}, the $1/T_1T$ values of ^{51}V for x = 1.00, 0.98 and 0.95 take constant values, which shows the Korringa law resulting from Pauli paramagnetic state.

1. Introduction

The spinel CuV_2S_4 exhibits CDW transitions at 92 K (T_{t1}) with a structural transformation from cubic to tetragonal phase and at 56 K (T_{t2}) with the transformation without changing tetragonal symmetry [1-4]. On the other hand, $CuIr_2S_4$ exhibits a temperature-induced metal-insulator transition (MIT) around T_{MI} = 226 K with a structural transformation from cubic to triclinic [5,6]. Simultaneously, at this MIT, a charge-ordering by Ir^{3+} and Ir^{4+} and a bond dimerization with spin-singlet state by Ir^{4+} take place [5]. The CDW and the MIT states are incompatible with each other. Therefore, a systematic study was made to investigate the concentration dependences of structural, electrical and magnetic properties for spinel-type $Cu(Ir_{1-x}V_x)_2S_4$ [5]. According to these results [5], a low substitution ($x \geq 0.95$) of Ir for V destroys the CDW transition. and similarly, a low substitution ($x \leq 0.05$) of V for Ir leads to disappearance of the step-like sharp MIT [6]. It was found that both transitions are very sensitive for the change of x in the vicinity of both x = 1.00 and 0.00. In order to investigate the physical properties of the CDW state on V rich side in $Cu(Ir_{1-x}V_x)_2S_4$ from a microscopic point of view, ^{51}V NMR measurements for $Cu(Ir_{1-x}V_x)_2S_4$ (x = 1.00, 0.98 and 0.95) have been carried out. The temperature dependences of the line width (W), Knight shift (K) and spin-lattice relaxation time (T_1) have been measured for these samples.

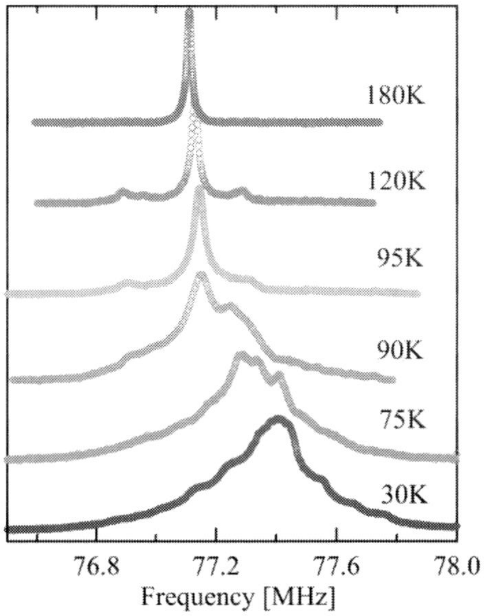

Figure 1. Temperature dependence of the spectra of ^{51}V NMR in CuV_2S_4.

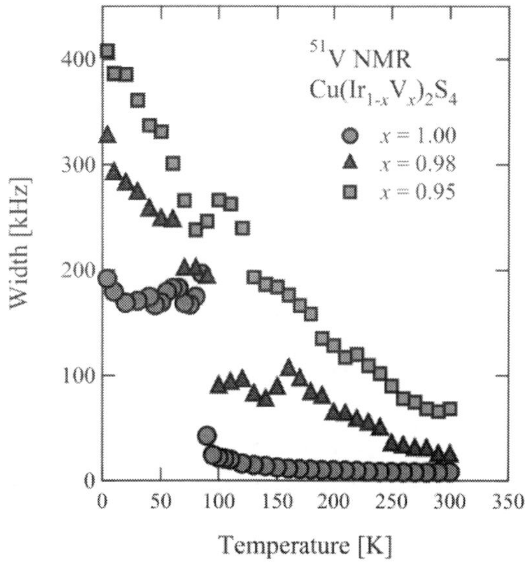

Figure 2. Temperature dependence of the line width (W) on ^{51}V NMR in $Cu(Ir_{1-x}V_x)_2S_4$ (x = 1.00, 0.98 and 0.95).

2. Experimental

The polycrystalline specimens were prepared by a solid state reaction method [5]. Mixtures of high-purity ne powders of Cu, Ir, V and S with normal stoichiometry were calcined in sealed quartz tubes at 850 °C for 10 days. Subsequently, the specimens were reground and pressed to form the pellets, and then were sintered at 850 °C for 2 days. The powdered specimens were prepared for NMR measurements. No secondary phase in the specimens was detected in the powder x-ray di raction pro le. A conventional pulse NMR spectrometer was utilized for the measurements of ^{51}V NMR. A magnetic eld was applied at about 6.9 T (about 78 MHz) by a superconducting magnet with the magnetic- eld homogeneity of 10^{-5}. ^{51}V-NMR shifts were obtained with gyro-magnetic ratio of 11.193 MHz/T. The line width and peak shift of the spectrum were derived from Fourier transform of echo envelope. Spin-lattice relaxation time T_1 was measured by means of the saturation recovery methods.

3. Results and Discussion

Figures 1 shows the spectra of ^{51}V NMR of CuV_2S_4 (x = 1.00) at several temperatures. The single narrow spectrum can be observed in high temperatures. As an example, the spectrum at 180 K is represented in Fig. 1. However, like the spectrum at 120 K in Fig. 1, the satellites like shoulder appears on both sides of the narrow central spectrum in the temperature region below about 130 K because of the appearance of short range CDW's as the precursory phenomenon for CDW transition. These satellite peaks appears due to a nuclear electrical quadruple interaction between ^{51}V nucleus and CDW charge modulation. Below CDW transition (T_{t1} = 92 K), the spectrum consists of superposition of the central resonance line and satellite ones resulting from incommensurate CDW charge modulation. In the spectrum, especially the central resonance line also occurs the line broadening by nuclear quadrupole interaction caused from the structural transformation from cubic to tetragonal phase due to CDW transition. When all ^{51}V nuclei are a ected by CDW charge modulation, the central line would be disappeared and the satellites would be grown up. However, the central resonance line is considerably remained even when

CDW state develops in the crystal at su ciently lower temperatures than T_{t1}. This may result from the imperfection of crystal. But the reason why the central line is remained is not clear. Below the second CDW transition (T_{t2} = 56 K), the resonance spectra shape is apparently di erent from that above T_{t2}.

Figure 2 shows temperature dependence of the line width (half width of half maximum) of ^{51}V NMR in $Cu(Ir_{1\ x}V_x)_2S_4$ (x = 1.00, 0.98, 0.95). For x = 1.00, the line width is 8.3 kHz at 300 K and slightly increases down to T_{t1}, while it shows an abrupt increase at T_{t1} without a temperature-hysteresis. The line width is about 170 kHz at 85 K below T_{t1} and then almost constant with decreasing temperature. Due to inhomogeneous line broadening, the line width increases with decreasing concentration x in high temperature region. For x = 0.98, the line width at 300 K, which is about 20 kHz, gradually increases with decreasing temperature, then abruptly changes at T_{t1}, and becomes about 200 kHz at 85 K. Below T_{t1}, the line width is increases with decreasing temperature. For x = 0.95, the line width, which is about 70 kHz at 300 K, increases more rapidly than that for x = 0.98 with decreasing temperature. At T_{t1}, the line width does not show the step-like anomaly, and more rapidly increases with decreasing temperature below T_{t1}. The line width of ^{51}V NMR is very sensible for T_{t1} transition, but is not sensitive for T_{t2} transition.

Figure 3 shows the temperature dependence of Knight shift (K) for ^{51}V NMR in $Cu(Ir_{1\ x}V_x)_2S_4$ between 4.2 and 300 K. For all samples, Knight shift in metallic phase above T_{t1} indicates the negative contribution to Knight shift from the core polarization of d conduction electrons. Knight shift for x = 1.00 and 0.98 samples represents a sudden change to a positive side by about 0.16 % without a temperature-hysteresis at T_{t1}. For x = 0.95 sample, however, it continuously changes to the positive side by about 0.15 % without a temperature-hysteresis from 130 to 80 K. These results show that the negative contribution to Knight shift from the core polarization of the d electrons disappears by the appearance of CDW state below T_{t1}. Below T_{t1}, Knight shift for x = 1.00 almost linearly shifts to the positive side. This phenomenon may be able to explain that the contribution of s electrons to Knight shift increases due to CDW transition. The rate of shifts to positive side is reduced when the concentration of x is decreased. This implies that the CDW state in x = 0.95 sample would be weaken. Therefore, Knight shift for x = 0.95 would keep almost constant below 50 K. The ^{51}V-Knight shift would be not sensitive for CDW transition at T_{t2}.

Spin-lattice relaxation time T_1 has been measured from 4.2 to 300 K. For x =1.00, the nuclear longitudinal magnetization recovery $f(t) = 1 \quad (M(t)/M_0)$ can be expressed by a single exponential type above T_{t1}. However, for x = 0.98 and 0.95, $f(t)$ cannot be expressed by a single exponential type above T_{t1} because of the distribution of T_1. The line width is narrow as the structure is cubic symmetry above T_{t1}. Therefore, the recovery can be analyzed by means of stretched exponential relaxation (SER) type as follows,

$$f(t) = y_0 \exp\left\{ \quad \left(\frac{t}{T_1}\right)^p \right\}. \tag{1}$$

The values of power p represent about 1.0 for x = 1.00, about 0.95 for x = 0.98 and about 0.85 for x = 0.95 above T_{t1}.

Below T_{t1}, as the structure is tetragonal symmetry, nuclear magnetic relaxation is a ected by nuclear quadrupole interaction because ^{51}V nuclei has a nuclear spin of 7/2. And also ^{51}V nuclei is in uenced by CDW charge modulation. Therefore, it can be expected that the nuclear magnetization recovery in the case of the transition +1/2 $\leftrightarrow$ -1/2 is analyzed by means of the following equation [7]:

$$f(t, T_1) = y_1 \left\{ \frac{1}{84} \exp\left(\quad \frac{t}{T_1} \right) + \frac{3}{44} \exp\left(\quad \frac{6t}{T_1} \right) + \frac{75}{364} \exp\left(\quad \frac{15t}{T_1} \right) + \frac{1225}{1716} \exp\left(\quad \frac{28t}{T_1} \right) \right\}. \tag{2}$$

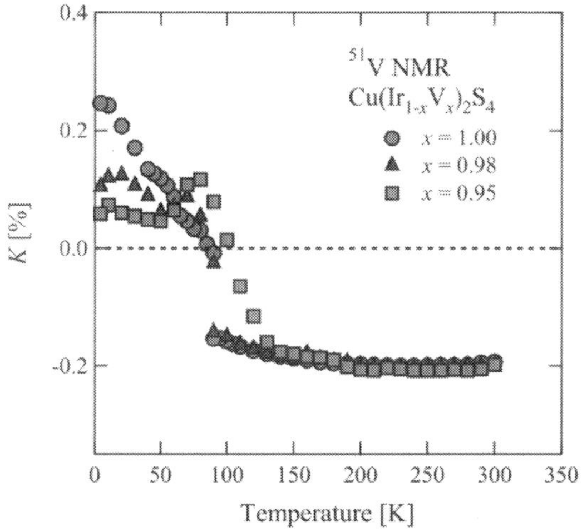

Figure 3. Temperature dependence of Knight shift (K) on ^{51}V NMR in $Cu(Ir_{1\ x}V_x)_2S_4$ (x = 1.00, 0.98 and 0.95).

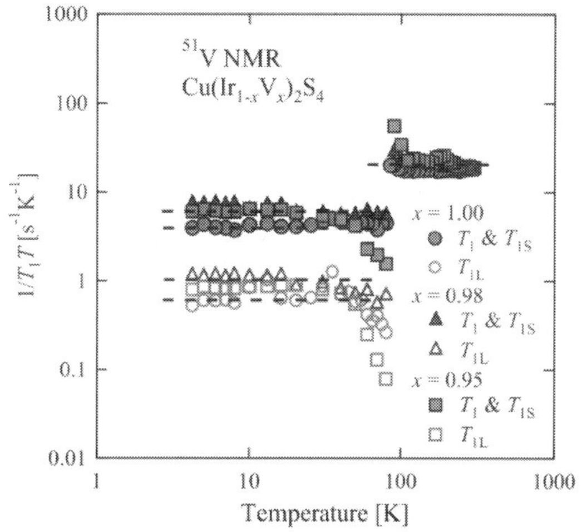

Figure 4. Temperature dependence of $1/T_1T$ on ^{51}V NMR in $Cu(Ir_{1\ x}V_x)_2S_4$ (x = 1.00, 0.98 and 0.95).

However, the recovery equation is slightly modi ed because T_1 is distributed. The recovery curve can be well explained by $g(t, T_1) = cf(t, T_{1S}) + (1\quad c)f(t, T_{1L})$, consisting of two components of short T_{1S} and long T_{1L}. The reason why T_1 is distributed in these materials below T_{t1} is not clear. The value of c is about 0.9 for $x = 1.00$, about 0.85 for $x = 0.98$ and about 0.8 for $x = 0.95$ below T_{t1}.

The temperature dependence of $1/T_1T$ from 4.2 to 300 K is shown in Fig. 4. In the metallic phase, $1/T_1T$ for all samples obeys the Korringa relation, which indicates T_1T = constant, as expected in a Pauri paramagnetic metallic state. Below T_{t1}, for $x = 1.00$ and 0.98 samples, $1/T_1T$ is almost constant. For $x = 0.95$, $1/T_1T$ becomes constant below 50 K. This means that $1/T_1T$ obeys Korringa relation resulting from Pauli paramagnetic state in the temperature region below T_{t1}. The appearance of Pauli paramagnetic state below T_{t1} would be ascribed to three dimensionality of this material.

In summary, for $x = 1.00$, the additional spectra of ^{51}V NMR due to the electrical quadruple interaction between ^{51}V nucleus and CDW charge modulation can be observed below about 130 K, which is induced by the appearance of short range CDW's as the precursory phenomenon for CDW transition. For $x = 1.00$, 0.98 and 0.95, Pauli paramagnetic metallic states can be also observed below CDW transition. From ^{51}V NMR results, in two CDW transitions (T_{t1} and T_{t2}), V sites are very sensible for T_{t1} transition, but are not sensitive for T_{t2} transition.

References

[1] Fleming R M, DiSalvo F J, Cava R J and Waszczak J V 1981 *Phys. Rev.* **24** 2850.
[2] Sekine T, Uchinokura K, Iimura H, Yoshizaki R and Matsuura E 1984 *Solid State Comm.* **51** 187.
[3] Okada H, Koyama K and Watanabe K 2004 *J. Phys. Soc. Jpn* **73** 3227.
[4] Okada H, Koyama K, Nakatake M, Shimada K, Namatame H, Taniguchi M and Watanabe K 2008 *J. Phys. Soc. Jpn* **77** 114711.
[5] Kawashima Y, Horibe N, Awaka J, Yamamoto H, Ebisu S and Nagata S 2007 *Physica* B **387** 208.
[6] Niki H, Okuda H, Okada Y, Higa K, Oshiro M, Fukuyoshi N, Mahoe R, Yogi M, Nakama T, Yagasaki K, Ebisu S and Nagata S 2011 *J. Phys.: Conf. Ser.* **273** 012139.
[7] Narath A 1967 *Phys. Rev.* **162** 320.

Spin-orbital state induced by strong spin-orbit coupling

Hiroaki Onishi
Advanced Science Research Center, Japan Atomic Energy Agency, Tokai, Ibaraki 319-1195, Japan

Abstract.
To clarify a crucial role of a spin-orbit coupling in the emergence of novel spin-orbital states in $5d$-electron compounds such as Sr_2IrO_4, we investigate ground state properties of a t_{2g}-orbital Hubbard model on a square lattice by Lanczos diagonalization. In the absence of the spin-orbit coupling, the ground state is spin singlet. When the spin-orbit coupling is strong enough, the ground state turns into a weak ferromagnetic state. The weak ferromagnetic state is a singlet state in terms of an effective total angular momentum. Regarding the orbital state, we nd the so-called complex orbital state, in which real xy, yz, and zx orbital states are mixed with complex coefficients.

1. Introduction

Novel quantum phenomenon driven by the interplay of strong spin-orbit coupling and Coulomb interaction has been a subject of significant study in the fields of condensed matter physics and materials science. Since spin and orbital degrees of freedom are entangled at a local level by the spin-orbit coupling, the total angular momentum provides a good description of ordered phases and fluctuation properties. On the other hand, we would envisage possible realization of devices with exotic functionalities that utilize cross correlation involving magnetic and electric degrees of freedom, in which magnetic behavior can be manipulated by an applied electric field, while electric behavior can be controlled by a magnetic field.

Recently, a $5d$ transition metal oxide Sr_2IrO_4, in which Ir^{4+} ions have five electrons with a low-spin $(t_{2g})^5$ state, has attracted growing attention as a candidate for a novel Mott insulator characterized by an effective total angular momentum $J_e = \frac{1}{2}$, where $\mathbf{J}_e = -\mathbf{L}+\mathbf{S}$, rather than spin commonly observed in $3d$ Mott insulators [1, 2]. The spin-orbit coupling in the t_{2g} manifold stabilizes the $J_e = \frac{1}{2}$ state. The emergent behavior of the $J_e = \frac{1}{2}$ Mott insulator has been explored in experiments such as angle resolved photoemission spectroscopy [1], optical conductivity [1, 3], resonant x-ray scattering [2], and resonant inelastic x-ray scattering [4]. An isostructural iridate Ba_2IrO_4 has also been reported as the $J_e = \frac{1}{2}$ Mott insulator [5]. Theoretical efforts have been devoted to clarify the characteristics of the $J_e = \frac{1}{2}$ Mott insulator on the basis of first-principles calculations [1, 6] and model calculations [7, 8, 9, 10].

The purpose of this paper is to gain an insight into the spin-orbital state under the spin-orbit coupling from a microscopic viewpoint. We investigate ground-state properties of a t_{2g}-orbital Hubbard model including the spin-orbit coupling by exploiting numerical techniques. We discuss a spin-orbit-induced ground-state transition from a spin singlet state to a singlet state in terms of the effective total angular momentum.

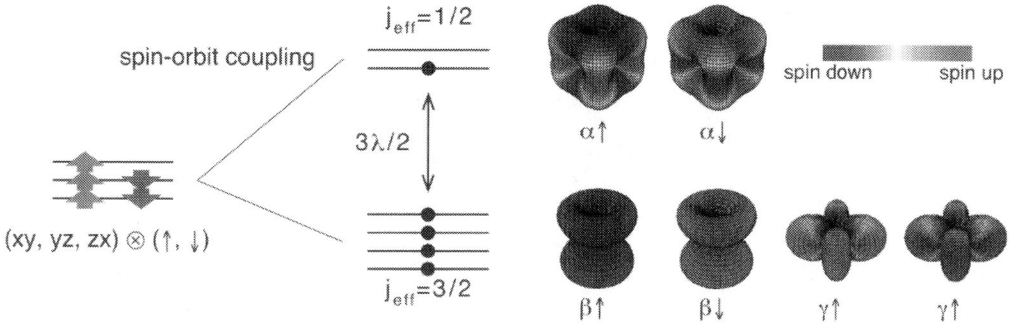

Figure 1. Local electron configuration and wavefunction of the t_{2g} orbitals under the spin-orbit coupling. To visualize the wavefunction, we draw the surface defined by $r=\sqrt{\sum_\sigma |\ (\ ,\ ,\ )|^2}$ in the polar coordinate with being real spin, while the color denotes the weight of spin up and down states.

2. Model and numerical method

We consider triply degenerate t_{2g} orbitals on a square lattice in the xy plane, and the electron number per site is five, corresponding to the low-spin state of Ir^{4+} ions. The t_{2g}-orbital Hubbard model is given by

$$\begin{aligned} H &= \sum_{\mathbf{i},\mathbf{a},\tau,\tau',\sigma} t^{\mathbf{a}}_{\tau\tau'}(d^\dagger_{\mathbf{i}\tau\sigma}d_{\mathbf{i}+\mathbf{a}\tau'\sigma}+\text{h.c.})+\lambda\sum_{\mathbf{i}}\mathbf{L}_{\mathbf{i}}\cdot\mathbf{S}_{\mathbf{i}}+U\sum_{\mathbf{i},\tau}\rho_{\mathbf{i}\tau\uparrow}\rho_{\mathbf{i}\tau\downarrow}+\frac{U'}{2}\sum_{\mathbf{i},\sigma,\sigma',\tau\neq\tau'}\rho_{\mathbf{i}\tau\sigma}\rho_{\mathbf{i}\tau'\sigma'} \\ &+\frac{J}{2}\sum_{\mathbf{i},\sigma,\sigma',\tau\neq\tau'} d^\dagger_{\mathbf{i}\tau\sigma}d^\dagger_{\mathbf{i}\tau'\sigma'}d_{\mathbf{i}\tau\sigma'}d_{\mathbf{i}\tau'\sigma}+\frac{J'}{2}\sum_{\mathbf{i},\sigma\neq\sigma',\tau\neq\tau'} d^\dagger_{\mathbf{i}\tau\sigma}d^\dagger_{\mathbf{i}\tau\sigma'}d_{\mathbf{i}\tau'\sigma'}d_{\mathbf{i}\tau'\sigma}, \qquad (1) \end{aligned}$$

where $d_{\mathbf{i}\tau\sigma}$ is an annihilation operator for an electron with spin $(=\uparrow,\downarrow)$ in orbital τ $(=xy, yz, zx)$ at site $\mathbf{i}$, $\rho_{\mathbf{i}\tau\sigma}=d^\dagger_{\mathbf{i}\tau\sigma}d_{\mathbf{i}\tau\sigma}$, and $\mathbf{S}_{\mathbf{i}}$ and $\mathbf{L}_{\mathbf{i}}$ represent spin and orbital angular momentum operators, respectively. The hopping amplitude is given by $t^{\mathbf{x}}_{xy,xy}=t^{\mathbf{x}}_{zx,zx}=t^{\mathbf{y}}_{xy,xy}=t^{\mathbf{y}}_{yz,yz}=t$, and zero for other combinations of orbitals. Hereafter, t is taken as the energy unit. λ denotes the spin-orbit coupling. U, U', J, and J' are intra-orbital Coulomb, inter-orbital Coulomb, exchange (Hund's rule coupling), and pair-hopping interactions, respectively. We assume that $U=U'+J+J'$ due to the rotational symmetry in the local orbital space, and $J=J'$ due to the reality of the xy, yz, and zx orbital functions.

We numerically investigate ground-state properties of the 2 2 four-site system by Lanczos diagonalization. Because of the three orbitals, the number of bases per site is $4^3=64$, and the matrix dimension of the Hamiltonian becomes huge as the system size increases. In general, the matrix dimension is reduced by decomposing the Hilbert space into a block-diagonal form by using symmetries of the Hamiltonian. In the present case, the total number of electrons is a good quantum number. We cannot use the z component of the total spin as a good quantum number, since the spin SU(2) symmetry is broken by the spin-orbit coupling. We have not utilized the lattice symmetry such as the translational symmetry. Thus the matrix dimension is $10,626$ for four sites, while it grows to $377,348,994$ for eight sites.

3. Results

First we briefly discuss the local electron configuration in the atomic limit. As shown in Fig. 1, the t_{2g} level is split by the spin-orbit coupling into a $j_{\rm e}=\frac{1}{2}$ doublet and a $j_{\rm e}=\frac{3}{2}$ quartet, in which the eigenstates are characterized by the effective total angular momentum $j_{\rm e}$. The

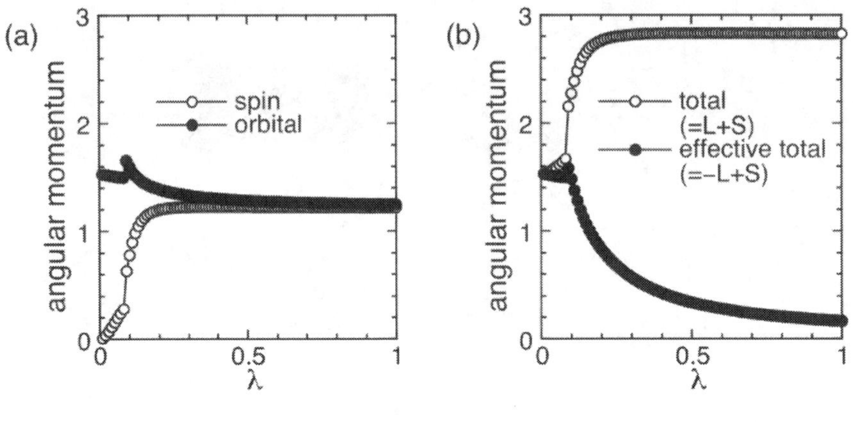

Figure 2. The magnitude of the sum total of angular momenta in the whole system at U'=10 and J=2. (a) Spin and orbital angular momenta. (b) Total and effective total angular momenta.

eigenstates are $|\ \ \rangle = \frac{1}{\sqrt{3}}|xy\ \ \rangle\ \ \frac{1}{\sqrt{3}}|yz\mp\rangle + \frac{i}{\sqrt{3}}|zx\mp\rangle$ with eigenenergy λ for the $j_{e}\ =\frac{1}{2}$ doublet, and $|\ \ \rangle = \frac{1}{\sqrt{2}}|yz\mp\rangle \mp \frac{i}{\sqrt{2}}|zx\mp\rangle$ and $|\ \ \rangle = \sqrt{\frac{2}{3}}|xy\ \ \rangle \mp \frac{1}{\sqrt{6}}|yz\mp\rangle - \frac{i}{\sqrt{6}}|zx\mp\rangle$ with eigenenergy $-\frac{\lambda}{2}$ for the $j_{e}\ =\frac{3}{2}$ quartet, where we introduce and orbitals to distinguish two Kramers doublets in the $j_{e}\ =\frac{3}{2}$ quartet, and pseudospins to label two states in each Kramers doublet. Note that due to the entanglement of spin and orbital states, the wavefunction exhibits anisotropic charge and spin distributions. Since we have five electrons for an Ir^{4+} ion, the lower $j_{e}\ =\frac{3}{2}$ quartet is fully occupied, while the upper $j_{e}\ =\frac{1}{2}$ doublet is half-filled.

Now we move on to the Lanczos results. In Fig. 2(a), we present the magnitude of sum total of spin and orbital angular momenta in the whole system, defined by $\langle(\sum_{\mathbf{i}}\mathbf{S}_{\mathbf{i}})^2\rangle = S_{\rm tot}(S_{\rm tot}+1)$ and $\langle(\sum_{\mathbf{i}}\mathbf{L}_{\mathbf{i}})^2\rangle = L_{\rm tot}(L_{\rm tot}+1)$, respectively. At λ=0, $S_{\rm tot}$ is found to be zero, indicating a spin singlet ground state. As λ increases, $S_{\rm tot}$ gradually increases and a sudden change occurs at a transition point, above which the ground state turns to a weak ferromagnetic state. Note that the induced spin moment is reduced from the maximum value $\frac{1}{2}\ \ 4$=2. We also find an abrupt change for $L_{\rm tot}$, implying that the orbital configuration is reorganized. Figure 2(b) represents the magnitude of sum total of total and effective total angular momenta in the whole system, defined by $\langle(\sum_{\mathbf{i}}\mathbf{J}_{\mathbf{i}})^2\rangle = J_{\rm tot}(J_{\rm tot}+1)$ and $\langle(\sum_{\mathbf{i}}\mathbf{J}_{\rm e\ ,\mathbf{i}})^2\rangle = J_{\rm e\ ,tot}(J_{\rm e\ ,tot}+1)$, respectively. $J_{\rm tot}$ is enhanced in the weak ferromagnetic state. In contrast, $J_{\rm e\ ,tot}$ decreases and approaches zero in the limit of large λ. Thus the weak ferromagnetic state is regarded as a singlet state in terms of the effective total angular momentum.

To clarify how the orbital state changes as λ varies, we measure the charge density in each of the t_{2g} orbitals, $n_{\tau} = \sum_{\sigma}\langle\rho_{\mathbf{i}\tau\sigma}\rangle$, in different two basis sets. Note that due to the translational invariance, the charge density is equivalent in all sites. In Fig. 3(a), we plot the charge density in the real xy, yz, and zx orbitals. At λ=0, we find that n_{xy}=1.5 and n_{yz}=n_{zx}=1.75, since the itinerancy of orbitals depends on the hopping direction and holes preferably occupy the xy orbitals rather than the yz and zx orbitals. As λ increases, n_{xy}, n_{yz}, and n_{zx} approach $\frac{5}{3}$, indicating that xy, yz, and zx orbital states are mixed with equal weight. Note that n_{yz} and n_{zx} are equivalent irrespective of λ. Transforming the basis set into the complex , , and orbitals, we can see the characteristics of the orbital state from the viewpoint of the effective total angular momentum. As shown in Fig. 3(b), we observe a sharp change at the transition point, implying again the reorganization of the orbital configuration. For large λ, the orbitals are singly occupied and relevant to the low-energy property, while the and orbitals are fully

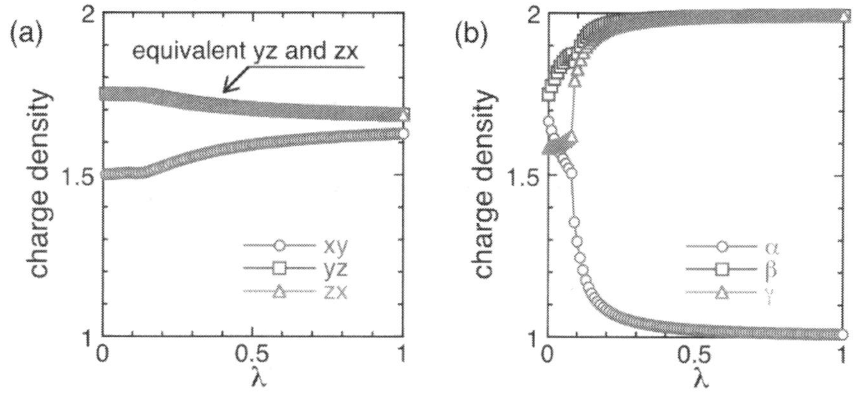

Figure 3. The charge density in each of the t_{2g} orbitals for different basis sets at U'=10 and J=2. (a) Real xy, yz, and zx orbitals. (b) Complex , , and orbitals.

occupied, consistent with the local electron configuration in the atomic limit, as shown in Fig. 1. This clearly indicates the emergence of a complex orbital state, in which spin and orbital states are entangled with complex number coefficients.

4. Summary

We have studied ground-state properties of the t_{2g}-orbital Hubbard model with the spin-orbit coupling by Lanczos diagonalization. We have found that due to the spin-orbit coupling, the ground state changes from the spin singlet to the singlet in terms of the effective total angular momentum. The complex orbital state characterized by $J_e = \frac{1}{2}$ emerges in the spin-orbit-induced state. To clarify novel magnetism in $5d$ transision metal oxides such as Sr_2IrO_4, an important issue is to understand the excitation dynamics under the strong spin-orbit coupling, which we will discuss elsewhere in future.

Acknowledgments

The author thanks G. Khaliullin, S. Maekawa, M. Mori, T. Shirakawa, H. Watanabe, and S. Yunoki for discussions. Part of numerical calculations were performed on the supercomputer at Japan Atomic Energy Agency. This work was supported by Grant-in-Aid for Scientific Research of Ministry of Education, Culture, Sports, Science, and Technology of Japan.

References

[1] Kim B J, Jun H, Moon S J, Kim J Y, Park B G, Leem C S, Yu J, Noh T W, Kim C, Oh S J, Park J H, Durairaj V, Cao G and Rotenberg E 2008 *Phys. Rev. Lett.* **101** 076402
[2] Kim B J, Ohsumi H, Komesu T, Sakai S, Morita T, Takagi H and Arima T 2009 *Science* **323** 1329
[3] Moon S J, Jin H, Choi W S, Lee J S, Seo S S A, Yu J, Cao G, Noh T W and Lee Y S 2009 *Phys. Rev.* B **80** 195110
[4] Ishii K, Jarrige I, Yoshida M, Ikeuchi K, Mizuki J, Ohashi K, Takayama T, Matsuno J and Takagi H 2011 *Phys. Rev.* B **83** 115121
[5] Okabe H, Isobe M, Takayama-Muromachi E, Koga A, Takeshita S, Hiraishi M, Miyazaki M, Kadono R, Miyake Y and Akimitsu J 2011 *Phys. Rev.* B **83** 155118
[6] Jin H, Jeong H, Ozaki T and Yu J 2009 *Phys. Rev.* B **80** 075112
[7] Jackeli G and Khaliullin G 2009 *Phys. Rev. Lett.* **102** 017205
[8] Watanabe H, Shirakawa T and Yunoki S 2010 *Phys. Rev. Lett.* **105** 216410
[9] Onishi H 2011 *J. Phys. Soc. Jpn.* **80** SA141
[10] Wang F and Senthil T 2011 *Phys. Rev. Lett* **106** 136402

Pressure-induced magnetic instability in Pd-Ni alloys

Gendo Oomi[1,2], Sadanori Iwai[2], Masashi Ohashi[3], Tomohito Nakano[4]
1) Department of Education and Creation Engineering, Kurume Institute of Technology, Kurume, Fukuoka 830-0052, Japan
2) Department of Physics, Kyushu University, Fukuoka 812-8581, Japan
3) Faculty of Environmental Design, Kanazawa University, Kanazawa 920-1192, Japan
4) Faculty of Engineering, Niigata University, Niigata 950-2181, Japan

E-mail:geomi@cc.kurume-it.ac.jp

Abstract. The electrical resistivity $\rho(T)$ of dilute Pd-Ni alloys has been measured at high pressure up to 3 GPa. It is found that the $\rho(T)$ of the ferromagnetic Pd-Ni alloy shows an anomalous temperature dependence near the critical pressure P_c, where the ferromagnetism disappears. The results are analysed in the framework of quantum critical behaviour induced by pressure. The effect of magnetic field on the $\rho(T)$ is also examined. Different behaviour of magnetoresistance against pressure was found depending on the alloy concentration and discussed in connection with an instability of ferromagnetism.

1. Introduction

There have been a lot of investigations of the quantum phase transition (QPT) for the intermetallic compounds including 4f or 5f rare earth elements, which is usually induced by changing control parameters such as concentration, magnetic field and pressure [1,2]. However a few reports on the QPT have been reported for the compounds or alloys with 3d transition metal elements [3,4]. It has been well known that Pd metal is an exchange-enhanced Pauli paramagnet and a ferromagnetism is easily induced by a small addition of magnetic elements such as Ni or Fe. A lot of experimental and theoretical investigations have been carried out for dilute Pd-Ni alloys [5,6]. Recently Nicklas et al. [7] and Aliev *et al.*[8] reported that non Fermi liquid (NFL) behaviors were observed near the critical concentration of $x_c = 0.025$ in dilute $Pd_{1-x}Ni_x$ alloys and near the critical saturation field H_s in Fe/Cr magnetic multilayers [8]. In their results characteristic temperature dependent electrical resistivity $\rho(T)$ was observed such as T^n with $n = 1$-1.5, which is largely different from the normal Fermi liquid (FL) behavior of $n = 2$. In order to get a deep insight into the new mechanism of QPT, it is worthwhile to explore the QPT in the wide range of materials. In the present work, the electrical resistivity $\rho(T)$ of dilute $Pd_{1-x}Ni_x$ alloys near the critical concentration($=x_C$) has been measured at high pressure up to 3 GPa and high magnetic field. The $\rho(T)$s are analysed by assuming the relation $\rho(T) = \rho_0 + A'T^n$, where ρ_0 is the residual resistivity, A' and n are constants.

2. Experimental

The polycrystalline samples of $Pd_{1-x}Ni_x$ alloys near x_c were prepared by arc melting the constituent metals of 6N Pd and 4N Ni. The alloys were annealed at 1000 °C for 5 days in an evacuated silica tube. High pressure was generated by using a piston-cylinder method up to 3 GPa. The electrical resistance was measured by means of standard four-probe method. The magnetic field was generated by using a superconducting magnet. The details of the high pressure apparatus were described previously[9].

3. Results and discussion

3.1. Effect of pressure on the temperature dependent electrical resistivity ρ(T) of Pd-Ni alloys

We have measured the $\rho(T)$ of three $Pd_{1-x}Ni_x$ alloys with $x = 0.03$, 0.027 and 0.025. In the following we indicate first the results for $x = 0.027$. Figure 1 illustrates the $\rho(T)$ at various pressures. All $\rho(T)$ curves show smooth temperature dependence without any crossing and discontinuities below room temperature. The pressure derivative of resistance is -5.2×10^{-2} GPa^{-1} at room temperature, which is much larger than those of Fe and Cr [10].

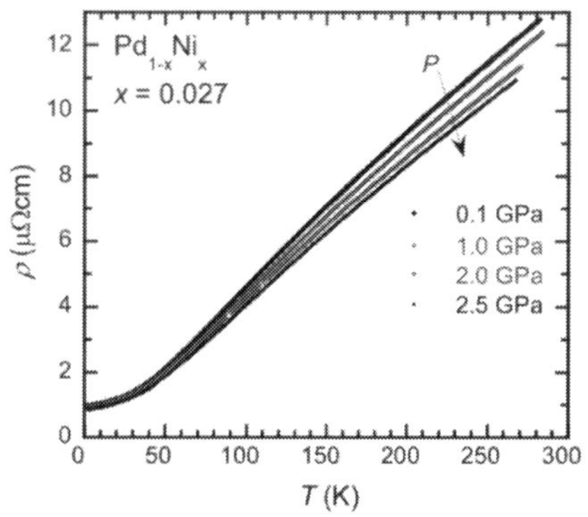

Figure 1. $\rho(T)$ of $x = 0.027$ at various pressures .

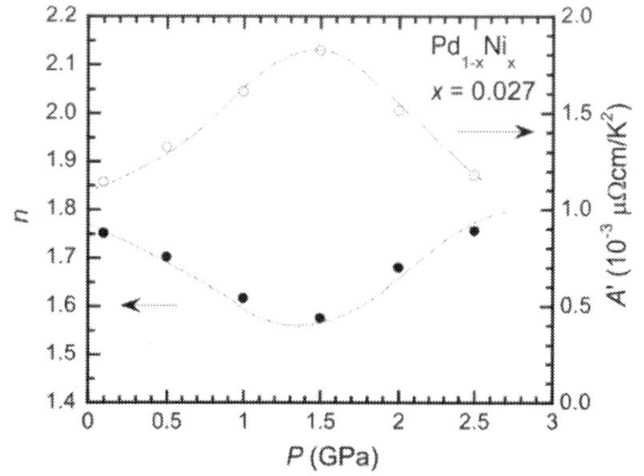

Figure 2. Temperature exponent n and the coefficient A' as a function of pressure.

The $\rho(T)$ at low temperature was fit to the following equation, $\rho(T) = \rho_0 + A'T^n$. Figure 2 shows the values of n and A' as a function of pressure. It is found that the n ($\simeq$1.75 at ambient pressure) decreases with increasing pressure below 1 GPa, shows a minimum near 1.5 GPa (=P_c) with n around 1.6, and increases above 1.8 GPa. Since the n is 2 for normal FL, the present result indicates that the electronic state of $x = 0.027$ deviates significantly from the normal FL, particularly around P_c. Furthermore the values of A' are found to show a maximum around P_c. Recently such behaviours have been observed for several intermetallic heavy Fermion (HF) compounds including f elements such as cerium and is called as non Fermi liquid (NFL) property[11,12]. The behaviours mentioned above are considered as a direct evidence of quantum critical point (QCP) and often discussed in connection with the occurrence of superconductivity. It was reported for $CePtSi_2$ that the values of n have a minimum near 1.4 GPa ($n \simeq 1$) and also the residual resistivity shows a maximum associated with superconductivity around $T_c = 0.12$ K[13].

On the other hand there have been a few reports for the QCP behavior in the alloys or compounds including d elements. The NFL behaviour for $Pd_{1-x}Ni_x$ alloys has been reported [7,14] around x_c ($\simeq$ 0.025), where n shows a minimum and A' has a maximum. In the present work, we have observed the same behaviour around P_c in x=0.027 alloy. Thus the present results as shown in Fig.2 indicate that the alloy shows a quantum critical behaviour at high pressure near P_c. From the temperature dependence of magnetic susceptibility of this alloy, the Curie temperature T_C is estimated to be about 5 K [7]. Since the T_C of ferromagnetic $x = 0.03$ ($T_C \sim 15$ K) decreases with increasing pressure having the rate -3K/GPa, the T_C of $x = 0.027$ may disappear around 1.7 GPa assuming the same pressure coefficient, which is in good agreement with P_c of $x = 0.027$.

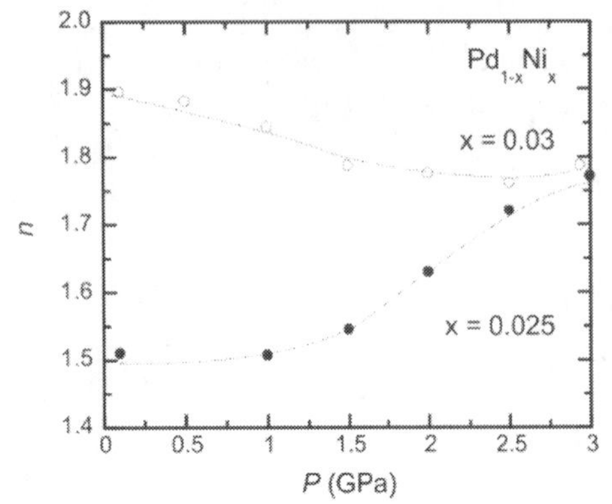

Figure 3 n of x= 0.03 and 0.025 as a function of pressure.

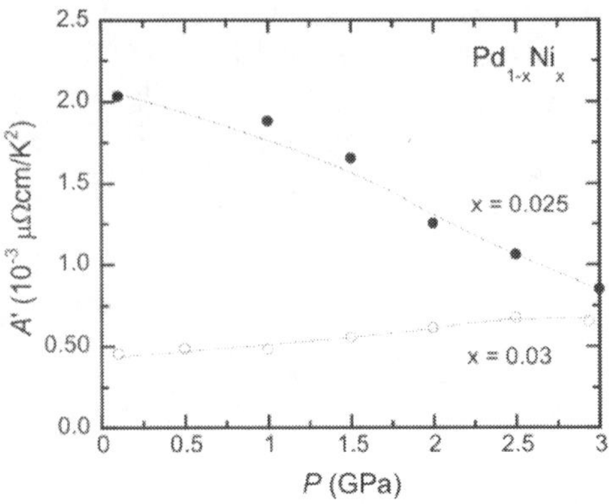

Figure 4 A' of x= 0.03 and 0.025 as a function of pressure.

Next we show the results for other concentrations, x = 0.03 and 0.025. The x = 0.025 alloy is paramagnetic, which is just on the critical concentration x_c, and shows a NFL behaviour even at ambient pressure. In fact, the $\rho(T)$ shows a temperature dependence with $n \simeq 1.5$ at ambient pressure. On the other hand, since the n of ferromagnetic x = 0.03 is 1.9 at ambient pressure, the electronic state is described by normal FL. The pressure dependence of n is shown in Fig. 3 for these two alloys. It is found that n decreases with increasing pressure for ferromagnetic x = 0.03 and smoothly increases for paramagnetic x = 0.025. These results imply that as pressure increases, both the FL state of x = 0.03 and the NFL state of x = 0.025 become unstable but the change in the magnitude of n of x = 0.027 by applying pressure is larger than that of x = 0.03. Considering that the n of x = 0.025 approaches 2 as pressure increases, the FL state of x = 0.025 is expected to recover above 3 GPa.
The values of A' are shown for x = 0.03 and 0.025 in Figure 4 as a function of pressure. There is a large difference among two alloys. A' increases smoothly with pressure for x = 0.03 but decreases significantly for x = 0.025. Since the ferromagnetic state of x = 0.03 is stable at ambient pressure, the change of A' is relatively small. But the change of A' of x = 0.025 is larger than that of x=0.03. This result implies that a magnetic quantum fluctuation in x = 0.025 is larger than that in x = 0.03. As the large fluctuation in x = 0.025 near QPT is suppressed by applying pressure, the FL state recovers at high pressure above 3 GPa.

3.2. Effect of magnetic field on the electronic state of x=0.027.

Figure 5 indicates the n as a function of magnetic fields $B(T)$ at 1.4 GPa, *i.e.*, near QPT. n increases rapidly below 0.3 T but above 1 T, the increasing rate tends to saturate and n is 1.9 at 2 T, which is the same as that of x = 0.03 showing normal FL at ambient pressure. In other words the NFL state at B=0 becomes unstable by applying magnetic field and the FL state recovers around 2T. Since the electronic state of x = 0.027 is considered to be dominated by large quantum fluctuation due to magnetic instability around 1.4 GPa, the present result indicates that this fluctuation is suppressed by applying magnetic field. This observation is consistent with the previous results[15]. Furthermore the A' at 1.4 GPa decreases with increasing B from 1.8 nΩcm/K^2 at 0 T to 0.5 nΩcm/K^2 at 2 T, which is also nearly the same as that of x = 0.03 at ambient pressure (FL state).
Figure 6 shows the magnetoresistance (MR) ratio(=($\rho(H)$- $\rho(H = 0)$)/ $\rho(H = 0)$) at T = 4.2 K and B = 6 T for the present alloys as a function of pressure. Positive MR ratios of 20-30% were found for the all samples. But the pressure dependence is different each other. The MR ratio of x = 0.025 decreases smoothly with pressure, but that of x = 0.027 decreases slightly below 1.5 GPa but above it a small discontinuous drop of about 2% is observed.

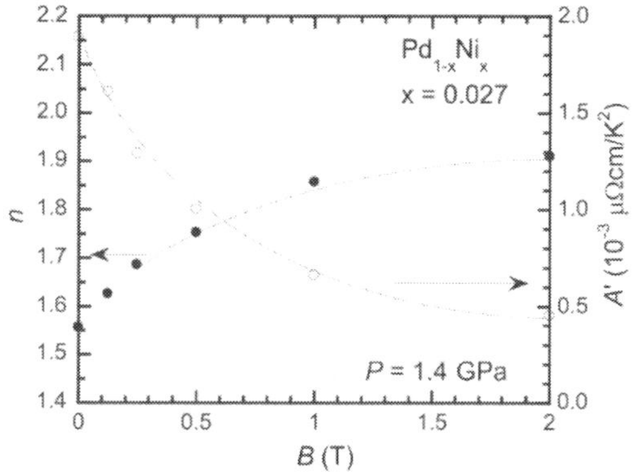

Figure 5. n and A' at 1.4 GPa as a function of magnetic field.

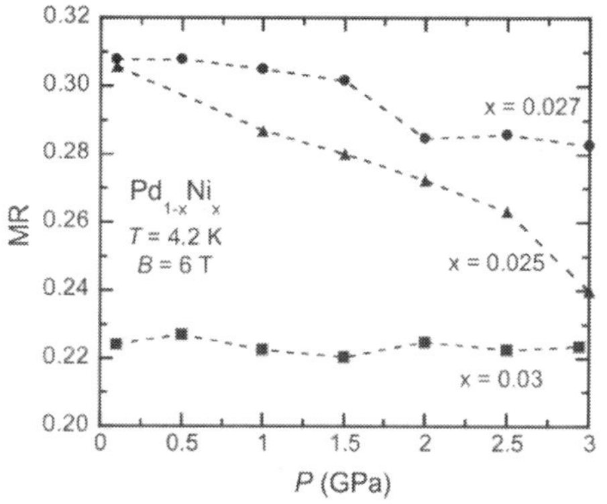

Figure 6 MR ratio at 4.2K of Pd1-xNix alloys as a function of pressure.

The MR ratio of $x = 0.03$ is almost constant below 3 GPa reflecting relatively stable ferromagnetism compared with $x = 0.027$ alloy. The large pressure dependence of $x = 0.025$ is due to the large quantum fluctuations. Since the MR ratio depends strongly on the characteristic temperature such as spin fluctuation or Kondo temperature, *i.e.*, the magnitude of A'[16], the small drop near P_c may be related to the QCP.

The authors would like to express their sincere thanks to Dr. M. J. Nicklas for his useful suggestions and showing them his PhD thesis.

References

[1] Y. Uwatoko *et al.*, The Handbook on the Physics and Chemistry of Rare Earths, vol.42 (to be published).
[2] see for example, Proc. of Novel Pressure-induced Phenomena in Condensed Matter Systems, ed. by T.Kagayama, M. Ohashi and Y.Uwatoko, J. Phys. Soc. Jpn, Suppl. A, 76 (2007).
[3] C. Thessieu, C. Pfleiderer, and J. Flouquet, Physica B, 239 (1997) 67.
[4] G. Oomi, Y. Fuchizaki, M. Ohashi,T. Nakano and Y. Uwatoko, J. Magn. Soc. Jpn. 34 (2010) 263.
[5] R. A. Beyerlein and D. Lazarus, Phys. Rev. B, 7 (1973) 511.
[6] T. Kato and J. Mathon, J. Phys. F:Metal Phys., 6 (1976) 221.
[7] M. Nicklas, M. Brando, G. Knebel, F. Mayr, W. Trinkl, and A. Loidl, Phys. Rev. Lett., 82 (1999) 4268.
[8] F. G. Aliev, V. V. Moshchalkov, and Y. Bruynseraede, Phys. Rev. Lett., 81 (1998) 5884.
[9] F. Honda, S. Kaji, M. Ohashi, G. Oomi, T. Eto and T. Kagayama, J. Phys. :Condens. Matter, 14 (2002)11501.
[10] P. W. Bridgman, The Physics of High Pressure, (Dover, New York,1970) 264.
[11]G. R .Stewart, Rev. Mod. Phys. 73 (2001) 797.
[12] H. Miyagawa, G. Oomi, M. Ohashi, I. Satoh, T. Komatsubara, M. Hedo and Y. Uwatoko, Phys. Rev. B, 78 (2008) 064403.
[13] T. Nakano, M. Ohashi, G. Oomi, K. Matsubayashi and Y. Uwatoko, Phys. Rev. B, 79 (2009) 172507.
[14] A. Tari and B. R. Coles, J. Phys. F: Metal Physics, 1 (1971) L69.
[15] M. J. Nicklas, PhD thesis, Univ. Augusburg, 2000.
[16] T. Kagayama, G. Oomi, R. Yagi, Y. Iye, Y. Onuki and T. Komatsubara, J. Phys. Soc. Jpn. 61 (1992) 2632.

Journal of Physics: Conference Series **391** (2012) 012104 doi:10.1088/1742-6596/391/1/012104

Ab initio study on magnetism and pressure-induced transitions in cubic MnGe

U. K. Rößler

IFW Dresden, Postfach 270116, D-01171 Dresden, Germany

E-mail: u.roessler@ifw-dresden.de

Abstract. State of the art density functional theory (DFT) calculations within the generalized gradient approximation (GGA) are reported for the (metastable) cubic MnGe with chiral B20 structure under applied pressure. It is found that the ambient pressure metallic (nominally) ferromagnetic high spin state (2 μ_B/ f.u.) in MnGe undergoes a transition into a low spin state (1 μ_B/ f.u.). From the calculations, the pressure at this first-order transition is about 18 GPa. The low moment state is halfmetallic, i.e., a small band-gap opens in the density of states for the majority spin electrons at the Fermi edge. Calculations for decreasing lattice parameter show that a fully metallic ferromagnetic low spin state should be recovered, before at still higher pressures the magnetic moment collapses and a non-spin polarized state arises. For the zero, low and high spin states, the equation of state has been determined. As the non-centrosymmetric crystal structure enables chiral Dzyaloshinskii-Moriya couplings, the magnetic order in MnGe is long-period modulated. The theoretical results suggest that half-metallic transport properties could be observed on the background of chiral complex spin-structures in MnGe under presssure or in strained thin films.

1. Introduction

The series of magnetic 3d transition metal – silicides with chiral B20 structure and their alloys, with MnSi as best investigated system, are attracting continuing attention, that derives from their chiral helimagnetism [1, 2, 3] and the possibility to reach a pressure-collapse of their magnetic order [4]. The magnetic properties of MnSi are those of a weakly ferromagnetic metallic system subject to strong spin-fluctuations. Chiral Dzyaloshinskii-Moriya couplings state [5] are responsible for the long-period helimagnetic ground-state [5, 2, 3]. The properties of the analogous germanides are less investigated, owing to their difficult synthesis, but the tendency towards magnetically ordered states is much stronger in this series. For applications and possible integration into spintronics devices, the higher transition temperatures of the germanide series appears promising [6, 7]. The magnetic germanide FeGe with the cubic B20 structure is a chiral helimagnet as MnSi, but the underlying electronic structure is closer to strong band ferromagnetism [8]. A quantum phase transition from the magnetically ordered state to an unconventional paramagnetic Non-Fermi-Liquid (NFL) state has been found at high pressures [9]. Cubic MnGe can be synthesized only under pressure [10, 11]. MnGe is known to undergo a magnetic transition, but information on magnetic order and phase diagram is incomplete. Experimental data report a magnetic ordering temperature 197 K, and from Curie-Weiss fits in paramagnetic state $\theta_C = 187$ K $\mu_{eff} = 3.0$ μ_B [10]. The experiments by Kanazawa et al. show a magnetic ordering temperature $T_N \simeq 170$ K. Magnetization curves at

5 K show technical saturation at magnetic fields $\mu_0 H_{c2} \geq 13$ T only with a moment of about 1.9 μ_B/f.u.[11]. Another anomaly with weak hystereses is seen in a field range $0.5 < \mu_0 H_{c1} < 2$ T for temperatures $T < 100$ K, which may indicate a helix reorientation transition or a transition from a low-field, effectively antiferromagnetic, modulated states into an intermediate possibly also modulated state. This magnetization process in applied magnetic field differs not only regarding the large magnitude of the characteristic fields, which should scale with the strengths of the chiral Dzyaloshinskii-Moriay coupling, from the usual properties of chiral cubic helimagnets. The temperature-dependence of the ferromagnetic saturation field $H_{c2}(T)$ is anomalous, in correspondence to a strong temperature dependence of the magnetic modulation length in the range 3 to 9 nm observed by small-angle neutron scattering experiments on polycrystals [11, 12]. An anomalous Hall effect contribution found by electrical transport measurements in polycrystalline bulk samples of MnGe at low temperatures [11] has been interpreted as indication of a complex spin texture such as a Skyrmion lattice, earlier predicted to exist in chiral ferromagnets [13, 14].

Further investigations on the basic magnetic properties of this compound are desirable to understand the origin of the anomalous magnetic phase diagram and transport data. Here, results from density-functional theory (DFT) calculations are reported on the magnetic properties of MnGe. The theoretical results allow to predict an interesting evolution of the basically ferromagnetic properties in this compound under pressure.

2. Method

Calculation have been performed using a full-potential local orbital method, as implemented in the FPLO code [15], with the generalized gradient approximation (GGA) parametrized by Perdew, Burke Ernzerhof[16]. For comparison also the local spin-density approximation (LSDA) from Perdew and Wang [17] has been used. In the calculations energies have been converged to better than 10^{-8} Ha / unit cell, and electron densities to better than 10^{-6} electrons /(a.u.)^3. Different magnetic states are found by the DFT calculations. Searches for these co-existing spin states have been done by fixed spin-calculations at various reduced lattice parameters compared to the experimentally reported $a = 4.795$ Å at RT [10]. Structure optimization of the B20 cells was performed for all calculated volumes and the different magnetic states with forces on the ions reduced to less than 10^-3 eV/Åusing the scalar relativistic option in the FPLO code. Reported results pertain to these calculations with the GGA, if not stated otherwise.

3. Results

The search for different magnetic states in MnGe resulted in the observation of a ferromagnetic high-spin (HS) state and a low-spin (LS) state with a spin-moment of only $1\mu_B$/ f.u. In the calculation using GGA, the HS state with a moment slightly larger than $2\mu_B$/ f.u. forms the energy minimum at zero pressure. For the LSDA, the LS state is the global minimum. Comparison with the experimental saturation magnetization suggests that the GGA provides the better theoretical description of magnetic properties in MnGe. The DFT results, as depicted in Fig. 1, also show that (i) the HS state is destabilized by reducing the volume of the unit cell, i.e. applying pressure; (ii) the existence range for the LS state is restricted to a certain range of lattice parameters; (iii) for low volumes the magnetic moment in the LS state is reduced, and (iv) this state finally collapses into a non-magnetic zero-spin (OS) state without spin splitting of the electronic bands. The dependence of the total energy on lattice parameter, Fig. 2, has been fitted by the Murnaghan formula to derive the equation of states (eos) for the three different magnetic states, the results are collected in Table 1.

The evolution of the band-structure is shown by the (spin-split) density of states (DOS) in different magnetic states in Fig. 3. The LS structure is half-metallic with a narrow gap in the bands for majority spin at the Fermi edge. For reduced lattice parameter a metallic state is

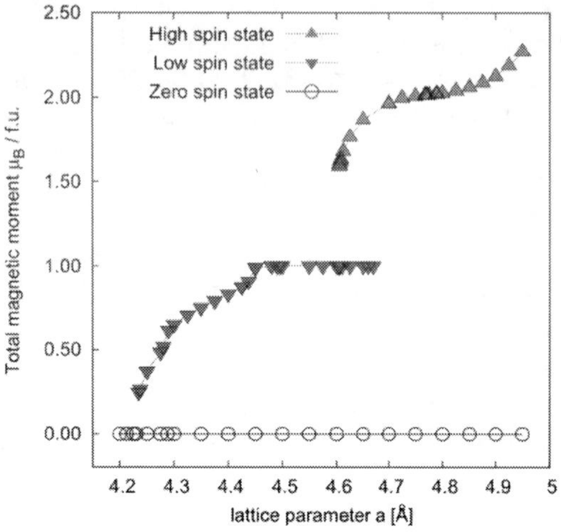

Figure 1. Magnetic moment vs lattice parameter for the three different magnetic states.

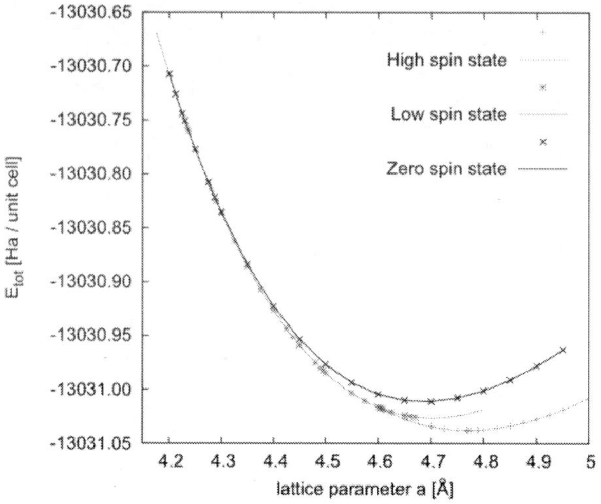

Figure 2. Total energy vs lattice parameter for the three different magnetic states. Symbols are results from the DFT calculations. Lines are fits to the Murnaghan formula.

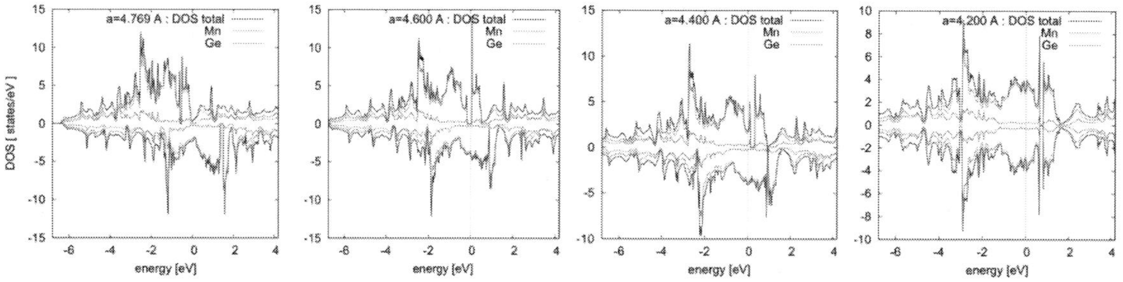

Figure 3. Density of states (DOS) with decreasing lattice constant: $a = 4.769$ Å in HS state close to equilibrium lattice constant at ambient pressure. $a = 4.600$ Å half-metallic LS state under pressure. $a = 4.400$ Å metallic LS state under higher pressure. $a = 4.200$ Å non-spin polarized 0S state at highest pressure.

found, as the majority spin DOS shifts upwards in energy. The LS state finally collapses and the spin-splitting of the band-structure is lost. From the eos results, the pressures for the various transitions can be estimated. The first-order transition from HS to LS state is expected at about 18 GPa, the metallic LS state sets in around 40 GPa, and the pressure collapse of the magnetic moment should occur as a first-order transition between LS and 0S state at or above 75 GPa. The transition pressures are estimates with relatively large uncertainty. This uncertainty is apparent already from the stabilization of the LS state in the LSDA calculations for ambient pressure.

4. Conclusions

The theoretical results suggest that MnGe would allow to investigate the effects of complex spin-textures on the transport in a half-metallic state, i.e. with only one spin character of the electrons at the Fermi edge. The half-metallic LS state should be realizable in experiments

Table 1. Results of the total energy calculations for the different spin states. Using GGA[16] and LSDA [17] functional. Average spin $< s >$ is quoted for the reachable state with lowest energy at lattice parameter a. Remaining columns are parameters from the Murnaghan fits: the equilibrium volume V_0 per atom, ΔE the total energy at this volume with respect to global energy minimum, and B_0, B_0' bulk modulus and its pressure derivative. Note, the LS state in GGA and the HS state in LSDA, the corresponding equilibrium states are not reachable in the calculations. The experimental lattice constant is $= a4.795$ Å at RT [10]. Arras et al. have calculated $a = 4.80$ Å in a GGA-DFT study [6].

		a [Å]	$< s >$ [μ_B/f.u.]	V_0 [Å^3/atom]	ΔE [meV/f.u.]	B_0 [GPa]	B_0'
GGA:	HS	4.77	2.01	13.56	0	136	5.0
	LS	4.67	1.00	12.99	77.4	187	3.6
	0S	4.70	0	12.89	179.8	176	4.3
LSDA:	HS	4.67	1.24	12.33	19.6	147	3.45
	LS	4.60	1.00	12.19	0	201	4.0
	0S	4.58	0	12.06	40.27	213	4.2

with standard high-pressure devices. It is also conceivable to shift the narrow band gap in the majority-spin bands through the Fermi edge by strains in epitaxial films.

Discussions with A.N. Bogdanov, K. Koepernik, M. Richter, Y. Tokura, and H. Wilhelm, and support by DFG project RO 2238/9-1, are gratefully acknowledged.

[1] Y. Ishikawa et al., Solid State Commun. 19, 525 (1976).
[2] P. Bak and M. H. Jensen, J. Phys.C: Solid State Phys. **13**, L881 (1980).
[3] O. Nakanishi et al., Solid State Commun. **35**, 995 (1980).
[4] J.D.Thompson *et al.*, Physica B **161**, 317 (1989).
[5] I.E. Dzyaloshinskii, Sov. Phys. JETP **19**, 960 (1964).
[6] E. Arras, D. Caliste, T. Deutsch, F. Lançon, P. Pochet, Phys. Rev. B **83**, 174103 (2011).
[7] A. Spiesser, V. L. Thanh, S. Bertaina, L. A. Michez, Appl. Phys. Lett. **99**, 121904 (2011)
[8] M. Neef *et al.*, Phys. Rev. B **80**, 035122 (2009).
[9] P. Pedrazzini *et al.*, Phys. Rev. Lett. **98**, 047204 (2007).
[10] H. Takizawa, T. Sato, T. Endo, M. Shimada, J. Solid State Chem. **73**, 40 (1988).
[11] N. Kanazawa, Y. Onose, T. Arima, D. Okuyama, K. Ohoyama, S. Wakimoto, K. Kakurai, S. Ishiwata, and Y. Tokura, Phys. Rev. Lett. **106**, 156603.2011 (2011).
[12] O.L. Makarova, A.V. Tsvyashchenko, I. Merebeau, G. Andre, L.N. Fomicheva, N. Ray, DMI 2011, Internation workshop "Spin chirality and Dzyaloshinskii-Moriya interaction", May 25-27, 2011, St. Petersburg, Russia, Abstract book, p.12
[13] A.N. Bogdanov and D.A. Yablonsky, Sov. Phys. JETP **68**, 101 (1989).
[14] A. Bogdanov, A. Hubert, J. Magn. Magn. Mater. **138**, 255 (1994), **195**, 182 (1999).
[15] K.Koepernik H.Eschrig, Phys. Rev. B **59**, 1743 (1999); [http://www.fplo.de].
[16] J. P. Perdew, K. Burke, M. Ernzerhof, Phys. Rev. Lett. **77**, 3865 (1996).
[17] J. P. Perdew, Y. Wang, Phys. Rev. B **45**, 13244 (1992).

Paramagnetic Resonance in the Cubic Helimagnet ϵ-FeGe

J. Sichelschmidt, M. Schmidt
Max Planck Institute for Chemical Physics of Solids, 01187 Dresden, Germany

H. Wilhelm
Diamond Light Source Ltd., Chilton, Didcot, OX11 ODE, United Kingdom

U.K. Rößler
Leibniz Institute for Solid State and Materials Research, 01187 Dresden, Germany

E-mail: Sichelschmidt@cpfs.mpg.de

Abstract. We have investigated the spin dynamics of the paramagnetic region of ϵ-FeGe by means of Electron Spin Resonance (ESR) spectroscopy. The high-quality single crystal showed a strong paramagnetic resonance line which is well-described by a Lorentzian shape above 287 K. The close relation of the linewidth to the magnetic susceptibility indicates a resonance of a strongly coupled $3d$ / conduction electron system.

1. Introduction

Investigations of ferromagnetic metals by the Electron Spin Resonance (ESR) technique provide valuable information on the spin dynamics in the paramagnetic regime. Important examples of 3d metals are $TiBe_2$ [1], $ZrZn_2$, $NbFe_2$ [2], and MnSi [3, 4] which allow an ESR investigation of the paramagnetic resonance without doping of ESR probe spins and at temperatures low enough to be easily accessible by the ESR technique (unlike the temperatures needed for Fe and Ni paramagnetic resonances [5]). They all have in common an ESR linewidth that is mainly determined by ferromagnetic correlations which considerably enhance the paramagnetic susceptibility. The observation of such correlation effects in ESR has been found to be relevant also for $4f$ heavy fermion metals [6, 7] and allows considerable insight into the nature of metallic magnetism.

Detailed magnetic measurements on the cubic helimagnet ϵ-FeGe (space group $P2_13$) revealed for temperatures below $T_C = 278.2$ K a rich variety of complex magnetically ordered phases as well as field dependent precursor phenomena above T_C [8]. Here, we report ESR investigations in the paramagnetic regime providing one additional example for the ESR observability of paramagnetic resonances in ferromagnetic metals without the need of doping ESR-active centers in the material.

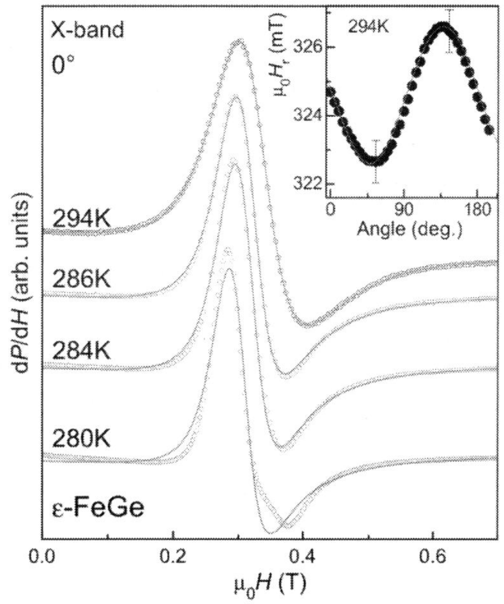

Figure 1. Four representative resonance lines at 9.4 GHz (X-band) for temperatures above magnetic ordering. Red lines show a Lorentzian shape of a paramagnetic resonance which provides an accurate spectra fitting for $T \gtrsim 287$ K only. Amplitudes were normalized for better illustration. Inset displays the resonance field H_r as a function of rotation angle around the [110] axis ($\perp \boldsymbol{H}$). Solid line shows uniaxial fit.

2. Experiment and Results

We measured the power P of a magnetic microwave field $\boldsymbol{h}$ absorbed by an high-quality ϵ-FeGe single crystal [8] as a function of a transverse external static magnetic field $\boldsymbol{H}$ ($\boldsymbol{H} \perp \boldsymbol{h}$). The crystal was mounted with its [110] axis nearly parallel to the $\boldsymbol{h}$-field which results in a $\approx 45°$ orientation of its [100] axis to the fields $\boldsymbol{H}$ and $\boldsymbol{h}$. The temperature was varied with a ^{4}He-flow-type cryostat. The paramagnetic resonance was recorded as the derivative dP/dH because a lock-in technique was used to enhance the signal-to-noise ratio. The parameters of the line (resonance field H_r, linewidth ΔH, intensity I^{ESR}) were evaluated at three microwave frequencies $\omega/2\pi = 1.1, 9.4, 34$ GHz (L-, X-, Q-band). They contain the static and dynamic magnetic properties of the resonating spin system which is characterized by the g-factor $g_{\mathrm{ESR}} \propto \omega/H_{\mathrm{res}}$, the relaxation rate $\Gamma \propto \Delta H$ and the intensity $I^{\mathrm{ESR}} \propto \chi^0$ (static magnetic susceptibility).

Figure 1 shows four examples of well-defined spectra of ϵ-FeGe for three temperatures above T_C. The spectra consist of strong resonance absorptions without any traces of weak and narrow lines which would indicate spurious Fe-containing phases. The lineshape could be well described by a metallic Lorentzian [9] for $T \gtrsim 287$ K. In the framework of a conduction electron spin resonance this metallic Lorentzian would correspond to a Dysonian lineshape [10] in the limit of a much faster relaxation of a conduction electron spin than its diffusion through the skin depth. Hence, from the shape, one cannot distinguish whether the resonance in ϵ-FeGe arises from a local or an itinerant spin probe. This distinction is possible, for instance, for the paramagnetic line of $Fe_{0.5}Co_{0.5}Si$ for which a Dysonian lineshape with a much faster spin-diffusion than spin-relaxation was used [11].

Below $T \approx 287$ K deviations from a Lorentzian shape as well as from a Dysonian shape

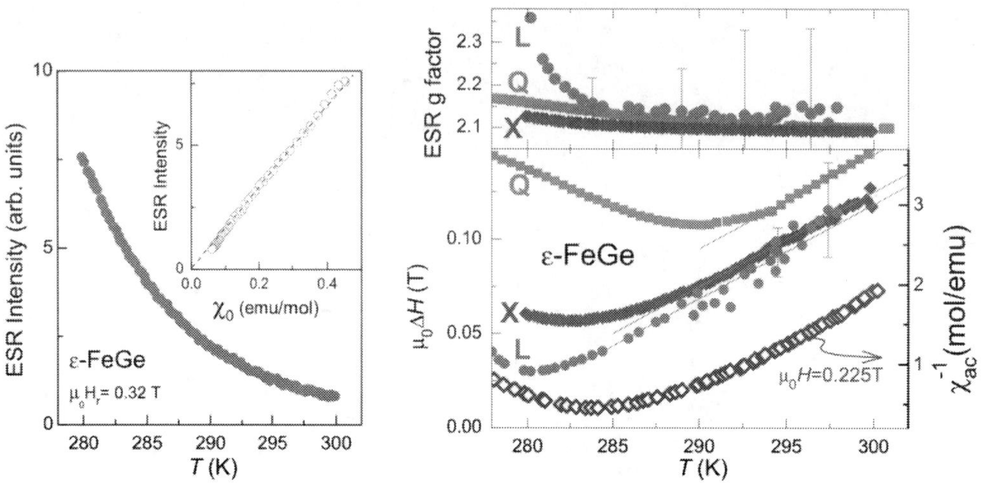

Figure 2. Left: Temperature dependence of ESR intensity and comparison with the static dc-susceptibility χ_0 at $\mu_0 H = 0.32$ T (inset). Right: ESR g factor and linewidth ΔH for L-, X-, and Q-band frequencies. The reciprocal ac-susceptibility χ_{ac}^{-1} (right axis) was measured at 0.225 T [8], close to the X-band resonance field (0.32 T). Solid lines emphasize a linear behavior as discussed in the main text.

gradually appear (compare red lines with data) and become pronounced below ≈ 284 K. These shape distortions are unusual for a purely paramagnetic resonance and occur in the same temperature region where the ac-susceptibility shows a maximum and several crossover lines have been found in the phase diagram above T_C [8]. In this precursor region short-range magnetic modulations build up which could be responsible for the observed line shape distortions. Upon entering the region of magnetic ordering a variety of ferromagnetic resonance modes appear at fields very similar to those reported earlier from polycrystalline FeGe [12]. Similar features where also reported for the resonance modes in the ordered states of MnSi and $Fe_xCo_{1-x}Si$ where a thorough investigation regarding lineshape and resonance field was performed [3, 4, 13, 14, 15]. These results might be applicable for ϵ-FeGe as well. In this paper, however, we refrain from presenting further details of the resonance in the ordered region of ϵ-FeGe and focus on the resonance in the paramagnetic regime instead.

Upon rotating the crystal around its [110] axis (which is aligned $\perp$ H) the resonance field varies with a 180°-periodicity as shown in the inset of Fig.1 for $T = 294$ K. A uniaxial fit (solid line) corresponds to $g_{45^\circ} = 2.107 \pm 0.004$ and $g_{135^\circ} = 2.082 \pm 0.004$. However, these g factors may not reflect the anisotropy caused by off-diagonal crystalline electric field parameters in a cubic crystal symmetry because then a structure with a 90°-periodicity would be likely for the applied rotation geometry [16]. Moreover, the amplitude of the resonance field variation decreases by 50% with increasing the temperature from 290 K to 300 K. Therefore, despite the almost spherical shape of the crystal, we suspect very weak demagnetization fields to contribute to the resonance field.

Figure 2 shows the temperature dependence of various resonance line parameters. The ESR intensity, obtained from doubly integrating the dP/dH-spectra, measures the static magnetic susceptibility of the resonating probe. This could nicely be confirmed with the linearity between the ESR intensity and the dc-susceptibility χ_0 shown for X-band in the inset.

The ESR g-factor and ESR linewidth ΔH were both obtained from Lorentzian line-fittings, keeping the dispersion-absorption ratio constant. The ESR g-factor shows a temperature dependence which is strongest for the smallest resonance field (i.e. for the L-band measurements). This behavior indicates the direct relation between the field dependent dc-susceptibility and the ESR g-factor which, hence, measures the magnetic molecular-field. The linewidth displays a *reciprocal* dependence to the ac-susceptibility, χ_{ac}. This behavior is a hallmark of a resonance of a strongly coupled 3d / conduction electron system [17]. Therefore, the linear behavior of ΔH with a slope $b = 5$ mT/K as shown by the solid lines is *not* indicating a Korringa relaxation mechanism as is often the case for the relaxation of diluted ESR probes in a metallic enviroment [17].

While a frequency dependence of the resonance for $T > 290$K is not visible in the g factor the linewidth shows a linear broadening with 0.8 ± 0.1 mT/GHz. This behavior is in agreement with the results for $Fe_{1-x}Co_xSi$ (≈ 1.2 mT/GHz) [11] but in contrast to the frequency independence of the linewidth in systems like $ZrZn_2$ [18]. There, the broadening mechanism is solely determined by the exchange enhanced relaxation of the conduction electron spins [19]. Hence, the frequency dependence of the linewidth in ϵ-FeGe indicates additional relaxation mechanisms.

3. Conclusion

In the paramagnetic regime of ϵ-FeGe, above temperatures of ≈ 287 K, the spin dynamics of the strongly coupled Fe-3d / conduction electron system can be characterized by a field-dependent, almost isotropic paramagnetic resonance. The deviations from a Lorentzian line shape observed for $T_C < T < 287$ K are ascribed to complex magnetic short-range modulations which evolve in this precursor region.

Acknowledgments

Support by Volkswagen foundation (I/84689) and valuable discussions with M. Baenitz and F. Garcia are acknowledged. We thank R. Koban and H. Rave for their technical assistance.

[1] Shaltiel D, Ioshpe D, Grayevsky A, Zevin V and Smith J L 1987 *Phys. Rev. B* **36** 4090
[2] Förster T, Sichelschmidt J, Grüner D, Brando M, Kimura N and Steglich F 2010 *J. Phys.: Conf. Ser.* **200** 012035
[3] Date M, Okuda K and Kadowaki K 1977 *J. Phys. Soc. Jap.* **42** 1555
[4] Demishev S V, Semeno A V, Bogach A V, Glushkov V V, Sluchanko N E, Samarin N A and Chernobrovkin A L 2011 *JETP Lett.* **93** 213
[5] Cooper B R and Keffer F 1962 *Phys. Rev.* **125** 896–900
[6] Krellner C, Förster T, Jeevan H, Geibel C and Sichelschmidt J 2008 *Phys. Rev. Lett.* **100** 066401
[7] Sichelschmidt J, Wykhoff J, Gruner T, Krellner C, Klingner C, Geibel C, Steglich F, Krug von Nidda H A, Zakharov D V, Loidl A and Fazlizhanov I 2010 *Phys. Rev. B* **81** 205116
[8] Wilhelm H, Baenitz M, Schmidt M, UK Rößler, Leonov A A and AN Bogdanov 2011 *Phys. Rev. Lett.* **107** 127203
[9] Wykhoff J, Sichelschmidt J, Lapertot G, Knebel G, Flouquet J, Fazlishanov I I, Krug von Nidda H A, Krellner C, Geibel C and Steglich F 2007 *Science Techn. Adv. Mat.* **8** 389
[10] Dyson F 1955 *Phys. Rev.* **98** 349
[11] Watanabe H, Tazuke Y Y and Nakajima H 1985 *J. Phys. Soc. Jap.* **54** 3978
[12] Haraldson S, Pettersson L and Bhagat S M 1978 *J. Magn. Res.* **32** 115 – 120
[13] Kataoka M 1987 *J. Phys. Soc. Jap.* **56** 3635
[14] Watanabe H 1989 *J. Phys. Soc. Jap.* **58** 1035
[15] Maleyev S V 2006 *Phys. Rev. B* **73** 174402
[16] Abragam A and Bleaney B 1970 *Electron Paramagnetic Resonance of Transition Ions* (Oxford: Clarendon Press)
[17] Barnes S E 1981 *Adv. Phys.* **30** 801
[18] Shaltiel D, Davidov D, Dublon G, Gabai A, Grill A and Valach E 1971 *J. Phys. Colloques* **32** C1–632
[19] Walsh W M, Knapp G S, L W Rupp J and Schmidt P H 1970 *J. Appl. Phys.* **41** 1081

International Conference on Strongly Correlated Electron Systems (SCES 2011) IOP Publishing
Journal of Physics: Conference Series **391** (2012) 012106 doi:10.1088/1742-6596/391/1/012106

Structural and magnetic properties of $La_{0.5}Ba_{0.5}CoO_{3-\delta}$ cobaltites

V Sikolenko[1,2], **V Efimov**[2], **I O Troyanchuk**[3], **D V Karpinsky**[3], **M V Bushinsky**[3], **D Sheptyakov**[4]

[1]Karlsruhe Institute of Technology, D-76131, Kaiserstr.12, Karlsruhe, Germany
[2]Joint Institute for Nuclear Research, Dubna, RU-141980 Russia [3]SSPA "Scientific-Practical Materials Research Centre of NAS of Belarus", P. Brovka str. 19, 220072 Minsk, Belarus [4]Paul Scherrer Institute, CH-5232, Villigen, PSI, Switzerland

E-mail: vadim.sikolenko@jinr.ru

Abstract. The magnetic and structural properties of non-stoichiometric $La_{0.5}Ba_{0.5}CoO_{3-\delta}$ cobaltites have been studied by neutron and synchrotron powder diffraction. The oxygen content decrease leads to transformation of the magnetic structure from ferromagnetic to G-type antiferromagnetic through the two-phase state. Both coexisting phases have a cubic crystal structure with different volume of the unit cell.

1. Introduction

Rare earth cobaltites with an $LnCoO_3$ perovskite structure and hole-doped $Ln_{1-x}A_xCoO_3$ (Ln = lanthanide, A = alkaline earth metal: Ca, Sr or Ba) attract much interest as they exhibit a variety of unusual magnetic and transport properties. However the nature of the ferromagnetic state in cobaltites has been a subject of debates for a long time [1]. Three main mechanisms were suggested for magnetic properties of mixed-valence cobaltites: the superexchange model based on the localized electron interaction via oxygen ion, the Zener double exchange via charge transfer and the itinerant electron ferromagnetism [1].

The structural studies performed on the cubic oxygen-stoichiometric perovskite $La_{0.5}Ba_{0.5}CoO_3$ have revealed the onset of a long-range tetragonal phase accompanying a para-ferromagnetic transition occurring at $T_C = 180$ K [4]. The tetragonal distortion has been discussed in terms of cooperative and static Jahn-Teller distortions of the CoO_6 octahedra. It was assumed that the Jahn-Teller effect is favored by the intermediate spin-state configuration, of the $Co^{3+}(d^6)$ and $Co^{4+}(d^5)$ ions derived from the measured ferromagnetic moment 1.9 μ_B per cobalt ion. However, the Sr-doped ferromagnetic cobaltites have approximately the same magnetic moment value and do not exhibit a structural transition at the Curie point. Moreover EXAFS and neutron diffraction studies do not reveal an appreciable local Jahn-Teller distortion in $La_{1-x}Sr_xCoO_3$ [6,7].
It was shown that the oxygen content in the Ln and Ba ordered perovskite systems $LnBaCo_2O_y$ can be varied in a wide range from y = 4.5 to y = 6 that leads to a change of the average oxidation state of the cobalt ions from 2+ to 3.5+. The reduction of the oxygen content from y = 3 down to y = 2.25 has been reported for $La_{0.5}Sr_{0.5}CoO_y$ perovskite [6]. This oxygen content corresponds to nominal Co^{2+} oxidative state. Considering these reports we have decided to investigate the effect of oxygen deficiency on structural and magnetic properties of $La_{0.5}Ba_{0.5}CoO_{3-\delta}$ cobaltites.

2. Experimental

Ceramic samples of $La_{0.5}Ba_{0.5}CoO_{3-\delta}$ compositions were prepared by a solid-state phase reaction method using high-purity $BaCO_3$, La_2O_3 and CoO taken in stoichiometric ratio and thoroughly mixed with a planetary ball mill (RETSCH PM-100). The synthesis was performed at 1200 °C for 10 hours in air, followed by cooling at rate of 50-300 °C/h down to 300 °C or quenching. Synchrotron powder diffraction (SPD) experiments ware carried out at the Swiss Light Source synchrotron of the Paul Scherrer Institute using MS beamline (wavelength 0.49 Å) The neutron powder diffraction (NPD) experiments were performed using a high-resolution diffractometer E9 in Helmholtz-Zentrum Berlin (HZB) and HRPT spectrometer in Paul Scherrer Institute. The neutron and X-ray powder diffraction data were analyzed using the Rietveld method incorporated into the FullProf software package [8].

3. Results and discussion

Figure 1 presents the temperature dependence of the magnetization of the $La_{0.5}Ba_{0.5}CoO_{2.87}$ sample. In the temperature range of 160 –170 K the M(T) dependencies exhibit a sharp drop in both the FC and ZFC magnetization confirming a transition into paramagnetic state. The temperature onset of the magnetic ordering ($T_C \sim 170$ K) is slightly lower than in stoichiometric $La_{0.5}Ba_{0.5}CoO_3$ in which the ferromagnetic-paramagnetic transition occurs near $T_C \sim 180$ K [2]. The estimated magnetic moment per Co ion is close to 1.1 μ_B at liquid helium temperature. This value is much lower compared with 1.9 μ_B/Co reported for stoichiometric $La_{0.5}Ba_{0.5}CoO_3$ [2].

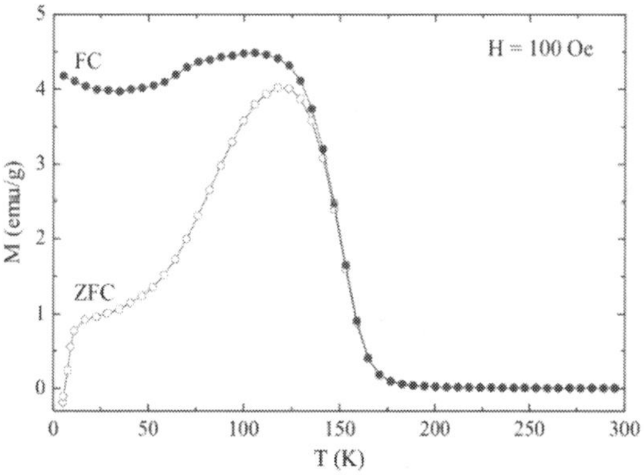

Figure 1. Temperature dependences of the dc magnetization in a field of 100 Oe for $La_{0.5}Ba_{0.5}CoO_{2.87}$.

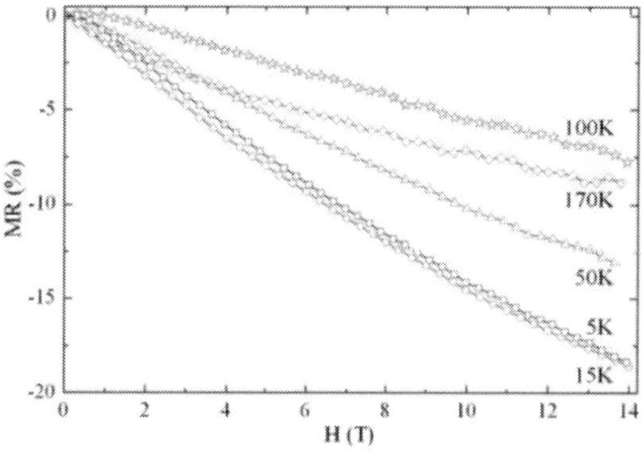

Figure 2. Field dependence of the magnetoresistance ratio MR=[ρ(H)–ρ(H=0)]/ ρ(H=0).100% for the ferromagnetic $La_{0.5}Ba_{0.5}CoO_{2.95}$

Figure 2 shows magnetoresistance ratio for the ferromagnetic sample $La_{0.5}Ba_{0.5}CoO_{2.95}$ as a function of magnetic field at various temperatures. The magnetoresistance exhibits a local maximum near Tc, and increases gradually with further cooling. At 5K we obtained the MR value of about 20% in the field of 14 T. The MR varies gradually with the field and does not show any tendency to saturation with the temperature decrease.

NPD patterns of $La_{0.5}Ba_{0.5}CoO_{2.87}$ were recorded at room temperature, 150, 120, 80 and 2 K, and for clarity only one of them are displayed in figure 3. The first pattern recorded at 293 K is characteristic of the high-temperature structural state which extends down to ~ 200 K as follows from the SPD study. At room temperature the Rietveld refinements lead to a simple cubic cell $a_p \times a_p \times a_p$ with the space group $Pm3m$ (global $\chi^2 = 1.76$). The observed, calculated and difference patterns of this sample recorded at 2K are displayed in figure 3. At 150 K we have noticed an asymmetric broadening of all the peaks which develops strongly upon cooling to 120 K. For the patterns recorded at 80 and 2 K we clearly observed a splitting of all the peaks at 2θ> 60° and an appearance of new small peaks strongly separated from the basic peaks described with $Pm3m$ space group.

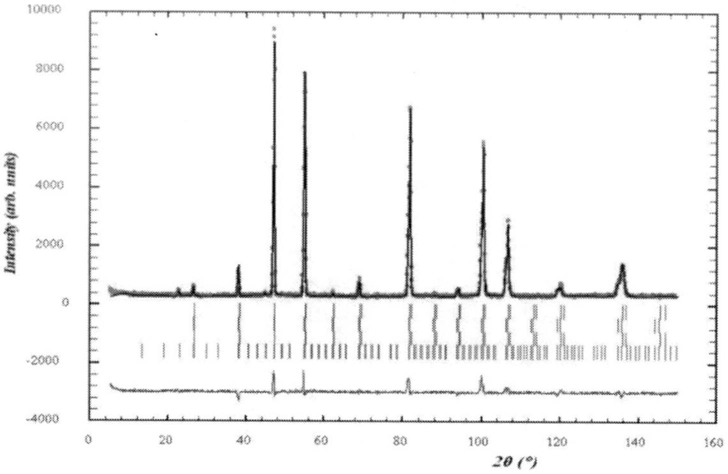

Figure 3 Neutron powder diffraction data collected from $La_{0.5}Ba_{0.5}CoO_{2.87}$ at 2K. Experimental curve and Rietveld refinement of crystal and magnetic structures are shown.

SPD experiments were performed in order to clarify a reason for NPD peaks splitting and appearance of the new peaks. The SPD spectra were recorded at every 5 K from 4 K up to room temperature using the same sample as for the NPD study. The volume ratio between the two phases at 2 K is approximately 1:2 where the major phase is closer to the high-temperature structural state. The amount of the minor phase, which has a larger unit cell decreases gradually upon warming up, and disappear completely above 200 K where the sample consists of one cubic phase. In contrast with NPD study we did not observe additional diffraction peaks associated with deviation from cubic symmetry of both low temperature phases, suggesting that additional peaks observed at low temperature in NPD patterns result from magnetic ordering. We suggest that both phases have a cubic unit cell at low temperature which can be described with $Pm3m$ space group. The Rietveld refinement in this model gave a satisfactory fit for the SPD pattern recorded at 4 K. In the temperature range where two phases coexist the unit cell parameters of both phases slightly increase with temperature. There is no evidence for intermediate state between phases with close unit cell parameters.

The Rietveld refinement of the NPD spectra acquired at 2 K in the two cubic phase model leads to satisfactory description of basic structural peaks for both the structural phases. The fit was further improved by taking into account magnetic neutron scattering. The additional peaks appearing at low temperature can be well indexed in $2a_p \times 2a_p \times 2a_p$ supercell where a_p is unit cell parameter for the minor phase with larger unit cell parameter. This means that the magnetic structure associated with the minor phase is G-type antiferromagnetic. The estimated magnetic moment for this phase is around $\pm 2\ \mu_B$/Co. The major phase is ferromagnetic with magnetic moment close to 1.6 μ_B/Co. The oxygen content of the minor phase was roughly estimated as smaller in compered with the major ones. Apparently phase separation is associated with different spin state of the cobalt ions in the ferromagnetic and ferromagnetic phases.

The neutron powder diffraction of strongly oxygen-deficient $La_{0.5}Ba_{0.5}CoO_{2.6}$ has been recorded at T = 2, 50, 100, 150 and 250K. Figure 4 presents the pattern at the T = 2 K, together with Rietveld refinement of crystal and magnetic structures. All data can be refined only with one phase in the frame of cubic $Pm3m$ space group, the magnetic structure is G-type antiferromagnetic.

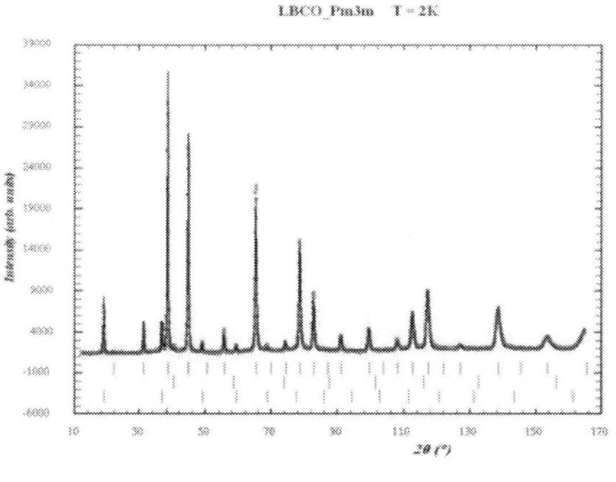

Figure 4. Neutron powder diffraction data for $La_{0.5}Ba_{0.5}CoO_{2.6}$ at 2 K. Experimental data and Rietveld refinement are shown.

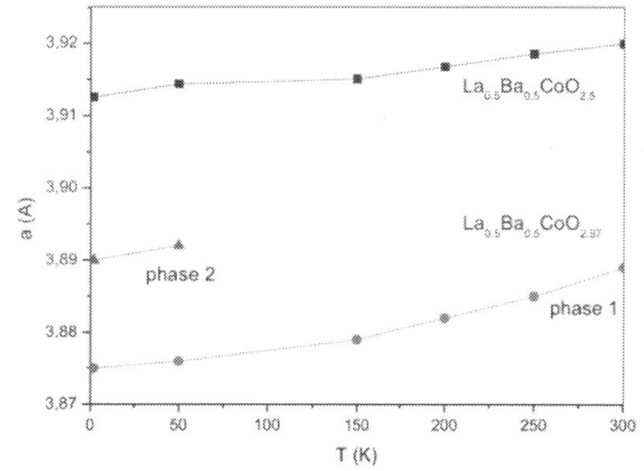

Figure 5. Temperature dependence of the unit cell parameters for $La_{0.5}Ba_{0.5}CoO_{2.6}$ and two phases of $La_{0.5}Ba_{0.5}CoO_{2.87}$

The temperature dependence of the cubic unit cell parameters for $La_{0.5}Ba_{0.5}CoO_{2.5}$ (upper curve) and $La_{0.5}Ba_{0.5}CoO_{2.87}$ (two lower curves) respectively is presented on the Figure 5. There is no evidence for the existence of intermediate state between phases with close unit cell parameters for $La_{0.5}Ba_{0.5}CoO_{2.87.}$

In conclusion, high-resolution neutron and synchrotron powder diffraction studies of the non-stoichiometric compounds $La_{0.5}Ba_{0.5}CoO_{3-\delta}$ show that the decrease of the oxygen content leads to a transformation of the ferromagnetic structure into G-type antiferromagnetic one via intermediate region of the compositions demonstrating a macroscopic phase separation into different structural and magnetic phases. Both antiferromagnetic and ferromagnetic phases have a cubic crystal structure with different volume of the unit cell. The amount of antiferromagnetic phase increases gradually with cooling, and the calculated ratio between antiferromagnetic and ferromagnetic phases at 2 K is about 1:2 for the $La_{0.5}Ba_{0.5}CoO_{2.87.}$ The estimated magnetic moment is around $\pm 2\ \mu_B$/Co for the antiferromagnetic and 1.6 μ_B/Co for the ferromagnetic phase. The strongly reduced $La_{0.5}Ba_{0.5}CoO_{2.6}$ is pure antiferromagnetic.

This work was supported by Belarussian Fund for Fundamental Research (grant T 11D-003) and by Russian Foundation for Basic Research (grant 09-02-01446-a).

References

[1] Goodenough J B 1971 *Mater. Res. Bull.* 6 p 967
[2] Senaris-Rodrigues H and Goodenough J B 1995 *J.Sol.Stat.Chem* 118 p 323
[3] Sazonov A et al 2005 *J.Phys.:Cond.Mat.* 17 p 4181
[4] Fauth P, Suard E, Caignaert V 2001 *Phys.Rev.B65* p 060401
[5] Sundaram N et al 2009 *Phys. Rev. Lett.* 102 p 026401
[6] Wang Z L and Yin J S 1998 *Philosophical Magazine Part B* 77 p 49
[7] Sikolenko V et al 2006 *Crystallography Reports* 51 p S67
[8] Rodrigues-Carvajal J 1993 *Physica B* 192 p 55

Metal-insulator transition and optical conductivity in $BaVS_3$

Arata Tanaka

Department of Quantum Matter, ADSM, Hiroshima University, Higashi-Hiroshima 739-8530, Japan

E-mail: atanaka@hiroshima-u.ac.jp

Abstract. Possible orbital orders below the metal-insulator transition temperature in quasi-one-dimensional $BaVS_3$ have been investigated in relation to the monoclinic lattice distortion in the insulating phase using a one-dimensional two-band Hubbard model. Orbital occupations in the ground state and optical conductivity have been calculated by means of exact diagonalization of finite size clusters. Depending on the size of crystal field splitting of the t_{2g} levels, two different orbital orders in the t_{2g} orbitals with the periodicity of four V ion sites along the c-axis are found to be stabilized with the lattice distortion. The on-site exchange interaction plays important role in the formation of these orbital orders.

1. Introduction

Quasi-one-dimensional $BaVS_3$ has the hexagonal perovskite structure, in which the face-sharing VS_6 octahedrons form linear chains of V ions along the c-axis. It exhibits three successive second order phase transitions, whose origin are still controversial. A hexagonal to orthorhombic structural transition occurs at $T_S = 240$ K, where the V ion chains become zigzag chains. A metal-insulator transition (MIT) at T_{MI} =70 K is accompanied by a monoclinic distortion doubling the unit cell with four different V ion sites along the chain. An incommensurate magnetic ordering occurs below T_X=30 K. The t_{2g} level of each of V ions is split into doubly degenerated e_g^π and single a_{1g} levels by a small trigonal field. Both the a_{1g} orbitals with one-dimensional character forming wide bands and the e_g^π orbitals with their lobes towards interchain directions forming narrow bands contribute the states near the Fermi level [1]. This t_{2g} orbital degree of freedom is considered to play important role in MIT. The formations of charge and orbital orders [2, 3, 4] and a spin-orbital liquid [5] have been proposed for the origin of MIT.

2. Model and Methods of Calculations

To elucidate the $3d$ electronic state of the V ion chain across MIT in connection with the t_{2g} orbital degeneracy and the monoclinic lattice distortion, a linear-chain two-band Hubbard model consists of eight sites with periodic boundary condition is considered. Since the degeneracy of the e_g^π level is lifted by the orthorhombic lattice distortion below T_S, the lowest two t_{2g} orbitals, the a_{1g} and one of the e_g^π orbitals, are taken into account in this model. The Hamiltonian of the

Table 1. V-V bond lengths R_l between sites l and $l+1$ at 40 K in Ref. [6] and hopping integrals $T_{l,a,a}$ for insulating phase calculated using eq. 2.

	$l=1,5$	$l=2,6$	$l=3,7$	$l=4,8$
R_l (Å)	2.896	2.941	2.753	2.806
$T_{l,a,a}$ (eV)	−0.526	−0.492	−0.660	−0.605

system is then written as

$$H = \frac{D}{2}\sum_l (n_{l,e} - n_{l,a}) + \sum_{l,m,m',\sigma} T_{l,m,m'}(c^\dagger_{l+1,m,\sigma}c_{l,m',\sigma} + c^\dagger_{l,m',\sigma}c_{l+1,m,\sigma})$$
$$+U\sum_{l,m} n_{l,m\uparrow}n_{l,m\downarrow} + U'\sum_l n_{l,a}n_{l,e} - J\sum_{l,\sigma'\sigma}(c^\dagger_{l,a,\sigma'}c_{l,a,\sigma}c^\dagger_{l,e,\sigma'}c_{l,e,\sigma} + c^\dagger_{l,a,\sigma}c_{l,e,\sigma'}c^\dagger_{l,a,\sigma'}c_{l,e,\sigma}), \quad (1)$$

where $n_{l,m,\sigma}$, $c^\dagger_{l,m,\sigma}$ and $c_{l,m,\sigma}$ are the number, creation and annihilation operators of the a_{1g} ($m = a$) or e^π_g ($m = e$) orbital with spin quantum number σ on site l, respectively. The first term describes the energy splitting of the e^π_g and a_{1g} levels with separation $D \equiv \varepsilon(e^\pi_g) - \varepsilon(a_{1g})$, the second term is for electron hopping between neighboring sites and the last three terms represent on-site Coulomb and exchange interactions ($U = U' + 2J$). The V ions are equally spaced along the chain for $T > T_{\rm MI}$ and in the monoclinic phases there are four different V-V bond lengths along the chain. This effect of the monoclinic lattice distortion due to the changes in the V-V bond lengths are included as the variations of hopping integrals $T_{l,a,a}$, which can be written as

$$T_{l,a,a} = \frac{1}{N_\pi}\left\{(dd\sigma) - \frac{2(pd\pi)^2}{3\Delta}\right\}, \quad (2)$$

where $(dd\sigma)$ and $(pd\pi)$ are the Slater-Koster parameters between V $3d$-$3d$ and V $3d$-S $3p$ orbitals of the neighboring sites, respectively, Δ denotes the charge transfer energy between the V $3d$ and S $3p$ orbitals and $N_\pi = 1 + 4(pd\pi)^2/\Delta^2$. With $(pd\pi)$ =0.478 eV, $\Delta = 2.0$ eV and the value of $(dd\sigma)$ after Harrison [7], which has bond length R dependence $(dd\sigma) \propto 1/R^5$, $T_{l,a,a}$'s for the metallic and insulating phases are calculated using the crystal structural data in Ref. [6] at 300 K and 40 K. For the metallic phase, $T_{l,a,a} = -0.566$ eV is adopted and those for the insulating phase are listed in table 1. The size of the other hopping integrals are small and the values adopted are $T_{l,e,e}$=0.05 eV and $T_{l,a,e} = -0.14$ eV; $U' = 2.0$ eV and $J = 0.65$ eV are assumed.

The optical conductivity has been calculated using the method in Ref. [8]. The optical conductivity with the electric polarization vector along the chain as a function of photon energy ω can be described as

$$\sigma_{zz}(\omega) = D_{\rm W}\delta(\omega) + \frac{e^2}{N\omega}{\rm Im}\left[\langle g|\hat{j}_z\frac{1}{H - E_g - \omega - {\rm i}\delta}\hat{j}_z|g\rangle\right], \quad (3)$$

where

$$\hat{j}_z = -{\rm i}\sum_{l,m,m',\sigma} T_{l,m,m'}(c^\dagger_{l+1,m,\sigma}c_{l,m',\sigma} - c^\dagger_{l,m',\sigma}c_{l+1,m,\sigma}) \quad (4)$$

and N denotes number of sites ($N = 8$) and E_g and $|g\rangle$ are the energy and wave function of the ground state. The Drude weight $D_{\rm W}$ can be obtained from the sum rule

$$\int_0^\infty \sigma_{zz}(\omega)\,d\omega = -\frac{\pi e^2}{4N}\sum_{l,m,m',\sigma} T_{l,m,m'}\langle g|(c^\dagger_{l+1,m,\sigma}c_{l,m',\sigma} + c^\dagger_{l,m',\sigma}c_{l+1,m,\sigma})|g\rangle. \quad (5)$$

The energy and wave function of the ground state are calculated using the Lanczos method and $\sigma_{zz}(\omega)$ in eq. 3 is obtained by the recursion method.

3. Results and Discussions

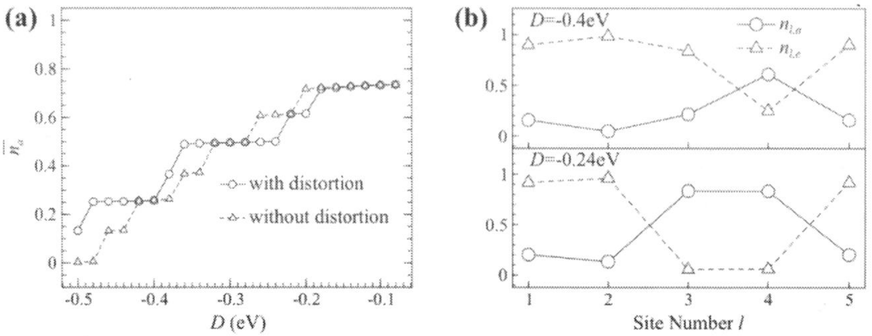

Figure 1. (a) a_{1g} orbital occupation per site as a function of D with (circles) and without (triangles) the effect of the monoclinic lattice distortion. (b) a_{1g} (circles) and e_g^π (triangles) orbital occupations of each site for $D = -0.4$ eV and $D = -0.24$ eV with monoclinic lattice distortion. The lines are drawn as guides for eyes.

Figure 1(a) shows the averaged occupation number of the a_{1g} orbital over sites $\overline{n}_a$ as a function of D with and without the monoclinic lattice distortion. The a_{1g} orbital occupation is increased steadily with increasing D without the lattice distortion. However, the state with $\overline{n}_a \sim 1/2$ and $\overline{n}_a \sim 1/4$ are particularly stabilized with the lattice distortion. To examine these two states more closely, the a_{1g} and e_g^π orbital occupations in each site are presented in Fig. 1(b) for $D = -0.24$ eV with $\overline{n}_a \sim 1/2$ and for $D = -0.4$ eV with $\overline{n}_a \sim 1/4$. The patterns of the orbital order are different between these two states. The alignment e_g^π-e_g^π-a_{1g}-a_{1g} is found with $D = -0.24$ eV, whereas e_g^π-e_g^π-e_g^π-a_{1g} with $D = -0.4$ eV. While there are large variations in the orbital occupations among the sites, the disproportionation of the charge $n_{l,a} + n_{l,e}$ among the sites is small as discussed in the previous work [4].

The a_{1g} orbital is preferably occupied at site 4, because of the strong a_{1g}-a_{1g} hybridization to the neighboring site owing to short V-V distance (see table 1). This can be regarded as the effective a_{1g}-e_g^π level splitting at site 4 is larger than the other sites and also implies that switching from the high-spin $S = 1$ $(a_{1g})^1(e_g^\pi)^1$ configuration to the low-spin $S = 0$ $(a_{1g})^2$ configuration occurs at smaller D than the other sites when the site is doubly occupied. The orbital and spin correlations between site 4 and the neighbors are strongly depend on the electronic configuration of the d^2 state of site 4 created by charge fluctuation. For small D, the high-spin $(a_{1g})^1(e_g^\pi)^1$ configuration at site 4 is preferred and this induces the a_{1g}-e_g^π orbital order with ferromagnetic spin correlation between site 4 and the neighbors. Similarly, the a_{1g}-a_{1g} orbital order with antiferromagnetic spin correlation are expected between site 4 and the neighbors for large D. In fact, the e_g^π-a_{1g}-e_g^π orbital alignment of sites 3 to 5 and the ferromagnetic spin correlation of site 4 and the neighbors $\langle \boldsymbol{S}_3 \cdot \boldsymbol{S}_4 \rangle \sim \langle \boldsymbol{S}_4 \cdot \boldsymbol{S}_5 \rangle \sim 0.19$ with $D = -0.4$ eV are contrasted with the a_{1g}-a_{1g} orbital alignment and the antiferromagnetic spin correlation $\langle \boldsymbol{S}_3 \cdot \boldsymbol{S}_4 \rangle \sim -0.45$ between sites 3 and 4 with $D = -0.24$ eV. Similar high- to low-spin state switching at doubly occupied sites at MIT in VO_2 and Ti_2O_3 has been discussed [9].

To examine the effects of the monoclinic lattice distortion on the electric transport properties, in Fig. 2(a), the Drude weight D_W as a function of D calculated with and without the distortion are shown. The unphysical negative D_W values are the result of the finite size effect and indicate insulating behavior of the system [8]. While such insulating behavior can be seen rather limited

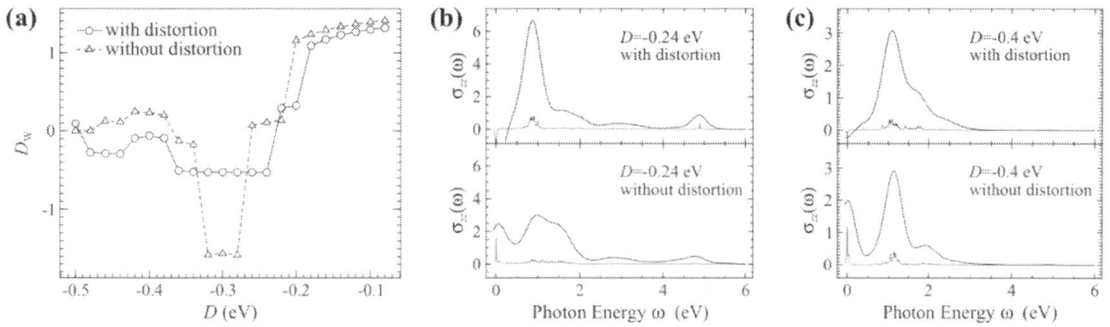

Figure 2. (a) Drude weight D_W as a function of D with (circles) and without (triangles) the effect of the monoclinic lattice distortion. (b) Optical conductivity spectra for $D = -0.24$ eV with (upper panel) and without (lower panel) the effect of the monoclinic lattice distortion. (c) The same to (b) but for $D = -0.4$ eV.

range of D where the value of $\bar{n}_a$ is $\sim 1/2$ without distortion, much wider ranges of D where the orbital orders e_g^π-e_g^π-a_{1g}-a_{1g} or e_g^π-e_g^π-e_g^π-a_{1g} occur are insulating with the distortion. The sign of D_W changes from positive to negative by including the effects of distortion in the ranges $-2.6\,\text{eV} \leq D \leq -2.4\,\text{eV}$, and $-4.8\,\text{eV} \leq D \leq -3.8\,\text{eV}$.

In Figs. 2(b) and 2(c), optical conductivity spectra calculated with $D = -0.24$ eV (b) and $D = -0.4$ eV (c) are shown. In the case of the spectra with $D = -0.24$ eV, the spectral weight distributed within $\omega < 2\,\text{eV}$ without the lattice distortion converges into a sharp peak located at $\omega \sim 0.8\,\text{eV}$ when the lattice distortion is considered. On the other hand, although large spectral weight transfer to the Drude peak can be seen, no such drastic change in the intensity of the peak at $\omega \sim 1.2\,\text{eV}$ is found for the spectra with $D = -0.4\,\text{eV}$. This is consistent with the fact that no clear change in the intensity of a peak centered around $\omega \sim 1.2\,\text{eV}$ is found across MIT in experimental spectra with the polarization vector parallel to the c-axis [10].

4. Conclusions

Possible orbital orders below T_{MI} and the optical conductivity are investigated as a function of the e_g^π-a_{1g} level separation. States having small Drude weights with orbital alignments e_g^π-e_g^π-a_{1g}-a_{1g} and e_g^π-e_g^π-e_g^π-a_{1g} are particularly stabilized when the effects of the monoclinic distortion is considered. The on-site exchange interaction plays important role in the formation of these orders. From these findings, it is inferred that the MIT is caused by the orbital order stabilized by the strong electron-lattice coupling.

References

[1] Mitrovic S, Fazekas P, Søndergaard C, Ariosa D, Barišić N, Berger H, Cloëtta D, Forró L, Höchst H, Kupčić I, Pavuna D and Margaritondo G 2007 *Phys. Rev. B* **75** 153103.
[2] Fagot S, Foury-Leylekian P, Ravy S, Pouget J P and Berger H 2003 *Phys. Rev. Lett.* **90** 196401.
[3] Radnoćzi K and Fazekas P 2006 *Physica B* **378—380** 663.
[4] Tanaka A 2006 *J. Magn. Magn. Mater.* **310** 898.
[5] Vernay F, Penc K, Fazekas P and Mila F 2004 *Phys. Rev. B* **70** 014428.
[6] Fagot S, Foury-Leylekian P, Ravy S, Pouget J P, Anne M, Popov G, Lobanov M V and Greenblatt M 2005 *Solid State Sciences* **7** 718.
[7] Harrison W A 1980 *Electronic Structure and Physical Properties of Solids* (San Francisco: Freeman).
[8] Dagotto E 1994 *Rev. Mod. Phys.* **66** 763.
[9] Tanaka A 2004 *J. Phys. Soc. Jpn.* **73** 152.
[10] Kézsmárki I, Mihály G, Gaál R, Barišić N, Akrap A, Berger H, Forró L, Homes C C and Mihály L 2006 *Phys. Rev. Lett.* **96** 186402.

Field-temperature phase diagram of superconductivity in Sr_2RuO_4-Ru under out-of-plane uniaxial pressure

H Taniguchi[1], S Kittaka[1,2], S Yonezawa[1], H Yaguchi[3] and Y Maeno[1]

[1] Department of Physics, Graduate School of Science, Kyoto University, Kyoto 606-8502, Japan

[3] Department of Physics, Faculty of Science and Technology, Tokyo University of Science, Chiba 278-8510, Japan

E-mail: `taniguchi@scphys.kyoto-u.ac.jp`

Abstract. One of the interesting issues on the spin-triplet superconductor Sr_2RuO_4 (T_c = 1.5 K) is the origin of the T_c enhancement up to about 3 K in the eutectic system Sr_2RuO_4-Ru. We have recently revealed a striking fact that T_c of pure Sr_2RuO_4 is also enhanced up to 3.2 K by uniaxial pressure along the c axis ($P_{\parallel c}$). When $P_{\parallel c}$ is applied to Sr_2RuO_4-Ru, there is a crossover at $P^*_{\parallel c} \sim 0.4$ GPa from behavior attributable to interfacial superconductivity (SC) to behavior similar to that in pure Sr_2RuO_4 under $P_{\parallel c}$. We focus on the field-temperature phase diagrams of Sr_2RuO_4-Ru above and below $P^*_{\parallel c}$. We revealed that the H-T curves of Sr_2RuO_4-Ru both above and below $P^*_{\parallel c}$ are concave-up. This fact suggests that the $P_{\parallel c}$-originated SC above $P^*_{\parallel c}$ as well as the interfacial SC below $P^*_{\parallel c}$ are granular-like. We also found that the H-T curves of these SCs do not scale with T_c's, which is possibly related to a difference in the spatial distribution of superconducting regions.

1. Introduction

Sr_2RuO_4 has been attracting much interest as a leading candidate of a spin-triplet chiral p-wave superconductor [1, 2]. One of the interesting issues for this compound is the enhancement of the superconducting transition temperature T_c in the eutectic system Sr_2RuO_4-Ru; the onset T_c of Sr_2RuO_4-Ru is $\sim$ 3 K, while T_c of pure Sr_2RuO_4 and Ru are 1.5 K and 0.49 K, respectively [3]. In the eutectic system, lamellae of Ru metal, whose typical dimension is $10 \times 10 \times 1$ μm^3, are embedded in a Sr_2RuO_4 single crystal with stripe patterns. The nonbulk nature of this SC was revealed by the absence of specific-heat anomaly around 3 K [4]. The zero-bias conductance peak observed above 1.5 K indicates that the interface between Sr_2RuO_4 and Ru is superconducting and that this SC is non s-wave [5, 6]. The anisotropic upper critical field similar to that of the 1.5-K SC in Sr_2RuO_4 indicates that it is the Sr_2RuO_4 region around the interfaces that becomes superconducting [7].

We have investigated uniaxial pressure effects on Sr_2RuO_4 and Sr_2RuO_4-Ru in order to clarify the mechanism of the T_c enhancement, because a possible origin of the T_c enhancement is anisotropic lattice distortion around the interfaces. We reported that T_c of "pure" Sr_2RuO_4

[2] Present address: Institute for Solid State Physics, University of Tokyo, Chiba 277-8581, Japan

is enhanced up to 3.2 K under pressure along the c axis ($P_{\parallel c}$) [8]. Here, "pure" means that the sample contains almost no Ru inclusions. The result strongly supports our scenario that the anisotropic lattice distortion plays an important role in the T_c enhancement. We also clarified that $P_{\parallel c}$ substantially enhances the shielding fraction of the 3-K SC in the eutectic system, based on DC magnetization measurements [9, 10]. Moreover, in Sr_2RuO_4-Ru under $P_{\parallel c}$, we found a crossover at $P^*_{\parallel c} \sim 0.4$ GPa from behavior attributable to interfacial SC to behavior similar to that in pure Sr_2RuO_4 under $P_{\parallel c}$ [8].

In this article, we report results of AC susceptibility measurements of Sr_2RuO_4-Ru under $P_{\parallel c}$ for several DC magnetic fields. We then compare the field-temperature (H-T) phase diagram of the $P_{\parallel c}$-originated SC above $P^*_{\parallel c}$ with that of the interfacial SC below $P^*_{\parallel c}$.

2. Experimental

Uniaxial pressure was applied along the c axis at room temperature using a piston-cylinder-type pressure cell (Fig. 1(a)). The room-temperature pressure value was confirmed to be in a reasonable agreement with low-temperature pressure value determined by the superconducting transitions of tin and lead [11]. When the screws at both ends are tightened, the dish-shaped springs hold the pressure. The pistons press a sample directly; no pressure medium is used. The inner parts and the outer body of the cell are made of Be-Cu and polybenzimidazole (hard plastic), respectively.

The single crystals used in this study were grown in the floating zone method with Ru self-flux [12]. The dimensions of the sample reported here are 1.70×1.46 mm^2 in the ab plane and 0.40 mm along the c axis. The top and bottom surfaces were polished to be parallel to each other. The side surfaces were covered with epoxy (Stycast 1266, Emerson-Cuming) to prevent a breakdown of the sample (Fig. 1(b)) [10]. A space between the epoxy and the inner wall of the cell allows the epoxy to spread freely under pressure.

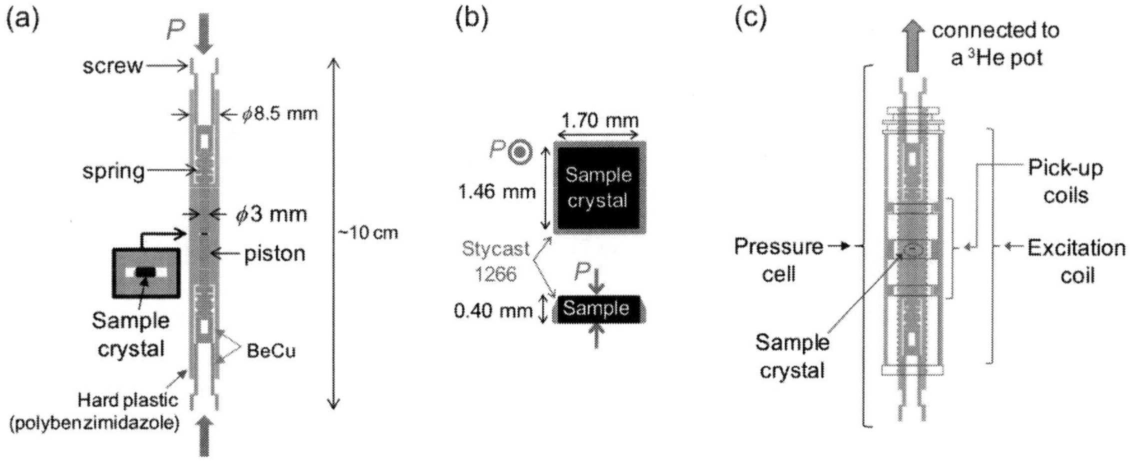

Figure 1. Schematic diagram of (a) the pressure cell, (b) the top and side views of the sample surrounded by Stycast 1266, and (c) the set-up for the AC susceptibility measurement.

AC susceptibility $\chi_{AC}(T) = \chi'(T) - i\chi''(T)$ was measured down to 0.3 K with a mutual inductance method using a lock-in amplifier (LIA) (Stanford Research Systems, SR830). The pressure cell with the susceptometer coils (Fig. 1(c)) were cooled down with a ^{3}He cryostat (Oxford Instruments, Heliox VL). Both the AC magnetic field H_{AC} and the DC magnetic field H_{DC} were applied parallel to the c axis. The accuracy of the field alignment was better than 5 degrees. χ_{AC} is related to the read-out voltage of the LIA ($V_{LIA} = V_x + iV_y$) as $\chi_{AC} = -iC_1V_{LIA}/fH^0_{AC} + C_2$. Here, H^0_{AC} and f are the amplitude and frequency of the applied AC magnetic field, and C_1 and C_2 are complex coefficients. The values of C_1 and C_2 were chosen so that $\chi'(4\ \mathrm{K}) = 0$ and $\chi'(0.3\ \mathrm{K}) = -1$ under $P_{\parallel c} = 0$ and $H_{DC} = 0$ for each sample; thus $|\chi'|$ corresponds to the AC shielding fraction. For the χ_{AC} curves under $P_{\parallel c} > 0$ or $H_{DC} > 0$, C_1 and C_2 determined under 0 GPa and 0 T were used.

3. Results and Discussion

Figure 2 represents the temperature dependence of χ_{AC} of Sr_2RuO_4-Ru under $P_{\parallel c} = 0$ and $P_{\parallel c} =$ 0.5 GPa $> P^*$. An enlarged view around the superconducting onset is shown in Fig. 2(b). The shielding fraction between 1.7 K and 3.0 K is enhanced, consistent with previous reports [9, 10]. The decrease of the volume fraction at lower temperatures is attributable to partial destruction of SC due to excess or inhomogenuity of distortion as well as to cracks in the sample [8]. Next, we measured χ_{AC} of the same sample at 0.5 GPa under several DC magnetic fields (Fig. 3). In the $P_{\parallel c}$-enhanced shielding region ($0 < |\chi'| < 0.05$), $\chi'(T)$ curves exihibit parallel shift to lower temperature with increasing field (Fig. 3(b)). This behavior is different from that of the interfacial SC below $P^*_{\parallel c}$; the interfacial SC does not exhibit parallel shift, as exemplified in the region of $0 < |\chi'| < 0.6$ in Fig. 5 of Ref. [13].

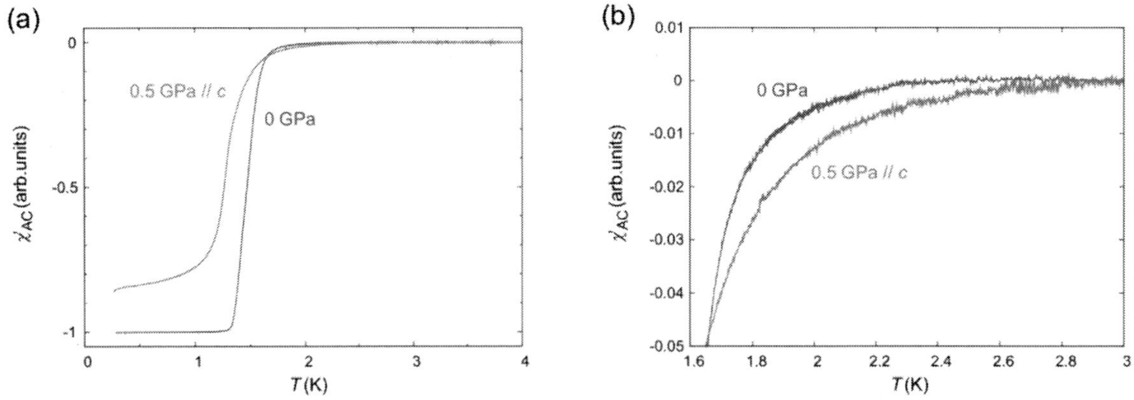

Figure 2. Zero-field temperature dependence of the real part of the AC susceptibility of Sr_2RuO_4-Ru under $P_{\parallel c} = 0$ and $P_{\parallel c} = 0.5$ GPa: (a) the entire behavior and (b) an enlarged view around the superconducting onset. These data were obtained with $\mu_0 H_{AC} = 2$ μT-rms and $f = 293$ Hz.

In order to illustrate the difference between the 3-K SC in Sr_2RuO_4-Ru above $P^*_{\parallel c}$ ($P_{\parallel c}$-originated SC) and that below $P^*_{\parallel c}$ (interfacial SC), we present in Fig. 4 the H-T phase diagram of these SCs. Figure 4 also contains the H-T phase diagram of the bulk SC of Sr_2RuO_4 at $P_{\parallel c} = 0$ [15]. All the H-T phase diagrams are determined by χ_{AC}. We define T_c in the following two ways: T_c^{cross} is defined as the temperature of the intersection of the extrapolation of the most rapidly changing part of χ' and that of the normal state χ' as shown in Fig. 3(b), and T_c^{onset} is defined as the temperature at which χ' reaches -0.005. We used the steepest slope in $0 < |\chi'| < 0.05$ to determine T_c^{cross} of $P_{\parallel c}$-originated SC, so that T_c^{cross} captures the characteristics of the $P_{\parallel c}$-originated SC. For the same reason, the steepest slope in $0 < |\chi'| < 0.6$ is used for T_c^{cross} of the interfacial SC.

We have revealed that (1) the H-T curves of the $P_{\parallel c}$-originated SC and the interfacial SC are concave up with decreasing temperature, while that of the bulk SC of Sr_2RuO_4 is convex up,

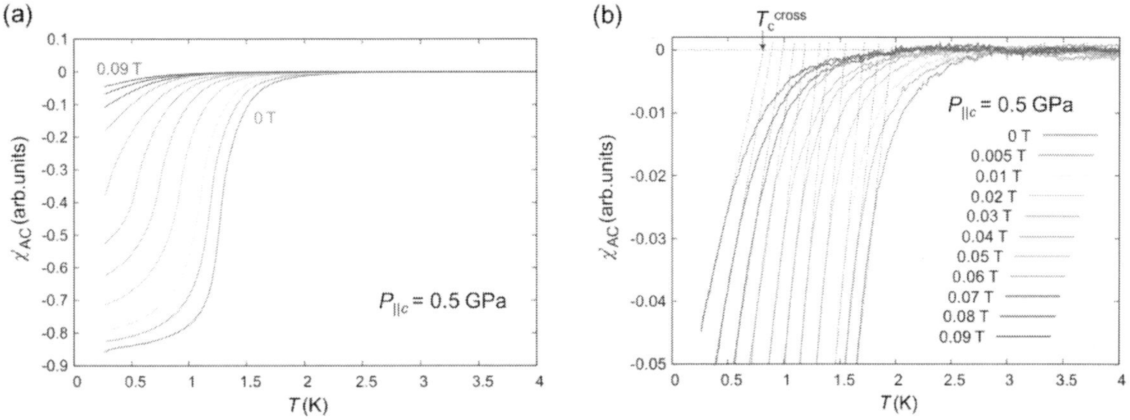

Figure 3. Temperature dependence of the real part of the AC susceptibility of Sr_2RuO_4-Ru at $P_{\parallel c}$= 0.5 GPa under different H_{DC}'s, measured with $\mu_0 H_{AC}$ = 2 μT-rms and f = 293 Hz: (a) the entire behavior and (b) an enlarged view of the $P_{\parallel c}$-enhanced shielding region.

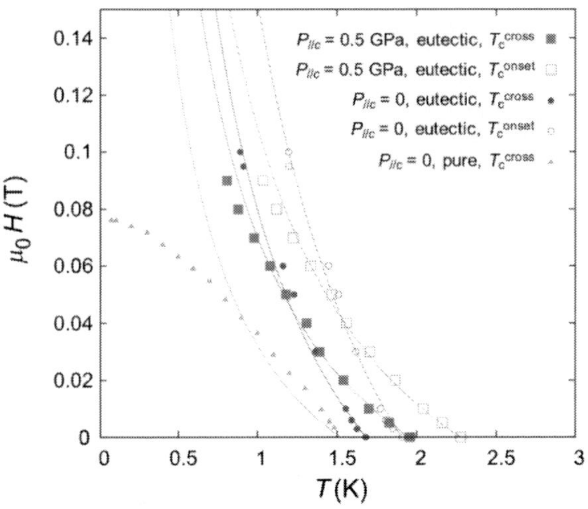

Figure 4. Comparison of H-T phase diagrams for $H \parallel c$ based on two different T_c definitions: T_c^{cross} and T_c^{onset} (see text). Solid and dashed lines indicate fitting curves with Eq. (1) for a model of an assembly of intergranular Josephson junctions [14]. These curves reproduce the concave-up behavior of the $P_{\parallel c}$-originated SC above $P^*_{\parallel c}$, as well as that of the interfacial SC below $P^*_{\parallel c}$.

and that (2) the slopes of the H-T curve of the $P_{\parallel c}$-originated SC are smaller than those of the interfacial SC resulting in a crossing at around 0.04 T. Note that these features are not affected by the definitions of T_c.

It is known that H-T curves of granular superconductors exhibit concave-up behavior with decreasing temperature. A calculation of the effective critical field H_j of an assembly of intergranular Josephson junctions revealed that H_j exhibits $1/T$ dependence in the vicinity of $T_c(H{=}0)$ [14, 16]. Therefore, in order to clarify the origin of the concave-up behavior for the $P_{\parallel c}$-originated SC, we fitted the equation

$$\mu_0 H_j(T) = a \left(\frac{1}{T} - \frac{1}{T_c(H=0)} \right) \quad (1)$$

to the data in the range $0.5T_c(H{=}0) < T < T_c(H{=}0)$. The well-fitted results shown in Fig. 4 imply that both the $P_{\parallel c}$-originated SC and the interfacial SC have granular-like features. The difference in the slope of the H-T curve is possibly related to certain changes in the spatial distribution of superconducting regions caused by pressure, such as an increase of intergranular distance or a change in the shape of superconducting grains themselves. We note, however, that possibilities of extrinsic origins such as cracks have not been excluded.

4. Conclusion

We measured the AC susceptibility of Sr_2RuO_4-Ru under uniaxial pressure along the c axis ($P_{\|c}$) in several DC magnetic fields. We revealed the field-temperature (H-T) phase diagrams of Sr_2RuO_4-Ru above and below $P^*_{\|c}$. We clarified that the H-T phase diagrams are concave up with decreasing temperature both in the $P_{\|c}$-originated SC above $P^*_{\|c}$ and in the interfacial SC below $P^*_{\|c}$, while the phase diagram is convex up in the bulk SC of Sr_2RuO_4. This result suggests that the $P_{\|c}$-originated SC is somewhat inhomogeneous as well as the interfacial SC. We also revealed that the slope of the H-T curve is smaller in the $P_{\|c}$-originated SC than in the interfacial SC. This is probably related to a spatial-distribution change of superconducting grains.

In the future, measurements in higher fields would be interesting, because a crossover from granular-like behavior to a bulk-like behavior of the $H_{\rm j}(T)$ curve is proposed in Ref. [8] when the field reaches the bulk H_{c2}. Another future direction is to try to reduce cracks in the sample caused by the pressure. Such cracks prevent us from more quantitative analyses. A reduction of cracks is probably achieved by controlling the amount of epoxy surrounding the sample.

Acknowledgments

This work is supported by a Grant-in-Aid for the Global COE program "The Next Generation of Physics, Spun from Universality and Emergence" from the MEXT of Japan, and by the "Topological Quantum Phenomena" (No. 22103002) Grant-in Aid for Scientific Research on Innovative Areas from the MEXT of Japan.

References

[1] Maeno Y, Hashimoto H, Yoshida K, Nishizaki S, Fujita T, Bednorz J G and Lichtenberg F 1994 *Nature* **372** 532
[2] Ishida K, Mukuda H, Kitaoka Y, Asayama K, Mao Z Q, Mori Y and Maeno Y 1998 *Nature* **396** 658
[3] Maeno Y, Ando T, Mori Y, Ohmichi E, Ikeda S, Nishizaki S and Nakatsuji S 1998 *Phys. Rev. Lett.* **81** 3765
[4] Yaguchi H, Wada M, Akima T, Maeno Y and Ishiguro T 2003 *Phys. Rev. B* **67** 214519
[5] Mao Z Q, Nelson K D, Jin R, Liu Y and Maeno Y 2001 *Phys. Rev. Lett.* **87** 037003
[6] Kawamura M, Yaguchi H, Kikugawa N, Maeno Y and Takayanagi H 2005 *J. Phys. Soc. Jpn.* **74** 531
[7] Ando T, Akima T, Mori Y and Maeno Y 1999 *J. Phys. Soc. Jpn.* **68** 1651
[8] Kittaka S, Taniguchi H, Yonezawa S, Yaguchi H and Maeno Y 2010 *Phys. Rev. B* **81** 180510(R)
[9] Yaguchi H, Kittaka S and Maeno Y 2009 *J. Phys.: Conf. Ser.* **150** 052285
[10] Kittaka S, Yaguchi H and Maeno Y 2009 *J. Phys. Soc. Jpn.* **78** 103705
[11] Smith T F and Chu C W 1967 *Phys. Rev.* **159** 353
[12] Mao Z Q, Maeno Y and Fukazawa H 2000 *Mater. Res. Bull.* **35** 1813
[13] Kittaka S, Nakamura T, Yaguchi H, Yonezawa S and Maeno Y 2009 *J. Phys. Soc. Jpn.* **78** 064703
[14] Gerber A, Grenet T, Cyrot M and Beille J 1992 *Phys. Rev. B* **45** 5099(R)
[15] Kittaka S, Nakamura T, Aono Y, Yonezawa S, Ishida K and Maeno Y 2009 *Phys. Rev. B* **80** 174514
[16] Chamaya E V, Tien C, Wur C S and Kumzerov Y A 1996 *Physica C* **269** 313

International Conference on Strongly Correlated Electron Systems (SCES 2011) IOP Publishing
Journal of Physics: Conference Series **391** (2012) 012109 doi:10.1088/1742-6596/391/1/012109

Anomalous electronic structures of transition-metal oxides with hollandite-type crystal structure

T Toriyama[1], **T Konishi**[2] **and Y Ohta**[1]

[1] Department of Physics, Chiba University, Chiba 263-8522, Japan
[2] Graduate School of Advanced Integration Science, Chiba University, Chiba 263-8522, Japan

E-mail: ohta@faculty.chiba-u.jp

Abstract. We make the electronic structure calculations of transition-metal oxides with a hollandite-type crystal structure $A_2M_8O_{16}$ using the generalized gradient approximation (GGA) in the density functional theory, where the Hubbard-type repulsive interaction is taken into account (GGA+U). We first examine the electronic structure of the 3d series (M=Ti, V, Cr, Mn) as well as the 4d series (M=Mo, Ru, Rh) to discuss their generic features. We then discuss in particular the origins of the metal-insulator transition observed in $K_2Cr_8O_{16}$ and quasi-one-dimensional electron conduction observed in $K_2Ru_8O_{16}$. We also consider the observed metal-insulator transition in $K_2V_8O_{16}$ where the effect of electron correlations may play an essential role.

1. Introduction

Hollandite materials have attracted considerable attention in recent years in the field of physics of strong electron correlations. The crystal structure of hollandites resembles that of the rutile structure but the single chains of MO_6 octahedra (M=transition element) in rutiles are replaced by the double chains of the edge-shared MO_6 octahedra in hollandites, resulting in a sparse structure with large tunnels, wherein a variety of cations A can be introduced as in the chemical formula $A_xM_8O_{16}$ with $0 \leq x \leq 2$.

We may point out that the material series with the 3d transition-metal elements M = Ti, V, Cr, and Mn, which are in a mixed valent state, show a variety of unusual electronic and magnetic properties: In Ti series, $K_xTi_8O_{16}$ has been reported to be either a dilute ferromagnetic semiconductor [1] or a paramagnetic metal [2], depending on the carrier concentration. In V series, $K_2V_8O_{16}$ shows a variety of physical properties associated with the metal-insulator transition [2, 3, 4, 5, 6, 7, 8, 9, 10]. In Cr series, $K_2Cr_8O_{16}$ has been reported to exhibit phase transitions from metallic paramagnet to ferromagnet and then to insulating ferromagnet by lowering temperatures [11, 12]. In Mn series, a variety of structural, electronic, and magnetic phase transitions have been reported as well [13, 14].

We may also point out that the material series with the 4d transition-metal elements M = Mo, Ru, and Rh show interesting physical properties: In Mo series, an Mo_4 clustering has been reported to occur in $Rb_{1.5}Mo_8O_{16}$ [15]. In Ru series, a quasi-one-dimensional electron conduction has been reported in KRu_4O_8, $RbRu_4O_8$, and $Cs_{0.8}Li_{0.2}Ru_4O_8$ [16, 17, 18], as well as in $BaRu_6O_{12}$ [19] with a similar crystal structure. In Rh series, a quasi-one-dimensional electron conduction has also been reported [20, 21, 22, 23].

In this paper, motivated by such materials development, we make the electronic structure calculations of the series of hollandite-type transition-metal oxides to clarify their basic electronic structures. We use the generalized gradient approximation (GGA) in the density functional theory, where the Hubbard-type repulsive interaction U is taken into account (GGA+U). We thereby examine the $3d$ series (M=Ti, V, Cr, Mn) as well as the $4d$ series (M=Mo, Ru, Rh) to discuss generic features in the electronic structure of hollandites first. Then, we in particular study the quasi-one-dimensional electron conduction observed in $K_2Ru_8O_{16}$ and $K_2Rh_8O_{16}$ by focusing on singularities in their calculated band structures and Fermi surfaces. The origins of the metal-insulator transition observed in $K_2Cr_8O_{16}$ is discussed in this respect. We also study the electronic structure of $K_2V_8O_{16}$ to consider the origins of the observed anomalous electronic states and metal-insulator transition and find that the effect of electron correlations plays an essential role in the metal-insulator transition of this material.

2. Method of calculation

We employ the computer code WIEN2k [24] based on the full-potential linearized augmented-plane-wave (FLAPW) method. We use the experimental crystal structure of each material observed at room temperatures. The symmetry of the lattice at high-temperature phase is body-centred-tetragonal ($I4/m$ space group). The primitive unit cell contains four M ions, one K ion, and eight O ions. All the M ions are crystallographically equivalent but there are two inequivalent O ions, i.e., O(1) inside the double chains and O(2) connecting the double chains. To improve the description for the electron correlations in the d orbitals, we use the rotationally invariant version of the GGA+U method [25, 28] as well. We use the codes VESTA [29] and XCrySDen [30] for graphical purposes.

3. Results of calculation

3.1. Generic features

Let us discuss some common features in hollandites first. The calculated densities of states show that there appear the t_{2g} and e_g bands separated by $10D_q$ of ~ 2 eV for the $3d$ series and ~ 3 eV for the $4d$ series and O $2p$ states well below the d bands. The hybridization between the d and $2p$ states are considerably strong. The Fermi level is located somewhere in the t_{2g} manifold, depending on the number of d electrons. The rigid-band approximation works well in the series considered at least in the high-energy scale of a few eV.

More precisely, there are 12 bands in the t_{2g} manifold, the character of which differs strongly between the topmost 4, the lowermost 4, and the middle 4 bands. The Fermi level is located in the topmost 4 bands in the Ru, Rh, and ferromagnetic Cr systems, in the lowermost 4 bands in the Ti, V, and Mo systems, and in the middle 4 bands in the Mn systems. Thus, the number of valence d electrons or the location of the Fermi level determines the basic electronic properties of hollandite-type transition-metal oxides.

3.2. $K_2Ru_8O_{16}$

The Ru ions are in the mixed valent state with the average valence of $Ru^{3.75+}$ ($d^{4.25}$), i.e., there are formally Ru^{4+} (d^4) and Ru^{3+} (d^5) in the 3 : 1 ratio. We find from the GGA+U calculations that, despite its structural complexity, the electronic state of $K_2Ru_8O_{16}$ near the Fermi level is surprisingly simple, consisting only of a single band coming predominantly from the $4d_{yz}$ and $4d_{zx}$ orbitals of Ru ions with strong admixture of the $2p_z$ orbitals of corner-shared O ions connecting the double RuO chains. The electronic band structure near the Fermi level is highly quasi-one-dimensional, exactly at half filling, and has a pair of nearly parallel sheetlike Fermi surfaces separated by π/c. The calculated results are consistent with observed quasi-one-dimensional electron conduction of this material [16]. These results establish that $K_2Ru_8O_{16}$

belongs to a class of materials for further studies of Tomonaga-Luttinger liquids [27]. See Ref. [18] for details.

3.3. $K_2Rh_8O_{16}$

There are some experimental reports on the hollandite rhodates: $Ba_{1.2}Rh_8O_{16}$ [20, 21], $(Ba,Bi)_{1.54}Rh_8O_{16}$ [22], and $Sr_{1.5}Rh_8O_{16}$ [23]. The formal oxidation state of Rh in these materials ranges between $Rh^{3.57+}$ ($d^{5.43}$) and $Rh^{3.7+}$ ($d^{5.3}$). The number of d electrons thus indicates that the second band counted from the top in the t_{2g} manifold crosses the Fermi level with the filling of electrons roughly between a half to three quarters if the second band is isolated (although of course more than one band can cross the Fermi level). Our preliminary electronic structurte calculations indicate that the band structure is not very simple but there is in fact a highly quasi-one-dimensional band, as expected from our results for $K_2Ru_8O_{16}$, which is consistent with the quasi-one-dimensional transport properties reported in $Ba_{1.2}Rh_8O_{16}$ [20, 21].

3.4. $K_2Cr_8O_{16}$

We point out that the above two materials have an interesting similarity with the ferromagnetic hollandite $K_2Cr_8O_{16}$ [11], where Cr ions are in the oxidation state $Cr^{3.75+}$ ($d^{2.25}$). The electronic structure calculations [12] have shown that the system is a fully spin-polarized half metal and that the Fermi level is located between the third and fourth bands counted from the top of the t_{2g} manifold for majority-spin electrons, or that both of the third and fourth bands cross the Fermi level to form a semimetallic (or compensated metallic) band structure. In fact, it has been found [12] that the Fermi surface of the fourth band is highly quasi-one-dimensional, as in the present $K_2Ru_8O_{16}$. The observed metal-insulator transition of the Cr system is therefore related to the nesting feature of the quasi-one-dimensional Fermi surfaces, details of which will be discussed elsewhere [31].

3.5. $K_2V_8O_{16}$

Such situations as above are in strong contrast to the hollandite vanadate $K_2V_8O_{16}$. The average valence of V ions is $V^{3.75+}$ ($d^{1.25}$), i.e., in the mixed valent state of $V^{3+} : V^{4+} = 3d^2 : 3d^1 = 1 : 3$, and therefore the Fermi level is located in the third band counted from the lowest in the t_{2g} manifold in the paramagnetic state. Our GGA+U calculations have shown [32] that the lowest four bands are predominantly of the d_{xy} character and show the nearly one-dimensional dispersion. The nesting of the Fermi surfaces may play some role in the observed metal-insulator transition but the effects of electron correlations, i.e., the charge and orbital ordering as well as the Mott localization, should rather be more important in $K_2V_8O_{16}$ than in other systems. In fact, depending on the values of U used, the calculated band structure and Fermi surfaces change largely, and the spin and orbital structure obtained in the low-temperature phase diplays a variety of patterns, which are not expected from the $2k_F$ singularity in the high-temperature phase [32].

3.6. Other materials

Besides the materials discussed above, we have also made the electronic structure calculations of other hollandite-type transition-metal oxides. We find that the calculated electronic structure of $K_2Ti_8O_{16}$ is consistent with the observed paramagnetic metallic state. We also find that the band structure of $K_2Mn_8O_{16}$ is complicated, i.e., the Fermi level is located somewhere in the middle 4 bands of the t_{2g} manifold.

4. Summary

We have made the electronic structure calculations of transition-metal oxides with the hollandite-type crystal structure using GGA and GGA+U in the density functional theory. We have examined the 3*d* series (M=Ti, V, Cr, Mn) as well as the 4*d* series (M=Mo, Ru, Rh) to discuss generic aspects in theelectronic structure of hollandites first. Then, we in particular study the origins of the metal-insulator transition observed in $K_2Cr_8O_{16}$ and quasi-one-dimensional electron conduction observed in $K_2Ru_8O_{16}$ by focusing on singularities in their calculated band structures and Fermi surfaces. We also study the electronic structure of $K_2V_8O_{16}$ to consider the origins of the observed anomalous electronic states and metal-insulator transition.

Acknowledgments
We would like to thank M Isobe, M Itoh, D I Khomskii, W Kobayashi, A Nakao, H Nakao, Y Shimizu, Y Ueda and T Yamauchi for useful discussions on general aspects of hollandite materials. This work was supported in part by a Grant-in-Aid for Scientific Research (No. 22540363) from the Ministry of Education, Culture, Sports, Science and Technology of Japan. A part of computations was carried out at the Research Center for Computational Science, Okazaki Research Facilities, Japan.

References
[1] Noami K, Muraoka Y, Wakita T, Hirai M, Kato Y, Muro T, Tamenori Y and Yokoya T 2010 *J. Appl. Phys.* **107** 073910.
[2] Isobe M, Koishi S, Yamazaki S, Yamaura J, Gotou H, Yagi T and Ueda Y 2009 *J. Phys. Soc. Jpn.* **78** 114713.
[3] Isobe M, Koishi S, Kouno N, Yamaura J, Yamauchi T, Ueda H, Gotou H, Yagi T and Ueda Y 2006 *J. Phys. Soc. Jpn.* **75** 073801.
[4] Waki T, Kato H, Kato M and Yoshimura K 2004 *J. Phys. Soc. Jpn.* **73** 275.
[5] Maignan A, Lebedev O I, Van Tendeloo G, Martin C and Hebert S 2010 *Phys. Rev.* B **82** 035122.
[6] Komarek A C 2009 *PhD thesis, Univ. Köln.*
[7] Shimizu Y, Okai K, Itoh M, Isobe M, Yamaura J, Yamauchi T and Ueda Y 2011 *Phys. Rev.* B **83** 155111.
[8] Ishige Y, Sudayama T, Wakisaka Y, Mizokawa T, Wadati H, Sawatzky G A, Regier T Z, Isobe M and Ueda Y 2011 *Phys. Rev.* B **83** 125112.
[9] Horiuchi S, Shirakawa T and Ohta Y 2008 *Phys. Rev.* B **77** 155120.
[10] Sakamaki M, Horiuchi S, Konishi T and Ohta Y 2008 *arXiv: 0811.4338.*
[11] Hasegawa K, Isobe M, Yamauchi T, Ueda H, Yamaura J I, Gotou H, Yagi T, Sato H and Ueda Y 2009 *Phys. Rev. Lett.* **103** 146403.
[12] Sakamaki M, Konishi T and Ohta Y 2009 *Phys. Rev.* B **80** 024416 (2009) and **82** 099903(E).
[13] Ishiwata S, Bos J W G, Huang Q and Cava R J 2006 *J. Phys.: Condens. Matter* **18** 3745.
[14] Sato H, Enoki T, Yamaura J I and Yamamoto N 1999 *Phys. Rev.* B **59** 12836.
[15] Ozawa T, Suzuki I and Sato H 2006 *J. Phys. Soc. Jpn.* **75** 014802.
[16] Kobayashi W 2009 *Phys. Rev.* B **79** 155116.
[17] Foo M L, Lee W, Siegrist T, Lawes G, Ramirez A P, Ong N P and Cava R J 2004 *Mater. Res. Bull.* **39** 1663.
[18] Toriyama T, Watanabe M, Konishi T and Ohta Y 2011 *Phys. Rev.* B **83** 195101.
[19] Mao Z Q, He T, Rosario M M, Nelson K D, Okuno D, Ueland B, Deac I G, Schiffer P, Liu Y and Cava R J 2003 *Phys. Rev. Lett.* **90** 186601.
[20] Pautrat A and Kobayashi W 2010 *Phys. Rev.* B **82** 115113.
[21] Kobayashi W, Hebert Pérez S O, Pelloquin D and Maignan A 2009 *Phys. Rev.* B **79** 085207.
[22] Klimczuk T, Lee W L, Zandbergen H W and Cava R J 2004 *Mater. Res. Bull.* **39** 1671.
[23] Plaisier J R, van Vliet A A C and Ijdo D J W 2001 *J. Alloys and Compounds* **314** 56.
[24] Blaha P, Schwarz K, Madsen G K H, Kvasnicka D and Luitz J 2002 *WIEN2k, An Augmented Plane Wave + Local Orbitals Program for Calculating Crystal Properties* (Technische Universität Wien, Austria).
[25] Anisimov V I, Solovyev I V, Korotin M A, Czyżyk M T and Sawatzky G A 1993 *Phys. Rev.* B **48** 16929.
[26] Djafri F, Canonne J, Abraham F and Thomas D 1985 *J. Less-Com. Met.* **109** 323.
[27] See, e.g., Giamarchi T 2004 *Quantum Physics in One Dimension* (Clarendon Press, Oxford).
[28] Liechtenstein A I, Anisimov V I and Zaanen J 1995 *Phys. Rev.* B **52** R5467.
[29] Momma K and Izumi F 2008 *J. Appl. Cryst.* **41** 653.
[30] Kokalj A 2003 *Comp. Mater. Sci.* **28** 155.
[31] Toriyama T *et al.* 2011 *preprint.*
[32] Toriyama T, Konishi T and Ohta Y 2011 *in preparation.*

Pressure effect on transport properties of $NdCo_2$

K Uchima[1], M Takeda[2], Y Takaesu[1,2], H G Flesch[2], M Hedo[2], T Nakama[2], K Yagasaki[2], Y Uwatoko[3] and A T Burkov[4]

[1]General Education, Okinawa Christian Junior College, Nishihara, Okinawa, 903-0207, Japan
[2]Faculty of Science, University of the Ryukyus, Nishihara, Okinawa, 903-0213, Japan
[3]Institute for Solid State Physics, University of Tokyo, Kashiwa, Chiba 277-8581, Japan
[4]A. F. Ioffe Physico-Technical Institute, Russian Academy of Sciences, Sankt-Petersburg, 194021, Russia

E-mail: uchima@ocjc.ac.jp

Abstract. Electrical resistivity ρ and thermopower S of the Laves phase compound of $NdCo_2$ has been investigated at temperatures from 2 K to 300 K. ρ has been measured under pressures up to 8 GPa and S has been measured under pressures up to 3 GPa. The magnetic transition temperature T_C obtained by ρ measurement decreases with increasing pressure. The temperature T_{mim} where the thermopower S takes minimum at high temperature region increases linearly with increasing pressure. The high-temperature minimum of S is associated with a sharp peak in density of states related mainly to the Co 3d-electron density. Since the width of an itinerant electronic band depends on the extent of the corresponding overlapping of 3d orbitals, the pressure variation of T_{min} can be attributed to the broadening of the peak width of 3d electron density of states.

1. Introduction

RCo_2 compounds with cubic Laves phase structure (R stays for rare earth elements) are known for magnetic instability of Co 3d-electron system [1, 2, 3]. When R is nonmagnetic (R=Y, Lu), the compounds are exchange-enhanced paramagnets exhibiting metamagnetic transition in external magnetic field of 70 T for YCo_2 [4] and 77 T for $LuCo_2$ [5]. In the case of magnetic RCo_2 compounds, long-range magnetic order of the itinerant 3d-electron subsystem is induced by exchange field of the ordered localized magnetic moments of the rare earths. Due to antiferromagnetic spin–spin coupling between 3d and 4f moments [6], the compounds with light R elements are ferromagnetic, whereas those with heavy R are ferrimagnetic. Accordingly, $NdCo_2$ is a ferromagnet with the Curie temperature of T_C 98 K. The magnetization of 4f moments in magnetically ordered state of $NdCo_2$ is parallel to the induced magnetization of Co 3d band. It has been found that electronic transport in RCo_2 compounds is strongly dependent on the state of Co 3d band and on 3d-4f interaction [3]. Resistivity of $NdCo_2$ shows a steep decrease at the magnetic transition temperature T_C, and reveals an anomaly at the spin reorientation temperature of T_s 40 K with decreasing temperature [7]. For thermopower of $NdCo_2$, the clear magnetic anomalies have been observed at T_C and at T_s. In order to investigate the pressure effect on transport properties of $NdCo_2$, we have measured electrical resistivity and thermopower at temperatures from 2 K to 300 K under pressure.

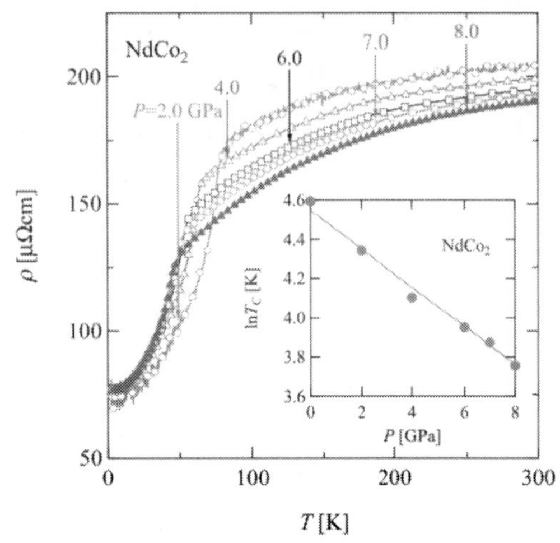

Figure 1. Temperature dependence of electrical resistivity of $NdCo_2$ under pressures up to 8 GPa. The inset shows the pressure dependence of T_C.

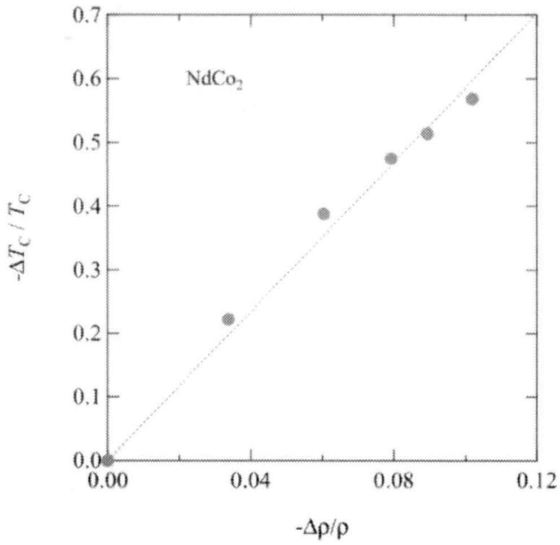

Figure 2. $\Delta T_C/T_C$ vs. Δ (300)/ (300) for $NdCo_2$. The dotted line is a guide for the eye.

2. Experimental procedures

The polycrystalline sample of $NdCo_2$ compound was prepared by melting the pure constituents (3N) in an arc furnace under a protective Ar atmosphere and were subsequently annealed in vacuum at 900 °C for one week. A ratio of R:Co = 1:1.93 for cast was chosen to avoid formation of magnetic RCo_3 phase. The cubic Laves phase structure was confirmed by X-ray diffraction measurement with lattice constant a = 7.305 Å. The electrical resistivity measurements were carried out by means of a standard four-probe dc method from 2 K to 300 K. The seesaw-heating method was used for thermopower measurements [8]. The electrical resistivity and thermopower S were measured simultaneously from 2 K to 300 K. A clamp-type piston cylinder pressure cell and cubic anvil cell were utilized for the measurements of and S under pressures up to 3 GPa and for up to 8 GPa, respectively.

3. Results and Discussion

Figure 1 shows the temperature dependence of of $NdCo_2$ under pressures up to 8 GPa. increases with increasing temperature and shows a knee around the Curie temperature T_C. T_C and T_r were determined as the temperature where d /dT takes maximum. As shown in Fig. 1, the application of pressure leads to decrease of T_C and of the resistivity at the high temperature region, while the resistivity at low temperatures increases with increasing pressures. Based on s-d model, T_C is given by

$$T_C = \frac{N_A \mu_B^2}{3k_B}\xi[I_{RR} + I_{Rd}^2 \quad _d(T_C)].$$

Here, N_A is Avogadro constant, ξ is the de Gennes factor, I_{RR} and I_{Rd} are the respective exchange interaction coefficients between the rare earths and between the rare earth and cobalt, and $_d$ is the magnetic susceptibility of Co-3d band [9], $_d$ $\mu_B^2 N_{\varepsilon_F}$ where N_{ε_F} is the 3-d electron density of state (DOS) at the Fermi level. The pressure variation of T_C is attributed to the pronounced pressure response of $_d$. From the result shown in the inset of Fig. 1, we obtained the pressure derivative of T_C for $NdCo_2$, $d\ln T_C/dP =$ 0.11, which is in good agreement with

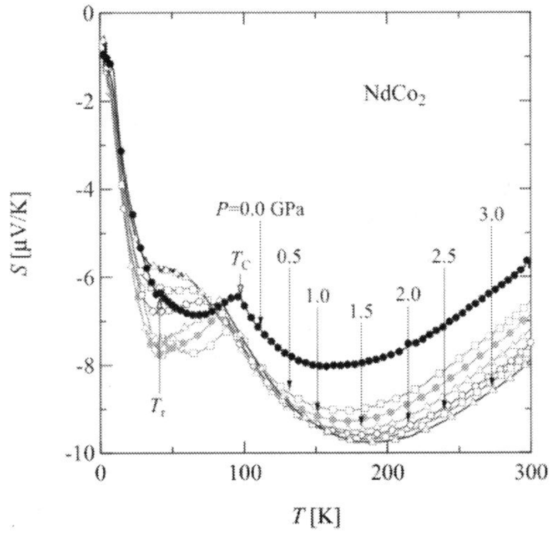

Figure 3. Temperature dependence of S of $NdCo_2$ at temperatures from 2 K to 300 K under pressures up to 3 GPa. The empty arrows indicate T_C and T_r at $P = 0$ GPa.

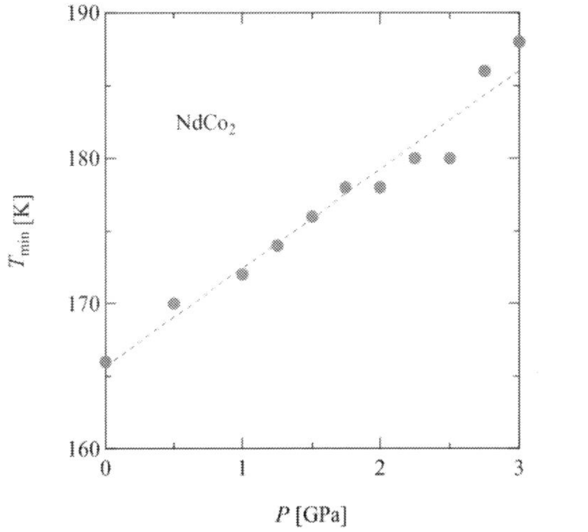

Figure 4. Pressure dependence of $S(T)$ minimum temperature T_{min}. The dashed line is a guide for the eye.

the value reported in Ref. [10]. On the other hand, the total resistivity $\ (T)$ of the magnetic RCo_2 compounds can be written as

$$(T) = \ _0 + \ _{ph}(T) + \ _m(T) + \ _{sf}(T).$$

Here, $_0$, $_{ph}$, $_m$, and $_{sf}$ denote the contributions due to impurities, phonon, 4f magnetic moments and Co-3d spin fluctuation scatterings, respectively. Within the scope of Mott's s-d scattering model, these all contributions are proportional to the 3d state density at Fermi level, so that $\quad N_{\varepsilon_F}$ [11]. At high temperatures, in paramagnetic state, the main effect of pressure on resistivity comes from decrease of N_{ε_F} due to increasing overlapping of 3d orbitals. Therefore, we expect that there a linear relation between the pressure variations of T_C and at high temperatures: $\mathrm{d}\ln T_C/\mathrm{d}P \quad \mathrm{d}\ln\ /\mathrm{d}P$. As can be seen in Fig. 2, the plot of $\Delta T_C/T_C$ vs. $\Delta\ (300)/\ (300)$ shows a good linear relation, which indicates that the pressure effect on T_C and at high temperatures is associated with the change of N_{ε_F}.

Figure 3 shows the temperature dependence of thermopower S of $NdCo_2$ in the temperature range from 2 K to 300 K under pressures up to 3 GPa. S is negative in the whole temperature range and under all applying pressure. At ambient pressure, S decreases linearly with increasing temperature, having two anomalies associated with the spin reorientation and the magnetic phase transition as indicated by the empty arrows, and reveals a broad minimum at $T_{min} \quad 160$ K. It was shown that T_{min} of the paramagnetic RCo_2 compounds is in good correlation with the electronic part of the specific heat which is a measure of DOS at the Fermi level. Since the width of 3d itinerant electronic band depends on the extent of the corresponding atomic orbital overlapping, it can be expected that the 3d peak width of RCo_2 compounds increases when lattice constant decreases. This assumption has been confirmed by very good correlation between the $S(T)$ minimum temperature T_{min} and lattice constants of RCo_2 compounds [3]. As shown in Fig. 4, T_{min} increases linearly with increasing pressure, $\mathrm{d}T_{min}/\mathrm{d}P \quad 10$ K/GPa. The variation of T_{min} against the lattice constant a, $\mathrm{d}T_{min}/\mathrm{d}a$, can be expressed as

$$\frac{\mathrm{d}T_{min}}{\mathrm{d}a} = \frac{3}{a\kappa}\frac{\mathrm{d}T_{min}}{\mathrm{d}P},$$

where κ is the compressibility of $NdCo_2$. With the values of $\kappa = 0.0107$ GPa ^{1} [14] and $a = 7.305$ Å, we can obtain $\mathrm{d}T_{\mathrm{min}}/\mathrm{d}a$ 410 K/Å, which is in agreement with the value of $\mathrm{d}T_{\mathrm{min}}/\mathrm{d}a$ 550 K/Å for the RCo_2 compounds [3]. This result indicates that the pressure variation of T_{min} of $NdCo_2$ can be attributed to the contraction of lattice, in other words, the broadening of the peak width of 3d electron density of states due to the application of external pressure.

In summary, the electrical resistivity and thermopower were measured from 2 K to 300 K under pressure. The decreases of T_{C} and of the high temperature resistivity with pressure are related to decrease of $N_{\varepsilon_{\mathrm{F}}}$ due to increasing overlapping of 3d orbitals. The temperature T_{mim} where the S takes minimum at high temperature region increases linearly with increasing pressure. The high-temperature minimum of S is associated with a sharp peak in the density of states related mainly to the Co 3d-electron. The pressure variation of T_{min} can be attributed to the broadening of the peak width of 3d electron density of states.

Acknowledgments
Part of this work was supported by a Grant-in-Aid for Scientific Research (No. 21540340) from Ministry of Education, Culture, Sports, Science and Technology of Japan, and by Russian Foundation for Basic Research under the Grant 09-02-01285a.

References
[1] Lemaire R 1966 *Cobalt* **32** 132; 1966 **33** 201
[2] Levitin R Z and Markosyan A S 1988 *Sov. Phys. - Usp.* **31** 730
[3] Gratz E, Resel R, Burkov A T, Bauer E, Markosyan A S and Galatanu A 1995 *J. Phys.: Condens. Matter* **7**6687
[4] Goto T, Fukamichi K, Sakakibara T and Komatsu H 1989 *Solid State Commun.* **72** 945
[5] Goto T, Sakakibara T, Murata K, Komatsu K and Fukamichi K 1990 *J. Magn. Magn. Mater.* **90/91** 700
[6] Cyrot M, Lavagna M 1979 *J. Appl. Phys.* **50(3)** 2333
[7] Gratz E, Resel and Nowotny H 1982 *J. Magn. Magn. Matter.* **29**127
[8] Resel R, Gratz E, Burkox A T, Nakama T, Higa M and Yagasaki K 1996 *Rev. Sci. Instrum.* **67**1970
[9] Bloch D and Lemaire R *Phys. Rev. B* **2** (1970) 2648–2650
[10] Hauser R, Bauer E and Gratz E *Phys. Rev. B* **57** (1998) 2904-2914 .11
[11] Burkov A T, Gratz E, Bauer E, and Resel R *J. Alloys Compd.* **198** (1993) 117-126
[12] Burkov A T, Zyuzin A Yu, Nakama T, Yagasaki K *Phys. Rev. B* **69** (2004) 144409-1-6.
[13] Yamada H, Inoue J, Terao K, Kanda S and Shimizu M *J.Phys. F: Met. Phys.* **14**1943.
[14] Klimker H and Rosen M *J. Magn. Magn. Matter.* **7**(1978) 361-364

Hydrostatic Pressure Study on 3-K Phase Superconductivity in Sr_2RuO_4-Ru Eutectic Crystals by AC Magnetic Susceptibility Measurements

Hiroshi Yaguchi, Hiromichi Watanabe and Akira Sakaue

Department of Physics, Faculty of Science and Technology, Tokyo University of Science, Noda 278-8510, Japan

Abstract.

We have investigated the effect of hydrostatic pressure on 3-K phase superconductivity in Sr_2RuO_4-Ru eutectic crystals by means of AC magnetic susceptibility measurements. We have found that the application of hydrostatic pressure suppresses the superconducting transition temperature T_c of the 3-K phase with a pressure coefficient of $dT_c/dP \approx -0.2$ K/GPa, similar to the case of the 1.5-K phase. We have also observed that the effect of hydrostatic pressure on the 3-K phase seems to be elastic whilst that of uniaxial pressure is plastic.

1. Introduction

Sr_2RuO_4 is the first layered perovskite superconductor without copper [1], isostructural to the cuprate high-temperature superconductor $La_{2-x}Ba_xCuO_4$. More importantly, Sr_2RuO_4 is known to be a spin-triplet superconductor [2]. The original superconducting phase in Sr_2RuO_4 occurs with a sharp transition at a T_c of 1.5K, called the 1.5-K phase. Besides this, the eutectic system, a two-phase composite structure of a single-crystalline Sr_2RuO_4 matrix and lamellar microdomains of ruthenium metal embedded in it, shows a broad superconducting transition with an enhanced onset of approximately 3 K, called the 3-K phase [3]. The 3-K phase transition is, on further cooling, followed by the superconducting transition to the original phase (the 1.5-K phase) in Sr_2RuO_4. (Figure 1 displays a SEM photograph of a polished surface of a Sr_2RuO_4-Ru eutectic crystal. Figure 2 shows typical AC magnetic susceptibility data for the 1.5-K phase in Sr_2RuO_4 and for the 3-K phase in Sr_2RuO_4-Ru.)

Whilst the origin of the 3-K phase still remains uncertain, 3-K phase superconductivity is non-bulk, and is believed to occur near Sr_2RuO_4 interfaces with Ru inclusions [4, 5, 6]. Sigrist and Monien's phenomenological theory [7] within the framework of Ginzburg Landau formalism, which assumes spin-triplet pairing similar to Sr_2RuO_4, successfully describes important aspects of the 3-K phase. Examples of experiments in support of the 3-K phase being unconventional superconductivity include tunnelling measurements on S/N junctions obtaining zero bias conductance peaks [8, 9, 10], and transport measurements on micro-fabricated samples observing hysteresis in the *I*-*V* characteristics attributable due to the chiral *p*-wave state [11].

Recent uniaxial-pressure studies on the 3-K phase have revealed that both in-plane pressure (parallel to the *ab*-plane) and out-of-plane pressure (parallel to the *c*-axis) enhance 3-K phase superconductivity in Sr_2RuO_4-Ru eutectic crystals, particularly in volume fraction [12, 13]. The enhancement by in-plane pressure is even greater than that by out-of-plane pressure. This fact probably supports the scenario that the rotation of the RuO_6 octahedra plays an important role in the occurrence of the 3-K phase[7] because in-plane pressure will affect the Ru-Ru lattice constant in the *ab*-plane directly, and the out-of-plane pressure will affect it indirectly.

In our recent hydrostatic pressure study, we used SQUID magnetometer (MPMS, Quantum Design) to measure the DC magnetisation at temperatures down to 1.8 K, and found that the shielding fraction or the apparent superconducting volume fraction associated with the 3-K phase is suppressed by the application of hydrostatic pressure [14]. In the present work, we have investigated the AC magnetic susceptibility of Sr_2RuO_4-Ru eutectic crystals under hydrostatic pressure at temperatures down to 1.2 K to obtain insight into the mechanism of 3-K phase superconductivity.

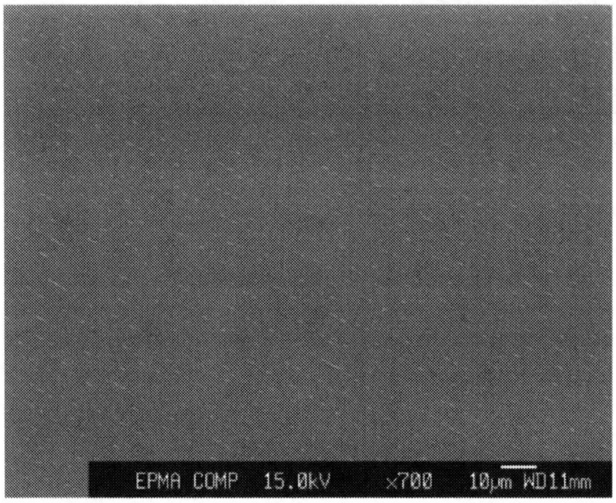

Figure 1. SEM image of a polished surface of a Sr_2RuO_4-Ru eutectic crystal. The darker part corresponds to Sr_2RuO_4 and the brighter part corresponds to Ru inclusions.

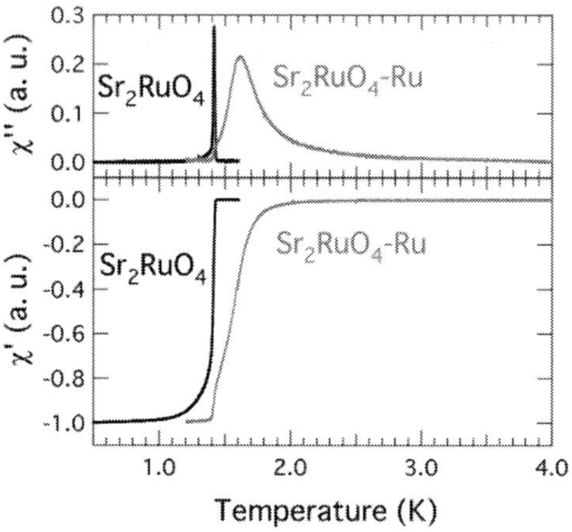

Figure 2. Example of AC susceptibility data for the 1.5 K phase in single-phase Sr_2RuO_4 and for the 3-K phase in the eutectic system Sr_2RuO_4-Ru.

2. Experimental

The samples used in the present study were Sr_2RuO_4-Ru eutectic crystals grown by a floating zone method [15], which were cut and polished such that approximate dimensions of the samples were 1 × 1 × 2 mm^3 with the longest dimension parallel or perpendicular to the *c*-axis. We used a mutual inductance method to measure AC susceptibility at temperatures down to 1.2 K with an AC measurement field of 2 μT-rms at a frequency of 777 Hz. Hydrostatic pressure was applied using a beryllium-copper clamp cell with Daphne 7373 oil (Idemitsu Co. Ltd.) as the pressure transmitting medium; pressures were determined from the pressure dependence of the superconducting critical temperature of tin [16].

3. Results and Discussion

Figure 3 shows the temperature dependence of the real and imaginary parts of the AC magnetic susceptibility of a Sr_2RuO_4-Ru eutectic crystal under hydrostatic pressure, with the AC field along the *c*-axis. These measurements were performed in the order of increasing applied pressure, and a subsequent measurement was done after the applied pressure was released. We have made a similar set of measurements on four eutectic crystals. All of the data in Fig. 3 are, for convenience, multiplied by a common factor such that the real part of the 0-GPa (before) trace tends to -1 at temperatures well below T_c. Similar to Fig. 2, rather broad changes associated with 3-K phase superconductivity are seen in both of the real and imaginary parts of the AC susceptibility, and those changes shift towards lower temperatures with increasing hydrostatic pressure; 3-K phase superconductivity is suppressed by the application of hydrostatic pressure. However, the shielding fraction, corresponding to the drop in the real part, hardly changes with pressure. These results enable the pressure coefficient to be determined as $\mathrm{d}T_c/\mathrm{d}P \approx -0.2$ K/GPa; the dissipation peak in the imaginary part associated with the 3-K phase has been used to define T_c of the 3-K phase. Measurements on the other samples also yield close values for the pressure coefficient. (We here note that hydrostatic pressure studies on Sr_2RuO_4 obtained the pressure coefficient of the superconducting transition temperature for the 1.5-K phase to be -0.2 K/GPa [17, 18, 19], which is very close to the value for the 3-K phase.)

Therefore, the basic relation in tetragonal symmetry

$$\frac{\mathrm{d}T_c}{\mathrm{d}P} = \frac{\mathrm{d}T_c}{\mathrm{d}P_{\|a}} + \frac{\mathrm{d}T_c}{\mathrm{d}P_{\|b}} + \frac{\mathrm{d}T_c}{\mathrm{d}P_{\|c}} = 2\times\frac{\mathrm{d}T_c}{\mathrm{d}P_{\|a}} + \frac{\mathrm{d}T_c}{\mathrm{d}P_{\|c}}, \quad (1)$$

constrains at least either $\mathrm{d}T_c/\mathrm{d}P_{\|a}$ or $\mathrm{d}T_c/\mathrm{d}P_{\|c}$ of the 3-K phase to be negative. Nevertheless, it has been reported that uni-axial pressure both parallel and perpendicular to the c-axis greatly enhances 3-K phase superconductivity in eutectic crystals [12, 13], showing an apparent disagreement with the present hydrostatic-pressure study.

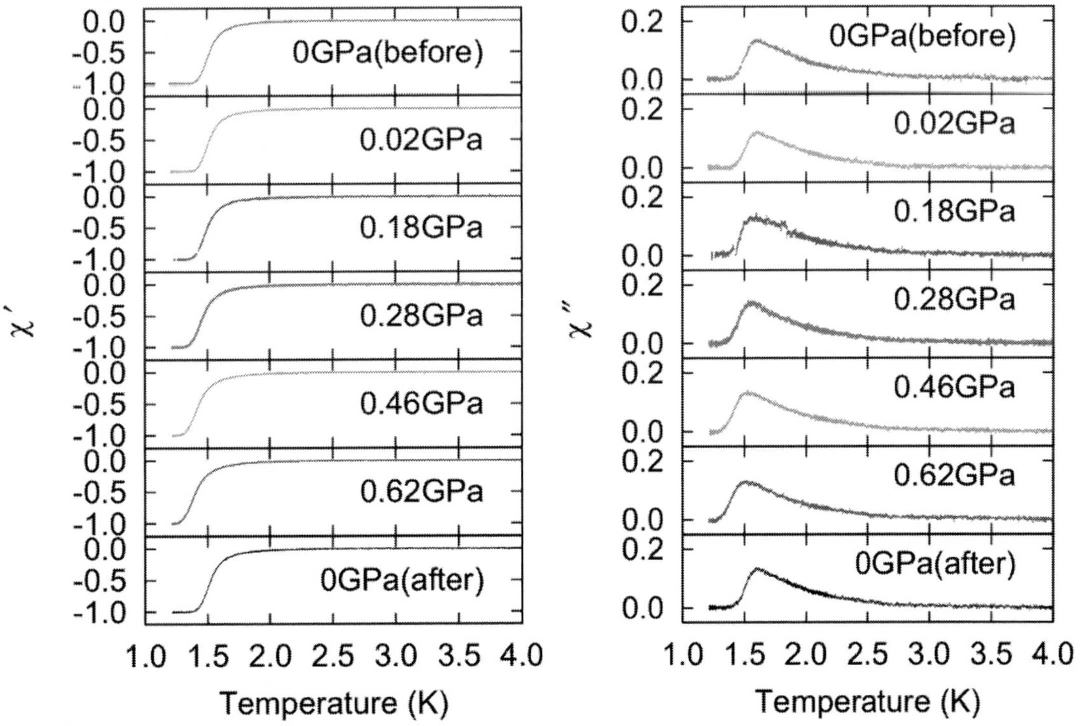

Figure 3. The real (left) and imaginary (right) parts of the AC susceptibility of a Sr_2RuO_4-Ru eutectic crystal at various hydrostatic pressures. The bottom trace of each panel was taken after the applied pressure was released.

The above apparent discrepancy is very likely to be due to the difference between the natures of hydrostatic and uni-axial pressure effects on this eutectic system. The effect of hydrostatic pressure is modest and appears to be elastic since the 0-GPa traces of "before" and "after" in Fig. 3 well coincide with each other. By contrast, the effect of uniaxial pressure has been reported to be dramatic and plastic; the effect remains even after the applied pressure is released [13]. These facts presumably support the idea that uniaxial pressure induces and/or promotes the rotation of the RuO_6 octahedra in Sr_2RuO_4-Ru eutectic crystals [12, 13, 20], leading to large enhancement of 3-K phase superconductivity, whilst hydrostatic pressure is unlikely to affect such rotation considerably.

4. Summary

In summary, we have found that hydrostatic pressure suppresses the superconducting transition temperature T_c of the 3-K phase in Sr_2RuO_4-Ru crystals, and have evaluated its pressure coefficient dT_c/dP to be approximately -0.2 K/GPa, very close to that of the 1.5-K phase. It has turned out that hydrostatic pressure effect, which suppresses the 3-K phase in a rather modest manner, presents a striking contrast with the uniaxial pressure effect largely enhancing the 3-K phase. Also the effect of hydrostatic pressure on the 3-K phase in Sr_2RuO_4-Ru eutectic crystals appears to be elastic, in contrast to uni-axial pressure causing plastic deformation to occur in eutectic crystals. Such elastic deformation will be related to the rotation of the RuO_6 octahedra in Sr_2RuO_4-Ru eutectic crystals.

Acknowledgements

We thank S. Kawada, S. Suzuki, S. Kittaka and Y. Maeno for their supports and discussions.

References

[1] Maeno Y, Hashimoto H, Yoshida K, Nishizaki S, Fujita T, Bednorz J G and Lichtenberg F 1994 *Nature* **372** 532
[2] Mackenzie A P and Maeno Y 2003 *Rev. Mod. Phys.* **75** 657
[3] Maeno Y, Ando T, Mori Y, Ohmichi E, Ikeda S, NishiZaki S and Nakatsuji S 1998 *Phys. Rev. Lett.* **81** 3765
[4] Ando T, Akima T, Mori Y and Maeno Y 1999 *J. Phys. Soc. Jpn.* **68** 1651
[5] Yaguchi H, Wada M, Akima T, Maeno Y and Ishiguro T 2003 *Phys. Rev. B* **67** 214519
[6] Ying Y A, Xin Y, Clouser B W, Hao E, Staley N E, Myers R J, Allard L F, Fobes D, Liu T, Mao Z Q and Liu Y 2009 *Phys. Rev. Lett.* **103** 247004
[7] Sigrist M and Monien H 2001 *J. Phys. Soc. Jpn.* **70** 2409
[8] Mao Z Q, Nelson K D, Jin R and Liu Y 2001 *Phys. Rev. Lett.* **87** 037003
[9] Kawamura M, Yaguchi H, Kikugawa N, Maeno Y and Takayanagi H 2005 *J. Phys. Soc. Jpn.* **74** 531
[10] Yaguchi H, Takizawa K, Kawamura M, Kikugawa N, Maeno Y, Meno T, Akazaki T, Semba K and Takayanagi H 2006 *J. Phys. Soc. Jpn* **75** 125001
[11] Kambara H, Kashiwaya S, Yaguchi H, Asano Y, Tanaka Y and Maeno Y 2008 *Phys. Rev. Lett.* **101** 267003
[12] Yaguchi H, Kittaka S and Maeno Y 2009 *J. Phys.: Conf. Seri.* **150** 052285
[13] Kittaka S, Yaguchi H and Maeno Y 2009 *J. Phys. Soc. Jpn* **78** 103705
[14] Yaguchi H, Kawada S, Watanabe H and Suzuki S 2010 *J. Phys. Soc. Jpn* **79** 125004
[15] Mao Z Q, Maeno Y and Fukazawa H 2000 *Mat. Res. Bull.* **35** 1813
[16] Smith T F and Chu C W 1967 *Phys. Rev.* **159** 353
[17] Shirakawa N, Murata K, Nishizaki S, Maeno Y and Fujita T 1997 *Phys. Rev. B* **56** 7890
[18] Forsythe D, Julian S R, Bergemann C, Pugh E, Steiner M J, Alireza P L, McMullan G J, Nakamura F, Haselwimmer R K W, Walker I R, Saxena S S, Lonzarich G G, Mackenzie A P, Mao Z Q and Maeno Y 2002 *Phys. Rev. Lett.* **89** 166402
[19] Svitelskiy O, Headley S, Tozer S W, Palm E C, Murphy T P, Shulyatev D and Suslov A V 2008 *Phys. Rev. B* **77** 052502
[20] Kittaka S, Taniguchi H, Yonezawa S, Yaguchi H and Maeno Y 2010 *Phys. Rev. B* **81** 180510(R)

International Conference on Strongly Correlated Electron Systems (SCES 2011) IOP Publishing
Journal of Physics: Conference Series **391** (2012) 012112 doi:10.1088/1742-6596/391/1/012112

Phase transition and magnetic structure of pyrochlore oxide $Cd_2Os_2O_7$

J Yamaura[1], H Ohsumi[2], K Sugimoto[3], S Tsutsui[3], Y Yoda[3], S Takeshita[2], A Tokuda[2], S Kitao[4], M Kurokuzu[4], M Seto[4], I Yamauchi[1], K Ohgushi[1], M Takigawa[1], T Arima[2,5], Z Hiroi[1]
[1]ISSP, University of Tokyo, Kashiwa, Chiba 277-8581, Japan
[2]RIKEN SPring-8 Center, Sayo, Hyogo 679-8148, Japan
[3]JASRI, SPring-8, Sayo, Hyogo 679-5198, Japan
[4]RRI, Kyoto University, Kumatori, Sennan, Osaka 590-0494, Japan
[5]GSFS, University of Tokyo, Kashiwa, Chiba 277-8561, Japan

E-mail: jyamaura@issp.u-tokyo.ac.jp

Abstract. The 5d transition metal oxide $Cd_2Os_2O_7$ having a pyrochlore lattice of Os atoms exhibits a metal-insulator transition with a magnetic ordering at 227 K, but the magnetic structure in the ordered state and the origin of the transition have not yet been specified. The commensurate magnetic reflection with the propagation vector $\boldsymbol{k}$ = (0, 0, 0) was first observed in the resonant X-ray scattering using a high-quality single crystal. We also clarified the only one kind of magnetic moment via the synchrotron radiation-based Mössbauer effect. We propose that a strong spin-orbit interaction is notably apparent in the spin arrangement and the insulating mechanism in this compound.

1. Introduction

Metal-insulator (MI) transition is one of the most dramatic phenomena of electrons in a solid. The strong electron correlation causes a localized insulator with a first-order type Mott transition, which is often accompanied by a simple collinear-type antiferromagnetic order [1]. On the other hand, the Fermi surface nesting leads to a different type of insulator with a CDW or SDW transition, in which either a nonmagnetic singlet state or an incommensurate magnetic state is stabilized. Moreover, the other type of MI transition is theoretically expected, which is the Slater transition induced by the Brillouin zone folding due to a magnetic order [2]. The magnetic order associated with the Slater transition can be a commensurate or incommensurate antiferromagnetic order and is accompanied by no lattice distortion. Because electron correlations are not crucial, the transition should appear in weakly correlated electron systems.

The 5d transition metal oxide $Cd_2Os_2O_7$ with a pyrochlore lattice of Os atoms exhibits an MI transition at T_{MI} = 226 K, accompanied by a magnetic ordering [3, 4]. The magnetic structure of the ordered state has not been determined, and the origin of the transition remains a mystery. A Slater mechanism and the formation of a SDW state were suggested in previous study [4, 5] but were questioned because of the lack of Fermi surface nesting in the calculated band structure [6, 7]. In this work, we investigated the MI transition of $Cd_2Os_2O_7$ by means of resonant X-ray scattering (RXS) and synchrotron radiation-based Mössbauer spectroscopy to unveil the magnetic ground state. These state-of-the-art techniques enable us to obtain element-selective signals from complex compounds [8, 9].

2. Experiments

A single crystal of $Cd_2Os_2O_7$ was synthesized from CdO and Os in a sealed quartz tube, which contained an appropriate amount of AgO for oxygen supply. The tube was placed in a furnace with a temperature gradient of 900-750 °C for a week. RXS experiments were carried out at the absorption edge of Os $L3$ (E = 10.874 keV) and Os $L2$ (E = 12.387 keV) using a six-circle diffractometer and an imaging plate at beamlines of SPring-8 BL19LXU and SPring-8 BL02B1, respectively [10, 11]. Synchrotron radiation-based Mössbauer spectroscopy was measured at 20 K in a beamline of SPring-8 BL09LXU, in which ^{189}Os enriched powder samples of $Cd_2Os_2O_7$ and Os were used for the transmitter and the scatter on the velocity transducer, respectively [9]. The radiation at 36.332 keV was chosen by a high-resolution Si(111)-Si(400) monochrometer to reduce the background, and the delayed emission was measured using a Si-APD detector.

3. Result and Discussions

Figure 1(a) shows the oscillation photograph on the imaging plate at the Os $L2$ edge at 100 K. The $00l$ ($l = 4n+2$) and $0kl$ ($k + l = 4n+2$) reflections, which violate the extinction rule of space group $Fd\text{-}3m$, are observed below T_{MI}. These forbidden reflections are not due to the lattice symmetry breaking but due to the magnetic or the anisotropic tensor susceptibility (ATS) scattering, as mentioned later.

Figure 1(b) shows the X-ray absorption spectra near the Os $L3$ edge and the energy dependence of the 006 reflection intensity at 10 K, which was obtained in the π-σ' polarization channel at the azimuthal angle $\varphi = -47°$. Note that the 006 reflection peak has a maximum value strikingly at the Os $L3$ edge depicted by the dotted line, which is evidence of the resonant effect.

The ATS intensity, which is derived from the local atomic environment, of the $00l$ ($l = 4n + 2$) forbidden reflection at the azimuthal angle φ in the π-σ' polarization channels are estimated using $I_{00l} = F^2\sin^2\theta\cos^2 2\varphi$, as shown by Dmitrienko [12]. Hence, the ATS contribution is negligible in the present experimental setting with $\varphi = -45°$ in the π-σ' channel. Thus, the observed 006 reflection is not due to Thomson scattering and ATS scattering but due to magnetic scattering. The magnetic peak was not observed with a powder sample in the previous neutron experiment [13] and was detected for the first time in the present experiment. Considering that the commensurate magnetic reflections represent the violations of the extinction rules with the d-glide and without the face-centered symmetry operations, we conclude that a commensurate magnetic structure appears with the propagation vector $\boldsymbol{k} = (0, 0, 0)$ below T_{MI}.

Figure 2 shows the temperature dependence of the 006 magnetic reflection intensity in the π-σ' channel at $\varphi = -47°$. The integrated intensity of the 006 reflection increases gradually below T_{MI}=227 K upon cooling, which can be fitted by a power law expression. The result reveals that this transition is the successive and the second-order type transition.

Figure 3 shows ^{189}Os Mössbauer spectrum of $Cd_2Os_2O_7$ measured at 20 K for 96 hours. The spectrum is preliminary estimated at the Isomer shift $I.S.$ = 0.7 mm/s relative to that of Os, the electric quadrupole splitting is e^2qQ = -6.2 mm/s, and the internal magnetic field is H_{int} = 28.8 T. This single spectrum reveals that the unique magnetic moment inhabits the Os atom.

Here, we discuss the origin of the MI transition in this compound. Usually, the on-site Coulomb energy U in 5d transition metal oxides is between 1-2 eV. Due to the hybridization of Os-5d and O-2p, the band width W in $Cd_2Os_2O_7$ is 2 eV, which represents an intermediate coupling region with $U \sim W$. Considering these points, Mandrus *et al.* has suggested that the MI transition of this compound is the Slater transition, which is a continuous phase transition in the weak coupling region, instead of a Mott transition. [4] However, the present result of the spin arrangement with $\boldsymbol{k} = (0, 0, 0)$ suggests that this transition is due to neither the Slater-type nor the SDW transitions accompanied by the band folding with the antiferromagnetic structure formation or the Fermi surface instability, respectively. Consequently, the band-gap formation of this compound is thought to arise from an unprecedented mechanism, that is presumably related to the unique spin arrangement within a primitive unit cell. Recently, Wan *et al.* provides a stabilization mechanism of the all-in/all-out spin arrangement in $R_2Ir_2O_7$, where the spin-orbit (SO) interaction plays an important role, from the LDA+U+SO band

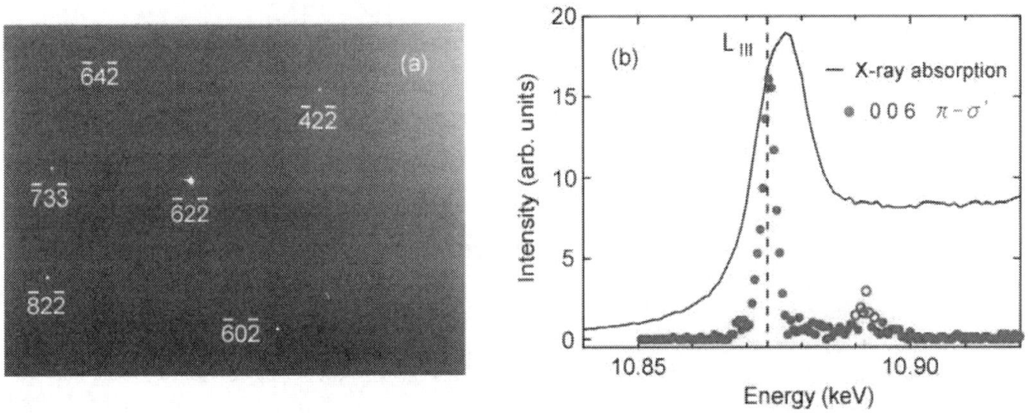

Figure 1. (a) Oscillation photograph at the Os *L*2 edge at 100 K. Arrow shows the reflection, which is forbidden in *Fd-3m*. (b) X-ray absorption spectra near the Os *L*3 edge at room temperature and the energy dependence of the 006 reflection intensity in the π-σ' polarization (φ = -47°) at 10 K.

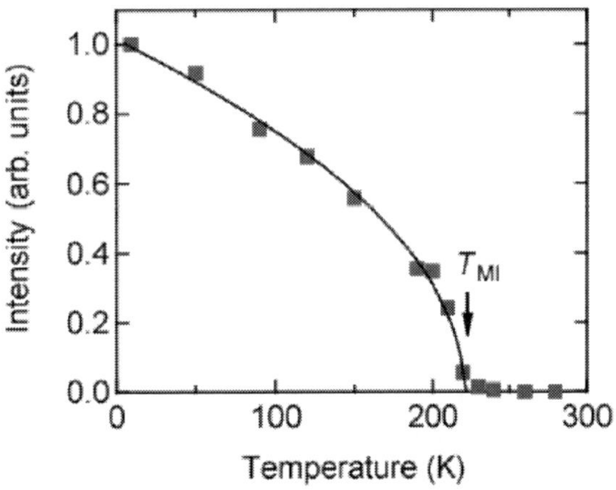

Figure 2. Temperature dependence of the intensity of 006 magnetic reflection in the π-σ' channel at φ = -47°, where the intensity is normalized by that at 10 K. The solid line represents the fitting by the power law expression for the second-order transition.

calculation (not yet experimentally confirmed) [14]. We also believe that the spin arrangement in the present compound is primarily related to the lattice symmetry and the SO interaction.

4. Summary

Pyrochlore oxide $Cd_2Os_2O_7$ exhibits the second-order type of metal-insulator transition at 227 K. The commensurate magnetic structure with the propagation vector $\boldsymbol{k}$ = (0, 0, 0) and the unique magnetic moment in the low-temperature phase was determined by the resonant X-ray scattering and the synchrotron-radiation based Mössbauer measurements. We propose that the feature of the 5d-electrons with the strong spin-orbit interaction notably emerges due to the spin arrangement and the insulating mechanism in this compound.

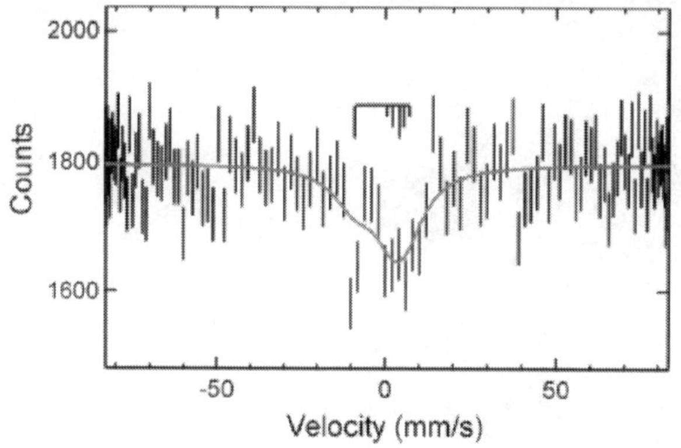

Figure 3. Mössbauer spectrum of the ^{189}Os enriched sample of $Cd_2Os_2O_7$ at 20 K. The solid line describes the fitting curve estimated by the Isomer shift, the electric quadrupole splitting and the internal magnetic field.

Acknowledgements

The synchrotron radiation experiments were performed at BL-02B1, BL-19XU and BL-09XU of SPring-8 with the approval of RIKEN (Proposal Nos. 2009B1200, 20090064 and 2011A1128). This work was supported by a Grant-in-Aid for Scientific Research on Priority Areas "Novel States of Matter Induced by Frustration" (19052003)

References

[1] Mott N F 1990 *Metal-Insulator Transitions* (Taylor and Francis, London/Philadelphia)
[2] Slater J C 1951 Phys. Rev. **82** 538
[3] Sleight A W, Gillson J L, Weither J F and Bindloss W 1974 *Solid State Commun.* **14** 357
[4] Mandrus D, Thompson J R, Gaal R, Forro L, Bryan J C, Chakoumakos B C, Woods L M, Sales B C, Fishman R S and Keppens V 2001 *Phys. Rev.* B **63** 195104
[5] Koda A, Kadono R, Ohishi K, Saha S R, Higemoto W, Yonezawa S, Muraoka Y and Hiroi Z 2007 *J. Phys. Soc. Jpn.* **76** 063703
[6] Singh D J, Blaha P, Schwarz K and Sofo J O 2002 *Phys. Rev.* B **65** 155109
[7] Harima H 2002 *J. Phys. Chem. Solids* **63** 1035
[8] Blume M 1985 *J. Appl. Phys.* **57** 3615
[9] Seto M, Masuda R, Higashitaniguchi S, Kitao S, Kobayashi Y, Inaba C, Mitsui T and Yoda Y 2009 *Phys. Rev. Lett.* **102** 217602.
[10] Yabashi M, Mochizuki T, Yamazaki H, Goto S, Ohashi H, Takeshita K, Ohata T, Matsushita T, Tamasaku K, Tanaka Y and Ishikawa T 2001 *Nucl. Instr. Meth. Phys. Res.* A **467-468** 678
[11] Sugimoto K, Ohsumi H, Aoyagi S, Nishibori E, Moriyoshi C, Kuroiwa Y, Sawa H and Takata M 2010 *AIP Conf. Proc.* **1234** 887
[12] Dmitrienko V E 1983 *Acta Crystallogr.* A **39** 29
[13] Reading J and Weller M T 2001 *J. Mat. Chem.* **11** 2373
[14] Wan X, Turner A M, Vishwanath A and Savrasov S Y 2011 *Phys. Rev.* B **83** 205101

Magnetic excitations in MnP

Shinichiro Yano[1], Jun Akimitsu[1], Shinich Itoh[2], Tetsuya Yokoo[2], Setsuo Satoh[2], Daichi Kawana[3] and Endoh Yasuo[2]

[1]Department of Physics and Mathematics, Aoyama Gakuin University, Sagamihara, Kanagawa 229-8558, Japan.
[2]Neutron Science Division, Institute of Materials Structure Science, High Energy Accelerator Research Organization, 305-0801, Japan
[3]Condensed Matter Research Center, Institute of Materials Structure Science, High Energy Accelerator Research Organization, 305-0801, Japan

E-mail: yanoshin@phys.aoyama.ac.jp

Abstract. We performed inelastic neutron scattering experiments on a three-dimensional magnetic system MnP, which is a ferromagnetic inter-metallic compound below T_C = 291 K and transitions into a screw state at T^* = 47 K. Using newly developed high resolution chopper spectrometer (HRC) and triple axis spectrometer LTAS(for E = 0-2 meV), 6G(for E = 2-8 meV) TAS1(for E = 10-35 meV), we succeeded in observing the excitations from low energy to Brillouin zone center. We assumed isotropic 6 exchange parameters, and calculated dispersion relation using Heisenberg model for 2-sub lattices. We confirm that nearest neighbor (n.n) exchange parameters J_{nn} are J > 0 (ferromagnetic interaction), the next nearest neighbor (n.n.n) exchange parameters are J < 0 (anti-ferromagnetic interaction), We conclude that the competition between n.n and n.n.n would be the cause of helimagnetic structure in MnP.

1. Introduction

Manganese phosphide MnP is a ferromagnetic intermetallic compound below T_C = 291K, and it transforms into a double helix proper screw state at T^* = 47K[1]. The crystal structure is a slightly distorted NiAs-type structure with the lattice parameters of a = 5.916Å, b = 5.260Å, c = 3.173Å at room temperature. In the ferromagnetic state, the easy-axis of the magnetization is the c-axis. In the helimagnetic state, the spin rotates in the b-c plane with a propagation vector 0.117a^* along the a-axis. The precise structure model of screw spiral state in MnP is proposed to be Elliptic or Bunching state[2,3].

Recently, Yamazaki *et.al.*[4] performed neutron diffraction experiment, and found that there is magnetic (δ,1,0) reflection in helimagnetic phase, and they could not find (1+δ,0,0) = (δ,1,0)+(1,-1,0). (the δ is the same period of magnetic propagation vector). They also reported the magnetic susceptibility measurement. From these experimental data, they concluded that anti-ferroic modulated magntic structure should exist in MnP.

The conventional theory for helimagnetism cannot explain why the complicated magnetic structure is stable and a ferromagnetic intermetallic compound transforms into a proper screw spiral state. In order to elucidate the mechanism of the spiral state, the information of spin wave in the whole Brillouin zone is crucially important. We focus our attention to the spin wave dispersion of the ferromagnetic state in MnP. Spin waves in MnP have already been measured in a small-q region below 150K by Tajima *et.al.*[5]. Adopting a Heisenberg model, they deduced

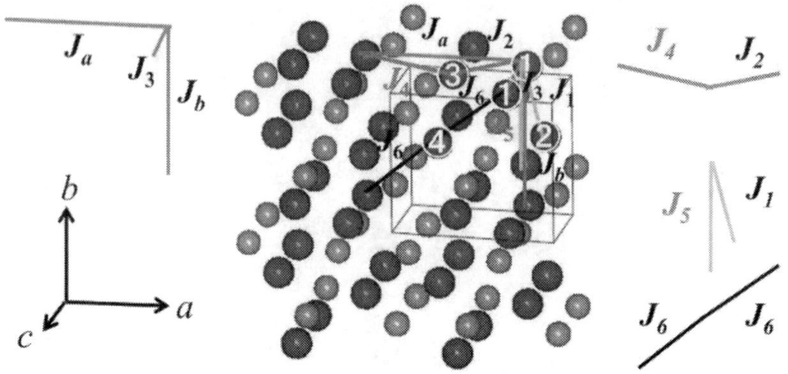

Figure 1. Purple and ash color atoms are Mn, P atoms respectively. Mn 4c cite in unit cell are Mn1$(-x,-y,-z)$,Mn2(x,y,z),Mn3$(-x+\frac{1}{2},y+\frac{1}{2},z)$,Mn4$(x+\frac{1}{2},-y+\frac{1}{2},-z)$.

J_m	J_1	J_2	J_3	J_4	J_5	J_6
r_m(Å)	2.70	2.82	3.18	3.92	4.17	4.27
z	2	2	2	2	4	4

Table 1. The difinition of exchange paramteres,J_1,J_2,J_3,J_4 are the nearest neighbor Mn1-Mn2,Mn1-Mn3,Mn1-Mn1,Mn1-Mn4. J_4,J_5 are the next nearest neighbor Mn1-Mn3,Mn1-Mn2. We show here the distance between Mn-Mn atoms and coordination number.

the nearest neighbor effective inter-planar interaction (J_1) and next-nearest neighbor one (J_2) along the a-axis and concluded that ferro-screw transition at 47K is caused by the slight change of the ratio of $(J_2)/(J_1)$ with temperature. Todate *et.al.*[6] also reported spin waves of MnP in ferromagnetic state along the three principal axes at relatively small-q region. Dispersion relation along the a-axis exhibits anomalous wave vector and temperature dependence, while the quadratic q dependence was observed along the b-axis and the c-axis. MnP is extensively studied so far, however, it was limited to be focused on relatively low energy because of limitation of experimental condition.

2. Experimental

2.1. Inelastic neutron scattering on HRC

We constructed the High Resolution Chopper Spectrometer (HRC) at the BL12 beamline in the Materials and Life Science Experimental Facility (MLF) of the Japan Proton Accelerator Research Complex (J-PARC) in collaboration with the High Energy Accelerator Research Organization (KEK) and the University of Tokyo [7]. The HRC delivers high-resolution and relatively high-energy neutrons for a wide range of studies on the dynamics of materials. We successfully installed the DAQ system on the HRC. A total of 251 PSDs and 2 monitors were mounted on the spectrometer. The event data of the detected neutrons are processed in the DAQ system and visualized in the form of the dynamic structure factor. We verified the data analysis process by visualizing the excitations in one- and three-dimensional single-crystal magnetic systems in inelastic neutron scattering experiments[8].We performed inelastic neutron scattering experiments in three-dimensional (3D) magnetic systems to demonstrate excitation spectra with the DAQ system on the HRC.Using HRC, we examined the wide range of $S(Q,\omega)$ with high energy resolution conditions (E_i = 30-350 meV, $\Delta E/E_i$ = 0.03-0.06), and succeeded in observing the excitations from low energy to nearly Brillouin zone center.

2.2. Inelastic neutron scattering on TAS's

We also performed the neutron inelastic scattering experiments at triple-axis spectrometer LTAS,TAS-1,6G JRR-3M reactor in JAEA(Tokai). Using triple axis spectrometer LTAS, we focus on relatively low energy magnetic excitation. Spin wave spectra were obtained by constant Q scan, and spectra were fit to Gaussian. We plot peak center and the error, the errors are half maximum full-width. We performed constant Q scan at E_f fix mode (E_f = 3.0 meV). We found that the linear dispersion relation contribution of phonon, and the energy gap at $q = 0$ in the ferromagnetic state at 54K was estimated to be (0.5±0.1) meV which is consistent value with the previously reported[5]. We performed the neutron inelastic scattering experiments at triple-axis spectrometer TOPAN (6G), for middle range energy measurement (E = 2-8 meV). The measurements are E_i fixed =13.5meV condition. Data are fitted to $\hbar\omega_a = Aq^3 + \Delta$, and determined to be A = 84.4meV Δ= 0.34meV which are also consistent values with the previously reported. We also performed the neutron inelastic scattering experiments at triple-axis spectrometer TAS-1, JRR-3M reactor in JAERI (Tokai) for high energy measurement (E = 10-35 meV). The measurements are done under E_f fixed = 14.7meV condition. The obtained data clearly showed that the other higher magnetic excitation exists in MnP. Using newly developed chopper spectrometer HRC and triple axis spectrometers LTAS, 6G and TAS-1, we succeeded in observing the two branches of magnetic excitations from low energy to Brillouin zone center.

3. Aanalysis and Discussion

In order to explain the 2 distinctive branch of magnetic excitation, we derive spin wave dispersion relation in the ferromagnetic state of MnP.We assumed on isotropic Heisenberg model has been taken acount of 6 exchange interactions between Mn-Mn, and calculation was made for 2-sub lattices.

$$\begin{aligned} H &= \sum_{nn'jj'} J_{nn'jj'} \vec{S_{nj}} \cdot \vec{S_{n'j'}} - D\sum_{nj}(S^z_{nj})^2 \\ &= \sum_{nn'jj'} J_{nn'jj'}[S^z_{nj}S^z_{n'j'} + \frac{1}{2}(S^+_{nj}S^-_{n'j'} + S^-_{nj}S^+_{n'j'})] + D\sum_{nj}(S^z_{nj})^2 \end{aligned} \tag{1}$$

,where n, n' denote unit cell (the number is N),j, j' denote sub lattice in each of unit cells(j, j'=1-4, the number of magnetic ion = $4N$).The definition of exchange parameter is discribed in Fig1 and Table1. Here, J_1, J_2, J_3 and J_6 are nearest neighbor interaction between Mn-Mn within the unit cell of MnP, and J_4, J_5 are next nearest neighbor interactions.D is the axial anisotropy constant whose easy axis is along the c-axis. We have four dispersion relations λ_{1-4}, however we found that λ_1 and λ_2 are spurious branches because dynamical structure factor becomes 0, so we have 2 magnetic branches. Obtained dispersion relations along the a-axis are shown in Fig2.

Using newly developed high resolution chopper spectrometer HRC and triple axis spectrometers LTAS, 6G and TAS-1, we succeeded in observing the excitations from low energy to Brillouin zone center. We assumed isotropic 6 exchange parameters, and calculated dispersion

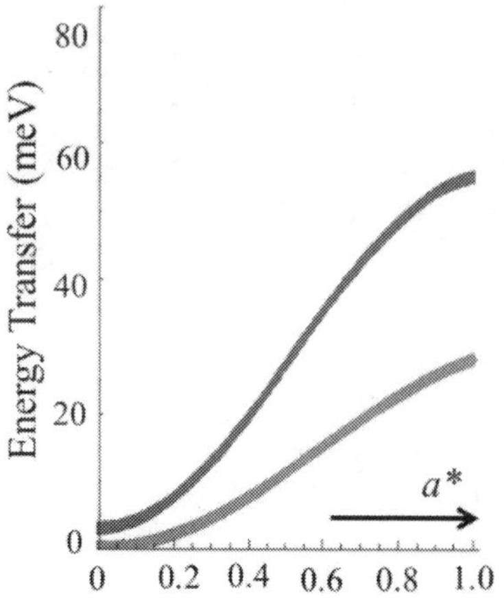

Figure 2. Calculated spin wave dispersion relation in MnP along the a-axis

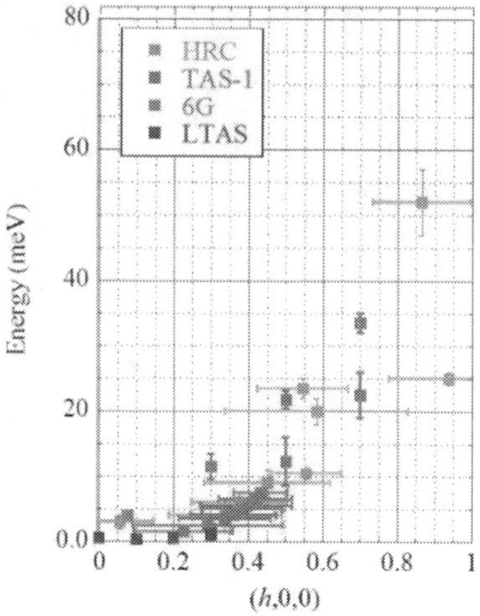

Figure 3. Obtained spin wave dispersion relation in MnP along the a-axis

J_m	J_1	J_2	J_3	J_4	J_5	J_6	D
E(meV)	11.30	1.67	8.99	**−5.27**	**−3.48**	2.55	0.24

Table 2. Determined exchange parameters using model fitting.

relation using Hisenberg model for 2-sub lattices. We determined exchange parameters J_m by model fitting to experimentaly obtained dipersion relation along the a-axis. We found that nearest neighbor (n.n) exchange parameters J_m are $J > 0$ (ferromagnetic interaction), the next nearest neighbor (n.n.n) exchange parameters are $J < 0$ (anti-ferromagnetic interaction), so that we concluded that the competition between n.n and n.n.n should be the cause of helimagnetic structure in MnP.

4. Acknowledgments

Authors wishing to acknowledge N.Metoki, K.Kakurai, S.Wakimoto and K.Iwasa for techical assistance for neutron scattering experiments at triple axis spectorometers LTAS, TAS-1 and 6G in JRR-3M. S.Y. was partially supported by the Iwanami Fujukai Foundation. Two of the authors (S.Y. and J.A.) were partially supported by a Gtant-in-Aid for Science Research and a High-Tech Research Center project for private universities with the matching fund subsidy from the Ministry of Education, Culture, Sports, Science and Technology, Japan

5. References

[1] G.P.Felcher, *J. Appl. Phys.* (1966) **37** 1056
[2] Hideki Obara, Yasuo Endoh, Yoshikazu Ishikawa and Takemi Komatsubara, *J.Phys.Soc.Jpn* **49** 928 (1980)
[3] R.M. Moon, *J.Appl.Phys.* **53** 1956 (1982)
[4] Yamazaki *et.al.*, arXiv:1104.3838v2
[5] K.Tajima,Y.Ishikawa and H.Obara,*J.Magn.and Magn.Mater.* **15-18** (1980) 373.
[6] K.Todate,K.Yamada,Y.Endoh and Y.Ishikawa,*J.Phys.Soc.Jpn.*, **56** (1987) 36.
[7] S. Itoh *et.al.*, *Nucl. Instr. and Meth. A* **631** (2011) 90.
[8] S. Yano *et.al.*, *Nucl. Instr. and Meth. A* **654** (2011) 421.

Low-temperature specific heat for ferromagnetic and antiferromagnetic orders in $CaRu_{1-x}Mn_xO_3$

M Yokoyama[1,†], S Nakano[1], S Someya[1], T Nakada[1], N Wada[1], H Kawanaka[2], H Bando[2], K Tenya[3], A Kondo[4] and K Kindo[4]

[1]Faculty of Science, Ibaraki University, Mito 310-8512, Japan
[2]National Institute of Advanced Industrial Science and Technology, Tsukuba 305-8568, Japan
[3]Faculty of Education, Shinshu University, Nagano 380-8544, Japan
[4]Institute for Solid State Physics, The University of Tokyo, Kashiwa 270-8581, Japan

E-mail: [†]makotti@mx.ibaraki.ac.jp

Abstract. Low-temperature specific heat of $CaRu_{1-x}Mn_xO_3$ was measured to clarify the role of d electrons in ferromagnetic and antiferromagnetic orders observed above $x = 0.2$. Specific heat divided by temperature C_p/T is found to roughly follow a T^2 function, and relatively large magnitudes of electronic specific heat coefficient γ were obtained in wide x range. In particular, γ is unchanged from the value at $x = 0$ (84 mJ/K^2 mol) in the paramagnetic state for $x \leq 0.1$, but linearly reduced with increasing x above $x = 0.2$. These features of γ strongly suggest that itinerant d electrons are tightly coupled with the evolution of magnetic orders in small and intermediate Mn concentrations.

1. Introduction

The relationship between anomalous charge transport and magnetism in the vicinity of metal-insulator transition is a longstanding subject in the physics of strongly correlated electron systems. The Ru-based metal $CaRuO_3$ is considered to have a paramagnetic ground state with itinerant characteristics of Ru 4d electrons [1,2]. In $CaRuO_3$, magnetization is found to be significantly enhanced at low temperatures [3,4], while no long-range magnetic order is observed in microscopic measurements such as neutron scattering and NMR [5–7]. In addition, electrical resistivity ρ shows an unusually large magnitude at high temperatures, corresponding to a very small mean-free pass of electron conduction comparable to lattice constants [8]. These features imply that $CaRuO_3$ involves instabilities for both magnetism and charge transport, which are enhanced by strong electronic correlation between itinerant Ru 4d electrons.

In the mixed compound $CaRu_{1-x}Mn_xO_3$ [9–14], it is revealed that a ferromagnetic (FM) order evolves above $x = 0.2$, and becomes stable in the intermediate x range. Both spontaneous magnetization M_0 and Curie temperature T_C show maximum values at $x \sim 0.7$ ($M_0 \sim 1\mu_B$/f.u. and $T_C = 170$ K), followed by the reductions of them with further increasing x. Furthermore, an antiferromagnetic (AFM) spin arrangement with the G-type structure is observed in the FM ordered phase. Finally, at $x = 1$ the FM order vanishes, and only the G-type AFM order appears below $T_N = 120$ K [15]. By doping Mn, the feature of ρ also changes from metallic to less conductive above $x \sim 0.1$, but the magnitude of ρ still remains in small values in the wide x range, that is, smaller than 0.1 Ω cm even at low temperatures for $x \leq 0.8$. At

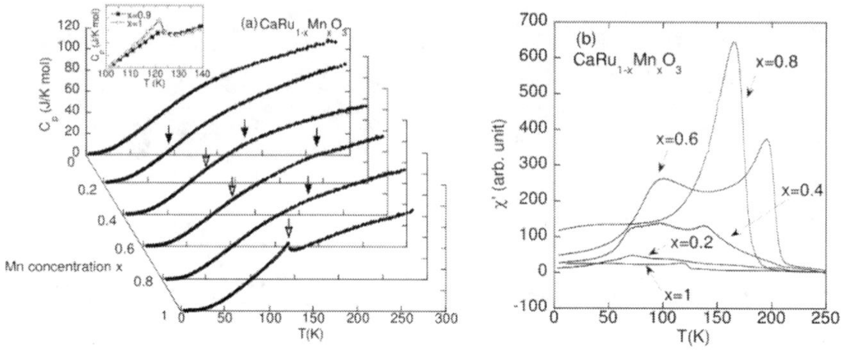

Figure 1. Temperature variations of (a) specific heat and (b) real part of ac susceptibility for $CaRu_{1-x}Mn_xO_3$. In (a), closed and open arrows indicate the onsets of FM and AFM orders, respectively, estimated from the peak position of the ac susceptibility. The types of magnetic order are determined from the previous reports on magnetization and neutron scattering measurements [10–14]. The inset of (a) shows an enlargement of C_p around the AFM transition temperatures for $x = 0.9$ and 1.

$x = 1$, in contrast, ρ dramatically increases with decreasing temperature, whose magnitude at low temperatures exceeds 10^6 Ω cm [16]. These properties indicate that doping Mn into $CaRuO_3$ modifies the itinerant electronic states due to strong correlations between Ru 4d and Mn 3d electrons, and it significantly affects the magnetic and transport properties. To clarify the variation of the d electronic states, we have performed the low-temperature specific-heat measurements for $CaRu_{1-x}Mn_xO_3$.

2. Experiment Details

Polycrystalline samples of $CaRu_{1-x}Mn_xO_3$ for $0 \leq x \leq 1$ were prepared by means of the conventional solid-state method. The mixture of appropriate amounts of $CaCO_3$, RuO_2 and MnO were first calcined at 800 °C for 24 hours. They were shaped into pellets after careful mixing, and then sintered at 1200 °C for 48 hours. By means of powder X-ray diffraction technique we checked that all the samples have a distorted perovskite structure (the $GaFeO_3$-type orthorhombic structure) [9–14] without any extrinsic phase. Specific heat C_p was measured between 3 K and 275 K with a thermal-relaxation method using both a commercial system (PPMS: Quantum Design) and a hand-made equipment, where we used plate-shaped samples with the mass of about 4 mg and 40 mg, respectively. The thermal-relaxation curves for all the measurements were well definitive, and the C_p data obtained from both equipments were consistent within the experimental accuracy. Ac-susceptibility was measured in the temperature range of 5-300 K to estimate the FM and AFM transition temperatures. Frequency and amplitude of the applied ac field were 180 Hz and ~ 0.5 Oe, respectively.

3. Results and Discussion

Figure 1(a) shows temperature variations of the specific heat $C_p(T)$ for $CaRu_{1-x}Mn_xO_3$. Clear jumps ascribed to the AFM transition are observed at 122.7 K for $x = 1$ and 125.1 K for $x = 0.9$ (see inset of Fig. 1(a)), but no distinct anomaly is seen in the $C_p(T)$ curves for $x \leq 0.8$. This contrasts with temperature dependence of the real part of ac susceptibility $\chi'(T)$ for $0.2 \leq x \leq 1$, where the FM and AFM transitions are indicated by the appearance of peaks (Fig. 1(b)). The transition temperatures estimated from $\chi'(T)$ are consistent with those determined by the dc magnetization and neutron scattering measurements [10–14]. The peaks seen in $\chi'(T)$ are somewhat broaden probably due to the distribution of the transition

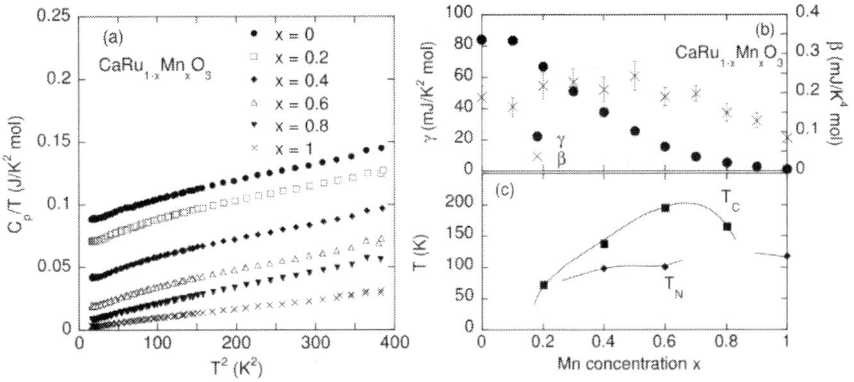

Figure 2. (a) Low-temperature specific heat divided by temperature plotted as a function of T^2, (b) electronic specific-heat coefficient γ and lattice contribution on specific heat β, and (c) T_C and T_N estimated from the $\chi'(T)$ data for $CaRu_{1-x}Mn_xO_3$. The definition of parameters γ and β is described in the text. The solid lines in (c) are guides to the eye [10, 11].

temperatures in the sample generated by doping. We consider this effect to be one of the reasons for the reduction of jumps in $C_p(T)$. On the other hand, it is interesting that similar reduction of anomaly is observed in the $C_p(T)$ data for Mn-doped $SrRuO_3$ [17]. $SrRuO_3$ is a ferromagnet with itinerant characteristics of Ru 4d electrons, and $C_p(T)$ exhibits a jump at $T_C = 162$ K. In the mixed compound $SrRu_{1-x}Mn_xO_3$, the anomaly at T_C in $C_p(T)$ is rapidly suppressed in the FM metallic phase for $x \leq 0.3$, but a shoulder-like anomaly again appears in $C_p(T)$ at the AFM transition temperatures in the strongly insulating region for x 0.4. As for $CaRu_{1-x}Mn_xO_3$, the electrical resistivity is much reduced for $x \leq 0.8$. The disappearance of anomaly in $C_p(T)$ may thus be related to the small entropy change associated with the magnetic transitions in the small-resistivity region, as well as the broadening of the onset due to the inhomogeneity.

Displayed in Fig. 2(a) is low-temperature specific heat divided by temperature C_p/T plotted as a function of T^2. C_p/T for entire x range are roughly proportional to T^2. The finite C_p/T values for $T \to 0$ except at $x = 1$ indicate that the Fermi-liquid excitation as well as phonon are mainly responsible for $C_p(T)$. However, C_p/T in the small and intermediate x range shows a very weak deviation from the T^2 function with a downward curvature below $T^2 \sim 100$ K^2 ($T \sim 10$ K). It may be ascribed to the spin fluctuation generated in the magnetically ordered phase. The C_p/T data are fitted with a function of $\gamma + \beta T^2$ to derive the electronic specific heat coefficient γ. In Fig. 2(b), we show x variations of γ and β obtained from the best fit in the temperature range between 3 K and 10 K. For comparison, we also plot the $x - T$ phase diagram obtained from the $\chi'(T)$ data in Fig. 2(c). The γ value for $x = 0$ is estimated to be 84 mJ/K^2 mol, which is consistent with that reported previously [18]. The magnitude of γ does not change in the paramagnetic phase for $x \leq 0.1$, but linearly decreases with increasing x in the magnetically ordered region (x 0.2). It is highly reduced and becomes ~ 5 mJ/K^2 mol or less for x 0.8, where T_C decreases through the maximum at $x \sim 0.7$. On the other hand, β is estimated to be 0.19 mJ/K^4 mol for $x = 0$ that yields the Debye temperature of $\Theta_D = 370$ K. It is reduced in Mn-rich region and finally becomes 0.085 mJ/K^4 mol ($\Theta_D = 490$ K) at $x = 1$. This feature seems to be consistent with the volume contraction in Mn-rich concentrations found by X-ray diffraction measurements [10, 12].

In $CaRu_{1-x}Mn_xO_3$, it is proposed that a double-exchange mechanism between Mn ions plays a crucial role in the evolution of the FM order in Mn-rich region, and in lower x range an interplay of double-exchange and superexchange interactions between d electrons located at Ru and Mn ions may yield the formation of inhomogeneous FM and AFM orders [10, 11]. On the other hand, the present C_p investigation revealed that γ shows relatively large values in small and intermediate x ranges, while it is strongly reduced in Mn-rich region. In particular, γ at $x = 0.4$ is estimated to be 38 mJ/K^2 mol, which is almost the same magnitude as that for the ferromagnetic metal $SrRuO_3$ (33 mJ/K^2 mol). The large γ values indicate that d electrons have itinerant characteristics in this x range. Furthermore, we observed that γ is nearly constant in the paramagnetic region, and it is then reduced when the FM order occurs above $x = 0.2$. Such a variation of γ cannot be simply attributed to an increase in the fraction of d electrons localized at Mn ions in the sample, but strongly suggests that the itinerant d electrons originating from both Ru and Mn are tightly coupled with the magnetic orders. This is supported by our recent magnetization and neutron-diffraction measurements for $x = 0.4$, where the magnetization shows typical features on fluctuation expected from itinerant electron spins, and the neutron-diffraction experiments revealed the occurrence of the long-range magnetic orders [14].

4. Summary

We performed C_p measurements for $CaRu_{1-x}Mn_xO_3$ ($0 \leq x \leq 1$). Low-temperature γ value is estimated to be 84 mJ/K^2 mol for pure $CaRuO_3$, and it is unchanged by doping Mn in the paramagnetic region for $x \leq 0.1$. In the magnetically ordered phase, γ is linearly reduced with increasing x for $0.2 \leq x \leq 0.6$, and then approaches $\sim$ 5 mJ/K^2 mol or less for x 0.8. The relatively large magnitude of γ in the small and intermediate Mn concentrations suggests that d electrons have itinerant characteristics contrary to localized ones expected in Mn-rich concentrations, and they play important roles in the magnetic orderings.

Acknowledgment

We thank Y. Nishihara and A. Yamaji for fruitful discussion. This work was carried out by the joint research in ISSP, the University of Tokyo, and partly supported by a Grant-in-Aid for Scientific Research on Innovative Areas "Heavy Electrons" (No.23102703) from the Ministry of Education, Culture, Sports, Science and Technology of Japan.

References

[1] Cao G, McCall S, Shepard M, Crow J E and Guertin R P, 1997 *Phys. Rev. B* **56** 321
[2] Kiyama T, Yoshimura K, Kosuge K, Michor H and Hilscher G, 1998 *J. Phys. Soc. Jpn.* **67** 307
[3] Kiyama T, Yoshimura K, Kosuge K, Mitamura H and Goto T, 1999 *J. Phys. Soc. Jpn.* **68** 3372
[4] Ferner I, Nowik I, Bradaric I and Gospodinov M, 2000 *Phys. Rev. B* **62** 11332
[5] Martinez J L, Prieto C, Rodriguez-Carvajal J, de Andres A, Vallet-Regi M and Gonzalez-Calbet J M, 1995 *J. Magn. Magn Mater.* **140-144** 179-180
[6] Mukuda H, Ishida K, Kitaoka Y, Asayama K, Kanno R and Takano M, 1999 *Phys. Rev. B* **60** 17
[7] Yoshimura K, Imai T, Kiyama T, Thurber K R, Hunt A W and Kosuge K, 1999 *Phys. Rev. Lett.* **83** 4397
[8] Cao G, Korneta O, Chikara S, Delong L E and Schlottmann P, 2008 *Solid State Commun.* **148** 305
[9] Sugiyama T and Tsuda N, 1999 *J. Phys. Soc. Jpn.* **68** 1306
[10] Shames A I, Rozenberg E, Martin C, Maignan A, Raveau B, André G and Gorodetsky G, 2004 *Phys. Rev. B* **70** 134433
[11] Markovich V, Auslender M, Fita I, Puzniak R, Martin C, Wisniewski A, Maignan A, Raveau B and Gorodetsky G, 2006 *Phys. Rev. B* **73** 014416
[12] Taniguchi T, Mizusaki S, Okada N, Nagata Y, Lai S H, Lan M D, Hiraoka N, Itou M, Sakurai Y, Ozawa T C, Noro Y and Samata H, 2008 *Phys. Rev. B* **77** 014406
[13] Mizusaki S, Taniguchi T, Okada N, Nagata Y, Hiraoka N, Itou M, Sakurai Y, Noro Y, Ozawa T C and Samata H, 2009 *J. Phys.: Condens. Matter* **21** 276003
[14] Kawanaka H, Yokoyama M, Noguchi A, Bando H and Nishihara Y, 2009 *J. Phys.: Condens. Matter* **21** 296002
[15] Wollan E O and Koehler W C, 1955 *Phys. Rev.* **100** 545
[16] Neumeier J J and Cohn J L, 2000 *Phys. Rev. B* **61** 14319
[17] Yamaji A, Yokoyama M, Nishihara N, Narumi Y and Kindo K, 2010 *J. Phys.: Conf. Ser.* **200** 012232.
[18] Kikugawa N, Balicas L and Mackenzie A P, 2009 *J. Phys. Soc. Jpn.* **78** 014701

Bulk-Sensitive Photoemission Spectroscopy of $TlFe_2Se_2$

Y Nishitani[1], A Okada[2], A Higashiya[3], A Irizawa[4], T Muro[5], S Horiguchi[6], H Sato[6], N Umeyama[7,8], N Miyakawa[8], S I Ikeda[7], H Kobori[2] and A Yamasaki[2]

[1] Graduate School of Natural Science, Konan University, Kobe 658-8501, Japan
[2] Faculty of Science and Engineering, Konan University, Kobe 658-8501, Japan
[3] Industrial Technology Center of Wakayama Prefecture, Wakayama 649-6261, Japan
[4] The Institute of Scientific and Industrial Research (ISIR), Osaka University, ibaraki, Osaka 567-0047, Japan
[5] Japan Synchrotron Research Institute (JASRI), Sayo, Hyogo 679-5198, Japan
[6] Faculty of Science and Engineering, Chuo University, Tokyo 112-8551, Japan
[7] National Institute of Advanced Industrial Science and Technology (AIST), Tsukuba, Ibaraki 305-8568, Japan
[8] Department of Applied Physics, Tokyo University of Science, Tokyo 162-8601, Japan

E-mail: yamasaki@konan-u.ac.jp

Abstract. We have investigated the electronic structures of the parent compound of an iron chalcogenide superconductor $TlFe_2Se_2$ by bulk-sensitive photoemission spectroscopy (PES). Valence-band PES demonstrates the energy gap opening at the Fermi level due to the shift of Fe 3*d* state to the higher binding energy side as compared to the calculated density of states, being consistent with the insulating behaviour in transport measurements.

1. Introduction

$TlFe_2Se_2$ has the same crystal structure as the parent compound of an Fe-based superconductor (SC), $BaFe_2As_2$. It is pointed out that the carrier doping makes this compound the SC. Similarly, the superconductivity was recently revealed in $(Tl,K)Fe_{2-x}Se_2$ [1]. Interestingly, most of parent compounds of Fe-based SCs are metals but $TlFe_2Se_2$ is not a metal but an insulator as well as cuprates. Therefore, it is expected that superconducting mechanism and electronic structures of the $TlFe_2Se_2$ system might be different from those of other Fe-based SC.

We have performed high-resolution bulk-sensitive soft-x-ray photoemission spectroscopy (PES) for the $TlFe_2Se_2$ to investigate bulk electronic structures.

2. Experimental

The $TlFe_2Se_2$ single crystal of which we performed the PES measurement was obtained under the following conditions. Powders of Tl_2Se (99.9%), Fe (99.9%) and Se (99.9%) were mixed in a ratio of Tl_2Se:Fe:Se = 1:4:4 and pressed into a pellet. The pellet was sealed in a quartz tube under 10^{-2} Pa of vacuum. Then, the tube was kept at 973 K and quenched to room temperature after the reaction [2]. The stoichiometric composition of grown samples (Tl : Fe : Se = 0.94-1.10 : 1.96-1.98 : 2) was determined by the x-ray fluorescence analysis with the fundamental parameter method.

Soft-x-ray PES was carried out at the undulator beamline BL-25SU in SPring-8 [3]. The highest total energy resolution ΔE (the full width at half maximum) is 100 meV evaluated by measuring the Fermi edge of Au. Clean surface of the *a-a* plane is obtained by fracturing the sample in ultrahigh vacuum (~10^{-8} Pa) at the measurement temperature (T=100 K) which is below the Neel Temperature (T_N=450 K).

3. Results and Discussion

Figure 1 shows Tl 4*f* core-level PES spectrum of $TlFe_2Se_2$. There are two peaks split by the spin-orbit interaction (Δ_{SO}=4.46 eV). The peak at the higher binding energy (E_B=122.6 eV) is attributed to Tl $4f_{5/2}$ component and another at the lower binding energy (E_B=118.2 eV) to Tl $4f_{7/2}$ component. As compared with the Tl metal [4], these peaks shift slightly toward the lower E_B side due to the chemical bonding of $TlFe_2Se_2$. In addition, there is no remarkable satellite peak in the spectrum even though very weak plasmon-satellite peaks are seen at E_B□136.2 and 140.6 eV (not shown here). These suggest that Tl atoms are almost isolated in this system. In other words, the electronic states of the Tl atom are hardly hybridized with those of the neighboring Se atoms and the Tl valence is monovalent. This is consistent with results of the valence-band PES and the band structure calculation as shown later.

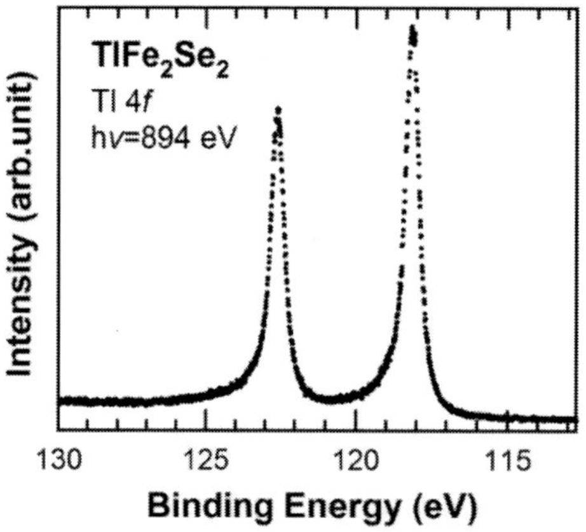

Figure 1. Tl 4*f* core-level PES spectrum of $TlFe_2Se_2$.

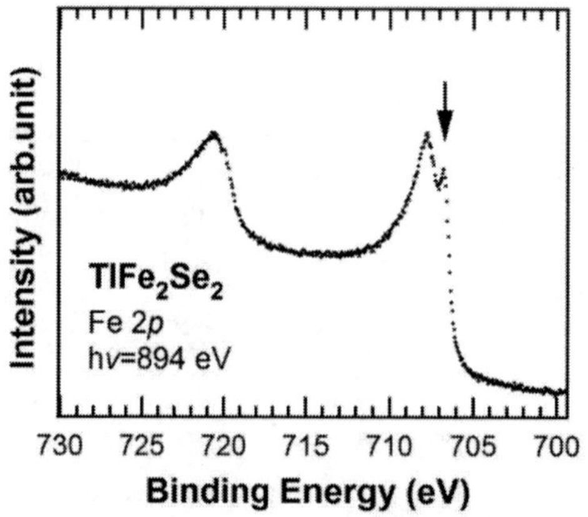

Figure 2. Fe 2*p* core-level PES spectrum of $TlFe_2Se_2$.

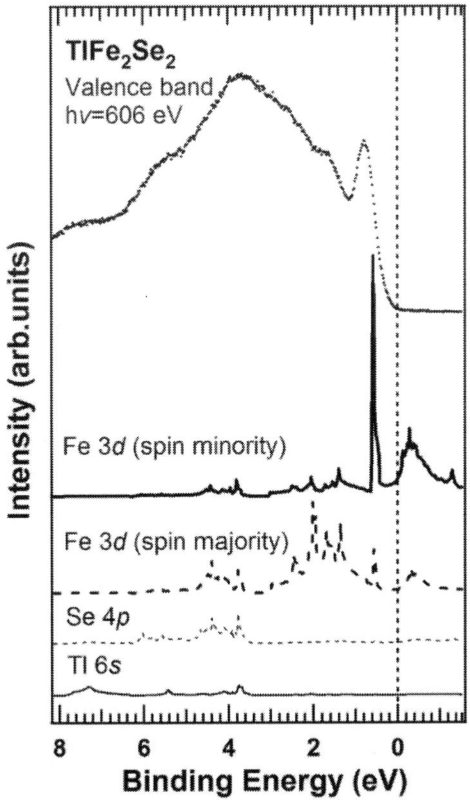

Figure 3. Valence-band PES spectrum of $TlFe_2Se_2$ and the DOSs calculated by D.J.Singh.

Figure 2 shows Fe 2*p* core-level PES spectrum of $TlFe_2Se_2$. There are two major components split by the spin-orbit interaction (Δ_{SO}=12.9 eV) as well as the Tl 4*f* core-level PES spectrum shown in Fig. 1. The peak at the higher binding energy (E_B=720.94 eV) is attributed to Fe $2p_{1/2}$ component and another at the lower E_B (E_B=708.04 eV) to Fe $2p_{3/2}$ component. The remarkable satellite peak is seen on the lower E_B side of the Fe $2p_{3/2}$ main peak (indicated by an arrow). The satellite peak has two possible origins. (1) It originates from the charge transfer from mainly Se 4*p* ligand orbitals to screen the core-hole potential. (2) A different electronic configuration from the Fe^{2+} ($|3d^6>$) state can give rise to another peak in the spectrum, which is realized by an additional electron doping from Tl atoms to conserve the charge neutrality of the crystal. In this energy region, Tl $4p_{1/2}$ peak (E_B□721 eV) should also appear. However, the photoionization cross section of Fe 2*p* state at hν~900 eV is 8 times larger than the that of Tl 4*p* one. Then, the photoemission from the Fe 2*p* states becomes predominant and the contribution of the Tl 4*p* states is negligible in this spectrum.

Figure 3 shows valence-band PES spectrum of $TlFe_2Se_2$ and the densities of states (DOSs) calculated by D. J. Singh [5]. These DOSs are obtained by the band structure calculation within the local spin-density approximation (LSDA) for the antiferromagnetic $TlFe_2Se_2$. The valence-band PES spectrum has some peak and shoulder structures. As compared with the calculated DOSs, it is revealed that the structures located at E_B□3.6 and 5.7 eV and at E_B□7.4 eV originate from Se 4*p* and Tl 6*s* electronic states, respectively. Meanwhile, Fe 3*d* minority- and majority-spin DOSs predominantly contribute to the spectrum near the Fermi level (E_F). The sharp peak just below E_F in the PES spectrum should correspond to the Fe 3*d* minority-spin state. The peak position shifts to the higher binding energy side compared to the calculated DOS. Nevertheless, an additional new structure is not observed in the vicinity of E_F as seen in the DOS. Instead, the spectral intensity rapidly decreases as the energy approaches E_F and becomes zero at the E_F. These indicate the gap formation at E_F, which is consistent with so-far reported other experimental results [2] although the calculation predicts the ground state of $TlFe_2Se_2$ below T_N (~450 K) is a metallic state.

4. Summary

We have carried out the high-resolution bulk-sensitive photoemission spectroscopy for the parent compound of an iron chalcogenide superconductor $TlFe_2Se_2$ using soft x-rays ($h\nu$=600-900 eV). The valence-band PES spectrum reproduces the overall characteristic structure of calculated DOSs. The spectral intensity, however, rapidly decreases near E_F and the energy gap is formed unlike the calculated result.

Acknowledgements

We would like to thank to D. J. Singh for providing the results of band structure calculations. The experiments were performed at BL-25SU in SPring-8 with the approval of the Japan Synchrotron Radiation Research Institute(JASRI) (Proposal No.2010A1244) under the support of Grant-in-Aid for Scientific Research for Young Scientists (B) (No.21740235 and No. 23740244) from the Ministry of Education, Culture, Sports, Science, and Technology, Japan, and Research Institute of Konan University.

5. References

[1] Fang M, Wang H, Dong C, Li Z, Feng C, Chen J and Yuan H Q 2010 *arXiv* 5236
[2] Horiguchi S, Sato H, Umeyama N, Hara Y, Miyakawa N, Takase K and Ikeda S I 2011 *Solid State Phenomena* **170** 47
[3] Yamasaki A, Matsui Y, Imada S, Takase K, Azuma H, Muro T, Kato Y, Higashiya A, Sekiyama A, Suga S, Yabashi M, Tamasaku K, Ishikawa T, Terashima K, Kobori H, Sugimura A, Umeyama N, Sato H, Hara Y, Miyagawa N and Ikeda S I 2010 *Phys. Rev. B* **82** 184511
[4] Moulder J F, Stickle W F, Sobol P E and Bomben K D 1978 *Handbook of x-Ray Photoelectron Spectroscopy* (Massachusetts : Perkin Elmer)
[5] Zhang L and Singh D J 2009 *Phys. Rev*. B **79** 094528

International Conference on Strongly Correlated Electron Systems (SCES 2011) IOP Publishing
Journal of Physics: Conference Series **391** (2012) 012116 doi:10.1088/1742-6596/391/1/012116

Low temperature thermal and electrical transport properties of $ZrZn_2$ in high magnetic field

Y Zou[1], M Sutherland[1], S Friedemann[1], S M Hayden[2], D Rothfuss[3], A Fleischmann[3], C Enss[3], F M Grosche[1]

[1] Cavendish Laboratory, J J Thomson Avenue, Cambridge CB30HE, UK
[2] H.H. Wills Physics Laboratory, Tyndall Avenue, Bristol BS8 1TL, UK
[1] Kirchhoff-Institute, Im Neuenheimer Feld 227, 69120 Heidelberg, Germany

E-mail: yz350@cam.ac.uk

Abstract. The low temperature electrical and thermal transport properties of the itinerant ferromagnet $ZrZn_2$ were investigated in order to explore the nature of the Fermi-liquid breakdown in this material. We have implemented electrical and thermal conductivity measurements down to temperatures of 100 mK and in high magnetic field. In zero field and above 2 K the electrical and effective thermal resistivities take a $T^{5/3}$ and T-linear form, respectively. These are the signatures of the marginal Fermi-liquid, predicted to occur close to a ferromagnetic quantum critical point by spin fluctuation theory. In contrast, we find that below 2 K and in external magnetic field the electrical resistivity assumes a quadratic temperature dependence, consistent with a return to conventional Fermi-liquid behaviour.

1. Introduction

Metals close to a ferromagnetic quantum critical point offer a comparatively clear and well-defined environment to investigate the breakdown of Landau's Fermi liquid theory. Since the interaction between quasiparticles is strongly enhanced by magnetic fluctuations, Fermi liquid theory can break down, leading to unconventional low temperature properties. The investigated material, $ZrZn_2$, undergoes the weakest form of a Fermi liquid breakdown, for which the quasiparticle scattering rate τ^{-1} is proportional to the excitation energy ϵ. This is the signature of the marginal Fermi liquid state, which separates the conventional Fermi liquid ($\tau \propto \epsilon^2$) from more exotic forms of Fermi liquid breakdown.

Standard theory based on low-frequency, long-wavelenth spin fluctuctions predicts the appearance of the marginal Fermi liquid state in a three dimensional system close to a ferromagnetic quantum critical point [1, 2, 3, 4]. The underlying quasiparticle scattering rate leads to characteristic temperature denpendences for key transport properties, such as a $T^{5/3}$ form for the electrial resistivity and a T-linear form for the effective thermal resistivity w.

2. Experimental details

The transition metal compound $ZrZn_2$ crystallizes in the cubic Laves structure C15. As a d-band itinerant ferromagnet it orders magnetically below $T_c = 28.5\,\mathrm{K}$ with a small ordered moment of $\mu \approx 0.17\mu_B$/f.u. in the limit $T \to 0$ and $B \to 0$ [5]. In contrast, the Curie-Weiss moment in the paramagnetic regime ($\mu_{\text{eff}} = 1.9\mu_B$/f.u.) is well above the ordered moment [1]. Furthermore,

Table 1: Temperature dependence of the transport properties

Transport Property	Fermi Liquid	Marginal Fermi Liquid
Electrical Resistivity ρ	$\rho = \rho_0 + AT^2$	$\rho = \rho_0 + aT^{5/3}$
Thermal Resistivity $w = L_0T/\kappa$	$w = w_0 + AT^2 + BT^2$	$w = w_0 + aT^{5/3} + bT$
$\delta = w - \rho$	$\delta = BT^2$	$\delta = bT$

ferromagnetism can be supressed by a hydrostatic pressure of 20kbar, which suggests that the system lies in the vicinity of a ferromagnetic quantum critical point [6, 7]. Inelastic neutron scattering shows highly dispersive spin density relaxation rates which supports the applicatibity of spin fluctuation theory [8].

We measured three single crystal samples with different residual resistivities of ρ_0 = 6.88 $\mu\Omega$cm (RRR=11), ρ_0 = 1.34 $\mu\Omega$cm (RRR=48) and ρ_0 = 0.96 $\mu\Omega$cm (RRR=68). They were cut into a typical geometry of 2 mm × 0.5 mm × 0.2 mm and contacted in a four wire geometry with a micro-spot-welding technique to give low contact resistance.

Resistivity measurements were carried out using a four terminal ac technique in a ^{4}He cryostat (PPMS) and an adiabatic demagnetisation refrigerator down to 100 mK. The measurements were repeated with several different excitation currents between 0.3 mA and 1 mA to ensure reliability.

Thermal conductivity was measured with a one-heater, two-thermometer method in a PPMS and a dilution refrigerator down to 100 mK. The setup was tested by comparing the electrical and thermal transport data in silver wire, obtaining the Wiedemann-Franz law. A temperature gradient of typically 3% – 5% was applied for the thermal conductivity measurement.

3. Experimental results

Figure 1 shows the resistivity of three $ZrZn_2$ samples with different impurity levels. A non Fermi-liquid behaviour of the form $\rho = \rho_0 + AT^{5/3}$ is clearly visible from 2 K to approximately

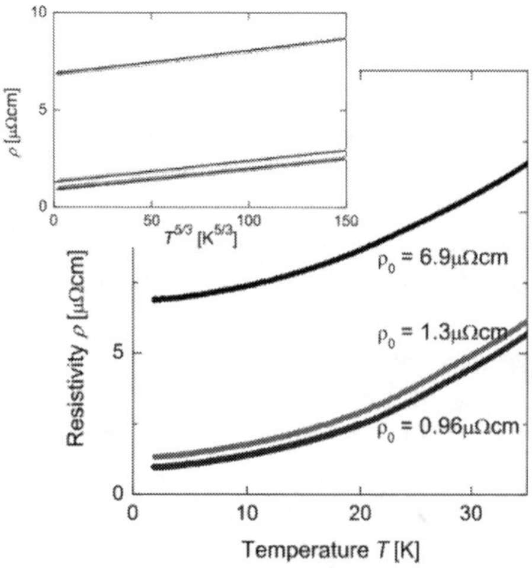

Figure 1: Main: Electrical resistivity of three different $ZrZn_2$ samples. Inset: Resistivity plotted versus $T^{5/3}$. The straight lines are guides to the eye.

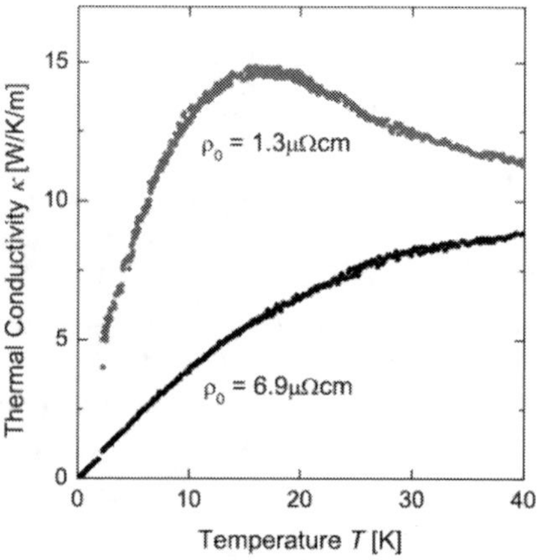

Figure 2: Thermal conductivity of two different $ZrZn_2$ samples.

16 K, in agreement with recent investigations [9, 10]. This indicates that the dominant scattering is caused by spin fluctuations in this temperature range. We can rule out phonon scattering as the origin of this unusual power law, since it would lead to a higher power in the temperature dependence of the resistivity. Above 16 K, however, the measured resistivity exceeds the $T^{5/3}$ temperature dependence, suggesting that phonon scattering becomes increasingly relevant.
In a system like $ZrZn_2$ the thermal conductivity at low temperature consists of contributions from electrons, phonons and magnons ($\kappa_{tot} = \kappa_{el} + \kappa_{ph} + \kappa_{mag}$). The magnon contribution is negligible due to the very small ordered moment in $ZrZn_2$. The phonon contribution can be isolated by comparing samples with different impurity levels, since κ_{ph} is not significantly affected by the presence of impurities. Figure 2 shows the thermal conductivity of two different samples. A characteristic peak appears at 15 K for the purer sample, but is strongly suppressed by impurity scattering in the second sample. The large difference in magnitude shows that the system is highly sensitive to elastic scattering from impurities. Additionally, we can conclude that the electronic contribution must dominate in this temperature range, which agrees with a detailed analysis of the phonon contribution. After isolating and subtracting the phononic part, we define the effective thermal resistivity w using the Wiedemann Franz law:

$$w = \frac{L_0 T}{\kappa_{el}}, \tag{1}$$

with the Lorentz number L_0. When comparing heat and charge transport, there is a crucial difference in the underlying scattering mechanics. While the electrical current is only reduced by large angle quasiparticle scattering, the heat current is also affected by inelastic small angle scattering. Thus, the difference between the thermal resistivity w and the electrical resistivity ρ, defined as

$$\delta = w - \rho, \tag{2}$$

reveals information about the latter process.

Figure 3 illustrates the linear temperature dependence in δ, which also indicates the dominant inelastic scattering from spin fluctuations. In contrast, phonon scattering would give rise to a T^3 term the effective thermal conductivity w. Both the electrical and the thermal transport

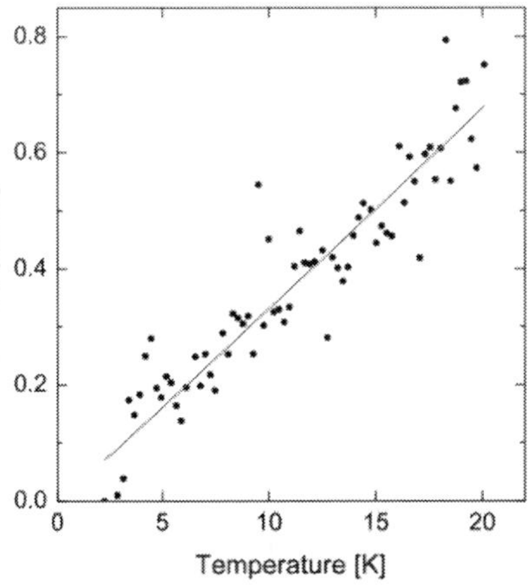

Figure 3: The difference $\delta = w - \rho$ of the thermal (w) and electrical resistivities (ρ) versus temperature for the sample with $\rho_0 = 6.88\,\mu\Omega$cm. The straight line is a power fit of the form $\delta = bT^c$ yielding the exponent $c = 1.03$.

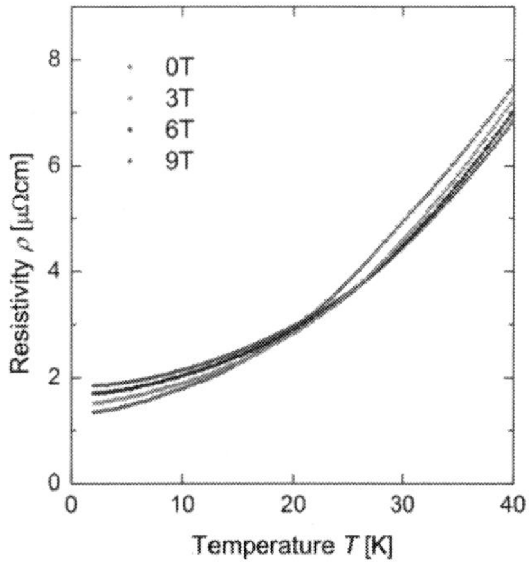

Figure 4: Electrical resistivity for the sample with $\rho_0 = 1.34\,\mu\Omega$cm in different magnetic fields.

measurements clearly show an unusual non Fermi-liquid behaviour in $ZrZn_2$ at low temperature. However, the data also illustrate why measurements on a sample with about an order of magnitude lower impurity level were required to achieve the high quality of thermal conductivity data evident in [6].

Figure 4 shows the temperature dependence of the resistivity in $ZrZn_2$ at different magnetic fields. At low temperature the resistivity increases with increasing field. Such a positive magnetoresistance at low T is frequently observed in clean metals. With increasing temperature, however, scattering from spin fluctuations becomes more significant in limiting the charge transport. The application of an external magnetic field suppresses the fluctuations which in turn leads to the negative magnetoresistance at high temperature.

For a detailed treatment of the resistivity exponent in magnetic field, resistivity data down to 100 mK are plotted in Figure 5. At high temperature above 10 K the resistivity in all fields assumes $T^{5/3}$ dependence in agreement with Figure 1. At low temperature, however, the resistivity recovers a T^2 dependence, characteristic for a Fermi liquid. The crossover temperature, at which the resistivity exponent changes from 2 to 5/3 is strongly field dependent. In zero field the T^2 dependence recovers below 2 K, whereas at 9 K the Fermi liquid behaviour is evident from the lowest temperature up to 7 K. The recovery of T^2 law and its strong field dependence can be explained by the suppression of spin fluctuations in magnetic field, as suggested by measurements of the electronic specific heat [11]. It also shows that $ZrZn_2$ is indeed a Fermi liquid in the low temperature limit, and suggests that disorder scattering is not responsible for the $T^{5/3}$ form observed at high temperature.

The above presented temperature denpendences in the transport data agree with the theoretical predictions of spin fluctuation theory and suggest that $ZrZn_2$ can be described as a marginal Fermi liquid over a wide range of temperature, field and pressure.

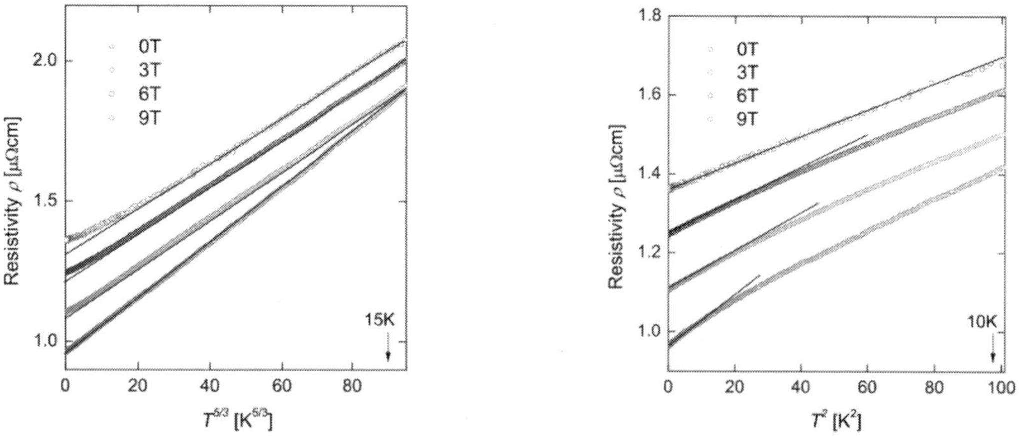

Figure 5: Electrical resistivity of $ZrZn_2$ in different magnetic fields. In the left diagram the resistivity is plotted versus T^2, in comparison the same data set is plotted vs. $T^{5/3}$ in the right diagram. The straight lines are fits to the data at high temperature (left) and low temperature (right), respectively.

References

[1] Moriya T 1985 *Spin Fluctuations in Itinerant Electron Magnetism* (Springer)
[2] Lonzarich G G 1997 *Electron* (Cambridge University Press)
[3] Hertz J 1976 *Physical Review B* **14** 1165
[4] Millis A 1993 *Physical Review B* **48** 7183–7196
[5] Jarlborg T, Freeman A J and Koelling D D 1981 *Journal of Magnetism and Magnetic Materials* **23** 291–298

[6] Smith R, Sutherland M, Lonzarich G G, Saxena S S, Kimura N, Takashima S, Nohara M and Takagi H 2008 *Nature* **455** 1220–1223
[7] Uhlarz M, Pfleiderer C, von Loehneysen H, Hayden S M and Lonzarich G G 2002 *Physica B: Condensed Matter* **312-313** 487–488
[8] Lonzarich G G, Bernhoeft N R and Paul D M 1989 *Physica B: Condensed Matter* **156-157** 699–705
[9] Sutherland M, Smith R P, Marcano N, Zou Y, Rowley S E, Grosche F M, Kimura N, Hayden S M, Takashima S, Nohara M and Takagi H 2012 *Phys. Rev. B* **85**(3) 035118
[10] Yelland E A, Yates S J C, Taylor O, Griffiths A, Hayden S M and Carrington A 2005 *Phys. Rev. B* **72**(18) 184436
[11] Pfleiderer C, Fait A, von Loehneysen H, Hayden S and Lonzarich G 2001 *Journal of Magnetism and Magnetic Materials* **226–230** 258–260

Quantitative Imaging of Flux Vortices in Superconductors

J.C. Loudon, C.J. Bowell and P.A. Midgley

Department of Materials Science and Metallurgy, University of Cambridge, Pembroke Street, Cambridge, CB2 3QZ.

Email: j.c.loudon@gmail.com

Abstract. Transmission electron microscopy offers the possibility of imaging flux vortices in superconductors quantitatively and at video rate. Here we use the technique to image flux vortices in MgB_2 and investigate their magnetic structure.

1. Introduction

Superconductors have zero electrical resistance and expel magnetic flux from their interiors (the Meissner effect). If a magnetic field is applied to an ideal (type I) superconductor no flux enters unless the field exceeds the critical field, H_c, at which the material ceases to be superconducting. However, in type II superconductors, the whole superconducting state is not destroyed at once but above the lower critical field, H_{c1}, magnetic flux penetrates the superconductor by flowing along channels called flux vortices. The vortices consist of a core where superconductivity is suppressed with a size given by the coherence length, ξ, surrounded by circulating supercurrents which persist over a distance called the penetration depth, Λ. The magnetic flux density associated with each vortex persists over the same distance, Λ. Each vortex carries a single quantum of magnetic flux given by $\Phi_0 = h/2e$ where h is Planck's constant and e is the electron charge. Vortices repel one another and so their arrangement in the absence of pinning is a 2-dimensional hexagonal lattice called an Abrikosov lattice [1]. The performance of almost all superconducting devices is determined by the behaviour of flux vortices as energy is dissipated when they move. It is important, therefore, to characterise their movement and magnetic structure.

Flux vortices can be imaged using transmission electron microscopy [2] due to the magnetic flux density from the vortices deflecting the electron beam and appear as black-white features in an out-of-focus image (Fig. 1). This is a unique imaging technique as it gives information on the internal structure of individual vortices, not just the stray fields, and allows imaging at video rate.

MgB_2 was investigated in this experiment. It was discovered to be a superconductor in 2001 and has a transition temperature, $T_c = 39$ K [3]. It has a hexagonal crystal structure (space group 191: *P6/mmm*, lattice parameters: $a = b = 3.086$ Å, $c = 3.542$ Å) consisting of alternating layers of magnesium and boron.

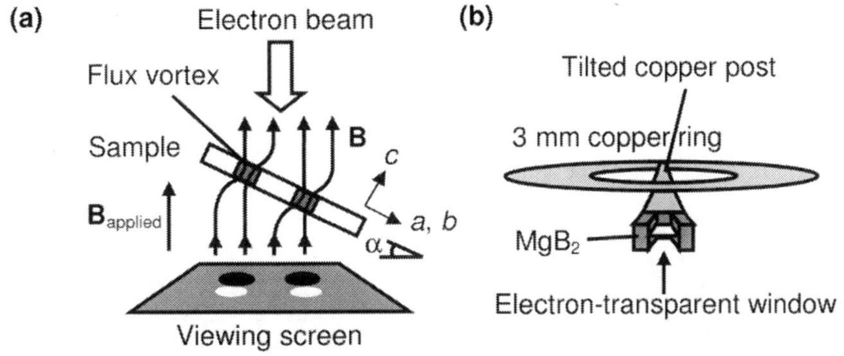

Figure 1. (a) Experimental arrangement for imaging flux vortices. The electrons are deflected by the component of the B-field normal to the electron beam so each vortex appears as a black-white feature in an out-of-focus image. (b) The specimen geometry (not to scale). The MgB_2 specimen was mounted to a copper post glued to a standard 3 mm diameter copper ring at an angle of 45°. An electron transparent window was then cut by focussed ion-beam milling.

2. Sample Preparation and Experimental Methods

MgB_2 single crystals were synthesised by Dr J. Karpinski as described in ref. [4] and were thinned to 250 nm in the c direction using focussed ion-beam (FIB) milling so they were electron-transparent. The sample needs to be tilted to a high angle (α in Fig. 1(a)) to maximise the electron beam deflection and so was attached to a tilted copper post as illustrated in Fig. 1(b) giving a tilt angle of $45 \pm 1°$. A liquid-helium cooled specimen stage with a base temperature of 10 K was used to cool the sample and energy-filtered images were recorded with a CCD camera using a Philips CM300 transmission electron microscope equipped with a field-emission gun operated at 300 kV. A magnetic field was applied to the sample by altering the setting of the twin lens in the microscope. Further details are given in ref. [5].

3. The Magnetic Structure of Flux Vortices

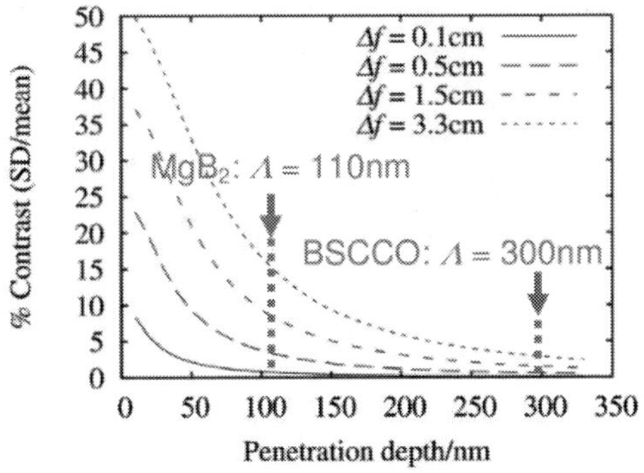

Figure 2. Simulations showing the variation of the contrast seen in out-of-focus images as a function of penetration depth, Λ, for a superconductor with a thickness l = 200 nm , tilted at an angle $\alpha = 45°$ for different defocus levels, Δf.

Information on the magnetic structure of flux vortices can be derived from the out-of-focus images. Simulations based on the modified London equation [6] shown in Fig. 2 demonstrate that the contrast of the images (defined here as the standard deviation divided by the mean) is a unique function of the penetration depth, Λ provided that the sample thickness is known. We note here that as the sample is tilted, two thicknesses are relevant: we denote l as the thickness measured normal to the plane of the specimen and t the thickness measured parallel to the electron beam. The two are related via $l = t \cos\alpha$. In principle, then, we can measure the penetration depth of individual vortices and if the magnetic structure of certain vortices is altered, say by pinning, it should be possible to identify which vortices have changed and in what way. This has been done in ref. [6] where some of the images of vortices appear stretched because the vortex cores are pinned on columnar defects. However, the changes to the magnetic structure of the vortices are inferred from a visual comparison between experimental and simulated images. Here we make a quantitative comparison.

The specimen thickness was found using energy-filtered imaging as explained in ref. [5]. This produced a thickness map of the specimen where the thickness (t) is given as a multiple of the inelastic mean-free path of the electrons, λ. The inelastic mean free path was then calibrated using electron holography [5], giving $\lambda = 152 \pm 2$ nm.

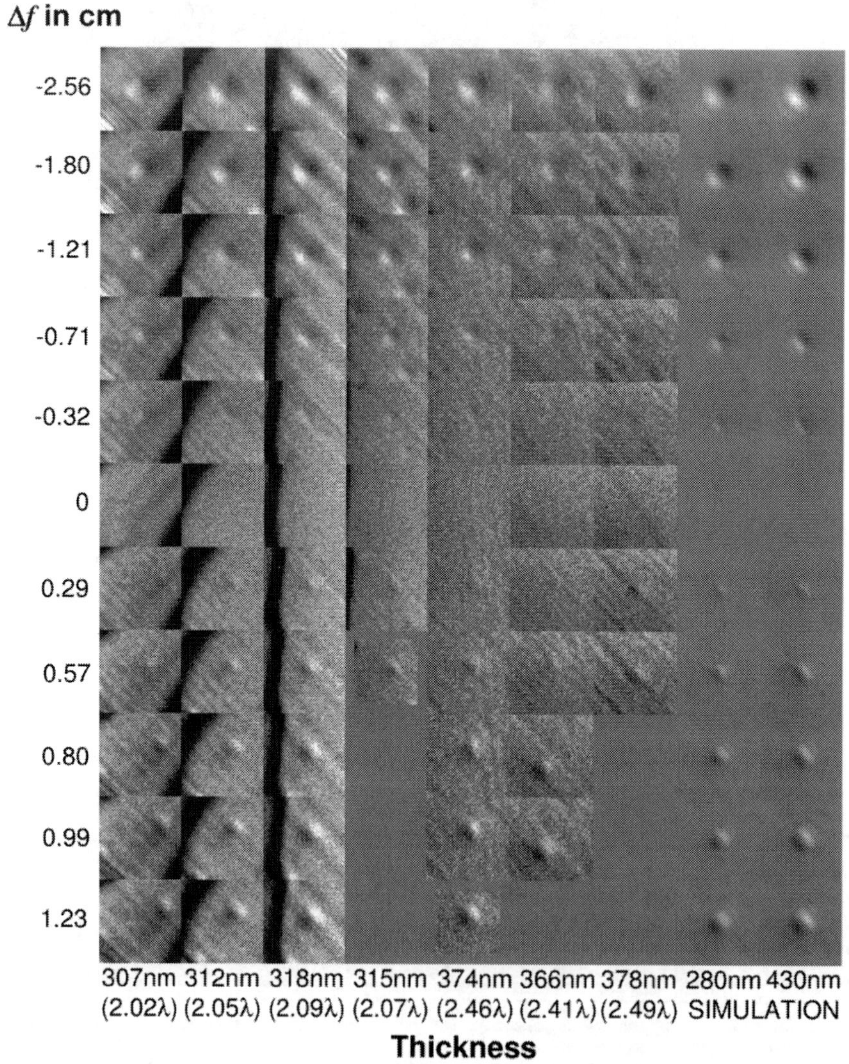

Figure 3. Flux vortices in MgB_2 taken at 11 K. Each column shows the same vortex taken at different defocus levels. The thickness (t) of the specimen along the electron beam direction is given at the bottom of each series in terms of λ, the inelastic mean-free-path and calibrated in nm using $\lambda = 152 \pm 2$ nm. Simulated images for thicknesses (t) of 280 nm and 430 nm with $\Lambda = 110$ nm [7] are shown on the right.

Fig. 3 shows images of vortices acquired from MgB_2 at 11 K at different defocus levels and simulations for 280 nm and 430 nm thicknesses assuming $\Lambda = 110$ nm [7] are shown for comparison in the two columns on the right. As discussed in ref. [8], a visual comparison between the images and simulations is not a very sensitive method of comparison. Even so, a comparison of the length of the vortex image along the black-white interface shows that the simulations for a 280 nm thickness compare better with the images than those for 430 nm. Note that altering the penetration depth in the simulations does not affect this length but only changes the contrast.

Simulations have shown [8] that a more sensitive comparison can be made by plotting the image contrast of the vortices versus the defocus. Fig 4(a) shows such a comparison for one vortex and the best fit for this vortex was with a thickness of 280 nm (the specimen thickness was 318 nm for this vortex) and a penetration depth of 110 nm. The data in such plots were more scattered than we are presently able to account for – the errorbars show the likely variation in the contrast measurement caused by the noise in the image – and they also tend to be assymmetric with positive defoci having larger contrast values than negative defoci, an effect which is difficult to understand.

Both the visual comparison and the contrast plots show that the data are best fit by a simulation of a vortex which is shorter than the specimen thickness. This indicates that there are dead layers of non-superconducting material of around 15–30 nm either side of the superconductor. These are likely caused by the ion-beam thinning of the specimen and electron microscopy experiments on semiconductors prepared using the same technique have reported a 30 nm thick dead layer [9].

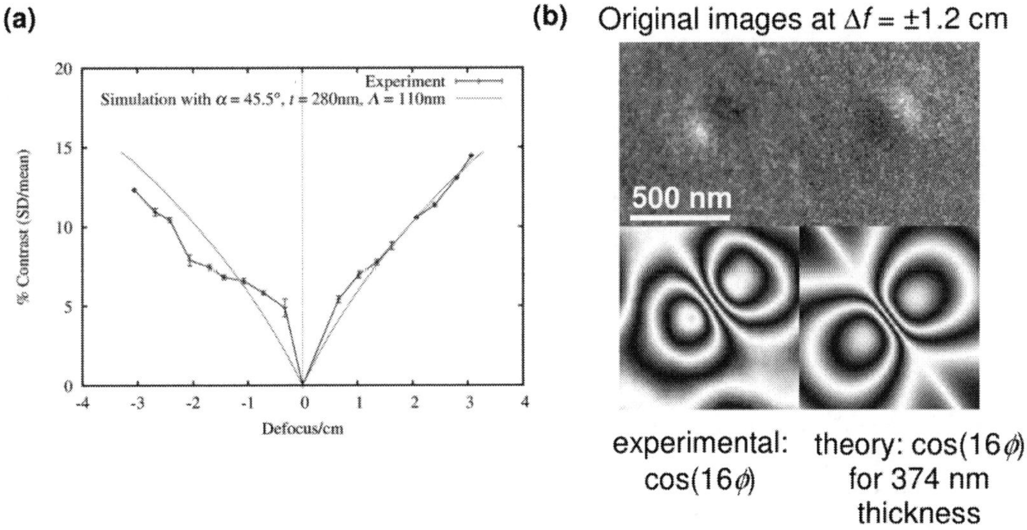

Figure 4. (a) A plot of contrast versus defocus for a single vortex determined experimentally is shown in black together with the contrast derived from simulated images based on the modified London equation with $t = 280$ nm and $\Lambda = 110$ nm, shown in red. (b) An alternative method is to reconstruct the phase of the electron wavefunction from the defocussed images using the transport of intensity equation [9]. The top two panels show the defocussed images and the lower left panel, the phase reconstructed from these. A simulation of the expected phase shift is shown in the lower right panel (see text for details).

Another method of comparison is to use the defocussed images to reconstruct the phase of the electron wavefunction as it exits the specimen using the transport of intensity equation [10]. From the phase, the component of flux density normal to the electron beam can be derived. Fig. 4(b) shows two vortex images equally disposed either side of focus on which the phase reconstruction was based. The images in the lower panel show the cosine of 16 times the phase which gives a contour map of the phase. The black lines can be thought of as field lines of the component of the B-field normal to the electron beam. It should be noted that the stray field in the reconstruction re-enters the superconductor which is unphysical. This is an artefact of the reconstruction which requires that the phase be a constant on the boundaries of the image. However, the simulation shown for comparsion has been performed with the same restriction so the two images are comparable. It can be seen that the field within the vortex is very similar indicating that $\Lambda = 110$ nm provides a good fit. Strangely, the fit is best for a simulation which matches the measured specimen thickness of 374 nm although the visual comparison in Fig. 2 showed that a thickeness of 280 nm was best.

4. Summary

Transmission electron microscopy is a unique technique for imaging flux vortices in superconductors as images can be acquired at video rate and the images give information on the internal magnetic structure of the vortices. This technique should help distinguish between different models of vortex structure and enable a better understanding of the effects of how vortices respond to pinning sites which is crucial for improving the performance of superconducting devices. We have acquired defocussed images from MgB_2 and were able to use these to obtain reasonable values for the penetration depth and the length of the vortex and to reconstruct the B-field within the specimen. Some difficulties remain: the data in plots of contrast versus defocus have more scatter than we can presently account for and show an asymmetry so that the contrast values at positive defocus are higher by some 3-5% than those for negative defocus. Further work to understand these effects is required.

Acknowledgments

The authors thank N.D. Zhigadlo and J. Karpinski, Laboratory for Solid State Physics, ETH Zurich, Switzerland for supplying the MgB_2 single crystals used in this experiment. This work was funded by the Royal Society and the EPSRC, grant number EP/E027903/1.

References

[1] Abrikosov AA 1957 *Sov. Phys. JETP* **5** 1174.
[2] Harada K, Matsuda M, Bonevich J, Igarashi M, Kondo S, Pozzi G, Kawabe U and Tonomura A 1992 *Nature* **360** 51.
[3] Nagamatsu J, Nakagawa N, Muranaka T, Zenitani Y and Akimatsu J 2001 *Nature* **410** 63.
[4] Karpinski J, Kazakov SM, Jun J, Angst M, Puzniak R, Wisniewski A and Bordet P 2003 *Physica C* **285** 42.
[5] Loudon JC, Bowell CJ, Zhigadlo ND, Karpinski J and Midgley PA 2012 *Physica C* **474** 18.
[6] Beleggia M, Pozzi G, Masuko J, Osakabe N, Harada K, Yoshida T, Kamimura O, Kasai H, Matsuda T and Tonomura A 2002 *Phys. Rev. B* **66** 174518.
[7] Manzano F, Carrington A, Hussey NE, Lee S, Yamamoto A and Tajima S 2002 *Phys. Rev. Lett.* **88** 047002.
[8] Loudon JC and Midgely PA 2009 *Ultramicroscopy* **109** 700.
[9] Twitchett AC, Dunin-Borkowski RE and Midgley PA 2002 *Phys. Rev. Lett.* 88 238302.
[10] Teague MR 1983 *J Opt. Soc. Am.* **73** 1434.

Journal of Physics: Conference Series **391** (2012) 012118 doi:10.1088/1742-6596/391/1/012118

The crystal structure of polycrystalline powder κ-(BEDT-TTF)$_2$Cu[N(CN)$_2$]Br studied by high energy x-ray diffraction

S Süllow[a], A U B Wolter[b], J Müller[c], M Lang[c], J E Daniels[d,e], V Honkimäki[e], T Buslaps[e]

[a]Institut für Physik der Kondensierten Materie, TU Braunschweig, Braunschweig, Germany
[b]Institut für Festkörper- und Werkstoffforschung, Dresden, Germany
[c]Physikalisches Institut, J.-W. Goethe Universität, SFB/TR 49, Frankfurt(M), Germany
[d]University of New South Wales, Sydney, Australia
[e]ID15, European Synchrotron Radiation Facility, Grenoble, France

E-mail: s.suellow@tu-bs.de

Abstract. We present a structural investigation of the organic charge-transfer salt superconductor κ-(BEDT-TTF)$_2$Cu[N(CN)$_2$]Br at a temperature of 90 K using high energy (90 keV) powder x-ray diffraction. With this approach we access the scattering profile over a wide range in Q-space (from $\sim$ 0.3 to 19 Å^{-1}). Overall, the observed scattering intensity is understood as a superposition of Bragg diffraction from a crystalline phase superimposed onto a large quasi-amorphous signal, indicative of short range order effects in our powder material. We discuss implications of these findings regarding the structural properties of this material.

In recent years, the organic charge-transfer salts κ-(BEDT-TTF)$_2$X with X = Cu(NCS)$_2$, Cu[N(CN)$_2$]Br and Cu[N(CN)$_2$]Cl stood in the focus of intensive research efforts [1, 2, 3, 4, 5]. In these studies, the prime topics of interest were their exotic superconducting and normal state properties, resembling to some degree the high-T_c cuprates. The materials X = Cu[N(CN)$_2$]Br, Cu[N(CN)$_2$]Cl crystallize in an orthorhombic crystal structure, space group $Pnma$ ($Z = 4$) [6]. The structure contains alternating layers of a conducting network of cations $(S_8C_{10}H_8)_2{}^+$ (bisethylenedithiotetrathiafulvalene, abbreviated BEDT-TTF) and insulating polymeric anion chains X$^-$. This layered crystal structure causes these materials to behave in a quasi-twodimensional fashion in terms of their electronic properties. Compositional modifications by replacing the anion X = Cu[N(CN)$_2$]Cl with Cu[N(CN)$_2$]Br or Cu(NCS)$_2$ tunes the ground state from an antiferromagnetic insulator into correlated metal states. Such behavior is assumed to reflect a bandwidth-controlled Mott transition [7].

The conceptual phase diagram presented in Ref. [7] simplifies the experimental situation, with spatial inhomogeneities and percolating states appearing close to the Mott transition in experimental studies (see for instance Refs. [8, 9]). The observation of inhomogeneous states opens up the question of the interplay of structural and electronic inhomogeneities, this in particular in view of the observation of glass-like transitions in the series of compounds [3, 10]. So far, in detailed structural investigations using x-ray laboratory techniques for instance on κ-(BEDT-TTF)$_2$Cu[N(CN)$_2$]Br (abbreviated κ-Br hereafter) the glass-like transition at $T_g \sim 77$ K could be identified in physical properties such as the temperature dependence of the lattice parameters (which at 100 K are $a = 12.884(5)$ Å, $b = 29.581(8)$ Å and $c = 8.484(4)$ Å [10]). Yet, these studies did not unambiguously reveal the microscopic nature of the glass-like transition. Only, the notion of a configurational freezing-out of the terminal ethylene groups of the BEDT-TTF molecule at T_g, as it was proposed previously [3, 11, 12, 13], appears to be inconsistent with the experimental data.

In this situation we decided to extend our structural investigation of the glass-like transition in κ-Br by using synchrotron-based x-ray techniques. This approach enables us to use high energy (90 keV) x-ray diffraction, which in turn allows to study the scattering profile of κ-Br over a wide parameter range of the scattering vector Q with high statistics. Here, we present first results of our study.

For the experiment we have used the beamline ID15B at the European Synchrotron Radiation Facility, which offers as operation modes monochromatic x-ray radiation at energies of 30, 60 or 90 keV (a description of the beamline set-up can be found in Ref. [14]). In our case, we have carried out a standard powder diffraction experiment at highest available energy, using an on-line image plate detector to record the Debye-Scherrer rings.

The powder sample κ-Br has been obtained from single crystals grown in Frankfurt as described in Ref. [4]. These crystals were crushed to powder by hand within a few minutes using a mortar and pestle. The powder was filled into a thin glass capillary, which was installed on the sample holder of a closed cycle cryostat and could be spinned within the horizontal plane. The sample holder of the cryostat (lowest achievable temperature ~ 20 K) was covered by a can with x-ray transparent windows. Size homogeneity and random arrangement of the powder was verified by checking the Debye-Scherrer diffraction patterns for an inhomogeneous intensity distribution. Since in a first experiment some intensity inhomogeneity was observed, the sample κ-Br was crushed another time for a few minutes by hand with mortar and pestle. Subsequently, the Debye-Scherrer ring intensity was evenly distributed.

For the experiment temperatures in the range 20 to 120 K were set and the Debye-Scherrer diffraction patterns for κ-Br were recorded by the image plate detector under continuous spinning of the capillary. By performing a numerical integration of the intensity measured at the detector along the trace of the diffraction cone with scattering vector Q we arrive at the common representation of the intensity distribution $I(Q)$ of a Debye-Scherrer experiment. As an example we present the data for κ-Br measured at 90 K in Fig. 1 (green line).

Subsequently, a diffraction pattern was taken for the same experimental geometry with only an empty glass capillary installed in the sample holder. Again, in the same way as for the filled capillary a numerical integration of the measured intensity was performed, with the resulting data included in Fig. 1 (red line). In a next step, the data for the empty capillary is subtracted from that of the filled one, in order to correct for scattering contributions from the experimental set-up. We include the corrected data in Fig. 1 (black line).

Clearly, and especially for $Q \leq 4$ Å^{-1}, in Fig. 1 a large number of Bragg peaks can be identified. The positions of all these peaks can be accounted for with the orthorhombic crystal structure of κ-Br using lattice parameters of $a = 12.91$ Å, $b = 29.82$ Å and $c = 8.524$ Å, which is in good agreement with the values reported in Ref. [10]. However, from Fig. 1 it is also apparent that even our data of κ-Br corrected for scattering from the set-up exhibits an additional and large quasi-amorphous scattering contribution. Without further information, the precise Q dependence of this contribution is not known. However, from Fig. 1 it can be estimated that for $Q \leq 4$ Å^{-1} the quasi-amorphous intensity in absolute counts is of the same order of magnitude as the intensity of well-defined Bragg peaks, while for larger Q values amorphous contributions dominate the scattering profile. As consequence, these large amorphous contributions prohibit to carry out a full structural refinement of our data. We can, however, model our data by separating out different contributions.

In Ref. [10] we have obtained detailed structural information on κ-Br. We use these positional

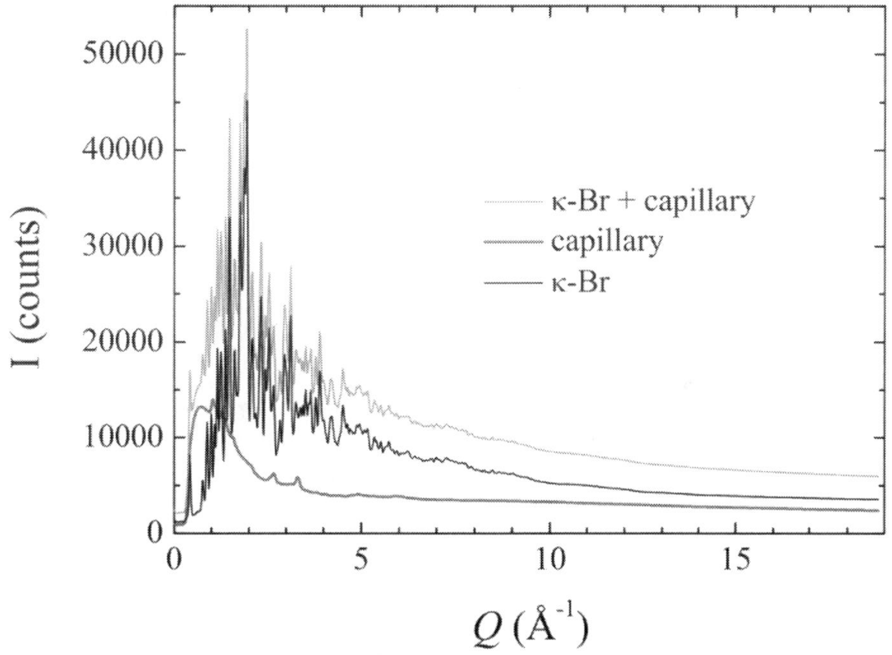

Figure 1. (Color online) Example of a measured diffraction pattern of κ-Br at 90 K: green line - capillary filled with κ-Br; red line - empty capillary; black line - resulting pattern after subtraction of background contribution; for details see text.

parameters from Ref. [10] together with the experimental lattice parameters to calculate the intensity distribution for perfectly crystalline material in a powder diffraction experiment. Next, we *ad-hoc* choose a Q-dependent profile, which - by adding it to the calculated powder diffraction profile of κ-Br - produces a reasonable description of our data. Our procedure is illustrated in Fig. 2, where we plot our data (+) together with the quasi-amorphous scattering profile (blue line) and the resulting fit of the data (black) for $Q \leq 4.2$ Å^{-1} (a.) and the full spectrum (b.).

From Fig. 2(a) it is apparent that for small Q values we can reproduce peak positions and almost quantitatively peak intensities for our powder sample κ-Br, if we assume a scattering profile with a broad quasi-amorphous shape (maximum of about 10000 counts in the range $Q \sim 2$ Å^{-1}), together with some broad peaks (up to 20000 counts close to 2 Å^{-1}). On top of this quasi-amorphous part sits our scattering intensity from a crystalline phase with maximum peak intensities up to 20000 counts. Moreover, the experimental data for larger Q values must almost entirely be attributed to a quasi-amorphous phase (see Fig. 2(b)). In other words, our data parameterization suggests that our powder sample κ-Br, produced from single crystalline material by crushing for a few minutes by hand using mortar and pestle, contains a significant amount of quasi-amorphous or strongly distorted crystalline phase.

It is well known that crushing eventually will cause a destruction of the crystalline lattice, and thus produce amorphous material (see Refs. [15, 16]). However, as indicated in Refs. [15, 16], in these cases crushing is performed by milling the material over a long period of time (a few to a thousand hours), rather than doing it by hand for a few minutes. This implies that the organic superconductors studied here exhibit an extraordinary structural sensitivity to metallurgical treatment such as crushing. This observation of course is in line with the general experience researchers have while working with these systems, with for instance reports of dependencies of physical properties on thermal treatments etc. [3, 4, 11, 12]. It suggests that in these cases the

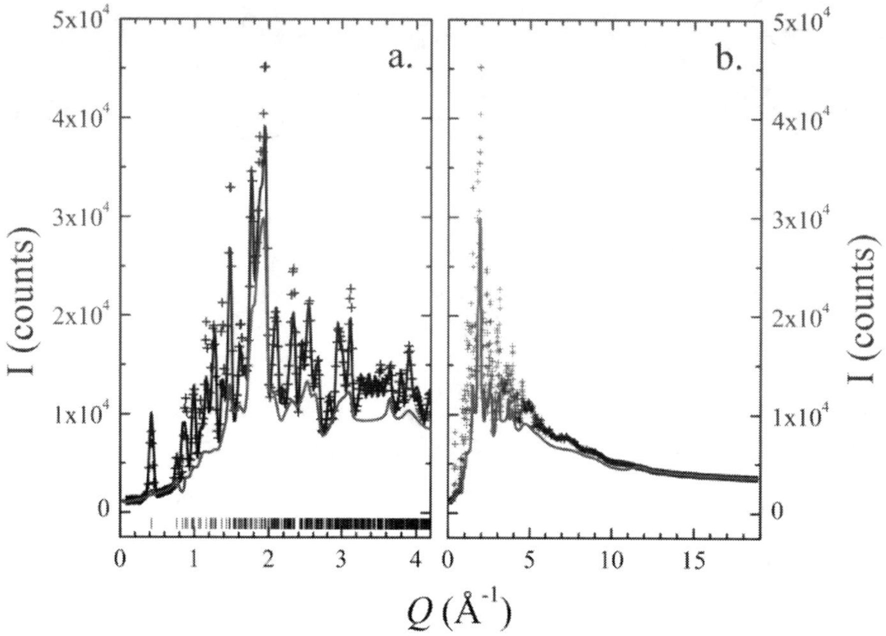

Figure 2. (Color online) Diffraction pattern of κ-Br at 90 K (+), together with the quasi-amorphous scattering profile (blue line) and the resulting data parameterization (black line, only plotted in (a.) for clarity) up to $Q \leq 4.2$ Å^{-1} (a.) and 19 Å^{-1} (b.); for details see text.

difference in sample treatment corresponds to a difference in the density of dislocations, lattice faults etc., which apparently are easy to be produce in these organic salts.

Regarding our present study, the next step will be to analyze the temperature dependence of our structural data. In particular, it will be worthwhile to determine if there are features to be observed for the Bragg peak intensities or amorphous scattering contributions of κ-Br, which reflect the glass-like transitions observed for single crystalline material.

This research was supported by the DFG under Grant No. SU229/8-2. We acknowledge fruitful discussions with R. Feyerherm.

References

[1] Lang M and Müller J 2004 *Organic Superconductors* in *The Physics of Superconductors - Vol.2*, K.-H. Bennemann, J.B. Ketterson (Eds.), Springer
[2] McKenzie R H 1997 *Science* **278** 820
[3] Müller J *et al.* 2002 *Phys. Rev. B* **65** 144521
[4] Strack C *et al.* 2005 *Phys. Rev. B* **72** 054511
[5] Powell B J and McKenzie R H 2006 *J. Phys. Condens. Matter* 18 R827
[6] X = $Cu(NCS)_2$ crystallizes in a closely related monoclinic structure; see Ref. [1].
[7] Kanoda K 1997 *Hyperfine Interact.* **104** 235
[8] Limelette P *et al.* 2003 *Phys. Rev. Lett.* **91** 016401
[9] Kagawa F, Miyagawa K and Kanoda K 2009 *Nature Phys.* **5** 880
[10] Wolter A U B *et al.* 2007 *Phys. Rev. B.* **75** 104512
[11] Su X *et al.* 1998 *Phys. Rev B* **57** R14056
[12] Tanatar M A *et al.* 1999 *Phys. Rev. B* **59** 3841
[13] Sasaki T *et al.* 2005 *J. Phys. Soc. Jpn.* **74** 2351
[14] see http://www.esrf.fr/UsersAndScience/Experiments/StructMaterials/ID15/
[15] Zhou G F and Bakker H 1995 *Phys. Rev. B* **52** 9437
[16] Leslie-Pelecky D L and Schalek R L 1999 *Phys. Rev. B* **59** 457

Theoretical modeling of the electronic structure and exchange interactions in Cu(II)Pc

Wei Wu[1,2],* A. J. Fisher[2], N. M. Harrison[1], Hai Wang[3], Zhenlin Wu[4],
Jules Gardener[5], Sandrine Heutz[4], Tim Jones[6], and Gabriel Aeppli[2]
Department of Chemistry and London Centre for Nanotechnology,
Imperial College London, South Kensington Campus, London, SW7 2AZ, U.K.[1]
UCL Department of Physics and Astronomy and London Centre for Nanotechnology,
University College London, Gower Street, London, WC1E 6BT, U.K.[2]
Department of Physics, Kuming University, P. R. China[3]
Department of Materials and London Centre for Nanotechnology,
Imperial College London, South Kensington Campus, London, SW7 2AZ, U.K.[4]
Department of Physics, Harvard University, MA 02138, USA[5] *and*
Department of Chemistry, Warwick University, Coventry, CV4 7AL, U.K.[6]

We calculate the electronic structure and exchange interactions in a copper(II)phthalocyanine (Cu(II)Pc) crystal as a one-dimensional molecular chain using hybrid exchange density functional theory (DFT). In addition, the intermolecular exchange interactions are also calculated in a molecular dimer using Green's function perturbation theory (GFPT) to illustrate the underlying physics. We find that the exchange interactions depend strongly on the stacking angle, but weakly on the sliding angle (defined in the text). The hybrid DFT calculations also provide an insight into the electronic structure of the Cu(II)Pc molecular chain and demonstrate that on-site electron correlations have a significant effect on the nature of the ground state, the band gap and magnetic excitations. The exchange interactions predicted by our DFT calculations and GFPT calculations agree qualitatively with the recent experimental results on newly found η-Cu(II)Pc and the previous results for the α- and β-phases. This work provides a reliable theoretical basis for the further application of Cu(II)Pc to molecular spintronics and organic-based quantum information processing.

PACS numbers: 75.50.Xx,71.15.Mb,71.70.Gm

I. INTRODUCTION

Magnetic organic materials [1–5] combining optical and magnetic functionality are promising candidates for novel spintronical devices [5–8] and quantum information processing (QIP) [9, 10]. The electronic structure of magnetic organic materials is also of fundamental interest and underlies both their magnetic and optical properties. The exchange interactions between molecular units largely determine the Neel temperature (T_N) or Curie temperature (T_C) for the transition from paramagnetic to ordered anti-ferromagnetic (AFM) or ferromagnetic (FM) phases. The theoretical prediction of the band gap of materials is particularly important for the understanding of optical properties.

Transition-metal phthalocyanines (TMPcs) are well known organic semiconductors [11] that can carry ions with unpaired electron spins in their molecular centres, e.g., copper(II)phthalocyanine (Cu(II)Pc, spin-$\frac{1}{2}$). In previous work we have reported the electronic structure of the Cu(II)Pc molecule and the variation of exchange interactions as a function of the molecular stacking angle in a molecular dimer [10, 12]. The calculations of the electronic structure of particular crystalline polymorphs of Cu(II)Pc have previously been reported by several groups [13–16]. However, in these works the spin structure of the material was not considered in detail, and there was no attempt to calculate the exchange interaction between copper spins; indeed, in [15] the spin nature of the ground state found is not explained and a non-spin-resolved band structure is presented. In ref.[16] it was found that the highest occupied state is of b_{1g} symmetry and derived from a Cu d-state based on pure DFT. Using a DFT+U treatment of the local correlations at the Cu site, the band gap was computed to be 1.5 eV and the splitting between the majority and minority spin states of the Cu was 2 eV [16].

The dominant exchange interactions occur within molecular chains along the stacking axis. The stacking angle and the *sliding angle* (defined in §III) together determine the stacking axis. The magnetic interactions between chains are very weak and presumably dipolar in origin, owing to the large distance (> 10 Å) between copper atoms in the neighbouring chains. The dominant exchange interactions in a Cu(II)Pc crystal may therefore be modeled using a one-dimensional molecular chain.

*Electronic address: wwu1@imperial.ac.uk

Recently a new stable form of Cu(II)Pc was reported, known as the η-phase [10] and possessing a wire-like morphology. Unlike the previously known α and β phases, the η-phase Cu(II)Pc displays a sliding angle significantly different from the plus phase and the cross phase as defined in ref.[12], implying the effect of the sliding angle as a structural parameter needs to be carefully considered. The possibility of the variation of the sliding angle opens up a very wide range of potential structures that can be prepared with comparable stability; indeed, many such structures are already known but their magnetic properties are not yet determined [17, 18]. Our aim in this paper is to assess systematically the importance of sliding angle for the magnetic interactions, and in doing so to provide a theoretical reference for the geometries, electronic structure, and magnetic properties of new morphologies of Cu(II)Pc (and possibly other transition metal phthalocyanines), which could be found in future experiments. We have therefore modeled the Cu(II)Pc one-dimensional chain for a wide range of stacking and sliding angles. Compared with our previous paper [12] we extend our modeling to a larger geometrical parameter space and extend the dimer model to a one-dimensional molecular chain with periodic boundary conditions — thereby isolating the important intra-chain magnetic interactions.

The investigation of the Cu(II)Pc one-dimensional chain falls into four parts. In §II we give the details of the computational methods and software. In §III we describe the important geometrical parameters, and in particular define the stacking and sliding angles. In §IV we present the exchange interactions calculated by density functional theory (DFT) and Green's function perturbation theory (GFPT), describe the electronic structure of the α-phase of Cu(II)Pc as a typical example, and discuss these results. Finally in §V we draw more general conclusions.

II. COMPUTATIONAL DETAILS

We carry out calculations for the Cu(II)Pc single molecule with a 6-31G basis set [19] in the Gaussian 03 code [20]. The total energies of molecular periodic chains were also calculated using DFT as implemented in the CRYSTAL 06 code [21]. The 6-31G basis set is designed for use in molecular studies and is inadequate to describe atomic orbitals of the copper atoms in the molecular crystal environment. In order to improve the description, we use a 6-311G Gaussian basis set to describe the atomic orbitals of the copper atoms. The outer d-shell gaussian function of each copper atom, with exponent 0.67 a_0^{-2}, is supplemented by a diffusion gaussian function with an exponent at 0.27 a_0^{-2} which is between one third and one half of the outer exponent. This successfully enhances the basis set while avoiding pseudo linear dependence. We use the standard molecular 6-31G basis set for other elements. The Monkhorst-Pack sampling [22] of reciprocal space is carried out choosing a grid of shrinking factor equal to eight. The truncation of the Coulomb and exchange series in direct space is controlled by setting the Gaussian overlap tolerance criteria to $10^{-6}, 10^{-6}, 10^{-6}, 10^{-6}$, and 10^{-12} [21]. The self-consistent field (SCF) procedure is converged to a tolerance of 10^{-6} a.u. per unit cell (p.u.c). To accelerate convergence of the SCF process, all calculations have been performed adopting a linear mixing of Fock matrices by 30 percent.

Electronic exchange and correlation are described using the B3LYP hybrid functional [23]. The advantages of B3LYP include a partial elimination of the self-interaction error and balancing the tendencies to delocalize and localize wave-functions by mixing Fock exchange with that from a generalized gradient approximation (GGA) exchange functional [23]. The broken-symmetry method [24] is used to localize opposite electron spins on each molecule in order to describe the AFM state. The exchange coupling, J in the one-dimensional Heisenberg spin chain [25] is defined here as,

$$\hat{H} = 2J \sum_i \hat{\vec{S}}_i \cdot \hat{\vec{S}}_{i+1}, \tag{1}$$

and determined by

$$J = (E_{\mathrm{FM}} - E_{\mathrm{AFM}})/2, \tag{2}$$

where E_{AFM} and E_{FM} are the total energies of a super-cell containing two molecules in which the spin configurations are anti-aligned and aligned respectively. DFT total-energy calculations have some intrinsic disadvantages for computing exchange interactions, e.g., a comparison between two very large numbers to get a small exchange splitting in an all-electron local basis set formalism. However, the performance of B3LYP as implemented in CRYSTAL has previously been shown to provide an accurate description of the electronic structure and magnetic properties for both inorganic and organic compounds [12, 26–28].

In addition to total energy calculations we use GFPT to compute exchange interaction in a Cu(II)Pc dimer in order to have a qualitative picture of the interaction mechanism. This has the advantage of avoiding the calculation of very small differences between large energies, at the cost of assuming a particular model for the exchange interaction. We adopt the formalism of indirect exchange in a perturbative way as described in the previous work [12]. Here we use

Gaussian 03 to compute the single-molecule electronic structure and the core Hamiltonian of a molecular dimer. The perturbative approximation to the exchange interaction is then computed as,

$$J = A(t_x^e t_x^h + t_y^e t_y^h), \tag{3}$$

where $t_{x,y}^e$ ($t_{x,y}^h$) is the electron (hole) hopping through the lowest unoccupied (highest occupied) e_{gx} and e_{gy} orbitals, and A is an intra-molecular parameter proportional to $\frac{J_p^2}{U^3}$ where J_p is the two-electron integral responsible for spin-polarization of ring states by the Cu spin, and U is the on-site Coulomb interaction. The magnitude of A is irrelevant if we are only interested in the trend of the exchange interaction in the variation of the geometry. In our perturbation theory calculations we therefore only compute the hopping integrals in eq.(3).

The three models adopted here (DFT calculations of a molecular dimer [12] and of a one-dimensional molecular chain, and GFPT calculations for a molecular dimer) provide complementary information about exchange interactions. The first two are used to analyze the electronic structure and to obtain a quantitative description of exchange interactions while the third reveals most clearly the physical mechanism responsible for the exchange interactions.

III. MOLECULAR GEOMETRY

The intra-molecular coordinates were determined by the optimization of an isolated Cu(II)Pc molecule. TMPc crystals are known to consist of parallel molecular planes along the stacking axis connecting copper atoms. In all our calculations we therefore assume that each molecule is in the same orientation as its neighbours along the molecular chain, with a constant inter-plane separation d. Let $\hat{\vec{r}}$ be the projection into the molecular plane of the vector joining the central Cu atoms of the two molecules. Then we define the stacking angle ϕ as the angle between the molecular plane and the stacking axis connecting the copper atoms, and the sliding angle ψ as the angle between $\hat{\vec{r}}$ and the X axis as shown in Fig.1. The X and Y axes are defined within a given molecular plane. Various crystal structures can therefore be generated by varying the inter-plane distance d, the stacking angle ϕ, and the sliding angle ψ. These two angles are related to the cartesian coordinates X and Y as $X = d \cot\phi \cos\psi$ and $Y = d \cot\phi \sin\psi$ (Fig.1).

It would be necessary to have a reliable description of Van der Waals forces to calculate the inter-plane distance d from first principles. However, widely used exchange-correlation functionals, including B3LYP, do not describe Van der Waals forces reliably. In the following calculations the inter-plane distance d is fixed at the experimentally determined value of 3.42 Å, obtained by X-ray diffraction (XRD) of Cu(II)Pc thin films [10, 29]. Calculations have been performed for stacking angles from 20° to 90° and sliding angles from 0° to 45° with 5° increments exploiting the D_{4h} symmetry of the Cu(II)Pc molecule. So far three phases of Cu(II)Pc molecular crystals have received extensive experimental study: the α-phase with stacking angle $\phi \simeq 63°$ and sliding angle $\psi \simeq 7°$, the β-phase with stacking angle $\phi \simeq 45°$ and sliding angle $\psi \simeq 43°$, and the η-phase with stacking angle $\phi \simeq 63°$ and sliding angle $\psi \simeq 27°$ [10].

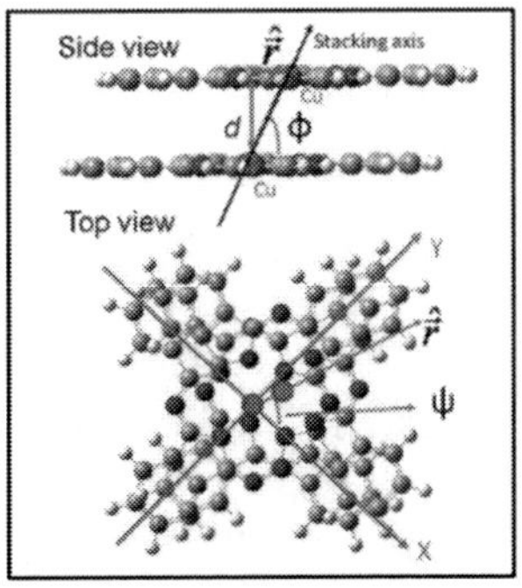

FIG. 1: (Color online.) Two neighboring Cu(II)Pc molecules from a side view and a top view to illustrate important geometrical parameters in the crystal structure. The atoms are color-coded, i.e., copper in red, carbon in grey, nitrogen in blue, and hydrogen in white. The inter-plane distance is labeled as d, the stacking angle ϕ, and the sliding angle ψ. The stacking axis is indicated by a black arrow connecting copper atoms. The vector $\vec{r}$, and the X and Y axes are used to define the sliding angle, i.e., $\cos\psi = \hat{\vec{r}} \cdot \hat{\vec{X}}$.

IV. RESULTS AND DISCUSSIONS

A. DFT calculations

In Fig.(2) the exchange interaction calculated using DFT in a one-dimensional Cu(II)Pc chain is shown as a function of X and Y. The exchange interaction depends strongly on stacking angle (distance from the centre of the plot) but weakly on sliding angle. The exchange interaction increases with the stacking angle in the chosen angular region and peaks with a value of 4.78 K at a stacking angle of 90°.

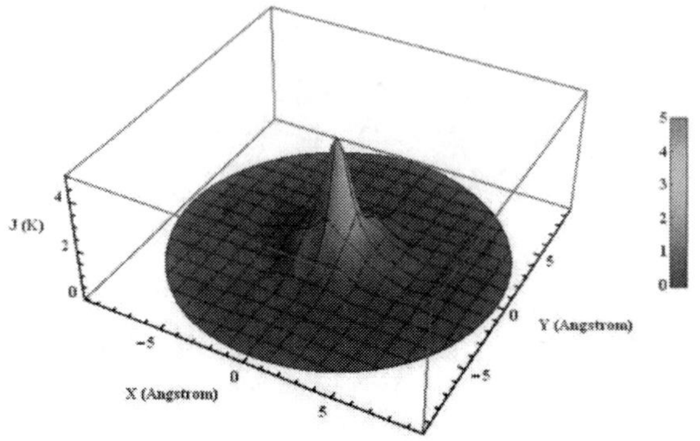

FIG. 2: (Color online.) The exchange interaction calculated using DFT as a function of the cartesian coordinates X and Y. Notice that exchange interactions strongly depend on stacking angles (and hence on the distance from the centre of the plot), but weakly on sliding angles, and peak at a stacking angle of 90° with 4.78 K.

The exchange interactions at the experimentally observed geometries [10] are $J_\alpha^{\rm DFT} = 1.30$ K, $J_\beta^{\rm DFT} = 0.14$ K, and $J_\eta^{\rm DFT} = 1.26$ K. These results are in qualitative agreement with magnetic measurements using superconducting quantum interference device (SQUID) magnetometry [3, 10], from which it has been deduced that $J_\alpha^{\rm exp} = 1.4$ K, $J_\beta^{\rm exp} = -0.15$ K, and $J_\eta^{\rm exp} = 0.5$ K. There is a quantitative agreement of the calculated result with the observed result for the α-phase, while the β-pase is paramagnetic ($J \simeq 0$) within the accuracy of DFT calculations, and also within the experimental uncertainty. There is some discrepancy in the η-phase, for which only single measurement is available [10]. We expect that basis set errors will largely cancel out when subtracting the energies in equation (2), leaving intrinsic limitations of hybrid DFT as the main source of error. The magnitude of exchange interactions is in the order of 1 K; this is much less then the DFT error in a single total energy, but it has been shown previously that the cancelation of errors in states differing just in their spin orientations can yield exchange constants in hybrid DFT accurate to about 1 K—about 9 orders of magnitude smaller than the individual total energies. In this context the agreement of the calculations with the experiments is seen to be remarkably good. These calculations also agree qualitatively with the results of the previous DFT calculations for the molecular dimers in these three phases [10, 12].

The total energy per unit cell of the α-phase is approximately 2 meV higher than that of the η−phase, and the total energies of both of them are much higher than the β−phase by 0.4 eV. This suggests the π-π electronic repulsion between rings might be important in the molecular interaction in Cu(II)Pc though Van der Waals forces also play a crucial role and are not fully accounted for in our calculations.

The electronic structures of the α-, the β-, and the η-phases are shown in Fig.(3), Fig.(4), and Fig.(5), respectively. In the band structures of all these three phases, k_x is oriented along the stacking axis of the molecular chain and the zero of energy for the DOS plots is chosen to be at mid-gap. 21 occupied and 10 unoccupied bands are plotted in the band structure from $k_x = 0$ to $k_x = 2\pi/a$ with $a = 9.56$Å in the AFM configurations (Fig.3-5(a)), while 22 (20) occupied bands and 9 (11) unoccupied bands of spin-up (spin-down) are plotted in the FM configurations (Fig.3-5(d)). The corresponding densities of states (DOS) are shown for the AFM (Fig.3-5(b)) and FM configurations (Fig.3-5(e)). Two valence bands are derived from the highest occupied molecular orbitals (HOMO), and four conduction bands from the lowest unoccupied molecular orbtals (LUMO) respectively of the Cu(II)Pc molecule. The band gap is about 2.0 eV and occurs between π and π^* states of the Pc ring. The band gap is significantly increased by our hybrid DFT calculations compared with the previous work in which on-site electron correlation is neglected [13–15], where the gap is significantly smaller (only about 1 eV). Furthermore in several previous calculations the lowest unoccupied state is a Cu d-state, in contrast to the molecular HOMO as we find. It is interesting that when on-site electron correlation is taken into account empirically at the DFT+U level with an on-site U about 5 eV [16], the band gap is opened up

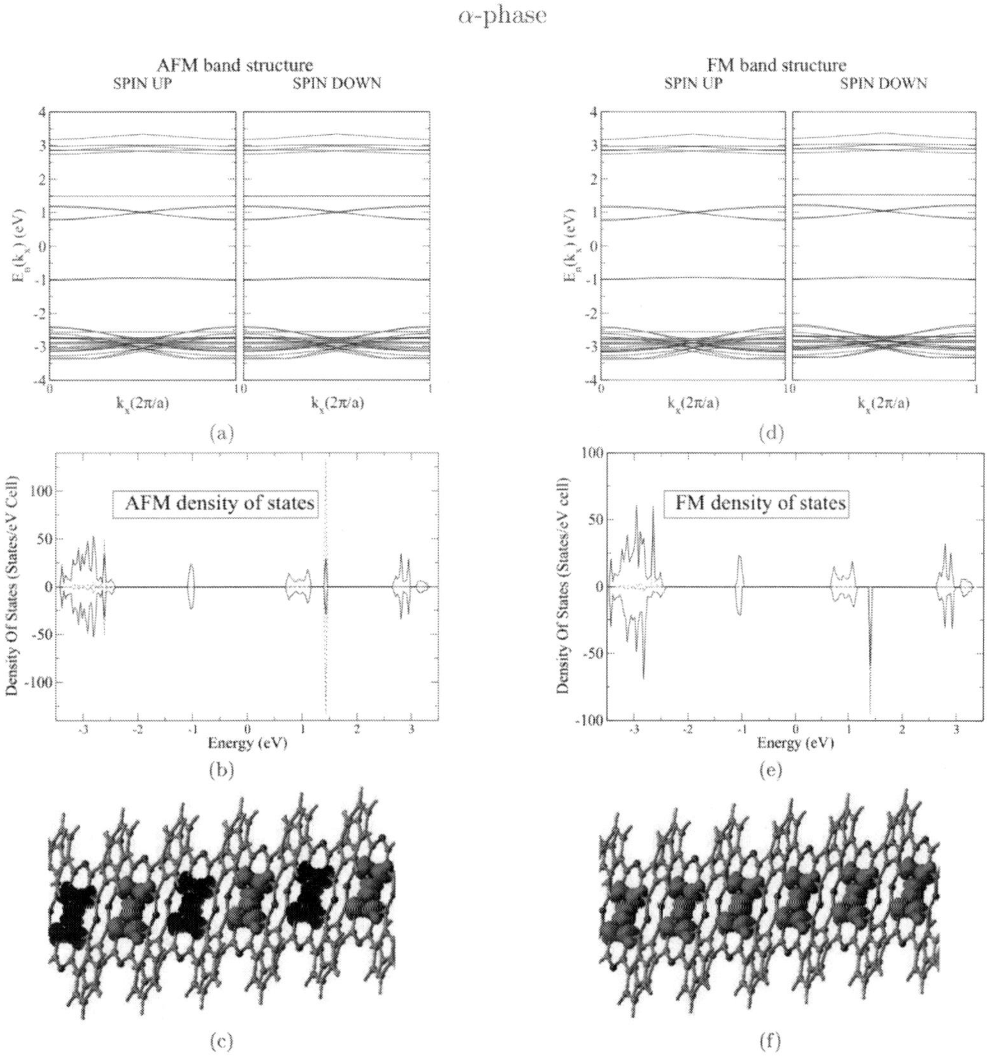

FIG. 3: (Color online.) The band structure, density of states, and spin densities of AFM (left column) and FM (right column) configurations for α-phase Cu(II)Pc are shown. In band structure the d-bands in spin-up are highlighted in red, and spin-down in blue. Notice that these bands are very narrow. In the density of states the spin-up is displayed as positive and spin-down as negative, and the projected density of states is scaled by a factor of 5.0 to illustrate the singly occupied orbitals. The spin densities show the spin configurations in different calculations where spin up is in red and spin down in blue.

but by somewhat less than in hybrid DFT (to approximately 1.5 eV). Optical absorption measurements for Cu(II)Pc crystals [11] show the wavelength of the Q absorption band arising from the HOMO-LUMO transition is about 650 nm ($\simeq$ 1.9 eV), in a good agreement with the band gap in our calculations.

The projected DOS (PDOS) onto two copper atoms in a supercell is also plotted along with the total DOS, colour-coded consistent with that in the spin densities, to illustrate the occupancy of the atomic orbitals on the copper atoms. The spin densities for the AFM configuration (FM configuration) are shown in Fig.3-5(c) (Fig.3-5(f)), respectively. The singly occupied bands can be found by comparing the projected density of states of the AFM and FM configurations. There are two peaks in the projected DOS on the left of the valence bands in spin-up and spin-down in AFM configuration (Fig.3-5(b)) while there are two peaks in the same energy position only in spin-up in the FM configuration (Fig.3-5(e)). So these peaks correspond to the singly occupied narrow bands. The spin densities of AFM

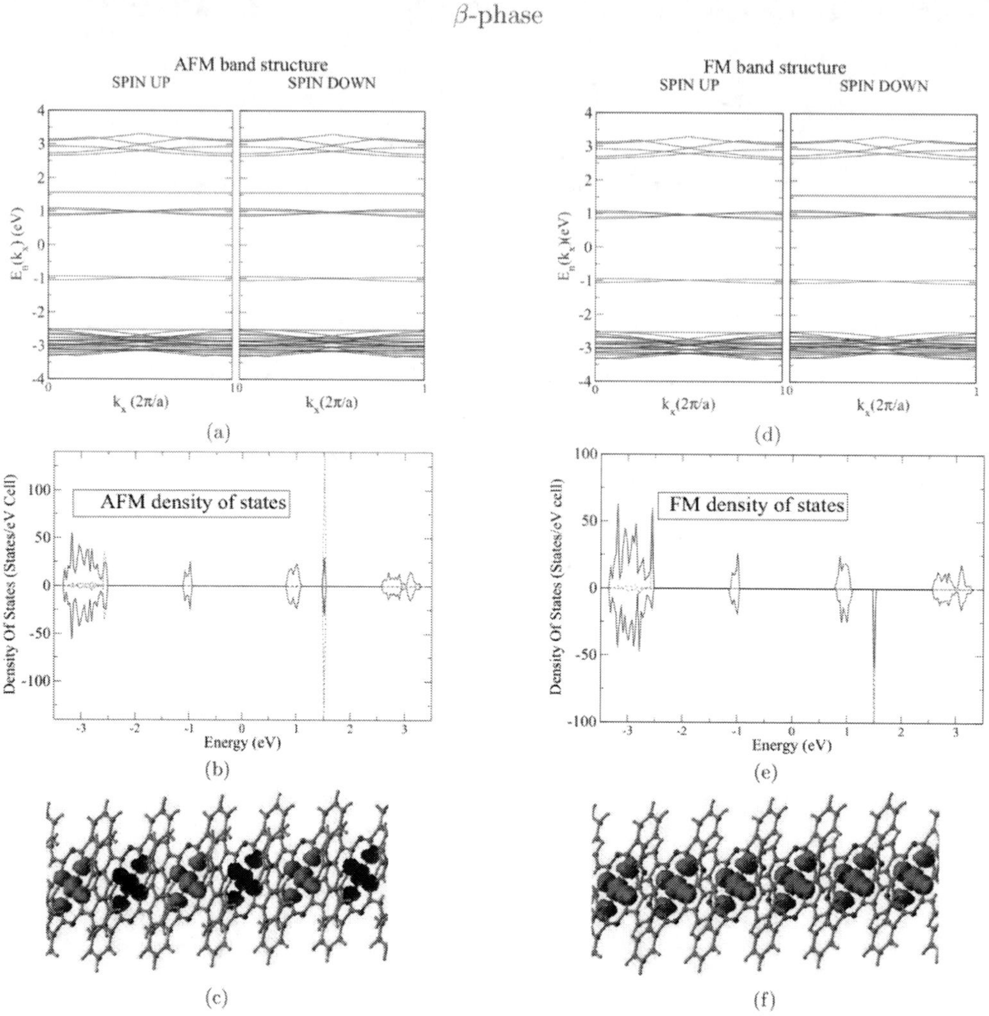

FIG. 4: (Color online.) The band structure, density of states, and spin densities of AFM (left column) and FM (right column) configurations for β-phase Cu(II)Pc are shown. In band structure the d-bands in spin-up are highlighted in red, and spin-down in blue. Notice that these bands are very narrow. In the density of states the spin-up is displayed as positive and spin-down as negative, and the projected density of states is scaled by a factor of 5.0 to illustrate the singly occupied orbitals. The spin densities show the spin configurations in different calculations where spin up is in red and spin down in blue.

(Fig.3-5(c)) and FM (Fig.3-5(f)) configurations both indicate the singly occupied orbital is $d_{x^2-y^2}$; this is in agreement with the results of our previous DFT calculations for the Cu(II)Pc single molecule and dimer. The Mulliken charges on the copper atoms are about $+0.9\ |e|$ for the AFM and FM configurations, and the Mulliken magnetic moments on two copper atoms for AFM (FM) are about $0.65\ \mu_B(0.65\ \mu_B)$ and $-0.65\ \mu_B(0.65\ \mu_B)$, respectively.

The electron hopping integral between copper atoms is very small as suggested by the small bandwidth of these singly occupied bands, and this further indicates that it is difficult for copper spins to interact directly. In particular, the narrow bandwidth suggests that the superexchange mechanism as defined in Ref. [31] will be suppressed, since it relies on virtual charge excitations of the Cu sites. However, the singly occupied band is energetically very close to a set of ligand bands, and spin polarization of the ligand due to copper spins is facile; this opens up the possibility of indirect exchange between the Cu sites involving polarization of the surrounding ligand spins. The gap between the upper and lower Hubbard bands is easily identified as about 4 eV in both the band structure and the DOS; the significantly greater splitting compared with the previous work [13–15] drives both the filled and empty d-states away

η-phase

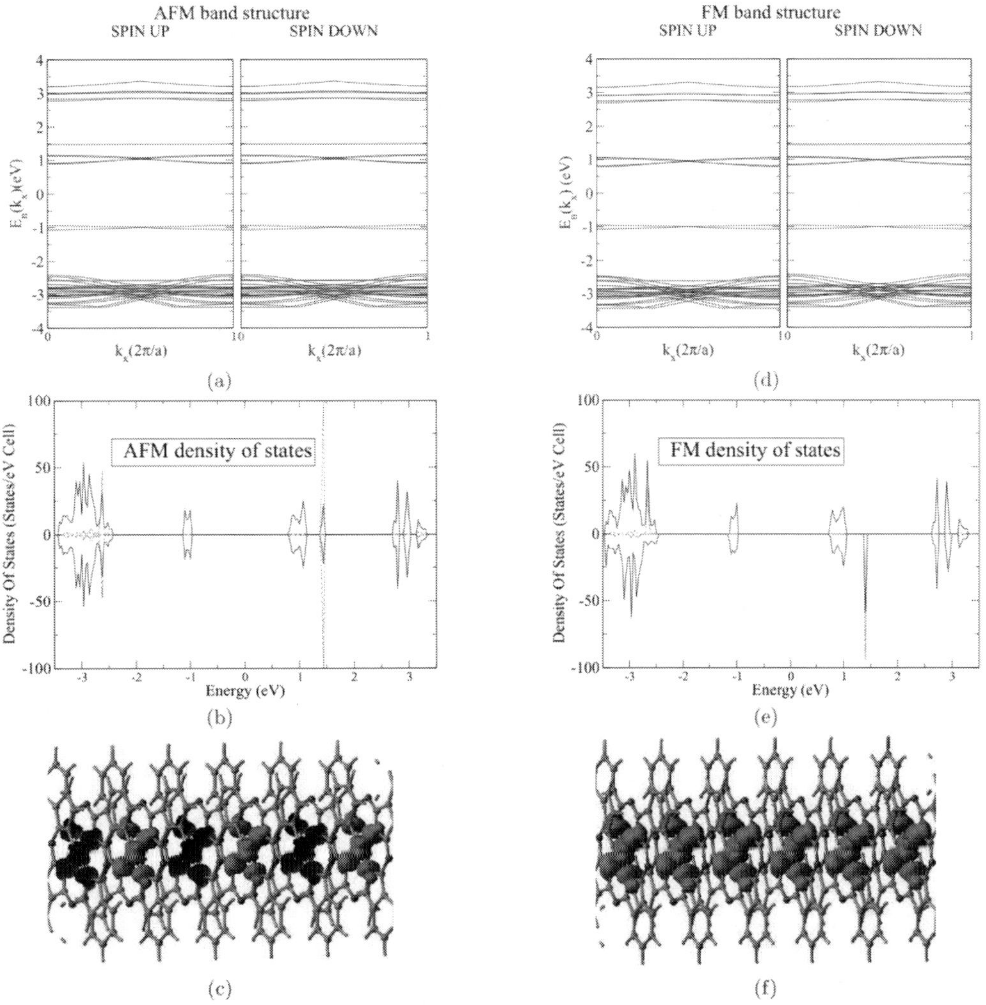

FIG. 5: (Color online.) The band structure, density of states, and spin densities of AFM (left column) and FM (right column) configurations for η-phase Cu(II)Pc are shown. In band structure the d-bands in spin-up are highlighted in red, and spin-down in blue. Notice that these bands are very narrow. In the density of states the spin-up is displayed as positive and spin-down as negative, and the projected density of states is scaled by a factor of 5.0 to illustrate the singly occupied orbitals. The spin densities show the spin configurations in different calculations where spin up is in red and spin down in blue.

from the Fermi energy, in qualitative agreement with DFT+U calculations [16].

The electronic structures of these three phases share many qualitative features in terms of the orbital occupancy, the band gap, the on-site Coulomb interaction on the copper site, etc as shown in Fig. (3-5). However, the transfer integrals are sensitive to the local geometry and thus the bandwidth varies significantly as the geometry changes. For example, LUMO bandwidths of α-phase ($\sim$ 0.5 eV), η-phase ($\sim$ 0.3 eV) and β-phase ($\sim$ 0.2 eV) differ significantly. This reflects the significant variation in the exchange; using perturbation theory, we now make this connection more directly.

B. Perturbation theory calculations

The variation of the exchange interactions in a dimer for the same set of geometrical parameters used in the DFT calculations has been estimated using GFPT, based on the model; of indirect exchange [30, 31] . The exchange interactions are once again strongly dependent on the stacking angle but weakly on the sliding angle, as in the DFT calculations for one-dimensional chain. The exchange interaction using GFPT peaks at a stacking angle 90° and is 12 times larger than that in the α-phase. The ratio of the exchange interaction of η-phase to α-phase is 0.4 in agreement with that observed [10]. In fact this GFPT result more closely resembles the trend observed experimentally than the DFT result for which the ratio is about 1. Furthermore, the oscillating behavior of the exchange interactions with the stacking angle is more obvious in perturbation theory than the DFT calculations. These comparisons suggest that GFPT based on indirect exchange indeed captures the dominant physical mechanism of the exchange interactions.

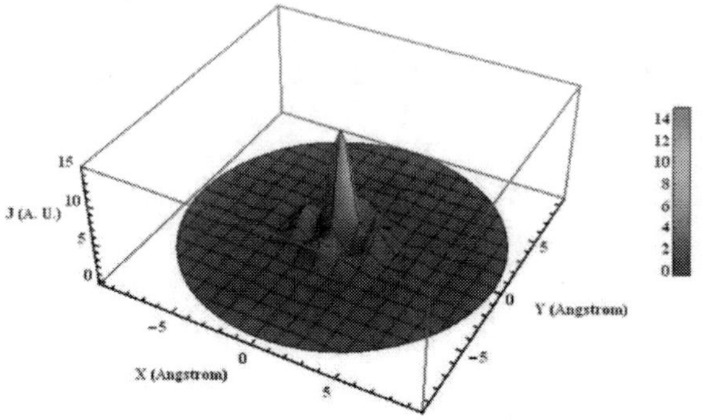

FIG. 6: (Color online.) Exchange interactions of Cu(II)Pc dimer calculated using perturbation theory as a function of the cartesian coordinates X and Y. Notice that exchange interactions strongly depend on stacking angles, but weakly on sliding angles.

V. CONCLUSIONS

We have calculated the exchange interaction in Cu(II)Pc for a variety of sliding and stacking angles which include the observed α-, β-, and η-phases both using DFT for one-dimension molecular chains and GFPT for molecular dimers. The qualitative trends of exchange interactions predicted by DFT and GFPT agree with those observed experimentally. Exchange interactions computed using DFT and GFPT are strongly dependent on stacking angles, but only weakly on sliding angles.

The DFT calculations give insight into the underlying electronic structure of the Cu(II)Pc molecule chain and its relationship to the magnetic properties. Valence and conduction bandwidths, the band gap, and the DOS are all important for understanding various optical and transport measurements in Cu(II)Pc. We find that the band gaps and d-d splittings are predicted to be significantly larger than in previous work [13–15] thanks to the use of the hybrid exchange density functional, leading to an improved description of the on-site correlation [16]. The DOS in our calculations is in qualitative agreement with the optical experiments shown in [10]. Our calculations have therefore shown a significant improvement in the treatment of the electronic structure of Cu(II)Pc crystals, without the need to introduce any material-specific parameters.

In particular the variation of the valence and conduction bandwidths between structures drives the computed (and experimentally observed) variation in the exchange constants. One important consequence is that, at least for this type of interaction mechanism, synthetic strategies aimed at maximizing the transition temperature should seek stacking angles as close to 90° as possible; another is that such geometries are also likely to give the highest charge-transport mobilities [32].

An important technical feature of our DFT calculations, employing a hybrid exchange-correlation functional, is the ability to describe localized unpaired d-electron and delocalized ligand states simultaneously; the DFT+U approach [16, 33] with a plane-wave basis, incorporates similar physics gives similar results provided U is chosen appropriately. A detailed comparison will be presented in a forthcoming paper [34]. Calculations of both types therefore provide a valuable basis for the further theoretical investigation of devices combining optics and magnetism or electronics and magnetism, e.g., magneto-resistant devices.

As shown by the GFPT results, the dominant magnetic interaction mechanism in Cu(II)Pc is indirect exchange [30, 31] which arises from the spin polarization of the ligand due to the central unpaired copper spin and the consequent delocalization of this spin polarization through the chain of delocalized π-conjugated ring systems formed by the stacked Cu(II)Pc molecules. Similar delocalization due to π-conjugation occurs in other organic systems, most notably in graphene [35]; it is therefore not surprising that the connection between spin and charge transport implied by our results has been noticed in other systems as far back as the early 1990s, where triplet energy transfer was attributed to double electron exchange [36] in a similar way to our work.

Acknowledgments

We wish to acknowledge the support of the UK Research Councils Basic Technology Programme under grant EP/F041349/1. We thank Gabriel Aeppli, Salahud Din, Sandrine Heutz, Chris Kay, Soumaya Mauthoor, Marshall Stoneham, Hai Wang, and Marc Warner for helpful discussions.

[1] Joel S. Miller, Arthur J. Epstein, William M. Reiff, Science, **240**, 40 (2000).
[2] James C. Ellenbogen and Kwan S. Kwok, Materials Today, **5**, 28 (2002).
[3] S. Heutz, C. Mitra, Wei Wu, A. J. Fisher, A. Kerridge, A.M. Stoneham, A. H. Harker, J. Gardener, H.-H. Tseng, T. S. Jones, C. Renner, and G. Aeppli, Adv. Mat., **19**, 3618 (2007).
[4] S. Pramanik, C.-G. Stefanita, S. Patibandla, S. Bandyopadhyay, K. Garre, N. Harth and M. Cahay, Nat. Nano., **2** 216 (2007).
[5] Z. H. Xiong, Di Wu, Z. Valy Vardeny, and Jing Shi, Nature, **427**, 821 (2004).
[6] V.N. Prigodin, N.P. Raju, K.I. Pokhodnya, J.S. Miller and A.J. Epstein, Adv. Mat., **14**, 1230 (2002).
[7] Stefano Sanvito, Nat. Mat. **6**, 803 (2007).
[8] Jung-Woo Yoo, Chia-Yi Chen, H.W. Jang, C.W. Bark, V. N. Prigodin, C. B. Eom, and A. J. Epstein, Nat. Mat., **9**, 638 (2010).
[9] Grigore A. Timco, Stefano Carretta, Filippo Troiani, Floriana Tuna, Robin J. Pritchard, Christopher A. Muryn, Eric J. L. McInnes, Alberto Ghirri, Andrea Candini, Paolo Santini, Giuseppe Amoretti, Marco Affronte, and Richard E. P. Winpenny, Nat. Nano. **4**, 173 (2009).
[10] Hai Wang, Soumaya Mauthoor, Salahud Din, Jules A. Gardener, Rio Chang, Marc Warner, Gabriel Aeppli, David W. McComb, Mary P. Ryan, Wei Wu, Andrew J. Fisher, Marshall Stoneham and Sandrine Heutz, ACS Nano, **4**, 3921 (2010).
[11] *The porphyrins* edited by David Dolphin, Academic Press, 1979.
[12] Wei Wu, A. Kerridge, A. H. Harker, and A. J. Fisher, Phy. Rev. B **77**, 184403 (2008).
[13] B. Bialek, In Gee Kim, and Jae Il Lee, Thin Solid Films **436**, 107 (2003).
[14] L. Lozzi, S. Santucci, S. La Rosa, B. Delley, and S. Picozzi, J. Chem. Phys. **121**, 1883 (2004).
[15] Yanting Yang, Yanming Yang, Fugen Wu, and Zhigang Wei, Solid State Communications **148**, 559 (2008).
[16] Gianluca Giovannetti, Geert Brocks, and Jeroen van den Brink, Phys. Rev. B **77**,035133 (2008).
[17] Peter Erk, Current Opinion in Solid State and Materials Science, **5**, 155 (2001).
[18] Peter Erk, Heidi Hengelsberg, Mairi F. Haddow and Richard van Gelder, Cryst. Eng. Comm., **6**, 474 (2004).
[19] R. Ditchfield, W. J. Hehre, and J. A. Pople, J. Chem. Phys., **54** 724 (1971).
[20] M. J. Frisch, *et al.*, Gaussian 03 (Gaussian, Inc., Pittsburgh, PA, 1998).
[21] R. Dovesi, V. R. Saunders, R. Roetti, R. Orlando, C. M. Zicovich-Wilson, F. Pascale, B. Civalleri, K. Doll, N. M. Harrison, I. J. Bush, P. DArco, and M. Llunell, CRYSTAL06 (CRYSTAL06 User's Manual. University of Torino, Torino, 2006).
[22] H. J. Monkhorst and J. D. Pack, Phys. Rev. B **13**, 5188 (1976).
[23] A. D. Becke, J. Chem. Phys. **98**, 5648 (1993).
[24] L. Noodleman, J. Chem. Phys. **74**, 5737 (1980).
[25] W. Heisenberg, Z. Phys. **49**, 619 (1928).
[26] F. Illas, I de P.R. Moreira, C. de Graaf and V. Barone, Theo. Chem. Acc, **104**, 265 (2000).
[27] J. Muscata, A. Wanderb, and N.M. Harrison, Chem. Phys. Lett. **342**, 397 (2001).
[28] Ling Ge, Barbara Montanari, John H. Jefferson, David G. Pettifor, Nicholas M. Harrison, and G. Andrew D. Briggs, Phys. Rev. B **77**, 235416 (2008).
[29] A. Hoshino, Y. Takenaka and H. Miyaji, Acta Cryst. **B59**, 393 (2003).
[30] M. A. Ruderman and C. Kittel, Phys. Rev. **96**, 99 (1954).
[31] P. W. Anderson, Phys, Rev, **115**, 2 (1959).
[32] W.J. Pietro, T.J. Marks and M.A. Ratner, J. Am. Chem. Soc. **107** 5387 (1985).
[33] A.I. Liechtenstein, V.I. Anisinov and J. Zaanen, Phys. Rev. B **52** R5467 (1995).
[34] N. Kearsey *et. al.*, in preparation.
[35] A. K. Geim and K. S. Novoselov, Nat. Mat. **6**, 183 (2007).
[36] Gerhard L. Closs, Mark D. Johnson, John R. Miller, Piotr Piotrowiak, J. Am. Chem. Soc. **111** 3751 (1989).

Superconductivity of heterofullerides $A_nB_mHg_xC_{60}$ (A=K,Rb,Cs; B=Be,Mg,B,Al,Ga,In; n=1, 2; m=1, 2; x<1)

V.A. Kulbachinskii[1], **B.M. Bulychev, R.A. Lunin, I.P. Kachan**
M.V. Lomonosov Moscow State University, 119991, GSP-2, Moscow, Russia

E-mail: kulb@mig.phys.msu.ru

Abstract. The fullerides $A_nM_mC_{60}$ (A=K, Rb, Cs; M= Be, Mg; B, Al, Ga. In; n=1, 2; m=1, 2) have been synthesized by exchange reactions of alkali metal fullerides with metal chlorides (groups 2 and 13), by method using liquid alloys of metals with mercury (amalgams), and by thermal decomposition of aluminum hydride. The samples are studied by X-ray diffraction, nuclear magnetic resonance, electron paramagnetic resonance, Raman scattering, and differential scanning calorimetry. It was found that the fulleride K_2GaC_{60} is superconductor with transition temperature T_c=22 K that exceeds T_c for K_3C_{60} (19 K). Fullerides with In and B are not supercomductors. The fullerides with composition $Cs_nHg_xC_{60}$ (n=2;3) are not superconductors and crystallize in orthorhombic lattices.

1. Introduction

The discovery of superconducting fullerides with alkaline metals in 1991 [1] leads to the intensive development of chemistry and physics of these compounds, primarily to the development of methods of synthesis, to the study of the electrical and structural properties, and to the definition of the spectral characteristics and many other features. One of the important conclusions was that superconductivity is observed in fullerides A_3C_{60} (A=K, Rb, Cs) with *fcc* lattice [2]. Later it was shown that superconducting and electrophysical properties of fullerides depend on method of synthesis, in particular on the nature of solvent. For example synthesis of Na_3C_{60} from simple substances in toluol leads to fulleride with triclinic lattice [3,4].

Synthesis of $K_3Hg_xC_{60}$ from amalgam gives fulleride with monoclinic lattice. The fulleride $K_3Hg_xC_{60}$ is not superconductor. Heterofulleride with composition $K_2Hg_xC_{60}$ (x<1) is a superconductor with transition temperature T_c=22 K. This value exceeds transition temperature for K_3C_{60} (T_c=19 K).Synthesis of Cs_3C_{60} gives amorphous or probably polymeric substances [4]. At the same time synthesis of this fulleride in liquid ammonia from simple substances gives superconducting fulleride with orthorhombic lattice in contradiction with previous results [5]. However, on important problem remains: in what composition of fullerides is the superconductivity possible and how to rise to the critical temperature T_c?

In the present study, the results of investigation of structural and superconducting properties of heterofullerides with metals of 1st, 2nd and 13th groups $A_nB_mHg_xC_{60}$ (A=K, Rb, Cs; M= Be, Mg, B, Al, Ga. In; n=1, 2; m=1, 2; x<1) are described.

[1] To whom any correspondence should be addressed.

2. Experimental

The fullerides have been synthesized by the exchange reactions of alkaline fullerides with anhydrous metal halides (i), by a method using liquid alloys of metals with mercury (amalgams) (ii) or by thermal decomposition of hydrides (iii).

2.1 Synthesis of amalgams and fullerides from amalgams

The calculated amounts of one or two metals (to within 0.02–0.15 g) were in a pyrex vial containing 6–7 g of mercury (i.e., a sufficient amount according to the phase diagrams for producing the liquid amalgams). The vial was then pumped down in a vacuum and sealed off. The metals were dissolved by heating a vial to 400 °C for 3 hours with periodic mixing of the contents. The vial with the amalgam and the reactor for synthesizing the fulleride were placed in a "Braun-M" vacuum cell. The vial was opened and its contents transferred to the reactor, into which the calculated amount of fullerite (0.15–0.25 g) and 20–30 mL of pure toluene had been placed. The reactor was extracted from the cell and its contents were frozen in liquid nitrogen; the entire system was pumped down in vacuum and unsealed, and the reactor was placed in a thermally controlled cabinet equipped with a mixing device for 15–20 days at a temperature of 100–110 °C. The progress of the reaction was monitored visually by the fading of the violet fullerite solution. Further operations for separating the reaction product were carried out in a fully sealed glass unit and included: separation of the excess mercury by decanting, washing the residue three or four times by recondensation of the solvent with cooling in liquid nitrogen, drying the residue at 90–100 °C, and packing it into vials for the electrical, spectral, and X-ray structural studies [4].

2.2 Synthesis of heterofullerides by exchange reactions in organic solvents

Fullerides were synthesized in the organic solvents in two stages. In the first stage, alkali metal fullerides with compositions A_nC_{60}, where A=Na,K,Rb and n=3–6 were obtained by direct reaction of an alkali metal with the fullerite solution in pure toluene using the method described in Refs. [4,6,7,8]. The second stage involved an exchange reaction of an alkali fulleride with anhydrous chlorides of the heterometals in pure tetrahydrofuran (THF) at temperatures below 80 °C in accordance with the reactions

$$A_4C_{60} + MCl_2 \xrightarrow{THF} A_2MC_{60} + 2ACl \quad (1)$$

$$A_5C_{60} + 2MCl_2 \xrightarrow{THF} AM_2C_{60} + 4ACl \quad (2)$$

$$A_5C_{60} + MCl_3 \xrightarrow{THF} A_2MC_{60} + 3ACl \quad (3)$$

where A-K, Rb, Cs; M- Be, Mg, B, Al, Ga, In.

In principle, this method makes it possible to obtain fullerides and heterofullerides of various metals here, thallium containing heterofullerides with different chemical compositions and atomic charge states [4].

2.3 Thermal decomposition of aluminum hydride

We used also a new method of thermal decomposition of aluminum hydride according to reaction

$$A_2C_{60}+AlH_3 \xrightarrow{толуол} A_2AlC_{60}+3/2\ H_2 \quad (4)$$

Aluminum hydride was synthesized by method described in Ref. [9]. Some parameters of samples are listed in table 1. The compositions of these fullerides calculated from the loaded components. As we show in table 3 our attempts to synthesize heterofullerides with cesium failed as it was previously [4]. According to X-ray and NMR data cesium heterofullerides synthesized at $T>20^oC$ are amorphous and polymerized. Superconductivity in cesium fullerides was found only in $RbCsTlC_{60}$ and $KCsTlC_{60}$ [10].

3. Results and discussion

X-ray data for some synthesized fullerides are summarized in table 1. Synthesized heterofullerides of potassium and rubidium are superconductors (fig. 1a). We observed that in general the superconducting transition temperature T_c increases with crystal lattice parameter a as seen in table 1. Raman scattering data shows that charge state of C_{60} molecule in fullerides is close to –3.

Journal of Physics: Conference Series **391** (2012) 012173 doi:10.1088/1742-6596/391/1/012173

Table 1. Some parameters of samples

№	Composition of initial mixture	Solvent	Composition of fulleride	T_c (K)	X-Ray data
368	K_4C_{60}+$BeCl_2$+C_{60}	THF	K_2BeC_{60}	13	*fcc, a*=14.268 Å +KCl
225	K_4C_{60}+$MgCl_2$+C_{60}	THF	K_2MgC_{60}	16	*fcc, a*=14.290 Å+KCl
942	{2K+Mg+$(Hg)_n$}+C_{60}	toluene	$K_2MgHg_xC_{60}$	13.6	*fcc, a*=14.280 Å
292	K_5C_{60}+2$MgCl_2$+C_{60}	THF	KMg_2C_{60}	15.5	*fcc, a*=14.256 Å +KCl
236	K_5C_{60}+$AlCl_3$+C_{60}	THF	K_2AlC_{60}	14.5	*fcc, a*=14.249 Å+KCl
20	K_2C_{60}+AlH_3	toluene	K_2AlC_{60}	14	*fcc, a*=14.282+<1%Al
941	{2K+Al+$(Hg)_n$}+C_{60}	toluene	$K_2AlHg_xC_{60}$	15.7	*fcc, a*=14,281 Å + second phase (~3 %)
931	{2K+Ga+$(Hg)_n$}+C_{60}	toluene	$K_2GaHg_xC_{60}$	20	*fcc, a*=14,297 Å +second phase
923	{2K+In+$(Hg)_n$}+C_{60}	toluene	$K_2InHg_xC_{60}$	-	*fcc, a*=14.281 Å +second phase
264	Rb_4C_{60}+$BeCl_2$+C_{60}	THF	Rb_2BeC_{60}	22	*fcc, a*=14.447 Å+RbCl
724	Rb_4C_{60}+$MgCl_2$+C_{60}	THF	Rb_2MgC_{60}	19	*fcc, a*=14.425Å+RbCl
982	{2Rb+Mg+$(Hg)_n$}+C_{60}	toluene	$Rb_2MgHg_xC_{60}$	21.8	*fcc, a*=14.447 Å +~5%M_2
12	Rb_5C_{60}+2$MgCl_2$+C_{60}	THF	$RbMg_2C_{60}$	-	*fcc, a*= 14.425 Å +RbCl
13	{Rb+2Mg+$(Hg)_n$}+C_{60}	toluene	$RbMg_2Hg_xC_{60}$	-	-
979	{2Rb+Al+$(Hg)_n$}+C_{60}	toluene	$Rb_2AlHg_xC_{60}$	24.8(11)	*fcc,a*=14.458 Å+M_2
929	{2Rb+Ga+$(Hg)_n$}+C_{60}	toluene	$Rb_2GaHg_xC_{60}$	25	*fcc,a*=14.439(3) (~90%) + *a*=9.055; *b*=10.057; *c*=14.190 Å (~10%)
922	{2Rb+In+$(Hg)_n$}+C_{60}	toluene	$Rb_2InHg_xC_{60}$	-	*a*=14.439(3) (~65%) + *a*=9.055; *b*=10.057; *c*=14.190 Å(~35%)
5	Cs_4C_{60}+$BeCl_2$+C_{60}	THF	Cs_2BeC_{60}	-	Amorphous + CsCl
690	Cs_4C_{60}+$MgCl_2$+C_{60}	THF	Cs_2MgC_{60}	-	Amorphous + CsCl
983	{2Cs+Mg+$(Hg)_n$} + C_{60}	toluene	$Cs_2MgHg_xC_{60}$	-	Amorphous
674	Cs_5C_{60}+$AlCl_3$+C_{60}	THF	Cs_2AlC_{60}	-	Amorphous +CsCl
980	{2Cs+Al+$(Hg)_n$}+ C_{60}	toluene	$Cs_2AlHg_xC_{60}$	-	Amorphous

Second phase M_2:- *a*=10.539; *b*=7.998; c=6.883 Å; β=103.21°; V=564.8 Å^3

The shift of line $A_g(2)$ in synthesized superconducting fullerides is almost the same as in a very well known superconductor K_3C_{60} (fig. 1b). For heterofullerides synthesized with metals of 13th group with composition $A_2M^{+3}C_{60}$ we may suppose the transfer of 5 electrons. Nevertheless as we can see in table1 only products of reaction of K_5C_{60} with $AlCl_3$ in THF and K_2C_{60} with AlH_3 in toluene are superconductors. Reactions of fullerite with ternary amalgams give superconducting potassium, rubidium, aluminum and gallium fullerides. All other substances with higher or less number of transferred electrons, for example Al_2C_{60}, $MAlC_{60}$, Al_3C_{60} etc., don't crystallize in *fcc* lattice and are not superconductors. At the same time indium fullerides have *fcc* lattice but they are not superconductors.

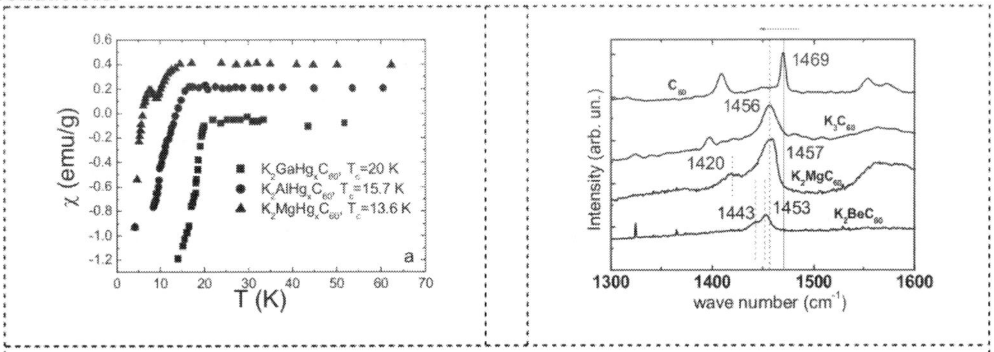

Figure 1. Temperature dependence of magnetic susceptibility for (a) $K_2GaHg_xC_{60}$, $K_2AlHg_xC_{60}$, and $K_2MgHg_xC_{60}$ and (b) shift of line $A_g(2)$ of Raman spectra in fullerite and synthesized fullerides.

As we see in table 1 critical temperature T_c for K_2AlC_{60} fullerides synthesized by three different methods is almost the same. The enhancement of T_c in fullerides synthesized from amalgam the most probably is due to intercalation of some amount of Hg in fulleride. The intercalation of Hg in fullerides also leads to the superconductivity in Rb-Al and K-Ga fullerides. We showed that small amount of Hg in fulleride change radically its properties [4,11]. Fullerides with composition K_2BC_{60} and Rb_2Bc_{60} are not superconductors. Fulleride $K_2InHg_xC_{60}$ is not a superconductor while $K_2Hg_xC_{60}$ is a superconductor with T_c=22 K [11]. Similar fulleride $Rb_2InHg_xC_{60}$ is not a superconductor while $Rb_2Hg_xC_{60}$ is a superconductor with T_c=25 K.

The work was supported by RFBR, grant number № 08-03-00237a.

References

[1] A.F. Hebard, M.J. Rosseinsky, R.C. Haddon, et. al., *Nature*, **350**, 600 (1991).
[2] O. Gunnarsson, *Reviews of Modern Physics*, **69**, 575 (1997).
[3] V.A. Kulbachinskii, B.M. Bulychev, V.G. Kytin, et. al., *Central Eur. J. Phys.* **8**,101 (2010)
[4] V.A. Kulbachinskii. B.M. Bulychev, et.al. *Low Temperature Physics*, **37**, 245 (2011).
[5] A.Y. Ganin, Y. Takjabayashi, P. Jeglič, et. al, *Nature,* **466,** 221 (2010).
[6] B.M. Bulychev, R.A. Lunin, et. al, *J. of Phys. and Chem. of Solids*, **65**, 337 (2004).
[7] O.A. Dityatyev, V.I. Privalov, B.M. Bulychev, et.al., *Mol. Mater.*, **13,** 281 (2000).
[8] B. M. Bulychev, R. A. Lunin, et.al, *Russ. Chem. Bull., Int. Ed.*, **53**, 1686 (2004).
[9] B. M. Bulychev, P. A. Storozhenko, V. N. Fokin, *Russ.Chem.Bull., Int.Ed.*, **58**, 1817 (2009).
[10] V.A.Kulbachinskii,B.M.Bulychev, V. G. Kytin, et.al, Hindawi Pub. Corp., *Advances in Condensed Matter Physics*, Volume 2008, Article ID 941372, (2008).
[11] B.M. Bulychev, V.A. Kulbachinskii, R.A. Lunin, V.G. Kytin and Yu.A. Velikodny, *Fullerenes, Nanotubes, and Carbon Nanostructures*, **18**, 381-385 (2010).

Calorimetric study of the superconducting and normal state properties of $Ca(Fe_{1-x}Co_x)_2As_2$

M. Abdel-Hafiez[1], **L. Harnagea**[1], **S. Singh**[2], **U. Stockert**[3], **S. Wurmehl**[1], **N. Leps**[4], **R. Klingeler**[4], **A. U. B. Wolter**[1], **B. Büchner**[1]

[1] Leibniz-Institute for Solid State and Materials Research, D-01171 Dresden, Germany
[2] Indian Institute of Science Education and Research (IISER), Maharashtra 411008, India
[3] MPI for Chemical Physics of Solids, D-01187 Dresden, Germany
[4] Kirchhoff Institute for Physics, University of Heidelberg, D-69120 Heidelberg, Germany

E-mail: `m.mohamed@ifw-dresden.de`

Abstract. We present a calorimetric study on single crystals of $Ca(Fe_{1-x}Co_x)_2As_2$ ($x = 0$, 0.032, 0.051, 0.056, 0.063, and 0.146). The combined first order spin-density wave/structural transition occurs in the parent $CaFe_2As_2$ compound at 168 K and gradually shifts to lower temperature for low doping levels ($x = 0.032$ and $x = 0.051$). It is completely suppressed upon higher doping $x \geq 0.056$. Simultaneously, superconductivity appears at lower temperature with a transition temperature around $T_c \sim 14.1$ K for $Ca(Fe_{0.937}Co_{0.063})_2As_2$. The phase diagram of $Ca(Fe_{0.937}Co_{0.063})_2As_2$ has been derived and the upper critical field is found to be $\mu_0 H_{c2}^{(C)} = 11.5$ T and $\mu_0 H_{c2}^{(ab)} = 19.4$ T for the c and ab directions, respectively.

1. Introduction

After the discovery of superconductivity in LaFeAsO (1111 compound) with a transition temperature $T_c \sim 26$ K [1], a lot of investigations have been undertaken finding an increase of T_c by replacing La with smaller size rare earth ions [2] reaching up to $T_c = 55$ K in $SmO_{1-x}F_xFeAs$ [3]. Simultaneously, several other families of Fe-based layered superconductors have been discovered, such as the $ThCr_2Si_2$ structure-type AFe_2As_2 (122 family, A = Ca, Sr, Ba, Eu). They adopt a structure analogous to oxypnictides with LaO layers replaced by layers of A. These series are of great interest because large single crystals can be grown [4] which show a structural transition from a tetragonal to an orthorhombic structure combined with an antiferromagnetic ground state due to a spin-density wave (SDW) formation [5]. Furthermore, the parent compound of the 122 series becomes superconducting upon electron and hole doping [6],[7]. The $CaFe_2As_2$ system is of particular interest because the application of a pressure of 0.69 GPa is sufficient to suppress the SDW/structural transition completely and to induce superconductivity with a T_c exceeding 10 K [8],[9] while for $SrFe_2As_2$ and $BaFe_2As_2$ higher pressure up to 4 GPa are needed [10]. We should also mention that the electron and hole doped $BaFe_2As_2$ and $SrFe_2As_2$ series of compounds have remained very much in focus, while studies on the analogous $CaFe_2As_2$ compounds are very scarce.

Since specific heat measurements can provide useful information about the electronic and the gap structural properties, it is important to explore the normal state and

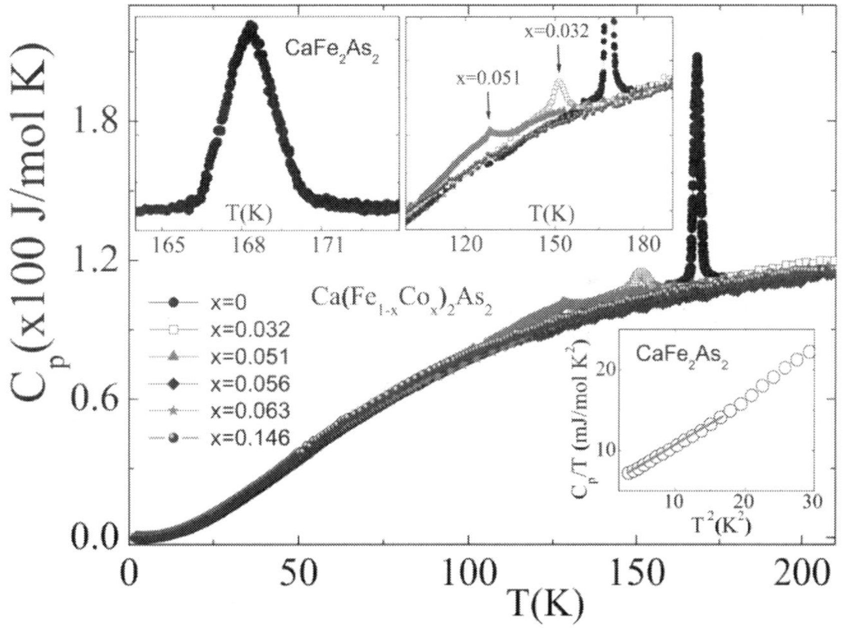

Figure 1. Temperature dependence of the specific heat of $Ca(Fe_{1-x}Co_x)_2As_2$ for $x = 0$, 0.032, 0.051, 0.056, 0.063, and 0.146 measured in zero magnetic field. The lower inset depicts the low temperature specific heat of the parent $CaFe_2As_2$ compound together with a fit to $C_p = \gamma T + \beta T^3$ shown as a red solid line. The upper insets show a zoom into the region around the phase transition for $CaFe_2As_2$ plotted as C_p vs. T and upon Co doping.

superconducting properties of $Ca(Fe_{1-x}Co_x)_2As_2$ single crystals using this technique. In this contribution, we present a first calorimetric study of both antiferromagnetic and superconducting $Ca(Fe_{1-x}Co_x)_2As_2$ for (x = 0, 0.032, 0.051, 0.056, 0.063 and 0.146) and report on the superconducting properties of the $Ca(Fe_{0.937}Co_{0.063})_2As_2$ single crystal.

2. Experimental

Single crystals of $Ca(Fe_{1-x}Co_x)_2As_2$ were grown by the high temperature solution growth method using Sn flux. All room-temperature processing (weighing, mixing, grinding and storage) of these materials was carried out in an Ar filled glove-box (O_2 and moisture level less than 0.1 ppm). The crystal growth and the physical properties of the $Ca(Fe_{1-x}Co_x)_2As_2$ series are presented in details in ref.[4],[11]-[15] and the x values are indicating the composition as measured by Energy-dispersive X-ray spectroscopy (EDX) [4]. The specific heat measurements have been performed in a Physical Property Measurement System (PPMS) using a relaxation technique. For the measurements $B \parallel ab$ a small copper block has been used to mount the sample on the specific heat puck. The heat capacity of the copper block was determined in a separate measurement and directly subtracted. A temperature rise of only 0.5 % was used for those measurements in the vicinity of the transition of the un-doped sample.

3. Results and discussions

Fig. 1 summarizes the temperature dependence of the zero-field specific heat at various Co-doping of $Ca(Fe_{1-x}Co_x)_2As_2$ between 1.8 and 220 K. In the parent compound ($x = 0$) a sharp first-order structural transition coincides with a SDW transition at $T_s = 168$ K (upon warming), which is in a good agreement with literature [4],[12],[16]. The transition width is estimated to

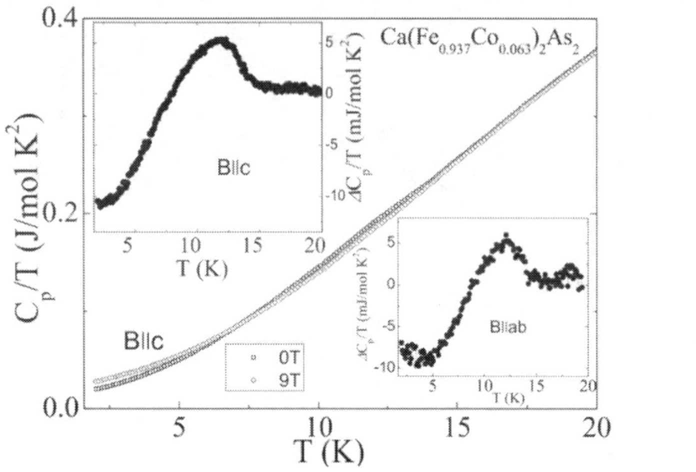

Figure 2. Specific heat measurements of $Ca(Fe_{0.937}Co_{0.063})_2As_2$ measured in zero and 9 T for $B \parallel c$. The upper inset shows the difference between the zero and 9 T data, $\Delta C_p/T$, which depicts a specific heat jump with a superconducting transition temperature of about 14.1 K for $B \parallel c$. The lower inset shows the difference between the zero and 9 T data, $\Delta C_p/T$, for $B \parallel ab$.

be about 4 K. To determine the Sommerfield coefficient γ of our $CaFe_2As_2$ single crystal, the low-temperature specific heat data are plotted as C_p/T as a function of T^2. These data can be fitted to a $C_p=\gamma T+\beta T^3$ power law, where γ and β are determined by the electronic and phononic contributions, respectively, as shown in the lower inset in Fig. 1. The γ value is found to be around 5.4 mJ/mol K^2, which is in agreement with values ranging between 4.7 - 8.2 mJ/mol K^2 in the literature [5],[16]. At present we do not understand this relatively large sample dependent variation in the measured values of γ. It is, however, conceivable that this difference is due to a different single crystal growth technique which is leading to differences in the crystal quality. The phononic coefficient β is found to be $\sim$ 0.508 mJ/mol K^4. Using the relation $\theta_D = (12\pi^4 RN/5\beta)^{1/3}$, where R is the molar gas constant and $N = 5$ is the number of atoms per formula unit, we obtain the Debye temperature $\theta_D = 267$ K.

Upon Co doping of the system the combined SDW/structural transition is shifted to lower temperature and becomes broader and considerably reduced in magnitude, with the transition being shifted to 148 K and 128 K for $x = 0.032$ and $x = 0.051$, respectively, as shown in the upper inset in Fig. 1. Contrary to the magnetization, resistivity and neutron diffraction studies [4],[15], in our specific heat studies no evident splitting of the structural and magnetic transition was found in our sample $x = 0.051$ and $x = 0.056$. For higher Co doping the transitions are completely suppressed and superconductivity evolves at low temperature. The transitions observed in the specific heat data are in general consistent with those found in the resistivity, magnetization and neutron diffraction [4],[11],[15] measurements performed in few of $Ca(Fe_{1-x}Co_x)_2As_2$ single crystals from the same batch like those used in the present study.

Figure 2 shows the temperature dependence of the heat capacity of $Ca(Fe_{0.937}Co_{0.063})_2As_2$ in zero and 9 T for $B \parallel c$. Taking into account the relatively broad superconducting transition in both resistivity and susceptibility measurements on the same single crystal as discussed in detail in [4], the apparent lack of the anomaly at T_c may be due to the relatively large transition width. This is strongly supported by specific-heat data on superconducting $CaFe_{1.94}Co_{0.06}As_2$ single crystals, where no superconducting transition has been observed either [17]. This probably reflects the presence of nanoscopic inhomogeneities intrinsic to the Ca-system which is beyond the detection limit of EDX or we can speculate that the broad nature of superconducting transitions is related to the extreme pressure sensitivity of this compound.

However, in order to obtain information about the electronic properties in the superconducting state, we subtracted our 9 T specific heat data from the zero-field data for

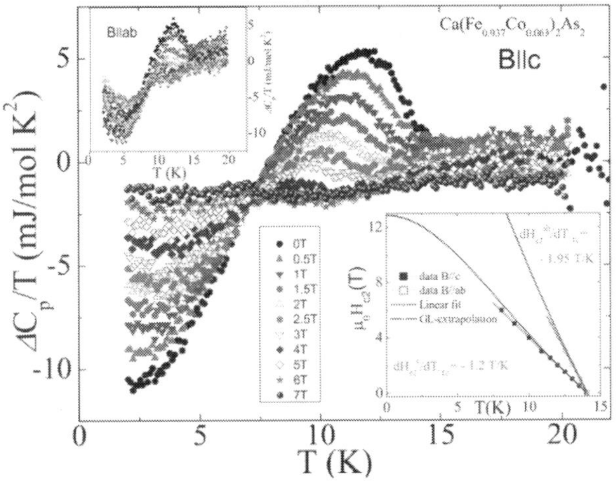

Figure 3. Specific heat data $\Delta C_p/T = C_p(B) - C_p(9T)$ of $Ca(Fe_{0.937}Co_{0.063})_2As_2$ in various magnetic fields for $B \parallel c$. The upper inset shows $\Delta C_p/T$ for $B \parallel ab$. The lower inset depicts the phase diagram estimating the upper critical fields, where the blue dotted lines show the theoretical curves based on the G-L theory and the red dotted lines represent linear fits with average slopes of $-\mu_0 dH_{c2}^{(c)}/dT = 1.2$ T/K and $-\mu_0 dH_{c2}^{(ab)}/dT = 1.95$ T/K for the c and ab directions, respectively.

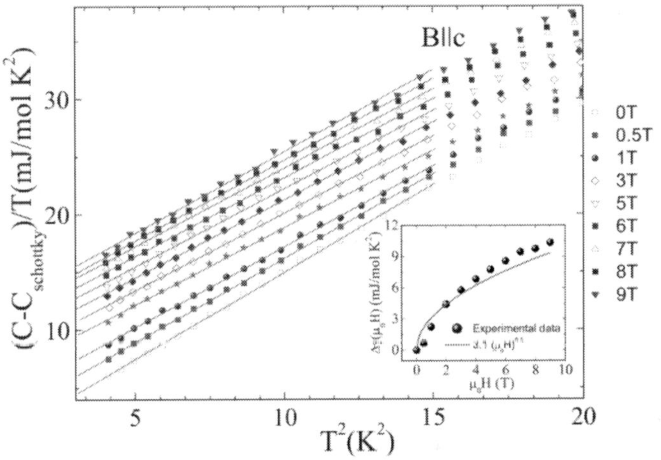

Figure 4. The electronic specific-heat coefficient C_p-$C_{Schottky}/T$ vs. T^2 of $Ca(Fe_{0.937}Co_{0.063})_2As_2$ for magnetic fields $B \parallel c$ up to 9 T. The data show a rather linear behavior is detected after subtracting the Schottky contribution (in the form $c_p = \gamma T + AT^{-3}$) to the specific heat at low temperatures. Solid red lines are represent linear fit in the form $C_p = \gamma T + \beta T^3$ (see table 1). The lower inset depicts the field dependence of $\Delta\gamma(H)$ for $B \parallel c$. The solid red line represents a fit to $\Delta\gamma(H) = A(\mu_0 H)^{0.5}$.

both directions, labeled as $\Delta C_p/T$ in the insets of Fig. 2. The extracted superconducting transition temperature is found to be 14.1 K, which is in agreement with resistivity and magnetization studies on single crystals from the same batch [4],[11]. Note, that this analysis for the estimation of the transition temperature T_c can be applied since the superconducting transition of $Ca(Fe_{0.937}Co_{0.063})_2As_2$ is strongly suppressed and shifted to lower temperaturesin in an external magetic field of 9T. The same procedure has been used to extract T_c for other fields up to 6 T (shown in Fig. 3) using the half jump height. From Fig. 3 it can be clearly seen that the superconducting transition is systematically shifted to lower temperature and reduced in height with increasing magnetic field. This behavior occurs for both field directions.

From the field dependence of T_c the phase diagram of $Ca(Fe_{0.937}Co_{0.063})_2As_2$ can be drawn, which is shown in the lower inset of Fig. 3 depicting the experimental data as well as the estimated upper critical field values H_{c2} for both directions. According to the Ginzburg-Landau (GL) equation [18], H_{c2} can be expressed as

$$H_{c2}(t) = H_{c2}(0)[\frac{1-t^2}{1+t^2}], \quad (1)$$

where t is the reduced temperature $t = T/T_c$. The upper critical field at $T = 0$ has been evaluated to be $\mu_0 H_{c2}^{(c)}(0) = 12.7$ T and $\mu_0 H_{c2}^{(ab)}(0) = 25.4$ T for the c and ab direction,

Table 1. Fitting parameters of the low temperature specific heat data of $CaFe_{0.937}Co_{0.063}As_2$ single crystals using $C_p=\gamma T+\beta T^3$ after subtracting the Schottky contribution.

μ_0(H) (T)	γ (mJ/mol K^2)	β (mJ/mol K^4)
0	0.945	1.51
0.5	1.63	1.46
1	3.19	1.4
2	5.49	1.34
3	6.74	1.34
4	7.8	1.34
5	8.73	1.35
6	9.57	1.36
7	10.41	1.36
8	10.71	1.4
9	11.3	1.4

respectively. Another possibility to extract the upper critical field is to consider the single-band Werthamer-Helfand-Hohenberg (WHH) model [19]:

$H_{c2}(0) = -0.69\, T_c(dH_{c2}/dT)_{T_c}$. Our data are perfectly described by a linear fit to the WHH model with an average slope $-\mu_0 dH_{c2}/dT$ = 1.2 T/K and 1.95 T/K yielding upper critical fields $\mu_0 H_{c2}(0)$ = 11.5 T and 19.4 T for the c and ab direction, respectively. Note, that resistivity studies on $CaFe_{1.94}Co_{0.06}As_2$ single crystals show similar values for the average slopes, i.e., $-\mu_0 dH_{c2}{}^{(c)}/dT$ = 1.4 T/K for B $\parallel$ c and $-\mu_0 dH_{c2}{}^{(ab)}/dT$ = 1.8 T/K for B $\parallel$ ab [17].

Figure 4 depicts the low temperature specific heat of $Ca(Fe_{0.937}Co_{0.063})_2As_2$ plotted as C_p-$C_{Schottky}$/T vs. T^2 in different applied magnetic fields for $B \parallel c$. One can see the roughly linear behavior after subtracting the Schottky contribution (in the form $C_p = \gamma T + AT^{-3}$) to the specific heat at low temperatures. A similar approach was successfully applied in the case of $Sr_3Ru_2O_7$ to subtract the Schottky contribution has been reported by *Zhou et al.*, [21]. It is clear that the magnetic field enhances the low-temperature specific heat continuously, indicating the increase of the quasiparticle density-of-states at the Fermi level induced by a magnetic field. A linear extrapolation of the low-T data to zero temperature yields the field dependence of the field-induced contribution (the values of the electronic and phnonic contributions are shown in table 1).

$$\Delta\gamma(H) = [C(T,H) - C(T,0)]/T. \tag{2}$$

The lower inset of Fig. 4 shows $\Delta\gamma(H)$ for $B \parallel c$, clearly indicating a non-linear behavior in magnetic fields. In fact, this non-linear behavior can be fitted by the simple equation $\Delta\gamma(H) = A(\mu_0 H)^{0.5}$ predicted for d-wave symmetry in the clean limit [20]. One can see that the theoretical curve can not describe the whole data. In the case of $Ba(Fe_{1-x}Co_x)_2As_2$ single crystal at (x = 0.045, 0.08, 0.103, and 0.105), the $\gamma(H)$ values of the low-temperature specific heat data can not be described by a single s-wave gap [22]. However, it is difficult to solely judge from specific heat data wether nodes exist or not and it should be noted that other pairing mechanisms can also result in a non-linear behavior of $\gamma(H)$.

4. Conclusions

In summary, we have studied the heat capacity of single crystals of $Ca(Fe_{1-x}Co_x)_2As_2$ ($x = 0$, 0.032, 0.051, 0.056, 0.063, and 0.146). Here, a combined magnetic and structural ordering has been observed for the parent compound evidenced by a first order transition around 168 K. For higher doping levels this transition shifts to lower temperatures, until it is completely suppressed for x = 0.056 and superconductivity appears around 14.1 K. From the low temperature specific heat data of the Co-doped superconducting $Ca(Fe_{0.937}Co_{0.063})_2As_2$ single crystal, the upper critical field has been determined using the WHH model. It is found to be 11.5 T and 19.4 T for $B \parallel c$ and $B \parallel ab$, respectively. A non-linear behavior of the field dependence of $\gamma(H)$ in the low-temperature region after substracting the Schottky contribution to the specific heat.

Acknowledgments

We thank M. Deutschmann, S. Müller-Litvanyi, R. Müller, J. Werner, S. Pichl, K. Leger and S. Gass for technical support. This project was supported by the DFG through BE1749/13.

References

[1] Kamihara Y, Watanabe T, Hirano M and Hosono H 2008 *J. Am. Chem. Soc.* **130** 3296.
[2] Chen G F, Wu D, Li G, Hu W Z, Dong J, Zheng P, Luo J L and Wang N L 2008 *Phys. Rev. Lett.* **100** 247002.
[3] Zhi-An R et al 2008 *Chinese Phys. Lett.* **25** 2215.
[4] Harnagea L et al 2011 *Phys. Rev. B* **83** 094523.
[5] Ni N, Nandi S, Kreyssig A, Goldman A I, Mun E D, Bud'ko S L and Canfield P C 2008 *Phys. Rev. B* **78** 014523.
[6] Rotter M, Tegel M and Johrendt D 2008 *Phys. Rev. Lett.* **101** 107006.
[7] Ni N, Thaler A, Kracher A, Yan J Q, Bud'ko S L and Canfield P C 2009, *Phys. Rev. B* **80**, 024511.
[8] Park T, Park E, Lee H, Klimczuk T, Bauer E D, Ronning F and Thompson J D 2008 *J. Phys.: Condens. Matter* **20** 322204.
[9] Baek S.-H, Lee H, Brown S E, Curro N J, Bauer E D, Ronning F, Park T, and Thompson J D 2009 *Phys. Rev. Lett.* **102** 227601.
[10] Alireza P L et al 2009 *J. Phys.: Condens. Matter* **21**, 012208.
[11] Klingeler R et al 2010 *Phys. Rev. B* **81** 024506.
[12] Kumar N et al 2009 *Phys. Rev. B* **80** 144524.
[13] Pramanik A K, Harnagea L, Singh S, Aswartham S, Behr G, Wurmehl S, Hess C, Klingeler R, and Büchner B 2010 *Phys. Rev. B* **82** 014503.
[14] Baek S.-H, Grafe H.-J, Harnagea L, Singh S, Wurmehl S, Büchner B *Pre-print-Condmat*/1103.0146.
[15] Prokes K, Mat'as S, Harnagea L, Singh S, Wurmehl S, Argyriou D N and Büchner B 2011 *Phys. Rev. B* **83** 104414.
[16] Ronning F, Klimczuk T, Bauer E D, Volz H and Thompson J D 2008 *J. Phys.: Cond. Matter* **20** 322201.
[17] Kumar N, Nagalakshmi R, Kulkarni R, Paulose P L, Nigam A K, Dhar S K and Thamizhavel A 2009 *Phys. Rev. B* **79** 012504.
[18] Woollam J A, Somoano R B and O'Connor P 1964 *Phy. Rev. Lett.* **32** 712.
[19] Werthamer N R, Helfand E and Hohenberg P C 1966 *Phys. Rev.* **147** 295.
[20] Volovik G E 1993 *JETP Lett.* **58** 469.
[21] Zhou Z X, McCall S, Alexander C S, Crow J E and Schlottmann P 2004 *Phys. Rev. B* **69** 140409(R).
[22] Gofryk K, Sefat A S, McGuire M A, Sales B C, Mandrus D, Thompson J D, Bauer E D and Ronning F 2010 *Phys. Rev. B.* **81** 184518.

Composite operator candidates for a study of the *p*-*d* model

A Avella[1,2,3]**, F Mancini**[1,3]**, F P Mancini**[1] **and E Plekhanov**[1,2,3]

[1] Dipartimento di Fisica "E.R. Caianiello", Università degli Studi di Salerno, 84084 Fisciano (SA), Italy

[2] CNR-SPIN, UoS di Salerno, 84084 Fisciano (SA), Italy

[3] Unità CNISM di Salerno, Università degli Studi di Salerno, 84084 Fisciano (SA), Italy

E-mail: avella@physics.unisa.it

Abstract. We study the two-dimensional *p*-*d* model by means of a four-pole approximation within the Composite Operator Method framework. Many results of numerical simulations have been correctly reproduced by considering as basic composite field a four-component spinor field, including the *p* field, the two Hubbard operators for the *d* field, and a composite operator describing the *p* field dressed by the spin excitations of the *d* field. In this manuscript, we propose an alternative choice for the latter field, which takes into account charge and pair fluctuations too. The results of this study show that near half-filling one can safely neglect the contribution of charge and pair fluctuations and safely use the previously proposed field.

1. Introduction

It is believed that real oxide cuprate materials can be adequately described by the two-dimensional *p*-*d* model [1]. The model describes CuO_2 layers, characterized by a strong intra-atomic Coulomb repulsion at Cu-sites and by a very highly correlation between the $3d$ electrons of Cu in the atomic orbital $3d_{x^2-y^2}$ and the $2p$ electrons of O in the atomic orbitals $2p_x$ and $2p_y$. Neglecting the Coulomb repulsion at O-sites and among the *p*- and *d*-electrons, as well as the O-O and Cu-Cu hopping terms, one can use a Bogoliubov transformation, to count out the antibonding degree of freedom of the *p*-electrons [2]. As a result, the Hamiltonian reads as

$$\begin{aligned} H = \sum_{\mathbf{i}} [(\varepsilon_d - \mu) d^\dagger(i) d(i) + (\varepsilon_p - \mu) p^\dagger(i) p(i) + U n_\uparrow^d(i) n_\downarrow^d(i)] \\ + 2t_{pd} \sum_{\mathbf{i}} \left[p^{\gamma\dagger}(i) d(i) + d^\dagger(i) p^\gamma(i) \right] . \end{aligned} \tag{1}$$

We use the Heisenberg representation, $i = (\mathbf{i}, t)$ with $\mathbf{i} = \mathbf{R}_i$ a Bravais vector in the CuO_2 plane pointing at a Cu site. $p(i)$ ($p^\dagger(i)$) and $d(i)$ ($d^\dagger(i)$) are annihilation (creation) operators of *p*- and *d*-electrons in the spinor notation and satisfy canonical anticommutation relations. μ is the chemical potential; ε_d and ε_p are the energy levels; $n_\sigma^d(i) = d_\sigma^\dagger(i) d_\sigma(i)$ is the charge density operator of *d*-electrons with spin σ. U describes the Coulomb repulsion at Cu sites. The short range hopping integral t_{pd} describes the hybridization between d and bonding p orbitals. For a generic operator $\Phi(i)$ we define:

$$\Phi^{\gamma}(i) = \sum_{\mathbf{j}} \gamma_{\mathbf{ij}} \Phi(j), \tag{2}$$

where

$$\gamma_{\mathbf{ij}} = \frac{a^2}{(2\pi)^2} \int_{\Omega_B} d^2\mathbf{k} e^{i\mathbf{k}\cdot(\mathbf{R}_i - \mathbf{R}_j)} \gamma(\mathbf{k}), \tag{3}$$

and

$$\gamma(\mathbf{k}) = \sqrt{1 - \alpha(\mathbf{k})}, \qquad \alpha(\mathbf{k}) = \frac{1}{2}[\cos(k_x a) + \cos(k_y a)]. \tag{4}$$

In the above equations, a is the lattice constant and Ω_B the volume of the unit cell in the reciprocal space. α corresponds to first neighbor projector, whereas γ has a more complex structure in result of the bonding - anti-bonding transformation.

As it has been illustrated in Refs. [3, 4], the study of highly interacting systems can be conveniently tackled by considering higher-order fields instead of the bare electronic operators $\{p, d\}$. These new operators, called composite operators, come out from the dynamics and include higher-order interactions among the original electronic fields. In the framework of the composite operator method [3, 4], one has first to identify the relevant composite fields. On the base of experimental and numerical results, we believe that one should use a four-component spinor field in order to capture sufficient physics to describe real oxide cuprate materials. The large covalence between oxygen and copper electrons leads to a large fluctuation of the energy of p-electrons at O sites, pointing out at the operator $p(i)$ - the bonding component for the p_σ electrons on oxygen site - as the first field. The splitting of the d-band into the lower and the upper Hubbard sub-bands - due to the strong intra-atomic Coulomb interaction at Cu sites - naturally leads to the choice of the Hubbard projection operators $\xi(i) = [1 - n^d(i)]d(i)$ and $\eta(i) = n^d(i)d(i)$.

While the choice of the first three fields is quite intuitive, it has proved a difficult task the right choice of the last field. Indeed, one would like to take into account quantum fluctuations associated with the Cu-O bonds. By taking higher-order time derivatives of the Hubbard fields, one may generate a hierarchy of composite operators. In particular the field π, defined as

$$\pi(i) = \frac{1}{2}\sigma^{\mu} n_{\mu}^{d}(i) p^{\gamma}(i) + \xi(i) p^{\gamma\dagger} \eta(i), \tag{5}$$

appears as a good candidate, since it describes p-electronic excitations accompanied by the nearest neighbor d-electron spin, charge and pair fluctuations. However, in a series of papers [5, 9, 10], we chose the fourth field as

$$p_s(i) = \sigma_k n_k^d(i) p^{\gamma}(i). \tag{6}$$

This field describes p-electronic excitations accompanied by the nearest neighbor d-electron spin fluctuations only. In the above equation, σ_k are the Pauli matrices, n_d is the particle number per site of d electrons, and $n_k^d(i) = d^{\dagger}(i)\sigma_k d(i)$ is the spin density operator of d-electrons. The reason for this choice is based on the experimental evidence that spin fluctuations are predominant over all other fluctuations in the region of physical interest (i.e., near half filling).

Thus, by considering as basic composite field the four-component spinor field:

$$\psi(i) = \begin{pmatrix} p(i) \\ \xi(i) \\ \eta(i) \\ p_s(i) \end{pmatrix} \tag{7}$$

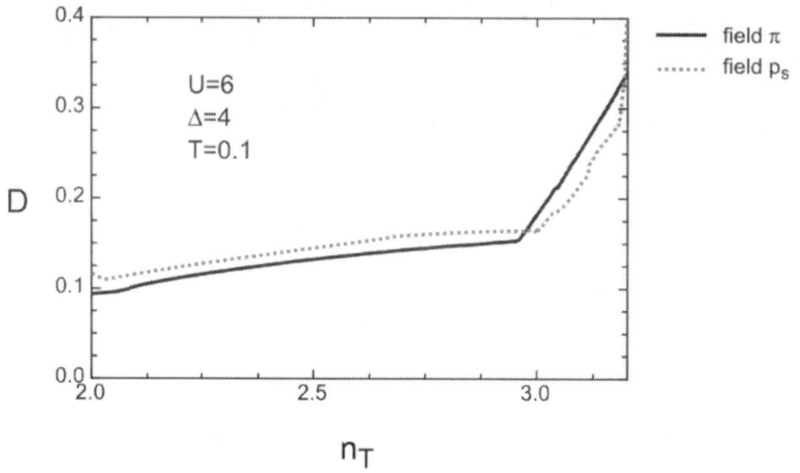

Figure 1. The double occupancy D as a function of n_T, for $U = 6$, $\Delta = 4$, and $T = 0.1$. The solid line is obtained by considering the field $\psi(i)$ as defined in eq. (7). The dotted line obtained by taking $\pi(i)$ instead of p_s.

the Hamiltonian (1) can been solved in the pole approximation [3, 4]. In this approximation, the field (7) is indeed an eigenoperator of the Hamiltonian (1) with eigenenergy $\varepsilon(\mathbf{k})$. As a consequence, a solution of the Hamiltonian can be obtained.

Within this framework, the retarded Green's function and the pertinent correlation functions turn out to be functions of a finite set of parameters, in terms of which one finds a solution of the model. The unknown parameters, which depend on the choice of the basis, can be determined by means of the Pauli principle and/or a decoupling procedure [3, 4]. We refer the interested reader to Ref. [5] for computational details. As it is shown in Ref. [5], this choice of the composite field $\psi(i)$ correctly reproduces many results of numerical simulations [6, 7, 8] and also qualitatively reproduces the band structure exhibited by the undoped single-layer La_2CUO_4 [9].

2. The field π: results and discussion

In this short manuscript, we would like to take a step back and estimate the error done by neglecting the charge and pair fluctuations in the choice of the fourth field. In particular, as fourth field we consider the field $\pi(i)$, as defined in Eq. (5). We recall that $\pi(i)$ is the higher-order non-local field generated by the dynamics of the Hubbard operators $\xi(i)$ and $\eta(i)$. We shall focus our attention on the comparison between the results obtained by the two different choices of the fourth basic field. The differences between the two solutions greatly depend on the value of the total filling n_T In particular, we will show that near half filling ($n_T = 3$) the spin fluctuations prevail over the charge and pair fluctuations, reducing, thus, the difference between the two solutions. In the following, we take $t_{pd} = 1$ as the unit of energy.

In order to test the adequacy of our original choice of the field p_s, we shall consider the double occupancy per site of d electrons D and the chemical potential μ. In Fig. 1, we report, as a function of the total number of electrons n_T, the double occupancy for $\Delta = 4$, $U = 6$, $T = 0.1$. As one can clearly see, the agreement between the two solution is good in the underdoped region, drifting apart in the overdoped region.

In Fig. 2, we show the dependence of n_T on the chemical potential for $U = 6$, $\Delta = 4$, and $T = 0.1$. The best agreement between COM predictions and QMC results [6], is clearly given by considering the field p_s. The comparison between the two choices leads to the conclusion

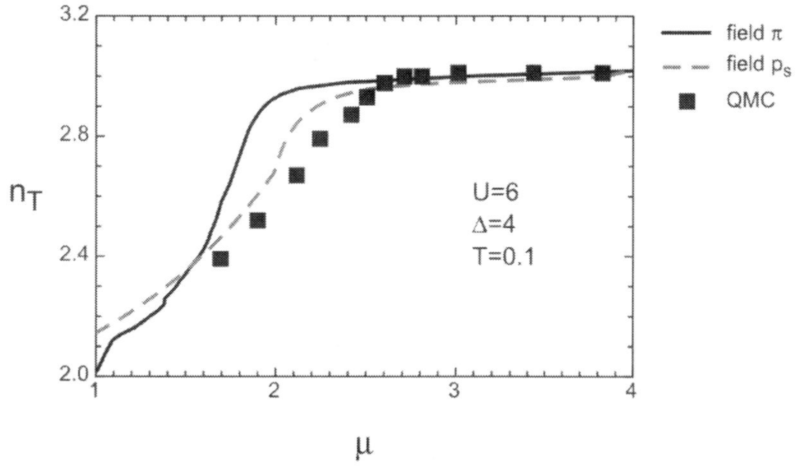

Figure 2. The n_T-μ plot for the values of external parameters $U = 6$, $\Delta = 4$ and $T = 0.1$. The squares are QMC data [6]. The solid line is COM prediction using the field p_s, the dotted line using the field π.

that considering also the charge and pair fluctuations in the definition of the fourth field does not dramatically improve the results obtained by considering only the field p_s, at least near half filling and in the underdoped region.

3. Conclusions

In summary, by means of a four-pole approximation, within the Composite Operator Method, we have studied the *p*-*d* model and, in particular, we focused our attention on the comparison between alternative choices of the basic composite field. We came to the conclusion that taking into account a wider set of fluctuations in an overall approximate framework does not naturally leads to more accurate results.

Acknowledgements

We acknowledge the CINECA award under the ISCRA initiative (project MP34dTMO), for the availability of high performance computing resources and support. A. A. thanks the Max-Planck-Institut FKF Stuttgart for hospitality and financial support.

References

[1] Emery V and Reiter G 1988 Phys. Rev. B **38** 4547.
[2] Matsumoto H and Tachiki M 1990 Suppl. Prog. Theor. Phys. **101** 353.
[3] Mancini F and Avella A 2004 Adv. Phys. **53** 537.
[4] Avella A and Mancini F 2011 *Strongly Correlated Systems: Theoretical Methods (Springer Series in Solid-State Sciences)* vol 171 (Berlin Heidelberg: Springer-Verlag) chap The Composite Operator Method (COM) p 103.
[5] Fiorentino V, Mancini F, Žaşinas E and Barabanov A F 2001 Phys. Rev. B **64** 214515.
[6] Dopf G, Muramatsu A and Hanke W 1990 Phys. Rev. B **41** 9264.
[7] Scalettar R T, Scalapino D J, Sugar R L and White S R 1991 Phys. Rev. B **44** 770.
[8] Dopf G *et al.* 1992 Phys. Rev. Lett. **68** 2082.
[9] Avella A, Mancini F, Mancini F P and Plekhanov E 2011 J. Phys. Chem. Sol. **72** 384.
[10] Avella A, Mancini F, Mancini F P and Plekhanov E 2011 J. Phys.: Conf. Series **273** 012091.

Journal of Physics: Conference Series **391** (2012) 012122 doi:10.1088/1742-6596/391/1/012122

High-resolution thermal expansion of isovalently substituted $BaFe_2(As_{1-x}P_x)_2$

A E Böhmer[1,2], P Burger[1,2], F Hardy[1], T Wolf[1], P Schweiss[1], R Fromknecht[1], H von Löhneysen[1,2], C Meingast[1], S Kasahara[3], T Terashima[3], T Shibauchi[4] and Y Matsuda[4]

[1] Karlsruhe Institute of Technology, IFP, 76021 Karlsruhe, Germany
[2] Karlsruhe Institute of Technology, Fakultät für Physik, 76131 Karlsruhe, Germany
[3] Research Center for Low Temperature and Materials Sciences, Kyoto University, Kyoto 606-8501, Japan
[4] Department of Physics, Kyoto University, Kyoto 606-8502, Japan

E-mail: anna.boehmer@kit.edu

Abstract. We have investigated the isovalently substituted system $BaFe_2(As_{1-x}P_x)_2$ by high-resolution thermal expansion using a home-built capacitive dilatometer. Accurate measurements succeeded despite the very small size of the available single crystals ($\sim 500 \times 500 \times 100\mu m^3$). Information on the uniaxial pressure derivatives of the transition temperatures is obtained using thermodynamic relations. In-plane and out-of-plane pressure derivatives have opposite sign, which demonstrates the sensitivity of the compound to uniaxial pressure. The structural and the superconducting transition always respond oppositely to uniaxial pressure, which signals their coupling and competition.

1. Introduction

In the intensively studied 122 family of iron-based superconductors, superconductivity can be induced by various substitutions, both charged [1, 2] and isovalent [3, 4] ("chemical pressure"), and by hydrostatic pressure [5]. Uniaxial pressure effects are expected to play a key role in these compounds due to their anisotropic crystal structures. Indeed, the uniaxial pressure derivatives of T_c were found to be large and anisotropic in $Ba(Fe_{1-x}Co_x)_2As_2$ by thermal expansion [6, 7]. Uniaxial pressure components are also expected to be important in the parent compound $BaFe_2As_2$ [8]. Recently, small high-quality single crystals of $BaFe_2(As_{1-x}P_x)_2$ have been grown [9]. Here, we study uniaxial pressure effects in $BaFe_2(As_{1-x}P_x)_2$ single crystals using high-resolution thermal-expansion measurements. Precise data could be obtained in spite of the extremely small (for dilatometry) crystal sizes ($\sim 500 \times 500 \times 100\,\mu m^3$). We discuss the uniaxial pressure derivatives of the transition temperatures as a function of substitution. Their pronounced anisotropy underlines the importance of hydrostaticity when performing high-pressure experiments. Our results are consistent with a coupling and competition between the orthorhombic SDW phase and superconductivity.

2. Experimental

High-quality single crystals of $BaFe_2(As_{1-x}P_x)_2$ were grown from stoichiometric mixtures of the components as described in Ref. [9]. The sample composition was determined by refinement of 4-

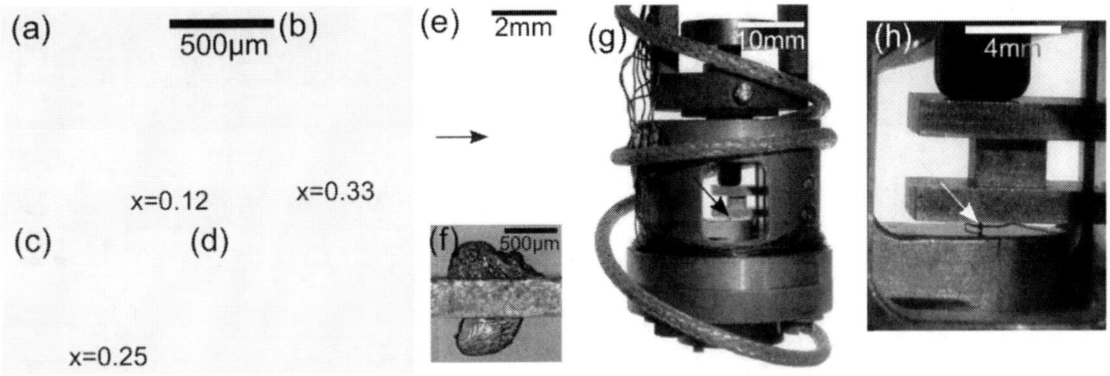

Figure 1. (a)-(c) Photographs of the single-crystalline $BaFe_2(As_{1-x}P_x)_2$ samples. (d) The sample of (c) glued to a thin wire. Panels (a)-(d) are on the same scale. (e)-(f) Top view and side view of the sample of (a) fixed into the sample holding spring. (g) Global view of the dilatometer with the capacitor at the bottom and the screw used to press a sample against the capacitor at the top. (h) Magnified view of the central part of (g). Samples are indicated by arrows. Note their tiny size when compared to the setup.

circle x-ray diffraction data. Uniaxial thermal-expansion coefficients were measured with a home-built capacitive dilatometer with a resolution of $0.1 - 0.01$ Å. In this setup, the sample is pressed against one plate of a plate-type capacitor with a force of ~ 0.2 N. When samples are inserted such that this pressure is directed along their tetragonal [110] direction, they will be detwinned [10] and thermal expansion along the shorter orthorhombic b-axis will be measured. Comparing this to the twinned case (when the dilatometer pressure is oriented along the tetragonal [100] direction) allows to estimate the response of the orthorhombic a-axis as well.

It is a special challenge to insert the sub-millimeter sized samples into the dilatometer for in-plane measurements and to control accurately the sample orientation, which is important because of the (de)twinning. Fig. 1 illustrates the setup and sample mounting. Images of the measured samples are shown in panels (a)-(c). In order to insert the samples into the dilatometer, they are either fixed to a sample-holding spring (e,f) or to a thin wire (d). Using these aids, the samples can be inserted and their orientation inside the dilatometer can be controlled. Panel (h) shows a sample inserted on a wire as an example. Note the small dimension of the sample compared to the dilatometer. Nevertheless, high-resolution thermal-expansion data could still be obtained.

3. Results and Discussion

Fig. 2 presents the measured uniaxial thermal-expansion coefficients $\alpha_i = (1/L_i)\, dL_i/dT$ (L is the sample length and "i" stands for the direction) of $BaFe_2(As_{1-x}P_x)_2$ (x = 0, 0.12, 0.25, 0.33). Thanks to the careful sample mounting, thermal expansion could be measured along all three independent crystallographic directions: along the c-axis, along the tetragonal [100] direction (twinned, a- and b-axis) and along [110] (detwinned, orthorhombic b-axis). The strong enhancement of the anomalies in α along the [110] direction demonstrates successful detwinning. The magneto-structural transition of the undoped compound [11] manifests as a sharp peak of the α_i's at T_{sm} = 135 K. T_{sm} decreases upon P substitution and it is found to split into two transitions for $x = 0.25$. By comparing measurements in the twinned and the detwinned configuration, we estimate that the structural transition is at $T_s = 56$ K and the SDW transition at $T_{SDW} = 49$ K.

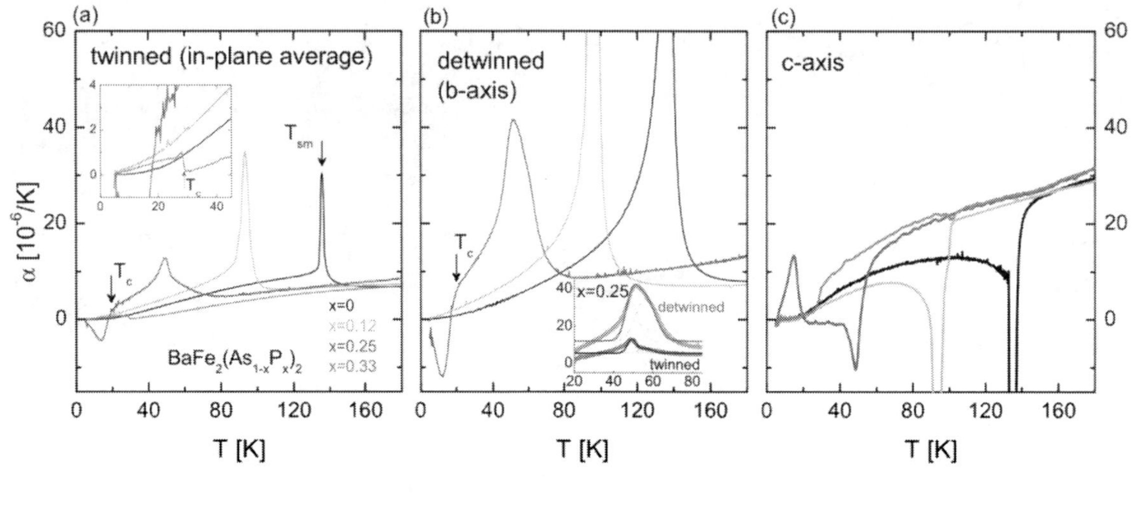

Figure 2. Measured uniaxial thermal-expansion coefficients of $BaFe_2(As_{1-x}P_x)_2$. (a) Twinned in-plane measurements that give an average of the a- and the b-axis. The inset shows an expanded view of the low-temperature region. (b) Measurements in the detwinned configuration for the (underdoped) samples that undergo the tetragonal-to-orthorhombic phase transition. The inset shows how the magneto-structural transition of the $x = 0.25$ sample can be decomposed into two transitions. (c) Measured thermal-expansion coefficients along the c-axis.

The samples with $x = 0.25$ and $x = 0.33$ show superconductivity below $T_c = 20$ K and $T_c = 30$ K, respectively. The $x = 0.25$ sample is underdoped and undergoes the superconducting transition below the structural and the magnetic transitions. For this underdoped sample, T_c manifests itself as a kink in the α_i's, and they reach a minimum (or maximum) somewhat below T_c. The superconducting anomaly changes its shape drastically on the overdoped side ($x = 0.33$). It is now step-like and of opposite sign as compared to the underdoped sample.

In general, thermal expansion allows to compute uniaxial pressure derivatives of transition temperatures dT_c/dp_i via the Ehrenfest relation $\frac{dT_c}{dp_i} = V_m \frac{\Delta\alpha_i}{\Delta C_p/T_c}$. Here, $\Delta\alpha_i$ and ΔC_p are the discontinuities in α and the specific heat C_p, respectively and V_m is the molar volume. ΔC_p is always positive. Hence, even in the absence of specific-heat data, the sign and the relative size of the anomalies in the α_i provide interesting information about the anisotropy of the dT_c/dp_i.

Due to the very small difference between T_s and T_{SDW} only a combined magneto-structural transition at T_{sm} will be considered. The analysis further assumes that the samples remain single-phase at all temperatures even if we cannot exclude a phase separation below T_c. It is readily seen from the in-plane measurements that T_{sm} would increase with an applied in-plane pressure. T_c on the underdoped (overdoped) side of the phase diagram would decrease (increase). The uniaxial pressure derivatives along the c-axis have opposite sign in all cases. Further, the effects cancel largely in volume. Interestingly, the anomalies at T_c and T_{sm} always have the same sign in $Ba(Fe_{1-x}Co_x)_2As_2$ [6, 7, 12] as in $BaFe_2(As_{1-x}P_x)_2$. Within the tetragonal phase (overdoped side) T_c may be enhanced by increasing the separation between FeAs planes and by reducing their area. Consistently, the ratio c/a increases as one approaches optimal doping from the overdoped side [9].

Comparison of measurements in the twinned (along [100]) and detwinned (along [110]) configurations further reveals a large ab-plane anisotropy. All the anomalies are more than twice larger in the detwinned configuration as compared to the twinned one. This means that a- and b-axis pressure derivatives are of opposite sign. The statement remains true even when the detwinning is incomplete or when "twinned" samples are partially detwinned. On the underdoped side, it is seen that T_c and T_{sm}, or the mechanisms that are at the origin of the respective phase transitions, are coupled and competing. When the structural distortion is favored, superconductivity is hindered and vice versa. For example, the structural transition is favored (disfavored) by b-axis (a-axis) pressure which also decreases (enhances) T_c. The results may also be summarized by identifying uniaxial pressure with a shift in the phase diagram: stress along the c-axis (and a-axis on the underdoped side) mimics a higher P-content while stress along the b-axis (in-plane axis on the overdoped side) mimics a lower P-content. P substitution and (uniaxial) pressure are thus closely linked, as has been found previously in phosphorus- [13, 14] and also cobalt- [15] doped $BaFe_2As_2$.

4. Summary

We succeeded in measuring the uniaxial thermal-expansion coefficients of very small single crystals of $BaFe_2(As_{1-x}P_x)_2$ for a number of substitution levels. Samples with $x = 0$, 0.12 and 0.25 undergo a magneto-structural transition. There is evidence for its splitting at the P content $x = 0.25$. Samples with $x = 0.25$ and $x = 0.33$ are superconducting. A pronounced anisotropy of the uniaxial pressure derivatives of all transition temperatures is found. In-plane and c-axis pressure derivatives are of opposite sign, as are a- and b-axis pressure derivatives. Further, the dT_c/dp_i's change sign close to optimal doping. Derivatives of the magneto-structural and superconducting transition temperatures, dT_{sm}/dp_i and dT_c/dp_i, are of opposite sign for all directions i which is consistent with a competition between the orthorhombic SDW phase and superconductivity.

Acknowledgments

This work is supported by the Deutsche Forschungsgemeinschaft through the priority program SPP1458.

References

[1] Sefat A S, Jin R, McGuire M A, Sales B C, Singh D J and Mandrus D 2008 *Phys. Rev. Lett.* **101** 117004
[2] Rotter M, Tegel M, Schellenberg I, Schappacher F M, Pöttgen R, Deisenhofer J, Günther A, Schrettle F, Loidl A and Johrendt D 2009 *New J. Phys.* **11** 025014
[3] Jiang S, Xing H, Xuan G, Wang C, Ren Z, Feng C, Dai J, Xu Z and Cao G 2008 *J. Phys.: Condens. Matter* **21** 382203
[4] Sharma S, Bharathia A, Chandra S, Reddy V R, Paulraj S, Satya A T, Sastry V S, Gupta A and Sundar C S 2010 Phys. Rev. B **81** 174512
[5] Alireza P L, Ko Y T C, Gillett J, Petrone C M, Cole J M, Lonzarich G G and Sebastian S E 2009 *J. Phys.: Condens. Matter* **21** 012208
[6] Hardy F, Adelmann P, Wolf T, v Löhneysen H and Meingast C 2009 *Phys. Rev. Lett.* **102** 187004
[7] Bud'ko S L, Ni N, Nandi S, Schmiedeshoff G M and Canfield P C 2009 *Phys. Rev.* B **79** 054525
[8] Yamazaki T, Takeshita N, Kobayashi R, Fukazawa H, Kohori Y, Kihou K, Lee C H, Kito H, Iyo A and Eisaki H 2010 *Phys. Rev.* B **81** 224511
[9] Kasahara S, Shibauchi T, Hashimoto K, Ikada K, Tonegawa S, Okazaki R, Shishido H, Ikeda H, Takeya H, Hirata K, Terashima T and Matsuda Y 2010 *Phys. Rev.* B **81** 184519
[10] Fisher I R, Degiorgi L and Shen Z X 2011 *Rep. Prog. Phys.* **74** 124506
[11] Rotter M, Tegel M, Johrendt D, Schellenberg I, Hermes W and Pöttgen R 2008 *Phys. Rev.* B **78** 020503
[12] Meingast C (unpublished)
[13] Goh S K, Nakai Y, Ishida K, Klintberg L E, Ihara Y, Kasahara S, Shibauchi T, Matsuda Y and Terashima T 2010 *Phys. Rev.* B **82** 094502
[14] Klintberg L E, Goh S K, Kasahara S, Nakai Y, Ishida K, Sutherland M, Shibauchi T, Matsuda Y and Terashima T 2010 *J. Phys. Soc. Jpn.* **79** 123706
[15] Drotziger S, Schweiss P, Grube K, Wolf T, Adelmann P, Meingast C and v Löhneysen H 2010 *J. Phys. Soc. Jpn.* **79** 124705

Coexistence of superconductivity and ferromagnetism in iron pnictides

Guang-Han Cao, Wen-He Jiao, Yong-Kang Luo, Zhi Ren, Shuai Jiang and Zhu-An Xu

Department of Physics and Center for Correlated Matter, Zhejiang University, Hangzhou 310027, China

E-mail: ghcao@zju.edu.cn

Abstract. Superconductivity (SC) and long-range ferromagnetism (FM) are mutually antagonistic, thus in general SC does not coexist with FM. In this mini-review, however, we will show that such a coexistence can inhabit several iron pnictide systems, including $EuFe_2(As_{1-x}P_x)_2$, $Eu(Fe_{1-x}Co_x)_2As_2$, $Eu(Fe_{1-x}Ru_x)_2As_2$, $CeFe(As_{1-x}P_x)O_{0.95}F_{0.05}$ and $Sr_2VFeAsO_3$. We will briefly discuss the possible reasons and consequences of the coexistence of 3d-electron SC and 4f-electron FM.

1. Introduction

Superconductivity (SC) and ferromagnetism are two intriguing collective phenomena in condensed matter physics. Conventional SC is characterized by spin-singlet s-wave Cooper pairing, in which time-reversal symmetry is conserved. However, ferromagnetism breaks time-reversal symmetry. This fundamental difference makes the two phenomena antagonistic and competitive. One may easily see this point by the fact that all the magnetic elements do not superconduct. Ginzburg[1] first pointed out that SC can be destroyed in external magnetic field by electromagnetic interaction. The incompatible nature of SC and FM was demonstrated by the experiments followed, which showed the competition of the two collective phenomena in (La,Gd) and (Ce,Pr)Ru_2 solid solutions.[2] The suppression of FM in the superconducting regime was explained by Anderson and Suhl[3] in terms of the Ruderman-Kittel-Kasuya-Yosida (RKKY) interactions by the end of 1950s. Judging from the energy scale, however, FM wins SC in most cases. Thus it was suggested that in the superconducting state pure ferromagnetism should be modified in the form of cryptoferromagnetic alignment for localized spins. It was not until late 1970s that coexistence of SC and FM was evidenced in $ErRh_4B_4$[4] and $Ho_{1.2}Mo_6S_8$[5], in narrow regimes of temperature and external field. In late 1990s, SC and weak ferromagnetism were observed in high-T_c rutheno-cuprates[6, 7]. In the examples above, SC and FM originate from obviously different electrons of different elements. However, there is a scenario that both SC and FM come from the same type of electrons, as in UGe_2[8] and URhGe,[9] where the SC emerges under the ferromagnetic background ($T_c < T_M$). It is widely believed that the spin structure of the Cooper pairs is in triplet states.

The recently discovered Fe-based superconductors[10] are characterized by multiband superconductivity as well as high transition temperatures. This feature makes it possible to see new phenomena including those on the interplay of SC and FM. The outstanding example comes

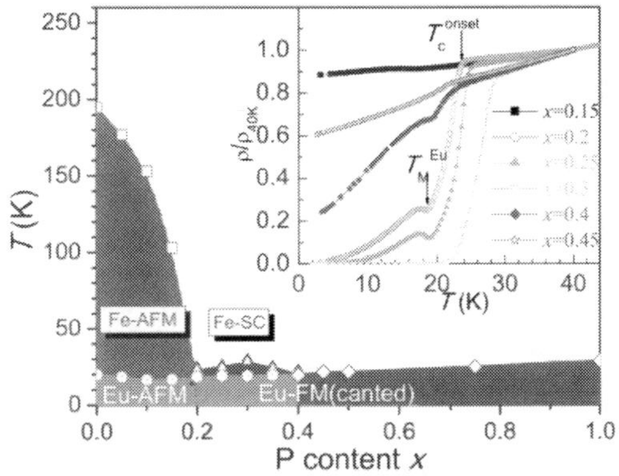

Figure 1. Superconducting and magnetic phase diagram of $EuFe_2(As_{1-x}P_x)_2$. The inset shows the temperature dependence of resistivity for $0.15 \leq x \leq 0.45$ (adapted from Ref. [29]).

from the Eu-based iron pnictides, $EuFe_2As_2$-related systems, in which the Eu^{2+} ions show large local magnetic moments with $J = S = 7/2$. By the chemical doping either at the Fe-site or at the As-site, coexistence of SC and FM (or with significant ferromagnetic components) has been observed[11, 12, 13] and confirmed by various kind of method[14, 15, 16, 17, 18, 19]. Evidence of additional coexistence of SC and FM was also observed in other iron pnictide systems, e.g., $SrFe_2As_2$ with internal strain[20] $CeFe(As_{1-x}P_x)O_{0.95}F_{0.05}$[21] and $Sr_2VFeAsO_3$[22], which suggests common mechanism that supports the coexistence.

2. $EuFe_2(As_{1-x}P_x)_2$ system

The two end members in $EuFe_2(As_{1-x}P_x)_2$ system crystallize in the same $ThCr_2Si_2$-type structure with space group $I4/mmm$, however, they show contrasting physical properties. It was confirmed that $EuFe_2As_2$ underwent two magnetic transitions at ~200 K and 20 K, respectively.[23] We first recognized the 200-K transition as a spin-density-wave (SDW) ordering associated with the iron sublattice, upon which superconductivity was predicted to realize through appropriate chemical doping.[23] Furthermore, we proposed an A-type antiferromagnetic (AFM) order for the Eu sublattice,[23, 24] which was then confirmed by magnetic resonant x-ray scattering[25] and neutron diffraction[26] experiments. On the other hand, the iron in $EuFe_2P_2$ is non-magnetic, and the Eu^{2+} moments order ferromagnetically at 29 K with the spin direction tilted toward the c-axis.[27, 28]

Although neither $EuFe_2As_2$ nor $EuFe_2P_2$ superconduct, their solid solutions with certain range of P concentration do. For instance, in $EuFe_2(As_{0.7}P_{0.3})_2$, SC transition occurs at T_c=26 K (see the inset of figure 1). Following the superconducting transition, surprisingly, the Eu sublattice becomes ferromagnetically ordered below 20 K.[11] Bulk nature for both SC and FM was confirmed by the heat capacity measurement, which demonstrates unambiguously the coexistence of SC and FM to the microscopic scale. The recent Mössbauer study[14, 19] shows the canting of the Eu spins toward the c-axis with the phosphorus doping. The spin canting angle and the ^{151}Eu hyperfine field "saturate" at x=0.2, suggesting ferromagnetism for the Eu sublattice. The spin canting at lower doping level (x = 0.12) was also evidenced by the anisotropic magnetization,[18] indicating that there is ferromagnetic component along the c-axis.

Figure 1 shows the superconducting and magnetic phase diagram in $EuFe_2(As_{1-x}P_x)_2$.[29] With the P doping, the Fe-AFM is suppressed rapidly, and then SC emerges for $x \geq 0.2$. This phenomenon is similar to that in the $BaFe_2(As_{1-x}P_x)_2$ system.[30] The difference is that the former has narrower superconducting window. Compared with the Fe sublattice, the magnetic transition temperature of the Eu sublattice changes very mildly. However, it changes from AFM to FM, accompanied with the tilting of Eu spin direction. SC coexists with FM for $0.2 \leq x \leq 0.4$. A recent paper based on single crystal investigations shows a similar phase diagram, however, the SC regime varies a little.[17] Despite of the discrepancy, it is clear that in the superconducting region the Eu sublattice at least has spontaneous ferromagnetic components because of the Eu^{2+} spins tilted along the same direction.

3. $Eu(Fe_{1-x}M_x)_2As_2$ (M=Co and Ru) system

As is known, Fe-site doping may also introduce SC. By doping with Ni, we observed ferromagnetism for the Eu^{2+} moments, though SC transition was not able to be seen.[31] However, evidence of SC was given for the Co doping.[12] As can be seen in figure 2, the resistivity of $Eu(Fe_{0.89}Co_{0.11})_2As_2$ crystals drops at ~21 K, suggestive of a SC transition. The corresponding SC diamagnetism can be detected only when the applied field is very small (see the inset of figure 2). There is a clear resistivity reentrance around 17 K, which was confirmed to be due to the magnetic ordering of Eu spins.

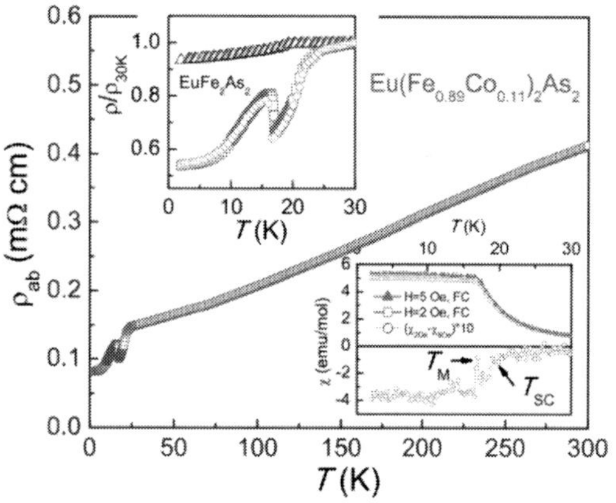

Figure 2. (Adapted from Ref. [12]) Resistivity and magnetic susceptibility (lower inset) for $Eu(Fe_{0.89}Co_{0.11})_2As_2$.

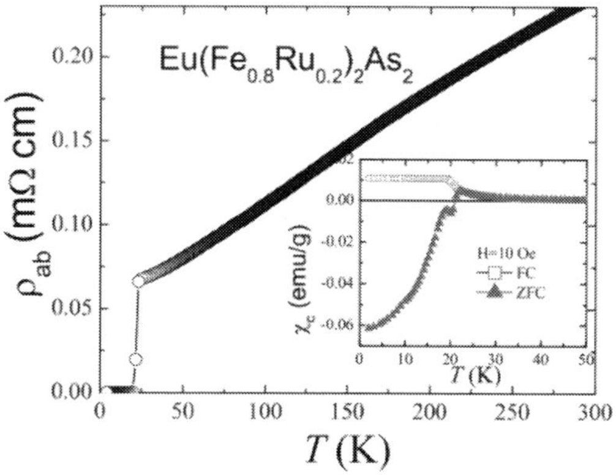

Figure 3. Temperature dependence of resistivity and magnetic susceptibility (inset) for $Eu(Fe_{0.8}Ru_{0.2})_2As_2$ crystals.

The Ru doping in $EuFe_2As_2$ gives stronger signal for SC, as shown in figure 3. Not only does the resistivity show sharp superconducting transition to zero, but also the magnetic susceptibility has strong diamagnetic signal for the zero-field-cooling mode (the step-like anomaly is due to the magnetic ordering of Eu spins). When changing the Ru content, interesting anisotropic SC was observed, which could be the result from either spontaneous vortices or a novel FFLO state.[13]

4. Discussion

Apart from the Eu-containing iron pnictide system, there are some other cases showing the evidence of coexistence of SC and FM. For example, In $CeFe(As_{1-x}P_x)O_{0.95}F_{0.05}$ system, a ferromagnetic transition (associated with Ce $4f$ electrons) occurs at ~5 K, meanwhile, SC transition happens at 13 K for x=0.5.[21] Another example is seen in $Sr_2VFeAsO_3$, where V $3d$ weak FM takes place at 55 K, followed by a SC transition at 24 K.[22] These facts suggest a common mechanism supporting the coexistence of SC and FM in iron pnictides.

Here we present a simple explanation. Since the superconducting upper critical fields H_{c2} in Fe-based superconductors are very high (*e.g.*, ~60 T for $BaFe_2As_2$[32]), which are probably higher than the exchange fields among the Eu spins, thus SC may survive in the presence of a strong ferromagnetic internal fields. On the other hand, the multiband feature of Fe $3d$ electrons allows the effective RKKY interactions among the Eu^{2+} spins even in the superconducting state. It is widely believed that d_{yz} and d_{zx} orbitals are most probably related to SC, and the d_{z^2} electrons are supposed to be responsible for mediating RKKY interactions. Therefore, both SC and FM could be satisfied simultaneously. It is our point that the multiband effect plays an important role to facilitate the coexistence of SC and FM.

Finally we would like to talk about the possible consequences of the coexistence. Just like the modulation of FM by SC, proposed by Anderson and Suhl[3], SC should also be modified by the internal exchange fields. Spontaneous vortex state and FFLO phase are the possible candidates.[13] In most cases the Fe-based SCs have the following characteristic: 1)pauli-limiting H_{c2} and large Maki parameters; 2)multi-Fermi pockets with small E_F (or E_F can be tuned to a small value); 3) quasi-two dimensionality. All these factors favor for the formation of FFLO state. Thus we expect a novel FFLO state at zero external fields in the SC/FM coexisted iron pnictide system.

Acknowledgments

This work is supported by the NSF of China (No. 10934005) and National Basic Research Program of China (Nos. 2010CB923003 and 2011CBA001103).

References

[1] V. L. Ginzburg 1956 *Zh. Eksp. Teor. Fiz.* **31** 202
[2] Matthais B T, Suhl H and Corenzwit E 1958 *Phys. Rev. Lett.* **1** 92
ibid 1958 *Phys. Rev. Lett.* **1** 449
[3] Anderson P W and Suhl H 1959 *Phys. Rev.* **116**, 898
[4] Fertig W A, Johnston D C, DeLong L E *et al* 1977 *Phys. Rev. Lett.* **38** 987
[5] Ishikawa M and Fischer O 1977 *Solid State Commun.* **23** 37
[6] Felner I, Asaf U, Levi Y and Millo O 1997 *Phys. Rev. B* **55** R3374
[7] Bernhard C, Tallon J L, Niedermayer Ch *et al* 1999 *Phys. Rev. B* **59** 14099
[8] Saxena S S, Agarwal P, Ahilan K *et al* 2000 *Nature* **406** 587
[9] Aoki D, Huxley A, Ressouche E *et al* 2000 *Nature* **413** 613
[10] Kamihara Y, Watanabe T, Hirano M and Hosono H 2008 *J. Am. Chem. Soc.* **130** 3296
[11] Ren Z, Tao Q, Shuai J, Feng C M, Wang C, Dai J H, Cao G H and Xu Z A 2009 *Phys. Rev. Lett.* **102** 137002
[12] Jiang S, Xing H, Xuan G F, Ren Z, Wang C, Xu Z A and Cao G H 2009 *Phys. Rev. B* **80** 184514
[13] Jiao W H, Tao Q, Bao J K, Sun Y L, Feng C M, Xu Z A, Nowik I, Felner I and Cao G H 2011 *EPL* **95** 67007
[14] Nowik I, Felner I, Ren Z, Cao G H and Xu Z A 2011 *J. Phys. Condens. Matter* **23** 065701
[15] Ahmed A, Itou M, Xu S G *et al* 2010 *Phys. Rev. Lett.* **105** 207003
[16] Wu D, Chanda G, Jeevan H S, Gegenwart P and Dressel M 2011 *Phys. Rev. B* **83** 100503(R)
[17] Jeevan H S, Kasinathen D, Rosner H and Gegenwart P 2011 *Phys. Rev. B* **83** 054511
[18] Zapf S, Wu D, Bogani L, Dressel M, Jeevan H S and Gegenwart P 2011 *Phys. Rev. B* **84** 140503(R)
[19] Munevar J, Micklitz H, Alzamora M *et al* 2011 *ArXiv:* 1105.1201
[20] Saha S R, Butch N P, Kirshenbaum K, Paglione J and Zavalij P Y 2009 *Phys. Rev. Lett.* **103** 037005
[21] Luo Y K, Han H, Jiang S *et al* 2011 *Phys. Rev. B* **83** 054501
[22] Cao G H, Ma Z F, Wang C *et al* 2010 *Phys. Rev. B* **82** 104518
[23] Ren Z, Zhu Z W, Jiang S *et al* 2008 *Phys. Rev. B* **78** 052501
[24] Jiang S, Luo Y K, Ren Z *et al* 2009 *New J. Phys.* **11** 025007
[25] Herrero-Martin J, Scagnoli V, Mazzoli C *et al* 2009 *Phys. Rev. B* **80** 134411
[26] Xiao Y, Su Y, Meven M *et al* 2009 *Phys. Rev. B* **80** 1742424
[27] Feng C, Ren Z, Xu S *et al* 2010 *Phys. Rev. B* **82** 094426
[28] Ryan D H, Cadogan J M, Xu S G, Xu Z A and Cao G H 2011 *Phys. Rev. B* **83** 132403
[29] Cao G H, Xu S G, Ren Z, Jiang S, Feng C M and Xu Z A 2011 *J. Phys. Condens. Matter* **23** 464204
[30] Jiang S, Xing H, Xuan G *et al* 2009 *J. Phys. Condens. Matter* **21** 382203
[31] Ren Z, Lin X, Tao Q, Jiang S, Zhu Z W, Wang C, Cao G H and Xu Z A 2009 *Phys. Rev. B* **79** 094426
[32] Yuan H Q, Singleton J, Balakirev F F *et al* 2009 *Nature* **457** 565

Paramagnetic Meissner effect at high fields in YCaBaCuO single crystal

F T Dias[1], V. N. Vieira[1], A. L. Falck[1], D. L. da Silva[1], P Pureur[2] and J Schaf[2]

[1]Instituto de Física e Matemática, Universidade Federal de Pelotas, RS, Brazil

[2]Instituto de Física, Universidade Federal do Rio Grande do Sul, Porto Alegre, RS, Brazil

e-mail: fabio.dias@ufpel.edu.br (Fábio Dias)

Abstract. We report on systematic magnetization experiments in an $Y_{1-x}Ca_xBa_2Cu_3O_{7-\delta}$ (x = 0.25 at%) single crystal. The magnetization experiments were made using a superconducting quantum interference device magnetometer (SQUID). Magnetic moments were measured as functions of the temperature according to the zero-field cooling (ZFC), field-cooled cooling (FCC), and field-cooled warming (FCW) prescriptions. The time-dependence of the FC magnetization at fixed magnetic fields was studied. Magnetic fields up to 50 kOe were applied and a paramagnetic response related to the superconducting state was observed when strong enough fields were applied parallel to the c axis. The magnitude of the high field paramagnetic moment (HFPME) increases when the field is augmented. The effect shows strong and anomalous time dependence, such that the paramagnetic moment increases as a function of the time. An $YBa_2Cu_3O_{7-\delta}$ single crystal exhibiting the same effect was used for comparison. We discuss our results in terms of the flux compression scenario into the sample modulated by Ca concentration.

1. Introduction

The Meissner paramagnetic effect (PME) is characterized by a paramagnetic response of the field cooled (FC) magnetization when a superconductor sample is field cooled through its critical temperature, T_c [1-7]. In the literate [1-7], the PME is distinguished between: the low-field PME (LFPME) [1-4] and the high-field PME (HFPME) [5-7]. The LFPME usually has its paramagnetic response gradually increased when $H \leq 10$ Oe are applied. On the other hand, when $H > 10$ Oe are applied the LFPME is rapidly suppressed and conventional Meissner magnetization response is observed. [1-4] There are successful theories explanations for the LFPME such as randomly oriented π junctions [8], flux compression [9] and the giant vortex state [10]. The HFPME shows some noticeable differences when compared to the LFPME [. For instance, the magnitude of the HFPME moment increases when strong magnetic fields ($H \geq 1$ kOe) are applied [5-7], its FC magnetization time-dependence is strong and anomalous. On the other hand, the theories approaches applied to the explanation of the LFPME are unsuccessful to explain the HFPME [5-7]. According to some authors [5-7] a detailed description of the relation between HFPME and pinning mechanisms still lacking since the existing flux-compressed scenarios do not take into consideration the crucial role of the pinning mechanism, nor the importance of vortex dynamics.

Motivated by the last considerations we report on field and time dependence of the FC magnetizations in a $YBa_2Cu_3O_{7-\delta}$ [YBaCuO] and a $Y_{1-x}Ca_xBa_2Cu_3O_{7-\delta}$ (x = 0.0025) [(YCa)BaCuO] single crystals with the proposal of study the role of the chemically introduced pinning mechanisms on the HFPME behavior contrasting these results with a no doped sample.

2. Experimental techniques and method of analysis

The single crystal samples of the studied systems were grown by the self-flux method [11]. The crystals were analyzed by X-ray diffraction [11]. The polarized light microscopy specifies that our single crystals are heavily twinned [11].

DC magnetization measurements were performed with a Quantum Design MPMS-XL SQUID magnetometer in DC mode. The SQUID signal and the sample centering were visually monitored during the experimental run. Magnetic fields from 1 to 50 kOe were applied perpendicular to c axis of the single crystals. Magnetic moments were measured as function of the temperature according to the zero field cooling (ZFC), field cooled cooling (FCC), and field-cooled warming (FCW) protocols. The time dependence of the FCC moment at a fixed temperature was studied up to a time of the 50.000s. All results were corrected for the corresponding demagnetization effect and sample holder signal contributions.

3. Results and discussion

Figure 1 displays the temperature evolution of the FCC (closed symbols) and the FCW (open symbols) DC magnetizations for YBaCuO and (YCa)BaCuO single crystals as function of applied field.

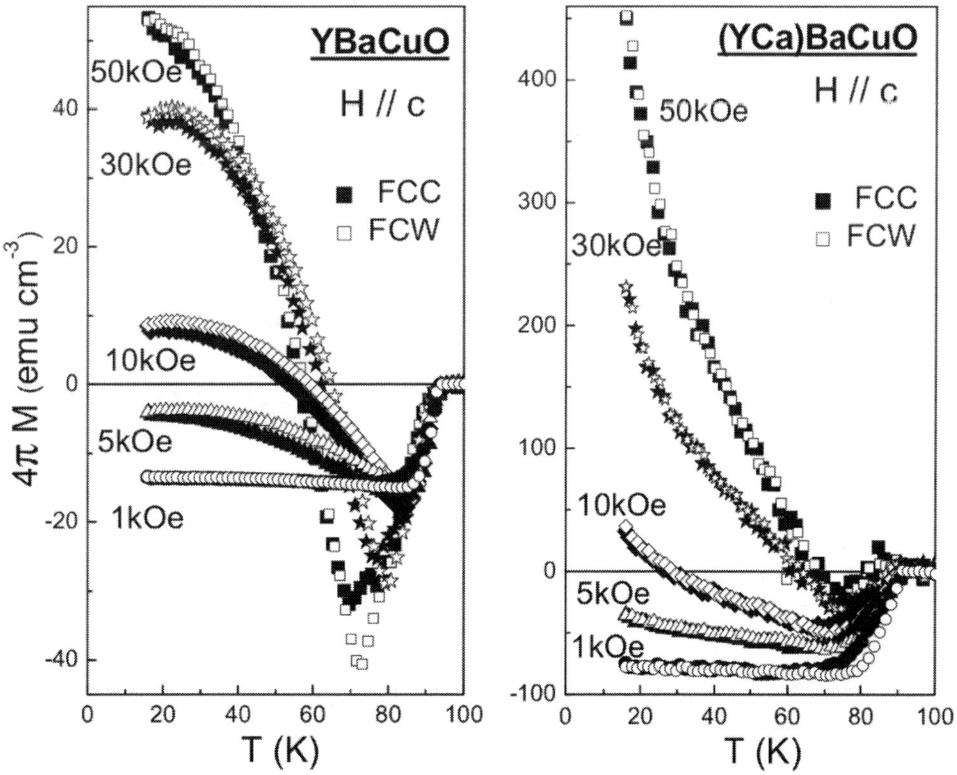

Figure 1. FCC and FCW magnetizations for our $YBa_2Cu_3O_{7-\delta}$ (left panel) and $Y_{1-x}Ca_xBa_2Cu_3O_{7-\delta}$ (x = 0.0025) (right panel) samples to the *H* // c. Measurements were performed using a cooling and a warming rate of the 3 K/min.

In both samples, for $H < 1$ kOe, not showed in figure 1, the LFPME is absent and FCC and FCW magnetizations decrease in the whole temperature range below T_c showing the usual Meissner behavior. However, when H ≥ 1 kOe the FCC and FCW magnetizations exhibit a diamagnetic "dip" in the temperature, T_d that became more pronounced in YBaCuO sample as H is augmented. On the other hand, for $T < T_d$ the FCC and FCW magnetic moments increase steadily as temperature is lowered and attain a positive value which essentially exceeds the magnetization values in the normal state when H ≥ 10 kOe are applied. This behavior is the signature of the HFPME and was observed in $YBa_2Cu_3O_{7-\delta}$ single crystals [6] and melted textured samples [7]. It is important to note that the FCC and FCW magnetizations for our samples show a weak irreversibility as well as the HFPME magnitude of the Ca doped sample shows no tendency to saturate as temperature decreases towards to zero and improves the magnitude of this effect as compared to the YBaCuO sample [6,7].

Figure 2 shows for both single crystals the FCC magnetization time-dependence as a function of the applied magnetic field. The magnetic moment time dependence was measured for approximately 50.000 s after field cooling the samples at 10 K/min from 100 K to 40 K. In the figure, M_0 represents the first measured value of the FCC magnetization time-dependence.

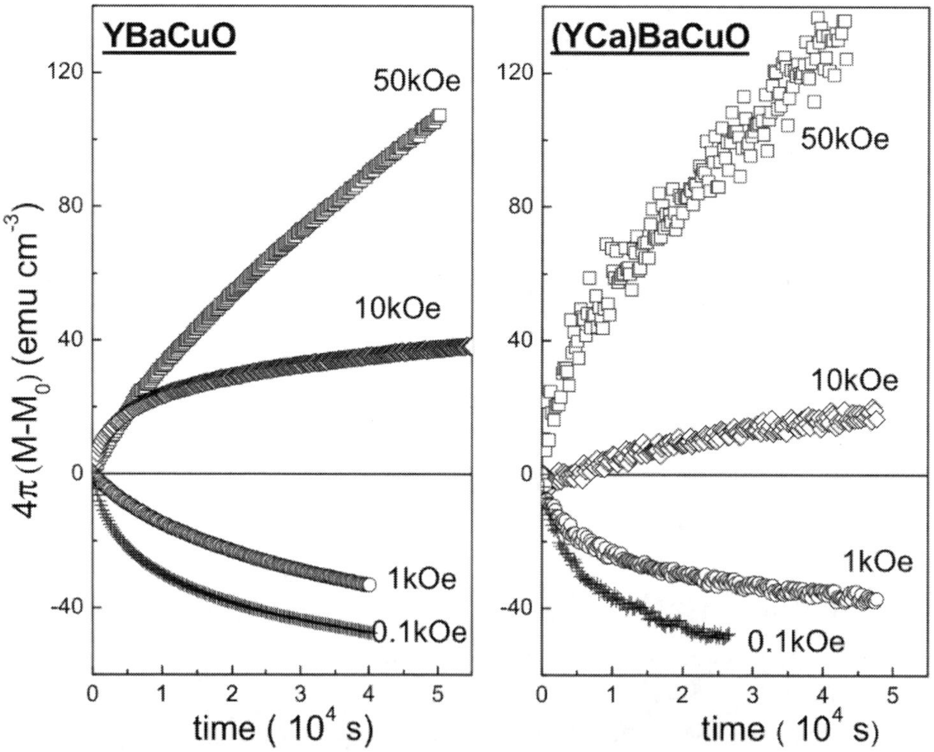

Figure 2. Time dependence of the FCC magnetization for our $YBa_2Cu_3O_{7-\delta}$ (left panel) and $Y_{1-x}Ca_xBa_2Cu_3O_{7-\delta}$ (x = 0.0025) (right panel) samples to the H // c. Measurements were performed at $T = 40$ K after a field cooling rate of the 10 K/min.

The behavior of the FCC moment time-dependence of the YBaCuO and the Y(Ca)BaCuO samples is very similar. While the applied magnetic field increases from 0.1 kOe to 50 kOe the FCC moment time-dependence monotonically proceeds from a diamagnetic to a paramagnetic performance. Remarkable in the results of Figure 2 is the fact that paramagnetic moments obtained after field cooling the samples relax monotonically to increasing positive values showing no time tendency to saturate.

The Ca doping acts as an efficient pinning centres in the microstructure $YBa_2Cu_3O_{7-\delta}$ [12,13]. Particularity, Ca substitutions give rise to formation of rows of oxygen vacancies with pronounced strain fields in the CuO_2 superconducting planes [12,13]. The defects thus formed seem to be analogous to array of normal nanodots in CuO_2 planes which are expected to have strong intragrain pinning interactions with vortices state structure.

HFPME results reported from Rykov [6] for an optimally doped and an overdoped YBaCuO single crystal display that the HFPME magnitude is suppressed in the overdoped sample. According to the authors, it is justified because HFPME magnitude is correlated to the structural disorder, like oxygen vacancies, that in the overdoped sample is smaller than optimally doped sample. The HFPME results showed in the Figure 1 are in agreement with this prediction once the HFPME magnitude is more intensity in Ca doped sample where the structural disorder character, enhanced by chemically substitution, is stronger than that one of the YBaCuO sample.

The behaviour of the FCC moment time-dependence, displayed in Figure 2, for our single crystals shows that as applied magnetic field is augmented more extra flux is allowed to penetrate the sample supporting an establishment of a flux compression scenario. A similar behaviour was observed in melted textured $YBa_2Cu_3O_{7-\delta}$ samples [7].

The weak irreversibility displayed by FCC and FCW magnetizations of our samples where stronger magnetic filed were applied as well as the enhancement of the HFPME by Ca doping suggest the establishment of a flux compressed state modulated by pinning as a probably responsible to the origin of the HFPME in our samples. In agreement with this scenario, we believe that in some regions of the samples, where pinning is strong, the vortex density can be depleted below that expected from equilibrium state. Consequently, this state opens a place for the admission of extra vortices into the sample, originating the HFPME.

Thus, it may be possible that pinning plays an important role to explain the HFPME in our samples. More studies are necessary in order to understand the correlated mechanism in both samples. The time dependence of FCC magnetization and the crucial role of the pinning mechanism in the YCaBaCuO single crystal are under current investigation.

Acknowledgments
The authors thank to the CNPq Universal Program, under contract N° (483100/2009-9), for partially financing this work.

References

[1] Svedlindh P, Niskanen K, Norling P, et al 1989 *Physica C* **4162-164** 1365
[2] Riedling S, Brauchle G, Lucht R, et al 1994 *Physical Review B* **49** 13283
[3] Sandim M J R, Stamopoulos D, Ghivelder L, et al 2010 *J. Supercond. Nov. Magn.* **23** 1533
[4] Pradip Das, Tomy C V, Takeya H, et al 2009 *J. of Physics: Conference Series* **150** 052041
[5] Prokhorov V G, Svetchnikov V L, Park J S, et al 2009 *J. Supercond. Sci.Technol.* **22** 045027
[6] Rykov A I, Tajima S and Kusmartsev F V, 1997 *Physical Review B* **55** 8557
[7] Dias F T, Pureur P, Rodrigues P, et al 2004 *Physical Review B* **70** 224519
[8] Sigrist M and Rice T M, 1992 *EJ. Phys. Soc. Jpn.***61** 4283
[9] Koshelev A E and Larkin A I, 1995 *Physical Review B* **52** 13559
[10] Moshchalkov V V, Qiu X G and Bruyndonex V, 1997 *Physical Review B* **55** 11793
[11] Vieira V N, Riegel I C and Schaf J, 2007 *Physical Review B* **76** 024518
[12] Huhtienen H, Awana V P S, Gupta A, et al 2007 *Supercond. Sci. Technol.* **20** 159
[13] Laval J Y and Orlova T S 2003 *Supercond. Sci. Technol.* **16** 1139

Polarization-dependent three-dimensional angle-resolved photoemission study on LiFeAs

T. Hajiri[1], R. Niwa[1], T. Ito[1,2], M. Matsunami[3,4], B. H. Min[5], S. Kimura[3,4] and Y. S. Kwon[5]
[1]Graduate School of Engineering, Nagoya University, Nagoya 464-8603, Japan
[2]Nagoya University Synchrotron Radiation Research Center, Nagoya University, Nagoya, 464-8603, Japan
[3]UVSOR Facility, Institute for Molecular Science, Okazaki 444-8585, Japan
[4]School of Physical Sciences, The Graduate University for Advanced Studies (SOKENDAI), Okazaki 444-8585, Japan
[5] Department of Emerging Materials Science, DGIST, Daegu 711-873, Republic of Korea

E-mail: t.ito@numse.nagoya-u.ac.jp, hajiri.tetsuya@f.mbox.nagoya-u.ac.jp

Abstract. We report the orbital character of the Fermi surfaces for the Fe-based superconductor LiFeAs using the polarization-dependent three-dimensional angle-resolved photoemission spectroscopy and the angular-dependent dipole selection rule analysis. The evaluated orbital character is well reproduced by the band calculation. The results demonstrate that the polarization-dependent ARPES combined with the selection rule analysis is useful to identify the orbital character of the Fe-based unconventional superconductivity.

1. Introduction

Since the discovery of iron-based superconductors in 2008, a new research field of superconductivity has been established [1]. The common feature of the new superconductors is the existence of iron pnictide layers, similar to CuO_2 planes of cuprate superconductors. The phase diagram of iron pnictides is also similar to that of cuprates, i.e., the superconductivity appears by an electron/hole doping to parent compounds [2]. Although intensive research works have been performed, the relation of antiferromagnetic fluctuation to superconductivity is still under debate.

From previous theoretical [3] and experimental [4] works, it has been reported that the high critical temperature (T_c) of iron pnictides has relevance to a Fermi surface (FS) nesting with a sign reversal of the order parameter. However, the determination of the FS nesting condition in iron pnictides is intrinsically difficult due to three-dimensionality of their FSs [5]. In addition, the different orbital character at the "multi" FSs has been proposed [6]. Therefore, the direct investigation of the orbital character forming three dimensional (3D) FS as well as the relation between the FS nesting and the superconductivity has been anticipated. In this point of view, the polarization-dependent angle-resolved photoemission spectroscopy (ARPES) is a preferable probe. Particularly, 3D-ARPES using linearly polarized synchrotron radiation is a powerful tool.

The title compound LiFeAs, categorized into the family of 111-type iron pnictides, is an ideal system to investigate the unconventional superconductivity as well as the orbital character at the multi FSs, because the superconductivity of the pristine LiFeAs emerges at T_c = 18 K without the antiferromagnetic and structural transitions, unlike the other 1111-type and 122-type iron pnictides. It should be noted that the FS nesting is expected to be small and the superconductivity is suggested to be assisted by ferromagnetic fluctuation [7].

In this work, we performed the polarization-dependent 3D-ARPES to determine the orbital character of band dispersions near the Fermi level (E_F) of LiFeAs. Furthermore, we analyse the orbital symmetry of the FSs by considering the angular-dependent dipole selection rule.

2. Experiment

High-quality single crystals of LiFeAs with T_c = 18 K were grown by a Bridgman method [8]. The polarization-dependent 3D-ARPES measurements were performed at the BL7U, namely "SAMRAI" beamline [9], of UVSOR-II, in the Institute for Molecular Science. At the beamline, the linearly [e.g., horizontally (E_P) and vertically (E_S)] polarized VUV light from the APPLE-II-type undulator is available. The experimental geometry of the present polarization-dependent ARPES is shown in Fig. 1 (a). The total energy and angular resolutions, and sample temperature were set to 6 ~ 15 meV, 0.17°, and 12 K, respectively. The experimental energy and angular resolutions are sufficiently high to detect orbital character of the multi FSs [10]. The clean surface was prepared by *in situ* cleaving under a vacuum of 8×10^{-9} Pa. The photon energies corresponding to high-symmetry k_z points were obtained by using normal-emission ARPES. The inner potential was 15.4 eV. The E_F of samples was referred to that of evaporated gold film. LDA band structure calculation was performed by using the full potential linearized augmented plane wave plus local orbital (LAPW+lo) method including spin-orbit coupling implemented in the Wien2k code [11].

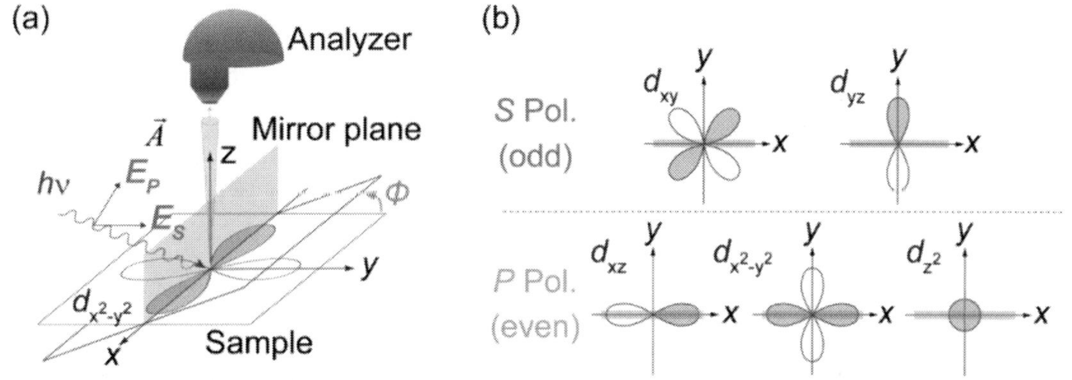

Figure 1. (a) Experimental geometry of the polarization-dependent ARPES. (b) The spatial symmetry of the 3*d* orbitals with respect to the mirror plane, which are selectively excited with each polarization (*S* or *P*) under normal-emission configuration. The mirror plane including the direction of incidence and emission is defined along the *x* axis as shown by bold lines.

3. Results & Discussion

Figure 2(a) shows the polarization-dependent 3D-ARPES images obtained near high-symmetry points. On the each images, the image of left (right) side represents the energy bands excited by the *S* (*P*) polarized light. We found that the FSs of LiFeAs mainly consist of the three and two hole-pockets at the Γ and Z points, respectively and the two electron-pockets at the M and A points. These findings are consistent with previous reports [12]. For all images, the prominent features as indicated by solid and dotted curves are clearly changed dependent on the polarization.

When the linear polarization dependence of the electronic band structure is identified, the dipole selection rule can be exploited to determine the orbital symmetry with respect to the mirror planes of the crystal surface. In the case of the normal emission, the final state $|f\rangle$ has an even symmetry with respect to the mirror plane [13]. Then the nonvanishing condition of the dipole transition $\langle f|\boldsymbol{A}\cdot\boldsymbol{p}|i\rangle$ is satisfied when the initial state $|i\rangle$ has the same symmetry (even/odd) as the dipole operator $\boldsymbol{A}\cdot\boldsymbol{p}$, where $\boldsymbol{A}$ and $\boldsymbol{p}$ are the vector potential of electromagnetic field and the momentum operator, respectively. For example, if $\boldsymbol{A}\cdot\boldsymbol{p}$ is even (odd) corresponding to the *P* (*S*) polarization in Fig. 1 (a), then the initial states with even (odd) symmetry should be reflected in ARPES spectra. In Fig. 1(b), we summarized the polarization-dependent sensitivity of Fe 3*d* orbitals under the normal-emission geometry [13].

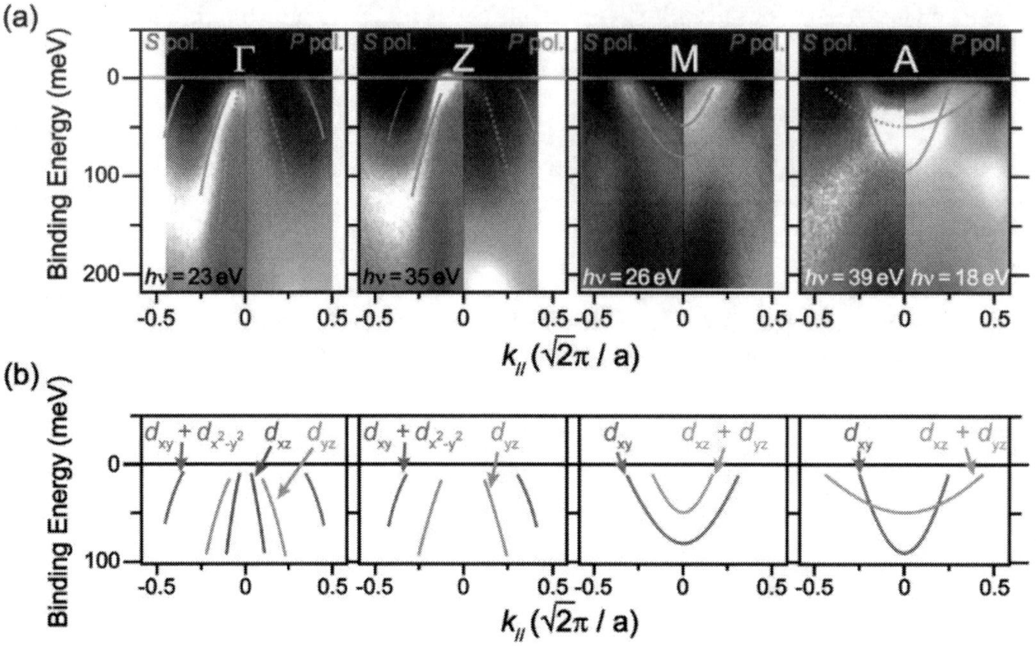

Figure 2. (a) Polarization-dependent ARPES images at some symmetry points. The left and right images in each panel are obtained by the *S* and *P* polarizations, respectively. (b) The schematic band dispersions and their orbital characters. See the text for detail.

Figure 2 (b) shows the resultant orbital character for each band near E_F estimated by using the present analysis. At Γ point ($\phi = 0°$), the emission with the *P* polarization provides the d_{xz} orbital band (the smallest hole pocket), while the d_{yz} orbital band (the medium hole pocket) is obtained by using the *S* polarization. The d_{yz} orbital band is also observed at Z point. The outer band is observed with both polarizations at Γ and Z points. It is attributed to the d_{xy} orbital band with the *S* polarization and the $d_{x^2-y^2}$ orbital band with the *P* polarization. Therefore the outer band at the Γ and Z points should be the combination of the the d_{xy} and $d_{x^2-y^2}$ orbitals. Since the intensity of d_{xy} orbital is expected to be smaller than that of d_{yz} orbital on the basis of the present analysis, the d_{xy} orbital is not clearly observed. Similarly, the intensity of $d_{x^2-y^2}$ orbital is also smaller than that of the d_{xz} orbital.

For M and A points, the electron band observed with *S* polarization is attributed to the d_{xy} orbital. Regarding *P* polarization, the d_{xz} orbital is naturally expected. In addition, considering the matrix element in $\phi = 30°$, the d_{yz} orbital should be also observed. As a result, the electron pocket observed with *P* polarization is attributed to $d_{xz} + d_{yz}$ orbitals. Compared to the band calculations for LiFeAs using LMTO method [12], our characterization of $d_{xz} + d_{yz}$ orbitals is consistent, but d_{xy} orbital is not so. On the other hand, our LDA calculations as shown in Fig. 3 give a good agreement even for the d_{xy} orbital.

4. Summary

We investigated the orbital character of band dispersions near E_F of LiFeAs by using the polarization-dependent 3D-ARPES. By applying the angular-dependent dipole selection rule analysis, we succeeded to clarify the orbital symmetry of the Fermi surfaces. The evaluated orbital character was well reproduced by the LDA band calculation, suggesting that the present analysis combined with the polarization-dependent 3D-ARPES is a powerful tool to clarify the relation between multi orbital 3D-FSs nesting and superconductivity.

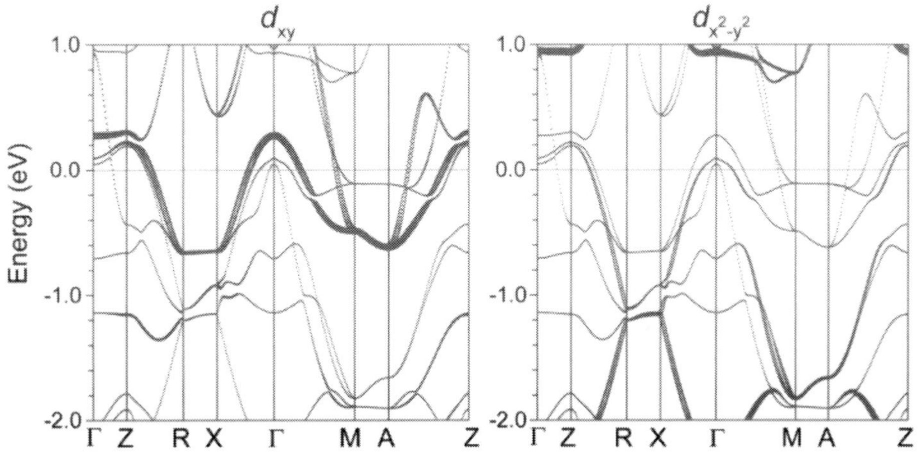

Figure 3. Character-dependent LDA band calculations for LiFeAs. The size of circles is proportional to the d_{xy} and $d_{x^2-y^2}$ partial weights.

Acknowledgments

Part of this work was supported by the Use-of-UVSOR Facility Program (BL7U, 2010) of the Institute for Molecular Science. The work at DGIST was partially supported by Basic Science Research Program through the NRF funded by the Ministry of Education, Science and Technology (2010-0007487).

References

[1] Kamihara Y, Watanabe T, Hirano M and Hosono H 2008 *J. Am. Chem. Soc.* **130** 3296.
[2] Mazin I I 2010 *Nature* **464** 183-186.
[3] Mazin I I, Singh D J, Johannes M D and Du M H 2008 *Phys. Rev. Lett.* **101**, 057003.
[4] Terashima K, Sekiba Y, Bowen J H, Nakayama K, Kawahara T, Sato T, Richard P, Xu Y-M, Li L J, Cao G H, Xu Z-A, Ding H, and Takahashi T 2007 *Proc. Natl. Acad. Sci. USA* **106** 7330.
[5] Vilmercati P, Fedorov A, Vobornik I, Manju U, Panaccione G, Goldoni A, Sefat A, McGuire M A, Sales B C, Jin R, Mandrus D, Singh D J and Mannella N 2009 *Phys. Rev. B* **79** 220503(R).
[6] Yoshida T, Nishi L, Ideta S, Fujimori A, Kubota M, Ono K, Kasahara S, Shibauchi T, Terashima T, Matsuda Y, Ikeda H and Arita R 2011 *Phys. Rev. Lett.* **106** 117001.
[7] Brydon P M R, Daghofer M, Timm C and Brink J. van den 2011 *Phy. Rev. B* **83** 060501(R).
[8] Song Y J, Ghim J S, Min B H, Kwon Y S, Jung M H and Rhyee J S 2010 *Appl. Phys. Lett.* **96** 212508.
[9] Kimura S, Ito T, Sakai M, Nakamura E, Kondo N, Horigome T, Hayashi K, Hosaka M, Katoh M, Goto T, Ejima T and Soda K 2010 *Rev. Sci. Instrum.* **81** 053104.
[10] Zhang Y, Chen F, He C, Zhou B, Xie B P, Fang C, Tsai W F, Chen X H, Hayashi H, Jiang J, Iwasawa H, Shimada K, Namatame H, Taniguchi M, Hu J P and Feng D L 2011 *Phys. Rev. B* **83** 054510.
[11] Blaha P, Schwarz K, Sorantin P and Trickey S B 1990 *Comput. Phys. Commun.* **59**, 399.
[12] Borisenko S V, Zabolotnyy V B, Evtushinsky D V, Kim T K, Morozov I V, Yaresko A N, Kordyuk A A, Behr G, Vasiliev A, Follath R and Buchner B 2010 *Phys. Rev. Lett.* **105** 067002.
[13] Hüfner S, *Photoelectron Spectroscopy* (Springer, Berlin, 2003).

International Conference on Strongly Correlated Electron Systems (SCES 2011) IOP Publishing
Journal of Physics: Conference Series **391** (2012) 012126 doi:10.1088/1742-6596/391/1/012126

Electric property of the iron pnictide oxide superconductor $(Fe_2As_2)(Ca_6(Al, Ti)_4O_y)$ under hydrostatic pressure

Akihiko Hisada[1], Naoki Fujiwara[2], Yoshiya Uwatoko[1,3], Hiraku Ogino[3,4], Kohji Kishio[3,4], Jun-ichi Shimoyama[3,4]

[1] Institute for Solid State Physics, The University of Tokyo, Kashiwanoha, Kashiwa, Chiba 277-8581, Japan

[2] Graduate School of Human and Environmental Studies, Kyoto University, Yoshida Nihonmatsu-cho, Sakyo-ku, Kyoto 606-8501, Japan

[3] TRiP, Japan Science and Technology Agency (JST), Sanban-cho Building 5, Sanban-cho, Chiyoda ku, Tokyo 102-0075, Japan

[4] Department of Applied Chemistry, The University of Tokyo, Tokyo 113-8656, Japan

E-mail: a-hisada@issp.u-tokyo.ac.jp

Abstract. We report the electrical resistivity measurements of $(Fe_2As_2)(Ca_6(Al, Ti)_4O_y)$ polycrystalline sample under hydrostatic pressure using a cubic-anvil apparatus. The temperature dependence of the normal state resistivity exhibits metallic behavior. The $\rho(T)$ curve is convex upward at low pressure, and it approaches T-linear dependence by applying high pressure. The superconductivity is suppressed by application of pressure. The onset temperature of superconductivity, which is 38 K at ambient pressure, decreases to 28 K at 8 GPa.

1. Introduction

Since the discovery of high-temperature superconductivity in $LaFeAsO_{1-x}F_x$ [1], several types of iron-based superconductors have been discovered. The iron-based superconductors are commonly composed of alternate stacking of quasi-two-dimensional iron pnictide (FePn) layers and blocking layers or ions. These compounds are categorized into the so-called 1111 systems RFeAsO (R = rare earth elements), 122 systems $AEFe_2As_2$ (AE = alkaline earth metals), 111 systems AFeAs (A = alkaline metal), and 11 systems FeSe and FeTe. The FePn interlayer distance for the 1111 systems is longer than that for other systems. For example, the typical distance for the 11, 111, 122 and 1111 systems is ~ 5.5 Å, ~ 6.4 Å, ~ 6.5 Å, and ~ 8.7 Å, respectively. Among the superconductors, the Sm-1111 system exhibits the highest superconducting transition temperature (T_C). Thus the FePn interlayer distance has been pointed out to be a key factor to raise T_C.

Recently, new superconducting (SC) families having perovskite-type oxide blocking layers, $(Fe_2As_2)(AE_{n+2}M_nO_y)$ (M = Sc, Cr, V, Al, (Mg, Ti), (Al, Ti), n = 2, 3, 4, and y ~ 3n) and $(Fe_2As_2)(AE_{n+1}M_nO_y)$ (M = Al, Sc, (Sc, Ti), (Mg, Ti), n = 2 ~ 5, and y ~ 3n-1), were found to have relatively higher T_C than the existing systems: T_C of them exceeds 30 K at ambient pressure [2,3]. Figure 1 shows the crystal structure of $(Fe_2As_2)(Ca_6(Al, Ti)_4O_y)$. The distance

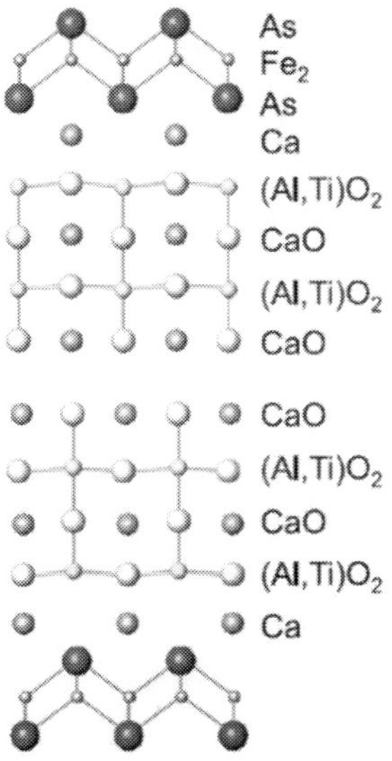

Figure 1. Crystal structure of $(Fe_2As_2)(Ca_6(Al, Ti)_4O_y)$ [2]. The perovskite-type (Al, Ti)O layers are sandwiched between the FeAs layers.

between the neighboring FeAs layers for $(Fe_2As_2)(Ca_6(Al, Ti)_4O_y)$ goes beyond ~ 22 Å, which is much longer than that for the 1111 systems. In these families, the distance can be systematically controlled by changing the numbers of the blocking layers. However, so far as these families are concerned, the distance is not directly related to T_C, because these families exhibit high T_C regardless of the numbers of the blocking layers.

Meanwhile, these families are characterized by large pnictogen height (h_{Pn}) from an Fe plane corresponding to a small a-axis unit length. Kuroki *et. al.* theoretically suggested that h_{Pn} is an important factor to raise T_C [4] and it is suggested that the optimal T_C is realized when $FePn_4$ forms a regular tetrahedron [5]. The lattice parameters can be controlled by applying pressure, thus the effect of pressure on these families is intriguing.

In this study, we focus on $(Fe_2As_2)(Ca_6(Al, Ti)_4O_y)$ having four (Al, Ti)O layers, because it contains less impurities than the families having two or three (Al, Ti)O layers. The onset and zero-resistivity temperatures are 39 and 25 K, respectively, at ambient pressure [2]. Moreover, an unusual successive phase transition from superconducting state to antiferromagnetic (AF) state was discovered from ^{75}As and ^{27}Al NMR studies [6]. The behavior of the AF transition under pressure is also notable.

2. Experimental procedure

Polycrystalline samples with a starting composition of $(Fe_2As_2)(Ca_6(Al_{0.33}Ti_{0.67})_4O_{11})$ were synthesized by solid-state reaction. Details of the preparations and characterizations are given in Ref. 2. Electrical resistivity (ρ) measurement under pressure was performed by using a cubic-anvil high-pressure apparatus, which consists of six tungsten-carbide (WC) anvils and a pyrophyllite gasket containing a teflon capsule. This apparatus has an advantage to producing highly hydrostatic pressure. As a pressure transmitting medium, we used glycerin. The pressure of the sample is calibrated by measuring resistivity anomalies of Bi, Te, and Sn, associated with their structural phase transitions at room temperature. Electrical resistivity was measured by a standard dc four-probe technique.

3. Results and Discussion

Figures 2(a) and 2(b) show the temperature dependence of the electrical resistivity under pressures up to 8 GPa. Electrical resistivity in the normal state shows metallic temperature

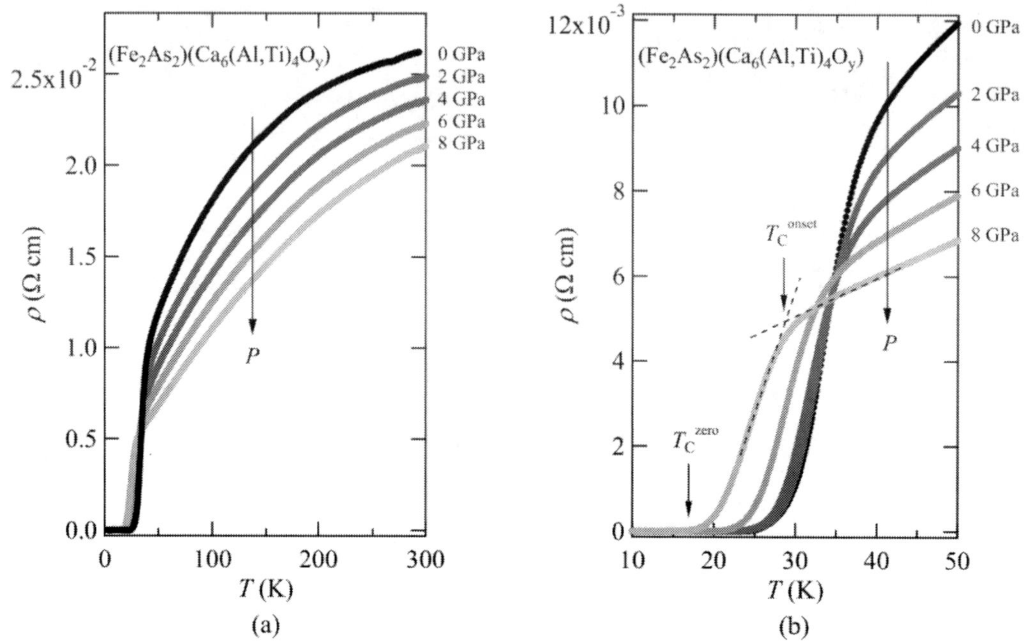

Figure 2. (a) Electrical resistivity (ρ) versus temperature (T) under different pressures. (b) Low temperature part of the $\rho(T)$ curve below 50 K. The arrows indicate the onset of superconducting transition temperature ($T_C{}^{onset}$) and the temperature of the zero resistivity ($T_C{}^{zero}$)

dependence and it monotonically decreases with increasing pressure. The $\rho(T)$ curve is convex upward at low pressure and it approaches T-linear dependence by applying high pressure. The onset of superconducting transition temperature ($T_C{}^{onset}$) and the temperature of the zero resistivity ($T_C{}^{zero}$) are shown in Figure 2(b). $T_C{}^{onset}$ and $T_C{}^{zero}$ are 38 and 23 K, respectively, at ambient pressure, which are in agreement with previous report [2]. They show small change below 4 GPa and they gradually decrease with increasing pressure above 6 GPa. At 8 GPa, $T_C{}^{onset}$ and $T_C{}^{zero}$ are 28 and 17 K, respectively. Any anomaly, which implies the magnetic ordering, was not found in the resistivity curve.

Figure 3 shows the pressure dependence of $T_C{}^{onset}$ and $T_C{}^{zero}$. They show the negative slope of about -2 K/GPa at high pressure. The pressure effect on T_C is associated with h_{Pn} and a-axis unit length. It is generally believed that the optimal T_C is realized when $FePn_4$ forms a regular tetrahedron. The optimal h_{Pn} is estimated to be between 1.34 and 1.38 Å. In $(Fe_2As_2)(Ca_5(Sc, Ti)_4O_y)$ having four (Sc, Ti)O layers, h_{Pn} and a-axis unit length have been estimated to be 1.4 and 3.902 Å, respectively [3]. In $(Fe_2As_2)(Ca_6(Al, Ti)_4O_y)$, a-axis unit length was estimated to be 3.815 Å, however, h_{Pn} has not been estimated. Taking account of the values for $(Fe_2As_2)(Ca_5(Sc, Ti)_4O_y)$, the h_{Pn} of $(Fe_2As_2)(Ca_6(Al, Ti)_4O_y)$ is more than 1.4 Å. For the compounds having h_{Pn} larger than the optimal value, pressure application causes the decrease in T_C because it causes reduction in a-axis unit length and thus the increase in h_{Pn} [7], that is, h_{Pn} deviates from the optimal value. This explanation seems valid for LiFeAs, in which h_{Pn} is about 1.5 Å [8, 9].

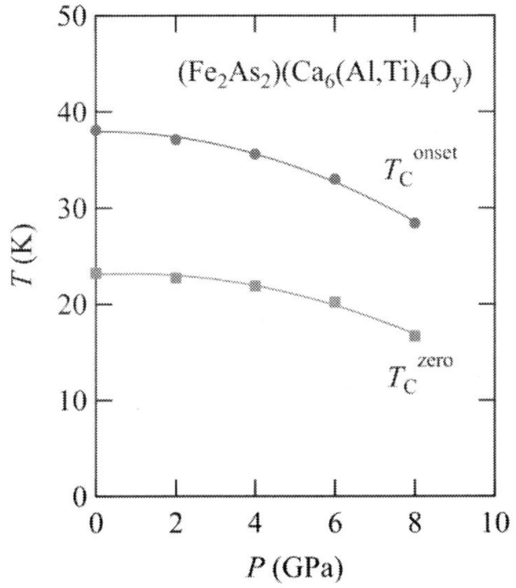

Figure 3. Pressure dependence of T_C^{onset} and T_C^{zero} for $(Fe_2As_2)(Ca_6(Al, Ti)_4O_y)$. The superconducting transition temperature decreases with increasing pressure.

4. Summary

We performed the electrical resistivity measurements of $(Fe_2As_2)(Ca_6(Al, Ti)_4O_y)$ polycrystalline sample under hydrostatic pressure using a cubic-anvil apparatus. The temperature dependence of the normal state resistivity shows metallic behavior, and the $\rho(T)$ curve, which is convex upward at low pressure, approaches T-linear dependence by applying high pressure. The superconductivity is suppressed by application of pressure. At ambient pressure, T_C^{onset} and T_C^{zero} are 38 and 23 K, respectively. They gradually decrease with increasing pressure, and they are 28 and 17 K, respectively, at 8 GPa. The pressure effect on T_C is associated with h_{Pn} and a-axis unit length.

Acknowledgement

We would like to thank K. Matsubayashi for fruitful discussions. This work was supported by KAKENHI 21340092.

References

[1] Kamihara Y, Watanabe T, Hirano M and Hosono H 2008 *J. Am. Chem. Soc.* **130** 3296.
[2] Ogino H, Machida K, Yamamoto A, Kishio K, Shimoyama J, Tohei T and Ikuhara Y 2010 *Supercond. Sci. Technol.* **23** 115005.
[3] Ogino H, Sato S, Kishio K, Shimoyama J, Tohei T and Ikuhara Y 2010 *Appl. Phys. Lett.* **97** 072506.
[4] Kuroki K, Usui H, Onari S, Arita R and Aoki H 2009 *Phys. Rev.* B **79** 224511.
[5] Lee C H, Iyo A, Eisaki H, Kito H, Fernandez-Diaz M T, Ito T, Kihou K, Matsuhata H, Braden M, and Yamada K 2008 *J. Phys. Soc. Jpn.* **77** 083704.
[6] Nakano T, Fujiwara N, Tsutsumi S, Ogino H, Kishio K and Simoyama J 2011 *Phys. Rev.* B **84** 060502.
[7] Ogino H, Sato S, Matsumura Y, Kawaguchi N, Ushiyama K, Katsura Y, Horii S, Kishio K, and Shimoyama J 2010 *Physica* C **470** S280.
[8] Pitcher M J, Parker D R, Adamson P, Herkelrath S J C, Boothroyd A T, Ibberson R M, Brunelli M and Clarke S J 2008 *Chem. Commun.* 5918.
[9] Mito M, Pitcher M J, Crichton W, Garbarino G, Baker P J, Blundell S J, Adamson P, Parker D R and Clarke S J 2009 *J. Am. Chem. Soc.* **131** 2986.

P-concentration dependence of the quasiparticle density of states in $BaFe_2(As_{1-x}P_x)_2$

Tetsuya Iye[1,2], Yusuke Nakai[1,2], Shunsaku Kitagawa[1,2], Kenji Ishida[1,2], Hiroaki Ikeda[1,2], Shigeru Kasahara[3], Takasada Shibauchi[1], Yuji Matsuda[1], and Takahito Terashima[3]

[1]Department of Physics, Graduate School of Science, Kyoto University, Kyoto 606-8502, Japan
[2]TRIP, JST, Sanban-cho Bldg., 5, Sanban-cho, Chiyoda, Tokyo 102-0075, Japan
[3]Research Center for Low Temperature and Materials Sciences, Kyoto University, Kyoto 606-8502, Japan

E-mail: tiye@scphys.kyoto-u.ac.jp

Abstract. ^{31}P-NMR measurements were performed in the iron pnictides $BaFe_2(As_{1-x}P_x)_2$ for $0.07 \leq x \leq 0.64$ in order to investigate evolution of electronic state with isovalent substitution. In general, the spin part of the Knight shift $K_{\rm spin}$ and $(T_1T)^{-1/2}$ are proportional to the quasiparticle density of states at the Fermi level $N(E_{\rm F})$ at high temperature where the Korringa relation holds. We found that $K_{\rm spin}$ and $(T_1T)^{-1/2}$ are almost unaffected by the P-substitution, indicating that $N(E_{\rm F})$ is nearly constant. These results are quantitatively consistent with our band calculation. We claim that the slight changes of $N(E_{\rm F})$ by P-substitution in $BaFe_2(As_{1-x}P_x)_2$ is not enough to explain the T_c variation ranging from 30 K to 0 K, and thus suggest that antiferromagnetic correlations related to the quantum critical character are essential for the high T_c superconductivity.

Despite a large number of experimental and theoretical researches on the physics of recently discovered iron-pnictides [1], the mechanism of high-T_c superconductivity is still under debate. Spin-fluctuation-mediated superconductivity which promotes the $s_\pm$-wave superconducting gap symmetry with a sign change between electron and hole bands is one of the most promising scenario [2, 3, 4, 5, 6], while orbital-fluctuation-mediated superconductivity whose gap symmetry is s_{++} with no sign change is considered to be another candidate [7]. In case of conventional phonon-mediated superconductivity, T_c is closely related to the quasiparticle density of states at the Fermi level $N(E_{\rm F})$ in the form of $T_c = 1.14\omega_{\rm D}e^{-2/gN(E_{\rm F})}$, where $\omega_{\rm D}$ is the characteristic frequency of collective excitations involved in pairing interaction and g is the appropriate dimensional coupling constant. Since there is a possibility that the changes in T_c by chemical substitution and/or physical pressure is attributed to the changes in $N(E_{\rm F})$, it should be examined with a proper system to discuss the nature of dominant pairing interaction.

In this sense, we studied $BaFe_2(As_{1-x}P_x)_2$ as such an ideal system. This is because the substitution of isovalent P for As is expected to introduce chemical pressure without doping charge carriers in contrast with hole-doping $(Ba_{1-x}K_x)Fe_2As_2$ and electron-doping $Ba(Fe_{1-x}Co_x)_2As_2$. $BaFe_2(As_{1-x}P_x)_2$ shows unconventional superconductivity at $T_c^{\rm max} = 31$ K ($x \sim 0.30$) [8, 9, 10, 11, 12] with line nodes in the superconducting gap [13, 14, 15, 17, 16, 18]. However, recent ARPES experiment revealed that P-substitution in $BaFe_2(As_{1-x}P_x)_2$ induced significant hole-doping [19]. This result suggests that the iso-valent picture in $BaFe_2(As_{1-x}P_x)_2$

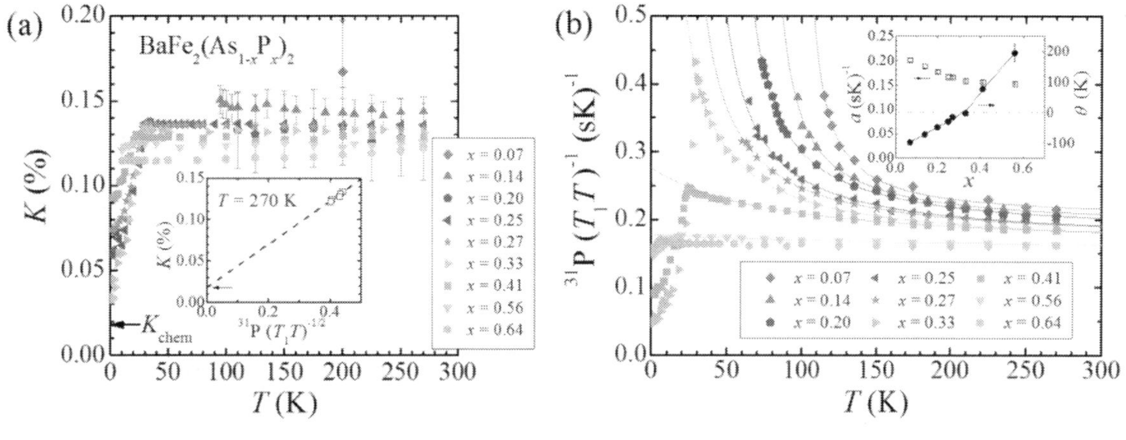

Figure 1. (Color online) (a) The temperature dependences of Knight shifts for $0.07 \le x \le 0.64$ in $BaFe_2(As_{1-x}P_x)_2$. The black arrow denotes the value of $K_{chem} = 0.018 \pm 0.019$ % determined by $(T_1T)^{-1/2} - K$ plot (inset). (b) The temperature dependences of $(T_1T)^{-1}$ for $0.07 \le x \le 0.64$ in $BaFe_2(As_{1-x}P_x)_2$. The solid lines are the fitting curves to $(T_1T)^{-1} = a + b(T+\theta)^{-1}$. The inset shows x-dependences of the fitting parameters a and θ, and the dotted line corresponds to 0 K in the right axis.

is oversimplified [19], and hence, it is important to investigate the variation of $N(E_F)$ experimentally. Our investigation clarified that $N(E_F)$ remains almost unchanged by P-substitution, while the characteristic temperature of magnetic anomaly θ deduced from $(T_1T)^{-1}$ largely changes with a sign change across an anticipated quantum critical point. These suggest that the change in antiferromagnetic correlations related to the quantum critical character is more important than the change in $N(E_F)$ for the T_c variation in the P-substituted system.

Samples of $BaFe_2(As_{1-x}P_x)_2$ for $0.07 \le x \le 0.64$, consisting of single crystals with average dimensions of $100 \times 100 \times 50$ μm^3 were prepared as described elsewhere [9]. ^{31}P-NMR spectra were obtained by sweeping frequency in a fixed magnetic field of 4.12 T at 200 K. The Knight shift K was determined from the peak frequency with respect to the reference material H_3PO_4 ($K = 0$). The ^{31}P nuclear spin-lattice relaxation rate T_1^{-1} was determined by fitting the time dependence of spin-echo intensity after saturation of nuclear magnetization to a theoretical $I = 1/2$ curve with a single component of T_1.

The resonance frequencies of ^{31}P-NMR spectra consisting of a single line for $0.07 \le x \le 0.64$ are temperature-independent in the normal state [10]. The temperature variation of K is shown in Fig.1(a). K in the normal state slightly decrease with increasing P-concentration x. In general, K is composed of the temperature-dependent spin shift K_{spin} and the temperature-independent chemical shift K_{chem}: $K(T) = K_{spin}(T) + K_{chem}$. K_{spin} is related to the uniform spin susceptibility $\chi(q=0)$ which is proportional to $N(E_F)$, while K_{chem} is unrelated to $N(E_F)$. Since no obvious AFM fluctuations were detected via $(T_1T)^{-1}$ at high temperatures as seen in Fig.1(b), it would be a good approximation to consider that $(T_1T)^{-1/2}$ is proportional to $N(E_F)$ at high temperatures, i.e., to assume that the usual Korringa relation holds. Based on the plot of $(T_1T)^{-1/2}$ against K at 270 K for different x which is shown in the inset of Fig.1(a), we estimated $K_{chem} = 0.018 \pm 0.019$ % from the intercept [10]. This value would be reasonable since K_{chem} for ^{31}P in many diamagnetic insulators is of the order of some hundreds of ppm [21]. Using the obtained K_{chem} with an assumption of x-independent K_{chem}, we deduced K_{spin} at 200 K for each x and plotted them in Fig.2(a). Since K_{spin} is proportional to $N(E_F)$, the

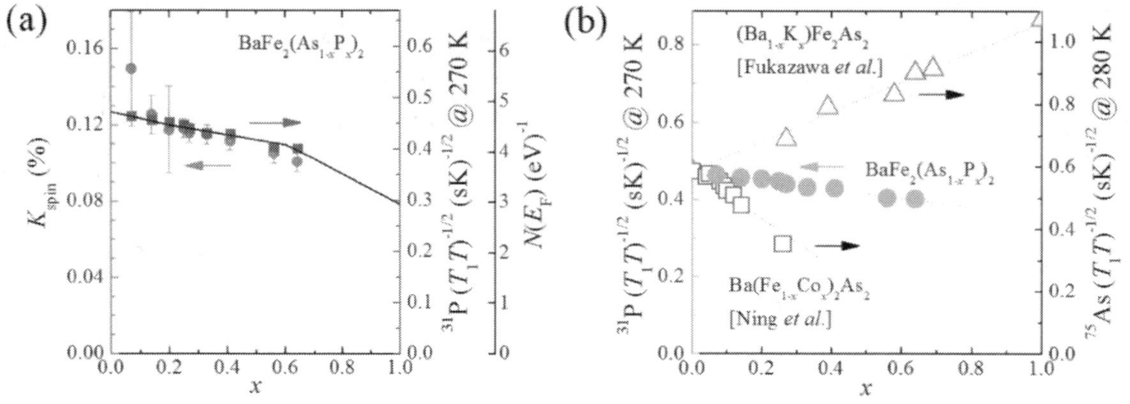

Figure 2. (Color online) (a) x-dependence of $K_{\rm spin}$ and ^{31}P $(T_1T)^{-1/2}$ at 270 K of $BaFe_2(As_{1-x}P_x)_2$ and x-evolution of $N(E_F)$ derived from the band calculation. (b) x-dependence of ^{31}P $(T_1T)^{-1/2}$ at 270 K of $BaFe_2(As_{1-x}P_x)_2$, ^{75}As $(T_1T)^{-1/2}$ at 280 K of $(Ba_{1-x}K_x)Fe_2As_2$ [23] and $Ba(Fe_{1-x}Co_x)_2As_2$ [24], all of which are proportional to $N(E_F)$. The dotted lines are guides to the eye. The vertical axis of ^{31}P $(T_1T)^{-1/2}$ is adjusted to unify the value of $BaFe_2As_2$ with that in the right axis.

x-dependence of $K_{\rm spin}$ suggests that the change in $N(E_F)$ is at most 16 % for superconducting $0.20 \leq x \leq 0.64$. This result is quantitatively consistent with the change in $N(E_F)$ deduced from $(T_1T)^{-1/2}$ and our band calculation. The calculations by local-density-approximation were performed for nonspin-polarized $BaFe_2As_2$ and $BaFe_2P_2$, using the WIEN2K package in the APW + local orbital basis [10]. We performed the local-density-approximation calculations for $BaFe_2As_2$ with linearly-interpolated $(a, c, z) = [a_0(1-x) + a_1x, c_0(1-x) + c_1x, z_0(1-x) + z_1x]$, where $a_{0(1)}$, $c_{0(1)}$, and $z_{0(1)}$ are the experimental values for the crystallographic parameters of $BaFe_2As_2(BaFe_2P_2)$ [9, 26]. The calculated $N(E_F)$ are plotted as a solid line in Fig.2(a), which agree with the experimental results quantitatively.

Low energy spin fluctuations are probed by measuring T_1^{-1}. AFM spin fluctuations are enhanced significantly with decreasing x from $x = 0.64$ where $(T_1T)^{-1}$ is almost constant. The measured $(T_1T)^{-1}$ can be well fit by the equation expected from the self-consistent renormalization (SCR) theory under the existence of two-dimensional (2D) AFM spin fluctuations [22]: $(T_1T)^{-1} = a + b(T+\theta)^{-1}$, which are shown as solid lines in Fig.1(b). The first term is the contribution originating from intraband scattering and is proportional to $N(E_F)^2$, the second term is the contribution of interband AFM spin fluctuation. x-dependence of a and θ are demonstrated in the inset of Fig.1(b). With increasing x, the parameter a decreases slightly whereas θ increases significantly with a sign change around $x \sim 0.33$ where quantum critical point is anticipated.

We plot the x-dependence of $(T_1T)^{-1/2}$ at high temperatures for $BaFe_2(As_{1-x}P_x)_2$, $(Ba_{1-x}K_x)Fe_2As_2$ [23], and $Ba(Fe_{1-x}Co_x)_2As_2$ [24] to compare the evolution of $N(E_F)$. The K- and Co-substitution induce significant changes in $N(E_F)$, while the P-substitution has little effects on $N(E_F)$. Note that the changes in $N(E_F)$ for superconducting dome region $0.2 \leq x \leq 1$ in $(Ba_{1-x}K_x)Fe_2As_2$ are $\sim 60\%$ increase and $0.04 \leq x \leq 0.14$ in $Ba(Fe_{1-x}Co_x)_2As_2$ are $\sim 20\%$ decrease from the value of $BaFe_2As_2$, whereas the change in $BaFe_2(As_{1-x}P_x)_2$ is only 11 % decrease. These results suggest that the $N(E_F)$ change is not a primary role for superconductivity in the "122" compounds. Rather, the AFM fluctuations with the quantum critical character are more important for superconductivity than the $N(E_F)$ change.

In summary, we have investigated the P-substitution effect to the electronic state in $BaFe_2As_2$ through ^{31}P-NMR. The quasiparticle density of states $N(E_F)$ detected via measuring the spin part of the Knight shifts K_{spin} and the nuclear spin-lattice relaxation rate T_1^{-1} show only slight decrease with increasing P-concentration x, quantitatively consistent with the results of the band calculation. Almost unaffected value of $N(E_F)$ by P-substitution is compared to more rapid changes in $N(E_F)$ caused by K and Co doping. Our experiments clarified that T_c in the "122" system is not mainly determined by the $N(E_F)$ change but by the AFM spin fluctuations.

Acknowledgements
We thank K. Kitagawa and Y. Maeno for experimental supports and valuable discussions. This work was supported by the Grants-in-Aid for Scientific Research on Innovative Areas "Heavy Electrons" (No. 20102006) from MEXT, for the GCOE Program "The Next Generation of Physics, Spun from Universality and Emergence" from MEXT, and for Scientific Research from JSPS.

References

[1] Kamihara Y, Watanabe T, Hirano M, and Hosono H 2008 *J. Am. Chem. Soc.* **130** 3296
[2] Mazin I I, Singh D J, Johannes M D, and Du M H 2008 *Phys. Rev. Lett.* **101** 057003
[3] Kuroki K, Onari S, Arita R, Usui H, Tanaka Y, Kontani H, and Aoki H 2008 *Phys. Rev. Lett.* **101** 087004
[4] Chubukov A V, Efremov D V, and Eremin I 2008 *Phys. Rev. B* **78** 134512
[5] Ikeda H 2008 *J. Phys. Soc. Jpn.* **77** 123707
[6] Graser S, Maier T A, Hirschfeld P J, and Scalapino D J 2009 *New J. Phys.* **11** 025016
[7] Kontani H, and Onari S 2010 *Phys. Rev. Lett.* **104** 157001
[8] Jiang S, Xing H, Xuan G, Wang C, Ren Z, Feng C, Dai J, Xu Z, and Cao G 2009 *J. Phys. Condens. Matter* **21** 382203
[9] Kasahara S, Shibauchi T, Hashimoto K, Ikada K, Tonegawa S, Okazaki R, Shishido H, Ikeda H, Takeya H, Hirata K, Terashima T, and Matsuda Y 2010 *Phys. Rev.* B **81** 184519
[10] Nakai Y, Iye T, Kitagawa S, Ishida K, Ikeda H, Kasahara S, Shishido H, Shibauchi T, Matsuda Y, and Terashima T 2010 *Phys. Rev. Lett.* **105** 107003
[11] Kim J S, Stewart G R, Kasahara S, Shibauchi T, Terashima T, and Matsuda Y 2011 *J. Phys. Condens. Matter* **23** 222201
[12] Ishikado M, Nagai Y, Kodama K, Kajimoto R, Nakamura M, Inamura Y, Wakimoto S, Nakamura H, Machida M, Suzuki K, Usui H, Kuroki K, Iyo A, Eisaki H, Arai M, and Shamoto S 2011 *Phys. Rev.* B **84** 144517
[13] Hashimoto K, Yamashita Y, Kasahara S, Senshu Y, Nakata N, Tonegawa S, Ikada K, Serafin A, Carrington A, Terashima T, Ikeda H, Shibauchi T, and Matsuda Y 2010 *Phys. Rev.* B **81** 220501(R)
[14] Nakai Y, Iye T, Kitagawa S, Ishida K, Kasahara S, Shibauchi T, Matsuda Y, and Terashima T 2010 *Phys. Rev.* B **81** 020503(R)
[15] Kim J S, Hirschfeld P J, Stewart G R, Kasahara S, Shibauchi T, Terashima T, and Matsuda Y 2010 *Phys. Rev.* B **81** 214507
[16] Suzuki K, Usui H, and Kuroki K 2011 *J. Phys. Soc. Jpn.* **80** 013710
[17] Yamashita M, Senshu Y, Shibauchi T, Kasahara S, Hashimoto K, Watanabe D, Ikeda H, Terashima T, Vekhter I, Vorontsov A B, and Matsuda Y 2011 *Phys. Rev.* B **84** 060507(R)
[18] Zhang Y, Ye Z R, Ge Q Q, Chen F, Jiang J, Xu M, Xie B P, and Feng D L *arXiv* :1109.0229
[19] Ye Z R, Zhang Y, Xu M, Ge Q Q, Chen F, Jiang J, Xie B P, Hu J P, and Feng D L *arXiv* :1105.5242
[20] Singh D J 2008 *Phys. Rev.* B **78** 094511
[21] Carter G C, Bennett L H, and Kahan D J 1977 *Metallic Shifts in NMR (Progress in Materials Science)* vol 20 (Oxford: Pergamon Press)
[22] Moriya T, Takahashi Y, and Ueda K 1990 *J. Phys. Soc. Jpn.* **59** 2905
[23] Hirano M, Yamada Y, Saito T, Nagashima R, Konishi T, Toriyama T, Ohta Y, Fukazawa H, Kohori Y, Furukawa Y, Kihou K, Lee C -H, Iyo A, and Eisaki H *arxiv* :1110.6081
[24] Ning F L, Ahilan K, Imai T, Sefat A S, McGuire M A, Sales B C, Mandrus D, Cheng P, Shen B, and Wen H -H 2010 *Phys. Rev. Lett.* **104** 037001
[25] Kuchinskii E Z, Nekrasov I A, and Sadovskii M V 2010 *JETP Lett.* **91** 518
[26] Rotter M, Tegel M, Johrendt D, Schellenberg I, Hermes W, and Pöttgen R 2008 *Phys. Rev.* B **78** 020503(R)

International Conference on Strongly Correlated Electron Systems (SCES 2011) IOP Publishing
Journal of Physics: Conference Series **391** (2012) 012128 doi:10.1088/1742-6596/391/1/012128

Quantized Massive Collective Modes and the Evolution Mechanism of Fermi Arc in High-Tc Cuprates

Ikuzo Kanazawa

Department of Physics, Tokyo Gakugei University, Koganeishi, Tokyo 184-8501, Japan

E-mail: kanazawa@u-gakugei.ac.jp

Abstract. We have discussed the origin of broad high-frequency spectra in Angle-resoluved photoemission spectroscopy(ARPES) in high-Tc cuprates by using quantized massive gauge fields. It is suggested that the emitted massive gauge fields, which are mediating Cooper pairing in high-Tc cuprates, decay into other excitations such as spin, charge, and phonon waves. The evolution mechanism of the Fermi arc with increasing hole-doping has been discussed, taking into account the restoration of spontaneous symmetry breaking.

1. Introduction

Since the discovery of high-Tc cuprates, the fundamental mechanism that gives rise to high-Tc superconductivity has been debated. Angle-resolved photoemission spectroscopy(ARPES) spectra in underdoped high-Tc cuprates are quite unusual [1,2]. A broad feature, " hump", at $100 \sim 300$ meV is present in photoemission spectra taken near $(\pi,0)$ in underdoped high-Tc cuprates. In addition, it has been observed that optical infrared-spectra have a broad feature at $200 \sim 400$ meV, whose intensity increases gradually with decreasing the temperature, in Bi-2212 cuprates [3]. We find that the various spectroscopies such as ARPES, the optical conductivity, and scanning tunneling spectroscopies are pointing to a unified picture of a broad high-frequency spectra of bosonic excitations in most of high-Tc cuprates [4]. In high-Tc cuprates, there is evidence that two aspect of the pseudogap, small pseudogap (SPG) and large pseudogap (LPG) [5]. The SPG, whose corresponding temperature T^*, is a characteristic energy comparable to the superconducting-state gap. The SPG developes progressively below a temperature T^* and then evolves into the fully coherent superconducting gap at Tc. It has been suggested that the SPG will be some kind of incoherent Cooper pairing or the real space electric organization [6,7]. The LPG corresponds to T_{mK} and $T_{m\chi}$, below which the Knight shift and the spin susceptibility start to decrease gradually, respectively. The magnitude of the LPG increases with decreasing hole concentration, and the temperature $T_{m\chi}$(or T_{mK}) above which the LPG closes also increases with decreasing hole concentration [5]. Recently Yazdani [8] has discussed visualizing Cooper pair formation of the atomic scale in high-Tc cuprates by using the scanning tunneling microscopy. For the optimal and overdoped high-Tc cuprates the onset of the gap is indeed due to pairing, which occurs locally at T_p. The T^* line on the overside is controlled by the largest pairing gap ($T_{p,max}$). Then it has been suggested that the high-energy (up to ~ 400 meV) hole-like excitations, which might corresponds to broad high-frequency spectra in

ARPES and the optical conductivity, of the normal state are a direct predictor of the strength of pairing. In this study, we discuss the origin of broad high-frequency spectra in ARPES of the normal state in high-Tc cuprates, and then present the evolution mechanism of the Fermi arc with increasing of hole-doping in high-Tc cuprates.

2. A model system and quantized massive collective gauge-fields

We have introduced quantized massive gauge fields around the doped hole as collective modes, which might contain effects of spin fluctuation, charge fluctuation, and phonons, from the viewpoint of adjoing interactions [10,12]. Taking into account that the symmetry in the undoped (2+1)-dimensional quantum antiferromagnet is invariant under local SU(2) [9], we assume that the perturbing gauge fields A_μ^a introduced by the hole has a local SU(2) symmetry. Since the distortion around the hole is due to strong many-body effects, the distortion effects around the hole is non-linear (Yang-Mills fields-like). Then SU(2) gauge fields A_μ^a are spontaneously broken through the Anderson-Higgs mechanism in a way similar to the breaking of the antiferromagnetic symmetry around the hole. After the symmetry breaking $\langle 0|\phi_a|0\rangle = \langle 0, 0, \mu(\hat{k}_{\rm F})\rangle$ of the Bose field ϕ_a, we can obtain the effective Lagrangian density L_{eff} [10].

$\hat{k}_{\rm F}$ is the vector of the Fermi momentum. Then $\langle 0|\phi_3|0\rangle$ can be regarded as a kind of the disorder parameter [11]. The value, $\mu(\hat{k}_{\rm F})$, of the symmetry breaking depends strongly on the direction of Fermi momentum, $\hat{k}_{\rm F}$, on the Fermi surface. Furthermore, the valve $\mu(\hat{k}_{\rm F})$, is much correlated to the gap-energy of the high energy pseudogap. If the value of the high-energy pseudogap is related to the strength of the antiferromagnetic short-range order [5], the distortion, which is induced by the doped hole, becomes larger as the hole is doped in the state of the larger gap-energy of the high-energy pseudogap. Since the value, $\mu(\hat{k}_{\rm F})$, means the strength of the distortion induced by the doped hole, the value, $\mu(\hat{k}_{\rm F})$, is higher around the hot spot. The effective Lagrangian describes two massive vector field A_μ^1 and A_μ^2, and one massless U(1) gauge field A_μ^3 [10]. Because masses of A_μ^1 and A_μ^2 are formed through the Higgs mechanism by introducing the hole, the fields A_μ^1 amd A_μ^2 exist around the hole within the length of $\sim 1/m_1$. The masses, m_1, of the gauge fields A_μ^1 and A_μ^2 induced by the hole depend strongly on an angle of Fermi momentum of the hole on the Fermi surface. The value, m_1, is higher in the case of the hole around the hot spot. It is thought that the interaction between two holes for the Cooper pair formation is derived from the exchange of the fields A_μ^a [17]. The attractive interaction between two holes at $k \sim (- \mid \hat{k}_{\rm F}^h \mid, 0)$ and $k \sim (\mid \hat{k}_{\rm F}^h \mid, 0)$ or between two holes at $k \sim (0, - \mid \hat{k}_{\rm F}^h \mid)$ and $k \sim (0, \mid \hat{k}_{\rm F}^h \mid)$ for the Cooper pair formation are the strongest ones in comparison with the other pairs. In addition it should be noticed that the quantized massive gauge fields induce strongly the short-range spin fluctuation [17]. Recently Yazdani [8] has found that the high-energy (up to $\sim$ 400 meV) hole-like excitations of the normal state are a direct predictor of the strength of Cooper pairing. We stress strongly that these high-energy hole-like excitations correspond to the complex excitations, which are composed of quantized massive gauge fields with the hole in the present formula.

As described previously [12], the effective interaction between two holes is like the asymptotic freedom, since the SU(2) Yang-Mills fields is more effective when the distance between the holes is shorter than $\sim 2/m_1$. For simplifying discussions, we parametrize $m_1(\hat{k}_F) = \mu(\hat{k}_F) \cdot g_4$ by $\hat{k}_F(\theta)$ and angle θ, with $\theta = 0$ corresponding to $\hat{k}$ along the zone diagonal and represents approximately as $m_1(\hat{k}_F) = \mu(\hat{k}_F) \cdot g_4 = \mu(\hat{k}_F^h) \cdot g_4 \cdot \sin^2(2\theta)$. Where $\hat{k}_F^h$ is the Fermi momentum at the hot spot. Recently the present author [13,14] has introduced the restoration temperature $T_{res}(\hat{k}_F)$ around the hot spot (π, 0), which corresponds to the temperature of LPG. Now we shall estimate some parameters by using experimental results of ARPES in triple-layered Bi2223 of Tc $\sim$ 108 K [15]. Taking into account of $m_1(\hat{k}_F^h) = \mu(\hat{k}_F^h) \cdot g_4 = (\sim 150 meV)$, we can estimate approximately to be $m_1(\hat{k}_F) = m_1(\hat{k}_F^h, \theta) = (150meV) \cdot \sin^2(2\theta)$. Now we shall consider the broad feature of ARPES spactra near (π,0) in underdoped high Tc cuprates from the present theoretical view. The creation of a photohole is more likely to produce massive collective gauge fields A_μ^1 and A_μ^2, whose effective mass is $\sim m_{1,eff}(\hat{k}_F, T)$. The ARPES spactra in high Tc cuprates are given approximately by

$$A(p,\varepsilon) \propto \frac{\mathrm{Im}\Sigma(p,\varepsilon)}{[\varepsilon - \varepsilon_k - \mathrm{Re}\Sigma(p,\varepsilon)]^2 + [\mathrm{Im}\Sigma(p,\varepsilon)]^2}, \tag{1}$$

$$\begin{aligned}\mathrm{Im}\Sigma(k,\varepsilon) \propto & -{g_2}^2 \int \mathrm{d}p_1 \int_{-\infty}^{\infty} \mathrm{d}\omega \int_{-\infty}^{\infty} \mathrm{d}\varepsilon_1 \\ & \times \frac{\mathrm{Im}D_R(k-p_1,\omega)A(p_1,\varepsilon_1)}{\omega + \varepsilon_1 - \varepsilon - i\delta} \\ & \times \left(\tanh\frac{\varepsilon_1}{2T} + \coth\frac{\omega}{2T}\right).\end{aligned} \tag{2}$$

$$\mathrm{Re}\Sigma(k,\varepsilon) \propto P\int [\mathrm{Im}\Sigma(k,\varepsilon)/(\varepsilon - \omega')]\mathrm{d}\omega'/\pi. \tag{3}$$

Where $D_R(k,\omega) \sim \{g_{\nu\mu}/\left(\omega^2 - (k^2 + m_1^2) + \prod\right)\}$ is the Fourier transform of the Green function, $\langle A_\mu^a A_\nu^a \rangle_{a=1,2}$, of the massive gauge fields A_μ^1 and A_μ^2 in the 't Hooft-Feynman gauge. $A(p_1, \varepsilon_1)$ is the hole spectral function. The loss peaks of ARPES spectra occur when $\mathrm{Im}\Sigma(k,\varepsilon)$ is large. This loss feature, caused by massive collective gauge fields, will have a broad energy distribution because of the momentum dependence of the massive collective mode spectrum as well as the recoil energy of the hole when a massive collective excitation is emitted. That is, the emitted massive gauge fields A_μ^1 and A_μ^2 of the massive collective modes decay into other excitations such as spin, charge, and phonon waves, with the lifetime $\sim \hbar/Im\Pi$. It is suspected that incoherent Cooper pairs begins to be created around the hot spot, whose angle $\theta = 45^o$ just below the temperature $T^*(\hat{k}_F) \sim \alpha T_{res}(\hat{k}_F) \propto \mu(\hat{k}_F) \sim \mu(\hat{k}_F^h) \cdot sin^2(2\theta)$ of small pseudogap (SPG). Where α is $\sim 1/3$ [5]. Then formation of incoherent Cooper pairs extends gradually into the lower angle as temperature decreases from T^* to Tc. Below Tc, coherent Cooper pairs are created. This might correspond to the reduction of the Fermi arc as temperature decreases from T^* to Tc. Hashimoto et al [16] have performed a temperature-dependent angle-integrated photoemission study of $La_{2-x}Sr_xCuO_4$ covering from lightly doped to heavily overdoped regions. They found that the pseudogap Δ^* and temperature T^* increase as hole concentration decreases. From the effective Lagrangian [10], we get the effective value, $\mu_{eff}(\hat{k}_F)$, from the sum of the polarization bubble by hole-electron popagator in the random phase approximation as follows,

$$\begin{aligned}\mu_{eff}(\hat{k}_F) & \sim \frac{\mu(\hat{k}_F)}{1 + \sum_q D_R(q,\omega) \sum_k g_2^2 G(k+q)G(-k)} \\ & = \frac{\mu(\hat{k}_F)}{1 + \sum_q \sum_k M_{qk}}\end{aligned} \tag{4}$$

The matrix, M_{qk}, is $D_R(q,\omega)g_2^2G(k+q)G(-k)$. Where $D_R \sim g_{\mu\nu}/(\omega^2 - (q^2 + m_1^2) + \Pi)$ is the Green function of massive gauge fields A_μ^1 and A_μ^2 in the 't Hooft-Feynman gauge. $G(k)$ is the Green function of the hole on the Fermi arc. When hole-doping increases, the matrix, M_{qk}, increases. As a result, the effective value, $\mu_{eff}(\hat{k}_F)$, decreases, and $T^*(\hat{k}_F)$ decreases. This is consistent with experimental results [16].

Journal of Physics: Conference Series **391** (2012) 012128 doi:10.1088/1742-6596/391/1/012128

3. Conclusion

The evolution mechanism of the Fermi arc have been discussed in high-Tc cuprates by using quantized massive gauge fields, taking into account the restoration of spontaneous symmetry breaking. The present theoretical-formula indicates that the large pseudogap (LPG) and the small pseudogap (SPG) decrease as hole-doping increases. It is suggested strongly that the emitted massive gauge fields, which are mediating Cooper pairing in high-Tc cuprates, decay into other excitations such as spin, charge, and phonon waves.

References

[1] Shen Z X, and Dessan D S 1995 *Phys. Rep.* **251** 1
[2] Shen Z X, and Schriefer J R 1997 *Phys. Rev. Lett.* **78** 1771
[3] Hwang J et al. 2004 *Nature* **427** 714
[4] Carbotte J R et al. 2011 *Rep. Prog. Phys.* **74** 066501
[5] Dipasupil R M et al., 2002 *J.Phys.Soc.Jpn.* **71** 1535
[6] Emery V J, and Kivelson S A 1995 *Nature* **374** 434
[7] Wise W D et al. 2008 *Nature Phys.* **4** 696
[8] Yazdani A 2009 *J. Phys. Condens Matters* **21** 164214
[9] Affleck I et al. 1988 *Phys. Rev.* B **38** 745
[10] Kanazawa I 2003 *J. Phys.* A **36** 9371
[11] Hooft G't 1978 *Nucl. Phys.* B **135** 1
[12] Kanazawa I 1991 *Physica* C **185-189** 1703
[13] Kanazawa I 2005 *J. Phys. Soc. Jpn. Suppl.* **74** 200
[14] Kanazawa I 2005 *J. Phys. Chem. Solids* **66** 1388
[15] Sato T et al. 2002 *Phys. Rev. Lett.* **89** 067005
[16] Hashimoto M et al. 2007 *Phys. Rev.* B **75** 140503
[17] Kanazawa I 2002 *Physica* C **378-381** 278

International Conference on Strongly Correlated Electron Systems (SCES 2011) IOP Publishing
Journal of Physics: Conference Series **391** (2012) 012129 doi:10.1088/1742-6596/391/1/012129

Crystal structure refinement of Cu_6O_8YCl by powder neutron diffraction

Kenji Kawashima[a], **Kazumasa Horigane**[b], **Kenji Ohoyama**[c], **and Jun Akimitsu**[a]

a Department of Physics and Mathematics, Aoyama Gakuin University, Fuchinobe 5-10-1, Kanagawa 252-0206, Japan
b WPI-AIMR, Tohoku University, Sendai 980-8577, Japan
c Institute for Materials Research, Tohoku University, Sendai 980-8577, Japan

E-mail: `kawaken@phys.aoyama.ac.jp`

Abstract. We report the crystal structure parameters of clathrate-type copper-oxide Cu_6O_8YCl, which belongs to the Cu_6O_8MX (M = cation, X = anion) family, investigated by the powder neutron diffraction technique and the Rietveld method. The crystal structure of Cu_6O_8YCl was identified to be a cubic structure with the space group $Fm\bar{3}m$. The Cl atom is located at the 4a site, which is the center position of the Cu_6O_8 cage, and the Y atom occupies the 4b site in the cuboid space. The Y and Cl atoms have anomalously large atomic displacement parameters (U_{iso}) at room temperature, which suggests that Y and Cl atoms respectively present in a cuboid and in a Cu_6O_8 cage exhibit intensive motion such as rattling, as recently observed for clathrate compounds.

1. Introduction

Research on compounds with a caged structure has been stimulated by interesting observations, such as the relationship between their physical properties and the motion of the icluded ions in the cage [1, 2, 3].

Members of the copper-oxide family of Cu_6O_8MX (M = cation, X = anion) compounds have a Cu_6O_8-cage in their crystal structure. $Cu_6O_8PbCl_{1-x}$ is a mineral known as Murdochite and its crystal structure and electrical resistivity have been studied previously[4, 5, 6]. Cu_6O_8MX (M = Tb, Dy, Ho, Er, Yb, Lu, Y, Sc, In, and Cu_2 and X = Cl and NO_3) compounds have been synthesized as candidates for a cubic-type copper-oxide superconductor, and their magnetic and electrical properties have been investigated previously [7, 8, 9, 10, 11, 12, 13].

Cu_6O_8MCl is crystallized in a cubic structure with a space group of $Fm\bar{3}m$ (No. 225). A unique structural feature throughout the Cu_6O_8MX family is the location of the X anions at the centers of oversized Cu_6O_8 cages. The M ions are located in the cuboid spaces between the Cu_6O_8 cages. The Cu_6O_8 cages share faces to form a three-dimensional network as shown in Fig. 1. Previous works have revealed that the Cu, O, M, and Cl ions mainly occupy the Wyckoff positions of 24d, 32f, 4b, and 4a, respectively, with $Fm\bar{3}m$ symmetry. However, the atomic positions of the Y ions in Cu_6O_8YCl are the Wyckoff position of 24e $(x, 0, 0)$ with $x = 0.0591$, which slightly deviates from the Wyckoff position of 4b (0,0,0), with $Fm\bar{3}m$ symmetry [14]. In this case, the Y ions partially occupy the 24e Wyckoff position with occupancy $g \sim 0.167$. This suggests that ions existing in an oversized void space vibrate anharmonically, like the rattling

Published under licence by IOP Publishing Ltd

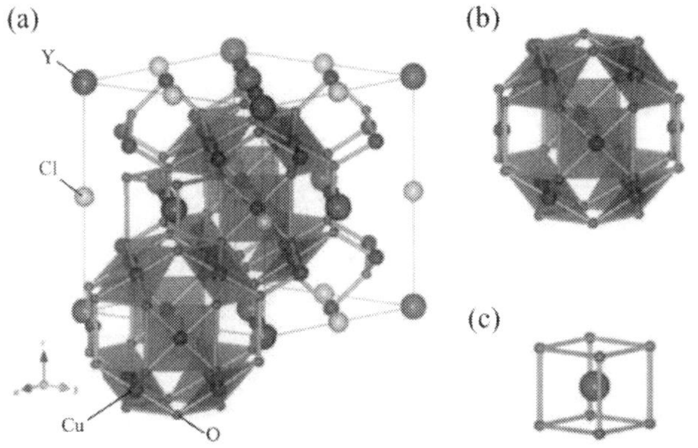

Figure 1. (Color online) The crystal structure of Cu_6O_8YCl ((a)) and enlarged views of the Cu_6O_8 cage ((b)) and the cuboid space including the Y ion ((c)). The solid line in Fig. 1(a) shows an unit cell of Cu_6O_8YCl. (The program VESTA was used.) [17].

phenomena observed in other cage-structured compounds. However, there is no additional information on the motion of the ions included in the cage.

In this paper, we report the results of the crystal structure refinement of Cu_6O_8YCl at room temperature using the powder neutron diffraction technique and the Rietveld method to obtain information on the motion of the included ions in a void space. In addition, we discuss the motion of the Y and Cl ions.

2. Experimental details

Polycrystalline samples of Cu_6O_8YCl were synthesized using the conventional solid state reaction. We used powdered $CuNO_3{\cdot}3H_2O$ (99.9 %), $CuCl_2{\cdot}2H_2O$ (99.9 %), and Y_2O_3 (99.9 %) as starting materials. Appropriate amounts of powders were placed in a glass beaker with distilled water and mixed together. The beaker was warmed to about 100 °C until all the water had evaporated. After evaporation, the powder was placed in an Al_2O_3 combustion boat. The temperature of the furnace was raised to 450 °C. After warming the system for stability, the boat was placed in the furnace and then left for 12 h in an O_2 atmosphere and quenched to room temperature. The color of the synthesized samples was dark black.

A neutron powder diffraction experiment was carried out using the *Kinken* powder diffractometer for high efficiency and high resolution measurements, HERMES, made at the Institute for Materials Research (IMR), Tohoku University, and installed in the guide hall of the JRR-3M reactor at the Japan Atomic Energy Agency (JAEA), Tokai [15]. The wavelength of the neutron beam was 1.8204(5) Å, monochromatized by Ge(331). The fine powder samples were sealed in a vanadium cylinder (ϕ10 mm) with helium gas. The neutron diffraction data were analyzed by the Rietveld method using the computer program RIETAN-FP [16]. Diffraction data were taken between $2\theta = 6°$ and 155° with an angle step of 0.1°.

3. Results and discussion

In accordance with an earlier report, the positions of the Cu, O, and Cl atoms were fixed at the 24d, 32f, and 4a sites, respectively. The Y atoms occupy the Wyckoff position 24e (x = 0.0591,0,0) with a site occupation factor g of 0.167 with $Fm\bar{3}m$ symmetry [14]. The M atoms occupy site 4b (0,0,0) as in other Cu_6O_8MCl compounds [8, 9]. In the data fitting, the site preference of framework atoms was tested by changing the Wyckoff position of the Y atom to examine the detailed crystal structure.

Figure 2(a) shows the final observed, calculated, and residual patterns for the refined powder neutron diffraction pattern of Cu_6O_8YCl at room temperature. The patterns of the calculated intensities are also shown in Fig. 2(b) and 2(c). The Cu, O, Cl, and Y atoms shown in Fig. 2(b)

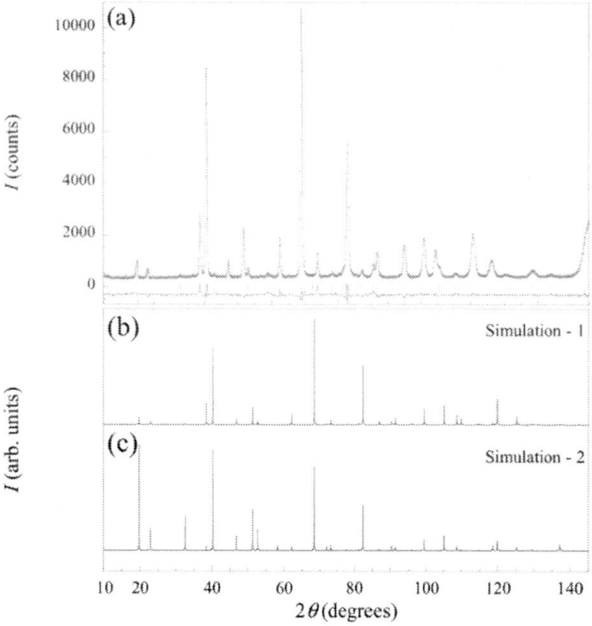

Figure 2. (Color online) Powder neutron diffraction patterns of Cu_6O_8YCl at room temperature (a) with simulation patterns (b) and (c), assuming that the Y occupy the 4b and 24e sites, respectively. In Fig. 2(a), the cross markers (exp) and the solid line (fit) represent the experimental and calculated patterns, respectively. The vertical markers represent the calculated Bragg angles. The lower solid line (diff) denotes the difference between the experimental and calculated intensities.

Table 1. Refined crystal parameters of Cu_6O_8YCl for the data in Fig. 2(a). The space group is $Fm\bar{3}m$ (No. 225) and $a = 9.256(1)$ Å. The R factors were $R_{wp} = 6.29\,\%$, $R_I = 3.53\,\%$, and $R_F = 2.86\,\%$.

atom	site	g	x	y	z	U_{iso} (10^{-2}Å^2)
Cu	24d	1	0	1/4	1/4	0.64(3)
O	32f	1	0.1437(1)	x	x	0.45(3)
Cl	4a	1	0.5	0.5	0.5	3.43(12)
Y	4b	1	0	0	0	3.20(15)

occupy the Wyckoff positions 24d, 32f, 4a, and 4b, respectively. On the other hand, the Cu, O, Cl, and Y atoms shown in Fig.2 (c) occupy the Wyckoff positions 24d, 32f, 4a, and 24e (x,0,0) with $x = 0.06$ and the site occupancy $g = 0.167$. Due to the powder neutron diffraction background, the lower angle part below 6.9° was not used in the analysis.

The shape of the experimentally observed diffraction intensity and peak positions in Fig. 2(a) correspond well with the simulation-1 data. The experimental diffraction pattern can be well fitted well assuming a cubic $Fm\bar{3}m$ structure. The calculated profiles obtained from simulation-1 and -2 data seem to agree well with the experimentally observed profiles. The convergence of the reliability gave $R_{wp} = 6.29\%$, $R_I = 3.53\%$, and $R_F = 2.86\%$ for the profile data used in Fig. 2(b), and $R_{wp} = 6.92\%$, $R_I = 4.96\%$, and $R_F = 3.62\%$ for the profile data used in Fig. 2(c). The reliability factor of the refinement result obtained by using the profile data of Fig. 2(b) is slightly better than that of the refinement result obtained using the profile data of the pattern in Fig. 2(c). However, the resulting isotropic temperature factors U_{iso} of the Y and Cl atoms obtained using the profile data of Fig. 2(c) have a large negative value ($\sim -4 \times 10^{-2}$ Å^2). This suggests that the structural parameters of Cu_6O_8YCl are in better agreement with the data of the simulation-1 pattern. Consequently, the Wyckoff positions of the Cu, O, Cl, and Y atoms are the 24d, 32f, 4a, and 4b sites, respectively.

The lattice constant a of Cu_6O_8YCl determined from the diffraction peaks is $a = 9.256(1)$ Å. The refinement crystal parameters are listed in Table 1. In a previous work, it was reported that the Y atoms partially the occupy 24e site which slightly deviates from the 4b (0,0,0) position, with $g \sim 0.167$. The most important finding in the present structural analysis is an unusually large atomic displacement parameter $U_{\rm iso}$ for the Y and Cl atoms, which is approximately five- (seven-) and five- (eight-) times greater than those of the Cu and O atoms that form the Cu_6O_8-cage. Similarly, large $U_{\rm iso}$ values have been observed in Si-clathrates (e.g. $U_{iso} = 3.3 \times 10^{-2}$ Å^2 for Ba atoms in the $Ba_8Ga_{16}Si_{30}$) [18]. The Y and Cl atoms exist in the center of cuboids and Cu_6O_8-cages, which have a large void space for the atoms. These facts suggest that both the Y and Cl atoms exhibit anomalous ion motion, such as rattling in an oversized void space.

In summary, we performed powder neutron diffraction measurements for copper-oxide Cu_6O_8YCl to obtain detailed information on the structure parameters and motion of the Y and Cl, which are in a void space in the crystal structure. The crystal structure of Cu_6O_8YCl is a cubic structure with the space group $Fm\bar{3}m$. The Y and Cl ions are located at the 4b and 4a sites, which respectively are located at the center of cuboids and the Cu_6O_8-cages. A large atomic displacement parameter was found for the Y and Cl ions. To clarify motion of the included ions at a low temperature and the relationship between these motions and the physical properties of the copper oxide, further research is required.

Acknowledge

This work was supported by the High-Tech Research Center Project for Private Universities from the Ministry of Education, Culture, Sports, Science and Technology (MEXT), Grant-in-Aid for Specially Promoted Research, and Grant-in-Aid for Scientific Research. The research was supported by Strategic International Collaborative Research Program (SICORP), Japan Science and Technology Agency.

[1] Y. Nagao, J. Yamaura, H. Ogusu, Y. Okamoto, and Z. Hiroi: J. Phys. Soc. Jpn. **78** (2009) 064702.
[2] E. D. Bauer, N. A. Frederick, P.-C. Ho, V. S. Zapf, and M. B. Maple: Phys. Rev. B **65** (2002) 100506.
[3] M. Udagawa, T. Hasegawa, Y. Takasu, N. Ogita, K. Suekuni, M. A. Avila, T. Takabatake, Y. Ishikawa, N. Takeda, Y. Nemoto, T. Goto, H. Sugawara, D. Kikuchi, H. Sato, C. Sekine, I. Shirotani, J. Yamaura, Y. Nagao, and Z. Hiroi: J. Phys. Soc. Jpn. **77** (2008) Suppl. A, 142E47
[4] E. Dubler, A. Vedani, and H. R. Oswald: Acta Cryst. C**39** (1983) 1143.
[5] Z. Medunić, P. Strobel, J. L. Hodeau, R. Argoud, D. Duževič, A. Božić, and D. Djurek: Mat. Res. Bull. **28** (1993) 741.
[6] B. Winkler, M. Chall, C. J. Pickard, V. Milman, and J. White: Acta Cryst. B**56** (2000) 22.
[7] I. Yazawa, R. Sugise, N. Terada, M. Jo, K. Oka, and H. Ihara: Jpn. J. Appl. Phys. **29** (1990) L1693.
[8] H. Hayakawa, E. Akiba, I. Yazawa, H. Ihara, and S. Ono: Jpn. J. Appl. Phys. **29** (1990) L1796.
[9] H. Hayakawa, E. Akiba, I. Yazawa, H. Ihara, and S. Ono: Jpn. J. Appl. Phys. **30** (1991) L1303.
[10] H. Hayakawa, R. Sugise, N. Terada, M. Jo, K. Oka, H. Hayakawa, and H. Ihara: Jpn. J. Appl. Phys. **29** (1990) L1480.
[11] Y. Sakata, S. Sakai, H. Imamura, and S. Tsuchiya: J. Mol. Catal. **88** (1994) 103.
[12] R. Sugise, K. Ohdan, T. Hamamoto, K. Kashiwagi, M. Shirai, I. Yazawa, and H. Ihara: Jpn. J. Appl. Phys. **32** (1993) L940.
[13] K. Bushida, I. Yazawa, G. Zouganel, T. Connard, N. Terada, K. Kaneko, M. Hirabayashi, and H. Ihara: Physica C **185-189** (1991) 2727.
[14] G. Zouganelis, K. Bushida, I. Yazawa, N. Terada, M. Jo, H. Hayakawa, and H. Ihara: Solid State Commun. **80** (1991) 709.
[15] K. Ohoyama, T. Kanouchi, K. Nemoto, M. Ohashi, T. Kajitani, and Y. Yamaguchi: Jpn. J. Appl. Phys. **37** (1998) 3319.
[16] F. Izumi and K. Momma, Sol. State Phenom., **130** (2007) 15.
[17] K. Momma and F. Izumi: Commission Crystallogr. Comput., IUCr Newslett. (2006) No. 7, 106.
[18] L. Qiu, I.P. Swainson, G.S. Nolas, and M.A. White: Phys. Rev. B **70** (2004) 035208.

Nuclear spin-lattice relaxation rate in the superconducting state on $La(Fe_{1-x}Zn_x)AsO_{0.85}$ studied by ^{75}As-NMR

S Kitagawa[1,2], Y Nakai[1,2], T Iye[1,2], K Ishida[1,2], Y F Guo[2,3], Y G Shi[2,3], K Yamaura[2,4] and E Takayama-Muromachi[2,3,4]

[1] Department of Physics, Graduate School of Science, Kyoto University, Kyoto 606-8502, Japan

[2] TRIP, JST, Sanban-cho bldg., 5, Sanban-cho, Chiyoda, Tokyo 102-0075, Japan

[3] International Center for Materials Nanoarchitectonics (MANA), National Institute for Materials Science, Tsukuba, Ibaraki 305-0044, Japan

[4] Superconducting Materials Center, National Institute for Materials Science, 1-1 Namiki, Tsukuba, 305-0044 Ibaraki, Japan

E-mail: `shunsaku@scphys.kyoto-u.ac.jp`

Abstract. ^{75}As nuclear spin-lattice relaxation rate $1/T_1$ was measured in Zn-free and Zn-substituted $La(Fe_{1-x}Zn_x)AsO_{0.85}$ ($x = 0$, and 0.05), T_c of which is 25 K and 0 K, respectively. Although the temperature dependence of $1/T_1T$ above T_c is the same between two samples, $1/T_1T$ of $x = 0$ shows $T^{3.0\pm0.1}$ dependence ($1/T_1 \propto T^4$) in the superconducting (SC) state, but $1/T_1T$ of non-SC $x = 0.05$ sample continuously decreases below 50 K. The normalized $1/T_1T$ $[(1/T_1T)_{x=0}/(1/T_1T)_{x=0.05}]$, which is related to the SC gap structure, is proportional to $T^{2.4\pm0.1}$ below T_c. These results indicate that the temperature dependence of $1/T_1T$ in the SC state can be affected by the temperature dependence of the normal-state $1/T_1T$.

Introduction

One of the most important issues in iron-based superconductors is to identify their superconducting(SC) gap structure and pairing symmetry experimentally. Until now, it has become clear that iron-based superconductors have a non-universal SC gap structure, i.e. various experiments suggest that RFeAs(O,F) (R:rare-earth elements) with the "1111" structure and (Ba,K)Fe_2As_2 with the "122" structure possess multi-SC gaps, but $BaFe_2(As,P)_2$ and $Ba(Fe,Co)_2As_2$ with heavily doping possess nodes in the SC gaps[1]. At present, the pairing symmetry and thus the SC mechanism are controversial issues, since the SC mechanism is closely related with the pairing symmetry. There are two candidates of the pairing symmetry for RFeAs(O,F); spin fluctuations mediated $s_{\pm}$-wave[2, 3, 4] and orbital fluctuations mediated s_{++}-wave[5].

In general, nuclear magnetic resonance (NMR) measurements give important information about the pairing symmetry. To determine the parity of SC pairing, the Knight-shift measurements give crucial information about the spin state of a SC pair. In addition, the measurements of nuclear spin-lattice relaxation rate $1/T_1$ give information about the SC gap structure. Existence of a significant coherence peak just below T_c, as well as an exponential temperature dependence of $1/T_1$ at low temperatures indicates that the SC gap is isotropic and finite. Temperature dependence of $1/T_1$ well below SC transition temperature T_c is related to low-energy quasiparticle excitations, and thus to a nodal structure.

In iron-based superconductors, there are many NMR studies reported in the SC state. Most Knight-shift measurements show that spin susceptibility decreases below T_c, indicative of a spin singlet pairing[6, 7]. Moreover, $1/T_1$ shows absence of a coherence peak just below T_c, suggestive of sign-changed SC gaps[8, 9] since a coherence factor of the sign-changed SC gaps vanishes when the coherence factor is integrated over the Brillouin zone. However, various temperature dependences of $1/T_1T$ below T_c have been reported; $1/T_1T \propto T^n$: $n \simeq 2 \sim 5$ has been observed[9, 8, 10, 11]. One of the reasons for this variety of temperature dependence of $1/T_1T$ in the SC state might be due to temperature dependence of $1/T_1T$ in the normal state. Temperature dependent normal-state is different with different carrier concentrations and affects the temperature exponent of $1/T_1T$ in the SC state. Indeed, the theoretical studies indicate that normal-state $1/T_1T$ is temperature dependent and is changed by doping due to the characteristic band dispersion around the Fermi energy[12, 13].

In this paper, we measured temperature dependence of $1/T_1T$ in Zn-free and Zn-substituted $La(Fe_{1-x}Zn_x)AsO_{0.85}$ ($x = 0$, and 0.05). Comparing with the SC $x = 0$ sample and the $x = 0.05$ sample in which superconductivity is suppressed by Zn-substitution, we found that $1/T_1T$ in the normal state is temperature dependent below 50 K and normalized SC-state $1/T_1T \propto T^{2.4\pm0.1}$ although $1/T_1T$ at $x = 0$ is proportional to $T^{3.0\pm0.1}$ below T_c. When we discuss the SC gap symmetry from $1/T_1T$, we should note that the temperature dependence of $1/T_1T$ in the SC state is affected by the normal-state $1/T_1T$, which is one of the reasons for various temperature dependences observed in the SC state.

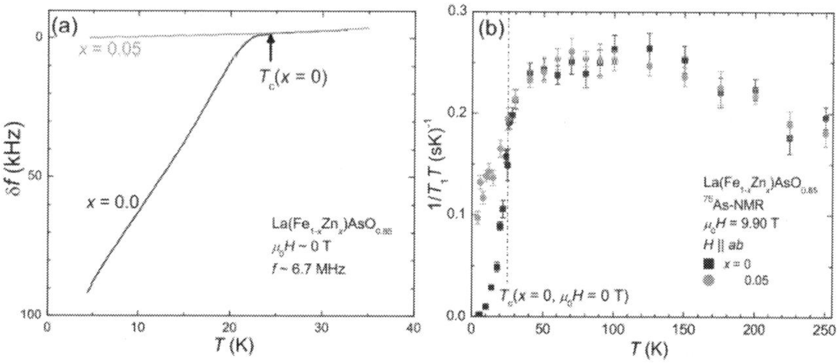

Figure 1. (Color online)(a) Meissner signals for $La(Fe_{1-x}Zn_x)AsO_{0.85}$ ($x = 0$, and 0.05) measured by an identical NMR coil. The arrow indicates T_c at x = 0. (b) Temperature dependence of $1/T_1T$ for $x = 0$, and 0.05. On cooling, $1/T_1T$ at $x = 0$ slightly increases down to 100 K, and decreases from $\sim$ 50 K. $1/T_1T$ in the normal state hardly changes by Zn substitution.

We performed ^{75}As-NMR measurements in polycrystalline samples of Zn-substituted $La(Fe_{1-x}Zn_x)AsO_{0.85}$ ($x = 0$, and 0.05), which were used in the previous reports[14]. T_c of the Zn-free $x = 0$ sample is 24 K, and the $x = 0.05$ sample does not show any Meissner signals as shown Fig. 1 (a), which is consistent with the previous reports[14, 15]. Conventional spin-echo technique was utilized for following NMR measurements.

Figure 1(b) shows the temperature dependence of $1/T_1T$ at 72.1 MHz ($\mu_0H = 9.90$ T) in $La(Fe_{1-x}Zn_x)AsO_{0.85}$ ($x = 0$, and 0.05). On cooling, $1/T_1T$ at $x = 0$ slightly increases down to 100 K, and decreases from $\sim$ 50 K above T_c. From comparing this result with previous NMR results in $LaFeAsO_{1-y}$ reported by Mukuda *et al.*[10], electron doping of our $x = 0$ sample is lower, since $1/T_1T$ at 100 K is larger than those in their underdoped $y \sim 0.2$ sample. $1/T_1T$ at $x = 0$, and 0.05 are almost the same above 24 K, suggesting that Zn substitution modifies neither carrier doping level nor Fermi surface properties.

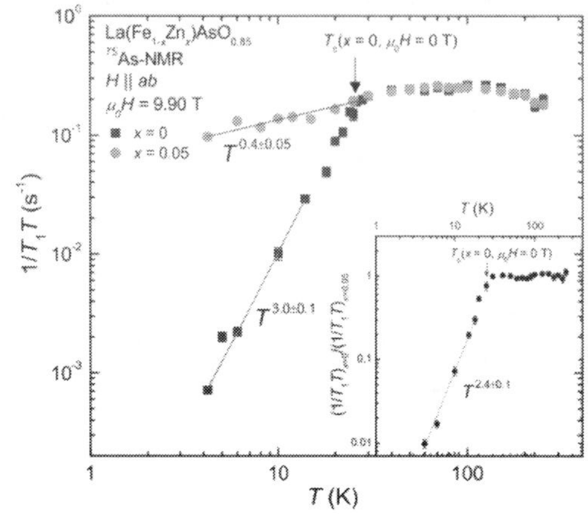

Figure 2. (Color online)(Main panel) Temperature dependence of $1/T_1T$ for $x = 0$, and 0.05. $1/T_1$ at $x = 0$ suddenly decreases due to opening SC gap and is proportional to $T^{3.0\pm0.1}$. On the other hand, $1/T_1$ at $x = 0.05$ is proportional to $T^{0.4\pm0.1}$ below 40 K although the $x = 0.05$ sample does not show superconductivity. Solid lines are corresponding to fitting lines. (Inset) Temperature dependence of normalized $1/T_1T$ $[(1/T_1T)_{x=0}/(1/T_1T)_{x=0.05}]$. Normalized $1/T_1T$ is proportional to $T^{2.4\pm0.1}$ below T_c. Solid line is corresponding to a fitting line.

Below 24 K, $1/T_1T$ in the Zn-free sample suddenly decreases due to opening SC gap, but $1/T_1T$ in the Zn-substituted sample gradually decreases below ~ 50 K. Since no Meissner signal is observed in the Zn-substituted sample, this decrease in $1/T_1T$ is not caused by superconductivity but can be ascribed to the characteristic band dispersion around the Fermi energy. In contrast, $1/T_1T$ of the Zn-free sample sharply decreases and is proportional to $T^{3.0\pm0.1}$ below 15 K down to 4.2 K, as shown in Fig. 2. In order to know the temperature dependence of $1/T_1T$ ascribed to the SC gap structure in the SC $x = 0$ sample, $1/T_1T$ of the $x = 0$ sample is divided by $1/T_1T$ of the non-SC $x = 0.05$ sample, which is shown in the inset of Fig. 2. Normalized $1/T_1T$ $[(1/T_1T)_{x=0}/(1/T_1T)_{x=0.05}]$ is proportional to $T^{2.4\pm0.1}$ below 20 K, which is close to the previous results ($1/T_1T \sim T^2$) in $LaFeAs(O_{0.89}F_{0.11})$ where $1/T_1T$ is constant in the temperature range between T_c and 50 K[9, 16]. This suggests that temperature dependence of $1/T_1T$ in the SC state can be affected by $1/T_1T$ in the normal state, which is estimated from a sample where superconductivity is suppressed by non-magnetic impurity or magnetic field.

Measurements of the temperature dependence of $1/T_1T$ at low temperatures help to determine which type of SC pairing is realized. For an s-wave superconductor, $1/T_1T$ shows activated temperature dependence, as expected from the existence of an isotropic gap. For a line-node superconductor, $N_s(E) \propto E$ for $E \ll \Delta_0$, leading to $1/T_1T \propto T^2$ where Δ_0 is a SC gap at 0 K. For a point-node superconductor, $N_s(E) \propto E^2$ for $E \ll \Delta_0$, leading to $1/T_1T \propto T^4$. These differences arise from quesiparticle excitations near the gap nodes. Then, normalized $1/T_1T$ $[(1/T_1T)_{\mathrm{SC}}/(1/T_1T)_{\mathrm{normal}}]$ can be described as follows:

$$\frac{1}{T_1T} \propto \begin{cases} \exp(-\Delta_0/k_{\mathrm{B}}T) & \text{for full gap,} \\ T^2 & \text{for line node gap,} \\ T^4 & \text{for point node gap.} \end{cases} \tag{1}$$

In the case of a simple $s_{\pm}$-wave superconductor, $1/T_1T$ follows $\exp(-\Delta_0/k_BT)$ at low temperatures, which is the same as in the case of a s_{++}-wave superconductor. On the other hand, in the case of an anisotropic SC gap model incorporated with impurities or a multi-SC gap effect[17, 18, 19], $1/T_1T$ approaches T^2 similar to behaviors in nodal gaps. It should be also noted that a development of SC gap just below T_c strongly affects the temperature variation of $1/T_1T$. Therefore, to distinguish type of SC pairing, we should check the temperature exponent of $1/T_1T$ well below T_c. Our experimental data can be explained by a line-node d-wave or an $s_{\pm}$-wave model with highly anisotropic SC gaps. However, in a superconductor with nodal SC gaps, residual density of states is easily introduced by tiny amount of impurities and $1/T_1T$ becomes constant at low temperatures. Considering that our samples are polycrystalline with oxygen deficiency, line node d-wave would be excluded. Then, an $s_{\pm}$-wave is the most promising pairing symmetry of $LaFeAsO_{0.85}$.

In summary, we have measured ^{75}As-$1/T_1T$ in Zn-free and Zn-substituted $La(Fe_{1-x}Zn_x)AsO_{0.85}$ ($x = 0$, and 0.05). $1/T_1$ at $x = 0$, and 0.05 are almost the same above T_c of $x = 0$, suggesting that the temperature dependent normal state are the same. $1/T_1T$ of the Zn-free sample sharply decreases and is proportional to $T^{3.0\pm0.1}$ below T_c down to 4.2 K, but $1/T_1T$ in the non-SC $x = 0.05$ sample shows gradual decrease below ~ 50 K. Normalized SC-state $1/T_1T$ $[(1/T_1T)_{x=0}/(1/T_1T)_{x=0.05}]$ is proportional to $T^{2.4\pm0.1}$, which is close to the previous reports. We suggest that temperature dependence of $1/T_1T$ in the normal state is one of the reasons that various temperature dependences of $1/T_1T$ have been observed in the SC state in La-based "1111" superconductors.

This work was supported by a Grant-in-Aid for the Global COE Program "The Next Generation of Physics, Spun from Universality and Emergence" from MEXT, and for Scientific Research from JSPS. This was also supported in part by "Funding Program for World-Leading Innovative R&D on Science and Technology (FIRST Program)".

[1] Paglione J and Greene R L 2010 *Nature Physics* **6** 645
[2] Mazin I I, Singh D J, Johannes M D and Du M H 2008 *Phys. Rev. Lett.* **101** 057003
[3] Kuroki K, Onari S, Arita R, Usui H, Tanaka Y, Kontani H and Aoki H 2008 *Phys. Rev. Lett.* **101** 087004
[4] Cvetkovic V and Tesanovic Z 2009 *Europhys. Lett.* **85** 37002
[5] Kontani H and Onari S 2010 *Phys. Rev. Lett.* **104** 157001
[6] Kawabata A, Lee S C, Moyoshi T, Kobayashi Y and Sato M 2008 *J. Phys. Soc. Jpn. Suppl. C* **77** 147
[7] Terasaki N, Mukuda H, Yashima M, Kitaoka Y, Miyazawa K, Shirage P M, Kito H, Eisaki H and Iyo A 2009 *J. Phys. Soc. Jpn.* **78** 013701
[8] Grafe H J, Paar D, Lang G, Curro N J, Behr G, Werner J, Hamann-Borrero J, Hess C, Leps N, Klingeler R and Buchner B 2008 *Phys. Rev. Lett.* **101** 047003
[9] Nakai Y, Ishida K, Kamihara Y, Hirano M and Hosono H 2008 *J. Phys. Soc. Jpn.* **77** 073701
[10] Mukuda H, Terasaki N, Yashima M, Nishimura H, Kitaoka Y and Iyo A 2009 *Physica C* **469** 559
[11] Sato M, Kobayashi Y, Lee S C, Takahashi H, Satomi E and Miurao Y 2010 *J. Phys. Soc. Jpn.* **79** 014710
[12] Ikeda H 2008 *J. Phys. Soc. Jpn.* **77** 123707
[13] Ikeda H, Arita R and Kuneš J 2010 *Phys. Rev. B* **82** 024508
[14] Kitagawa S, Nakai Y, TIye, Ishida K, Kamihara Y, Hirano M and Hosono H 2010 *Phys. Rev. B* **81** 212502
[15] Guo Y F, Shi Y G, Yu S, Belik A A, Matsushita Y, Tanaka M, Katsuya Y, Kobayashi K, Nowik I, Felner I, Awana V P S, Yamaura K and Takayama-Muromachi E 2010 *Phys. Rev. B* **82** 054506
[16] Nakai Y, Kitagawa S, Ishida K, Kamihara Y, Hirano M and Hosono H 2009 *New J. Phys.* **11** 045004
[17] Parker D, Dolgov O V, Korshunov M M, Golubov A A and Mazin I I 2008 *Phys. Rev. B* **78** 134524
[18] Nagai Y, Hayashi N, Nakai N, Nakamura H, Okumura M and Machida M 2008 *New J. Phys.* **10** 103026
[19] Kinouchi H, Mukuda H, Yashima M, Kitaoka Y, Shirage P M, Eisaki H and Iyo A 2011 *Phys. Rev. Lett.* **107** 047002

Variation of superconductivity in the pseudoternary layer systems $Sr(T_{1-x}T'_x)_2Ge_2$ (T, T' = transition metals)

H C Ku[1], J W Wang[1], I A Chen[1], Y B You[1], and Y Y Hsu[2]

[1] Department of Physics, National Tsing Hua University, Hsinchu, Taiwan, Republic of China
[2] Department of Physics, National Taiwan Normal University, Taipei, Taiwan, Republic of China

E-mail: hcku@phys.nthu.edu.tw

Abstract. Variation of superconducting transition temperature T_c for the pseudoternary $Sr(T_{1-x}T'_x)_2Ge_2$ layer systems (T, T' = Co, Ni, Rh, Pd; $0 \leq x \leq 1$) are reported. For the system $Sr(Pd_{1-x}Rh_x)_2Ge_2$, T_c decreases from 3.12 K for $x = 0$, to 2.92 K for $x = 0.1$, to 2.50 K for $x = 0.2$, and extrapolated to 0 K for $x = 0.65$. No T_c is expected for $SrRh_2Ge_2$. For the system $Sr(Pd_{1-x}Co_x)_2Ge_2$, T_c decreases sharply to 2.80 K for $x = 0.01$, 2.58 K for $x = 0.03$, and extrapolates to 0 K for $x = 0.15$. No T_c is also expected for $SrCo_2Ge_2$. In these systems, stronger T_c suppression for Co($3d^7$) and Rh($4d^7$) were due to lower density of states in $d_{xz,yz}$ bands with one less conduction electron per transition metal. Weaker T_c suppression for isoelectronic Ni is due to shift from Pd-$4d^8$ to Ni-$3d^8$ band. T_c decreases monotonically from 3.12 K for $SrPd_2Ge_2$ to 0.92 K for $SrNi_2Ge_2$. The lower T_c of the present electron-overdoped (nd^7 or nd^8) compounds are due to dispersive 3D-like $nd_{xz,yz}$ conduction bands with weaker electron correlation, as compared to the less-electron-doped ($3d^{6.1}$) 22 K superconductor $BaFe_{1.8}Co_{0.2}As_2$ or the hole-doped ($3d^{5.9}$) 38 K superconductor $Ba_{0.6}K_{0.4}Fe_2As_2$ with additional less dispersive 2D-like $3d_{xy}$ band contribution and stronger electron correlation.

1. Introduction

The recent discovery of high superconducting transition temperatures T_c up to 55 K in $LaFeAs(O_{1-x}F_x)$ has generated profound interest in FeAs-based systems.[1] For the AFe_2As_2 (122) (A = Ca, Sr, or Ba) layer system with the $ThCr_2Si_2$-type body-centered-tetragonal structure, the $(FeAs)^{1-}$ layers are separated by A^{2+} layers. The parent compound $BaFe_2As_2$ is a poor metal with an antiferromagnetic spin density wave (SDW) transition at $T_N = 140$ K. T_c up to 38 K were reported in the hole-doped $(Ba_{0.6}K_{0.4})Fe_2As_2$ and $(Sr_{0.6}K_{0.4})Fe_2As_2$, where A^{2+} sites are partially substituted by K^{1+} ions,[2, 3] and T_c up to 22 K were reported by the electron-doped $BaFe_{1.8}Co_{0.2}As_2$, where Fe($3d^6$) are partially substituted by Co($3d^7$).[4]

Superconductivity was also discovered in the arsenic-free ternary germanide system SrT_2Ge_2 (T = transition metals) with the same $BaFe_2As_2$-type (122) structure where $(PdGe)^{1-}$ layers are separated by Sr^{2+} layers.[5, 6, 7, 8, 9] T_c up to 3.12 K was reported for $SrPd_2Ge_2$.[5, 8, 9] Lower $T_c = 0.92$ K for $SrNi_2Ge_2$ was recently reported by our group.[7, 9]

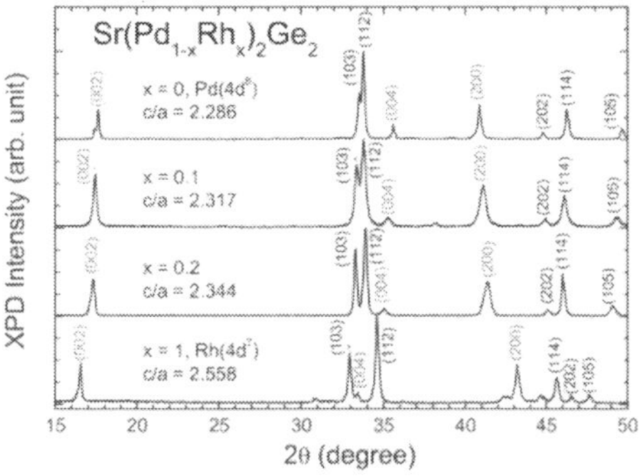

Figure 1. X-ray powder diffraction (XPD) patterns for the tetragonal pseudoternary $Sr(Pd_{1-x}Rh_x)_2Ge_2$ system (x = 0, 0.1, 0.2, 1).

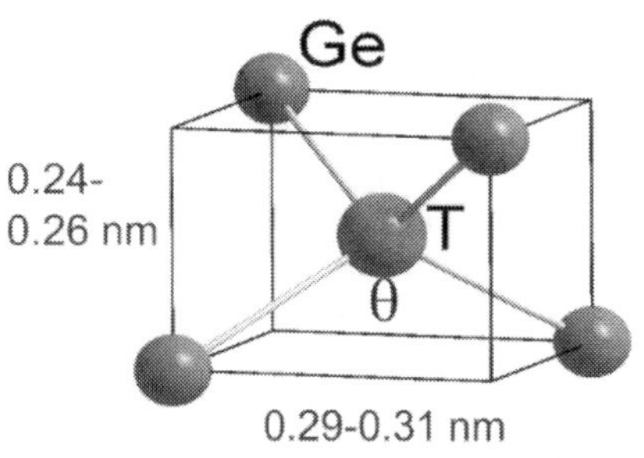

Figure 2. The c-axis squeezed TGe_4 tetrahedron in the T-Ge layer (T = Co, Ni, Rh, Pd). The bond angle θ(Ge-T-Ge) = 118°–122° > 109.3°.

2. Experiment

The pseudoternary $Sr(T_{1-x}T'_x)_2Ge_2$ samples (T, T' = Co, Ni, Rh, Pd; $0 \leq x \leq 1$) were prepared by two-step arc-melting under argon atmosphere.[9] The x-ray powder diffraction data were collected by a Rigaku Rotaflex 18-kW rotating anode diffractometer with graphite monochromatized Cu-K_α radiation. The magnetic susceptibility data was collected with a Quantum Design 1-T μ-metal shielded $MPMS_2$ superconducting quantum interference device (SQUID) magnetometer down to 2 K.

3. Results and Discussion

The x-ray powder diffraction (XPD) patterns for four pseudoternary $Sr(Pd_{1-x}Rh_x)_2Ge_2$ samples (x = 0, 0.1, 0.2, 1) are shown collectively in Fig. 1. All samples crystallized with the $BaFe_2As_2$-type (122) structure of space group $I4/mmm$ (Z = 2), where the $(TGe)^{1-}$ layers are separated by Sr^{2+} layers. Within the layer, the TGe_4 tetrahedron as shown in Fig. 2 was squeezed along the c axis where T at (4a)(0,1/2,1/4) and Ge at (4e):(0,0,z) with internal coordinate z = 0.370 for T = Pd and 0.368 for Rh.[9]

The T-T bond length d(T-T) = $a/\sqrt{2}$ decreases with reducing Pd contents, from d(Pd-Pd) = 0.313 nm to d(Rh-Rh) = 0.296 nm. The long T-T bond length indicates weak direct electron T-T hopping within the T-Ge layer. An anomalous c/a ratio increases from 1.0104 nm/0.4420 nm = 2.286 for T = Pd($4d^8$) to 1.0724 nm/0.4193 nm = 2.558 for Rh($4d^7$) due to the squeezed TGe_4 tetrahedron along c axis, but is still smaller than c/a = 1.2980 nm/0.3964 nm = 3.275 for the 22 K superconductor $BaFe_{1.8}Co_{0.2}As_2$.[4]

All bond angles θ(Ge-T-Ge) in the squeezed tetrahedron along c axis of the pseudoternary systems $Sr(Pd_{1-x}T_x)_2Ge_2$ (T = Co, Ni, Rh) are larger than 109.3° for unsqueezed ideal tetrahedron. Bond angle of 122.5° for Pd with eight $4d$-electrons is similar to 122.4° for the isoelectronic Ni with eight $3d$-electrons, but is greater than 118.2° for Co with seven $3d$-electrons and 116.1° for Rh with seven $4d$-electrons.[9]

The five degenerate nd orbitals were splitted into up-lying t_{2g} triplet and lower-lying e_g doublet in the undistorted tetrahedral crystal field with bond angle of 109.3°. In the c-axis squeezed TGe_4 tetrahedral crystal field, the t_{2g} triplet are further splitted into one up-lying doublet $d_{xz,yz}$ and one lower-lying d_{xy} level. The e_g doublet is splitted into $d_{3z^2-r^2}$ and $d_{x^2-y^2}$ levels. The conduction bands are formed by $d_{xz,yz}$ and d_{xy} orbitals with light electron-like bands from up-lying $d_{xz,yz}$ doublet and heavy hole-like band from lower lying d_{xy} level. With eight

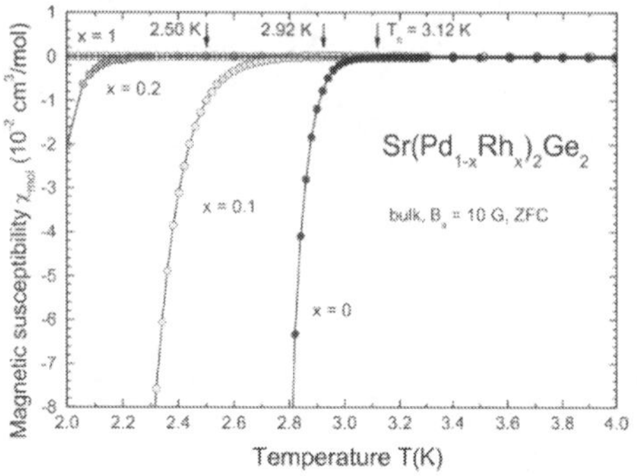

Figure 3. Zero-field-cooled (ZFC) molar magnetic susceptibility for the pseudoternary $Sr(Pd_{1-x}Rh_x)_2Ge_2$ superconducting samples ($x = 0, 0.1, 0.2, 1$). No T_c down to 2 K was observed for $SrRh_2Ge_2$.

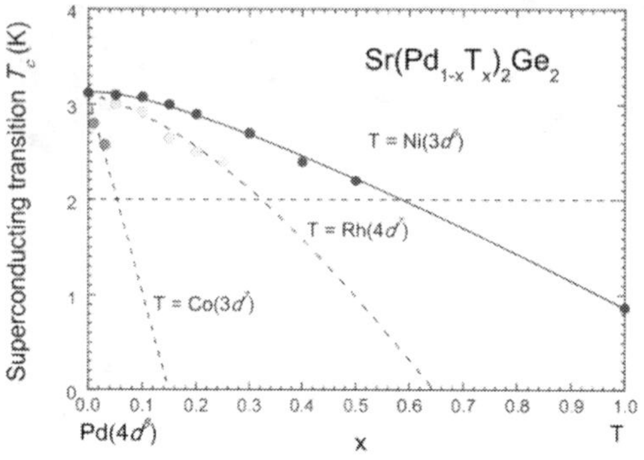

Figure 4. Variation of superconducting transition temperature T_c for the pseudoternary $Sr(Pd_{1-x}T_x)_2Ge_2$ layer system (T = Co, Ni, Rh). No superconductivity is expected for ternary $SrCo_2Ge_2$ and $SrRh_2Ge_2$.

$4d$ electrons in T = Pd, the d_{xz} and d_{yz} bands are close to half-filled, with some hole pockets in lower d_{xy} band. With seven $4d$ electrons in T = Rh, the d_{xz} and d_{yz} bands are less than half-filled, with some hole pockets in lower d_{xy} band.

The variation of superconducting T_c for the pseudoternary $Sr(Pd_{1-x}Rh_x)_2Ge_2$ system is shown in Figure 3. T_c onset decreases from 3.12 K for $x = 0$, to 2.92 K for $x = 0.1$, to 2.50 K for $x = 0.2$, and no T_c down to 2 K can be detected for $x = 1$. Variation of T_c for three pseudoternary systems $Sr(Pd_{1-x}T_x)_2Ge_2$ (T = Co, Ni, Rh) are shown collectively in Fig. 4. For T = Rh, T_c extrapolated to 0 K for $x = 0.65$ and no T_c is expected for $SrRh_2Ge_2$. For T = Co, T_c decreases sharply to 2.80 K for $x = 0.01$, and 2.58 K for $x = 0.03$, and extrapolates to 0 K for x = 0.15. No T_c is also expected for $SrCo_2Ge_2$. In these systems, stronger T_c suppression for Co($3d^7$) and Rh($4d^7$) were observed due to low DOS in $d_{xz,yz}$ bands. Weaker T_c suppression for isoelectronic Ni($3d^8$) is due to shift from $4d$ to $3d$ band. T_c decreases monotonically from 3.12 K for $SrPd_2Ge_2$ with Pd($4d^8$) to 0.92 K for $SrNi_2Ge_2$ with Ni($3d^8$).

The LDA band structure calculation for $SrPd_2Ge_2$ suggests higher dispersion energy bands across Fermi energy E_F with multi-sheet Fermi surface, low total DOS at E_F with partial DOS contributions from Pd-$4d$ and Ge-$4p$.[6] However, a small but non-zero on-site Coulomb repulsion U ($U \sim 0.5$ band width W) is probably needed in the LDA+U calculations for this electron-correlated system to get a more realistic band structure. The lower T_c of this electron-overdoped ($4d^8$) compound is probably due to more dispersive 3D-like $3d_{xz,yz}$ conduction bands, as compared with the less-electron-doped ($3d^{6.1}$) 22 K superconductor $BaFe_{1.8}Co_{0.2}As_2$ or the hole-doped ($3d^{5.9}$) 38 K superconductor $Ba_{0.6}K_{0.4}Fe_2As_2$ with less dispersive 2D-like $3d_{xy}$ conduction band.[2, 3, 4]

The x-ray powder diffraction (XPD) patterns for five pseudoternary $Sr(Ni_{1-x}Co_x)_2Ge_2$ samples ($x = 0, 0.25, 0.5, 0.75, 1$) are shown collectively in Fig. 5. Figure 6 presents variation of T_c and bond lengths d(T-Ge), d(T-T) of the squeezed TGe$_4$ tetrahedron in the $(T\mathrm{Ge})^{1-}$ layer for the combined systems $Sr(Pd_{1-x}Ni_x)_2Ge_2$ and $Sr(Ni_{1-x}Co_x)_2Ge_2$. The d(T-T) bond length decreases from 0.313 nm for T = Pd to 0.288 nm for Co, and enhances direct hopping in basal plane thus suppresses the T_c. The T-Ge bond length d(T-Ge) $= [a^2/4+(z-1/4)^2c^2]^{1/2}$ decreases with decreasing transition metal T size, from 0.252 nm for Pd to 0.237 nm for Co. The short T-Ge bond length indicates strong hybridization between T-nd and Ge-$4p$ orbitals, with delocalize, itinerant electron hopping through this channel. The symmetry of superconductivity for this low

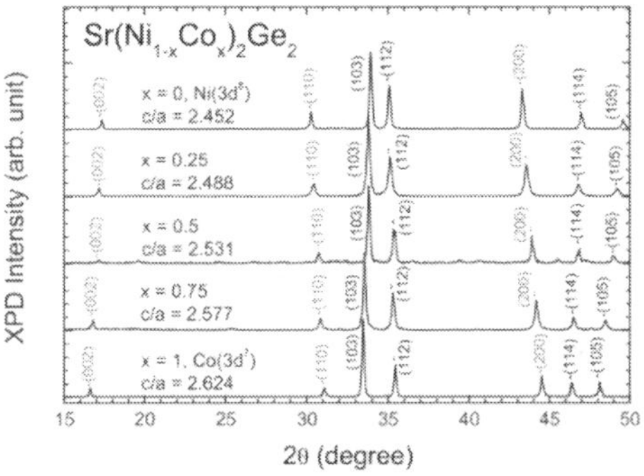

Figure 5. X-ray powder diffraction (XPD) patterns for the pseudoternary $Sr(Ni_{1-x}Co_x)_2Ge_2$ tetragonal system ($x = 0$, 0.25, 0.5, 0.75, 1).

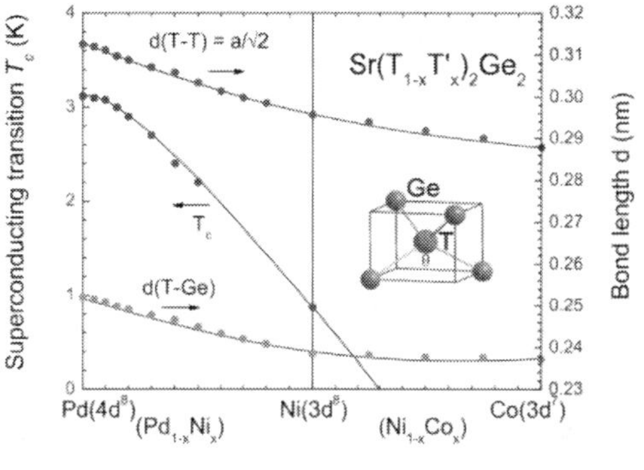

Figure 6. Variation of T_c and bond lengths d(T-Ge), d(T-T) in the T-Ge layer for the $Sr(Pd_{1-x}Ni_x)_2Ge_2$ and $Sr(Ni_{1-x}Co_x)_2Ge_2$ systems.

T_c system is probably a multi-band s-wave symmetry and the mechanism of superconductivity may require structural related electron-correlation with phonon mediation.

4. Conclusion

Pseudoternary $Sr(T_{1-x}T'_x)_2Ge_2$ system (T, T' = Co, Ni, Rh, Pd) is a tetragonal layer system where nd orbitals of transition metal T are split by the squeezed TGe_4 tetrahedral crystal field. At normal state, the system is weakly electron correlated and close to Fermi liquid regime. The lower T_c of the electron-overdoped $SrPd_2Ge_2$ and $SrNi_2Ge_2$ compounds are the results of close to half-filled dispersive 3D-like $nd_{xz,yz}$ conduction bands. No superconductivity is expected for $SrRh_2Ge_2$ and $SrCo_2Ge_2$ compounds with lower density of states in these bands due to one less electron per transition metal.

Acknowledgments

Supported by NSC of ROC Grant No. 98-2811-M-007-040 and 99-2811-M-007-095.

References

[1] Kamihara Y, Watanabe T, Hirano M, and Hosono H 2008 *J. Am. Chem. Soc.* **130** 3296
[2] Rotter M, Tegel M and Johrendt D 2008 *Phys. Rev. Lett.* **101** 107006
[3] Sasmal K, Lv B, Lorenz B, Guloy A M, Chen F, Xue Y-Y, and Chu C-W 2008 *Phys. Rev. Lett.* **101** 107007
[4] Sefat A S, Jin R, McGuire M A, Sales B C, Singh D J and Mandrus D 2008 *Phys. Rev. Lett.* **101** 117004
[5] Fujii H and Sato A 2009 *Phys. Rev. B* **79** 224522
[6] Shein I R and Ivanovskii A L 2010 *Physica B* **405** 3213
[7] Yang C D, Hsu H C, Tseng W Y, Chen H C, Ku H C, Ou M N, Chen Y Y, and Hsu Y Y 2011 *J. Phys.: Conf. Ser.* **273** 012089
[8] Sung N H, Rhyee J-S and Cho B K 2011 *Phys. Rev. B* **83** 094511.
[9] Wang J W, Chen I A, You Y B, Ku H C, and Hsu Y Y 2012 *Phys. Rev. B* **85** 024538

Effects of Pressure and Magnetic Field on the Pressure-Induced Superconductivity in $EuFe_2As_2$

N Kurita[1,2], M Kimata[3], K Kodama[1,4], A Harada[1], M Tomita[1], H S Suzuki[1], T Matsumoto[1], K Murata[5], S Uji[1,2,4] and T Terashima[1,2]

[1]National Institute for Materials Science, Tsukuba, Ibaraki 305-0003, Japan
[2]JST, Transformative Research-Project on Iron Pnictides (TRIP), Chiyoda, Tokyo 102-0075, Japan
[3]Institute for Solid State Physics, The University of Tokyo, Kashiwanoha, Kashiwa, Chiba 277-8581, Japan
[4]Graduate School of Pure and Applied Sciences, University of Tsukuba, Ibaraki 305-0003, Japan
[5]Department of Physics, Graduate School of Science, Osaka City University, Osaka 558-8585, Japan

E-mail: KURITA.Nobuyuki@nims.go.jp

Abstract.
We have performed electrical resistivity measurements of $EuFe_2As_2$ under hydrostatic pressures (P) up to 3.2 GPa and magnetic fields up to 15 T. In zero field, the superconducting (SC) ground state with a sharp transition to zero resistivity, indicative of bulk superconductivity, emerges at $T_c \sim 30$ K in a pressure range from 2.5 GPa to $\sim$3.0 GPa. Under applied field of 15 T, the SC transition is suppressed to lower temperatures with increasing pressure, suggesting that the application of pressure reduces the upper critical field B_{c2} for the P-induced superconductivity. Meanwhile, the antiferromagnetic (AF) order of the Eu^{2+} moments survives up to 3.2 GPa, the highest pressure in the experiments, with no significant change in the AF ordering temperature and the magnetic field effect.

1. Introduction

Since the discovery of superconductivity in LaFeAs(O,F) with $T_c = 26$ K [1], a number of experimental and theoretical researches have been performed on a group of materials with stacked Fe-pnictide (or Fe-chalcogenide) layers [2]. In the RFeAsO (R = rare earth) and AFe_2As_2 (A = alkaline earth or Eu) systems, high-T_c superconductivity can be driven with suppression of the antiferromagnetic (AF) order of Fe moments at T_0, by doping [1] or application of pressure (P) [3]. It is, therefore, of importance to elucidate the interplay between superconductivity and magnetism, which could lead to a better understanding for the high-T_c mechanism in Fe-based systems. For this purpose, application of pressure is an excellent probe for systematically tuning ground states, without a random potential, generally produced by elemental substitutions.

In recent works, we have revealed that $EuFe_2As_2$ exhibits P-induced superconductivity with $T_c \sim 30$ K, at the critical pressure (P_c) of 2.5 - 2.7 GPa where $T_0 \to 0$ and the bulk superconductivity emerges [4–8]. $EuFe_2As_2$ is a unique compound in that the AF order of the Eu^{2+} moments at $T_N \sim 20$ K [9–12] is less sensitive to applied pressure [4, 8, 13, 14], and is detectable even inside the bulk superconducting (SC) phase [4, 14]. This leads to the peculiar temperature dependence of the upper critical field B_{c2} for the P-induced superconductivity [4, 7].

Here, we present the evolution of the P-induced superconductivity in $EuFe_2As_2$ as a function of pressure and magnetic field, via the high-pressure and high-field resistivity measurements.

2. Experimental Details

Single crystals of $EuFe_2As_2$ were grown by Bridgman method from a stoichiometric mixture of the constituent elements. The crystal examined in this study is sample #3 (Residual resistivity ratio = 7, $P_c \sim 2.5$ GPa) as used in Ref. [8]. High-pressure resistivity measurements up to 3.2 GPa were performed in an Oxford ^{4}He system equipped with a 17 - T SC magnet, using a hybrid-type piston cylinder pressure device [15]. The resistivity was measured by a four-probe method with an ac current $I = 0.3$ mA under applied fields (B) up to 15 T for $B \parallel I \parallel ab$. To generate hydrostatic pressure, Daphne 7474 (Idemitsu Kosan), which remains in a liquid state up to 3.7 GPa at room temperature [16], was used as a pressure-transmitting medium. The generated pressure was determined at 4.2 K from the resistance change of a calibrated Manganin wire [4]. In this study, B denotes an externally applied field, and the magnetization within a sample (up to ~ 0.9 T [5]) is neglected.

3. Results and Discussions

Figure 1 shows the temperature dependence of the resistivity, $\rho(T)$, in $EuFe_2As_2$ at several pressures from 2.4 GPa to 3.2 GPa under zero applied field. At 2.4 GPa ($< P_c$), the $\rho(T)$ curve shows a partial SC transition below the onset temperature of 31 K, with the reentrant-like hump around 20 K due to the AF order of the Eu^{2+} moments [4, 13, 14]. As was revealed by ac-susceptibility and ac-calorimetry measurements for $P < P_c$ [4, 14], the partial SC transition with nonzero resistivity is probably filamentary. At 2.5 GPa ($\sim P_c$), one can see a sharp SC transition to zero resistivity, indicative of bulk superconductivity [4, 14]. With increasing pressure, the SC transition gradually shifts to lower temperatures. Judging from the sharp SC transitions to zero-resistivity, the bulk SC phase exists in narrow pressures of 2.5 GPa to ~ 3.0 GPa. At 3.1 GPa,

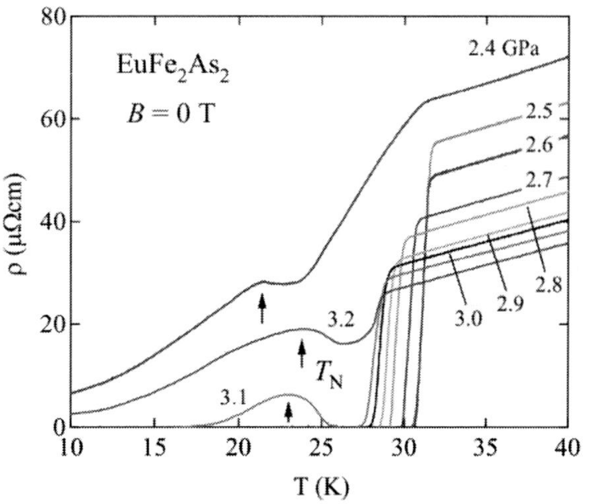

Figure 1. (Color online) Temperature dependence of the resistivity, $\rho(T)$, in $EuFe_2As_2$ at several pressures from 2.4 GPa to 3.2 GPa under zero applied field. Arrows indicate the AF ordering temperature, T_N, of the Eu^{2+} moments.

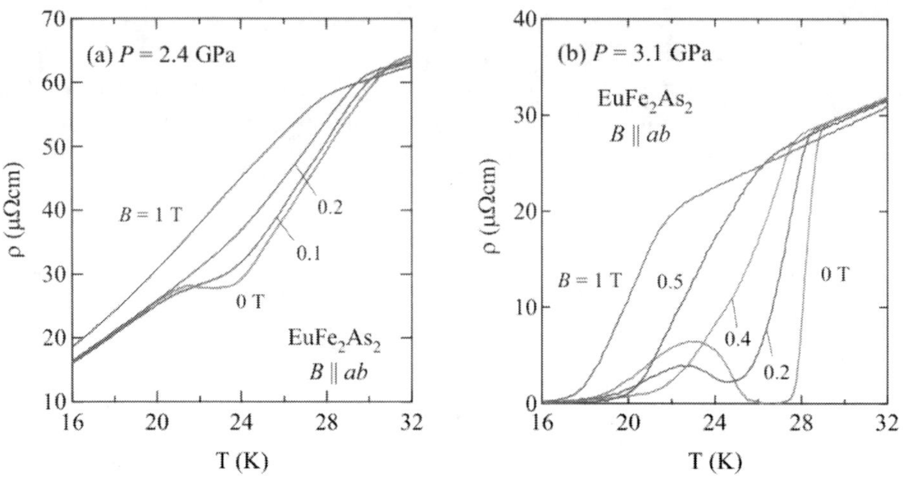

Figure 2. (Color online) $\rho(T)$ data in $EuFe_2As_2$ at (a) 2.4 GPa and (b) 3.1 GPa under several magnetic fields up to 1 T.

with decreasing temperature, $\rho(T)$ exhibits a slightly broader transition to zero resistivity, followed by a humplike behavior, and then returning to zero resistivity again. At 3.2 GPa, the $\rho(T)$ curve with the humplike behavior no longer reaches zero down to the lowest temperature of 1.6 K. It is noted that the resistive hump is observed at similar temperatures at 2.4, 3.1, and 3.2 GPa as indicated by arrows. In addition, as shown in Figs. 2 (a) and (b), the resistive hump at 2.4 and 3.1 GPa is rapidly suppressed under applied field less than 1 T, similar to the case at $P = 2.6$ GPa [4]. These results are consistent with Ref. [14] which demonstrates that the AF order of the Eu^{2+} moments remains up to ~ 8 GPa with only a moderate increase in T_N.

Figure 3 shows the magnetic field effect on the $\rho(T)$ data in $EuFe_2As_2$ at $P = 2.6$ GPa and 2.9 GPa. At 2.6 GPa, the SC transition is observed at ~ 10 K under the highest measured field of

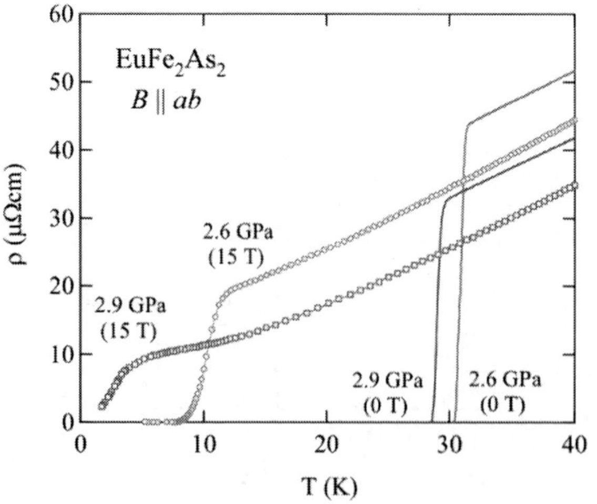

Figure 3. (Color online) ρ vs T of $EuFe_2As_2$ at $P = 2.6$ GPa and 2.9 GPa under $B = 0$ T and 15 T.

15 T. This is consistent with the $B_{c2}-T_c$ phase diagram at 2.5 GPa which we recently established using a 27-T resistive magnet [7]. The $B_{c2}(T=0)$ value for $B \parallel ab$ at 2.5 GPa is obtained to be 20 T or 25 T by the zero resistivity or the onset of resistive transitions, respectively [7]. At the higher pressure of 2.9 GPa, the SC transition under 15 T is suppressed to lower temperatures and longer approaches zero resistivity down to 1.6 K. This could indicate that the $B_{c2}(0)$ value attains its maximum at $P \sim P_c$, followed by a decrease as further pressure is applied. Details of this work including the pressure evolution of the $B_{c2}-T_c$ phase diagram will be presented elsewhere [17].

4. Conclusions

We have investigated pressure and magnetic field effects on the P-induced superconductivity in $EuFe_2As_2$, via the high-pressure and high-field resistivity measurements. In zero field, a sharp SC transition to zero resistivity, indicating bulk superconductivity, appears in a limited pressure range of 2.5 GPa ($\sim P_c$) to $\sim$3.0 GPa. Under applied field of 15 T, application of pressure suppresses the SC transition to lower temperatures. This suggests that $B_{c2}(0)$ attains its maximum at $P \sim P_c$ and decreases with increasing pressure. On the other hand, the AF order of the Eu^{2+} moments persists up to the highest measured pressure of 3.2 GPa with the moderate increase in T_N and the similar magnetic field effect.

Acknowledgment

This work was partially supported by a Grant-in-Aid for Young Scientists (No. 23740279) from the Ministry of Education, Culture, Sports, Science and Technology, Japan.

References

[1] Kamihara Y, Watanabe T, Hirano M and Hosono H 2008 *J. Am. Chem. Soc.* **130** 3296
[2] For recent reviews, see, Ishida K, Nakai Y and Hosono H 2009 *J. Phys. Soc. Jpn.* **78** 062001; Johnston D C 2010 *Adv. Phys.* **59**, 803, and references therein.
[3] Alireza P L, Ko Y T C, Gillett J, Petrone C M, Cole J M, Lonzarich G G and Sebastian S E 2009 *J. Phys.: Condens. Matter* **21** 012208
[4] Terashima T, Kimata M, Satsukawa H, Harada A, Hazama K, Uji S, Suzuki H S, Matsumoto T and Murata K 2009 *J. Phys. Soc. Jpn.* **78** 083701; **78** 118001
[5] Terashima T, Kurita N, Kikkawa A, Suzuki H S, Matsumoto T, Murata K, and Uji S, 2010 *J. Phys. Soc. Jpn.* **79** 103706
[6] Kurita N, Kimata M, Kodama K, Harada A, Tomita M, Suzuki H S, Matsumoto T, Murata K, Uji S and Terashima T 2011 *J. Phys.: Conf. Ser.* **273** 012098
[7] Kurita N, Kimata M, Kodama K, Harada A, Tomita M, Suzuki H S, Matsumoto T, Murata K, Uji S and Terashima T 2011 *Phys. Rev. B* **83** 100501 (R)
[8] Kurita N, Kimata M, Kodama K, Harada A, Tomita M, Suzuki H S, Matsumoto T, Murata K, Uji S and Terashima T 2011 *Phys. Rev. B* **83** 214513
[9] Ren Z, Zhu Z W, Jiang S, Xu X F, Tao Q, Wang C, Feng C M, Cao G H and Xu Z A 2008 *Phys. Rev. B* **78** 052501
[10] Jeevan H S, Hossain Z, Kasinathan D, Rosner H, Geibel C and Gegenwart P 2008 *Phys. Rev. B* **78** 052502
[11] Jiang S, Luo Y, Ren Z, Zhu Z, Wang C, Xu X, Tao Q, Cao G and Xu Z 2009 *New J. Phys.* **11** 025007
[12] Xiao Y, Su Y, Meven M, Mittal R, Kumar C M N, Chatterji T, Price S, Persson J, Kumar N, Dhar S K, Thamizhavel A and Brueckel Th 2009 *Phys. Rev. B* **80** 174424
[13] Miclea C F, Nicklas M, Jeevan H S, Kasinathan D, Hossain Z, Rosner H, Gegenwart P, Geibel C and Steglich F 2009 *Phys. Rev. B* **79** 212509
[14] Matsubayashi K, Munakata K, Isobe M, Katayama N, Ohgushi K, Ueda Y, Kawamura N, Mizumaki M, Ishimatsu N, Hedo M, Umehara I and Uwatoko Y 2011 *Phys. Rev. B* **84** 024502
[15] Uwatoko Y, Hedo M, Kurita N, Koeda M, Abliz M and Matsumoto T 2003 *Physica C* **329-333** 1658
[16] Murata K, Yokogawa K, Yoshino H, Klotz S, Munsch P, Irizawa A, Nishiyama M, Iizuka K, Nanba T, Okada T, Shiraga Y and Aoyama S 2008 *Rev. Sci. Instrum.* **79** 085101
[17] Kurita N *et. al.*, unpublished (in preparation).

Journal of Physics: Conference Series **391** (2012) 012133 doi:10.1088/1742-6596/391/1/012133

Upper Critical Field of the Stoichiometric Fe-based Superconductor LiFeAs

N Kurita[1,2], K Kitagawa[2,3], K Matsubayashi[2,3], A Kismarahardja[4], E S Choi[4], J S Brooks[4], Y Uwatoko[2,3], S Uji[1,2], and T Terashima[1,2]

[1]National Institute for Materials Science, Tsukuba, Ibaraki 305-0003, Japan
[2]JST, Transformative Research-Project on Iron Pnictides (TRIP), Chiyoda, Tokyo 102-0075, Japan
[3]Institute for Solid State Physics, The University of Tokyo, Kashiwanoha, Kashiwa, Chiba 277-8581, Japan
[4]National High Magnetic Field Laboratory, Florida State University, Tallahassee, FL 32310, USA

E-mail: KURITA.Nobuyuki@nims.go.jp

Abstract.
We have investigated the upper critical field B_{c2} of the stoichiometric Fe-based superconductor LiFeAs, via the magnetic torque measurements in dc-magnetic fields up to 35 T and at temperatures down to 0.3 K. B_{c2} at 0.3 K is obtained to be 26.4 T and 15.5 T for the applied field $B_a \parallel ab$ and $B_a \parallel c$, respectively, indicating nearly-isotropic superconductivity. The detailed analyses show that, while B_{c2} for $B_a \parallel c$ is limited by the orbital effect, the spin-paramagnetic effect is evident in the temperature dependence of B_{c2} for $B_a \parallel ab$.

1. Introduction

The discovery of the Fe-based superconductors (SCs) [1] has triggered intensive research efforts, not only due to the high transition temperature T_c (up to 54 - 56 K in some compounds), but the possible unconventional pairing state, a variety of the FeAs layered crystal structures, etc. [2].

As to the upper critical field B_{c2} of Fe-based SCs, thus far, there are few reports on the systematic investigations to address the issue of the pair-breaking mechanism. This is mainly due to the fact that high-T_c SCs including Fe-based systems generally have extremely high B_{c2}. In many cases, accordingly, the low temperature behavior is extrapolated from the high temperature data around T_c. However, such an extrapolation, generally assuming an orbitally limited case [3], could lead to misleading conclusions for the Fe-based systems where a spin-paramagnetic effect [4] may play an important role. For better understanding of the pair-breaking mechanism of the high-T_c superconductivity, therefore, it is of importance to determine B_{c2} experimentally in a wide temperature range from T_c down to $T \rightarrow 0$.

Recently, we have performed magnetic torque measurements of the stoichiometric Fe-based superconductor LiFeAs [5, 6] using a single crystal with T_c 16 K, and have first reported on the whole temperature dependence of B_{c2} and its anisotropy [8]. Here, we present more detailed data which supplement the results of Ref. [8].

2. Experimental Details

LiFeAs single crystals with the residual resistivity ratio of 45 were grown by a self-flux method as described in Ref [7]. The samples were carefully handled so as not to expose them to air directly. The bulk superconductivity was confirmed from the clear diamagnetic transition of the dc magnetic susceptibility below T_c 16 K [8]. The magnetic torque measurements were carried out using a static resistive magnet in fields up to 35 T and at temperatures down to 0.3 K, by means of a piezoresistive microcantilever. The specific heat C was measured in a physical property measurement system (PPMS; Quantum Design). The nuclear magnetic resonance (NMR) spin-lattice relaxation time T_1 was measured for the ^{75}As NMR lines. The crystalline axes were aligned within 1 ° by using a two-axis goniometric stage during the NMR experiment.

3. Results and Discussions

Figures 1(a) and (b) show the torque signals in LiFeAs as a function of the applied field B_a for $B_a \parallel ab$ and $B_a \parallel c$, respectively, at several temperatures down to 0.3 K. As indicated by arrows, highly hysteretic behavior, which is more significant at low temperatures, is observed between the field-up and -down sweeps. The hysteresis of the torque response, the irreversible curve, appears in the superconducting mixed state when the pinning force is strong enough to trap the flux lines. Note that the shapes of the small loops near B_{c2} for $B_a \parallel ab$ and $B_a \parallel c$ are

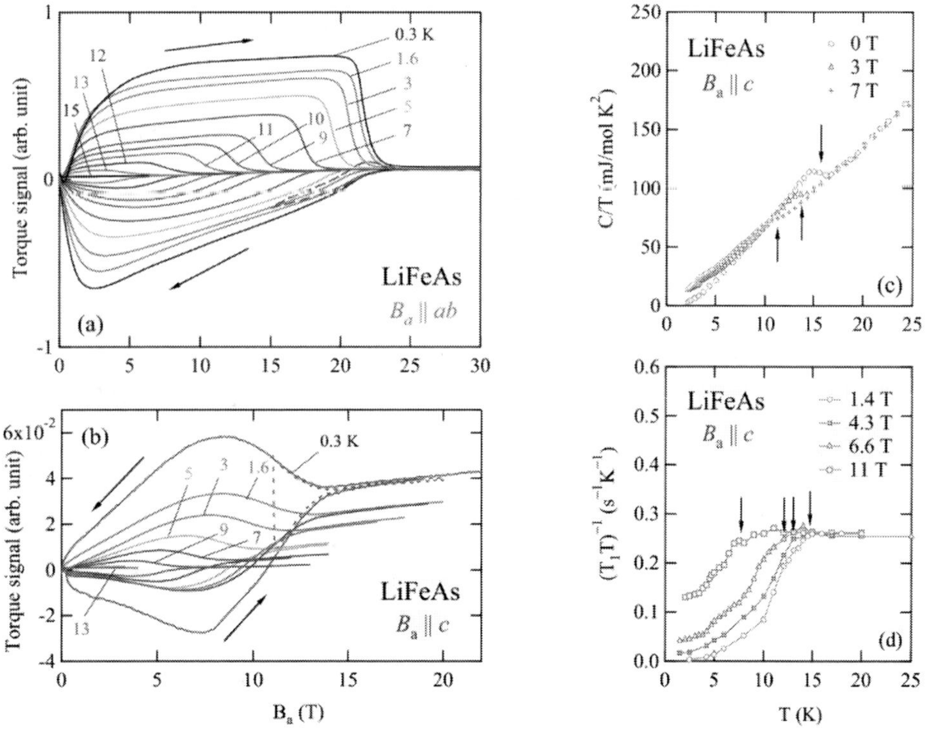

Figure 1. (Color online) Magnetic torque signals vs B_a of LiFeAs at several temperatures down to 0.3 K for (a) $B_a \parallel ab$ and (b) $B_a \parallel c$. The solid curves represent the data obtained in field-up and -down sweeps between $B_a = 0$ and above B_{c2}, where the arrows in (a) and (b) indicate the sweep directions. Small-loop measurements around B_{c2}, as shown by the dashed curves at $T = 0.3$ K, were performed at each temperature. (c) C/T vs T and (d) $1/(T_1T)$ vs T of LiFeAs in several fields for $B_a \parallel c$. The arrows in (c) and (d) indicate B_{c2} at each field.

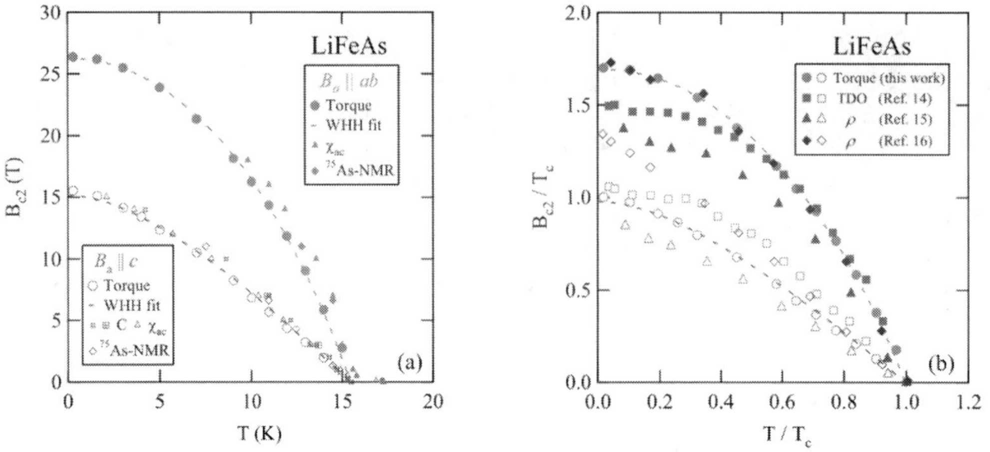

Figure 2. (Color online) (a) B_{c2} vs T of LiFeAs for $B_a \parallel ab$ and $B_a \parallel c$, deduced from the magnetic torque, C, ^{75}As-NMR, and $_{\rm ac}$ measurements. The dashed curves indicate fits to the data based on the WHH theory (see text). (b) Scaled B_{c2}/T_c vs T/T_c of LiFeAs. For comparison, data obtained from TDO [14] and [15, 16] are also shown.

quite different. It could be possible that the pinning mechanism is different for the two field orientations, although the detail remains to be clarified.

From the magnetic torque data, we define the upper critical field B_{c2} as the field where the irreversibility disappears. To accurately determine B_{c2} by minimizing a drift of the torque signals, we made the small-loop measurements around B_{c2} at each temperature, as shown by the dashed curves in (a) and (b) for $T = 0.3$ K, in addition to the large-loop ones between $B_a = 0$ and above B_{c2}. The field defined here is, indeed, the irreversibility field, but we can safely assume that it is practically identical to B_{c2} for this compound because thus determined B_{c2} values are in good agreement with those determined from other bulk property measurements; specific heat (C), spin-lattice relaxation time of ^{75}As NMR (T_1), and ac-susceptibility ($_{\rm ac}$) [see Fig. 2(a)]. From the temperature dependence of C/T, T_c is obtained by the entropy balance construction as indicated by arrows in Fig. 1(c). From the temperature dependence of $1/(T_1T)$, T_c is determined as the temperature where there is a sudden decrease in $1/(T_1T)$ as indicated by arrows in Fig. 1(d).

Figure 2(a) displays B_{c2} vs T of LiFeAs for $B_a \parallel ab$ and $B_a \parallel c$, determined by the magnetic torque, C, ^{75}As-NMR, and $_{\rm ac}$ [9] measurements using samples from the same batch. As indicated by the dashed curves, the data can be well fitted using a Werthamer-Helfand-Hohenberg (WHH) formula containing the spin-paramagnetic and orbital pair-breaking effects [3]. From the best fits, the Maki parameter α and spin-orbit scattering parameter $_{\rm so}$ are obtained to be (α, $_{\rm so}$) = (2.30, 0.51) for $B_a \parallel ab$ and (0.75, ∞) for $B_a \parallel c$, where $T_c = 15.5$ K is fixed. The orbital critical fields B^*_{c2} at $T = 0$ are estimated to be 47.2 and 15.3 T for $B_a \parallel ab$ and $B_a \parallel c$, respectively, from $B^*_{c2}(0) = 0.69\, T_c\, {\rm d}H_{c2}/{\rm d}T|_{T=T_c}$ [3]. The Pauli-Clogston paramagnetic limit $B_{\rm po} = 1.84\, T_c$ [4] is 28.5 T.

It is interesting to note that $B^{*c}_{c2}(0) < B_{\rm po} < B^{*ab}_{c2}(0)$. This relation results in the orbitally limited B_{c2} for $B_a \parallel c$, as indicated by $^{c}_{\rm so} = \infty$, and strongly spin-paramagnetically limited B_{c2} for $B_a \parallel ab$, as indicated by that $B^{ab}_{c2}(0.3\,{\rm K}) = 26.4\,{\rm T} \ll B^{*ab}_{c2}(0)$, and effectively reduces the B_{c2} anisotropy at low temperatures. A similar trend that the weaker orbital effect for $B_a \parallel ab$ is partly compensated for by the spin-paramagnetic effect to yield a reduced anisotropy is clearly seen in a stoichiometric superconductor KFe_2As_2 with low T_c of 2.8 K [10], as well as in the high-T_c systems such as "122"-type (Ba,K)Fe_2As_2 [11] and "11"-type FeSe [12].

As temperature decreases, the anisotropy parameter Γ, defined as $\Gamma = B_{c2}^{ab} / B_{c2}^{c}$, decreases from 3.1 at T_c (from the WHH fit) to 1.7 at 0.3 K. The low-temperature variation of Γ is similar to those observed in the "122" and "11" systems [11, 12], but the value of Γ in LiFeAs is slightly larger. The slightly larger Γ of LiFeAs than those of the "122" and "11" systems might be due to the more two-dimensionality of LiFeAs [13].

Soon after the publication of our study [8], three groups independently reported on the full B_{c2} T phase diagram of LiFeAs using different techniques: tunnel diode resonator (TDO) [14], electrical resistivity (pulse magnet) [15], and (static magnet) [16]. In Fig. 2(b), we show a comparison of the scaled B_{c2}/T_c vs T/T_c. Note that our results are qualitatively consistent with others in terms of that the pair-breaking for $B_a \parallel c$ is mostly caused by the orbital effect while a spin-paramagnetic effect plays an important role for $B_a \parallel ab$. The recognizable discrepancies may be explained by a combination of several factors such as the manner to define the B_{c2} value, the sample quality which affects B_{c2}, and so on.

4. Conclusions

To conclude, we have performed high-field magnetic torque measurements of the LiFeAs single crystal, and have determined the full temperature dependence of B_{c2} down to 0.3 K. The detailed analyses show that, while B_{c2} for $B_a \parallel c$ is limited by the orbital effect, the weaker orbital effect for $B_a \parallel ab$ due to the mass anisotropy is partly compensated for by the spin-paramagnetic effect leading to a reduced anisotropy at low temperatures.

Acknowledgment

We would like to thank M. Imai for the experimental supports, and M. Takigawa for the useful discussion. The magnetic torque measurements were performed at the National High Magnetic Field Laboratory, which is supported by National Science Foundation Cooperative Agreement No. DMR-0654118, the State of Florida, and the U.S. Department of Energy. A. K. was supported by NSF-DMR 0602859.

References

[1] Kamihara Y, Watanabe T, Hirano M and Hosono H 2008 *J. Am. Chem. Soc.* **130** 3296
[2] For recent reviews, see, Ishida K, Nakai Y and Hosono H 2009 *J. Phys. Soc. Jpn.* **78** 062001; Johnston D C 2010 *Adv. Phys.* **59**, 803, and references therein.
[3] Werthamer N R, Helfand E and Hohenberg P C 1966 *Phys. Rev.* **147** 295
[4] Chandrasekhar B S 1962 *Appl. Phys. Lett.* **1** 7; Clogston A M 1962 *Phys. Rev. Lett.* **9** 266
[5] Wang X C, Liu Q Q, Lv Y X, Gao W B, Yang L X, Yu R C, Li F Y and Jin C Q 2008 *Solid State Commun.* **148** 538
[6] Tapp J H, Tang Z, Lv B, Sasmal K, Lorenz B, Chu P C W and Guloy A M 2008 *Phys. Rev.* **B 79** 060505
[7] Imai Y, H. Takahashi, K. Kitagawa, K. Matsubayashi, N. Nakai, Y. Nagai, Y. Uwatoko, M. Machida, and A. Maeda 2011 *J. Phys. Soc. Jpn.* **80** 013704
[8] Kurita N, Kitagawa K, Matsubayashi K, Kismarahardja A, Choi E S, Brooks J S, Uwatoko Y, Uji S and Terashima T 2011 *J. Phys. Soc. Jpn.* **80** 013706
[9] Matsubayashi K and Kitagawa K, private communications.
[10] Terashima T, Kimata M, Satsukawa H, Harada A, Hazama K, Uji S, Harima H, Chen G F, Luo J L and Wang N L 2009 *J. Phys. Soc. Jpn.* **78** 063702
[11] Yuan H Q, Singleton J, Balakirev F F, Baily S A, Chen G F, Luo J L. and Wang N L 2009 *Nature* **457** 565
[12] Fang M, Yang J, Balakirev F F, Kohama Y, Singleton J, Qian B, Mao Z Q, Wang H and Yuan H Q 2010 *Phys. Rev.* **B 81** 020509(R).
[13] Singh D J 2008 *Phys. Rev.* **B 78** 094511.
[14] Cho K, Kim H, Tanatar M A, Song Y J, Kwon Y S, Coniglio W A, Agosta C C, Gurevich A and Prozorov R 2011 *Phys. Rev. B* **83** 060502(R)
[15] Zhang J L, Jiao L, Balakirev F F, Wang X C, Jin C Q and Yuan H Q 2011 *Phys. Rev. B* **83** 174506
[16] Khim S, Lee B, Kim J W, Choi E S, Stewart G R and Kim K H 2011 *Phys. Rev. B* **84** 104502

Determination of the Nature of the Structural Phase Transitions in 122 Pnictide Systems

J.C. Loudon[1], C.J. Bowell[1], J. Gillett[2], S.E. Sebastian[2] and P.A. Midgley[1]

[1] Department of Materials Science and Metallurgy, University of Cambridge, Pembroke Street, Cambridge, CB2 3QZ.
[2] Department of Physics, University of Cambridge, Cavendish Laboratory, J. J. Thomson Avenue, Cambridge, CB3 0HE.

Email: j.c.loudon@gmail.com

Abstract. Transmission electron microscopy was used to investigate the tetragonal-to-orthorhombic phase transitions in $SrFe_2As_2$ and $Ba(Fe_{0.985}Co_{0.015})_2As_2$ which are members of the recently discovered iron-based superconductors. We use a method which measures the spatially-resolved order parameter and show that whereas the transition in $SrFe_2As_2$ is a first-order martensitic phase change, $Ba(Fe_{0.985}Co_{0.015})_2As_2$ shows a continuous change of the order parameter as the phase transition takes place. This would normally indicate a second order phase transition but the situation seems to be more complicated as, during the transition, different regions of the sample had different values of the order parameter: something which should only happen in a first order change.

1. Introduction

$Sr(Fe_{1-x}Co_x)_2As_2$ and $Ba(Fe_{1-x}Co_x)_2As_2$ are iron-based superconductors, discovered in 2008 [1,2]. Superconductivity occurs for a range of cobalt concentrations as shown by the phase diagram in Fig. 1. The non-superconducting compositions show a paramagnetic-to-antiferromagnetic transition accompanied by a change of crystal structure from tetragonal to orthorhombic. It has been postulated that the superconducting phase is the result of a quantum critical point [3] and this implies that the tetragonal to orthorhombic transition should be second order close to the superconducting phase. However, determining the order of the transition is no easy matter. For example, both Tegel *et al.* [4] and Jesche *et al.* [5] used x-ray diffraction to determine the order of the phase transition in $SrFe_2As_2$ and the first group found it was second order and the second group first order. We believe that part of the reason for such disagreements is the macroscopic nature of the measurements used and so introduce a technique to measure the spatially-resolved order parameter on a scale of tens of nanometres using transmission electron microscopy. Here we examine the phase transitions in $SrFe_2As_2$ and $Ba(Fe_{0.985}Co_{0.015})_2As_2$.

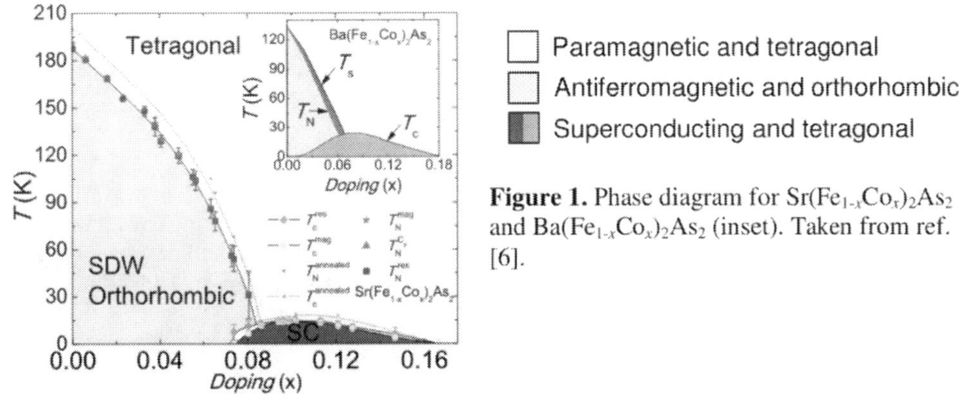

Figure 1. Phase diagram for $Sr(Fe_{1-x}Co_x)_2As_2$ and $Ba(Fe_{1-x}Co_x)_2As_2$ (inset). Taken from ref. [6].

2. The Phase Transition

The room temperature tetragonal unit cell of $Sr(Fe,Co)_2As_2$ or $Ba(Fe,Co)_2As_2$ is shown in Fig. 2(a). Both compounds have space group *I4/mmm* with $SrFe_2As_2$ having lattice parameters $a_T = b_T = 3.92$ Å and $c_T = 12.36$ Å and $BaFe_2As_2$, $a_T = b_T = 3.96$ Å, $c_T = 12.99$ Å. Fig. 2(b) shows that in the orthorhombic phase, the right angle between a_T and b_T is lost (dashed lines). The orthorhombic phase has space group *Fmmm* with lattice parameters $a_O = 5.58$ Å, $b_O = 5.52$ Å, $c_O = 12.30$ Å for $SrFe_2As_2$ at 90 K and $a_O = 5.62$, $b_O = 5.57$ Å, $c_O = 12.94$ Å for $BaFe_2As_2$ at 5 K [4,7].

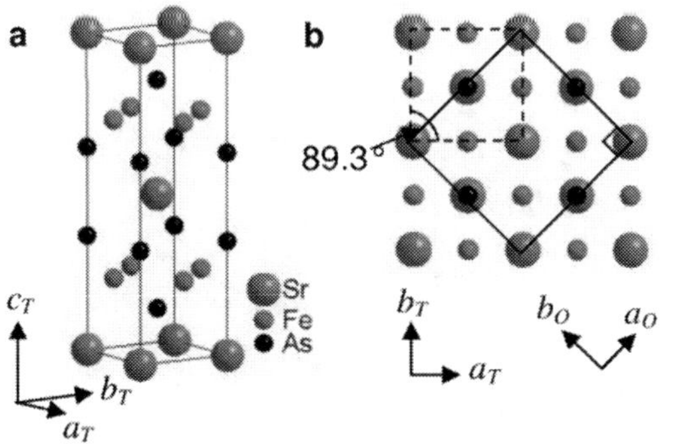

Figure 2. (a) Tetragonal and (b) Orthorhombic $Sr(Fe,Co)_2As_2$ or $Ba(Fe,Co)_2As_2$ viewed down [001] showing the relationship between the tetragonal (dashed line) and orthorhombic unit cells (solid lines). (Data taken from ref. 4.)

To track the progress of the transition we define the order parameter, Q, such that $Q = \frac{a_O - b_O}{a_O}$.

3. Sample Preparation and Experimental Methods

We used single crystals of $SrFe_2As_2$ and $Ba(Fe_{0.985}Co_{0.015})_2As_2$ synthesised as described in ref [6] and these were argon ion-thinned to electron transparency as described in ref. [8]. The local order parameter was measured using large-angle convergent beam electron diffraction (LACBED) which produces an image with dark contours running across it which correspond to specific crystallographic planes (Fig. 1(a)). Twins form in the orthorhombic phase and the change in orientation of the crystal planes on moving from one twin to the next gives a broken contour (Fig. 1(b)). The spacing between the breaks in the LACBED contours is proportional to the order parameter, Q. LACBED patterns were recorded as videos on warming and cooling through the phase transition. Further details are given in ref. [8].

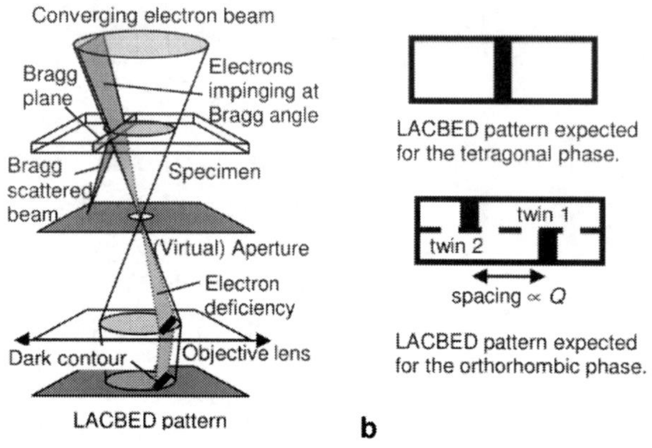

Figure 3. (a) Schematic illustrating the formation of large-angle convergent beam electron diffraction (LACBED) patterns. (b) LACBED patterns consist of dark contours corresponding to crystallographic planes. In the tetragonal phase, these contours are continuous but in the orthorhombic phase, twinning causes the contour to be broken and the spacing between the breaks is proportional to the order parameter.

4. Results

The results of the investigation of the phase transition in $SrFe_2As_2$ are reported fully in ref. [8] but we summarise our findings here. The phase transition takes place by the growth of the orthorhombic twins along their length. At the end of the twins is a needle tip which is surrounded by transformation dislocations and it is the passage of these dislocations through the material which effects the transformation. The orthorhombic twins move across the specimen one at a time in a series of jumps so that the local order parameter changes faster than the time between video frames (1/25 s). The phase transition can be described as a first-order Martensitic transition.

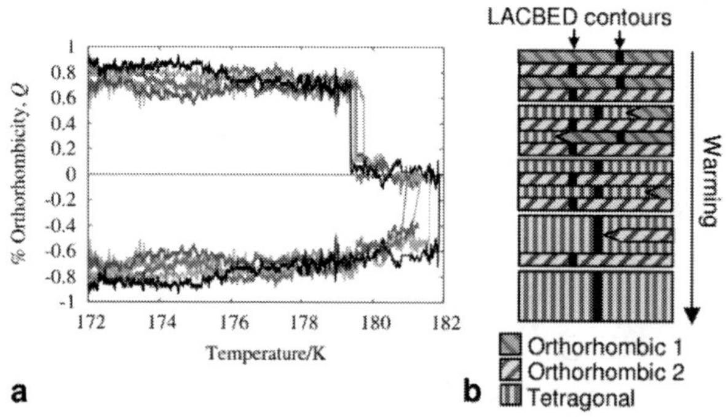

Figure 4. (a) The order parameter, Q, versus temperature. The region of interest contained two twin types to which we give opposite signs. Each colour represents a different pair of twins. (b) Schematic illustration showing how the phase transition took place. As the specimen was warmed through the transition, first one set of orthorhombic twins withdrew leaving behind the tetragonal phase. About 2 K higher the other twin variant withdrew. On cooling the process was reversed.

The transition in $Ba(Fe_{0.985}Co_{0.015})_2As_2$ was very different so that rather than jumping abruptly to zero as the specimen was warmed, the order parameter varied continuously as shown in Fig. 5(a). This suggests a second order phase transition but it appears the situation is more complicated as some of the video frames such as that shown in Fig. 5(b) clearly show twins with different values of the order parameter coexisting at the same temperature: something which should only happen in a first order change.

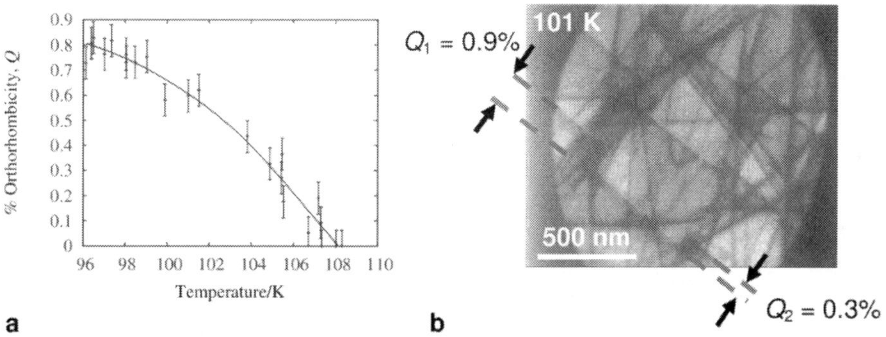

Figure 5. (a) The order parameter, Q, versus temperature from one of the twins in the region of interest. (b) A frame from the video of LACBED patterns showing two different values of the order parameter, 0.3% and 0.9% coexisting at the same temperature.

5. Summary

Transmission electron microscopy was used to investigate the tetragonal-to-orthorhombic phase transitions in $SrFe_2As_2$ and $Ba(Fe_{0.985}Co_{0.015})_2As_2$, by mapping the spatially resolved order parameter. This showed that the transition in $SrFe_2As_2$ was a first-order martensitic phase change whereas $Ba(Fe_{0.985}Co_{0.015})_2As_2$ showed a continuous change of the order parameter as the phase transition took place. This would normally indicate a second order phase transition but the situation seems to be more complicated than this as, during the transition, different regions of the sample had different values of the order parameter: something which should only happen in a first order change.

Acknowledgments
This work was funded by the Royal Society and the EPSRC, grant number EP/E027903/1.

References

[1] Kamihara Y, Watanabe T, Hirano M and Hosono H *J. Am. Chem. Soc.* 2008 **130** 3296.
[2] Sasmal K, Lv B, Lorenz B, Guloy AM, Chen F, Xue Y-Y and Chu C-W *Phys. Rev. Lett.* 2008 **101** 107007.
[3] Ning FL, Ahilan K, Imai T, Sefat AS, Jin RY, McGuire MA, Sales BC and Mandrus D *J. Phys Soc. Jpn.* 2009 **78** 013711.
[4] Tegel M, Rotter M, Weiß V, Schappacher FM, Pöttgen and Johrendt D *J. Phys.: Condens. Matter.* 2008 **20** 452201.
[5] Jesche A, Caroca-Canales N, Rosner H, Borrmann H, Ormeci A, Kasinathan D, Klauss HH, Luetkens H, Khasanov R, Amato A, Hoser A, Kaneko K, Krellner C and Geibel C *Phys. Rev. B* 2008 **78** 180594(R).
[6] Gillett J, Das SD, Syers P, Ming AKT, Espeso JI, Petrone CM, Sebastian SE 2010 *Preprint* arXiv: 1005.1330.
[7] Huang Q, Qiu Y, Bao W, Green MA, Lynn JW, Gasparovic YC, Wu T, Wu G, Chen XH *Phys. Rev. Lett.* 2008 **101** 257003.
[8] Loudon JC, Bowell CJ, Gillett J, Sebastian SE and Midgley PA *Phys. Rev. B* 2010 **81** 214111.

International Conference on Strongly Correlated Electron Systems (SCES 2011) IOP Publishing
Journal of Physics: Conference Series **391** (2012) 012135 doi:10.1088/1742-6596/391/1/012135

Competing orders in the phase diagram of cuprate superconductors

S.G. Magalhaes and E. V. L. de Mello

[1] Instituto de Física, Universidade Federal Fluminense, 24210-346 Niteroi,RJ, Brazil

E-mail: *sgmagal@gmail.com

Abstract.

We perform self-consistent calculations in the framework of the Bogoliubov-deGennes theory with a model Hubbard Hamiltonian with strong local correlations. The calculations are performed in a charge density disordered system to study the competition of charge inhomogeneity, spin disorder and superconductivity These calculations are appropriate to describe the cuprate superconductors' phase diagram with the competition of different phases. We find that, in the presence of charge disorder the spin density wave (SDW) order occurs at higher temperatures above the opening of the superconducting gap. This is in opposition to the homogeneous systems where the SDW phase appears in the low doping compounds in the superconducting phase. These finding provides an explanation to the non-Fermi liquid behavior of the cuprates normal phase.

1. Introduction

The origin of the superconducting gap associated with the superconducting state and the relation to the pseudogap above the transition temperature (T_c) remains one of the central questions in high-T_c research[1, 2]. The normal phase properties differs completely from the BCS superconductors. It is a matter of intense debate whether this problem is connected with the charge inhomogeneities observed in many different experiments[3, 4, 5, 6, 7, 8, 9, 10].

In order to obtain a unified interpretation to all of these observations we studied an electronic phase separation (EPS) transition that generates regions of low and high densities. Such a transition may be driven by the lower free energy of undoped antiferromagnetic (AF) regions[11] (intrinsic) or by the out of plane dopants[12] (extrinsic origin).

In this context we perform the study of the the superconducting (SC) and antiferromagnetic (AF) order in a model t-t'-U-V Hamiltonian commonly used to describe the cuprates[13, 14] in an electronic disordered system. Chen and Ting[14] have studied this model for different doping and temperature and found that the AF is stronger at low doping and high temperatures. Self-consistent calculations in the Bogoliubov-deGennes (BdG) framework indicates that the AF order persists above the opening of the SC order parameter at very high temperatures. This is used to interpret the normal pseudogap phase, since this type of spin density wave will cause an additional scattering to the electrons and will change the Fermi liquid properties.

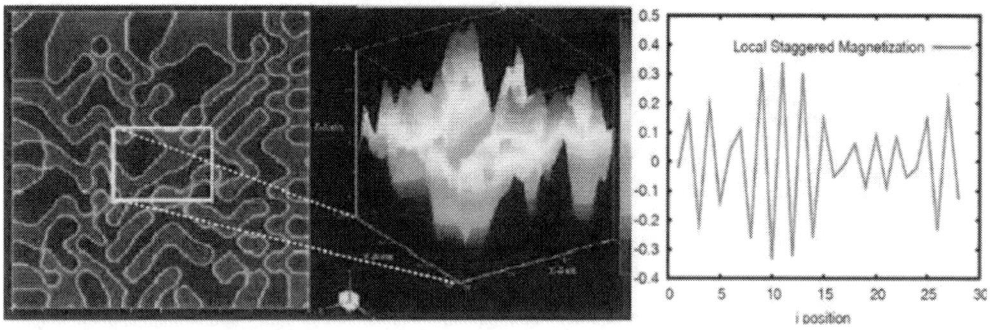

Figure 1. Typical phase separation calculation on a 105x105 square lattice with inhomogeneous densities represented by the color code[13, 11]. The BdG d-wave SC amplitude calculations are performed in a smaller 28x28 square lattice as drawn in the figure. The results are shown in the middle panel and yields $7 < \Delta_\delta(\mathbf{r}_i) < 13$meV. On the right panel we show the local staggered magnetization M_i along 28 points in row performed by the same calculation.

2. The model

The t-t'-U-V Hamiltonian in a square lattice i s defined as

$$\begin{aligned} H &= -\sum_{\{ij\}\sigma} t_{ij} c^\dagger_{i\sigma} c_{j\sigma} + \sum_{i\sigma} (\mu) n_{i\sigma} \\ &\quad + U \sum_i n_{i\uparrow} n_{i\downarrow} - \frac{V}{2} \sum_{\langle ij\rangle\sigma\sigma'} n_{i\sigma} n_{j\sigma'}, \end{aligned} \tag{1}$$

where $c^\dagger_{i\sigma}(c_{i\sigma})$ is the usual fermionic creation (annihilation) operator at site $\mathbf{r}_i$, with lattice parameter a=1 and spin $\sigma\{\uparrow\downarrow\}$. $n_{i\sigma} = c^\dagger_{i\sigma} c_{i\sigma}$ is the density operator, and t_{ij} is the hopping between sites i and j. U is the magnitude of the on-site repulsion, and V is the magnitude of the nearest-neighbor attractive interaction. μ is the chemical potential. To apply the Bogoliubov-deGennes (BdG) approach we define the pairing amplitude $\Delta_\delta(\mathbf{r}_i) = V\langle c_{i\downarrow} c_{i+\delta\uparrow}\rangle$, where δ=$\pm\hat{\mathbf{x}}$,$\pm\hat{\mathbf{y}}$ are nearest neighbors unit vectors for a square lattice. The resulting effective Hamiltonian H_{eff} is given by

$$\begin{aligned} H_{eff} &= -\sum_{i\delta\sigma} t_{i,i+\delta} c^\dagger_{i\sigma} c_{i+\delta\sigma} + \sum_{i\sigma} (\tilde{\mu}_i) n_{i\sigma} \\ &\quad + \sum_{i\delta} [\Delta^*_\delta(\mathbf{r}_i) c_{i\downarrow} c_{i+\delta\uparrow} + \Delta_\delta(\mathbf{r}_i) c^\dagger_{i+\delta\uparrow} c^\dagger_{i\downarrow}] \end{aligned} \tag{2}$$

where $\tilde{\mu}_i$=μ-$U/2\langle n_i\rangle$ is the local chemical potential, which incorporates the site dependent Hartree shift $\frac{U}{2}\langle n_i\rangle$ and the granular charge inhomogeneity of the system. The usual BdG equations for

$$\begin{pmatrix} K & \Delta \\ \Delta^* & -K^* \end{pmatrix} \begin{pmatrix} u_{n\sigma}(\mathbf{r}_i) \\ v_{n\sigma}(\mathbf{r}_i) \end{pmatrix} = E_n \begin{pmatrix} u_{n\sigma}(\mathbf{r}_i) \\ v_{n\sigma}(\mathbf{r}_i) \end{pmatrix} \tag{3}$$

with

$$\begin{aligned} K u_{n\sigma}(\mathbf{r}_i) &= -\sum_\delta t_{i,i+\delta} u_{n\sigma}(\mathbf{r}_i + \delta) + (\tilde{\mu}_i) u_{n\sigma}(\mathbf{r}_i) \\ \Delta u_{n\sigma}(\mathbf{r}_i) &= \sum_\delta \Delta_\delta(\mathbf{r}_i) u_{n\sigma}(\mathbf{r}_i + \delta) + \Delta_U(\mathbf{r}_i) u_{n\sigma}(\mathbf{r}_i), \end{aligned} \tag{4}$$

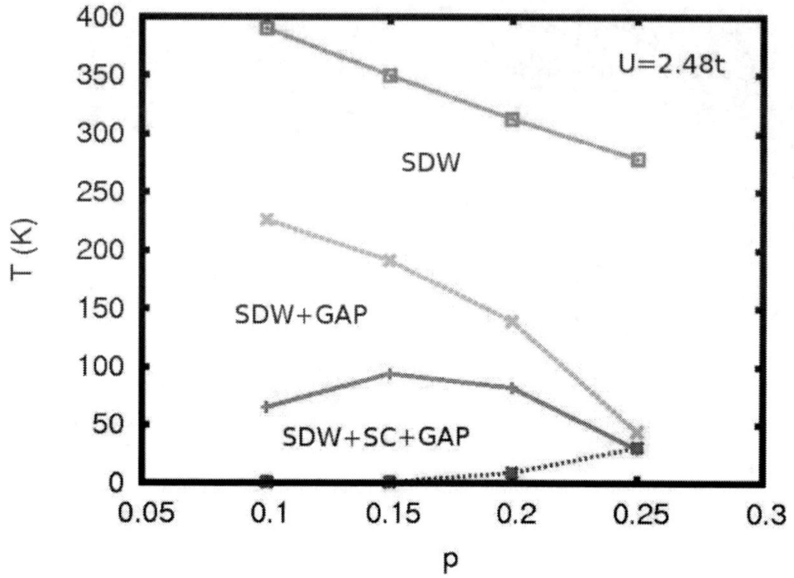

Figure 2. The calculated phase diagram showing the regions where the local staggered magnetization is different than zero and we call the SDW phase region. It starts at low temperature (blue curve) and ends at high temperature shown by the magenta line (high pseudogap line). The d-wave SC gap (GAP) onset temperature is shown by the green line (or low pseudogap line). The superconducting phase (SC) is the onset of phase coherence among the islands shown in Fig.1 (left panel).

and similar equations for $v_n(\mathbf{r}_i)$. These equations give the quasiparticle eigenenergies $E_n(\geq 0)$. The BdG equations are solved self-consistently together with the $d_{x^2-y^2}$ pairing amplitudes

$$\Delta_\delta(\mathbf{r}_i) = -\frac{V}{2}\sum_n [u_{n\sigma}(\mathbf{r}_i)v_n^*(\mathbf{r}_i+\delta) + v_{n\sigma}^*(\mathbf{r}_i)u_{n\sigma}(\mathbf{r}_i+\delta)]\tanh\frac{E_n}{2k_BT}. \quad (5)$$

and the inhomogeneous input (as shown in the left panel of Fig.1) hole density is given by

$$n(\mathbf{x}_i)_\sigma = 1 - 2\sum_n [|u_{n\sigma}(\mathbf{x}_i)|^2 f_n + |v_{n\sigma}(\mathbf{x}_i)|^2(1-f_n)], \quad (6)$$

The staggered magnetization $M_i = (-1)^i(n_{\mathbf{x}_i\uparrow} - n_{\mathbf{x}_i\downarrow})$.

We solved the above model self-consistently for the d-wave pairing amplitude $\Delta_\delta(\mathbf{r}_i)$ for many values of doping level p and temperature T, in order to study the phase diagram of cuprates. We used parameters close to the previous calculations of Chen and Ting for a homogeneous system, namely, U=2.48t, t=1 (0.15 eV) and nearest neighbors $t_{ij} = 0.25t$. V varies from $1.4t$ at n=0.90 to $V = 1.14t$ at n=0.75. This change in the two-body attractive potential is due to a phase segregation of the holes which forms islands of low and high densities. The holes get trapped in these grains, they loose kinetic energy what is a process that favors the pair attraction as discussed previously[13, 11, 15, 16]. Typical solutions of the local d-wave gap $\Delta_\delta(\mathbf{r}_i)$ are shown in the middle panel of Fig.(1).

3. Results and Conclusions

The EPS transition generates inhomogeneity in the (granular) local density n_i that was simulated by the Cahn-Hilliard or time dependent Ginzburg-Landau approach[13, 11, 15, 16]. The result is a system with grains or islands with characterized by low and high charge densities. In the BdG calculations we keep this granular structure fixed and use also the possibility that the attractive potential V is scaled by the degree of disorder. In general, the self-consistent calculations yield non-zero values of the d-wave SC gap at temperatures well above T_c. This is shown by the green line in Fig.(2) or the onset of the $SDW + GAP$ phase region. For instance, for $p = 0.1$, the local gaps $\Delta_\delta(\mathbf{r}_i)$ vanish near $T = 230K$. In this way the resistive transition at T_c is estimated by the Josephson coupling among these regions[15, 16].

As mentioned, Chen and Ting[14] studied this model on homogeneous systems and found that the spin disorder or SDW phase starts near the optimum value using $U = 2.44t$ and $|V| = t$ at all doping and vanishes in the overdoped region. They identified the onset of SDW phase with the pseudogap temperature. We essentially generalized their calculations for the case of a system with disorder in the local charge density. In both the homogeneous and disordered system the local staggered magnetization M_i depends strongly on the choice of parameters. The main purpose here is to show that the local SDW persists above T_c at very high temperatures, close to the room temperature. The additional scattering from this spin disordered phase change the Fermi liquid properties and it is a possible interpretation of the cuprates normal phase. In this way we identify the high pseudogap line with the onset of staggered magnetization and the low pseudogap with the the onset of superconducting amplitudes Δ_d. This last case is in agreement with various measurements connected with some superconducting properties like, for instance, the Nernst effect[17, 18].

Our results show a phase diagram more in agreement with the cuprates phase diagram[2] than that one calculated by Chen and Ting[14]. The difference is due to the charge disorder introduced as input in the self-consistent calculations. Therefore, our conclusion is that this local charge and spin disorder are important ingredients to study the different phases and crossover lines in the transport properties and phase diagram of cuprate superconductors.

4. Acknowledgments

This work was partially supported by the Brazilian agencies Faperj and CNPq.

5. References

[1] T. Timusk and B. Statt, Rep. Prog. Phys., **62**, 61 (1999).
[2] J.L. Tallon and J.W. Loram, Physica C **349**, 53 (2001).
[3] J.M.Tranquada, et al Nature (London),**375**, 561 (1995).
[4] P. M. Singer, A. W. Hunt, and T. Imai, Phys. Rev. Lett. **88**, 47602 (2002).
[5] S. H. Pan et al, Nature (London), 413, 282-285 (2001).
[6] K. McElroy, et al cond-mat/0404005 an Phys. Rev. Lett. **94**, 197005 (2005).
[7] Kenjiro K. Gomes et al, Nature **447**, 569 (2007).
[8] Abhay N. Pasupathy et al, Science **320**, 196 (2008).
[9] Takuya Kato et al, J. Phys. Soc. Jpn., **77**, 054710 (2008).
[10] Michela Fratini et al, Nature, **466**, 841 (2010).
[11] E.V.L. de Mello, et al J. Phys.: Condens. Matter **21**, 235701 (2009).
[12] Rinat Ofer, and Amit Keren, Phys. Rev. **B80** 224521 (2009).
[13] E.V.L. de Mello et al, Phys. Rev. **B70**, 224517 (2004).
[14] Hong-Yi Chen and C. S. Ting, Phys. Rev. **71** 132505R (2005).
[15] E. V. L. de Mello and R.B. Kasal, J. Supercond. Nov. Magn., **24**, 1123 (2011).
[16] E. V. L. de Mello and R.B. Kasal, Physica C 472, 60, (2012).
[17] Yayu Wang, Lu Li, and N. P. Ong, Phys. Rev. **B73**, 024510 (2006).
[18] E. V. L. de Mello, and D. N. Dias, J. Phys. C.M. **19**, 086218 (2007).

Isotope exponent in disordered underdoped and overdoped La214

S H Naqib* and R S Islam

Department of Physics, University of Rajshahi, Rajshahi-6205, Bangladesh

*E-mail: salehnaqib@yahoo.com

Abstract

The effect of oxygen isotope substitution on T_c has been analyzed for heavily underdoped and overdoped $La_{2-x}Sr_xCu_{1-y}Zn_yO_4$ compounds with different Zn contents in the CuO_2 plane. The effect of Zn on the isotope exponent, α, was more pronounced in the case of the underdoped ($x = 0.09$) compounds compared to the overdoped ($x = 0.22$) ones. The variation of α with in-plane disorder can be described quite well within a model based solely on impurity induced Cooper pair-breaking in the case of the underdoped compounds. This model, on the other hand, fails to describe the behavior of $\alpha(y)$ for the overdoped samples, even though Zn still breaks pairs quite effectively at this hole (Sr) content. We discuss the implications of these finding in details by considering the Zn induced magnetism, stripe correlations, and possible changes in the superconducting order parameter as number of charge carriers in the CuO_2 plane, p ($\equiv x$), is varied.

PACS: 74.72.Dn; 74.25.Kc; 74.62.Dh

Keywords: La214; isotope exponent; effects of disorder

1. Introduction

The physical mechanism responsible for superconductivity in high-T_c cuprates remains a mystery still after a quarter of a century since their discovery. It is widely believed that the normal and superconducting (SC) state properties of these materials are non-Fermi liquid like [1, 2] and the conventional BCS theory [3] fails to describe the various features shown below the SC transition temperature. Investigation of the isotope effect, which played a major role in the development of the BCS theory within the framework of electron-phonon coupling [4], is highly confusing in the case of cuprates [5 – 7]. The possible role played by phonons on the appearance of superconductivity in cuprates is a matter of strong debate [5 – 10]. In these strongly correlated electronic systems, the isotope exponent (IE), α, depends largely on the number of added holes in the CuO_2 plane, p. $\alpha(p)$ increases in the underdoped (UD) side and can significantly exceed the canonical BCS value of 0.5 near the so-called 1/8th anomaly [10, 11]. The value of IE near the optimum doping, on the other hand, is extremely small and stays almost the same (or increases slightly) in the overdoped (OD) region [6, 10]. The problem in interpretation of the IE is complicated because of the various electronic correlations present in cuprates. The IE is affected both by spin/charge ordering (stripe correlations) and pseudogap (PG) in the quasiparticle energy spectrum [10]. How these correlations are related to superconductivity itself is not well-understood yet. Besides, pair-breaking due to impurities and the symmetry of the order parameter also play significant roles [9, 12]. There is renewed interest in possible role of lattice on the quasiparticle dynamics in cuprates following the isotope dependent ARPES results [13]. A clear understanding of the hole content and disorder dependences of the IE is important to clarify various issues regarding carrier pairing and to reveal the possible roles of the underlying electronic correlations present in high-T_c cuprates.

In this study we have analyzed the oxygen ($^{16}O \rightarrow {}^{18}O$) isotope effect on the T_c of $La_{2-x}Sr_xCu_{1-y}Zn_yO_4$ (Zn-La214) compounds. Two different Sr contents, $x = 0.09$ (UD) and $x = 0.22$ (OD), are used. La214 is a particularly interesting system as strong stripe correlations in both spin and charge channels have been well established [14]. Formation of charge/spin stripes lead to a strong coupling between electronic and lattice degrees of freedom and IE is expected to get enhanced. The UD compounds used in this study has a substantial PG, but PG is absent for the OD compounds. The existence and the extent of the PG can increase the IE appreciably [10]. For the Zn-free La214, we have found the oxygen isotope exponent (OIE) substantially higher in the UD compound compared to the OD one. Zn substitution further increases OIE systematically for both the UD and the OD compounds but the rate of increment is much higher for the UD ones, even though Zn suppresses T_c very effectively in the OD compounds. The variation of α in the UD compounds with disorder can be described well using a model based solely on pair-breaking [9, 12], but this model fails completely to describe the behavior of $\alpha(y)$ for the OD compounds. We discuss the possible implications of these in the subsequent sections.

2. Experimental samples

Polycrystalline, single-phase, sintered compounds with almost identical density were used. OIE results obtained for heavily underdoped $La_{1.91}Sr_{0.09}Cu_{1-y}Zn_yO_4$ (with Zn contents y = 0.00, 0.005, and 0.01) and heavily overdoped $La_{1.78}Sr_{0.22}Cu_{1-y}Zn_yO_4$ (with y = 0.00, 0.01, and 0.02) are reported in this paper. Details of sample preparation, characterization, and isotopic substitution procedure can be found in ref. [15]. Almost identical value of the room-temperature thermopower, S[290 K], showed that the p-values did not change appreciably with Zn substitution for fixed value of Sr content [16]. T_c, on the other hand, was suppressed rapidly by Zn substitution. The initial rate of suppression was strongly dependent on the hole content, dT_c/dy ~ - 15 K/%Zn and - 8 K/%Zn for the UD and the OD compounds, respectively.

The level of oxygen isotope enrichment was determined from the weight changes of the ^{16}O and ^{18}O exchanged samples. Raman spectroscopy was also used [15] as a means to double-check if samples were successfully isotope-exchanged, as any isotopic substitution will induce shifts in the associated Raman lines. The levels of isotope substitution found from the weight changes (Raman shifts) were: 99% (98%) for $La_{1.91}Sr_{0.09}CuO_4$; 99% (96%) for $La_{1.91}Sr_{0.09}Cu_{0.995}Zn_{0.005}O_4$; 97% (93%) for $La_{1.91}Sr_{0.09}Cu_{0.99}Zn_{0.01}O_4$ and 100% (96%) for $La_{1.78}Sr_{0.22}CuO_4$; 93% (93%) for $La_{1.78}Sr_{0.22}Cu_{0.99}Zn_{0.01}O_4$; 94% (92%) for $La_{1.78}Sr_{0.22}Cu_{0.98}Zn_{0.02}O_4$. These values are accurate within 5%. To check the reliability of the IE data shown here, we have back-exchanged ($^{18}O \rightarrow {}^{16}O$) two of the randomly selected compounds with different Sr contents and have found highly reproducible results.

3. Experimental results

IE was obtained from the DC magnetization measurements using a SQUID magnetometer (*Quantum Design MPMS2*). The field-cooled mode was employed using a DC magnetic field of 150 Oe parallel to the longest sample dimension. The *M-T* curves for the 9% and 22% Sr compounds with different Zn contents are shown in Fig. 1a and 1b, respectively. The back-exchanged data for a representative pair of samples is shown in Fig. 2, showing that the observed isotope effect is not due to the differences in grain size, demagnetization factor and remanent magnetization of the ^{16}O and ^{18}O sample pairs. T_c is taken at the point of intersection between the line extrapolated from the steepest part of the magnetization curve and the horizontal line of almost zero magnetization in the normal state [8, 17]. An example is shown in the inset of Fig. 1a. The isotope exponent was calculated from the expression $\alpha = - m\Delta T_c/T_c\Delta m$. We show the effects of hole content and in-plane disorder on the IE in Fig. 3. The IE for the pure compound, α_p, is almost three times larger (0.271) for the UD one compared to the OD sample (0.0939).

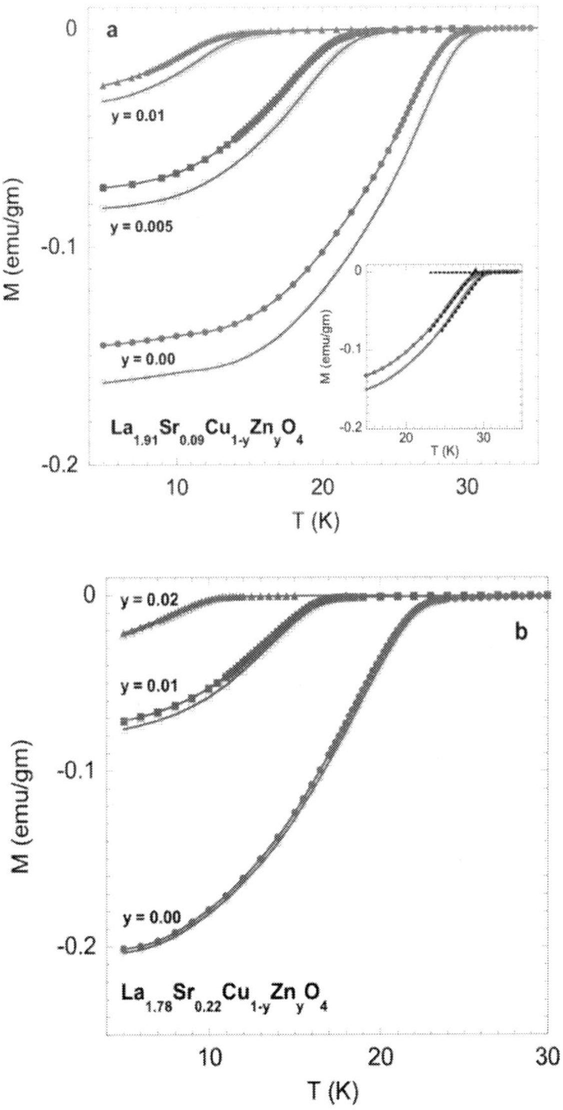

Figure 1 (color online): *M-T* data for (a) UD Zn-La214 and (b) OD Zn-La214. The open symbols are for the samples with ^{16}O and the filled ones are for the ones with ^{18}O. The sample compositions are shown in the plots. The inset of (a) shows the method used to determine T_c values (see text for details).

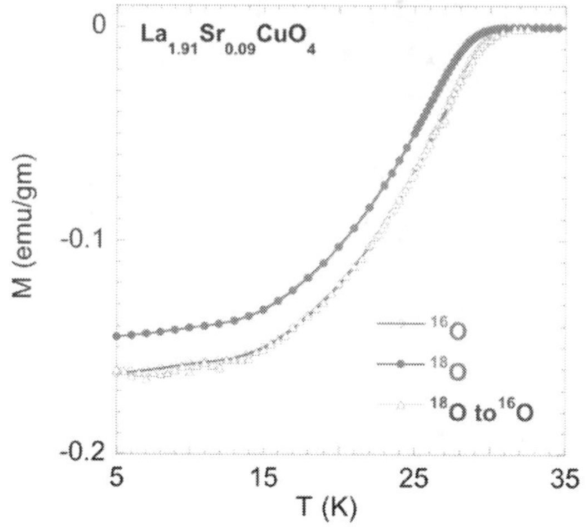

Figure 2 (color online): *M-T* plots for isotope exchanged and back-exchanged pure UD La214.

4. Analysis of the IE data

If it is assumed that the same pairing mechanism is at play in the UD and in the OD cuprates (a reasonable assumption), then the main point of concern should be, what are the differences between the UD and the OD samples? And how do those differences affect T_c? The major differences between the UD and OD samples are (a) presence of a large PG in the former and its absence in the later [17, 18] and (b) proximity of the UD compound to the 1/8th *static* stripe anomaly [14]. There could be others, somewhat less firmly established, electronic band structure related issues and the questions of non-adiabacity and mixed pairing schemes (coexistence of phonon and non-phonon mechanisms) that can also affect the values of the IE [9] as hole content varies.

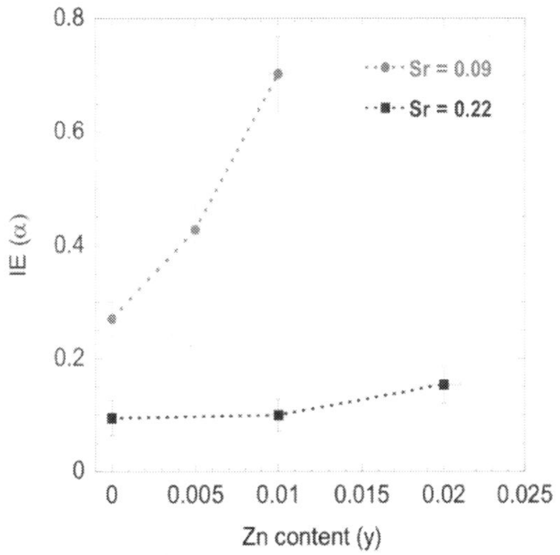

Figure 3 (color online): The IE versus Zn content plot. Sr concentrations are shown. The dashed lines are drawn only as guide to the eyes.

In case the PG reveals itself as a competing correlation, then the spectral weight removed due to its presence will be unavailable to the SC condensate and as a result T_c would be reduced. The extent of this reduction depends on the relative magnitudes of the PG and the SC gap *i.e.* the ratio E_g/Δ, here E_g is the characteristic PG energy and Δ is the maximum spectral gap at $T = 0$K [19]. Thus, an isotope effect in Δ will necessarily generate sizeable isotope effect in T_c and will show a diverging trend as $T_c \rightarrow 0$. Since PG increases with decreasing p (T_c decreases in the UD side), an increase in α_p (the IE of the in-plane disorder free compound) is expected. Williams *et al.* [19] and Pringle *et al.* [20] developed a model where the variation of α_p with p in pure compounds follows from the p-dependence of the E_g. In this scenario the IE in the SC gap (α_Δ) is assumed to be a constant, around 0.06, independent of the hole content and the observed variation in the IE in T_c with hole content is due to the variation in the PG energy scale. In the heavily OD compounds, where PG is absent, $\alpha_p \equiv \alpha_\Delta$ [19, 20]. The origin and significance of this small *intrinsic* p-independent value of isotope exponent is unclear though.

It is generally agreed upon that static stripe is detrimental to superconductivity and the lattice plays an important role in the stabilization of the stripe order. Therefore, an enhanced IE in the vicinity of the 1/8th doping is likely. There are indications that PG and stripe ordering originates from different physical mechanisms [9, 21]. In this situation the contribution to the IE originating from the PG and the stripe correlations should be additive and the significantly larger value (compared to the OD one) of α_p for the UD compound might be qualitatively accounted for.

Several earlier studies [9, 12] have analyzed the variation of the IE in disordered cuprates considering only the role played by pair-breaking due to intrinsic or extrinsic disorder. The general trend being, as T_c is decreased due to impurity scattering, IE increases. It is important to realize that both PG and stripe correlations can be incorporated directly to such pair-breaking schemes. In the unitary limit, PG determines the impurity scattering rate ([22], stripe ordering on the other hand, reduces both T_c and the superfluid density around the $1/8^{th}$ doping [23]. The combined effect of static stripe ordering and in-plane impurity is complex as in-plane disorder can change the strength of stripe correlations [21]. An important consequence of the magnetic pair-breaking model for the IE is that α can be expressed as a universal function $\alpha(T_c)$ [9]. This follows directly from the definition of the isotope exponent and the Abrikosov-Gorko'v (AG) equation [24], given by

$$\ln(\frac{T_c}{T_{c0}}) = -\kappa[\psi(\frac{1}{2}+\gamma) - \psi(\frac{1}{2})] \quad (1)$$

Here T_{c0} is the SC transition temperature of the pure compound and κ (≤ 1) is a constant depending on the symmetry of the SC order parameter and the nature of the pair-breaking mechanism. $\gamma = \Gamma/2\pi T_c$, where Γ is the pair-breaking scattering rate proportional to the impurity concentration and ψ is the *di-gamma* function. Consequently, one gets [9]

$$\frac{\alpha}{\alpha_p} = \frac{1}{1-\kappa\psi'(\frac{1}{2}+\gamma)\gamma} \quad (2)$$

ψ' is the first derivative of ψ. Eqs. 1 and 2 are also valid for non-magnetic potential scattering for *d*-wave superconductors [25]. Therefore, irrespective of the nature of the pair-breaking impurity scatterer, the IE in cuprates can be expressed in terms of the experimentally measurable parameters T_c and α_p. Here α_p denotes the IE of the pure (*e.g.*, Zn free, in this study) compound. This universal dependence of $\alpha(T_c)$ is shown in Fig. 4. In the same plot, the IE data for all the $La_{2-x}Sr_xCu_{1-y}Zn_yO_4$ compounds are exhibited. $\alpha(y)$ values for the UD compounds agree remarkably well with the universal curve whereas $\alpha(y)$ for the OD compounds falls significantly below this trend.

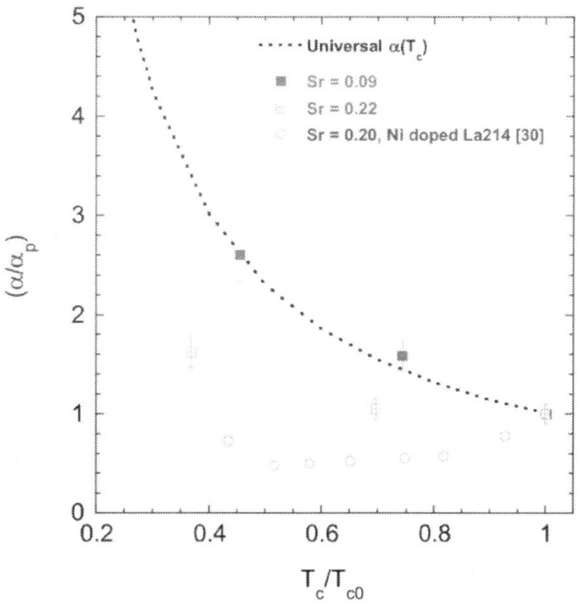

Figure 4 (color online): α/α_p versus T_c/T_{c0} for $La_{1.91}Sr_{0.09}Cu_{1-y}Zn_yO_4$ and $La_{1.78}Sr_{0.22}Cu_{1-y}Zn_yO_4$. The dashed curve shows the universal $\alpha(T_c)$ based on pair-breaking model (using Eq. 2, with $\kappa = 1$). For comparison, similar data for Ni substituted OD $La_{1.80}Sr_{0.20}Cu_{1-y}Ni_yO_4$ [30] compounds are also shown.

5. Discussion and conclusions

The results shown in Fig. 4 are surprising in several respects. Compared to the UD compounds, the physics of the OD ones are largely free from possible complications due to a sizeable PG and the relevance to the stripe ordering. Zn induced enhancement of magnetic susceptibility found in the UD compounds is also greatly reduced in the OD compounds [26]. As in the case for the UD region, the rate of suppression of T_c with Zn in the OD side can be modeled equally well within the framework of unitarity scattering in *d*-wave superconductors [22]. To illustrate this point explicitly, we show the experimental $T_c(x, y)$ data together with the fits to Eq. 1 in Fig. 5. Within this context, there is apparently no reason why the simple pair-breaking model would fail to describe the $\alpha(y)$ behavior for the OD $La_{1.78}Sr_{0.22}Cu_{1-y}Zn_yO_4$ compounds. This discrepancy is the central result of the present paper.

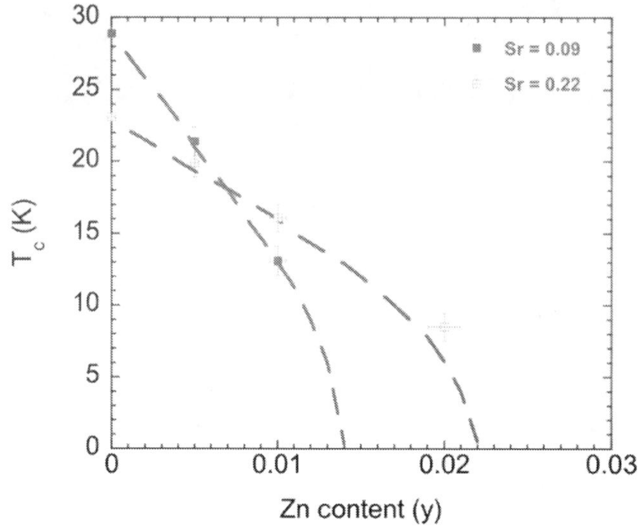

Figure 5 (color online): $T_c(x,y)$ for $La_{1-x}Sr_xCu_{1-y}Zn_yO_4$ compounds. The dashed lines show fits to the pair-breaking model given by Eq. 1.

It is interesting to note the dependence of α/α_p on the value of κ (Eq. 2). In *s*-wave superconductors with non-magnetic impurities, $\kappa = 0$, as a result there is no enhancement in the IE due to pair-breaking. For anisotropic *s*-wave or mixed (e.g., dx^2-y^2 + *s*) type order parameter, the value of $\kappa < 1$ and the enhancement of α/α_p could be significantly lower with increasing impurity concentration. In fact such a hole content dependent change in the SC order parameter was found in OD Y(Ca)123 [27], where it was found that the SC order parameter is dx^2-y^2 type for optimally and UD Y(Ca)123 but it consists of a significant *s* wave component as one moves into the OD region (orthorhombic structure). Therefore, for OD Y(Ca)123 it would not be surprising if α/α_p exhibits a weaker impurity content (or T_c/T_{c0}) dependence. Unfortunately, this line of reasoning does not help in understanding the behavior of the IE in the UD and OD La214. Unlike in Y123, the OD La214 becomes tetragonal. The low-*T* orthogonal phase does not exist in OD La214 for $x > 0.20$ [28]. The failure of pair-braking model to describe the IE in disordered (Ni and Fe doped) moderately OD ($x = 0.20$) La214 has been reported before [29, 30]. In ref. [29] the authors tried to describe the IE of Fe substituted La214 using the stripe scenario. This type of analysis is questionable since stripe correlations are not expected to play a large role in the OD compound. Also, Fe can change the in-plane hole content in addition to contributing in pair-breaking. Surprisingly the change in the IE with Ni doping was found to be minimal in the OD La214, and in fact α/α_p

decreased very gradually with increasing Ni content [29, 30] (see Fig. 4). It is interesting to note the contrasting effects of iso-valent Zn and Ni on the IE in OD La214. Both Zn and Ni have the same qualitative effect on T_c and the observed behavior can be modeled using the AG pair-breaking model. Even though the rate of suppression of T_c with Ni is weaker than that due to Zn. On the other hand, unlike Zn, Ni substitution has almost no effect on the IE in the OD La214 [30]. The only significant difference between Zn and Ni lies in their electronic spin configurations. Zn^{2+} is non-magnetic ($s = 0$, $3d^{10}$), while Ni^{2+} carries $s = 1$ due to its $3d^8$ electronic configuration. Therefore, the observed difference in the effect of these iso-valent impurities on the IE might be connected with spin physics. A complete understanding of the detailed mechanism is clearly lacking at this moment. Effects of Zn on the optimally doped (OPD, $x = 0.15$) La214 and Fe on moderately OD La214 were also studied earlier in ref. [30], where the authors tried the explain values of α/α_p by invoking to a peak in the EDOS for the OPD compounds (van Hove singularity (vHS) scenario [31]). In the case of heterovalent paramagnetic Fe doped $La_{1.80}Sr_{0.20}Cu_{1-y}Ni_yO_4$ compounds, the authors considered a shift in the position of the singularity with the level of Fe doping. Such an analysis is debatable because the ARPES results [32] showed that the Fermi level crosses the peak in the EDOS at $p \sim 0.225$, unlike the central assumption of the vHS, which asserts that the peak is located at the optimum doping.

In summary, the present work reports the IE of UD and OD La214 with different amount of in-plane iso-valent disorder. The evolution of $\alpha(y)$ for the UD compounds can be described well within the pair-breaking model for *d*-wave superconductors. The gradual increment of $\alpha(y)$ for the OD compounds, on the other hand, does not agree with the pair-breaking calculations, even though significant pair-breaking is present in these samples due to Zn. We have taken into account of the various other possible mechanisms and examined the role played by different electronic correlations that can affect the IE in cuprates, but none of these schemes provide with a completely satisfactory explanation for the experimentally observed OIE in La214 with different levels of disorder and hole content. The OIE data for the relatively *uncomplicated* OD La214 seem to pose a greater theoretical challenge.

Acknowledgements

The authors acknowledge the Commonwealth Commission, UK, Trinity College, University of Cambridge, UK, the Industrial Research Limited and MacDiarmid Institute for Advanced Materials and Nanotechnology, Wellington, New Zealand, for financial support and for the experimental facilities.

References

[1] Lee P A 2008 *Rep. Prog. Phys.* **71** 012501.

[2] Damascelli A, Hussain Z and Shen Z -X 2003 *Rev. Mod. Phys.* **75** 473.

[3] Bardeen J, Cooper L N and Schrieffer J R 1957 *Phys. Rev.* **106** 162 and *Phys. Rev.* **108** 1175.

[4] Frohlich H 1950 *Phys. Rev.* **79** 845.

[5] Schneider T and Keller H 2001 *Phys. Rev. Lett.* **86** 4899.

[6] Crawford M K, Kunchur M N, Farneth W E, McCarron III E M and Poon S J 1990 *Phys. Rev.* B **41** 282.

[7] Zhao G -M, Hunt M B, Keller H and Muller K A 1997 *Nature* **385** 236.

[8] Zhao G -M, H. Keller H, and Conder K 2001 *J. Phys. Cond. Matt.* **13** 569.

[9] Kresin V Z, Andreas B, Wolf S A and Ovchinnikov Y N 1997 *Phys. Rev.* B **56** 107 and Bill A, Kresin V Z and Wolf S A *arXiv:cond-mat/9801222v1*.

[10] Tallon J L, Islam R S, Storey J, Williams G V M and Cooper J R 2005 *Phys. Rev. Lett.* **94** 237002.

[11] Yang G Y, Zhang J D, Yang R L and Chen X H 2007 *Phys. Rev.* B **75** 212503.

[12] Harshman D R, Dow J D and Fiory A D 2008 *Phys. Rev.* B **77** 024523.

[13] Gweon G -H, Sasagawa T, Zhou S Y, Graf J, Takagi H, Lee D -H, and Lanzara A 2004 *Nature* **430** 187.

[14] Tranquada J M, Sternlib B J, Axe J D, Nakamura Y and Uchida S 1995 Nature **375** 561

[15] Naqib S H and Islam R S 2011 *Physica* C **471** 242.

[16] Naqib S H, Cooper J R, Tallon J L and Panagopoulos C 2003 *Physica* C **387** 365 and Tallon J L, Cooper J R, DeSilva P S I P N, Williams G V M and Loram J W 1995 *Phys. Rev. Lett.* **75** 4114.

[17] Naqib S H, Cooper J R, Tallon J L, Islam R S and Chakalov R A 2005 *Phys. Rev.* B **71** 054502.

[18] Naqib S H and Islam R S 2008 *Supercond. Sci. Technol.* **21** 105017.

[19] Williams G V M, Tallon J L, Quilty J W, Trodahl H J and Flower N E 1998 *Phys. Rev. Lett.* **80** 377.

[20] Pringle D J, Williams G V M and Tallon J L 2000 *Phys. Rev.* B **62** 12527.

[21] Islam R S and Naqib S H 2008 *Supercond. Sci. Technol.* **21** 125020.

[22] Tallon J L, Bernhard C, Williams G V M and Loram J W 1997 *Phys. Rev. Lett.* **79** 5294.

[23] Panagopoulos C, Xiang T, Anukool W, Cooper J R, Wang W S, Chu C W 2003 *Phys. Rev.* B **67** 220502.

[24] Abrikosov A A and Gor'kov L P 1961 *JETP* **12** 1243.

[25] Alexandrov A S and Zhao G -M 2009 *Phys. Rev.* B **80** 136501.

[26] Islam R S, Cooper J R, Loram J W and Naqib S H 2010 *Phys. Rev.* B **81** 054511.

[27] Yeh N -C, Chen C -T, Hammerl G, Mannhart J, Schmehl A, Schneider C W, Schulz R R, Tajima S, Yoshida K, Garrigus D and Strasik M 2001 *Phys. Rev. Lett.* **87** 087003.

[28] Torrence J B, Bezinge A, Nazzal A A I, Huang T C, Parkin S S P, Keane D T, LaPlaca S J, Horn P M and Held G A 1989 *Phys. Rev.* B **40** 8872.

[29] Mierzynska M and Wysokinsi K I 2003 *Phys. Rev.* B **68** 014504.

[30] Babushkina N, Zhernov A P, Inyuskhin A V, Kulagina T N, Ozhogin V I and Molchanova L V 1996 *Physica* C **272** 257.

[31] Tsuei C C, Newns D M, Chi C C and Pattnaik P C 1990 *Phys. Rev Lett.* **65** 2724.

[32] Kaminski A, Rosenkranz S, Fretwell H M, Norman M R, Randeria M, Campuzano J C, Park J M, Li Z Z and Raffy H 2006 *Phys. Rev.* B **73** 174511.

Journal of Physics: Conference Series **391** (2012) 012137 doi:10.1088/1742-6596/391/1/012137

Inverse-photoemission spectroscopy of iron-based superconductors $NdFeAsO_{1-\delta}$ and $Ba(Fe_{1-x}Co_x)_2As_2$

H Sato[1,4], Y Utsumi[2], O Morimoto[1], Y Nakashima[2], A Ino[2,4], Y Aiura[3,4], A Iyo[3,4], H Kito[3,4], K Miyazawa[3], P M Shirage[3], C H Lee[3,4], K Kihou[3], H Eisaki[3,4], H Namatame[1] and M Taniguchi[1,2]

[1] Hiroshima Synchrotron Radiation Center, Hiroshima University, Kagamiyama 2-313, Higashi-Hiroshima 739-0046, Japan

[2] Graduate School of Science, Hiroshima University, Kagamiyama 1-3-1, Higashi-Hiroshima 739-8526, Japan

[3] National Institute of Advanced Industrial Science and Technology, Ibaraki 305-8568, Japan

[4] Japan Science and Technology Agency, Transformative Research-Project on Iron Pnictide (JST-TRIP), Chiyoda 102-0075, Japan

E-mail: jinjin@hiroshima-u.ac.jp

Abstract. Unoccupied electronic structure of iron-based superconductors $NdFeAsO_{0.7}$ (T_c=51 K), $BaFe_2As_2$ and $Ba(Fe_{0.89}Co_{0.11})_2As_2$ (T_c=23 K) has been investigated by means of inverse-photoemission spectroscopy (IPES). The unoccupied Fe 3d states are observed around 1 eV above the Fermi level for all compounds. The Fe 3p-3d resonant IPES suggests that the unoccupied Fe 3d states of $Ba(Fe_{1-x}Co_x)_2As_2$ have more localized character compared with those of $NdFeAsO_{0.7}$. The unoccupied Nd 4f states of $NdFeAsO_{0.7}$ are located around 5 and 7 eV, and the Ba 5d and Ba 4f states of $Ba(Fe_{1-x}Co_x)_2As_2$ are located around 5 and 12 eV, respectively.

1. Introduction

The recent discovery of new superconductivity in the Fe-pnictide $LaFeAsO_{1-x}F_x$ with the critical temperature of T_c=26 K [1] has generated great interest because of its high T_c except for the cuprates. For RFeAsO (R: rare-earth metals), so-called 1111 system, $SmFeAsO_{1-x}F_x$ exhibits the highest T_c of 56 K [2], and for AFe_2As_2 (A: alkali and alkali-earth metals), so-called 122 system, the highest T_c of 38 K has been reported in $(Ba_{1-x}K_x)Fe_2As_2$ [3]. The FeAs-based superconductors are commonly composed of the FeAs layers and electronic structure of the FeAs layer is considered to play an essential role to superconductivity. The tremendous studies by means of angle-resolved photoemission spectroscopy have been carried out, so far, for, for example, $NdFeAsO_{1-x}F_x$ [4], $BaFe_2As_2$ [5] and $Ba(Fe_{1-x}Co_x)_2As_2$ [6], in order to reveal detailed feature of the Fermi surfaces with the hole pockets around the Γ point and electron pockets around the M point.

In this paper, we report whole feature of unoccupied electronic structure of $NdFeAsO_{1-\delta}$ and $Ba(Fe_{1-x}Co_x)_2As_2$ investigated by means of inverse-photoemission spectroscopy (IPES). Experimental information on unoccupied density of states (DOS) is important because it is directly related to the excited states. In addition, theoretical unoccupied DOS derived from the band-structure calculations is known to be often located at the lower energy compared with experimentally determined DOS [7]. In particular, in this study, by tuning incident electron energy (E_k) to the Fe 3p-3d excitation region, the Fe 3d structure near the Fermi level (E_F) is successfully speci ed. The IPES results suggest that the unoccupied Fe 3d states of $Ba(Fe_{1\ x}Co_x)_2As_2$ have more localized character compared with those of $NdFeAsO_{1\ \delta}$.

2. Experimental

IPES experiments were carried out at Hiroshima Synchrotron Radiation Center (HSRC) using the self-developed IPES spectrometer equipped with a low-energy electron gun, a non-periodic spherical grating and a one-dimensional photon detector [9]. The total energy resolution was 0.5 eV at E_k=50 eV. The energy of the IPES spectra is referred to E_F, determined from the Fermi edge of the IPES spectra of Au lm. All experiments were carried out at 26 K [10].

The polycrystals of $NdFeAsO_{0.7}$ (Nd1111), and single crystals of $BaFe_2As_2$ (Ba122) and $Ba(Fe_{0.89}Co_{0.11})_2As_2$ (Co-Ba122) were grown by the high-pressure synthesis method [11] and ux method [12], respectively. The oxygen de ciency for Nd1111 and the Co content for Co-Ba122 presented here are estimated values from the starting materials. The actual T_c-values of the grown samples were 51 K for Nd1111 and 23 K for Co-Ba122. Clean surfaces were *in situ* prepared by fracturing for Nd1111 and cleaving for (Co-)Ba122 at 26 K.

3. Results and discussion

Figure 1 shows IPES spectra of Nd1111 measured in the Nd 4d-4f excitation region (E_k=120 138 eV). We notice that the feature of the spectra strongly depends on E_k and the spectra exhibit prominent resonance above E_k=122 eV. In the spectrum at E_k=120 eV below the Nd 4d-4f excitation, two broad structures are observed around 2 and 7 eV. With increasing E_k to 122 eV, the structure around 5 eV, a dip region in the spectrum at E_k=120 eV, starts to be gradually enhanced and becomes a remarkable peak at E_k=124 eV. With further increasing E_k to 126 eV, a new component begins to be observed as a shoulder around 7 eV. The IPES spectra reach maximum in intensity at E_k=130 eV, and turn to decrease in intensity at E_k=134 and 138 eV. Since only the cross section of the Nd 4f inverse-photoemission resonantly increases in the Nd 4d-4f excitation region, the enhanced structures with the two components around 5 and 7 eV are attributed to the unoccupied Nd 4f states.

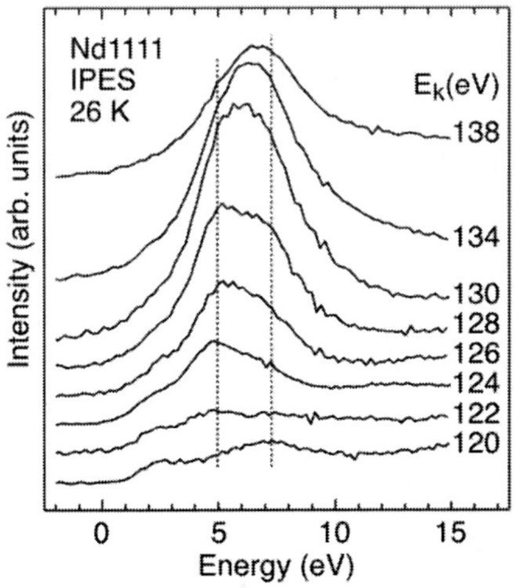

Figure 1. A series of IPES spectra of Nd1111 measured in the Nd 4d-4f excitation region.

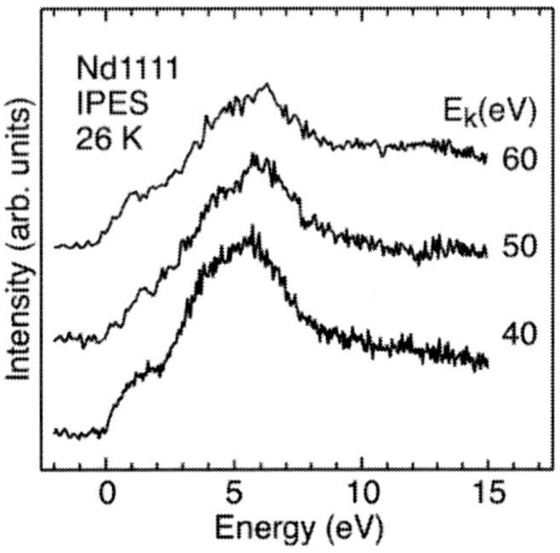

Figure 2. IPES spectra of Nd1111 measured at E_k=40, 50 and 60 eV.

Figure 2 shows the IPES spectra of Nd1111 measured at E_k=40, 50 and 60 eV. Note that the Fe 3p-3d excitation takes place between E_k=50 and 60 eV. In contrast to the Nd 4d-4f resonant IPES results (Fig. 1), the resonance enhancement is not observed between E_k=50 and 60 eV and the Fe 3d states can not be speci ed experimentally. The whole feature of the spectra is essentially unchanged up to E_k=100 eV. We nd a weak structure near E_F and a broad structure centered at 6 eV in the spectra at E_k=40 eV. The higher-lying one is derived from the unoccupied Nd 4f states. According to the band-structure calculation [13], the theoretical Fe 3d DOS of LaFeAsO exists from -5 to 2 eV region across E_F, and have a prominent and narrow peak at 1 eV in the conduction bands. From the comparison, the structure near E_F is attributed to the unoccupied Fe 3d states. Almost no resonance behavior of this structure in the Fe 3p-3d excitation region suggests that the unoccupied Fe 3d states are rather delocalized, in consistent with no satellite structure of the Fe 2p photoemission spectra of $LaFeAsO_{1-x}F_x$ [14].

Figure 3 shows the IPES spectra of Ba122 measured at E_k=40, 50 and 60 eV. The IPES spectra show a weak structure near E_F, a broad structure around 5 eV and a remarkable peak around 12 eV. In comparison with the IPES spectra of Nd1111 in Fig. 2, the remarkable peak is related to the Ba ion. The feature of the spectra of Co-Ba122 is similar to that of Ba122. It should be noticed that the structure near E_F is enhanced between E_k=50 and 60 eV, in contrast to the IPES results of Nd1111. This enhancement experimentally indicates that this structure is attributed to the unoccupied Fe 3d states and suggests the unoccupied Fe 3d states of Ba122 have a relatively localized character compared with those of Nd1111. Since the Fe 3d states of the FeAs layer is expected to be almost unchanged between Nd1111 and Ba122, the IPES results of Ba122 also support the assignment of the weak structure near E_F of the Nd1111 to the unoccupied Fe 3d states. Regarding the other structure, with the aid of the band-structure calculations [15], the broad structure around 5 eV, the most clearly observed at E_k=40 eV, is assigned to the unoccupied Ba 5d states and the remarkable peak around 12 eV to the unoccupied Ba 4f states.

Finally, the IPES spectra of Nd1111, Ba122 and Co-Ba122 measured at E_k=60 eV are compared in Fig. 4. The Fe 3d structures are observed near E_F for all compounds. Even taking into account the Nd 4f contribution to the spectra around 6 eV, the intensity of the Fe 3d structure of Nd1111 is smaller than those of (Co-)Ba122. The Fe 3d structure of Nd1111

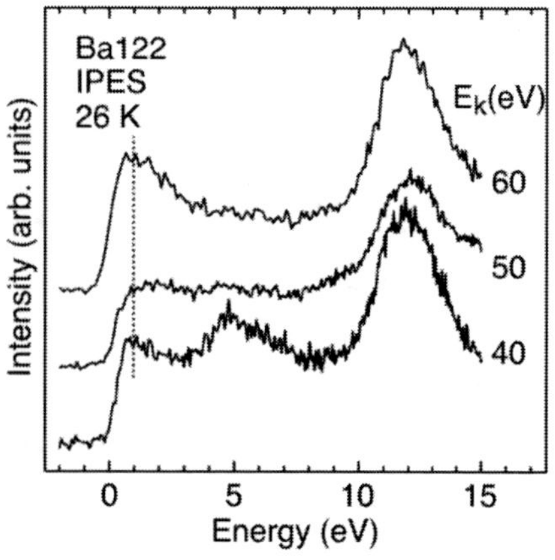

Figure 3. IPES spectra of Ba122 measured at E_k=40, 50 ans 60 eV.

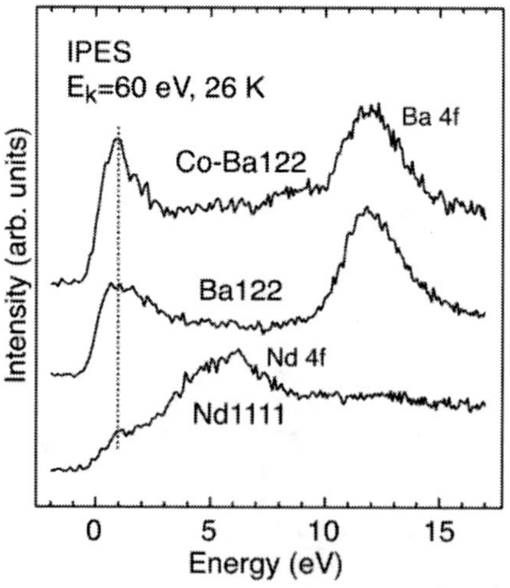

Figure 4. Comparisons of IPES spectra of Nd1111, Ba122 and Co-Ba122 measured at E_k=60 eV.

exhibits almost no resonance in the Fe 3p-3d excitation region, while those of (Co-)Ba122 exhibit clear resonance. The experimental results suggest that the unoccupied Fe 3d states of (Co-)Ba122 are relatively localized compared with those of Nd1111. This increase of the localized character of the Fe 3d states from 1111 to 122 system qualitatively correspond to the recent indications by Miyake *et al.*; the electron correlation becomes strong in order from 1111 (LaFePO and LaFeAsO) to 11 (FeSe and FeTe) through 122 ($BaFe_2As_2$) systems and, in this order, the compounds progressively lose covalent character of the chemical bonding between Fe and the pnictogen/chalcogen elements and gain stronger ionic character [16]. The IPES results are also qualitatively consistent with the smaller Fe moment of 0.63 μ_B for La1111 [17] than that of 0.87 μ_B for Ba122 [18, 19].

4. Summary

The unoccupied electronic structures of Nd1111 and (Co-)Ba122 have been investigated by means of IPES. The unoccupied Fe 3d states are located around 1 eV for all compounds and are relatively localized for (Co-)Ba122 compared with Nd1111. The unoccupied Nd 4f states of Nd1111 locate around 5 and 7 eV, and the Ba 5d and Ba 4f states of (Co-)Ba122 around 5 and 12 eV, respectively.

Acknowledgments

The authors are grateful to Prof. T Oguchi for the band-structure calculations of Ba122 taking into account the Ba 4f states. This work was supported by JST-TRIP. The IPES experiments have been done under the approval of HSRC (Proposal Nos. 08-A-48 and 09-A-60).

References

[1] Kamihara Y, Watanabe T, Hirano M and Hosono H 2008 *J. Am. Chem. Soc.* **130** 3296
[2] Ren Z A, Lu W, Yang J, Yi W, Shen X L, Li Z C, Che G C, Dong X L, Sun L L, Zhou A and Zhao Z X 2008 *Chin. Phys. Lett.* **25** 2215
[3] Rotter M, Tegel M and Johrendt D 2008 *Phys. Rev. Lett.* **101** 107006
[4] Kondo T, Santander-Syro A F, Copie O, Liu C, Tillman M E, Mun E D, Schmalian J, Bud'ko S L, Tanatar M A, Canfield P C and Kaminski A 2008 *Phys. Rev. Lett.* **101** 147003
[5] For example, Shimojima T, Ishizaka K, Ishida Y, Katayama N, Ohgushi K, Kiss T, Okawa M, Togashi T, Wang X -Y, Chen C -T, Watanabe S, Kadota R, Oguchi T, Chainani A and Shin S 2010 *Phys. Rev. Lett.* **104** 057002
[6] For example, Sudayama T, Wakisaka Y, Takubo K, Morinaga R, Sato T J, Arita M, Namatame H, Taniguchi M and Mizokawa T 2010 *Phys. Rev. Lett.* **104** 177002
[7] For example, Sato H, Tamura M, Happo N, Mihara T, Taniguchi M, Mizokawa T, Fujimori A and Ueda Y 1994 *Solid State Commun.* **92** 921
[8] Sato H, Kotsugi T, Senba S, Namatame H and Taniguchi M 1998 *J. Synchrotron Radiat.* **5** 772
[9] Although the measurement temperature is below T_c for $NdFeAsO_{0.7}$ and above T_c for $Ba(Fe_{0.89}Co_{0.11})_2As_2$, we assume that the IPES spectra do not depend on temperature significantly within the present experimental energy resolution.
[10] Kito H, Eisaki H and Iyo A 2008 *J. Phys. Soc. Jpn.* **77** 063707
[11] Nakajima Y, Taen T and Tamegami T 2009 *J. Phys. Soc. Jpn.* **78** 023702
[12] Ishibashi S, Terakura K and Hosono H 2008 *J. Phys. Soc. Jpn.* **77** 053709
[13] Malaeb W, Yoshida T, Kataoka T, Fujimori A, Kubota M, Ono K, Usui H, Kuroki K, Arita R, Aoki H, Kamihara Y, Hirano M and Hosono H 2008 *J. Phys. Soc. Jpn.* **77** 093714
[14] Oguchi T, private communication.
[15] Miyake T, Nakamura K, Arita R and Imada M 2010 *J. Phys. Soc. Jpn.* **79** 044705
[16] Qureshi N, Drees Y, Werner J, Wurmehl S, Hess C, Klingeler R, Buchner B, Fernandez-Diaz M T and Braden M 2010 *Phys. Rev. B.* **82** 184521
[17] Huang Q, Qiu Y, Wei Bao, Green M A, Lynn J W, Gasparovic Y C, Wu T, Wu G and Chen X H 2008 *Phys. Rev. Lett.* **101** 257003
[18] The Fe moment of Nd1111 is estimated to be $\sim$0.9 μ_B comparable to Ba122, which is reinforced by the Nd moment of $\sim$1.55 μ_B [Qiu Y, Wei Bao, Huang Q, Yildirim T, Simmons J M, Green M A, Lynn J W, Gasparovic Y C, Li J, Wu T, Wu G and Chen X H 2010 *Phys. Rev. Lett.* **101** 257002].

Study of the Two-Gap Superconductivity in GdO(F)FeAs by ScS-Andreev Spectroscopy

T.E. Shanygina[1,2], Ya.G. Ponomarev[2], S.A. Kuzmichev[2], M.G. Mikheev[2], S.N. Tchesnokov[2], O.E. Omel'yanovsky[1], A.V. Sadakov[1], Yu.F. Eltsev[1], V.M. Pudalov[1], A.S. Usol'tsev[1,2], E.P. Khlybov[3], and L.F. Kulikova[3]

[1] P.N. Lebedev Physical Institute, 119991 Moscow, Russia
[2] M.V. Lomonosov Moscow State University, 119991 Moscow, Russia
[3] Institute for High Pressure Physics, 142190 Troitsk, Russia

E-mail: tatiana.shanygina@gmail.com

Abstract. Current-voltage characteristics and dynamic conductance of the superconductor - constriction - superconductor junctions in GdO(F)FeAs polycrystalline samples with critical temperatures $T_C^{\rm local} = 46 \div 53\,$K were investigated. Two superconducting gaps, the large $\Delta_{\rm L} = 10.5 \pm 2$ meV, and the small one $\Delta_{\rm S} = 2.3 \pm 0.4\,$meV were clearly observed at $T = 4.2\,$K. The $2\Delta_{\rm L}/k_{\rm B}T_C^{\rm local} = 5.5 \pm 1$ ratio gives support to the strong coupling mechanism which is responsible for the high T_C value. Temperature dependence of the large gap $\Delta_{\rm L}(T)$ indicates the presence of intrinsic proximity effect (in k−space) between two superconducting condensates.

Novel Fe-based superconductors [1] are now in the focus of intensive experimental research. Some of their features such as layered crystal structure and spatial separation of superconducting layers are similar to those of cuprates and MgB_2. The stoichiometric 1111-family compounds were shown to be antiferromagnetic metals with spin density wave (SDW) ground state [2, 3]. Partial oxygen deficiency or fluorine substitution for oxygen induces superconductivity in the Fe-As layers. Rare-earth elements replacement also affects the critical temperature T_C. In particular, T_C for Gd-1111 oxypnictides can be lowered by replacement of Y for Gd [4] or increased by introducing Th instead of Gd [5]. The $T_C = 56\,$K reported in $Gd_{0.8}Th_{0.2}FeAs$ compound is today the highest one for Fe-based superconductors.

Band-structure calculations showed the total density of states $N(0)$ at the Fermi level to be formed mainly by Fe $3d$-states [6, 7, 8]. A large Fe-isotope effect [9] and the correlation between T_C and $N(0)$ for different iron-based superconductors [10] give support to the phonon-mediated coupling in these compounds. The theoretically calculated Fermi surface for 1111-system compounds [11, 12, 13, 14] comprises cylinder-like hole sheets centered at the Γ point and quasi-two-dimensional (2D) electron sheets at the M point of the reduced Brillouin zone.

Here we present the superconducting gap measurements in GdO(F)FeAs by Andreev spectroscopy of ScS-contacts (superconductor - constriction - superconductor) using a "break-junction" technique.

Nearly optimally doped Gd-1111 samples were prepared by high pressure synthesis described in detail in [15, 16]. Superconducting properties of the samples were tested by measuring temperature dependences of the AC-magnetic susceptibility and resistance $R(T)$. Both showed a sharp superconducting transition at $T_C^{\rm bulk} = (52 \div 53)\,{\rm K}$ (the bulk value of the critical temperature was defined at the maximum of $dR(T)/dT$-curve). The two sets of polycrystalline GdO(F)FeAs samples were studied: "El"-series ($GdO_{0.88}F_{0.12}FeAs$, $T_C^{\rm bulk} = 53 \pm 1\,{\rm K}$) and "Kh"-series ($GdO_{0.91}F_{0.09}FeAs$, $T_C^{\rm bulk} = 52 \pm 1\,{\rm K}$). Figure 1 shows typical temperature dependences of resistance for Kh-series samples.

In order to determine superconducting gap values for GdO(F)FeAs, we used (i) Andreev spectroscopy [17] of individual Sharvin-type [18] ScS-contacts and (ii) intrinsic Andreev spectroscopy (intrinsic multiple Andreev reflections effect (IMARE) [19], which usually exists due to the presence of steps and terraces at clean cryogenic clefts). ScS-Andreev nanocontacts in polycrystalline samples were prepared by the "break-junction" technique [20, 21]. The samples of the typical size $1.5 \times 0.5 \times 0.1\,{\rm mm}^3$ were attached to the flexible sample holder with two current and two potential leads by In-Ga alloy that is liquid at room temperature. The holder with the sample was cooled down to 4.2 K, then a microcrack in the sample was generated by precise bending the sample holder.

The resulting nano-contact is a mechanical contact of two clean cryogenically cleaved surfaces separated by a constriction; this geometry excludes impurity presence at the cryogenic clefts. The $I(V)$-, $dI(V)/dV$-characteristics of the contact were measured using a standard AC-modulation technique [22, 23].

According to the theory by Kümmel *et al.* [24] for the SnS (superconductor - normal metal - superconductor) Andreev-type contact, the main features of its current-voltage characteristics (CVC) are the excess current at low bias voltages and the subharmonic gap structure (SGS). The latter shows series of dynamic conductance dips at bias voltages

$$V_n = \frac{2\Delta}{en}, \; n = 1, 2, \ldots \tag{1}$$

due to the multiple Andreev reflections effect. For a multigap superconductor, several independent SGSs corresponding to each gap should be observed. The CVCs studied in this work are typical for the clean classical SnS-contact (with a constriction acting as a normal metal) with excess-current at low bias voltage, therefore, we believe, the theoretical model by Kümmel *et al.* [24] is applicable to our break-junctions. Strictly speaking, the sharpest SGS are intrinsic of Andreev contacts of the highest quality with a small diameter a which is less than the quasiparticles mean free path l (i.e., in the ballistic limit) [18]. In such a case one may observe a large number of gap peculiarities, thus facilitating the gap energy measurements. In the case $a \approx l$, only few SGS minima would contribute to the dynamic conductance spectra, which in most cases corresponds to the studied GdO(F)FeAs ScS-contacts.

In the framework of the "break-junction" technique and due to the layered structure of GdO(F)FeAs, the so-called stacks of Andreev contacts may be formed on cryogenic clefts; this type of junctions, the $S - c - \ldots - S - c - S$-array, enables one to observe IMARE similar to that observed earlier in Bi-2201 [25]. For such an array, bias voltage for any peculiarities caused by the bulk properties at $dI(V)/dV$-characteristic scales with number of sequentially connected contacts N as compared with the single junction case. On the contrary, this is not the case for such CVC features that originate from any defects influence on the cleaved surfaces. One can determine N by collating dynamic conductance spectra of Andreev arrays with different number of junctions in a stack and normalizing them to a single contact characteristics. The array contacts are a factor of N less sensitive to surface defects (which otherwise would seriously affect the properties of superconductor [26]). Thus, by making CVC measurements on the stack contacts, one can determine the *true bulk gap* values much more accurately.

Figure 2 shows dynamic conductance of two ScS-Andreev contacts: single junction El3d1, lower black line, and an array Kh3d4, upper gray line. After normalizing the $dI(V)/dV$-characteristic of the array ($N = 2$ for this case) to that for the single junction, we achieve a good

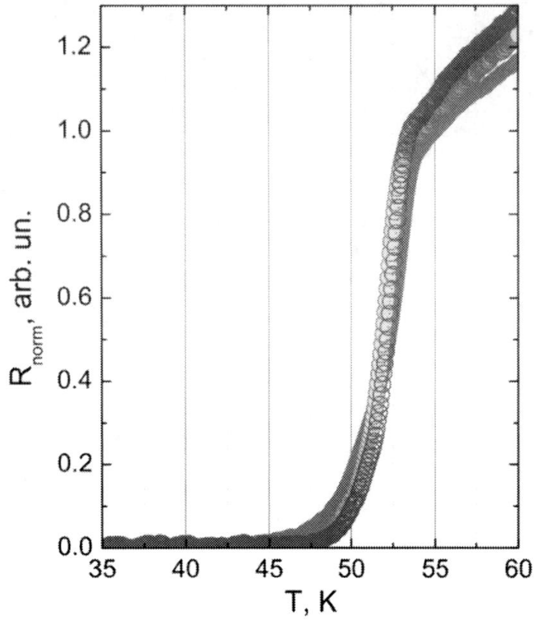

Figure 1. Normalized temperature dependence of resistance for polycrystalline $GdO_{0.91}F_{0.09}FeAs$ samples Kh1 (solid yellow circles), Kh2 (red line) and Kh3 (open blue circles) measured prior a microcrack formation. The bulk critical temperatures $T_C^{bulk} = 52 \pm 1\,K$.

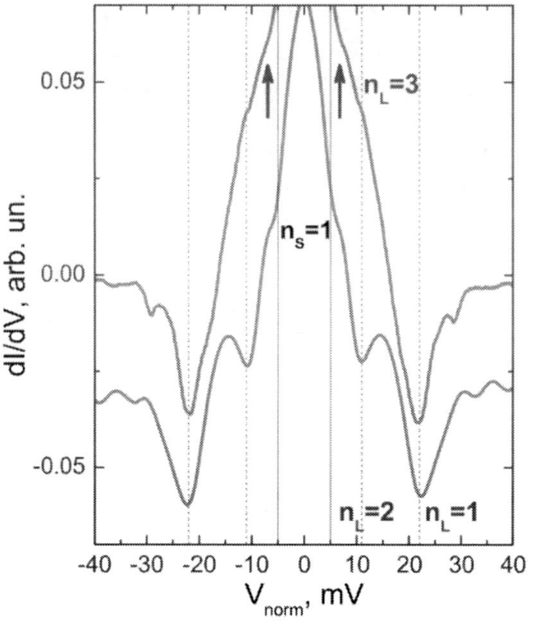

Figure 2. dI/dV-curves for the single-junction ScS-contact El1d6 (lower yellow line) and 2-junctions Kh3d4 array (normalized to a single junction; upper red line). $T = 4.2\,K$. SGS dips defining $\Delta_L \approx 11\,meV$ and $\Delta_S \approx 2.5\,meV$ are marked with n_L and n_S labels, respectively.

coincidence of the spectra peculiarities. Such a procedure was done for all ScS-arrays, with N as a fitting parameter. The normalized dynamic conductance shows a set of dips at bias voltages $V \approx 22\,mV$ and $11\,mV$ (marked by dashed vertical lines). The minimum at $V \approx 7.3\,mV$ is seen only at the Kh3d4-curve (marked by arrows) for stack contact. These three peculiarities may be associated with the SGS minima $n_L = 1, 2, 3$, which leads to the estimated local value of the large superconducting gap $\Delta_L \approx 11\,meV$. The minima at $V \approx \pm 5\,mV$ at the El1d6-curve (marked by solid vertical lines) don't correspond to the expected bias voltage $V \approx 7.3\,mV$ for the 3rd harmonic $n_L = 3$ of the large gap, and, hence, indicates the presence of the small gap $\Delta_S \approx 2.5\,meV$. The satellite minima above the $n_L = 1$ dips may originate from excitation of Leggett collective modes and require additional studies.

For some contacts (such as El3 sample, single ScS-Andreev junction #d1), one can see the SGS originating from the small gap, but not only a single peculiarity. Figure 3 shows such series comprising $n_S = 1, 2, 3$ at $V \approx (4.4,\ 2.2,\ 1.3)\,mV$ (vertical dashed lines). Using eq. (1), we directly find the small superconducting gap value $\Delta_S \approx 2.2\,meV$. The additional fine structure seen in Fig. 3 at $V \approx 5\,mV$ may be caused by the Δ_S order parameter anisotropy and is subject of further studies. As the small gap SGS is located near zero bias voltages, it becomes smeared by the dramatic increasing of the dynamic conductance at the dI/dV-curve. Hence, the background subtraction or taking the second derivative of the CVC helps one to distinguish the small gap peculiarities, as was done to the spectrum at the Fig.3.

By a precise mechanical readjusting of the contact, we were able to observe characteristics of several ScS break-junctions within the same low-temperature experiment; the independent

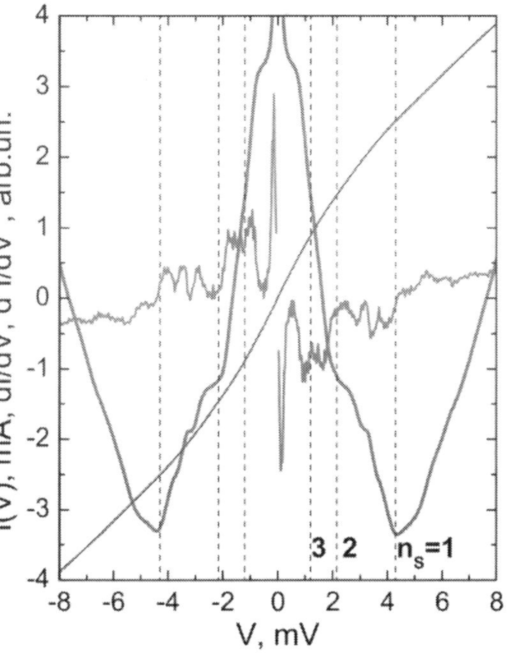

Figure 3. $I(V)$- (blue line), dI/dV- (red line) and d^2I/dV^2-characteristics (yellow line) for single ScS-contact El3d1. $T = 4.2\ K$. The SGS dips associated with small gap $\Delta_S \approx 2.2\ meV$ are marked by dashed vertical lines. Linear background is subtracted.

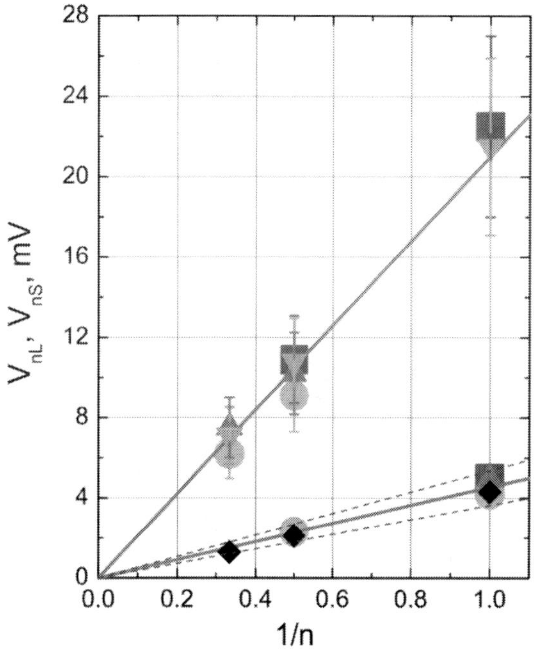

Figure 4. Normalized to a single junction SGS bias voltages V_{nL}, V_{nS} versus $1/n$ for several ScS-contacts ($T_C = 46 \div 53\ K$) investigated. The solid lines indicate the averaged gap values $\Delta_L = 10.5 \pm 2\ meV$ and $\Delta_S = 2.3 \pm 0.4\ meV$ at $T = 4.2\ K$.

local gap measurements improve the statistics of the results. To get values of $\Delta_{L,S}$ and their scatter, in accordance with eq. (1), we plotted SGS minima positions V_n as a function of their inverse number $1/n$; this line should start at the $(0,0)$ point. Our experimental data measured at $T = 4.2$ K are summarized on Figure 4 that shows the results for three samples with a similar T_C^{bulk} (Kh3d4 contact - down triangles, El1d6 - squares, El3d1 - rhombs, El3d2 - circles, and El3d13 - up triangles). All the data agree with each other and confirm the presence of the two distinct superconducting gaps. The averaged gap values obtained at $T = 4.2$ K [16, 27] are:

$$\Delta_L = 10.5 \pm 2\,\text{meV},\ \Delta_S = 2.3 \pm 0.4\,\text{meV}$$

for the critical temperatures $T_C^{local} = (46 \div 53)$ K. The close agreement of the data for various samples shows the reproducibility of the measured gaps magnitude.

IMARE spectroscopy technique also facilitates measurements of temperature dependence of the superconducting gaps. The subharmonic gap structure of the dynamic conductance of symmetrical SnS-contacts remains rather well-pronounced up to T_C. Importantly, to define the gap value no additional fitting is required [24]: at any temperatures up to T_C the gap Δ can be obtained straightforward using the expression (1). We note, that in case of alternative technique of the Andreev spectroscopy of SN-junctions, and for a two-gap superconductor, one needs to fit the resulting conductance curves using 7 fitting parameters.

Figure 5 shows the dynamic conductance of ScS-array (sample Kh3, contact #d4, see Fig. 2) measured in the range $4.2\ K \leq T \leq T_C$. The curves are shifted vertically for clarity. The

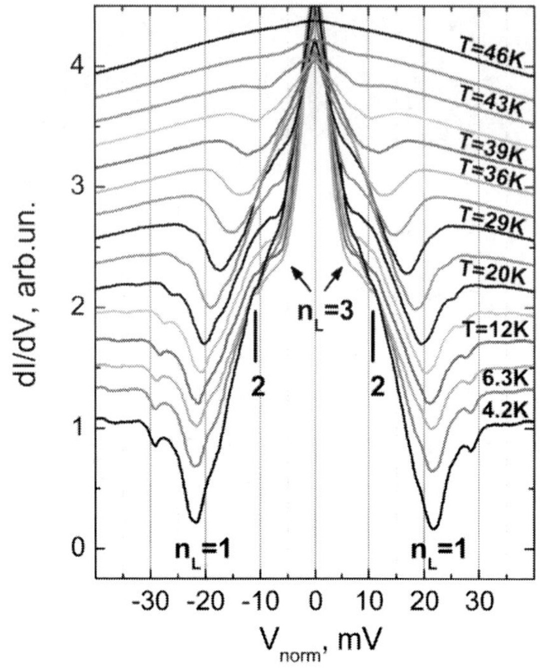

Figure 5. The temperature affecting on the SGS (black labels) at normalized dynamic conductance of 2 ScS-junctions array Kh3d4 ($4.2\,K \leq T \leq T_C$). $\Delta_L(4.2\,K) \approx 11\,meV$. Local critical temperature $T_C^{local} \approx 46\,K$. The curves were shifted along the vertical scale for the sake of clarity.

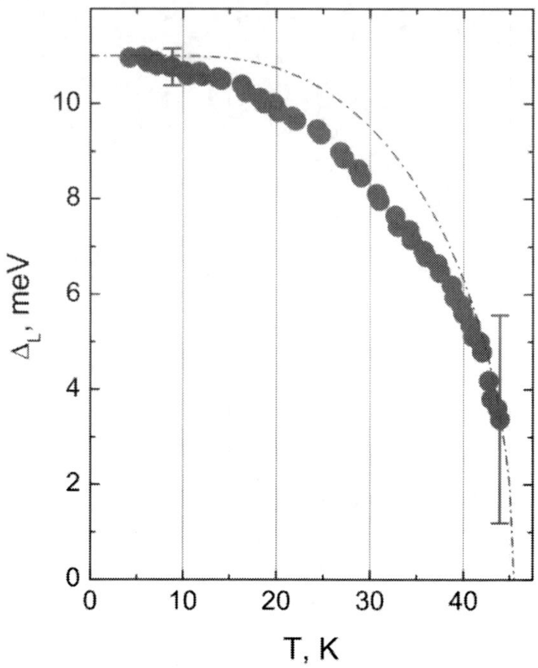

Figure 6. Temperature dependence of the large gap $\Delta_L(T)$ (circles) plotted under the data of Figure 5. $T_C^{local} \approx 46\,K$. Single-gap BCS-like behavior (dash-dotted line) is presented for comparison. Error bars show typical temperature smearing for Cooper pairs at two representative temperatures.

peculiarities are washed out at $T_C^{local} \approx 46\,K$. The "local" critical temperature is referred to as the temperature of the transition to the normal state in the contact area (with diameter $a \approx l$). It is worthy to note that the T_C^{local} can differ from the T_C^{bulk}, the averaged characteristic measured by non-local methods. The coincidence of the local T_C^{local} and the averaged T_C^{bulk} may be observed only in the ideal clean and homogeneous sample. By taking the $\Delta_L \approx 11$ meV value (see Fig. 2) and $T_C^{local} \approx 46$ K from Figure 5, for the Kh3d4 array we obtain an estimate for the BCS ratio, $2\Delta_L/k_BT_C^{local} \approx 5.5$; the ratio would however be reduced to 4.8 if we use the averaged $T_C^{bulk} = 53 \pm 1$ K value, as in Refs. [16, 27]. The former value 5.5 is close to that for $LaO_{0.9}F_{0.1}FeAs$ (a lower-T_C analogue of Gd-1111) [28]. At the same time, the measured $2\Delta_L/k_BT_C^{local}$-ratio exceeds the standard single-gap BCS value 3.52 for the weak coupling limit. Presumably, the $2\Delta_L/k_BT_C$ value suggests the strong coupling in the condensates with a large gap value. As for the small gap, our data leads to $2\Delta_S/k_BT_C = 1.1 \div 1.3$ which is less than the BCS value 3.52.

Figure 6 shows the temperature dependence of the large gap $\Delta_L(T)$ obtained from our IMARE measurements (see Fig. 5) within the range $4.2\,K \leq T \leq T_C^{local}$. Similar measurements of the temperature dependence for the small gap represent a harder task, possibly require higher quality of samples, and have not been accomplished yet. The $\Delta_L(T)$ dependence lies noticeably below the standard single-gap BCS-like curve (dash-dotted line in Figure 6). Note that the temperature smearing (see error bars plotted for the last data point) allows to resolve the gap value up to ≈44.5 K. The deviation in the $\Delta_L(T)$ dependence could not be attributed to surface gap influence (i.e., it is not due to the real space proximity effect), because it is obtained from Andreev reflection measurements in the *array* contact, where the surface effects are reduced. Moreover, the $\Delta_L(T=0)$ energy coincides with the value for the single ScS-contacts (see Fig. 2, 4). Were the real-space proximity effect be responsible for the deviation, the $\Delta_L(T=0)$ would differ; our experiment, however, shows this is not the case. Such behavior resembles the $\Delta_\sigma(T)$-dependence for MgB_2 [29, 30, 31] and $\Delta_L(T)$ for FeSe [23]. Theoretical studies [32, 33, 34] explain such a deviation by interband coupling effect. By analogy, we assume that similar deviation for Gd-1111 may arise from nonzero interband coupling. The reduced BCS-ratio for the small gap also supports this conclusion and suggests that the "weak" superconductivity in the low gap condensate may be induced by k-space (internal) proximity effect between two superconducting condensates [35]. According to Refs. [32, 33, 34], the observed shape of the $\Delta(T)$ curve (see Figure 6) is typical for the "driving" condensate in the presence of the second (driven) condensate with a small gap.

The existence of the large superconducting gap ($2\Delta_L/k_BT_C^{bulk} > 4$) in the 1111-family compounds with critical temperatures $40\,K \leq T_C \leq 53\,K$ was confirmed by tunneling spectroscopy ("break-junction" technique) [36, 37], point-contact Andreev reflection (PCAR) spectroscopy of SN-junctions [38, 39, 40, 41, 42, 43, 44], and angle-resolved photoemission spectroscopy (ARPES) [45]. To the best of our knowledge, there are no other data available for Gd-1111 to compare with. Thus, comparing our results for GdO(F)FeAs with the aforementioned data on other 1111-superconductors, we find rather good agreement with $2\Delta_L/k_BT_C$ values determined by "break-junction" technique [36, 37] and in some PCAR-measurements [43, 44].

In conclusion, we studied the dynamic conductance of ScS-contacts in $GdO_{1-x}F_xFeAs$ ($x = 0.09, 0.12$) polycrystalline samples ($T_C^{local} = 46 \div 53\,K$) by ScS Andreev spectroscopy, in the temperature range $4.2\,K \leq T \leq T_C$. We detected two superconducting gaps, the large $\Delta_L = 10.5 \pm 2\,meV$, and the small $\Delta_S = 2.3 \pm 0.4\,meV$. The observed temperature dependence of the large gap $\Delta_L(T)$ suggests the existence of k-space proximity effect between two superconducting condensates and, therefore, nonzero interband interaction. The estimated $2\Delta_L/k_BT_C^{local} = 5.5 \pm 1$ ratio exceeds the BCS limit 3.52 and indicates the strong coupling regime in the "driving" condensate.

The work was supported by grants from RFBR, and the Russian Ministry of Education and Science. We also thank A. Bianconi and A. Kordyuk for appreciation and valuable discussions.

References

[1] Kamihara Y *et al.* 2008 *J. Am. Chem. Soc.* **130** 3296
[2] Klauss H H *et al.* 2008 *Phys. Rev. Lett.* **101** 077005
[3] Luetkens H *et al.* 2008 *Phys. Rev. Lett.* **101** 097009
[4] Kadowaki K *et al.* 2009 *J. Phys.: Conf. Ser.* **83** 67006
[5] Wang C *et al.* 2008 *Europhys. Lett.* **150** 052088
[6] Nekrasov I A *et al.* 2008 *JETP Lett.* **87** 560
[7] Eschrig H and Koepernik K 2009 *Phys. Rev.* B **80** 104503
[8] Miyake T *et al.* 2010 *J. Phys. Soc. Jpn.* **79** 044705
[9] Liu R H *et al.* 2009 *Nature* **459** 64
[10] Sadovskii M V *et al.* 2010 *JETP Lett.* **91** 518
[11] Singh D J and Du M H 2008 *Phys. Rev. Lett.* **100** 237003
[12] Singh D J 2009 *Physica C* **469** 418
[13] Nekrasov I A *et al.* 2008 *JETP Lett.* **88** 144
[14] Coldea A I *et al.* 2008 *Phys. Rev. Lett.* **101** 216402
[15] Khlybov E P *et al.* 2009 *JETP Lett.* **90** 387
[16] Shanygina T E *et al.* 2011 *JETP Lett.* **93** 94
[17] Andreev A F 1964 *Sov. Phys. JETP* **19** 1228
[18] Sharvin Yu V 1965 *Zh. Eksp. Teor. Fiz.* **48** 984
[19] Nakamura H *et al.* 2009 *J. Phys. Soc. Jpn.* **78** 123712
[20] Moreland J and Ekin J W 1985 *J. Appl. Phys.* **58** 3888

[21] Müller J *et al.* 1992 *Physica C* **191** 485
[22] Ponomarev Ya G and Rakhmanina A V 1970 *Prib. Tehn. Eksp.* **5** 120
[23] Ponomarev Ya G *et al.* 2011 *Zh. Eksp. Teor. Fiz.* **140** 527
[24] Kümmel R *et al.* 1990 *Phys. Rev.* B **42** 3992
[25] Ponomarev Ya G *et al.* 2000 *Inst. Phys. Conf. Ser.* **167** 241
[26] van Heumen E *et al.* 2011 *Phys. Rev. Lett.* **106** 027002
[27] Pudalov V M *et al.* 2011 *Physics-Uspekhi* **54** 648
[28] Ponomarev Ya G *et al.* 2009 *Phys. Rev.* B **79** 224517
[29] Ponomarev Ya G *et al.* 2004 *Solid State Comm.* **129** 85
[30] Ponomarev Ya G *et al.* 2004 *JETP Lett.* **79** 484
[31] Kuzmichev S A *et al.* 2012 *Solid State Comm.* **152** 119
[32] Moskalenko V A 1959 *Phys. Met. Metall.* **4** 503
[33] Suhl H *et al.* 1959 *Phys. Rev. Lett.* **12** 552
[34] Nicol E J and Carbotte J P 2005 *Phys. Rev.* B **71** 054501
[35] Yanson I K *et al.* 2003 *Phys. Rev.* B **67** 024517
[36] Ekino T *et al.* 2009 *Physica C* **470** S358
[37] Sugimoto A *et al.* 2010 *Physica C* **470** 1070
[38] Daghero D *et al.* 2009 *Phys. Rev.* B **80** 060502(R)
[39] Gonnelli R S *et al.* 2009 *Physica C* **469** 512
[40] Wang Y L *et al.* 2009 *Supercon. Sci. Technol.* **22** 015018
[41] Miyakawa N *et al.* 2010 *J. Supercond. Novel Magn.* **23** 575
[42] Tanaka M and Shimada D 2010 *J. Supercond. Novel Magn.* **24** 1491
[43] Yates K A *et al.* 2009 *New J. Phys.* **11** 025015
[44] Samuely P *et al.* 2009 *Supercon. Sci. Technol.* **22** 014003
[45] Kondo T *et al.* 2008 *Phys. Rev. Lett.* **101** 147003

International Conference on Strongly Correlated Electron Systems (SCES 2011) IOP Publishing
Journal of Physics: Conference Series **391** (2012) 012139 doi:10.1088/1742-6596/391/1/012139

Magnetic spectrum of the $J_1 - J_2$ model for pnictides

M. B. Silva Neto*, **E. C. Marino**

Instituto de Física, Universidade Federal do Rio de Janeiro, Caixa Postal 68528, Rio de Janeiro, Brazil

E-mail: *mbsn@if.ufrj.br

C. M. S. da Conceicao

Departamento de Física Teórica, Universidade do Estado do Rio de Janeiro, CEP 20550-013, Rio de Janeiro, Brazil

Abstract. We calculate the magnetic excitation spectrum of the $J_1 - J_2$ model for iron pnictides within the framework of the nonlinear sigma model. We find, in addition to the usual acoustic spin-wave branch, a second, optical spin-wave branch, not captured by previous theoretical studies. We also find that the spin-wave velocity has a planar directional anisotropy, resulting from the collinear/striped structure of the antiferromagnetic ground state. Finally, we include single ion anisotropy, which fully gaps all branches, and we discuss the relevance of the modified, multi-branched, fully gapped character of the spectrum to the understanding of neutron scattering results for $SrFe_2As_2$

1. Introduction

The discovery of superconductivity above 50 K in $RFeAsO_{1-x}F_x$ (R = La, Ce, Sm, etc.) [1] brought the new iron-based pnictide compounds, including the MFe_2As_2 (M = Ba, Sr, etc.) family [2], into the class of the so called high-temperature superconductors. Like cuprates, pnictides also exhibit a layered spin density wave (SDW) ordered structure in the parent compound, which gives room to superconductivity upon doping [3]. From the theoretical point of view, the magnetism of the parent compounds could be, in principle, described in terms of localized magnetic moments at the Fe positions, interacting antiferromagnetically both via a superexchange J_1 between nearest Fe neighbors, Fe−Fe, and a second superexchange J_2 between As bridged next-to-nearest Fe neighbors, Fe−As−Fe [4]. Although an itinerant description in terms of band electrons, Fermi surface nesting, and spin-density-wave (SDW) instability, is also a good candidate for the understanding of the magnetism in iron pnictides [5, 6], in the following we shall adopt an effective description in terms of local moments.

2. The Model

Let us start by considering the $J_1 - J_2$ spin-Hamiltonian

$$\hat{H} = J_1 \sum_{\langle i,j \rangle} \hat{\mathbf{S}}_i \cdot \hat{\mathbf{S}}_j + J_2 \sum_{\langle\langle i,j \rangle\rangle} \hat{\mathbf{S}}_i \cdot \hat{\mathbf{S}}_j, \tag{1}$$

where $J_1 > 0$ and $J_2 > 0$ are, respectively, the antiferromagnetic superexchanges between nearest-neighbors, $\langle i,j \rangle$, and next-to-nearest neighbors, $\langle\langle i,j \rangle\rangle$, spins $\hat{\mathbf{S}}_i$ on a two dimensional square lattice. The existence of two superexchanges J_1 and J_2 in the spin Hamiltonian (1) renders the antiferromagnetism collinear, with wave vectors at $(\pi, 0)$ and/or $(0, \pi)$ [7]. This is in agreement with recent inelastic neutron scattering experiments in (Sr,Ba)Fe_2As_2 [7, 8], which exhibit clear peaks at the antiferromagnetic zone center $Q = (1, 0, 1)$, for energies between $5-15$ meV. Such wave vector degeneracy in the classical ground state of the Hamiltonian (1) gives rise to an extra Ising symmetry [9, 10, 11], which is broken at a different temperature than the AF ordering one.

Previous linear spin wave studies of the $J_1 - J_2$ model have considered, as a starting point, a classical ground state with a *two sublattice structure*, and have obtained that the quantized magnetic spectrum would be composed by a single gapless spin-wave branch [12], with $\hbar\omega(\mathbf{k}) \sim |\mathbf{k}|$ for small wave vector $\mathbf{k}$. The existence of such gapless mode reflects the full rotational invariance of the spin Hamiltonian (1) and is in agreement with Goldstone's theorem. We shall refer to this branch as the *acoustic branch*. As we shall soon see, the actual *four sublattice structure* of the AF/Ising broken classical ground state, with two coupled, interpenetrating Néel ordered states, gives rise to a second, optical spin-wave branch.

For $J_2 \simeq 2J_1$, as the experiments suggest, the ground state is a collinear antiferromagnet, composed by two coupled, interpenetrating Néel ordered states. Thus, we can span spins in a coherent basis for each of these Néel states, which we shall label as A and B, and we write $\hat{\mathbf{S}}_i^A = S\mathbf{\Omega}_i^A = S\left[e^{i\mathbf{Q}\cdot\mathbf{x}_i}\mathbf{n}_A\sqrt{1-\left(\frac{\mathbf{L}_A}{\bar{s}}\right)^2} + \frac{\mathbf{L}_A}{\bar{s}}\right]$, $\hat{\mathbf{S}}_i^B = S\mathbf{\Omega}_i^B = S\left[e^{i\mathbf{Q}\cdot\mathbf{x}_i}\mathbf{n}_B\sqrt{1-\left(\frac{\mathbf{L}_B}{\bar{s}}\right)^2} + \frac{\mathbf{L}_B}{\bar{s}}\right]$, where $\mathbf{Q} = (\pi/a, \pi/a)$ is the ordering wave vector for each Néel state, $\mathbf{n}_{A,B}$ and $\mathbf{L}_{A,B}$ are, respectively, the staggered and uniform components of the spins belonging to the two states, and $\bar{s} = S/a^d$ is the density of spin in the unit cell. We use the fixed length constraint $\mathbf{n}_{A,B}^2 = 1$.

After integrating out the uniform (fast) components $\mathbf{L}_A$ and $\mathbf{L}_B$ of the spins we arrive at the action that describes the low-energy, long-wavelength fluctuations of the staggered (slow) order parameter $\mathbf{n}_{A,B}$ (as usual we use $\beta = 1/k_BT$)

$$\begin{aligned} \mathcal{S}(\mathbf{n}_A, \mathbf{n}_B) &= \frac{\rho_s}{2\hbar}\int_0^{\hbar\beta} d\tau \int d^2\mathbf{x} \left\{ \left(|\nabla\mathbf{n}_A|^2 + |\nabla\mathbf{n}_B|^2\right) + \frac{1}{c_0^2}\left(|\partial_\tau\mathbf{n}_A|^2 + |\partial_\tau\mathbf{n}_B|^2\right) \right. \\ &+ \gamma\left(\mathbf{n}_A\cdot\partial_x\partial_y\mathbf{n}_B + \mathbf{n}_B\cdot\partial_x\partial_y\mathbf{n}_A\right) + \eta\,\mathbf{n}_A\cdot\mathbf{n}_B \\ &- \left.\frac{1}{c_1^2}\left[(\mathbf{n}_A\cdot\mathbf{n}_B)(\partial_\tau\mathbf{n}_A\cdot\partial_\tau\mathbf{n}_B) - (\mathbf{n}_A\cdot\partial_\tau\mathbf{n}_B)(\mathbf{n}_B\cdot\partial_\tau\mathbf{n}_A)\right]\right\}. \end{aligned} \quad (2)$$

The first line contains the action for two nonlinear sigma models corresponding to the order parameters $\mathbf{n}_A, \mathbf{n}_B$. The second line contains their coupling, through γ, within the AF/Ising broken phase, where the term $\mathbf{n}_A \cdot \mathbf{n}_B$ is allowed (this model was derived explicitly for the $(\pi, 0)$ magnetic configuration, which breaks explicitly the Ising symmetry). For the Ising symmetric part of the phase diagram, such term would have been absent, but integration over fluctuations would give rise, instead, to a term like $(\mathbf{n}_A \cdot \mathbf{n}_B)^2$, which is Ising invariant [9]. Finally, the fourth line contains interacting dynamical terms that also arise from the integration over the uniform components.

All couplings in (2) are expressed in terms of the original parameters. The spin stiffness is $\rho_s = 2J_2S^2a^{2-d}$, the spin-wave velocity equals $c_0 = 2\sqrt{2}Sa\sqrt{4J_2^2 - J_1^2}/\hbar$, we will also have a contribution from $c_1 = c_0(J_2/J_1)$, while the couplings between $\mathbf{n}_A$ and $\mathbf{n}_B$ are

$$\gamma = \frac{2J_1}{J_2}\left(1 + \frac{J_1^2/4}{4J_2^2 - J_1^2}\right) \quad \text{and} \quad \eta = \frac{2}{a^2}\left(\frac{J_1}{J_2}\right)\frac{J_1^2}{4J_2^2 - J_1^2}. \quad (3)$$

The bare values of the parameters described above are, as expected, higher than the measured values. For example, for $a = 5.695$ Å, $J_1 = 20$ meV,and $J_2 = 40$ meV, we find $\hbar c_0 = 1.2$ eV Å, while the typical values are actually around 0.25 eV Å. For a theory like the one described by Eq. (2), which is highly interacting, a full renormalization procedure is required to reduce the values of all couplings. For now, and for the purposes of comparison with experiments in this letter, we shall use the already established values for some of these constants.

3. The Magnetic Spectrum

To find the magnetic excitation spectrum we look at the poles of the staggered spectral function, $\mathcal{A}(\mathbf{k},\omega)$. Within the Green's function formalism, these are obtained from the imaginary part of the retarded Green's function, $G_{ret}(\mathbf{k},\omega)$, for transverse staggered fluctuations, usually denoted by $\mathcal{A}(\mathbf{k},\omega) = -(1/\pi)\lim_{\delta\to 0}\mathcal{I}m[G_{ret}(\mathbf{k},\mathrm{i}\omega_n\to\omega+\mathrm{i}\delta)]$, where $\omega_n = 2\pi n/\beta$ are the Matsubara frequencies. Thus, we consider a configuration such as $\mathbf{n}_A = (\sigma_A,\pi_A^x,\pi_A^y)$ and $\mathbf{n}_B = (\sigma_B,\pi_B^x,\pi_B^y)$, in which case the coefficient of the quadratic part in the transverse fluctuations, $(\pi_{A,B}^{x,y})^T\, G^{-1}\, \pi_{A,B}^{x,y}$, is written, in reciprocal space, as

$$G^{-1} = \begin{bmatrix} \varepsilon_0(\mathbf{k},\omega_n) & b\sigma_B\omega_n & -\varepsilon_1(\mathbf{k},\omega_n) & -b\sigma_B\omega_n \\ -b\sigma_B\omega_n & \varepsilon_0(\mathbf{k},\omega_n) & b\sigma_A\omega_n & -\varepsilon_1(\mathbf{k},\omega_n) \\ -\varepsilon_1(\mathbf{k},\omega_n) & -b\sigma_A\omega_n & \varepsilon_0(\mathbf{k},\omega_n) & b\sigma_A\omega_n \\ b\sigma_B\omega_n & -\varepsilon_1(\mathbf{k},\omega_n) & -b\sigma_A\omega_n & \varepsilon_0(\mathbf{k},\omega_n) \end{bmatrix},$$

where

$$\varepsilon_0(\mathbf{k},\omega_n) = \mathbf{k}^2 + \frac{1}{c_0^2}\omega_n^2 + \frac{2\mathrm{i}\hbar\lambda_{A,B}}{\rho_s}, \tag{4}$$

$$\varepsilon_1(\mathbf{k},\omega_n) = \gamma\, k_x k_y + \frac{\sigma_A\sigma_B}{2c_1^2}\omega_n^2 - \frac{\eta}{2}. \tag{5}$$

Here $\lambda_{A,B}$ are the Lagrange multipliers for the fixed length constraint, while σ_A and σ_B are the renormalized order parameters of the two Néel ordered states.

The structure of the magnetic spectrum then follows from the zeroes of the determinant of the transverse propagator matrix, $\det G = 0$, or simply

$$(\varepsilon_0^2 - \varepsilon_1^2)^2 - b^2(\sigma_A+\sigma_B)^2\omega^2(\varepsilon_0-\varepsilon_1)^2 = 0, \tag{6}$$

with ε_0 and ε_1 given by Eqs. (4) and (5), after the replacement $\mathrm{i}\omega_n \to \omega$. The condition for stationary action determines the saddle point to be at $\mathrm{i}\lambda_A = \mathrm{i}\lambda_B = \eta\rho_s/4\hbar$ and $\sigma_A = -\sigma_B = \sigma$. The first set of poles correspond to

$$\hbar\omega(\mathbf{k}) = \hbar\, c_{SW}\sqrt{\mathbf{k}^2 + \gamma k_x k_y}, \tag{7}$$

where $\frac{1}{c_{SW}^2} = \left(\frac{1}{c_0^2} - \frac{\sigma^2}{2c_1^2}\right)$. This is the usual acoustic spin-wave branch also found in linear spin wave theory [12], which is gapless at the zone center, $\hbar\omega(0) = 0$, in agreement with Goldstone's theorem. Notice that the observable spin-wave velocity c_{SW} *increases* by a factor $\sqrt{2+\gamma}$ as one moves from the $k_y = 0$ (or $k_x = 0$) line to the $k_x = k_y$ direction (along the spin stripes), showing that the spin-wave dispersion has a *planar directional anisotropy*, see Fig. 1, a consequence of the collinear character of the antiferromagnetic order.

The second set of poles correspond to

$$\hbar\Omega(\mathbf{k}) = \hbar\, c_{op}\sqrt{\mathbf{k}^2 - \gamma k_x k_y + \eta}, \tag{8}$$

where $\frac{1}{c_{op}^2} = \left(\frac{1}{c_0^2} + \frac{\sigma^2}{2c_1^2}\right)$. Here, instead, we find an optical gap, $\hbar\Omega(0) = \Delta_{op} \neq 0$, at the antiferromagnetic zone center $(1,0,1)$, given by

$$\Delta_{op} = \hbar\, c_{op}\sqrt{\eta}, \tag{9}$$

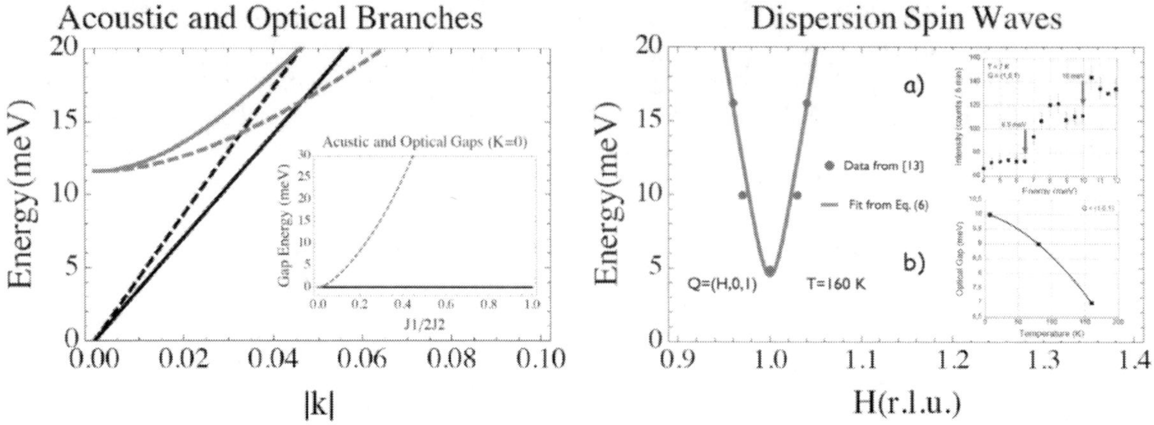

Figure 1. (Color online): Left.: Acoustic (lower, black) and Optical (upper, red) spin-wave branches for, as a function of $|k|$, for two different directions, $k_y = 0$ (solid lines) and $k_x = k_y$ (dashed lines). Inset: Acoustic and Optical gaps as a function of $J_1/2J_2$. Notice that while the acoustic gap is always zero, as imposed by Goldstone's theorem, the optical gap hardens as J_1 increases, diverging for $J_1 = 2J_2$. Right.: Comparison between our theoretical predictions and results from neutron diffraction from Ref. [7], where the two magnon gaps are evident.

which corresponds to long lived optical excitations. Notice now that the optical spin-wave velocity along the $k_x = k_y$ line *decreases* with respect to the $k_y = 0$ (or $k_x = 0$) directions, see Fig. 1, exactly the opposite from the case for the acoustic branch. The optical gap hardens as J_1 increases, see inset in Fig. 1, and eventually diverges at $J_1 = 2J_2$, when the classical configuration changes and the action (2) has to be modified.

4. Comparison with Experiments

In Fig. 1 we show our results for the magnon dispersion of the acoustic and optical modes, as a function of momentum. As we can clearly see, the dispersion is anisotropic in both cases. We also compare our results to neuron diffraction experiments from Ref. [7] in the right panel of Fig. 1, where clearly a second magnon gaps is observed at higher energies, which becomes softer as the temperature is increased.

Acknowledgments

The authors acknowledge fruitful discussions with A. H. Castro Neto and M. El Massalami.

[1] Y. Kamihara, *et al.*, J. Am. Chem. Soc. **130**, 3296 (2008).
[2] M. Rotter, M. Tegel, and D. Johrendt, *Physical Review Letters* **101**, 107006 (2008).
[3] M. R. Norman, Physics **1**, 21 (2008).
[4] F. Ma, Zhong-Yi Lu, and T. Xiang, *Physical Review B* **78**, 224517 (2008).
[5] V. Cvetkovic and Z. Tesanovic, Eur. Phys. Lett. **85**, 37002 (2009).
[6] A. V. Chubukov, D. V. Efremov, and I. Eremin, *Physical Review B* **78**, 134512 (2008); J. Knolle, *et al.*, preprint arXiv:1002.1668.
[7] J. Zhao *et al.*, *Physical Review Letters* **101**, 167203 (2008).
[8] R. A. Ewings, *et al.*, *Physical Review B* **78**, 220501(R) (2008).
[9] P. Chandra, P. Coleman, and A. I. Larkin, *Physical Review Letters* **64**, 88 (1990).
[10] C. Xu, M. Mueller, and S. Sachdev, *Physical Review B* **78**, 020501(R) (2008).
[11] C. Fang *et al.*, *Physical Review B* **77**, 224509 (2008).
[12] Dao-Xin Yao and E. Carlson, *Physical Review B* **78**, 052507 (2008).

Gaussian and critical fluctuations in the electrical conductivity of $YBa_2Cu_3O_{7-\delta}$ with chemically introduced disorder

V N Vieira[1], F T Dias[1], D L da Silva[1], P Pureur[2] and J Schaf[2]

[1]Instituto de Física e Matemática, Universidade Federal de Pelotas, RS, Brazil

[2]Instituto de Física, Universidade Federal do Rio Grande do Sul, Porto Alegre, RS, Brazil

e-mail: vdnvieira@gmail.com (Valdemar Vieira)

Abstract. We have studied experimentally by electrical magnetoconductivity measurements the influence of the site disorder, introduced by chemical doping, on the superconducting paring transition of the $YBa_2Cu_3O_{7-\delta}$ superconductor. The measurements were performed in $YBa_2Cu_3O_{7-\delta}$ and $YBa_{1.9}Sr_{0.1}Cu_3O_{7-\delta}$ single crystals with low magnetic fields applied either parallel or perpendicular to the superconducting Cu-O atomic planes. The in-plane electrical conductivity results show the systematic occurrence of tree-dimensional Gaussian (3D), tree-dimensional XY (3DXY) and beyond tree-dimensional XY fluctuation regimes, as the temperature is decreasing toward to the T_c of our samples. We propose that the observed beyond tree-dimensional XY fluctuation regime is precursory to a weak first-order superconducting transition driven by antiferromagnetic excitations related to the pseudo gap phenomenon. The experimental results shows that chemical introduced disorder, by Sr doping, does not hardly affect the stability of the beyond tree-dimensional XY fluctuation regimes in the $YBa_2Cu_3O_{7-\delta}$ superconductor when $H \leq 600$Oe are applied.

1. Introduction

The nature of superconducting state of the high T_c superconducting cuprates (HTSC) is characterized by a short coherence length of Ginsburg-Landau (GL) order parameter. In this scenario, structural disorder, like oxygen vacancies, which are present in the most pure crystals of HTSC, affects the GL order parameter and the nature of superconducting transition of these materials. On the other hand, scientific investigations focus on the effects from point like structural disorder on the superconducting paring transition of the HTSC is scarce [1-3].

An alternative way to introduce point like disorder into a superconducting system is to partially substitute a chemical element of the compound structure. For instance, the partial substitution of Ba for Sr, in $YBa_2Cu_3O_{7-\delta}$ increases the disorder in the structure of this material [4,5]. It is evidenced by a dramatic increased NQR linewith [4]. The Sr substitution also induces a variation of structural parameters, causing a charge distribution. Bond valence sum calculations and nuclear quadrupole resonance measurement indicate a transfer of holes from oxygen to copper atoms in the CuO_2 planes and from copper to oxygen in the chains [4]. The linear decrease of T_c as Sr doping level arises may be due to the increased site disorder [5]. The doping with Sr are responsible for increases of $J_C(H,T)$ and even the development of a second peak [6,7] and enhancement of superconducting granularity in this system [1-3].

In this communication, we report on field-dependent fluctuation conductivity measurements in homogeneous $YBa_2Cu_3O_{7-\delta}$ [YBCO] and in the doped $YBa_{1.9}Sr_{0.1}Cu_3O_{7-\delta}$ [YB(Sr)CO] single crystals. We focus on the investigation of the effects from chemically introduced disorder on the electrical fluctuations conductivity regimes in the vicinity T_c of the $YBa_2Cu_3O_{7-\delta}$ superconductor.

2. Experimental techniques and method of analysis

The single crystal samples of the studied systems were grown by the self-flux method [3]. The crystals were analyzed by X-ray diffraction. The obtained lattice parameters are in agreement with values reported in the literature for well oxygenated samples [2,4].

In-plane resistivity versus temperature measurements were performed with a low current-low frequency AC technique using a lock-in amplifier as a null detector. Constant magnetic fields in the range 0 – 600 Oe were applied in the configuration H parallel to ab-planes (H // ab) and H parallel to c-axis (H // c). During the measurements the temperature was swept (down) very slowly (0.05 K/min) and measured with a Pt thermometer having a resolution better than 2.10^{-3} K. The resistive data points, $\rho(T)$ are spaced closely enough in order to allow a precise numerical calculation of the temperature derivative of the resistivity, $d\rho/dT$, for details see figure 1 [2,3,8].

We separate the contributions of thermal fluctuations to the measured conductivity by using a method based on the Kouvel-Fischer analysis of critical phenomena [8]. Their contribution can be estimated by subtracting from extrapolated linear resistivity, defined by the data well above T_C, the effectively measured resistivity. The possibly experimental uncertainty inherent to the adoption to this method [9] is minimized by the high density of the experimental $\rho(T)$ data recorded at the temperature scale correspondent to the normal state resistivity of our samples [3,8].

Assuming that excess electric conductivity $\Delta\sigma(T)$ diverges when the temperature approaches T_c from above according to the power law

$$\Delta\sigma(T) = A(T - T_c)^{-\lambda}, \tag{1}$$

where A is a critical amplitude and λ is a critical exponent. Then we calculate the temperature derivative of $\ln[\Delta\sigma(T)]$ and plot $[\chi_\sigma(T)]^{-1}$, obtaining

$$[\chi_\sigma(T)]^{-1} = \frac{1}{\lambda}(T - T_c). \tag{2}$$

The existence of asymptotically critical fluctuation regimes becomes evident if a linear behavior of the data over considerable temperature intervals occurs in this plot and its temperature axis extrapolation yields the best evaluation of the paring critical temperature, T_C, see figure 2.

3. Results and discussion

The upper panels of figure 1 displays the superconducting resistive transition of the YBCO and YB(Sr)CuO samples to the H // ab configuration. In order to make a closer examination of these superconducting resistive transitions we displayed in the lower panels of figure 1 its respective resistive temperature derivatives, $d\rho/dT$. It can be observed that $d\rho/dT$ data of the YBCO crystal is characterized by a single sharp peak that is barely affect by magnetic field. In contrast, the $d\rho/dT$ data of the YB(Sr)CO crystal is characterized by a single sharp peak followed by a satellite or a hump, at lower temperature scale. The satellite or hump is strong affected by the magnetic field. Similar behavior of $\rho(T)$ and $d\rho/dT$ was observed to the H // c configuration.

These features, described in the last paragraph, characterizes the YBCO sample as a homogeneous superconductor and the YB(Sr)CO sample as a granular superconductor. The YB(Sr)CO granularity can be a result of random distribution of oxygen rich superconducting regions that possibly are embedded in a weaker superconducting or no superconducting background material probably resulting from the Sr chemical doping.

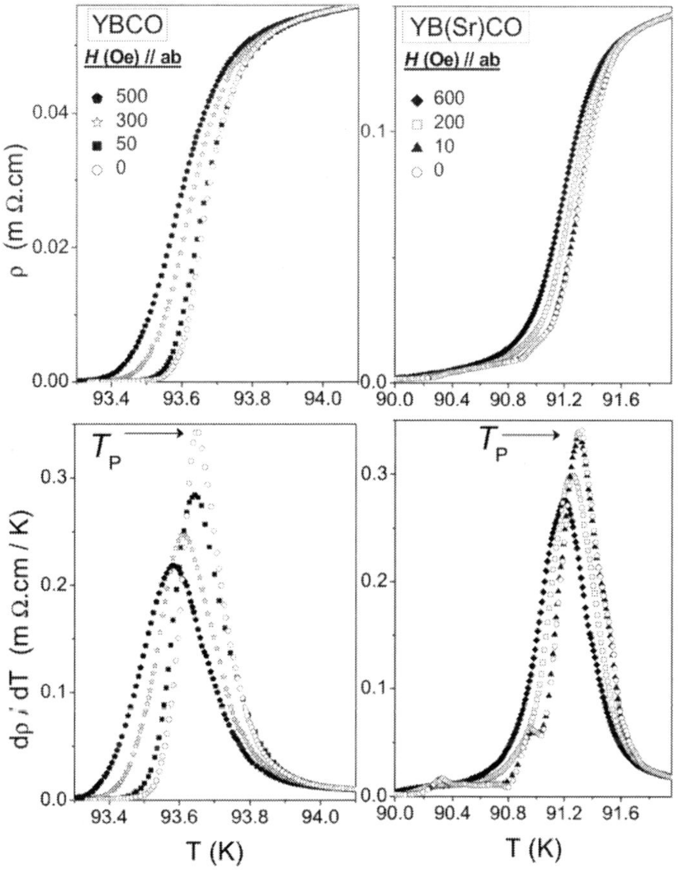

Figure 1. The characteristic resistive transitions (upper panels) and temperature derivatives (lower panels) of the $YBa_2Cu_3O_{7-\delta}$ [YBCO] and $YBa_{1.9}Sr_{0.1}Cu_3O_{7-\delta}$ [YB(Sr)CuO] single crystals to H // ab .

The upper temperature peak, T_P, in figure 1, marks the occurrence of the first step in the resistive transition of a granular superconductor. T_P is well to correspond closely but not exactly to the pairing transition temperature T_C within the superconducting grains, see figure 2. The second stage is related to superconducting grain coupling process and the establishment of a coherence transition where the zero resistence temperature, T_0, not represented in the figure 1, takes place only when grain coupling diverges and extends over the whole sample. In particular, the flux flow contributions on the superconducting magnetoresistivity transition of our samples are relevant at this second stage ($T < T_C$) and are specially intensified to the H // c configuration, but is important to remark that the focus of this paper is to study the magnetic field stability of the critical fluctuation regimes to $T > T_C$.

In figure 2 is shown a representative $[\chi_\sigma(T)]^{-1}$ result for the pure YBCO and YB(Sr)CO single crystals when no magnetic field is applied. As the temperature approaches T_c from above, a sequence of linear temperature behaviours may be seen for both samples. Farther from T_c, the regimes characterized by the exponents $\lambda_G = 0.50 \pm 0.03$, in the YBCO sample, and $\lambda_G = 0.49 \pm 0.03$, in the YB(Sr)CO sample, are due to 3D-Gaussian fluctuations. Decreasing the temperature, a crossover occurs to a genuine critical region at the Ginzburg temperature T_G of our samples [3,8]. Two power law regimes for the fluctuation conductivity are identified within the critical interval. Just below the Gaussian region, the regimes labeled by the exponents $\lambda_{XY} = 0.34 \pm 0.02$, in the YBCO sample, and $\lambda_{XY} = 0.30 \pm 0.03$, in the YB(Sr)CO sample, are characteristic of 3D-XY-E scaling [3,8]. The asymptotic fluctuation regimes corresponds to $\lambda_A = 0.17 \pm 0.02$, in the YBCO sample, and $\lambda_A = 0.17 \pm 0.03$, in the YB(Sr)CO sample. Its respective temperature axis extrapolation provides the $T_C = 93.72$ K for YBCO and $T_C = 91.52$ K for YB(Sr)CO. The overall behaviour depicted in figure 1 reproduces rather well previous observations in an YBCO crystal [3,8].

The fluctuation conductivity in superconductors is expected to diverge at T_c as [8]

$$\lambda = \nu (z - 1 - \eta), \tag{3}$$

where ν is the critical exponent for the order parameter correlation length, z is the dynamical exponent and η is a small exponent related to the order parameter correlation function. In the mean-field 3D-Gaussian region, $\nu = 0.5$, $z = 2$ and $\eta = 0$. Then, from equation (3) one obtains $\lambda = 0.5$. For the 3D-XY-E model, $\nu = 0.67$, $\eta \approx 0.0$ and $z = 1.5$. Substituting these values in equation (3) one calculates $\lambda \approx 0.33$, which is in agreement with the experimentally observed λ_{XY}. The full dynamic 3D-XY-E scaling was observed in YBCO by several authors [3,8,10].

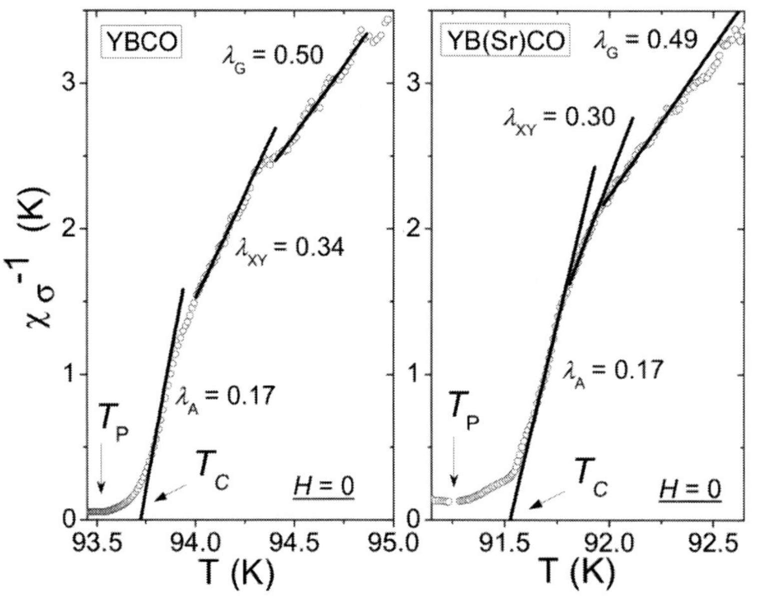

Figure 2. Plots of the $[\chi_\sigma(T)]^{-1}$ data for our $YBa_2Cu_3O_{7-\delta}$ [YBCO] and $YBa_{1.9}Sr_{0.1}Cu_3O_{7-\delta}$ [YB(Sr)CuO] samples. The λ_A and λ_{XY} exponents corresponds to critical fluctuations regimes. The λ_G labels a 3D Gaussian regime. The straight lines are fittings to equation (2).

Figure 3 magnifies the $[\chi_\sigma(T)]^{-1}$ data in the narrow fluctuation regime beyond 3D-XY for all the studied systems. Measurements were performed in the presence of the quoted magnetic fields for *H* // c and *H* // ab. The results in YBCO and YB(Sr)CO samples show that the external field tends to destabilize the super-critical regime so that the 3D-XY-E scaling is recovered around H = 600 Oe. The destabilization effect is more accentuated in H // c than H // ab field configurations.

The low values found for the conductivity critical exponent in the asymptotic regime immediately above T_c raises the possibility that the superconducting transition in YBCO is weakly first-order. A first-order and discontinuous transition would be characterized by the exponent $\lambda = 0$. Thus, the observed regimes beyond 3D-XY-E may be interpreted as precursor to the ultimate first-order transition in this HTSC. An interesting possibility is that the first-order transition is approached within a slowing down of the dynamics. Equation (3) shows that $\lambda = 0$ is reached when $z \approx 1$. Thus, considering that the static exponents ν and η keep the values of the 3D-XY model, the asymptotic conductivity exponents $\lambda_{XY} \approx 0.33$, $\lambda_A \approx 0.17$ are obtained when we substitute in equation (3) the sequence $z = 3/2$ and $5/4$, respectively. If this picture is valid, one might expect to observe still smaller exponents in the asymptotic conductivity regime of YBCO.

The mechanism that drives the superconducting transition in YBCO first-order is unknown. In a classical paper, Halperin, Lubensky and Ma, see reference [11], proposed that a superconducting transition may be driven first-order because of fluctuations of the electromagnetic field. However, the small magnitude of the blackbody radiation is too small to render the effect experimentally observable in an extreme type-II superconductor as YBCO. Recently, the influence of antiferromagnetic (AF) fluctuations was proposed by Ferreira, Continentino and Marino [12] as an alternative mechanism to render a superconducting transition weakly first-order. These authors studied the coupling between a two-component superconducting order-parameter and a metallic AF field within the Ginzburg-Landau

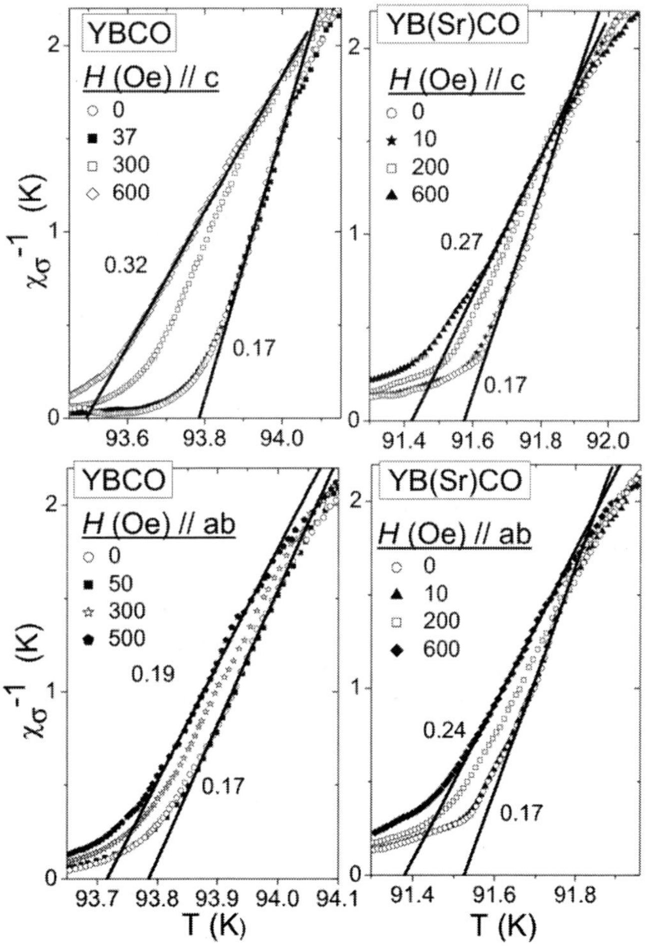

Figure 3. Plots of the $[\chi_\sigma(T)]^{-1}$ data for our single-crystal samples, showing the evolution of the asymptotic critical regime under several magnetic fields applied parallel and perpendicular to c-axis. The exponent values label fits to equation (2).

theory including quantum corrections to one-loop order. In particular, they consider the case where two separate but in close proximity quantum critical points (QCP) occur along the carrier concentration axis. Due to proximity of the AF ground state, the superconducting QCP is shifted towards the magnetic QCP so that AF spin fluctuations may strongly affect the superconducting state. In particular, the magnetic fluctuations may drive the superconducting transition weakly first-order, as much as the fluctuations of the electromagnetic field in reference [11]. Calculations in reference [12] strictly applies to $T = 0$. However, in non-zero temperatures AF fluctuations related to the so-called pseudogap phenomenon are believed to play a relevant role in the normal phase of the HTSC [13]. Thus, one may conceive that a tendency to first-order behaviour occurs in the superconducting transition of the HTSC. This might explain the observation of the critical regimes beyond 3D-XY-E in the fluctuation conductivity of our samples.

On the other hand, from figure (3), it is evident that the fluctuations regimes dynamics observed above the paring transition in our YB(Sr)CO sample evolve with applied magnetic filed quite similarly as observed in our YBCO sample. This observed behavior suggests that chemical introduced disorder effects, like charge transfer [4] and chemical pressure [4], introduced by Sr doping, are irrelevant to the fluctuations regimes in YBCO.

In summary, we present fluctuation conductivity measurements in pure and doped YBCO single crystals that reveal the occurrence of critical regimes beyond 3D-XY-E that we attribute to the proximity of a weak first-order transition induced by AF excitations related to the pseudogap phenomenon and that chemical introduced disorder, by Sr doping, does not affect the stability of the 3D Gaussian and critical fluctuations regimes in the vicinity T_c of the $YBa_2Cu_3O_{7-\delta}$ superconductor.

Acknowledgments
The authors thank to the FAPERGS and Cnpq agencies to the PRONEM 03/2011 program, under contract N° (11/2042-7), for partially financing this work.

References

[1] Menegotto Costa R, Mendonça Ferreira L, Vieira V N, et al 2007 *Eur. Phys. J. B* **58** 107
[2] Vieira V N, Pureur P and Schaf J 2001 *Physica C* **353** 241
[3] Vieira V N, Pureur P and Schaf J 2002 *Physical Review B* **66** 224506-1
[4] Licci F, Gauzzi A, Marezio M et al 1998 *Physical Review B* **58** 15208
[5] Attfield J, Kharlanov A and McAllister J 1998 *Nature* **394** 157
[6] Saito K, Nissen H U, Beeli C et al 1998 *Physical Review B* **58** 6645
[7] Vieira V N and Schaf J 2004 *Physica C* **408-410** 533
[8] Menegotto Costa R, Pureur P, Gusmão et al 2001 *Physical Review B* **64** 214513
[9] Alloul H, Rullier-Albenque F, Vignolle B et al 2010 *EPL* **91** 37005
[10] Holm W E, Eltsev Y and Rapp Ö 1995 *Physical Review B* **51** 11992
[11] Halperin B I, Lubensky T C and Ma S K 1974 *Phys. Rev. Lett.* **32** 292
[12] Ferreira A S, Continentino M A and Marino E C 2004 *Solid State Communications* **130** 321
[13] Hinkov V, Bourges P, Pailhes S et al 2007 *Nature Physics* **3** 780

International Conference on Strongly Correlated Electron Systems (SCES 2011) IOP Publishing
Journal of Physics: Conference Series **391** (2012) 012141 doi:10.1088/1742-6596/391/1/012141

Electronic structures of the FeSe superconductor studied by high-energy photoelectron spectroscopy

A Yamasaki[1], Y Matsui[1], S Imada[2], K Takase[3], H Azuma[3], T Muro[4], Y Kato[4], A Sekiyama[5], S Suga[5], A Higashiya[6], M Yabashi[4], K Tamasaku[7], T Ishikawa[7], K Terashima[2], H Kobori[1], A Sugimura[1], N Umeyama[8,9], H Sato[10], Y Hara[11], N Miyakawa[8] and S I Ikeda[9]

[1]Faculty of Science and Engineering, Konan University, Kobe 658-8501, Japan
[2]College of Science and Engineering, Ritsumeikan University, Kusatsu 525-8577, Japan
[3]College of Science and Technology, Nihon University, Tokyo 101-8308, Japan
[4]Japan Synchrotron Research Institute, Sayo, Hyogo 679-5198, Japan
[5]Graduate School of Engineering Science, Osaka University, Osaka 560-8531, Japan
[6]Industrial Technology Center of Wakayama Prefecture, Wakayama 649-6261, Japan
[7]RIKEN SPring-8 Center, Sayo, Hyogo 679-5148, Japan
[8]Department of Applied Physics, Tokyo University of Science, Tokyo 162-8601, Japan
[9]Nanoelectronics Research Institute, AIST, Tsukuba 305-8568, Japan
[10]Faculty of Science and Engineering, Chuo University, Tokyo 112-8551, Japan
[11]Ibaraki National College of Technology, Ibaraki 312-8508, Japan

E-mail: yamasaki@konan-u.ac.jp

Abstract.
The bulk electronic structures of an iron-based superconductor FeSe have been investigated by photoelectron and photoabsorption spectroscopies (PES and XAS) with use of high-energy photons (h $\simeq$ 0.7-0.9 and 8 keV). The core-level PES and XAS combined with the configuration interaction cluster model calculation reveal the highly coherent nature of the Fe $3d$ electrons which is realized by the hybridization between Fe $3d$ and Se $4p$ states.

1. Introduction

In a simple Fe-Se binary compound FeSe which consists of a two dimensional Fe plane in Fe_2Se_2 layer, superconductivity has been discovered below 27 K under 1.5 GPa [1]. The appearance of superconductivity in the FeSe system indicates Fe_2X_2 (X=P, As, and Se) layer is essential for the superconductivity in Fe-based superconductors. The density functional study has pointed out that FeSe is not a conventional electron-phonon superconductor, being similar to $LaFeAsO_{1\ x}F_x$ system [2]. We have performed the high-energy spectroscopies, that is, soft-x-ray and hard-x-ray photoelectron spectroscopies (SXPES and HAXPES) and soft-x-ray absorption spectroscopy (SXAS) for Fe $2p$ core level to reveal the electronic structures of the FeSe superconductor. Core-level spectroscopy is well-known to be a powerful method to probe the valence electronic structure through the Coulomb interaction between the created core hole and valence electrons.

2. Experimental

For measurements, tetragonal FeSe superconductor [FeSe(Tet): $T_c^{\text{zero}} \simeq 6$ K at ambient pressure] is used, together with hexagonal FeSe [FeSe(Hex)] (likely, Fe_7Se_8) as a reference material. Details of the sample preparation processes and their characteristics have been reported elsewhere [3]. SXPES was carried out at the twin-helical undulator SX beamline BL25SU in SPring-8 using GAMMADATA-SCIENTA SES-200 spectrometer. HAXPES was performed at the beamline BL19LXU in SPring-8 with MB Scienti c A1-HE spectrometer. SXAS spectra were measured at the BL25SU by the total electron yield method in which sample currents were directly measured while scanning photon energy. Photon energies for both the PES and XAS spectra were calibrated with respect to the Fermi edge of an evaporated gold thin lm. Clean surfaces of samples were obtained by fracturing the samples *in situ* in UHV (4×10 ^{8} Pa) at the measurement temperature ($T \simeq 20$ K).

3. Results and discussion

Fe $2p$ core-level PES spectra of FeSe (Tet) and FeSe (Hex) are shown in Fig. 1 (only the Fe $2p_{3/2}$ part is illustrated). The SXPES spectrum of FeSe (Hex) has a broad peak structure with a weak shoulder on the lower binding energy side. The broad spectral shape derived from the many nal states with di erent energy eigenvalues, that is, the ne multiplet structure indicates the localized character of Fe $3d$ electrons in FeSe (Hex). Meanwhile, the signi cantly sharp peak at the lowest binding energy is seen in both the SXPES and HAXPES spectra of FeSe (Tet). The sharp peak is more prominent in the HAXPES spectrum, which indicates that the peak originates from the bulk electronic state because of much higher bulk sensitivity in HAXPES than in SXPES (the estimated inelastic mean-free paths are about 9.9 nm and 1.7 nm for the HAXPES with h = 7896 eV and the SXPES with h = 900 eV, respectively [4]). The peak suggests FeSe (Tet) has highly coherent valence electrons probably related to the superconducting nature in this system.

Figure 2 shows Fe $2p$-$3d$ core-level XAS spectra of FeSe (Tet) and FeSe (Hex). The spectrum of FeSe (Tet) has a single peak and one shoulder structure in the $2p_{3/2}$ part (left hand side of the spectrum). Meanwhile, three shoulders in addition to a sharp peak are seen in the spectrum of FeSe (Hex). In order to understand the origin of their structures, we have performed multiplet calculations. The required parameters, such as Slater integrals and spin-orbit coupling constants are obtained by Cowan's Hartree-Fock program [7]. Some of major calculated spectra are also shown in Fig. 2. First, let us consider the basis state consisting of the single $|3d^6\rangle$ con guration in the initial state. Without any anisotropic crystal elds, that is, under the spherical eld the calculated spectrum which has a dull peak structure with two or more shoulders do not reproduce both the spectra of FeSe (Tet) and FeSe (Hex). Then, turning on the octahedral (O_h) crystal eld with the splitting energy Δ_{CF} of 1.3 eV, we have obtained a very similar calculated spectrum to the experimental one for FeSe (Hex). The XAS spectrum of FeSe (Tet), however, cannot be reproduced by spectra with any intensities of the tetrahedral (T_d) crystal eld. Instead of the crystal eld e ect, the hybridization between the Fe $3d$ and ligand (Se $4p$) orbitals is essential to reproduce the spectral shape. We have carried out the con guration interaction (CI) cluster model calculation using the XTLS 9.0 program [5, 6]. As one can see, the spectrum with the basis state consisting of the $|3d^6\rangle$ and $|3d^7\underline{L}\rangle$ con gurations which arises from the orbital mixing, reproduces the experimental spectrum of FeSe (Tet). The optimized values for the parameters U_{dd}, Δ, and $pd\sigma$ are 5.5 eV, 3.5 eV, and -1.1 eV, respectively (for details of these parameters, see the references), which lead to the Fe valence of +1.9. Other parameters are obtained by applying the following relations, $U_{dd}/U_{dc} = 0.8$, $pd\sigma/pd\pi =$ 2.2, T $= (2\sqrt{6}/3)pd\pi$, and $T_\pi = [\{(2\sqrt{3}/3)pd\sigma\}^2 + \{(2\sqrt{2}/3)pd\pi\}^2]^{1/2}$ [8]. These obtained values for the parameters are similar to those in a moderately correlated electron system $Fe_{1/3}TiS_2$. These results suggest that the FeSe (Hex) has well-localized Fe $3d$ electrons under the e ective

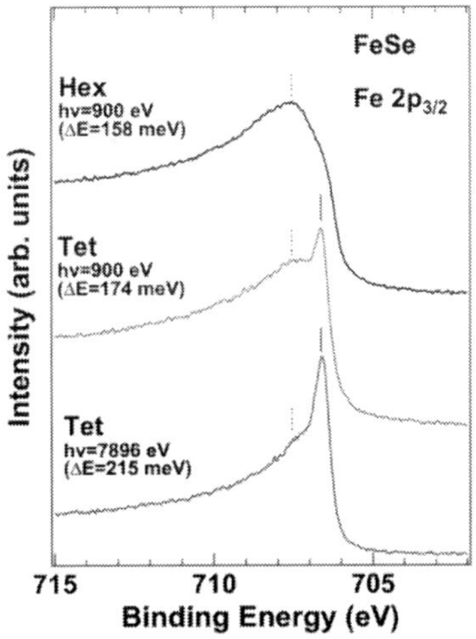

Figure 1. Fe $2p_{3/2}$ core-level PES spectra of FeSe (Tet) and FeSe (Hex).

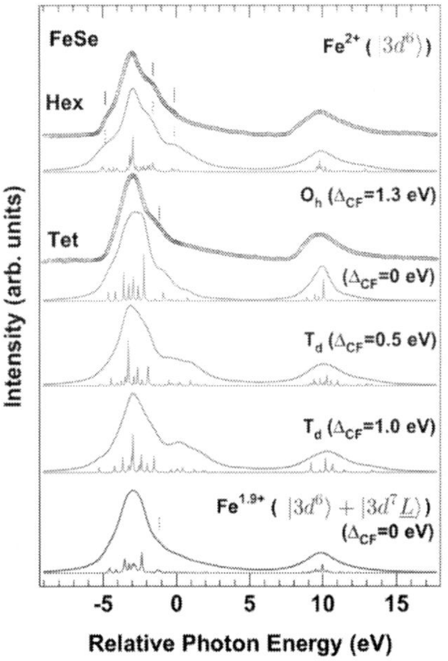

Figure 2. Fe $2p$-$3d$ core-level XAS spectra of FeSe (Tet) and FeSe (Hex). Spectra obtained by atomic and CI cluster model multiplet calculations with and without Gaussian and Lorentzian broadenings (broadening widths [FWHM] of $\gamma_G = 100$ meV, $\gamma_L(j = 3/2) = 2.0$ eV, and $\gamma_L(j = 1/2) = 3.0$ eV) are also shown.

octahedral crystal eld, being consistent with the result of the Fe $2p$ core-level PES. In contrast, the hybridization between Fe $3d$ and Se $4p$ states is revealed to be important rather than the crystal eld e ect for describing the electronic structure of FeSe (Tet), which leads to the highly coherent character of the Fe $3d$ electrons observed in Fe $2p$ core-level PES.

4. Summary

We have investigated bulk electronic states of the FeSe superconductor with the tetragonal crystal structure. The high-energy photoelectron spectroscopy using the hard x ray reveals the highly coherent nature of the Fe $3d$ electrons which is realized by the hybridization between Fe $3d$ and Se $4p$ states.

Acknowledgments

We would like to thank K. Sugimoto, K. Mima, R. Yamaguchi, Y. Miyata for supporting experiments and D. J. Singh for providing results of band structure calculations, A Tanaka for providing multiplet calculation program. The experiments were performed at SPring-8 with the approval of the Japan Synchrotron Radiation Research Institute (JASRI) (Proposal No. 2008B1149 and No. 2009A1122) under the support of Grant-in-Aid for Scienti c Research for Young Scientists (B) (No. 21740235 and No. 23740244) from the Ministry of Education, Culture, Sports, Science, and Technology, Japan, and Research Foundation for Opto-Science and Technology, the Sumitomo Foundation, Research Institute of Konan University, and the Hirao Taro Foundation of the Konan University Association.

References

[1] Mizuguchi Y, Tomioka F, Tsuda S, Yamaguchi T and Takano Y 2008 *Appl. Phys. Lett.* **93** 152505
[2] Subedi A, Zhang L, Singh D J and Du M H 2008 *Phys. Rev.* B **78** 134514
[3] Yamasaki A *et al.* 2010 *Phys. Rev.* B **82** 184511
[4] Tanuma S, Powell C J and Penn D R 1994 *Surf. Interface anal.* **21** 165
[5] Tanaka A, Jo T and Sawatzky G A 1992 *J. Phys. Soc. Jpn.* **61** 2636
[6] Yamasaki A *et al.* 2007 *Phys. Rev. Lett.* **98** 156402
[7] Cowan R D 1981 *The Theory of Atomic Structure and Spectra* (Berkeley, University of California Press)
[8] Bocquet A E, Mizokawa T, Saitoh T, Namatame H and Fujimori A 1992 *Phys. Rev.* B **46** 3771

Journal of Physics: Conference Series **391** (2012) 012142 doi:10.1088/1742-6596/391/1/012142

A dynamical probe of superfluidity in one-dimension: The adiabatic quantum pump

Roberta Citro

Dipartimento di Fisica "E. R. Caianiello", Università degli Studi di Salerno, and Institute CNR-SPIN, UO Salerno, Via Ponte don Melillo, I-84084 Fisciano (Sa), Italy

E-mail: citro@sa.infn.it

Abstract. Quantum pumping as a dynamical probe of the superfluid response of a one-dimensional (1D) quantum fluid is discussed. It is shown that a spatially periodic potential, oscillating adiabatically in time with frequency ω_0, acts as a quantum pump inducing a continuous momentum current from broken spatiotemporal symmetries of the driven potential. The momentum current generated by the pump is strongly affected by the interactions. It has a power-law dependence on the frequency and the temperature with the exponent determined by the interaction. It depends on the phase difference between two umklapp terms of the drive, providing indication for the effect of quantum phase slips on the decay of superflow. Application of the results in understanding the superfluid properties of helium confined in nanometer-size pores or of laser-cooled atoms is briefly discussed.

1. Introduction

Dynamical effects and quantum transport phenomena are fundamental problems in physics. An intriguing example is the quantum pumping, where in the absence of a biased force a directed current of particles is produced along a periodic structure[1]. An adiabatic quantum pump is a device that generates a dc current by a periodic slow variation of some system characteristic, the variation being slow enough so that the system remains close to its ground state throughout the pumping cycle. From a fundamental physics standpoint, this mechanism represents a macroscopic quantum phenomenon reminiscent of superconductivity where current flows without dissipation. While quantum pumping has been mainly studied in solid state systems, as in quantum dots[2, 3], here we discuss quantum pumping in 1D quantum fluids as a dynamical probe of its superfluid response. Our pump is made of a spatially periodic potential, oscillating adiabatically in time with frequency ω_0 on a one-dimensional quantum fluid described by a Tomonaga-Luttinger liquid. It induces a continuous momentum current from broken spatiotemporal symmetries of the driven potential which indicates the effect of quantum phase slips on the degree of dissipation of the superflow. The zero temperature analysis recently appeared[4] while the present study at finite temperature can be relevant in understanding the superfluid properties of helium confined in nanometer-size pores or of laser-cooled atoms in confined potentials.

2. The model and the pump

The proposed pump consists of a 1D quantum fluid exposed to a spatially periodic optical lattice potential oscillating wave like with frequency ω_0 and momentum q_0. The low-energy properties of the fluid are described by a Tomonaga-Luttinger liquid, and the pumping is carried out at small ω_0, staying this way in the neighborhood of the Luttinger liquid fixed point. The low-temperature and frequency degrees of freedom of the system are described by two collective canonical conjugate fields, $\theta(x,t)$ and $\partial_x\phi(x,t)/\pi$, which describe the phase and the density fluctuations, respectively [5, 6] with Hamiltonian:

$$H = \frac{\hbar}{2\pi}\int dx[\frac{v_s}{K}(\nabla\phi(x))^2 + \frac{v_s K}{\hbar^2}(\pi\Pi(x))^2]. \tag{1}$$

($\hbar = 1$ hereafter). This Hamiltonian is a standard sound wave one in which the fluctuations of the phase $\phi(x)$ represent the phonon modes of the density wave

$$\rho(x) = [\rho_0 - \frac{1}{\pi}\nabla\phi(x)]\sum_{p=-\infty}^{\infty} e^{i2p(\pi\rho_0 x - \phi(x))}, \tag{2}$$

ρ_0 being the average particle density and v_s is the sound velocity. The Luttinger parameters v_s and K used in (1) are related by the compressibility of the fluid $K/(\hbar\pi v_s\rho_0^2)$.

When the interaction goes to zero, K goes to infinity, while $K = 1$ for infinitely strong hard-core interactions (Tonks-Girardeau limit)[7]. The situation $K < 1$ can be instead realized in Bose-gases with long-range interactions, e.g. in dipolar gases.

An external potential $V(x)$ which is periodic (with period l and period wavevector $Q = \frac{2\pi}{l}$) acting on a length L of the system is considered. Such potential couples to the density (2) and leading terms of such coupling are of the form: $H_{n,m} \sim g_{n,m}\int dx e^{ik_{n,m}x}e^{i2n\phi(x)}$, the so called umklapp terms, where $k_{n,m} = n2\pi\rho_0 - mQ$ represents the momentum transfer. A commensurability between the boson density and the imposed periodicity implies $k_{n,m} = 0$ for some n, m. Thus the smallest $k_{n,m}$ provides us with a measure of the incommensurability bewteen the 1D fluid density and the external potential. We then allow the external periodic potential to oscillate with frequency ω_0 and propagate with characteristic momenta $\{q\}$ centered around a value q_0, $V(x) \to V(x,t) = \sum_q A_q\cos(\omega_0 t - qx)V(x)$, i.e. it acts as a pump $H^{ext}(t) = H^{pump}$. Close to the Luttinger fixed point, the umklapp terms, which are irrelevant operators in the renormalization group sense, will have time and phase dependent coupling constant of the form $g_{n,m}(t) = g_{n,m}e^{i\omega_0 t - \varphi_{n,m}}$. The momenta $\{q\}$ in the driving potential break the reflection symmetry and this is reflected by the umklapp phases $\varphi_{n,m}$ that for weak periodic potential are given by $\varphi_{n,m} \sim nq_0/\omega_0$. When reflection symmetry is present $\varphi_{n,m} = 0$ and no momentum current will be generated in the system.

The bosonized version of the pump Hamiltonian reads:

$$H^{pump} = \sum_{n,m} H_{n,m}(t) = \sum_{n,m}\frac{g_{n,m}}{(2\pi\rho_0^{-1})^2}\int dx\{e^{i(\omega_0 t - \varphi_{n,m})}e^{ik_{n,m}x}e^{i(2n\phi(x))} + h.c.\}, \tag{3}$$

where the sum runs over n, m integers (m is the order of commensurability, e.g. $m = 1$ corresponds to one particle per site). Let us note that such irrelevant terms describe the effect of quantum phase slips responsible for the decay of the superflow. We wish to study the effect of the oscillating terms on the momentum current operator that in bosonized version is given by $Mj = M\frac{1}{\pi}\partial_t\phi = \Pi(x)$, where $j(x,t)$ is the particle current operator. In the following, we shall consider the oscillating lattice as a perturbation around the Luttinger liquid fixed point and compute the current perturbatively, by the Keldysh technique[8]. Explicit calculations show that to get a continuous current out of the pumping potential at least two umklapp operators with a nonzero phase difference are required[9].

By using the bosonized expression of H^{pump}, the pumped current to second order can be calculated as:

$$j^{pump}(\omega_0, T) = Kv_s \sum_{n,n'} \sum_{m,m'} n_- (\rho_0 L)^{-n_-^2 K} \mathcal{A}_{n_\pm,m_\pm} I(\omega_0, k_{n_+,m_+}, T) \frac{\sin(k_{n_-,m_-}\frac{L}{2})}{k_{n_+,m_+}} \quad (4)$$

where $k_{n_\pm,m_\pm} = (\frac{k_{n,m} \pm k_{n',m'}}{2})$, $n_\pm = n \pm n'$ and similarly for $m_\pm$, and,

$$\mathcal{A}_{n_\pm,m_\pm} = \frac{g_{n,m}}{(2\pi\rho_0^{-1})^2} \frac{g_{n',m'}}{(2\pi\rho_0^{-1})^2} \sin \varphi_{n,m}^{n',m'} \quad (5)$$

is the area enclosed in a pumping cycle by the periodic parameters $g_{n,m}(t)$ and $g_{n',m'}(t)$, $\varphi_{n,m}^{n',m'}$ is the phase difference between two umklapp operators.

Using the finite temperature expression for the correlation functions of the boson operators[6] the expression for $I(\omega_0, k_{n_+,m_+}, T)$ reads:

$$I(\omega_0, k_{n_+,m_+}, T) = Sgn(\omega_0) \sinh(\frac{\omega_0}{\pi T})$$

$$\left(\frac{2\pi T}{\rho_0 v_s}\right)^{2K^{nn'}-1} \sin(\pi K^{nn'}) B(-\frac{is_+}{2\pi} + \frac{K^{nn'}}{2}; 1 - K^{nn'}) B(-\frac{is_-}{2\pi} + \frac{K^{nn'}}{2}; 1 - K^{nn'}), \quad (6)$$

where $s_\pm = \frac{(\omega_0 \pm v_s k_{n_+,m_+})}{2T}$; $B(x, y) = \Gamma(x)\Gamma(y)/\Gamma(x + y)$ is the Euler beta function; $K^{nn'} = nn'K$.

When we consider incommensurate fillings, $k_{n_+,m_+} \neq 0$, two interesting regimes occur, $T \ll \omega_0$, or $T \gg \omega_0$. In the first case, we get the zero temperature result, i.e.

$$I^{T=0}(\omega_0, k_{n_+,m_+}) \simeq$$

$$Sgn(\omega_0) \sin(\pi K^{nn'}) \Gamma^2(1 - K^{nn'}) \left(\frac{\rho_0^{-1}}{2v_s}\right)^{2K^{nn'}-2} ||\omega_0| - |k_{n_+,m_+}||^{2K^{nn'}-1} \theta(|\omega_0| - |v_s k_{n_+,m_+}|) \quad (7)$$

For $T \gg \omega_0$ and ω_0 not too small compared to $v_s k_{n_+,m_+}$ one has:

$$I(\omega_0, k_{n_+,m_+}, T) \simeq$$

$$\sin^2(\pi K^{nn'}) \Gamma^2(1 - K^{nn'}) \left(\frac{\rho_0^{-1}}{2v_s}\right)^{2K^{nn'}-2} ||\omega_0| - |k_{n_+,m_+}||^{2K^{nn'}-1} e^{-\frac{v_s k_{n_+,m_+}}{2T}} \sinh(\frac{\omega_0}{\pi T}) \quad (8)$$

where the exponential factor describes the suppression of processes involving momentum transfer k_{n_+,m_+}. When $\omega_0 \to 0$ at low-temperature the exponential factor in (8) prevails and the processes with the smallest k_{n_+,m_+} are favored, i.e. the pumped current is suppressed and thus the system supports a superflow. At some commensurate point $k_{n_0,m_0} \sim 0$ and temperature not too low, we have to balance algebraic and exponential suppression in (8) and in the limit $\omega_0 \ll T$, the dominant contribution to the d.c. current will be proportional to $\left(\frac{2\pi\rho_0^{-1}T}{v_s}\right)^{2K_{n_0 n_0'}-1} \frac{\omega_0}{T} Sgn(\omega_0)$. By taking the first commensurate point and the Tonks- Girardeau limit (hard-core bosons) where $K = 1$, one recovers the same behavior as that of a non-interacting fermion gas $I \simeq \omega_0$[6]. With interactions present, the current behaves as a power-law of the temperature with an exponent depending on the interactions, indicating a strong renormalization of the scattering process due to phase slips. The Fig. 1 (left panel) shows the low-frequency behavior of the pumped

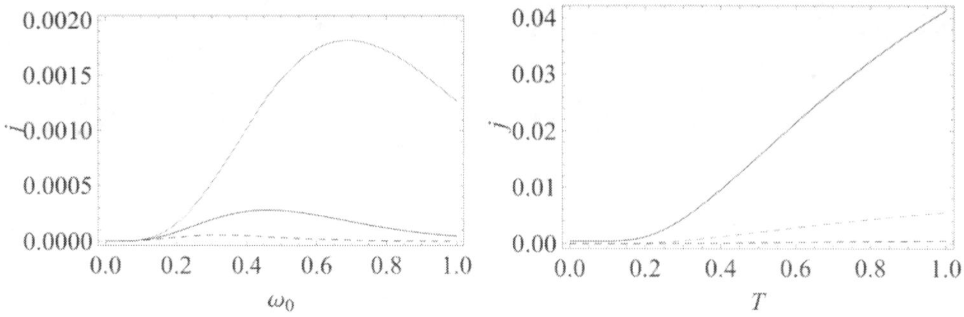

Figure 1. (Color online) Low frequency behavior of the pumped current in arbitrary units having taken into account umklapp terms $g_{1,0}, g_{20}, g_{21}$. Left panel: The dashed line is for $T = 0.07$, the straight (black) line for T=0.1 and the straight (red) for T=0.2, the Luttinger parameter is $K = 1.5$, the system length (in units of $1/k_{20}$) is $L = 1.$; ω_0 is in units of $v_s k_{20}$. Right panel: The straight line is for $K = 1.1$, the dashed (red) line for K=1.4 and the dashed (black) for K=1.8, the system length is $L = 1.$ and $\omega_0 = 0.2$, data in this panel have been magnified by a factor 10^2.

momentum current (PMC) at finite temperature T, taking into account few umklapp terms. The pumped current is an increasing function of T and has a characteristic power-law behavior (interaction dependent) at very small ω_0. In the right panel, the PMC is plotted as a function of T for different values of the Luttinger parameter. The figure clearly shows an onset temperature below which the pumped current is a constant, indicating the dissipative contribution from phase-slips. As K increases the pumped current is decreases and this is in agreement with the expectation that decreasing the interaction tends to increase the superfluid response of the system. Let us note that the PMC is never zero, unless ω_0 is zero, and the constant values goes to zero only for $L \to \infty$ at increasing K. Thus superfluidity could be observable in the dynamical sense in 1d for sufficiently long systems. A similar conclusion was drawn recently[10] by the analysis of the dynamical momentum response of a Tomonaga-Luttinger liquid subject to a periodic potential. Experiments of torsional oscillator in nanometer-sized pores filled with ^4He[11] offer a way to detect the dynamical aspects of superfluidity. In conclusion, we have shown that a quantum pump made of a periodic potential oscillating with a finite frequency on a one dimensional quantum fluid could be used to probe the superfluid behavior at finite temperature.

Acknowledgments
R.C. acknowledges enlightening discussions with N. Andrei.

References
[1] Thouless D J, 1983 *Phys. Rev.* B **10** 6083.
[2] Shilton J, 1996 *J. Phys.: Condens. Matter* **8** L337.
[3] Kemppinen A, Meschke M, Möottönen M, Averin D V and Pekola J P, 2009 *Eur. Phys. Journ.* **172**, 311.
[4] Citro R, Andrei N, 2011 *Phys. Rev. A* **83**, 015601.
[5] Haldane F D M, 1981 *Phys. Rev. Lett.***47** 1840.
[6] Giamarchi T, 2004 *Quantum Physics in One Dimension* (Oxford: Oxford University Press).
[7] Girardeau M, 1960 *J. Math. Phys.* **1**, 516.
[8] Rammer J, 1986 *Rev. Mod. Phys.* **58**, 323.
[9] Citro R, Andrei N, Niu Q 2003 *Phys. Rev. B* **68**, 165312.
[10] Eggel T, Cazalilla M A, Oshikawa M, arXiv:1104.0175.
[11] Toda R *et al.*, 2007 *Phys. rev. Lett.* **99**, 255301.

Journal of Physics: Conference Series **391** (2012) 012143 doi:10.1088/1742-6596/391/1/012143

Dynamic density-density correlations in interacting Bose gases on optical lattices

S. Ejima[1], **H. Fehske**[1], **and F. Gebhard**[2]

Institut für Physik, Ernst-Moritz-Arndt-Universität Greifswald, 17489 Greifswald, Germany
Fachbereich Physik, Philipps-Universität Marburg, 35032 Marburg, Germany

E-mail: `<ejima,fehske>@physik.uni-greifswald.de`

Abstract. In order to identify possible experimental signatures of the superfluid to Mott-insulator quantum phase transition we calculate the charge structure factor $S(k,\omega)$ for the one-dimensional Bose-Hubbard model using the dynamical density-matrix renormalisation group (DDMRG) technique. Particularly we analyse the behaviour of $S(k,\omega)$ by varying—at zero temperature—the Coulomb interaction strength within the first Mott lobe. For strong interactions, in the Mott-insulator phase, we demonstrate that the DDMRG results are well reproduced by a strong-coupling expansion, just as the quasi-particle dispersion. In the superfluid phase we determine the linear excitation spectrum near $k = 0$. In one dimension, the amplitude mode is absent which mean-field theory suggests for higher dimensions.

Recent experimental realisations of optical lattices make it possible to investigate the properties of ultracold dilute atoms in a new regime of strong correlations [1, 2]. Tuning the strength of the laser field, the effective interactions between the atoms can be tuned to be stronger than their kinetic energy. The competition between the Coulomb energy and the kinetic energy may even drive a quantum phase transition between superfluid (SF) and Mott insulating (MI) phases. The Bose–Hubbard model (BHM) captures the essential physics of this problem. In previous work [3] we studied the photoemission spectra of the one-dimensional (1D) BHM both in the MI and SF phases. In this contribution we extend our preceding investigations of the dynamical properties of the BHM by analysing the dynamical structure factor $S(k,\omega)$. Analytically, $S(k,\omega)$ has been determined by mean-field approaches [4, 5] and, numerically, for rather small 1D systems, by exact diagonalisation [6] or, at finite temperatures, by QMC [7]. Here, we carry out large-scale dynamical density-matrix renormalisation group (DDMRG) [8] calculations in order to examine the density-density correlations of N interacting Bose particles on L sites at zero temperature. Moreover, fixing $\rho = N/L = 1$, we set up a perturbation theory whose results can be compared with the DDMRG data in the strong-coupling limiting case.

The Hamiltonian of the 1D BHM reads

$$\hat{\mathcal{H}} = -t\sum_j(\hat{b}_j^\dagger\hat{b}_{j+1} + \hat{b}_j\hat{b}_{j+1}^\dagger) + \frac{U}{2}\sum_j \hat{n}_j(\hat{n}_j - 1)\,, \qquad (1)$$

where $\hat{b}_j^\dagger$ and $\hat{b}_j$ are the creation and annihilation operators for bosons on site j, $\hat{n}_j = \hat{b}_j^\dagger\hat{b}_j$ is the boson number operator on site j. The tunnel amplitude between nearest neighbour lattice sites is denoted by t, and U gives the on-site Coulomb repulsion of the Bose particles. For the DDMRG treatment of (1) we use periodic boundary conditions, take into account $n_b = 5$ bosons per site and keep up to $m = 500$ density-matrix eigenstates, ensuring that the discarded weight is always smaller than 3×10^{-5}.

The dynamical charge structure factor is defined as

$$S(k,\omega) = \sum_n |\langle\psi_n|\hat{n}_k|\psi_0\rangle|^2 \delta(\omega - \omega_n), \tag{2}$$

where $|\psi_0\rangle$ and $|\psi_n\rangle$ denote the ground and n-th excited state, respectively, and $\omega_n = E_n - E_0$ gives the corresponding excitation energy. $S(k,\omega)$ characterises the density-density response of the Bose gas and is directly accessible in experiment, for instance, by Bragg spectroscopy [9].

From lowest-order strong-coupling perturbation theory we find for $\rho = N/L = 1$ $(N, L \to \infty)$

$$S^{(1)}(k,\omega > 0) = \left(\frac{4t\sin(k/2)}{U}\right)^2 \int_{-\pi}^{\pi} \frac{\mathrm{d}k}{\pi} \sin^2 k\, \delta(\omega - U + 2t\cos k\sqrt{5 + 4\cos k}). \tag{3}$$

Then, for $U - 2\sqrt{5+4\cos k} \le \omega \le U + 2\sqrt{5 + 4\cos k}$, we have in units of t

$$S^{(1)}(k,\omega > 0) = \left[\frac{4\sin(k/2)}{U}\right]^2 \frac{\sqrt{20 + 16\cos k - (\omega - U)^2}}{2\pi(5 + 4\cos k)}, \tag{4}$$

where $0 \le k < 2\pi$. Since DDMRG provides $S(k,\omega)$ with a finite broadening η, it is useful to convolve the strong-coupling result $S^{(1)}(k,\omega)$ with the Lorentz distribution [10]: $S_\eta(k,\omega) = \int_{-\infty}^{\infty} \mathrm{d}\omega' S(k,\omega')\eta/\left[\pi\left[(\omega-\omega')^2 + \eta^2\right]\right]$. For the 1D BHM, from Eq. (4), we thus get

$$S_\eta^{(1)}(k,\omega > 0) = \left(\frac{4\sin(k/2)}{U}\right)^2 \frac{2}{\pi^2} \int_{-1}^{1} \mathrm{d}\lambda \frac{\eta\sqrt{1-\lambda^2}}{(\omega - U + 2\lambda\sqrt{5 + 4\cos k})^2 + \eta^2}. \tag{5}$$

Figure 1 illustrates the frequency and momentum dependencies of $S(k,\omega)$. Peaks in $S(k,\omega)$ assign charge excitations. For $k \simeq 0$, $S(k,\omega)$ exhibits a maximum around $\omega \approx U$ that can be attributed to excitations across the Mott gap. Within strong-coupling theory (4) this signature stays at $\omega \approx U$ for all k but becomes more pronounced as k reaches the Brillouin zone boundary.

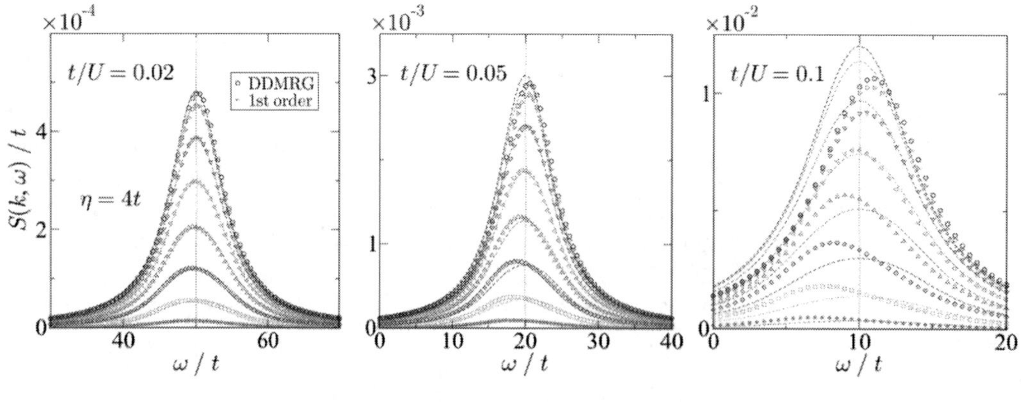

Figure 1. Frequency dependence of the dynamical charge structure factor $S(k,\omega)$ at various momenta $k = 2\pi j/L$ (for $j = 1, 2, \ldots, L/2$ from bottom to top). DDMRG data (symbols) were obtained for $L = 16$; $S_\eta^{(1)}(k,\omega)$ (lines) give the corresponding first-order strong-coupling expansion results (for the same value of $\eta = 4t$). The dashed line indicates the point $\omega = U$.

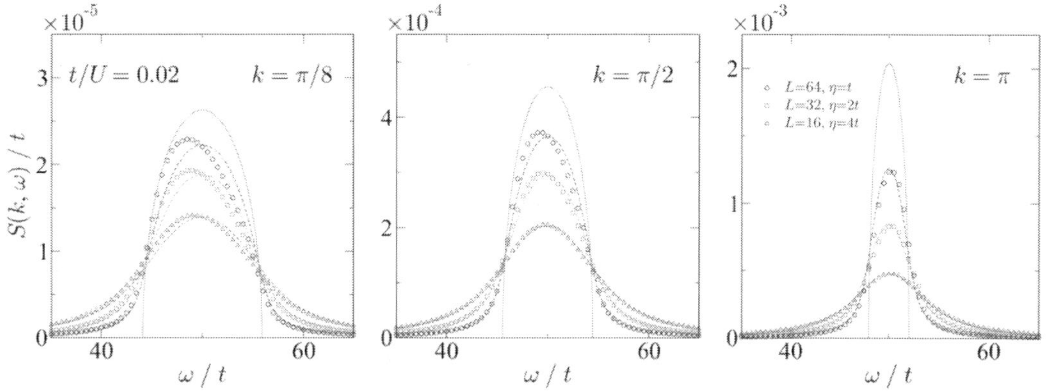

Figure 2. η-dependence of the dynamical structure factor $S(k,\omega)$. DDMRG data (symbols) are compared with the corresponding strong-coupling results (dashed lines) at various momenta k. The solid green line shows the unconvolved result of the strong-coupling expansion.

Our DDMRG data corroborate this prediction (cf. the left panel of Fig. 1 depicting the results for $t/U = 0.02$). Naturally, as U gets smaller, differences between the numerical DDMRG and the analytical strong-coupling results emerge. Most importantly, the position of the maximum in $S(k,\omega)$ varies with k: it is shifted to smaller (larger) frequencies for k values near the band centre (Brillouin zone boundary); see middle and right-hand panels of Fig. 1. We have to include higher-order t/U-corrections to reproduce this feature analytically. Note that the maximum in $S(k,\omega)$ amplifies as $k \to \pi$.

Next, we investigate the dependences of $S(k,\omega)$ on the broadening η and the system size L to scrutinise whether the DDMRG data "converge" to the unconvolved strong-coupling result (4) as $\eta \to 0$ and $L \to \infty$. Figure 2—showing $S(k,\omega)$ in the MI phase at $k = \pi/8$ (left), $\pi/2$ (middle) and π (right panel) for L = 16, 32, 64 η = $4t$, $2t$, t, respectively—demonstrates that this is indeed the case. Firstly, for $t/U = 0.02$, we see that the DDMRG results are in a satisfactory accord with $S_\eta^{(1)}(k,\omega)$, where as a matter of course for smaller system sizes a larger value of η has to be used to achieve good agreement. Secondly, $S(k,\omega)$ calculated by DDMRG approaches the curve given by (4) for increasing system size L and decreasing broadening η.

Finally, we look at the changes in the dynamical density-density response as the system crosses the MI-SF quantum phase transition with decreasing Coulomb interaction strength. Figure 3 shows the intensity distribution of $S(k,\omega)$ in the MI and SF phases as well as in the vicinity of the Kosterlitz–Thouless (KT) transition point, where the charge excitation gap closes. For large U, the spectral weight is mainly concentrated around $\omega = U$ in the region $\pi/2 \leq k \leq \pi$, in agreement with the strong-coupling prediction. As U weakens in the MI phase, the distribution of the spectral weight broadens. At the same time, the maximum value of $S(k,\omega)$ acquires a sizable dispersion (see upper-row panels). As the system reaches the MI-SF transition point we observe a significant redistribution of spectral weight to lower k values and, most notably, the excitation gap closes (see lower panels). In our previous work, we evaluated the scaling of the Tomonaga–Luttinger liquid parameter and determined the KT transition point to be located at $t/U = 0.305(1)$ [3]. In the SF phase spectra for $k \gtrsim 0$ (cf. the panel for $t/U = 0.4$), we observe an almost linear dispersion of the $S(k,\omega)$, which—in accordance with bosonization [11] and with Bogoliubov theory [12] —can be taken as a signature of the Bose condensation process. Our 1D DDMRG BHM data are unsuggestive of two distinct (gapless sound and massive) modes in the SF phase. These two phases are found in mean-field theory and may appear for dimensions $d \geq 2$ where a true condensate exists in the SF ground state.

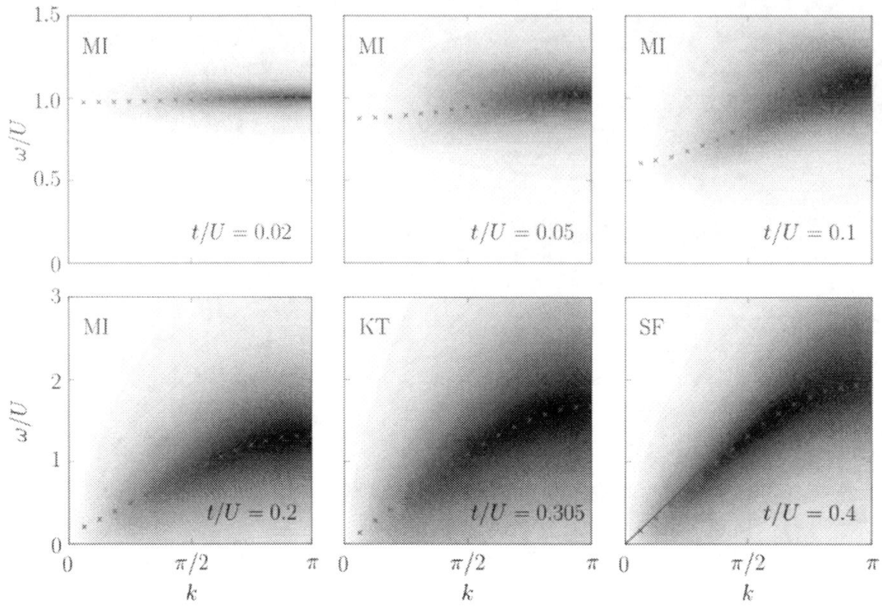

Figure 3. Intensity of the dynamical structure factor $S(k,\omega)$ in the k–ω plane for different t/U. DDMRG data were obtained for $L = 32$ system, using $\eta = 2t$ ($\eta = t$) in the MI phase and at the KT transition point (in the SF phase). The red crosses mark the positions of the maximum in each $k = 2\pi j/L$-sector. The green line for $t/U = 0.4$ marks the field-theory/Bogoliubov result $\omega(k) = \sqrt{\epsilon_0(k)[\epsilon_0(k) + 2U]}$ with $\epsilon_0(k) = -2t(\cos k - 1)$.

To summarise, we have determined the dynamical structure factor $S(k,\omega)$ for the 1D BHM with particle density $\rho = N/L = 1$ by means of unbiased numerical DDMRG calculations. As discussed for photoemission spectra previously [3], $S(k,\omega)$ agrees with the first-order perturbation theory result in the Mott insulator phase for $U \gg t$. Naturally, as the Coulomb interaction is lowered, noticeable deviations appear between both approaches, in particular the DDMRG $S(k,\omega)$ becomes dispersive and we find a substantial redistribution of spectral weight into the small k-sector. In this regime, higher-order corrections have to be taken into account in our analytical treatment of the BHM. Approaching the SF state, the charge excitations gap closes and the maximum in $S(k,\omega)$ exhibits a linear dispersion. The quantum phase transition between MI and SF phases is located at about $t/U \simeq 0.3$ and found to be of KT type.

Acknowledgements. SE and HF acknowledge funding by DFG through SFB 652.

References

[1] Greiner M, Mandel O, Esslinger T, Hänsch T W and Bloch I 2002 *Nature* **415** 39
[2] Bloch I, Dalibard J and Zwerger W 2008 *Rev. Mod. Phys.* **80** 885
[3] Ejima S, Fehske H and Gebhard F 2011 *Europhysics Letters* **93** 3002
[4] Oosten D van *et al.* 2005 *Phys. Rev. A* **71** 021601(R)
[5] Huber S D, Altman E, Büchler H P and Blatter G 2007 *Phys. Rev. B* **75** 085106
[6] Roth R and Burnett K 2004 *J. Phys. B* **37** 3893
[7] Pippan P, Evertz H G and Hohenadler M 2009 *Phys. Rev. A* **80** 033612
[8] White S R 1992 *Phys. Rev. Lett.* **69** 2863; Jeckelmann E 2002 *Phys. Rev. B* **66** 045114
[9] Stenger J *et al.* 1999 *Phys. Rev. Lett.* **82** 4569
[10] Nishimoto S and Jeckelmann E 2004 *J. Phys.:Condens. Matter* **16** 613
[11] Cazalilla M A, Citro R, Giamarchi T, Orignac E and Rigol M, 2011 *Rev. Mod. Phys.* **83** 1405
[12] Bogoliubov N N 1947 *J. Phys. (USSR)* **11** 23

Dynamical properties of ultracold fermions with attractive interactions in an optical lattice

Akihisa Koga[1], **Philipp Werner**[2]

[1]Department of Physics, Tokyo Institute of Technology, Tokyo 152-8551, Japan
[2]Theoretische Physik, ETH Zurich, 8093 Zürich, Switzerland

E-mail: `koga@phys.titech.ac.jp`

Abstract. We study dynamical properties of ultracold fermions with attractive interactions by means of dynamical mean-field theory and a continuous-time quantum Monte Carlo method. By calculating the pair potential and the density of states, we discuss the stability of the superfluid state. We also show that when holes are doped into the fermionic optical lattice system with intermediate interaction strength, the quasi-particle peak smears and a gap structure instead appears in the dilute system.

Ultracold atomic systems have attracted wide-spread interest in the physics community since the demonstration of Bose-Einstein condensation (BEC) in a Rb atom system [1]. Interesting topics in the field are the nature of the superfluid state in fermionic systems, including the BCS-BEC crossover [2, 3, 4], and the recently observed pseudogap behavior [5, 6]. These experimental observations stimulate theoretical investigations on the superfluid state and related phenomena in ultracold atomic systems.

Two-component Fermi gas systems have been studied theoretically in much detail and it has been clarified that a pseudogap phenomenon indeed appears in the BCS-BEC crossover region above the critical temperature [7, 8, 9, 10]. On the other hand, in fermionic optical lattice systems, such dynamical properties at finite temperatures have not been discussed. In our previous paper, we have studied the attractive Hubbard model to clarify how the superfluid state is stabilized and how the gap structure appears at finite temperatures [11]. It is also instructive to systematically study the doping dependence of dynamical properties. This topic was beyond the scope of our previous paper, but it may be interesting to investigate how the introduction of the lattice potential changes low energy properties of the Fermi gas system.

To clarify this, we consider the infinite-dimensional attractive Hubbard model and discuss how the particle density affects dynamical properties at low temperatures. The model Hamiltonian is given as $\hat{\mathcal{H}} = \sum_{ij,\sigma}(-t_{ij} - \mu\delta_{ij})\, c_{i\sigma}^\dagger c_{j\sigma} - U\sum_i n_{i\uparrow}n_{i\downarrow}$, where $c_{i\sigma}$ ($c_{i\sigma}^\dagger$) annihilates (creates) a fermion on the ith site with spin $\sigma(=\uparrow,\downarrow)$, and $n_{i\sigma} = c_{i\sigma}^\dagger c_{i\sigma}$. U is the onsite attractive interaction, t_{ij} is the transfer integral between sites, and μ is the chemical potential. The low-energy properties have been studied in one dimension [17, 18, 19, 20], two dimensions [21, 22, 23, 24] and higher dimensions [11, 25, 26, 27, 28, 29]. It is known that the superfluid ground state is always realized in two and higher dimensions. Here, we focus on the BCS-BEC crossover region to study how dynamical properties are affected by the particle density and temperature. To this end, we make use of DMFT [12, 13, 14, 15]. In DMFT, the original lattice model is mapped to an effective impurity model, where local particle correlations are taken into account precisely.

The lattice Green's function is obtained via a self-consistency condition imposed on the impurity problem. This treatment is formally exact in the limit of infinite spatial dimensions. The self-consistency condition of our model [30] is given by $\hat{G}_0^{-1}(i\omega_n) = i\omega_n\hat{\sigma}_0 + \mu\hat{\sigma}_z - \left(\frac{D}{4}\right)^2 \hat{\sigma}_z\hat{G}(i\omega_n)\hat{\sigma}_z$, where $\hat{G}_0(i\omega_n)$ $[\hat{G}(i\omega_n)]$ is the non-interacting [interacting] Green's function for the impurity model, $\omega_n = (2n+1)\pi T$ is the Matsubara frequency, and T is the temperature. $\hat{\sigma}_0$ is the identity matrix and $\hat{\sigma}_z$ is the z-component of the Pauli matrix. We have used the semi-circular density of states, $\rho_0(x) = \frac{2}{\pi D}\sqrt{1-\left(\frac{x}{D}\right)^2}$, where D is the half bandwidth.

In the DMFT framework, an impurity solver is necessary to obtain the local Green's function from the effective impurity model. Here we use the CTQMC technique [16], which has successfully been applied to a wide variety of models such as the Hubbard model [11, 31, 32, 33, 34], the periodic Anderson model [35, 36], the Kondo lattice model [37], and the Holstein-Hubbard model [38, 39]. In this paper, using a CTQMC method in the hybridization expansion formulation [31, 32, 40] extended to the Nambu formalism [11], we directly compute how the superfluid state is realized in the system.

To discuss the stability of the superfluid state, we first calculate the pair potential $\Delta = \langle c_{i\uparrow}c_{i\downarrow}\rangle$ in the system with $U = 2D$ and $n = 0.5, 0.25$ and 0.125, where $n = \sum_\sigma \langle n_{i\sigma}\rangle/2$, as shown in Fig. 1 (a). At high temperatures, the pair potential is zero and the normal state is realized.

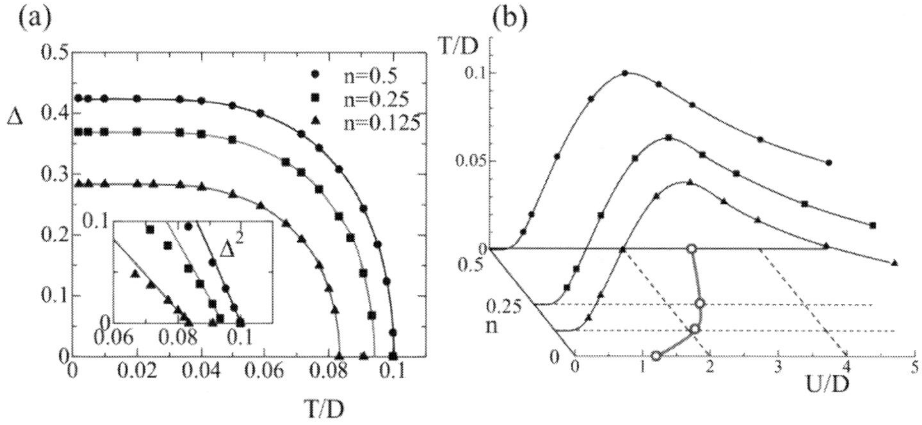

Figure 1. (a) Pair potential as a function of temperature in the system with $U = 2D$. (b) Phase diagram of the attractive Hubbard model. Solid circles, squares, and triangles represent results for the system at half, quarter, and one-eighth filling. Open circles represent the phase boundary at zero temperature where the first-order pairing transition occurs [29].

As the temperature is decreased, a phase transition occurs to the superfluid state at a critical temperature T_c, where the pair potential is induced. We find that the decrease of the particle density reduces the critical temperature. By examining the critical behavior $\Delta \sim |(T-T_c)/D|^\beta$ with the exponent $\beta = 1/2$, we determine the critical temperatures $T_c \sim 0.10D$ ($n = 0.5$), $0.094D$ ($n = 0.25$), and $0.084D$ ($n = 0.125$).

By performing similar calculations, we obtain the phase diagram, as shown in Fig. 1 (b). In the small U case, weakly coupled Cooper pairs are formed and the BCS-type superfluid state is realized. In the large U case, the strong attraction tightly couples the fermions, and hence a BEC-type superfluid state is realized. In this case, the superfluid critical temperature is scaled by the effective hopping for paired bosons $\sim t^2/U$. The BCS-BEC crossover, which may be characterized by the maximum of the critical temperature, occurs in the intermediate region ($U \sim 2D$). Fig. 1 (b) shows that the decrease in the particle density results in only a small shift of the BCS-BEC crossover region, at least for $n \geq 0.125$. Nevertheless, a drastic change appears in the density of states around the BCS-BEC crossover at $T = T_c$. The density of

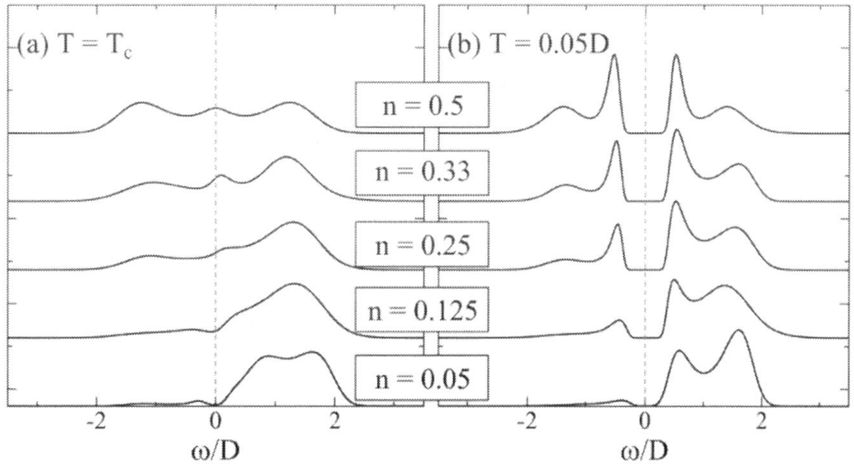

Figure 2. Density of states of the attractive Hubbard model with $U = 2D$ at $T = T_c$ (a) and $T = 0.05$ (b) when $n = 0.5, 0.33, 0.25, 0.125$ and 0.05.

states is calculated by means of the maximum entropy method [41, 42, 43], using the classic algorithm and a Gaussian default model. We find in Fig. 2 (a) a quasi-particle peak near the Fermi level for $n \geq 0.25$, while a gap structure appears in the case $n \leq 0.125$. This may originate from the pairing transition, which is realized only in the normal state. It has already been clarified that when the system is restricted to be paramagnetic, the pair transition as a function of the attractive interaction takes place between the heavy metallic Fermi liquid state and the insulating bound pairs state [29]. Note that this transition occurs in systems with arbitrary fillings, in contrast to the Mott transition in the repulsive Hubbard model. The zero-temperature phase boundary is shown by the open circles in Fig. 1 (b). It is found that the phase boundary crosses the $U = 2D$ line around $n \sim 0.125$. At the finite temperature we consider here ($T = T_c$), the crossover behavior between the metallic and insulating states appears at a slightly smaller interaction, compared to the transition point at $T = 0$. Therefore, in the system with $U = 2D$, the decrease of the particle density induces a change from heavy metallic behavior to insulating behavior. Once the system is in the superfluid state ($T = 0.05D$), the pairing transition does not affect the low energy properties. Instead, the superfluid gap opens in the vicinity of the Fermi level, as shown in Fig. 2 (b).

We comment on the difference of low energy properties between the dilute lattice system and the Fermi gas system. It has already been clarified that in the Fermi gas system, a pseudogap behavior appears around the BCS-BEC crossover region [7, 8, 9, 10]. In the lattice model, either a metallic behavior or a gap behavior appears, depending on the interaction strength and the particle density. Therefore, a competition between these incompatible behaviors should appear in dynamical properties when the lattice potential is gradually introduced in the Fermi gas system.

We have investigated the attractive Hubbard model in infinite spatial dimensions. By combining dynamical mean-field theory with the strong-coupling version of the continuous-time quantum Monte Carlo method, we have directly dealt with the superfluid phase in the system. By calculating the pair potential and the density of states of the system at different band fillings, we have clarified that a gap behavior indeed appears in the density of states when the system is far away from half filling.

Acknowledgment
This work was partly supported by the Grant-in-Aid for Scientific Research 20740194 (A.K.) and the Global COE Program "Nanoscience and Quantum Physics" from the Ministry of Education, Culture, Sports, Science and Technology (MEXT) of Japan. PW acknowledges support from SNF Grant PP0022-118866. The simulations have been performed using some of the ALPS libraries [44].

References
[1] Anderson M H, Ensher J R, Matthews M R, Wieman C E, and Cornell E A, 1995 *Science* **269** 198
[2] Jochim S, et al. 2003 *Science* **302** 2101
[3] Zwierlein M W, et al. 2003 *Phys. Rev. Lett.* **91** 250401
[4] Bourdel T, et al. 2004 *Phys. Rev. Lett.* **93** 050401
[5] Gaebler J P, Stewart J T, Drake T E, Jin D S, Perali A, Pieri P, and Strinati G C 2010 *Nat. Phys.* **6** 569
[6] Feld M, Frölich B, Vogt E, Koshorreck M and Köhl M 2011 *Nature* **480** 75
[7] Chen Q and Levin K 2009 *Phys. Rev. Lett.* **102** 190402
[8] Tsuchiya S, Watanabe R, and Ohashi Y 2009 *Phys. Rev. A* **80** 033613
[9] Chien C C, Guo H, He Y, and Levin K 2010 *Phys. Rev. A* **81** 023622
[10] Su S Q, Sheehy D E, Moreno J, and Jarrell M 2010 *Phys. Rev. A* **81** 051604
[11] Koga A and Werner P 2011 *Phys. Rev. A* **84** 023638
[12] Metzner W and Vollhardt D 1989 *Phys. Rev. Lett.* **62** 324
[13] Müller-Hartmann E 1989 *Z. Phys. B* **74** 507
[14] Georges A, Kotliar G, Krauth W and Rozenberg M J 1996 *Rev. Mod. Phys.* **68** 13
[15] Pruschke T, Jarrell M, and Freericks J K 1995 *Adv. Phys.* **44** 187
[16] Gull E, Millis A J, Rubtsov A N, Lichtenstein A I, Troyer M, and Werner P 2011 *Rev. Mod. Phys.* **83** 349
[17] Lieb E H and Wu F Y 1968 *Phys. Rev. Lett.* **20** 1445
[18] Shiba H 1972 *Prog. Theor. Phys.* **48** 2171
[19] Machida M, Yamada S, Ohashi Y, and Matsumoto H 2006 *Phys. Rev. A* **74** 053621
[20] Fujihara Y, Koga A, and Kawakami N 2009 *Phys. Rev. A* **79** 013610
[21] Moreo A and Scalapino D J 1991 *Phys. Rev. Lett.* **66** 946
[22] Paiva T, dos Santos R R, Scalettar R T, and Denteneer P J H 2004 *Phys. Rev. B* **69** 184501
[23] Dupuis N 2004 *Phys. Rev. B* **70** 134502
[24] Koga A, Higashiyama T, Inaba K, Suga S, and Kawakami N 2008 *J. Phys. Soc. Jpn.* **77** 073602
[25] dos Santos R R 1994 *Phys. Rev. B* **50** 635
[26] Freericks J K, Jarrell M and Scalapino D J 1993 *Phys. Rev. B* **48** 6302
[27] Keller M, Metzner W, and Schollwöck U 2001 *Phys. Rev. Lett.* **86** 4612
[28] Koga A and Werner P 2010 *J. Phys. Soc. Jpn.* **79** 064401
[29] Capone M, Castellani C, and Grilli M 2002 *Phys. Rev. Lett.* **88** 126403
[30] Georges A, Kotliar G, and Krauth W 1993 *Z. Phys. B* **92** 313
[31] Werner P, Comanac A, de'Medici L, Troyer M, and Millis A J 2006 *Phys. Rev. Lett.* **97** 076405
[32] Werner P and Millis A J 2007 *Phys. Rev. B* **75** 085108
[33] Werner P and Millis A J 2006 *Phys. Rev. B* **74** 155107
[34] Werner P and Millis A J 2007 *Phys. Rev. Lett.* **99** 126405
[35] Luitz D J and Assaad F F 2010 *Phys. Rev. B* **81** 024509
[36] Koga A and Werner P 2010 *J. Phys. Soc. Jpn.* **79** 114401
[37] Otsuki J *et al.* 2007 *J. Phys. Soc. Jpn.* **76** 114707
[38] Assaad F F and Lang T C 2007 *Phys. Rev. B* **76** 035116
[39] Werner P and Millis A J 2007 *Phys. Rev. Lett.* **99** 146404
[40] Boehnke L, Hafermann H, Ferrero M, Lechermann F, and Parcollet O 2011 *Phys. Rev. B* **84** 075145
[41] Gull S F, in *Maximum Entropy and Bayesian Methods in Science and Engineering*, ed. G. J. Erickson and C. R. Smith (Kluwer Academic, Dordrecht, 1988) p. 53; Skilling J (Kluwer Academic, Dordrecht, 1989) p. 45; Gull S F, *ibid.* p. 53.
[42] Silver R N, *et al.* 1990 *Phys. Rev. B* **41** 2380; Gubernatis J E, *et al.* 1991 *Phys. Rev. B* **44** 6011
[43] Press W F, Teukolsky S A, Vetterling W T and Flannery B R, *Numerical Recipes* (Cambridge University Press, Cambridge, 1992) p. 809.
[44] Albuquerque A F *et al.* 2007 *J. Magn. Magn. Mater.* **310** 1187

Journal of Physics: Conference Series **391** (2012) 012145 doi:10.1088/1742-6596/391/1/012145

Long range magnetic order in azurite from μSR

M Kraken[1], J Engelke[1], S Süllow[1], F J Litterst[1], M Lang[2], C Baines[3] and H Luetkens[3]

[1] Institut für Physik der kondensierten Materie, TU Braunschweig, D-38106 Braunschweig, Germany
[2] Physikalisches Institut, J.W. Goethe Universität, D-60438 Frankfurt (Main), Germany
[3] SμS, Paul Scherrer Institut, CH-5232 Villigen PSI, Switzerland

E-mail: m.kraken@tu-bs.de

Abstract. We report a Muon Spin Rotation (μSR) study on single crystalline azurite. The measurements revealed clear evidence for long range magnetic order below the Néel temperature at $T_N = 1.9$ K. The spectra in the magnetically ordered regime were fitted with three different muon sites, which can be related to muons close to Cu^{2+} monomer and dimer ions, respectively. The spontaneous muon spin rotation frequencies revealed the development of a magnetic order parameter. Below 0.4 K, transverse damping data indicated a further change in the local magnetic environment of the Cu^{2+} monomer ions.

1. Introduction

The natural mineral azurite $Cu_3(CO_3)_2(OH)_2$ has raised increased interest after the observation of a 1/3 magnetization plateau under external magnetic fields [1, 2]. Recently, azurite has successfully been described using an effective generalized spin-$\frac{1}{2}$ diamond chain model [3].
Under zero magnetic field, dimers of Cu^{2+} ions form $S = 0$ singlets below about 18 K and azurite remains paramagnetic down to 1.9 K, due to the effective 1D magnetic coupling. Below $T_N \approx 1.9$ K, indications for an antiferromagnetic state were found from heat capacity, susceptibility and proton NMR [4, 5, 6].
Azurite crystallizes in the space group $P2_1$ with the parameters $a = 4.99$ Å, $b = 5.82$ Å, and $c = 10.33$ Å with $\beta = 92.21°$ [7]. As shown in figure 1, the chain structure consists of alternating Cu^{2+} dimer and monomer ions along the b axis. Every Cu^{2+} ion is surrounded by four O^{2-} ions on the apices of a rectangle with a distance of ≈ 2 Å to the Cu^{2+} for each O^{2-} ion, the different chains are connected through CO_3 groups (not shown in figure 1) and the H^+ ion is located close to the O^{2-} ion between the monomer and dimer spins. From neutron scattering, the magnetic moments of each Cu^{2+} monomer and dimer ion is $0.68\mu_B$ and $0.26\mu_B$, respectively [7].
Especially in view of these small magnetic moments, μSR is an excellent method to investigate the magnetic properties of azurite. From earlier μSR measurements [8] static order was suggested to occur below T_N, however, no spontaneous muon spin rotation signals have been detected. Recently, we found evidence for static order from the vanishing of the TF (transverse field) μSR signal below T_N for a powder specimen [9]. Here, we present first data on a single crystal sample from the same mineral source. In contrast to the data from the powder sample, we were able to trace spontaneous muon spin rotation frequencies with a temperature dependence following a magnetization curve. We therefore can definitely prove the occurrence of long

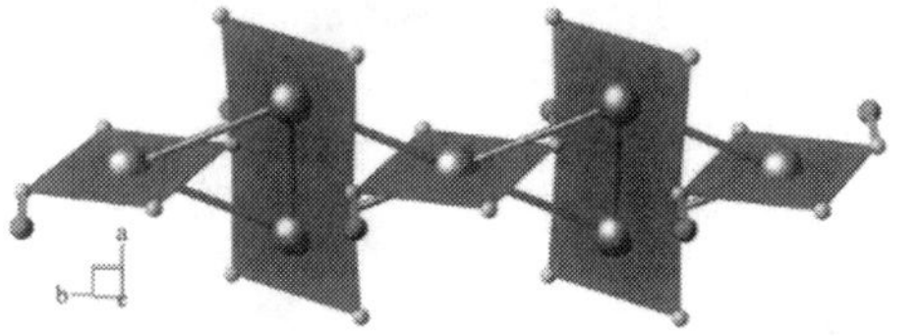

Figure 1. The crystal structure of azurite. The blue spheres represent the Cu^{2+} ions, the pink ones the O^{2-} ions and the red spheres the H^{+} ions. The carbon atoms are not shown in this picture.

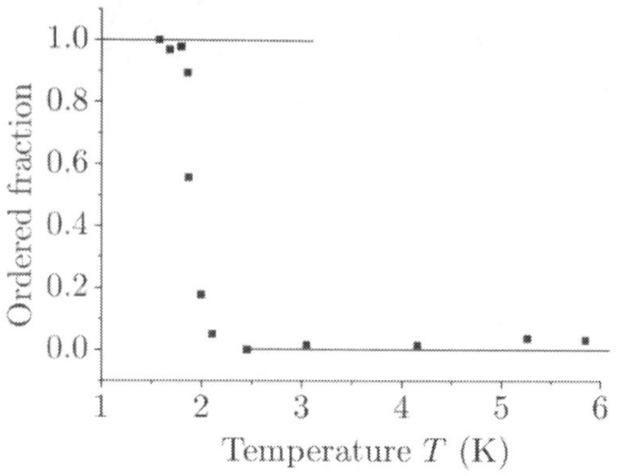

Figure 2. Temperature dependence of the magnetically ordered volume fraction from TF μSR measurements in an external field of 5 mT.

range magnetic order in an azurite single crystal by means of μSR measurements.

2. Experimental

The measurements were performed at the SμS of the Paul Scherrer Institut in Villigen, Switzerland between 19 mK and 6 K using the LTF and GPS facilities [10]. For an introduction to the technique of μSR we refer to [11].
The used azurite single crystal is the same as previously studied by neutron scattering [12]. Its quality has been proven by various experimental methods [7, 9, 12, 13].

3. Results and Discussion

From the asymmetry of the TF μSR (5mT) signal between 1.6 K and 6 K, we gain information about the magnetically ordered volume fraction by fitting the spectra with one weakly damped and one strongly damped cosine function (at GPS, there is no background signal from sample holder, etc.). The weakly damped signal can be attributed to the non-ordered fraction of the sample, while the strongly damped signal can be attributed to the magnetically ordered fraction. The amount of the magnetically ordered volume fraction is plotted in figure 2. It is zero well above T_N, increases strongly between 2.0 K and 1.8 K and saturates below. We determine $T_N = (1.9 \pm 0.1)$ K from these measurements, which is in good agreement with the reported values [4, 6]. The transition is sharper than that from earlier μSR measurements on polycrystalline material [9].
We examined the magnetically ordered regime below T_N by performing ZF (zero field) measurements. In figure 5 we show the ZF spectrum for $T = 0.019$ K. For the fit, we used a function $A_{ZF}(t)$ that describes three different muon sites in a magnetically ordered environment and an additional signal from the non-magnetic sample holder for the case of the dilution refrigerator used for the low temperature measurements in LTF

$$A_{ZF}(t) = \sum_{i=1}^{3} \left(A_{T,i} \cos\left(\omega_{\mu,i} t + \phi_i\right) \exp(-\lambda_{T,i} t) + A_{L,i} \exp(-\lambda_{L,i} t)\right) + A_{BG}. \quad (1)$$

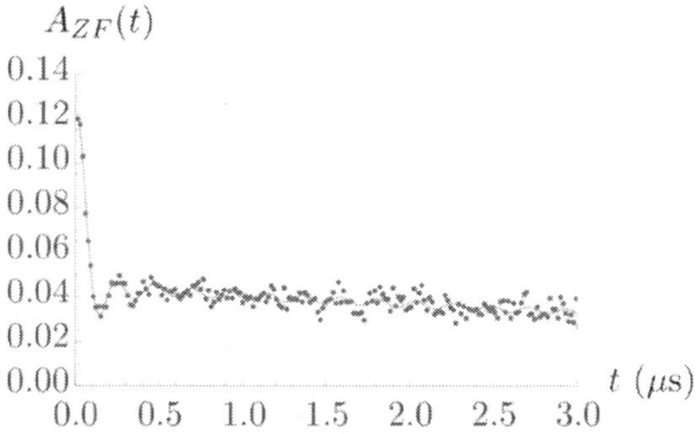

Figure 3. Zero field μSR spectrum at $T = 0.019\,\mathrm{K}$ (background subtracted).

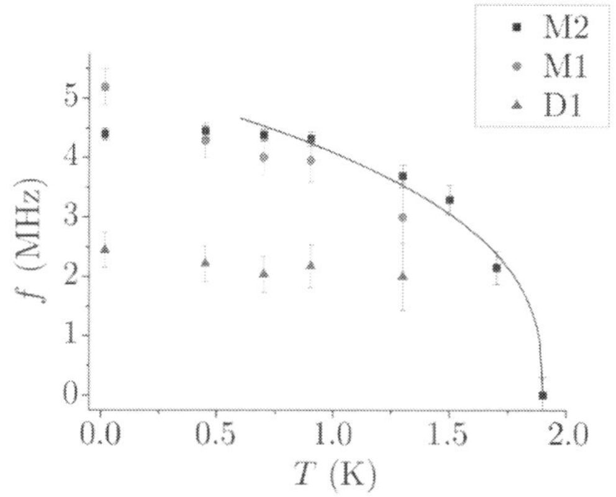

Figure 4. Muon spin rotation frequencies f versus temperature for $T \leq T_N$. The fit was performed for M2 in the region between 0.7 K and 1.9 K with a fixed $T_N = 1.9\,\mathrm{K}$.

The values of the spontaneous muon spin rotation frequencies f at $T = 0.45\,\mathrm{K}$ for the three different sites are 2.2 MHz, 4.3 MHz and 4.5 MHz with clearly distinct transverse dampings. With the assumption, that the positively charged anti-muons used in these measurements come to rest close to the O^{2-} ions, we expect that the measured values of the spontaneous muon spin rotation frequencies f scale with the magnetic moments of the nearest Cu^{2+} ions, since each O^{2-} ion has the same distance to its nearest dimer/monomer ion. An influence from the next nearest Cu^{2+} ions will result only in slight differences of the observed frequency values as well as different values for the transverse damping of the different sites. With the knowledge of the magnetic moments of the monomer and dimer ions from neutron scattering [7], we tentatively attribute the signal with the smaller frequency to a muon site close to the dimer ions (D1) and the two signals with the larger frequencies to muon sites close to the monomer ions (M1 and M2, respectively). The values of the μSR frequencies correspond to magnetic moments at Cu^{2+} of a few $\frac{1}{10}\mu_B$ (using a rule of thumbs after [14]). These values and their ratio match the values from neutron scattering experiments described above.

The transverse damping λ_T of the signals D1 and M1 is approximately $9\,\mu\mathrm{s}^{-1}$ while λ_T of the signal M2 is of the order of $0.2\,\mu\mathrm{s}^{-1}$. This different behavior can be understood when we attribute the signal M2 to the muon site close to the O^{2-} ion next to the H^+ ion, since the influence of the coupled dimer ions at this site is probably negligible. Therefore this signal displays only the interaction of the muon with the monomer ion. Without the influence of the next nearest neighbors, the damping of the signal is strongly decreased.

The development of the spontaneous muon spin rotation frequencies f with temperature is shown in figure 4. The values for M1 and D1 carry a large error due to the strong damping of these signals, we therefore consider only signal M2. The temperature dependent frequency $f(T)$, which is proportional to the sublattice magnetization of the system, is fitted using

$$f(T) \sim \left(1 - \frac{T}{T_N}\right)^{\nu}, \qquad (2)$$

with $T_N = 1.9\,\mathrm{K}$ and $\nu = 0.36 \pm 0.06$. The value of ν is typical for magnetization curves. The big uncertainty in ν arises from the unavailability of well resolved spectra close to T_N. Therefore, unambiguous assignment of the critical exponent from these data is not possible.

Below 0.4 K, we detect strong deviations from the systematic temperature dependence of transverse dampings λ_T and of the frequencies f, for the sites M1 and M2 as can be seen

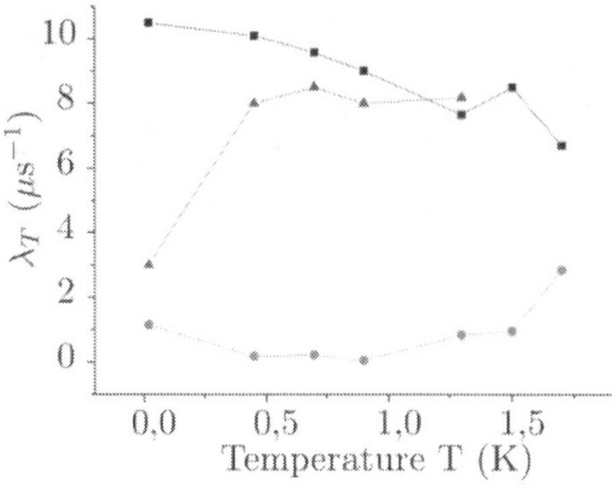

Figure 5. The development of the transversal damping versus temperature for the three signals D1 (■), M1 (▲) and M2 (•). Above 1.3 K, the signals D1 and M1 cannot be distinguished due to the small muon spin rotation frequencies close to T_N.

in figures 4 and 5. This temperature range deserves further investigations in view that there are indications for an additional phase transition below 0.5 K from ultrasonic measurements [13] and neutron scattering [7]. Whether there occurs indeed a phase transition or only a spin reorientation is still unclear.

4. Conclusion

We have studied a single crystal of azurite by μSR and identified the magnetic transition at $T_N = (1.9 \pm 0.1)$ K. The transition is sharper than that observed on polycrystalline material by μSR [9]. A more detailed examination of this behavior will be presented elsewhere. Below T_N, spontaneous muon spin precession is observed giving clear evidence for long range magnetic order. The ZF μSR spectra in this regime, were fitted with three different muon sites, whose signals could be tentatively attributed to the different dimer and monomer ions, respectively. From these data, we were able to extract the sublattice magnetization as a function of temperature. Below 0.4 K, indications of an additional transition, which affects mainly the Cu^{2+} monomer ions, were observed.

Acknowledgments

M.K. and J.E. are grateful for travel support by European Community. M.L. acknowledges support by SFB/TR 49.

References

[1] Kikuchi H *et al.* 2005 *Phys. Rev. Letters* **94** 227201
[2] Gu B and Su G 2006 *Phys. Rev. Letters* **97** 089701
[3] Jeschke H *et al.* 2011 *Phys. Rev. Letters* **106** 217201
[4] Forstat H, Taylor G and King B R 1959 *J. Chem. Phys.* **31** 929
[5] Garber M and Wagner R 1960 *Physica* **26** 777
[6] Spence R D and Ewing R D 1958 *Phys. Rev.* **112**(5) 1544–1545
[7] Rule K C *et al.* 2011 *Phys. Rev. B* **83** 104401
[8] Kikuchi H *et al.* 2005 *Prog. Theor. Phys. Suppl.* **159** 1
[9] Gibson M C R *et al.* 2010 *Phys. Rev. B* **81** 140406
[10] URL `http://lmu.web.psi.ch/`
[11] Brewer J H 1994 *Encyclopedia of Applied Physics (VCH Publisher)* **11** 23
[12] Rule K C *et al.* 2008 *Phys. Rev. Letters* **100** 117202
[13] Cong P T *et al.* 2010 *J. Phys.: Conf. Ser.* **200** 012226
[14] MacLaughlin D E *et al.* 1988 *Phys. Rev. B* **37** 3153

Transport through a single Anderson impurity coupled to one normal and two superconducting leads

Akira Oguri[1] **and Yoichi Tanaka**[2]

[1] Department of Physics, Osaka City University, Osaka 558-8585, Japan
[2] Condensed Matter Theory Laboratory, RIKEN, Saitama 351-0198, Japan

Abstract. We study the interplay between the Kondo and Andreev-Josephson e ects in a quantum dot coupled to one normal and two superconducting (SC) leads. In the large gap limit, the low-energy states of this system can be described exactly by a local Fermi liquid for the interacting Bogoliubov particles. The phase shift and the renormalized parameters for the Bogoliubov particles vary depending on the Josephson phase ϕ between the two SC leads. We explore the precise features of a crossover that occurs between the Kondo singlet and local Cooper-pairing states as ϕ varies, using the numerical renormalization group approach.

1. Introduction

The Kondo e ect in quantum dots (QD) connected to superconducting (SC) leads has been studied intensively in these years, experimentally [1–4] and theoretically [5–15]. The competition between the strong Coulomb repulsion and superconductivity gives rise to the quantum phase transition (QPT) between a nonmagnetic singlet and magnetic doublet ground states. It also causes the Andreev scattering at an interface between the QD and a normal lead, and a ects substantially transport properties near the QPT, or the corresponding crossover region.

In the present work, we consider a single QD coupled to one normal and two SC leads as illustrated in Fig. 1, and study the interplay of the Kondo, Andreev, and Josephson e ects, by using the Wilson numerical renormalization group (NRG) approach. We show that three phase variables, the phase shift δ for the interacting Bogoliubov particles, the Bogoliubov angle Θ_B at the interface between the QD and normal lead, and the Josephson phase ϕ between the two SC leads play a central role, and determine the low-temperature properties in the large SC gap limit.

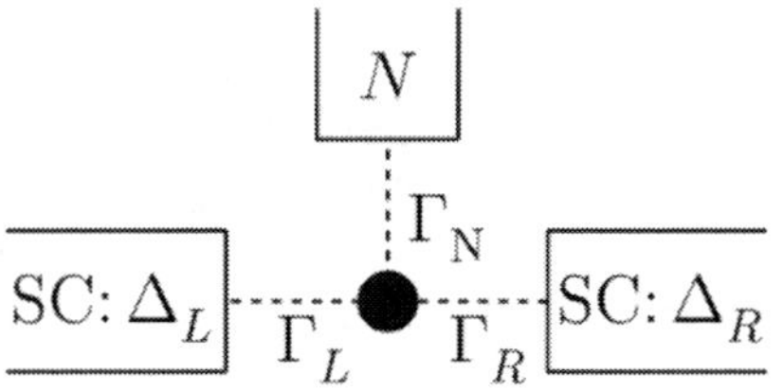

Figure 1. Anderson impurity (•) coupled to one normal (N) and two superconducting leads (L,R): $\Gamma_\nu \equiv \pi\ v_\nu^2$ with the density of states of the lead ($= L, R, N$), and v_ν the tunneling matrix element. The SC gaps $\Delta_{L/R} = |\Delta_{L/R}|\, e^{i\ \ L/R}$ give rise to the Andreev scattering and Josephson e ect.

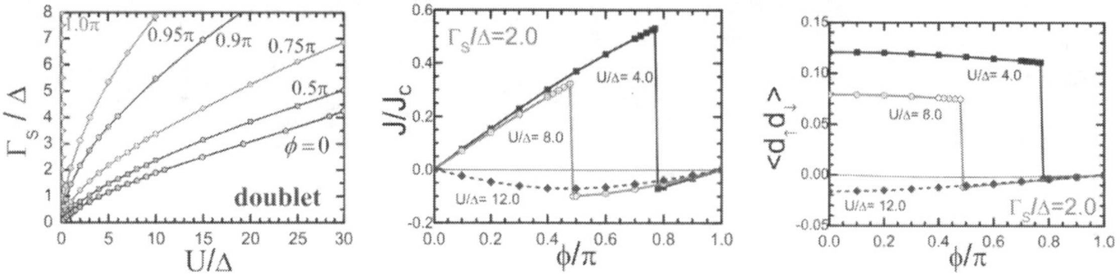

Figure 2. The left panel shows the phase boundary between the non-magnetic singlet (upper side) and magnetic doulbet (lower side) ground states in the U vs Γ_S plane for several values of $\phi \equiv \theta_R \;\; \theta_L$. The middle and right panels show the Josephson current J and $\langle d_\uparrow d_\downarrow \rangle$ as functions of ϕ for $\Gamma_S = 2.0\Delta$ for several U. The critical current is given by $J_C = e\Delta/\hbar$ for nite Δ.

2. Model

We start with a single impurity Anderson model of the form,

$$H = \xi_d \left(n_d \quad 1\right) + \frac{U}{2}\left(n_d \quad 1\right)^2 + \sum_{\nu=N,L,R} \sum v_\nu \left(\psi^\dagger_{\nu,} \, d \; + d^\dagger \, \psi_{\nu,} \right) + H_{\rm lead} + H_{\rm BCS} \,, \qquad (1)$$

$$H_{\rm lead} = \sum_{\nu=N,L,R} \sum_{k,} \epsilon_k \, c^\dagger_{\nu,k} \; c_{\nu,k} \,, \qquad H_{\rm BCS} = \sum_{\alpha=L,R} \sum_{k} \left(\Delta_\alpha \, c^\dagger_{\alpha,k\uparrow} \, c^\dagger_{\alpha, \;\; k\downarrow} + {\rm H.c.}\right) . \qquad (2)$$

Here, $\xi_d \equiv \epsilon_d + U/2$, U the Coulomb interaction, $n_d \equiv \sum d^\dagger d$, and $d^\dagger$ ($c^\dagger_{\nu,k}$) the creation operator for an electron with energy ϵ_d (ϵ_k) and spin σ in the quantum dot (lead). The QD and leads are coupled via v_ν, and $\psi_{\nu,} \;\; \equiv \sum_k c_{\nu,k} \; /\sqrt{\mathcal{N}_\nu}$ with $\mathcal{N}_\nu$ the normalization factor one-particle states in the leads. We assume that $\Gamma_\nu \equiv \pi \;\; v_\nu^2$ and the density of states are independent of the frequency ω. The complex superconducting gap $\Delta_{L/R} = |\Delta_{L/R}| e^{i \;\; L/R}$ gives rise to the Josephson current between the two SC leads for nite $\phi \equiv \theta_R \;\; \theta_L$. Since this Hamiltonian still has a number parameters to be explored, in this report we concentrate on the half- lled case $\epsilon_d = \;\; U/2$, assuming a symmetric junction with $\Gamma_L = \Gamma_R$ ($\equiv \Gamma_S/2$) and $|\Delta_L| = |\Delta_R|$ ($\equiv \Delta$).

3. Two SC terminal case without the normal lead $\Gamma_N = 0$

We rst of all consider a simpler $\Gamma_N = 0$ case, where the QD is coupled only to the two SC leads, in order to see the competition between the Kondo and Josephson e ects [10–14]. Figure 2 shows the NRG results for the ground state properties. We see that the region of the magnetic doublet state, which emerges for large U/Δ and small Γ_S/Δ, expands in the parameter space as the Josephson phase ϕ increases. This indicates that ϕ disturbs the Kondo screening. At the QPT, the Josephson current and the SC correlation $\langle d_\uparrow d_\downarrow \rangle$ at the impurity site vary discontinuously, and take small negative values in the magnetic doublet phase for nite Δ.

4. Low-energy properties of the QD connected to one normal and two SC leads

We next consider the three terminal case, $\Gamma_N \neq 0$, with the additional normal lead. Speci cally, we concentrate on the $\Delta \to \infty$ limit, where Δ is much larger than the other energy scales $\Delta \gg \max(\Gamma_S, \Gamma_N, U, |\epsilon_d|)$, and the quasi-particle excitations in the continuum energy region above the SC gap are projected out. Nevertheless, the essential physics of the low-energy transport is still preserved in this limit, and a static pair potential Δ_d is induced into the impurity site owing to the SC proximity e ects [7, 10],

$$\Delta_d \equiv \Gamma_R e^{i \;\; R} + \Gamma_L e^{i \;\; L} = |\Delta_d| \, e^{i \;\; d} \,, \qquad |\Delta_d| = \Gamma_S \sqrt{1 \quad \mathcal{T}_0 \sin^2 (\phi/2)} \,, \qquad (3)$$

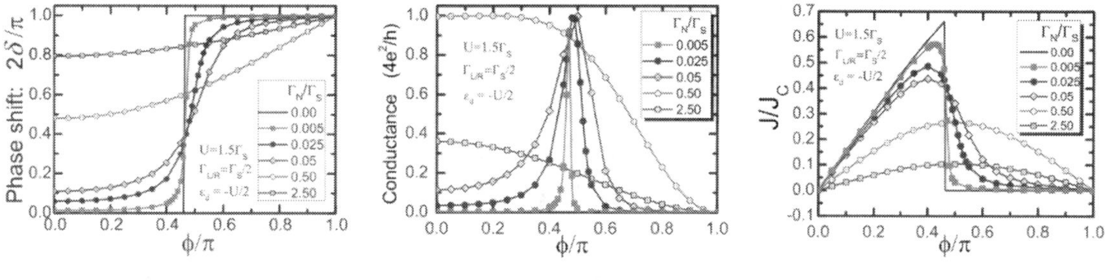

Figure 3. NRG results for the phase shift δ for the Bogoliubov particles (left), the conductance g_{NS} between the QD and normal lead (middle), and the Josephson current J (right), are plotted vs ϕ for several Γ_N in the large gap limit. The other parameters are chosen such that $U = 1.5\Gamma_S$, $\Gamma_L = \Gamma_R$ $(= \Gamma_S/2)$, and $\epsilon_d = -U/2$. Here, $J_C = e\Gamma_S/\hbar$ since $\Delta \to \infty$.

where $\mathcal{T}_0 = 4\Gamma_R\Gamma_L/(\Gamma_R + \Gamma_L)^2$. Carrying out the Bogoliubov transform with the angle Θ_B,

$$\cos\Theta_B = \frac{\xi_d}{E_A}, \qquad \sin\Theta_B = \frac{|\Delta_d|}{E_A}, \qquad E_A \equiv \sqrt{\xi_d^2 + |\Delta_d|^2}, \tag{4}$$

the Hamiltonian can be mapped onto the asymmetric Anderson model for the Bogoliubov particles, the Green's function for which takes the form $G(\omega) = [\omega - E_A + i\Gamma_N - \Sigma(\omega)]^{-1}$ [15]. The phase shift δ, defined by $G(0) = -|G(0)|e^{i\delta}$, determines the ground-state properties

$$\langle n_d \rangle - 1 = (2\delta/\pi - 1)\cos\Theta_B\,, \qquad \langle d_\downarrow d_\uparrow \rangle = \frac{1}{2}(2\delta/\pi - 1)\, e^{i\phi_d} \sin\Theta_B\,. \tag{5}$$

The conductance g_{NS} between the QD and normal lead, and the Josephson current J between the two SC leads can be determined by the three phase variables δ, Θ_B and ϕ,

$$g_{NS} = \frac{4e^2}{h}\sin^2\Theta_B \sin^2 2\delta, \qquad J = \frac{e\Gamma_S}{\hbar}\,\frac{\mathcal{T}_0\,|2\delta/\pi - 1|\,\sin\Theta_B\,\sin\phi}{2\sqrt{1 - \mathcal{T}_0 \sin^2(\phi/2)}}\,. \tag{6}$$

Furthermore, the renormalization factor $Z = [1 - \partial\Sigma/\partial\omega]^{-1}$, the position of the Andreev level $\widetilde{E}_A = \pm Z[E_A + \Sigma(0)]$, and the Wilson ratio R for the Bogoliubov particles can also be deduced from the local Fermi-liquid behavior, using the NRG.

The results are plotted *vs* ϕ in Figs. 3 and 4 for several values of Γ_N/Γ_S. The Coulomb interaction is chosen to be $U = 1.5\Gamma_S$, and for this value of U the QPT occurs in the $\Gamma_N \to 0$ limit at $\phi \simeq 0.46\pi$, where $E_A = U/2$. Note that $\mathcal{T}_0 = 1$ and $\xi_d = 0$ in the present case. The gapless excitations near the normal lead change the sharp QPT into a continuous crossover between the two different singlet states. We see in Fig. 3 that for $\phi \lesssim 0.46\pi$ the ground state is a local Cooper pairing consisting of a linear combination of the empty and doubly occupied impurity states with a small δ. On the other side, for $\phi \gtrsim 0.46\pi$, the ground state is a Kondo singlet with $\delta \simeq \pi/2$. At the transient region, the conductance g_{NS} has a peak, and the Josephson current J decreases rapidly to zero. Note that a weak current, which flows in the opposite direction as seen in Fig. 2 for the magnetic doublet state with finite Δ, is absent in the large gap limit. These features that are caused by the crossover are smeared as Γ_N increases.

Figure 4 shows the results for the renormalized parameters. We see for small Γ_N $(= 0.05\Gamma_S)$ that the parameters are strongly renormalized in the Kondo-singlet region for $\phi \gtrsim 0.46\pi$, where $Z \ll 1.0$, $R \simeq 2.0$, and the couple of the Andreev peaks $\widetilde{E}_A$ stay close to the Fermi level $\omega = 0$. In contrast, in the local Cooper-pairing region for $\phi \lesssim 0.46\pi$, the parameters are almost *not* renormalized $Z \simeq 1.0$, $R \simeq 1.1$, and the Andreev peaks situate away from the Fermi level. When the coupling Γ_N between the QD and normal lead is large, these two singlet states cannot be distinguished clearly.

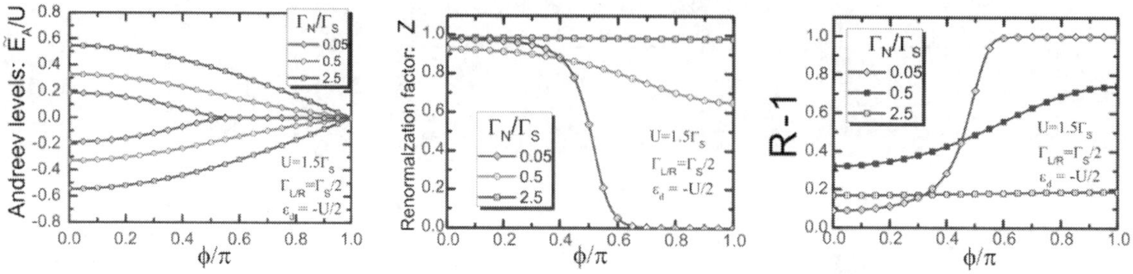

Figure 4. NRG results for the level position of renormalized Andreev resonance $\widetilde{E}_A$ (left), the renormalization factor Z (middle) and $R - 1$ (right), where R is the Wilson ratio, are plotted *vs* ϕ in the large gap limit for $U = 1.5\Gamma_S$ and $\Gamma_L = \Gamma_R$ $(= \Gamma_S/2)$ for several Γ_N.

5. Summary

We have studied the crossover between the Kondo singlet and the local Copper pairing, which consists of the linear combination of the empty and doubly occupied impurity states, in the quantum dot coupled to the one normal and two SC leads. In this three terminal geometry the normalized parameters, which characterize the Fermi-liquid behavior of the Bogoliubov particles, vary depending on the Josephson phase ϕ. Therefore, the crossover can occur at nite Josephson phase ϕ_C. We calculated the phase shift δ and the renormalized parameters with the NRG approach, and observed numerically that the Andreev conductance between the dot and normal lead has a peak near the crossover region at $\phi \simeq \phi_C$. The Josephson current between the the two SC leads decreases rapidly in the strongly renormalized Kondo-singlet ground state for $\phi \gtrsim \phi_C$, where the Andreev states $\widetilde{E}_A$ stay close to the Fermi level. In the local Cooper-paring ground state at the opposite side $\phi \lesssim \phi_C$, the local Fermi-liquid parameters are not renormalized so much.

Acknowledgments

We would like to thank J. Bauer and N. Kawakami for discussions. This work is supported by the JSPS Grant-in-Aid for Scienti c Research C (No. 23540375). Y.T. was supported by Special Postdoctoral Researchers Program of RIKEN. Numerical computation was partly carried out at Yukawa Institute Computer Facility.

References

[1] M. R. Graber, T. Nussbaumer, W. Belzig, and C. Sconenberger, Nanotechnology **15**, S479 (2004).
[2] A. Eichler, M. Weiss, S. Oberholzer, C. Schonenberger, A. Levy Yeyati, J. C. Cuevas, and A. Martın-Rodero, Phys. Rev. Lett. **99**, 126602 (2007).
[3] T. Sand-Jespersen, J. Paaske, B. M. Andersen, K. Grove-Rasmussen, H. I. Jørgensen, M. Aagesen, C. B. Sørensen, P. E. Lindelof, K. Flensberg, and J. Nygård, Phys. Rev. Lett. **99**, 126603 (2007).
[4] R. S. Deacon, Yoichi Tanaka, A. Oiwa, R. Sakano, K. Yoshida, K. Shibata, K. Hirakawa, and S. Tarucha, Phys. Rev. B **81**, 121308 (2010).
[5] R. Fazio and R. Raimondi, Phys. Rev. Lett. **80**, 2913 (1998); **82**, 4950(E) (1999).
[6] A. A. Clerk, V. Ambegaokar, and S. Hershfield, Phys. Rev. B **61**, 3555 (2000).
[7] J. C. Cuevas, A. L. Yeyati, and A. Martın-Rodero, Phys. Rev. B **63**, 094515 (2001).
[8] Y. Avishai, A. Golub, and A. D. Zaikin, Phys. Rev. B **63**, 134515 (2001).
[9] M. Governale, M. G. Pala and J. Konig, Phys. Rev. B **77**, 134513 (2008).
[10] Yoshihide Tanaka, A. Oguri, and A. C. Hewson, New J. Phys. **9**, 115 (2007); **10**, 029801(E) (2008).
[11] J. Bauer, A. Oguri, and A. C. Hewson, J. Phys.: Condes. Mat. **19**, 486211 (2007).
[12] C. Karrasch, A. Oguri, and V. Meden, Phys. Rev. B **77**, 024517 (2008).
[13] T. Hecht, A. Weichselbaum, J. von Delft, R. Bulla, J. Phys.: Condes. **20**, 275213 (2008).
[14] T. Meng, S. Florens, and P. Simon, Phys. Rev. B **79** 224521 (2009).
[15] Yoichi Tanaka, N. Kawakami, and A. Oguri, J. Phys. Soc. Jpn. **76**, 074701 (2007); **77**, 098001(E) (2008).

Variational cluster approach to the extended Falicov-Kimball model: A BCS-BEC crossover in the Excitonic insulators

K Seki[1], T Kaneko[1], S Yamaki[1], R Eder[2] and Y Ohta[1]

[1]Department of Physics, Chiba University, Chiba 263-8522, Japan
[2]Karlsruhe Institute of Technology, Institut für Festkörperphysik, D-76021 Karlsruhe, Germany

E-mail: seki-kazuhiro@graduate.chiba-u.jp

Abstract. We study the Coulomb-interaction induced spontaneous symmetry breaking of the excitonic insulator state in the two-dimensional extended Falicov-Kimball model. Using the variational cluster approximation, we evaluate the order parameter, single-particle excitation gap, momentum distribution functions, and anomalous Green's function as a function of U at zero temperature. We nd that in the weak-to-intermediate couplinb regime, the Fermi surface is well-de ned and the calculated results can be understood in the close correspondence with the BCS theory, whereas in the strong-coupling regime, the Fermi surface is ill-de ned and the results are consistent with the picture of BEC.

1. Introduction

Realization of excitonic insulators (EI) in the proximity of the semimetal-semiconductor transition was suggested about half a century ago [1, 2]. Because of the weak screening of the Coulomb attraction between electrons and holes due to a small number of carriers, the electrons and holes may spontaneously form the bound states (excitons), giving rise to the EI state. Recently, $TaNi_2Se_5$ has been studied by the angle-resolved photoemission spectroscopy measurements and it was reported that the valence-band top is extremely flat, and the material can be a new candidate for EI of the bound pairs between Ni $3d$-Se $4p$ holes and Ta $5d$ electrons [3]. From the theoretical point of view, the extended Falicov-Kimball model (EFKM) has been studied extensively as a simple lattice model to describe EI by means of various numerical and theoretical approaches. The ground-state phase diagram of EFKM, on one hand, was determined by the constrained path Monte Carlo (CPMC) [4, 5]. On the other hand, the excitation properties of the EFKM are still of great interest. Projector-based remormalization method (PRM) calculation on the one-dimensional EFKM [6] reported incoherent parts of the single-particle excitation spectra, which are related to dissociation of the excitons. Detailed studies on the dynamical excitonic susceptibility at finite temperature calculated by use of the PRM [7] and slave boson technique [8] confirmed that tightly bound excitons exist even above the critical temperature T_c of the exciton condensation. The results strongly support the existence of so-called excitonic halo above T_c suggested by Bronold and Fehske [9], where tightly bound excitons exist without condensation, leading to a scenario of the Bose-Einstein condensation (BEC) of preformed excitons in the strong Coulomb interaction regime.

In this paper, we study the BCS-BEC crossover [10] of the EI state of the EFKM defined on the two-dimensional (2D) square lattice as a function of the Coulomb interaction strength U. We employ the variational cluster approximation (VCA) [11] based on the self-energy functional theory [12] at zero temperature. The cluster perturbation theory (CPT) [13] is used to calculate the single-particle and anomalous Green's functions. Details will be presented elsewhere [14].

2. Model and Method

The Hamiltonian of EFKM reads

$$\mathcal{H} = -\sum_{=c,f}\left(t \sum_{\langle ij\rangle}(\ ^{\dagger}_{i} \ _{j} + \mathrm{H.c.}) + (\ - \ )\sum_{i} n_{i} \right) + U\sum_{i} n_{ic}n_{if} \qquad (1)$$

where $_{i}$ denotes the annihilation operator of an electron on the -orbital at site i and $n_{i} = \ ^{\dagger}_{i} \ _{i}$. t is the hopping integral between the neighboring sites of the 2D square lattice and is the on-site energy level of the -orbitals. U is the inter-orbital Coulomb repulsion between electrons. The chemical potential is determined so as to maintain the average particle density n at half filling $n = 1$. Throughout the paper, we set $\hbar = k_B = 1$. We use $t_c = 1$ as the unit of energy and we focus on the band parameters with values $_{c} = 0$, $_{f} = -1$, and $t_f = -0.3$.

We employ VCA in order to study the EI state of EFKM at zero temperature, which enables us to take into account the effects of electron correlations. VCA is based on the variational principle for the grand potential as a functional of the self-energy [12, 15]. The trial self-energy for the variational method is generated from the exact self-energy of the disconnected finite-size clusters, so-called reference system. The Hamiltonian of the reference system is defined as $\mathcal{H}_{\mathrm{ref}} = \mathcal{H} + \mathcal{H}_{\mathrm{pair}} + \mathcal{H}_{\mathrm{local}}$, $\mathcal{H}_{\mathrm{pair}} = \ '\sum_i (c_i^{\dagger} f_i + \mathrm{H.c.})$, $\mathcal{H}_{\mathrm{local}} = \ '\sum_i (n_{ic} + n_{if})$, where the Weiss field for the on-site electron-hole pairing $'$ and the orbital-independent potential $'$ are the variational parameters optimized based on the variational principle. $'$ is introduced in order to calculate the average particle density n correctly [16]. The single-particle Green's function for -electron $\mathcal{G}$ $(\mathbf{k},\omega)$ and anomalous Green's function $\mathcal{G}^{cf}(\mathbf{k},\omega)$ are calculated by CPT [13]. An 8-site (16-orbital) cluster is used as a reference system, thus the effects of static and dynamical electron correlation within the cluster size are taken into account. Details of VCA can be found in Ref. [17].

3. Results of calculation

We first calculate the U dependence of the order parameter for the exciton condensation $2 \ = \frac{U}{L}\sum_i \langle c_i^{\dagger} f_i + \mathrm{H.c.}\rangle$ and single-particle excitation gap $_{\mathrm{gap}} = \ ^{+} - \ ^{-}$, where L is the number of the lattice sites, $\langle \cdots \rangle$ denotes the ground-state expectation value, and $^{+(-)}$ is the upper (lower) bound of the chemical potential. The factor 2 for the order parameter is introduced in order to compare with the single-particle gap, by analogy with the BCS mean-field theory. Calculated results are shown in Fig. 1.

First, we can see from the results that there is the lower (upper) bound of the Coulomb interaction strength U_{c1} (U_{c2}) for the EI state. The obtained values are $(U_{\mathrm{c1}}, U_{\mathrm{c2}}) = (0.65, 6.6)$. The existence of the upper bound U_{c2} seems to contradict to the case of the s-wave superconducting state in the attractive Hubbard model, which has no U_{c2} [18]. What happened at $U = U_{\mathrm{c2}}$ is that the Hartree potential makes the f-band fully occupied and c-band empty, so that there are no Coulomb interactions between c-electrons and f-holes and thus the system becomes simply a band insulator above U_{c2}.

Then, in the weak-to-intermediate coupling regime ($U \lesssim 5$), both the order parameter and single-particle gap increase with increasing U with the relation $2| \ | \simeq \ _{\mathrm{gap}}$. This result is consistent with the relation $2| \ | = \ _{\mathrm{gap}}$ of the BCS mean-field theory. In the strong-coupling

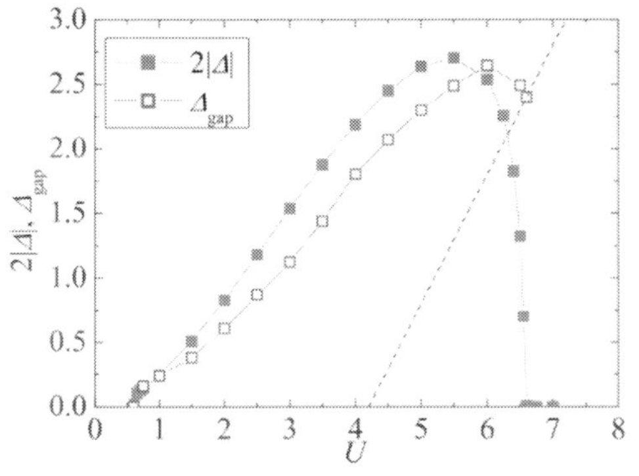

Figure 1. U dependence of the order parameter 2| | (filled symbol) and the single-particle gap $_{\rm gap}$ (open symbol). The dashed straight line indicates the single-particle gap in the normal state.

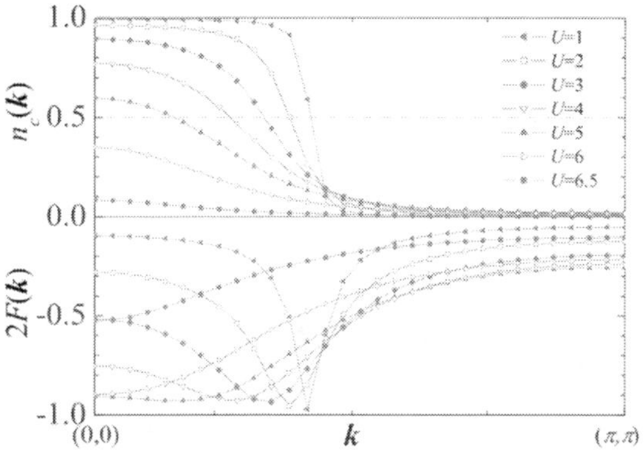

Figure 2. The momentum distribution functions $n_c(\mathbf{k})$ and $2F(\mathbf{k})$ along $\mathbf{k}$ = $(0,0) \to (\ ,\ )$ for various U. The f-electron momentum distribution functions $n_f(\mathbf{k})$ = $1 - n_c(\mathbf{k})$ are not shown.

regime ($U \gtrsim 5$), the order parameter rapidly decreases with increasing U but the single-particle gap remains large. If we can assume that the energy scale of the single-particle gap $_{\rm gap}$ and order parameter 2| | may correspond to that of the characteristic temperature for the exciton formation ($T_{\rm ex}$) and critical temperature for the condensation of excitons ($T_{\rm c}$), respectively, then the two temperatures should be comparable ($T_{\rm ex} \simeq T_{\rm c}$) in the weak-coupling (BCS) regime but may be well separated ($T_{\rm ex}$ $T_{\rm c}$) in the strong-coupling (BEC) regime. The BCS-BEC crossover may then be expected in this model although our calculations are done at zero temperature.

We next calculate the c-electron and anomalous momentum distribution functions defined as $n_c(\mathbf{k}) = -\frac{1}{\ } \int_{-\infty}^{\mu} d\omega \Im \mathcal{G}^{cc}(\mathbf{k}, \omega + i0^+)$, $F(\mathbf{k}) = -\frac{1}{\ } \int_{-\infty}^{\mu} d\omega \Im \mathcal{G}^{cf}(\mathbf{k}, \omega + i0^+)$, respectively. The results are shown in Fig. 2. Here we define, by analogy with the BCS mean-field theory, the "Fermi momentum" $\mathbf{k}_{\rm F}$ as $n_c(\mathbf{k}_{\rm F}) = n_f(\mathbf{k}_{\rm F}) = 0.5$. We can see that, in the weak-coupling regime, $n_c(\mathbf{k})$ sharply drops across the $\mathbf{k}_{\rm F}$ and $F(\mathbf{k})$ is peaked at $\mathbf{k}_{\rm F}$. With increasing U, $\mathbf{k}_{\rm F}$ approaches to $(0,0)$ because the Hartree potential for c-electrons reduces the c-electron density. Accordingly, $F(\mathbf{k})$ becomes broad in the momentum space, indicating that the radius of electron-hole pairs becomes small in real space. When U reaches the crossover regime ($U \sim 5$), we have no $\mathbf{k}_{\rm F}$ and $|F(\mathbf{k})|$ decreases in all momentum with increasing U. This behavior is consistent with the rapid decrease of | | with increasing U in the strong-coupling regime (see Fig. 1).

We also calculate the anomalous spectral function $F(\mathbf{k}, \omega) = -\frac{1}{\ } \Im \mathcal{G}^{cf}(\mathbf{k}, \omega + i\ )$ with $= 0.1$ at $U = 2$, 5, and 6.5. The results are shown in Fig. 3. At $U = 2$, $F(\mathbf{k}, \omega)$ shows sharp peak near $\mathbf{k}_{\rm F}$ and its intensity rapidly decreases as the momentum goes away from $\mathbf{k}_{\rm F}$ and frequency goes away from the Fermi energy . At $U = 5$, the incoherent continua appear in the spectral function. We can see that both the single particle gap and intensity of $F(\mathbf{k}, \omega)$ are large. At $U = 6.5$, $F(\mathbf{k}, \omega)$ shows a semiconductor-like dispersion mainly due to the Hartree potential. The frequency dependence of the intensity of $F(\mathbf{k}, \omega)$ is weaker than that at $U = 2$ and 5.

4. Summary

We have analyzed the EI state of EFKM using VCA as a function of Coulomb interaction strength U. In the weak-to-intermediate coupling regime ($U \lesssim 5$), the order parameter and single-particle gap $_{\rm gap}$ behave like the BCS theory, i.e., $_{\rm gap} \simeq 2|\ |$, the Fermi momentum $\mathbf{k}_{\rm F}$ can be defined from the momentum distribution function, and the anomalous Green's function

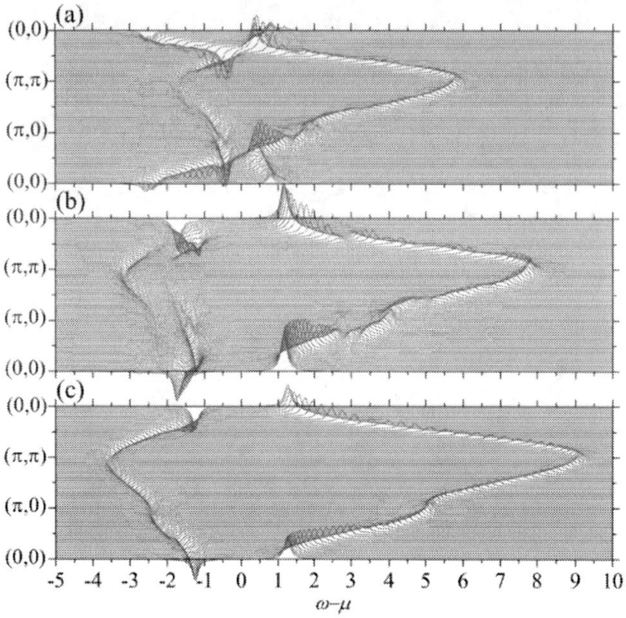

Figure 3. Calculated anomalous spectral function $F(\mathbf{k}, \omega)$ at (a) $U = 2$, (b) $U = 5$, and (c) $U = 6.5$.

is peaked near Fermi energy at $\mathbf{k}_F$. In the crossover regime ($U \sim 5$), both and $_{gap}$ take large values and they are close to the maximum. Correspondingly, $\mathbf{k}_F$ becomes ill-defined. In the strong-coupling regime ($U \gtrsim 5$), rapidly decreases with increasing U, while $_{gap}$ remains open, $\mathbf{k}_F$ is ill-defined, and the frequency and momentum dependence of the intensity of the anomalous Green's function becomes weak. Thus, the Fermi surface plays no role in the strong coupling regime, which is consistent with the picture of BEC.

Acknowledgment

We thank H Fehske, T Mizokawa, and S Ejima for useful discussions. KS acknowledges support from JSPS Research Fellowship for Young Scientists. This work was supported in part by Kakenhi Grant No. 22540363 of Japan. A part of computations was done at Research Center for Computational Science, Okazaki, Japan.

References

[1] Mott N F 1961 *Phil. Mag.* **6** 287.
[2] Jérome D, Rice T M and Kohn W 1967 *Phys. Rev.* **158** 462.
[3] Wakisaka Y, Sudayama T, Takubo K, Mizokawa T, Arita M, Namatame H, Taniguchi M, Katayama N, Nohara M and Takagi H 2009 *Phys. Rev. Lett.* **103** 026402.
[4] Batista C D, Gubernatis J E, Bonča J and Lin H Q 2004 *Phys. Rev. Lett.* **92** 187601.
[5] Batista C D 2002 *Phys. Rev. Lett.* **89** 166403.
[6] Phan V N, Becker K W and Fehske H 2010 *Phys. Rev.* B **81** 205117.
[7] Phan V N, Fehske H and Becker K W 2011 *EPL* **95** 17006.
[8] Zenker B, Ihle D, Bronold F X and Fehske H 2010 *Phys. Rev.* B **81** 115122.
[9] Bronold F X and Fehske H 2006 *Phys. Rev.* B **74** 165107.
[10] Nozières P and Schmitt-Rink S 1985 *J. Low Temp. Phys.* **59** 195.
[11] Potthoff M, Aichhorn M and Dahnken D 2003 *Phys. Rev. Lett.* **91** 206402.
[12] Potthoff M 2003 *Eur. Phys. J.* B **32** 429; **36**, 335.
[13] Sénéchal D, Perez D, and Pioro-Ladriere M 2000 *Phys. Rev. Lett.* **84** 522.
[14] Seki K, Eder R and Ohta Y *in preparation.*
[15] Luttinger J M and Ward J C 1960 *Phys. Rev.* **118** 1417.
[16] Aichhorn M, Arrigoni E, M. Potthoff M and Hanke W 2006 *Phys. Rev.* B **74** 024508.
[17] Sénéchal D 2008 *arXiv* 0806:2690v2.
[18] Toschi A, Capone M and Castellani C 2005 *Phys. Rev.* B **72** 235118.

Spin and charge orderings in the atomic limit of the *U*-*V*-*J* model

F Mancini[1,2], E Plekhanov[1,2,3], and G Sica[1]

[1] Dipartimento di Fisica "E.R. Caianiello", Università degli Studi di Salerno, 84084 Fisciano (SA), Italy
[2] Unità CNISM di Salerno, Università degli Studi di Salerno, 84084 Fisciano (SA), Italy
[3] CNR-SPIN, UoS di Salerno, 84084 Fisciano (SA), Italy

E-mail: `plekhanoff@physics.unisa.it`

Abstract. In this paper we study a generalization of the 1D Hubbard model in the atomic limit by considering density-density and Ising-type spin-spin nearest neighbor (NN) interactions, parameterized by V and J, respectively. We present the $T = 0$ phase diagram for both ferro ($J > 0$) and anti-ferro ($J < 0$) coupling obtained in the narrow-band limit by means of an extension to zero-temperature of the transfer-matrix method. Based on the values of the Hamiltonian parameters, we identify a number of phases that involve orderings of the double occupancy, NN density and spin correlations.

1. Introduction

The Hubbard model (HM) is undoubtedly one of the most studied model in condensed matter physics. In one spatial dimension (1D), the exact Bethe solution for the pure HM is known [1]. During its long history, several extensions to the HM have been proposed. Among them, the inclusion of non-local charge density and/or spin exchange interactions appears to be the most natural choice. Density correlations correspond to an effective finite-range Coulomb interaction (repulsive or attractive) as opposed to the on-site Hubbard U. On the other hand, the spin-spin interaction is believed to be responsible for the interplay between various spin orderings (SDW or ferromagnetism) with strong (triplet) superconducting correlations in a variety of compounds: Sr_2RuO_4 [2], UGe_2 [3], URhGe [4], $ZrZn_2$ [5] and $(TMTSF)_2X$ [6] family. Any extension of the HM breaks the integrability, so that either approximate analytic (*e.g.* bosonization) or numerical finite-system methods (*e.g.* Density Matrix Renormalization Group or Quantum Monte Carlo) should be used even in 1D. Bosonization works well in the limit of large band width and predicts several interesting effects like coexistence of spin density wave correlations and (triplet) superconducting ones (for spin exchange interactions [7, 8]) or a transition towards Mott-insulating phases at commensurate fillings (for charge interactions [9]). On the other hand, in the opposite limit of narrow-band width (atomic limit), the extended Hubbard model can be exactly solved by means of any of the two equivalent methods - Transfer Matrix (TM) technique [10] or Composite Operators Method [11, 12]. In the present work we employ a $T \to 0$ extrapolation of the TM method in order to derive the zero-temperature phase diagram of the extended HM with both nearest-neighbor (NN) charge and spin interactions in the narrow-band limit.

2. Results

The Hamiltonian of the U-V-J system reads as follows:

$$H = \sum_i [-\mu n(i) + UD(i)] + V \sum_i n(i)n(i+1) - J \sum_i S^z(i) S^z(i+1). \tag{1}$$

Here $n(i) \equiv n_\uparrow(i) + n_\downarrow(i)$ denotes the electron density operator at site i, $D(i) \equiv n_\uparrow(i) n_\downarrow(i)$ is the double occupancy at site i while $S^z(i) \equiv (n_\uparrow(i) - n_\downarrow(i))/2$ denotes the z-component of the spin at site i. The Hamiltonian (1) contains only NN term, hence the TM will be a 4×4 matrix. It is easy to check that the TM elements can be calculated through the matrix elements of an auxiliary matrix Z: $T_{ij} = \exp(-\beta Z_{ij})$, where Z is defined as follows:

$$Z = \begin{pmatrix} 0 & -\frac{\mu}{2} & -\frac{\mu}{2} & \frac{U}{2} - \mu \\ -\frac{\mu}{2} & V - \frac{J}{4} - \mu & V + \frac{J}{4} - \mu & \frac{U}{2} + 2V - \frac{3}{2}\mu \\ -\frac{\mu}{2} & V + \frac{J}{4} - \mu & V - \frac{J}{4} - \mu & \frac{U}{2} + 2V - \frac{3}{2}\mu \\ \frac{U}{2} - \mu & \frac{U}{2} + 2V - \frac{3}{2}\mu & \frac{U}{2} + 2V - \frac{3}{2}\mu & U + 4V - 2\mu \end{pmatrix}. \tag{2}$$

In our longer paper [13] we show how to reconstruct the $T = 0$ phase diagram from the TM matrix elements. Here we apply this method to the U-V-J model. We first identify all different matrix elements (we call them energy scales). In the present case these are:

$$\begin{aligned} F^0 &= 0 \\ F^{1/2} &= -\frac{\mu}{2} \\ F^1_1 &= \frac{U}{2} - \mu;\ F^1_2 = V - \frac{J}{4} - \mu;\ F^1_3 = V + \frac{J}{4} - \mu \\ F^{3/2} &= \frac{U}{2} + 2V - \frac{3}{2}\mu \\ F^2 &= U + 4V - 2\mu. \end{aligned} \tag{3}$$

In (3) we have already sorted the energy scales based on the values of particle number (given by the superscript of F) allowed for a given configuration. In particular, the three energy terms at $n = 1$ can be written as a unique expression as follows:

$$F^1 = A - \mu, \tag{4}$$

where:

$$A \equiv \min\left(\frac{U}{2}, V - \frac{|J|}{4}\right) = \begin{cases} \frac{U}{2}, & \frac{U}{2} < V - \frac{|J|}{4} \\ V - \frac{|J|}{4}, & \frac{U}{2} > V - \frac{|J|}{4}. \end{cases} \tag{5}$$

We choose J as energy scale and explore the phase diagram in the $V-U$ plane. As μ increases, the particle number in the system increases as well. Therefore, for each value of n ($n = 0, \frac{1}{2}, 1, \frac{3}{2}, 2$) we can establish the ranges of the chemical potential, within which μ can change without changing n. Such ranges can be summarized as follows:

$$\begin{aligned} &n = 0: \ \mu < \min\left(0, A, x, \tfrac{2}{3}x\right) && n = 1: \ \max(A, 2A) < \mu < 2x - \max(A, 2A) \\ &n = \tfrac{1}{2}: \ 0 < \mu < \min\left(2A, x, \tfrac{4}{3}x\right) && n = \tfrac{3}{2}: \ \max\left(\tfrac{2}{3}x, x, 2(x - A)\right) < \mu < 2x \\ &n = 2: \ \max\left(x, \tfrac{4}{3}x, 2x - A, 2x\right) < \mu, \end{aligned} \tag{6}$$

where we introduced $x \equiv \frac{1}{2}(U + 4V)$. Except for the cases $n = 0$ and $n = 2$, which are always possible, the other cases might be incompatible with a given choice of U, V, J. This occurs when the upper-bound of μ becomes lower than the lower-bound. This condition determines the existence of the jump at a given n in the dependence $\mu(n)$. Such conditions are:

$$\begin{aligned} n = \tfrac{1}{2} &\quad x > 0 \wedge A > 0 \\ n = 1 &\quad 0 < 2A < x \text{ or } A < 0 \wedge A < x \\ n = \tfrac{3}{2} &\quad x > 0 \wedge A > 0. \end{aligned} \tag{7}$$

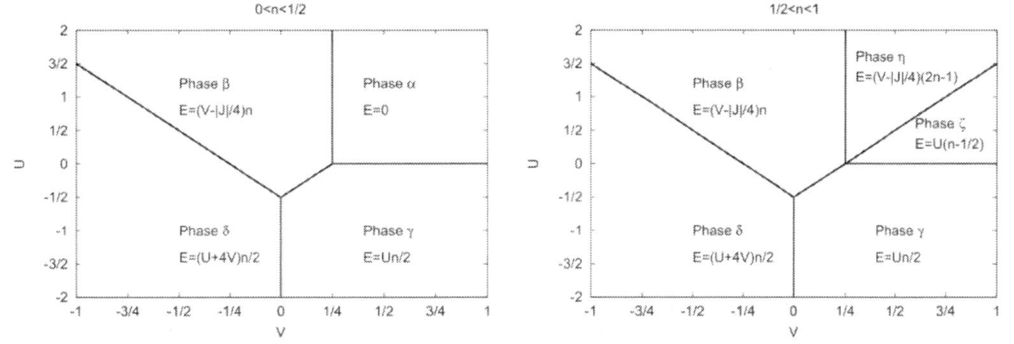

Figure 1. Left panel: Phase diagram in the case $0 < n < \frac{1}{2}$. Right panel: Phase diagram in the case $\frac{1}{2} < n < 1$.

The conditions for the existence of jumps at $n = \frac{1}{2}$ and $n = \frac{3}{2}$ coincide as these values of n are related by the particle-hole relation. Moreover, depending on the values of the Hamiltonian parameters, we can distinguish four cases in the conditions (6):

$$\text{(i) } x > 0 \wedge A > 0 \quad \text{(ii) } x > 0 \wedge A < 0 \quad \text{(iii) } x < 0 \wedge A > x \quad \text{(iv) } x < 0 \wedge A < x.$$

We can identify three zones in the $V - U$ plane depending on how many jumps there are in the dependence $\mu(n)$:

(i) $V > \frac{|J|}{4} \wedge U > 0$

(ii) (a) $\frac{U}{2} > -V - \frac{|J|}{4} \wedge \frac{U}{2} > V - \frac{|J|}{4} \wedge V < \frac{|J|}{4}$ (b) $U < 0 \wedge V > 0 \wedge \frac{U}{2} < V - \frac{|J|}{4}$

(iii) $V < 0 \wedge U < -2V - \frac{|J|}{2}$.

The zone i) is characterized by the presence of jumps at all "allowed" values of n: $n = 0, \frac{1}{2}, 1, \frac{3}{2}, 2$, while in the zone ii) the jumps are only at $n = 0, 1, 2$. Finally, in the zone iii), the only jumps present are at $n = 0, 2$. The dependence $\mu(n)$ in the three zones can be summarized as follows:

i)

$n=0$	$n=\frac{1}{2}$	$n=1$	$n=\frac{3}{2}$	$n=2$
$\mu<0$	$0<\mu<2A$	$2A<\mu<U+4V-2A$	$U+4V-2A<\mu<U+4V$	$U+4V<\mu$

ii)

$n=0$	$n=1$	$n=2$
$\mu<A$	$A<\mu<U+4V-A$	$U+4V-A<\mu$

iii)

$n=0$	$n=2$
$\mu<\frac{U}{2}+2V$	$\frac{U}{2}+2V<\mu$

We are now in a position to transform from canonical (n fixed) to grand canonical (μ fixed) ensemble. In the zone i) and in the range $0 < n < \frac{1}{2}$, $\mu = 0$ and the free-energy per site (which is a linear function of n) goes from $F_1 = 0$ to $F_2 = -\mu/2|_{\mu=0} = 0$. Therefore, in the above interval F is constantly zero and hence the internal energy per site, which is defined as $E = \frac{1}{N}\left[\langle H\rangle + \mu\sum_i\langle n(i)\rangle\right] = F + \mu n$, is zero as well. Analogously, we can determine the behavior of the internal energy as a function of n in the whole range. These findings for the zones i) and ii) are summarized in Tables 1 and 2. In the zone iii), it is easy to verify that $E = (U + 4V)\frac{n}{2}$ for all $0 < n < 2$.

Table 1. Internal energy as a function of particle concentration $E(n)$ for the Zone i).

$0<n<\frac{1}{2}$	$\frac{1}{2}<n<1$	$1<n<\frac{3}{2}$	$\frac{3}{2}<n<2$
$E=0$	$E=A(2n-1)$	$E=(U+4V)(n-1)+A(3-2n)$	$E=(U+4V)(n-1)$

Table 2. Internal energy as a function of particle concentration $E(n)$ for the Zone ii).

$0<n<1$	$1<n<2$
$E=An$	$E=(U+4V)(n-1)+A(2-n)$

By definition, A takes two different forms depending on whether $\frac{U}{2}$ is greater or not than $V-\frac{|J|}{4}$. This means that everywhere A appears, the two cases should be considered and, therefore, the zones i) and ii) will be further divided in two by the line $\frac{U}{2}=V-\frac{|J|}{4}$. Thanks to the particle-hole symmetry enjoyed by the system, we can consider the phase diagrams just for $n\leqslant 1$. Two cases can be distinguished: $0<n<\frac{1}{2}$ and $\frac{1}{2}<n<1$, together with the particular cases $n=\frac{1}{2}$ and $n=1$. The two phase diagrams are depicted in Fig. 1. We proceed now to the description of the different phases appearing in the whole phase diagram.

Phase α is located in the range $V>\frac{|J|}{4}\wedge U>0$ and $0<n<\frac{1}{2}$. The internal energy is zero because the particles are are far apart to activate any of the Hamiltonian terms. The chemical potential vanishes, $\mu=0$. When $n>\frac{1}{2}$, the former Phase α gives rise to two new phases (η and ζ).

Phase β is located in the range $\frac{U}{2}>|V|-\frac{|J|}{4}\wedge V<\frac{|J|}{4}$. Energy: $E=(V-\frac{|J|}{4})n$ and thus in this phase the singly occupied sites at NN distance interact via density-density and spin-spin interactions. The spin-spin correlation function takes the value:

$$\langle S^z(i)S^z(i+1)\rangle=-\frac{\partial E}{\partial J}=\frac{\mathrm{sign}(J)n}{4}. \tag{8}$$

Hence, if $J<0$, the system is dominated by antiferromagnetic correlations, while if $J>0$ the correlations are ferromagnetic. This phase exists in the range $0<n<1$. Chemical potential $\mu=V-\frac{|J|}{4}$, NN charge-density correlations $\langle n(i)n(i+1)\rangle=n$. This phase exhibits spontaneous magnetization, uniform in case $J>0$, or staggered in case $J<0$.

Phase γ is located in the range $V>0\wedge U<0\wedge\frac{U}{2}<V-\frac{|J|}{4}$. Energy is $E=\frac{U}{2}n$, and therefore in this phase the doubly occupied sites are separated by at least one empty site. This phase exists in the range $0<n<1$. Chemical potential $\mu=\frac{U}{2}$. The only non-zero correlation function is $\langle D(i)\rangle=n/2$.

Phase δ is located in the range $V<0\wedge\frac{U}{2}<-V-\frac{|J|}{4}$ for all $0<n<2$. The energy: $E=(U+4V)\frac{n}{2}$ and thus, in this phase, the doubly occupied sites are placed at the NN distance. Chemical potential $\mu=\frac{U}{2}+2V$. The only non-zero correlation functions are $\langle D(i)\rangle=n/2$, $\langle n(i)n(i+1)\rangle=2n$ and $\langle D(i)D(i+1)\rangle=n/2$.

Phase η is located in the range $V>\frac{|J|}{4}\wedge\frac{U}{2}>V-\frac{|J|}{4}$ and $\frac{1}{2}<n<1$. The energy $E=(V-\frac{|J|}{4})(2n-1)$. This is the second phase with magnetic correlations and is similar to the adjacent Phase β in the sense that in both phases density-density and spin-spin correlations are present, although the expressions for the correlation functions are different: $\langle n(i)n(i+1)\rangle=2n-1$, $\langle S^z(i)S^z(i+1)\rangle=\frac{\mathrm{sign}(J)}{4}(2n-1)$. Chemical potential $\mu=2(V-\frac{|J|}{4})$. There also exists spontaneous magnetization, uniform in case $J>0$, or staggered in case $J<0$.

The fact that in both the Phases β and η the energy depends only on $|J|$ reflects the invariance of the system under a sublattice rotation.

Phase ζ is located in the range $\frac{U}{2} > V - \frac{|J|}{4} \wedge U > 0$ and $\frac{1}{2} < n < 1$. The energy $E = U(n - \frac{1}{2})$. This phase is similar to the Phase γ in the sense explained above. Chemical potential $\mu = U$, while the only non-vanishing correlation function is $\langle D(i) \rangle = n - 1/2$.

The phase boundaries for the phases at $n > 1$ are identical to those at $n < 1$ and the expressions for the internal energy can be obtained from those at $n < 1$ by invoking the particle-hole symmetry:

$$E(2 - n) = E(n) + (U + 4V)(1 - n), \tag{9}$$

while for the chemical potential the particle-hole relation states:

$$\mu(2 - n) = (U + 4V) - \mu(n). \tag{10}$$

3. Conclusions

In this work we present an original method for extrapolating the TM technique at $T = 0$ and use it to obtain the exact $T = 0$ phase diagram of the extended Hubbard U-V-J model in the narrow-band limit. Depending on the values of the Hamiltonian parameters, the orderings involve double occupancy, NN density and spin correlations. We note that the phase diagram boundaries do not depend on the sign of J, although the spin correlations do. The spin correlations appear to be rather fragile in our model: the Ising term in the Hamiltonian acts only if the NN sites are singly occupied. However, in the same conditions the V term is also active, while the U term tends to create either doubly occupied or empty sites (depending on the sign of U). That is why, when either V is large and positive or U is large and negative the spin correlations are completely suppressed. The introduction of J affects only the phases (β and η) where the sites are singly occupied. In these phases, for $J = 0$ [14, 12], the spins are not ordered while, if $J \neq 0$, a ferromagnetic or antiferromagnetic order is established depending on the sign of J.

4. Acknowledgements

We acknowledge the CINECA award under the ISCRA initiative (project BSMOSCES), for the availability of high performance computing resources and support.

References

[1] Bethe H 1931 *Z. Phys.* **71** 205
[2] Mackenzie A P and Maeno Y 2003 *Rev. Mod. Phys.* **75** 657
[3] Saxena S S *et al.* 2000 *Nature* **406** 587
[4] Aoki D *et al.* 2000 *Nature* **413** 613
[5] Pfeiderer C *et al.* 2001 *Nature* **412** 58
[6] Ishiguro T, Yamaji K and Saito G 2001 *Organic Superconductors (Springer Series in Solid-State Sciences)*
[7] Dziurzik C, Japaridze G, Schadschneider A and Zittartz J 2004 *Eur. Phys. J. B* **37** 453
[8] Dziurzik C, Japaridze G I, Schadschneider A, Titvinidze I and Zittartz J 2006 *Eur. Phys. J. B* **51** 41
[9] Giamarchi T 2004 *Quantum Physics in One Dimension* (Oxford University Press) ISBN 0198525001
[10] Baxter R J 2007 *Exactly Solved Models in Statistical Mechanics* (Dover Publications) ISBN 0486462714
[11] Mancini F and Avella A 2004 *Adv. Phys.* **53** 537
[12] Mancini F 2005 *Eur. Phys. J. B* **47** 527
[13] Mancini F, Plekhanov E and Sica G *in preparation*
[14] Mancini F and Mancini F P 2008 *Phys. Rev. E* **77** 061120

A Numerical Study of the Superconducting Proximity Effect in Topological Surface States

Roland Grein[1,2], **Jens Michelsen**[1], **Matthias Eschrig**[2]

[1] Institut für Theoretische Festkörperphysik and DFG-Center for Functional Nanostructures, Karlsruhe Institute of Technology, D-76128 Karlsruhe, Germany

[2] SEPnet and Hubbard Theory Consortium, Department of Physics, Royal Holloway, University of London, Egham, Surrey TW20 0EX, United Kingdom

E-mail: `roland.grein@kit.edu`

Abstract. We study the superconducting proximity effect induced in the surface states of the 3-d topological insulator Bi_2Se_3 by a singlet, s-wave superconductor deposited on its surface. To this effect, the $\mathbf{k}\cdot\mathbf{p}$-Hamiltonian of Bi_2Se_3 and the BCS-Hamiltonian are mapped onto tight-binding chains which we couple through a transfer-Hamiltonian at the interface. We then employ the Recursive Green's Function technique to obtain the local spectral function and infer the dispersion of the interface-states from it. In agreement with earlier microscopic studies of this problem, we find that the Fu-Kane model is a reasonable approximation at energies $\epsilon \ll \Delta_{\rm SC}$. However, for energies close to the SC bulk gap, the Fu-Kane model is expected to break down. Indeed, our numerical calculations show strong modifications of the interface-state dispersion for $\epsilon \gtrsim \Delta_{\rm SC}$. We find that the proximity effect can be strong enough to induce a gap in the surface state that is comparable to the superconducting gap. An analysis of the spatial profile of the states shows that their weight shifts towards the SC as the coupling strength increases. We conclude that an intermediate coupling is ideal for realising the Fu-Kane scenario.

1. Introduction

The research effort devoted to topological insulators has grown rapidly since their initial discovery in 2007 [1] and is presently developing in a multitude of directions [2, 3], one of them being the superconducting proximity effect. It was argued that in superconducting heterostructures involving 3-dimensional topological insulators, the physics of the interface region can effectively be described in terms of a topological superconductor, with Majorana-Fermions at line-junctions and vortex-cores [4]. These excitations obey non-abelian statistics and are thus highly relevant to the field of topological quantum computation [5]. Apart from proximity structures, topological superconductivity can also be realised in non-centrosymmetric materials, which feature spin-current carrying surface states [6]. Such surface states are characteristic for topological materials, where the electronic structure is characterised by non-trivial topological invariants.

In their initial work, Fu and Kane described the proximity effect in a topological insulator by simply including a singlet, s-wave pair-potential in the effective Hamiltonian of the 2-dimensional surface state [4]:

$$\mathcal{H}_{\rm surface} = \begin{pmatrix} H_0(\mathbf{k}) & \Delta i\sigma_2 \\ -\Delta i\sigma_2 & -H_0^*(-\mathbf{k}) \end{pmatrix}, \quad H_0(\mathbf{k}) = v_{\rm F}(k_x\sigma_1 + k_y\sigma_2) - \mu. \tag{1}$$

v_F, Δ and μ can be considered as free parameters that we will use to fit the dispersion of this model to our numerical results.

While the proximity effect is known to induce pairing correlations which suppress the density of states and, under certain circumstances, opens up a gap in the quasiparticle spectrum of an adjacent material, it does not induce a pair potential as such [7]. The latter would require a finite pairing interaction on the normal side, as can easily be seen from the gap-equation. Thus, there was no obvious answer to the question whether this model can be justified from a microscopic theory of the proximity effect, and a first investigation was provided by Stanescu et al. [8]. Their work was based on a tight-binding model describing the whole structure within a transfer-Hamiltonian approach. The authors concluded that the Fu-Kane model is a valid low-energy theory and that the inclusion of a pair-potential in the surface state Hamiltonian can be derived from integrating out the superconducting part of the structure and including it as a self-energy within a Green's function formalism. Since the relevant states are localised at the interface, this self-energy formally appears as an effective order parameter in the surface state Hamiltonian [9]. A similiar study has also been performed for the edge-states of 2D TIs [10].

A different approach was presented a little later by Lababidi et al. [11], using a numerical method to solve the Bogoliubov-de-Gennes equations for a TI-SC heterostructure, where the $\mathbf{k} \cdot \mathbf{p}$-Hamiltonian for Bi_2Se_3 [12, 13] was used to model the TI. For technical reasons, the SC-part of the structure had to be described by a two-band model, with a superconducting conduction band and a split-off valence band, where the conduction band is assumed to be identical to one of the two orbitals of the $\mathbf{k} \cdot \mathbf{p}$-Hamiltonian. The interface is modelled by a step-function dependence of the Hamiltonian parameters. The authors claim that this model would account for the strong-coupling limit of the proximity effect, while the approach of Stanescu et al. was limited to weakly transmitting interfaces. They also reach the conclusion that the Fu-Kane model is a reasonable low-energy approximation, but do not endeavour to extract a surface state Hamiltonian from their calculation. They do, however, perform a self-consistent calculation of the SC-gap and find a suppression in proximity to the interface.

We here propose a different approach to this problem, which can directly be applied to material specific $\mathbf{k} \cdot \mathbf{p}$- or tight-binding-Hamiltonians obtained from fits to ab-initio band structure calculations. Thus, the starting point is similar to the work of Lababidi et al., but we then proceed by solving the problem with a combination of the finite differences and recursive Green's function methods. This has proven to be convenient tool for transport calculations in low-dimensional conductors [14]. By exploiting that translational invariance is only broken in one direction, we show that it is also a very efficient method for solving the problem discussed here. As opposed to the approach of Stanescu et al., this allows us to treat an infinite system, where potential finite size effects are absent. This is not important if the energy gaps on both sides of the interface are of the order of the band gap, and the interface-states are thus strongly localised. This assumption was made by Stanescu et al. However, a realistic SC-gap is in fact a lot smaller than the band-gap of the TI. The interface-states hence extend into the SC on the scale of the coherence length. In our model, there is also no need to maintain the same Hilbert-space dimensionality on both sides of the interface in the continuum models - which appears to be the reason for the inclusion of an additional valence band on the SC-side in the work of Lababidi et al. We show that the proximity induced superconducting gap in the surface states can in principle become comparable to the SC-bulk gap, and that the Fu-Kane model is a sensible low-energy theory if the coupling is not too strong.

2. The model

We start from the continuum model of the $\mathbf{k} \cdot \mathbf{p}$-Hamiltonian on the TI-side and a Bogoliubov-de-Gennes Hamiltonian with a single parabolic band on the SC-side. We consider an infinite

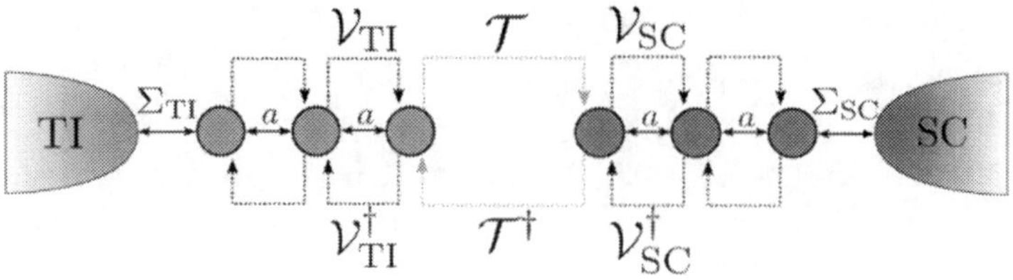

Figure 1. Tight-binding model of the TI-SC heterostructure. A central device, consisting of a one-dimensional tight-binding chain is connected to semi-infinite leads via self-energies that are added to the on-site terms at the respective final sites.

system and the interface is assumed to lie in the $z = 0$-plane. We exploit the translational invariance in the x-y-directions – which implies the conservation of parallel momentum, $k_{||}$, at the interface – by Fourier-transforming the Hamiltonians to real-space coordinates only in the z-direction ($k_z \mapsto -i\partial_z$). Using the method of finite differences [15], we then map the model onto a 1-dimensional tight-binding Hamiltonian which depends parametrically on $k_{||}$. The coupling between the two parts of the system is modelled by a transfer-Hamiltonian of the form:

$$\mathcal{H}_{\rm T} = \mathcal{T}|N+1\rangle\langle N| + \mathcal{T}^\dagger|N\rangle\langle N+1|, \tag{2}$$

coupling the final site N of the TI to the first site $N+1$ of the SC via the matrix element:

$$\mathcal{T} = \frac{t}{2}\left(\mathcal{P}\mathcal{V}_{\rm TI} + \mathcal{V}_{\rm SC}\mathcal{P}\right). \tag{3}$$

Here, $\mathcal{V}_{\rm TI,SC}$ are the respective hopping elements of the TI and the SC derived from the finite differences discretisation (see Fig. 1). $\mathcal{P}$ is a projector connecting the single band of the SC to the two orbitals of the $\mathbf{k}\cdot\mathbf{p}$-Hamiltonian. This allows for a greater flexibility in the description of orbital-coupling compared to the approach of Lababidi et al. The inclusion of a second 'dummy' band on the SC side is also not necessary, since we are not forced to maintain the same Hilbert-space dimensionality on both sides of the interface. We found that coupling the SC-band to the TI-orbital with positive parity ('+'-orbital in Ref. [12]) is most effective. We hence consider a $\mathcal{P}$ here, where the matrix elements between the '+'-orbital and the SC-band are equal to 1 and all other elements zero. The parameter t allows us to tune the coupling strength. In the limit $t \to 0$, the two subsystems are decoupled and the surface state dispersion of a vacuum boundary is recovered. It agrees with the dispersion obtained directly from the continuum model. For $t \to \infty$, the proximity effect is fully absent as well, but the dispersion is slightly modified compared to the vacuum boundary solution. To investigate the spectral properties of the system, we calculate the local spectral function, which is obtained with a recursive Green's function approach [16]. This technique allows us to describe an infinite tight-binding (TB) system by coupling a finite tight-binding chain to semi-infinite leads on both sides of the interface. The devision of the infinite TB-chain into a 'central device' and 'lead' regions is arbitrary, since the procedure of integrating out the leads is exact and not an approximation. The only requirement is that the leads are homogenous, i.e., the interface must be part of the central device (for the case of a homogeneous order order parameter in the SC; otherwise the region of varying order parameter must be included in the central region). The Green's function of the central device coupled to the leads is then formally defined as:

$$\mathcal{G}(\mathbf{k}_{||},\epsilon) = (\epsilon + i\eta - \mathcal{H}(\mathbf{k}_{||}) - \Sigma_{\rm SC}(\mathbf{k}_{||},\epsilon) - \Sigma_{\rm TI}(\mathbf{k}_{||},\epsilon))^{-1}, \tag{4}$$

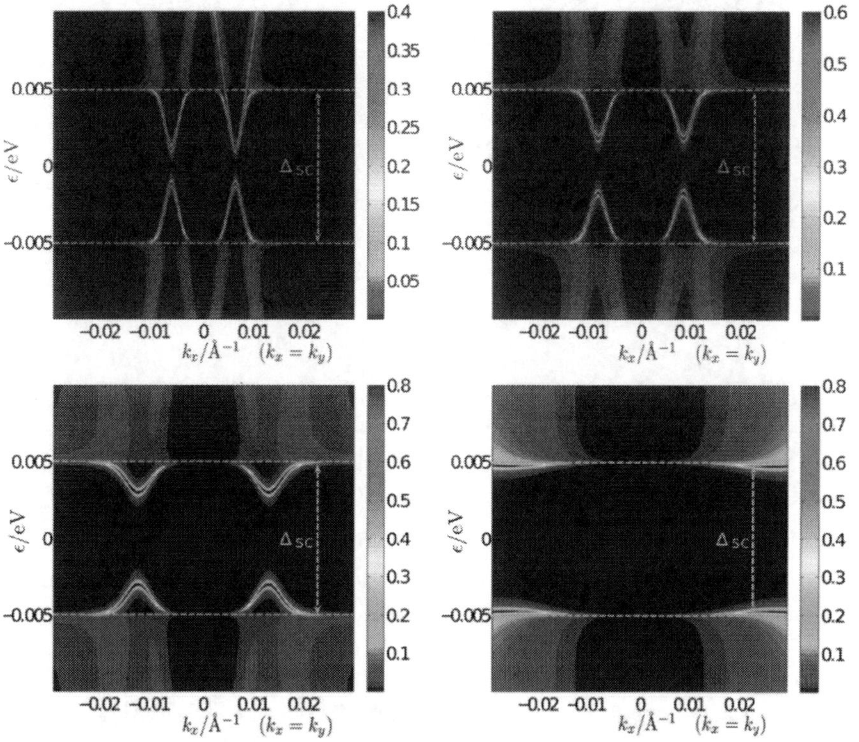

Figure 2. The spectral function $A_N(\mathbf{k}_{||}, \epsilon)$ as a function of momentum and energy along the cut $k_x = k_y$ for $t = 0.92$ (upper-left), $t = 0.9$ (upper-right), $t = 0.88$ (lower-left) and $t = 0.86$ (lower-right). The upper-left plot also shows a tentative fit to one branch of the Fu-Kane dispersion (pink line).

where $\Sigma_{\rm SC}$ and $\Sigma_{\rm TI}$ are the self-energies of the SC and TI respectively, and $\mathcal{H}$ is the tight-binding, Bogoliubov-de-Gennes (BdG) Hamiltonian of the central device:

$$\mathcal{H}(\mathbf{k}_{||}) = \mathcal{H}_{\rm SC}(\mathbf{k}_{||}) + \mathcal{H}_{\rm TI}(\mathbf{k}_{||}) + \mathcal{H}_{\rm T}(\mathbf{k}_{||}). \tag{5}$$

The local spectral function is given by:

$$A_n(\mathbf{k}_{||}, \epsilon) = -\frac{1}{2\pi a}{\rm Im\ Tr}[\langle n|\mathcal{G}(\mathbf{k}_{||})|n\rangle]. \tag{6}$$

a is the lattice spacing of the TB-model. Note that we take the full trace of $\mathcal{G}$, i.e., the Gor'kov Green's function, to map out the whole dispersion of the BdG-Hamiltonian. This implies the presence of hole-like branches in the dispersions we plot. We use $a = 0.05$ Å and $\eta = 10^{-4}$ eV in the following.

3. Results

We use the parameters provided in Ref. [13] for the Bi_2Se_3 $\mathbf{k}\cdot\mathbf{p}$-Hamiltonian, but the value of C_0 is shifted so that the chemical potential lies at the crossing point of the surface-state dispersion. The chemical potential of Bi_2Se_3 usually lies above the bottom of the conduction band, i.e., the material is in fact not an insulator [17, 18]. However, chemical doping [18] or gate-control [19] have been reported to allow for tuning the chemical potential. For the superconductor model,

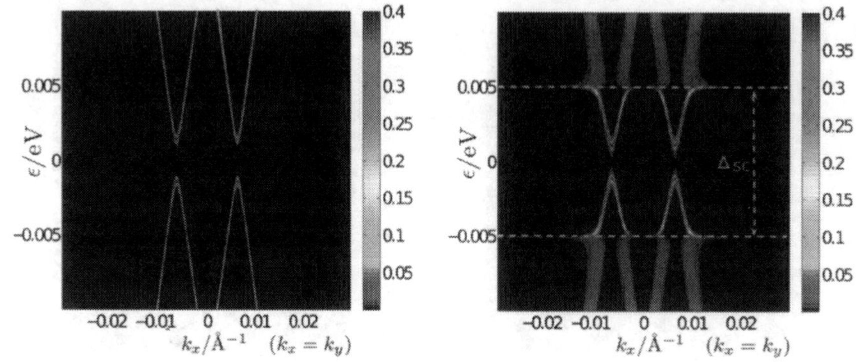

Figure 3. The right plot is the same as the upper-left one in Fig. 2 ($t = 0.92$). The left plot shows the same calculation with the assumption$\Sigma_{SC}(\mathbf{k}_{||}, \epsilon) = \Sigma_{SC}(\mathbf{k}_{||} = 0, \epsilon = 0)$. The indicated fit to the Fu-Kane dispersion matches very well in the energy range shown here and is the same as the one depicted in Fig. 2.

we assume a bulk-gap of $\Delta_{SC} = 5$ meV, a Fermi-energy $E_F = 0.4$ eV, and an effective mass equal to the free electron mass $\hbar^2/2m = 3.81$ eVÅ^2. For other values of m, we obtain qualitatively the same results. To analyse the surface-state dispersion, we plot the spectral function $A_N(\epsilon, \mathbf{k}_{||})$, N being the site of the TI next to the interface, in Fig. 2 for different values of the interface parameter t and on an energy scale comparable to the SC-gap. The band-gap of the TI is a lot larger than the SC-gap (about 0.3-0.35 eV [17, 18]). Spectral weight from the bulk states of the TI is outside the energy window shown in Fig. 2. Well-defined interface states in the TI only exist for energies smaller than the SC bulk-gap. Above this energy, the surface states of the TI hybridise with the continuum of itinerant SC-states, which results in the broadening of the spectral function above Δ_{SC}. A tentative fit to one branch of the Fu-Kane model is indicated in the upper-left plot. We observe that the avoided crossing at the chemical potential of the interface states is shifted away from the Γ-point. This is consistent with the shift of μ in the Fu-Kane model reported by Lababidi et al. The gap induced in the dispersion of interface states by the superconducting proximity effect is in principle only limited by the SC-gap itself. This strong-coupling limit would, however, require a self-consistent calculation of the SC-gap, which is feasible within our approach, but we have not done it yet. Note also that the dispersion is strongly modified at the energy of the SC-gap and that the Fu-Kane model breaks down in this energy range. In the range of very strong coupling, which we reach for about $t = 0.86 - 0.84$ with our choice of parameters, the interface states merge into the bulk-DOS of the SC, which leaks into the TI on a length scale given by the band-gap. In this case, there are basically no quasiparticle excitations anymore that can be considered interface states. For even smaller values of t the proximity effect becomes weaker again. The calculation of these dispersions is highly effective in our approach, as considering a single site coupled to the semi-infinite leads on the left and right side is sufficient to obtain $A_N(\epsilon, \mathbf{k}_{||})$. When a self-consistent calculation of the SC order parameter is included, the TB-chain must be extended into the superconductor until the point where Δ_{SC} recovers its bulk-value.

While the agreement of the interface state dispersion is a necessary condition for proving the validity of the Fu-Kane model, it is, in principle, not a sufficient one. To complete this prove, an effective Hamiltonian for the interface states must be extracted from the microscopic calculation. This can be achieved by formulating the proximity of the superconductor in terms of a self-energy, i.e. coupling the final TI-site directly to the SC-lead, and then approximating $\Sigma_{SC}(\mathbf{k}_{||}, \epsilon) \approx \Sigma_{SC}(\mathbf{k}_{||} = 0, \epsilon = 0)$. This is illustrated in Fig. 3, were we show that the dispersion

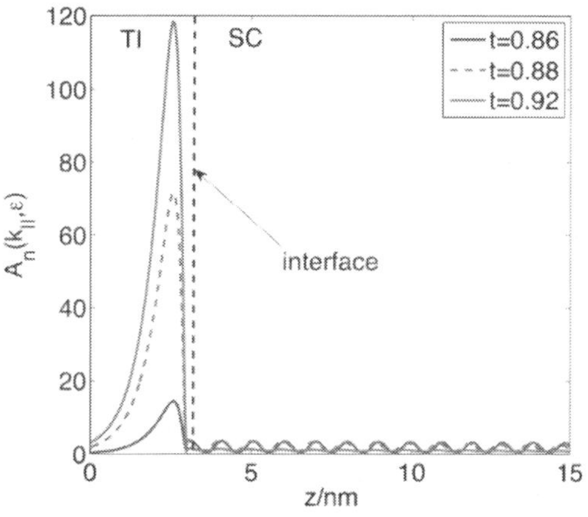

Figure 4. Spatial profile of the states at the bottom of the surface-state dispersion for $t = 0.92, 0.88, 0.86$. Specifically, $k_x = k_y = 0.0064$ Å^{-1} (red), 0.013 Å^{-1} (green), 0.0275 Å^{-1} (blue); and $\epsilon = 1.13$ meV (red), $= 3.03$meV (green), $= 4.74$ meV (blue). The interface is located at $z = 3$ nm.

thus obtained agrees very well with the Fu-Kane dispersion. The off-diagonal components in particle-hole space of the self-energy $\Sigma_{\rm SC}$ do then indeed have the structure of a singlet, s-wave gap. A surface-state Hamiltonian could now be obtained by considering a *finite* TI tight-binding model, were the self energy $\Sigma_{\rm SC}(\mathbf{k}_{||} = 0, \epsilon = 0)$ is added to the site at the interface. This TB-Hamiltonian can then be diagonalised exactly and projected onto the interface states, which would provide the interface state Hamiltonian.

The spatial profile of the states can also be inferred from the spectral function, as shown in Fig. 4. For weak coupling, the interface states are mainly localised on the TI-side, but as the coupling increases, their weight shifts towards the SC. The decay length on either side of the interface is governed by the energy $\Delta_{\rm SC,TI} - \epsilon$, where $\Delta_{\rm TI,SC}$ is the respective bulk energy gap, and ϵ is the energy of the state. The band gap $\Delta_{\rm TI}$ is of the order 0.1eV, while the SC gap is typically two orders of magnitude smaller. Thus, we hardly see any decay on the superconducting side on the interatomic length-scale shown in Fig. 3, in particular when the coupling is strong and the band bottom of the interface state dispersion is close to the SC-gap.

4. Conclusions

We presented an efficient numerical approach to the proximity effect in topological surface states, based on the recursive Green's function technique within the transfer-Hamiltonian approach. The underlying tight-binding model can be derived from ab-initio bandstructure calculations and thus allows for material specific calculations in a straight-forward manner. We showed that the only fundamental limitation to the proximity induced gap in the surface states of the TI is the SC energy-gap itself, but an intermediate coupling regime is most desirable to achieve a superconducting interface state which is well localised and reasonably described by the Fu-Kane model at small energies. We showed that in the case of weak transmission, approximating the self-energy by its zero energy, zero momentum value reproduced the Fu-Kane dispersion to a high precision and argue that this can likely be used to extract an effective interface-Hamiltonian from our calculation.

Acknowledgements
RG acknowledges financial support from the Karlsruhe House of Young Scientists (KHYS) and the South East Physics Network (SEPnet).

References

[1] Koenig M, Wiedmann S, Bruene C, Roth A, Buhmann H, Molenkamp L W, Qi X L and Zhang S C 2007 *Science* **318** 766–70
[2] Hasan M Z and Kane C L 2010 *Rev. Mod. Phys.* **82** 3045–67
[3] Qi X L and Zhang S C 2011 *Rev. Mod. Phys.* **83** 1057–110
[4] Fu L and Kane C L 2008 *Phys. Rev. Lett.* **100** 096407
[5] Nayak C, Simon S H, Stern A, Freedman M and Das Sarma S 2008 *Rev. Mod. Phys.* **80** 1083–159
[6] Vorontsov A B, Vekhter I and Eschrig M 2008 *Phys. Rev. Lett.* **101** 127003
[7] De Gennes P G 1964 *Rev. Mod. Phys.* **36** 225–37
[8] Stanescu T D, Sau J D, Lutchyn R M and Das Sarma S 2010 *Physical Review* B **81** 241310
[9] Sau J D, Lutchyn R M, Tewari S and Das Sarma S 2010 *Physical Review B* **82** 094522
[10] Black-Schaffer A M 2011 *Physical Review B* **83** 060504
[11] Lababidi M and Zhao E 2011 *Physical Review* B **83** 184511
[12] Zhang H, Liu C X, Qi X L, Dai X, Fang Z and Zhang S C 2009 *Nat. Phys.* **5** 438–42
[13] Liu C X, Qi X L, Zhang H, Dai X, Fang Z and Zhang S C 2010 *Phys. Rev.* B **82** 045122
[14] Ferry D and Goodnick S 1997 *Transport in nanostructures* (Cambridge University Press)
[15] Datta S 1995 *Electronic Transport in Mesoscopic Systems* (Cambridge University Press)
[16] Metalidis G and Bruno P 2005 *Physical Review* B **72** 235304
[17] Xia Y, Qian D, Hsieh D, Wray L, Pal A, Lin H, Bansil A, Grauer D, Hor Y S, Cava R J and Hasan M Z 2009 *Nature Physics* **5** 398–402
[18] Hsieh D, Xia Y, Qian D, Wray L, Dil J H, Meier F, Osterwalder J, Patthey L, Checkelsky J G, Ong N P, Fedorov A V, Lin H, Bansil A, Grauer D, Hor Y S, Cava R J and Hasan M Z 2009 *Nature* **460** 1101–U59
[19] Chen J, Qin H J, Yang F, Liu J, Guan T, Qu F M, Zhang G H, Shi J R, Xie X C, Yang C L, Wu K H, Li Y Q and Lu L 2010 *Phys. Rev. Lett.* **105**

Journal of Physics: Conference Series **391** (2012) 012150 doi:10.1088/1742-6596/391/1/012150

Majorana fermions in topological superconductors with spin-orbit interaction

Masatoshi Sato

Institute for Solid State Physics, University of Tokyo, Kashiwanoha 5-1-5, Kashiwa, Chiba, 277-8581, Japan

E-mail: `msato@issp.u-tokyo.ac.jp`

Abstract. When a topological order is realized in a certain class of superconductors, it supports Majorana fermions on the edge or in the vortex core. Vortices with Majorana fermion modes are neither fermions nor bosons but non-Abelian anyons, obeying the non-Abelian statistics for which the exchange operations of particles are not commutative. Because of this remarkable feature, Majorana fermions plays an important role for the realization of fault-tolerant topological quantum computation.

Here I would like to report our recent work on Majorana fermions and non-Abelian anyons in topological superconductors with spin-orbit interaction [1, 2, 3, 4, 5, 6]. Until recently, only spin-triplet superconducting states, such as such $\nu = 5/2$ fractional quantum Hall states or Sr_2RuO_4, had been known to exhibit Majorana fermions [7, 8, 9, 10, 11]. However, now it is revealed that spin-singlet superconductors also support Majorana fermions if we take into account the spin-orbit interaction. While a possible realization of non-Abelian anyon in an s-wave superconducting state was first discussed in the context of cosmology [1], by using a zero mode solution in a vortex by Jackiw and Rossi [12] (or its non-gauged version [13]), now there are many examples of such systems in condensed matter physics. In particular, a scheme using the spin-orbit interaction and Zeeman one, which was proposed in [2, 3, 4], is powerful to realize Majorana fermions in various systems. It is shown that Majorana fermions are realized in full-gapped s-wave superfluids/superconductors [3, 4, 14, 15]. Also, Majorana fermions is shown to possible even in nodal superconductors such as high-T_c cuprates [5].

We also show that a peculiar Majorana fermion is obtained in a certain non-centrosymmetric superconductor [6, 16, 17]. In sharp contrast to ordinary Majorana fermions, the novel Majorana fermion has a flat dispersion and preserves the time-reversal invariance.

1. What is Majorana fermion

First, I would like to explain briefly what a Majorana fermion is. The definition of a Majorana fermion is very simple. A Majorana fermion is a Dirac fermion which satisfies the Majorana condition. Its Hamiltonian is given by the Dirac Hamiltonian,

$$\mathcal{H}(\boldsymbol{k}) = \boldsymbol{\sigma} \cdot \boldsymbol{k} \quad (2\text{D}) \quad \text{or} \quad \mathcal{H}(k_x) = ck_x \quad (1\text{D}), \tag{1}$$

and its wavefunction Ψ satisfies the Majorana condition

$$\Psi = C\Psi^*, \tag{2}$$

with a constant matrix C. The latter equation implies that the particle Ψ is identified with its anti-particle Ψ^*. In other words, the Majorana fermion is its own anti-particle. Originally, the

Majorana fermion was conceived as an elementary particle like a neutrino, but now it has been known that the Majorana fermion can be realized in topological superconductors [18].

A representative example of superconductors with Majorana fermion is a two-dimensional chiral p-wave superconductor [7]. The chiral superconductor is analogous to the quantum Hall state. Like a quantum Hall state, its bulk wave function has a non-zero Chern number and a one-dimensional Dirac fermion is realized as a topologically protected edge state. Furthermore, the Dirac fermion naturally satisfies the Majorana condition due to the superconductivity: In a superconductor, an electron is scattered into its anti-particle, or vice versa, by forming or destroying a Cooper pair. Thus, in the background of many Cooper pairs, a particle excitation cannot be distinguished from its anti-particle, thus it satisfies the Majorana condition naturally.

In a two-dimensional chiral p-wave superconductor, there also exist a Majorana fermion in a vortex. In two dimensions, a vortex can be considered as a small hole in the superconductor [7]. Thus as an edge state of the small hole, one obtains a topologically stable zero mode in a vortex. The zero mode γ_0 also satisfies the Majorana condition, which is given by

$$\gamma_0^\dagger = \gamma_0. \tag{3}$$

At first sight, the Majorana condition (3) looks problematic. For an ordinary zero mode, $\gamma_0^\dagger$ is considered as the creation operator for the zero mode, but with the condition (3), this gives rise to contradiction, *i.e*, the creation operator $\gamma_0^\dagger$ becomes the annihilation operator γ_0 at the same time. Fortunately, the apparent contradiction can be avoided if we consider a pair of vortices [7, 8]. From the zero modes $\gamma_0^{(i)}$ $(i = 1, 2)$ for a pair of vortices 1 and 2, one can construct a well-defined creation operator $\gamma^\dagger$,

$$\gamma^\dagger = \frac{\gamma_0^{(1)} + i\gamma_0^{(2)}}{\sqrt{2}}. \tag{4}$$

The creation operator $\gamma^\dagger$ is not the same as its conjugate annihilation operator γ and it satisfies the standard anti-commutation relation,

$$\left\{\gamma^\dagger, \gamma\right\} = 1. \tag{5}$$

Here one should note that spatially separate vortices are needed to define the creation operator. The nonlocal definition gives rise to a nonlocal quantum correlation between vortices, which changes the statistical properties of the vortices drastically. Actually the vortices obey a new kind of statistics called the non-Abelian anyon statistics [8]. Unlike ordinary fermions/bosons or Abelian anyons, one can create many degenerate states and manipulate them by exchanging non-Abelian anyons. The non-Abelian anyons are of particular interest since they can be considered as qubits for fault-tolerant quantum computers [19].

From the reason above, the two-dimensional chiral p-wave superconductor is very interesting. However, at the same time, the chiral p-wave superconductor is very special. 1) The chiral p-wave superconductor supports spin-triplet Cooper pairs. At present, however, only a few such as ^{3}He superfluids or Sr_2RuO_4 have been widely accepted as spin-singlet superconductors. 2) The chiral p-wave superconductor is a full gapped but unconventional superconductor, while most unconventional superconductors including the high T_c cuprate have nodes in their gap functions. 3) Finally, the chiral p-wave superconductor breaks the time-reversal symmetry. Thus a natural question arises: Which property is essential to realize the Majorana fermion ? In the following, I will show that the answer is none of them: Majorana fermions are possible in spin-singlet superconductors. Furthermore, nodal superconductors also can support Majorana fermions. And even for time-reversal invariant system, we may have a peculiar time-reversal invariant Majorana fermion. Thus, Majorana fermions can be obtained without any special property of the chiral p-wave superconductivity. However, as I will show below, we find that the spin-orbit interaction is indispensable to realize Majorana fermions instead.

2. Majorana fermion in spin-singlet superconductors

Here I will show that Majorana fermion and non-Abelian anyon are possible even for spin-singlet s-wave superconducting states [1, 20, 3, 4, 14, 15]. This possibility was first considered in Ref.[1].

The basic idea of Ref.[1] is to consider a two dimensional Dirac fermion coupled with an s-wave condensate Φ,

$$\mathcal{H} = \begin{pmatrix} -i\sigma_i\partial_i & \Phi^* \\ \Phi & -i\sigma_i\partial_i \end{pmatrix} \tag{6}$$

where σ_i $(i = 1, 2)$ is the Pauli matrices. From the analysis of Jackiw and Rossi [12] (and its non-gauged version [13]), it has been known that there exist a zero mode in the vortex configuration given by $\Phi = \Phi_0 f(r)e^{i\theta}$. Therefore, imposing the Majorana condition on the system, one has a Majorana zero mode in a vortex, which implies that the vortices are non-Abelian anyons [1].

When this idea was considered in 2003, no condensed matter system supporting a single Dirac fermion was known. Thus the above idea was applied to a cosmic string called axion string [1]. After the publication of Ref.[1], however, it was found that a single Dirac fermion is naturally realized on a surface of a topological insulator. Actually, Fu and Kane showed, independently of Ref.[1], that a similar Hamiltonian is realized in the interface between a topological insulator and an s-wave superconductor [20]. They also showed that the vortex obeys non-Abelian anyon statistics in the same reason mentioned above. Furthermore, the idea was generalized to unconventional superconductors [21]. Here one should note that the spin-orbit interaction is very important since the spin-orbit interaction is a key gradient to to realize topological insulators.

In the above scheme of non-Abelian anyons, we need a special system which supports a two dimensional Dirac fermion. However, if one consider an two dimensional s-wave superconductor with Rashba spin-orbit interaction, one can realize a Majorana fermion from electrons with a usual parabolic dispersion [3, 4]. The model Hamiltonian considered in Refs.[3, 4] is

$$\mathcal{H}(\boldsymbol{k}) = \begin{pmatrix} \varepsilon_{\boldsymbol{k}} - h\sigma_z + \boldsymbol{g}_{\boldsymbol{k}}\cdot\boldsymbol{\sigma} & i\Delta\sigma_y \\ -i\Delta\sigma_y & -\varepsilon_{\boldsymbol{k}} + h\sigma_z + \boldsymbol{g}_{\boldsymbol{k}}\cdot\boldsymbol{\sigma}^* \end{pmatrix} \tag{7}$$

where $\boldsymbol{k} = (k_x, k_y)$ is the momentum in the xy plane, $\varepsilon_{\boldsymbol{k}}$ the electron energy measured from the Fermi energy, Δ the s-wave gap function, and $\boldsymbol{g}_{\boldsymbol{k}} \sim \lambda(k_y, -k_x)$ the Rashba spin-orbit interaction. We also consider the Zeeman magnetic field $h = \mu_{\rm B}H_z$ in the z-direction. Contrary to the Hamiltonian (6), we consider here that $\varepsilon_{\boldsymbol{k}}$ is parabolic around $\boldsymbol{k} = 0$ as a usual electron. To see the basic idea of Ref.[3], perform the following unitary transformation

$$\mathcal{H}^{\rm D}(\boldsymbol{k}) = D\mathcal{H}(\boldsymbol{k})D^\dagger, \quad D = \frac{1}{\sqrt{2}}\begin{pmatrix} 1 & i\sigma_y \\ i\sigma_y & 1 \end{pmatrix}, \tag{8}$$

which leads to the dual Hamiltonian $\mathcal{H}^{\rm D}$,

$$\mathcal{H}^{\rm D}(\boldsymbol{k}) = \begin{pmatrix} \Delta - h\sigma_z & -i\varepsilon_{\boldsymbol{k}}\sigma_y - i\boldsymbol{g}_{\boldsymbol{k}}\cdot\boldsymbol{\sigma}\sigma_y \\ i\varepsilon_{\boldsymbol{k}}\sigma_y + i\boldsymbol{g}_{\boldsymbol{k}}\sigma_y\cdot\boldsymbol{\sigma} & -\Delta + h\sigma_z \end{pmatrix}. \tag{9}$$

Interestingly, the spin-triplet gap function $-i\boldsymbol{g}_{\boldsymbol{k}}\cdot\boldsymbol{\sigma}\sigma_y$ is induced in the dual Hamiltonian, by the spin-orbit interaction of the original Hamiltonian. This implies that the original Hamiltonian $\mathcal{H}(\boldsymbol{k})$ also has a property similar to the spin-triplet superconductor in the presence of the spin-orbit interaction. Actually, we can show that gapless Majorana edge state appears like a two dimensional chiral p-wave superconductor if the Zeeman magnetic field exceeds the critical value [3, 4],

$$h > \sqrt{\Delta^2 + \mu^2}. \tag{10}$$

One can also show that a vortex in the system support a Majorana zero mode, so it becomes a non-Abelian anyon in this parameter region [3, 4, 14]. The bulk wave function has a non-zero Chern number in this parameter region [3, 4], thus the state is topologically equivalent to a chiral p-wave superconductor.

Here one might notice that a rather strong magnetic field, which is larger than the gap function Δ, is needed to satisfy the condition (10). Under such a strong magnetic field, the orbital depairing effect may break the superconductivity, although the Pauli depairing effect is strongly suppressed due to the Rashba spin-orbit interaction [22, 23]. This difficulty, however, has been resolved by various systems, (i)s-wave superfluids in neutral ultracold fermionic atoms with lager generated Rashba spin-orbit interaction [3, 4], (ii) heterostructure semiconductor devices [14, 15], and (iii) heavy fermion systems [4].

Recently, a similar idea was applied to nanowire systems [9, 24], and it was found that the same condition (10) is required to obtain Majorana fermions at the boundary of the semiconductor nanowire on an s-wave superconductor [25, 26, 27].

3. Majorana fermion in nodal superconductors

Now we show that Majorana fermions are possible even in nodal superconductors [5]. The model we consider is a two-dimensional d-wave superconductor with the Rashba spin-orbit interaction and Zeeman term,

$$\mathcal{H}(\boldsymbol{k}) = \begin{pmatrix} \varepsilon_{\boldsymbol{k}} - h\sigma_z + \boldsymbol{g}_{\boldsymbol{k}} \cdot \boldsymbol{\sigma} & i\Delta(\boldsymbol{k})\sigma_y \\ -i\Delta(\boldsymbol{k})\sigma_y & -\varepsilon_{\boldsymbol{k}} + h\sigma_z + \boldsymbol{g}_{\boldsymbol{k}} \cdot \boldsymbol{\sigma}^* \end{pmatrix}. \tag{11}$$

The Hamiltonian is almost the same as Eq.(7), but instead of an s-wave gap function Δ, we consider the d-wave gap functions,

$$\Delta(\boldsymbol{k}) \sim \Delta_0 k_x k_y, \quad \text{or} \quad \Delta(\boldsymbol{k}) \sim \Delta_0(k_x^2 - k_y^2). \tag{12}$$

Gap-nodes exist in both d-wave pairings.

To understand what happens, we can use the dual transformation again. In a manner similar to (9), the spin-triplet gap function $-i\boldsymbol{g}_{\boldsymbol{k}} \cdot \boldsymbol{\sigma}\sigma_y$ is induced in the dual Hamiltonian by the Rashba spin-orbit interaction. Therefore, one can expect that the original Hamiltonian is topologically similar to the spin-triplet superconductor. Indeed, although there are bulk nodes in the system, a gapless Majorana edge state appears like a chiral p-wave superconductor if the Zeeman magnetic field satisfies

$$h > \mu. \tag{13}$$

We can also show that there exist a single Majorana zero mode in a vortex in this parameter region, thus the vortex obeys non-Abelian anyon statistics [5]. On contrary to the s-wave case (see (10)), the condition (13) does not depend on the magnitude of the gap function Δ_0. Therefore, by choosing a system with small μ, the Majorana fermion can be realized in a weak magnetic field. The topological phase in nodal superconductors can be characterized by the parity of the Chern number [5].

4. Time-reversal invariant Majorana fermions

Finally, I would like to show the existence of Majorana fermion in two dimensional time-reversal invariant superconductors [6, 16, 17]. Usually, the time-reversal breaking is necessary to realize Majorana fermion in two dimensional superconductors. This is because the Majorana fermion on the edge has a linear dispersion

$$E = ck_y, \tag{14}$$

and it already breaks the time-reversal invariance. However, if the coefficient c vanishes, and the dispersion becomes flat,

$$E = 0 \tag{15}$$

then the Majorana fermion can be obtained even in time-reversal invariant systems. In Ref.[6], we have shown that such a Majorana fermion with flat dispersion is realized in a special kind of two dimensional noncentrosymmetric superconductors.

We found that the spin-orbit interaction is indispensable to realize the dispersionless Majorana fermion in our model. Without the spin-orbit interaction, the single branch of the edge state with flat dispersion disappears, and no Majorana fermion is obtained [6, 16].

5. Summary

We show that Majorana fermions can be realized in various two dimensional superconductors other than chiral p-wave ones, if we take into account the spin-orbit interaction. In particular, s-wave superconductors or d-wave superconductors in high T_c cuprate can support Majorana fermions and non-Abelian anyons. From the stableness of s-wave pairing against disorder or a large superconducting gap of high T_c cuprate, the experimental detection of non-Abelian anyons in these schemes is expected to be feasible in comparison with that in chiral p-wave superconductors.

Acknowledgments

The author thanks S. Fujimoto, Y. Takahashi, Y. Tanaka, K. Yada, T. Yokoyama, and Y. Mizuno for collaboration on which this work is based. This work was supported by the Grant-in Aid for Scientific Research from MEXT of Japan "Topological Quantum Phenomena" No. 22103005 and No.22540383.

[1] M. Sato, Phys. Lett. B **575**, 126 (2003).
[2] M. Sato and S. Fujimoto, Phys. Rev. B **79**, 094504 (2009).
[3] M. Sato, Y. Takahashi, and S. Fujimoto, Phys. Rev. Lett. **103**, 020401 (2009)
[4] M. Sato, Y. Takahashi, and S. Fujimoto, Phys. Rev. B **82**, 134521 (2010) (Editor's suggestion).
[5] M. Sato and S. Fujimoto, Phys. Rev. Lett. **105**, 217001 (2010).
[6] Y. Tanaka, Y. Mizuno, T. Yokoyama, K. Yada, and M. Sato, Phys. Rev. Lett. **105**, 097002 (2010).
[7] N. Read and D. Green, Phys. Rev. B **61**, 10267 (2000).
[8] D. A. Ivanov, Phys. Rev. Lett. **86**, 268 (2001).
[9] A.Y.Kitaev, Physics-Uspekhi **44**, 131 (2001).
[10] M. Sato, Phys. Rev. B **79**, 214526 (2009).
[11] M. Sato, Phys. Rev. B **81**, 220504(R) (2010).
[12] R. Jackiw and P. Rossi, Nucl. Phys. B **190**, 681 (1981).
[13] C. G. Callan, Jr. and J. A. Harvey, Nucl. Phys. B **250**, 427 (1985).
[14] J. D. Sau et. al., Phys. Rev. Lett. **104**, 040501 (2010).
[15] J. Alicea, Phys. Rev. B **81**, 125318 (2010).
[16] K. Yada, M. Sato, Y. Tanaka, and T. Yokoyama, Phys. Rev. B **83**, 064505 (2011).
[17] M. Sato, K. Yada, Y. Tanaka, and T. Yokoyama, Phys. Rev. B **83**, 224511 (2011).
[18] For recent reviews, X. -L. Qi and S. C. Zhang, arXiv:1008.2026; Y. Tanaka, M. Sato, and N. Nagaosa, arXiv:1105.4700.
[19] M. Freedman, A. Kitaev, M. Larsen, and Z. Wang, Bull. Am. Math. Soc. **40**, 31 (2003).
[20] L. Fu and C. Kane, Phys. Rev. Lett. **100**, 096407 (2008).
[21] J. Linder, Y. Tanaka, T. Yokoyama, A. Sudbo, and N. Nagaosa, Phys. Rev. Lett. **104**, 067001 (2010).
[22] P. A. Frigeri, D. F. Agterberg, A. Koga, and M. Sigrist, Phys. Rev. Lett. **92**, 097001 (2004).
[23] S. Fujimoto, J. Phys. Soc. Jpn. **76**, 051008 (2007).
[24] A. C. Potter and P. A. Lee, Phys. Rev. Lett. **105**, 227003 (2010).
[25] Y. Oreg, G. Refael, and F. von Oppen, Phys. Rev. Lett. **105**, 177002 (2010).
[26] R. M. Lutchyn, T. Stanescu, and S. Das Sarma, Phys. Rev. Lett. **106**, 127001 (2011).
[27] J. Alicea, Y. Oreg, G. Rafael, F. von Oppen, and M. P. A. Fisher, Nat. Phys. **7**, 412 (2011).

A 4-pole approach to the Hubbard model within the Composite Operator Method

Adolfo Avella[1,2,3], Ferdinando Mancini[1,3] and Gerardo Sica[1]

[1] Dipartimento di Fisica "E.R. Caianiello", Università degli Studi di Salerno, I-84084 Fisciano (SA), Italy

[2] CNR-SPIN, UoS di Salerno, I-84084 Fisciano (SA), Italy

[3] Unità CNISM di Salerno, Università degli Studi di Salerno, I-84084 Fisciano (SA), Italy

E-mail: avella@sa.infn.it

Abstract. The Hubbard model is studied within the Composite Operator Method framework in a 4-pole approximation. The operatorial basis is chosen according to both the hierarchy of equations of motion and the exact solution of the model reduced to the minimal cluster (2 sites) where all Hamiltonian terms are still active. Such a recipe amounts to include into the basis not only the two Hubbard operators, which usually take care of the scale of energy related to the on-site Coulomb repulsion U, but also two other non-local operators describing the Hubbard operators dressed by charge, spin and pair fluctuations on the nearest-neighboring sites and addressing the dynamically generated scale of energy related to the exchange $J = 4t^2/U$. The calculation framework is outlined and a very first comparison to numerical simulations, showing very good qualitative and quantitative agreement, is reported. As reference, the results in the 2-pole approximation are also reported and the differences discussed.

1. Introduction

Strongly correlated systems in solid state physics and condensed matter theory pose such a difficult, but inspiring, challenge to be by far one of the research fields with the higher number of active scholars. Among all other open quests in the field, the one for an analytical calculation framework capable to catch the most relevant features of the many model Hamiltonians proposed in the literature is surely very exciting and fundamental to any advancement in the understanding of these systems. The Composite Operator Method (COM) [1, 2] has proved to be quite efficient in capturing the anomalous unconventional features [3, 4] of the Hubbard model [5, 6], which is the prototype of all strongly correlated systems in absence of extrinsic magnetic defects. COM is based on three main ingredients: the use of a basis, for the relevant Green's functions, consisting of composite fields (that is, fields obtained by the composition of canonical electronic fields) embedding as much as possible of the interactions under study; the use of Algebra Constraints [1, 2] to impose relevant symmetries - secure the correct Hilbert space as well as to determine the unknowns of the theory; the choice of an appropriate (with respect to the specific case under study) treatment of the residual self-energy [1, 2] in order to describe the interactions among the composite field adopted in the basis.

In this short manuscript, we outline a calculation framework where the Composite Operator Method is applied to the two-dimensional Hubbard model: a basis of four composite fields is chosen, the unknowns of the theory are fixed by means of Algebra Constraints as well as by exploiting the knowledge of the exact solution on two sites, the residual self-energy is neglected so that the approximation used falls into the 4-pole class. A very first comparison to numerical simulation data (specifically quantum Monte Carlo ones) for the double occupancy is reported and analysed. 2-pole approximation results, obtained within the same COM framework, are also reported in order to establish the main differences and the quantitative and qualitative improvements obtained by augmenting the number of poles.

2. Formulation

2.1. The Hamiltonian

The Hamiltonian of the 2-dimensional Hubbard model reads as

$$H = \sum_{\mathbf{ij}} \left(-4t\alpha_{\mathbf{ij}} - \mu\delta_{\mathbf{ij}}\right) c^\dagger(i)c(j) + U\sum_{\mathbf{i}} n_\uparrow(i)n_\downarrow(i) \tag{1}$$

where $\mathbf{i}$ is a vector of the lattice; t is the hopping and the energy unit; $\alpha_{\mathbf{ij}}$ is the projector on the nearest-neighbour sites; μ is the chemical potential; U is the strength of the on-site Coulomb repulsion; the electronic operators $c(i)$ and $c^\dagger(i)$, as well as all other fermionic operators, are expressed in the spinorial notation and in the Heisenberg picture ($i = (\mathbf{i}, t)$); $n_\sigma(i) = c_\sigma^\dagger(i)c_\sigma(i)$ is the electronic number operator of an electron of spin σ.

2.2. The equations of motion

The electronic operator $c(i)$ obeys the following equation of motion

$$\mathrm{i}\frac{\partial}{\partial t}c(i) = -\mu c(i) - 4tc^\alpha(i) + U\eta(i) \tag{2}$$

Hereafter, given a generic operator $\Phi(i)$, we will use the notation $\Phi^\kappa(i) = \sum_{\mathbf{j}} \kappa_{\mathbf{ij}}\Phi(\mathbf{j}, t)$ for any projector κ (e.g. $\kappa = \alpha$, α^2, ...).

The Hubbard operators $\xi(i) = c(i) - \eta(i)$ and $\eta(i) = n(i)c(i)$ ($n(i) = c^\dagger(i)c(i)$) satisfy the following equations of motion

$$\mathrm{i}\frac{\partial}{\partial t}\xi(i) = -\mu\xi(i) - 4tc^\alpha(i) - 4t\left(\xi_s(i) + \eta_s(i)\right) \tag{3}$$

$$\mathrm{i}\frac{\partial}{\partial t}\eta(i) = (U - \mu)\,\eta(i) + 4t\left(\xi_s(i) + \eta_s(i)\right) \tag{4}$$

Keeping only the 2-site terms ($\delta_{\mathbf{ij}}$ and $\alpha_{\mathbf{ij}}$) and forcing a particle-hole symmetric closure, the operators $\xi_s(i) = \frac{1}{2}\sigma^\mu n_\mu(i)\xi^\alpha(i) + \eta^{\dagger\alpha}(i)c(i)c(i)$ and $\eta_s(i) = \frac{1}{2}\sigma^\mu n_\mu(i)\eta^\alpha(i) + \xi^{\dagger\alpha}(i)c(i)c(i)$ enter the following approximated equations of motion

$$\mathrm{i}\frac{\partial}{\partial t}\xi_s(i) \approx -\mu\xi_s(i) + 8t\eta^{\alpha^2}(i) + 4t\xi_s^\alpha(i) + 8t\eta_s^\alpha(i) \tag{5}$$

$$\mathrm{i}\frac{\partial}{\partial t}\eta_s(i) \approx (U - \mu)\,\eta_s(i) + 4t\eta^{\alpha^2}(i) + 4t\xi_s^\alpha(i) \tag{6}$$

where $n_\mu(i) = c^\dagger(i)\sigma_\mu c(i)$, $\sigma^\mu = (-\mathbf{1}, \vec{\sigma})$ and $\sigma_\mu = (\mathbf{1}, \vec{\sigma})$, where $\vec{\sigma}$ are the Pauli matrices.

2.3. The basis

Given the above reported equations (3-6) of motion and the full knowledge of the exact solution, within the COM framework [7], of the corresponding two-site system, we have chosen as operatorial basis for our fermionic retarded Green's function $G^{\mathcal{R}}(i,j) = \langle \mathcal{R}\{\psi(i), \psi^\dagger(j)\}\rangle$ the following one

$$\psi(i) = \begin{pmatrix} \xi(i) \\ \eta(i) \\ \xi_s(i) \\ \eta_s(i) \end{pmatrix} \tag{7}$$

The first two fields describe the Hubbard subbands and, consequently, the scale of energy of U. The second two fields, together with the first two, give a fermionic closed basis [2] for the Hubbard model on two sites [7] and, in such a case, they open up the possibility to access and exactly describe the virtual exchange energy $J = 4t^2/U$.

This basis, once the equations (5-6) are enforced, closes the following equation of motion

$$\mathrm{i}\frac{\partial}{\partial t}\psi(i) = [\psi(i), H] \cong \sum_{\mathbf{j}} \varepsilon(\mathbf{i},\mathbf{j})\psi(\mathbf{j},t) \tag{8}$$

where $\varepsilon(\mathbf{i},\mathbf{j})$, named energy matrix, has entries that can be just read in equations (3-6). Following this route, we have neglected residual terms in the current $(\mathrm{i}\frac{\partial}{\partial t}\psi(i))$ of the chosen basis that would lead to the appearance of higher-order composite operators.

2.4. *The normalization matrix*

The equation of motion (8) leads to the following expression of the Green's function $G^{\mathcal{R}}(i,j)$

$$G^{\mathcal{R}}(\mathbf{k},\omega) = \frac{1}{\omega - \varepsilon(\mathbf{k}) + \mathrm{i}\delta} I(\mathbf{k}) \tag{9}$$

where $I(\mathbf{i},\mathbf{j}) = \langle\{\psi(\mathbf{i},t), \psi^\dagger(\mathbf{j},t)\}\rangle$ is the normalization matrix, whose entries have the following expressions

$$I_{\xi\xi}(\mathbf{k}) = 1 - \frac{n}{2} \tag{10}$$
$$I_{\xi\eta}(\mathbf{k}) = 0 \tag{11}$$
$$I_{\eta\eta}(\mathbf{k}) = \frac{n}{2} \tag{12}$$
$$I_{\xi\xi_s}(\mathbf{k}) = \Delta + \alpha(\mathbf{k})\left(p - \frac{n}{2}\right) \tag{13}$$
$$I_{\xi\eta_s}(\mathbf{k}) = 0 \tag{14}$$
$$I_{\eta\xi_s}(\mathbf{k}) = 0 \tag{15}$$
$$I_{\eta\eta_s}(\mathbf{k}) = -\Delta - \alpha(\mathbf{k})p \tag{16}$$
$$I_{\xi_s\xi_s}(\mathbf{k}) = I^{(0)}_{\xi_s\xi_s} + I^{(1)}_{\xi_s\xi_s}\alpha(\mathbf{k}) + I^{(2)}_{\xi_s\xi_s}\beta(\mathbf{k}) + I^{(3)}_{\xi_s\xi_s}\eta(\mathbf{k}) \tag{17}$$
$$I_{\xi_s\eta_s}(\mathbf{k}) = I^{(0)}_{\xi_s\eta_s} + I^{(1)}_{\xi_s\eta_s}\alpha(\mathbf{k}) \tag{18}$$
$$I_{\eta_s\eta_s}(\mathbf{k}) = I^{(0)}_{\eta_s\eta_s} + I^{(1)}_{\eta_s\eta_s}\alpha(\mathbf{k}) + I^{(2)}_{\eta_s\eta_s}\beta(\mathbf{k}) + I^{(3)}_{\eta_s\eta_s}\eta(\mathbf{k}) \tag{19}$$

where $\Delta = C^{\alpha}_{11} - C^{\alpha}_{22}$, $p = \frac{1}{4}\langle n^{\alpha}_{\mu}(i)n_{\mu}(i)\rangle - \langle[c_\uparrow(i)c_\downarrow(i)]^{\alpha}\, c^\dagger_\downarrow(i)c^\dagger_\uparrow(i)\rangle$ and $C^{\alpha}_{ab} = \left\langle \psi^{\alpha}_a(i)\psi^\dagger_b(i)\right\rangle$. $\beta(\mathbf{k})$ and $\eta(\mathbf{k})$ are the projectors on the second-neighbour sites along the diagonals and the main directions of the lattice, respectively.

$I^{(0)}_{\xi_s\xi_s}$, $I^{(1)}_{\xi_s\xi_s}$, $I^{(2)}_{\xi_s\xi_s}$, $I^{(3)}_{\xi_s\xi_s}$, $I^{(0)}_{\xi_s\eta_s}$, $I^{(1)}_{\xi_s\eta_s}$, $I^{(0)}_{\eta_s\eta_s}$, $I^{(1)}_{\eta_s\eta_s}$ $I^{(2)}_{\eta_s\eta_s}$ and $I^{(3)}_{\eta_s\eta_s}$ have so lengthy and complex forms (very many correlation functions of higher-order appear) that we were forced to reduce them to treatable expressions (where no new higher-order correlation functions are present) by selecting only two-site correlation functions and mode-coupling non-nearest-neighbor charge, spin, and pair correlations function.

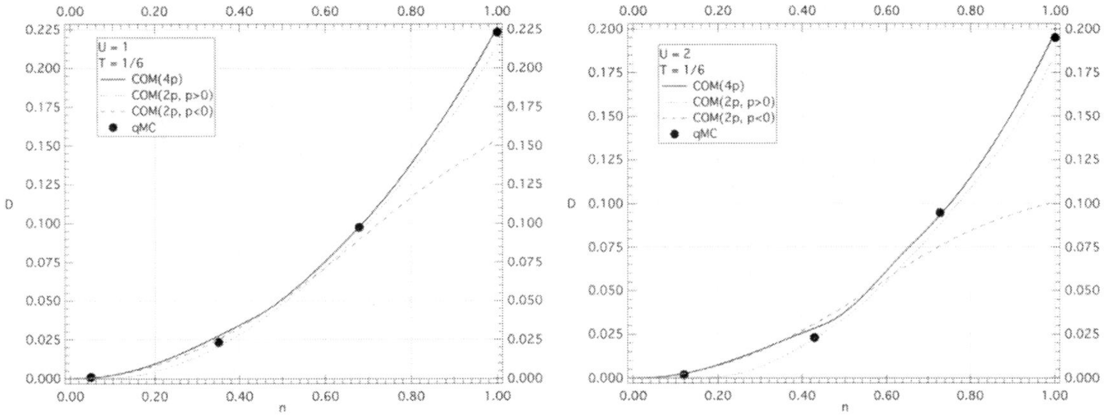

Figure 1. Double occupation D for $T = 1/6$ as function of doping n for two different values of the Coulomb repulsion U ($U = 1, 2$). The quantum Monte Carlo (qMC) data refer to a 12×12 cluster [8]. The COM results in the 2-pole approximation for both $p < 0$ and $p > 0$ solutions are also reported for reference and comparison.

2.5. *Self-consistency*

The only self-consistent parameters not connected to correlation functions straightforwardly attainable by the Green's function exploiting the fluctuation-dissipation theorem in a self-consistent manner are μ and p. They will be fixed by the following two equations

$$n = 2\left(1 - C_{11} - C_{22}\right) \tag{20}$$

$$C_{12} = 0 \tag{21}$$

where $C_{ab} = \left\langle \psi_a(i)\psi_b^\dagger(i)\right\rangle$. These two equations should be added to all others coming from the fluctuation-dissipation theorem. The first equation trivially fixes the number of particles in a self-consistent way and directly springs from the definition of the chemical potential. The second equation is an Algebra Constraint [1, 2] coming from the Pauli exclusion principle and can be easily traced back to $c_\sigma(\mathbf{i})c_\sigma(\mathbf{i}) = 0$. Such a constraint will ensure that no spurious same-spin double-occupancy of any site can take place securing the fundamental connection between charge configurations and spin ones at the basis of the exchange processes that rule low-energy physics of the model.

3. Results

In fig. 1, we present the double occupancy $D = \langle n_\uparrow(i)n_\downarrow(i)\rangle = I_{\eta\eta} - C_{22}$ for $T = 1/6$ as function of doping n for two different values of the Coulomb repulsion U ($U = 1, 2$). The quantum Monte Carlo (qMC) data refer to a 12×12 cluster [8]. The COM results in the 2-pole approximation [1] for both $p < 0$ and $p > 0$ solutions are also reported for reference and comparison. We can see that the 4-pole approximation results improve upon the 2-pole ones as they better reproduce the numerical data both quantitatively and qualitatively. In particular, the 4-pole approximation results are closer to the numerical points in average and, much more relevant, they manage to interpolate between the $p < 0$ and $p > 0$ 2-pole solutions. As a matter of fact, the 4-pole solution avoids the extended zero-double-occupancy region in filling featured by the 2-pole $p > 0$ solution, but not by the $p < 0$ one, but also the sudden down-turn featured by the 2-pole $p < 0$ solution, but not by the $p > 0$ one. Accordingly, the 4-pole solution seems to capture the best of both 2-pole solutions.

4. Conclusions

In this very short manuscript, we have outlined a 4-pole approximation for the two-dimensional Hubbard model within the Composite Operator Method framework. The choices related to the basis, the closure of the equations of motion, the determination of the normalization matrix and of the energy matrix have been discussed and their main aspects illustrated in some detail. The results of such a 4-pole approximation, as regards a very fundamental local quantity, the double occupancy, have been compared with those of numerical simulation, i.e. quantum Monte Carlo, and of the corresponding 2-pole approximation [1] (for both flavours available: $p < 0$ and $p > 0$). The 4-pole approximation performs better than both 2-pole ones and seems to nicely interpolate between them retaining only the better aspects of both. This result is quite encouraging and we are planning to extend the present 4-pole analysis to the study of single-particle and thermodynamics properties of the Hubbard model.

Acknowledgments

A. A. thanks the Max-Planck-Institut FKF Stuttgart for hospitality and financial support.

References

[1] Mancini F and Avella A 2004 *Adv. Phys.* **53** 537
[2] Avella A and Mancini F 2011 *Strongly Correlated Systems: Theoretical Methods* (*Springer Series in Solid-State Sciences* vol 171) (Berlin Heidelberg: Springer-Verlag) chap The Composite Operator Method (COM), p 103
[3] Odashima S, Avella A and Mancini F 2005 *Phys. Rev. B* **72** 205121
[4] Avella A and Mancini F 2007 *Phys. Rev. B* **75** 134518
[5] Hubbard J 1963 *Proc. Roy. Soc. A* **276** 238
[6] Anderson P W 1987 *Science* **235** 1196
[7] Avella A, Mancini F and Saikawa T 2003 *Eur. Phys. J. B* **36** 445
[8] Moreo A, Scalapino D J, Sugar R L, White S R and Bickers N E 1990 *Phys. Rev. B* **41** 2313

International Conference on Strongly Correlated Electron Systems (SCES 2011) IOP Publishing
Journal of Physics: Conference Series **391** (2012) 012152 doi:10.1088/1742-6596/391/1/012152

Metal-insulator transition in the Edwards model

H. Fehske[1], S. Ejima[1], G. Wellein[2] and A. R. Bishop[3]

[1]Institut für Physik, Ernst-Moritz-Arndt-Universität Greifswald, 17489 Greifswald, Germany
[2]RRZE, Friedrich-Alexander-Universität Erlangen-Nürnberg, 91058 Erlangen, Germany
[3]Los Alamos National Laboratory, Los Alamos, New Mexico 87545, U.S.

E-mail: `<fehske,ejima>@physik.uni-greifswald.de`

Abstract. To understand how charge transport is affected by a background medium and vice versa we study a two-channel transport model which captures this interplay via a novel, effective fermion-boson coupling. By means of (dynamical) DMRG we prove that this model exhibits a metal-insulator transition at half-filling, where the metal typifies a repulsive Luttinger liquid and the insulator constitutes a charge density wave. The quantum phase transition point is determined consistently from the calculated photoemission spectra, the scaling of the Luttinger liquid exponent, the charge excitation gap, and the entanglement entropy.

The way a system evolves from a metallic to an insulating state is one of the most fundamental problems in solid state theory. Electron-electron and electron-phonon interactions are the driving forces behind metal-insulator transitions (MITs) in the majority of cases. For example, the Mott-Hubbard MIT [1] is caused by strong Coulomb correlations, whereas the Peierls MIT [2] is triggered by the coupling to vibrational excitations of the crystal. Theoretically the MIT problem can be addressed by the investigation of generic Hamiltonians for interacting electrons and phonons such as Hubbard or Holstein models [3]. In one dimension (1D), these models exhibit a MIT at half-filling, where on the insulating side of the MIT a spin-density-wave (SDW) or a charge-density-wave (CDW) broken-symmetry ground state appears, respectively. On the metallic side, near the MIT, charge transport then takes place within a strongly correlated "background" that anticipates the developing SDW, respectively CDW, order. Since the particles responsible for charge transport and the background order phenomena are the same, the problem is very complex.

A path forward might be the construction of simplified transport models, which capture the basic mechanisms of quantum transport in a background medium in a rather effective way. Along this line a novel quantum transport model has been proposed recently [4],

$$\mathcal{H} = -t_b \sum_{\langle i,j \rangle} f_j^\dagger f_i (b_i^\dagger + b_j) - \lambda \sum_i (b_i^\dagger + b_i) + \omega_0 \sum_i b_i^\dagger b_i \,. \tag{1}$$

This so-called Edwards model mimics the correlations inherent to a spinfull fermionic many-particle systems by a boson affected hopping of spinless particles (see Fig. 1). For the half-filled band case, the model describes a repulsive Tomonaga-Luttinger liquid (TLL), provided the excitations of the background are energetically inexpensive ($\omega_0 < \omega_{0,c}$) or will readily relax ($\lambda > \lambda_c(\omega_0)$). This defines the fluctuation dominated regime. By contrast, strong background correlations, which will develop for large ω_0 and small $\lambda \ll t_b$ tend to immobilize the charge carriers and may even drive a MIT by establishing CDW long-range order [5].

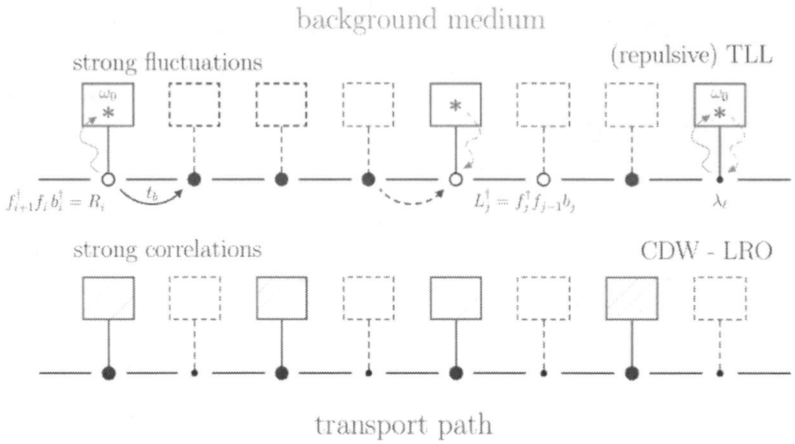

Figure 1. The Edwards model (1) describes a very general situation: As a charge carrier (•) moves along a 1D transport path it creates an excitation with energy ω_0 (∗) in the background at the site it leaves or annihilates an existing excitation at the site it enters. The background medium may represent, e.g., a magnetically, orbitally or charge ordered lattice. One assumes that the (de)excitation of the background can be parameterized as a bosonic degree of freedom. Any distortion of the background can heal by quantum fluctuations. Accordingly the λ-term allows for spontaneous boson creation and annihilation processes.

In the present work, we employ density-matrix renormalization group (DMRG) and dynamical DMRG methods [6] to analyse the ground-state properties of the Edwards model and the charge carrier dynamics for the limiting case of high-energy background fluctuations.

Let us start with the discussion of the photoemission (PE) spectra. The single-particle spectral function probed by angle-resolved [inverse] PE reads

$$A(k,\omega) = A^-(k,\omega) + A^+(k,\omega)\,, \quad \text{with} \quad A^\pm(k,\omega) = \sum_n |\langle\psi_n^\pm|f_k^\pm|\psi_0\rangle|^2\,\delta[\omega \mp \omega^\pm]\,. \tag{2}$$

Here $A^-(k,\omega)$ $[A^+(k,\omega)]$ is associated with the emission [injection] of an electron with wave vector k, i.e. $f_k^- = f_k$ and $f_k^+ = f_k^\dagger$. $|\psi_0\rangle$ is the ground state of a N–site system in the N_f–particle sector, while $|\psi_n^\pm\rangle$ denote the n-th excited states in the $N_f \pm 1$-particle sectors with excitation energies $\omega_n^\pm = E_n^\pm - E_0$. For the half-filled Edwards model we have $N_f = N/2$.

Figure 2 shows $A(k,\omega)$ for a stiff background, i.e. the distortions induced by particle hopping are energetically costly. In this regime the bosons will strongly affect particle transport: The quasiparticle mass is sizeably enhanced and a renormalized band structure appears but—if λ is large enough—the system remains metallic, as can be seen from the finite spectral weight at the Fermi energy E_F (left panel). As the system's ability for relaxation decreases, i.e., at fixed ω_0, λ falls below a certain critical value, a gap opens in the single-particle spectrum at $k_F = \pi/2$ (middle panel). Evidently the system has become an insulator. We note the internal feedback mechanism: The collective boson excitations originate from the motion of the charge carriers and have to persist long enough to finally inhibit particle transport, thereby completely changing the nature of the many-particle ground state. The collective boson-particle dynamics leads to an asymmetric band structure for $k \le k_F$ and $k \ge k_F$ (see inset). While the induced hole probed by PE can only move coherently by a six-step process with three bosons first excited and afterwards consumed, an additional electron can easily move by a two-step process even if strong CDW correlations exist in the background [5]. We note that the (I)PE spectra exhibit weak

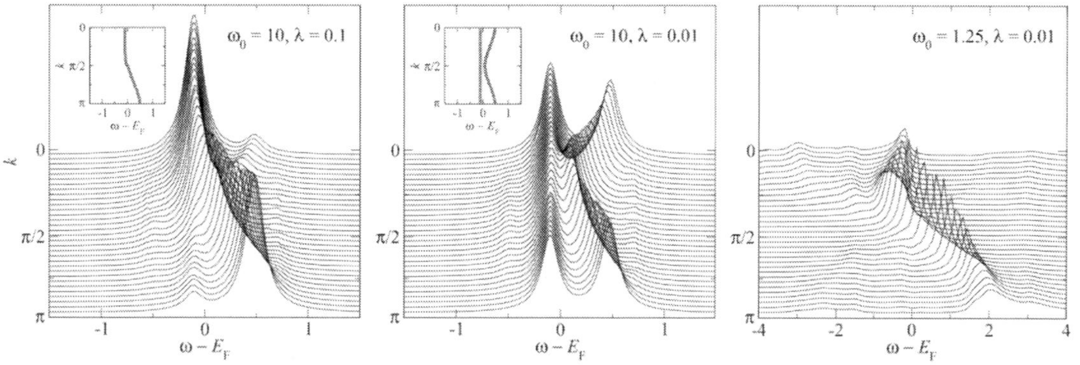

Figure 2. Line-shape of the single-particle spectral $A(k,\omega)$ in the half-filled band sector of the 1D Edwards model. The insets shows the dispersion of the absorption/emission maximum. For the numerics, we consider an $N = 32$–site chain with open boundary conditions (BC) and map a boson site, containing 2^{n_b} states, to n_b pseudosites. We take up to 4 pseudosites, keep $m = 500$ density-matrix eigenstates, and use a broadening $\eta = 0.1$. All energies are given in units of t_b.

signals around the bare boson energies $\pm\omega_0$ (not shown). The interrelation of charge dynamics and background fluctuation becomes apparent again, if we decrease ω_0 keeping λ fixed (right panel). Now the fluctuations overcome the correlations and the system returns to a metallic state which is different in nature, however, from the state we started with: $A(k,\omega)$ shows sharp absorption features near k_F only and is "overdamped" at the Brillouin zone boundaries, where the spectrum is dominated by bosonic excitations.

In order to determine more precisely the phase boundary between the metallic and insulating ground states, typifying a Tomonaga-Luttinger liquid (TLL) and a CDW, respectively, we analyse the limiting ($N \to \infty$) behaviour of the TLL charge exponent

$$K_\rho = \pi \lim_{q\to 0} \frac{S_c(q)}{q}, \quad \text{with} \quad S_c(q) = \tfrac{1}{N}\sum_{i,j} e^{iq(j-k)}\langle (n_j - \tfrac{1}{2})(n_k - \tfrac{1}{2})\rangle, \quad q = \frac{2\pi}{N}, \tag{3}$$

as well as those of the single-particle (charge) gap, $\Delta_c(N) = E_0^+ + E_0^- - 2E_0$, and monitor the finite-size scaling of the entanglement entropy difference [7]

$$\Delta S_N = S_N(N/2) - S_N(N/2-1) = -\frac{c^*}{3}\ln\cos\left[\frac{\pi}{N}\right], \tag{4}$$

where $S_N(l) = -\mathrm{Tr}[\rho_l \ln \rho_l] = \frac{c^*}{3}\ln\left[\frac{N}{\pi}\sin\left(\frac{\pi l}{N}\right)\right] + s_1$. We expect that the TLL charge exponent decreases from $K_\rho = 1$, as λ is lowered, and finally reaches $1/2$ at the MIT point, if the transition is of Kosterlitz-Thouless type [8, 9]. The central charge c^* should scale to unity in the metallic TLL regime [10].

Figure 3 demonstrates that the $N \to \infty$ extrapolated K_ρ indeed becomes $1/2$ at some critical value, where $\lambda_c^{-1}(\omega_0 = 10) \simeq 5.89$, indicating the MIT. In the metallic phase we find a repulsive particle interaction, $K_\rho \le 1$. Our DMRG results point towards an exponential opening of the charge gap entering the insulating state, which corroborates the Kosterlitz-Thouless transition scenario. Note that the CDW state of the Edwards model is a few boson state, in contrast to the Peierls CDW phase of the Holstein model [5]. That means the MIT in the Edwards model is driven by strong correlations, as for the Mott-Hubbard transition. To extract the central charge c^* we use the entanglement entropy difference, Eq. (4), rather than directly exploiting $S_N(l)$.

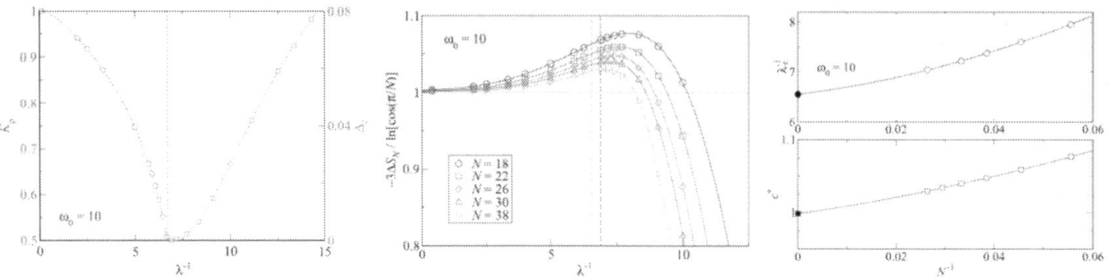

Figure 3. Left panel: $N \to \infty$ extrapolated value of the TLL parameter K_ρ, respectively of the charge gap Δ_c, as a function of λ^{-1} for $\omega_0 = 10$ (open BC). Middle panel: Entanglement entropy differences ΔS_N for different system sizes (periodic BC). The red dashed-dotted line gives the MIT transition point in reasonable agreement with the value obtained from K_ρ (black dashed line). Right panels: Critical value of λ_c^{-1} (filled circle, top panel) and central charge $c^* \simeq 1$ (filled square, bottom panel), both extrapolated from the maxima of ΔS_N. Here we use $m = 2000$, $n_b = 2$, and ensure a discarded weight less than 10^{-10}.

For a model with spinless fermions this is advantageous because we can work with a fixed system size, thereby avoiding antiperiodic BC that give rise to complex phase factors [7]. As can be seen from the middle panel of Fig. 3, for $\lambda^{-1} < \lambda_c^{-1}$, the rescaled quantity $-3\Delta S_N / \ln[\cos(\pi/N)]$ extrapolates to unity as $N \to \infty$. This opens an alternative route to detect the MIT point. We find that the $\lambda_c^{-1}(\omega_0)$ determined by extrapolating the maximum of ΔS_N, i.e. in a completely different manner, matches the critical value obtained from K_ρ surprisingly well. Simultaneously, indeed $c^* \to 1$ (see right panel).

To summarise, we have studied the spectral and ground-state properties of the 1D Edwards fermion-boson transport model by large-scale (dynamical) DMRG numerics. We showed that strong correlations within the background medium will not only affect the charge-carrier's dynamics by enhancing the quasiparticle mass but may even trigger a metal-insulator quantum phase transition. The MIT transition point has been determined in good agreement both from the TLL charge exponent and the entanglement entropy difference. We stress that to date only a very small number of microscopic model exists which have been rigorously shown to exhibit a MIT.

Acknowledgements. The authors would like to thank A. Alvermann, D. M. Edwards, G. Hager and S. Nishimoto for valuable discussions, and the RRZE for providing computer resources. This work is supported by DFG SFB 652 and, at Los Alamos, by CINT and the USDoE.

References

[1] Mott N F 1990 *Metal–Insulator Transitions* (London: Taylor & Francis)
[2] Peierls R 1955 *Quantum theory of solids* (Oxford: Oxford University Press)
[3] Hubbard J 1963 *Proc. Roy. Soc. London, Ser. A* **276** 238; Holstein T 1959 *Ann. Phys. (N.Y.)* **8** 325
[4] Edwards D M 2006 *Physica B* **378-380** 133; Alvermann A, Edwards D M and Fehske H 2007 *Phys. Rev. Lett.* **98** 056602
[5] Wellein G, Fehske H, Alvermann A and Edwards D M 2008 *Phys. Rev. Lett.* **101** 136402; Ejima S, Hager G and Fehske H 2009 *Phys. Rev. Lett.* **102** 106404; Ejima S and Fehske H 2009 *Phys. Rev. B* **80** 155101
[6] White S R 1992 *Phys. Rev. Lett.* **69** 2863; Jeckelmann E 2002 *Phys. Rev. B* **66** 045114
[7] Nishimoto S 2011, private communication
[8] Kosterlitz J M and Thouless D J 1973 *J. Phys. C* **6** 1181
[9] Bursill R J, McKenzie R H and Murray D W 1998 *Phys. Rev. Lett.* **80** 5607; Ejima S, Gebhard F and Nishimoto S *EPL* **70** 492; Ejima S and Fehske H 2009 *EPL* **87** 27001
[10] Calabrese P and Cardy J 2004 *J. Stat. Mech.* P06002; Laflorencie N, Sørensen E S, Chang M S and Affleck I 2006 *Phys. Rev. Lett.* **96** 100603

Journal of Physics: Conference Series **391** (2012) 012153 doi:10.1088/1742-6596/391/1/012153

Investigation of BCS gap equation of (d+id) hole doped cuprates

Partha Goswami

Physics Department, D.B.College(University of Delhi),Kalkaji,Delhi-110019, India

physicsgoswami@gmail.com

Abstract. We consider a (d + i d) cuprate superconductor and model the functional dependence of the pairing interactions $V(\mathbf{k},\mathbf{k}') = (V_{x2-y2}(\mathbf{k},\mathbf{k}')+V_{xy}(\mathbf{k},\mathbf{k}'))$ required for d+id ordering in the pseudo-gap(PG) phase by a function of the form $V_{trial} = [(V_{x2-y2}(k_F, k_F) + V_{xy}(k_F,k_F))\, F(\varphi,\varphi')]$, where $V_{x2-y2}(\mathbf{k},\mathbf{k}') = V_1 (\cos k_xa - \cos k_ya)(\cos k'_xa - \cos k'_ya)$, $V_{xy}(\mathbf{k}, \mathbf{k}') = V_2 \sin(k_xa)\sin(k_ya)\sin(k'_xa)\sin(k'_ya)$, V_1 and V_2 ($V_1 > V_2$) are the coupling strengths, k_F is the Fermi momentum, $\varphi = \arc(\tan(k_y/k_x))$, and (k_x,k_y) belong to the first Brillouin zone (BZ). We further assume that an attractive interaction $-|U_1| (\cos k_xa - \cos k_ya)(\cos k'_xa - \cos k'_ya)$, where U_1 is a model parameter, is responsible for d-wave superconductivity(DSC).Within the BCS framework, for $V_2 << V_1$, we show that the resultant zero-temperature gap $\Delta_0(0)$ is an increasing function of $g(k_F)$.(~ $((D/2)(|U_1| + V_1))$ where the quantity D is the density of energy states).; the solutions are possible if $|U_1| \approx V_1$. The exercise underscores the fact that the unconventional superconductivity in the hole-doped cuprates may definitely be described within the BCS framework.

1. Introduction

The high temperature superconductivity in hole-doped cuprates, such as $YBa_2Cu_3O_{7-\delta}$, is derived by doping the parent two-dimensional Mott insulators with a large anti-ferromagnetic exchange interaction. It is now an established view [1] that the superconducting (SC) gap in such systems have robust d-wave symmetry while the pseudo-gap (PG) is of d+id variety[2,3]. The under-doped metallic region between $0.10 \le \delta \le 0.16$ for this system has attracted much attention because of the anomalous temperature dependence (a non-Fermi liquid behavior) of the electronic specific heat, the magnetic susceptibility, etc.. In $YBa_2Cu_3O_{7-\delta}$, upon hole doping, the anti-ferromagnetism is destroyed at the hole density $\delta \sim 0.10$ and the superconductivity is optimized at $\delta \sim 0.16$. Further doping leads to decreasing T_c and more or less conventional Fermi-liquid behavior. Regarding the origin of the PG and its relation with superconductivity (SC), however, there are two different views. In one, the PG is regarded as a superconducting precursor state (SPS) involving incoherent electron-electron pairings above T_c [4,5,6,7, 8,9] with the particle-hole symmetry preserved whereas the other regard the PG, distinct from SC, to be an ordered state with particle-hole asymmetry and both the phases compete[10,11,12,13].Chakravarty et al.[11] had put forward interpretation of this gap in terms of a hidden long-range order, viz. d-density wave(DDW) order which corresponds to spontaneous currents along the bonds of a square lattice for the ordering wave vector $\mathbf{Q}=(\pm\pi,\pm\pi)$. Although the currents violate the microscopic time reversal symmetry

(MTRS), there is no macroscopic violation. This happens as the DDW state preserves the combined effect of MTRS and translation by a lattice spacing. The net result is that the staggered magnetic flux produced by these currents is zero on the macroscopic scale. Our view regarding the origin of the PG is, however, centered around the paradigm that PG corresponds to d+id-density wave (chiral DDW or CDDW)ordering[2,3] at the anti-ferromagnetic wave vector $\mathbf{Q} = (\pi,\pi)$. Many theorists and experimentalists subscribe to this view [2,3,14]. Accordingly, we model the effective fermion pairing interaction $V(\mathbf{k},\mathbf{k}')$ in the singlet pairing channel by suitable function of the form $V_{trial} = [(V_{x2-y2}(k_F, k_F) + V_{xy}(k_F,k_F)) F(\varphi,\varphi')]$, where $V_{x2-y2}(\mathbf{k},\mathbf{k}') = V_1 (\cos k_xa - \cos k_ya)(\cos k'_xa - \cos k'_ya)$, $V_{xy}(\mathbf{k}, \mathbf{k}') = V_2 \sin(k_xa)\sin(k_ya)\sin(k'_xa)\sin(k'_ya)$, V_1 and V_2 are the coupling strengths, k_F is the Fermi momentum, $\varphi = \mathrm{arc}(\tan(k_y/k_x))$, and (k_x,k_y) belong to the first Brillouin zone (BZ). As regards d-wave superconductivity (DSC), we model the effective two-particle pairing interaction $U(\mathbf{k},\mathbf{k}')$ in the singlet pairing channel by suitable function of the form $-|U_1|(\cos k_xa - \cos k_ya)(\cos k'_xa - \cos k'_ya)$ where U_1 is the coupling strength (model parameter). We assume implicitly that this unconventional superconductivity is initiated by the strongly coupled bosonic modes, such as those corresponding to the electron spin fluctuations (proximity to an anti-ferromagnetic phase raises the possibility of spin-fluctuation-mediated pairing).

In our scheme the Fermi level is pinned at the Van Hove singularity [15]of the normal state dispersion $\varepsilon_\mathbf{k}$ involving the first, the second, and the third neighbor hopping plus a constant term. All energies are expressed in units of the first neighbor hopping. With all these paraphernalia, there are only two energy gaps $D_\mathbf{k} = (-\chi_\mathbf{k}+i\Delta_\mathbf{k})$ and $\Delta_\mathbf{k}^{(sc)}$ corresponding to PG and DSC, respectively, and two distinct quasi-particle dynamics in our formulation of the problem. The second-neighbor hopping in the dispersion, which is known to be important for cuprates [16]and frustrates the kinetic energy of electrons, leads to Fermi surface sheets being not connected by $\mathbf{Q} = (\pi,\pi)$ (non-nesting property). One will notice that, the two gaps in the excitation spectrum are distinct and do not merge into one `quadrature' gap $(D_{eff}(\mathbf{k}))^2 = [|D_\mathbf{k}|^2 + \Delta_\mathbf{k}^{(sc)2}]$) if the nesting property, $\varepsilon_\mathbf{k} = -\varepsilon_{\mathbf{k}+\mathbf{Q}}$, of the dispersion is absent. We have solved the coupled equations for $(D_\mathbf{k}, D_\mathbf{k}^\dagger)$ and $\Delta_\mathbf{k}^{(sc)}$ assuming $V_2 << V_1$ and the specific form for the interactions $U(\mathbf{k},\mathbf{k}')$ and $V(\mathbf{k},\mathbf{k}')$ given above together with the equation to determine the chemical potential (μ) obtainable from the Luttinger rule. The noteworthy outcome of the exercise is the condition $|U_1| \approx V_1$ for the superconductivity in the hole-doped cuprates. Since V_1 has the coulombic origin [11], the superconductivity in the hole-doped cuprates is likely to have the same origin.

In the present communication we derive (see section 2) an expression for single-particle excitation spectrum in CDDW+DSC state. In section 3 we set up the equations for $D_\mathbf{k}$, $D_\mathbf{k}^\dagger$ and $\Delta_\mathbf{k}^{(sc)}$. We solve these equations together with the equation to determine the chemical potential (μ) of the fermion number in a sample case. All the numerical values are expressed in the units of the first neighbor hopping. We show in section 4 that the gap equations have reasonable solutions within the BCS framework even for the unconventional superconductivity.

2.CDDW Hamiltonian and single-particle excitation spectrum

We consider the usual[2,3,11] tight-binding energy dispersion $\varepsilon_{\mathbf{k}} = \varepsilon_{\mathbf{k}}^{(1)} + \varepsilon_{\mathbf{k}}^{(2)} + \varepsilon_{\mathbf{k}}^{(3)} + 4t'$, $\varepsilon_{\mathbf{k}}^{(1)} = -2t\,(c_x + c_y)$, $\varepsilon_{\mathbf{k}}^{(2)} = 4t'\,c_x c_y$, $\varepsilon_{\mathbf{k}}^{(3)} = -2t''\,(c'_x + c'_y)$, $c_j = \cos k_j a$, $c'_j = \cos 2k_j a$ (j = x,y), and 'a' is the lattice constant. The quantity $\varepsilon_{\mathbf{k}}$ involves t, t′, t′′ which are the hopping elements between nearest, next-nearest (NN) and NNN neighbors, respectively. The energy $\varepsilon^{(1)}(\mathbf{k})$ satisfies the perfect nesting condition $\varepsilon^{(1)}(\mathbf{k+Q}) = -\varepsilon^{(1)}(\mathbf{k})$ with anti-ferromagnetic wave vector $\mathbf{Q} = (\pm\pi,\pm\pi)$. For the hole-doped materials, $t' > 0$ (for the electron-doped materials $t' < 0$), and, in all cases, $t' < (t/2)$. For example, typical values are $t \sim 0.2$ eV, $(t'/t) \sim 0.4$, and $(t''/t) \sim 0.01$[15]. In the second-quantized notation, the Hamiltonian to deal with the d+id density-wave (CDDW) order at the wave vector **Q** plus the d-wave superconductivity can be expressed as

$$H = \sum_{\mathbf{k},\sigma} \varepsilon_{\mathbf{k}} d^{\dagger}_{\mathbf{k},\sigma} d_{\mathbf{k},\sigma} + \sum_{\mathbf{k},\sigma} \varepsilon_{\mathbf{k+Q}} d^{\dagger}_{\mathbf{k+Q},\sigma} d_{\mathbf{k+Q},\sigma} + \sum_{\mathbf{k},\sigma} D_{\mathbf{k}}\, d^{\dagger}_{\mathbf{k},\sigma} d_{\mathbf{k+Q},\sigma}$$

$$+\sum_{\mathbf{k},\sigma} D_{\mathbf{k}}^{\dagger}\; d^{\dagger}_{\mathbf{k+Q},\sigma} d_{\mathbf{k},\sigma} + \sum_{\mathbf{k},\sigma} \Delta_{\mathbf{k}}^{\dagger(sc)} d_{-\mathbf{k},-\sigma} d_{\mathbf{k},\sigma} + \sum_{\mathbf{k},\sigma} \Delta_{\mathbf{k}}^{(sc)} d^{\dagger}_{\mathbf{k},\sigma}\; d^{\dagger}_{-\mathbf{k},-\sigma} \qquad (1)$$

where $D_{\mathbf{k}} = (-\chi_{\mathbf{k}} + i\Delta_{\mathbf{k}}) \equiv -\sum_{\mathbf{k}',\sigma} V(\mathbf{k},\mathbf{k}')\, \langle d^{\dagger}_{\mathbf{k}'+\mathbf{Q},\sigma} d_{\mathbf{k}',\sigma} \rangle$ and $D_{\mathbf{k}}^{\dagger} = (-\chi_{\mathbf{k}} - i\Delta_{\mathbf{k}}) \equiv -\sum_{\mathbf{k}',\sigma} V(\mathbf{k},\mathbf{k}')\, \langle d^{\dagger}_{\mathbf{k}',\sigma} d_{\mathbf{k}'+\mathbf{Q},\sigma} \rangle$. The gap function $\Delta_{\mathbf{k}}^{\dagger(sc)} \equiv \sum_{\mathbf{k}',\sigma} U(\mathbf{k},\mathbf{k}') \langle d^{\dagger}_{\mathbf{k},\sigma} d^{\dagger}_{-\mathbf{k},-\sigma} \rangle = -\Delta^{\dagger}_{\mathbf{k+Q}}{}^{(sc)}$. In (1), $d_{\mathbf{k},\sigma}$, with $\sigma = \pm 1$, corresponds to the fermion annihilation operator for the single-particle state $(\mathbf{k},\sigma)$. In the pure d-wave case, $\Delta_{\mathbf{k}}^{(sc)} = \Delta^{\dagger}_{\mathbf{k}}{}^{(sc)} = (\Delta_0^{(SC)}(T))(\cos k_x a - \cos k_y a)$. The quantities $(\chi_{\mathbf{k}}, \Delta_{\mathbf{k}})$ are given by $\chi_{\mathbf{k}} = -\chi_0 \sin(k_x a)\sin(k_y a)$,and $\Delta_{\mathbf{k}} = (\Delta_0^{(PG)}(T))(\cos k_x a - \cos k_y a)$.The conical brackets stand for the thermal average calculated with $H - \mu\check{N}$ where μ is the chemical potential of the fermion number, and $\check{N}$ is the total number operator. Upon ignoring the third neighbor hopping term in the dispersion $\varepsilon_{\mathbf{k}}$ above, we find that the dispersion typically has two in-equivalent saddle points(van Hove singularities(vHs)) at $(\pi,0)$ and $(0,\pi)$ in the first Brillouin Zone. Upon assuming that for fillings such that the Fermi curve lies close to the singularities, the majority of states participating in the pairing formation will come from regions in the vicinity of these saddle points. It must be noted that the CDDW + DSC ordering leads to pining of the Fermi level close to, but not precisely at the vHs.

At this point we introduce few thermal averages determined by $H - \mu\check{N}$,viz. $G_{\sigma}(\mathbf{k},\tau) = -\langle T\{d_{\mathbf{k},\sigma}(\tau) d^{\dagger}_{\mathbf{k},\sigma}(0)\}\rangle$, $\Gamma_{\sigma}(\mathbf{k},\tau) = -\langle T\{ d^{\dagger}_{-\mathbf{k},-\sigma}(\tau)\, d^{\dagger}_{\mathbf{k},\sigma}(0)\}\rangle$, $G'_{\sigma}(\mathbf{k},\tau) = -\langle T\{d_{\mathbf{k+Q},\sigma}(\tau)\, d^{\dagger}_{\mathbf{k},\sigma}(0)\}\rangle$, $\Gamma'_{\sigma}(\mathbf{k},\tau) = -\langle T\{ d^{\dagger}_{-\mathbf{k}-\mathbf{Q},-\sigma}(\tau)\, d^{\dagger}_{\mathbf{k},\sigma}(0)\}\rangle$,and $G''_{\sigma}(\mathbf{k},\tau) = -\langle T\{ d_{\mathbf{k},\sigma}(\tau)\, d^{\dagger}_{\mathbf{k+Q},\sigma}(0)\}\rangle$.Here T is the time-ordering operator which arranges other operators from right to left in the ascending order of imag-inary time τ. The first step of the scheme involves the calculation of (imaginary) time evolution of the operators $d_{\mathbf{k},\sigma}(\tau)$ where, in units such that $\hbar = 1$, $d_{\mathbf{k},\sigma}(\tau) = \exp(H\tau) d_{\mathbf{k},\sigma}(0) \exp(-H\tau)$. We obtain, for example, $\partial\, d_{\mathbf{k},\sigma} = -(\varepsilon_{\mathbf{k}} - \mu)\, d_{\mathbf{k},\sigma} - \Delta_{\mathbf{k}}^{(sc)}\; d^{\dagger}_{-\mathbf{k},-\sigma} - D_{\mathbf{k}}\, d_{\mathbf{k+Q},\sigma}$, $\partial\, d^{\dagger}_{-\mathbf{k},-\sigma} = (\varepsilon_{\mathbf{k}} - \mu)\, d^{\dagger}_{-\mathbf{k},-\sigma} - \Delta^{\dagger}_{\mathbf{k}}{}^{(sc)} d_{\mathbf{k},\sigma} + D_{\mathbf{k}}^{\dagger}\; d^{\dagger}_{-\mathbf{k}-\mathbf{Q},-\sigma}$ and so on. Here $\partial \equiv (\partial/\partial\tau)$ and the argument part has been dropped in writing the operators $d_{\mathbf{k},\sigma}(\tau)$ and their derivative. As the next step, we find that the equations of motion of the averages are given by

$$\partial G_{\sigma}(\mathbf{k},\tau) = -(\varepsilon_{\mathbf{k}} - \mu)\, G_{\sigma}(\mathbf{k},\tau) - \Delta_{\mathbf{k}}^{(sc)} \Gamma_{\sigma}(\mathbf{k},\tau) - D_{\mathbf{k}}\, G'_{\sigma}(\mathbf{k},\tau) - \delta(\tau),$$

$$\partial \Gamma_{\sigma}(\mathbf{k},\tau) = (\varepsilon_{\mathbf{k}} - \mu)\; \Gamma_{\sigma}(\mathbf{k},\tau) - \Delta_{\mathbf{k}}^{(sc)} G_{\sigma}(k,\tau) + D_{\mathbf{k}}^{\dagger}\; \Gamma'_{\sigma}(\mathbf{k},\tau),$$

$$\partial G'_{\sigma}(\mathbf{k},\tau) = -(\varepsilon_{\mathbf{k+Q}} - \mu)\, G'_{\sigma}(\mathbf{k},\tau) - D_{\mathbf{k}}^{\dagger}\; G_{\sigma}(\mathbf{k},\tau) + \Delta_{\mathbf{k}}^{(sc)} \Gamma'_{\sigma}(k,\tau),$$

$$\partial \Gamma'_{\sigma}(\mathbf{k},\tau) = (\varepsilon_{\mathbf{k+Q}} - \mu)\ \Gamma'_{\sigma}(\mathbf{k},\tau) + D_{\mathbf{k}}\ \Gamma_{\sigma}(\mathbf{k},\tau) + \Delta_{\mathbf{k}}^{(sc)} G'_{\sigma}(\mathbf{k},\tau)\ . \qquad (2)$$

The final step is the calculation of the Fourier coefficients $G_\sigma(\mathbf{k},\omega_n) = {}_0\int^\beta d\tau\ e^{i\omega_n\tau}\ G_\sigma(\mathbf{k},\tau)$ (where the Matsubara frequencies are $\omega_n = [(2n+1)\pi / \beta]$ with $n = 0,\pm1,\pm2,\ldots.$) of these temperature Green's functions. Here $\beta = (k_B T)^{-1}$ .We refrain from writing explicitly the equations to determine these coefficients as this is a trivial exercise in view of (2). Upon solving the equations we obtain $G_\sigma(\mathbf{k},\omega_n) = (\Delta_1/\Delta)$, $\Gamma_\sigma(\mathbf{k},\omega_n) = (\Delta_2/\Delta)$, $G'_\sigma(\mathbf{k},\omega_n) = (\Delta_3/\Delta)$, $\Gamma'_\sigma(\mathbf{k}, \omega_n) = (\Delta_4/\Delta)$, and $G''_\sigma(\mathbf{k},\omega_n) = (\Delta_5/\Delta)$where

$$\Delta = (i\omega_n)^4 - (i\omega_n)^2 (\xi_k^2 + \xi_{k+Q}^2 + 2\,D_{eff}(\mathbf{k})^2) + \{\xi_k^2\,\xi_{k+Q}^2 - 2\xi_k\xi_{k+Q}\,|D_k|^2 + \Delta_k^{(sc)2}(\xi_k^2 + \xi_{k+Q}^2)$$

$$+\Delta_k^{(sc)4} + |D_k|^4 - (D_k^\dagger)^2\,\Delta_k^{(sc)2} - (D_k)^2\,\Delta_k^{(sc)2}\},$$

$$\Delta_1 = (i\omega_n + \xi_k)((i\omega_n)^2 - \xi_{k+Q}^2 - \Delta_k^{(sc)\,2}) - |D_k|^2\ (i\omega_n - \xi_{k+Q}),$$

$$\Delta_2 = \Delta_k^{(sc)}\{(i\omega_n)^2 - \xi_{k+Q}^2 - \Delta_k^{(sc)2} + (D_k^\dagger)^2\},$$

$$\Delta_3 = D_k^\dagger\,(i\omega_n + \xi_k)(i\omega_n + \xi_{k+Q}) - D_k\,(-\Delta_k^{(sc)2} + (D_k^\dagger)^2),$$

$$\Delta_4 = D_k^\dagger\,\Delta_k^{(sc)}\,(i\omega_n + \xi_k) + D_k\,\Delta_k^{(sc)}\,(i\omega_n - \xi_{k+Q}),$$

$$\Delta_5 = D_k\,(i\omega_n + \xi_k)(i\omega_n + \xi_{k+Q}) - D_k^\dagger\,(-\Delta_k^{(sc)2} + D_k^2), \qquad (3)$$

$\xi_k = \varepsilon_k - \mu$, and $D_{eff}(\mathbf{k})^2 = (\,|D_k|^2 + \Delta_k^{(sc)2})$. The denominator Δ of the Fourier coefficients yields the single particle excitation spectrum E_k ($T<T_c$): E_k $(T<T_c) = \pm E^{(U,L)}(\mathbf{k})$,$E_k^{(U,L)2} = (\frac{1}{2})[(\xi_k^2 + \xi_{k+Q}^2 + 2D_{eff}(\mathbf{k})^2) \pm p(\mathbf{k})]$, and $p(\mathbf{k}) = [(\xi_k^2 - \xi_{k+Q}^2)^2 + 4|D_k|^2(\xi_k + \xi_{k+Q})^2 + 4\{(D_k^\dagger)^2 + (D_k)^2 - 2|D_k|^2\}\Delta_k^{(sc)2}]^{1/2}$. The excitations in cuprates are thus demonstrably Bogoliubov quasi-particles in the DSC phase. One may notice that, the two gaps in the excitation spectrum are distinct and do not merge into one `quadrature' gap $(D_{eff}(\mathbf{k})^2 = [|D_k|^2 + \Delta_k^{(sc)2}]^{1/2})$ even if the dispersion is nested ($\varepsilon_k = -\varepsilon_{k+Q}$). The reason being the complex d-wave order parameter D_k describing the PG state breaks the time-reversal symmetry of the normal state. The time-reversal operator $\hat{O}$ transforms the order parameter to its complex conjugate: $\hat{O}\,D_k = D^\dagger_k$. If the time-reversal symmetry is preserved, D_k and $D^\dagger_k$ are identical to within a common spatially independent phase. If, however, the time-reversal symmetry is broken, the two states are distinct albeit with the same free energy. The bands $E_k^{(U,L)}$ above the Fermi energy are the reflected ones of those below the Fermi energy, i.e.$(-E_k^{(U,L)})$. For $V_2 << V_1$,the band-maxima in $E_k^{(U)}$ occur at the points($\pm\pi$, $\pm\pi$) and (0,0) whereas those in $E_k^{(L)}$ occur at the points($\pm\pi/2$, $\pm\pi/2$). On the other hand, the band-maxima in $(-E_k^{(U)})$ occur at the anti-nodal points($\pm\pi$,0) and (0,$\pm\pi$) while the band-minima in $(-E_k^{(L)})$ occur at the nodal points($\pm\pi/2$, $\pm\pi/2$). Since the Bogoluibov quasi-particles here with energy above the Fermi sea do not behave the same as those below the Fermi energy, we have clear particle-hole asymmetry in the CDDW+DSC state.

3. Equations for $D_k, D_k^\dagger$ and $\Delta_k^{(sc)}$

For the perfectly nested dispersion, since $p(\mathbf{k}) = 4i\Delta_k\,\Delta_k^{(sc)}$, one may write the Fourier coefficient $G_\sigma(\mathbf{k},\omega_n)$ as

$$G_\sigma(\mathbf{k},\omega_n) = \sum_{j=\pm1} \{U^{(j)}{}_{\mathbf{k}}{}^2(i\omega_n + j\,(\acute{\varepsilon}_{\mathbf{k}} + i\,\acute{\iota}_{\mathbf{k}}{}^{-1}))^{-1} + V^{(j)}{}_{\mathbf{k}}{}^2(i\omega_n + j\,(\acute{\varepsilon}_{\mathbf{k}} - i\,\acute{\iota}_{\mathbf{k}}{}^{-1}))^{-1}\}, \quad (4a)$$

$$\acute{\varepsilon}_{\mathbf{k}} = R_{\mathbf{k}}{}^{1/2}\cos(\theta_{\mathbf{k}}/2),\ \acute{\iota}_{\mathbf{k}}{}^{-1} = R_{\mathbf{k}}{}^{1/2}\sin(\theta_{\mathbf{k}}/2),\ R_{\mathbf{k}} = [(\xi_{\mathbf{k}}{}^2 + D_{eff}(\mathbf{k})^2)^2 + 4\Delta_{\mathbf{k}}{}^2\Delta_{\mathbf{k}}{}^{(sc)2}]^{1/2}, \quad (4b)$$

$$\tan^2(\theta_{\mathbf{k}}) = (4\Delta_{\mathbf{k}}{}^2\Delta_{\mathbf{k}}{}^{(sc)2})/\,(\xi_{\mathbf{k}}{}^2 + D_{eff}(\mathbf{k})^2)^2, \quad (4c)$$

$$U^{(j)}{}_{\mathbf{k}}{}^2 = (1/4)[1-(j\xi_{\mathbf{k}}/\,R_{\mathbf{k}}{}^{1/2}\exp(i\theta_{\mathbf{k}}/2))],\ V^{(j)}{}_{\mathbf{k}}{}^2 = (1/4)[1-(j\xi_{\mathbf{k}}/\,R_{\mathbf{k}}{}^{1/2}\exp(-i\theta_{\mathbf{k}}/2))], \quad (4d)$$

In 4(d) we have used the fact that for the nested dispersion $E_{\mathbf{k}}{}^{(U,L)\,2} = [\xi_{\mathbf{k}}{}^2 + D_{eff}(\mathbf{k})^2 \pm 2i\Delta_{\mathbf{k}}\,\Delta_{\mathbf{k}}{}^{(sc)}]$. The sum of the coherence factors $\sum_{j=\pm1}\{U^{(j)}{}_{\mathbf{k}}{}^2 + V^{(j)}{}_{\mathbf{k}}{}^2\} = 1$ but the factors are complex unlike the usual Fermi liquid(FL) picture. We shall, however, presently focus our attention on the imperfect nesting(and when $p(\mathbf{k})$ is likely to be real) and revisit the near-nested situation while seeking approximate solutions of the gap equations analytically in section 4. For the imperfect nesting case we have

$$G_\sigma(\mathbf{k},\omega_n) = \sum_{j=\pm1}\{u^{(j)}{}_{\mathbf{k}}{}^2(i\omega_n + jE_{\mathbf{k}}{}^{(U)})^{-1} + v^{(j)}{}_{\mathbf{k}}{}^2(i\omega_n + jE_{\mathbf{k}}{}^{(L)})^{-1}\}, \quad (5a)$$

$$u^{(j)}{}_{\mathbf{k}}{}^2 = (1/2)[(1-j\xi_{\mathbf{k}}/\,E_{\mathbf{k}}{}^{(U)}) \times\{(E_{\mathbf{k}}{}^{(U)2} - \xi'_{\mathbf{k+Q}}{}^2)\,/\,q(\boldsymbol{k})\} - \{(1 + j\xi_{\mathbf{k+Q}}/\,E_{\mathbf{k}}{}^{(U)}) \times (|D_{\mathbf{k}}|^2/\,q(\mathbf{k})\}], \quad (5b)$$

$$v^{(j)}{}_{\mathbf{k}}{}^2 = (-1/2)[(1-j\xi_{\mathbf{k}}/\,E_{\mathbf{k}}{}^{(L)}) \times\{(E_{\mathbf{k}}{}^{(L)2} - \xi'_{\mathbf{k+Q}}{}^2)\,/\,q(\boldsymbol{k})\} - \{(1 + j\xi_{\mathbf{k+Q}}/\,E_{\mathbf{k}}{}^{(L)})) \times (|D_{\mathbf{k}}|^2/\,q(\mathbf{k})\}], \quad (5c)$$

where $\xi'_{\mathbf{k+Q}}{}^2 = (\xi_{\mathbf{k+Q}}{}^2 + \Delta_{\mathbf{k}}{}^{(sc)2})$,and $q(\mathbf{k}) = (E_{\mathbf{k}}{}^{(U)2} - E_{\mathbf{k}}{}^{(L)2})$.We find that the coherence factors $u^{(\pm)}{}_{\mathbf{k}}{}^2$ and $v^{(\pm)}{}_{\mathbf{k}}{}^2$ satisfy the sum rule $\sum_{j=\pm1}\{u^{(j)}{}_{\mathbf{k}}{}^2 + v^{(j)}{}_{\mathbf{k}}{}^2\} = 1$. The equation to determine the chemical potential(μ), according to the Luttinger rule, is now given by

$$(1+p) = (N_s/2)^{-1}\sum_{\mathbf{k}',\,j=\pm1}[u^{(j)}{}_{\mathbf{k}'}{}^2\times(\exp(-j\beta\,E_{\mathbf{k}'}{}^{(U)})+1)^{-1} + v^{(j)}{}_{\mathbf{k}'}{}^2\times(\exp(-j\beta E_{\mathbf{k}'}{}^{(L)})+1)^{-1}] \quad (6)$$

where p is the hole-doping level, and N_s is the number of unit cells in the **k**-space. The **k**–summations in these equations will get replaced by the integration $\int d(\mathbf{k}a)\ \rho_{Fermi}(\mathbf{k}) \rightarrow \int_{-\pi}^{+\pi} d(k_xa)/2\pi \int_{-\pi}^{+\pi} (d(k_ya)/2\pi\ \rho_{Fermi}(\mathbf{k})$ where $\rho_{Fermi}(\mathbf{k})$ stands for the Fermi energy density of states (DOS). The dimensionless density of states $\rho(\mathbf{k},\omega) \equiv (-1/2\pi^2\rho_0)\,\mathrm{Im}G^{(R)}(\mathbf{k},\omega)$ (where we take a broad band-width, say, $(10t_1) \sim 1.65$ eV which gives $\rho_0 = (10t_1)^{-1} \sim 0.6\ (\mathrm{eV})^{-1}$)while the single-particle spectral function in the spin-σ channel is given by $A_\sigma(\mathbf{k},\omega) = (-\pi^{-1})\ \mathrm{Im}\ G_\sigma{}^{(R)}(\mathbf{k},\omega)$, where $G_\sigma{}^{(R)}(\mathbf{k},\omega)$ is the retarded Green's function given by $G^R{}_\sigma(k,\omega) = \int_{-\infty}^{\infty} (d\omega'/2\pi)\{\zeta_\sigma(k,\omega')/\,(\omega-\omega' + i\,0^+)\}$and $\zeta_\sigma(k,\omega) = (1/\,i)\,\{G_\sigma(k, z = i\omega_n)\,\big|_{z=\omega-i0+} - G_\sigma(k, z = i\omega_n)\,\big|_{z=\omega+i0+}\}$. Upon using the result $(x \pm i\,0^+)^{-1} = [P(x^{-1}) \pm (1/i)\ \pi\ \delta\ (x)]$,where P represents a Cauchy's principal value,the spectral function A(**k**,ω in the CDDW+DSC phase is given by a sum of δ functions at the quasi-particle energies: $A(\mathbf{k},\omega) = 2\pi\sum_{j=\pm1}[u^{(j)}{}_{\mathbf{k}}{}^2\,\delta(\omega + jE_{\mathbf{k}}{}^{(U)}) + v^{(j)}{}_{\mathbf{k}}{}^2\,\delta(\omega + jE_{\mathbf{k}}{}^{(L)})]$. We simply replace the δ functions by Lorentzians with an assumed intrinsic life-time broadening $(\gamma/t_1) \sim 0.1$. Thus, from the expression for spectral density, putting put $\omega = 0$, we obtain $\rho_{Fermi}(\mathbf{k}) = (1/2\pi\rho_0)\sum_{j=(\pm)}[u^{(j)}{}_{\mathbf{k}}{}^2(\gamma/t_1)\times[E_{\mathbf{k}}{}^{(U)2} + (\gamma/t_1)^2]^{-1} + v^{(j)}{}_{\mathbf{k}}{}^2(\gamma/t_1)\times[E_{\mathbf{k}}{}^{(L)2} + (\gamma/t_1)^2]^{-1}$.

This leads to the usual Fermi arc picture on the first Brillouin zone. We shall, however, approximate this by a constant in section4 while seeking solutions of the gap equations analytically.

The remaining Fourier coefficients $G'_\sigma(k,\omega_n)$ and $\Gamma_\sigma(\mathbf{k},\omega_n)$which correspond to the CDDW gap and the DSC gap, respectively, are given by

$$G'_\sigma(k,\omega_n)=$$

$$\sum_{j=\pm1}(-j/2E_{\mathbf{k}}^{(U)})\times[\{ D_{\mathbf{k}}^{\dagger}(E_{\mathbf{k}}^{(U)}-j\,\xi_{\mathbf{k}})\times(E_{\mathbf{k}}^{(U)}-j\xi_{\mathbf{k+Q}})-D_{\mathbf{k}}(-\Delta_{\mathbf{k}}^{(sc)2}+(D_{\mathbf{k}}^{\dagger})^2)\}/q(\mathbf{k})]$$

$$\times(i\omega_n+jE_{\mathbf{k}}^{(U)})^{-1}$$

$$+\sum_{j=\pm1}(j/2E_{\mathbf{k}}^{(L)})\times[\{ D_{\mathbf{k}}^{\dagger}(E_{\mathbf{k}}^{(L)}-j\xi_{\mathbf{k}})\times(E_{\mathbf{k}}^{(L)}-j\xi_{\mathbf{k+Q}})-D_{\mathbf{k}}(-\Delta_{\mathbf{k}}^{(sc)2}+(D_{\mathbf{k}}^{\dagger})^2)\}/q(\mathbf{k})]$$

$$\times(i\omega_n+jE_{\mathbf{k}}^{(L)})^{-1},$$

$$\Gamma_\sigma(\mathbf{k},\omega_n)=\sum_{j=\pm1}(-j\Delta_{\mathbf{k}}^{(sc)}/2E_{\mathbf{k}}^{(U)})[(E_{\mathbf{k}}^{(U)2}-\xi_{\mathbf{k+Q}}^{\ 2}-\Delta_{\mathbf{k}}^{(sc)2}+D_{\mathbf{k}}^{\dagger 2}))/q(\mathbf{k})](i\omega_n+j\,E_{\mathbf{k}}^{(U)})^{-1}$$

$$+\sum_{j=\pm1}(j\Delta_{\mathbf{k}}^{(sc)}/2E_{\mathbf{k}}^{(L)})[(E_{\mathbf{k}}^{(L)2}-\xi_{\mathbf{k+Q}}^{\ 2}-\Delta_{\mathbf{k}}^{(sc)2}+(D_{\mathbf{k}}^{\dagger})^2))/q(\mathbf{k})](i\omega_n+jE_{\mathbf{k}}^{(L)})^{-1}. \quad (7)$$

We note that the Fourier coefficient $G'_\sigma(\mathbf{k},\omega_n)$ corresponds to $D^{\dagger}_{\mathbf{k}}$ while the the coefficient $G''_\sigma(\mathbf{k},\omega_n)$ corresponds to $D_{\mathbf{k}}$ .Upon replacing $D_{\mathbf{k}}$ and $D^{\dagger}_{\mathbf{k}}$, respectively, by $(-\chi_{\mathbf{k}}+i\Delta_{\mathbf{k}})$ and $(-\chi_{\mathbf{k}}-i\Delta_{\mathbf{k}})$, the three Fourier coefficients $G'_\sigma(k,\omega_n)$, $G''_\sigma(\mathbf{k},\omega_n)$ and $\Gamma_\sigma(\mathbf{k},\omega_n)$ lead to the coupled equations for CDDW gaps ($D^{\dagger}_{\mathbf{k}}$, $D_{\mathbf{k}}$) and DSC gap($\Delta_{\mathbf{k}}^{(sc)}$). With $G'_\sigma(k,\omega_n)$ and $G''_\sigma(k,\omega_n)$, in view of the definitions $D_{\mathbf{k}}^{\dagger}\equiv-\sum_{\mathbf{k}',\sigma}V(\mathbf{k},\mathbf{k}')\langle d^{\dagger}_{\mathbf{k}',\sigma}d_{\mathbf{k}'+\mathbf{Q},\sigma}\rangle$ and $D_{\mathbf{k}}\equiv-\sum_{\mathbf{k}',\sigma}V(\mathbf{k},\mathbf{k}')\langle d^{\dagger}_{\mathbf{k}'+\mathbf{Q},\sigma}d_{\mathbf{k}',\sigma}\rangle$, in the zero-temperature limit, we obtain

$$\chi_{\mathbf{k}}=\sum_{\mathbf{k},\sigma}V(\mathbf{k},\mathbf{k}')[\chi_{\mathbf{k}'}(1+a(\mathbf{k}'))/\{2(E_{\mathbf{k}'}^{(U)}+E_{\mathbf{k}'}^{(L)})\}], \quad (8a)$$

$$\Delta_{\mathbf{k}}=\sum_{\mathbf{k},\sigma}V(\mathbf{k},\mathbf{k}')[\Delta_{\mathbf{k}'}(1+b(\mathbf{k}'))/\{2(E_{\mathbf{k}'}^{(U)}+E_{\mathbf{k}'}^{(L)})\}], \quad (8b)$$

while with $\Gamma_\sigma(\mathbf{k},\omega_n)$ we obtain

$$\Delta_{\mathbf{k}}^{(sc)}=-\sum_{\mathbf{k}',\sigma}U(\mathbf{k},\mathbf{k}')[\Delta_{\mathbf{k}'}^{(sc)}(1+c(\mathbf{k}'))/\{2(E_{\mathbf{k}'}^{(U)}+E_{\mathbf{k}'}^{(L)})\}]. \quad (9)$$

where $a(\mathbf{k})\equiv(\xi_{\mathbf{k}}^2+|D_{\mathbf{k}}|^2-\Delta_{\mathbf{k}}^{(sc)2})/(E_{\mathbf{k}}^{(U)}E_{\mathbf{k}}^{(L)})$, $b(\mathbf{k})\equiv(\xi_{\mathbf{k}}^2+|D_{\mathbf{k}}|^2+\Delta_{\mathbf{k}}^{(sc)2})/(E_{\mathbf{k}}^{(U)}E_{\mathbf{k}}^{(L)})$, and $c(\mathbf{k})\equiv(\xi_{\mathbf{k+Q}}^2+D_{\mathrm{eff}}(\mathbf{k})^2+2\chi_{\mathbf{k}}D_{\mathbf{k}}^{\dagger})/(E_{\mathbf{k}}^{(U)}E_{\mathbf{k}}^{(L)})$. In the zero-temperature limit when $\chi_{\mathbf{k}}\ll\Delta_{\mathbf{k}}$ (or $V_1\gg V_2$) and the nesting of the Fermi surface is near-perfect, Eq.(8b) may be written as $\Delta_{\mathbf{k}}\approx\left(\frac{1}{2}\right)\sum_{\mathbf{k}',\sigma}V(\mathbf{k},\mathbf{k}')\,\{\Delta_{\mathbf{k}'}/(\xi_{\mathbf{k}'}^{\ 2}+\Delta_{\mathbf{k}'}^{\ 2}+\Delta_{\mathbf{k}'}^{(sc)2})^{1/2}\}$. For $V(\mathbf{k},\mathbf{k}')\approx V_1(\cos k_xa-\cos k_ya)(\cos k'_xa-\cos k'_ya)$, the equation assumes a simple form

$$1\approx(V_1/2)\sum_{\mathbf{k}}[(\cos k_xa-\cos k_ya)^2/\{\xi_{\mathbf{k}}^2+(\Delta_0^{(PG)}(0)^2+\Delta_0^{(SC)}(0)^2)(\cos k_xa-\cos k_ya)^2\}^{1/2}]. \quad (10a)$$

Equation (9), on the other hand, yields $\Delta_{\mathbf{k}}^{(SC)} = -\left(\frac{1}{2}\right)\sum_{\mathbf{k}',\sigma} U(\mathbf{k},\mathbf{k}')\, \Delta_{\mathbf{k}'}^{(SC)} \times [\surd(\xi_{\mathbf{k}'}^{2}+\Delta_{\mathbf{k}'}^{2} +\Delta_{\mathbf{k}'}^{(sc)2})]^{-1}$ which is formally similar to the weak coupling BCS gap equation. With an appropriate attractive interactions $U(\mathbf{k},\mathbf{k}') = -|U_1|$ (cos $k_x a$–cos$k_y a$) (cos $k'_x a$–cos$k'_y a$) where U_1 is the coupling strength, we find that the equation assumes a simple form

$$1 = (|U_1|/2)\sum_{\mathbf{k}}[(\cos k_xa-\cos k_ya)^2/\{\xi_{\mathbf{k}}^2 +(\Delta_0^{(PG)}(0)^2+\Delta_0^{(SC)}(0)^2)(\cos k_xa-\cos k_ya)^2\}^{1/2}].(10b)$$

The approximate solutions of the Eqs.(10), once obtained, may be substituted in (8a) to obtain χ_0. The gap trio thus obtained may constitute a good starting point for an iterative scheme to solve Eqs.(8) and (9). For example, for the doping level 9.94% and temperature T= 60 K, solving these equations simultaneously we have found $\mu/t = -0.0018$, $\Delta_0^{(PG)}/t = 0.0240$, $\chi_0/t = 0.0005$ and $\Delta_0^{(SC)}/t = 0.0150$ for $(t'/t) \sim 0.3913$, and $(t''/t) \sim 0.0005$.

4. Unconventional superconductivity and BCS framework

Upon combining the two gap equations given by (10) we obtain

$$1=\left(\frac{1}{2}\right)(|U_1|+V_1)\sum_{\mathbf{k}}[(\cos k_xa-\cos k_ya)^2\times\{\xi_{\mathbf{k}}^2+(\Delta_0^{(PG)}(0)^2+\Delta_0^{(SC)}(0)^2)(\cos k_xa-\cos k_ya)^2\}^{-1/2}]. \quad (11)$$

We also obtain $|U_1|\approx V_1$ as the sum $\sum_{\mathbf{k}}[(\cos k_xa-\cos k_ya)^2\times\{\xi_{\mathbf{k}}^2 + (\Delta_0^{(PG)}(0)^2 + \Delta_0^{(SC)}(0)^2)\times(\cos(k_xa)-\cos(k_ya))^2\}^{-1/2}]\neq 0$. Since V_1 has its origin in the coulombic interaction[9], the condition $|U_1|\approx V_1$ obtained hints at the possibility of the superconductivity in the hole-doped cuprates initiated by the same interaction. Upon modeling the functional dependence of the pairing interactions and order parameter as $U(\mathbf{k},\mathbf{k}') \to -|U_1|\cos(2\varphi)\cos(2\varphi')$, $V(\mathbf{k},\mathbf{k}')\to V_1\cos(2\varphi)\cos(2\varphi')$, and squared order parameter $\Delta_{\mathbf{k}}(T=0)^2 = (\Delta_0^{(PG)}(T=0)^2 +\Delta_0^{(SC)}(T=0)^2)(\cos k_xa-\cos k_ya)^2\to\Delta_0(0)^2\cos^2(2\varphi)$, where $\varphi \equiv \arctan(k_y/k_x)$ and (k_x,k_y) belong to the first Brillouin zone (BZ), we may write the gap equation (11) as

$$1 \approx g(k_F)\,{}_0\!\int^{2\pi}(d\varphi/2\pi)\times({}_0\!\int^{+1}d\xi\;[\cos^2(2\varphi)/\surd(\xi^2+(\Delta_0(0)/\hbar\omega_c)^2\cos^2(2\varphi))])$$

$$= g(k_F)\,{}_0\!\int^{2\pi}(d\varphi/2\pi)\cos^2(2\varphi)\times\log|\{\surd((\Delta_0(0)/\hbar\omega_c)^2\cos^2(2\varphi)+1)+1\}/\{(\Delta_0(0)/\hbar\omega_c)\times\cos(2\varphi)\}| \quad (12)$$

where the dimensionless quantity $g(k_F) = (D/2)(|U_1|+V_1)$ and $\Delta_0(T=0) = (\Delta_0^{(PG)}(T=0)^2 +\Delta_0^{(SC)}(T=0)^2)^{1/2}$. In writing energy integrations in the gap equation, we have assumed an arbitrary energy cut-off '$\hbar\omega_c$' less than the Fermi energy. This facilitates integration over a length larger than k_F^{-1}. The quantity D with dimensions $(\text{Energy})^{-1}$, which will be assumed to be a constant, is the density of energy states. In the weak coupling limit $g(k_F) << 1$, for the special situation $\varphi\to 0$, we obtain $(\Delta_0(0)/\hbar\omega_c) \approx 2\exp(-1/g(k_F))$. This is reminiscent of the corresponding result of the BCS theory for the 'conventional superconductors'.

For the general situation, we adopt the simple strategy of assigning numerical values to $(\Delta_0(0)/\hbar\omega_c)$ and performing the integration in (12), by discretizing the integral, to obtain $g(k_F)$. The 2-D graph in Figure 1 displays the outcome. We find that the quantity $(\Delta_0(0)/\hbar\omega_c)$ is an increasing function of $g(k_F)$. Now at all temperatures above T = 0 K, there is a finite possibility of finding electrons in the non-superconducting state. At finite temperature the occupation of the excited one electron state $E_k = \sqrt{(\xi_k^2+\Delta_k(T)^2)}$ obeys the Fermi statistics with the Fermi distribution $(\exp\beta E_k+1)^{-1}$. Equation (12) is therefore replaced by

$$1 \approx \quad g(k_F)\,{}_0\!\int^{2\pi}(d\varphi/2\pi) \times ({}_0\!\int^{+1}d\xi\ [\cos^2(2\varphi)/\sqrt{(\xi^2+(\Delta_0/\hbar\omega_c)^2\cos^2(2\varphi))}])$$

$$\times(1-2\,f\,(T,\xi,(\Delta_0/\hbar\omega_c),\varphi)) \qquad (13)$$

where $f\,(T,\xi,(\Delta_0/\hbar\omega_c),\varphi) = (\exp\{\beta\hbar\omega_c(\sqrt{(\xi^2+(\Delta_0/\hbar\omega_c)^2\cos^2(2\varphi))})\}+1)^{-1}$. We assume that $\Delta_0(T{=}0)^2 = (\Delta_0^{(PG)}(T{=}0)^2 + \Delta_0^{(SC)}(T{=}0)^2) \approx \Delta_0^{(SC)}(T{=}0)^2$. In that case one may say that the factor of 2 multiplying the Fermi function appears because either one of the states **k** or **−k** may be occupied. One can derive the equation for the critical temperature T_c readily from above:

$$2 = \ g(k_F)\,({}_0\!\int^{+1}d\xi\ \ (1/\xi)\tanh(\xi\,\hbar\omega_c/2k_B T_c). \qquad (14)$$

Equation (14) immediately gives $k_B T_c = 0.57\ \hbar\omega_c \exp(-1/\ g(k_F))$. With the aid of data available(for $g(k_F)$ =1.0(strong coupling) we have $(\Delta_0/\hbar\omega_c) = 0.5$) from Figure 1, we then obtain the ratio $(2\Delta_0/k_B T_c) \approx 4.8$. This value is quite different from the experimental value[17] where T_c is approximately related both to the gap(G) and to the extension of Fermi arcs by $2G\ (4\alpha/\pi) = 4.3\ k_B T_c$ where the angle α spans the half-Fermi arc as measured from the nodal direction. Though the quantitative aspect of our results may be artefact of the approximations made, qualitatively the exercise underscores the fact that reasonable solution for the order parameter amplitude $\Delta_0(T{=}0)$ for a unconventional superconductor is available within the BCS framework.

References

[1] A. Damascelli, Z. Hussain, and Z.-X.Shen, Rev. Mod.Phys. 75, 473 (2003).
[2] P. Kotetes, and G.Varelogiannis, Phys. Rev. B 78, 2205-09(R) (2008); P. Kotetes, and G. Varelogiannis, Phys. Rev. Lett. 104, 106404 (2010).
[3] C. Zhang, S. Tewari, V.M. Yakovenko, and S. Das Sarma, Phys. Rev.B **78**, 174508 (2008); S. Tewari, C. Zhang, V. M. Yakovenko, and S. Das Sarma, Phys. Rev. Lett. **100,** 217004 (2008).
[4] M. Franz, Z. Tesˇanovic´ and O. Vafek ,Phys. Rev. B 66, 054535 (2002).
[5] Y. Wang, L. Li, and N. P. Ong, Phys. Rev. B 73, 024510(2006).
[6] L. Li, Y. Wang, S. Komiya, S. Ono, Y. Ando, G.D. Gu,and N.P. Ong, Phys. Rev. B 81, 054510 (2010).
[7] A. Kanigel, U. Chatterjee, M. Randeria, M. R. Norman, G. Koren, K. Kadowaki, and J. C. Campuzano, Phys. Rev. Lett. 101, 137002 (2008).
[8] O. Yuli, I. Asulin, Y. Kalcheim, G. Koren, and O. Millo , Phys. Rev. Lett. 103, 197003 (2009).
[9] A. Levchenko, M. R. Norman, and A. A. Varlamov, Phys. Rev. B 83, 020506(R) (2011).
[10] S. H¨ufner, M.A. Hossain, A. Damascelli, and G.A. Sawastzky, Rep. Prog. Phys. 71, 062501 (2008).

[11]S. Chakravarty,R. B. Laughlin, D. K. Morr, and C. Nayak, Phys. Rev.B **63**, 94503 (2001);I. Dimov, P. Goswami, Xun Jia, and S. Chakravarty, Phys. Rev.B **78**, 134529 (2008); S. Chakravarty , Rep. Prog. Phys. **74** 022501 (2011).
[12] J.-X. Li, C.Q.Wu and D. H. Lee, Phys. Rev. B 74, 184515 (2006).
[13] A. Levchenko, T. Micklitz, M. R. Norman, and I. Paul, Phys. Rev. B 82, 060502(R) (2010).
[14] Jing Xia, Elizabeth Schemm, G. Deutscher, S.A. Kivelson, D.A.Bonn,W.N.Hardy, R.Liang, W. Siemons, G. Koster, M. M. Fejer, and A. Kapitulnik,Phys. Rev. Lett. 100, 127002 (2008).
[15] A. Hackl,and M. Vojta, Phys. Rev. B 80, 220514(R) (2009).
[16] The experimental analysis [K. Tanaka, T. Yoshida, A. Fujimori, D.H. Lu, Z.-X. Shen, X.-J. Zhou, H. Eisaki, Z. Hussain, S. Uchida, Y. Aiura, K. Ono, T. Sugaya, T. Mizuno, I. Terasaki.Phys. Rev. B 70 , 092503(2004)] shows that some differences among the different families of cuprate superconductors is strongly correlated with second neighbor hopping.
[17] A. Pushp, C. V. Parker, A. N. Pasupathy, K. K. Gomes,S: Ono, J. Wen, Z. Xu, G. Gu, A. Yazdani, Science 324,1689 (2009).

Figure and (figure) caption

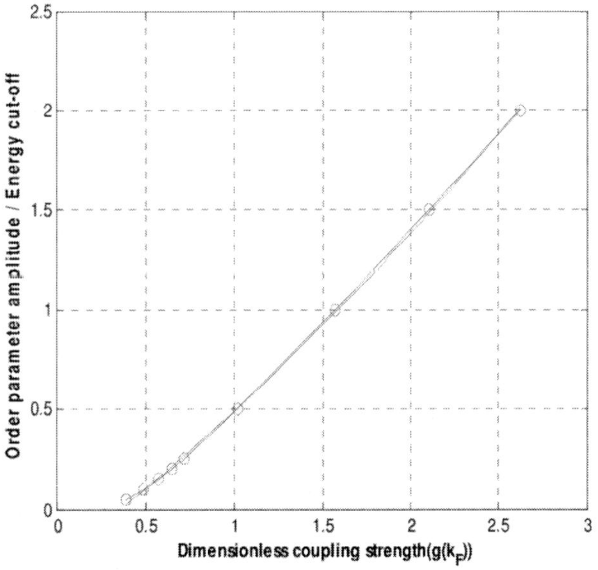

Figure 1. The 2-D plot of ($\Delta_0/\hbar\omega_c$) as a function of the dimensionless coupling strength $g(k_F)$ and second degree polynomial fit.

Microscopic picture on antiferromagnetic order induced by paramagnetic pair-breaking in d-wave superconductor

Y Hatakeyama and R Ikeda

Department of Physics, Kyoto University, Kyoto 606-8502, Japan

E-mail: `hatakeyama@scphys.kyoto-u.ac.jp`

Abstract. We theoretically investigate possible antiferromagnetic (AFM) orders in Fulde-Ferrell-Larkin-Ovchinnikov (FFLO) superconducting phase with $d_{x^2\ y^2}$-wave paring induced by the paramagnetic pair-breaking effect. The AFM $\boldsymbol{Q}$ vector whose incommensurate part $\boldsymbol{q}_{\rm AFM}$ is parallel to [1,1,0] direction and independent of the eld strength is obtained with the dispersion relation which have a remarkably in ected Fermi surface and the largest DOS along [1,1,0] direction. It is shown that, in a eld parallel to [1,1,0], $\boldsymbol{q}_{\rm AFM}$ perpendicular to the eld direction is stabilized as a result of the coupling between $\boldsymbol{Q}$ vector and the wavevector of FFLO modulation. These results are consistent with the observations in the high- eld low-temperature phase of $CeCoIn_5$.

1. Introduction

The discovery of antiferromagnetic (AFM) order in the high field corner of the d-wave superconducting (SC) phase of $CeCoIn_5$ [1] has led to a tremendous interest in the genuine picture of the high-field and low-temperature (HFLT) superconducting phase of this compound in fields parallel to the basal plane, which has been identified earlier [2, 3] with the long-sought Fulde-Ferrell-Larkin-Ovchinnikov (FFLO) phase. The detected field-induced AFM order is peculiar in the sense that it is absent outside the HFLT phase, and this feature has been discussed theoretically from different viewpoints so far [4, 5, 6]. Besides this, doping [7] and NMR [8] experiments have given strong support on the picture [6] identifying the HFLT phase with a FFLO state.

The incommensurate $\boldsymbol{Q}$ vector of the AFM order observed in the neutron scattering experiments [1, 9] has exhibited the following three features: (i) the incommensurate part $\boldsymbol{q}_{\rm AFM}$ of the $\boldsymbol{Q}$ vector ($\boldsymbol{q}_{\rm AFM} = \boldsymbol{Q} - (\ \ ,\ \ ,\ \ )$) is fixed in [1,1,0] or [1,-1,0] direction regardless of the direction of a field $\boldsymbol{H}$ (Note that these directions are equivalent due to the tetragonal crystal symmetry of $CeCoIn_5$); (ii) $\boldsymbol{q}_{\rm AFM}$ is independent of the field strength; (iii) A $\boldsymbol{q}_{\rm AFM}$ perpendicular to $\boldsymbol{H} \parallel [1,1,0]$ is preferred. However, these features have not been successfully explained so far.

In this work, we extend our theory [6, 10] identifying the HFLT phase of $CeCoIn_5$ with the coexistent phase of AFM and FFLO orders in a $d_{x^2-y^2}$-wave superconductor with strong paramagnetic pair-breaking (PPB) in order to study the detailed structures of the AFM order. It is found that the $\boldsymbol{Q}$ vector of the AFM order is determined by the nesting vector of the Fermi surface, and that $\boldsymbol{q}_{\rm AFM}$ parallel to [1,1,0] (or [1,-1,0]) direction reflects the fact that the principal

Fermi surface of $CeCoIn_5$ is remarkably inflected and has the largest density of state along [1,1,0] direction. It is also found that $\boldsymbol{q}_{\rm AFM}$ is independent of the field strength. Moreover, it is shown that, for $\boldsymbol{H} \parallel [1,1,0]$, $\boldsymbol{q}_{\rm AFM}$ perpendicular to $\boldsymbol{H}$ is stabilized by the coupling between $\boldsymbol{Q}$ vector of the AFM order and the wavevector of the FFLO modulation parallel to $\boldsymbol{H}$. Therefore, the features of the observed $\boldsymbol{Q}$ vector of the AFM order in $CeCoIn_5$ are explained by the picture that the HFLT phase of $CeCoIn_5$ is the coexistent phase of AFM and FFLO orders induced by strong PPB.

2. Model and calculation

We start with a weak-coupling BCS-like Hamiltonian including the effect of PPB and the mean fields of FFLO SC and AFM orders:

$$\mathcal{H} = \mathcal{H}_{\rm kin} + \mathcal{H}_{\rm SC} + \mathcal{H}_{\rm AFM} \tag{1}$$

$$\mathcal{H}_{\rm kin} = \sum_{\boldsymbol{k},\sigma} [\ _{\boldsymbol{k}} - \ _{\rm B}H\ ]\, c^{\dagger}_{\boldsymbol{k},\sigma} c_{\boldsymbol{k},\sigma} \tag{2}$$

$$\mathcal{H}_{\rm SC} = \sum_{\boldsymbol{q}} \left\{ \frac{1}{g} |\ _{\boldsymbol{q}}|^2 - \frac{1}{2} \sum_{\boldsymbol{k},\ ,\beta} \left[\ _{\boldsymbol{q}} w_{\boldsymbol{k}} (i\ _y)\ _{,\beta} c^{\dagger}_{\boldsymbol{k},\ } c^{\dagger}_{-\boldsymbol{k}+\boldsymbol{q},\beta} + \text{h.c.} \right] \right\} \tag{3}$$

$$\mathcal{H}_{\rm AFM} = \sum_{\boldsymbol{Q}} \left\{ \frac{1}{U} |m_{\boldsymbol{Q}}|^2 - \frac{1}{2} \sum_{\boldsymbol{k},\ ,\beta} \left[m_{\boldsymbol{Q}} (\ _x)\ _{,\beta} c^{\dagger}_{\boldsymbol{k}+\boldsymbol{Q},\ } c_{\boldsymbol{k},\beta} + \text{h.c.} \right] \right\}. \tag{4}$$

Here, $\mathcal{H}_{\rm kin}$, $\mathcal{H}_{\rm SC}$, and $\mathcal{H}_{\rm AFM}$ are the kinetic energy term with the PPB effect, SC-interaction term, and AFM-interaction one of the Hamiltonian, respectively. Further, $\ _{\boldsymbol{k}}$ is the dispersion relation of quasiparticles measured from the Fermi energy, $\boldsymbol{H}$ is the magnetic field parallel to the basal plane, $\ _{\boldsymbol{q}}$ is a FFLO SC order parameter with momentum $\boldsymbol{q}$, and $m_{\boldsymbol{Q}}$ is an AFM staggered moment with wavevector $\boldsymbol{Q}$ of the AFM modulation. On the basis of observations in $CeCoIn_5$ [1], we have assumed that the AFM staggered moment is perpendicular to the basal plane and the pairing symmetry of SC order is $d_{x^2-y^2}$ [11]: $w_{\boldsymbol{k}} = \cos(k_x a) - \cos(k_y a)$ where a is the lattice spacing of the basal plane. Based on the recent NMR measurement [8] suggesting the existence of nodal planes of SC order parameter in the HFLT phase, the FFLO modulation of the SC order parameter have been assumed to be of Larkin-Ovchinnikov (LO) type: $\ (\boldsymbol{r}) = \sqrt{2}\ \ \cos(\boldsymbol{q}_{\rm FFLO} \cdot \boldsymbol{r})$, where $\boldsymbol{q}_{\rm FFLO}$ is the wavevector of the FFLO modulation.

From the Hamiltonian (1), we have calculated a AFM ordering in FFLO state in Pauli-limit and have found the most optimal $\boldsymbol{Q}$-vector of the AFM order. The dispersion relation of 2D tight-binding model has been used in order to incorporate the tetragonal crystal symmetry of $CeCoIn_5$:

$$\ _{\boldsymbol{k}} = \ + 2t_1(\cos(k_x a) + \cos(k_y a)) + 4t_2 \cos(k_x a)\cos(k_y a) + 2t_3(\cos(2k_x a) + \cos(2k_y a)), \tag{5}$$

where $\ $ is the chemical potential, and t_1, t_2, t_3 are the hopping integrals. In order to investigate the AFM ordering in a FFLO state, we have focused on the $|m_{\boldsymbol{Q}}|^2$ term in the free energy. Since the coefficient of the $|m_{\boldsymbol{Q}}|^2$ term is proportional to the inverse of the AFM susceptibility, the most optimal $\boldsymbol{Q}$ vector has been found by the numerical minimization of the coefficient of the $|m_{\boldsymbol{Q}}|^2$ term with respect to $\boldsymbol{Q}$. The detail of our calculation is discussed in [10].

Moreover, we have investigated the coupling between the $\boldsymbol{Q}$ vector of the AFM order and the wavevector $\boldsymbol{q}_{\rm FFLO}$ of the FFLO modulation. The dispersion relation has been assumed to have a quasi-2D cylindrical Fermi surface and satisfy the following relation for the nesting condition: $\ _{\boldsymbol{k}+\boldsymbol{Q}} = -\ _{\boldsymbol{k}} - \boldsymbol{v}_{\boldsymbol{k}} \cdot \boldsymbol{q}_{\rm AFM} - \ _{\rm IC} T_c$, where $\boldsymbol{v}_{\boldsymbol{k}}$ is the Fermi velocity, $\ _{\rm IC}$ is the parameter representing the deviation from the perfect nesting, and T_c is the SC transition temperature.

We have calculated the free energy expanded in powers of and $m = m_{\boldsymbol{Q}}$:

$$F(\;, m, \boldsymbol{q}_{\rm FFLO}, \boldsymbol{q}_{\rm AFM}) = f_{\Delta^2}(\boldsymbol{q}_{\rm FFLO}) + f_{\Delta^4}(\boldsymbol{q}_{\rm FFLO}) + f_{\Delta^6} + f_{m^2}(\boldsymbol{q}_{\rm AFM}) + f_{m^4} + f_{\Delta^2 m^2}(\boldsymbol{q}_{\rm FFLO}, \boldsymbol{q}_{\rm AFM}) \quad (6)$$

In the calculation of these terms, the orbital depairing effect and the spatial modulation of in a FFLO state are incorporated using the familiar quasi-classical approximation for the orbital depairing [3, 12]. We have focused on the m^2 part of the expanded free energy (6): $F_m = f_{m^2} + f_{\Delta^2 m^2}$. We have examined the dependence of F_m on the angle between $\boldsymbol{q}_{\rm AFM}$ and $\boldsymbol{q}_{\rm FFLO}$ in order to find the most optimal .

3. Results

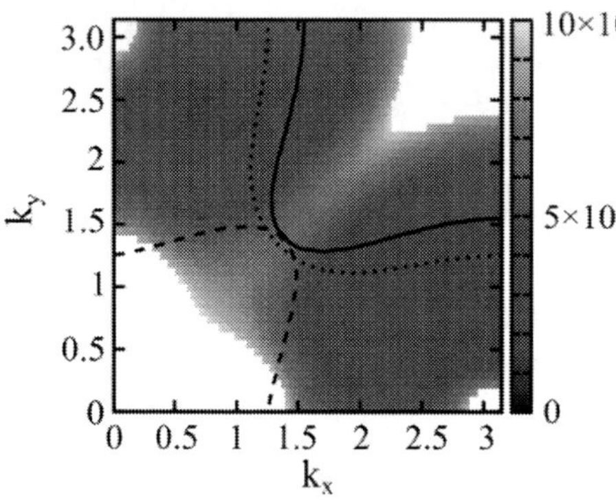

Figure 1. Fermi surface (solid curve) and the density of states (colormap) of the dispersion relation (5) with the parameter values $t_1 = 20T_c$, $/t_1 = 1.85, t_2/t_1 = -1.25, t_3/t_1 = 0.65$. The broken curve is the Fermi surface shifted by the nesting vector ($+0.37$, $+0.37$) in k-space, and the dotted curve is the Fermi surface obtained by using the result of the band calculation[13].

We have numerically investigated the most optimal $\boldsymbol{q}_{\rm AFM}$. It is found that $\boldsymbol{q}_{\rm AFM}$ is determined by the nesting vector of the Fermi surface in this model. For example, from the solid and the broken curves in Fig. 1, we can see that the nesting vector of the Fermi surface depicted in Fig 1 is ($+0.37$, $+0.37$), and the $\boldsymbol{Q}$ vector obtained from the numerical minimization coincides with the nesting vector. As shown below, the $\boldsymbol{Q}$ vector consistent with the observation [1] can be reproduced in this model; therefore it is suggested that the origin of the incommensurate AFM order in the HFLT phase of $\rm CeCoIn_5$ consists in the electronic band structure.

We have examined how the direction of the most optimal $\boldsymbol{q}_{\rm AFM}$ depends on the dispersion relation. In the simplest case $t_3 = 0$, where the nesting condition of the Fermi surface is relatively kept, $\boldsymbol{q}_{\rm AFM}$ parallel to [1,0,0] direction is stabilized, which is inconsistent with the experimental $\boldsymbol{q}_{\rm AFM}$. On the other hand, in the case where t_3 is finite, it is possible to have a remarkably inflected Fermi surface and the largest DOS along [1,1,0] direction, as depicted in Fig. 1. These conditions of the Fermi surface correspond to the band 14-electron one obtained from the band calculation [14]. When these conditions are satisfied, we obtain $\boldsymbol{q}_{\rm AFM}$ parallel to [1,1,0] direction, which is consistent with the observed feature (i) of the $\boldsymbol{Q}$ vector indicated in Sec. 1.

We have investigated the temperature and the field dependence of $q_{\rm AFM}$. It is found that $q_{\rm AFM}$ is almost independent of the temperature and the field strength, as shown in Fig. 2. This results reproduce the observed feature (ii) of the $\boldsymbol{Q}$ vector. The reason for the field-independent $q_{\rm AFM}$ is that the nesting vector, and thus $q_{\rm AFM}$ are not affected by the Zeeman splitting of the

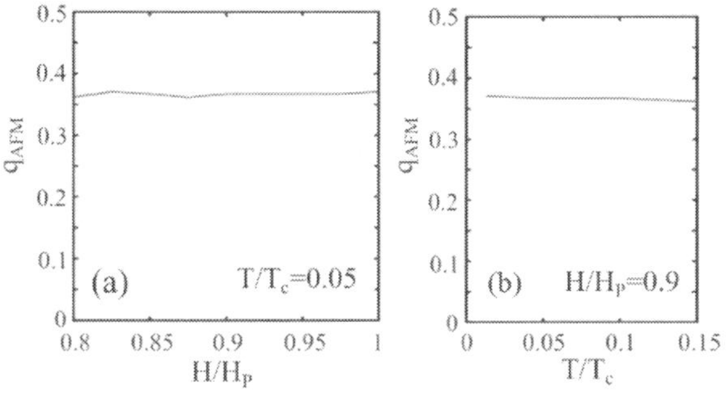

Figure 2. (a) Field dependence and (b) tempareture dependence of $q_{\rm AFM}$ obtained by a numerical minimization. The same dispersion relation as Fig. 1 is used.

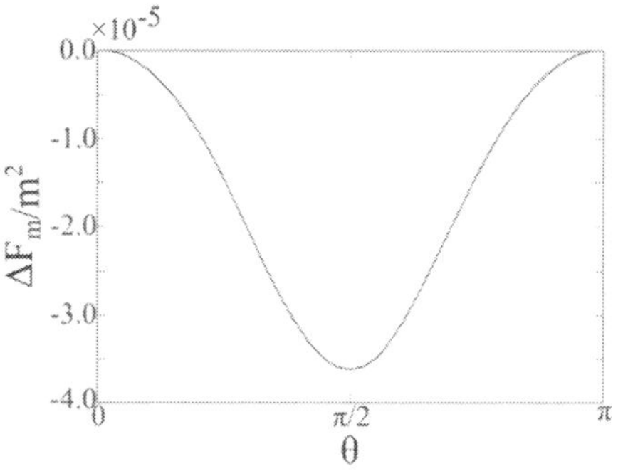

Figure 3. Coupling energy F_m of the AFM modulation and the FFLO modulation as a function of the angle between $\boldsymbol{q}_{\rm AFM}$ and $\boldsymbol{q}_{\rm FFLO}$.

Fermi surface when the AFM staggered moment m is perpendicular to the field. In this case, the nesting between the up-spin Fermi surface and the down-spin Fermi surface is relevant for the AFM ordering. Because the volume changes of these Fermi surfaces induced by the Zeeman splitting cancels out, the nesting vector is not affected by the Zeeman splitting.

We have examined the interaction between the AFM $\boldsymbol{Q}$ vector and the wavevector $\boldsymbol{q}_{\rm FFLO}$ of the FFLO modulation. Figure 3 shows the coupling energy F_m as a function of , where $F_m = F_m( ) - F_m(0)$. The parameter values used in our calculation are, $_M = 1.2$, $_{\rm IC} = 0.4, T/T_c = 0.1, H/H_{\rm orb} = 0.2$, where $_M = \sqrt{2}H_{\rm orb}/H_{\rm P}$ is Maki parameter, $H_{\rm orb}$ and $H_{\rm P}$ are the depairing fields in the orbital limit and in Pauli-limit, respectively. From Fig. 3, we can see that F_m is minimized when $\boldsymbol{q}_{\rm AFM}$ is perpendicular to $\boldsymbol{q}_{\rm FFLO}$. The experiments such as the ultrasound measurement [15] and NMR measurement [8] are also consistent with the FFLO order in which $\boldsymbol{q}_{\rm FFLO}$ is parallel to $\boldsymbol{H}$. Therefore, this result means that $\boldsymbol{q}_{\rm AFM}$ perpendicular to $\boldsymbol{H}$ is stabilized, which is consistent with the observed feature (iii) of the $\boldsymbol{Q}$ vector.

Acknowledgments

This work was supported by the Grant-in-Aid for the Global COE Program "The Next Generation of Physics, Spun from Universality and Emergence" from the Ministry of Education, Culture, Sports, Science and Technology (MEXT) of Japan.

References

[1] Kenzelmann M *et al.* 2008 *Science* **321** 1652
[2] Bianchi A, Movshovich R, Capan C, Pagliuso P G and Sarrao J L 2003 *Phys. Rev. Lett.* **91** 187004
[3] Adachi H and Ikeda R 2003 *Phys. Rev.* B **68** 184510
[4] Aperis A, Varelogiannis G and Littlewood P B 2010 *Phys. Rev. Lett.* **104** 216403
[5] Yanase Y and Sigrist M 2011 *J. Phys.: Condens. Matter* **23** 094219
[6] Ikeda R, Hatakeyama Y and Aoyama K 2010 *Phys. Rev.* B **82** 060510
[7] Tokiwa Y *et al.* 2010 *Phys. Rev.* B **82** 220502; Ikeda R 2010 *Phys. Rev.* B **81** 060510(R)
[8] Kumagai K, Shishido H, Shibauchi T and Matsuda Y 2011 *Phys. Rev. Lett.* **106** 137004
[9] Kenzelmann M *et al.* 2010 *Phys. Rev. Lett.* **104** 127001
[10] Hatakeyama Y and Ikeda R 2011 *Phys. Rev.* B **83** 224518
[11] Izawa K *et al.* 2001 *Phys. Rev. Lett.* **87** 057002; Hiasa N and Ikeda R 2008 *Phys. Rev. Lett.* **101** 027001
[12] Ikeda R 2007 *Phys. Rev. B* **76** 134504
[13] Tanaka K, Ikeda H, Nisikawa Y and Yamada K 2006 *J. Phys. Soc. Jpn.* **75** 024713
[14] Onuki Y *et al.* 2004 *J. Phys. Soc. Jpn.* **73** 769
[15] Watanabe T *et al.* 2004 *Phys. Rev. B* **70** 020506; Ikeda R 2007 *Phys. Rev.* B **76** 054517

International Conference on Strongly Correlated Electron Systems (SCES 2011) IOP Publishing
Journal of Physics: Conference Series **391** (2012) 012155 doi:10.1088/1742-6596/391/1/012155

Mott Insulator in Two-Channel Kondo Lattice

S Hoshino[1,†], **J Otsuki**[1,2] **and Y Kuramoto**[1]

[1]Department of Physics, Tohoku University, Sendai 980-8578, Japan

[2]Theoretical Physics III, Center for Electronic Correlation and Magnetism, Institute of Physics, University of Augsburg, Augsburg 86135, Germany

E-mail: †`hoshino@cmpt.phys.tohoku.ac.jp`

Abstract. The Mott insulator is found in the two-channel Kondo lattice by using the dynamical mean- eld theory combined with continuous-time quantum Monte Carlo method. At quarter lling of two conduction bands, the paramagnetic state shows metal-insulator crossover with increasing the coupling constant at low temperatures. The insulating state is explained by the strong coupling limit, where the system is effectively interpreted as the half- lled Hubbard model. It is shown that the present Mott insulating state is well described by the Hubbard I approximation.

1. Introduction

In Pr- and U-based systems with even number of f electrons per site, the ground state can be a non-Kramers doublet [1]. When this doublet strongly couples with conduction electrons, exotic behaviors are expected as in URu_2Si_2 [2, 3] and $PrAg_2In$ [4, 5]. Recently non-Kramers doublet systems attract renewed attention due to the discovery of $_3$ doublet compounds such as $PrIr_2Zn_{20}$ [6, 7] and PrV_2Al_{20} [8]. Theoretically, the simplest description of the non-Kramers doublets coupled with conduction electrons is given by the two-channel Kondo lattice (2ch KL) [9]. It is desirable to reveal basic properties of this fundamental model, since it is not restricted to a particular compound.

The 2ch KL is described by the following Hamiltonian:

$$\mathcal{H} = \sum_{\boldsymbol{k}\ \sigma} (\varepsilon_{\boldsymbol{k}} - \ ) c^{\dagger}_{\boldsymbol{k}\ \sigma} c_{\boldsymbol{k}\ \sigma} + J \sum_{i} \boldsymbol{S}_i \cdot \boldsymbol{s}_{ci} \ , \tag{1}$$

where $c_{\boldsymbol{k}\ \sigma}$ is the annihilation operator of conduction electrons with channel and spin . The operators $\boldsymbol{S}_i$ and $\boldsymbol{s}_{ci} = \frac{1}{2}\sum_{\sigma\sigma'} c^{\dagger}_{i\ \sigma} \boldsymbol{\sigma}_{\sigma\sigma'} c_{i\ \sigma'}$ are the localized and conduction spins at site i, where $\boldsymbol{\sigma}$ is a spin-1/2 Pauli matrix. The 2ch KL forms ordered states in the wide parameter region. At half filling, which corresponds to the case with two conduction electrons per site, antiferromagnetic order is found [10]. The channel symmetry breaking is also reported if suppression of staggered ordering is assumed [11]. At quarter filling, an instability toward the antiferro-channel order is pointed out [1, 12]. The antiferro-channel order actually realizes in infinite dimensions [13]. On the other hand, the paramagnetic state of the 2ch KL has been identified as an incoherent metal [9]. In this paper, we show that the paramagnetic state at quarter filling becomes the Mott insulator [14] in strong coupling region. The Mott insulating state is well described by the Hubbard I approximation [15] as will be demonstrated in the following.

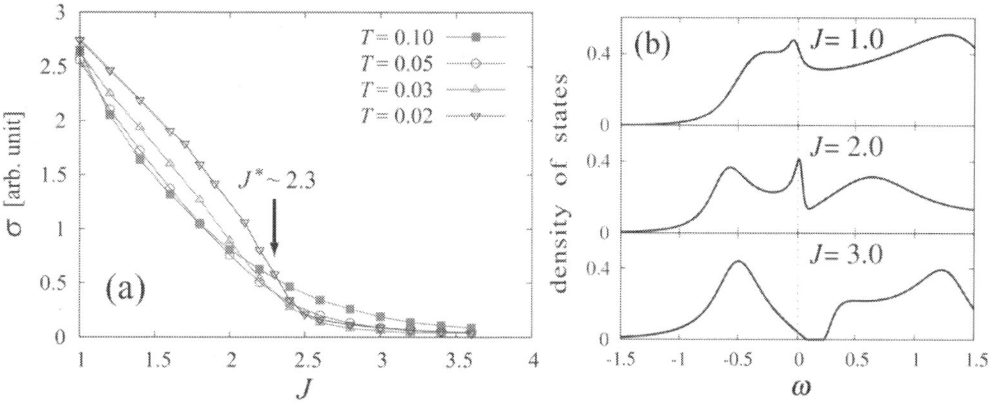

Figure 1. J dependence of the (a) electrical conductivity at quarter filling, and (b) density of states at $T = 0.02$ with several values of J.

We analyze the 2ch KL by using the dynamical mean-field theory (DMFT) [16] combined with the continuous-time quantum Monte Carlo method (CT-QMC) [17, 18]. We take the semi-circular density of states given by $\rho_0(\omega) = N^{-1}\sum_{\boldsymbol{k}} \ (\omega - \varepsilon_{\boldsymbol{k}}) = (2/\ \ D)\sqrt{1-(\omega/D)^2}$ where N is the number of sites, and D is a half of the band width with $D = 1$ as a unit of energy. For analytic continuation from imaginary axis onto real one, we have employed the Pade approximation [19]. In this paper, we concentrate mainly on the paramagnetic state at quarter filling, and do not allow for the staggered ordering. The suppression of the staggered order is expected in realistic situations because of *e.g.* substantial nearest neighbor hopping or geometrical frustration. Note that the present discussion is valid only at finite temperatures, since the paramagnetic state cannot be the ground state because of the residual entropy.

2. Metal-Insulator Crossover

In this section, we discuss the metal-insulator crossover in the 2ch KL at quarter filling. Figure 1(a) shows the J dependence of the electrical conductivity calculated from the current-current correlation function [16]. At $T = 0.10$, the conductivity smoothly decreases with increasing J. On the other hand, the result at $T = 0.02$ shows the characteristic behavior near $J \ \sim 2.3$. The conductivity steeply decreases around this point, and the system tends to become an insulator when $J > J$. Although we could not go to further low temperatures because of the increasing computing time, the change in conductivity seems to become steeper at J .

The change in the electronic states is clearly seen in the density of states as shown in Fig. 1(b). In the cases with $J = 1.0, 2.0 (< J\ )$, the value at $\omega = 0$ is finite and the system is metallic. However, the density of states with $J = 3.0 (> J\ )$ forms the gap structure near the Fermi level. Thus, the the electronic state changes from metal to insulator near the crossover point $J \sim J$. On the other hand, it is also characteristic that the sharp peak develops at the Fermi level when $J \lesssim J$. These behaviors are very similar to that of the Mott transition in the half-filled Hubbard model [14].

The origin of the insulating state can be understood in terms of the strong coupling limit. Figure 2(a) shows the schematic illustration of the quarter-filled 2ch KL in $J \to \infty$. In this limit, the electron at each site does not have spin degrees of freedom owing to the formation of the Kondo singlet with the localized spin, and has only the channel degrees of freedom. If we regard the channel $\ = 1$ as the spin 'up' state and $\ = 2$ as 'down' state, the system is mapped onto the half-filled Hubbard model which shows the Mott physics. Such situation is schematically drawn in Fig. 2(b). In the 2ch KL, the repulsive interaction between conduction electrons is

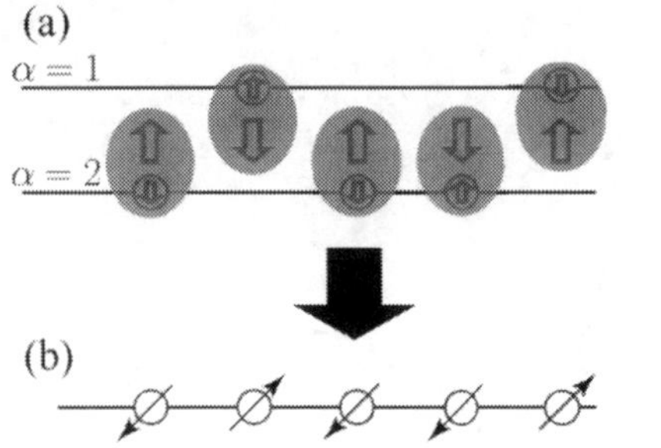

Figure 2. Schematic illustrations of (a) quarter-filled 2ch KL in the strong coupling limit and (b) half-filled Hubbard model.

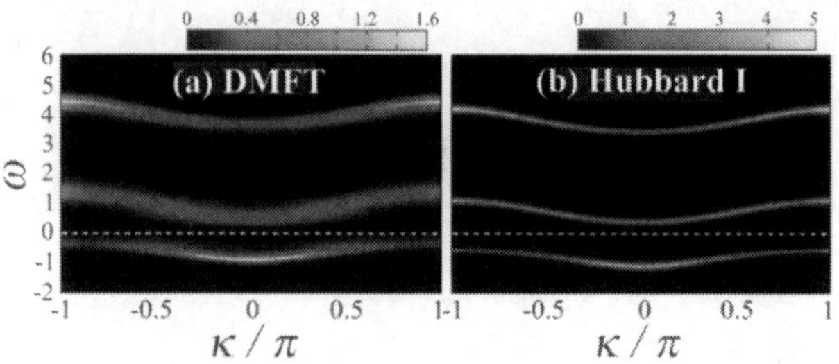

Figure 3. Single-particle spectrum for the quarter-filled 2ch KL with $J = 3.0$ and $T = 0.02$ obtained by the (a) DMFT and (b) Hubbard I Approximation.

caused by the Kondo exchange J instead of the Coulomb interaction. Although the present mapping is valid only in $J \to \infty$, the system at low temperatures can show the properties in this limit in wide parameter region, since the effective exchange interaction is scaled to strong coupling by the Kondo effect.

We note that the Mott insulator realizes also in the half-filled 2ch KL in the strong coupling region. This is understood in a manner similar to the quarter-filled case by considering the limit $J \to \infty$. The difference is that the electrons in the strong coupling limit have only the spin degrees of freedom instead of channel.

3. Single-Particle Spectrum and Hubbard I Approximation

Let us consider the single-particle spectrum in the Mott insulating state. The spectrum is derived from the Green function as

$$A(\boldsymbol{k},\omega) = -\mathrm{Im}\, G(\boldsymbol{k},\omega+\mathrm{i}\ )/\ , \tag{2}$$

where is a positive infinitesimal. In the present condition, the spectrum does not depend on the spin and channel. Since the self energy is local in the DMFT, the wave-vector dependence of the spectrum enters only through $\varepsilon_{\boldsymbol{k}}$. We introduce the parameter defined by $\varepsilon_{\boldsymbol{k}} = -D\cos\ $, and draw the spectrum as if the system were in one dimension.

Figure 3(a) shows the single-particle spectrum of the 2ch KL with $J = 3.0$ and $T = 0.02$ obtained by the DMFT. We can see the formation of the gap at the Fermi level, which is consistent with Fig. 1(b). The bands below and above the Fermi level are due to the emergence of lower and upper Hubbard bands. On the other hand, we observe another band in the high-energy region near $\omega \sim 4$, which does not appear in the Hubbard model.

Now we show that the spectrum in the Mott insulating states are well explained by the Hubbard I approximation [15]. Let us consider the model in atomic limit:

$$\mathcal{H}_{\mathrm{atom}} = -\ \ \sum_{\sigma} c^{\dagger}{}_{\sigma} c\ {}_{\sigma} + J \sum \boldsymbol{S}\cdot\boldsymbol{s}_{\mathrm{c}} \tag{3}$$

Here only the local operators appear. We define the Green function in the atomic limit by

$$G_{\mathrm{atom}}(\tau) = -\langle T_{\tau} c\ {}_{\sigma}(\tau) c^{\dagger}{}_{\sigma}\rangle. \tag{4}$$

In the Hubbard I approximation, the Green function in the original Hamiltonian (1) is approximated as

$$G(\boldsymbol{k},\mathrm{i}\varepsilon_n)^{-1} = G_{\mathrm{atom}}(\mathrm{i}\varepsilon_n)^{-1} - \varepsilon_{\boldsymbol{k}}, \tag{5}$$

where $\varepsilon_n = (2n+1)\ T$ is the fermionic Matsubara frequency. Thus we obtain the single-particle properties in the 2ch KL. We note that the Hubbard I approximation is reasonable only in the strong coupling regime. At $T = 0$, the quarter-filled situation is realized by the condition $\ = -3J/4$. When we plot the spectrum, the infinitesimal constant is replaced by $0.05D$.

The single-particle spectrum obtained by the Hubbard I approximation is shown in Fig. 3(b). The bands below and above the Fermi level originate from removal and addition processes of local electrons, respectively. Since there are two final states in the electron-addition process in the atomic limit of the quarter filled 2ch KL, we observe two bands above the Fermi level. Compared with Fig. 3(a), the Hubbard I approximation well reproduces the spectrum obtained by the DMFT not only qualitatively but quantitatively. However, we should mention that the Hubbard I approximation fails in metallic states away from integer filling, since the correct Fermi volume is not reproduced. This approximation can describe only the Mott insulator at integer filling.

4. Summary

We have analyzed the paramagnetic phase of the 2ch KL at quarter filling by using the DMFT+CT-QMC method. The system shows the metal-insulator crossover by tuning strength of the Kondo coupling. The resultant insulating state is regarded as the Mott insulator, which is understood in the strong coupling limit. The electronic structure of the Mott insulating state is well described by the Hubbard I approximation.

One of the author (SH) is supported by Research Fellowships of the Japan Society for the Promotion of Science for Young Scientists. This work is partly supported by a Grant-in-Aid for Scientific Research on Innovative Areas "Heavy Electrons" (No 20102008) of MEXT, Japan.

References

[1] D. L. Cox and A. Zawadowski: Adv. Phys. **47** (1998) 599.
[2] Y. Kuramoto, H. Kusunose and A. Kiss: J. Phys. Soc. Jpn. **78** (2009) 072001.
[3] J. A. Mydosh and P. M. Oppeneer: arXiv:1107.0258.
[4] A. Yatskar, W. P. Beyermann, R. Movshovich, P. C. Can eld, A. Panchula and S. L. Bud'ko: Physica B **230-232** (1997) 46.
[5] O. Suzuki, H. S. Suzuki, H. Kitazawa, G. Kido, T. Ueno, T. Yamaguchi, Y. Nemoto and T. Goto: J. Phys. Soc. Jpn. **75** (2006) 013704.
[6] T. Onimaru, K. T. Matsumoto, Y. F. Inoue, K. Umeo, Y. Saiga, Y. Matsushita, R. Tamura, K. Nishimoto, I. Ishii, T. Suzuki and T. Takabatake: J. Phys. Soc. Jpn **79** (2010) 033704.
[7] T. Onimaru, K. T. Matsumoto, Y. F. Inoue, K. Umeo, T. Sakakibara, Y. Karaki, M. Kubota and T. Takabatake: Phys. Rev. Lett. **1061** (2011) 177001.
[8] A. Sakai and S. Nakatsuji: J. Phys. Soc. Jpn. **80** (2011) 063701.
[9] M. Jarrell, H.-B. Pang, D. L. Cox and K.-H. Luk: Phys. Rev. Lett. **77** (1996) 1612.
[10] M. Jarrell, H. Pang and D. L. Cox: Phys. Rev. Lett. **78** (1997) 1996.
[11] R. Nourafkan and N. Nafari: J. Phys. Condens. Matter **20** (2008) 255231.
[12] T. Schauerte, D. L. Cox, R. M. Noack, P. G. J. van Dongen and C. D. Batista: Phys. Rev. Lett. **94** (2005) 147201.
[13] S. Hoshino, J. Otsuki and Y. Kuramoto: J. Phys. Soc. Jpn. Suppl. A **80** (2011) SA135.
[14] M. Imada, A. Fujimori and Y. Tokura: Rev. Mod. Phys. **70** (1998) 1039.
[15] J. Hubbard: Proc. Roy. Soc. A **276** (1963) 238.
[16] A. Georges, G. Kotliar, W. Krauth and M. J. Rozenberg: Rev. Mod. Phys. **68** (1996) 13.
[17] J. Otsuki, H. Kusunose, P. Werner and Y. Kuramoto: J. Phys. Soc. Jpn **76** (2007) 114707.
[18] E. Gull, A. J. Millis, A. I. Lichtenstein, A. N. Rubtsov, M. Troyer and P. Werner: Rev. Mod. Phys. **83** (2011) 349.
[19] H. J. Vidberg and J. W. Serene: J. Low Temp. Phys. **29** (1977) 179.

Quantum disordered ground state for the frustrated square lattice

A Kalz, A Honecker, S Fuchs and T Pruschke
Institut für Theoretische Physik, Georg-August-Universität Göttingen, Friedrich-Hund-Platz 1, 37077 Göttingen, Germany
E-mail: kalz@theorie.physik.uni-goettingen.de

Abstract. We present Quantum Monte-Carlo simulations of an exchange-anisotropic spin-1/2 Heisenberg model on a square lattice with nearest and next-nearest neighbor interactions. The ground state phase diagram shows two classical magnetically ordered phases for dominating antiferromagnetic S^z-interactions and for large quantum fluctuations a ferromagnetic order in the x-y plane. In between a finite region is detected where neither classical nor quantum mechanical order, e.g. long-ranged dimer correlations, are found.

1. Introduction

The investigation of frustrated spin models has become a rather active field in the past decades [1–3]. In particular the appearance of quantum critical behavior and emergence of new quantum phases in these systems has connected this field of low temperature physics directly with the research on high-T_C superconductors and topological insulators [4–8, and references therein]. The frustration induced suppression of classical order especially for systems in two dimensions and the corresponding enhancement of quantum fluctuations yield an interesting competition between ordered ground states – like staggered dimer phases – and quantum disordered states as the spin liquid state [9–11].

The analysis of such frustrated models with Quantum Monte-Carlo (QMC) simulations is often complicated due to the sign problem which limits the ability of the algorithm for competing quantum fluctuations. This is one reason why in the present work we choose the quantum fluctuations to be non-frustrating (i.e., the S^{xy}-interactions are ferromagnetic). In addition such a spin model can be mapped exactly onto a model of hard-core bosons with repulsive interactions [12, 13]. These bosonic models can be analyzed in optical lattices [14] and also frustrated systems have been realized recently [15].

In the following section we will introduce the model and describe the expected ground states before explaining our QMC results in Sec. 3. We conclude with a short discussion in Sec. 4.

2. Model

The model is given by summing over all interactions on nearest neighbor (NN) and next-nearest neighbor (NNN) bonds between spin-1/2 operators on a $L \times L$ square lattice:

$$H = \sum_{\mathrm{NN}} t_1(S_i^+ S_j^- + \mathrm{H.c.}) + V_1\, S_i^z S_j^z + \sum_{\mathrm{NNN}} t_2(S_i^+ S_j^- + \mathrm{H.c.}) + V_2\, S_i^z S_j^z\,. \quad (1)$$

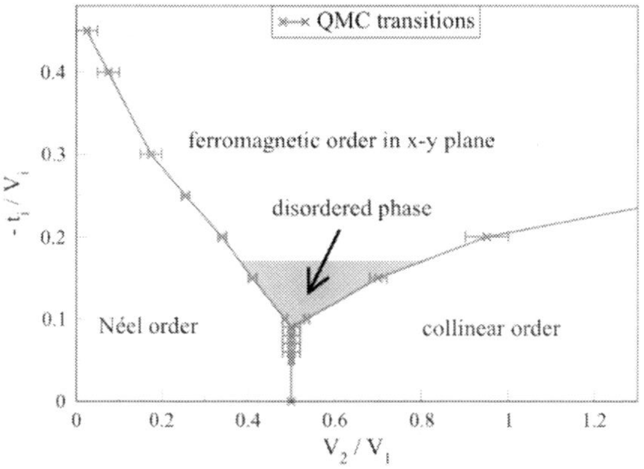

Figure 1. Ground state phase diagram of the anisotropic Heisenberg model on the frustrated square lattice (data from our previous work [11]). For small fluctuations $|t_k|$ the classical antiferromagnetic phases survive and for large values a ferromagnetic order in the x-y plane is stabilized. For the intermediate region we find no finite signal of any order parameter.

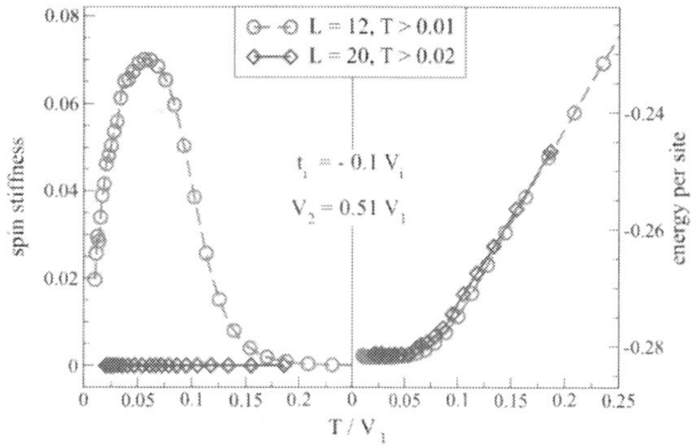

Figure 2. Left: The spin stiffness inside the disordered region at $t_k = -0.1\ V_k$ and $V_2 = 0.51\ V_1$ only shows a small signal for an intermediate temperature on the 12×12 lattice. For a larger lattice no finite signal is found. **Right:** The energies are saturated to their ground state value for $T < 0.05\ V_1$.

The $t_k < 0$ couple the S^{xy}-direction ferromagnetically and the $V_k > 0$ are chosen antiferromagnetic and, hence, introduce frustration in the model. The nearest neighbor S^z-coupling favors a Néel state which has the lowest energy in the classical Ising case ($t_k = 0$) for values $V_2 < V_1/2$ and competes with a collinear configuration, i.e., parallel spins in one direction and antiparallel spins in the other direction, at the critical point $V_2 = V_1/2$. Classically the ground state is highly degenerate at this point [16, 17]. For large quantum fluctuations ($|t_k| \approx 1/2$) a ferromagnetic alignment of all spins in the x-y plane is expected. This corresponds to a superfluid state of the above mentioned hard-core bosons [18, 19].

In the vicinity of the critical point $V_2 = V_1/2$ and for non-zero fluctuations the emergence of quantum ordered phases is possible since quantum fluctuations play a crucial role in low dimensional systems. This was discussed for similar models, e.g. in [9]. These phases consist of quantum mechanically entangled states coupling two or more spins on neighboring sites. Examples are staggered or columnar configurations of dimers on nearest-neighbor bonds. However, for the present model these dimers are not given by the spin singlet state of two spins but due to the exchange anisotropy by the $S^z = 0$ triplet state.

3. Results

The phase diagram of the model (1) is given in Fig. 1 in terms of the strength of fluctuations $-t_1/V_1 = -t_2/V_2$ and frustration V_2/V_1. The phase transition lines were derived in Ref. [11] from finite-temperature QMC simulations using an implementation of the *Stochastic Series Expansion* [20, 21] which is part of the ALPS project [22–24]. Furthermore, an exchange Monte-Carlo update [25–28] was used to overcome thermalization problems in the vicinity of the critical point $V_2 = V_1/2$.

To detect the different magnetic phases we calculated the structure factor which is given by the Fourier transformed spin-spin correlation function for the wave vectors $\vec{q} = (\pi, \pi)$ (Néel order) and $\vec{q} = (\pi, 0), (0, \pi)$ (collinear order). The ferromagnetic phase – corresponding to superfluid order in an analogue bosonic model – can be detected by calculating the spin stiffness

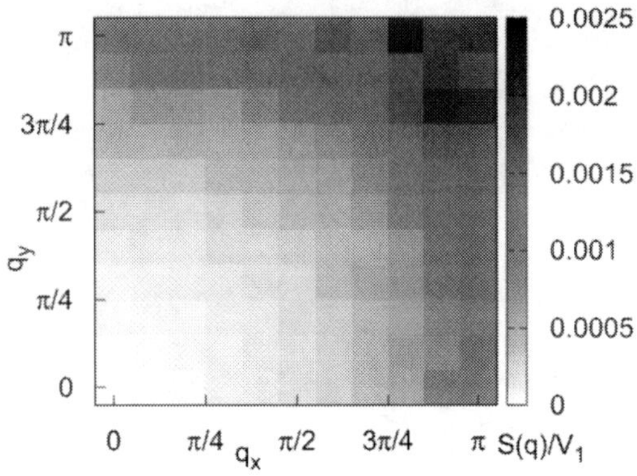

Figure 3. For $t_k = -0.1\ V_k$ and $V_2 = 0.51\ V_1$ the complete structure factor of a 20×20 lattice is shown and only a vanishingly small signal at $\vec{q} = (\pi, \pi)$ (Néel type) is identified – probably due to finite-size effects and the finite temperature ($T = 0.02\ V_1$).

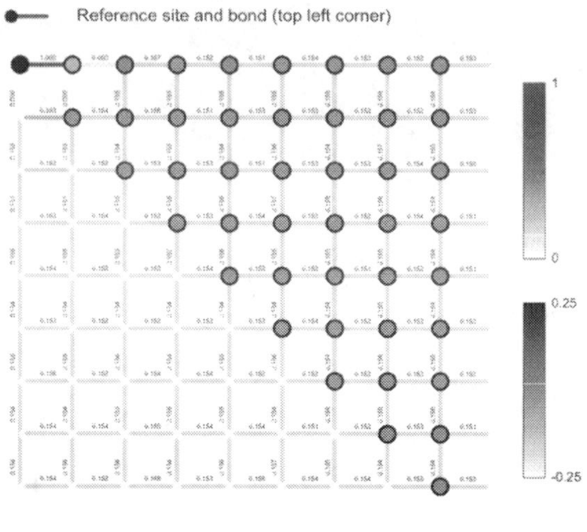

Figure 4. The correlation of spins and dimers with respect to the top left site and adjacent bond are shown in a color coded lattice for $t_k = -0.12\ V_k$ and $V_2 = 0.5\ V_1$ at $T = 0.01\ V_1$. No sign of long-ranged order is visible neither in the spin-spin correlation function nor in any configuration of parallel dimers or plaquettes.

(or superfluid density). In the QMC simulations the estimator for this observable is connected with the winding number [29].

As expected the classical antiferromagnetic states survive for small $|t_k|$ and a direct transition between both states is found. However, above the critical point we find for $|t_k| \geq 0.1\ V_k$ an opening of a region without classical order in the phase diagram. We performed careful calculations of the spin stiffness in this region and found it to be suppressed for lattice sizes $L \geq 16$ and low temperatures $T < 0.05\ V_1$ (Fig. 2). Only for larger values of $|t_k| > 0.175\ V_k$ the ferromagnetic order parameter shows a finite and size-independent signal [11].

For a set of parameters in the shaded region of the phase diagram in Fig. 1 we calculated further observables. In Fig. 3 we show as an example the complete structure factor of a 20×20 lattice in a color-coded map. Apart from a small enhancement of the values for $\vec{q} \approx (\pi, \pi)$ which occurs due to finite-size effects and thermal fluctuations ($T = 0.02\ V_1$) we find no evidence for any magnetic order in the S^z-components of commensurate or incommensurate type.

Since no classical order is found, the emergence of quantum ordered phases as the columnar or staggered configuration of dimers was also tested. Therefore, we implemented an estimator for the correlation of four spins residing on two different nearest-neighbor bonds [30]. The result of such a calculation is shown in Fig. 4 for a 16×16 lattice and parameters inside the unidentified region. The correlation of parallel and perpendicular dimers on nearest-neighbor bonds is plotted in a color-coded lattice which reflects the relative distance between the top left dimer and a second dimer. The strength of the correlation is normalized to the value of the self-correlation of the reference bond. We also include the results of spin-spin correlation functions on the lattice sites. We cannot identify any long-ranged order in the dimer or spin correlations and, hence, conclude that for a small region of the phase diagram the system is in a quantum disordered ground state.

4. Discussion

We performed QMC simulations for an anisotropic Heisenberg model on a square lattice with competing interactions. The phase diagram was derived and a finite region without classical order was identified. An additional analysis of higher order correlation functions also ruled out the appearance of dimer ordered phases. In our recent work [11] we also presented exact diagonalization results for a set of parameters inside the disordered region and found a finite gap in the energy spectrum which is compatible with the rapid decay of the spin-spin correlation functions shown in Fig. 4. We conclude from our simulations that we have found a quantum mechanically disordered ground state with a finite gap which can be referred to as a gapped spin liquid.

Investigations of a similar model on the honeycomb lattice with additional third-nearest neighbor interactions are in progress. This model is known to exhibit emergent quantum phases for the exchange-isotropic case [31–33].

Acknowledgments

We acknowledge financial support by the Deutsche Forschungsgemeinschaft under grant No. HO 2325/4-2 and through SFB 602.

References

[1] Richter J, Schulenburg J and Honecker A 2004 *Quantum magnetism* (*Lecture Notes in Physics* vol 645) ed Schollwöck U, Richter J, Farnell D J J and Bishop R F (Springer) chap 2, p 85
[2] Diep H T (ed) 2005 *Frustrated Spin Systems* (World Scientific)
[3] Lacroix C, Mendels P and Mila F (eds) 2011 *Introduction to Frustrated Magnetism* 1st ed (*Springer Series in Solid-State Sciences* vol 164) (Springer)
[4] Ihle D and Kasner M 1990 *Phys. Rev. B* **42** 4760–4762
[5] Manousakis E 1991 *Rev. Mod. Phys.* **63** 1–62
[6] Nori F, Gagliano E and Bacci S 1992 *Phys. Rev. Lett.* **68** 240–243
[7] Bergman D L, Wu C and Balents L 2008 *Phys. Rev. B* **78** 125104
[8] Guo H M and Franz M 2009 *Phys. Rev. B* **80** 113102
[9] Balents L, Bartosch L, Burkov A, Sachdev S and Sengupta K 2005 *Phys. Rev. B* **71** 144508
[10] Sachdev S 2008 *Nature Physics* **4** 173–185
[11] Kalz A, Honecker A, Fuchs S and Pruschke T 2011 *Phys. Rev. B* **83** 174519
[12] Batrouni G G and Scalettar R T 2000 *Phys. Rev. Lett.* **84** 1599–1602
[13] Hébert F, Batrouni G G, Scalettar R T, Schmid G, Troyer M and Dorneich A 2001 *Phys. Rev. B* **65** 014513
[14] Lewenstein M, Sanpera A, Ahufinger V, Damski B, Sen A and Sen U 2007 *Advances in Physics* **56** 243–379
[15] Jo G, Guzman J, Thomas C, Hosur P, Vishwanath A and Stamper-Kurn D 2011 *cond-mat:1109.1591*
[16] Kalz A, Honecker A, Fuchs S and Pruschke T 2008 *Eur. Phys. J. B* **65** 533–537
[17] Kalz A, Honecker A, Fuchs S and Pruschke T 2009 *J. Phys.: Conf. Ser.* **145** 012051
[18] Bloch F 1932 *Z. Phys.* **74** 295–335
[19] Matsubara T and Matsuda H 1956 *Prog. Theor. Phys.* **16** 569–582
[20] Sandvik A W and Kurkijärvi J 1991 *Phys. Rev. B* **43** 5950–5961
[21] Syljuåsen O F and Sandvik A W 2002 *Phys. Rev. E* **66** 046701
[22] Alet F, Wessel S and Troyer M 2005 *Phys. Rev. E* **71** 036706
[23] Alet F *et al* 2005 *J. Phys. Soc. Jpn.* **74S** 30
[24] Albuquerque A F *et al* 2007 *J. Magn. Magn. Mat.* **310** 1187
[25] Hukushima K and Nemoto K 1996 *J. Phys. Soc. Jpn.* **65** 1604–1608
[26] Hansmann U H E 1997 *Chem. Phys. Lett.* **281** 140–150
[27] Katzgraber H G, Trebst S, Huse D A and Troyer M 2006 *J. Stat. Mech.: Theory and Experiment* P03018
[28] Melko R G 2007 *J. Phys.: Cond. Matt.* **19** 145203
[29] Pollock E L and Ceperley D M 1987 *Phys. Rev. B* **36** 8343–8352
[30] Sandvik A W 1992 *J. Phys. A* **25** 3667–3682
[31] Fouet J, Sindzingre P and Lhuillier C 2001 *Eur. Phys. J. B* **20** 241–254
[32] Cabra D, Lamas C and Rosales H 2010 *Phys. Rev. B* **83** 094506
[33] Albuquerque A F, Schwandt D, Hetényi B, Capponi S, Mambrini M and Läuchli A M 2011 *Phys. Rev. B* **84** 024406

Mott Transitions in Three-Orbital Hubbard Model at Fractional Band Filling

Tomoko Kita[1], Takuma Ohashi[2] and Norio Kawakami[1]

[1] Department of Physics, Kyoto University, Kyoto 606-0852, Japan

[2] Department of Physics, Osaka University, Toyonaka, Osaka 560-0043, Japan

E-mail: `kita@scphys.kyoto-u.ac.jp`

Abstract. We investigate the Mott transitions in the three-orbital Hubbard model by means of the dynamical mean field theory with continuous-time quantum Monte Carlo simulations. We obtain the phase diagram determined in the orbital level splitting versus Hubbard interaction plane in the case of two electrons per site. It is also found that the introduction of holes into the orbital selective Mott insulating phase induces the metallic state with heavy quasiparticles.

Strongly correlated electron systems with multiorbital degrees of freedom have been studied extensively. In these systems, orbital fluctuations enhanced by electron correlations induce a variety of intriguing phenomena, such as magnetic and/or orbital ordering, metal-insulator transition, etc. The orbital selective Mott transition (OSMT), where some of the orbitals become localized while the others still remain itinerant, is one of the interesting problems induced by the interplay of multiorbital effects and strong electron correlations [1, 2].

The dynamical mean field theory (DMFT) has been applied to the two-orbital Hubbard model with different bandwidths, and it has been clarified that the Hund's coupling plays an important role in the OSMT [3–6]. Recently, the OSMT has also been addressed in the system with the same bandwidths [7, 8]. In these studies, it has been suggested that the OSMT occurs due to the difference not in bandwidths but in orbital degeneracy. The OSMT may occur in a three-orbital Hubbard model when the orbital degeneracy is lifted into doubly degenerate orbitals and a single orbital. In the doubly degenerate orbitals, a metallic state is expected to be further stabilized by orbital fluctuations than in a non-degenerate orbital, which results in separate transitions to the Mott insulating phase. However, the filling dependence of the orbital selective Mott insulator (OSMI) phase has not been sufficiently understood yet [7, 9], which may be important in real materials.

In this study, we investigate the Mott transitions in the three-orbital Hubbard model with particular focus on the filling dependence of the insulating phases in the case of two electrons per site. We clarify that the hole-doping effect on the OSMI phase induces the metallic state with the heavy quasiparticle peak.

For the purpose mentioned above, we consider the three-orbital Hubbard model defined by

$$\begin{aligned} H = & \ t \sum_{<i,j>,\alpha,} c^{\dagger}_{i\alpha} \ c_{j\alpha} \ + \sum_{i,\alpha,} (\ \ \mu + \Delta_{\alpha}) n_{i\alpha} \\ & + U \sum_{i,\alpha} n_{i\alpha\uparrow} n_{i\alpha\downarrow} + \sum_{i,\alpha \neq \alpha', \ , \ '} (U' \quad \delta \ \ {}'J) n_{i\alpha} \ n_{i\alpha'} \ ', \end{aligned} \tag{1}$$

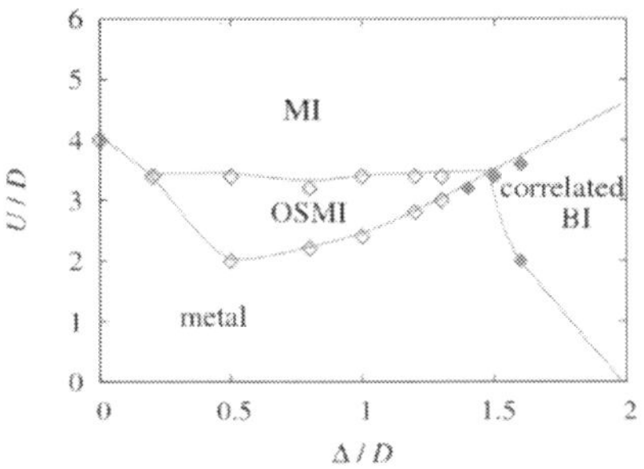

Figure 1. Δ-U phase diagram for the three-orbital Hubbard model with $J/U = 0.15$ and $n = 2.0$ at $T/D = 0.02$. The characteristic interaction strength of the first-order transition (crossover) is denoted by filled (open) diamonds. We only show U_{c_2} for the first-order transition. Solid lines are to guide the eye.

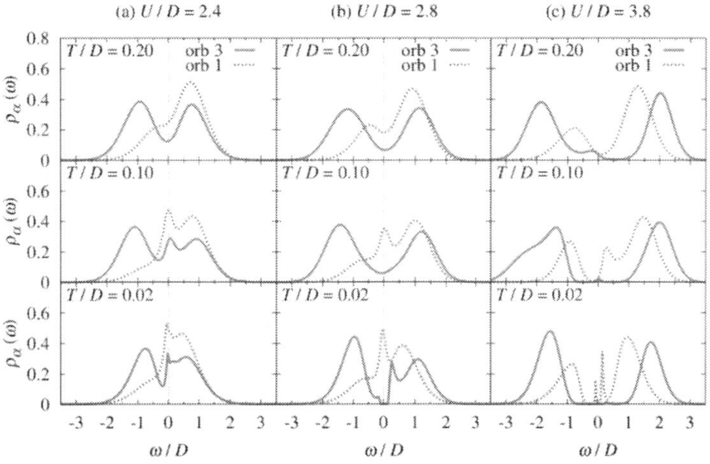

Figure 2. Density of states for the upper orbital 1 and the lower orbital 3 with $J/U = 0.15$ and $\Delta/D = 0.8$ at $n = 2.0$. Density of states are shown for three phases, (a) the metal, (b) the OSMI, and (c) the MI, with changing temperature T.

where $c_{i\alpha}^{(\dagger)}$ is an annihilation (creation) operator of an electron with spin σ $(=\uparrow, \downarrow)$ and orbital α $(= 1, 2, 3)$ at the ith site, and $n_{i\alpha} \;\; = c_{i\alpha}^{\dagger} \; c_{i\alpha}$ is the number operator. Here, t is the nearest-neighbor hopping integral, μ is the chemical potential, Δ_α is the energy level for orbital α. In the interaction terms, $U(U')$ is the intraorbital (interorbital) Coulomb interaction, and J is the Hund's coupling. We impose the condition $U = U' + 2J$ as usual. Orbital level splitting is given by Δ_α and we set $\Delta_1 = \Delta_2 = 0$, $\Delta_3 = \quad \Delta$ so that the energy level of doubly degenerate orbitals 1 and 2 is higher than that of the single orbital 3.

To investigate the Mott transition in this model, we use the single-site DMFT [10]. In the framework of DMFT, the lattice model is mapped onto an effective impurity model, where local electron correlations are taken into account precisely. The lattice Green's function is obtained via a self-consistency condition imposed on the impurity problem. In this paper, we use a semicircular density of states (DOS), $\;(\omega) = (2/\pi D)\sqrt{1 \quad (\omega/D)^2}$, which corresponds to the infinite-coordination Bethe lattice. We choose the energy unit as $D = 1$ (half bandwidth). To solve the effective impurity problem, we use the hybridization expansion continuous-time quantum Monte Carlo (CT-QMC) method [11, 12]. This method allows us to access the strong interaction regime in a wide range of temperature especially down to very low temperature [13–15]. Actually, calculations have been done successfully in previous DMFT + CT-QMC studies on the three-orbital Hubbard model [16–19].

In the model (1), it is known that the OSMT occurs when the electron filling per site is fixed at $n = 2.0$ [7, 8, 20]. Figure 1 shows the Δ-U phase diagram for $J/U = 0.15$ and $n = 2.0$ at low temperature $T/D = 0.02$. Note that for small Hund's coupling, e.g., $J/U = 0.05$, we have shown that the OSMI phase does not appear down to $T/D = 0.02$ at $n = 2.0$ in Ref. [20]. Therefore, we here choose a typical strength of Hund's coupling $J/U = 0.15$ which stabilizes the OSMI phase. In the phase diagram, the critical interaction of the first-order transition (crossover) is denoted by filled (open) diamonds, which is determined by the U-dependence of the double occupancy; the first order transition point is defined as the position where the double occupancy jumps and the crossover point is defined as the inflection point in the double occupancy curve. The crossover behavior is due to the finite-temperature effects and all the phase boundaries shown

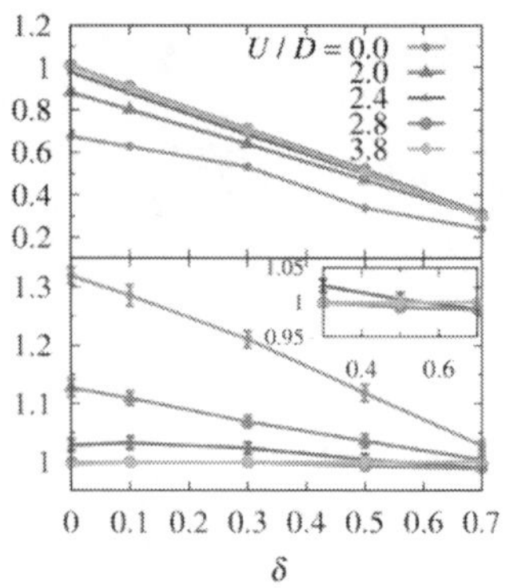

Figure 3. Electron number in the upper orbitals 1 and 2, $n_1 + n_2$ (top panel), and in the lower orbital 3, n_3 (bottom panel), as a function of doping δ. Other parameters are set as $J/U = 0.15$, $\Delta/D = 0.8$ and $T/D = 0.02$.

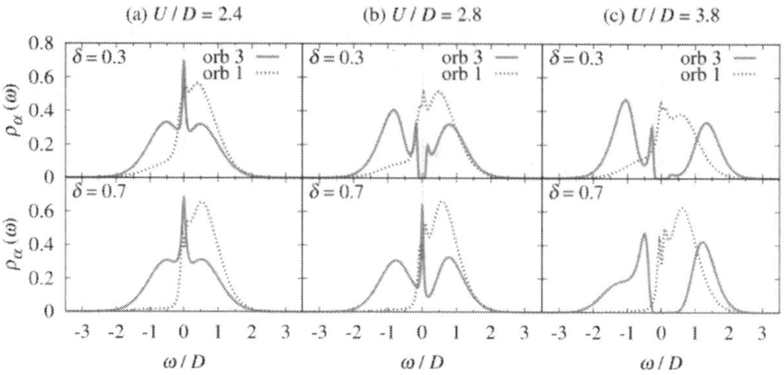

Figure 4. Density of states for the upper orbital 1 and the lower orbital 3 for several values of Coulomb interaction U and hole-doping δ. Other parameters are set as $J/U = 0.15$, $\Delta/D = 0.8$ and $T/D = 0.02$.

in the figure should be of first order at zero temperature. In the strong interaction regime, two kinds of insulators appear. One is the Mott insulator (MI) and the other is the correlated band insulator (BI). In a small Δ region, the Hund's coupling dominates the effect of Δ, inducing the MI with the electron number in each orbital $(n_1 + n_2, n_3) = (1, 1)$, where n_α denotes the electron number in orbital α. For a large Δ region, another insulating phase is stabilized, which is adiabatically connected to a BI specified by $(n_1 + n_2, n_3) = (0, 2)$. This insulating phase is referred to as the correlated BI. It is seen that the OSMI phase emerges in the MI region with intermediate interactions. For finite Δ $(0.2 < \Delta/D < 1.5)$, we find that the transition (or crossover) occurs twice. The Mott transition in the lower orbital occurs first at the smaller interaction, where the upper orbital is still metallic. This discrepancy in critical interaction strengths is due to the difference in orbital degeneracy.

To see how the quasiparticles evolve with changing temperature, we show the orbital-dependent DOS. We calculate the DOS in orbital α, $\;_\alpha(\omega)$, by applying the maximum entropy method [21] to the imaginary-time QMC data. In Fig. 2, we show $\;_\alpha(\omega)$ for several choices of T in three phases, the metal, the OSMI, and the MI. In the metallic phase (Fig. 2 (a)), the peaks around the Fermi level appear in both orbitals when temperature decreases. It is seen that at low temperature, the DOS for both orbitals form quasiparticle states around the Mott transition. With increasing U, the system enters the OSMI phase, as shown in Fig. 2 (b). The DOS for the lower orbital 3 opens a gap at low temperature while the DOS for the upper orbital 1 is still metallic with the quasiparticle peak. In the MI phase (Fig. 2 (c)), the DOS for both orbitals open gaps.

We discuss the hole-doping effect to clarify the stability of the OSMI phase and the quasiparticle behavior. Here, we introduce the parameter $\delta = 2.0 \quad n$. In Fig. 3, we show the electron number in each orbital as a function of δ. It is seen that the electron number in the upper orbitals $n_1 + n_2$ decreases with increasing δ for all U. On the other hand, in the lower orbital, n_3 remains 1 in the insulating phases $(U/D \geq 2.8)$ for small δ; most of the holes are doped in the upper orbitals due to correlation effects. In Fig. 4, we also show the orbital-dependent DOS to clarify the quasiparticle behavior. For $U/D = 2.8$, the undoped system is in the OSMI phase, as shown in the bottom panel of Fig. 2 (b). With increasing δ, the quasiparticle peaks in the upper orbitals (1 and 2) are suppressed, which is caused by the gradual decrease of $n_1 + n_2$. In the lower orbital 3, we find the filling-control Mott transition. The orbital 3 keeps insulating for small δ and becomes metallic at not infinitesimal but finite critical doping. As shown in Fig. 4 (b), $_3(\omega)$ is insulating for $\delta = 0.3$ and becomes metallic for $\delta = 0.7$. In this metallic phase, the heavy quasiparticle state is formed and sharp peak appears. These results are consistent with the behavior of the particle number. For $U/D = 3.8$, where the system is in the MI phase for $\delta = 0$, the gap in the upper orbital closes for infinitesimal doping, because $n_1 + n_2$ decreases and the filling becomes incommensurate one. On the other hand, the lower orbital 3 remains insulating for $\delta \leq 0.7$. Therefore, in the strong U regime, the system changes from the MI to the OSMI with hole doping.

In summary, we have studied the Mott transitions in the three-orbital Hubbard model with particular focus on the filling dependence of insulating states at $n = 2.0$. We have clarified that the OSMI phase appears in the doped system by doping holes into the MI phase at $n = 2.0$. It has also been found that the hole-doping effect on the OSMI phase at $n = 2.0$ induces the transition to the metal with the formation of the heavy quasiparticle state near the filling-control Mott transition.

Acknowledgments

The authors thank S. Sakai and K. Inaba for valuable discussions. This research is granted by the Japan Society for the Promotion of Science (JSPS) through the "Funding Program for World-Leading Innovative R&D on Science and Technology (FIRST Program)", initiated by the Council for Science and Technology Policy (CSTP). T.O. is supported by the Grant-in-Aid for Scientific Research [Grant nos. 21740232, 20104010] and the Next Generation Super Computing Project "Nanoscience Program" from the MEXT of Japan. N.K. is supported by the Grant-in-Aid for Scientific Research [Grant nos. 21540359, 20102008].

References

[1] Anisimov V, Nekrasov I, Kondakov D, Rice T and Sigrist M 2002 *The European Physical Journal B - Condensed Matter and Complex Systems* **25**(2) 191–201
[2] Koga A, Kawakami N, Rice T M and Sigrist M 2004 *Phys. Rev. Lett.* **92** 216402
[3] Koga A, Inaba K and Kawakami N 2005 *Progress of Theoretical Physics Supplement* **160** 253–273
[4] Knecht C, Blümer N and van Dongen P G J 2005 *Phys. Rev. B* **72** 081103
[5] Arita R and Held K 2005 *Phys. Rev. B* **72** 201102
[6] Inaba K and Koga A 2006 *Phys. Rev. B* **73** 155106
[7] de' Medici L, Hassan S R, Capone M and Dai X 2009 *Phys. Rev. Lett.* **102** 126401
[8] de' Medici L 2011 *Phys. Rev. B* **83** 205112
[9] Werner P and Millis A J 2007 *Phys. Rev. Lett.* **99** 126405
[10] Georges A, Kotliar G, Krauth W and Rozenberg M J 1996 *Rev. Mod. Phys.* **68** 13
[11] Werner P, Comanac A, de' Medici L, Troyer M and Millis A J 2006 *Phys. Rev. Lett.* **97** 076405
[12] Werner P and Millis A J 2006 *Phys. Rev. B* **74** 155107
[13] Haule K 2007 *Phys. Rev. B* **75** 155113
[14] Gull E, Werner P, Millis A and Troyer M 2007 *Phys. Rev. B* **76** 235123
[15] Gull E, Millis A J, Lichtenstein A I, Rubtsov A N, Troyer M and Werner P 2011 *Rev. Mod. Phys.* **83** 349–404
[16] Werner P, Gull E, Troyer M and Millis A J 2008 *Phys. Rev. Lett.* **101** 166405
[17] Werner P, Gull E and Millis A J 2009 *Phys. Rev. B* **79** 115119
[18] Chan C K, Werner P and Millis A J 2009 *Phys. Rev. B* **80** 235114
[19] Läuchli A M and Werner P 2009 *Phys. Rev. B* **80** 235117
[20] Kita T, Ohashi T and Kawakami N 2011 *Phys. Rev. B* **84** 195130
[21] Jarrell M and Gubernatis J E 1996 *Physics Reports* **269** 133 – 195

Conditions of ferromagnetic ordering in a model with Anderson-Hubbard centers: effective Hamiltonian approach

O Kramar, L Didukh, and Yu Skorenkyy

Ternopil National Technical University, Physics Department, 56 Ruska Str., Ternopil, Ukraine

E-mail: kramar@tu.edu.te.ua

Abstract. A model of strongly correlated electron system in which magnetic impurity levels are hybridized with conduction band has been considered. The effective Hamiltonian has been constructed for the case of strong Coulomb correlation on basis of configurational representation of Hamiltonian with Hubbard X-operators describing the localized spin subsystem. Criteria for the ferromagnetic ordering stabilization have been found in partial case of partially filled band for arbitrary temperatures and Curie temperature has been calculated. For the partial case of weak effective exchange the formula for the Curie temperature reproducing the well-established results has been calculated analytically. The electron concentration region favorable for ferromagnetic ordering is determined by hybridization through effective exchange integral.

1. Introduction

For description of unique properties of heavy-fermion compounds [1] the Anderson model [2] and its generalizations (see [3]) are widely used. One of the first generalizations of the single impurity Anderson model for the case of periodically spaced Anderson-Hubbard centers has been proposed in papers [4, 5]. It has been proven in the framework of the periodic Anderson model (PAM) that in the regime of strong intra-atomic interaction of localized magnetic moments the indirect (through the conduction band) exchange interaction occurs, which is proportional to the fourth order of the hybridization parameter $V(i\mathbf{k})$ and also the indirect hopping of the current carriers takes place, being dependent on the hybridization parameter squared. Recent years have seen a renewed interest to the mechanisms of ferromagnetism stabilization in the periodic Anderson model [6]-[10]. New class of systems to be modeled by the periodic Anderson model, namely organic ferromagnets [11] have been synthesized and investigated. In this work the effective Hamiltonian of PAM is obtained by canonical transformation method. It allows the derivation of the Hamiltonians for partial cases in controllable manner. A partial case of the model is studied and the mechanisms of ferromagnetic ordering are analyzed and concentration region favorable for ferromagnetism is found.

2. Configurational representation of Hamiltonian

Following papers [4, 12], we start from the model of Anderson-Hubbard material which describe localized ($d-$) subsystem hybridized with conduction ($c-$) band. Coulomb and exchange interactions within the localized subsystem are the most conveniently described in the configurational representation of Hubbard X-operators.

$$
\begin{aligned}
H &= H_0 + H_{\tilde{h}} + H_{\tilde{d}} + H_{2\tilde{h}} + H_{2\tilde{d}} + H_{\tilde{h}\tilde{d}}, \qquad (1)\\
H_0 &= (E_d-\mu)\sum_i\left(X_i^{\uparrow}+X_i^{\downarrow}+2X_i^2\right)+U\sum_i X_i^2+\sum_{\mathbf{k}\sigma}\varepsilon_{\mathbf{k}}c_{\mathbf{k}\sigma}^{+}c_{\mathbf{k}\sigma}\\
&- \sum_{ij\sigma}{}' \frac{J(ij)}{2}\left((X_i^{\sigma}+X_i^2)(X_j^{\sigma}+X_j^2)+X_i^{\sigma\bar{\sigma}}X_j^{\bar{\sigma}\sigma}\right),\\
H_{\tilde{h}} &= \sum_{i\mathbf{k}\sigma}(\eta_\sigma V(\mathbf{k}i)c_{\mathbf{k}\sigma}^{+}X_i^{0\sigma}+h.c.), \quad H_{2\tilde{h}} = 2\sum_{ij\mathbf{k}}(V(ij\mathbf{k},-\mathbf{k})X_i^{\uparrow 0}X_i^{\downarrow 0}c_{\mathbf{k}\uparrow}c_{-\mathbf{k}\downarrow}+h.c.),\\
H_{\tilde{d}} &= \sum_{i\mathbf{k}\sigma}(V(i\mathbf{k})c_{\mathbf{k}\sigma}^{+}X_i^{\sigma 2}+h.c.), \quad H_{2\tilde{d}} = 2\sum_{ij\mathbf{k}}(V(ij\mathbf{k},-\mathbf{k})X_i^{2\uparrow}X_i^{2\downarrow}c_{-\mathbf{k}\uparrow}c_{\mathbf{k}\downarrow}+h.c.),\\
H_{\tilde{h}\tilde{d}} &= 2\sum_{ij\mathbf{k}}(V(ij\mathbf{k},-\mathbf{k})(X_i^{\downarrow 0}X_i^{2\downarrow}c_{-\mathbf{k}\uparrow}c_{\mathbf{k}\downarrow}-X_i^{2\uparrow 0}X_i^{\uparrow 0}c_{-\mathbf{k}\uparrow}c_{\mathbf{k}\downarrow})+h.c.).
\end{aligned}
$$

Here operator X_i^{kl} describes transition of site i from state $|l\rangle$ to state $|k\rangle$, $c_{\mathbf{k}\sigma}^{+}(c_{\mathbf{k}\sigma})$ are creation (annihilation) operators for band electrons. Energy parameters of the model are the chemical potential μ, the energy of intra-site Coulomb repulsion of electrons U, the direct inter-site exchange interaction $J(ij)$, hybridization parameters $V(i\mathbf{k})$ and $V(ij\mathbf{k},-\mathbf{k})$. The peculiarity of the above model is the use of configurational representation which enables us to apply quite simple procedure for energy spectrum calculation, as strong intra-atomic correlation term is diagonal in this representation. As a consequence, we separate the hybridization processes of band states and the lower localized level located at E_d and hybridization of band states and the upper localized level located at E_d+U, though they are characterized by equal integrals in the present model. In the case when $E_d+U-E_F >> E_F-E_d$ (or opposite case) one can neglect corresponding translation processes in the Hamiltonian. These conclusions are in accordance with estimation of the hybridization matrix elements in the model of heavy fermions (see monograph [13]). The X−operator representation of the Anderson-like Hamiltonian is also suitable for mathematical treatment within Green function method.

3. Canonical transformation and effective Hamiltonian

According to the methodology of effective Hamiltonian derivation we assume that parameters characterizing the relative value of hybridization are small enough and perform Schrieffer-Wolff transformation [14] which excludes the terms of the first order in hybridization parameters $V(i\mathbf{k})$ and $V(ij\mathbf{k},-\mathbf{k})$. The resulting effective Hamiltonian has the form

$$
\tilde{H} = \tilde{H}_0 + H_1' + H_{cd}, \qquad (2)
$$

$$
\begin{aligned}
\tilde{H}_0 &= H_0 + \sum_{ij\sigma}{}' t_0(ij)X_i^{\sigma 0}X_j^{0\sigma} + \sum_{ij\sigma}{}' t_2(ij)X_i^{2\sigma}X_j^{\sigma 2} -\\
&- \frac{1}{2}\sum_{ij\sigma}{}' \tilde{J}(ij)\left(X_i^{\sigma}X_j^{\bar{\sigma}} - X_i^{\sigma\bar{\sigma}}X_j^{\bar{\sigma}\sigma} - X_i^0X_j^2\right),\\
H_1' &= \sum_{ij}{}' t_{02}(ij)\left(X_i^{2\downarrow}X_j^{0\uparrow} - X_i^{2\uparrow}X_j^{0\downarrow}\right) + h.c.,\\
H_{cd} &= \sum_{i\mathbf{k}\mathbf{k}'} J_1(i\mathbf{k}\mathbf{k}'i)(c_{\mathbf{k}\uparrow}^{+}c_{\mathbf{k}'\uparrow}(X_i^{\uparrow}+X_i^0) + c_{\mathbf{k}\downarrow}^{+}c_{\mathbf{k}'\downarrow}(X_i^{\downarrow}+X_i^0) -\\
&- c_{\mathbf{k}\downarrow}^{+}c_{\mathbf{k}'\uparrow}X_i^{\uparrow\downarrow} - c_{\mathbf{k}\downarrow}^{+}c_{\mathbf{k}'\uparrow}X_i^{\downarrow\uparrow}) + \sum_{i\mathbf{k}\mathbf{k}'} J_2(i\mathbf{k}\mathbf{k}'i)(c_{\mathbf{k}\uparrow}^{+}c_{\mathbf{k}'\uparrow}(X_i^{\downarrow}+X_i^2) +\\
&+ c_{\mathbf{k}\downarrow}^{+}c_{\mathbf{k}'\downarrow}(X_i^{\uparrow}+X_i^2) + c_{\mathbf{k}\downarrow}^{+}c_{\mathbf{k}'\uparrow}X_i^{\uparrow\downarrow} + c_{\mathbf{k}\downarrow}^{+}c_{\mathbf{k}'\uparrow}X_i^{\downarrow\uparrow}).
\end{aligned}
$$

In the above formulae $t_0(ij)$, $t_2(ij)$, $t_{02}(ij)$ are the integrals of indirect hopping through the sites with localized electrons (cation subsystem in transition metal compounds, quantum dots, etc), $J_1(i\mathbf{kk}'i)$ and $J_2(i\mathbf{kk}'i)$ are hybridization exchange integrals,

$$t_0(ij) = \sum_{\mathbf{k}<\mathbf{k_F}} \frac{V(i\mathbf{k})V(\mathbf{k}j)}{E_d - \epsilon_{\mathbf{k}}}, \quad t_2(ij) = \sum_{\mathbf{k}>\mathbf{k_F}} \frac{V(i\mathbf{k})V(\mathbf{k}j)}{E_d + U - \epsilon_{\mathbf{k}}}, \quad t_{02}(ij) = \frac{t_0(ij) + t_2(ij)}{2},$$

$$J_1(i\mathbf{kk}'i) = \frac{V(i\mathbf{k})V(\mathbf{k}'i)}{2}\left(\frac{1}{\epsilon_{\mathbf{k}} - E_d} + \frac{1}{\epsilon_{\mathbf{k}'} - E_d}\right),$$

$$J_2(i\mathbf{kk}'i) = \frac{V(i\mathbf{k})V(\mathbf{k}'i)}{2}\left(\frac{1}{\epsilon_{\mathbf{k}} - E_d - U} + \frac{1}{\epsilon_{\mathbf{k}'} - E_d - U}\right).$$

The magnitudes of these parameters can essentially renormalize the bare band hopping integral and enhance localization effects.

We restrict ourselves to the case of total (band and localized) electron concentration $n < 1$. In this case for sufficiently large U a hybridization with the upper localized level is negligible. Thus, in the case of strong correlation $U >> w_d$ (here w_d is $d-$band halfwidth) and $n < 1$ the effective Hamiltonian of localized electron subsystem takes the form:

$$H_{eff} = (E_d - \mu)\sum_i \left(X_i^{\uparrow} + X_i^{\downarrow}\right) + \sum_{ij\sigma} t_0(ij)X_i^{\sigma 0}X_j^{0\sigma} - \frac{J_{eff}}{2}{\sum_{ij\sigma}}' \left(X_i^{\sigma\sigma}X_j^{\sigma\sigma} + X_i^{\sigma\bar{\sigma}}X_j^{\bar{\sigma}\sigma}\right). \quad (3)$$

The used canonical transformation operator maps the general PAM onto an effective model with hybridization-renormalized hopping integral $t_0(ij)$ and effective exchange J_{eff} of ferromagnetic type determined by both hybridization of localized and band states and strong intra-atomic Coulomb correlation. In distinction from standard $t - J$ Hamiltonian the hopping amplitude is substantially suppressed as it represents an indirect hopping here. In other respects the Hamiltonian (3) is very simple and allows for analytical calculations. The simplest analytical approach of choice is the decoupling of equation of motions, similar to the first step of work [15].

4. Condition of ferromagnetic ordering

The energy spectrum obtained within projection procedure [16] in the Green function method $E_{\mathbf{k}} = -\mu - t_{0\mathbf{k}} + zn_\sigma J_{eff}$ allows us to calculate the mean numbers of spin-up and spin-down electrons. Equation for system magnetization has the form

$$\exp\left(\frac{zJ_{eff}m}{\Theta}\right) = \frac{\sinh\left(\frac{2(1-n)}{2-n-m}\frac{w_d}{\Theta}\right)\sinh\left(\frac{w_d}{\Theta} - \frac{w_d}{\Theta}\frac{2(1-n)}{2-n+m}\right)}{\sinh\left(\frac{2(1-n)}{2-n+m}\frac{w_d}{\Theta}\right)\sinh\left(\frac{w_d}{\Theta} - \frac{w_d}{\Theta}\frac{2(1-n)}{2-n-m}\right)}. \quad (4)$$

At $n \to 0$ the above equation reproduces the corresponding molecular field equation. The above equation allows us to obtain the equation for Curie temperature. Let us take $\Theta_C \ll w_d$. Then

$$\frac{\Theta_C}{w_d} = \frac{(1-n)}{(2-n)\ln\frac{2-n}{m_0}}, \quad m_0 = \left[(2-n)^2 - \frac{4(1-n)w_d}{zJ_{eff}}\right]^{\frac{1}{2}}, \quad (5)$$

where m_0 is the ground state system magnetization. If the band is less than half-filled, one has $\Theta_C = zJ_{eff}/2$ from eq. (5), in agreement with the above considerations. It is interesting to note that in eq. (5) Curie temperature value is proportional to the conduction band width, though ferromagnetic ordering is stabilized by exchange mechanism. From figs. 1,2 one can see that the ferromagnetic ordering is quite stable for electron concentrations $n > 0.5$, in agreement with the results of paper [7]. The electron concentration region favorable for ferromagnetic ordering

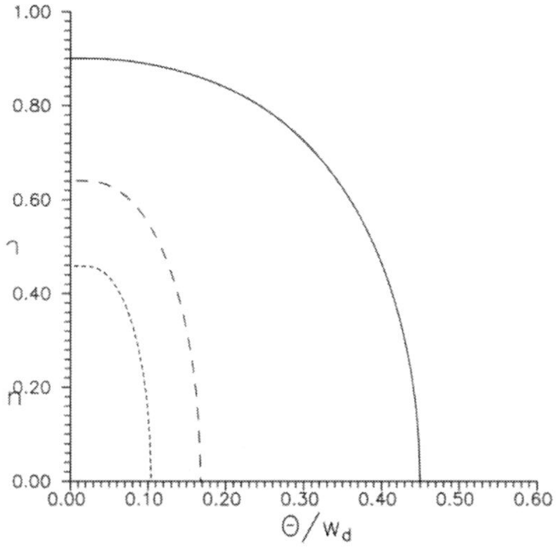

Figure 1. Temperature dependence of magnetization at $n = 0.9$. $zJ_{eff}/w_d = 1$ for solid curve, $zJ_{eff}/w_d = 0.5$ for dashed curve, $zJ_{eff}/w = 0.4$ for dotted curve.

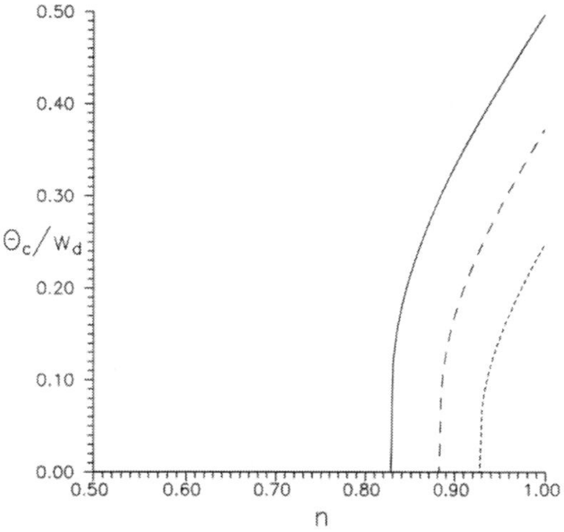

Figure 2. Concentration dependence of Curie temperature. $zJ_{eff}/w_d = 1$ for solid curve, $zJ_{eff}/w_d = 0.75$ for dashed curve, $zJ_{eff}/w = 0.5$ for dotted curve.

depends essentially on the position of localized level with respect to band center (this result agrees with [7, 9, 10]) and is considerably widened by hybridization through effective exchange integral. For the considered variant of PAM in which hopping amplitudes are renormalized by hybridization, assumption of paper [7] about the translational mechanism of ferromagnetic order stabilization is questionable, though taking into account the peculiarities of unperturbed electron density of states (e.g., its asymmetry) may change the situation in favor of scenario of paper [7]. Still, in the present model the driving force of the ordering here is definitely the effective exchange. One more important factor to be taken into account in the model to allow its direct application to real material is the orbital degeneracy of localized states [9].

References

[1] N. Grewe, F. Steglich, *Handbook on the Physics and Chemistry of Rare Earth* , Vol. 14, 343, Elsevier Science Publishers B.V. 1991.
[2] Anderson P.W. 1961 *Phys. Rev.* **124**, 41.
[3] Hewson A.C. *The Kondo Problem to Heavy Fermions, Cambridge University Press*, New York, N.Y., 1993.
[4] Didukh L.D., Stasyuk I.V. 1968 *Fiz. Metal. Metaloved.* **26**, 582 [in Russian].
[5] Didukh L.D., Stasyuk I.V. 1968 *Ukr. Fiz. Zhurn.* **13**, 1774 [in Russian].
[6] Meyer D., Nolting W., Reddy G.G., Ramakanth A. 1998 *phys. stat. sol. (b)* **208**, 473.
[7] Meyer D., Nolting W. 2000 *Eur. Phys. Journ.* **18**, 385.
[8] Meyer D., Nolting W. 2000 *Phys. Rev. B.* **62**, 5657.
[9] Unjong Yu, Byczuk K., Vollhardt D. 2008 *Phys. Rev. B.* **78**, 205118.
[10] Majidi M.A. et al 2007 Charge density wave driven ferromagnetism in the periodic Anderson model Preprint cond-mat.str-el/0710.5937.
[11] Suwa Yu., Arita R., Kuroki K. and Aoki H. 2010 *Phys. Rev. B.* **82**, 235127.
[12] Didukh L.D., Didukh V.D., and Stasyuk I.V. 1975 *Ukr. Fiz. Zhurn.* **20**, 97 [in Russian].
[13] Fazekas P. *Lecture notes on electron correlation and magnetism*, Singapore, World Scientific Publishing, 1999.
[14] Schrieffer J.R., Wolff P.A. 1966 *Phys. Rev.* **149**, 491.
[15] Fransson J., Eriksson O., and Sandalov I. 2002 *Physical Review Letters* **88**, 226601.
[16] Didukh L. 2000 *Acta Physica Polonica (B).* **31**, 3097.

Mass enhancement in an extended periodic Anderson model with valence fluctuations

Katsunori Kubo

Advanced Science Research Center, Japan Atomic Energy Agency, Tokai, Ibaraki 319-1195, Japan

E-mail: `kubo.katsunori@jaea.go.jp`

Abstract. We study the mass enhancement in an extended periodic Anderson model with the Coulomb interaction U_{cf} between conduction and f-electrons by the Gutzwiller method. In the Kondo regime, where the number of f electrons n_f per site is almost one, the mass enhancement factor becomes large as in the ordinary periodic Anderson model without U_{cf}. In the intermediate-valence regime, we nd that the mass enhancement factor becomes large for a large U_{cf}. As a result, the effective mass can vary nonmonotonically as a function of n_f, which may be relevant to the experimental observations of $CeCu_2Si_2$ under pressure.

1. Introduction

In recent years, a possibility of valence fluctuation mediated superconductivity has been discussed [1–4]. Such valence fluctuations are inferred from the observed rapid change in the effective mass. For example, the effective mass deduced from specific heat measurement in $CeCu_2Si_2$ changes rapidly at around the pressure where the superconducting transition temperature becomes maximum [5]. The rapid change in the effective mass indicates a rapid change in the valence according to the following relation [6]:

$$\frac{m}{m} = \frac{1 - n_f/2}{1 - n_f}, \tag{1}$$

where m is the free-electron mass, m is the effective mass, and n_f is the number of f electrons per site. However, this relation is derived for the periodic Anderson model, in which a sharp valence change does not occur. Thus, it is not clear whether we can apply this relation to a system with large valence fluctuations. In addition, the effective mass varies nonmonotonically in $CeCu_2Si_2$ under pressure [5]. In $CeCu_2Si_2$, recent X-ray absorption experiment shows that n_f decreases monotonically under pressure [7]. Equation (1) is a monotonic function of n_f, and the nonmonotonic variation in the effective mass of $CeCu_2Si_2$ cannot be explained by this relation. Thus, in this study, we extend this relation to a model in which a sharp valence change can occur.

2. Model and Method

We study an extended periodic Anderson model with the Coulomb interaction U_{cf} between conduction and f electrons. The Hamiltonian is given by [8]

$$H = \sum_{\boldsymbol{k}\sigma} {}_{\boldsymbol{k}} c^\dagger_{\boldsymbol{k}\sigma} c_{\boldsymbol{k}\sigma} + {}_{f} \sum_{i\sigma} n_{fi\sigma} - V \sum_{\boldsymbol{k}\sigma} (f^\dagger_{\boldsymbol{k}\sigma} c_{\boldsymbol{k}\sigma} + c^\dagger_{\boldsymbol{k}\sigma} f_{\boldsymbol{k}\sigma}) + U \sum_{i} n_{fi\uparrow} n_{fi\downarrow} + U_{cf} \sum_{i\sigma\sigma'} n_{ci\sigma} n_{fi\sigma'}, \quad (2)$$

where $c_{\boldsymbol{k}\sigma}$ and $f_{\boldsymbol{k}\sigma}$ are the annihilation operators of conduction and f electrons, respectively, with the momentum $\boldsymbol{k}$ and the spin . $n_{ci\sigma}$ and $n_{fi\sigma}$ are the number operators at site i with of the conduction and f electrons, respectively. ${}_{\boldsymbol{k}}$ is the kinetic energy of the conduction electron, ${}_{f}$ is the energy level of the f electron, V denotes hybridization, and U is the Coulomb interaction between f electrons. In the following, we set the energy level of the conduction band as the origin of energy, i.e., $\sum_{\boldsymbol{k}} {}_{\boldsymbol{k}} = 0$. We set $U \to \infty$, since the onsite Coulomb interaction between well-localized f electrons is large.

In this study, we employ the Gutzwiller method [9]. The Gutzwiller method has already applied to the ordinary periodic Anderson model without U_{cf} [6, 10]. In this study, we extend the method developed by Fazekas and Brandow [10] to the extended periodic Anderson model. The details of the method and some results have been reported in our previous papers [11, 12]; here, we report some new results.

In the Gutzwiller method, we consider the variational wave function given by

$$| \ \rangle = P_{ff} P_{cf} | \ \rangle, \quad (3)$$

where

$$P_{ff} = \prod_{i} [1 - n_{fi\uparrow} n_{fi\downarrow}] \quad (4)$$

excludes the double occupancy of f electrons at the same site, and

$$P_{cf} = \prod_{i\sigma\sigma'} [1 - (1-g) n_{ci\sigma} n_{fi\sigma'}] \quad (5)$$

is introduced to deal with the onsite correlation between conduction and f electrons [13]. g is a variational parameter. The one-electron part of the wave function is given by

$$| \ \rangle = \prod_{k<k_{\mathrm{F}},\sigma} [c^\dagger_{\boldsymbol{k}\sigma} + a(\boldsymbol{k}) f^\dagger_{\boldsymbol{k}\sigma}] |0\rangle, \quad (6)$$

where k_{F} is the Fermi momentum for the free conduction band without f electrons, $|0\rangle$ denotes vacuum, and $a(\boldsymbol{k})$ is determined variationally. Here, we have assumed that the total number n of electrons per site is less than 2 and consider only the lower hybridized band.

We evaluate the expectation value of energy of the variational wave function by applying Gutzwiller approximation. Then, we determine the variational parameters, g and $a(\boldsymbol{k})$, so that the energy becomes minimum. By using the optimized wave function, we can evaluate physical quantities.

3. Result

Before showing the calculated results, we discuss a state at a very large U_{cf}. For a very large U_{cf}, conduction and f electrons avoid each other. Then, at a moderate ${}_{f}$, each site is occupied by two conduction electrons or one f electron, that is, $n_c/2 + n_f \simeq 1$, where n_c is the number

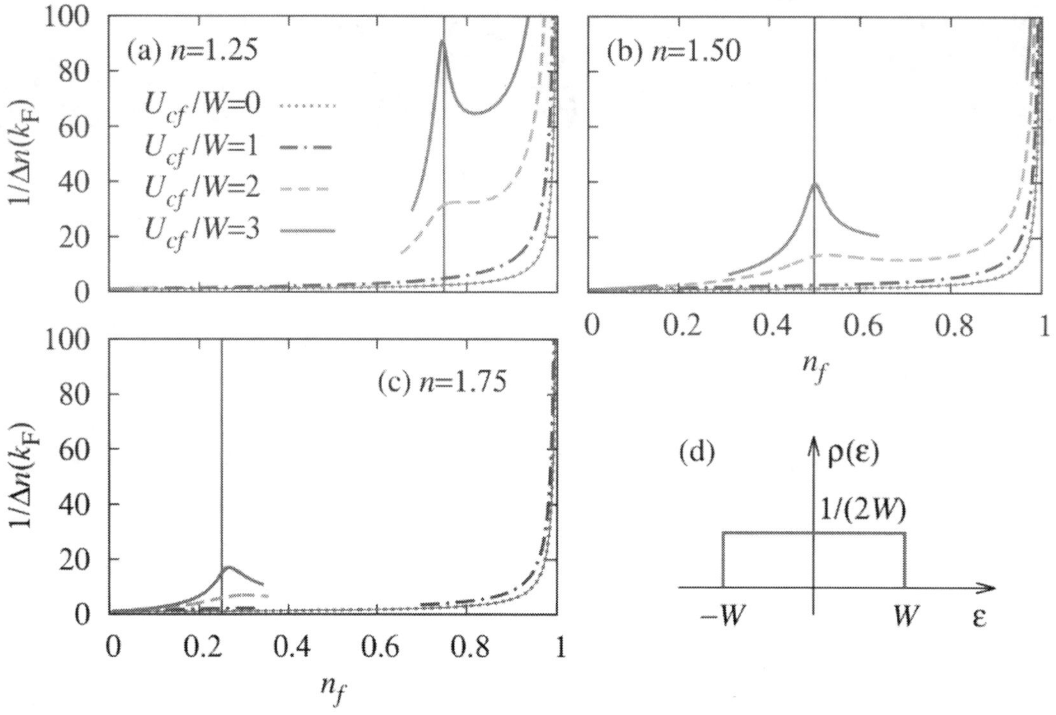

Figure 1. Mass enhancement factor $1/\ n(k_{\rm F})$ as functions of n_f for (a) $n = 1.25$, (b) $n = 1.50$, and (c) $n = 1.75$. $U_{cf}/W = 0$ (dotted lines), 1 (dash-dotted lines), 2 (dashed lines), and 3 (solid lines). The thin lines are $(1-n_f/2)/(1-n_f)$. The vertical lines indicate $n_f = 2-n$. (d) Density of states per spin of the conduction band.

of conduction electrons per site. The total number of electrons is $n = n_c + n_f$ and we obtain $n_f \simeq 2 - n$. In such a state, both the conduction and f electrons tend to be localized, and we expect a large effective mass at $n_f \simeq 2 - n$.

Now, we show the calculated mass enhancement factor. In the calculation, we consider a simple density of states per spin for the conduction band [shown in Fig. 1(d)]: $\rho(\ ) = 1/(2W)$ for $-W \quad W$; otherwise $\rho(\ ) = 0$. In this study, we define the inverse of the jump $\ n(k_{\rm F})$ in the momentum distribution function of electrons at the Fermi level as the mass enhancement factor. Figure 1 shows the mass enhancement factor $1/\ n(k_{\rm F})$ as functions of n_f for $n = 1.25$, 1.50, and 1.75. For a large U_{cf}, n_f jumps by changing $_f$, that is, a first-order valence transition takes place. We can recognize the first-order transitions by the terminations of lines in Fig. 1. For example, n_f jumps from 0.65 to 0.96 for $n = 1.50$ and $U_{cf}/W = 3$. The thin lines are the renormalization factor given by Eq. (1), which is derived for the ordinary periodic Anderson model without U_{cf} with $g = 1$. The $U_{cf} = 0$ data are almost overlapping with these thin lines. However, we note that $g \neq 1$ even for $U_{cf} = 0$ in the present theory. In the Kondo regime, i.e., $n_f \simeq 1$, the mass enhancement factor becomes large, irrespective of U_{cf}, as in the ordinary periodic Anderson model. By increasing U_{cf} from zero, $1/\ n(k_{\rm F})$ deviates from the relation Eq (1). In particular, $1/\ n(k_{\rm F})$ is enhanced by the effect of U_{cf} in the intermediate-valence regime $n_f \simeq 2-n$ as expected. As a result, the mass enhancement factor varies nonmonotonically as a function of n_f for a large U_{cf}.

4. Summary

To summarize, we have studied the mass enhancement in the extended periodic Anderson model with the Coulomb interaction U_{cf} between conduction and f electrons. We find that the mass enhancement factor becomes large in the intermediate-valence regime by the effect of U_{cf}. This enhancement results in the nonmonotonic variation of the mass enhancement factor as a function of n_f. This nonmonotonic variation may be relevant to $CeCu_2Si_2$ under pressure.

Acknowledgments

This work is supported by a Grant-in-Aid for Young Scientists (B) from the Japan Society for the Promotion of Science.

References

[1] Miyake K, Narikiyo O and Onishi Y 1999 *Physica B* **259–261** 676
[2] Onishi Y and Miyake K 2000 *J. Phys. Soc. Jpn.* **69** 3955
[3] Watanabe S, Imada M and Miyake K 2006 *J. Phys. Soc. Jpn.* **75** 043710
[4] Sugibayashi T, Saiga Y and Hirashima D S 2008 *J. Phys. Soc. Jpn.* **77** 024716
[5] Holmes A T, Jaccard D and Miyake K 2004 *Phys. Rev. B* **69** 024508
[6] Rice T M and Ueda K 1986 *Phys. Rev. B* **34** 6420
[7] Rueff J P, Raymond S, Taguchi M, Sikora M, Itié J P, Baudelet F, Braithwaite D, Knebel G and Jaccard D 2011 *Phys. Rev. Lett.* **106** 186405
[8] Gonçalves da Silva C E T and Falicov L M 1975 *Solid State Commun.* **17** 1521
[9] Gutzwiller M C 1965 *Phys. Rev.* **137** A1726
[10] Fazekas P and Brandow B H 1987 *Phys. Scr.* **36** 809
[11] Kubo K 2011 *J. Phys. Soc. Jpn.* **80** 063706
[12] Kubo K 2011 *Preprint* arXiv:1108.3132
[13] Onishi Y and Miyake K 2000 *Physica B* **281–282** 191

Interplay between magnetic correlation and evolution of Fermi liquid in the periodic Anderson model

T Mutou[1], E Nagira[1] and H Kusunose[2]

[1] Interdisciplinary Faculty of Science and Engineering, Shimane University, Matsue 690-8504, Japan

[2] Department of Physics, Ehime University, Matsuyama 790-8577, Japan

E-mail: tmutou@riko.shimane-u.ac.jp

Abstract. A simple e ective scheme to improve the self energy obtained by the dynamical mean field theory is proposed, in which a feedback of magnetic fluctuations is taken into account. We demonstrate e ectiveness of the scheme for the two-dimensional periodic Anderson model by investigating the e ect of the magnetic fluctuation in the formation of heavy quasiparticles. It is found that the spectral intensity near the Fermi level is strongly suppressed by the antiferromagnetic fluctuation slightly above the magnetic instability.

1. Introduction

In the heavy-fermion systems, the Kondo e ect leading to a coherent Fermi-liquid state and the inter-site interaction for a magnetic order often compete with each other. The heavy quasiparticle emerges as a consequence of the strong local uctuations, and simultaneously the evolution is a ected by developing magnetic uctuations. As a result, the interplay between nonlocal magnetic uctuations and the evolution of the Fermi liquid has attracted considerable interest to elucidate the origin of the characteristic phenomena observed in the heavy fermion systems, especially so-called the quantum critical behaviors. However, a useful scheme to treat e ects of the magnetic uctuations and the strong local correlations on the same footing is still under development. In the present study, we propose a simple e ective scheme to treat the strong local correlations and the nonlocal magnetic uctuations based on the dynamical mean eld theory (DMFT). We demonstrate e ectiveness of the scheme for the periodic Anderson model by showing the e ect of the antiferromagnetic spin uctuations on the formation of the heavy quasiparticles.

2. Model and Method

In order to demonstrate our scheme, let us consider the periodic Anderson model on a two-dimensional square lattice,

$$\mathcal{H} = \sum_{\boldsymbol{k}} \{(\varepsilon_{\boldsymbol{k}} \quad \mu) c^\dagger_{\boldsymbol{k}} c_{\boldsymbol{k}} + (\varepsilon^f_{\boldsymbol{k}} \quad \mu) f^\dagger_{\boldsymbol{k}} f_{\boldsymbol{k}} + V(c^\dagger_{\boldsymbol{k}} f_{\boldsymbol{k}} + f^\dagger_{\boldsymbol{k}} c_{\boldsymbol{k}})\} + U \sum_i f^\dagger_{i\uparrow} f_{i\uparrow} f^\dagger_{i\downarrow} f_{i\downarrow},$$

$$\varepsilon_{\boldsymbol{k}} = \quad 2(\cos k_x + \cos k_y), \qquad \varepsilon^f_{\boldsymbol{k}} = \varepsilon^f + \alpha \varepsilon_{\boldsymbol{k}},$$

where ε^f denotes the f level and α is a dispersiveness coe cient of the f electron, and the nearest-neighbor hopping on the square lattice is assumed (the hopping energy of the conduction electron and a lattice constant is set to be unity). In the present model, we introduce the energy dispersion of the f electron so that the metallic ground state can be realized in the model under the particle-hole symmetric condition between the f level ε^f, the on-site Coulomb energy U, and the chemical potential μ: $\varepsilon^f + U/2 = \mu = 0$.

To treat the strong local correlation between f electrons, we employ the DMFT, i.e., the local approximation for the self energy [1]. As an impurity solver in the DMFT framework, we adopt the iterative perturbation theory (IPT) [1, 2]. It is well known that the IPT is accurate to calculate the Green's function at wide temperature range in the presence of the particle-hole symmetry [3]. We calculate the local Green's function of f electrons $G_{ff}(z)$ by the DMFT with the IPT,

$$G_{ff}(z) = \frac{1}{N}\sum_{\boldsymbol{k}} G_{ff}(\boldsymbol{k}, z), \quad G_{ff}(\boldsymbol{k}, z) = \left(z - \varepsilon^f_{\boldsymbol{k}} - \Sigma(z) - \frac{V^2}{z - \varepsilon_{\boldsymbol{k}}}\right)^{-1},$$

where z denotes either the Matsubara frequency or the real frequency, and N represents a total number of the lattice points (hereafter a paramagnetic phase is considered and spin indices are omitted). In the IPT, the self energy is approximated by the second-order contribution in terms of the Weiss function $\mathcal{G}(z)$,

$$\Sigma(i\varepsilon_n) \simeq U^2 T \sum_m \bar{\chi}_0(i\omega_m)\mathcal{G}(i\varepsilon_n + i\omega_m), \quad \bar{\chi}_0(i\omega_m) \equiv -T\sum_n \mathcal{G}(i\varepsilon_n)\mathcal{G}(i\varepsilon_n + i\omega_m),$$
$$\mathcal{G}(z)^{-1} = G_{ff}(z)^{-1} + \Sigma(z).$$

After the self-consistent solution of the Green's function is obtained in the IPT-DMFT framework, we take account of the magnetic fluctuations as follows. First, we construct the (longitudinal) spin susceptibility $\chi^{ff}_{\mathrm{S}}(\boldsymbol{q}, z)$ by the RPA-like procedure with use of the renormalized repulsion U^*,

$$\chi^{ff}_{\mathrm{S}}(\boldsymbol{q}, z) = \frac{\chi^{ff}_0(\boldsymbol{q}, z)}{1 - U^* \chi^{ff}_0(\boldsymbol{q}, z)}, \qquad \chi^{ff}_0(\boldsymbol{q}, i\omega_m) = -\frac{1}{N}\sum_{\boldsymbol{k}} T \sum_n G_{ff}(\boldsymbol{k}, i\varepsilon_n) G_{ff}(\boldsymbol{k}+\boldsymbol{q}, i\varepsilon_n + i\omega_m).$$

Next, we improve the DMFT self energy by taking account of the spin fluctuations as

$$\Sigma(\boldsymbol{k}, i\varepsilon_n) = \Sigma(i\varepsilon_n) + \frac{UU^*}{2}\frac{1}{N}\sum_{\boldsymbol{q}} T \sum_m \left(\chi^{ff}_{\mathrm{S}}(\boldsymbol{q}, i\omega_m) - \chi^{ff}_{\mathrm{S}}(i\omega_m)\right) G_{ff}(\boldsymbol{k}+\boldsymbol{q}, i\varepsilon_n + i\omega_m), \quad (1)$$

where we have subtracted the local contributions in the second term, since it is already taken into account in the DMFT self energy, $\Sigma(i\epsilon_n)$. $\chi^{ff}_{\mathrm{S}}(z)$ is the local counterpart of the susceptibility defined by

$$\chi^{ff}_{\mathrm{S}}(z) \equiv \frac{\chi^{ff}_0(z)}{1 - U^* \chi^{ff}_0(z)}, \qquad \chi^{ff}_0(i\omega_m) = -T\sum_n G_{ff}(i\varepsilon_n) G_{ff}(i\varepsilon_n + i\omega_m).$$

In fact, the $\boldsymbol{k}$-summation of $\Sigma(\boldsymbol{k}, i\varepsilon_n)$ in the above expression (1) gives the local self energy though the present framework is phenomenologically constructed. In the present framework, U^* is the phenomenological parameter, corresponding to the local and instantaneous approximation of the four-point vertex function in the longitudinal particle-hole channel [4]. Here, we have neglected the charge fluctuations because the charge fluctuations are strongly suppressed in our parameter range in question. Finally, we obtain the final expression of the spectral function $\rho_f(\boldsymbol{k}, \varepsilon)$ with including the spin fluctuations,

$$\rho_f(\boldsymbol{k}, \varepsilon) = -\frac{1}{\pi}\mathrm{Im}\left(z - \varepsilon^f_{\boldsymbol{k}} - \Sigma(\boldsymbol{k}, z) - \frac{V^2}{z - \varepsilon_{\boldsymbol{k}}}\right)^{-1}\Bigg|_{z=\varepsilon+i\eta}.$$

We have used $U = 8$, $V^2 = 3$, and $\alpha = 0.3$ in the present work. In the computation, we have solved the self-consistent equation in the DMFT scheme and the dynamical spin susceptibilities in the rst Brillouin zone with equally spaced mesh of 256×256 and 2048 in the real frequency range, [-15,15]. Note that since all the above equations can be analytically continued explicitly, we have performed the computations in the real-frequency domain with the damping factor, $\eta = 0.01$.

3. Results

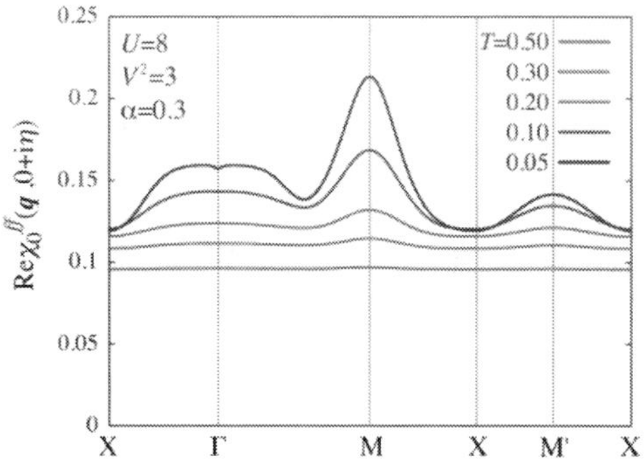

Figure 1. The f-electron static spin susceptibility (proper part) along the high symmetry line at $T = 0.50$, 0.30, 0.20, 0.10, and 0.05.

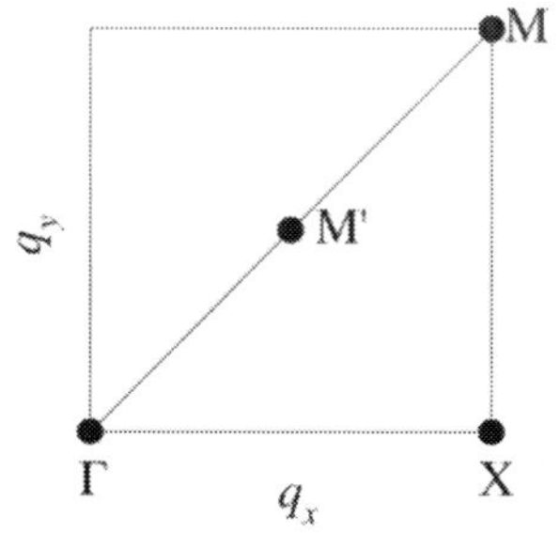

Figure 2. The high symmetry points of the rst Brillouin zone.

At rst, we show the T dependence of a proper part of the f-electron static spin susceptibility, Re $_0^{ff}(\boldsymbol{q}, 0+i\eta)$, along the high symmetry line of the rst Brillouin zone in Fig. 1 (The de nition of the high symmetry points is shown in Fig. 2). The peak of Re $_0^{ff}(\boldsymbol{q}, 0 + i\eta)$ is always located at $\boldsymbol{q} = (\pi, \pi) \equiv \boldsymbol{Q}$ (M point) at all temperatures due to the perfect nesting across the hybridized bands, indicating that the antiferromagnetic uctuation is dominant in the present system. In order to demonstrate the e ect of the critical spin uctuations, we choose the phenomenological parameter U^* so that the magnetic instability occurs at slightly below the lowest temperature in the present computation, i.e., $T = 0.05$. By the instability condition, 1 U^*Re $_0^{ff}(\boldsymbol{Q}, 0+i\eta) = 0$, we have chosen $U^* = 4.68$. In principle, the renormalized vertex U^* does depend on the temperature, however, we have neglected it for simplicity.

In the results with the spin uctuation (denoted as DMFT+SF), the self energy is improved according to the expression (1). As shown in Fig. 3, the spectra of the f-electron density of states with or without the spin uctuations exhibit considerable di erences in approaching to the magnetic instability in spite of no essential di erences at high temperatures. At temperatures below the characteristic temperature of the renormalized Fermi energy, the quasiparticle peak develops at the Fermi level in the spectrum without the spin uctuation (Fig. 3(b)). On the other hand, in the spectrum with the spin uctuation (Fig. 3(a)), the dip structure appears at the center of the quasiparticle peak, yielding the development of the pseudogap due to the critical spin uctuations.

Next, we show the quasiparticle dispersion in the intensity map of the f-electron spectral function in Fig. 4. The f-electron spectral function without the spin uctuation exhibits strong intensities corresponding to the dispersion of the heavy quasiparticle band near the Fermi level. The critical antiferromagnetic uctuations strongly in uence the electronic states near the Fermi

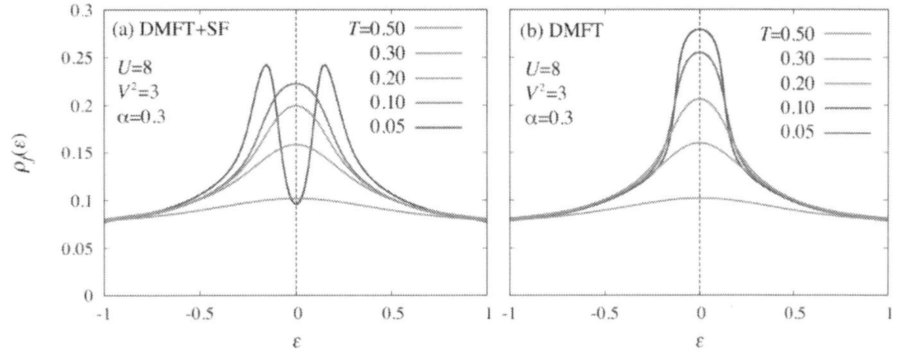

Figure 3. The density of states of f electron with the spin uctuation (a), and without it (b) at $T = 0.50$, 0.30, 0.20, 0.10, and 0.05.

level, leading to considerable suppression of the spectral intensity near the Fermi level as shown in Fig. 4(a).

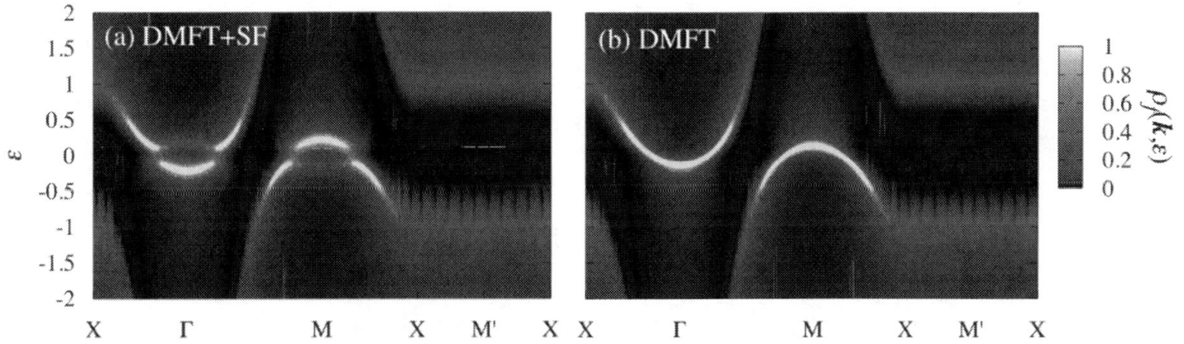

Figure 4. The intensity map of the spectral function of f electron with the spin uctuation (a), and without it (b) at $T = 0.05$ in the vicinity of the antiferromagnetic instability.

4. Summary

We have proposed the simple scheme to treat the nonlocal magnetic uctuations beyond the DMFT. The application to the periodic Anderson model on a square lattice clearly demonstrates that the evolution of the heavy quasiparticle is strongly suppressed by the critical antiferromagnetic spin uctuations, yielding the pesudogap behavior in the f-electron spectrum near the Fermi level. The application to more realistic systems is left for future investigations.

References

[1] Georges A, Kotliar G, Krauth W and Rozenberg M J 1996 *Rev. Mod. Phys.* **68** 13
[2] Georges A and Kotliar G 1992 *Phys. Rev.* B **45** 6479
[3] Zhang X Y, Rozenberg M J and Kotliar G 1993 *Phys. Rev. Lett.* **70** 1666
[4] Kusunose H 2006 *J. Phys. Soc. Jpn.* **75** 054713

Excitation spectra of the one-dimensional double-exchange model: An exact solution for one mobile electron

K Nakano[1], **R Eder**[2] **and Y Ohta**[1]

[1] Department of Physics, Chiba University, Chiba 263-8522, Japan
[2] Karlsruhe Institute of Technology, Institute for Solid State Physics, 76021 Karlsruhe, Germany

E-mail: kyohei-1985@graduate.chiba-u.jp

Abstract. We study the one-dimensional double-exchange model with L localized spins and one mobile electron. We solve the Schrödinger equation analytically and obtain the energies and wave functions for all the eigenstates with spin $S = (L-1)/2$ exactly. As an application, we compute the single-particle Green's function. We show that, for vanishing exchange interactions between localized spins, the single-particle spectrum is entirely incoherent and the lowest band has an infinite band mass, i.e., the single electron is localized due to its interaction with the spin excitations. The analysis of the wave function gives us a particularly simple ground state.

1. Introduction

Half-metallic ferromagnets [1] offer a unique opportunity for studying the electronic states of strongly correlated electron systems. Here, only the majority-spin (minority-spin) electrons form the Fermi surface with a gapped minority-spin (majority-spin) band and can couple with excitations of the spin (and possibly other) degrees of freedom of the system. In-gap states (or so-called nonquasiparticle states) in such half-metallic ferromagnets have attracted considerable attention in physics of strong electron correlations [1]. The states appear in the band gap of say spin-down band just above the Fermi level of say spin-up band due to the effects of electron correlations (see Fig. 1) and affect physical properties of the system. The origin of the in-gap states is beyond the one-electron band theory and their nature cannot be described by the Landau Fermi-liquid theory. So far, the appearance of the states has been understood in connected with spin polaron processes [1, 2, 3].

In this paper, we study such in-gap states using the simplest model for the half-metallic ferromagnets, which is constructed as follows. Suppose a double-exchange ferromagnet [4], where there are well localized electrons in the orbitals with energy $-\varepsilon$ and on-site Coulomb repulsion U, which are coupled, via the Heisenberg-type ferromagnetic exchange interaction J, with conduction electrons in the noninteracting tight-binding band with hopping parameter t (see Fig. 1). Hereafter, we use the simplest double-exchange model [5, 6] defined in the limit of $\varepsilon = U/2 \to \infty$ as given below, which retains the essential physics on the in-gap states.

We focus on this model in the case of one conduction electron because we can obtain an analytically exact solution for the energies and wave functions for all the eigenstates of the

Journal of Physics: Conference Series **391** (2012) 012161 doi:10.1088/1742-6596/391/1/012161

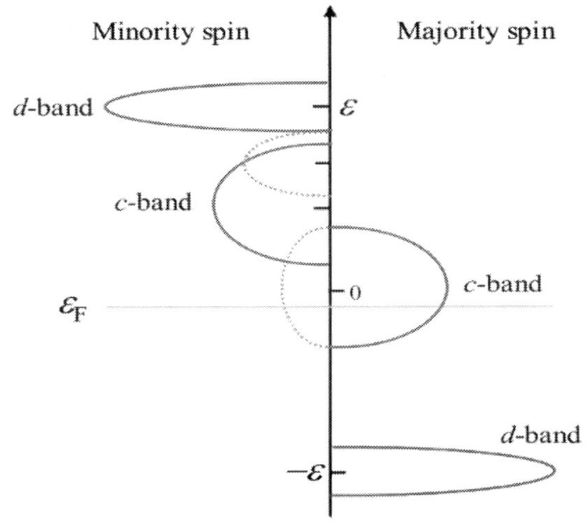

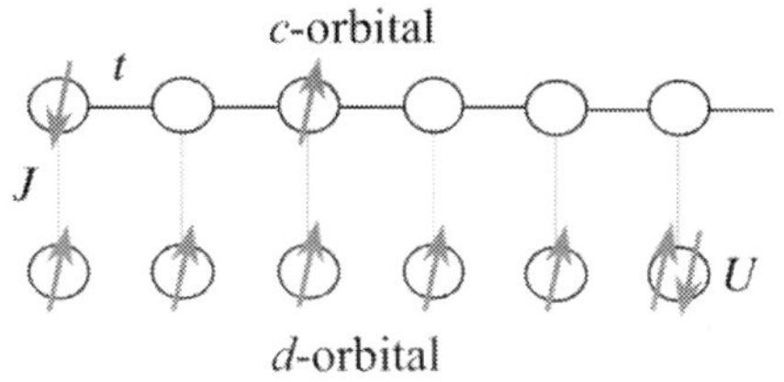

Figure 1. (Left) Schematic representation of the single-particle density of states of the half-metallic ferromagnet (solid line). The in-gap states caused by the electron correlation is also shown by the dotted line. (Right) Schematic representation of the double-exchange model.

Hamiltonian. We will thereby show that the single-particle Green's function is entirely incoherent and the lowest band has an infinite band mass, i.e., the single electron is localized due to its interaction with the spin excitations.

2. Model

The double-exchange model is defined by the Hamiltonian

$$\hat{H} = \hat{H}_0 + \hat{H}_J \tag{1}$$

$$\hat{H}_0 = -t \sum_{\langle ij\rangle\sigma} \left(\hat{c}^\dagger_{i\sigma}\hat{c}_{j\sigma} + \text{H.c.} \right) \tag{2}$$

$$\hat{H}_J = -J \sum_i \hat{\mathbf{s}}_i \cdot \hat{\mathbf{S}}_i \tag{3}$$

where $\hat{c}_{i\sigma}$ ($\hat{c}^\dagger_{i\sigma}$) is the annihilation (creation) operator of an electron with spin σ in the conduction orbital at site i. $\hat{\mathbf{S}}_i$ is the spin operator for the localized spin at site i, $\hat{\mathbf{s}}_i$ the spin operator of an electron in the conduction orbital at site i defined as $\hat{\mathbf{s}}_i = (1/2)\sum_{\alpha\beta} \hat{c}^\dagger_{i\alpha}\sigma_{\alpha\beta}\hat{c}_{i\beta}$ with the Pauli spin matrix σ. $\hat{H}_0$ represents the hopping of electrons between the conduction orbitals on nearest-neighbor sites $\langle ij\rangle$ and $\hat{H}_J$ represents the ferromagnetic ($J > 0$) exchange interaction between an electron in the conduction orbital and the localized spin $\hat{\mathbf{S}}_i$ on the same site i. We define the number operator of conduction electrons at site i as $\hat{n}_i = \sum_\sigma \hat{c}^\dagger_{i\sigma}\hat{c}_{i\sigma}$. We assume the system of L unit cells, where each cell contains a conduction and a localized orbital. The ground state of the double-exchange model is known to be fully spin-polarized when there is a small number of conduction electrons [5, 6]. Without conduction electrons, the ground state is 2^L-fold degenerate, one of which is the fully spin-polarized state. We take $t = 1$ as the unit of energy unless otherwise stated.

3. Results of calculations

3.1. Single-particle spectra

To see the in-gap states, we first calculate the single-particle Green's function for a single electron which is added to the fully spin-polarized (spin-up) state. We follow the method of Refs. [7, 8, 9, 10] to obtain the exact expression for the Green's function when the added electron is spin-down. The ground state $|\mathrm{FM}\rangle$ is assumed to be with a full spin-up polarization without conduction electrons. The result obtained reads

$$G(k,\omega)=\frac{1}{G_0^{-1}(k,\omega-\frac{J}{4})-\Sigma(\omega)} \tag{4}$$

where G_0 is the noninteracting Green's function and $\Sigma(\omega)$ is the momentum-independent self-energy. The spectrum consists of two components: an incoherent continuum in the range $[-2t-\frac{J}{4},2t-\frac{J}{4}]$ and a single δ-function-like peak at higher energy. The dispersive band corresponds to a spin polaron, i.e., the propagating electron is dressed by a magnon. This dressing leads to a modified dispersion with a bandwidth that strongly depends on J and a significant reduction of the spectral weight. The single-particle spectrum at low-energies therefore is entirely incoherent and the lowest energy for each momentum is dispersionless so that the effective mass is infinite and the electron is localized due to its interaction with the spin excitations.

3.2. Wave function

In our model, the Dyson's infinite sequence of equations terminates in the second order and thus the result is exact. This occurs because the addition of a spin-down electron yields only two types of states due to the spin rotational symmetry of the model; it is within the sector of the Hilbert space where the z-component of the total spin is $(L-1)/2$. Hence, we can write the general form of the wave function as:

$$|\Psi(k,\omega)\rangle=\left[\alpha(k,\omega)\hat{c}^{\dagger}_{k\downarrow}+\frac{1}{\sqrt{L}}\sum_q\beta(k,q,\omega)\hat{c}^{\dagger}_{k-q\uparrow}\hat{S}^{-}_{q}\right]|\mathrm{FM}\rangle. \tag{5}$$

The first term is the spin polaron state described by the usual Fermi liquid theory. The second term is the superposition of spin-up electron excitations and virtual magnons. Inserting Eq. (5) into the Schrödinger equation, we obtain the three equations for the coefficients:

$$\begin{aligned}\omega-\epsilon_k-\frac{J}{4}\quad\alpha(k,\omega)&=-\frac{J\gamma(k,\omega)}{2}\\ \omega-\epsilon_{k-q}+\frac{J}{4}\quad\beta(k,q,\omega)&=\frac{J}{2}\left[\gamma(k,\omega)-\alpha(k,\omega)\right]\\ \gamma(k,\omega)&=\frac{1}{L}\sum_q\beta(k,q,\omega).\end{aligned} \tag{6}$$

Adding the normalization condition to the above, we have four equaitions, which we can solve exactly. The solution (see Fig. 3) gives us the ground state properties with the polaron and magnon weight:

$$\alpha(k,\omega)\sim\frac{1}{\sqrt{L}} \tag{7}$$

$$\beta(k,q,\omega)\sim\sqrt{L}\delta_{k,q}. \tag{8}$$

The relation (7) means that we can neglect the polaron-state contributions to the ground state in thermodynamic limit. According to (8), the magnon momentum should be the same as the conduction electron momentrum. This fact implies that the conduction electron minimizes its kinetic energy by transferring their momentum to a magnon and accumulate around zero momentum. As a result, the ground state is infinitely degenerate, and the effective mass becomes infinite. This degeneracy is lifted by adding the direct ferromagnetic exchange interaction between localized spins. Details of the calculations will be given elsewhere [11].

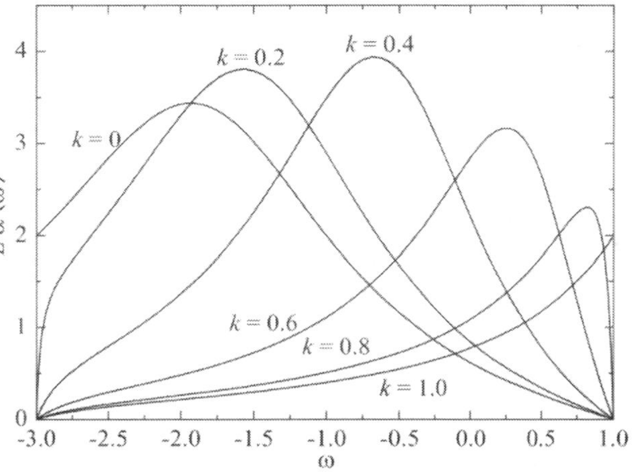

Figure 2. Calculated spin polaron spectral weight $L\alpha(k,\omega)^2$.

4. Summary

We have studied the one-dimensional double-exchange model with one conduction electron. We have obtained the analytically exact solution for the energies and wave functions for all the eigenstates of the Hamiltonian and have calculated the single-particle spectral function exactly. We have shown that the single-particle Green's function at low energy is entirely incoherent and the electron has an infinite band mass, i.e., the single electron is localized due to its interaction with the spin excitations. We hope that the results presented in this paper will shed some light on the origin of the in-gap states in the half-metallic ferromagnets. Further details will be given elsewhere [11].

Acknowledgments

This work was supported in part by a Grant-in-Aid for Scientific Research (No. 22540363) from the Ministry of Education, Culture, Sports, Science and Technology of Japan. A part of computations was carried out at the Research Center for Computational Science, Okazaki Research Facilities, Japan.

References

[1] Katsnelson M I, Irkhin V Y, Chioncel L, Lichtenstein A I and de Groot R A 2008 *Rev. Mod. Phys.* **80** 315.
[2] Edwards D M and Hertz J A 1973 *J. Phys. F: Met. Phys.* **3** 2191.
[3] Irkhin V Y and Katsnelson M I 2006 *Phys. Rev. B* **73** 104429.
[4] Zener C 1951 *Phys. Rev.* **82** 403.
[5] Yunoki S, Hu J, Malvezzi A L, Moreo A, Furukawa N and E. Dagotto 1998 *Phys. Rev. Lett.* **80** 845.
[6] Dagotto E, Yunoki S, Malvezzi A L, Moreo A, Hu J, Capponi S, Poilblanc D and Furukawa N 1998 *Phys. Rev. B* **58** 6414.
[7] Shastry B S and Mattis D C 1981 *Phys. Rev. B* **24** 5340.
[8] Nolting M, Reddy G G, Ramakanth A, Meyer D 2001 *Phys. Rev. B* **64** 155109.
[9] Berciu M and Sawatzky G A 2009 *Phys. Rev. B* **79** 195116.
[10] Berciu M 2009 *Physics* **2** 55.
[11] Nakano K, Eder R and Ohta Y, 2011 in preparation.

Journal of Physics: Conference Series **391** (2012) 012162 doi:10.1088/1742-6596/391/1/012162

Non-Fermi liquid transport and "universal" ratios in quantum Griffiths phases

David Nozadze and Thomas Vojta

Department of Physics, Missouri University of Science & Technology, Rolla, MO 65409, USA

E-mail: dn9z2@mst.edu, vojtat@mst.edu.

Abstract. We use the semi-classical Boltzmann equation to investigate transport properties such as electrical resistivity, thermal resistivity, thermopower, and the Peltier coefficient of disordered metals close to an antiferromagnetic quantum phase transition. In the quantum Griffiths phase, the electrons are scattered by spin-fluctuations in the rare regions. This leads to singular temperature dependencies not just at the quantum critical point, but in the entire Griffiths phase. We show that the resulting non-universal power-laws in transport properties are controlled by the same Griffiths exponent λ which governs the thermodynamics. λ takes the value zero at the quantum critical point and increases throughout the Griffiths phase. We also study some of the "universal" ratios commonly used to characterize Fermi-liquid behavior.

1. Introduction

In recent years, quantum phase transitions [1] have attracted a lot of attention in condensed matter theory. Quantum phase transitions happen at absolute zero temperature. They are induced by the change of an external parameter (pressure, chemical composition, and magnetic field) and are driven by quantum fluctuations. A number of metallic systems show strong deviations from conventional Landau Fermi-liquid properties when they are tuned through a quantum critical point [2].

Significant attention has also been attracted by phase transitions in the presence of quenched disorder. It has become clear that at low temperatures, strongly correlated materials can show a surprising sensitivity to imperfections and randomness. The interplay between large-scale quantum fluctuations and random fluctuations due to disorder leads to exotic phenomena such as quantum Griffiths singularities [3, 4, 5], infinite randomness critical points featuring exponential instead of power-law scaling [6, 7] and the smearing of the phase transition [8]. The Griffiths effects are caused by large spatial regions (rare regions) that are devoid of impurities and can show local order even if the bulk system is in the disordered phase. The locally ordered rare regions are not static but retain their quantum dynamics. Griffiths showed that these rare regions can lead to a singularity in the free energy in a whole parameter region which is now known as the Griffiths phase. In a *quantum* Griffiths phase, the rare-region low-energy density of states follows a power law, $\rho(\epsilon) \propto \epsilon^{\lambda-1}$, where λ is the non-universal Griffiths exponent. λ varies systematically within the Griffiths phase and vanishes at the critical point. This kind of density of states leads to power-law dependencies of several observables on the temperature T, including specific heat, $C \sim T^{\lambda}$, and magnetic susceptibility, $\chi \sim T^{\lambda-1}$. The zero-temperature magnetization-field curve behaves as $M \sim H^{\lambda}$ (for reviews, see Refs. [9, 10]). While the thermodynamics of quantum Griffiths phases is comparatively well understood, much less is known about transport properties.

We showed in Ref. [11] that the rare-region contributions to electrical resistivity, thermal resistivity, thermopower, and the Peltier coefficient in the quantum Griffiths phase associated with an antiferromagnetic quantum phase transition are characterized by non-universal power-laws in T which are controlled by the same Griffiths exponent λ which also governs the thermodynamics. Here, we summarize these results. We then investigate the behaviors in the antiferromagnetic quantum Griffiths phase of some of the "universal" ratios commonly used to characterize Fermi-Liquid behavior in metals. Our paper is organized as follows. In Sec. 2, we briefly discuss the model and method of solution. In Sec. 3, we then summarize the derivations of the transport properties and find that the scattering of the electrons by spin-fluctuations in the rare regions leads to singular temperature dependencies not just at the quantum critical point but in the entire antiferromagnetic quantum Griffiths phase. In Sec. 4, we present the behaviors of various "universal" ratios. Finally, we conclude in Sec. 5.

2. Model

The transport properties of the itinerant antiferromagnetic systems we are interested in can be described by a two-band model consisting of s and d electrons [12, 13]. The Hamiltonian has the form

$$H = H_s + H_d + H_{s-d}\,, \tag{1}$$

where H_s and H_d are the Hamiltonians of s and d electrons, respectively. Only the s electrons contribute to the transport properties. They are scattered by the spin-fluctuations of the d electrons which are assumed to be in the antiferromagnetic quantum Griffiths phase. The contribution to the resistivity from the scattering by the spin-fluctuations stems from the $s-d$ exchange interaction term of the Hamiltonian

$$H_{s-d} = g \int d\mathbf{r}\; \mathbf{s}(\mathbf{r}) \cdot \mathbf{S}(\mathbf{r})\,, \tag{2}$$

where g is the coupling between s and d electrons. $\mathbf{s}$ and $\mathbf{S}$ are the spin densities of the s and d electrons, respectively.

In three space dimensions, the clean critical theory is above its upper critical dimension [14, 15]; moreover the concept of quasiparticles is still (marginally) well defined. Therefore, transport properties can be treated within a semi-classical Boltzmann approach in the Griffiths phase (note that the method may break down sufficiently close to the actual quantum critical point [11]). The linearized Boltzmann equation in the presence of a temperature gradient ∇T, and an electric field $\mathbf{E}$ but zero magnetic field can be written as [16]

$$-\mathbf{v_k}\frac{\partial f^0_{\mathbf{k}}}{\partial T}\nabla T - \mathbf{v_k}\frac{\partial f^0_{\mathbf{k}}}{\partial \varepsilon_{\mathbf{k}}}\mathbf{E} = \left.\frac{\partial f_{\mathbf{k}}}{\partial t}\right|_{\text{scatt}}, \tag{3}$$

where $f^0_{\mathbf{k}}$ is the equilibrium Fermi-Dirac distribution function. The first and second terms describe changes of the electron distribution function $f_{\mathbf{k}}$ due to diffusion and electric field $\mathbf{E}$, respectively. The last one is the collision term which comprises elastic impurity scattering as well as magnetic scattering due to the spin fluctuations. In the following, we focus on the magnetic part of the scattering. Let the stationary solution of the Boltzmann equation be $f_{\mathbf{k}} = f^0_{\mathbf{k}} - \Phi_{\mathbf{k}}(\partial f^0_{\mathbf{k}}/\partial \varepsilon_{\mathbf{k}})$, where $\Phi_{\mathbf{k}}$ is a measure of the deviation of the electron distribution from equilibrium. Then the linearized scattering term due to the spin-fluctuations has the form [13, 17]

$$\left.\frac{\partial f_{\mathbf{k}}}{\partial t}\right|_{\text{scatt}} = \frac{2g^2}{T}\sum_{\mathbf{k}'} f^0_{\mathbf{k}'}(1-f^0_{\mathbf{k}})n(\varepsilon_{\mathbf{k}}-\varepsilon_{\mathbf{k}'})\text{Im}\chi(\mathbf{k}-\mathbf{k}',\varepsilon_{\mathbf{k}}-\varepsilon_{\mathbf{k}'})(\Phi_{\mathbf{k}}-\Phi_{\mathbf{k}'})$$

$$= \frac{1}{T}\sum_{\mathbf{k}'} \mathcal{P}_{\mathbf{k}'}(\varepsilon_{\mathbf{k}} - \varepsilon_{\mathbf{k}'})(\Phi_{\mathbf{k}} - \Phi_{\mathbf{k}'}), \tag{4}$$

where the second line defines the transition rates $\mathcal{P}_{\mathbf{k}}$. The symbol $n(\varepsilon_{\mathbf{k}} - \varepsilon_{\mathbf{k}'})$ denotes the Bose-Einstein distribution function, and χ is the total dynamical susceptibility of the spin-fluctuations (d electrons). In the above equation, the spin-fluctuations are assumed to be in equilibrium. This approximation is valid if the system can lose momentum efficiently by Umklapp or impurity scattering.

Quantum Griffiths effects in disordered metallic systems are realized in both Heisenberg magnets [18] and Ising magnets. In the latter case, they occur in a transient temperature range where the damping is unimportant [19]. In the following, we consider both cases.

In order to study the rare-region contribution to the transport properties in the Griffiths phase, we need to find the total rare-region dynamical susceptibility. To get it, we can simply sum over the susceptibilities of the individual clusters (rare regions). The imaginary part of the dynamical susceptibility of a single cluster of characteristic energy ϵ of a disordered itinerant quantum Heisenberg antiferromagnet in the quantum Griffiths phase is given by [20]

$$\mathrm{Im}\chi_{\mathrm{cl}}(\mathbf{q},\omega;\epsilon) = \frac{\mu^2\gamma\omega}{\epsilon^2(T)+\gamma^2\omega^2}F_\epsilon^2(\mathbf{q}), \tag{5}$$

where μ is the moment of the cluster and γ is the damping coefficient which results from the coupling of the spin-fluctuations and the electrons. $\epsilon(T)$ plays the role of the local distance from criticality. For high temperatures $\gamma T \gg \epsilon$, $\epsilon(T) \approx T$ and for low temperatures $\gamma T \ll \epsilon$, $\epsilon(T) \approx \epsilon$. $F_\epsilon(\mathbf{q})$ is the form factor of the cluster which encodes the spatial magnetization profile.

The imaginary part of the dynamical susceptibility of a single cluster in random quantum Ising models is given by [19]

$$\mathrm{Im}\chi_{\mathrm{cl}}(\mathbf{q},\omega;\epsilon) = \pi\frac{\mu^2}{4}\tanh\left(\frac{\epsilon}{2T}\right)[\delta(\epsilon-\omega)-\delta(\epsilon+\omega)]F_\epsilon^2(\mathbf{q}). \tag{6}$$

The precise functional form of $F_\epsilon(\mathbf{q})$ is not known, but we can find it approximately by analyzing the Fourier transform of a typical local magnetization profile of the rare region. We find $F_\epsilon^2(\mathbf{q}) = X[(\mathbf{q}-\mathbf{Q})^3\log(\epsilon^{-1})]$, where X and $\mathbf{Q}$ are a scaling function and the ordering wave vector, respectively [11].

To estimate the total rare-region susceptibility, we integrate over all rare regions using the rare-region density of states $\rho(\epsilon)$,

$$\mathrm{Im}\chi(\mathbf{q},\omega) = \int_0^\Lambda d\epsilon\rho(\epsilon)\mathrm{Im}\chi_{\mathrm{cl}}(\mathbf{q},\omega;\epsilon), \tag{7}$$

where Λ is an energy cut-off. In the Heisenberg case we find that the rare-region contribution to the zero-temperature susceptibility in the quantum Griffiths phase can be expressed as (up to logarithmic corrections)

$$\mathrm{Im}\chi(\mathbf{q},\omega) \propto |\omega|^{\lambda-1}\mathrm{sgn}(\omega)\,\mathrm{X}[(\mathbf{q}-\mathbf{Q})^3\log(\omega^{-1})]. \tag{8}$$

The rare-region susceptibility of the random quantum Ising model has the same structure as Eq.(8) [19].

3. Transport properties

3.1. Electrical resistivity

In order to calculate the electrical resistivity we consider Ziman's variational principle [16]. The resistivity ρ is given as the minimum of a functional of $\Phi_{\mathbf{k}}$. [1]

$$\rho[\Phi_{\mathbf{k}}] = \min\left[\frac{1}{2\mathrm{T}} \frac{\int\int(\Phi_{\mathbf{k}} - \Phi_{\mathbf{k}'})^2 \Gamma_{\mathbf{k}}^{\mathbf{k}'} \mathrm{d}\mathbf{k}\mathrm{d}\mathbf{k}'}{(\int \mathrm{v}_{\mathbf{k}} \Phi_{\mathbf{k}} \frac{\partial f_{\mathbf{k}}^0}{\partial \varepsilon_{\mathbf{k}}} \mathrm{d}\mathbf{k})^2}\right], \tag{9}$$

where

$$\Gamma_{\mathbf{k}}^{\mathbf{k}'} = \int_0^\infty d\omega\, \mathcal{P}_{\mathbf{k}'}(\omega) \delta(\varepsilon_{\mathbf{k}'} - \varepsilon_{\mathbf{k}} + \omega), \tag{10}$$

with $\mathcal{P}_{\mathbf{k}'}(\omega)$ defined in Eq.(4).

By making a suitable ansatz for the distribution function $\Phi_{\mathbf{k}}$ in the functional (9), we can find the resistivity. Close to the antiferromagnetic quantum phase transition, the susceptibility is strongly peaked around the ordering wave vector $\mathbf{Q}$. This leads to anisotropic scattering processes. However, in the presence of a high concentration of impurities, the low temperature resistivity is dominated by elastic impurity scattering which is isotropic. The isotropic scattering redistributes the electrons over the Fermi surface. This allows us to make the standard ansatz

$$\Phi_{\mathbf{k}} \propto \mathbf{n} \cdot \mathbf{k}, \tag{11}$$

where $\mathbf{n}$ is a unit vector parallel to the electric field. Note that any constant prefactor in $\Phi_{\mathbf{k}}$ is unimportant because it drops out of the resistivity functional (9). Then, following the calculations for the electrical resistivity outlined in Ref. [16], we obtain

$$\Delta\rho \propto T \int d^3\mathbf{q} \frac{(\mathbf{n}\cdot\mathbf{q})^2}{q} \int_0^\infty d\omega \frac{\partial n(\omega)}{\partial T} \mathrm{Im}\chi(\mathbf{q},\omega). \tag{12}$$

Inserting the susceptibility (8), the temperature dependence of the electrical resistivity due to the spin-fluctuations in the Griffiths phase is given by

$$\Delta\rho \propto T^\lambda, \tag{13}$$

up to logarithmic corrections. Thus, the temperature-dependence of the electrical resistivity follows a non-universal power-law governed by the Griffiths exponent λ.

3.2. Other transport properties

In the same way, we study the thermal resistivity W (i.e., the inverse of the thermal conductivity κ). We find that the rare-region contribution to the thermal resistivity in the antiferromagnetic quantum Griffiths phase has the form

$$\Delta W \propto T^{\lambda-1}. \tag{14}$$

The Seebeck effect, the existence of an electric field $\mathbf{E}$ in a metal subject to a thermal gradient ∇T, is characterized by the thermopower S which is defined by the relation $\mathbf{E} = S\ \nabla T$. To calculate the thermopower, we analyze the Boltzmann equation (3) in the presence of both ∇T and $\mathbf{E}$. Elastic impurity scattering leads to the usual linear temperature dependence $S_{\mathrm{imp}} \propto T$

[1] We set Plank's constant, electron's charge and Boltzmann constant $\hbar = e = k_B = 1$ in what follows.

while the contribution due to the magnetic scattering by the rare regions in the Griffiths phase reads

$$\Delta S \propto T^{\lambda+1} . \tag{15}$$

Another transport coefficient called the Peltier coefficient Π characterizes the flow of a thermal current in a metal in the absence of a thermal gradient. It is related to the thermopower by $\Pi = S\, T$. Correspondingly, the rare-region contribution to the Peltier coefficient has the form

$$\Delta\Pi \propto T^{\lambda+2} . \tag{16}$$

In analogy to the resistivity (13), the results for ΔW, ΔS, and $\Delta\Pi$ are given up to logarithmic corrections only.

4. "Universal" ratios in quantum Griffiths phases

In this section, we investigate the behaviors in quantum Griffiths phases of some of the "universal" ratios commonly used to characterize Fermi-liquid behavior. Two of the ratios, the Wilson ratio $\chi T/C$ [21, 22] and the Grüneisen parameter [23], the ratio between thermal expansion coefficient β and specific heat C, only involve thermodynamics. They have been discussed before. In the Wilson ratio, the Griffiths-phase power-laws of C and χ cancel, leading to a logarithmic temperature dependence [24]

$$R_W = \chi T/C \propto [\log(1/T)]^2 . \tag{17}$$

The Grüneisen parameter in the quantum Griffiths phase also diverges logarithmically with vanishing temperature T [25]

$$\Gamma = \beta/C \propto \log(1/T) . \tag{18}$$

The knowledge of the transport properties (Sec. 3) allows us to study further ratios. According to the Wiedemann-Franz law [26], the ratio of the electrical resistivity divided by the temperature and the thermal resistivity, the so-called Lorenz number $L = \rho/WT$, takes the universal value L_0 within Fermi-liquid theory. In the Griffiths phase, dominant impurity scattering leads to the universal value L_0 while the rare-region contributions to the Wiedemann-Franz law give a subleading contribution

$$\Delta L \propto T^{\lambda} . \tag{19}$$

Let us now consider the quasi-universal ratio $q = S/C$ of the Seebeck coefficient to the specific heat [27]. Due to the spin-fluctuations in the Griffiths phase, this ratio strongly deviates from its usual temperature-independent behavior and is given by

$$q \propto T^{1-\lambda} . \tag{20}$$

Another "universal" ratio is the Kadowaki-Woods ratio, normally defined as AT^2/C^2 [28], where A is the coefficient of the T^2-term of the resistivity. To evaluate it in the Griffiths phase where the leading temperature dependence of the resistivity is not quadratic, we rewrite the definition as $(\rho - \rho_{\mathrm{imp}})/C^2$. Then, we find that the Kadowaki-Woods ratio behaves as

$$R_{KW} = (\rho - \rho_{\mathrm{imp}})/C^2 \propto T^{-\lambda} \tag{21}$$

in the antiferromagnetic quantum Griffiths phase, in contrast to the temperature-independent Fermi-liquid behavior.

5. Conclusions

We have studied the transport properties in antiferromagnetic quantum Griffiths phases. The rare-region contributions to electrical resistivity, thermal resistivity, thermopower, and the Peltier coefficient are characterized by non-universal power-laws in T which are controlled by the Griffiths exponent λ. We have also investigated the behaviors of some "universal" ratios. In antiferromagnetic quantum Griffiths phases, the Kadowaki-Woods ratio and the ratio between Seebeck coefficient and specific heat show strong deviations from the usual Fermi-liquid behavior. In contrast, the Wiedemann-Franz law is fulfilled to leading order in T, but the spin-fluctuations in the rare regions lead to a subleading non-universal power-law in T controlled by the Griffiths exponent. The two ratios involving thermodynamics only, the Wilson ratio and the Grüneisen ratio, show only logarithmic deviations from Fermi-liquid behavior.

Our results have been obtained using the semi-classical Boltzmann equation approach. As discussed in Ref. [11], this approach is valid in the Griffiths phase and may break down sufficiently close to the quantum critical point. We also emphasize that our results have been derived for antiferromagnetic quantum Griffiths phases and may not be valid for ferromagnetic systems. First results on the thermodynamics of ferromagnetic quantum Griffiths phases [29] suggest that the Griffiths singularities take slightly different functional forms than in antiferromagnets. A detailed study of the transport properties remains a task for the future.

Acknowledgments

This work has been supported by the NSF under Grant No. DMR-0906566.

References

[1] Sachdev S 1999 *Quantum Phase Transitions* (Cambridge, England: Cambridge University Press)
[2] Löhneysen H V, Pietrus T, Portisch G, Schlager H G, Schröder A, Sieck M and Trappmann T 1994 *Phys. Rev. Lett.* **72** 3262
[3] Griffiths R B 1969 *Phys. Rev. Lett.* **23** 17
[4] Thill M and Huse D A 1995 *Physica A* **214** 321
[5] Rieger H and Young A P 1996 *Phys. Rev. B* **54** 3328
[6] Fisher D S 1992 *Phys. Rev. Lett.* **69** 534
[7] Fisher D S 1995 *Phys. Rev. B* **51** 6411
[8] Vojta T 2003 *Phys. Rev. Lett.* **90** 107202
[9] Vojta T 2006 *J. Phys. A* **39** R143
[10] Vojta T 2010 *J. Low Temp. Phys.* **161** 299
[11] Nozadze D and Vojta T 2011 *Europhys. Lett.* **95** 57010
[12] Mills D L and Lederer P 1966 *J. Phys. Chem. Solids* **27** 1805
[13] Ueda K and Moriya T 1975 *J. Phys. Soc. Jpn.* **39** 605
[14] Hertz J 1976 *Phys. Rev. B* **14** 1165
[15] Millis A J 1993 *Phys. Rev. B* **48** 7183
[16] Ziman J 1960 *Electrons and Phonons* (Oxford: Clarendon)
[17] Hlubina R and Rice T M 1995 *Phys. Rev. B* **51** 9253
[18] Vojta T and Schmalian J 2005 *Phys. Rev. B* **72** 045438
[19] Castro Neto A H and Jones B A 2000 *Phys. Rev. B* **62** 14975
[20] Vojta T, Kotabage C and Hoyos J A 2009 *Phys. Rev. B* **79** 024401
[21] Wilson K G 1975 *Rev. Mod. Phys.* **47** 773
[22] Stewart G R 1984 *Rev. Mod. Phys.* **56** 755
[23] Grüneisen E 1912 *Ann. Phys.* **344** 257
[24] Miranda E and Dobrosavljević V 2005 *Rep. Prog. Phys.* **68** 2337
[25] Vojta T 2009 *Phys. Rev. B* **80** 041101(R)
[26] Franz R and Wiedemann G 1853 *Ann. Phys.* **165** 497
[27] Behnia K, Jaccard D and Flouquet J 2004 *J. Phys.: Condens. Matter* **16** 5187
[28] Kadowaki K and Woods S B 1986 *Solid St. Commun.* **58** 507
[29] Nozadze D and Vojta T 2012 *Phys. Rev. B* **85** 174202

International Conference on Strongly Correlated Electron Systems (SCES 2011) IOP Publishing
Journal of Physics: Conference Series **391** (2012) 012163 doi:10.1088/1742-6596/391/1/012163

Spectral functions in a two-velocities Tomonaga-Luttinger Liquid

E Orignac[1], **M Tsuchiizu**[2] **and Y Suzumura**[2]

[1] Laboratoire de Physique de l'École Normale Suérieure de Lyon, CNRS UMR 5672, 46 Allée d'Italie, F-69364 Lyon Cedex 07, France

[2] Department of Physics, Nagoya University, Nagoya 464-8602, Japan

E-mail: `Edmond.Orignac@ens-lyon.fr`

Abstract. We obtain the spectral functions of a two-velocities Tomonaga-Luttinger liquid at $T = 0$K in terms of the Appell hypergeometric functions F_1 and F_2. In the case of SU(N) spin symmetry, the spectral functions can be expressed with the Gauss hypergeometric functions. We discuss the singularities and thresholds of the spectral functions for both the SU(N) invariant case and the general case.

1. Introduction

The Tomonaga-Luttinger liquid (TLL) [1] is a key paradigm of the theory of strongly correlated fermions in one dimension. In the case of fermions with spin, the property of spin-charge separation is expected, where fermion excitations separate into spin ( ) excitations and charge (ρ) excitations propagating with respective velocities u_σ and u_ρ. The technique of angle-resolved photoemission spectroscopy has been used to measure the spectral functions of electrons in quasi-one dimensional conductors such as $BaVS_3$ [2], TTF-TCNQ [3, 4], $(TMTSF)_2X$ [5], and $LiMO_6O_{17}$ [6]. Self-assembled one-dimensional gold metallic chains of atoms on semiconductor surfaces have also been considered as potential candidates for the observation of TLL and spin-charge separation [7, 8, 9, 10, 11]. Other candidate systems to detect spin-charge separation are ultracold fermionic gases [12] in optical lattices [13] or magnetic traps [14]. In these systems, however, the internal degrees of freedom may possess a higher symmetry than SU(2) [15] or may result from a mixture of spinless fermionic atoms with different masses [16] leading to a reduced U(1)   U(1) symmetry as if a magnetic field was present. The accurate calculation of fermion spectral functions beyond the case of SU(2) symmetry is thus necessary for testing quantitatively the TLL theory in such systems. Previously, the asymptotic behavior of spectral functions has been predicted [17, 18] for energies $h\omega$ close to $\ hu_{\rho,\sigma}q$. The zero temperature spectral functions have been calculated in Ref. [1] for the SU(2) invariant case, and finite temperature spectral functions have been studied numerically [19].

2. Results

In a TLL with two components (charge ρ and spin) with fully broken U(1)   U(1) symmetry, calling $\ _s(x) = \sum_{r=\pm} e^{irk_{F,s}x}\ \ _{r,s}(x)$ the fermion annihilation operator with (pseudo)-spin s the

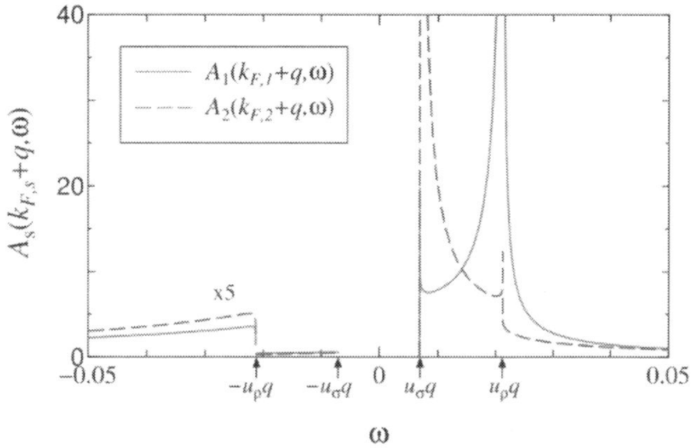

Figure 1. The spectral function for the case U(1) U(1) symmetry, with $q = 0.01$. The parameters in determining the spectral function are the bare velocities u_s for different (pseudo)-spin ($s = 1, 2$) and the mutual interaction g [20]. In this figure, we set $u_1 = 2$, $u_2 = 1$, and $g = 3$, where the renormalized velocities are given by $u_\rho \simeq 2.13$ and $u_\sigma \simeq 0.69$.

zero-temperature fermion Green's function is obtained from the correlators:

$$\langle \;_{r,s}(x,t)\; ^{\dagger}_{r,s}(0,0)\rangle = \frac{1}{2} \prod_{\beta=\rho,\sigma} \left[\frac{}{+i(u_\beta t - rx)}\right]^{\;_{s,\beta}} \left[\frac{}{+i(u_\beta t + rx)}\right]^{\;'_{s,\beta}}, \tag{1}$$

where u_β is the velocity of mode and the exponents $\nu_{s,\beta}$, $\nu'_{s,\beta}$ can be obtained from bosonization techniques [20]. The spectral function $A_s(a, \omega)$ is obtained by taking the Fourier transform of the retarded Green's function (1). The Fourier transform can be obtained in closed form using Feynman identities [21] following Ref. [22]. Following that route, in Ref. [20], we derived expressions of the spectral functions $A_s(k_F + q, \omega)$ in terms of Appell F_1 and F_2 hypergeometric functions [23]. We also found that $A_s(k_{F,s} + q, \omega) = 0$ for $|\omega| < u_\sigma q$. The typical behavior of the spectral function is shown in Fig. 1. We find, in $A_1(k_{F,1} + q, \omega)$, the large weight at $\omega \simeq u_\rho q$ while small weight at $\omega \simeq u_\sigma q$. In contrast, large weight can be seen at $\omega \simeq u_\sigma q$ and small weight at $\omega \simeq u_\rho q$ in $A_2(k_{F,2} + q, \omega)$.

Next we consider the model with SU(N) symmetry [24, 25]. The Hamiltonian reads:

$$\begin{aligned} H &= H_\rho + H_\sigma, \\ H_\rho &= \int dx \left[u_\rho K_\rho (\;\Pi_\rho)^2 + \frac{u_\rho}{K_\rho}(\partial_x \;_\rho)^2 \right], \\ H_\sigma &= \frac{2\;\, u_\sigma}{(N+1)} \sum_{A=1}^{N^2-1} \int dx \left[T_R^A(x) T_R^A(x) + T_L^A(x) T_L^A(x) \right], \end{aligned} \tag{2}$$

where $T^A_{R,L}$ are the SU(N)$_1$ currents [1] and K_ρ is the charge Luttinger exponent. The expression of Eq. (1) is still valid and the exponents are given by $\nu_{s,\rho} = (2\;_\rho + 1)/N$, $\nu_{s,\sigma} = (N-1)/N$, $\nu'_{s,\rho} = (2\;_\rho)/N$, and $\nu'_{s,\sigma} = 0$ [25], where $\;_\rho = (K_\rho + 1/K_\rho - 2)/8$. Since these exponents are spin-independent, we redefine $\nu_\rho \;\; \nu_{s,\rho}$, $\nu_\sigma \;\; \nu_{s,\sigma}$, and $\nu'_\rho \;\; \nu'_{s,\rho}$. The spectral function for the SU(N) case is given by

$$A_s(k_F + q, \omega)|_{u_\sigma q < \omega < u_\rho q} = \frac{(\;/\;\;u)^{\;-1}}{(\nu_\rho + \nu'_\rho)\;\;(\nu_\sigma)} \frac{(\omega - u_\sigma q)^{\;\rho + \;'_\rho - 1}(-\omega + u_\rho q)^{\;\sigma + \;'_\rho - 1}}{(\omega + u_\rho q)^{\;'_\rho}}$$

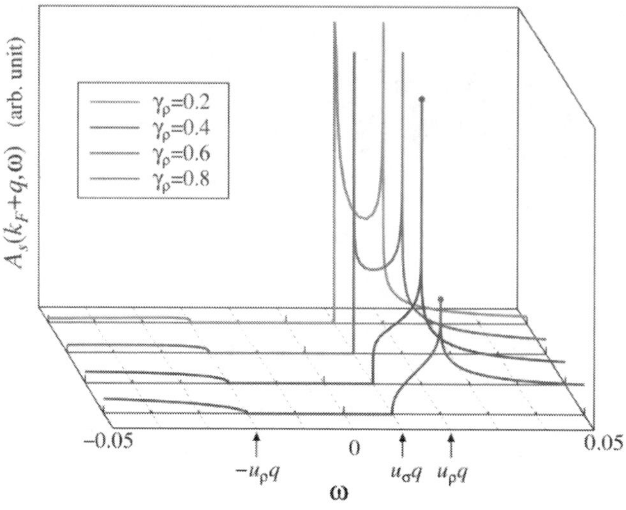

Figure 2. The evolution of the spectral function with $_\rho = 0.2, 0.4, 0.6, 0.8$ (from top to bottom) for the case of SU(3) symmetry. For $_\rho < 1/2$ power-law divergences are obtained for both $\omega = u_\rho q$ and $\omega = u_\sigma q$. For $_\rho > 1/2$, both divergences are replaced by cusps. The dots represent the finite values at $\omega = u_\sigma q$.

$$ {}_2F_1\left(\nu-1, \nu'_\rho; \nu_\rho+\nu'_\rho; \frac{2u_\rho\,(\omega-u_\sigma q)}{u\,(\omega+u_\rho q)}\right), \tag{3}$$

for $u_\sigma q < \omega < u_\rho q$, and

$$A_s(k_F+q,\omega)|_{|\omega|>u_\rho q} = \frac{(\ /2u_\rho)^{\ -1}}{(\nu_\rho+\nu_\sigma)\ \ (\nu'_\rho)} \frac{|\omega-u_\rho q|^{\ _\sigma+\ '_\rho-1}|\omega+u_\rho q|^{\ _\rho+\ _\sigma-1}}{|\omega-u_\sigma q|^{\ _\sigma}}$$

$$ {}_2F_1\left(\nu-1, \nu_\sigma; \nu_\rho+\nu_\sigma; \frac{u|\omega+u_\rho q|}{2u_\rho|\omega-u_\sigma q|}\right), \tag{4}$$

for $|\omega| > u_\rho q$. Here ${}_2F_1(\ ,\ ;\ ;x)$ is the Gauss hypergeometric function, and $\nu\ \ (\nu_\rho+\nu_\sigma+\nu'_\rho)$. For $-u_\rho q < \omega < +u_\sigma q$ the spectral function vanishes. It is worthwhile to note that Eqs. (3) and (4) can also be derived from the results of Ref. [20] using known limits of Appell hypergeometric functions in terms of Gauss hypergeometric functions. We also note that by taking $N = 2$, these expressions reproduce the formula for the SU(2) case given in Eq. (19.27) in Ref. [1]. From Eqs. (3)–(4), a power-law divergence $A_s(k_F+q,\omega) \sim |\omega-u_\rho q|^{\ _\sigma+\ '_\rho-1} = |\omega-u_\rho q|^{(2\gamma_\rho-1)/N}$ of the spectral function for $\omega = u_\rho q$ is expected when $_\rho < 1/2$. For $_\rho > 1/2$ a cusp is obtained instead of a divergence. When $\omega \to u_\sigma q$, Eq. (3) predicts a power-law divergence $A_s(k_F+q,\omega) \sim (\omega-u_\sigma q)^{\ _\rho+\ '_\rho-1} = (\omega-u_\sigma q)^{(4\gamma_\rho+1-N)/N}$. For $_\rho > (N-1)/4$ instead, the spectral function goes to zero as $\omega \to u_\sigma q$. In contrast with the $N = 2$ case, for $N\ \ 4$, the power-law divergence at $\omega = u_\sigma q$ is the last to disappear. For $N = 3$, both divergences disappear for $_\rho = 1/2$. Finally, for $\omega \to -u_\rho q$, the spectral function vanishes as $A_s(k_F+q,\omega) \sim (u_\rho q+\omega)^{\ _\rho+\ _\sigma-1} = (u_\rho q+\omega)^{2\gamma_\rho/N}$. We have represented the evolution of the spectral function as a function of $_\rho$ for $N = 3$ on Fig. 2.

3. Summary and discussions

In the present paper, we have derived expressions of the spectral functions of a two-velocities TLL in terms of Appell hypergeometric functions. For the model with SU(N) spin symmetry, the spectral functions can be expressed with the Gauss hypergeometric functions. We have analyzed the power-law singularities of the spectral functions for both cases.

Here we briefly discuss the experimental findings by the angle-resolved photoemission spectroscopy (ARPES) measurements for gold atom chains on silicon surfaces [7, 8, 9, 10, 11]. Initially, the ARPES measurement revealed a double-band structure and the spin-charge separation was suggested for its origin [7, 8, 9]. However, recent *spin-resolved* ARPES experiments on this system have shown that the double-band nature originates in the spin splitting caused by the Rashba effect [11]. The present analysis shown in Fig. 1 is relevant to the *spin-resolved* ARPES measurements. The results of Fig. 1 suggest that the spectral functions A_1 and A_2 exhibit strongest peaks at different frequencies. This behavior is consistent with the spin-splitting situation observed in Ref. [11]. From the present analysis, we infer that, in addition to the largest peak at $\omega = u_\rho q$ in $A_1(k_{F,1}+q,\omega)$ ($\omega = u_\sigma q$ in $A_2(k_{F,2}+q,\omega)$), another weaker cusp or shoulder structure can also be seen at $\omega = u_\sigma q$ ($\omega = u_\rho q$) due to the correlation effects in one dimension.

References

[1] Gogolin A O, Nersesyan A A and Tsvelik A M 1999 *Bosonization and Strongly Correlated Systems* (Cambridge, UK: Cambridge University Press)
[2] Nakamura M, Seikiyama A, Namatame H, Fujimori A, Yoshiara H, Ohtani T, Misu A and Takano M 1994 *Phys. Rev. B* **49** 16191
[3] Claessen R, Sing M, Schwingenschlogl U, Blaha P, Dressel M and Jacobsen C S 2002 *Phys. Rev. Lett.* **88** 96402
[4] Ito T, Chainani A, Haruna T, Kanai K, Yokoya T, Shin S and Kato R 2005 *Phys. Rev. Lett.* **95** 246402
[5] Vescoli V, Zwick F, Henderson W, DeGiorgi L, Grioni M, Gruner G and Montgomery L K 2000 *Eur. Phys. J. B* **13** 503
[6] Wang F, Mo S K, Allen J W, Kim H D, He J, Jin R, Mandrus D, Sekiyama A, Tsunekawa M and Suga S 2006 *Phys. Rev. B* **74** 113107
[7] Losio R, Altmann K N, Kirakosian A, Lin J -L, Petrovykh D Y and Himpsel F J 2001 *Phys. Rev. Lett.* **86** 4632
[8] Himpsel F J, Altmann K N, Bennewitz R, Crain J N, Kirakosian A, Lin J -L and McChesney J L 2001 *J. Phys. Condens. Matter* **13** 11097
[9] Crain J N, McChesney J L, Zheng F, Gallagher M C, Snijders P C, Bissen M, Gundelach C, Erwin S C and Himpsel F J, 2004 *Phys. Rev. B* **69** 125401
[10] Oncel N 2008 *J. Phys. Condens. Matter* **20** 393001
[11] Okuda T, Miyamaoto K, Takeichi Y, Miyahara H, Ogawa M, Harasawa A, Kimura A, Matsuda I, Kakizaki A, Shishidou T and Oguchi T 2010 *Phys. Rev. B* **82** 161410
[12] Recati A, Fedichev P O, Zwerger W and Zoller P 2003 *Phys. Rev. Lett.* **90** 020401
[13] Stöferle T, Moritz H, Schori C, Köhl M and Esslinger T 2004 *Phys. Rev. Lett.* **92** 130403
[14] Bouchoule I, Van Druten N J and Westbrook C I 2011 *Atom chips and one-dimensional Bose gases* in *Atom chips* edited by J. Reichel and V. Vuletic (Weinhheim: Wiley-VCH Verlag) chapter 11
[15] Taie S, Takasu Y, Sugawa S, Yamazaki R, Tsujimoto T, Murakami R and Takahashi Y 2010 *Phys. Rev. Lett.* **105** 190401
[16] Wille E, Spiegelhalder F M, Kerner G, Naik D, Trenkwalder A, Hendl G, Schreck F, Grimm R, Tiecke T G, Walraven J T M, Kokkelmans S J J M F, Tiesinga E and Julienne P S 2008 *Phys. Rev. Lett.* **100** 053201
[17] Voit J 1993 *Phys. Rev. B* **47** 6740
[18] Meden V and Schönhammer K 1992 *Phys. Rev. B* **46** 15753
[19] Nakamura N and Suzumura Y 1997 *Prog. Theor. Phys.* **98** 29
[20] Orignac E, Tsuchiizu M and Suzumura Y 2011 *Phys. Rev. B* **84** 165128
[21] Le Bellac M 1992 *Quantum and Statistical Field Theory* (Oxford, UK: Oxford University Press)
[22] Iucci A, Fiete G A and Giamarchi T 2007 *Phys. Rev. B* **75** 205116
[23] Erdélyi A, Magnus W, Oberhettinger F and Tricomi F G 1953 *Higher transcendental functions* vol 1 (New York: McGraw-Hill)
[24] Kawakami N 1993 *Phys. Rev. B* **47** 2928
[25] Assaraf R, Azaria P, Caffarel M and Lecheminant P 1999 *Phys. Rev. B* **60** 2299

Momentum-Dependent Variational Approach to Correlated Electron System

M A R Patoary* and Y Kakehashi**

Department of Physics and Earth Sciences, Faculty of Science, University of the Ryukyus, 1 Senbaru, Nishihara, Okinawa, 903-0213, Japan

E-mail: `*k108609@eve.u-ryukyu.ac.jp,**yok@sci.u-ryukyu.ac.jp`

Abstract. We present here the momentum dependent local-ansatz wavefunction approach (MLA) with the best choice of the self-consistent variational parameters to describe the ground-state properties of the correlated electron systems in solids. With use of the self-consistent variational scheme, we performed the numerical calculations for the half-filled band as well as non-half-filled band Hubbard model on the hypercubic lattice in infinite dimensions. The ground-state energy in the MLA is lower than those of the local-ansatz approach (LA) and the Gutzwiller approach (GA) in weak and intermediate Coulomb interaction regimes. The double occupation number is suppressed as compared with the LA. We observe the distinct momentum dependence of the momentum distribution functions which is qualitatively different from those of the LA and the GA.

The variational theory has been developed to describe the ground state properties of correlated electron system in solids [1]. To characterize the correlated electrons in the variational approach a minimum basis set is constructed by applying one-particle, two-particle, and higher-order particle operators onto the Hartree-Fock wavefunction. In principle, their amplitudes are chosen to be best. Gutzwiller [2, 3, 4] described a simple trial wavefunction to clarify the role of electron correlations in metallic ferromagnetism. It reduces the amplitudes of doubly occupied states on the local orbitals by making use of a projection operator $\Pi_i(1 - gn_{i\uparrow}n_{i\downarrow})$ onto the Hartree-Fock wavefunction $|\phi_0\rangle$; $|\Psi_{\text{GA}}\rangle = \left[\prod_i(1 - gn_{i\uparrow}n_{i\downarrow})\right]|\phi_0\rangle$. Here $n_{i\sigma}$ is the number operator for electrons on site i with spin σ, variational parameter g controls the double occupations on sites.

Stollhoff and Fulde [5, 6, 7] proposed a method called the local-ansatz approach (LA), which is simpler than the GA in treatment. The LA wavefunction takes into account the states created by local two-particle operators such as the residual Coulomb interactions $\{O_i\} = \{\delta n_{i\uparrow}\delta n_{i\downarrow}\}$; $|\Psi_{\text{LA}}\rangle = \left[\prod_i(1-\eta_{\text{LA}}O_i)\right]|\phi_0\rangle$. Here $\delta n_{i\sigma} = n_{i\sigma} - \langle n_{i\sigma}\rangle_0$, $\langle n_{i\sigma}\rangle_0$ being the average electron number on site i with spin σ in the Hartree-Fock approximation. The amplitudes η_{LA} is determined variationally.

The Hilbert space of the LA expanded by the local operators $\{O_i\}$ is not sufficient to characterize exactly the weakly correlated region; it does not reduce to the second-order perturbation theory in the weak Coulomb interaction limit. The same difficulty arises also in the GA. To overcome the difficulty in the weak Coulomb interaction regime and to improve the correlated states in the intermediate Coulomb interaction regime, we introduce in this paper a wavefunction with momentum-dependent variational parameters, and demonstrate that the new approach much improves the LA as well as the GA in those regimes. We call our new scheme the momentum-dependent local ansatz approach(MLA).

To introduce the MLA wavefunction, we expand the wavefunction with respect to the Coulomb interaction strength. We find that each coefficient of the two-particle excited states is momentum independent. In order to obtain the best local operators, we replace the momentum-independent coefficients with the momentum dependent variational parameters resulting in a new set of local operators $\{\tilde{O}_i\}$. That is, we define the local operators $\tilde{O}_i$ with the momentum-dependent variational parameter $\eta_{k'_2k_2k'_1k_1}$ as follows:

$$\tilde{O}_i = \sum_{k_1k'_1k_2k'_2} \langle k'_1|i\rangle\langle i|k_1\rangle\langle k'_2|i\rangle\langle i|k_2\rangle \; \eta_{k'_2k_2k'_1k_1} \delta(a^{\dagger}_{k'_2\downarrow}a_{k_2\downarrow})\delta(a^{\dagger}_{k'_1\uparrow}a_{k_1\uparrow}) \, . \tag{1}$$

Here $\langle i|k\rangle = \exp(-ik\cdot R_i)/\sqrt{N}$ is an overlap integral between the localized orbital and the Bloch state with momentum k, R_i denotes the atomic position, and N is the number of sites. $a^{\dagger}_{k\sigma}$ ($a_{k\sigma}$) denotes the creation (annihilation) operator for an electron with momentum k and spin σ, and $\delta(a^{\dagger}_{k'\sigma}a_{k\sigma}) = a^{\dagger}_{k'\sigma}a_{k\sigma} - \langle a^{\dagger}_{k'\sigma}a_{k\sigma}\rangle_0$. We then construct the MLA wavefunction from the flexible set of local operators $\{\tilde{O}_i\}$ as

$$|\Psi_{\rm MLA}\rangle = \prod_i (1-\tilde{O}_i)|\phi_0\rangle. \tag{2}$$

The best local basis set is chosen by controlling the variational parameters in the momentum space, which leads to the best wavefunction.

We calculate the ground-state energy using the MLA wavefunction within a single-site approximation (SSA) [8]. The correlation energy per atom is then given by

$$\epsilon_c = \frac{-\langle\tilde{O}_i^{\dagger}\tilde{H}\rangle_0 - \langle\tilde{H}\tilde{O}_i\rangle_0 + \langle\tilde{O}_i^{\dagger}\tilde{H}\tilde{O}_i\rangle_0}{1+\langle\tilde{O}_i^{\dagger}\tilde{O}_i\rangle} \tag{3}$$

Here $\tilde{H} = H - \langle H\rangle_0$. Minimizing the energy, we obtain a self-consistent equation with variational parameters. It is difficult to solve the self-consistent equation exactly. Because of this, we obtained an approximate solution as follows, which interpolates between the weak Coulomb interaction limit and the atomic limit.

$$\eta_{k'_2k_2k'_1k_1}(\tilde{\eta},\epsilon_c) = \frac{U\tilde{\eta}}{\Delta E_{k'_2k_2k'_1k_1} - \epsilon_c} \, . \tag{4}$$

Here $\tilde{\eta} = [1-\eta(1-2\langle n_{i\uparrow}\rangle_0)(1-2\langle n_{i\downarrow}\rangle_0)]$, the two-particle excitation energy $\Delta E_{k'_2k_2k'_1k_1}$ is given by :

$$\Delta E_{k'_2k_2k'_1k_1} = \epsilon_{k'_2\downarrow} - \epsilon_{k_2\downarrow} + \epsilon_{k'_1\uparrow} - \epsilon_{k_1\uparrow} \, . \tag{5}$$

In the preliminary version of our theory [8], we made use of the variational parameter in the LA; $\tilde{\eta}_{\rm LA} = [1-\eta_{\rm LA}(1-2\langle n_{i\uparrow}\rangle_0)(1-2\langle n_{i\downarrow}\rangle_0)]$ where $\eta_{\rm LA}$ is the variational parameter in the LA and for simplicity we might also adopt the correlation energy in the LA for ϵ_c in Eq. (4). In this paper we determine $\tilde{\eta}$ variationally. According to the variational principle, the ground-state energy E_0 satisfies the the following inequality.

$$\langle E_0\rangle \leq E(\{\eta^{*}_{k'_2k_2k'_1k_1}\}) \leq E(\{\eta_{k'_2k_2k'_1k_1}(\tilde{\eta},\epsilon_c)\}) \, . \tag{6}$$

The above relation implies that the best $\tilde{\eta}$ is again determined from the stationary condition of the trial energy $E(\{\eta_{k'_2k_2k'_1k_1}(\tilde{\eta},\epsilon_c)\})$. Because ϵ_c should satisfy the stationary condition $\delta\epsilon_c = 0$ for the value $\tilde{\eta}^*$, $\tilde{\eta}^*$ is determined by the following condition

$$\left[\frac{\partial\epsilon_c(\{\eta_{k'_2k_2k'_1k_1}(\tilde{\eta},\epsilon_c)\})}{\partial\tilde{\eta}}\right]_{\epsilon_c} = 0 \, . \tag{7}$$

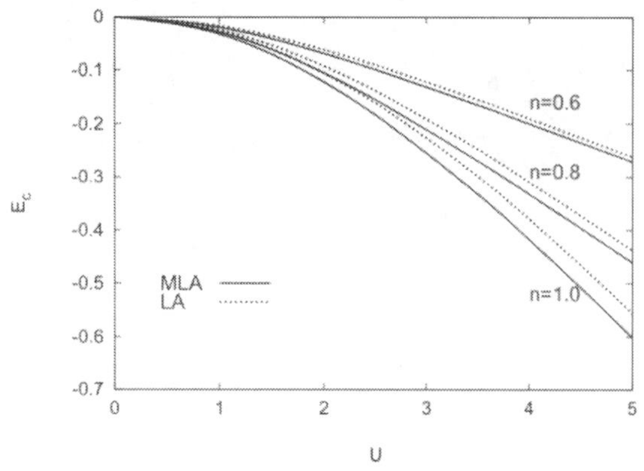

Figure 1. The correlation energies E_c vs. Coulomb interaction energy parameter U for electron number $n = 1.0, 0.8, 0.6$ in the MLA (solid curve) and the LA (dotted curve).

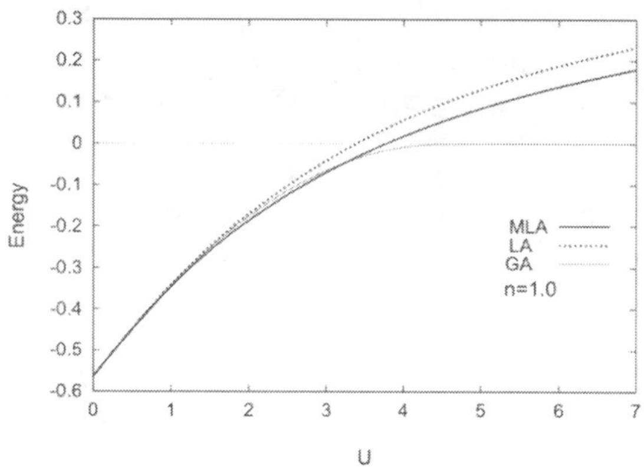

Figure 2. The energy vs. Coulomb interaction energy U curves in the MLA (solid curve), LA (dotted curve),and the GA (thin solid curve) for electron number $n = 1.0$.

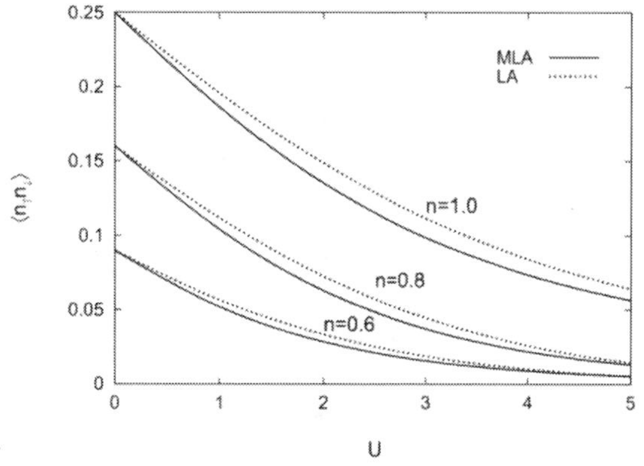

Figure 3. The double occupation number $\langle n_\uparrow n_\downarrow \rangle$ vs. Coulomb interaction energy U curves in the MLA (solid curve) and the LA (dotted curve) for electron number $n = 1.0, 0.8, 0.6$.

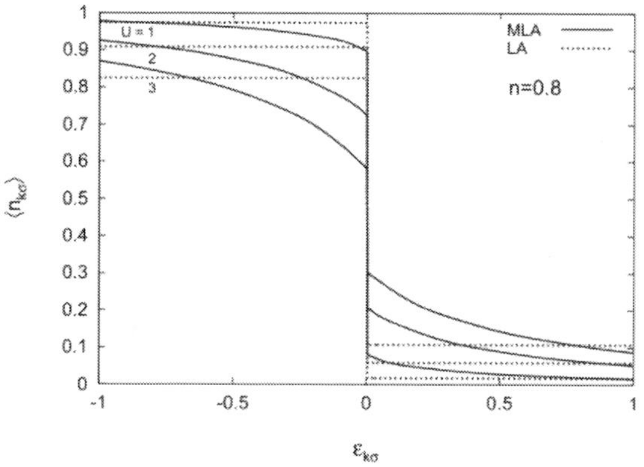

Figure 4. The momentum distribution as a function of energy ϵ_k for various Coulomb interaction energy parameters U with electron number $n = 0.8$. The MLA: solid curves, the LA: dotted curve.

Therefore, we obtained the solution choosing the best variational parameter $\tilde{\eta}$ using the above variational scheme .

To verify the validity of the MLA, we have performed the numerical calculations for the half-filled band as well as non-half-filled band ($n = 0.6, 0.8$) in the Hubbard model [9, 10, 11] on the hypercubic lattice in infinite dimensions where the SSA works best [12]. In this case, the density of states (DOS) for non-interacting system is given by $\rho(\epsilon) = (1/\sqrt{\pi}) \exp(-\epsilon^2)$ [13]. The energy unit is chosen to be $\int d\epsilon \rho(\epsilon) \epsilon^2 = 1/2$.

We present in Fig. 1 the results of correlation energy per atom vs Coulomb interaction U. We observe that, the correlation energy for the MLA is lower than that of the LA over all Coulomb interaction energy parameters U irrespective of electron number. Thus the MLA improves the LA. Note that, with increasing U the correlation corrections increase as U^2 for small U and cancel the Hartree-Fock energy loss being linear in U for large U. Therefore, the magnitude of the correlation energy $|\epsilon_c|$ tends to increase with increasing U. The ground state energy vs Coulomb interaction energy U curves for half-filling case are shown in Fig. 2. We observe that the energy for the MLA is lower than those of the LA and the GA in the weak and intermediate Coulomb interaction regimes U.

Figure 3 shows the double occupation number $\langle n_{\uparrow} n_{\downarrow} \rangle$ as a function of Coulomb interaction energy U for the half-filled and the non-half-filled cases ($n = 0.8, 0.6$). The double occupancy is the same for both LA and MLA in the uncorrelated limit. In order to reduce the loss of Coulomb energy, the double occupancy decreases with increasing Coulomb interaction U. We find that the MLA wavefunction more reduces the double occupancy as compared with that of the LA.

We have also investigated the momentum-distribution functions which are shown in Fig 4. It shows a qualitative change between the LA and the MLA; the distribution functions in the LA are constant below and above the Fermi level irrespective of U, while the MLA curves show a monotonical decrease of the distribution with increasing ϵ_k. It indicates a distinct momentum dependence of $\langle n_{k\sigma} \rangle$ via energy ϵ_k, which is qualitatively different from both the LA and the GA.

In summary, we have presented the variational scheme of the MLA which self-consistently determines both the variational amplitude $\tilde{\eta}$ and the correlation energy ϵ_c making use of variational principles. We demonstrated that the theory much improves the standard variational methods such as the LA and the GA in the weak and intermediate regimes. The correlation energy as well as the ground-state energy in the MLA is lower than both the LA and the GA in those regimes. The double occupation number is suppressed as compared with the LA. Especially, we find that calculated momentum distribution functions show a distinct momentum dependence, which is qualitatively different from those of the LA and the GA.

References

[1] Fulde P 1995 *Electron Correlations in Molecules and Solids* (Springer Verlag Pub., Berlin)
[2] Gutzwiller M C 1963 Phys. Rev. Lett. **10** 159
[3] Gutzwiller M C 1964 Phys. Rev. **134** A923
[4] Gutzwiller M C 1965 Phys. Rev. **137** A1726
[5] Stollhoff G and Fulde P 1977 Z. Phys. B **26** 257
[6] Stollhoff G and Fulde P 1978 Z. Phys. B **29** 231
[7] Stollhoff G and Fulde P 1980 J. Chem. Phys. **73** 4548
[8] Kakehashi Y 2008 J. Phys. Soc. Jpn. **77** 114702
[9] Hubbard J 1963 Proc. Roy. Soc. (London) **A276** 238
[10] Hubbard J 1964 Proc. Roy. Soc. (London) **A277** 237
[11] Hubbard J 1964 Proc. Roy. Soc. (London) **A281** 401
[12] Kakehashi Y 2004 Adv. Phys. **53** 497
[13] Metzner W and Vollhardt D 1989 Phys. Rev. Lett. **62** 324

Energy spectrum and metal-insulator transition in system with Anderson-Hubbard centers

Yu Skorenkyy, O Kramar and L Didukh
Ternopil National Technical University, Physics Department,
56 Ruska Str., Ternopil, Ukraine

E-mail: skorenkyy@tstu.edu.ua

Abstract. Energy spectrum of a model of narrow-band metal into which the periodically spaced Anderson-Hubbard centers are introduced has been studied. Hybridization with conduction band results in the indirect exchange interaction which is different from the interactions between localized magnetic moments and strong on-site Coulomb interaction. To study effect of the lattice deformation under the external pressure (or the self-contraction of the lattice) on electrical properties of the system, the phonon term and elastic energy have been taken into account. The equilibrium values of lattice strain and chemical potential have been calculated self-consistently for non-zero temperatures. Within the Green function approach, the energy spectrum has been calculated as function of model parameters, temperature and external pressure. Our results show that there exists a threshold value of the external pressure above which energy gap value decreases rapidly with temperature and system becomes a correlated metal.

1. Introduction

In recent years the metal-insulator transition problem has been intensively studied in the framework of periodic Anderson model [1-4]. To large extent, these studies are based on successful description of Mott-Hubbard localization in Hubbard model, especially within the framework of dynamical mean field theory (see [5-7] and references therein). Methods developed for investigation of Hubbard model, namely dynamical mean field theory [1, 2], Gutzwiller variational method [3, 4], exact diagonalization [4], spectral density approximation and modified alloy analogy [8] have been applied to clarify the role of hybridization in destabilization of Mott insulator. It has been found that the width of the Mott gap is renormalized by hybridization effects [2], and the increase of temperature leads to the gap closure [1], however the pressure-temperature phase diagram has not been built so far. To complement the results cited above, we have considered a model of the periodically spaced Anderson-Hubbard centers hybridized with conduction band which takes into account the lattice elastic energy also and allows to describe the effect of the external pressure on energy spectrum of the system.

2. The model Hamiltonian

We start from the model of Anderson-Hubbard material which generalizes the models [9, 10] and take into account the peculiarities of correlation effects in narrow energy bands. The Hamiltonian contains terms describing localized (d) subsystem and band (s) subsystem as well as their hybridization

$$
\begin{aligned}
H = & -\mu \sum_{i\sigma} \left(c_{i\sigma}^{+} c_{i\sigma} + d_{i\sigma}^{+} d_{i\sigma} \right) + E_d \sum_{i\sigma} d_{i\sigma}^{+} d_{i\sigma} + E_b(u) \sum_{i\sigma} c_{i\sigma}^{+} c_{i\sigma} + U \sum_{i} n_{i\uparrow} n_{i\downarrow} + \\
& + H_s + H_{sd} + \sum_{\vec{q}f} \hbar \omega_f(\vec{q}) b_{\vec{q}f}^{+} b_{\vec{q}f} + \frac{1}{2} N V_0 C \bar{u}^2 ,
\end{aligned} \tag{1}
$$

$$
H_s = \sum_{ij\sigma} t_{ij}(u) c_{i\sigma}^{+} c_{i\sigma} ,
$$

$$
H_{sd} = V(u) \sum_{i\sigma} \left(c_{i\sigma}^{+} d_{i\sigma} + d_{i\sigma}^{+} c_{j\sigma} \right),
$$

where $d_{i\sigma}^{+}$, $d_{i\sigma}$ are creation and annihilation operators for spin σ electron on i^{th} center in localized (d) state; $c_{i\sigma}^{+}$, $c_{i\sigma}$ are creation and annihilation operators of the band electron, $b_{\vec{q}f}$ operators describe phonon subsystem, $n_{i\sigma} = d_{i\sigma}^{+} d_{i\sigma}$, μ is chemical potential. Principal energy parameters of the model are intra-atomic Hubbard repulsion energy U, interatomic hopping amplitude t_{ij} and hybridization parameter V, though positions of localized level E_{d}, conduction band center E_{b} and elastic constant of the lattice C are also important. The parameters of hybridization of band and localized states $V(u) = V - gu$ and electron hopping $t_{ij}(u) = t_{ij}\left(1 + \frac{BV_0}{2w} u\right)$ are renormalized due to the lattice strain u (values of parameters g, B, V_0, C depend on the narrow-band compound [11]). Electron-phonon interactions play an important role in dynamics of strongly correlated electron systems, as revealed by studies of Hubbard-Holstein model (see [12, 13] and references therein). Here, we study equilibrium properties and focus on the hybridization effect, however, as the system energy is minimized with respect to lattice parameters, the lattice elastic energy must be taken into account. The role of electron-phonon coupling in the hybridized system described by the Hamiltonian (1) will be studied elsewhere. The model Hamiltonian (1) takes into account basic processes and interactions in a narrow non-degenerate band, namely electron hoppings (H_s term), intra-site Coulomb repulsion (the fifth sum), interatomic exchange (the fourth sum), the hybridization of band and localized states (H_{sd} term). The terms "localized" and "band" used here can have different sense depending on the peculiarities of the material under consideration. If a transition metal is studied then the localized subsystem are *3d*-electrons and band subsystem is formed by *s-p*-electrons. For the case of narrow band oxides, *3d* electrons form the localized subsystem and band states correspond to both *3d* electrons of transition metal and *2p* of oxygen subsystem, in rare earth compounds one has localized *f*-electrons and band *s-p-d*-electrons.

3. Lattice self-contraction and metal-insulator transition

We use the Green function method for calculation of single-electron energy spectrum. The equations of motion for localized electron $\langle\langle d_{p\uparrow} | d_{p'\uparrow}^{+} \rangle\rangle$ and band electron $\langle\langle c_{p\uparrow} | c_{p'\uparrow}^{+} \rangle\rangle$ Green functions involve higher-order Green functions. We apply a projection procedure [14]

$$
\left[c_{p\uparrow} ; H_{sd} \right] = \sum_{i} \varepsilon_{pj} d_{j\uparrow} ; \tag{2}
$$

$$
\left[d_{p\uparrow} ; H_{sd} \right] = \sum_{j} \xi_{pj} c_{j\uparrow} , \tag{3}
$$

$$
\langle\langle n_{p\downarrow} d_{p\uparrow} | d_{p'\uparrow}^{+} \rangle\rangle \cong \langle n_{p\downarrow} \rangle \langle\langle d_{p\uparrow} | d_{p'\uparrow}^{+} \rangle\rangle \tag{4}
$$

to break off the chain of equations and analogous decouplings in the equations for functions $\langle\langle c_{p\uparrow} | d_{p'\uparrow}^{+} \rangle\rangle$ and $\langle\langle d_{p\uparrow} | c_{p'\uparrow}^{+} \rangle\rangle$. Non-operator quantities ε_{pj}, ξ_{pj} are found to be functions of electron density and doubly occupied states concentration. Solving the equations with respect to band

and localized electrons Green functions we obtain the energy spectrum for the region close to metal-insulator transition

$$E_{1,2} = -\mu + \frac{E_d + E_b(u)}{2} + \frac{U\langle n_{p\downarrow}\rangle}{2} + \frac{t_{\vec{k}}(u)}{2} \mp \frac{1}{2}\sqrt{\left(E_d - E_b(u) + U\langle n_{p\downarrow}\rangle - t_{\vec{k}}(u)\right)^2 + 4(V(u))^2} \qquad (5)$$

for localized electrons and standard band spectrum for itinerant ones. The spectrum (5), calculated using simple projection procedure, is equivalent to the corresponding results of Gutzwiller variational approach [3, 4]. The distinctive feature of the present result and results of papers [3, 4] from the results of paper [8] is the presence of hybridization gap even in the formal limit of U=0.

Both the bandwidth and band center position cal be changed by the external pressure application. The equilibrium value of the lattice strain can be found from a minimum condition for Gibbs function

$$G = F + PV = F + NPV_0(1+\bar{u}), \qquad (6)$$

as

$$\bar{u} = -\frac{1}{V_0 C}\left(\frac{S}{N}\sum_{\vec{k}\sigma}\langle c^+_{\vec{k}\sigma}c_{\vec{k}\sigma}\rangle + \frac{BV_0}{2w}\frac{1}{N}\sum_{\vec{k}\sigma} t_{\vec{k}}\langle c^+_{\vec{k}\sigma}c_{\vec{k}\sigma}\rangle\right) - \frac{PV_0}{CV_0}. \qquad (7)$$

To find $\langle c^+_{\vec{k}\sigma}c_{\vec{k}\sigma}\rangle$, determined by the spectral function of band electrons, we first calculate the chemical potential from the condition $\sum_i\left(\langle d^+_{i\sigma}d_{i\sigma}\rangle + \langle c^+_{i\sigma}c_{i\sigma}\rangle\right) = nN$. For the transition to be initiated the equilibrium value of lattice strain has to be equal $\bar{u} = -(W - w_c)/(S - 0.5BV_0) < 0$.

After numerical calculation with model rectangular density of states at non-zero temperature we obtain from equation (7) the monotonic dependence of equilibrium lattice strain on temperature. As lattice contracts at the increase of temperature, it induces the metallization of the system. Following the paper [11] we take parameter values $1/CV_0 = 0.05eV^{-1}$; $BV_0 = -3eV$, S=0.4 eV; W=2,2 eV; w=2 eV; U=5 eV and obtain the pressure-temperature phase diagram shown in figure 1.

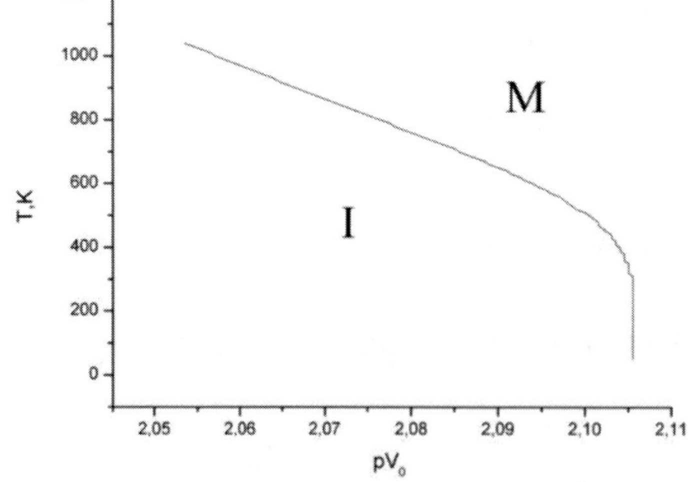

Figure 1. The pressure-temperature phase diagram of metal-insulator transition in the model.

Metal-insulator transition in the model is accompanied by abrupt change of doublon concentration $\langle n_{i\sigma} n_{i\bar{\sigma}} \rangle$. For strong enough Coulomb intra-atomic interactions this quantity is shown to depend weakly on electron-phonon coupling [12] but is very sensitive to the bandwidth-to-interaction parameter rate [5, 6, 14]. Together with the fact that in the vicinity of metal-insulator transition the intra-atomic Hubbard repulsion suppresses the Holstein electron-phonon coupling [13] allows us to argue that the phase diagram presented in figure 1 is qualitatively correct. In distinction from results obtained in the limit of infinite lattice coordination for periodic Anderson model [1,2] and Hubbard model [5-7], we do not observe quasiparticle peaks on position of chemical potential, but the qualitative picture of metal-insulator transition remains valid.

4. Conclusions

In the simple model of narrow band compound with Anderson-Hubbard centers taking into account lattice elastic energy a destabilization of the low temperature insulator phase can be easily described. The increase of temperature leads to the transition from insulating to metallic state, contrary to the Hubbard model at integer filling. Qualitatively, the obtained temperature dependence of the hybridization gap agrees with the corresponding result of dynamical mean field theory for periodic Anderson model [1] differs from the corresponding temperature dependence for Hubbard model [7]. No matter how weak, the hybridization of band and localized states provide a sufficient mechanism for localization effects to dominate in wide temperature-pressure ranges. Insulator or correlated metal phases in the phase diagram and the Coulomb repulsion-to-bandwidth or localized level energy-to-bandwith ratios, where both the bandwidth and localized levels position can be substantially changed by the external pressure or chemical substitution can in principle be estimated from the experimental data on the basis of the considered model. There also exists a threshold value of the external pressure, above which the system is metallic to the lowest temperature.

References

1. Schork T and Blawid S 1997 *Phys. Rev. B* **56** 6559.
2. Sordi G, Amaricci A and Rozenberg M J 2007 *Phys. Rev. Lett.* **99** 196403.
3. Itai K and Fazekas P 1996 *Phys. Rev. B* **54** R752.
4. Hagymasi I, Itai K and Soloyom J 2011 Periodic Anderson model with correlated conduction electrons: variational and exact diagonalization study *Preprint* cond-mat.str-el/1106.4999.
5. Rozenberg M J, Chitra R, Kotliar G 1999 *Phys. Rev. Lett.* **83** 3498.
6. Kotliar G, Lange E, Rozenberg M J 2000 *Phys. Rev. Lett.* **84** 5180.
7. Bulla R, Costi T A, Vollhardt D 2001 *Phys. Rev. B* **64** 045103.
8. Meyer D and Nolting W 2000 *Eur. Phys. Journ.* **18** 385.
9. Didukh L and Stasyuk I 1968 *Fizika metallov i metallovedenie* **26** 582 [in Russian].
10. Didukh L and Stasyuk I 1975 *Ukrainian Physical Journal* **20** 97 [in Russian].
11. Grygorchuk R and Stasyuk I 1980 *Ukrainian Physical Journal* **25** 404 [in Russian].
12. Bauer J and Hewson A C 2010 *Phys. Rev. B* **81** 235113.
13. Capone M, Castellani C and Grilli M 2010 *Advances in Condensed Matter Physics* **2010** 920860.
14. Skorenkyy Yu, Didukh L, Kramar O and Dovhopyaty Yu 2007 *Acta Physica Polonica A* **111** 635.

Perturbative approach to three-component fermionic atoms in optical lattices

Sei-ichiro Suga[1] **and Kensuke Inaba**[2,3]

[1] Department of Materials Science and Chemistry, University of Hyogo, Himeji 671-2280, Japan
[2] NTT Basic Research Laboratories, NTT Corporation, Atsugi 243-0198, Japan
[3] JST, CREST, Sanbancho, Chiyoda-ku, Tokyo 102-0075, Japan

E-mail: suga@eng.u-hyogo.ac.jp

Abstract. We develop a perturbative approach combined with the dynamical mean- eld theory for investigating the Mott transition in repulsively interacting three-component fermionic atoms in optical lattices. We show that this method captures the essentials of the correlation effects and describes well the crossover between the Fermi liquid and the paired Mott insulator.

The high controllability of cold atoms in optical lattices enables us to investigate quantum many-body effects that cannot be observed in condensed matter physics [1, 2, 3]. Recently, degenerate gases of three-component (color degree of freedom) ^{6}Li fermions have been successfully created and their novel features have been revealed [4, 5]. For three-component repulsive fermionic atoms in optical lattices by using the self-energy functional approach (SFA) we showed that a paired Mott insulator (PMI) appears at half filling, although the average atom number per site is non-integer $3/2$ [6, 7]. When two of the three repulsions between different two-color atoms are stronger than the other, effective pairs of weakly repulsing atoms are formed to avoid the stronger two repulsions. Accordingly, in a PMI the effective particle number at a site becomes an integer [7]. In this system, other phase transitions accompanied by symmetry breaking are possible such as a color density wave [6, 8, 9] and a color selective antiferromagnet [6]. For attractively interacting systems, it was shown that a color superfluid appears [10, 11]. To investigate these characteristic phase transitions, we require another method complementary to the SFA . A Feynman diagrammatic approach combined with the dynamical mean-field theory (DMFT) can be an efficient method. This method is called the iterated perturbation theory (IPT), and it has been applied successfully to various types of phase transitions [12, 13]. The diagrammatic approaches have provided us with rich knowledge on the scattering processes relevant to the various phase transitions. Furthermore, the IPT consumes few computational resources. Therefore, the IPT helps us to carry out systematic calculations that capture the essentials of the phase transitions.

In this paper, we develop the IPT by modifying straightforward perturbative calculations. We apply this method to the phase transition/crossover between a Fermi liquid (FL) and a PMI. We compare the results with those obtained with unperturbative calculations using the SFA [6, 7]. We discuss the validity of our approach.

The low-energy properties of repulsively interacting three-component fermionic atoms in optical lattices are well described by the following Hubbard-type Hamiltonian:

$$\hat{\mathcal{H}} = -t \sum_{\langle i,j \rangle} \sum_{=1}^{3} \hat{a}_i^\dagger \, \hat{a}_j \; - \sum_{i} \sum_{=1}^{3} \quad \hat{n}_i \; + \frac{1}{2} \sum_{i} \sum_{\neq \beta} U_{\;\beta} \hat{n}_i \; \hat{n}_{i\beta}, \tag{1}$$

where the subscript $\langle i, j \rangle$ is the summation over the nearest-neighbor sites, and $\hat{a}_i^\dagger$ ($\hat{a}_i$) and $\hat{n}_i$ are creation (annihilation) and number operators of a fermion with color at the ith site. Here, t denotes the hopping integral and is the chemical potential for an atom with color . By choosing adequately, atoms with each color achieve a balanced filling $n(= n \; = \langle \hat{n}_i \; \rangle)$. We set U_{12} U and $U_{23} = U_{31}$ U', and introduce an anisotropy ratio R $U/U'(< 1)$. We neglect the confinement potential for the first approximation.

To investigate the phase transition driven by the local correlation effects, we consider an infinite dimensional system, which is a nontrivial limit for discussing the Mott insulator [12]. According to the DMFT [12] the self-energy $(i\omega_n)$ is purely local and independent of the wave number $\boldsymbol{q}$. The self-energy $(i\omega_n)$ is obtained from the effective impurity model, whose Hamiltonian is

$$\hat{\mathcal{H}}_{\rm imp} = \sum_{\boldsymbol{q}} \sum_{=1}^{3} \;{}_{\boldsymbol{q}} \hat{c}_{\boldsymbol{q}}^\dagger \; \hat{c}_{\boldsymbol{q}} \; - \sum_{=1}^{3} \quad \hat{n}_d \; + \sum_{\boldsymbol{q}} \sum_{=1}^{3} \left[V_{\boldsymbol{q}} \; \hat{c}_{\boldsymbol{q}}^\dagger \; \hat{d} \; + \mathbf{c.c.} \right] + \frac{1}{2} \sum_{\neq \beta} U_{\;\beta} \hat{n}_d \; \hat{n}_{d\beta}, \tag{2}$$

where $\hat{n}_d \; = \; \hat{d}^\dagger \hat{d}$. We introduce the cavity Green's function $g_c \; (i\omega_n)$ corresponding to the non-interacting impurity Green's function, which satisfies the Dyson equation $G_{\rm imp,} \; (i\omega_n) \; = \; [g_c^{-1}(i\omega_n) - \quad (i\omega_n)]^{-1}$. The impurity Green's function $G_{\rm imp,} \; (i\omega_n)$ has to be equal to the local Green's function $G_{\rm loc,} \; (i\omega_n)[= \sum_{\boldsymbol{q}} G \; (i\omega_n, \boldsymbol{q})]$. In this way, the DMFT self-consistent loop is completed. We approximate the self-energy from the first- and second-order contributions. The first-order contributions are given by ${}_1^{(1)}(= {}_2^{(1)}) \; = \; Un_2 + U'n_3$ and ${}_3^{(1)} \; = \; U'n_1 + U'n_2$. The second-order contributions are given by ${}_1^{(2)}(i\omega_n)(= \; {}_2(i\omega_n)) \; = \; \sum_l [U'^2 g_{c1}(i\omega_n + i\nu_l)\chi_3(i\nu_l) + U^2 g_{c1}(i\omega_n + i\nu_l)\chi_2(i\nu_l)]$ and ${}_3^{(2)}(i\omega_n) = \sum_l [2U'^2 g_{c3}(i\omega_n + i\nu_l)\chi_2(i\nu_l)]$, where $\chi \; (i\nu_l) = -\sum_n g_c \; (i\omega_n + i\nu_l) g_c \; (i\omega_n)$.

We modify the self-energy as $\; = \; {}^{(1)} + A \quad {}^{(2)}$ by introducing the prefactor A . This kind of the prefactor was successfully introduced for investigating the crossover from BCS superconductivity to Bose-Einstein condensation in the attractive Hubbard model [13]. To determine the prefactor A , we perform a momentum expansion of $G_{\rm imp,} \; (\omega)$ in the power of $1/\omega$ for $\omega \to \infty$,

$$\begin{aligned} G_{\rm imp,} \; (\omega) \;\; &= \;\; \frac{\langle \{ \hat{d} \; , \hat{d}^\dagger \; \} \rangle}{\omega} + \frac{\langle \{ [\hat{d} \; , \hat{\mathcal{H}}_{\rm imp}], \hat{d}^\dagger \; \} \rangle}{\omega^2} + \frac{\langle \{ [[\hat{d} \; , \hat{\mathcal{H}}_{\rm imp}], \hat{\mathcal{H}}_{\rm imp}], \hat{d}^\dagger \; \} \rangle}{\omega^3} + \cdots \\ &= \;\; \frac{1}{\omega} + \left(- \quad + \sum_{\beta \neq} n_\beta U_{\;\beta} \right) \frac{1}{\omega^2} \\ &+ \left({}^2 + \sum_{\beta \neq} U_{\;\beta} (U_{\;\beta} - 2 \quad) n_\beta + 2 U_{\;\beta} U_{\;\gamma} D_{\beta\gamma} + \sum_{\mathbf{q}} |V_{\mathbf{q}} \; |^2 \right) \frac{1}{\omega^3} + \cdots, \end{aligned} \tag{3}$$

where $[\; , \;]$ and $\{ \; , \; \}$ denote the commutator and anticommutator, respectively, and $D_{\;\beta} \; = \; \langle n \; n_\beta \rangle$ is the double occupancy obtained from $D_{\;\beta} \; = \; -\sum_\gamma {\rm tr} [G_{{\rm imp},\gamma} \partial \;\; {}_\gamma / \partial U_{\;\beta}]$. From the Dyson equation $(i\omega_n) \; = \; g_c^{-1}(i\omega_n) - G_{\rm imp,}^{-1} \; (i\omega_n)$ and Eq. (3), we obtain an exact expression for the self-energy $(i\omega_n)$. For instance, we obtain the exact expression as ${}_3(\omega) = [2U'^2 n_1(1 - n_1) + 2U'^2(D_{12} - n_1 n_2)]/\omega + \cdots$, while the perturbative expression is

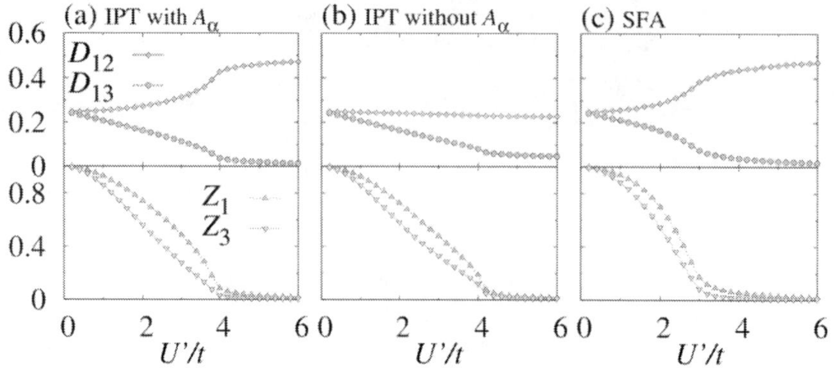

Figure 1. (Color Online) Renormalization factor Z and double occupancy $D_{\ \beta}$ calculated (a) by including the prefactor A , (b) without including it ($A \ = 1$), and (c) by using the SFA as functions of U'/t for $R = 0.1$ and $T/t = 0.1$ at half filling.

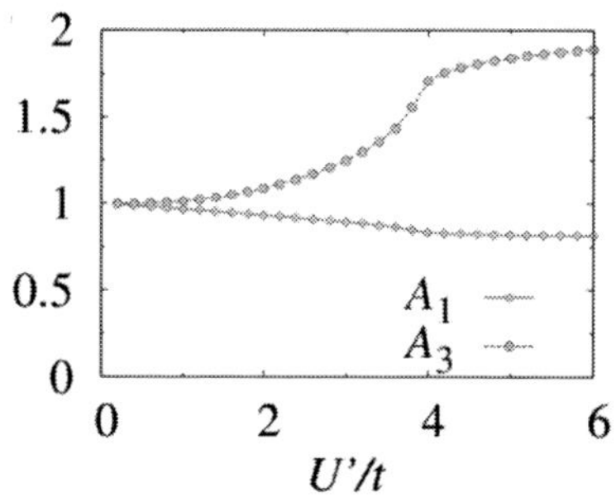

Figure 2. (Color Online) (a) Prefactor A as a function of U'/t for $U/t = 3$, $R = 0.1$, and $T/t = 0.01$ at half filling.

$^{(2)}_{3}(\omega) = 2U'^2 n_{c1}(1 - n_{c1})/\omega + \cdots$. Comparing the two expressions, we obtain the prefactors A as

$$A_1 = A_2 = \frac{U^2 n_2(1 - n_2) + U'^2 n_3(1 - n_3) + 2UU'(D_{23} - n_2 n_3)}{U^2 n_{c2}(1 - n_{c2}) + U'^2 n_{c3}(1 - n_{c3})}, \quad (4)$$

$$A_3 = \frac{n_1(1 - n_1) + D_{12} - n_1 n_2}{n_{c1}(1 - n_{c1})}, \quad (5)$$

where $n_c \ = \sum_n g \ (i\omega_n)e^{i\omega_n 0}$.

On the basis of this IPT, we calculate the renormalization factor Z $1/\{1 - \mathrm{Im}[\ (i \ T) - (-i \ T)]/2 \ T\}$ [14] and the double occupancy $D_{\ \beta}$. For noninteracting atoms, we use a semicircular density of states, $\rho_0(x) = \sqrt{4t^2 - x^2}/(2 \ t^2)$. We also perform unperturbative calculations using the SFA. We compare both results and discuss the validity of this IPT.

In Fig. 1, we show the U'/t dependence of $D_{\ \beta}$ and Z for $R = 0.1$ and $T/t = 0.1$ at half filling. Figure 1(a), (b), and (c) show the results obtained by including the prefactor A , without it ($A \ = 1$), and by using the SFA, respectively. Since we set U_{12} U and $U_{23} = U_{31}$ U', $Z_1 = Z_2$ and $D_{23} = D_{31}$. As U'/t increases, Z decreases and approaches zero asymptotically, indicating that there is a crossover from the FL to the Mott insulator . By extrapolating Z from small U'/t, we find that Z_1 and Z_3 become zero at the same U'/t. The crossover points thus obtained are evaluated to be $U'/t = 4.0$, 4.1, and 2.8 in Fig. 1(a), (b), and (c), respectively. In Fig. 1(a) and (c), D_{12} (D_{13}) increases (decreases) towards 0.5 (zero), as U'/t increases. The results show that paired color-1 and -2 atoms are formed in the Mott insulating phase. This is consistent with the fact that the renormalization effects of color-3 atoms are stronger than those of color-1 and -2 atoms ($Z_3 < Z_1$). In Fig. 1(a) and (c), the U'/t dependences of Z and D_{β} are qualitatively the same. In particular, the double occupancies well describe the situation for the appearance of the PMI, in that the stronger U' is avoided. On the other hand, in Fig. 1(b) both D_{12} and D_{13} decrease and maintain finite values with increasing U'/t. It is difficult to picture the Mott insulating state based on this result. Therefore, we argue that the modified self-energy that results from the inclusion of A captures the essentials of the correlation effects and well describes the PMI transition/crossover.

In Fig. 2, we show the U'/t dependence of the prefactor A for $R = 0.1$ and $T/t = 0.1$ at half filling. As U'/t increases, A_3 increases and A_1 decreases. The results indicate that the effective interaction of color-1 and -2 atoms decreases, while that of color-3 atoms increases. These features occur because the color-1 and -2 atoms form effective pairs to avoid the strong repulsions from color-3 atoms.

In summary, we have shown that our modified IPT captures the essentials of the correlation effects and well describes the FL-PMI transition/crossover. Close to the PMI transition point, we expect superfluid pair fluctuations to be strongly enhanced. Therefore, the superfluid transition can occur in spite of the repulsively interacting atoms. The present IPT is efficient for studying the superfluid transition in repulsively interacting three-component fermionic atoms in optical lattices. The results will be published elsewhere [15]. We finally remark on a possible application of the present IPT method to translational symmetry breaking states such as the density wave state. We can analyze these states by dividing the lattice into sublattices and then slightly modifying our IPT formulation.

We thank T. Ohashi for useful comments and valuable discussions. This work was supported by Grants-in-Aid for Scientific Research (No. 23540467) from JSPS and (No. 21104514) from MEXT.

References

[1] Greiner M and Fölling S 2008 *Nature* **453** 1
[2] Bloch I 2005 *Nat. Phys.* **1** 23
[3] Jaksch D and Zoller P 2005 *Annals of Physics* **315** 52
[4] Ottenstein T B, Lompe T, Kohnen M, Wenz A N and Jochim S 2008 *Phys. Rev. Lett.* **101** 203202
[5] Huckans J H, Williams J R, Hazlett E L, Stites R W and O'Hara K M 2009 *Phys. Rev. Lett.* **102** 165302
[6] Miyatake S, Inaba K and Suga S 2010 *Phys. Rev. A* **81** 021603(R)
[7] Inaba K, Miyatake S and Suga S 2010 *Phys. Rev. A* **82** 051602(R)
[8] Honerkamp C and Hofstetter W 2004, *Phys. Rev. Lett.* **92** 170403
[9] Rapp A and Rosch A 2011, *Phys. Rev. A* **92** 053605
[10] Rapp A, Zarand G, Honerkamp C and Hofstetter W 2007 *Phys. Rev. Lett.* **98** 160405
[11] Inaba K and Suga S 2009 *Phys. Rev. A* **80** 041602(R)
[12] Georges A, Kotliar G, Krauth W and Rozenberg M J 1996 *Rev. Mod. Phys.* **68** 13
[13] Garg A, Krishnamurthy H R and Randeria M 2005 *Phys. Rev. B* **72** 024517
[14] We use this de nition of the renormalization factor at nite temperatures instead of that at $T = 0$, $Z \equiv 1/(1 - \partial\Sigma\ /\partial\omega)|_{\omega=0}$.
[15] Inaba K and Suga S 2012 *Phys. Rev. Lett. in press*

The chiral charge density wave transition in *1T*-$TiSe_2$

Jasper van Wezel
Materials Science Division, Argonne National Laboratory, Argonne, IL 60439, USA
E-mail: jvanwezel@anl.gov

Abstract. It was recently suggested that the formation of charge density wave order in *1T*-$TiSe_2$ is accompanied by the emergence of orbital order, which renders the combined state chiral. Here, we analyse the transition between the chiral and non-chiral phases of $TiSe_2$ in terms of a Ginzburg-Landau description. We find that the phase transition is expected to be of second order, and that the consequences of the formation of chiral charge order should be observable in a variety of experimental settings.

1. Introduction

In the layered transition metal dichalcogenide *1T*-$TiSe_2$, the formation of charge order was recently suggested to give rise to a chiral pattern of electronic density [1–3]. The spiral order consists of three charge density wave components, which are shifted in phase with respect to each other. As a result, the dominant propagation vector of the charge ordered state is rotated as one goes from one atomic layer to the next. Similar chiral phases are common in magnetically ordered materials, where the order parameter is a vector whose rotation around a propagation direction straightforwardly yields a spiral magnetization [4]. In the case of charge order however, it is not *a priori* clear how the scalar electronic density can be molded into a helical configuration.

This apparent paradox may be resolved by the realization that the formation of chiral charge order in $TiSe_2$ is accompanied by the simultaneous emergence of orbital order. The coupling between electronic density, orbital occupation and lattice deformation yield a chiral pattern in the combined charge and orbital ordered state [3]. Here, we discuss the transition from the non-chiral into the chiral phase in terms of a Ginzburg-Landau expansion of the free energy. We also discuss various experiments in which the transition into the combined state of chiral charge and orbital order may be observed.

2. Chiral charge and orbital order in $TiSe_2$

The quasi-two dimensional, layered material titanium-diselenide is either a semimetal or a semiconductor with a small indirect (positive or negative) band gap of the order of 150 meV or less [5, 6]. The valence band is derived predominantly from Se-$4p$ orbitals, and has its maximum at the center of the first Brillouin zone, as indicated in figure 1. The conduction band has minima at the centers of the three inequivalent zone boundaries, and consist predominantly of Ti-$3d$ electrons. Although the electron pockets are poorly nested [7], $TiSe_2$ forms a commensurate 2x2x2 charge density wave at 202 K, accompanied by a periodic lattice distortion.

In spite of many theoretical and experimental investigations, it is still unclear whether the emergence of charge order in $TiSe_2$ is driven by a variant of the Jahn-Teller effect [7–10], the

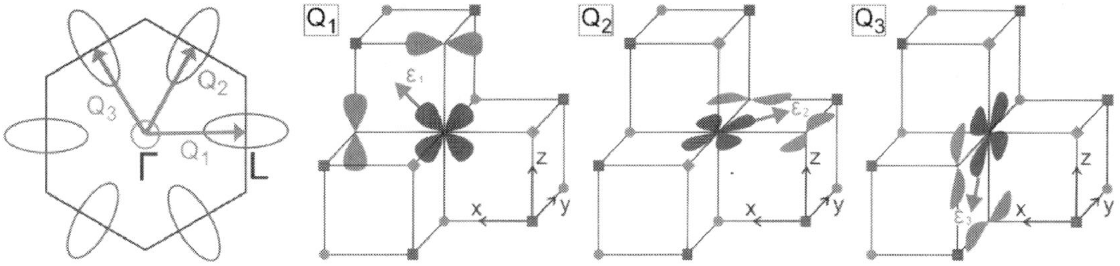

Figure 1. Left: Schematic indication of the Fermi surface pockets in the first Brillouin zone of $TiSe_2$. The three propagation vectors of the charge density modulations connect the central Se hole pocket with the Ti electron pockets at the L points. **Right**: The orbitals involved in the charge transfer processes of each of the three charge density wave components. The layer of Ti atoms (squares) is sandwiched between two layers of Se atoms (circles and diamonds). The resulting polarization directions of the corresponding displacement waves are indicated.

formation of an excitonic insulator [11–14], or a combined mechanism involving cooperation between excitons and phonons [15–17]. Regardless of the relative importance of phonons or excitons, it is clear that the emergence of charge order in $TiSe_2$ is associated with the transfer of electronic weight from the central Se pocket to the Ti pockets at the Brillouin zone boundaries.

2.1. Orbitals and atomic displacement

The charge density wave in $TiSe_2$ consists of three component waves, with three distinct propagation vectors, which connect the central Se pocket in figure 1 to the three inequivalent Ti pockets. The electrons in each of these pockets have a specific orbital character, which can be determined for example from a tight binding fit to the band structure [3]. As indicated in figure 1, the electrons in the Ti pockets are predominantly of the t_{2g} type, and their orientation rotates as one goes from one pocket to the next. Likewise, the orientations of the Se-p orbitals rotates as one goes around the central hole pocket.

Because the charge order is due to the transfer of charge between a specific Ti-d orbital and its neighboring Se-p orbitals, the atomic displacements $\boldsymbol{\delta}\mathbf{x} \propto \boldsymbol{\epsilon}\cos(\mathbf{Q}\cdot\mathbf{x})$ associated with the charge density modulations along $\mathbf{Q}$ will be confined to the plane containing these orbitals, as indicated in figure 1. We thus find that the polarizations of the atomic displacement waves accompanying the three components of the charge density wave in $TiSe_2$, are all transversal.

The total displacement pattern in the charge ordered phase consists of the superposition of three differently polarized components with three distinct propagation vectors. Having found both the propagation vectors and the polarization vectors, we still need to consider the possibility of nonzero relative phase differences between the superposed components of the displacement wave: $\boldsymbol{\delta}\mathbf{x} \propto \sum_i \boldsymbol{\epsilon}_i \cos(\mathbf{Q}_i\cdot\mathbf{x}+\varphi_i)$. These phase differences cannot be found from the electronic band structure alone. They instead depend on the interactions between charge density wave components, which are mediated both by Coulomb interactions and coupling to the lattice.

2.2. Relative phases

To describe the emergence of relative phases, we start from the expression for the Landau free energy of a charge density modulation $\alpha = \Re\{\psi_1+\psi_2+\psi_3\}$ consisting of three components $\psi_j(\mathbf{x}) = \psi_0 e^{i\mathbf{Q}_j\mathbf{x}+\varphi_j}$, where $\mathbf{Q}_j$ are given by the propagation vectors of the charge ordered state: $F = \int d\vec{x}\ \ a(x)\alpha^2 + b(x)\alpha^3 + c(x)\alpha^4 + d(x)\left[|\psi_1\psi_2|^2 + |\psi_2\psi_3|^2 + |\psi_3\psi_1|^2\right]$. The spatial variation of the coefficients reflects the periodicity of the lattice as well as the internal structure of the unit cell. This can be achieved by expanding the coefficients according to

$a = a_0 + a_1 \sum_i e^{i\mathbf{G}_i \mathbf{x}} + a_1\gamma \sum_{i,j} e^{i\mathbf{G}_i(\mathbf{x}+\mathbf{R}_j)} + ..$, where $\mathbf{G}_i$ are reciprocal lattice vectors, and $\mathbf{R}_j$ connect a Titanium atom to its Selenium neighbors in the same unit cell [18]. Notice that γ may in general be different from one because it depends on the specific values of the electron-phonon coupling on the Ti and Se sites. Inserting this form of the coefficients into the free energy, the integrals can be evaluated directly [3]. The final expression is then found to be:

$$F = \frac{3}{2}a_0\psi_0^2 + \frac{1}{2}a_1\psi_0^2(1-\gamma)\sum_j \cos(2\varphi_j) + \frac{3}{8}(15c_0 + 8d_0)\psi_0^4 + \frac{3}{4}c_2\psi_0^4 \sum_j \cos(2\varphi_j - 2\varphi_{j+1}). \quad (1)$$

Minimizing this expression with respect to the phase variables yields two possible solutions. The first is the well known non-chiral triple-**Q** charge density wave with all phases equal to $\pi/2$. There is however also a second solution, which has $\varphi_1 = \pi/2$, $\varphi_2 = -\varphi_3$, and $\varphi_3 = \pm 1/2\cos^{-1}[(3c_2\psi_0^2 - 2a_1(1-\gamma))/(6c_2\psi_0^2)]$. This solution corresponds to the recently observed chiral state [1–3]. It has the lowest overall energy if the argument of the cosine is larger than -1, as can be seen from the comparison of the Landau free energy expressions in the presence of these phase values:

$$\begin{aligned} F_{non-chiral} &= \frac{3}{2}a_0\psi_0^2 + \frac{3}{8}(15c_0 + 8d_0)\psi_0^4 - \frac{3}{2}a_1\psi_0^2(1-\gamma) + \frac{9}{4}c_2\psi_0^4 \\ F_{chiral} &= \frac{3}{2}a_0\psi_0^2 + \frac{3}{8}(15c_0 + 8d_0)\psi_0^4 - \frac{a_1^2(1-\gamma)^2}{6c_2} - \frac{9}{8}c_2\psi_0^4. \end{aligned} \quad (2)$$

Since the chiral state requires at least a finite value of the order parameter amplitude ψ_0, a phase diagram is implied where the uniform, high temperature state first goes into the usual non-chiral triple-**Q** structure at a critical temperature T_C, and only then enters into the chiral phase at a lower temperature T_{Ch}. The initial triple-**Q** state is entered when ψ_0 first becomes non-zero. Assuming that all temperature dependence is contained in the coefficient a_0, we thus write $a_0 = \tilde{a}(T - T_C) + a_1(1-\gamma)$. Using this definition, and minimizing the Landau free energy with respect to ψ_0, the order parameter amplitude acquires the temperature dependence $\psi_0^2 = 2\tilde{a}(T_C - T)/(15c_0 + 8d_0 + 6c_2)$. The chiral state emerges when $9c_2\psi_0^2 = 2a_1(1-\gamma)$, so that finally we find its transition temperature to be $T_{Ch} = T_C - a_1(1-\gamma)(15c_0 + 8d_0 + 6c_2)/(9\tilde{a}c_2)$.

From equation (2), it is straightforward to see that both the free energy itself and its first derivative with respect to temperature are continuous across T_{Ch}. The second derivative however, which is proportional to the specific heat, has a jump-like discontinuity:

$$C_V = \begin{cases} 3\tilde{a}^2T/[15c_0 + 8d_0 + 6c_2] & T > T_{Ch} \\ 3\tilde{a}^2T/[15c_0 + 8d_0 - 3c_2] & T < T_{Ch} \end{cases} \quad (3)$$

The transition is therefore found to be of second order.

3. Observing the transition

The phase transition between the non-chiral and the chiral charge ordered states predicted to occur at T_{Ch} has not been experimentally observed yet. The recent observation of the chiral phase in $TiSe_2$ employed low-temperature scanning tunneling microscopy (STM), and found the chiral phase to be present throughout the studied temperature range [1]. The transition temperature must thus lie above 84 K, but below $T_C = 202$ K. Besides the extension of the STM study to higher temperatures, a variety of alternative experimental probes can be used to observe the transition between the two types of charge order.

It has already been shown that optical reflectometry for example is sensitive to the breakdown of rotational symmetry associated with the formation of chiral order [1]. The triple-**Q** mode either shares the symmetry of the lattice, or breaks it down to a threefold symmetry. The chiral phase on the other hand is at most twofold symmetric. This difference can be observed in the varying intensity of reflected light with changing polarization.

Alternatively, one can study the distortions of the atomic lattice directly in diffraction experiments. The space group of the high temperature phase of $TiSe_2$ is lowered from $P\bar{3}m1$ to $P\bar{3}c1$ as the triple-**Q** pattern sets in. The chiral phase breaks the symmetry down further, so that only a P2 space group remains. This change in symmetry is accompanied by an alteration of the reflection conditions. Whereas the triple-**Q** state has forbidden reflections at $[h, -h, 0, 2l]$, the chiral state has no such constraints.

Finally, one can also make use of the fact that the formation of chiral charge order in $TiSe_2$ coincides with the appearance of an ordered structure in the orbital occupations. Because the different components of the displacement wave transfer charge between different sets of orbitals, and because one component dominates in each atomic layer, the orbital order is inevitable. The orbital-specific nature of resonant inelastic X-ray scattering (RIXS) experiments may be used to probe this orbital order in the chiral state.

4. Conclusions

The recently observed chirality in the layered dichalcogenide *1T*-$TiSe_2$ emerges from the simultaneous formation of orbital and charge density wave order. The three components of the charge density wave involve three distinct sets of orbitals, giving rise to three distinct polarization directions for their associated displacement waves. Superposing the components with non-zero relative phase differences then yields the observed chiral pattern. The origin of these phase differences can be understood from a Landau expansion of the free energy, in which the competition between local Coulomb interactions and elastic energy yields two possible phases for the charge order in $TiSe_2$. At the initial charge ordering temperature, the phase differences are strictly zero, and the charge density wave is non-chiral. At a second, lower temperature another phase transition is encountered and the phase differences become non-zero, yielding a chiral charge ordered state. The transition into the chiral state can be observed using a variety of experimental probes, including scanning tunneling microscopy, optical reflectometry, diffraction and inelastic X-ray scattering.

Acknowledgments

This work was supported by the US DOE, Office of Science, Contract No. DE-AC02-06CH11357.

References

[1] Ishioka J, Liu Y H *et al.* 2010 *Phys. Rev. Lett.* **105** 176401
[2] van Wezel J and Littlewood P B 2010 *Physics* **3** 87
[3] van Wezel J 2011 *preprint* cond.mat/1106.1930
[4] Overhauser A 1962 *Phys. Rev.* **128** 1437
[5] Di Salvo F and Waszczak J 1978 *Phys. Rev. B* **17** 3801
[6] Zunger A and Freeman A J 1978 *Phys. Rev. B* **17** 1839
[7] Rossnagel K, Kipp L and Skibowski M 2002 *Phys. Rev. B* **65** 235101
[8] Whangbo M H and Canadell E 1992 *J. Am. Chem. Soc.* **114** 9587
[9] Hughes H P 1977 *J. Phys. C* **10** L319
[10] Motizuki K, Suzuki N *et al.* 1981 *Solid State Comm.* **40** 995
[11] Wilson J 1977 *Solid State Comm.* **22** 551
[12] Li G, Hu W *et al.* 2007 *Phys. Rev. Lett.* **99** 027404
[13] Cercellier H, Monney C *et al.* 2007 *Phys. Rev. Lett.* **99** 146403
[14] Monney C, Cercellier H *et al.* 2009 *Phys. Rev. B* **79** 45116
[15] van Wezel J, Nahai-Williamson P and Saxena S S 2010 *Europhys. Lett.* **89** 47004
[16] van Wezel J, Nahai-Williamson P and Saxena S S 2010 *Phys. Rev. B* **81** 165109
[17] van Wezel J, Nahai-Williamson P and Saxena S S 2010 *Phys. Status Solidi b* **247** 592
[18] McMillan W L 1975 *Phys. Rev. B* **12** 1187

Dispersion of orbital excitations in 2D quantum antiferromagnets

Krzysztof Wohlfeld[1], Maria Daghofer[1], Giniyat Khaliullin[2], Jeroen van den Brink[1]

[1]IFW Dresden, P. O. Box 27 01 16, D-01171 Dresden, Germany
[2]Max-Planck-Institut für Festkörperforschung, Heisenbergstrasse 1, D-70569 Stuttgart, Germany

E-mail: k.wohlfeld@ifw-dresden.de

Abstract. We map the problem of the orbital excitation (orbiton) in a 2D antiferromagnetic and ferroorbital ground state onto a problem of a hole in 2D antiferromagnet. The orbiton turns out to be coupled to magnons and can only be mobile on a strongly renormalized scale by dressing with magnetic excitations. We show that this leads to a dispersion relation reflecting the two-site unit cell of the antiferromagnetic background, in contrast to the predictions based on a mean-field approximation and linear orbital-wave theory.

1. Introduction

Calculating collective excitations in nonfrustrated but interacting spin systems may be considered as one of the few extremely successful stories in solving the strongly correlated electron systems. Independently on whether the approach which is used to calculate such excitations stems from the linear spin wave theory (for ordered states) or the Bethe Ansatz solutions (for quantum disordered states in 1D), a remarkable qualitative and quite often also quantitative agreement with a number of experimental results is usually achieved, see e.g. [1].

On the other hand, collective excitations in interacting *orbital* systems have attracted much less attention of the community, primarily due to experimental problems in detecting the collective orbital excitations (orbitons) [2]. *A priori*, excitations in the orbital degree of freedom of Mott insulators should behave in a similar manner as those of the spin sector: in both cases the large Coulomb repulsion in the Mott insulators 'freezes' the charge degrees of freedom and largely reduces the effective low energy Hilbert space so that instead of electrons it is given by spin or orbital (pseudospin) degrees of freedom. In place of the 'spin-wave theory' for magnetic excitations, one would then use 'orbital-wave theory' [3].

It has recently been shown [4], however, that this orbital-wave picture does not hold in many realistic systems. Here, we prove that in a generic 2D Kugel-Khomskii (KK) spin-orbital model [5] with the antiferromagnetic (AF) and ferroorbital (FO) ground state the linear orbital wave theory breaks down and one needs to explicitly take care of the coupling between spins and orbitals. Furthermore, we show that for realistic parameters of the KK Hamiltonian, the dispersion of the coherent excitation changes qualitatively and reflects the two-site unit cell of the antiferromagnetic background, in contrast to expectations based on linear orbital-wave theory.

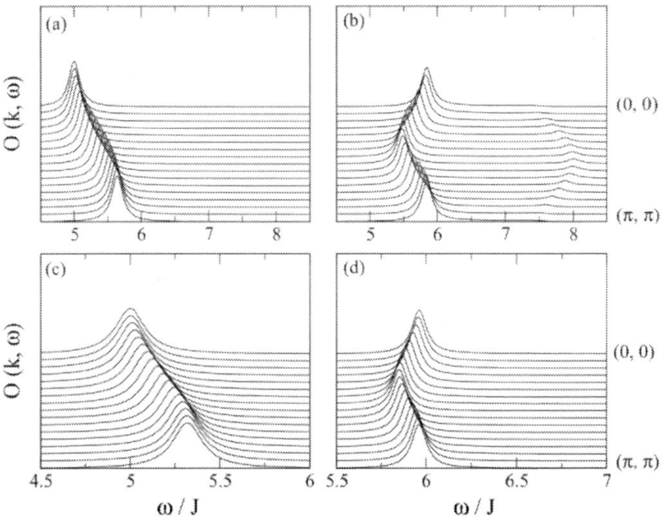

Figure 1. Spectral function for the orbital excitation $O(\mathbf{k},\omega)$ [Eq. (2)] calculated in the linear orbital wave picture [(a) and (c)] and using the exact mapping onto the effective t-J model [(b) and (d)]. Upper (lower) panels for equivalent (inequivalent) values of the hoppings between the two orbitals, see text. Note the different scales on upper and lower panels and that the incoherent spectrum in panel (d) is extremely small and not visible at this scale. Parameters: $E_z = 5J$ and broadening $\eta = 0.05J$.

2. Model and results

The Hamiltonian of the generic 2D KK model that we want to study describes a two-orbital system with orbital-conserving hopping and reads

$$H = J\sum_{\langle \mathbf{i},\mathbf{j}\rangle}\left(\mathbf{S_i}\cdot\mathbf{S_j}+\frac{1}{4}\right)\left(\mathbf{T_i}\cdot\mathbf{T_j}+\frac{1}{4}\right)+E_z\sum_{\mathbf{i}}T^z_{\mathbf{i}}, \tag{1}$$

where $\mathbf{S}$ ($\mathbf{T}$) are the spin (pseudospin) operators which fulfill the SU(2) algebra and J is the superexchange constant which describes the interaction of $S = 1/2$ spins and $T = 1/2$ (orbital) pseudospins. Besides, E_z is the crystal field which is implicitly assumed to be so large that in the ground state $|\phi\rangle$ of H only one type of orbital is occupied (the FO order of orbitals) and consequently there is long-range AF order in temperature $T = 0$.

The purpose of this study is to calculate the orbital excitations from the ground state $|\phi\rangle$ of Hamiltonian (1), as defined by the spectral function $O(\mathbf{k},\omega)$:

$$O(\mathbf{k},\omega) = \frac{1}{\pi}\lim_{\eta\to 0}\Im\left\langle \phi\left| T^-_{\mathbf{k}}\frac{1}{\omega+E_\phi-H-i\eta}T^+_{\mathbf{k}}\right|\phi\right\rangle, \tag{2}$$

where E_ϕ is the ground state energy and $T^+_{\mathbf{k}}$ is the momentum-dependent orbital raising operator.

In the 'standard' orbital-wave approach [6, 7], this spectral function can be easily calculated (see, e.g., Ref. [4]) by mean-field decoupling of spins and orbitals and by: (i) introducing the Holstein-Primakoff bosons for orbital pseudospins (cf. Ref. [8]) and (ii) neglecting interactions between bosons. We then obtain:

$$O(\mathbf{k},\omega) = \delta\left[\omega - E_z + \frac{1}{2}zJ_{\rm OW}(1-\gamma_{\mathbf{k}})\right], \tag{3}$$

with $z = 4$ is the coordination number in 2D and $\gamma_{\mathbf{k}} = (\cos k_x + \cos k_y)/2$ is the lattice structure factor in 2D. Here $J_{\rm OW} = 2J\langle\tilde{\phi}|\mathbf{S_i}\cdot\mathbf{S_j}+\frac{1}{4}|\tilde{\phi}\rangle \simeq -0.16J$ where $|\tilde{\phi}\rangle$ is the 2D AF, i.e. spin-only part of ground state $|\phi\rangle$. The spectrum consists of a single quasiparticle peak with a relatively large dispersion, cf. Fig. 1(a) for the spectrum along the high symmetry direction $(0,0)\to(\pi,\pi)$ in the Brillouin zone.

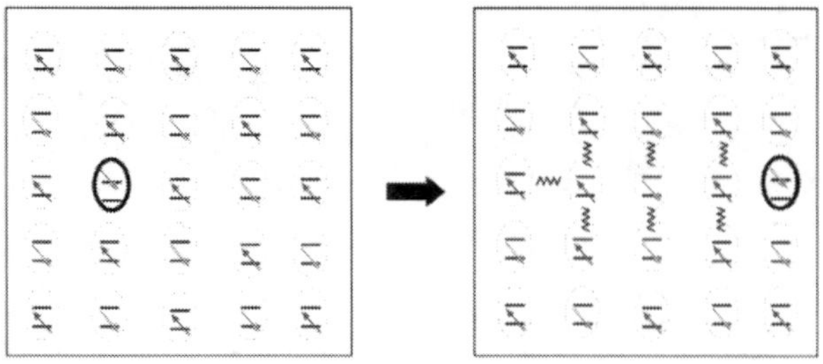

Figure 2. Artist's view of the defects (springs) in the spin background (up and down arrows) created by the moving orbiton (bold oval). Only if these defects are healed by quantum fluctuations the orbiton can move coherently – otherwise, the orbiton is trapped in a string-like potential, see text.

However, it was shown in detail in Ref. [4] that the above mean-field decoupling of spins and orbitals does not lead to a reliable description of orbital dynamics in a coupled spin-orbital system. Therefore, following Ref. [4] and using *inter alia* the Jordan-Wigner transformation [9] for spin and orbital degrees of freedom,[1] we map the above problem of a single orbiton in the AF and FO ground state Eq. (2) onto a a problem of a single hole in the AF ground state

$$O(\mathbf{k},\omega)=\frac{1}{\pi}\lim_{\eta\to 0}\Im\left\langle\tilde{\phi}\left|p^{\dagger}_{\mathbf{k}\uparrow}\frac{1}{\omega+E_{\tilde{\phi}}-\tilde{H}-E_z-i\eta}p_{\mathbf{k}\uparrow}\right|\tilde{\phi}\right\rangle, \tag{4}$$

described by an effective t–J Hamiltonian

$$\tilde{H}=-t\sum_{\langle\mathbf{i},\mathbf{j}\rangle,\sigma}(p^{\dagger}_{\mathbf{i}\sigma}p_{\mathbf{j}\sigma}+h.c.)+J\sum_{\langle\mathbf{i},\mathbf{j}\rangle}\left(\mathbf{S_i}\cdot\mathbf{S_j}+\frac{1}{4}n_{\mathbf{i}}n_{\mathbf{j}}\right). \tag{5}$$

Here the electron operators $p_{\mathbf{i}\sigma}$ ($p^{\dagger}_{\mathbf{i}\sigma}$) act in the restricted Hilbert space without double occupancies and create (annihilate) holes in the 2D AF ground state $|\tilde{\phi}\rangle$. Besides, $n_{\mathbf{i}}=\sum_{\sigma}n_{\mathbf{i}p\sigma}$ and the effective hopping parameter t describing the motion of orbital excitation is given by $t\equiv J/2$.

The spectrum Eq. (4) was calculated using the self-consistent Born approximation on 32×32 sites (which gives results in very good agreement with the exact diagonalization studies, cf. Ref. [10]) and consists of a large quasiparticle peak and a very small incoherent spectrum, cf. Fig. 1(b). The bandwidth W of the coherent orbiton peak, as obtained from Fig. 1(b) or from perturbative calculations cf. Ref. [10], is $W\simeq 2t^2/J\simeq J/2$ (for $t=J/2$) and thus somewhat smaller than the mean-field one for which $W\simeq 0.64J$. In more striking contrast to the mean-field results, the dispersion has a minimum at $(\pi/2,\pi/2)$ instead of $(0,0)$, as a result of the coupling between orbitons and the AF 'background'.

In many realistic cases, the spin-orbital model (1) has to be modified to account for the different hoppings for electrons in the two active orbitals. We have verified that the effective hopping t is then no longer $t=J/2$ but instead $t=2t_tt_2/U$ where t_1 and t_2 denote the bare hopping of an electron in orbital 1 and 2 and U is the on-site Hubbard repulsion. Furthermore, quite often the hopping of the occupied orbital (defined below as orbital 1) is larger, as the gain in hybridization energy contributes to its energy gain with respect to the empty orbital, i.e., $t_1>t_2$. For example for copper oxides with an occupied x^2-y^2 orbital and empty t_{2g} orbital, $t_2\approx t_1/2$ and thus $t=J/4$ since $J=4t_1^2/U$. In the orbital-wave picture, the dispersion width is then likewise reduced by a factor of two to $W=0.32J$, but the effect is stronger in the interacting t-J model, where we find $W\simeq 0.125J$, cf. Fig. 1(c)-(d).

[1] See also supplementary material in Ref. [4] which discusses the validity of this mapping in the 2D case.

3. Discussion

Let us now try to understand the physical origin of these large differences between the orbiton motion obtained in the mean-field approximation and in a far more exact way (i.e. via mapping onto the effective t–J model), cf. Fig. 2. As in the mean-field case the orbiton moves freely and does not excite any spin excitations, the whole effect of the spin background is taken into account by renormalizing the superexchange constant J into its reduced value $J_{\rm OW}$. This stays in contrast with the result obtained from the mapping onto the t–J model: now the orbiton is effectively a 'hole' in the AF spin background which, when moving, flips the spins of the 2D AF, cf. Fig. 2. In fact, the further the orbiton moves the more spin flips (defects in 2D AF) it excites which leads to the so-called string-like potential acting on the orbiton, cf. Fig. 2. The orbiton can then move coherently only due to the presence of quantum spin fluctuations in the ground state which 'heal' these defects, 'attach' to the orbiton and jointly move as an orbiton quasiparticle on a renormalized scale (cf. Ref. [11, 10] with the hole in 2D AF moving effectively as a spin polaron). We also note that this result is entirely different from that in the 1D case [4], where the orbital flip excitation fractionalizes into free 'spinon' and 'orbiton' quasiparticles, similar to decay of a doped hole into spinon and holon.

4. Conclusions

In summary, we showed that a single orbital excitation in the 2D AF and FO ground state moves similarly to a hole in the 2D AF state. Therefore, it can only be mobile by coupling to the spin fluctuations which leads to a minimum in the orbiton dispersion for the $(\pi/2, \pi/2)$ point in the Brillouin zone and to the very strong renormalization of the orbiton bandwidth. In particular, in the realistic case which can correspond to the *dd* excitations in copper oxides, the orbiton bandwidth is of the order of $\sim 0.1J$ which explains why typically such *dd* excitations in the AF ground state are nearly dispersionless and can be well modelled using quantum chemistry calculations on small clusters [12].

Acknowledgments

We acknowledge support of the Alexander von Humboldt Foundation (K.W.) and the DFG Emmy Noether Program (M.D.).

References

[1] Coldea R *et al.* 2001 *Phys. Rev. Lett.* **86** 5377
Walters A C *et al.* 2009 *Nat. Phys.* **5** 867
[2] Grüninger M *et al.* 2002 *Nature* **418** 39
Ulrich C *et al.* 2009 *Phys. Rev. Lett.* **103** 107205
[3] van den Brink J, Stekelenburg W, Khomskii D I, Sawatzky G A and Kugel K I 1998 *Phys. Rev.* B **58** 10276
van den Brink J, Horsch P, Mack F and Oleś A M 1999 *Phys. Rev.* B **59** 6795
[4] Wohlfeld K, Daghofer M, Nishimoto S, Khaliullin G and van den Brink J 2011 *Phys. Rev. Lett.* **107** 147201
[5] Kugel K I and Khomskii D I 1982 *Sov. Phys. Usp.* **25** 231
[6] Khaliullin G and Oudovenko V 1997 *Phys. Rev.* B **56** 14243
[7] Oleś A M, Khaliullin G, Horsch P and Feiner L F 2005 *Phys. Rev.* B **72** 214431
[8] Wohlfeld K, Oleś AM and Horsch P 2009 *Phys. Rev.* B **79** 224433
[9] Jordan P and Wigner E 1928 *Z. Physik* **47** 631
[10] Martinez G and Horsch P 1991 *Phys. Rev.* B **44** 317
[11] Schmitt-Rink S, Varma C M and Ruckenstein A E 1988 *Phys. Rev. Lett.* **60** 2793
[12] Hozoi L, Siurakshina L, Fulde P and van den Brink J 2011 *Sci. Rep.* **1** 65

Journal of Physics: Conference Series **391** (2012) 012169 doi:10.1088/1742-6596/391/1/012169

Hubbard model on the anisotropic kagomé lattice: Mott transition and ferrimagnetism

A Yamada[1], K Seki[1], R Eder[2], and Y Ohta[1]

[1] Department of Physics, Chiba University, Chiba 263-8522, Japan
[2]Karlsruhe Institut of Technology, Institut fur Festkorperphysik, 76021 Karlsruhe, Germany

E-mail: `atsushi@physics.s.chiba-u.ac.jp`

Abstract. The Mott transition and ferrimagnetism are studied in the Hubbard model on the anisotropic kagome lattice using the variational cluster approximation, and the phase diagram at half-filling and zero temperature is examined. We find that the ferrimagnetic phase rapidly grows as the geometric frustration is relaxed, and the Mott-insulating phase disappears in moderately frustrated region, indicating that the ferrimagnetic fluctuations stemming from the relaxation of the geometric frustration is enhanced by the electron correlations. In the metallic phase, heavy fermion behavior is observed and mass enhancement factor is computed. Enhancement of e ective spatial anisotropy by the electron correlations is also confirmed in moderately frustrated region, and its e ect on the heavy fermion behavior is examined.

1. Introduction

Electron systems with geometric frustration play an important role to reveal the nature of the Mott transition and physics related to it, e.g., heavy fermion behavior and spin liquids. When spatial anisotropy is introduced in systems with geometric frustration, the interplay between the spin uctuations and Mott transition appears as a new feature and provides unique phenomena which take place neither in the unfrustrated nor fully frustrated systems. As for the kagomé lattice, the issues related to the anisotropy have been considered only recently using the cellular dynamical mean eld theory[1], where the Mott transition point was analyzed and enhancement of spatial anisotropy and spin correlations were observed. Such enhancement may give rise to the extension of the ordered (ferrimagnetic) phase. Therefore, if the Mott transition itself persists without being veiled by the ferrimagnetic phase remains to be examined. Also, the e ect of the enhanced anisotropy on the heavy fermion behavior is worth being studied. We investigate the ferrimagnetism and Mott transition on the anisotropic kagomé lattice using the variational cluster approximation (VCA)[2, 3, 4], which is formulated based on a rigorous variational principle and exactly takes into account the short-range correlations. We study the phase diagram at zero temperature and half- lling. We show that, in moderately frustrated region, the ferrimagnetic phase rapidly grows down to the metal-insulator phase boundary, and the Mott insulator (MI) phase disappears. In the metallic phase, heavy fermion behavior is observed and the mass enhancement of the quasiparticle is computed. E ective spatial anisotropy becomes also larger due to the electron correlations, in agreement with the previous study.[1] This e ect gives rise to an enhancement of the anisotropy of the e ective masses of the quasiparticles.

2. The model and formalism

The Hamiltonian of the Hubbard model on the anisotropic kagomé lattice (see Fig. 1) reads

$$H = \quad \sum_{i,j,} t_{ij} c_i^{\dagger} c_j + U \sum_i n_{i\uparrow} n_{i\downarrow} \quad \mu \sum_{i,} n_i \, , \tag{1}$$

where $t_{ij} = t$ between the sites 1 and 2, 3 and $t_{ij} = t'$ between the sites 2 and 3, U is the on-site Coulomb repulsion, and μ is the chemical potential. The energy unit is set as $t = 1$ hereafter.

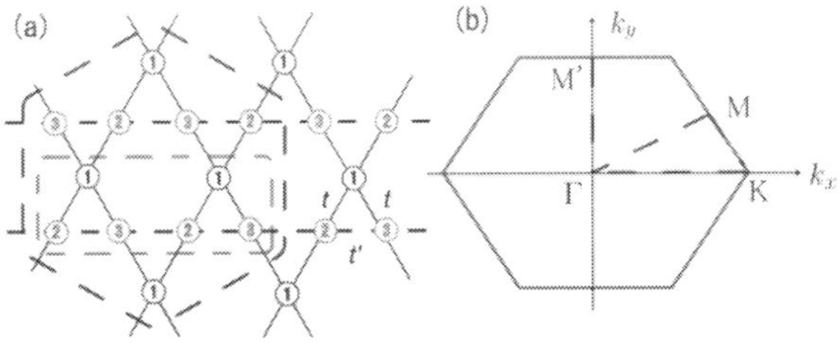

Figure 1. (a) Anisotropic kagomé lattice. In our lattice geometry, the three sites 1, 2, and 3 form an equilateral triangle of the unit length and the dashed lines are along the x direction. Inside the hexagon (dotted line) and square (dash-dotted line) are the 12- and 6-site clusters, respectively, which will be used in our analysis. (b) The rst Brillouin zone of the anisotropic kagomé lattice.

In our analysis, we use the 6- and 12-site clusters in Fig. 1(a) to set up the cluster Hamiltonian, and compute the density of states, double occupancy and ferrimagnetic order parameter $M = \frac{1}{3}\sum_{a=1}^{3} \mathrm{sign}(a)(\langle n_{a\uparrow}\rangle \quad \langle n_{a\downarrow}\rangle)$ with $\mathrm{sign}(a) = \quad 1$ for the site 1 and $\mathrm{sign}(a) = 1$ for the sites 2 and 3. Paramagnetic states with a gap at Fermi level is considered as the Mott insulating states.

3. Results of calculation

In Fig. 2, we show the phase diagram at zero temperature and half- lling obtained using the 12-site cluster. In this region of t', the ferrimagnetic phase is an insulator since there is a gap, and the transition between the ferrimagnetic and paramagnetic (including MI) phases is a level crossing (rst order). The critical interaction strength U_{MI} separating the MI and metallic phases is slightly smaller than the noninteracting band width W, where $W = 6$ at $t' = 1$ and $W = 4\sqrt{2} \simeq 5.66$ at $t' = 0$. The behavior of our U_{MI} according to the relaxation of the frustration is qualitatively consistent with the previous results[1], though our values for U_{MI} are relatively small compared to those in Ref. [1]. At present the origin of these discrepancies are not clear to us. As for the Mott transition, we could not nd out the Mott insulator and paramagnetic metal coexisting region of U at half- lling within our two controlling parameters μ and μ'. Detailed analysis of the Mott gap and the double occupancy in Ref. [6] also shows that there are no indication of the discontinuity around the Mott transition, suggesting that the transition is continuous. In Fig. 2, we show the double occupancy near the Mott transition. In Ref. [1] this transition is reported to be rst order. First order Mott transitions are obtained in other models in the variational cluster approach with bath degrees of freedom and treating the hybridization between the bath sites and cluster sites as a variational parameter.[7, 8] In these analyses, the coexisting metal and insulator solutions, leading to the rst order transition, di er by the value of these hybridization parameters, and these situations will be similar to the case of the Ref. [1]. Our analysis does not have bath degrees of freedom and technically this will be the origin of the di erence. It remains to be clari ed which is the correct picture.

4. Heavy fermion behavior

In Fig. 3 we show the spectral weight function (ω, k) calculated using the 12-site cluster for solutions corresponding to (a), (b) the ferrimagnetic phase (up and down spin parts are plotted separately), (c) the MI phase, and (d) the metallic phase at $t' = 0.75$. In (a) and (b), the mean- eld spin-density-wave (SDW) dispersion is also included (solid red lines) to see its general

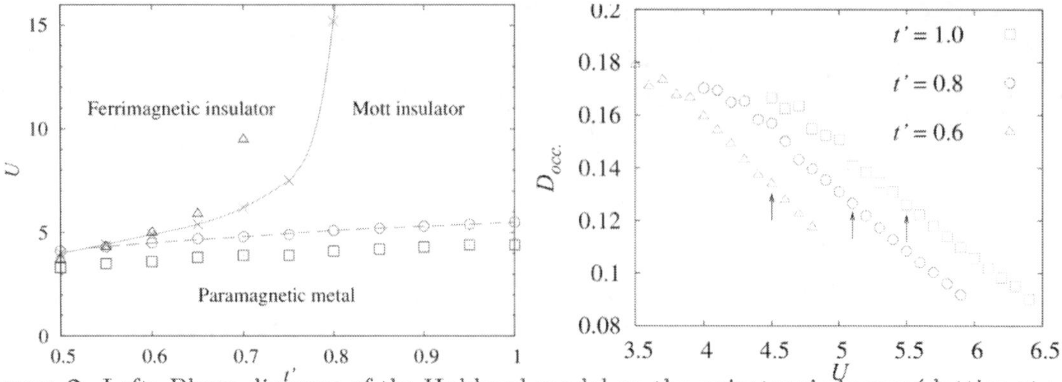

Figure 2. Left: Phase diagram of the Hubbard model on the anisotropic kagomé lattice at zero temperature and half- lling as a function of t' and U obtained by VCA, where the 12-site cluster is used for the crosses and circles and the 6-site cluster is used for the squares and triangles, respectively. Right: The double occupancy $D_{occ.}$ as a function of U for $t' = 1.0$, 0.8, and 0.6. The 12-site cluster is used. The three arrows indicate the Mott transition points.

features, where $M = 0.92$ in the mean- eld solution while $M = 0.72$ in VCA. In (d), the noninteracting band structure is also plotted (solid red lines) for comparison. In (d), we notice that the spectral function is consistent with the Fermi liquid state and the interacting bands slightly shrink toward the Fermi surface, leading to heavy fermion behavior. To study it in detail, we consider well below the MI transition line where Fermi liquid natures are con rmed from the behavior of the spectral function, and compute the mass enhancement factor m^*/m along the x and y directions in k space, where the x direction corresponds to the direction of t' hopping in real space. We also compute the ratio of the Fermi momenta in the x and y

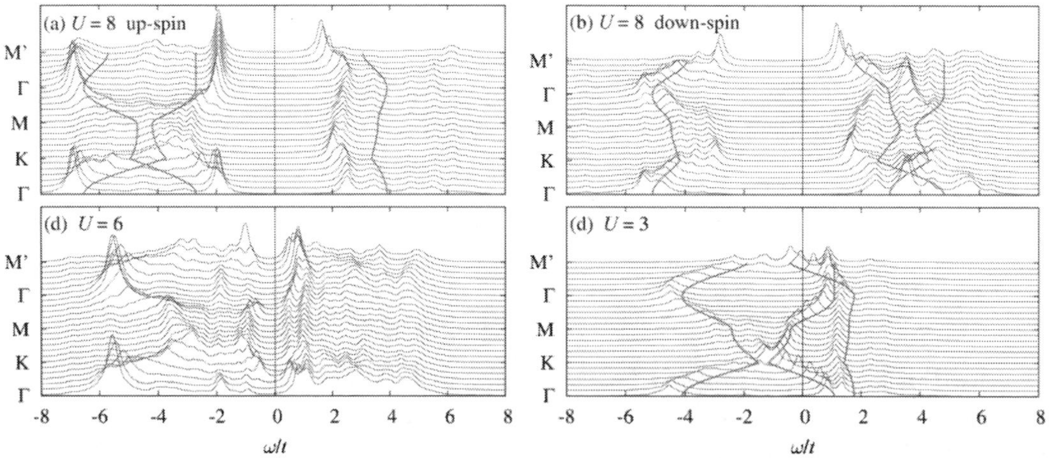

Figure 3. Spectral density at $t' = 0.75$ for (a), (b) $U = 8$ (ferrimagnetic state), (c) $U = 6$ (MI state), and (d) $U = 3$ (metallic state) along the dashed line in Fig. 1(b). The Lorentzian broadening with $\eta = 0.15t$ is used in all the cases. In (a) and (b), the solid red lines are the mean- eld SDW dispersion for the same values of U, t' and μ. In (d), the solid red lines are the noninteracting band structure. In (a), (b), and (c) the peaks are scaled by 5 compared to (d).

directions, $r = k_{yF}/k_{xF}$. As t' decreases, the noninteracting Fermi surface slightly shrinks in the x direction and slightly evolves in the y direction[9], so r is a measure of the anisotropy including the e ect of the electron correlations. we show in Fig. 4 r and m^*/m in the x and

y directions as functions of U for $t' = 1.0$, 0.8, and 0.6 in our lattice geometry for the 12-site cluster, to see general features about the e ect of the electron correlations on these quantities. The rapid growth of r at $t' = 0.6$ indicates that the e ective anisotropy is enhanced due to the electron correlations in moderately frustrated region. The analysis of the e ective anisotropy was also done in Ref. [1] by considering the renormalization of the hoping parameters and our results are qualitatively consistent with their analysis. As is shown in Fig. 4(b) (d), the heavy fermion behavior is observed in all cases, and appears largely in moderately frustrated region.

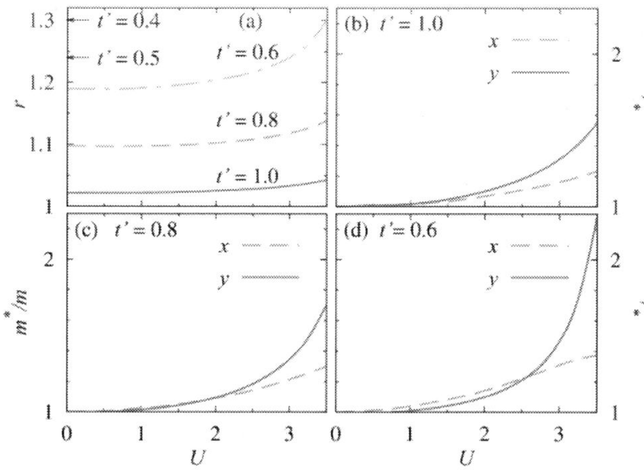

Figure 4. (a) Ratio r as a function of U for $t' = 1.0$, 0.8, and 0.6. The two arrows indicate the value of r for noninteracting band at $t' = 0.5$ ($r = 1.24$) and at $t' = 0.4$ ($r = 1.30$). (b) (d) Mass enhancement factor m^*/m in the x and y directions as functions of U for (b) $t' = 1.0$, (c) $t' = 0.8$, and (d) $t' = 0.6$. The 12-site cluster is used.

5. Summary

In summary we have investigated the ferrimagnetism and Mott transition on the anisotropic the kagomé lattice using VCA. The phase diagram at zero temperature and half- lling is determined. The ferrimagnetic phase rapidly grows in moderately frustrated region and the MI phase disappears there. In the metallic phase, heavy fermion behavior is studied and the mass enhancement is computed. Enhancement of spatial anisotropy due to the electron correlations is also observed for moderately frustrated region and its e ect on the heavy fermion behavior is discussed. Thus, the interplay between the spin correlations and Mott transition is quantitatively studied above and below the metal-insulator transition.

Acknowledgments

One of us (A.Y.) would like to thank N Fukui, K Kurasawa, H Mikami, and H Nakada for useful discussions on numerical analysis. This work was supported in part by Kakenhi Grant No. 22540363 of Japan. A part of computations was done at Research Center for Computational Science, Okazaki, Japan.

References

[1] Furukawa Y, Ohashi T, Koyama K, and Kawakami N, Phys. Rev. B **82**, 161101 (2010).
[2] Senechal D *et al.*, Phys. Rev. Lett. **84**, 522 (2000); Phys. Rev. B **66**, 075129 (2002).
[3] Pottho M *et al.*, Phys. Rev. Lett. **91** 206402 (2003); Dahnken C*et al.*, Phys. Rev. B **70**, 2451100 (2004).
[4] Pottho M, Eur. Phys. J. B **32**, 429 (2003).
[5] Aichhorn M, Arrigoni E, Pottho M, and Hanke W, Phys. Rev. B **74**, 024508 (2006).
[6] Yamada A, Seki K, Eder R, and Ohta Y, Phys. Rev. B **83**, 195127 (2011).
[7] Pottho M, Eur. Phys. J. B **36**, 335 (2003).
[8] Balzer M, Kyung M, Senechal D, Tremblay A M, Pottho M, Eur. Phys. Lett. **85**, 17002 (2009).
[9] Imai Y, Kawakami N, and Tsunetsugu H, Phys. Rev. B **68**, 195103 (2003).

Journal of Physics: Conference Series **391** (2012) 012170 doi:10.1088/1742-6596/391/1/012170

Study of Charge-Density-Wave Instability in Heavy Electron Systems

Tsuneya Yoshida and Norio Kawakami
Department of Physics, Kyoto University, Kyoto, 606-8502, Japan
E-mail: yoshida@scphys.kyoto-u.ac.jp

Abstract. In this article an extended periodic Anderson model, including correlations between conduction and localized electrons, is studied with dynamical mean eld theory and continuous-time quantum Monte Carlo method. In particular, we clarify how the antiferromagnetic phase, observed in the ordinary periodic Anderson model, changes into the charge-density-wave phase which emerges in a strong coupling region.

1. Introduction

Heavy fermion behavior can be found in lanthanide or actinide compounds where interesting phenomena are observed such as unconventional superconductivity, metamagnetism, Kondo insulators and so on. In heavy fermion systems, the effective mass of electrons can be thousands times heavier than that of free electrons and the hybridization between itinerant orbital and strongly correlated localized orbital plays an essential role in description of the system. To study these systems, the periodic Anderson model, which has the above feature, is often used.

In some compounds, however, correlations between conduction (c-) electrons and localized (f-) electrons are also important. In $CePd_2Si_2$ and $CeIn_3$, antiferromagnetism is suppressed with increasing pressure and the superconducting phase appears near the antiferromagnetic (AF) quantum critical point (QCP)[1]. In these compounds, a mechanism forming Copper pairs is related to AF fluctuations. On the other hands, the pairing mechanism in $CeCu_2Si_2$ and $CeCu_2Ge_2$ cannot be understood only in terms of AF fluctuations. In $CeCu_2Si_2$ and $CeCu_2Ge_2$, with increasing pressure, the AF phase is suppressed and the superconducting phase appears. However, the transition temperature of the superconducting phase has a maximum far from AF-QCP. Moreover, in $CeCu_2(Si_{0.9}Ge_{0.1})_2$, two superconducting domes are found in the P-T phase diagram[2]. To understand these phenomena, Miyake and his collaborators have extended the periodic Anderson model by taking into account local interaction between c- and f-electrons and proposed that a mechanism for the unconventional superconducting phases is related to valence fluctuations[3]. Numerical calculations, such as density matrix renormalization and dynamical mean field theory (DMFT) with exact diagonalization, also indicate that the c-f correlations cause a valence transition (VT). Another example showing the importance of the c-f correlation is $Yb_{0.38}In_{0.62}Cu_4$. This compound exhibits a metamagnetic behavior accompanied by volume expansion with lowering temperature. Motivated by this phenomenon, Watanabe *et al.* have proposed that the metamagnetism is caused by the VT which is induced by the cooperation of c-f correlation and magnetic field[4].

From these facts, the c-f electron correlation has received much attention and many groups have studied this effect intensively. Recently, it has been proposed that c-f correlations also enhance fluctuations of charge density wave (CDW) in the weak coupling region[6] and its instability is confirmed in the strong coupling region by our previous study[7]. However, the relation between this phase and the AF phase, stabilized in the ordinary periodic Anderson model, is unclear.

Thus, in this article, we study how the AF phase changes into the CDW phase with increasing c-f correlation. We analyze this system by using DMFT[8] combined with continuous time-quantum Monte Carlo method, which enables us to study the system with high accuracy[9, 10]. This paper is organized as follows. First, the model and method are explained in the next section. We discuss the effect of interorbital correlation on heavy fermion systems in sec. 3, which is followed by a brief summary in sec. 4.

2. Model and Methods

We consider the extended periodic Anderson model including the c-f correlations. The Hamiltonian is given by

$$H_{\mathrm{ePAM}} = -t \sum_{<i,j>,\sigma} c^{\dagger}_{i\sigma} c_{j\sigma} + H_{\mathrm{loc}} \tag{1}$$

$$H_{\mathrm{loc}} = {}_f \sum_{i,\sigma} n^f_{i,\sigma} + V \sum_{i,\sigma} (c^{\dagger}_{i\sigma} f_{i\sigma} + \mathrm{h.c.}) + U_{ff} \sum_i n^f_{i\uparrow} n^f_{i\downarrow} + U_{cf} \sum_{i\sigma,\sigma'} n^c_{i\sigma} n^f_{i\sigma'}, \tag{2}$$

where $c_{i\sigma}$ ($f_{i\sigma}$) is annihilation operator of c-(f-) electron at site i in spin state $\ \ =\uparrow, \downarrow$. The c-electrons are itinerant with hopping integral t while f-electrons are localized with energy $\ _f$. The hybridization between c- and f-electrons at the same sites is represented by V. Correlations between f-electrons (between c-and f-electrons) are represented by U_{ff} (U_{cf}).

To analyze this model, DMFT is employed. In DMFT, the lattice problem is mapped onto an effective impurity problem which has to be solved self-consistently. For simplicity, we assume the lattice structure to the Bethe lattice, and this lattice structure simplifies the self-consistent equation to the following equation[8],

$$\left[g^{-1}_{\sigma}(i\omega)\right]_{cc} = i\omega + \ \ - (\frac{D}{2})^2 \left[G_{\sigma}(i\omega)\right]_{cc}, \tag{3}$$

where D is the half-band width of the conduction band. The $g_{\sigma}(i\omega)$ is the Green's function of the effective bath, and $G_{\sigma}(i\omega)$ is the full Green's function in the impurity model. Long-range ordered phases are treated with sublattice method. The order parameter of the AF (CDW) phase is defined by $m_{c(f)} = \frac{1}{2}(n^{c(f)}_{\uparrow} - n^{c(f)}_{\downarrow})$ ($\rho = \frac{1}{2}\sum_{\sigma} |(n^c_{\sigma} - n^f_{\sigma})|$). In the following, we set $D = 1$ as energy unit and fix the temperature at $T = 0.02$. For simplicity, we analyze the particle hole symmetric case. Namely we set $\ \ = U_{cf}$ and $\ _f = -U_{ff}/2$, where   is chemical potential.

3. Correlation effects between conduction and localized electrons

Let us first analyze the order parameters. The order parameters for the AF and CDW phase are shown in Fig. 1. The AF order parameter at $U_{cf} = 0.5$ is plotted in Fig. 1(a). As seen in this figure, c- and f-electrons are magnetized by RKKY interaction. Note that a paramagnetic phase in $V \lesssim 0.3$ is different from the paramagnetic phase in $V \gtrsim 0.425$. In $V \lesssim 0.3$, the magnetic moments are suppressed by temperature, while in $V \gtrsim 0.425$, the moments are suppressed by the Kondo effect. This behavior is qualitatively the same as in the ordinary periodic Anderson model, i.e. $U_{cf} = 0$ case. However, in the strong U_{cf} region, this antiferromagnetic phase is replaced by the CDW phase. In Fig. 1(b), the order parameters for each phase at $V = 0.1$ are plotted as a function of the c-f correlation.

In Fig. 1 (b), we can observe how the system changes its character under strong c-f correlation. As seen from this figure, if the correlation is weak, the system has no long range order. Note, however, that this paramagnetic phase is due to temperature effects, and the AF phase is considered to be stabilized in the lower temperature region. If c-f correlation becomes stronger, the AF phase is stabilized even at $T = 0.02$. With increasing c-f correlation further, the AF phase is suppressed and the CDW phase emerges in the strongly correlated region. Sketches of these phases are shown in Figs. 1(c) and (d). The origin of this CDW phase may be the competition between f-electron interaction and hybridization. It is confirmed that strong hybridization suppresses the CDW phase. As mentioned above, the AF phase is affected by the temperature particularly in the region $V/D \lesssim 0.25$. The detailed discussions on the temperature effects will be given elsewhere[11].

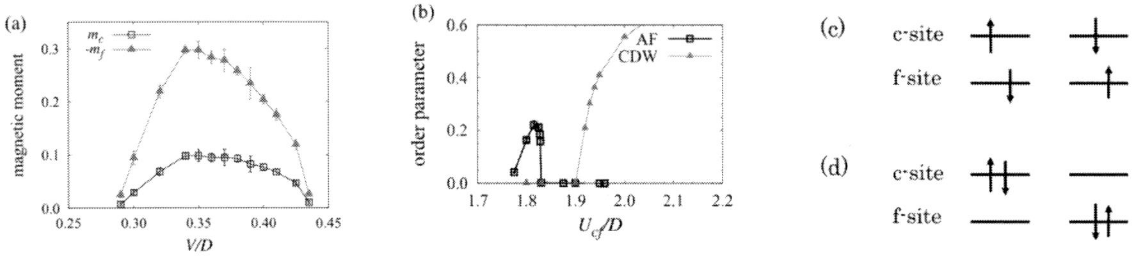

Figure 1. (a) Antiferromagnetic moments as a function of hybridization at $U_{cf} = 0.5$. (b) Order parameters of the AF and CDW phases as a function of intraorbital interaction at $V = 0.1$. (c) and (d) Sketches of the AF and CDW phase respectively.

For further understanding of this system in the strong c-f region, we observe the local spin correlation of each orbital (S^2, $= c, f$) and c-f spin correlation. These quantities reflect the CDW instability. Let us consider the hybridization dependence, from which we can find some interesting phenomena in local spin correlations. The results for $U_{cf} = 1$, and 2 are plotted in Fig. 2. For U_{cf}=2, the CDW phase is stabilized in $V < 0.105$. In this case, we can see a rapid change in $\langle S_c \cdot S_f \rangle$ due to CDW instability. Namely, in $V < 0.105$, $\langle S_c \cdot S_f \rangle$ is rapidly decreased with decreasing V. In the $U_{cf} = 1$ case, the c-f spin correlation exhibits qualitatively the same V dependence as in the $U_{cf} = 1$ case except for the influence of CDW transition.

On the other hand, we can see that the behaviors of the other quantities for the $U_{cf} = 2$ case are qualitatively different from the $U_{cf} = 1$ case, even if the system is in a paramagnetic phase. In Fig. 2 (b), we can find that in the $U_{cf} = 2$ case, local spin correlation of c-electrons increases with increasing V, while in the $U_{cf} = 1$ case, the local spin correlation decreases around $V = 0.3$ with decreasing V. Local spin correlation of f-electrons ($\langle S_f^2 \rangle$) also exhibits similar V dependence. These qualitatively different behaviors for $U_{cf} = 1$ and 2 are attributed to the difference of charge configuration between the ground states at $V = 0$. Namely, in the case of $U_{cf} = 2$, the system has CDW order at $V = 0$ (sketch of this state is shown in Fig. 1(d)). In the case of $U_{cf} = 1$, the ground state at $V = 0$ is a paramagnetic state whose charge configuration is drawn in Fig. 1(c). Therefore, if $U_{cf} = 2$ and V is small, the occupancy of an f-orbital is nearly two at c-electron rich sites. Increasing V, the f-electrons become easy to hop to c-orbital. On the other hand, in the $U_{cf} = 1$ case, one electron occupies each site, and the increase in V leads to the increase in f-electron double occupancy.

In the analysis of how the system behaves with increasing V, we have observed that the f-electron double occupancy exhibits qualitatively different behaviors for $U_{cf} = 1$ and 2. These

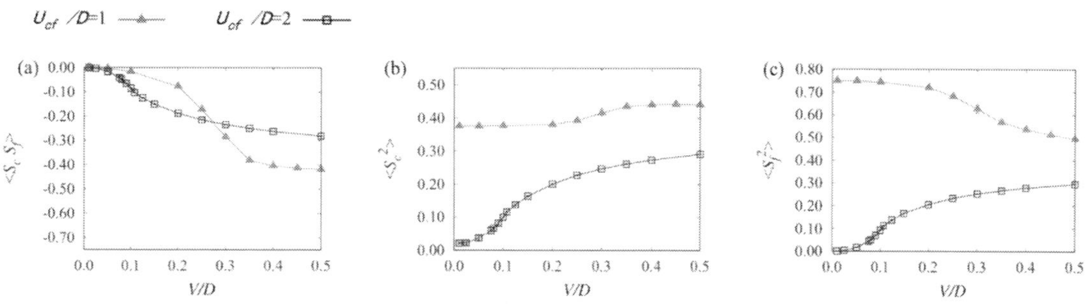

Figure 2. V-dependence of the physical quantities for $U_{cf} = 1$ and 2: (a) c-f-spin correlation, (b) ((c)) local spin correlation of c- (f-) electron $\langle S_c^2 \rangle$ ($\langle S_f^2 \rangle$). Note that in the $U_{cf} = 2$ case, the CDW phase is stabilized in $V < 0.105$.

interesting behaviors are attributed to the difference of charge configuration between the ground states at $V = 0$. The CDW phase studied in this paper emerges as a result of competition of the hybridization, the f-f interaction and the c-f interaction. Thus, for further understanding of the CDW phase, a more systematic study is necessary, which is currently in progress.

4. Summary

In this paper, we have studied AF and CDW instabilities in the extended periodic Anderson model with c-f correlations. The c-f spin correlation and the local spin correlation for each orbital have been calculated as a function of hybridization. From the behavior of local spin correlation, it has been found that the system exhibits qualitatively different behaviors depending on c-f correlation even if the system is in the paramagnetic phase. We have also elucidated how the AF phase, emerging in the ordinary periodic Anderson model, changes into the CDW phase in the presence of strong c-f correlation.

To conclude this paper, a brief comment on a VT is in order. It should be naturally expected that the interorbital correlation can induce a VT. In our study, however, the VT has not been found. This is due to our choice of parameters. Namely, we have restricted ourselves to the particle-hole symmetric case, where the deep f-electron level suppresses the valence fluctuations and thus the condition for the VT is not satisfied. We will extend our analysis to the valence fluctuation regime to incorporate the possibility of VT in our treatment.

This work is partly supported by JSPS through its FIRST Program, KAKENHI (Nos. 21740232, 20104010), and the Global COE Program "The Next Generation of Physics, Spun from Universality and Emergence" from MEXT of Japan.

References

[1] Mathur N D, Grosche F M, Julian S R, Walker I R,Freye D M, Haselwimmer R K W, and Lonzarich G G 1998 *Nature* **394** 39.
[2] Yuan H Q, Grosche F M, Deppe M, Geibel C, Sparn G, and Steglich F 2003 *Sience* **302** 2104.
[3] Onishi Y and Miyake K 2000 *J. Phys. Soc. Jpn.* **69** 3955.
[4] Watanabe S, Tsuruta A, Miyake K, and Flouquet J 2008 *Phys. Rev. Lett.***100** 236401.
[5] Sugibayashi T, and Hirashima D S 2006 *J. Phys. Soc. Jpn. Suppl.* **75** 244-6.
[6] Sugibayashi T, Saiga Y, and Hirashima D S 2008 *J. Phys. Soc. Jpn.* **77** 024716.
[7] Yoshida T, Ohashi T, and Kawakami N 2011 *J. Phys. Soc. Jpn* **80** 064710
[8] Georges A, Kotliar G, Krauth W and Rozenberg M J 1996 *Rev. Mod. Phys.* **68** 13-125.
[9] Werner P, Comanac A, de'Medich L, Troyer M and Millis A J 2006 *Phys. Rev. Lett.* **97** 76405.
[10] Haule K, 2007 *Phys. Rev. B* **75** 155113.
[11] Yoshida T. and Kawakami N. in preparation.

Multiple Lifshitz transitions driven by short-range antiferromagnetic correlations in the two-dimensional Kondo lattice model

Yu Liu and Guang-Ming Zhang

State Key Laboratory of Low-Dimensional Quantum Physics and Department of Physics, Tsinghua University, Beijing, 100084, China

E-mail: gmzhang@tsinghua.edu.cn

Abstract. With a mean field approach, the heavy Fermi liquid in the two-dimensional Kondo lattice model is carefully considered in the presence of short-range antiferromagnetic correlations. As the ratio of the local Heisenberg superexchange coupling to the Kondo coupling increases, the Fermi surface structure changes dramatically. From the analysis of the ground state energy density, multiple Lifshitz type phase transitions occur at zero temperature.

1. Introduction

Quantum phase transitions have attracted much interest in studying correlated electron systems. An electronic phase transition associated with the change of Fermi surface (FS) topology, the so-called Lifshitz transition[1], can be induced without any spontaneous symmetry breaking and local order parameter. The Kondo lattice model is a prototype model and believed to capture the basic physics of heavy fermion materials. The huge mass enhancement of the quasiparticles can be attributed to the coherent superposition of individual Kondo screening clouds, and the resulting metallic state is characterized by a *large* FS with the Luttinger volume containing both conduction electrons and localized moments. Competing with the Kondo singlet formation, the localized spins indirectly interact with each other via magnetic polarization of the conduction electrons – the Ruderman-Kittel-Kasuya-Yosida interaction. Such an interaction dominates at low values of the Kondo exchange coupling and is the driving force for the antiferromagnetic (AFM) long-range order and quantum phase transitions[2, 3]. So far most of investigations focus on the possible FS reconstruction around the magnetic quantum critical point[4, 5, 6, 7, 8]. However, the FS topology in the paramagnetic heavy fermi liquid phase may also be drastically changed by the short-range AFM spin correlations between the localized spins, leading to the Lifshitz phase transitions[9]. The nature of such a quantum phase transition has not been thoroughly explored yet.

In this paper, we consider the two-dimensional Kondo lattice model with the Heisenberg AFM superexchange coupling between localized spins. By introducing uniform short-range AFM valence-bond and Kondo screening parameters, a fermionic mean-field theory is derived and carefully re-examined. Away from half-filling, at the conduction electron density $n_c = 0.85$, for example, the possible changes of FS topology in the paramagnetic heavy Fermi liquid phase are considered carefully as increasing the short-range AFM spin correlations.

2. Mean field theory

The model Hamiltonian defined on a square lattice is given by

$$H=\sum_{\mathbf{k},\sigma}\epsilon_{\mathbf{k}}c_{\mathbf{k}\sigma}^{\dagger}c_{\mathbf{k}\sigma}+J_K\sum_i \mathbf{S}_i\cdot\mathbf{s}_i+J_H\sum_{\langle ij\rangle}\mathbf{S}_i\cdot\mathbf{S}_j. \quad (1)$$

The spin-1/2 operators of the local magnetic moments have the fermionic representation $\mathbf{S}_i=\frac{1}{2}\sum_{\sigma\sigma'}f_{i\sigma}^{\dagger}\tau_{\sigma\sigma'}f_{i\sigma'}$ with a local constraint $\sum_\sigma f_{i\sigma}^{\dagger}f_{i\sigma}=1$, where τ is the Pauli matrices. Following the large-N fermionic approach[10], the Kondo spin exchange and Heisenberg superexchange terms can be expressed up to a chemical potential shift as

$$\mathbf{S}_i\cdot\mathbf{S}_j=-\frac{1}{2}\sum_{\sigma\sigma'}f_{i\sigma}^{\dagger}f_{j\sigma}f_{j\sigma'}^{\dagger}f_{i\sigma'},\qquad \mathbf{S}_i\cdot\mathbf{s}_j=-\frac{1}{2}\sum_{\sigma\sigma'}f_{i\sigma}^{\dagger}c_{j\sigma}c_{j\sigma'}^{\dagger}f_{i\sigma'}, \quad (2)$$

then uniform short-range AFM valence bond and Kondo screening order parameters can be introduced as

$$\chi=-\sum_\sigma\left\langle f_{i\sigma}^{\dagger}f_{i+l\sigma}\right\rangle,\qquad V=\sum_\sigma\left\langle c_{i\sigma}^{\dagger}f_{i\sigma}\right\rangle. \quad (3)$$

To avoid the incidental degeneracy of the conduction electron band on a square lattice, we choose $\epsilon_{\mathbf{k}}=-2t\left(\cos k_x+\cos k_y\right)+4t'\cos k_x\cos k_y-\mu$, where t and t' are the first and second nearest neighbor hoping matrix elements, respectively, while μ is the chemical potential, which should be determined self-consistently by the density of the conduction electrons n_c. Under the uniform mean-field approximation, the f-spinons form a very narrow band with the dispersion $\chi_{\mathbf{k}}=J_H\chi\left(\cos k_x+\cos k_y\right)+\lambda$, where λ is the Lagrangian multiplier to be used to impose the local constraint on average.

Thus the corresponding mean-field Hamiltonian reads

$$H=\sum_{\mathbf{k}\sigma}\begin{pmatrix} c_{\mathbf{k}\sigma}^{\dagger} & f_{\mathbf{k}\sigma}^{\dagger}\end{pmatrix}\begin{pmatrix}\epsilon_{\mathbf{k}} & -\frac{1}{2}J_KV\\ -\frac{1}{2}J_KV & \chi_{\mathbf{k}}\end{pmatrix}\begin{pmatrix}c_{\mathbf{k}\sigma}\\ f_{\mathbf{k}\sigma}\end{pmatrix}+E_0, \quad (4)$$

with $E_0=N\left(-\lambda+J_H\chi^2+J_KV^2/2\right)$. The quasiparticle excitation spectra can be easily obtained

$$E_{\mathbf{k}}^{\pm}=\frac{1}{2}\left[\left(\epsilon_{\mathbf{k}}+\chi_{\mathbf{k}}\right)\pm W_{\mathbf{k}}\right], \quad (5)$$

which implies that the conduction electron band $\epsilon_{\mathbf{k}}$ has a finite hybridization with the spinon band $\chi_{\mathbf{k}}$. Here $W_{\mathbf{k}}=\sqrt{\left(\varepsilon_{\mathbf{k}}-\chi_{\mathbf{k}}\right)^2+\left(J_KV\right)^2}$. Accordingly, the ground-state energy density can be evaluated as

$$\varepsilon_g=\frac{2}{N}\sum_{\mathbf{k},\pm}E_{\mathbf{k}}^{\pm}\theta\left(-E_{\mathbf{k}}^{\pm}\right)-\lambda+J_H\chi^2+\frac{1}{2}J_KV^2, \quad (6)$$

where $\theta(-E_{\mathbf{k}})$ is the theta function. Then the self-consistent equations for the mean-field variables χ, V, and λ and the chemical potential μ can be deduced from the relations

$$\frac{\partial\varepsilon_g}{\partial\chi}=0,\quad \frac{\partial\varepsilon_g}{\partial V}=0,\quad \frac{\partial\varepsilon_g}{\partial\lambda}=0,\quad n_c=-\frac{\partial\varepsilon_g}{\partial\mu}. \quad (7)$$

3. Results and Discussions

In the following, we will assume that $t'/t=0.1$ and $n_c=0.85$. When the self-consistent equations are carefully evaluated, we find that the mean-field AFM order parameter χ is *always* positive in the range $0<J_H/J_K\leq 3$ so that the resulting state is a stable paramagnetic metal.

In Fig.1, as the strength of the AFM spin fluctuations grows up, we present the evolution of the band structure of the renormalized heavy quasiparticles around the Fermi level in the direction $(0,0) \to (\pi,\pi) \to (\pi,0) \longrightarrow (0,0)$ of the first Brillouin zone. Fig.1a to Fig.1h correspond to $J_K/t = 2.0$ and $x = J_H/J_K = 0$, 0.11, 0.12, 0.234, 0.235, 1.2, 2.76, and 2.77, respectively. At the critical values $x = 0.1181$, the point (π,π) changes from the local maximum to local minimum of the quasiparticle band, while at $x = 0.2341$, the separated band around the momentum $(\pi,\pi/2)$ starts to move away from the Fermi level. As the ratio of the coupling strengths increases, the separated band moves to the momentum $(\pi,0)$, and reaches the Fermi level again at $x = 2.77$.

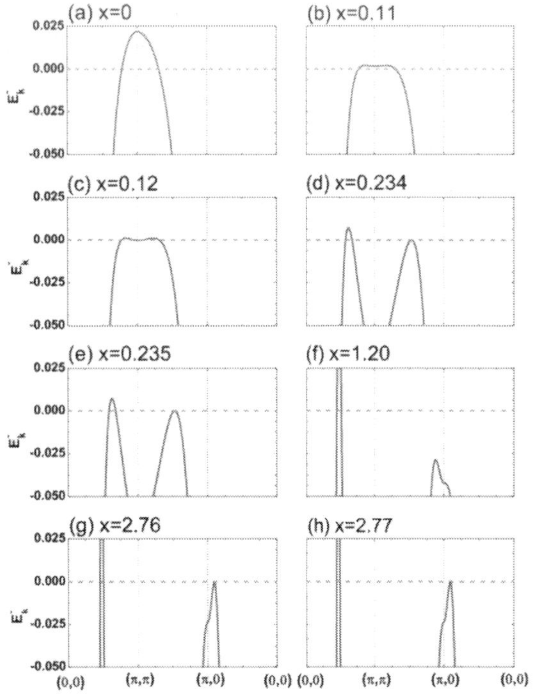

Figure 1. The lower renormalized quasiparticle band in the direction $(0,0) \to (\pi,\pi) \to (\pi,0) \to (0,0)$ as increasing the strength of the AFM spin fluctuations.

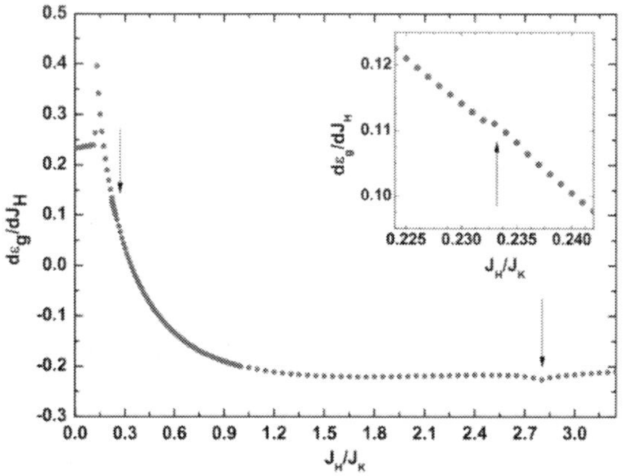

Figure 2. The first-order derivative of the ground state energy as a function of the parameter $x = J_H/J_K$. There are three singular points, $x_{c,1} = 0.1181$, $x_{c,2} = 0.2341$, and $x_{c,3} = 2.77$. The first one corresponds to a first order transition, while the latter two correspond to two second order phase transitions, respectively.

From the quantum phase transition aspects, we calculate the ground state energy density ε_g and its first-order derivative with respect to the ratio of the coupling parameters x. The numerical results are displayed in Fig.2. We find that there are three non-analytical points. ε_g is finite and continuous in the parameter range $0 < x < 3$, However, its first-order derivative has a large jump at $x_{1c} = 0.1181$, corresponding to a first-order quantum phase transition. Moreover, two small kinks appear at $x_{2c} = 0.2341$ and $x_{3c} = 2.77$ in the first-order derivative, which correspond to the jumps in the second-order derivative of ε_g. So x_{2c} and x_{3c} denote two second-order quantum phase transitions.

Once the renormalized quasiparticle band structure is available, the corresponding FS can be easily obtained. In Fig.3a-3g, the center of the FS is shifted from $(0,0)$ to (π,π). For a fixed $J_K/t = 2.0$, we can see that the FS is a hole-like circle around (π,π) for the parameter range $0 \le x \le 0.11$. At $x_{1c} = 0.1181$, the topology of the FS starts to change: a small circle emerges in the center of the deformed large square FS. As x is further increased, both circles expand

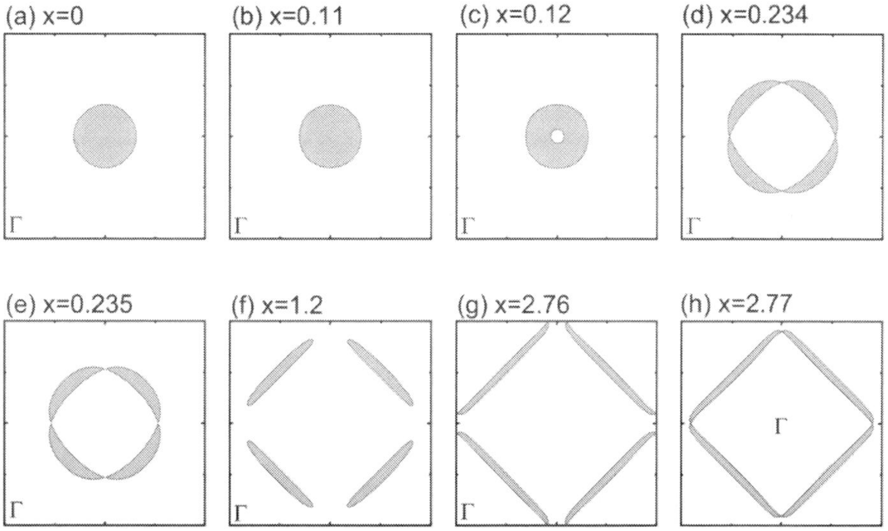

Figure 3. The structure of the FS changes as increasing the strength of the AFM spin fluctuations. The shaded area represents the hole-like FS. The center of the FS is shifted at (π, π) in (a)-(g).

and the small one is deformed into a rotated square. Up to $x_{2c} = 0.2341$, the two deformed circles intersect each other and then decompose into four Fermi pockets. When x increases to $x_{3c} = 2.77$, the FS pockets are reconnected again and form two closed square-like circles. Three critical values correspond to three different quantum phase transitions, which belong to the category of Lifshitz phase transitions.

To some extent our present mean-field theory captures the heavy-fermion liquid physics of the Kondo-Heisenberg lattice systems, especially the Fermi surface evolution of the renormalized heavy quasiparticles as the short-range AFM spin correlations between the localized magnetic moments are gradually increased. In order to put the present results on a more solid ground, further investigation beyond the mean-field theory is certainly needed.

Acknowledgments. The authors would like to acknowledge the support of NSF of China.

References

[1] I. M. Lifshitz, Sov. Phys. JETP **11**, 1130 (1960).
[2] S. Doniach, Physica B & C **91**, 231 (1977).
[3] G. M. Zhang, Q. Gu, and L. Yu, Phys. Rev. B **62** , 69 (2000); G. M. Zhang and L. Yu, Phys. Rev. B **62**, 67 (2000).
[4] Q. Si, S. Rabello, K. Ingersent, and J. Smith, Nature **413**, 804 (2001).
[5] T. Senthil, M. Vojta, and S. Sachdev, Phys. Rev. B **69**, 035111 (2004).
[6] H. Watanabe and M. Ogata, Phys. Rev. Lett. **99**, 136401 (2007).
[7] L. C. Martin and F. F. Assaad, Phys. Rev. Lett. **101**, 066404 (2008); L. C. Matrin, M. Berex, and F. F. Assaad, Phys. Rev. B **82**, 245105 (2010).
[8] T. Grover and T. Senthil, Phys. Rev. B **81**, 205102 (2010).
[9] G. M. Zhang, Y. H. Su, and L. Yu, Phys. Rev. B **83**, 033102 (2011).
[10] P. Coleman and N. Andrei, J. Phys.: Condens. Matter **1**, 4057 (1989).

Application of the underscreened Kondo lattice model to neptunium compounds

Christopher Thomas[1,2], Acirete S da Rosa Simões[1], J R Iglesias[1], C Lacroix[3*] and B Coqblin[4]

[1] Instituto de Física, Universidade Federal do Rio Grande do Sul, 91501-970 Porto Alegre, Brazil

[2] International Institute of Physics, Universidade Federal do Rio Grande do Norte, 59078-400 Natal-RN, Brazil

[3] Institut Néel, CNRS et Université Joseph Fourier , BP 166, 38042 Grenoble Cedex 9, France

[4] Laboratoire de Physique des Solides, CNRS, Université Paris-Sud, 91405 Orsay, France

E-mail: *••••••••••••••••••••••••••••••

Abstract. The coexistence of Kondo effect and ferromagnetic order has been observed in many uranium and neptunium compounds such as UTe or Np_2PdGa_3. This coexistence can be described within the underscreened Anderson lattice model with two f-electrons and $S = 1$ spins on each site. After performing the Schrieffer-Wolff transformation on this model, we have obtained an effective Hamiltonian with a f-band term in addition to the Kondo interaction for $S = 1$ spins. The results indicate a coexistence of Kondo effect and ferromagnetic order, with different relative values of the Kondo T_K and Curie T_C temperatures. We emphasize here especially the case $T_K < T_C$ where there is a Kondo behavior below T_C and a clear decrease of the magnetization below T_K. Such a behavior has been observed in the magnetization curves of $NpNiSi_2$ at low temperatures.

1. Introduction

It is now well established that cerium and uranium compounds have a different Kondo behavior. First, in the case of cerium and ytterbium systems, the Kondo effect and the magnetic order are provided by the $s - f$ exchange interaction between the localized $S = 1/2$ spins of the $4f$-electrons and the conduction electrons. As the conduction electrons may completely screen the local magnetic moment, there is a strong competition between the Kondo effect and the magnetic ordering, which has given the well known Doniach diagram [1] and relatively small Néel or Curie temperatures, typically of order 5-10 K. On the other hand, uranium and neptunium compounds, like UTe [2] or Np_2PdGa_3 [3], have larger localized spins, as for example $S = 1$; these spins are partially screened (underscreened) by the conduction electrons and it results a coexistence between the underscreened Kondo effect and a ferromagnetic ordering, yielding relatively large Curie temperatures T_C of order 50-100 K. Moreover, the $5f$ electrons in actinide compounds are less localized than the $4f$ electrons in rare-earth systems.

We have recently developed the underscreened Kondo lattice model [4, 5], where ferromagnetic order and Kondo behavior coexist, as observed in actinide compounds like UTe (with T_C=102 K) [2] or $UCu_{0.9}Sb_2$ (with T_C=113 K) [6], but we extend the parameter space to include the case where $T_K < T_C$ that seems to be particularly suitable to describe the magnetization curves of neptunium compounds like $NpNiSi_2$ (with T_C=51.5 K) [7] or Np_2PdGa_3 (with T_C=62.5 K) [3].

2. The Underscreened Kondo Lattice model

Starting from the underscreened Anderson lattice model (UAL) with two f-electrons in different orbitals, we have recently derived an effective hamiltonian, using the Schrieffer-Wolff (SW) transformation [5]. We first briefly summarize the main results of ref. [5]: supposing that the two f-electrons are strongly coupled in the triplet $S = 1$ configuration through Hund's rule. After performing the SW transformation, and keeping terms with f-operators in different sites (usually neglected), the effective hamiltonian is given by:

$$H = \sum_{\mathbf{k}\sigma} \varepsilon_{\mathbf{k}} n^c_{\mathbf{k}\sigma} + \sum_{i\sigma\alpha} E_o n^{f_\alpha}_{i\sigma} + J_K \sum_i \mathbf{s}_i \cdot \mathbf{S}_i + J_H \sum_{\langle ij \rangle} \mathbf{S}_i \cdot \mathbf{S}_j + H_f \qquad (1)$$

In this hamiltonian, $\varepsilon_{\mathbf{k}}$ is the energy of the conduction band, the second term of the Hamiltonian accounts for the energy E_o of the two $5f$ localized electrons (the indices $\alpha = 1, 2$ correspond to the two f-orbitals) of the $5f^2$ configuration of the actinide atoms. In the third term $J_K > 0$ is the Kondo coupling (proportional to the square of the hybridization) between the localized spin $\mathbf{S}_i$ and the spin $\mathbf{s}_i$ of a conduction electron at the same site derived with the SW transformation, J_H is an *ad hoc* interaction between nearest-neighboring spins [4, 5], which we consider as ferromagnetic ($J_H < 0$) and H_f is the effective f-band term, which is also derived from the SW transformation. The presence of a finite f-band width improves the model which can, therefore, describe a small delocalization of the $5f$ electrons [5]. Finally, the intersite exchange is considered here as an independent parameter, though RKKY can also be obtained from the starting UAL.

We have previously studied the underscreened Kondo lattice model (UKL) which is similar to equation (1), but without the f-band term [4]. Coexistence of ferromagnetism and Kondo effect was already obtained in the UKL model, but inclusion of the last term in equation (1) gives a better description of the actinide compounds in which $5f$ electrons are not completely localized.

Because of the triplet state of the $5f$ electrons, the Kondo effect cannot lead to a complete screening of the localized $S = 1$ spins, and the ferromagnetic exchange between the partially screened spins may lead to the formation of a ferromagnetic order. In the present paper, we will study in detail the case where the Kondo temperature is smaller than the Curie temperature.

We have used the mean field approximation which has showed to be a very powerful tool to describe the competition between Kondo effect and magnetic ordering in the Kondo lattice [8, 9, 10]. In this approximation, the Kondo effect is described by the Kondo parameters:

$$\lambda_{i\sigma} = \sum_\alpha \lambda^\alpha_{i\sigma} = \sum_\alpha \langle c^\dagger_{i\sigma} f^\alpha_{i\sigma} \rangle \qquad (2)$$

which couple conduction and f electrons at the same site and that we assume site-independent, i.e. $\lambda_\sigma = \lambda_{i\sigma}$.

Moreover, in order to describe the magnetic properties of the system, we introduce also the magnetization operators for f- and c- electrons :

$$M^f_i = S^z_i = \tfrac{1}{2}(n^f_{i\uparrow} - n^f_{i\downarrow}) \text{ and } m^c_i = \sigma^z_i = \tfrac{1}{2}(n^c_{i\uparrow} - n^c_{i\downarrow})$$

The ferromagnetic phase is described by non-zero values of $\langle M^f \rangle$ and $\langle m^c \rangle$, while non-zero λ_σ describe the Kondo effect and the formation of the heavy-fermion state. In the ferromagnetic phase, λ_σ depends on σ. The detailed calculations can be found in ref. [4, 5].

The last term of equation (1) is an effective $5f$ band, which is derived from the starting UAL model, within SW transformation. The half f-band width is spin-dependent and is given in mean field approximation by $W_f = A_\sigma$, where:

$$A_\sigma = -\frac{J_K P}{16}\left[7 + 3(\langle M^f\rangle)^2 + 6\sigma\langle M^f\rangle\right] . \quad (3)$$

where $\sigma = \pm 1/2$.

The effective bandwidth depends on the Kondo coupling $J_K > 0$ and the f-magnetization. P is here an adjustable parameter, which can be related to the parameters of the starting UAL model. This description has the advantage of describing the beginning of the delocalization of $5f$ electrons which are clearly less localized then the $4f$ electrons of rare-earths.

We have obtained a phase diagram giving the two temperatures T_K and T_C versus J_K for fixed values of J_H. For small J_K values, the Kondo temperature is zero and the Curie temperature increases slowly. At a critical value of J_K, T_K increases rapidly and becomes larger than T_C. When T_K and T_C are of the same order of magnitude, the Curie temperature reaches a maximum [5].

This theoretical result can account for the experimental behavior of the Curie temperature T_C of the monochalcogenide UTe under pressure, which is firstly increasing with pressure up to roughly 10 GPa and then decreases above this pressure [11, 12]. It is well known that pressure makes increase the Kondo coupling parameter J_K in uranium compounds and our theoretical result can well explain the pressure dependence of the Curie temperature T_C of UTe [5]. Thus, we propose that UTe can be described by the UAL with T_K and T_C of the same order of magnitude.

3. The case $T_K < T_C$: Comparison with experiments and conclusions

The case $T_K < T_C$ has been more recently studied and we present here briefly some results [13] showing that the $S = 1$ case is very different from what is obtained for the $S = 1/2$ Doniach diagram. The left panel of figure 1 shows the temperature dependence of the mean field parameters for this case. When decreasing the temperature below T_C, we can see that M^f and m^c increase with opposite signs (due to the positive sign of J_K), while the value of both λ_σ remain zero below T_C. However, at a lower temperature corresponding to the Kondo temperature, the two λ_σ become non zero, and there is a sudden drop of the magnetization parameters.

Thus, with decreasing temperature, there is first a ferromagnetic state between T_C and T_K and then a mixed Kondo-ferromagnetic state below T_K with a clear decrease of the ferromagnetic magnetization. In the ferromagnetic phase, the magnetic moment is almost saturated, reaching a value close to 1. In the Kondo-ferromagnetic phase, the f-moment is partially screened, and the f-moment is reduced to a value close to 1/2. This is in agreement with the description given by Nozières and Blandin [14] of the underscreened Kondo effect: the conduction electrons cannot completely screen a $S = 1$ localized spin, and the system is described below T_K by an effective $S = 1/2$ spin, which can order due to the intersite exchange.

This model applies well to the compound $NpNiSi_2$: the right panel of figure 1 shows the magnetization curves of this compound [7]. The magnetization obtained at a small magnetic field B=0.05 T as a function of temperature shows the onset of ferromagnetism at T_C=51.5 K. A sharp collapse of the magnetization is observed at T^* of order 12 K. There are at least three arguments in favor of the fact that this collapse is due to the Kondo effect: (i) the magnetic resistivity exhibits a *LogT* dependence above T_C; (ii) the deduced electronic specific heat coefficient at low temperatures (below T^*) is large: γ=133 mJ/mol K^2; (iii) the saturated moment $\mu_{sat} = 1.07\ \mu_B$ at 5 K is rather small, as shown in figure 1. This small value may be explained by a partial polarization of conduction electrons with spins opposite to those of localized electrons, exactly like in the underscreened Kondo Lattice case [7].

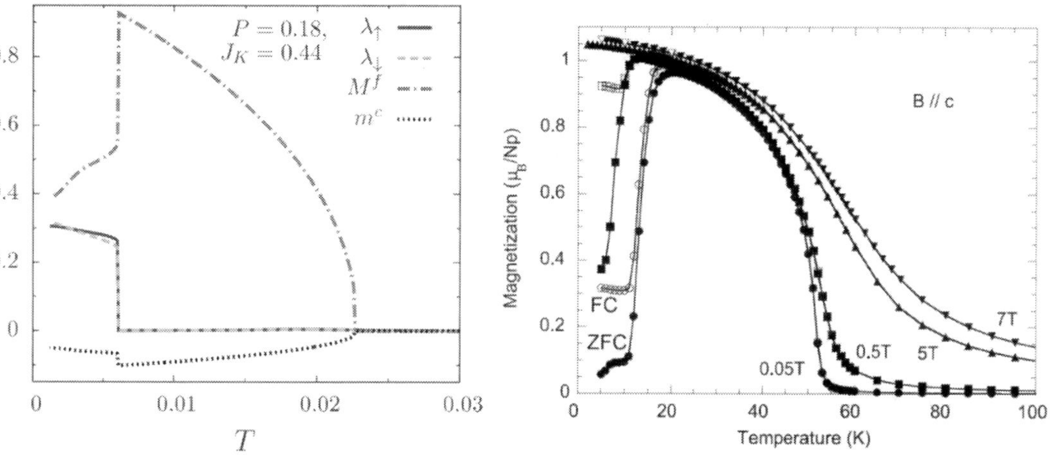

Figure 1. Left: Calculated magnetizations and Kondo parameters for $T_K < T_C$. The details of the calculations can be found in refs. [5, 13]. Right: Magnetization curves of $NpNiSi_2$ compound [7].

Thus, our model can give a good explanation of the magnetization curves of the $NpNiSi_2$ compound, if we take the Kondo temperature T_K equal to T^* of order 12 K, which is clearly much smaller than the Curie temperature T_C=51.5 K.

There is another ferromagnetic and Kondo neptunium compound, namely Np_2PdGa_3, which has been recently studied experimentally [3]. The magnetic resistivity, which follows a $LogT$ decrease above the Curie temperature $T_C = 62.5$ K and the electronic specific heat constant, which is equal to γ=120 mJ/mol K^2 at very low temperatures, clearly indicate a Kondo behavior in addition to the ferromagnetic one [3]. The magnetization curves show a broad maximum below T_C, which has been interpreted as the possible development of a ferromagnetic order influenced by a narrow domain wall or strong magnetocrystalline anisotropy and the two ZFC- and FC-magnetization curves have a different behavior at low temperatures [3]. However, a more probable explanation comes again from the Kondo effect which reduces the magnetization and the estimated Kondo temperature is here 35 K, which is also smaller than T_C [3].

In conclusion, we have shown that the UAL model for $T_K < T_C$ gives a good description of the magnetization of $NpNiSi_2$ at low temperatures. However, we remark that we have applied the UAL model to neptunium compounds, although the configuration of Np atoms is different from our starting configuration $5f^2$ appropriate to uranium compounds and moreover the crystalline field effect plays probably an important role [3]. Finally, the present underscreened Kondo lattice model accounts well for the results obtained for $T_K < T_C$ and further experimental studies are necessary for a more detailed understanding of the behavior of the Kondo-ferromagnetism coexistence below the Kondo temperature.

Acknowledgments

CT and JRI acknowledge financial support from Brazilian agencies CNPq and CAPES.

4. References

[1] Doniach S 1976, Proceedings of the *Valence Instabilities and Related Narrow-Band Phenomena*, ed R D Parks (New York: Plenum Press), p 168
[2] Schoenes J, Vogt O, Lohle J, Hulliger F and Mattenberger K 1996 *Phys. Rev.* B **53** 14987
[3] Tran V H, Griveau J C, Eloirdi R, Miiller W and Colineau E 2010 *Phys. Rev.* B **82** 094407
[4] Perkins N B, Nunez-Regueiro M D, Iglesias J R and Coqblin B 2007 *Phys. Rev.* B **76** 125101

[5] Thomas C, Simões A S R, Iglesias J R, Lacroix C, Perkins N B and Coqblin B 2011 *Phys. Rev.* B **83** 014415
[6] Bukowski Z, Troc R, Stepien-Damm J, Sulkowski C and Tran V H 2005 *J. Alloys Compd.* **403** 65
[7] Colineau E, Wastin F, Sanchez J P and Rebizant J 2008 *J. Phys.: Condens. Matter* **20** 075207
[8] Lacroix C and Cyrot M 1979 *Phys. Rev.* B **20** 1969
[9] Coleman P and Andrei N 1989 *J. Phys.: Condens. Matter* **1** 4057
[10] Iglesias J R, Lacroix C and Coqblin B 1997 *Phys. Rev.* B **56** 11820
[11] Link P, Benedict U, Wittig J and Wuhl H 1992 *J. Phys.: Condens. Matter* **4** 5585
[12] Cornelius A L, Schilling J S, Vogt O, Mattenberger K and Benedict U 1996 *J. Mag. Mag. Mater.* **161** 169
[13] Thomas C, Ph. D. Thesis, U.F.R.G.S., Porto Alegre, Brazil (May 2011)
[14] Nozières P and Blandin A 1980 *J. de Physique* **41** 193

Lithium nanoparticles in lithium fluorite crystals

E M Ibragimova, M A Mussaeva, M U Kalanov, N M Mukhamedshina and V N Sandalov

Institute of nuclear physics, Tashkent, Uzbekistan

E-mail: ibragimova@inp.uz

Abstract. Using intensive optical absorption techniques we studied formation of lithium nanoparticles in LiF crystals under ^{60}Co-γ-irradiation in the dose range of 10^5–10^9 R at 300 K. At 10^6 R the F-center absorption band 4.95 eV grows up D>3 and splits into a symmetric doublet of narrow resonances similar to s^1-metal nanoparticles surface plasmons. The maximal concentration $3.3 \cdot 10^{17}$ cm^{-3} corresponds to ~14 nm distance between F-centers when they form F-F-pair and then M-center. Absorption band of divacancies 2.75 eV splits into triplet at >10^7 R, when Li_n nanoparticles 3.26 eV are formed. Twinned LiF lattice provides the defects ordering and assembling of Li_n nanorods with the size of 8 nm related with the peak electric conductivity at 240-280 K. PACS: *71.70.-d ; 73.20Mf ; 73.22.-f*

1. Introduction

Early experiments on radiation induced defect production in ionic LiF crystal with the widest gap (>14 eV) revealed besides point electron defects, such as anion vacancies (*F*-center, 252 nm) and divacancies (*M*-center, 445-450 nm), also ionic defects, such as molecular fluorine interstitials (*H*- center, 340-380 nm), vacancy aggregates (F_n-centers, 450, 518 nm), colloidal precursors (*U*-center, 495 nm) and colloids of various nature and sizes, such as impurity colloids (*X*-center, 275 nm) or intrinsic Li_n ones (440, 530, 1000 nm) [1]. Aside the strong absorption of *F*-center an unknown band at 218 nm and a colloid center band at 265-285 nm were found [2]. After irradiation of LiF with thermal neutron fluency 3×10^{17} cm^{-2}, plate-like Li particles of 5 nm size were discovered with transmission electron microscope, which grew to 100 nm after the fluency 10^{18} cm^{-2} and following heating. Besides a prominent *X*-band at 280 nm, which belongs to impurity Mg colloids, a weak absorption band at 550 nm was also attributed to lithium colloids. The impurity colloids were formed easier at dislocations [3-5]. The defects induced at high fluencies were found to form an ordered substructure [6], which is very important.

Basing on Mie theory, describing extinction of small metal particles in transparent dielectric matrix, Li-colloid extinction in LiF was calculated at the sizes of spherical particles taken much less than the incident light wavelength λ as shown in figures of [7]. The most intensive and narrow peak at 458 nm was attributed to colloids with a size <5nm (curve 0). When the size grows within 10-20 nm, the peak moves to 480 nm, its width broadens and its magnitude decreases in two times. If the colloid size increases within 25-60 nm, there appear doublet

bands characterized by a weak magnitude and broad width, with the both maxima moving towards a larger wavelength. The model of a colloid precursor as Li or impurity anion in an anion site was developed to explain W-center absorption at 270-275 nm [8]. Later the energy states of Li-colloid nucleus as *F*-center with interstitial *Li* atom in a fluorine vacancy were calculated [9]. Taking into account the exchange interaction between the electron trapped at *F*-center and *2s*-electron of Li resulted in transformations of *1s* level of *F*-center into singlet *1S +3S* terms of *F(Li)*-center, and *2p* level into triplet *1P + 3P* terms. Thus, in the related absorption spectrum the singlet transition $1s \rightarrow 2p$ of *F*-center (250 nm) transforms into doublet: $1S \rightarrow 1P$ 5.81 eV (213 nm) and $3S \rightarrow 3P$ 4.53 eV (274 nm). The optical spectrometers available at that time made it difficult to observe the absorption band at 213 nm aside the intensive band 250 nm of *F*-centers, so only 270 nm could be seen and 250 nm should disappear to confirm this model.
Within the last decade the interest to metal nanoparticles in liquid has grown greatly because of their exciting properties, especially plasmonic resonant oscillations of their free electrons [10-12]. According to classic Mie theory, the surface plasmon frequency $\omega_s = \upsilon_f / R$, where υ_f is Fermi electron velocity and *R* is a spherical nanocluster size, and the extinction line width $\Gamma \sim a/R$ [10]. Experiments confirm the line broadening for smaller Ag clusters, but the ω depends mainly on the permittivity of dielectric matrix ε and the cluster concentration. By Mie-Gans theory, noble metal nanorods support two distinct plasmon modes: a longitudinal one moving toward red side and a transverse one to UV-side of spectra, with their frequencies dependent on the nanoparticle a/b aspect ratio [11]. This behavior differs from that for spheres of various sizes in [7]. Recently there appeared some articles on observation of nanosized metal colloids in a thin layer of LiF under fast Au ion irradiation that produced also nanosized ion tracks in the layer [13]. Broad complex absorption band at 2.3 - 3.3 eV was attributed to intrinsic Li colloids and a broader one at 4.4 - 4.6 eV to impurity Mn colloids. Since ion irradiation is very inhomogeneous, optical absorption of lithium colloids with growing size and probably shape should be studied thoroughly again with the use of homogeneous irradiation and fast reading highly luminous spectrometers and should be discussed as plasmon oscillations rather than just electron transitions in non-interacting point defect centers. Calculations in [7-9,11,12] should be tested experimentally.
The goal of this work was to study the formation of Li nanoparticles by assembling point defects in LiF crystal lattice under gamma-irradiation with the use of X-ray and optical spectroscopy.

2. Experiment

Available detector LiF crystals of the optical quality were taken for the experiment. The samples of 1 mm thickness were cleaved from a large crystal in (100) crystallographic planes so as to obtain a flat optical surface. With the use of X-ray fluorescent analysis at ^{109}Cd source the element composition of LiF single crystals was determined.

Samples were irradiated at the ^{60}Co radioisotope source of gamma-rays (1.17+1.32 MeV) in the wide dose range 10^5-10^9 R at the rate of 526 R/s at 320 K in dry air. Defects were accumulated at 30 steps so as to hit the particular dose when charge exchange interaction between neighboring defects becomes strong and the spectral band should split, or one type of defects (F_n) transforms into another *F(Li)*-center [9]. So the task was to estimate *F-F* distances.

Since radiation defect absorption bands are known to be wide, with the intensity growing fast with the dose above the limited optical density D=2 for the most double-beam spectrophotometers [1-9], primary task was to choose the proper optical spectrometer to measure spectral dependence of optical extinction (absorption + scattering) above 2 up to 5 so as to see intensive excitonic and plasmonic peaks. It was a single-beam spectrophotometer SF-56A (LOMO, Russia) in the range of 190-1100 nm with fast reading Si-photodiode array and the pc code FORT, the admitted absolute error less than ±1 %. Here we used the mode of the maximal slit opening to 6 nm that ensured registration of D up to 5 due to significant increasing the intensity of incident light beam. Optical spectra were measured 5 min. after each irradiation step so as to detect unstable optical centers (M, R_2, V_k , Li-colloid and *H*-interstitial [1]).

According to Smakula-Dexter formula $n_F = 9.48{\cdot}10^{15}D_F$ adapted for LiF (n=1.35) [13,14], the number N of color–centers per cm^2 was determined at each dose, and volume concentrations were calculated both for *F*–centers (250 nm) and *M*–centers (445 nm).

X-ray diffraction techniques were used for structure and phase analysis at nanoscale. For this task a serial diffraction meter DRON-3M (Russia) was upgraded for narrowing the apparatus function and increasing of the beam intensity. The improved collimation system provided two times increase in the signal-to-background ratio at small and medium scattering angles. It allowed us to select the low angle scattering background and weak narrow selective reflections, broadening and splitting of the basis lattice reflections into doublets and triplets, which determine the character of local stresses and twinning, respectively [15]. Nanoparticle sizes were evaluated by Selyakov-Scherrer formula with an experimental error below 10 %.

DC Electric conductivity was measured with 2-probe technique at 25 Volt within 80-400 K.

Journal of Physics: Conference Series **391** (2012) 012172 doi:10.1088/1742-6596/391/1/012172

3. Results and discussion

3.1. Element composition of crystals

The most prominent cation impurity was found to be Cu 0.02 wt% i.e. $6{\cdot}10^{17}$ cm^{-3} (the rest Ti, V, Cr, Mn, As, Au are below 0.01%). Since Cu^+ has the ion radius of 0.091 nm, which is larger than that of Li^+ 0.073 nm, it causes the local lattice deformation around, which can be compensated with an adjacent anion vacancy. The only anion impurity was Br^- < 0.01%, but due to a larger difference in ionic radii i.e. 0.182 nm vs 0.119 nm for F^- ion it causes a stronger local lattice deformation, than Cu^+, that can be compensated with an anion vacancy too. Therefore in non-irradiated sample there is a weak semi-band at the spectral edge 190 nm [1-6]. And under ionizing gamma-irradiation we expect charging of these vacancies i.e. creation of *F*- centers near the impurities, the concentration of which should saturate maximum at $6{\cdot}10^{17}$ cm^{-3}.

3.2. X-ray diffraction analysis of structure and phase

The powders for structure studies were made by mechanical abrasion of small LiF single crystals. The obtained grain sizes were within 14 - 21 μm. Part of the samples were irradiated with ^{60}Co-gamma-source to doses 10^7, 10^8 and 10^9 R in the dry air at 300 K. Measurements of the same mass samples were done at 300 K. Figure 1 (left) shows X-diffraction pattern on the non-irradiated LiF powder, which consists of an inelastic background and 6 selective reflections of different intensities over it, which belong to LiF lattice. The basic singlet 200-reflection with d/n

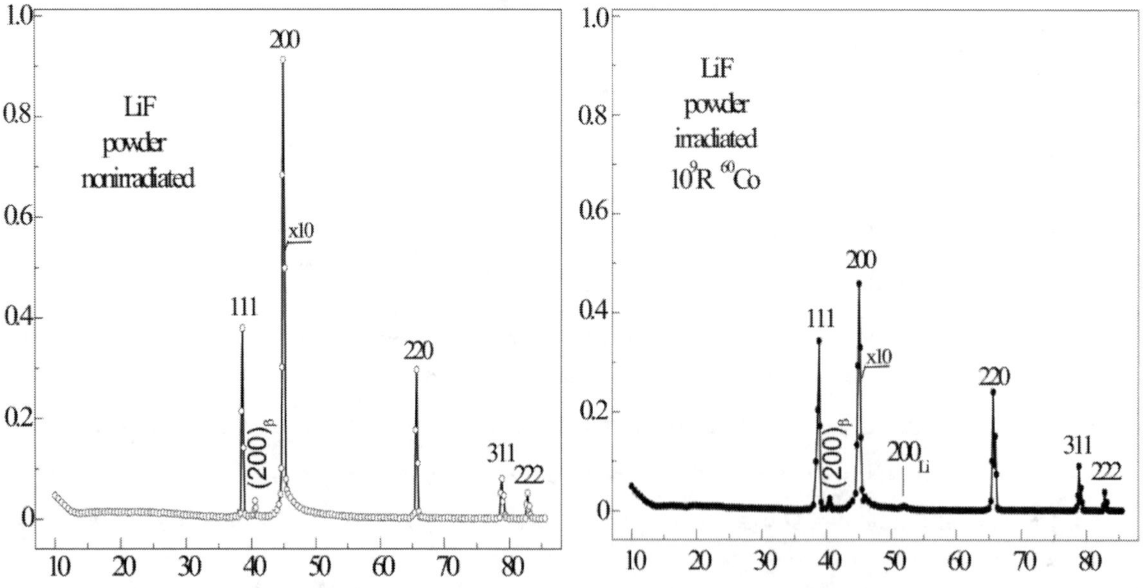

Figure 1. X-ray diffraction patterns of non-irradiated and dry gamma-irradiated LiF powder samples.

spacing 0.2011 nm is the most intensive ($I\times10^4$ pulse·sec^{-1}). Besides at small angles 2Θ=15–32° very weak broad diffuse scattering is due to structure fragments with non-saturated bonds distributed chaotically in subsurface layers of grains. Figure 1 (right) shows the pattern for LiF irradiated to the maximal dose 10^9 R. Comparing to non-irradiated one, all the selective peaks of the lattice have widened and got lower intensities: that of (200) has decreased in 2 times, the rest by 6-26 %. Such intensity losses are due to radiation induced "evaporation" of fluorine (i.e. radiolysis). Besides, the non-elastic background has got a non-even shape with weak steps, wells and hills, indicative of F emission preferably from the subsurface sites with a high potential, where self-assembling of Li metal clusters (then Li lattice) takes less energy.

The diffraction peaks of metallic Li particles are expected to occur at 2Θ = 32–40°, 48-56° and 68–75°. However the spectrum shows no signal in the non-irradiated sample. Since the volume of subsurface layers is a negligible part of the total bulk, the reflection intensity for metal Li lattice must be very weak, let alone its small atom number 3. Nevertheless such a selective reflection (200) with d/n = 0.1758 nm from Li crystal particles could be observed experimentally at 52° only in the heavily irradiated (10^9 R) LiF sample, as seen in Fig.1 right. Having the same orientation (200) and cubic symmetry as matrix LiF crystal, radiation induced Li nanocrystals form coherent metal-dielectric interfaces. The characteristic size of Li nanocrystals estimated by Selyakov-Scherrer expression is 8 nm, which confirms the calculated values [7,11].

Twinning of LiF lattice with (100) orientation occurs within (111) plane [6,15]. Nanotwin length was estimated from splitting of the basic reflection (200) into triplet as 100 nm in non-irradiated and 20-30 nm in irradiated crystals. In the case of wet irradiation the diffuse band at 26 grad is responsible for particles with an average size < 5 nm, while a weak narrow peak at 57 grad is attributed to LiOH nanocrystals oriented at (112) with a length of 28 nm [15].

3.2. Optical extinction spectra.

We found recently [16], that the absorption band at 250 nm (*F*-center) increases after gamma-irradiation in the dose range of 0.09 - 0.64 MR up to the optical density *D*=3, while its width narrows from 0.89 to 0.72 eV. It is characteristic for spherical nanoparticles [11]. Fig.2 left shows splitting of the band 250 nm into symmetric doublet of 240 and 260 nm with intensities *D*>3. At concentration $3\cdot10^{17}$ cm^{-3} (less than Cu) the distance between neighboring *F*-centers becomes 14 nm, which is less than the size of Li exciton 20 nm [1]. The reduced line width of the doublet means a lower rate of the plasmon decay, and the extinction is dominated by absorption [11].

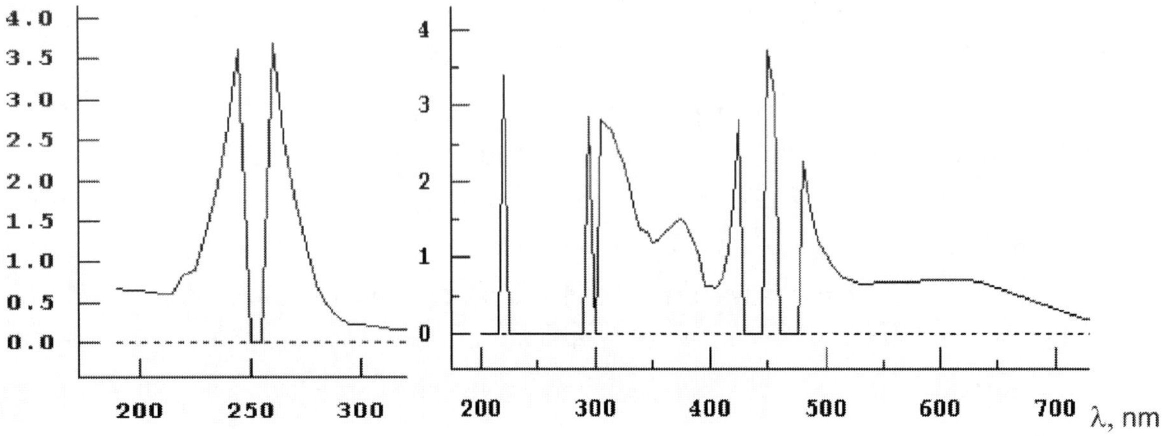

Figure 2. Optical extinction $D(\lambda)$ of 1 mm thick LiF (100) after gamma-irradiation to 0.73 MR (left) and 20 MR (right).

There also appears a weak band at 220 nm which was ascribed either to self-localized excitons or to *X*-centers [1-5]. The following irradiation with the small dose steps caused moving of the doublet components away each from other with descending intensity of the long-wave component, as it can be seen in Fig.2 right and calculated in [7]. At the dose of 7.5 MR the wavelengths of the doublet components reach the values of 220 and 270 nm that confirm well the *F*- band splitting calculated in [9]. Besides, simultaneously there emerged well known band at 445 nm attributed to *M*-centers and hardly resolved band about 460 nm, which is ascribed by various authors either to F_4 aggregate centers or Li colloids <5 nm [1,7]. Our measurements showed that divacancy based *M*-centers appear only after splitting of 250 nm band due to transformation of *F*-center into *F-F*-pair by the model of [9]. It is reasonable to assume that the observed behavior of the doublet with the dose grows (i.e increase in splitting) is due to shortening of the distance between *F*-centers in pair and the related increase in the exchange interaction. Fig. 2 right also shows formation of F_3 (320 and 375 nm) and F_4 , F_5 aggregates (500-700 nm) at the following dose increase to 20 MR. After growing to D>3 the band at 445 nm splits asymmetrically and unstably, when the *M-M*-center distance shortens to 22 nm. Since the aggregates may gain various configurations of different stability [11], the measurements were repeated many times within a few hours, days. The broad weak band about 620 nm turned out unstable and annealed in 3 hours at 300 K. It may be responsible for the dominating light scattering as $1/\lambda^4$ from large nanoparticles of 50 nm as calculated in [7], and may be attributed to an unstable linear *F-F-F-F* configuration which transforms into more stable square of 4*F*-centers, where Li colloid is formed. F_3-centers grow quite slow due to their non-stability.

Considering *F*-aggregate centers as places for Li colloid generation, it seems reasonable to implement the ratio of a longitudinal plasmon wavelength λ and the related length of a metal nanorod *L*. In the case, when only dipole oscillations contribute to the optical density *D*, the ratio is $\lambda=10L$, so the observed peaks correspond to *L* changes within 27-30 nm, which is in a good agreement with the estimation of the twin length from X-ray diffraction measurements (fig.1).

3.3. Electric conductivity

Thermally activated direct electric current (DC) of non-irradiated and irradiated LiF samples was measured with 2-probe technique at applied 25 V as shown in Figure 3 (1kGy= 0.1 MR).

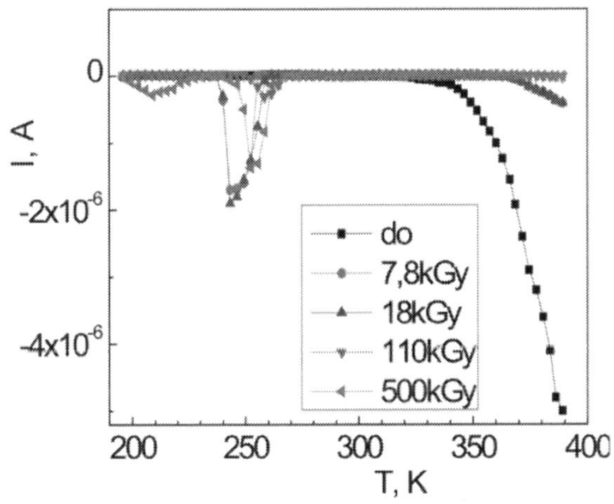

Figure 3. Temperature dependence of DC at 25 V in LiF samples irradiated to indicated doses (left) and repeated 1-2-3-measurements after the dose 1MGy (right).

Prior to irradiation there occurred only ion conductivity at T> 350 K. Irradiation induced peak conductivity around 250 K means delocalization of charge carriers from traps, their depth increasing with the dose growth, that make a 3D-conducting cluster. When the Li nanoparticle size *L* grows within 10-50 nm, the coulomb barrier for single electron transition decreases.

4. Conclusions

Mie theory based Radchenko [7] calculations of extinction dependent Li-colloid size and Lobanov's [9] model of *F(Li)* doublet absorption were confirmed well in our optical experiments. Broad singlet absorption bands of isolated *F*- and *M*-centers grow up to D~3.5 at 10 times different doses then split into doublet and triplet due to *F-F* and *M-M*-pair centers formation below the critical distances 14 and 22 nm, respectively, when their exchange interaction and oscillator strength is maximal. The maximal concentration of the pair centers is 2 times less than that of Cu impurity. So Li colloids are formed around Cu at the existing anion vacancies. Irradiation dose determines the size of Li-nano-particles >8 nm by self-assembling from the point defects and their arrangement into nanostructure of super-lattice type by changeably twinning. Optical density D and wavelength λ of longitudinal and transverse plasmon resonances depend on Li-nanorod aspect size and charge, and hence the matrix permittivity ε. Free electrons are confined in nano-Li in LiF matrix at the coherent metal-insulator interface oriented in (200) plane. When Li nano-crystals form 3D- cluster, electric conductivity emerges at 240-280 K possible due to single electron transport through the interface.

Non-linear properties of gamma-irradiated LiF (passive Q-switch mcs-ps for YAG-Nd) [1,15] come from Li-nanorods ordered with twin nanostructure rather than *M- or* F_2 centers

Other applications of LiF with nano-Li sublattice are possible, such as plasmon-enhanced resonant optical forces and also nanophotonic devices.

Acknowledgments. The researches are supported by national grant FA-F2-F068.

References

[1] Kotomin EA and Popov AI 2007 *Radiation effects in solids* (Amsterdam: Springer) Chap 7

[2] Kubo K 1961 *J. Phys. Soc. Jap.* **16** 2294

[3] Farge Y, Lambert M and Guinier A 1966 *J. Phys. Chem. Sol.* **27** 499

[4] Shwartz KK, Lyushina AF and Vitol AY 1969 *Solid State Phys.* **11** 1885

[5] Jain U and Lidiard AB 1977 *Philos. Mag.* **35** 245

[6] Kolontsova EV, Telegina IV and Zefirova VL 1977 *Radiation effects in solids* (Kiev: Naukova Dumka) pp 102, 158

[7] Radchenko IS 1969 *Solid State Phys.* **11** 1829

[8] Politov NG and Vorozheykina L F 1970 *Solid State Phys.* **12** 343

[9] Lobanov BD *et al* 1995 *Solid State Phys.* **37** 2545

[10] Kreibig U 1997 *Handbook of optical properties. Vol.2. Optics of small particles, interface and surface* eds. R Hummel, P Wibmann.. (Boca Raton, FL: CRC Press) p 145

[11] Wei A 2004 *Plasmonic nanomaterials.* In "Nanoparticles building blocks for nanotechnology" ed. V Rotello (Springer) p 173-200

[12] Gao Guozhong 2004 *Nanostructures & nanomaterials* (London: ICP) p 345

[13] Lushchik A et al 2007 *Phys.Rev. B* **76** 054114

[14] Perez A, Balanzat E and Dural J 1990 *Phys.Rev. B* **41** 3943

[15] Mussaeva MA, Ibragimova EM, Kalanov MU and Muminov MI 2006 *Solid State Phys.* **48** 2170

[16] Mussaeva MA, Ibragimova EM and Buzrikov S 2011 *Doklady Akademii Nauk Uzbekistana* # **1** 22

Supplementary data attachments to enhance the online versions of the article consist of data files, tables of extra information and extra figures.

XDA of LiF powder. β-component of the main peak (200) is seen at 40.7°. The rest selective peaks also belong to LiF structure: (111) d/n = 0.2327 nm at 2Θ = 38.7°, (220) 0.1224 nm 65.7°, (311) 0.1213 nm 78.9° and (222) 0.1162 nm 83.1°.

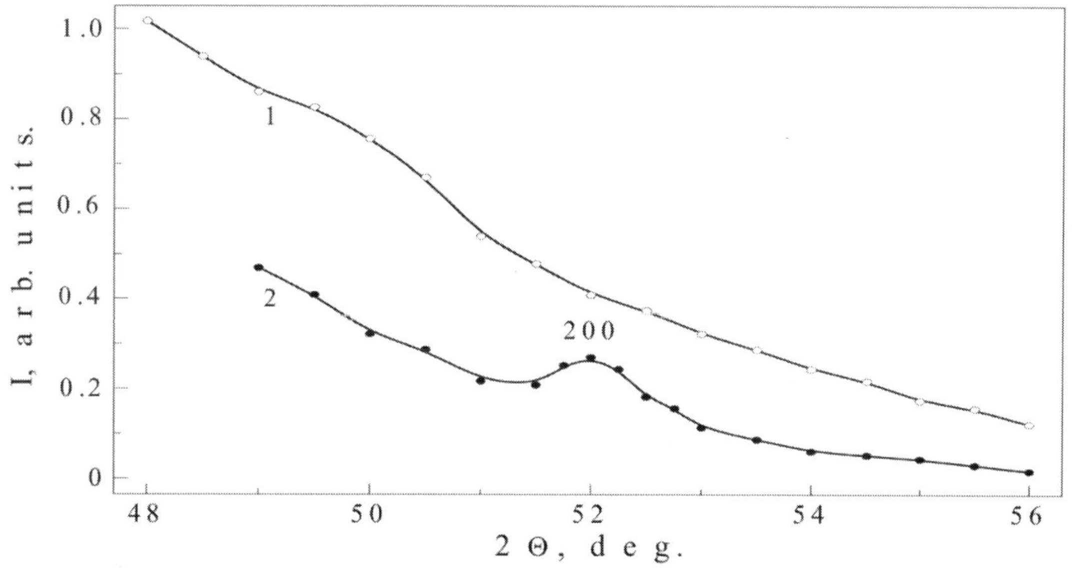

Figure 1-s. X-ray diffraction fragment of (200) peak shoulder : 1- non-irradiated, 2- irradiated LiF powder.

Optical reflection spectra are not applicable in this case, since expected nanoparticle sizes must be much less than incident wavelengths. So optical absorption of 1 mm thick LiF plate with (100) orientation was measured prior and after gamma-irradiation to the indicated doses. Figures 2 A,B,C were published in our paper in : Reports of Uzbekistan Academy of Sciences 2011, # 1.

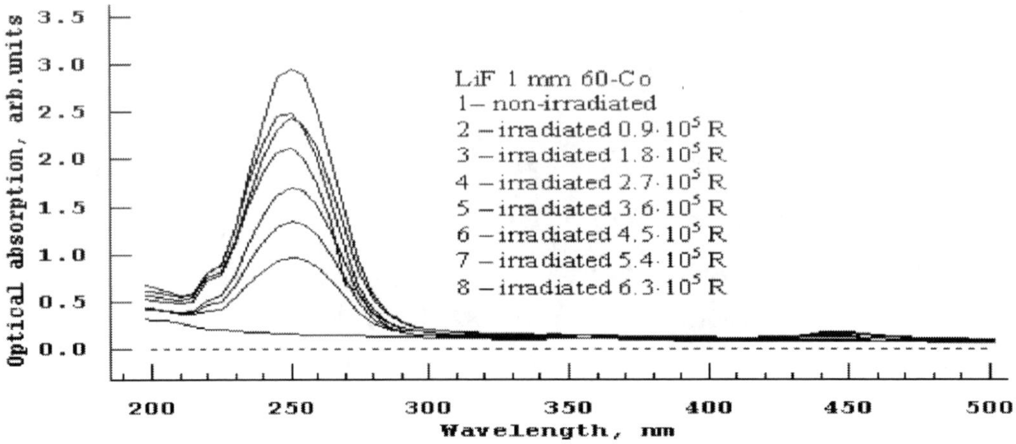

A

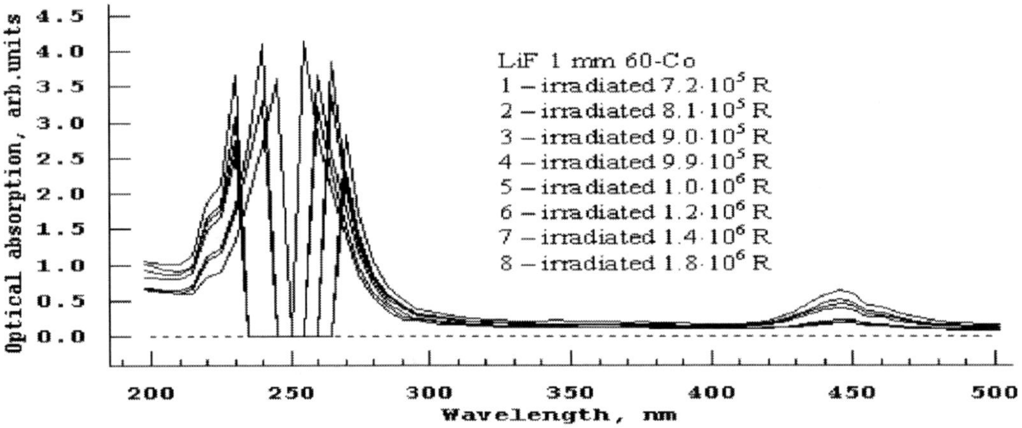

B

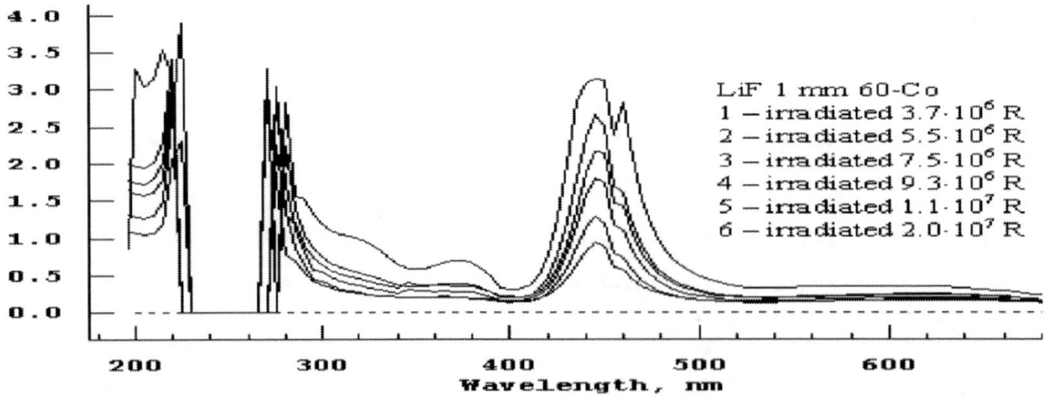

C

Figure 2-s. Optical absorption spectra. **A**: the lowest curve – non-irradiated, the magnitude of band at 250 nm grows with increasing the irradiation dose to 0.63 MR. **B**: the band splitting increases with the dose growth to 1.8 MR. Band at 450 nm is attributed to M-centers. **C:** The bands at 210 nm and 450 nm grow up and begin splitting at the dose 20 MR, when laser generation becomes effective. Table shows calculated concentrations N_F of F-centers

Dose 10^5 R	N_F (250 nm) 10^{17} cm^{-3}	d between F–F centers nm		Dose, 10^5 R	N_M (445 nm) 10^{16} cm^{-3}	d between M–M centers nm
0	0	–		8,1	0,64	54,3
0,9	0,97	21,7		9,9	1,14	44,8
1,8	1,33	19,5		18	1,61	40,0
2,7	1,63	18,2		37	2,24	35,8
3,6	1,97	17,2		55	2,92	32,9
4,5	2,25	16,4		75	3,71	30,4
5,4	2,26	16,3		93	5,69	26,4
6,3	2,74	15,4		110	6,67	25,1
7,3	3,33	14,5		200	9,64	22,2

AUTHOR INDEX

AUTHOR INDEX

AUTHOR INDEX

AUTHOR INDEX

AUTHOR INDEX

AUTHOR INDEX

AUTHOR INDEX